About the cover:

The molecule on the apple is anthocyanin. It is responsible for an apple skin's red color. Anthocyanins are found in many colorful fruits. This molecule's highly conjugated system of alternating double and single bonds allows it to absorb visible light. They have recently become useful in dye-sensitized solar cells. The highly conjugated molecule next to the yellow maple leaf is xanthophyll. It is responsible for the leaf's yellow color. You can learn more details on how this works in chapter 19.

The apple photo on the cover is by jmiltenburg at Morguefile.com.

The yellow maple leaf photo on the cover is by dhester at Morguefile.com.

About the author:

Dr. Chad Wallace serves as Dean of the School of Science & Engineering at Anderson University (Indiana) and Professor of Chemistry. He has taught organic chemistry for over two decades and noticed his students often struggled in class because they focused entirely too much on trivial, mundane organic chemistry facts and missed the most important concepts. To solve this problem, he has written this organic chemistry textbook, which uses easy-to-understand language and distills organic chemistry to its most basic and important concepts.

Dr. Wallace enjoys traveling. He has led trips with his students and/or family where they have served in Honduras, Nicaragua, Peru, Guatemala, and Kenya.

Organic Chemistry

Chad E. Wallace, Ph. D.

Volume 1. Seventh printing.

Table of Contents

List of Figures

To the Student.

Organic chemistry. It has the reputation of being one of the toughest courses offered in college, right? It is definitely not the easiest course, but one student might find their time in organic chemistry class to be fun and fulfilling while another student hates it and thinks it does not make any sense. What is the difference between these two students? Is it innate intelligence? Although it helps to be smart, I have seen very bright students get a D or F in the class. Is it how hard they work or how much they study? Working hard obviously helps. Students who do not put in a sufficient amount of time studying and practicing problems usually do not do very well. But, I have seen students who spend hours trying to memorize everything in the text, yet they still don't get that A they're after.

After years of teaching organic chemistry, I've tried to analyze what makes some students succeed and others struggle. I've come to find that the students who end up not passing the course, those students who really struggle, never learn the more important ideas. They never learn those few, central ideas that occur throughout the entire course. Imagine trying to learn higher mathematics without knowing your multiplication tables. Imagine trying to write an essay without knowing the alphabet.

Why do these students fail to learn these most important ideas? In class, I would always stop, tell the class that this is a very important idea, and tell them that if they do not understand this concept, they will probably fail this class, yet they still did not learn it. Why is this? I think the reason is that there is so much information in a typical organic chemistry text; students have a difficult time sorting out the most important concepts from the other, less important details.

I have tried to distill organic chemistry into its most important concepts. I really focus on what your professor may call pushing electrons. I must also confess that in this text, I have explained some concepts using language that I have developed over the years that I find helpful in understanding organic chemistry, but you will not find that language in other texts. I have decided to include that language to make the organic chemistry easier to understand.

You should read through this text with a pencil in hand. It is designed for you to write in it. Every time you encounter an exercise, write your answers down. If you are not sure of an answer, guess. But, the important thing is to write something down for each problem. The answers to the in-chapter exercises are found at the end of the book. Do not check your answers until you have attempted all of the problems. If you find you have made a mistake, go back and correct your answer. It is OK to make mistakes. Learn from those mistakes.

Do not fall into the trap of reading this book without a pencil and writing down your answers. You need the practice. Attempt to answer EVERY question. You become good at organic chemistry by doing problems. You will also build confidence by seeing that you do know how to do the exercises. Also, do not check the answer key before you try the problems on your own. It is easy to fool yourself by looking at the answers and thinking, "Yeah, I knew how to do that."

There is no replacement for reading your organic chemistry text. There is also no replacement for going to class. To be successful, go to class, read this text, reread this text, and then do every problem you find. Good luck!

Chapter 1. Periodic Table and Trends

1.1 Periodic table

It is convenient to know where on the periodic table the few most common elements encountered in organic chemistry are located. You won't always have a periodic table in front of you. We want to know the element locations in the table because we want to be able to compare the electronegativities of atoms (how much they attract electrons to them) and their sizes. Both will be important for us to analyze, predict, and understand organic reactions.

Most of organic chemistry deals with the elements hydrogen (H), carbon (C), oxygen (O), sulfur (S), nitrogen (N), and the halogens (fluorine (F), chlorine (Cl), bromine (Br), and iodine (I)). Therefore, for our purposes, we don't need to memorize too much of the periodic table, just the first couple of horizontal rows and one vertical column.

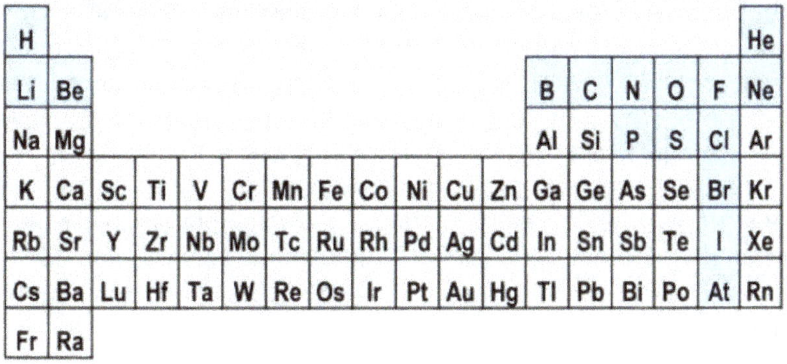

Figure 1.1 Common Organic Chemistry elements

So, let's begin memorizing where these elements are on the periodic table.

Most people know that hydrogen (H) is the first element in the periodic table. It is number 1 and in the top left corner. Helium (He) is the top right.

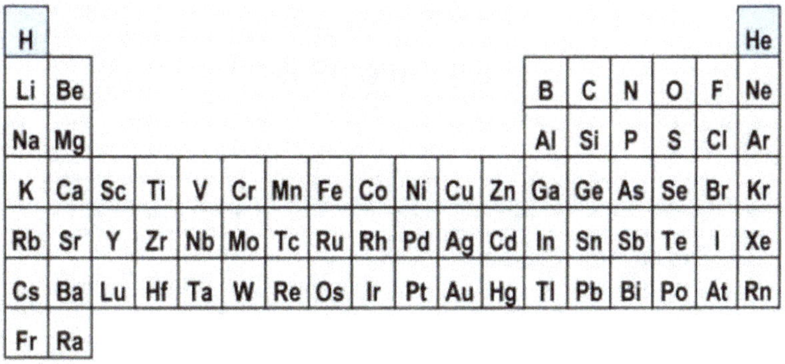

Figure 1.2 Hydrogen and Helium

Hydrogen, Helium. H He. It's so easy, it makes you want to laugh, "Ha ha." But, instead of "Ha Ha", we say "H He." H He. H He. So, to write the first row of the periodic table, just remember to laugh, H He.

Periodic Table and Trends

You can take the symbols for all of the elements of the second row (lithium (Li), beryllium (Be), boron (B), carbon (C), nitrogen (N), oxygen (O), fluorine (F), and neon (Ne)), squeeze them together, and make the nonsensical phrase "LiBe BCNOFNe" (pronounced Lih-bee Bik-noff-nee).

H																	He
Li	Be											B	C	N	O	F	Ne
Na	Mg											Al	Si	P	S	Cl	Ar
K	Ca	Sc	Ti	V	Cr	Mn	Fe	Co	Ni	Cu	Zn	Ga	Ge	As	Se	Br	Kr
Rb	Sr	Y	Zr	Nb	Mo	Tc	Ru	Rh	Pd	Ag	Cd	In	Sn	Sb	Te	I	Xe
Cs	Ba	Lu	Hf	Ta	W	Re	Os	Ir	Pt	Au	Hg	Tl	Pb	Bi	Po	At	Rn
Fr	Ra																

Figure 1.3 Period 2 Elements

The two elements on the left side of row two, lithium (Li) and beryllium (Be) make "LiBe" (pronounced Lih-bee). On the right side of row 2, boron (B), carbon (C), nitrogen (N), oxygen (O), fluorine (F), and neon (Ne) make "BCNOFNe". "BC" is pronounced Bik. "NOF" is pronounced Noff. Ne is pronounced Nee. Put it all together and you get "LiBe BCNOFNe" (Lih-bee Bik-noff-nee). Say it out loud five times. If you have difficulty remembering "LiBe BCNOFNe", perhaps it would help to sing it. If it is helpful, I've included one way to sing it.

"Hakuna matata" means "no worries." Hakuna matata is a Swahili phrase made popular by the 1994 Disney animated movie, "The Lion King." A meerkat, named Timon, and a warthog, named Pumbaa sing a song about Hakuna matata to their new friend, Simba. This is a catchy little tune. If you don't know this song, take a minute and look it up online so you too can have the tune in your head.

Instead of "Hakuna matata," sing, "LiBe BCNOFNe." Because in chemistry, if you remember LiBe BCNOFNe, you'll have "no worries" when it comes to the periodic table.

LiBe BCNOFNe! What a wonderful phrase.
LiBe BCNOFNe! Ain't no passing craze.
It means no worries for the rest of our days.
It's our problem-free philosophy.
LiBe BCNOFNe!

1.1 Write H He, LiBe BCNOFNe six times below.

_______________ _______________ _______________

_______________ _______________ _______________

1.2 Fill in the missing elements on this periodic table.

Na	Mg											Al	Si	P	S	Cl	Ar
K	Ca	Sc	Ti	V	Cr	Mn	Fe	Co	Ni	Cu	Zn	Ga	Ge	As	Se	Br	Kr
Rb	Sr	Y	Zr	Nb	Mo	Tc	Ru	Rh	Pd	Ag	Cd	In	Sn	Sb	Te	I	Xe
Cs	Ba	Lu	Hf	Ta	W	Re	Os	Ir	Pt	Au	Hg	Tl	Pb	Bi	Po	At	Rn
Fr	Ra																

Soon, when you encounter the question, "which is farther to the right in the periodic table, carbon (C) or oxygen (O)?" You'll know it is oxygen. Remember, LiBe BCNOFNe and have no worries!

The next to last vertical column of the periodic table is called group 17, or the halogens.

H																	He
Li	Be											B	C	N	O	F	Ne
Na	Mg											Al	Si	P	S	Cl	Ar
K	Ca	Sc	Ti	V	Cr	Mn	Fe	Co	Ni	Cu	Zn	Ga	Ge	As	Se	Br	Kr
Rb	Sr	Y	Zr	Nb	Mo	Tc	Ru	Rh	Pd	Ag	Cd	In	Sn	Sb	Te	I	Xe
Cs	Ba	Lu	Hf	Ta	W	Re	Os	Ir	Pt	Au	Hg	Tl	Pb	Bi	Po	At	Rn
Fr	Ra																

Figure 1.4 The Halogens

The halogens are commonly found in organic chemistry. It is sometimes important to know where they are in the periodic table. Fluorine (F) is at the top followed by chlorine (Cl), bromine (Br), iodine (I), and astatine (At). Notice that Br is bromine, not B. B is the symbol for boron. To understand the reactivity of some types of organic compounds, you will need to know where F, Cl, Br, and I are. But, unless you decide to become an astatine chemist, I doubt you will encounter it again. Who would, though? Astatine is radioactive!

Photo by katmystiry at Morguefile.com

There are several mnemonics used to memorize the halogens. The one that sticks in my mind is a little mean, so I apologize. But, it is effective. I remember the order of the halogens F, Cl, Br, I, At by the mnemonic, "**F**at **Cl**owns **Br**ing **I**n **At**tention". Feel free to make up your own. Go back up and take a look at the halogen column in the periodic table above with this new mnemonic in mind.

1.3 Say your halogen mnemonic out loud while you write the halogens in order, F Cl Br I At six times below.

________________ ________________ ________________

________________ ________________ ________________

1.4 Write the name of each element next to its atomic symbol.

H ____________ He ____________ Li ____________

C ____________ N ____________ O ____________

F ____________ Cl ____________ Br ____________

1.5 Fill in the missing elements (the halogens) in the periodic table below.

H																	He
Li	Be											B	C	N	O		Ne
Na	Mg											Al	Si	P	S		Ar
K	Ca	Sc	Ti	V	Cr	Mn	Fe	Co	Ni	Cu	Zn	Ga	Ge	As	Se		Kr
Rb	Sr	Y	Zr	Nb	Mo	Tc	Ru	Rh	Pd	Ag	Cd	In	Sn	Sb	Te		Xe
Cs	Ba	Lu	Hf	Ta	W	Re	Os	Ir	Pt	Au	Hg	Tl	Pb	Bi	Po		Rn
Fr	Ra																

There are two final elements you may want to memorize. Under nitrogen (N) and oxygen (O), we find phosphorus (P) and sulfur (S).

H																	He
Li	Be											B	C	N	O	F	Ne
Na	Mg											Al	Si	P	S	Cl	Ar
K	Ca	Sc	Ti	V	Cr	Mn	Fe	Co	Ni	Cu	Zn	Ga	Ge	As	Se	Br	Kr
Rb	Sr	Y	Zr	Nb	Mo	Tc	Ru	Rh	Pd	Ag	Cd	In	Sn	Sb	Te	I	Xe
Cs	Ba	Lu	Hf	Ta	W	Re	Os	Ir	Pt	Au	Hg	Tl	Pb	Bi	Po	At	Rn
Fr	Ra																

Figure 1.5 Phosphorus and Sulfur

I am not a fan of the vegetable peas. So, I remember "NO PeaS" to remember that N and O are above P and S.

You might also encounter the elements sodium (Na), magnesium (Mg), potassium (K), zinc (Zn), copper (Cu) in this text, but we won't worry about memorizing where those elements are. You should just be able to recognize that they are all metals.

1.6 Fill in the missing elements in the periodic table below.

Na	Mg											Al	Si				Ar
K	Ca	Sc	Ti	V	Cr	Mn	Fe	Co	Ni	Cu	Zn	Ga	Ge	As	Se		Kr
Rb	Sr	Y	Zr	Nb	Mo	Tc	Ru	Rh	Pd	Ag	Cd	In	Sn	Sb	Te		Xe
Cs	Ba	Lu	Hf	Ta	W	Re	Os	Ir	Pt	Au	Hg	Tl	Pb	Bi	Po		Rn
Fr	Ra																

1.2 Periodic trends

1.2.1 The periodic table in organic chemistry

The periodic table is intimidating to beginning chemistry students. The fact is, it is pretty important. There is no need to memorize the entire thing (although it can be done fairly easily with a few choice mnemonics). The reason it is convenient to know the placement of some of the elements on the periodic table is we can quickly get information, like electronegativity and atomic size, from some simple periodic trends.

1.2.2 Electronegativity

It's always a competition. In the world of atomic bonds, it's a tug-of-war. Two atoms are engaged in a battle for the negative electrons that are between them making the bond(s). Sometimes, the atoms are equally matched in the tug-of-war. For instance, in the hydrogen molecule, H-H, both hydrogen atoms have the same electronegativity and are equally matched so they equally share the two negative electrons in the bond. We call this a nonpolar bond. But, they are not always equally matched. Sometimes, one of the atoms is a little "stronger" and begins to win the tug-of-war. This "stronger" atom pulls the negative electrons a little closer to itself. Below, we have part of an organic compound, the carbonyl. In it, a carbon atom and an oxygen atom are in a tug-of-war over the negative electrons in the bonds between them.

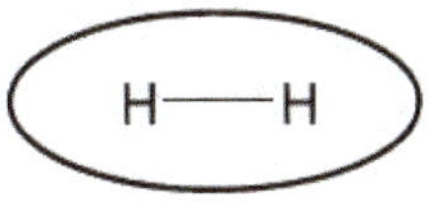

Nonpolar

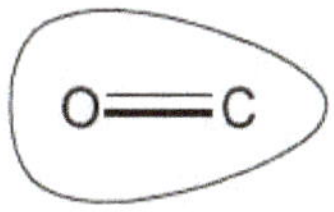

Polar

In this example, O is a little "stronger" or pulls the electrons more than C. In chemistry jargon, oxygen (O) is more electronegative than carbon (C). In simple terms, *electronegativity* is a measurement of the strength of an atom to attract to itself electrons in a bond. Because O is more electronegative than C, O has a little more negative electron density around it than does C. Unlike H-H, which is nonpolar, we have an unequal distribution of electron density. Therefore, we have one atom with more of a negative charge on it and one with more of a positive charge on it.

1.7 Analyze this carbonyl bond.

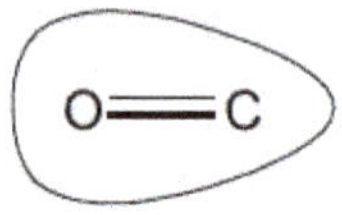

The atom that has a more positive charge on it is (O or C)? _________

The atom that has a more negative charge on it is (O or C)? _________

1.8 For the C=O bond below, circle the partial positive charge (δ^+) over the partially positive charged atom and circle the partial negative charge (δ^-) over the partially negatively charged atom.

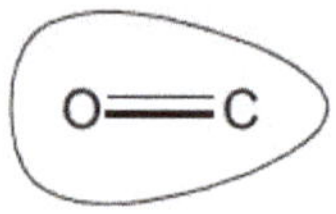

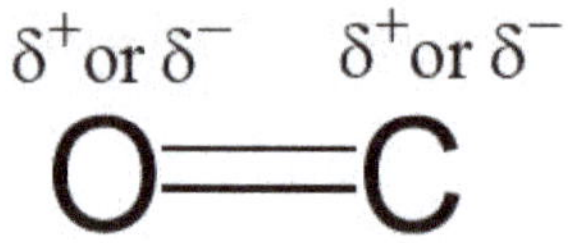

Since negative charges often attack positive charges in organic chemistry, knowing which atoms are more positive and which are more negative will help us out.

If a negatively charged molecule is going to attack C=O, which atom do you think it would attack? The answer is the same as the atom it is most attracted to. Remember the old adage, opposites attract.

1.9 Finish drawing an arrow from our negative pair of electrons below to the atom in C=O it would attack. I've started the arrow for you at the negative electrons.

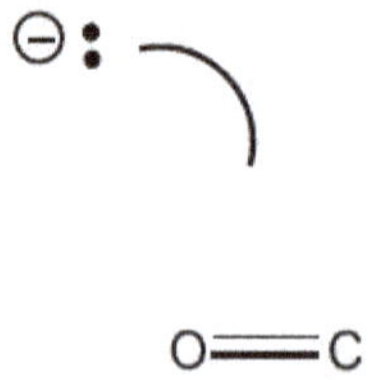

If you drew the arrow attacking the more positive atom, C, you were correct. Opposites attract. Negatives attack positives in organic chemistry. If you understand this, you are a long way towards understanding organic chemistry. Quite often, if you don't know how a reaction will proceed, this will bail you out. Reactions often occur at the most negative spot on one molecule with the most positive spot on another molecule. Often, it is our job to find those spots. That, my friends, is one reason why we need to understand electronegativity. Negative attacks positive. Opposites attract. Beautiful. Easy. But, which atoms are more electronegative?

There are different ways to measure electronegativity. Linus Pauling developed the most common method used. Pauling died in 1994 at the age of 93 after winning two Nobel Prizes. He won one in chemistry (1954) and one in peace (1962) for advocating the end of nuclear testing. Pauling's scale gives the most electronegative element, fluorine (F), an electronegativity value of 4.0. All other elements are less than this.

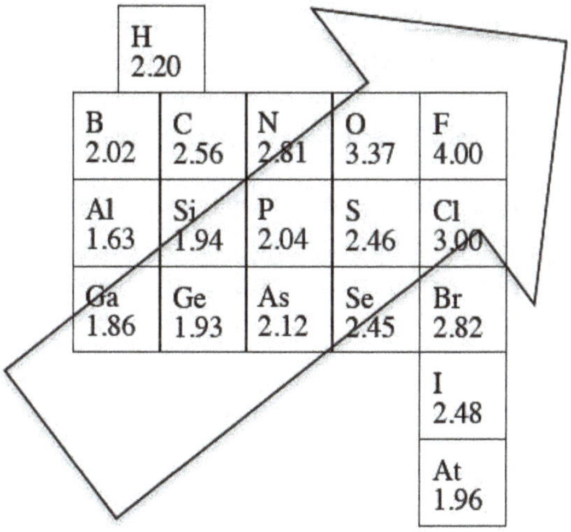

Figure 1.6 Electronegativity chart

The closer an element is to fluorine on the periodic table, the more electronegative it is. That means it is "stronger" in the tug-of-war for the negative electrons in the bond and therefore has a partial negative charge. As we go towards the upper right of the periodic table (closer to fluorine) atoms become more electronegative. See the numbers in the chart above. Hydrogen has an electronegativity of 2.2, so it is slightly less electronegative than carbon (C) and more electronegative than boron (B). Notice that most of the halogens are also more electronegative than carbon. A bond between two atoms that has one atom that is more negative and one atom that is more positive is called a polar bond. We say that atom has a dipole moment. We can easily show which atom is more negative by drawing a dipole moment arrow (+→) above the bond. The head of the arrow points towards the more negative atom, and the plus part of the arrow is, of course, above the more positive atom.

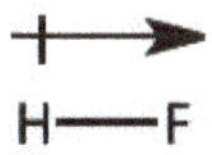

1.10 Above each bond below, draw dipole moment arrows pointing the appropriate directions. Remember, the arrow points towards the atom with a greater density of negative electrons around it!

B-O C-N F-H C=O C-Li H-C H-O Cl-C C-F P-Cl S-O B-H

Sometimes in organic chemistry, you will encounter a situation where an atom has a negative charge on it. Some atoms hate to have a negative charge on them, while others are happier with the negative charge. One thing that dictates how happy an atom is with a negative charge is its electronegativity. More electronegative atoms are more likely to have a negative charge.

1.11 Rank the following ions from 1-6 (1 is the best) for how happy, or stable, the following negatively charged anions are. Li^- B^- C^- N^- O^- F^-

___ ___ ___ ___ ___ ___

1.2.3 Atomic Size

Electronegativity is one reason we need to know the periodic table. Another reason is the size of atoms. When a chemist is talking about the size of an atom, they are referring to the size of the electron cloud surrounding the nucleus. The general trend is atoms get smaller as you go to the right in a horizontal row of the periodic table. For instance, as we move from nitrogen to oxygen, we add one more electron to the outermost shell (the 2s) and one more proton to the nucleus. Since we added the electron to the same outer shell, we might expect oxygen to be the same size as nitrogen. But, since oxygen has one more positive proton in the nucleus, it pulls more on the negative electrons in the outermost shell than nitrogen does making it ever so slightly smaller.

As we go down a row in the periodic table, we add more positive protons to the nucleus once again. But, as we go down the periodic table, we are adding the negative electrons to entirely new shells, so the atoms are bigger. This effect is quite a bit bigger than when we go horizontally on the table. The general trend is atoms get bigger, much bigger, as we go down the periodic table.

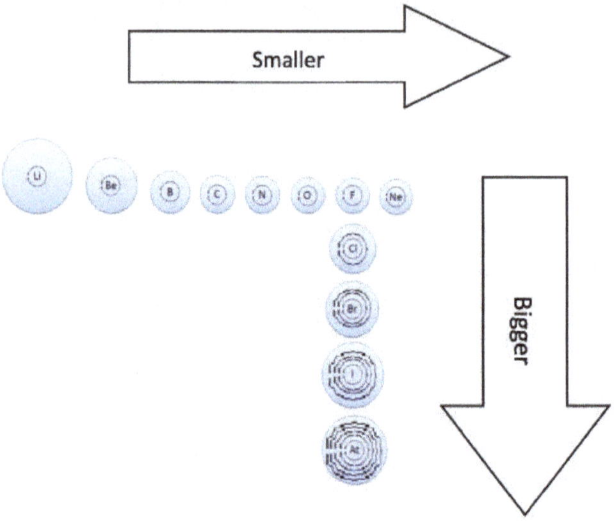

Figure 1.7 Atomic Size chart

1.12 Of each pair of atoms, circle the one that is bigger. Remember to have no worries!

a) C or O b) Be or F c) O or N d) N or C

1.13 Of each pair of atoms, circle the one that is bigger.

a) Cl or Br b) I or Cl c) F or Cl

d) Br or F e) P or N f) O or S

The size of an atom can help us predict how stable it is with a negative charge on it. In nature, positive or negative charges that are localized to small areas are bad. The more we can spread out a charge on an atom, the happier the ion will be. A negative charge can spread out more over big atoms than it can on small atoms. So, the bigger the atom, the happier it would be with a negative charge.

For instance, Br^- is more stable than F^- because Br^- is larger.

1.14 For each pair of anions, circle the one that is happier (or more stable).

a) Cl^- or Br^- b) I^- or Cl^- c) F^- or Cl^-

One might think that we have two competing ideas when determining the stability of a negative ion. Br^- is larger than F^-, so it is more stable, but F is more electronegative than Br so one might think that F^- would be more stable. In fact, we find that being a bigger atom is more important than the electronegativity of an atom when determining anion stability. It is really important to spread out the negative charge as much as we can. Nature abhors small, localized charges.

We use electronegativity when we are comparing two anions in the same horizontal row of the periodic table and we use atomic size when we are comparing two anions in the same vertical column of the periodic table.

1.15 For each pair of anions, circle the one that is happier (or more stable).

a) N^- or O^- b) O^- or S^- c) O^- or F^-

1.3 Problems

1.16 Strictly using your memory and the mnemonics you have learned, fill in the missing elements in the periodic table below. Practice until you can easily do this.

Na	Mg											Al	Si				Ar
K	Ca	Sc	Ti	V	Cr	Mn	Fe	Co	Ni	Cu	Zn	Ga	Ge	As	Se		Kr
Rb	Sr	Y	Zr	Nb	Mo	Tc	Ru	Rh	Pd	Ag	Cd	In	Sn	Sb	Te		Xe
Cs	Ba	Lu	Hf	Ta	W	Re	Os	Ir	Pt	Au	Hg	Tl	Pb	Bi	Po		Rn
Fr	Ra																

1.17 Draw in appropriate dipole moment arrows above the following bonds. Make sure they point the correct direction.

O=C C≡N F-C Br-C O-C C-Cl

1.18 Identify which member of each pair has the more stable anion.

a) Cl^- vs. Br^- b) O^- vs. F^- c) I^- vs. F- d) O^- vs. C^-

1.19 Identify the larger anion in each pair.

a) I^- vs. Br^- b) I^- vs. Cl^- c) Br^- vs. Cl^- d) O^- vs. S^-

Chapter 2. Bonding and Structure

2.1 Electron waves

The world really is strange. It is stranger than most people realize. Back in your Junior High Science class, you were probably told about the duality of light, where light is both a particle and a wave. This wave/particle duality is not only true about light, but it is true about most of nature. This is the strangeness that we explore in a field called quantum mechanics.

For now, I want us to focus on the wave/particle duality of electrons. One way to think about electrons is that they are little particles orbiting the nucleus. This is a fairly good model that many of us have in our minds, the Bohr planetary model. But, the interesting thing is that electrons are also waves! Let's focus on two very simple waves. Think of a guitar string. If I pluck a guitar string, it vibrates up and down.

If I combine these, the string vibrates so fast that it looks something like this.

This simple vibrational shape is spherical in nature and leads us to the shape of the simplest electron orbital, the s orbital.

Figure 2.1 The s orbital

If I put my finger down in the middle of the guitar string and pluck it, I get a wave where when one side is up, the other side is down. A moment later, it switches which side is up and which is down.

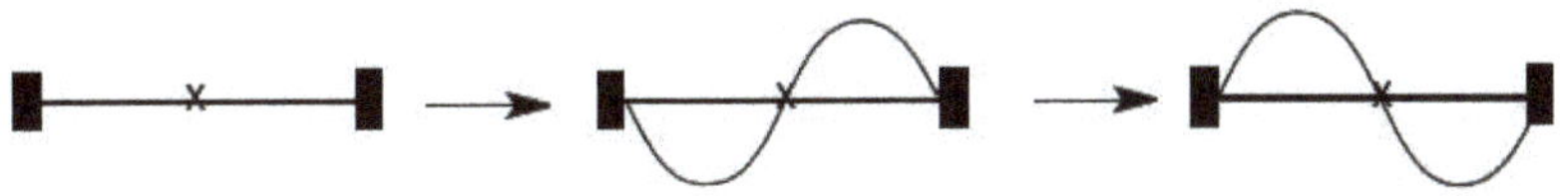

When the string vibrates fast, it looks something like this.

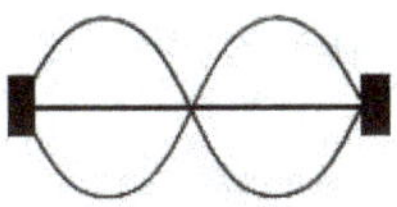

There is one spot in the center where there is no wave. We say that it has one node in the center. This vibrational shape has two lobes of equal size. It leads us to the shape of another electron orbital, the p orbital.

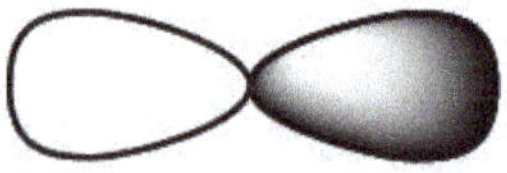

Figure 2.2 The p orbital

These orbital waves can be described with mathematical functions—just like all waves can. We call this mathematical function for the wave the wavefunction which is often represented by the greek letter psi (Ψ). These mathematical functions can be added together or subtracted. Intuitively, you understand this about waves.

For example, a child on a swing set moves forward and backward. This motion is a wave-like motion. When children are learning to swing on a swing set, they are taught to kick their legs out when they are moving forward and bend at the knees and put their feet down when they are moving backward. This motion of their feet is also a wave. They are taught to do this so that their feet wavefunction can be added to the wavefunction of the swing. *Constructive interference* is when two waves overlap to make a larger wave. When done correctly, the child's feet motion and the swinging motion constructively interfere to increase the amplitude of the wave, and their swinging, so they go higher and higher.

Photo by GaborfromHungary at Morguefile.com

Many children get confused when learning this, though. If instead, they kicked their feet straight out when they were going backward and bent their knees putting their feet down when they were going forward, their feet wave would be opposite the swing's wave. *Destructive interference* is when two waves are out of phase with each other making smaller waves. Children who swing this way will not be swinging for long. When the wavefunctions destructively interfere, they are subtracted from each other and get canceled out.

Two electron orbitals can add constructively or destructively because they are waves. This can happen when orbitals overlap on the same atom (to make hybrid orbitals) or when orbitals on two different atoms overlap when the atoms come close together (to make molecular orbitals or bonds).

2.2 Atomic orbitals

A hydrogen atom (atomic number 1) has only one electron surrounding one positive proton in the nucleus. This lone electron is in a round orbital called a 1s orbital and is in the first and only shell of electrons of the hydrogen atom. The outermost electrons of an atom are called its valence electrons. Since this first shell is also the outermost shell of electrons, this 1s electron is the *only* valence electron for hydrogen.

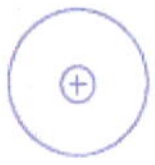

Figure 2.3 Hydrogen atom, one shell, 1s orbital

The first shell is small and can only hold two electrons in it. Helium (atomic number 2) fills up this first shell with its two electrons, both in the 1s orbital. Another shell of electrons outside of this first shell is needed for the other important organic chemistry atoms like carbon, nitrogen and oxygen because they contain more than two electrons. Therefore, these other important organic chemistry atoms have two shells of electrons.

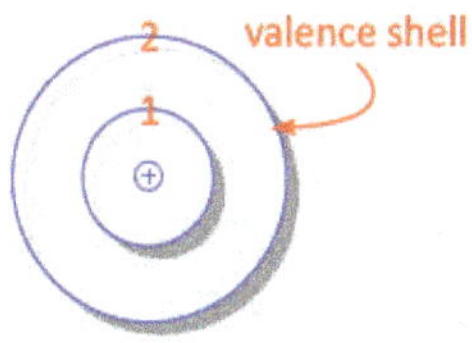

Figure 2.4 The two electron shells of common organic atoms

For these atoms, the first, innermost shell of electrons is made up of two electrons in a round orbital called a 1s orbital. These two electrons are not very interesting because they are tucked inside the atom and don't interact with other atoms. The second shell is farther away from the positively charged nucleus. It is more interesting because it contains the outermost electrons of the atom. These electrons are the ones that will bump into other atoms to form bonds. This outermost electron shell is called the valence shell and its electrons are the valence electrons. It is also made up of two electrons in a round orbital called a 2s orbital. It is the same shape as the 1s orbital, but just a little bigger as the electrons are a little farther from the nucleus in the second shell.

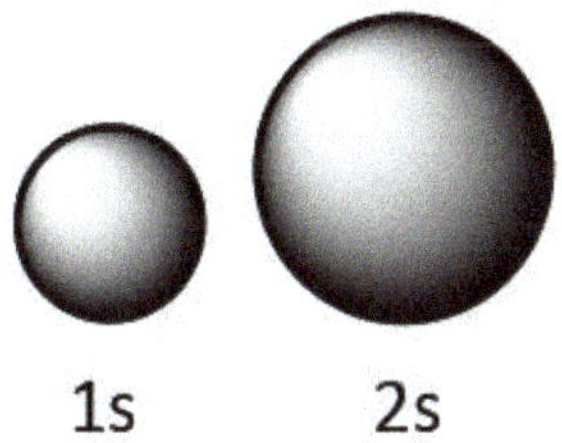

The round 2s orbital is not the only type of electron orbital in the 2nd shell. There are also three double-lobed 2p orbitals in this valence shell. These three p orbitals are all rotated 90° from each other.

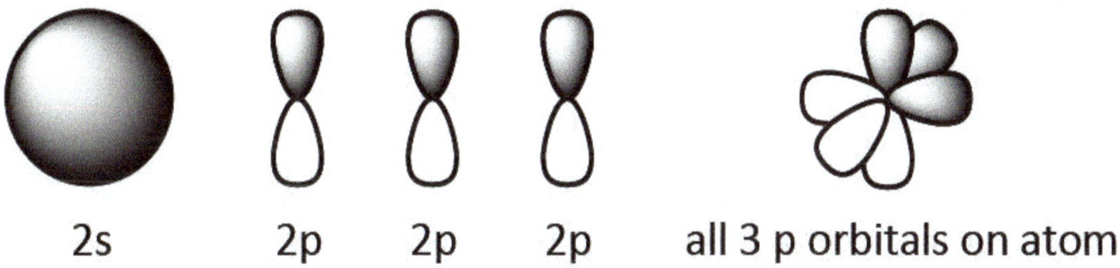

Figure 2.5 Atomic orbitals

They all have the number "2" in them because they make up the second shell – that is, the meatiest parts of these orbitals are all about the same distance away from the nucleus, where the 2nd shell is located.

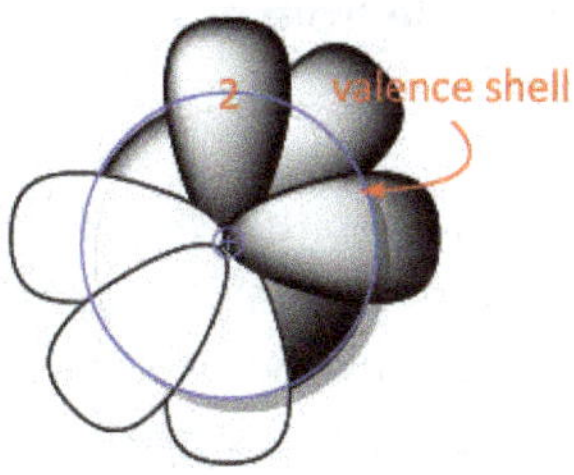

Figure 2.6 Showing how 2s and 2p orbitals make the valence (2nd) shell

This is what the atomic orbitals of carbon atom look like. To make bonds between atoms, the outermost valence electrons of one atom must overlap with the outermost valence electrons of another atom to make bonds. If the atomic orbitals described above were the only electron orbitals that atoms contain, we would expect that we could only bond atoms with a carbon atom at 90° and 180° angles. This is not the case. We often see bond angles of 180°, 120°, and 109.5°, but not 90°. Let's look at methane (CH_4) if only atomic orbitals can overlap compared to its true, tetrahedral shape.

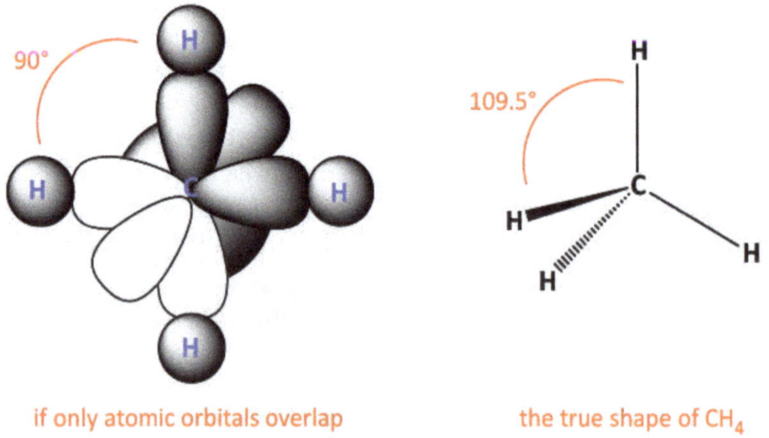

Figure 2.7 CH_4 if only atomic orbitals overlap vs. its true shape

Something else must be going on to get these other bond angles. Because all electron orbitals are waves, we can get these other bonding angles when the s and p atomic orbitals that are on the same atom mix together to form new oblong electron orbitals that are called *hybrid orbitals.* A hybrid is the offspring of two different species. So, hybrid orbitals are the offspring of two different types of orbitals on the same atom. It is important that you realize we are mixing together electron waves on *the same* atom because mixing together orbitals that are on *different* atoms is something else entirely, something we call making bonds.

2.3 Hybrid orbitals

What are they?

This is how we mix atomic orbitals on the same atom together to make hybrid orbitals. We mix together the outside, valence electrons. The number of atomic orbitals that mix together is the exact same number of hybrid orbitals that are created, always. So, if two atomic orbitals (2s and 2p) are mixed together, that will make two hybrid orbitals. Those hybrid orbitals would be called sp hybrid orbitals because they are made from one s and one p orbital. If three atomic orbitals (2s, 2p, 2p) are mixed together, that will make three hybrid orbitals. Those hybrid orbitals would be called sp^2 hybrid orbitals because they are made from one s and two p orbitals. If four atom orbitals (2s, 2p, 2p, 2p) are mixed together, that will make four hybrid orbitals. Those hybrid orbitals would be called sp^3 hybrid orbitals because they are made from one s and three p orbitals.

Their shapes

Like charges repel each other. Since these hybrid orbitals all contain negative electrons, they repel each other. This means the hybrid orbitals that are formed get as far away from each other as possible. If all four valence atomic orbitals (2s, 2p, 2p, 2p) are mixed, four sp^3 hybrid orbitals are made. When these four orbitals get as far away from each other as they can, they form a tetrahedral shape with an angle of 109.5° between them. This is what happened in our methane, CH_4, example earlier. If three of the four valence atomic orbitals are mixed (2s, 2p, 2p), then three sp^2 hybrid orbitals are made. When these three orbitals try to get as far away from each other as they can, they form a trigonal planar shape with an angle of 120° between them. If only two of the four valence atomic orbitals mix (2s, 2p) then two sp hybrid orbitals are made. When these two hybrid orbitals try to get as far away from each other as they can, they form a linear shape with an angle of 180° between them.

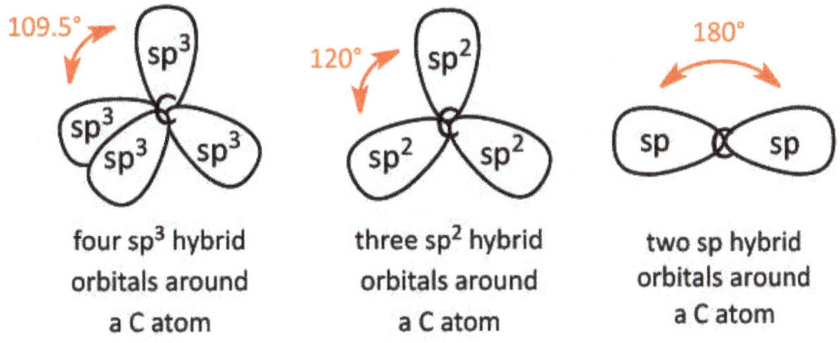

Figure 2.8 Geometry of hybrid orbitals

The unhybridized p orbitals

We are almost there, but this is not the complete picture. Most of the important organic chemistry atoms (outside of hydrogen) start with one s and three p valence atomic orbitals—four total. So, if we start with four atomic orbitals, we must end up with four total orbitals in the end, always. We must not forget about the atomic orbitals that are *not* mixed together to make the hybrids. When we make sp^2 hybrid orbitals, we mix three out of the four orbitals, the s, and two p orbitals to make three sp^2 hybrid orbitals. There is a p atomic orbital left over, untouched, that is not mixed at all! It is unchanged. We call it an unhybridized orbital because, well, it's not hybridized. It is still on the atom as a regular old p orbital. When we make sp hybrid orbitals, we mix two out of the four atomic orbitals, the s and p to make two sp hybrid orbitals. There are two p orbitals that remain untouched or unhybridized.

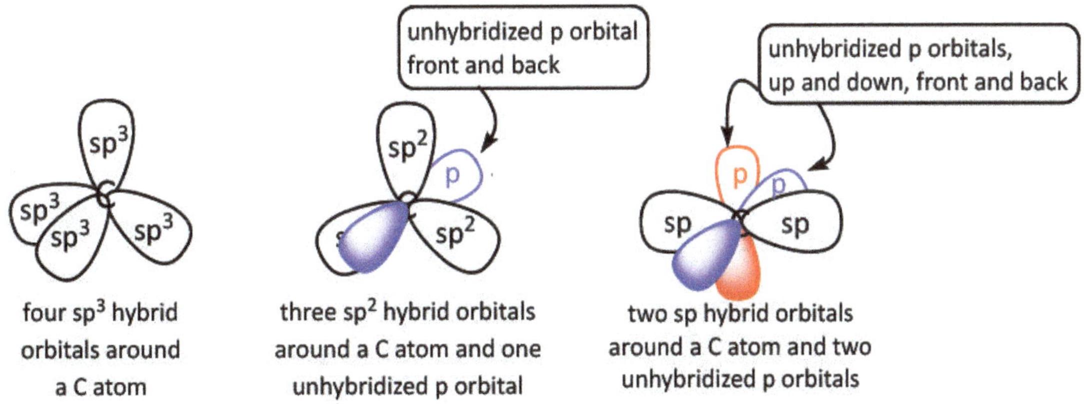

Figure 2.9 The unhybridized p orbitals with hybrid orbitals

2.3.2 What is the hybridization of an atom?

How does one determine what the hybridization (sp, sp^2, or sp^3) is for an atom? You need to count how many electron groups are on the atom. Lone pairs of electrons, even though they don't point to an atom, count as electron groups since they do take up space. They must be in a hybrid orbital. Additionally, single bonds, double bonds and triple bonds each count as only one electron group since they each point to one atom. Alternatively, you can count the atoms at the ends of those single, double, or triple bonds.

We can determine the hybridization for an atom after we have counted the number of electron groups on it. If *two* electron groups are on an atom, it needs two hybrid orbitals – one for each electron group. Such an atom has sp hybridization because sp hybridization is when one s and one p orbital overlap to make *two* sp hybrid orbitals. If *three* electron groups are on an atom, it needs three hybrid orbitals. Such an atom has sp^2 hybridization because sp^2 hybridization is when one s and two p atomic orbitals mix to make *three* sp^2 hybrid orbitals. If *four* electron groups are on an atom, the atom has sp^3 hybridization because sp^3 hybridization is when one s and three p atomic orbitals mix to make *four* sp^3 hybrid orbitals.

Let's look at an example, the hybridization of the carbon and oxygen atoms of formaldehyde. Let's analyze the oxygen atom first. We notice that the oxygen atom has two lone pairs of electrons (1 and 2). We always need to remember that electron pairs take up space and need to go into a hybrid orbital. The oxygen atom also has one atom attached (the carbon on the double bond labeled 3). This is a total of three electron groups, so the oxygen atom needs three hybrid orbitals. It is therefore sp^2 hybridized (one s and two p orbitals gives three sp^2 hybrid orbitals). When we analyze the carbon atom's hybridization, we notice there are no electron pairs of electrons on the carbon, but there are three atoms attached to it. These three atoms are the oxygen atom at the end of the double bond (1) and the two hydrogen atoms (2 and 3). With three electron groups on the carbon atom, it is also sp^2 hybridized.

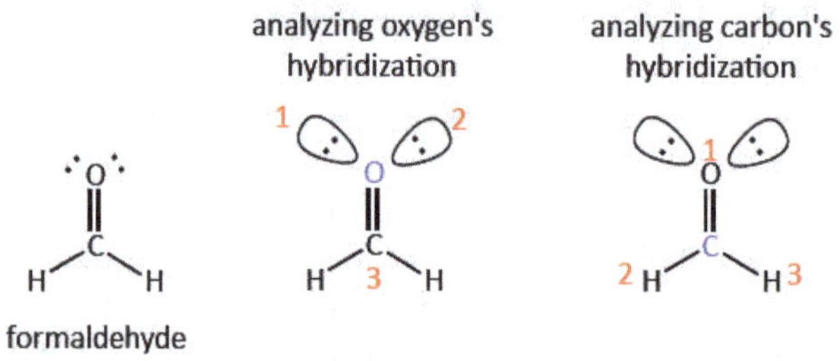

Figure 2.10 Hybridization of atoms in formaldehyde

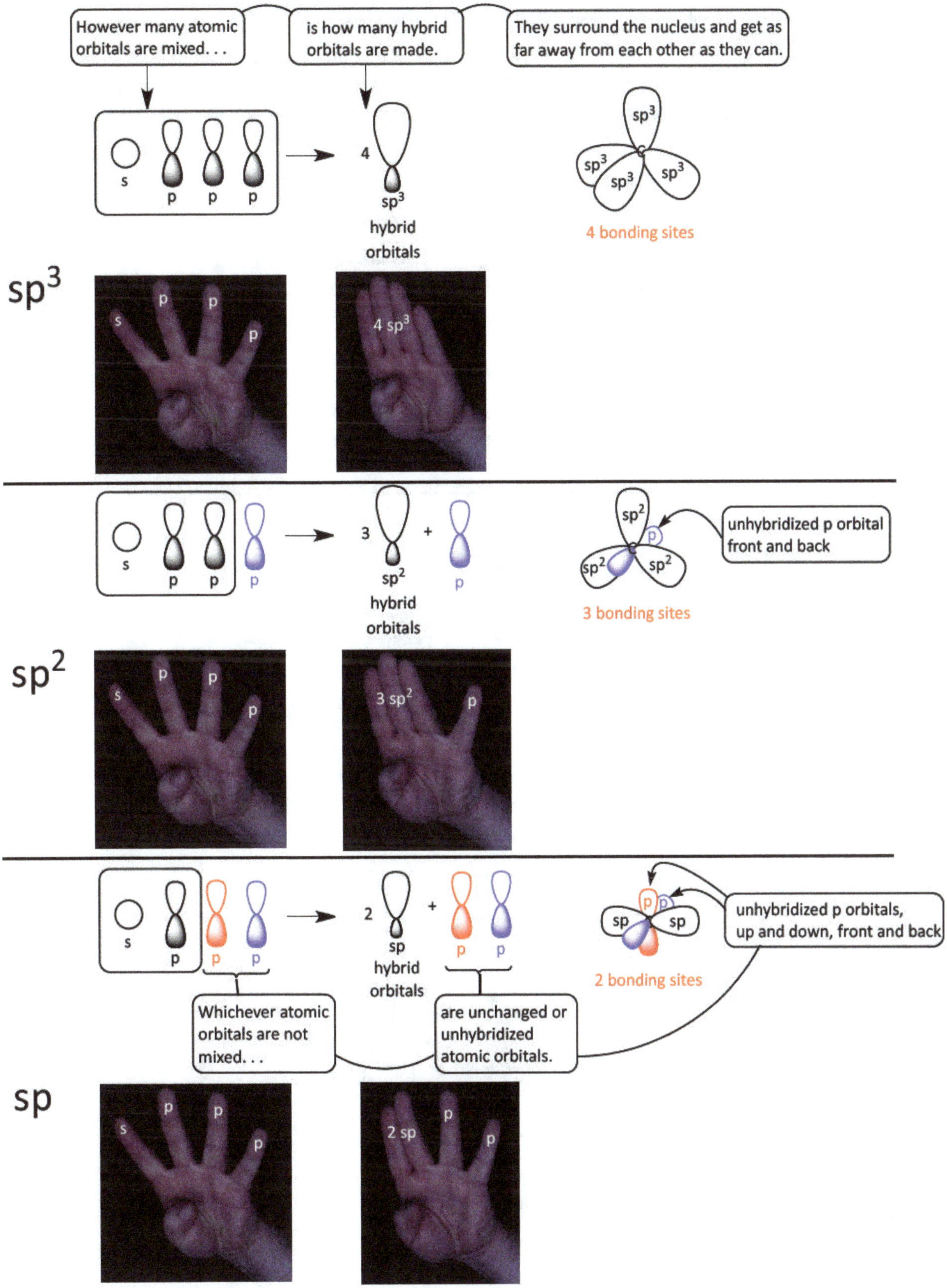

Figure 2.11 Hybrid orbitals

2.3.3 Hybridization of atoms in some common bonding patterns

For neutral carbon atoms in most organic chemistry molecules, if it contains only single bonds, it is sp^3 hybridized. If a carbon atom contains one double bond, it is sp^2 hybridized. If a carbon atom contains one triple bond, it is sp hybridized. Below are common hybridizations and show how to count the electron groups.

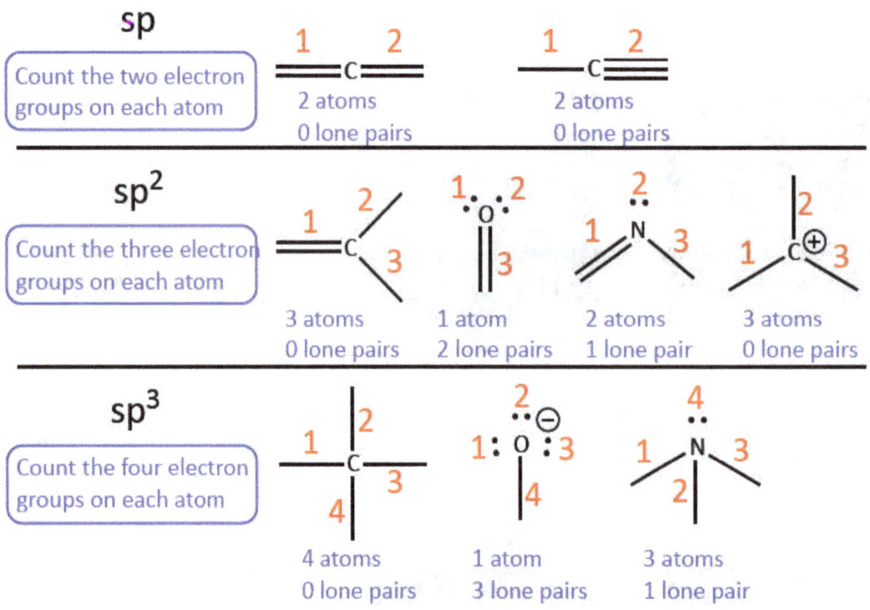

Figure 2.12 Common hybridization bonding patterns

2.1 Identify the hybridization on the central atoms of the amino acid, histidine.

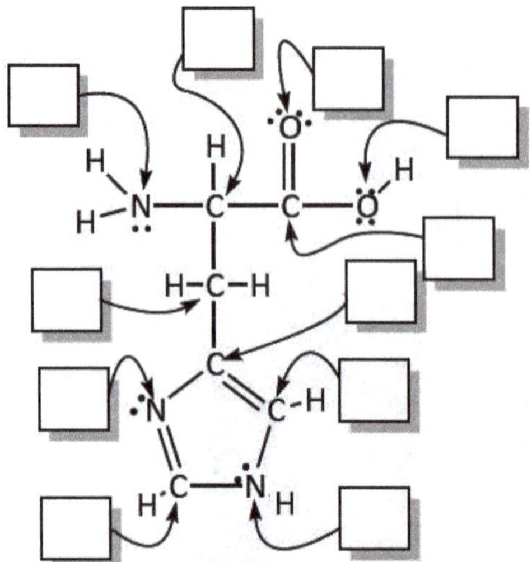

Histidine

2.4 Molecular orbitals

2.4.1 Sigma and Pi Bonds

When two atoms come together, the electron orbitals on the atoms can overlap. This constructive overlap of electron density between two atoms makes bonding molecular orbitals, also called bonds. There are two main types of bonds we will deal with in organic chemistry, sigma and pi bonds. A sigma bond is directly between two atomic nuclei. Or, a more technical way to describe it is the electron density lies on the internuclear axis. A pi bond has electron density above and below the internuclear axis.

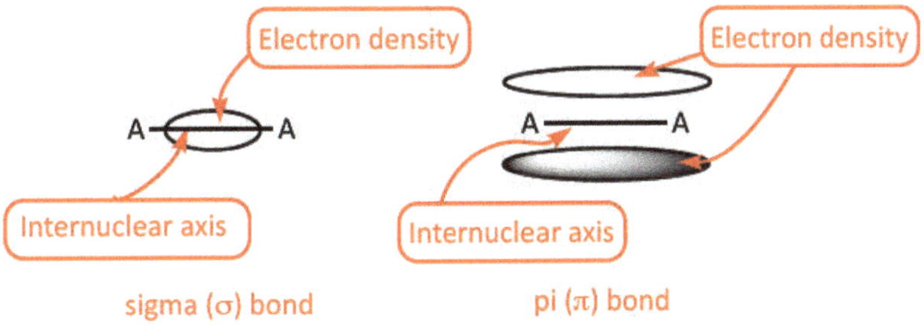

Figure 2.13 Sigma and pi-bonds

A mnemonic to help you learn these at first is to think of the first, single bond between two atoms as a snake, the sigma snake. This can help you think of a sigma bond. The second bond is split into two regions, above and below the internuclear axis. A pi bond may remind you of a *pie* crust.

Figure 2.14 Sigma and pi-bond mnemonics

We've learned that hybridization is a mixing of atomic orbitals on one atom. We need to look at how two or more atoms come together to make bonds and form molecules. We mix hybrid and/or atomic orbitals to make these molecular orbitals. These bonds that we form between atoms are called molecular bonds.

> *Atomic orbitals* are s and p orbitals on individual atoms.
>
> *Hybrid orbitals* are sp, sp^2, s p^3 orbitals that are mixed from the s and p atomic orbitals on the same atom.
>
> *Molecular orbitals* are a mixing of atomic and hybrid orbitals between different atoms to make bonds between those atoms. This is how we make molecules.

2.4.2 Alkane bonding

Let's look at how some simple molecules are put together. Let's begin with propane.

$$H_3C—CH_2—CH_3$$

Let's redraw it in 3D. We see that each of the carbon atoms has four bonds and is sp^3 hybridized.

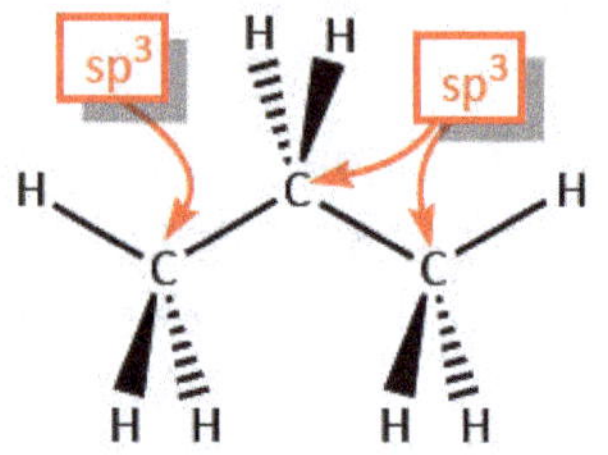

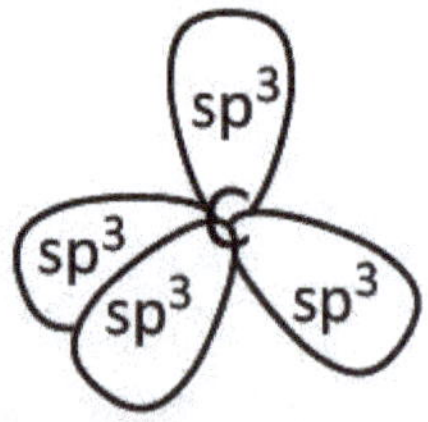

First, let's draw the central carbon atom with its four sp^3 hybrid orbitals around it. We draw it with the bottom two sp^3 hybrid orbitals pointing towards the bottom right and bottom left so they are pointing towards where the other carbon atoms will be drawn. Notice, when we draw the hybrid orbitals, we typically leave off the small back lobes on each one. This makes it easier to draw and makes it less complicated.

Now, let's draw the other, outside, carbon atoms with four sp^3 hybrid orbitals around each one.

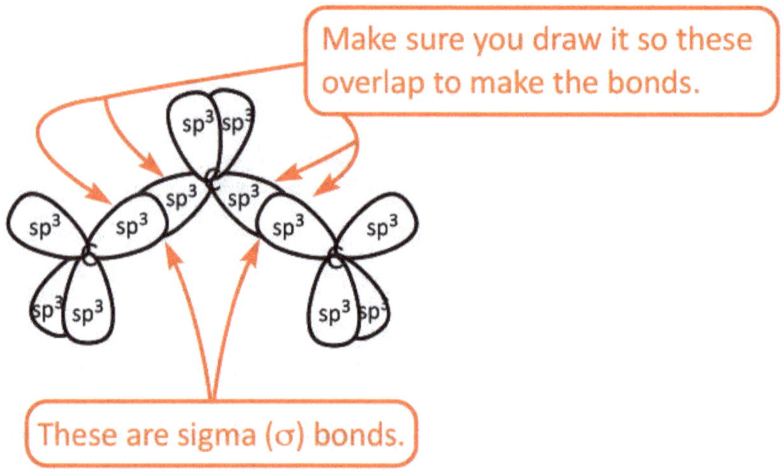

A sigma bond (σ-bond) is made by two hybrid orbitals overlapping directly between the nuclei of two atoms.

Finally, we draw in the hydrogen atoms. A hydrogen atom cannot make hybrid orbitals because it is just one 1s orbital. They are simple to draw. We draw the spherical 1s orbitals for each hydrogen atom. Notice, they overlap with the hybrid orbitals on the carbon atoms.

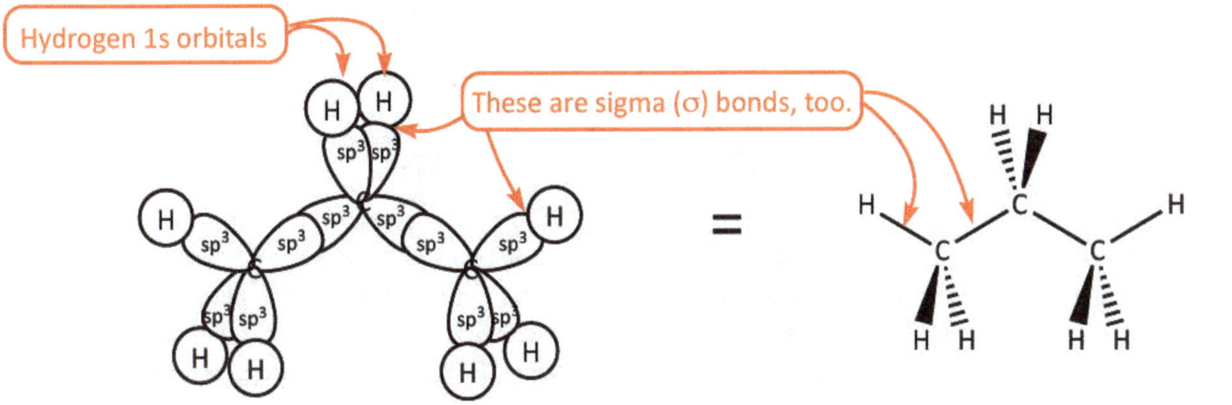

Figure 2.15 Orbital representation for propane

2.4.3 Alkene bonding

Let's look at something with a double bond in it, ethene.

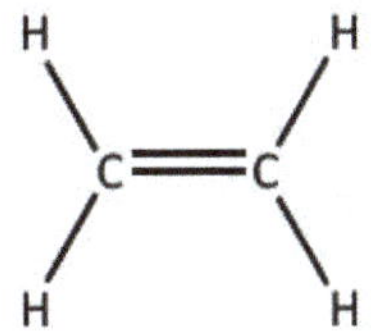

We identify the hybridization of the atoms.

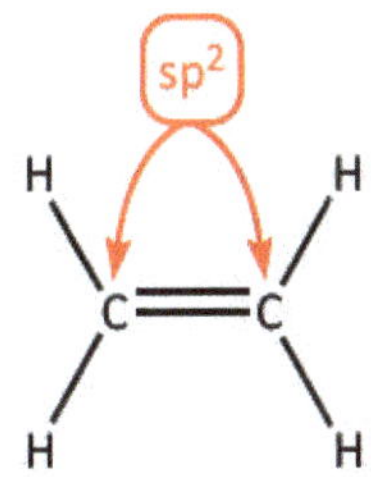

Then, we draw in the sp^2 hybrid orbitals around the carbon atoms making sure we arrange them so they overlap between the carbon atoms. If they do not overlap, they do not make a bond.

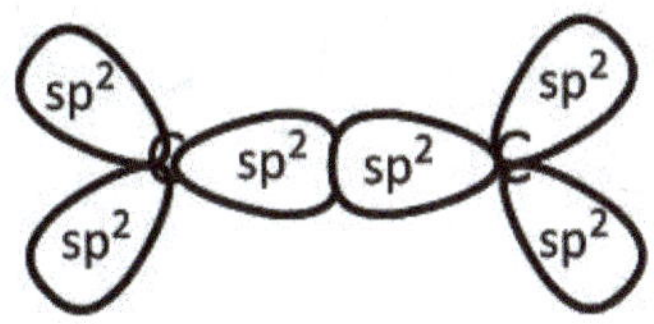

2.4.4 Pi bonds

But, whenever we have sp^2 hybridization, there is one atomic p-orbital on each carbon that remains unhybridized. If we draw those in, we see that they can overlap above and below the carbon-carbon internuclear axis. They make a pi bond (π-bond).

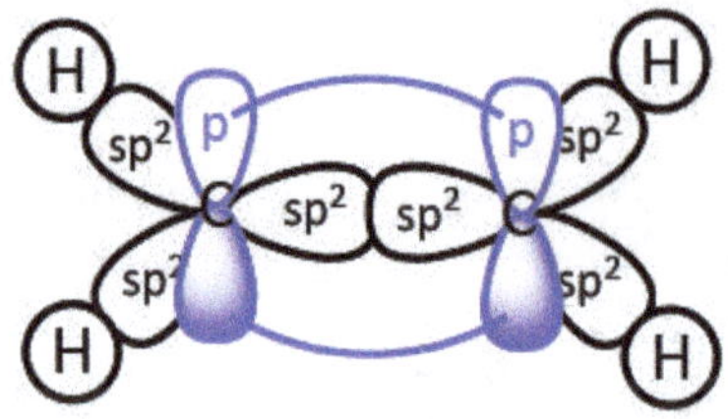

Notice, the σ-bond is the one that lies directly between the carbon nuclei. Notice the π-bond is above and below the internuclear axis (between the carbon nuclei). Half of the π-bond is above the plane and the other half is below the plane. These two parts together make one π-bond. So, there is one σ-bond and one π-bond between the carbon atoms. This is how we get a double bond. A double bond consists of one σ-bond and one π-bond.

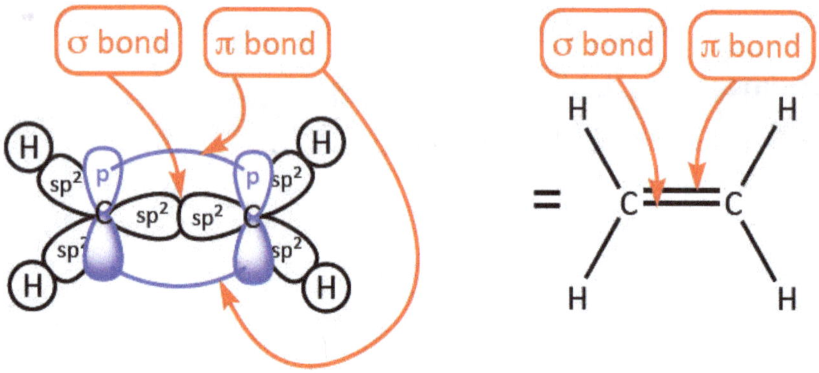

Figure 2.16 Orbital representation for ethene

2.4.5 Alkyne bonding

For acetylene, H-C≡C-H, the carbons are sp hybridized. There are two unhybridized p-orbitals on each carbon that can overlap to make two π-bonds.

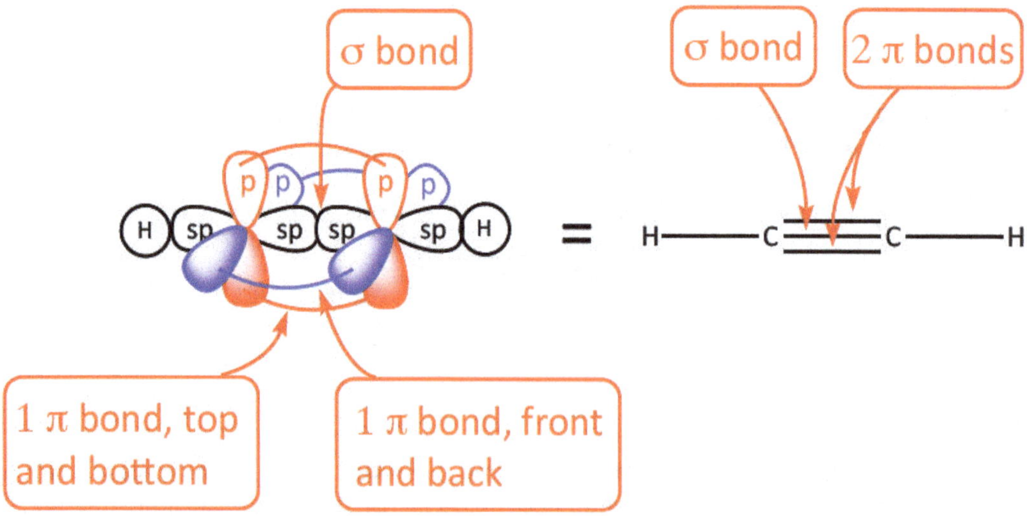

Figure 2.17 Orbital representation for ethyne

Important point

The first bond (the single bond) is a σ-bond. The second and third bonds of a multiple bond (double or triple bond) are π-bonds.

2.2 Identify the number of sigma bonds and pi bonds at the indicated bonding sites.

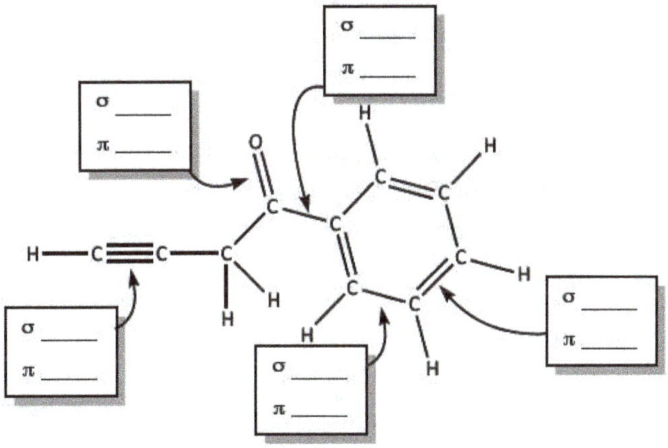

2.3 Draw an orbital representation for the cyanide ion, :C≡N:$^-$.

2.5 Lewis Structures

Electrons. Chemistry is all about the electrons, those negative charges floating around on the outside of atoms. In fact, the most important electrons in chemistry are those that are on the very outside of the atom. That makes sense because when two atoms approach each other to react, those are the first ones that will bump into each other. These special, outside electrons are called "valence electrons." The number of valence electrons that surround an atom may be determined by its placement in the periodic table. Hydrogen has one valence electron, carbon has four, nitrogen and phosphorus have five, oxygen and sulfur have six, and the halogens have seven.

1	2											3	4	5	6	7	8
H																	He
Li	Be											B	C	N	O	F	Ne
Na	Mg											Al	Si	P	S	Cl	Ar
K	Ca	Sc	Ti	V	Cr	Mn	Fe	Co	Ni	Cu	Zn	Ga	Ge	As	Se	Br	Kr
Rb	Sr	Y	Zr	Nb	Mo	Tc	Ru	Rh	Pd	Ag	Cd	In	Sn	Sb	Te	I	Xe
Cs	Ba	Lu	Hf	Ta	W	Re	Os	Ir	Pt	Au	Hg	Tl	Pb	Bi	Po	At	Rn
Fr	Ra																

Figure 2.18 Periodic table column numbers showing the number of valence electrons

Bonding and Structure

The normal Lewis dot structures for these atoms have the valence electrons represented as dots around the atom. One electron is placed on each of the four sides of the atom before they are paired.

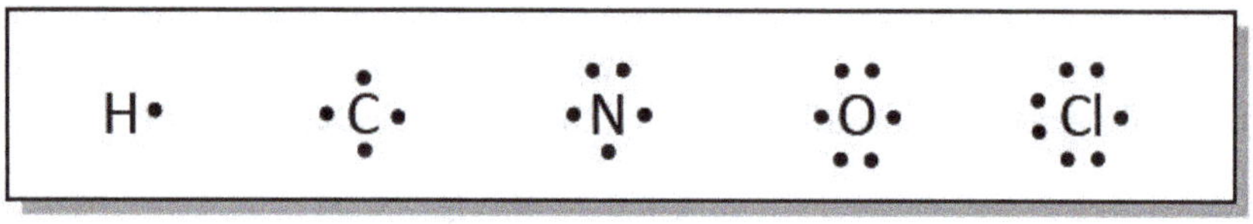

Figure 2.19 Valence electrons on organic chemistry elements

The normal bonding pattern for these atoms is with a bond at each site that has only one electron. It will be useful to know the normal bonding patterns of common atoms used in organic chemistry. This is important to aid us in drawing compounds as well as helping us recognize sites in molecules that may be the most reactive.

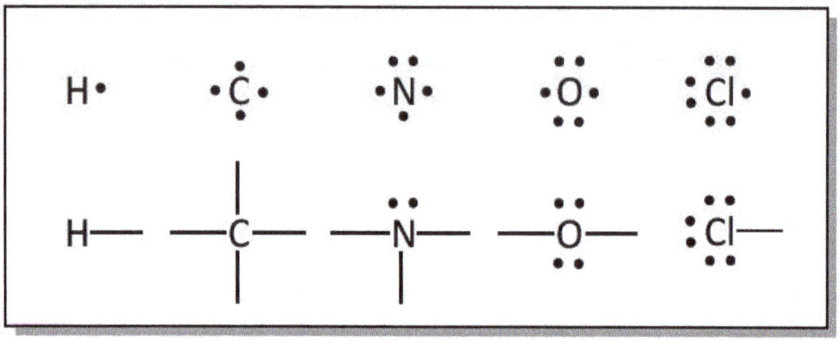

Figure 2.20 Number of bonds and lone pairs for organic chemistry atoms

Single bonds are not the only types of bonds you will see in organic chemistry. Double (=) and triple (≡) bonds are also commonly found. Because carbon, nitrogen, and oxygen normally form more than one bond, there is more than one normal bonding pattern for each of them.

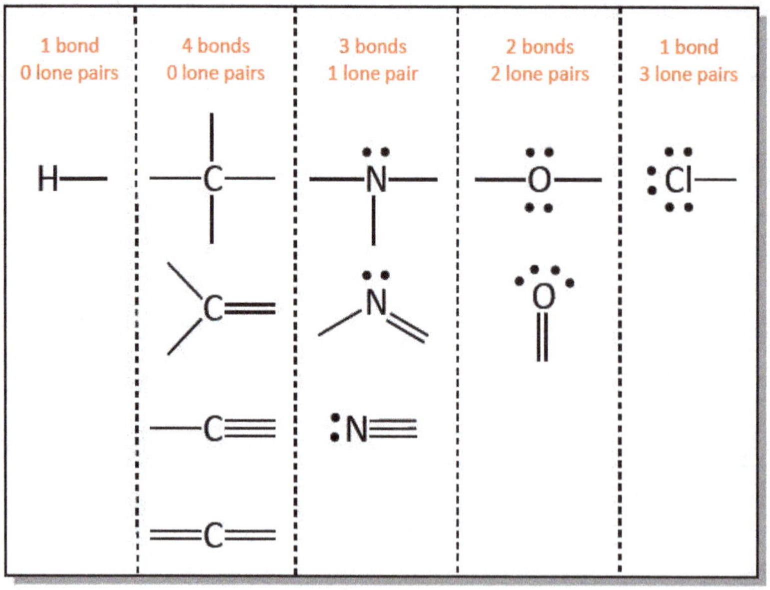

Figure 2.21 Common bonding patterns for organic chemistry atoms

There are four different bonding patterns for carbon. This is one of the reasons so many carbon-based, organic, compounds exist.

2.4 In each of the following compounds, the atoms are drawn, but the bonds and the lone pairs of electrons are missing. Using the normal bonding patterns for these atoms, fill in the bonds and the lone pairs of electrons. I've given a few hints in the paragraph after each problem to help you out. Feel free to check your answers at the end of the chapter AFTER you've given each one your best attempt.

a)

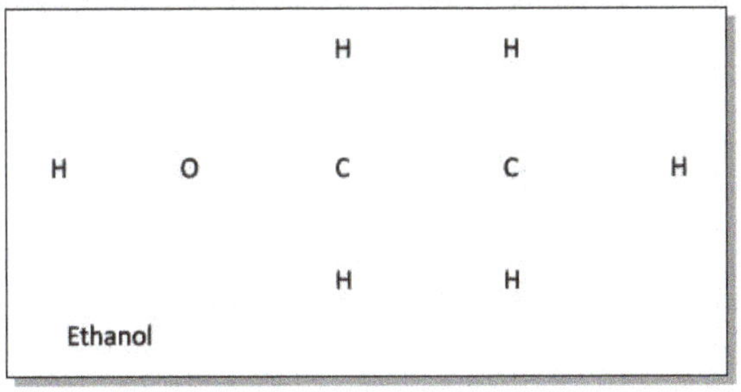

The best way to do problems like this, start off by drawing single bonds to connect all of the atoms.

Then, double check to make sure each atom has the number of bonds it normally has. If more bonds are needed for atoms to be "normal" then consider making double or triple bonds. In problem a, after all of the single bonds are drawn, every atom is "happy". The oxygen atom normally has two bonds and two lone pairs of electrons. To complete our structure, two sets of electron pairs need to be drawn on the oxygen atom.

b)

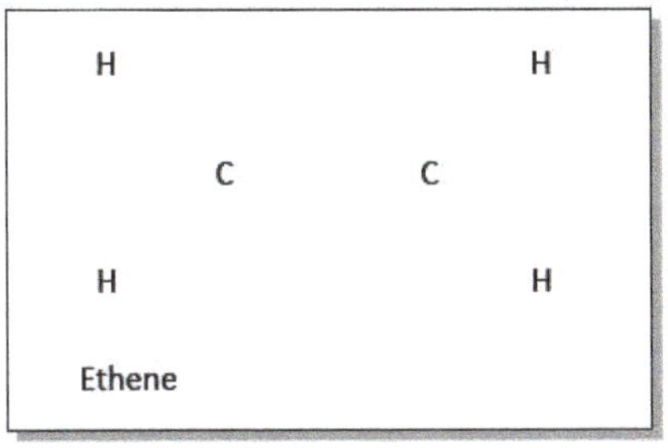

After drawing all of the single bonds in problem b, we notice that each carbon atom only has three bonds. Since each carbon atom wants four bonds, we need to make a double bond between the two carbon atoms.

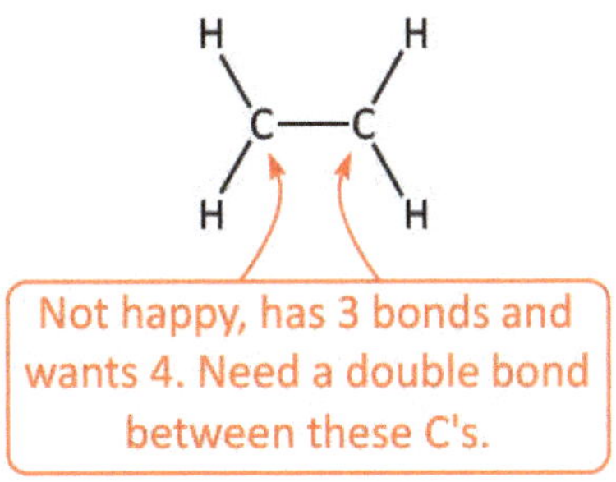

No electron pairs need to be drawn because each carbon atom usually has four bonds and no lone pairs of electrons.

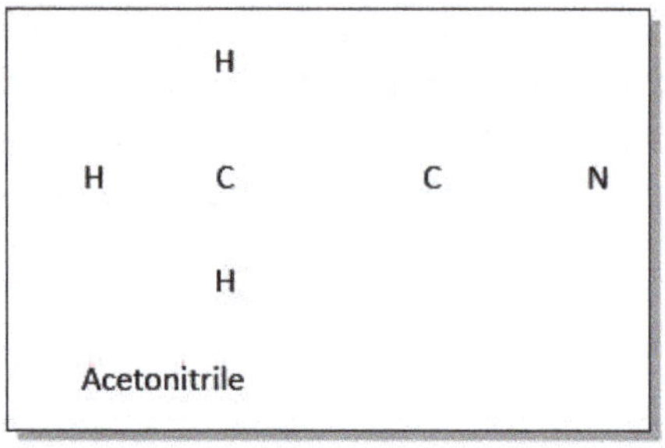
Acetonitrile

c)

After drawing in all of the single bonds in problem c, we notice that the hydrogens and the first carbon are happy, but the second carbon atom and the nitrogen atom are not. We need to draw in more bonds, but make sure you don't draw a double bond between the two carbons, or that will make the first carbon have 5 bonds! That is a mortal sin to an organic chemistry professor! So, draw in a double bond between the second C and the N. They're still not happy. Can you see what we need to do to fix this problem?

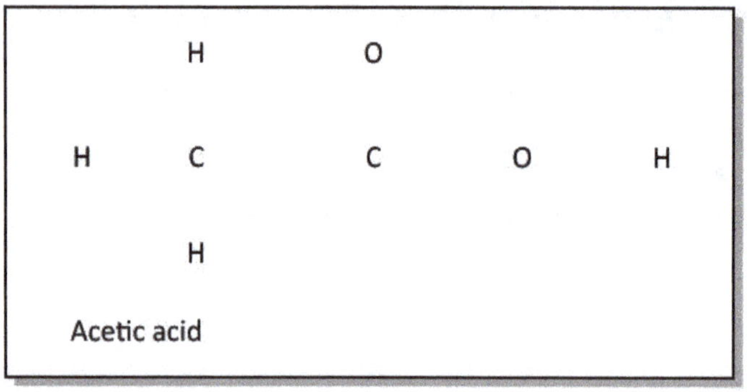
Acetic acid

d)

After drawing in all of the single bonds, we find that all of the atoms are happy except the second carbon and the top oxygen. This is a pretty easy fix by drawing in a double bond.

Try these next few problems on your own.

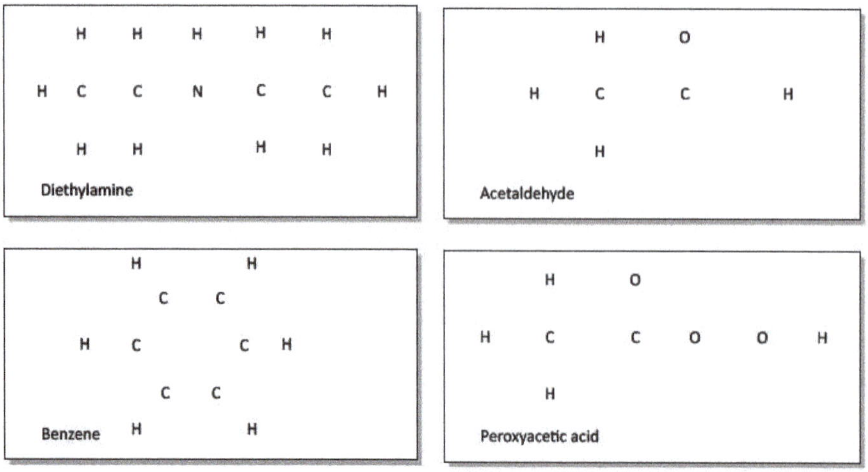
Diethylamine

Acetaldehyde

Benzene

Peroxyacetic acid

e)

2.6 Formal charges

An atom will have a negative charge if it contains an extra negative electron or two over what it normally has. An atom will have a positive charge if it contains one or two fewer negative electrons than it normally has. Chemists have agreed on a set of rules, or an accounting of electrons, to communicate how many electrons an atom has. This accounting is the atom's **formal charge.** Understanding formal charges will be useful to you later in this course. It is one way we can identify the reactive sites on molecules. It is crucial to understand formal charges in order to draw good resonance structures. Resonance structures will help you predict how a reaction will proceed.

2.5 Consider the following oxygen atom, . This oxygen atom contains _____ negative electrons around it. Oxygen's placement in the periodic table tells us that it normally contains ____ negative electrons around it. Since this oxygen atom contains one more negative electron compared to normal, it has a -1 formal charge.

This oxygen atom, . This oxygen atom contains _____ negative electrons around it. Normally, oxygen contains ____ negative electrons around it. Since this oxygen atom contains one fewer negative electron compared to normal, it has a +1 formal charge. That's pretty simple, right?

2.6 What are the formal charges of the following species? Two of the answers have been filled in for you as a self-check.

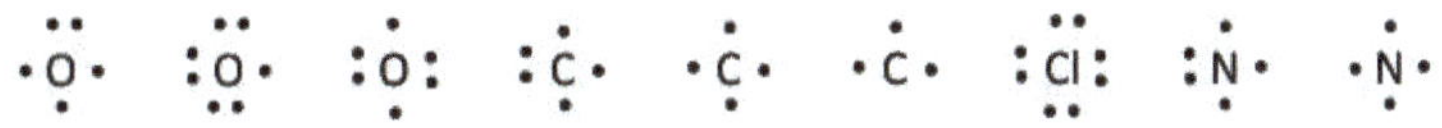

+1 ___ ___ ___ ___ ___ ___ 0 ___

Usually, you will not be assigning formal charges to nonbonded atoms like the previous exercise. Instead, you will need to analyze atoms in molecules. When you analyze atoms in molecules, the trick is to remember which electrons are available to the atom. Every nonbonded electron is available to the atom. For every two electrons available in a bond, usually represented by a line (—), one-half of those electrons are available to be counted on the atom.

Let's look at the nitrogen atom in ammonia, NH_3.

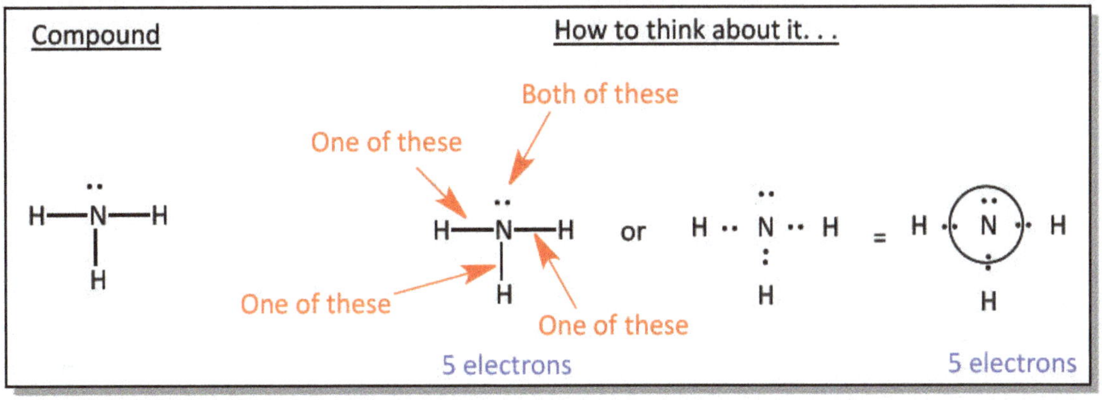

Figure 2.22 Bonding of ammonia

Normal nitrogen has five valence electrons. The nitrogen atom in ammonia has five valence electrons. Since it is normal, the nitrogen atom in ammonia has a zero formal charge. Zero formal charges are normally not written on atoms.

The negatively charged amide ion (NH_2^-) has a nitrogen atom in a different electron environment.

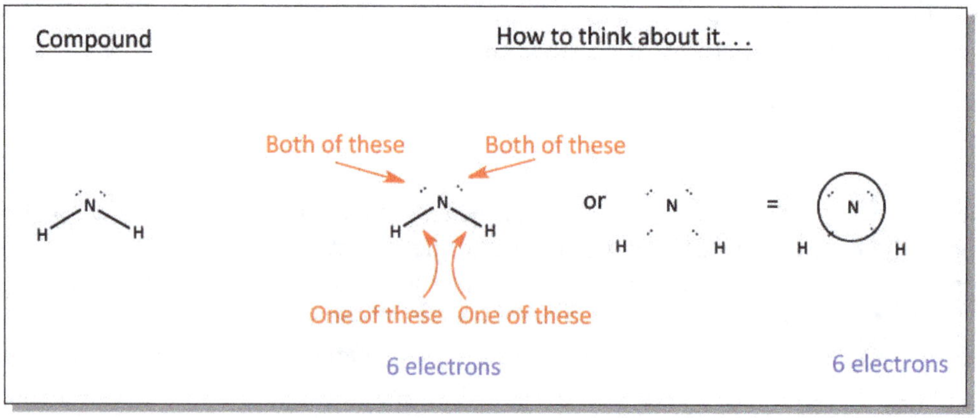

Figure 2.23 Understanding formal charges in amide

Normal nitrogen has five valence electrons. The nitrogen atom in the amide anion has six valence electrons. Since it has one extra negative electron compared to what is normal, the nitrogen atom in amide has a -1 formal charge. Since this is a non-zero formal charge, it must be written on the nitrogen atom. If the formal charge is +1 or -1, typically the "1" is omitted and only the charge is written.

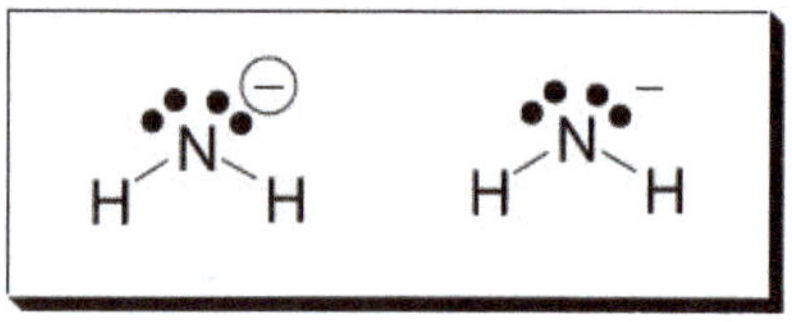

Let's analyze the nitrogen atom in the ammonium cation, NH_4^+.

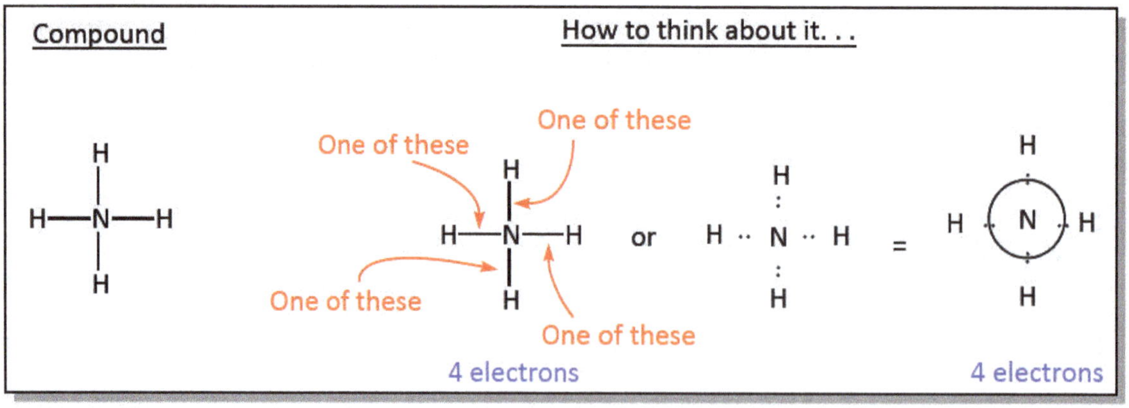

Figure 2.24 Understanding formal charges in the ammonium cation

Normal nitrogen has five valence electrons. The nitrogen atom in ammonium has four valence electrons. Since it has one fewer negative electron compared to what is normal, the nitrogen atom in ammonium has a +1 formal charge. Since this is a non-zero formal charge, it must be written on the nitrogen atom.

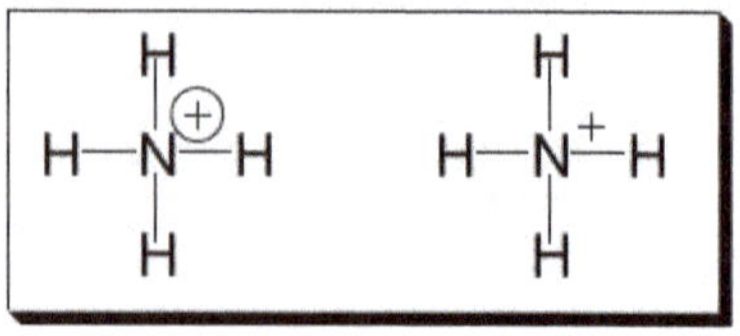

2.7 Identify the formal charges on the non-hydrogen atoms in the following structures. Write the formal charge on the atoms. If the formal charge is zero, do not write it.

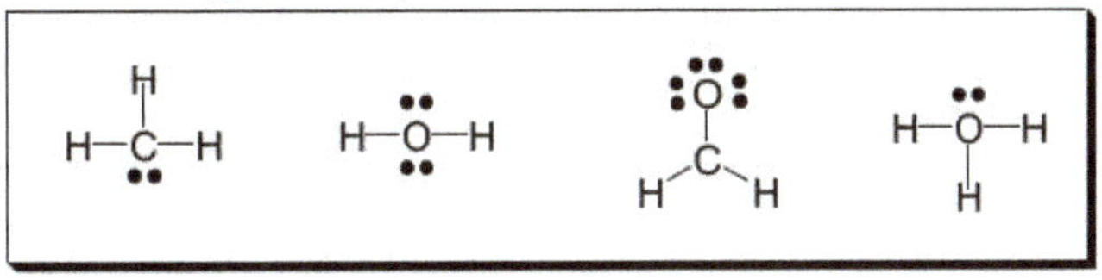

If you don't like to waste your time, it is important to be able to recognize when atoms are normal, with a zero formal charge, or not. Because if we have a molecule we are inspecting and it has dozens of atoms in it, we don't want to think about and analyze each of those atoms. This inspection of every atom would be a great waste of time. It is imperative that we be able to recognize normal bonding patterns. If we can quickly identify "normal" atoms, we can stop and analyze only the abnormal atoms. Take a moment to review the normal bonding patterns of atoms.

2.8 In the following molecules, try to quickly recognize the "normal" atoms with a zero formal charge, and then move on to the next atom. For the atoms that look abnormal, with a non-zero formal charge, stop and calculate its formal charge and write it on the atom.

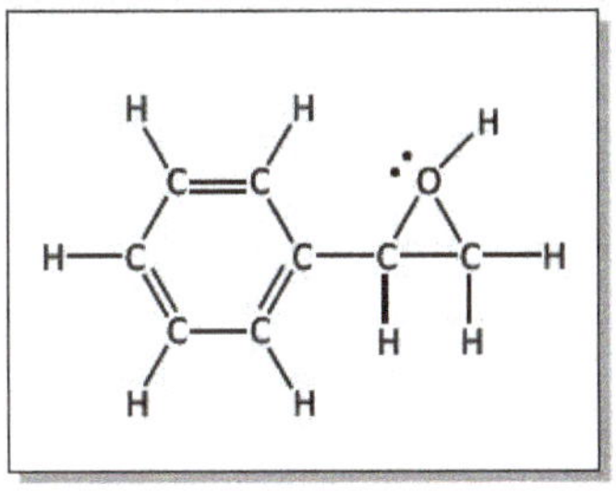

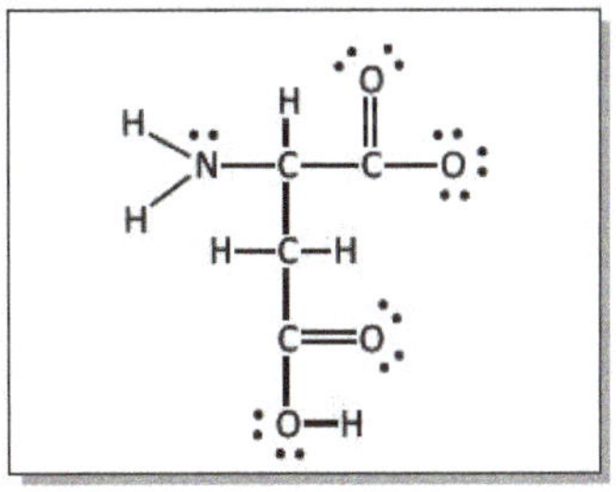

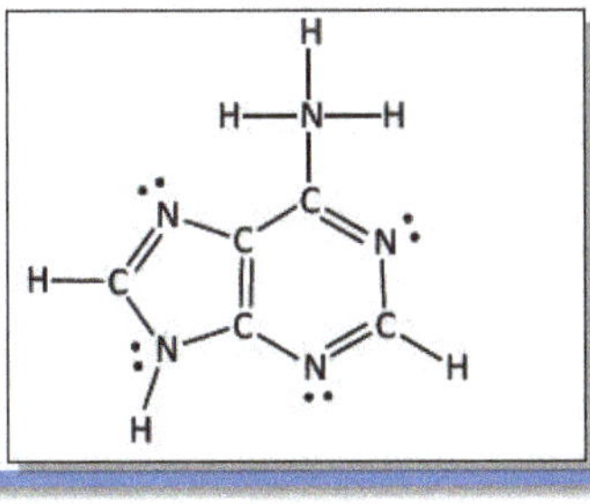

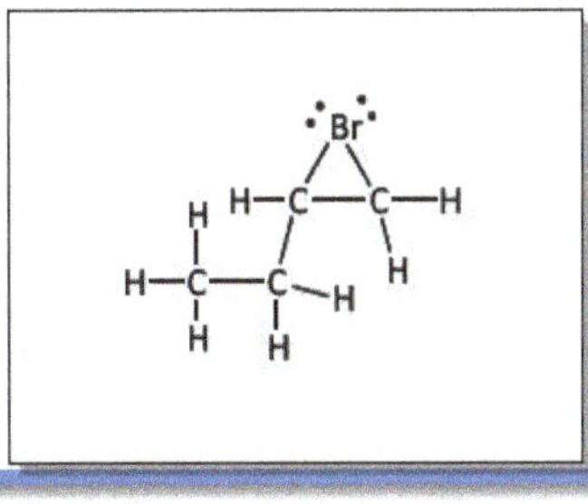

2.7 Drawing compounds

2.7.1 Condensed Structural Formula

Go back and review the normal, common bonding pattern for each atom. With these normal bonding patterns in mind, let's look at how organic compounds are typically drawn.

Let's consider the following alcohol. All atoms have a normal bonding pattern. Notice every carbon atom has four bonds, the oxygen atom has two bonds and two lone pairs and each hydrogen atom has one bond.

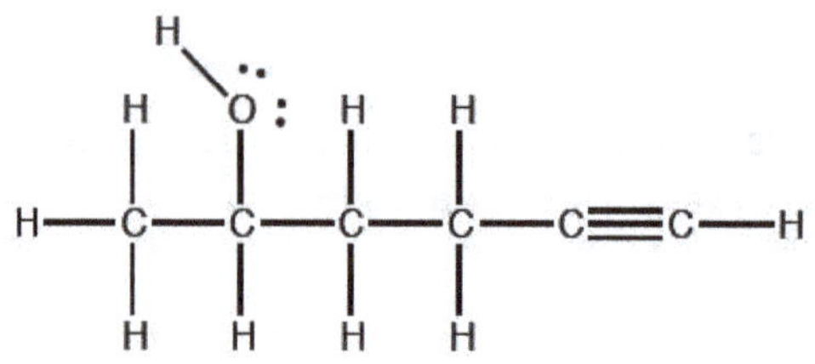

This drawing is perfectly acceptable. The problem with it is that it takes too long to draw in all of those hydrogen atoms. A quicker way to draw this compound is to not draw every bond to every hydrogen atom. A *condensed structural formula* is a drawing of a compound where the carbon-hydrogen bonds are not explicitly drawn. These may be drawn several ways.

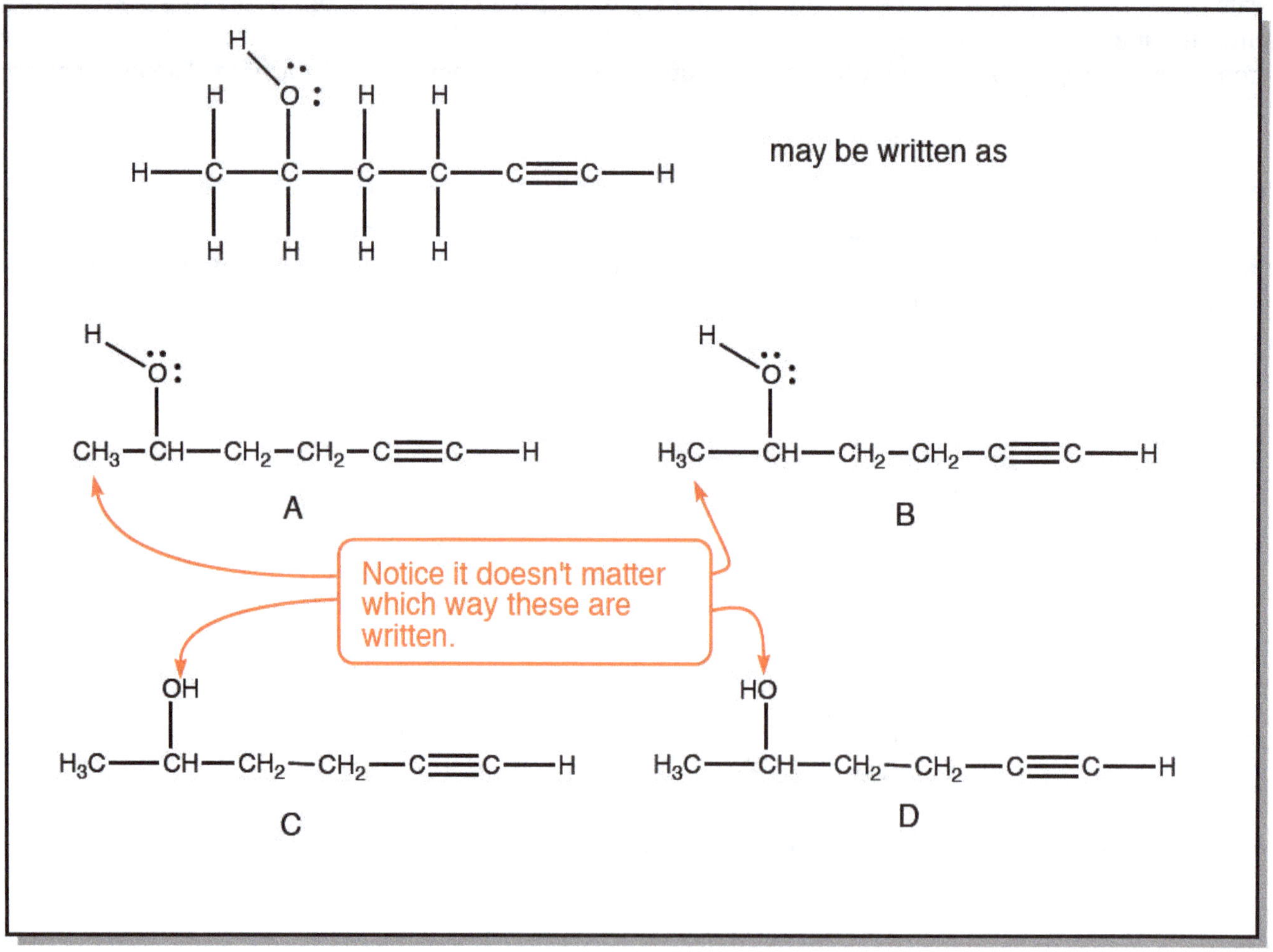

Figure 2.25 Drawing Condensed Structural Formulas

In compound A, all of the carbon-hydrogen bonds have been removed. The hydrogen atoms are usually written on the right of the carbon atom they are attached to. The exception to drawing the hydrogen atoms on the right of the carbon is shown in compound B. The carbon atom on the very left is drawn with the hydrogen atoms on the left. This may be done as it helps to show that the first and second carbon atoms are both bonded together. Compounds C and D show other ways to draw the OH. All of these are perfectly acceptable condensed structural formulas. I show them so you don't worry when you see subtle differences drawn.

2.9 Draw each of the following structures as a condensed structural formula.

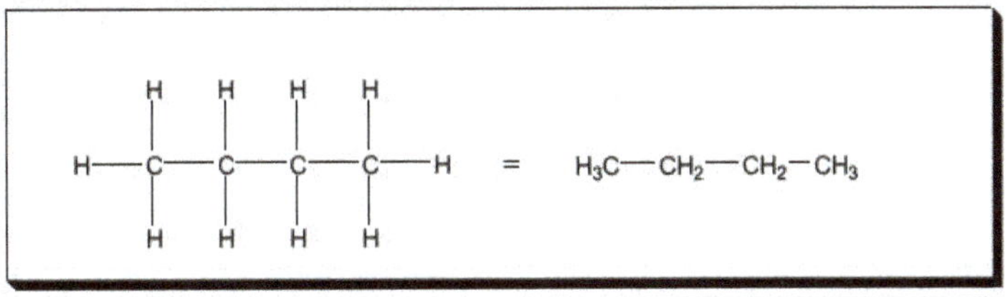

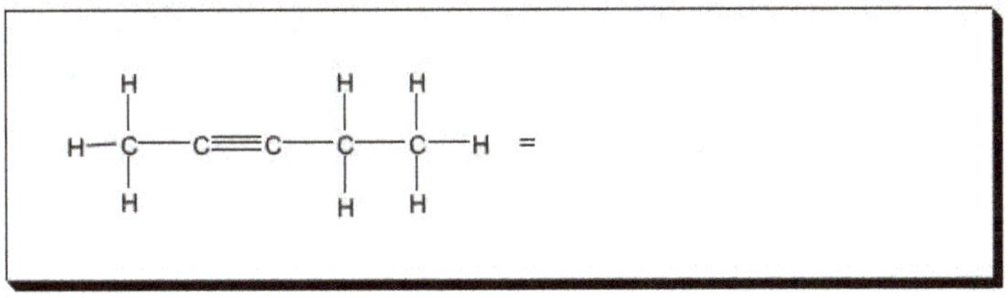

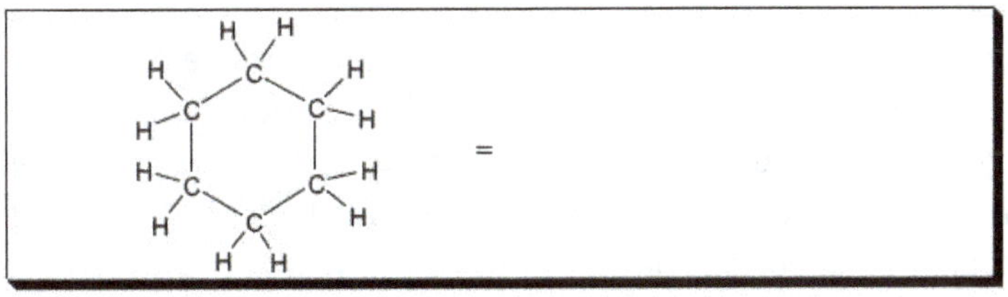

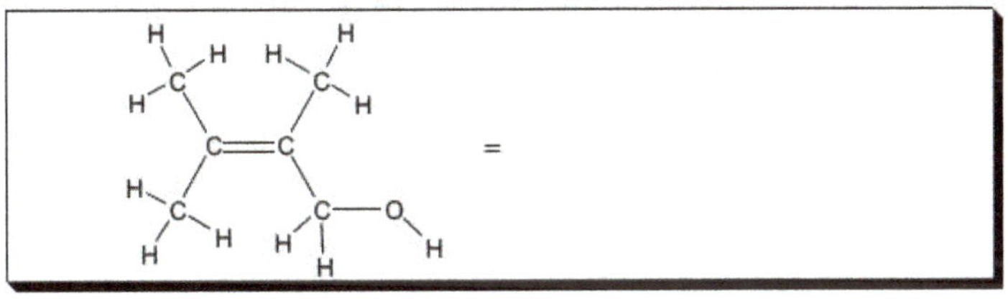

Instead of drawing out every bond, we can abbreviate the condensed structural formula if we have repeating groups. For instance, in the hexane, there are four methylene, CH_2, groups in the middle of the molecule. We can show this repeating group by putting it in parentheses and using a subscript to tell how many of them there are.

$$CH_3{-}CH_2{-}CH_2{-}CH_2{-}CH_2{-}CH_3 = CH_3(CH_2)_4(CH_3)$$

In this next molecule, there are two methyl, CH_3, groups on the same carbon atom. We can again use parentheses and a subscript to show this. There are three repeating methylene, CH_2, groups.

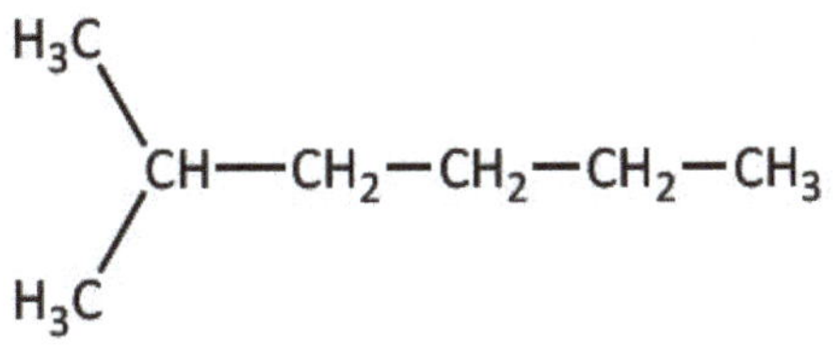

$= (CH_3)_2CH(CH_2)_3CH_3$

And, in this next molecule, there are three methyl groups, CH_3, on one carbon atom. There are four repeating methylene, CH_2, groups.

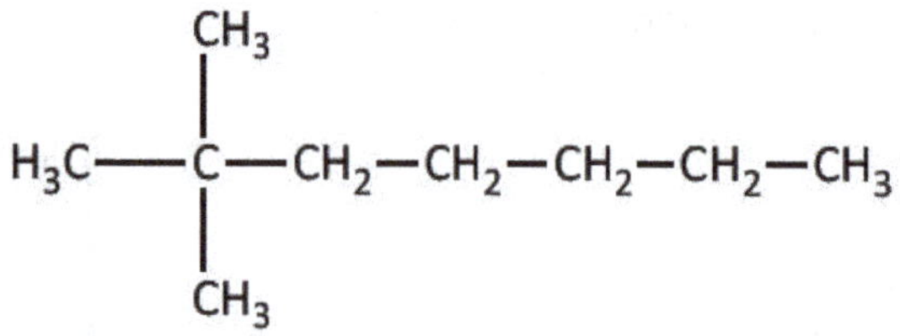

$= (CH_3)_3C(CH_2)_4CH_3$

2.10 Draw the Lewis structure (every bond and atom) for the following molecules.

a) $CH_3(CH_2)_3C(CH_3)_3$ b) $CH_3CH_2CH(CH_3)CH_2CH_3$ c) $C(CH_3)_4$ d) $CH_3(CH_2)_3CH_3$

2.7.2 Line Angle Formula

Chemists are smart, so they found a way to avoid drawing in all of those hydrogen atoms. If every practicing organic chemist knows that each carbon atom has four bonds, we could just not draw the H atoms at all. If we did not draw in the hydrogen atoms, and a chemist saw that the carbon atom only had two bonds drawn in, they would know that the carbon atom needs two more bonds to get to four, so it must have two hydrogen atoms attached. Remember, carbon atoms typically have four bonds. If you ever draw five bonds on a carbon atom, you might see your professor get angry. Trust me. You don't want to see your professor get angry. It is not pretty. It's funny what upsets an organic chemistry professor, isn't it?

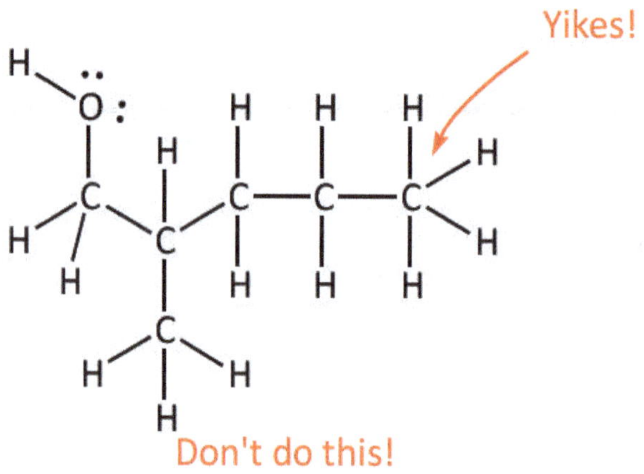

2.11 Draw in all of the missing hydrogen atoms on the following carbons.

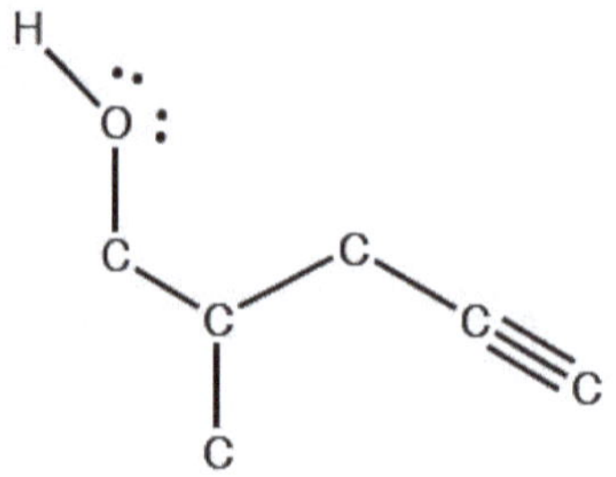

It's no fun drawing in all of those hydrogen atoms, is it? So, make sure you get good at drawing them using the other suggestions in this chapter! In fact, why draw in carbon atoms at all? Organic chemists have found an even easier-to-write notation. This notation is the line-angle formula or the line-angle structure. In the line-angle structure, a carbon atom is assumed every time a line ends, two lines come together at an angle, or a single bond meets a double or triple bond. So, instead of the above structure, chemists may write the following to save time.

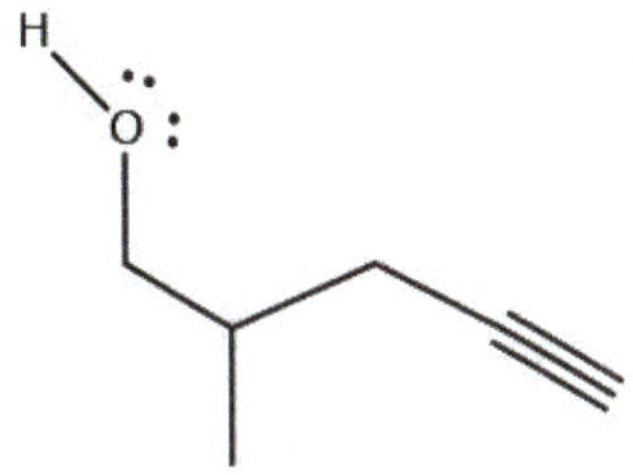

Notice that the hydrogen atom on the oxygen is still drawn in. It is customary to draw the hydrogen atoms that appear on all non-carbon atoms, also called *heteroatoms*. Let's make sure you can see where all of the carbon atoms are. Carbon atoms are located at the following sites.

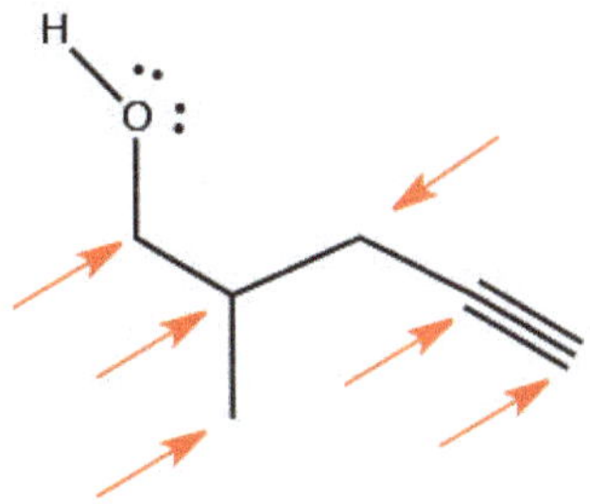

2.12 Count the number of carbon atoms in the following compounds.

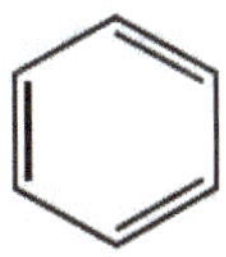

6
___

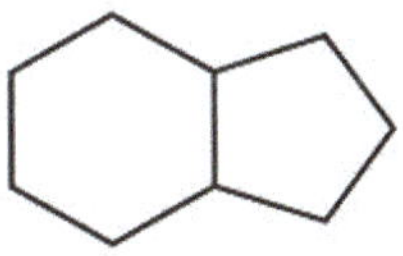

___

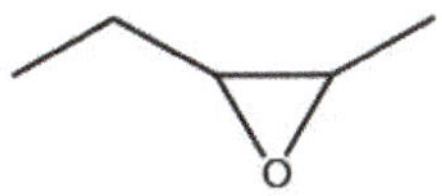

___

___

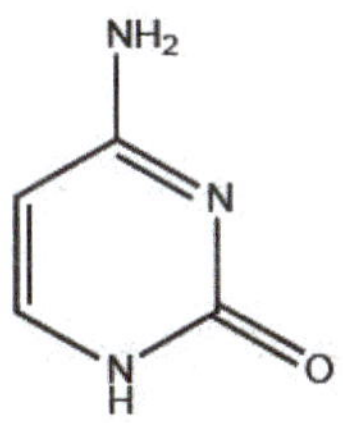

___

There is one more timesaving feature we can use to draw this compound. If practicing chemists know the normal bonding patterns, they can also figure out how many lone pairs of electrons are on atoms like nitrogen or oxygen. Normal oxygen has two bond and two lone pairs of electrons. So, our compound may be drawn like this.

Bonding and Structure

2.13 Fill in the requested information.

How many hydrogen atoms are here? ____

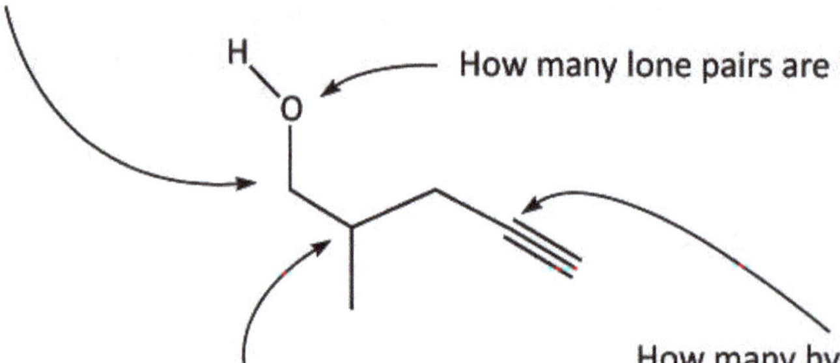

How many lone pairs are here? ____

How many hydrogen atoms are here? ____

How many hydrogen atoms are here? ____

Most textbooks do not draw in the lone pairs of electrons. This is a rather unfortunate practice because many chemical reactions occur with these lone pairs. Oh well, we must remember to be diligent remembering how many lone pairs are on these heteroatoms. I like to draw them in so I do not forget they are there.

2.14 How many lone pairs of electrons are on the following heteroatoms?

a) 

b) 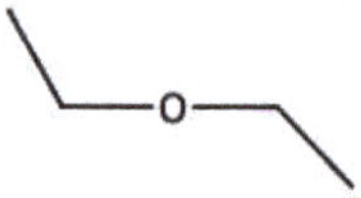

c) $H_3C—C\equiv N$ d) $H_3C—Br$

2.15 Identify the number of hydrogen atoms attached to each *carbon* atom.

a) 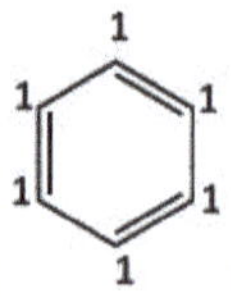

b) 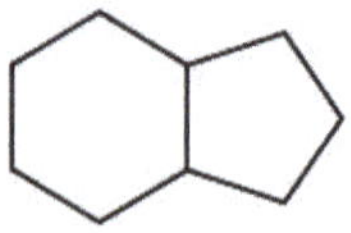
c) 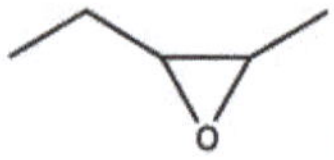

d) 
e) 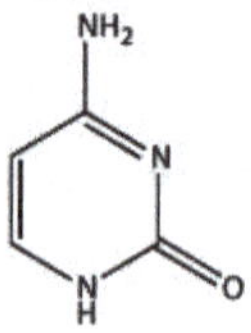

To review, these three structures are all equivalent.

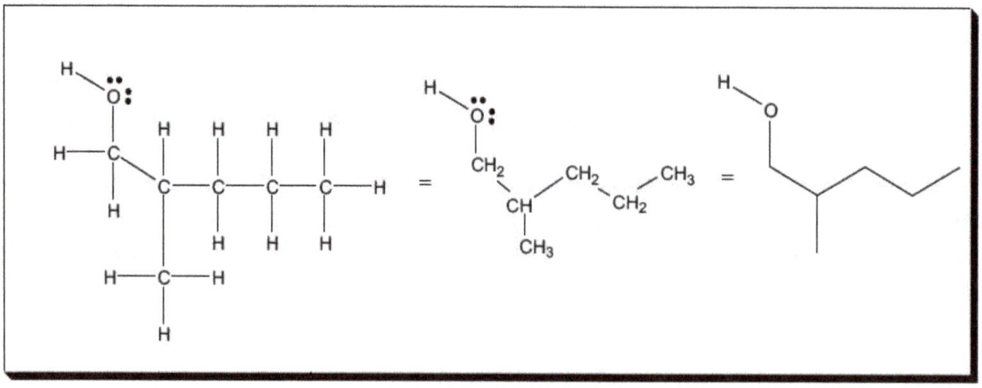

There are a few other things you should know about drawing compounds. One is that single bonds on these compounds can rotate around. That means that the following drawings are all for the same compound. It does not matter which way you draw it.

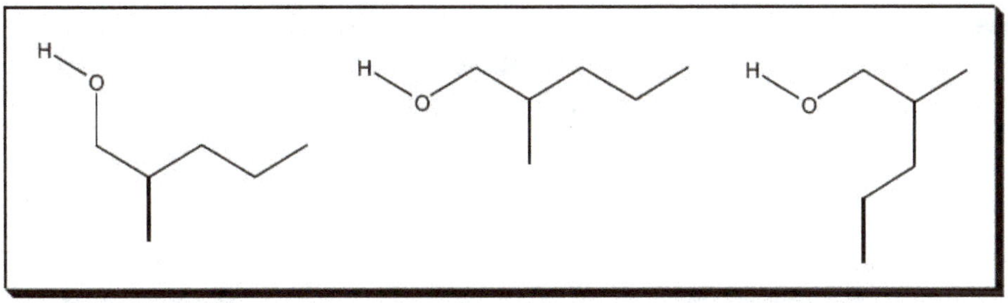

But, since the bonds all contain negative electrons, they repel each other. The bonds want to get as far away from each other as possible. Therefore, you would ***not*** want to draw this compound this way.

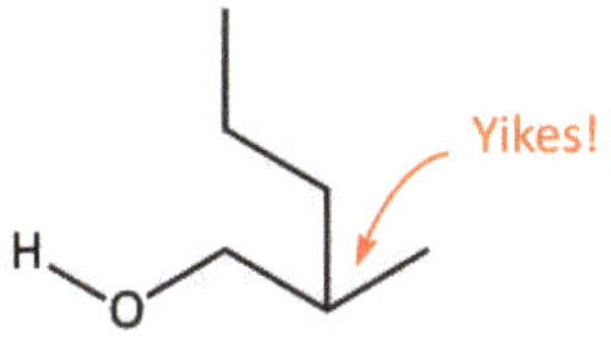

2.16 Draw the following compounds in the shorthand, line-angle notation.

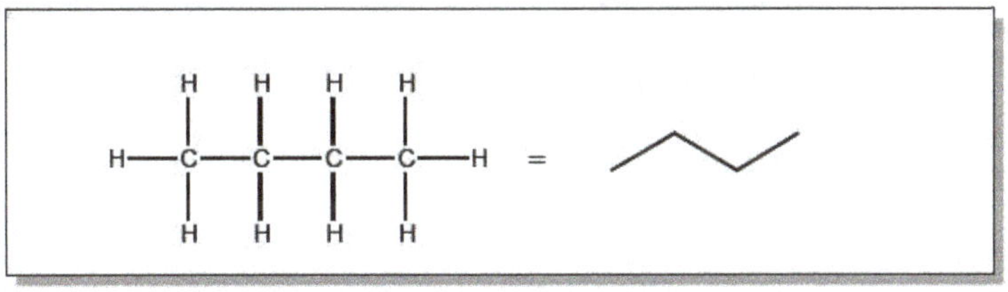

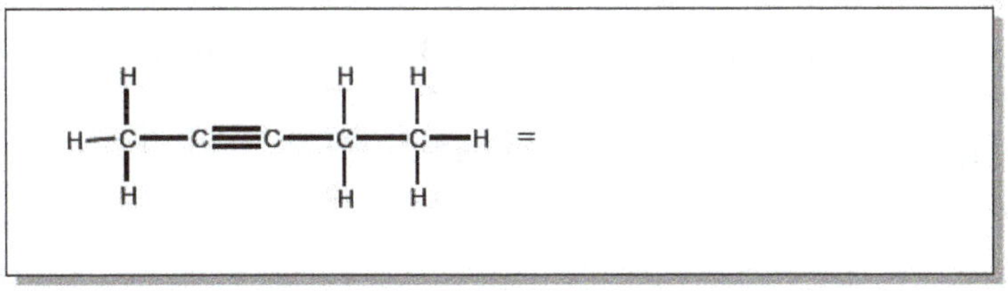

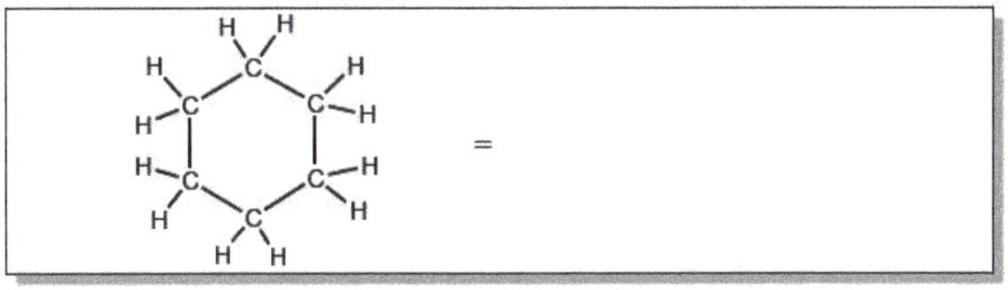

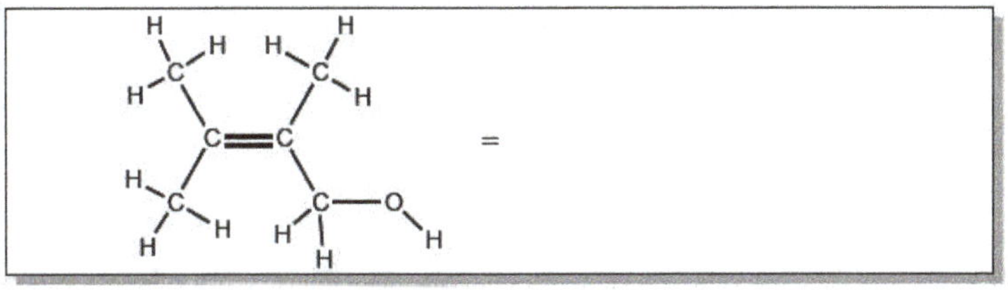

2.8 Geometry

Now that we can draw organic compounds, let's take a closer look at common geometries we will encounter. The simple theory that helps us understand the shapes of molecules is the *valence shell electron pair repulsion* theory, *VSEPR* (pronounced "Vesper") *theory*. The idea is that all of the electron groups on an atom (single bonds, double bonds, triple bonds, lone pairs of electrons) are all negatively charged groups. Since like charges repel each other, these groups get as far away from each other as they can. The farthest two groups can get away from each other on an atom is 180°, a linear shape.

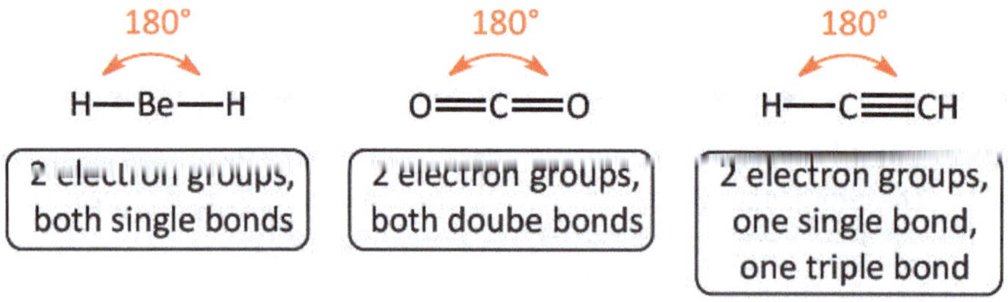

Figure 2.26 Linear molecular geometry

The farthest three groups can get away from each other on an atom is 120°, a trigonal planar shape.

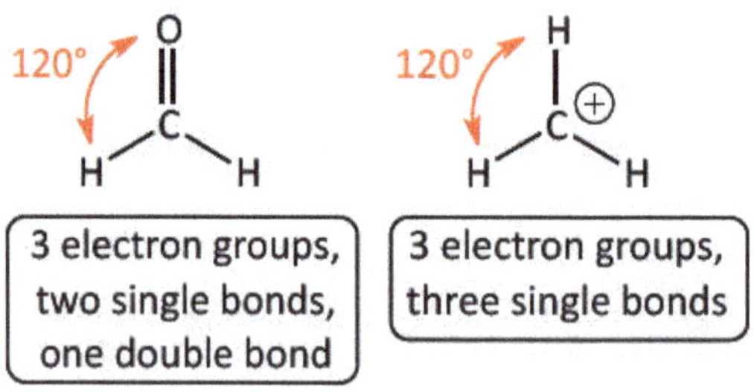

Figure 2.27 Trigonal planar molecular geometry

The farthest four groups can get away from each other is 109.5°, a tetrahedral shape. We use wedges and dashes to help us draw these shapes. A regular line means that atom lies in the plane of the paper. A wedge indicates an atom is coming towards you, in front of the page. A dashed line indicates an atom is going away from you, behind the page.

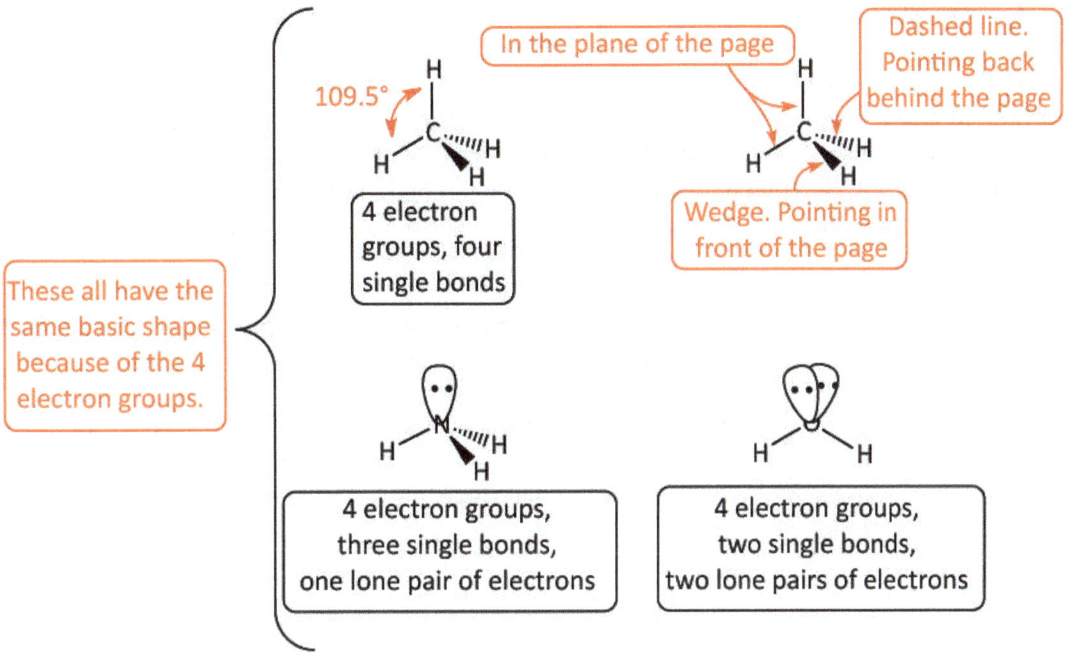

Figure 2.28 Tetrahedral molecular geometry

Even though these all have the same basic shape, we call them different things. Whenever we name a shape, we treat lone pairs of electrons like they are invisible. You can eve cover them up with your hand and ignore them. We just look at the atoms and name the shape we see. On a side note, notice "trigonal pyramid" is the name of a pyramid with a triangle as its base. This is different than a pyramid with a square pyramid as its base (named obviously "square pyramid").

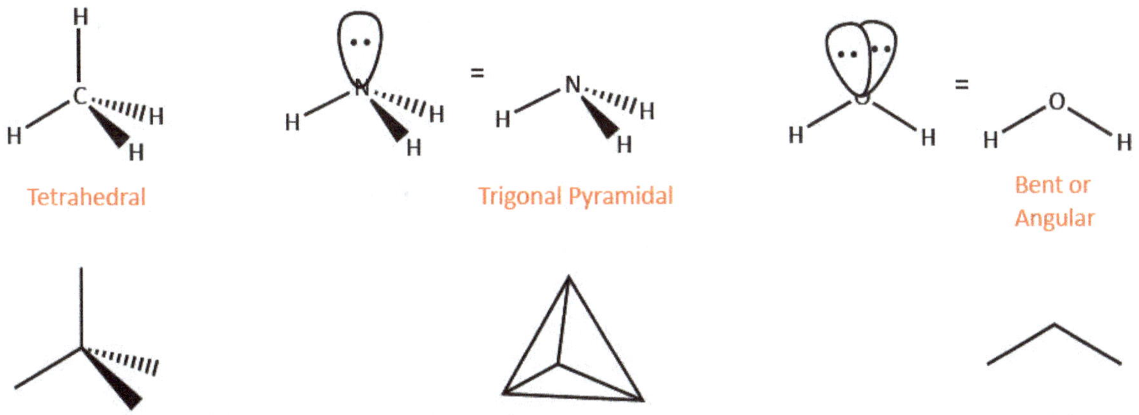

2.17 State the geometric shape of the molecule with the following bond angles between atoms (assume no lone pairs of electrons).

a) 109° b) 120° c) 180°

2.18 Name the geometric shape of the following molecules.

a) CCl_4 b) PH_3 c) $H_2C{=}O$ d) $O{=}C{=}O$ e) F_2O f) $AlCl_3$

2.9 Dipole Moment

Once we know the geometry of a molecule, we can use that information, along with what we know about electronegativity to determine if a molecule is polar or nonpolar. If a molecule is polar, it has one side of the molecule that is more positive and one side that is more negative. The charges are lopsided. Having a positive and negative side means we have two poles, or a dipole. The strength of this positive-negative separation is called the dipole moment of the molecule. The positive and negative sites are balanced and symmetrical if a molecule is nonpolar.

To determine if a molecule is polar or nonpolar, we look at the molecule's shape and the electronegativity of the atoms at each bond. We also remember that lone pairs of electrons are very negative spots.

Let's look at some nonpolar molecules first. They all have a dipole moment of zero. There are some more positive and some more negative spots in the molecules, but overall nonpolar molecules have a symmetrical distribution of charge. Notice the symmetry of the following nonpolar molecules. They all have a mirror plane of symmetry or an inversion point of symmetry.

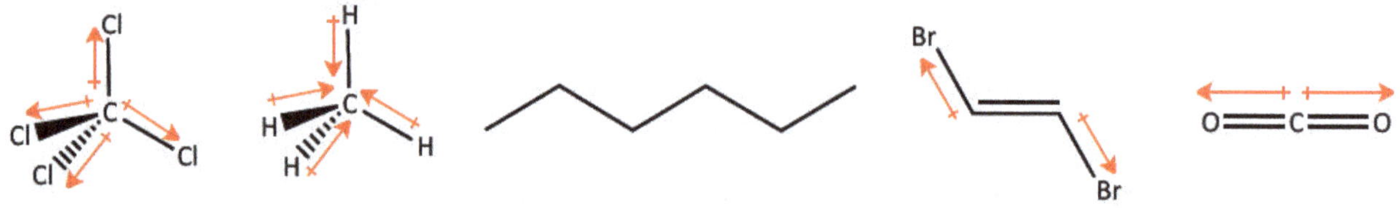

Figure 2.29 Nonpolar molecules

Let's look at some polar molecules. These all have a dipole moment. The dipole moment of each bond is shown with the red dipole arrows. The overall dipole moment for the molecule is shown with the blue arrow. The vector sum of the individual bond dipole moments is the overall molecular dipole moment. Another way to think about that is if you pulled or pushed individually in the direction of all of the little red arrows, how would the molecule move? This direction, the blue arrow, is the overall molecular dipole moment.

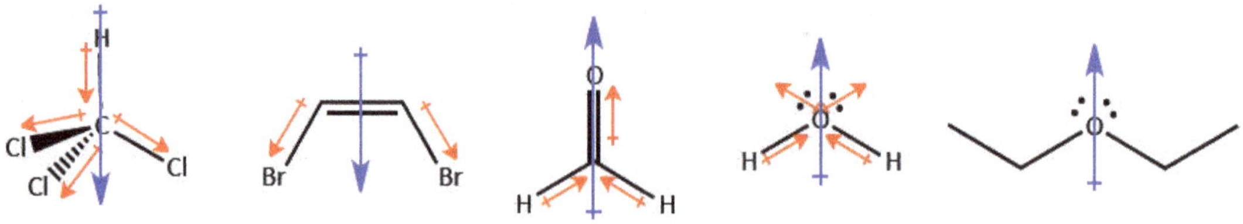

Figure 2.30 Polar molecules

2.19 Circle the following compounds that are polar (have a dipole moment).

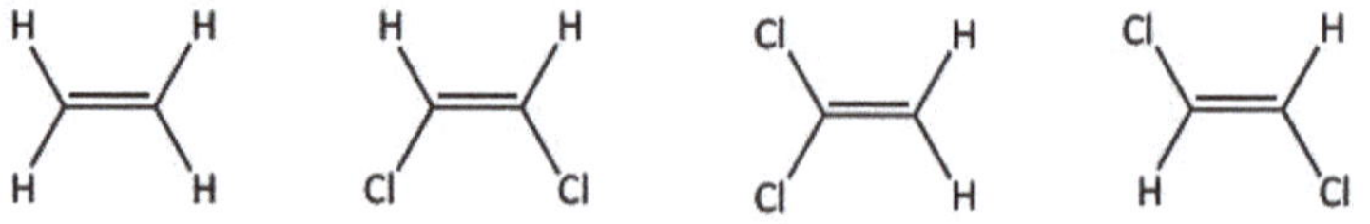

$H_3C—OH$ $CHCl_3$ CCl_4 CH_2Cl_2

2.20 Indicate whether each of the following molecules contain polar bonds? Which molecules are polar overall?

a) Cl_2 b) HCl c) H_2O d) CH_4 e) $CHBr_3$ f) CH_3OCH_3

2.10 Solubility

But, why do we care?

Some very practical chemistry comes out of knowing the polarity of molecules. For instance, we can begin to predict what solvent is best to dissolve something. Polar solvents tend to dissolve polar compounds and nonpolar solvents dissolve nonpolar compounds. One can remember this by remembering the phrase "Like dissolves like." For instance, dirt and grease are nonpolar. So, good spot removing cleaners are nonpolar solvents like carbon tetrachloride or hexane.

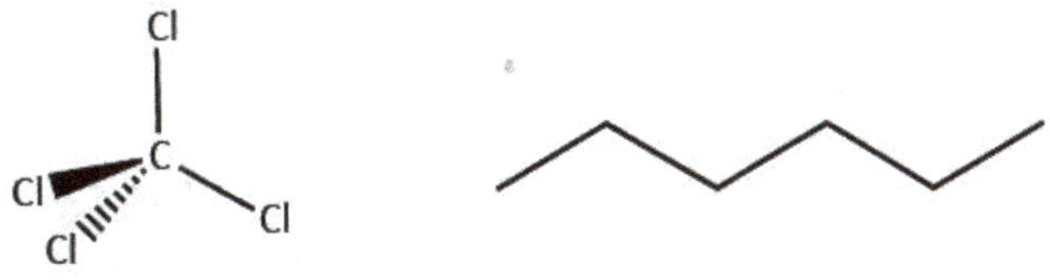

Carbon tetrachloride Hexane

Figure 2.31 Some nonpolar solvents

More polar compounds, like glucose, and ionic compounds like sodium chloride (table salt) are more soluble in more polar solvents like water. Glucose and salt do not dissolve in nonpolar solvents at all!

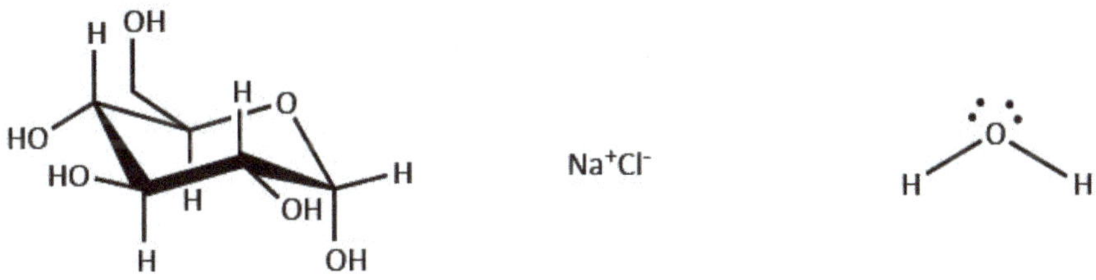

Glucose Table salt Water

Figure 2.32 Water soluble compounds

2.21 Which compound is more soluble in water?

a) CH_3CH_2OH or $CH_3CH_2CH_2CH_3$

b) $CH_3CH_2OCH_2CH_3$ or $(CH_3)_2NH$

2.11 Intermolecular forces

If you are ever asked, the rest of your life, something about a compound's boiling point, your first thought should be about this section. Imagine you have a beaker full of some compound. The zillions of molecules of this compound are swimming around as a liquid in this beaker. If this liquid is going to boil, you begin heating up the liquid until the molecules start flying away. The boiling point of the liquid is that temperature. The more the molecules stick together to each other; will you need to heat up the liquid **more or less** to get them to fly away? Of course, you need to heat the liquid up **more** to get it to boil when the molecules stick together more. If the molecules weakly stick together, the boiling point will be lower. It will not take as much heat to get them to fly away.

Here is another question. Between solid, liquid, and gas, which one has molecules that stick the most together? Of course, it is the solid. The molecules stick together very well and do not move around each other. In the liquid, the molecules stick together pretty well, but they can slide past each other. As a gas, the molecules do not stick together very well at all. The attractive forces or bonds within one molecule are called intramolecular forces. *Intra* means *within*. The attractive forces between molecules are called intermolecular forces. *Inter* means *between*.

So, the real question is, how do we know how well the molecules stick together? The answer is to look at the intermolecular forces involved. Intermolecular forces are simply the forces between molecules that hold them together. These forces are weaker than intramolecular forces or regular atom-atom bonds. There are three intermolecular forces to consider. All of them are related to dipole moments.

2.11.1 Dipole-Dipole interaction

When a molecule has a dipole moment, it has a positive end and a negative end. Since opposite charges attract, the positive end of a molecule is attracted to the negative end of another molecule. This intermolecular force is called a dipole-dipole interaction.

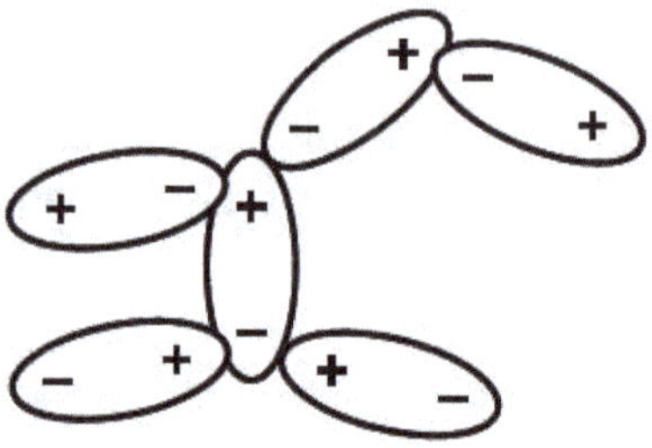

Figure 2.33 Dipole-dipole interaction

2.11.2 Hydrogen Bonding

Hydrogen bonding is an intermolecular force that is a special, very strong case of dipole-dipole interaction. It is the strongest of the intermolecular forces. For hydrogen bonding to occur, there needs to be a very positive site and a very negative site to stick together. The very positive site involves a hydrogen atom on a very electronegative atom like nitrogen, oxygen, or fluorine. Organic compounds do not contain H-F bonds, so we usually only consider H-N or H-O bonds. Because the hydrogen is on a very electronegative atom, the N or O win the tug-of-war for the negative electrons in the bond and make the hydrogen atom a very small, concentrated, positive spot. This is a special positive location that can bond to negative spots on other molecules.

In hydrogen bonding, the negative site is a lone pair of electrons on nitrogen, oxygen, or fluorine. Since N, O, and F are the most electronegative atoms, they are strongly pulling negative electrons towards themselves. They have a very negative charge, especially with the very large, juicy lone pair of electrons on them.

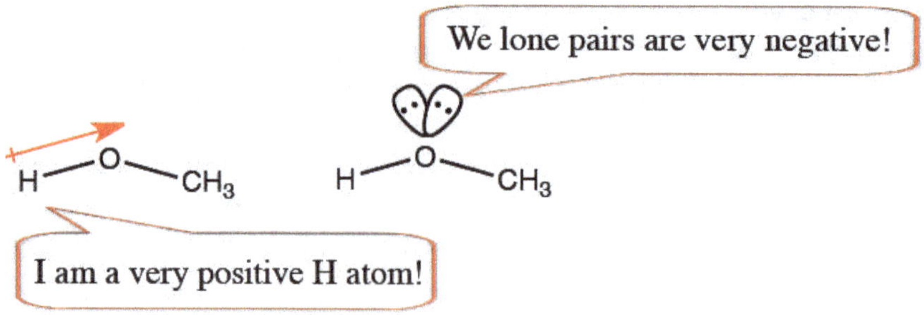

These very negative spots and these very positive spots stick together strongly.

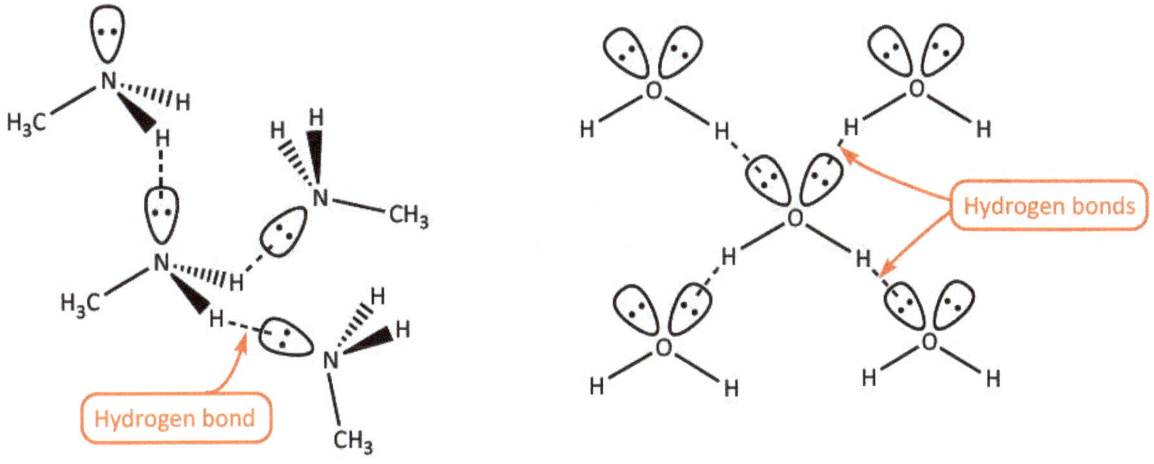

Methylamine

Water

Figure 2.34 Hydrogen bonding examples

2.22 Rank the following compounds from the highest boiling point (1) to the lowest boiling point (3).

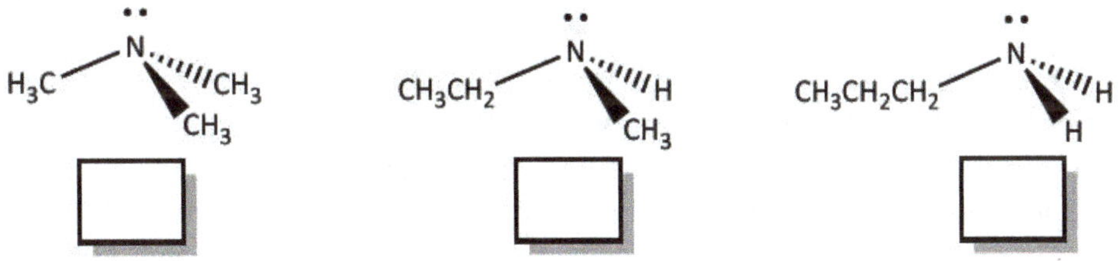

2.11.3 London Dispersion Forces

The weakest of the intermolecular forces are the London dispersion forces. We find these in non-polar compounds. "Molecule 1" is a nonpolar molecule with the electrons evenly distributed throughout.

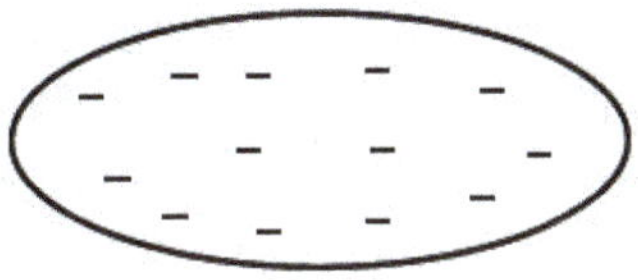

Molecule 1

Sometimes, when two nonpolar molecules are near each other, one of them will get a temporary dipole moment because some of the electrons will slosh more towards one end of it. This only lasts for a brief time before they will slosh back and the temporary dipole is lost. But, when the negative electrons slosh more to one side, one end of the molecule is temporarily negative and one side is temporarily positive.

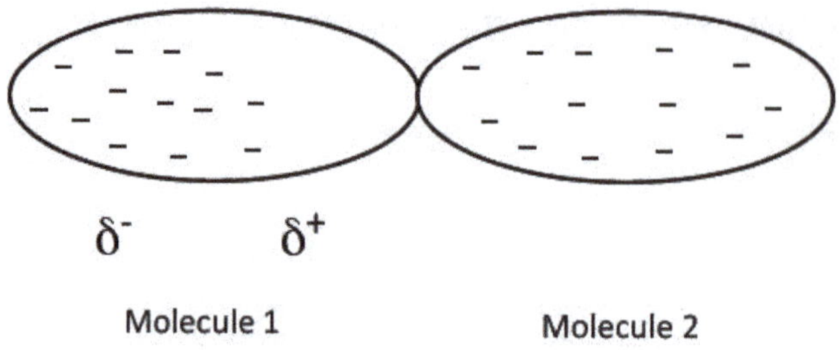

Since we have a partially positive side in this molecule for a brief moment, what will happen to the negative electrons in Molecule 2? The negative electrons in Molecule 2 are attracted to the positive spot on Molecule 1. Molecule 2 has an induced dipole. The two molecules then stick together in a weak dipole-dipole interaction.

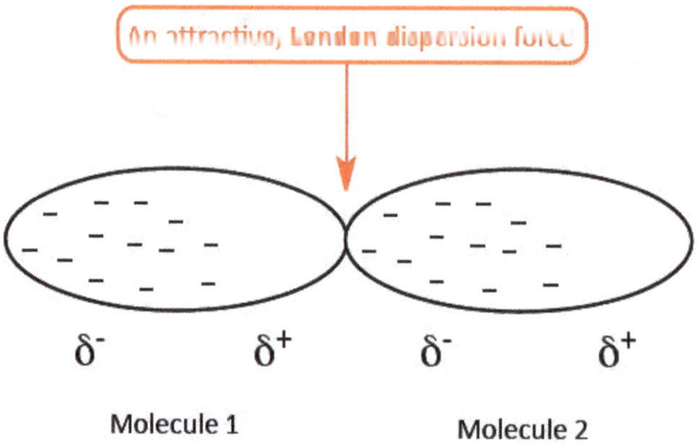

Figure 2.35 London Dispersion interaction

A brief moment later, the electrons will slosh back, and we'll lose this attraction. This is why London forces tend to be weak. Electrons are more likely to slosh around to make a dipole with the more electrons the nonpolar molecule has. Therefore, the London dispersion forces are greater the larger surface area a molecule has. Compare the diatomic molecules of chlorine, bromine, and iodine. These are all perfectly symmetrical nonpolar compounds. Yet, at room temperature, chlorine is a gas, bromine is a liquid, and iodine is a solid. So, the molecules must be sticking together the most for iodine and the least for chlorine. How do we explain this? I_2 has many more electrons (106) than Br_2 (70) and Cl_2 (34). It also has a larger surface area. So, we would predict the London dispersion forces are greatest for iodine.

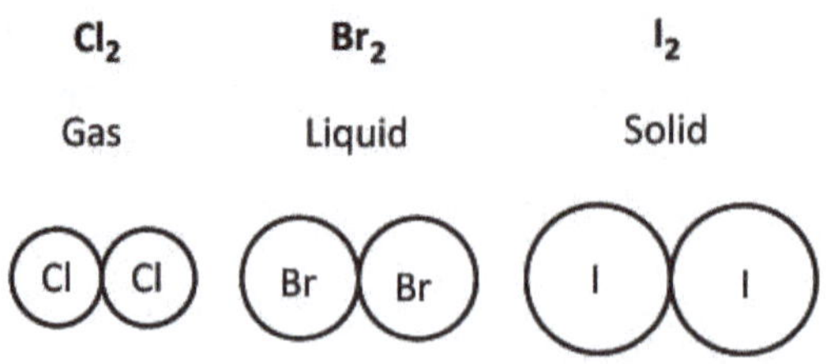

Figure 2.36 London forces in diatomic halides

In organic chemistry, we can make similar arguments. The following alkanes are all nonpolar. Long, straight molecules have a greater surface area than ones that have many branches coming off of it. The more branches we have, the more spherical the molecule behaves and the smaller surface area it has. So, London Dispersion forces are greatest for longer chains. Longer chains stick together better and have higher boiling points. So, more branched alkanes have lower boiling points.

Less branched ______________________ **Highly branched**

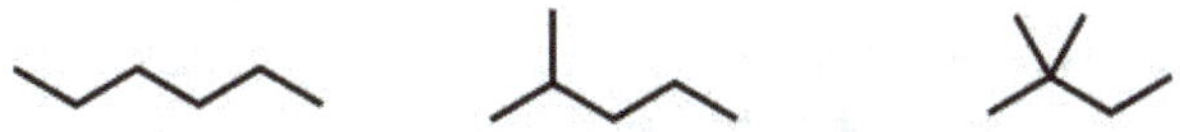

n-hexane	2-methylpentane	2,2-dimethylbutane
b.p.= 69°	b.p.= 60°	b.p.= 50°

Figure 2.37 London Forces in branched alkanes

2.23 Rank the following compounds from the highest boiling point (1) to the lowest boiling point (4).

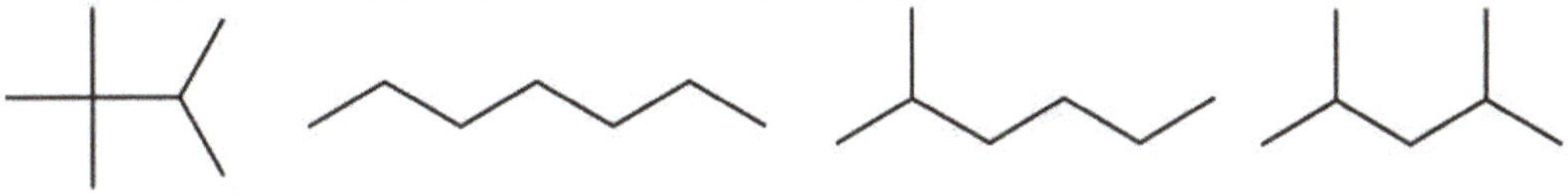

☐ ☐ ☐ ☐

2.24 Rank the following compounds from the highest boiling point (1) to the lowest boiling point (4).

H_3C—OH H_2O

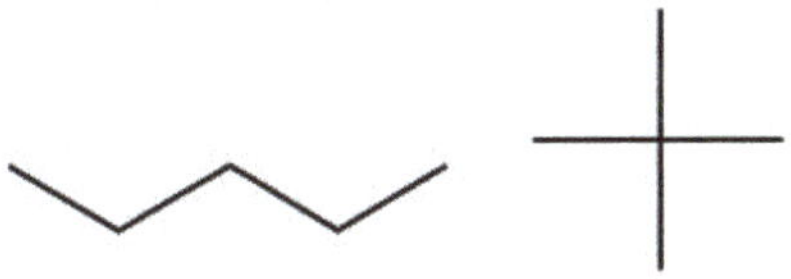

☐ ☐ ☐ ☐

2.12 Types of organic compounds

Organic chemistry is an interesting mixture of understanding concepts and memorization. It's time to memorize some things. We need to begin to be able to describe compounds so we can discuss them throughout this course. Many organic molecules contain reactive parts we call *functional groups*.

First, just like learning any new foreign language, you need to learn to count to ten. This is how we count in organic chemistry.

1 = meth	6 = hex
2 = eth	7 = hept
3 = prop	8 = oct
4 = but	9 = non
5 = pent	10 = dec

Let's look at various types of compounds.

2.12.1 The Hydrocarbons

Hydrocarbons are compounds that are made up only of hydrogen and carbon atoms.

Bonding and Structure

Alkanes

Alkanes are singly bonded carbon atoms, C-C. Alkane names typically have the *–ane* suffix. Knowing how to count in organic chemistry and the suffix, we can begin to name some compounds. We count the number of carbon atoms in the chain to get the prefix, and since they are all singly bonded, we use the alkane *–ane* suffix.

CH_4	$H_3C—CH_3$	$H_3C—CH_2—CH_3$	$H_3C—CH_2—CH_2—CH_3$
methane	ethane	propane	butane

2.25 Write the names of the straight-chain alkanes that contain 5- to 10-carbon atoms.

5: ____________________ 6: ____________________ 7: ____________________

8: ____________________ 9: ____________________ 10: ____________________

Cycloalkanes

If an alkane is in a ring, it is a cycloalkane. We use the prefix *cyclo-* to name it.

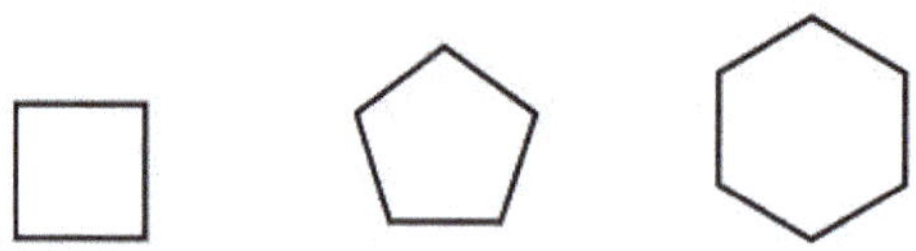

cyclobutane cyclopentane cyclohexane

2.26 Name these compounds.

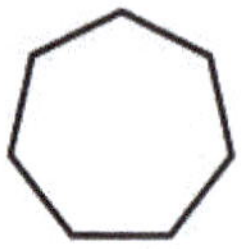

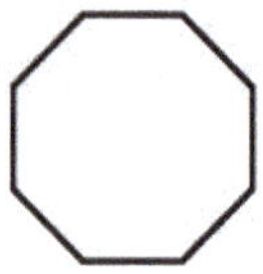

____________________ ____________________ ____________________

Alkyl groups

When we are naming a piece of an alkane as a group on a more important part of a structure, we call that piece of alkane an alkyl group. We name alkyl groups with the *–yl* suffix.

$—CH_3$	$—CH_2CH_3$	$—CH_2CH_2CH_3$	$—CH_2CH_2CH_2CH_3$
methyl (Me)	ethyl (Et)	propyl	butyl

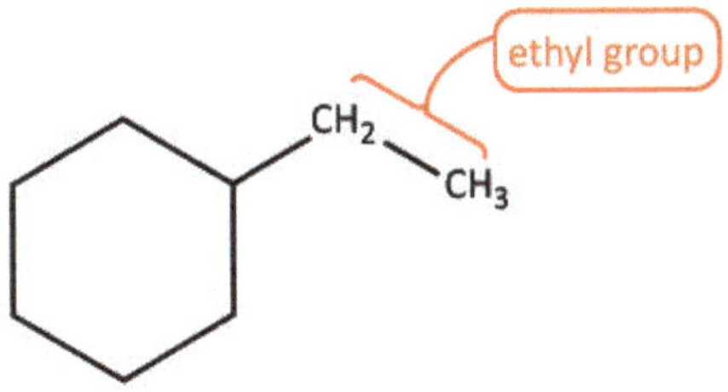

ethylcyclohexane

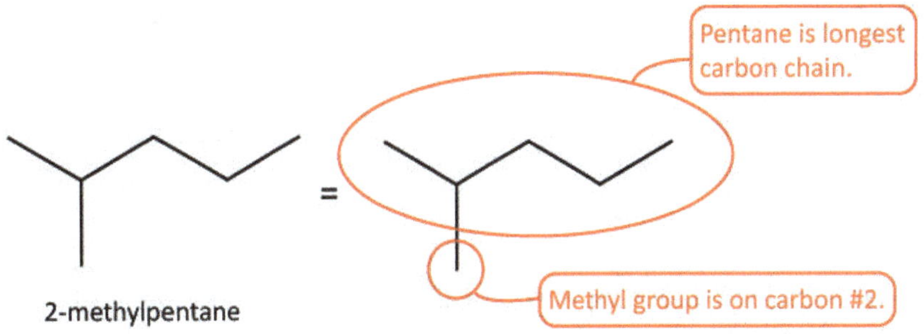

2-methylpentane

If we want to write a generic alkyl group, because it is not important to the discussion we are having, we use the symbol –R. This stands for the "Rest" of the molecule.

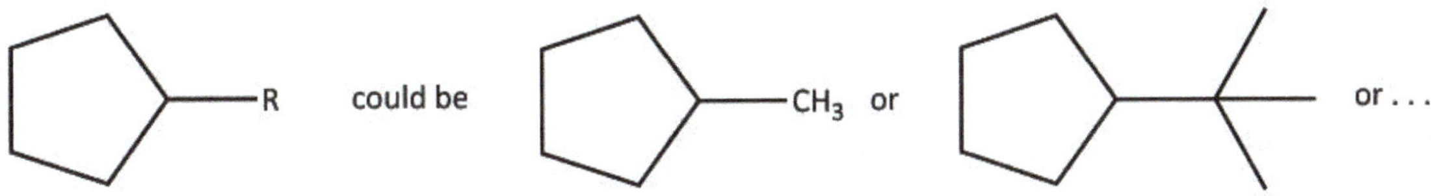

Alkenes

Alkenes are hydrocarbons that contain a double bond between carbon atoms, C=C. Alkenes have the *–ene* suffix.

$H_2C{=}CH_2$	$H_2C{=}CH{-}CH_3$	$H_2C{=}CH{-}CH_2{-}CH_3$
ethene	propene	butene

How about this compound that has four carbons and a double bond?

$$CH_3{-}CH{=}CH{-}CH_3$$

It should also be called some sort of butene, but we need a different name since the double bond is in the middle and not on the end. We have to tell which position the double bond is located, so we call it 2-butene.

$$CH_3{-}CH{=}CH{-}CH_3$$

2-butene

And, once again, if the alkene is in a ring, we use the prefix *cyclo-*.

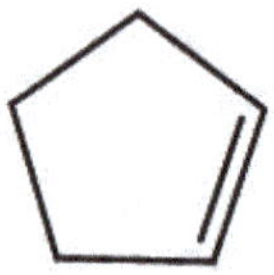

cyclopentene

Cis/Trans

Alkene double bonds cannot rotate around because they are made using one π-bond locking them into place. This means that we will sometimes get some isomers because of the double bond.

If we draw out an alkene compound, we see that the C=C has four groups on it, two on the left side and two on the right side of the double bond.

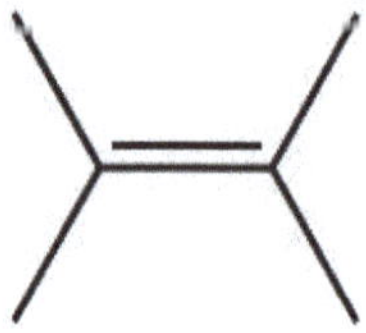

If we draw out 2-butene, we have one hydrogen atom and one methyl group to put on each side of the double bond. We could put the methyl groups on top-top, bottom-bottom, top-bottom, bottom-top.

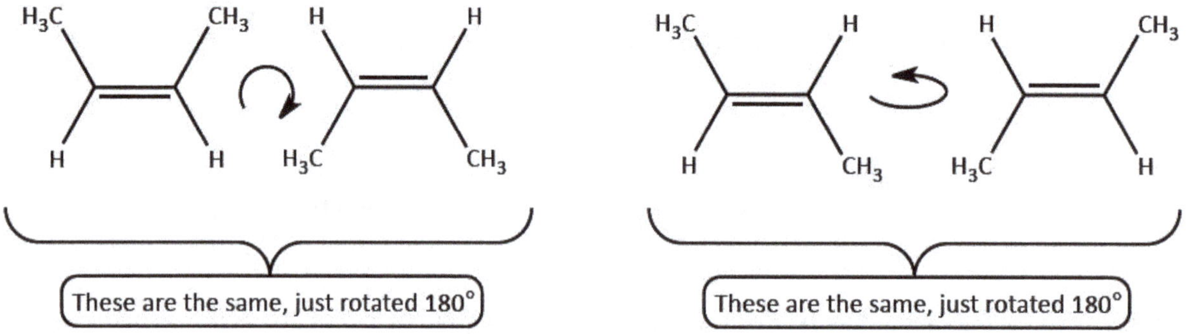

We see that we really only have two different structures here, not four. One of the structures has the CH_3 groups on the same side of the double bond, while the other has the CH_3 groups on the opposite side of the double bond. We use the prefix, *cis-*, to identify the same side and the prefix, *trans-*, to identify the opposite side.

Cis is Latin for "On this side of". So, if I am reading cisatlantic literature, I'm reading literature from the same side of the Atlantic Ocean as myself. If I'm reading transatlantic literature, I'm reading literature from the other side of the Atlantic Ocean. (*Cis* = same, *trans* = opposite).

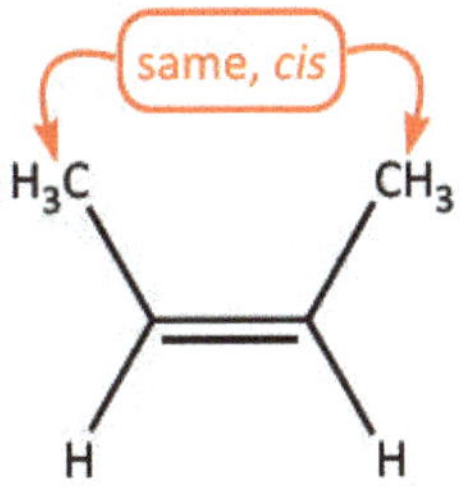

CH_3 groups on the same side

cis-2-butene

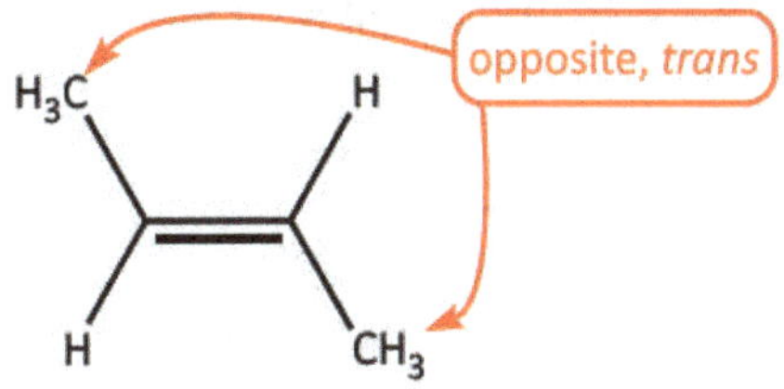

CH_3 groups on the opposite side

trans-2-butene

Not every double bond has *cis-trans* isomers, though. If both groups on the left side or both groups on the right side are the same, we cannot have *cis* or *trans* compounds.

2.27 Circle the compounds below that exhibit *cis-trans* isomerism.

$H_2C{=}CHCH_2CH_2CH_3$ $CH_3CH_2CH{=}CHCH_2CH_3$ $Cl_2C{=}CHCH_2CH_2CH_3$ $ClHC{=}CHCl$

Alkynes

Alkynes are hydrocarbons that contain a triple bond between carbon atoms, C≡C. Alkynes have the *–yne* suffix.

$H{-}C{\equiv}C{-}H$ ethyne

$H{-}C{\equiv}C{-}CH_3$ propyne

$H{-}C{\equiv}C{-}CH_2{-}CH_3$ butyne

$H_3C{-}C{\equiv}C{-}CH_3$ 2-butyne

Aromatic Compounds

Although they look like cycloalkenes, aromatic rings have special properties. We'll learn about these later, but for now, we want to realize that they are named differently. The simplest aromatic ring is benzene.

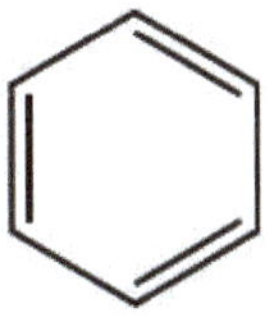

benzene

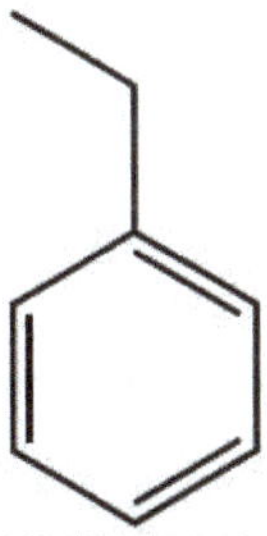

ethylbenzene

A *phenyl* group (abbreviated, Ph) is a benzene that is a group on another, more important chain.

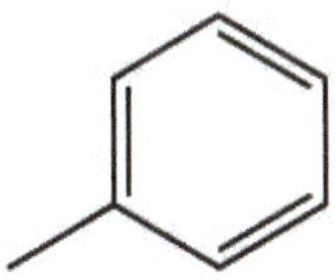

phenyl (Ph)

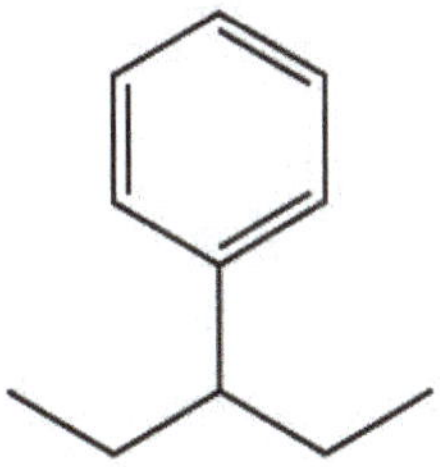

3-phenylpentane

=

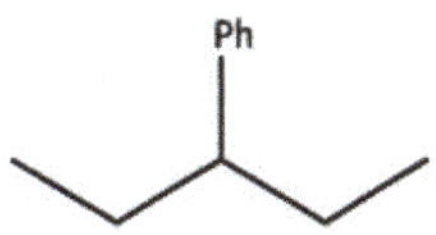

3-phenylpentane

Much like we use -R for a generic alkyl group, we use –Ar for a generic aromatic, or aryl, group. We use this when we want to indicate an aryl group is on a structure that we consider more important.

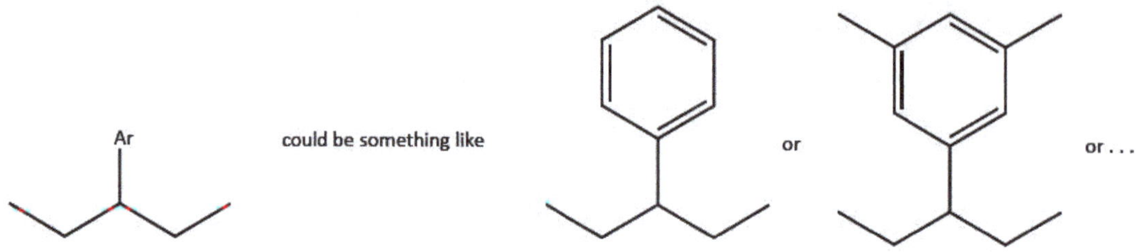

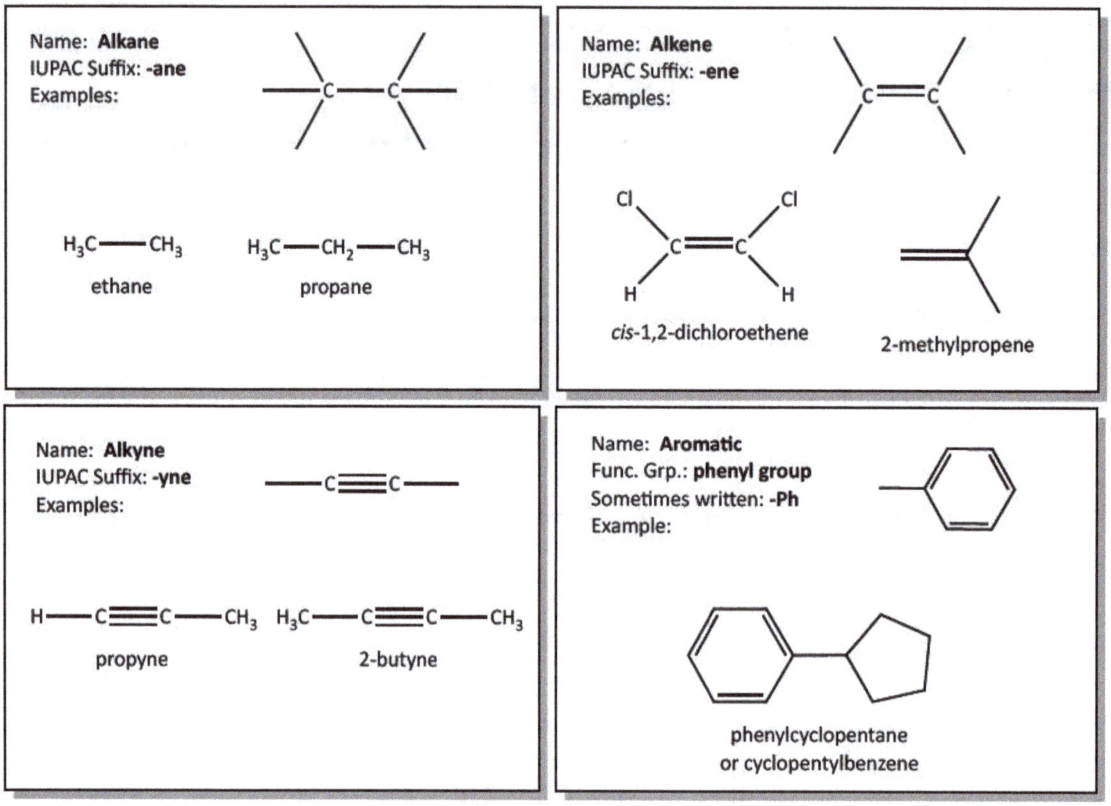

Figure 2.38 Hydrocarbons

2.12.2 Oxygen-containing compounds

Single bonds to oxygen

There are several types of functional groups that contain carbon-oxygen single bonds. Among these are alcohols, phenols, ethers, and epoxides. Alcohols are compounds with an –OH (hydroxyl) group on them. Phenols are a combination of phenyl groups and alcohols, hence the name. Ethers are compounds with an oxygen atom surrounded by two carbon atoms. An epoxide is an ether that is a 3-membered ring.

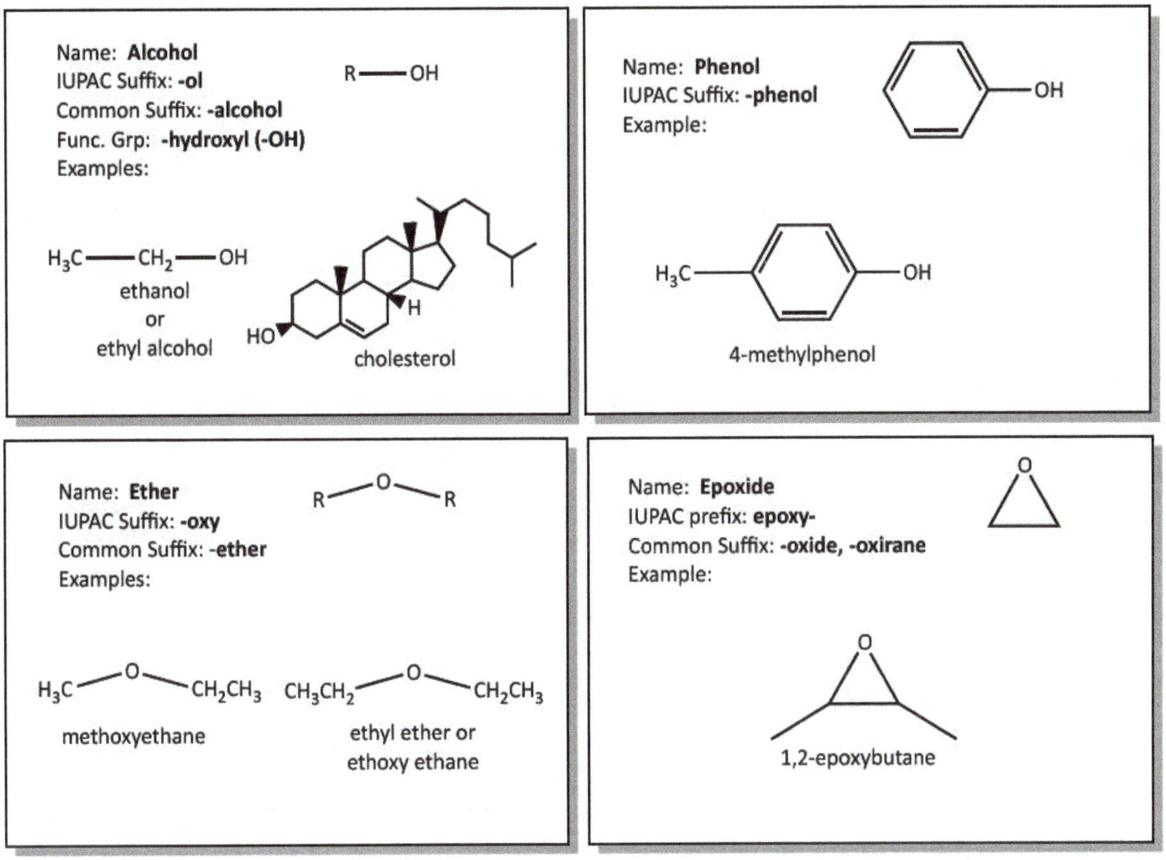

Figure 2.39 C-O single bonds

Carbonyl compounds

A carbonyl compound contains a C=O functional group. There are several of these carbonyl compounds.

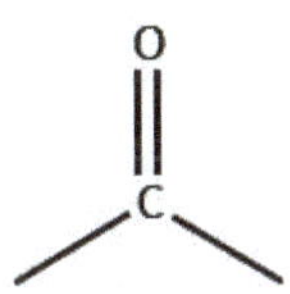

carbonyl

Aldehydes, ketones, and esters are often fruity smelling compounds. Carboxylic acids are acidic compounds. They are a part of amino acids.

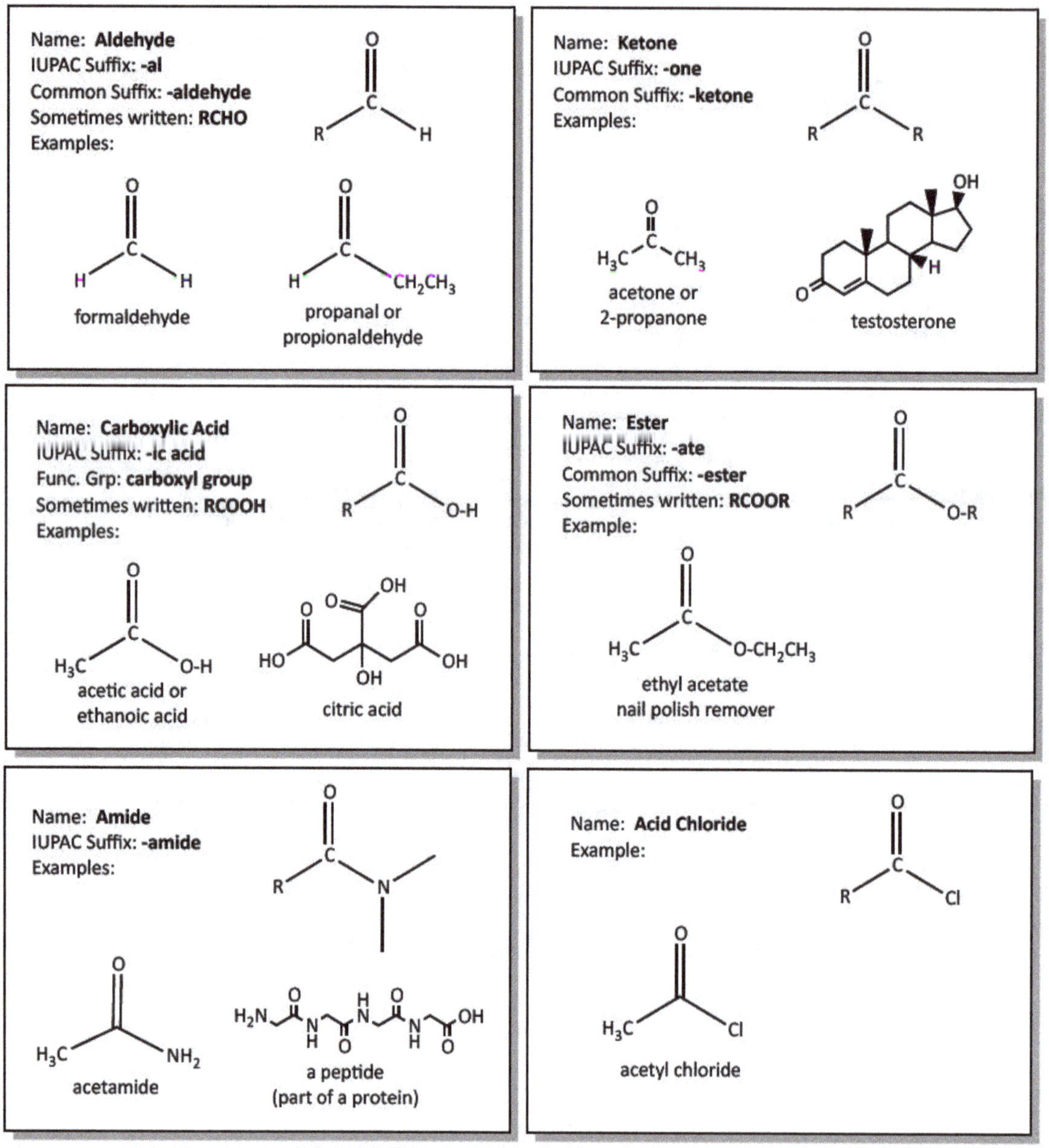

Figure 2.40 Carbonyl compounds

2.28 Draw the Lewis structures for the following compounds. Make sure you use the normal bonding patterns for each atom.

a) CH_3CHO b) CH_3CH_2COOH c) $CH_3CH_2COCH_3$ d) $CH_3CO_2CH_3$ e) CH_3COCl f) $CH_3CH_2COOCH_2CH_3$ g) $CH_3CONHCH_3$ h) CH_3CO_2H i) $CH_3CON(CH_3)_2$ j) $CH_3CH_2OCH_3$ k) CH_3CH_2OH l) $CH_3CHOHCH_3$

2.12.3 Nitrogen-containing compounds

Amines sometimes have a fishy smell. They are also usually basic compounds. Amines are also a part of amino acids. When the nitrogen of an amine has one carbon atom attached, it is a primary amine. When two carbon atoms are attached directly to the nitrogen atom, it is a secondary amine. When three carbon atoms are directly attached to the nitrogen atom, it is a tertiary amine. Nitro compounds tend to be explosive. You have probably heard of nitroglycerine or trinitrotoluene (TNT).

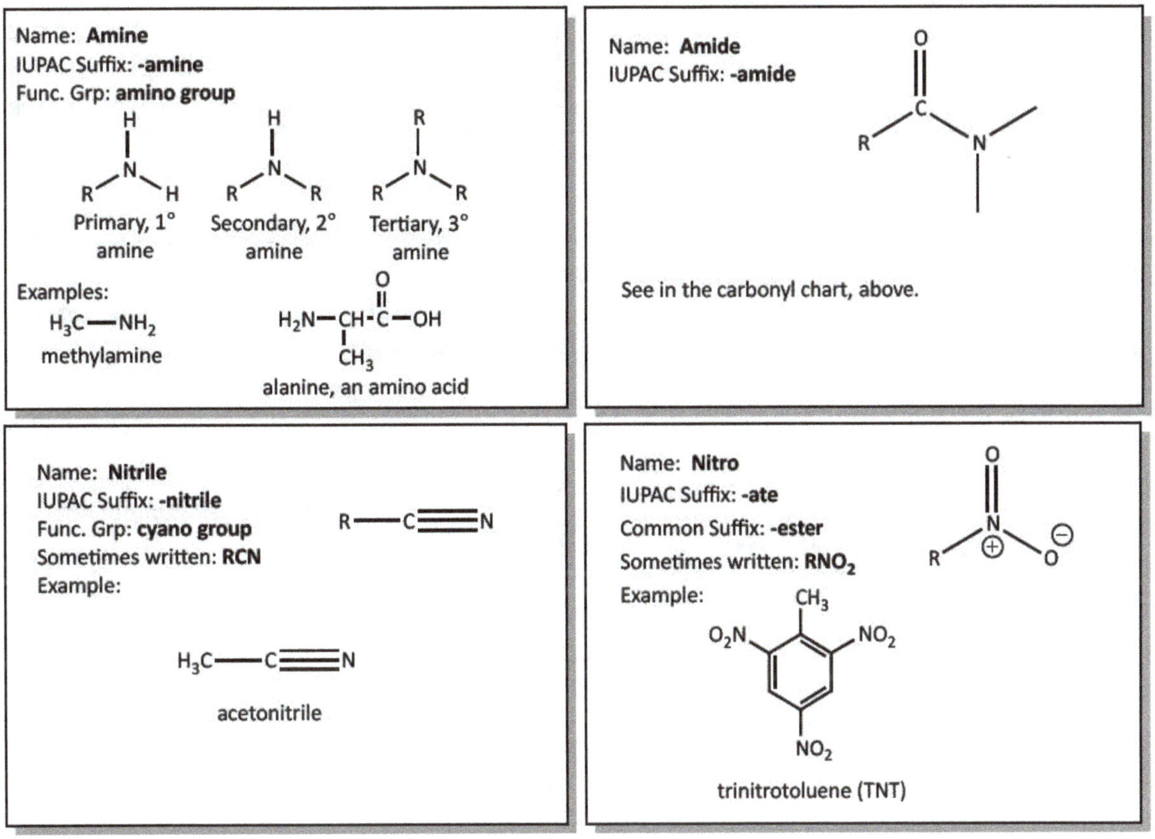

Figure 2.41 Nitrogen-containing compounds

2.12.4 Sulfur-containing compounds

Since sulfur is right below oxygen in the periodic table (the same group), sulfur forms some compounds similar to oxygen compounds. You'll notice that thiol is an analog of alcohol and sulfide or thioether is an analog of ether. Thiols are found in many foul-smelling compounds. Thiols are responsible for the pungent smells of skunks.

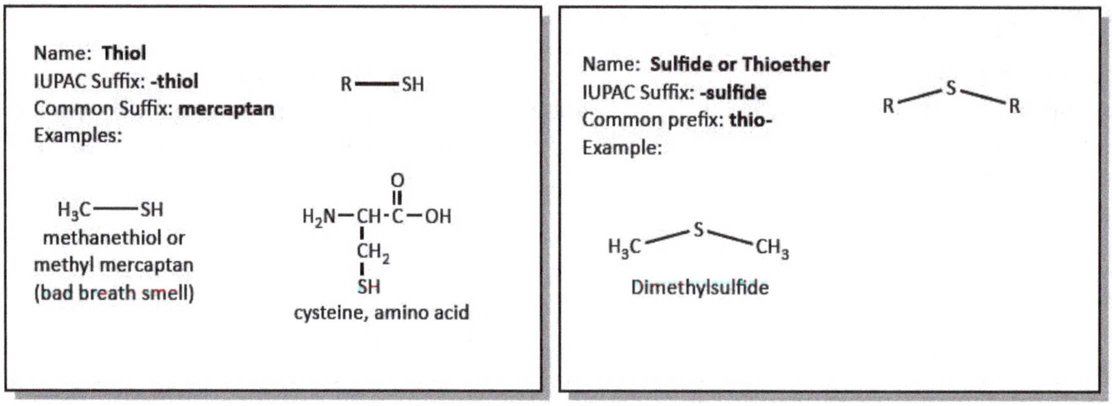

Figure 2.42 Sulfur-containing compounds

2.29 Identify the circled functional groups in the following molecule.
Possibilities: Alcohol, amine, nitro, nitrile, amide, ether, epoxide, ester, aldehyde, ketone, carboxylic acid, acid chloride, thiol, disulfide, thioether, alkane, alkene, alkyne, phenyl, phenol.

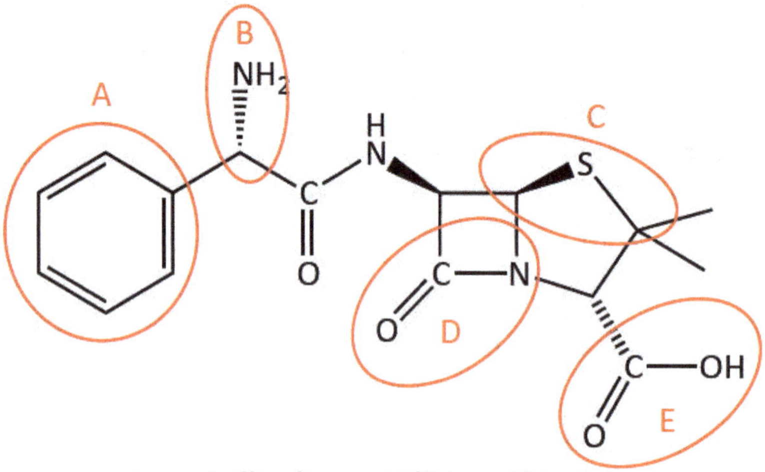

Amoxicillin (a penicillin antibiotic)

A ______________________

B ______________________

C ______________________

D ______________________

E ______________________

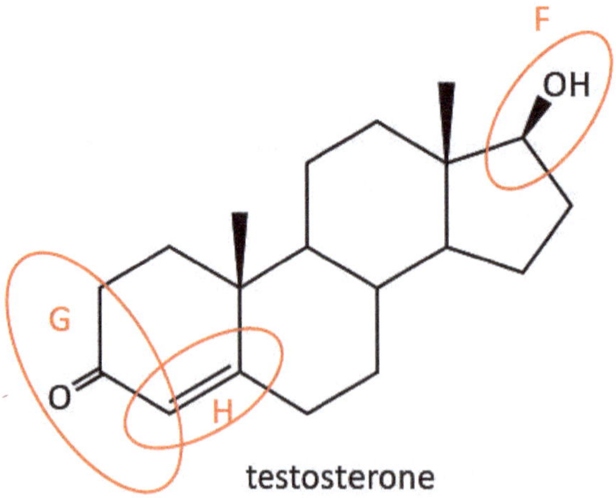

testosterone

F ______________________

G ______________________

H ______________________

2.13 Functional Group Study Cards

On the next few pages are to help you study your functional groups. Copy these pages and cut out each card and study them. Carry them around with you until you know them inside and out. Memorize these right now. Don't delay.

Name: **Alkane**
IUPAC Suffix: **-ane**
Examples:

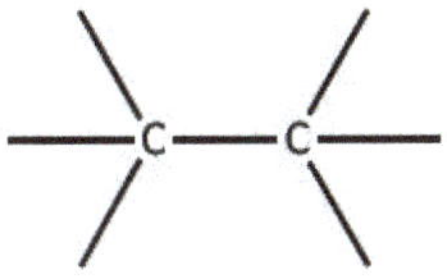

$H_3C—CH_3$
ethane

$H_3C—CH_2—CH_3$
propane

Name: **Alkene**
IUPAC Suffix: **-ene**
Examples:

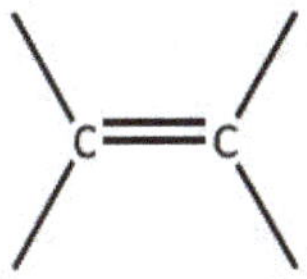

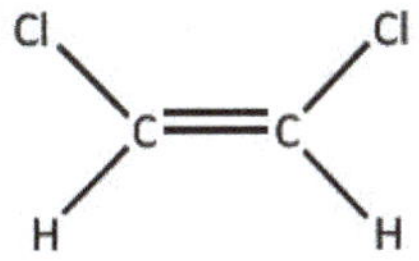

cis-1,2-dichloroethene

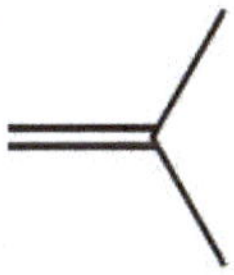

2-methylpropene

Name: **Alkyne**
IUPAC Suffix: **-yne**
Examples:

$—C\equiv C—$

$H—C\equiv C—CH_3$
propyne

$H_3C—C\equiv C—CH_3$
2-butyne

Name: **Aromatic**
Func. Grp.: **phenyl group**
Sometimes written: **-Ph**
Example:

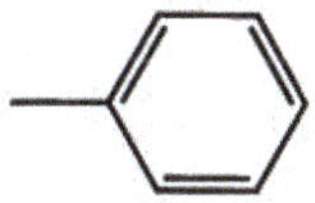

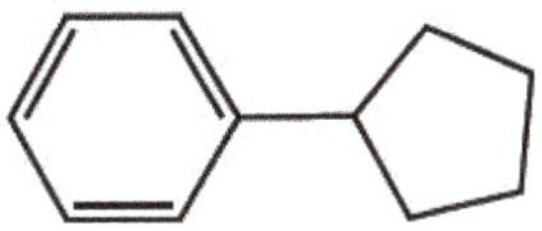

phenylcyclopentane
or cyclopentylbenzene

Name: **Alcohol**
IUPAC Suffix: **-ol**
Common Suffix: **-alcohol**
Func. Grp: **-hydroxyl (-OH)**
Examples:

$H_3C—CH_2—OH$
ethanol
or
ethyl alcohol

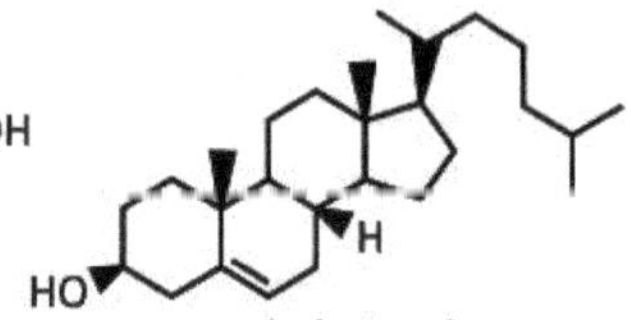

cholesterol

Name: **Phenol**
IUPAC Suffix: **-phenol**
Example:

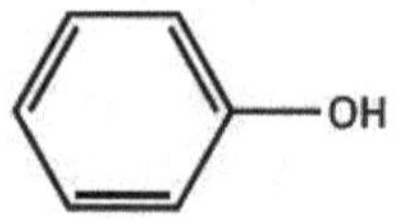

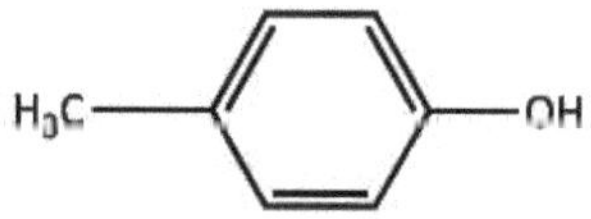

4-methylphenol

Name: **Ether**
IUPAC Suffix: **-oxy**
Common Suffix: **-ether**
Examples:

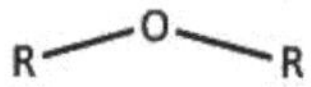

$H_3C—O—CH_2CH_3$
methoxyethane

$CH_3CH_2—O—CH_2CH_3$
ethyl ether or
ethoxy ethane

Name: **Epoxide**
IUPAC prefix: **epoxy-**
Common Suffix: **-oxide, -oxirane**
Example:

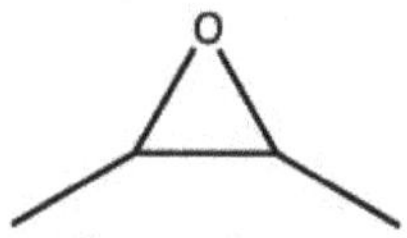

1,2-epoxybutane

Name: **Aldehyde**
IUPAC Suffix: **-al**
Common Suffix: **-aldehyde**
Sometimes written: **RCHO**
Examples:

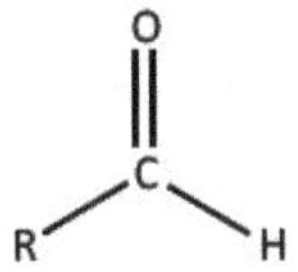

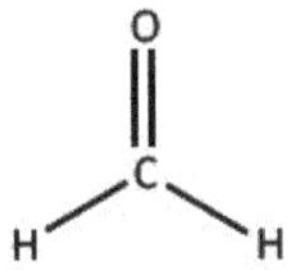

formaldehyde

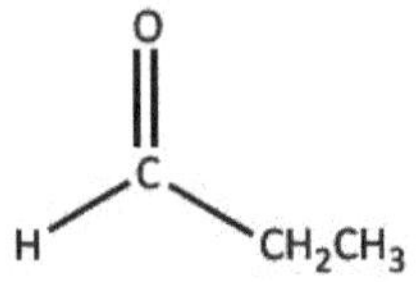

propanal or
propionaldehyde

Name: **Ketone**
IUPAC Suffix: **-one**
Common Suffix: **-ketone**
Examples:

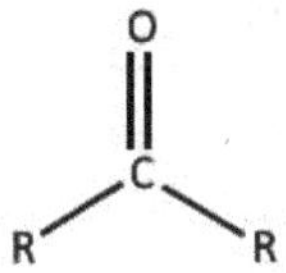

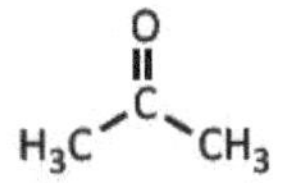

acetone or
2-propanone

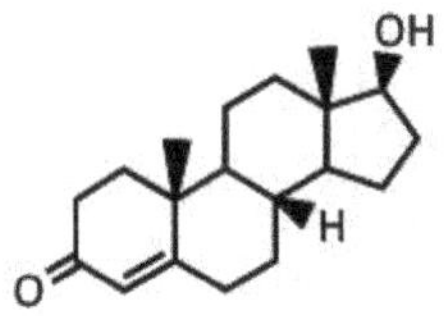

testosterone

Name: **Carboxylic Acid**
IUPAC Suffix: **-ic acid**
Func. Grp: **carboxyl group**
Sometimes written: **RCOOH**
Examples:

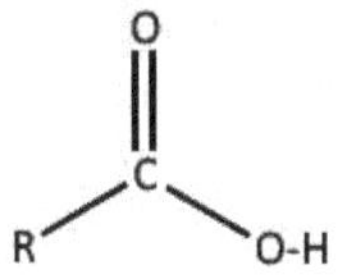

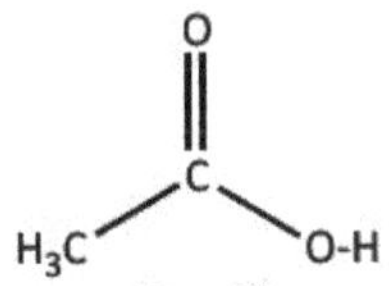

acetic acid or
ethanoic acid

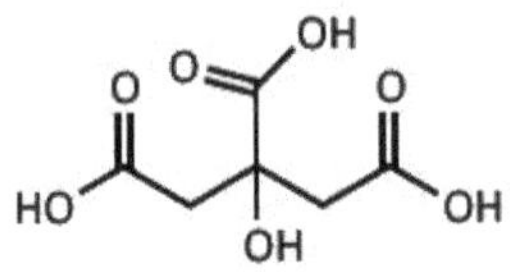

citric acid

Name: **Ester**
IUPAC Suffix: **-ate**
Common Suffix: **-ester**
Sometimes written: **RCOOR**
Example:

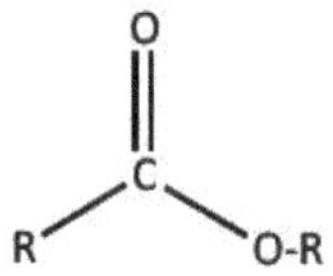

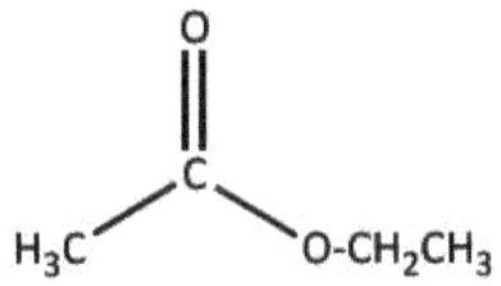

ethyl acetate
nail polish remover

Name: **Amide**
IUPAC Suffix: **-amide**
Examples:

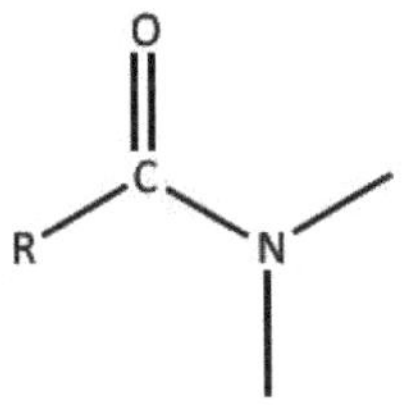

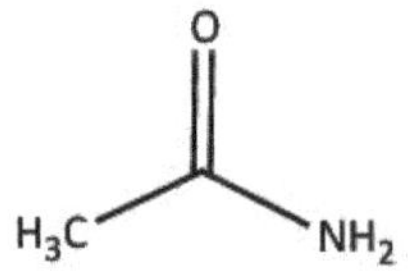

acetamide

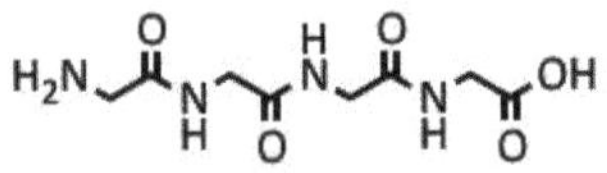

a peptide
(part of a protein)

Name: **Acid Chloride**
Example:

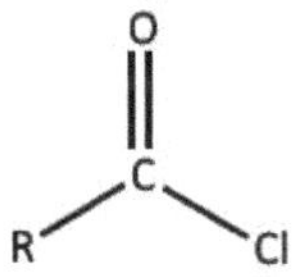

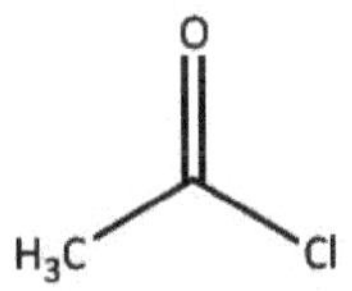

acetyl chloride

Bonding and Structure

Name: **Amine**
IUPAC Suffix: **-amine**
Func. Grp: **amino group**

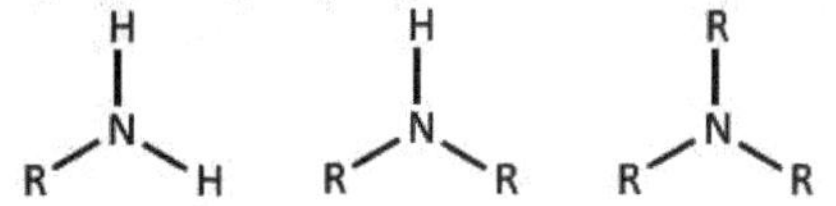

Primary, 1° amine — Secondary, 2° amine — Tertiary, 3° amine

Examples:

$H_3C—NH_2$
methylamine

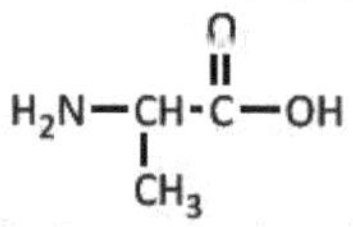

alanine, an amino acid

Name: **Amide**
IUPAC Suffix: **-amide**

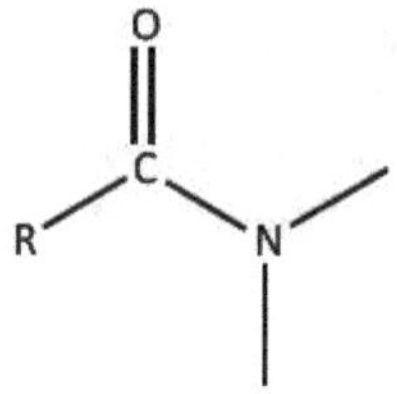

See in the carbonyl chart, above.

Name: **Nitrile**
IUPAC Suffix: **-nitrile**
Func. Grp: **cyano group**
Sometimes written: **RCN**
Example:

$R—C\equiv N$

$H_3C—C\equiv N$

acetonitrile

Name: **Nitro**
IUPAC Suffix: **-ate**
Common Suffix: **-ester**
Sometimes written: **RNO_2**
Example:

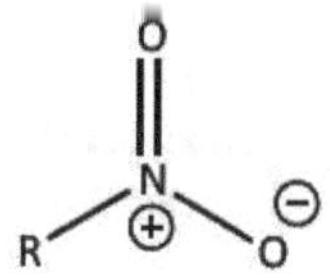

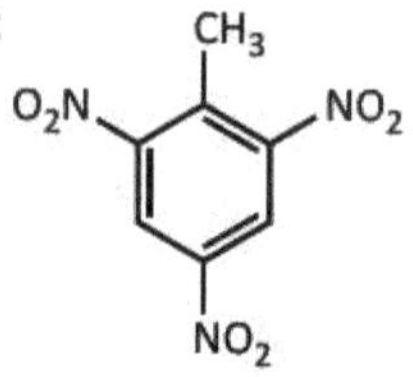

trinitrotoluene (TNT)

Name: **Thiol**
IUPAC Suffix: **-thiol**
Common Suffix: **mercaptan**
Examples:

R—SH

$H_3C—SH$
methanethiol or
methyl mercaptan
(bad breath smell)

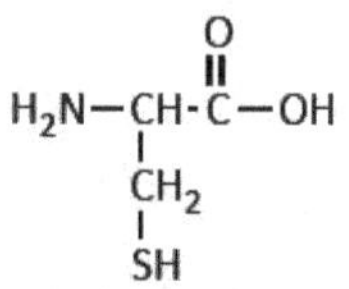

cysteine, amino acid

Name: **Sulfide or Thioether**
IUPAC Suffix: **-sulfide**
Common prefix: **thio-**
Example:

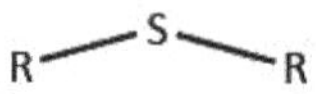

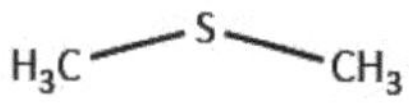

Dimethylsulfide

2.14 Problems

2.30 Identify the number of sigma and pi bonds in each of the following marked locations.

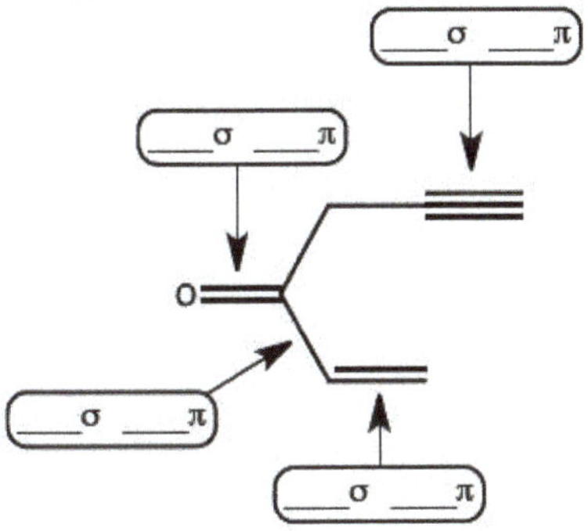

2.31 Homobatrachotoxin is a toxic alkaloid found in poison-dart frogs. If you put your finger in your mouth after handling such a frog, it tastes like a 9-volt battery. Identify the circled functional groups in the following molecule.

Possibilities: Alcohol, amine, nitro, nitrile, amide, ether, epoxide, ester, aldehyde, ketone, carboxylic acid, acid chloride, thiol, disulfide, thioether, alkane, alkene, alkyne, phenyl, phenol.

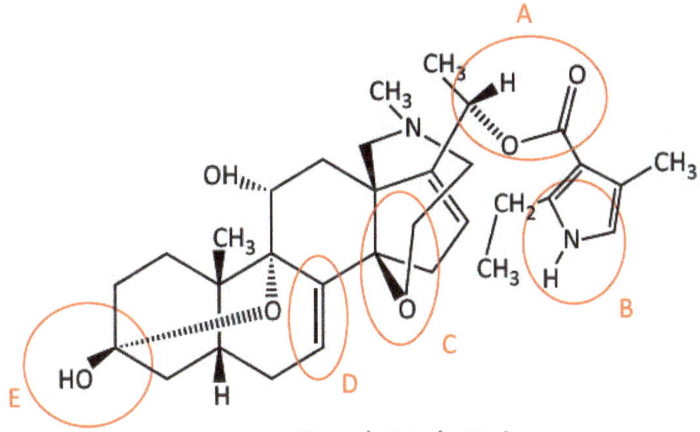

A ______________

B ______________

C ______________

D ______________

E ______________

Homobatrachotoxin

2.32 Write any nonzero formal charges on the following totally fictitious and crazy molecules.

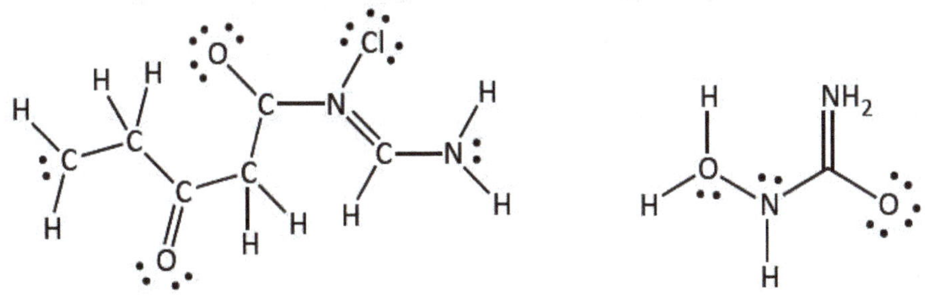

2.33 Draw Lewis structures for the following compounds/ions:

a) $CH_3CO_2CH_3$ b) acetic acid, CH_3COOH c) methanol, CH_3OH d) $CH_3OCH_2CH_3$

e) acetonitrile, CH_3CN f) $CH_3NHCOCH_3$ g) $(CH_3)_2CHCOOH$ h) $CH_3COCH_2COOCH_3$

i) $(CH_3)_3COH$ j) CH_3CHO k) $NCCH_2CH_2CHO$

Bonding and Structure

2.34 Draw an acceptable line-angle structure for the compounds shown below.

a) 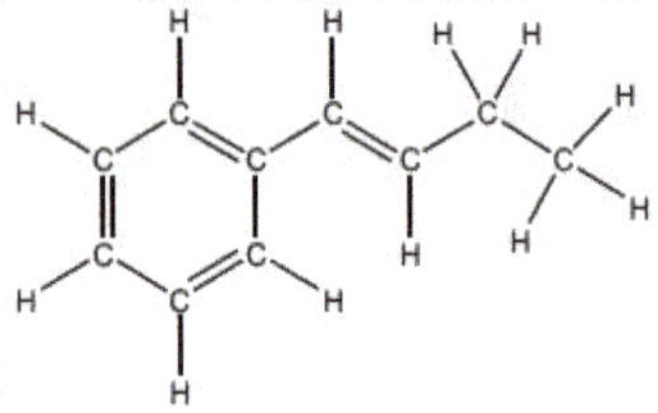 b) $CH_3COCH_2COOCH_3$ c) $(CH_3)_2CHCOOH$ d) $NCCH_2CH_2CHO$

2.35 Draw the Lewis structure (all atoms and lone pairs of electrons) for the following line-angle structures.

a) 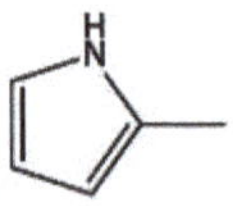 b) 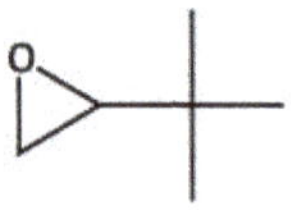 c) 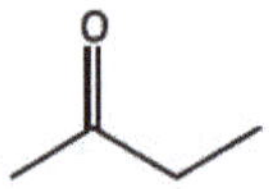 d) 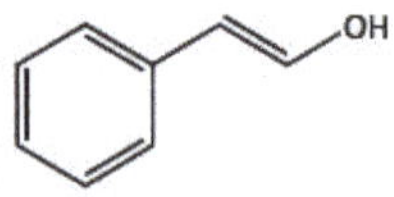

2.36 a) Draw any isomer of C_5H_7ClO that contains a ketone. (There are many correct answers.) b) Draw any isomer of C_5H_7ClO that does not contain a ketone. (Again, there are many correct answers.)

2.37 The structure of Vitamin C is shown here. Refer to this structure to answer the following questions.

a) Vitamin C contains _____ sp^2 hybridized oxygen atom.
b) Vitamin C contains _____ sp^2 hybridized carbon atoms.
c) Vitamin C contains _____ hydroxyl groups.
d) Vitamin C contains _____ sp^3 hybridized carbon atoms.
e) Vitamin C contains _____ aldehyde groups.
f) Vitamin C contains _____ esters groups.
g) Vitamin C contains _____ alkene groups.
h) Vitamin C contains _____ pi bonds.

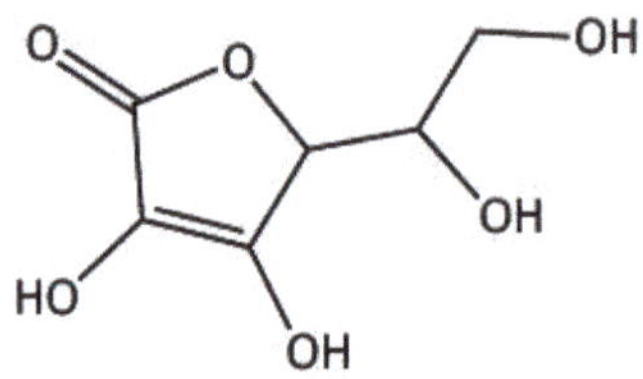

2.38 Refer to the following molecule to answer these questions.

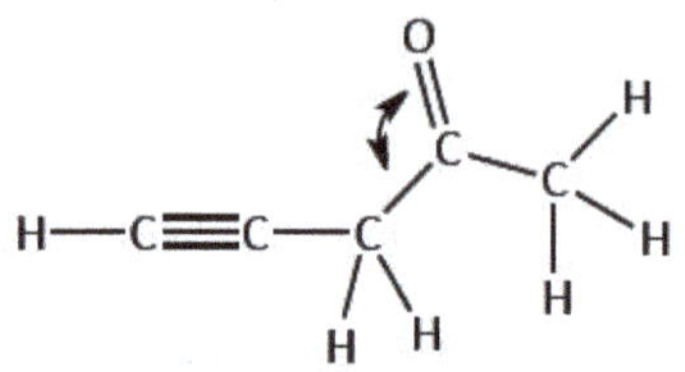

a) The molecule shown below contains _______ pi bonds and ________ sigma bonds.
b) The hybridization of the oxygen atom is _____ and the O-C-C bond angle marked with an arrow is _____°.
c) This molecule contains which two of the following functional groups? (Alcohol, amine, nitrile, amide, ether, ester, aldehyde, ketone, carboxylic acid, thiol, disulfide, thioether, alkene, alkyne, phenyl, phenol)

2.39 Draw an example of the following:

a) an ether	b) a carboxylic acid	c) an amide	d) a nitrile	e) a phenol	f) an ester
g) a ketone	h) an aldehyde	i) an alcohol	j) an alkyne	k) an amine	l) an alkene

2.40 Calculate the molecular formulas for each of the following.

a) 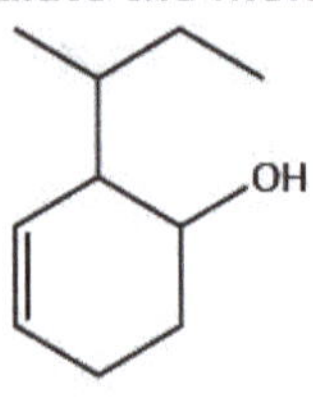 b) 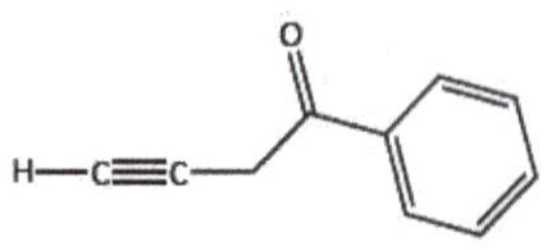 c) 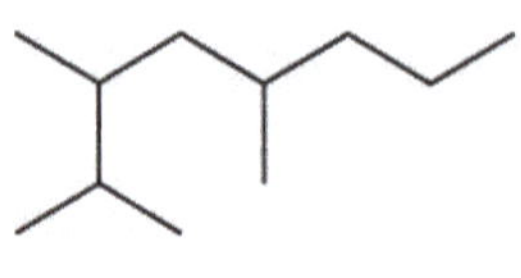

2.41 Which of the following molecules would have a net dipole moment? (Circle them)

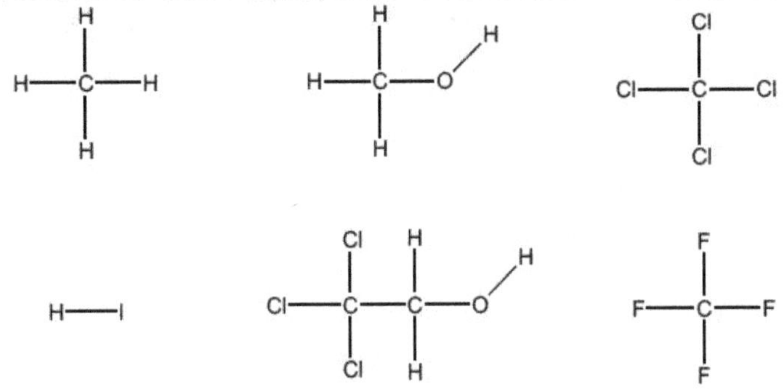

2.42 Identify the circled functional groups in the following molecules.

Possibilities: Alcohol, amine, nitro, nitrile, amide, ether, epoxide, ester, aldehyde, ketone, carboxylic acid, acid chloride, thiol, disulfide, thioether, alkane, alkene, alkyne, phenyl, phenol.

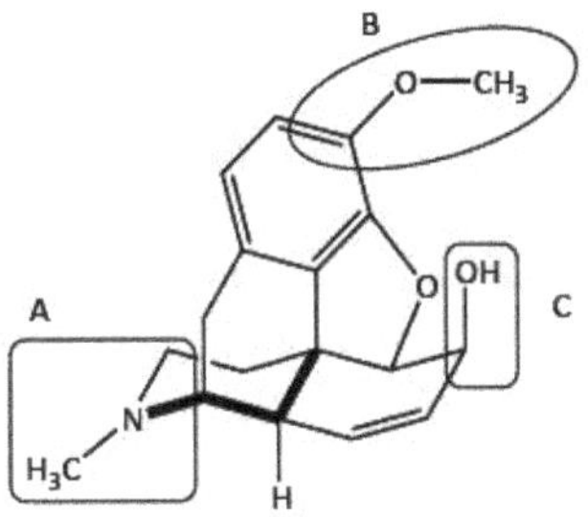

morphine

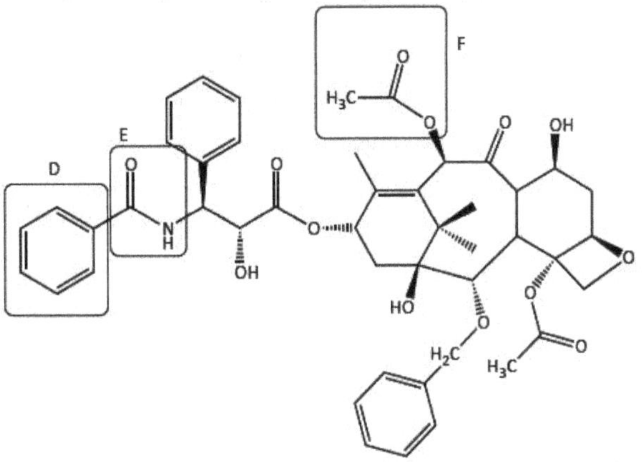

Taxol (a cancer fighting drug)

A ____________________

B ____________________

C ____________________

D ____________________

E ____________________

F ____________________

2.43 Which compound is more soluble in water? CH_3OCH_3 or CH_3CH_2OH? Briefly explain your choice.

2.44 The molecule shown below contains __________ pi bonds and __________ sigma bonds.

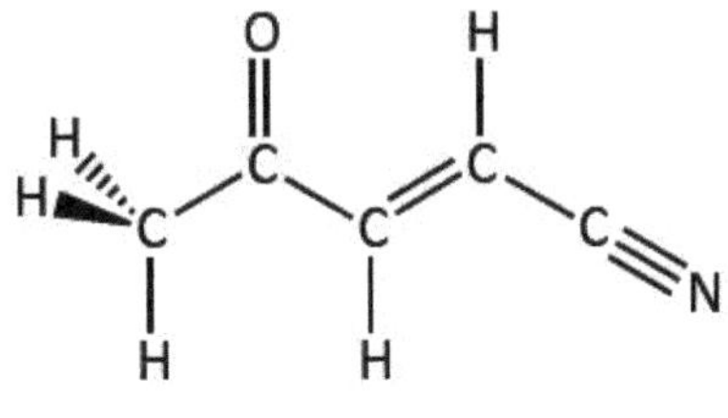

Bonding and Structure

2.45 Which two functional groups from the following list contain a triple bond? (Alcohol, amine, nitrile, amide, ether, ester, aldehyde, ketone, carboxylic acid, acid anhydride, thiol, disulfide, thioether, alkene, alkyne, phenyl)

2.46 Label the formal charges on the nitrogen and oxygen atoms in the following molecule.

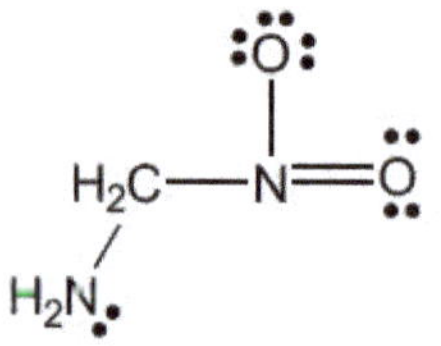

2.47 Label the hybridization of each of the nitrogen atoms in adenine (a DNA base).

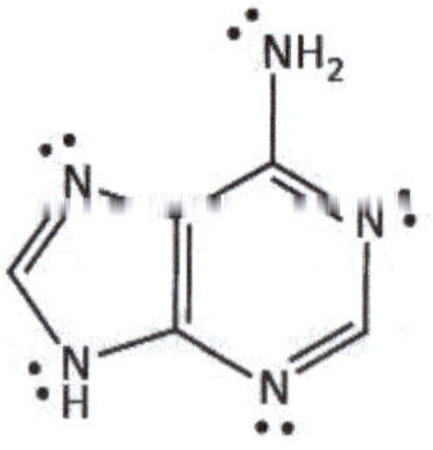

Adenine

2.48 Identify the circled functional groups in the following molecules.

Possibilities: Alcohol, amine, nitro, nitrile, amide, ether, epoxide, ester, aldehyde, ketone, carboxylic acid, acid chloride, thiol, disulfide, thioether, alkane, alkene, alkyne, phenyl, phenol.

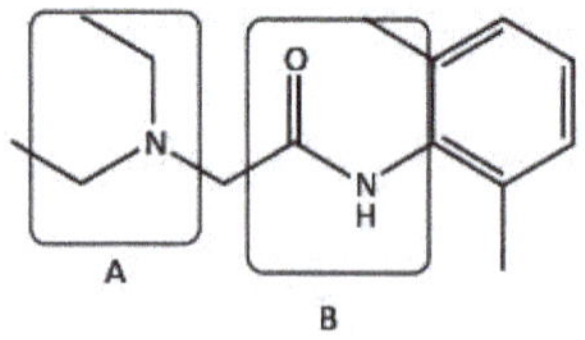

Lidocaine
(anesthetic)

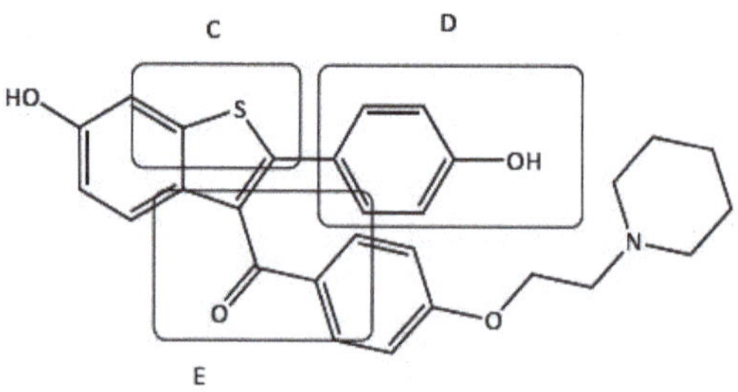

Raloxifene
(osteoporosis treatment)

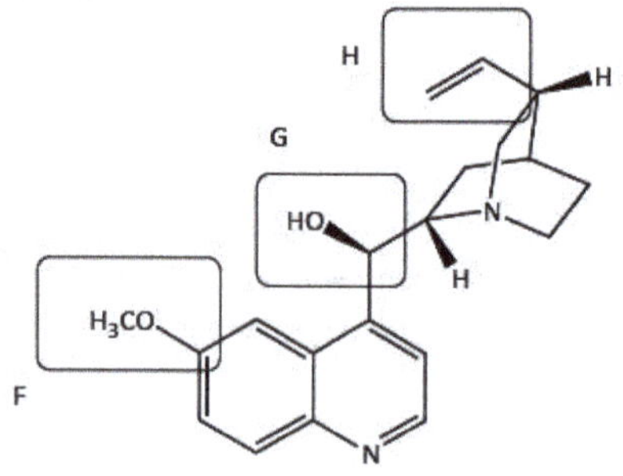

Quiniine
(anti-malarial)

A ______________________

B ______________________

C ______________________

D ______________________

E ______________________

F ______________________

G ______________________

H ______________________

2.49 For the following molecule (azidomethane), fill in any non-zero formal charges on the atoms.

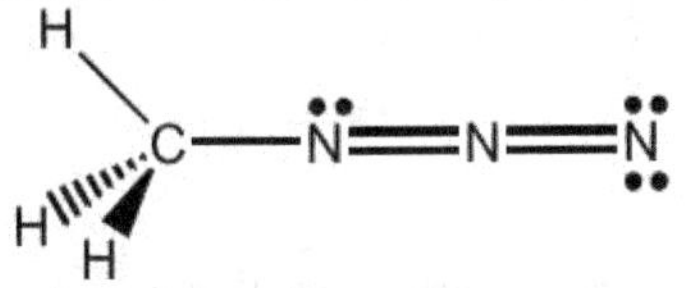

Draw a diagram showing all of the atomic orbitals for azidomethane. Indicate which atomic orbitals overlap to form sigma bonds, which ones overlap to form pi bonds, and which ones hold lone pairs of electrons. I've drawn out the orbitals around N to get you started. (Note: For sake of clarity, please don't draw the small back lobes of the hybrid orbitals).

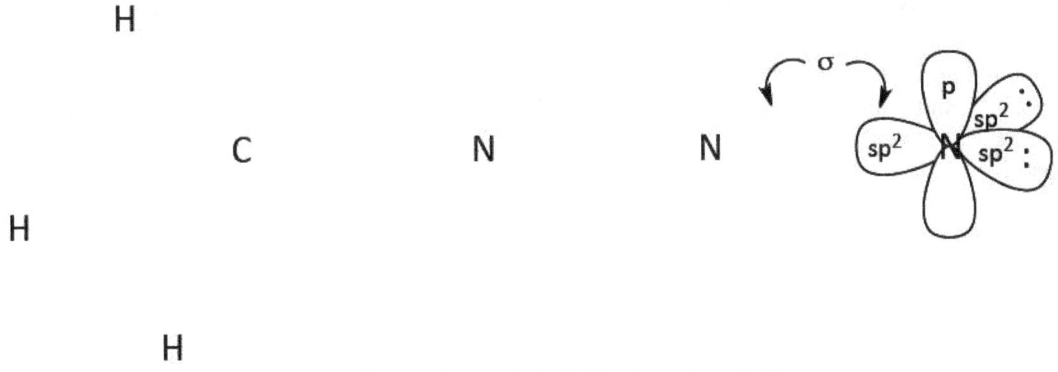

2.50 Draw the line-angle formula for $CH_3CH_2C(CH_3)_2CH_2CH(CH_2CH_3)CH_2CH(CH_3)_2$.

2.51 Predict the dipole moment of each of the following bonds.
a) C-O b) C-N c) C-Cl d) S-O e) C-F

2.52 Indicate the hybridization, the geometry, and the bond angle for the most central atom in each of the following compounds
a) $CH_3CH_2CH_3$ b) CH_3OCH_3 c) $CH_3C{\equiv}N$ d) $CH_2{=}CH{-}CH_3$ e) $CH_2{=}C{=}CH_2$ f) O=C=O

2.53 Two compounds have the formula Br-CH=CH-Br. a) Draw them both. b) Of these two isomers, one is polar and one is nonpolar. Identify the one that is polar. c) Then, explain why only one compound with the formula $Br_2C{=}CH_2$ exists.

2.54 Sketch the line-angle structures for the following oxygen-containing molecules and identify each according to its functional group. (alcohol, ether, ketone, aldehyde, carboxylic acid, phenol, or ester)
a) $Cl{-}CH_2{-}CHO$ b) $CH_3CH_2{-}O{-}CH_3$ c) $(CH_3)_2CHCOOH$ d) $(CH_3)_3CCH_2CO_2CH_3$
e) $CH_3CH_2COCH_3$ f) $CH_3CHOHCH_2CH_3$ g) $HO_2CCH_2CH_3$ h) $(CH_3)_2CHCO_2H$

2.55 Draw the Lewis structure for each compound and then identify its functional group.

a) 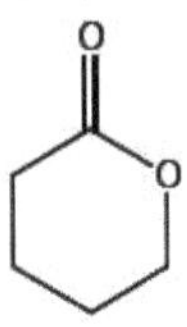 b) $HC{\equiv}C{-}CH_3$ c) 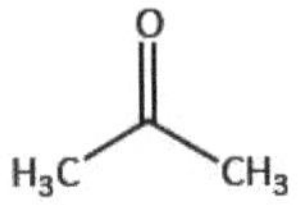 d) 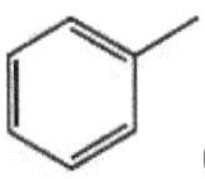 e) $H_3C{-}CHO$ f) $H_3C{-}COOH$ g) 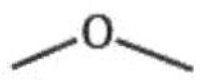

h)  i) 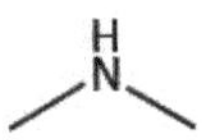 j) 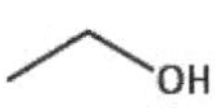 k) 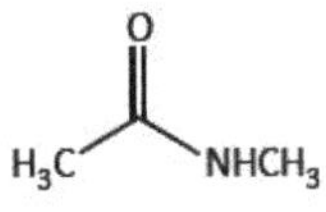 l)  m) CH_3SCH_3

Bonding and Structure

2.56 For each of the following compounds identify the atoms with pi bonds between them. Identify the hybridization of these atoms. Sketch the orbitals that overlap to make sigma and pi bonds between these atoms.

a) 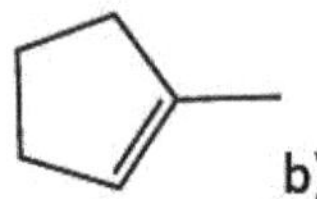 b) :N≡C—CH_3 c) 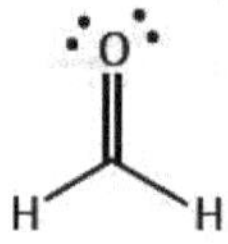

2.57 Which of the following pure compounds would have hydrogen bonding between the molecules? Circle them. Underline those that can hydrogen bond to water?

a) CH_3-NH_2 b) CH_3OCH_3 c) CH_2=CH-CH_2CH_3 d) $(CH_3)N$ e) CH_3OH f) CH_3CH_2CHO
g) CH_3CO_2H f) $CH_3CO_2CH_3$ h) $CH_3CONHCH_2CH_3$ i) CH_3COCH_3 j) $CH_3CON(CH_3)_2$

2.58 For each of the following pairs of compounds, identify the one with the higher boiling point.

a) CH_3CH_2OH and CH_3OCH_3 b) $CH_3(CH_2)_7CH_3$ and $CH_3(CH_2)_7CH_2OH$ c) CH_3NHCH_3 and $CH_3CH_2NH_2$
d) $HOCH_2CH_2CH_2CH_2OH$ and $CH_3CH_2CH_2CH_2OH$ e) $CH_3(CH_2)_7CH_3$ and $CH_3(CH_2)_3O(CH_2)_3CH_3$
f) $CH_3(CH_2)_6CH_3$ and $(CH_3CH_2)_4C$

Chapter 3. Mass Spectrometry & Infrared Spectroscopy

3.1 Mass Spectrometry

Mass spectrometry (mass spec is a method to measure the molecular weights of molecules and parts of molecules. This is a wonderful marvel of technology. If we want to identify a compound, perhaps a new molecule we made in the lab, mass spectrometry is a great first step in identifying that compound. Mass spectrometry can give us the molecular weight of a molecule. Mass spectrometry also breaks molecules into pieces and measures the molecular weights of those pieces. Mass spectrometry can also quickly tell us whether a molecule contains nitrogen, sulfur, or halogen atoms. These are all very useful in helping us identify and learn more about a molecule.

3.1.1 How it works

There are several methods by which mass spectrometry can be done. Electron impact (EI) is one of the simplest.

A beam of high-energy electrons strikes a sample. This beam of electrons breaks the molecule apart. The molecule and its fragments are ionized. The molecular weights of the ionized molecule and its fragments are measured and graphed. This graph is called the *mass spectrum*.

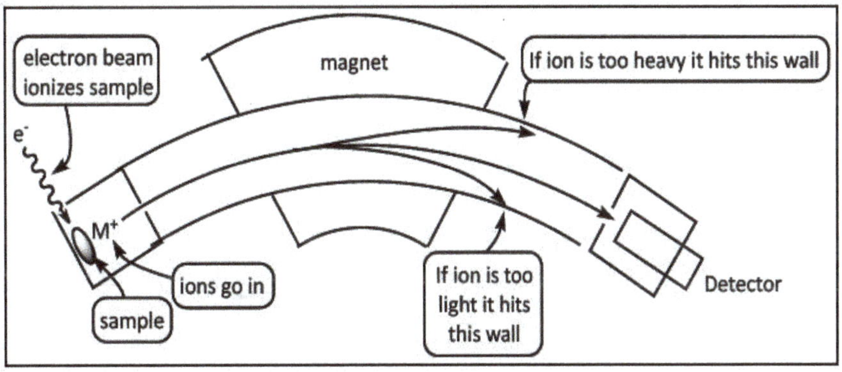

Figure 3.1 Generic Mass Spectrometer

Let's look at propane. An electron beam slams into propane and knocks an electron away making what is called the **molecular ion** (M^+), a radical cation. It is a radical because it has an unpaired electron and it is a cation because it has a positive charge. The mass spectrometer measures the molecular weights of ions only. The ions enter into a vacuum tube and flies towards the detector. A magnetic field bends the ion. Heavy ions have too much inertia and are not bent enough to reach the detector. They hit the wall. Light ions have too little inertia and are bent too much by the magnet to reach the detector. They hit the wall. Ions that have a certain mass for that magnetic field are bent in just the right way to reach the detector. Once an ion hits the detector, it records that is has been hit.

The magnetic field strength can be varied. If it is increased, it allows the heavier ions to bend more and hit the detector. If it is lessened, it allows the lighter ions to hit the detector. Before a mass spectrum is obtained, the instrument is calibrated with molecules of different masses. The magnetic field strength for the various molecular weights is carefully measured. Then, when an unknown sample is run, the magnetic field strength is varied. When the detector records a hit, knowing the magnetic field strength for that detector hit gives the molecular weight of that ion.

What is reported on the mass spectrum is actually a mass-to-charge ratio (m/z), the mass (m) divided by the charge (z). Usually, the charge, z, is +1, so what is recorded on the mass spectrum is the molecular weight of the ion. If an ion happens to have +2 charge, the magnet would bend it twice as much resulting in a peak on the mass spectrum at ½ the molecular weight of the ion.

3.1.2 Fragments

Not only is the M^+ molecular weight measured, but the molecules also break into ion fragments whose weights can be measured. The mass spectrum graphs how many of each ion of different m/z hits the detector. Don't worry about identifying every peak in a mass spectrum. Some strange reactions occur in the mass spectrometer. For our purposes, it is best to just look at the simpler ions made. Remember that a carbon atom has an atomic mass of 12 and a hydrogen atom has an atomic mass of 1.

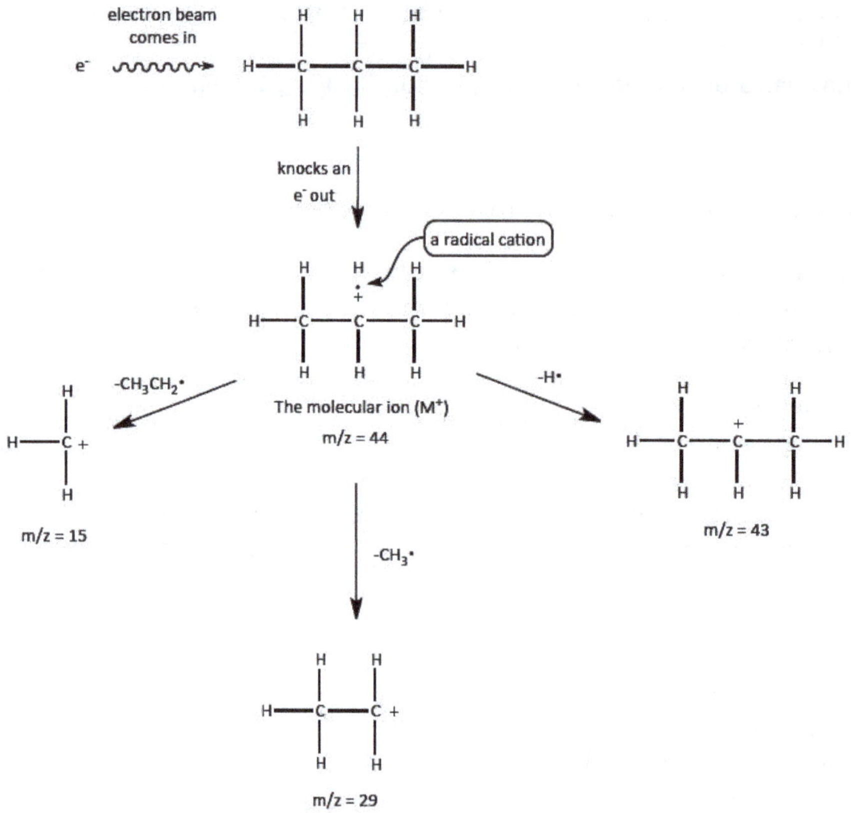

Figure 3.2 Mass Spec Fragmentation of Propane

3.1.3 The Mass Spectrum

The *molecular ion peak* of a mass spectrum is the peak that represents the molecular weight of the entire molecule being studied. It is usually the peak farthest to the right in the mass spectrum, the heaviest peak. The weights of the fragments also appear. In propane's mass spectrum, ion fragments that weigh 29 are the most abundant. More of those fragments hit the detector than any other. The *base peak* is the largest, tallest peak in the mass spectrum. For propane, the base peak is at 29 m/z.

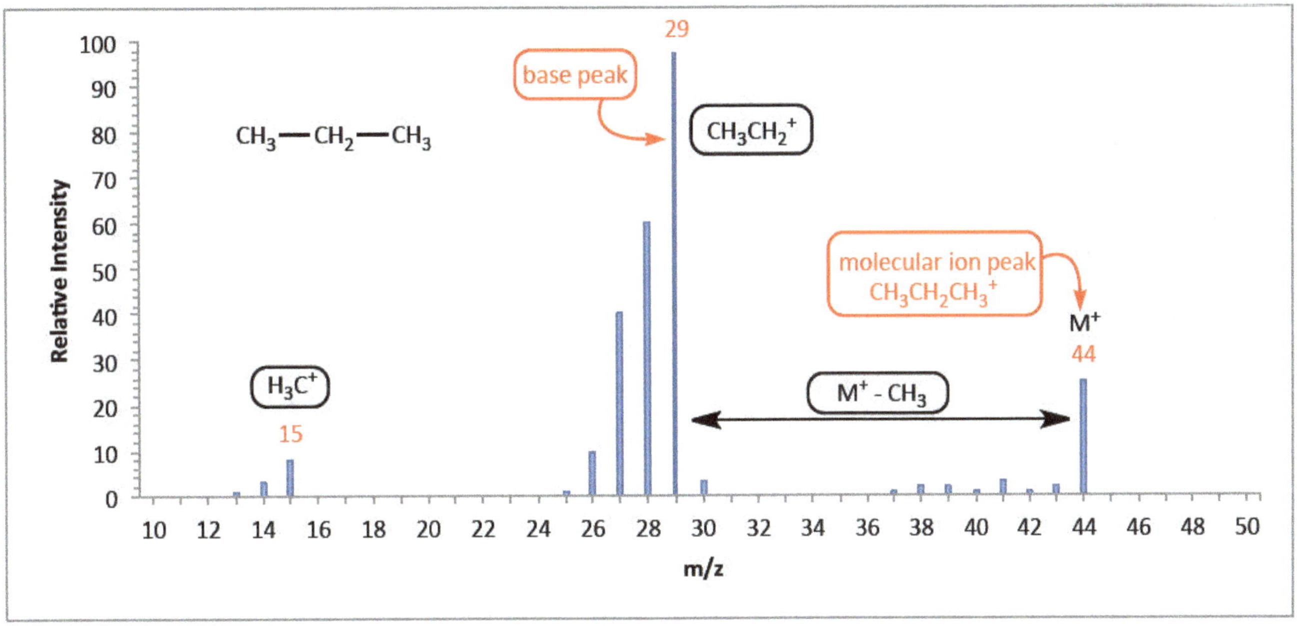

Figure 3.3 Mass Spectrum of Propane

3.1.4 Isotopes

It is possible to easily see if a compound contains a special atom, a bromine, chlorine, or sulfur atom. We look to the right side of the spectrum and at the area for the molecular ion peak (M^+). An M+2 peak is a peak that is two amu higher than the M^+ peak. If the molecule being studied contains one of these special atoms, there will be a larger than usual M+2 peak. These M+2 peaks exist because these special atoms have two major isotopes that are two amu apart from each other.

Bromine

The periodic table says that the atomic mass of bromine is 79.9amu. No bromine atom weighs 79.9. Bromine atoms either weigh 79 or 81. We call these various forms of an atom with different weights, isotopes. They differ by the number of neutrons in their nucleus. In nature, there is about an equal mixture of the two (with just a little more of the 79 isotope). We average those weights of the isotopes found in nature to come up with the atomic mass found on the periodic table of 79.9.

Since bromine atoms weigh either 79 or 81, what would we expect for the molecular weight we would see in a mass spectrum for bromomethane, CH_3Br? There are two forms of CH_3Br. One bromine isotope weighs 79 and one weighs 81. Since carbon weighs 12 and hydrogen weighs 1, CH_3 weighs 15. Therefore, the molecular mass of $CH_3{}^{79}Br$ is 94 while the molecular mass of $CH_3{}^{81}Br$ is 96. Since the two isotopes of bromine are found in nearly equal amounts, we see these two peaks in the mass spectrum in nearly equal amounts. The peak at 94 is the M+ and the peak at 96 is the M+2.

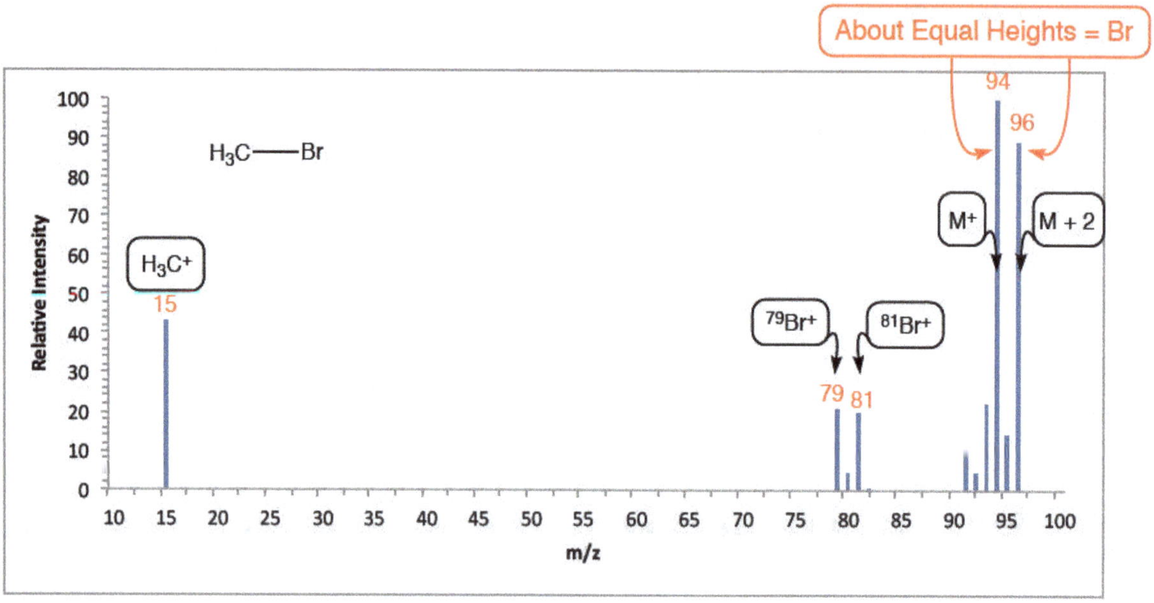

Figure 3.4 Mass spectrum of bromomethane

We often look at the M+ peak in a mass spectrum to see if there is an M+2 peak. If we see an M+ and an M+2 peak with equal heights, it indicates a bromine atom is present. This is a fingerprint for bromine.

You might wonder about being confused and calling the peak farthest to the right the M+ peak and thinking that there is another fragment peak two amu less (an M-2 peak). It would be quite strange to lose something that weighs two amu off of a molecule. So, we know that we are nearly always dealing with an M+ and an M+2 isotope peak.

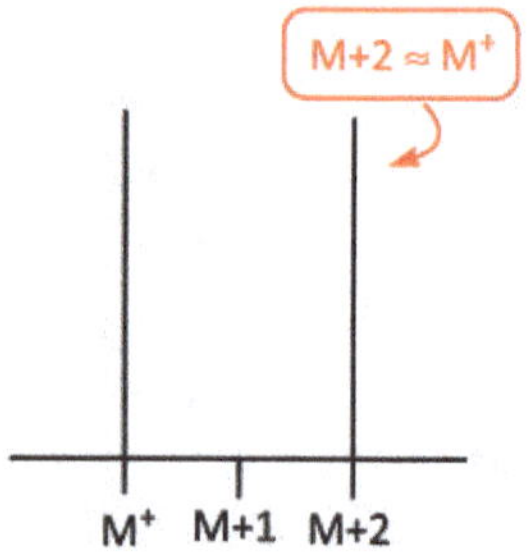

Figure 3.5 Bromine pattern

Chlorine

Chlorine's atomic weight on the periodic table is 35.5 amu. Chlorine has two isotopes, ^{35}Cl and ^{37}Cl. 75% of all chlorine naturally found is ^{35}Cl while the other 25% is ^{37}Cl.

Therefore, the M+2 peak for a compound that contains chlorine is about 1/3 the height of the M+ peak.

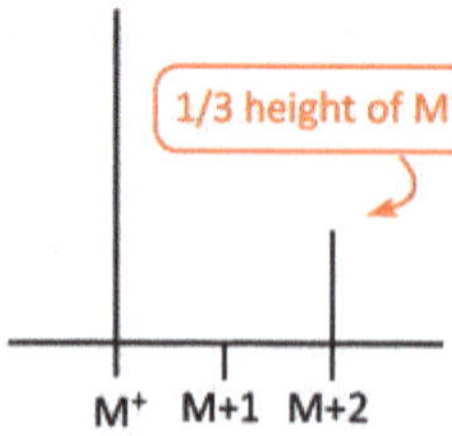

Figure 3.6 Chlorine pattern

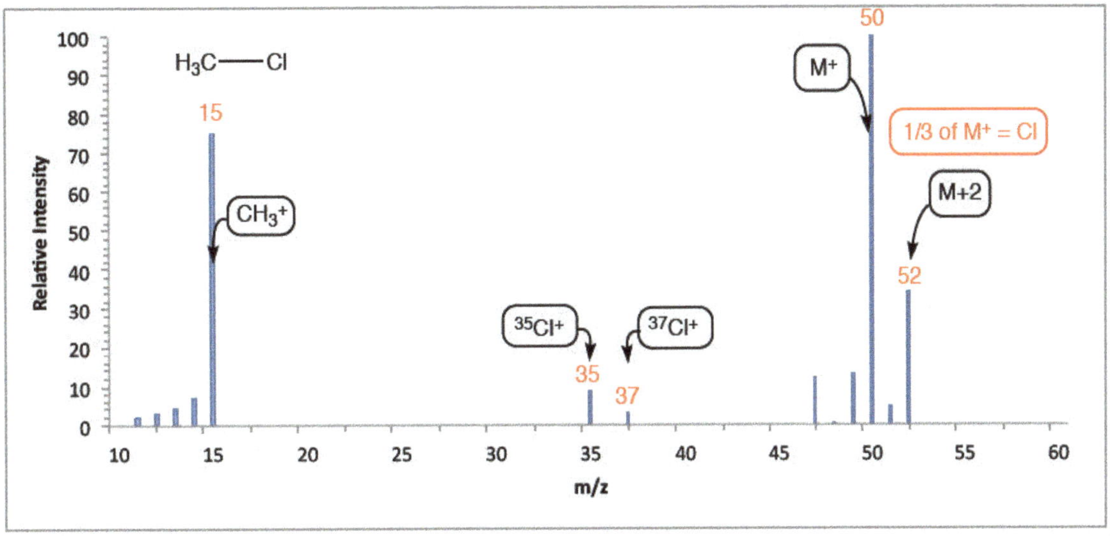

Figure 3.7 Mass Spectrum of Chloromethane

Iodine

Iodine only has one isotope. All iodine atoms are iodine-127. So, in a mass spectrum for a compound that contains iodine, we often see a peak at 127. This is the molecular weight for iodine. A peak of M^+-127 (loss of an iodine atom) is also seen.

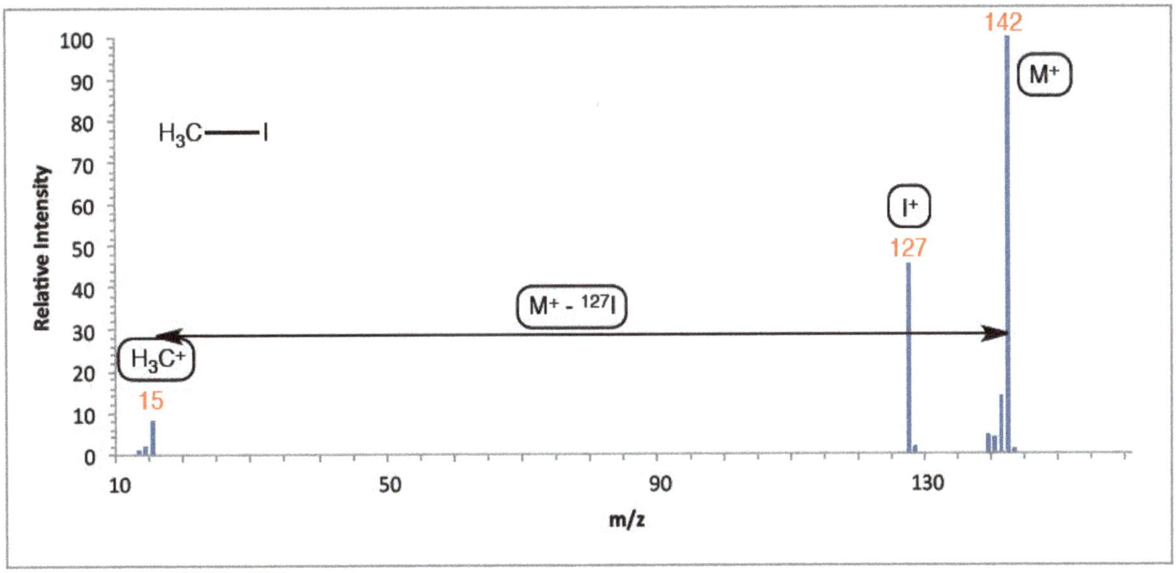

Figure 3.8 Mass Spectrum of Iodomethane

Sulfur

Sulfur has two main isotopes, sulfur-32 and sulfur-34. The natural abundance for sulfur-34 is 4.2% of sulfur-32. So, in a mass spectrum, the M+2 peak is usually about 4% of the M+ peak. Usually, if there is no special isotope, the M+2 peak is non-existent. Look for an M+2 peak that is larger than usual to indicate the presence of sulfur.

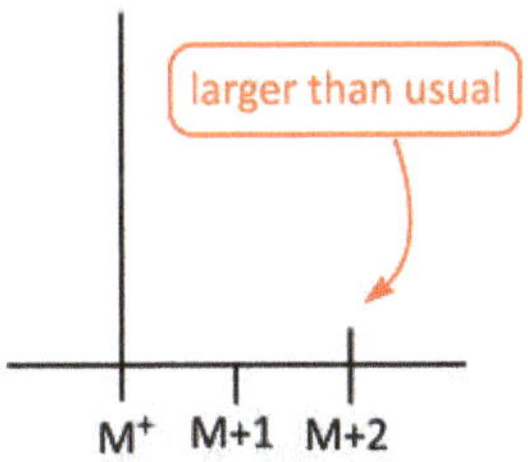

Figure 3.9 Sulfur pattern

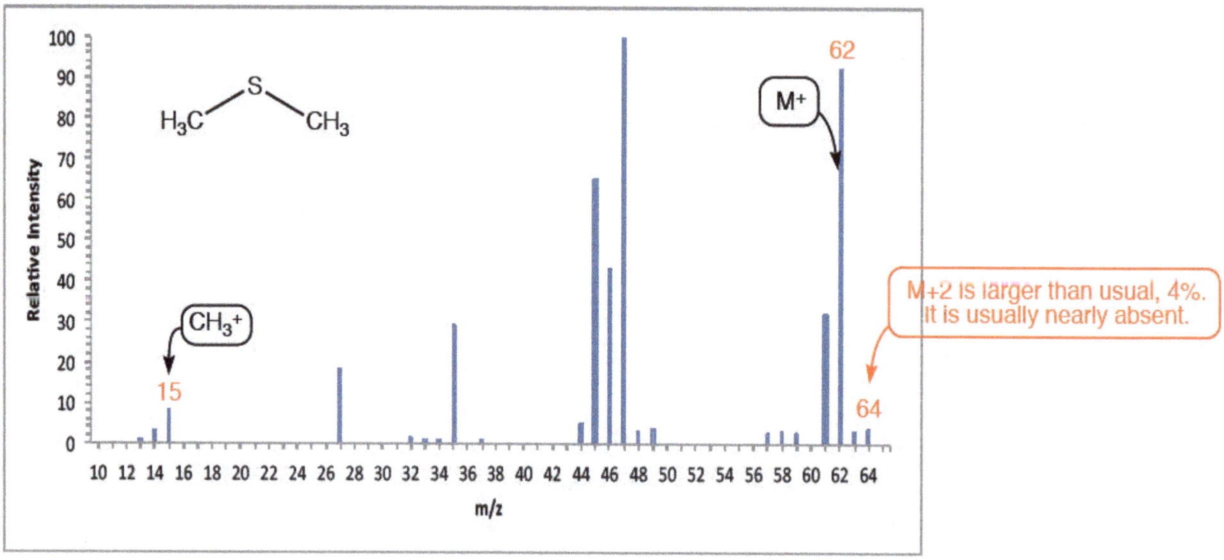

Figure 3.10 Mass spectrum of dimethyl sulfide

Nitrogen

Nitrogen atoms do not have M+2 isotope peaks to help us identify them. But, they do give a clue when they are present. All of the M^+ peaks you have seen so far have something in common; they all have an even number m/z. This is remarkable. But, nitrogen compounds are different. If a molecule has an odd number of nitrogen atoms in it, it has an *odd* molecular mass. So, a typical nitrogen-containing compound with one nitrogen atom will be odd, but if there are two nitrogen atoms present, it is even. Below is the mass spectrum for methylamine, CH_3NH_2.

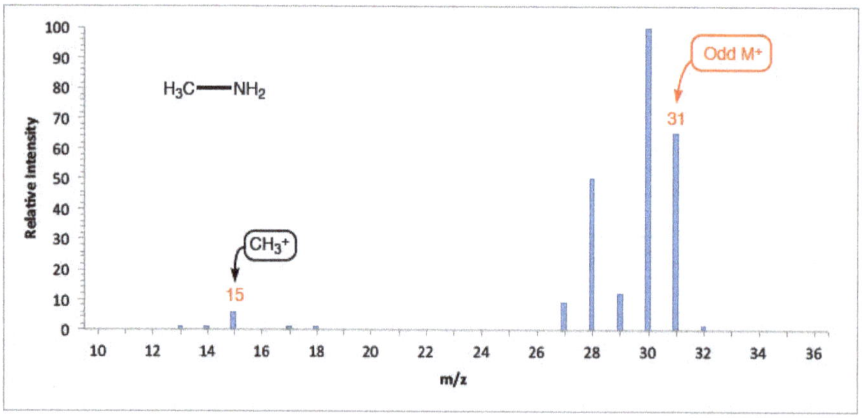

Figure 3.11 Mass Spectrum of methylamine

3.1 Identify the atom the following mass spectrum data indicates.

a) The M+2 peak is 1/3 the height of the M^+ peak.
b) The M^+ peak is an odd number.
c) The M+2 peak is nearly equal the height of the M^+ peak.
d) There is a large peak at 127 and M^+ - 127 is also prominent.
e) The M+2 peak is slightly larger than usual, about 4% of the M^+ peak.

3.1.5 M+1 peaks

The special isotope peaks are all at M+2. There is usually a very small peak at M+1. This peak mostly arises because even though most carbon is carbon-12, about 1% of carbon is carbon-13. Don't let this very small peak at M+1 confuse you. It is not sulfur! Usually, the M+2 peak is missing for compounds that contain only C, H, and O atoms. If the M+2 peak is suspiciously large, look for the presence of a special atom in the molecule.

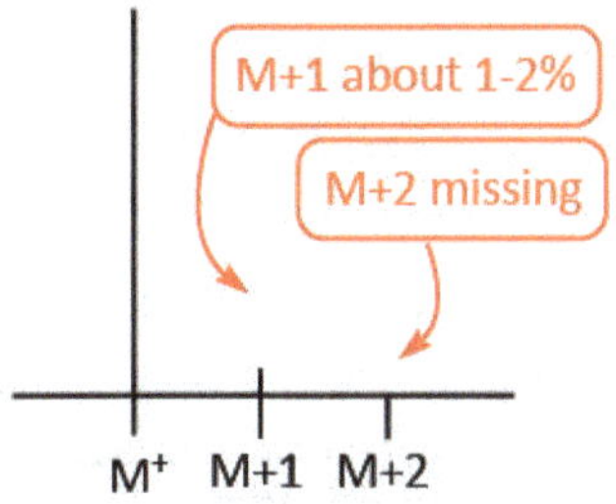

Figure 3.12 Typical M+1 pattern with only C, H, and O atoms in molecule

3.1.6 Degrees of Unsaturation

A compound's molecular formula gives some insight into its structure. For a hydrocarbon, every extra π-bond or ring formed decreased the molecular formula by two hydrogen atoms. Each of these π-bonds or rings is called a **degree of unsaturation** or an **element of unsaturation**.

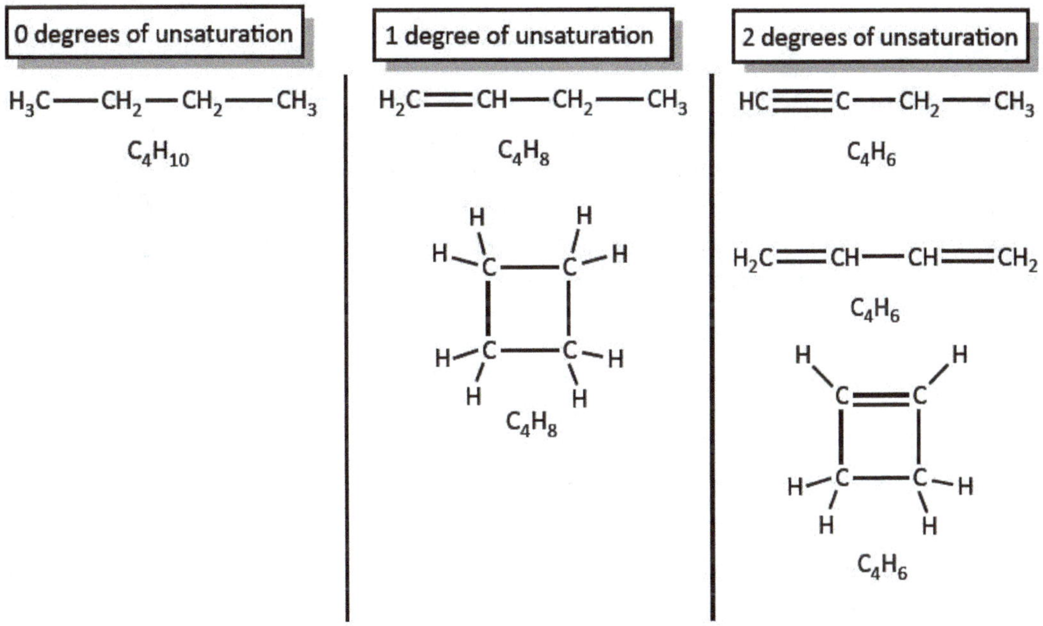

Figure 3.13 Examples for Number of Degrees of Unsaturation

So, if we know a compound has the formula C_4H_8, we could know that the compound has one degree of unsaturation, either a π-bond or a ring. We can determine the degrees of unsaturation from a molecular formula by using the following formula. Take a couple of minutes and memorize this formula now.

$$\#\ Degrees\ of\ Unsaturation = \frac{(2C + 2 - H)}{2}$$

C = # of carbon atoms
H = # of hydrogen atoms

Figure 3.14 Formula to calculate degrees of unsaturation

A compound normally has 2C+2 hydrogen atoms. So, our formula is taking the normal number of hydrogen atoms that should be present in a saturated compound and subtracts the number of hydrogen atoms that are actually there. Two hydrogen atoms are removed for each unsaturation. That is why we divide by two. Oxygen atoms do not count in the formula. Nitrogen atoms count as ½ a carbon atom. Halogens count as one H atom. This is because halogen atoms and hydrogen atoms both usually only have one bond. The number of degrees of unsaturation should always be a whole number. If you ever get 0.5 of an unsaturation, you did something wrong.

3.2 How many degrees of unsaturation are in the compounds with the following formulas?

C_6H_6O $\qquad$ C_7H_4ONCl _____

C_3H_7OBr _____ $\qquad$ C_7H_8 _____

3.3 Is propene the only compound with the formula C_3H_6?

3.2 Infrared Spectroscopy

3.2.1 Nature of light

Most chemistry instruments use light as a means to gather information about chemical compounds. Some light we can see, other light is invisible. There are gamma rays, X-rays, ultraviolet light, visible light with its colors of the rainbow, infrared light, microwaves, AM/FM radio waves. All of these are light waves and travel at the speed of light ($c = 3X10^8$ m/s) in a vacuum. They all vary by wavelength and frequency of the light. A light wave can be described by its wavelength (λ) and frequency (ν).

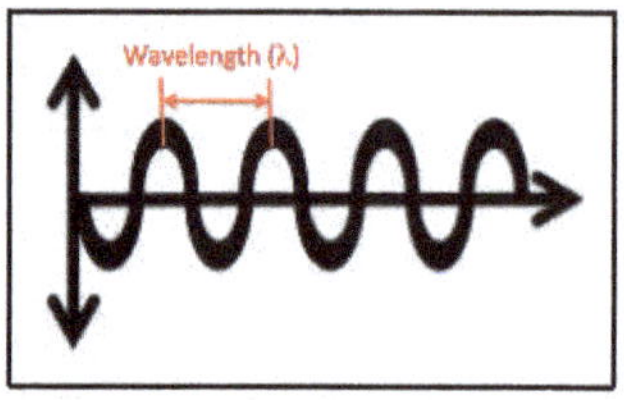

Figure 3.15 Wavelength

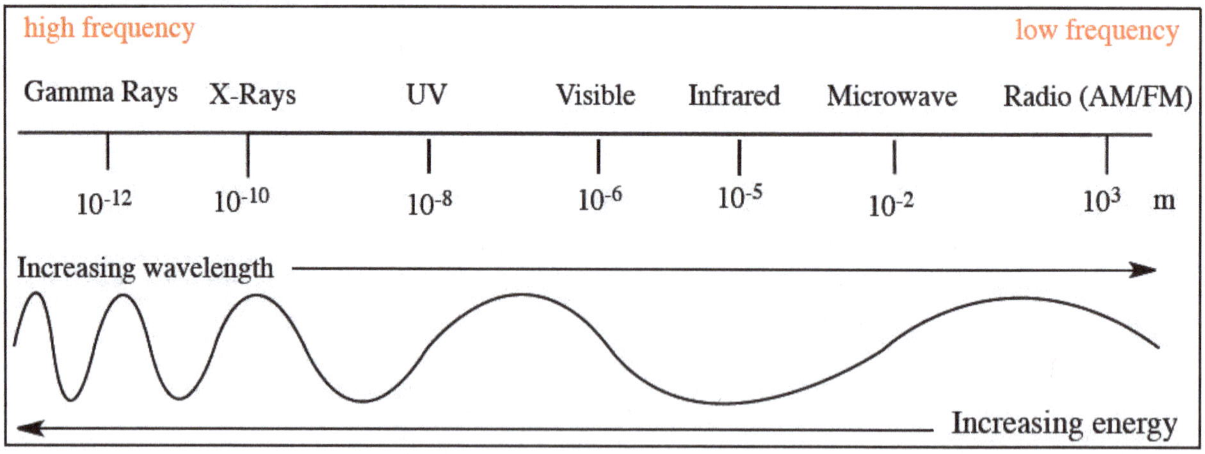

Figure 3.16 Various Wavelengths of the Light Spectrum

The distance between the peaks of the waves is the wavelength of the light. This is a distance, so it is usually measured in a unit like meters or nanometers. Since all types of light travel at the speed of light, we could also count how many waves cross a point in one second. This number of waves that pass by in one second is called the frequency. It is written as a #/second. Sometimes the unit is written as s^{-1} which is also called Hz (Hertz). Wavelength and frequency are inversely related to each other. The longer the wavelength, the lower the frequency. This is because, with long wavelengths, fewer waves can pass by in one second. The product of the wavelength and the frequency is equal to the speed of light, c. The speed of light in a vacuum is $3X10^8$ m/s.

$$c = \lambda\nu$$

The energy of a light wave is directly proportional to the frequency. Planck's constant (h = $6.63X10^{-34}$Js) is the constant of proportionality. So, the higher the frequency of a light wave, the greater its energy is.

$$E = h\nu$$

3.2.2 Springs

If a spring is pulled and released, it vibrates at differing resonance frequencies depending on the strength of the spring (the spring constant) and the masses of the weights attached to it. A stiff spring (a high spring constant) vibrates faster. A loose spring (a low spring constant) vibrates slower. If a spring has high masses attached, it vibrates slowly. If it has low masses attached, it vibrates faster. A molecular bond can be thought of as a tiny spring with two weights (atoms) on it. So, molecular bonds vibrate very fast with very light atoms like hydrogens on it. Bonds with heavier atoms, C, O, or N vibrate slower. Stronger triple bonds vibrate faster than weaker double bonds which vibrate faster than even weaker single bonds.

3.2.3 Infrared light (IR)

Infrared light is below visible red light in energy. *Infra* is Latin for "below". IR light has a longer wavelength than visible light. Infrared light is heat light. Infrared goggles can see the heat emanating from people at night. The infrared part of the spectrum of the light coming from the sun is what warms us. The reason infrared light is the heat light is that infrared light is the correct frequency of light to cause molecules to vibrate. The vibration of molecules gives us heat. For molecules, when the correct vibrational frequency of infrared light hits a molecule and matches the resonance vibrational frequency of the bond, the bond begins to vibrate.

In infrared spectroscopy, the light waves are described by their wavenumber ($\bar{\nu}$). Wavenumbers are the inverse of the light's wavelength, 1/ λ. Usually, this is a measurement of how many waves are in one centimeter of space. It is written as #/cm or cm^{-1} (inverse centimeters or reciprocal centimeters). The higher the wavenumber of light, the more waves fit into one centimeter. The higher the wavenumber of light, the shorter the wavelength of the light must be. The higher the wavenumber of light, the greater its frequency must be. The higher the wavenumber of light, the greater the energy light wave has. Read this paragraph again, slowly. Make sure you understand why each of the statements is true.

A typical infrared spectrum is 4000-400 cm^{-1}. A molecular bond vibrates when the correct frequency of infrared light hits it. Below is the general area for various bond vibrations in the IR spectrum.

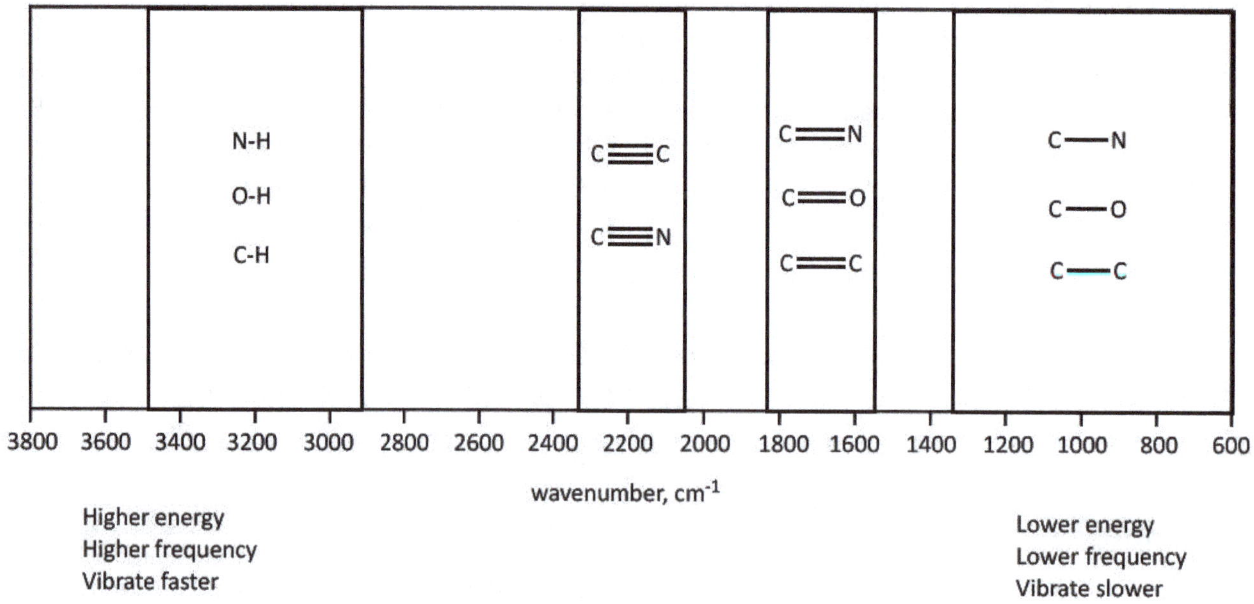

Figure 3.17 General areas for bond vibrations

3.2.4 Fingerprint region

The area of an infrared spectrum below 1400 cm^{-1} is called the fingerprint region. It is so named because this area is unique for every compound (its fingerprint). Even though there are some identifiable peaks in this area, there are extra peaks that are difficult to identify. Therefore, we spend most of our time identifying the peaks above 1400 cm^{-1} and then look down into the fingerprint region secondarily.

Stretching of the bond causes most of the peaks we will identify. The "extra" peaks that are in the fingerprint region that we usually don't bother identifying are bending, rocking, wagging, scissoring, and twisting. There is more than one way for the molecule to vibrate and absorb infrared radiation. Do not worry about trying to memorize these.

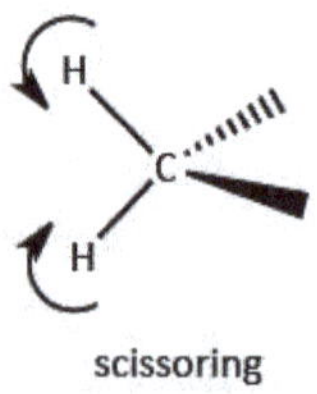

Figure 3.18 Example of Scissoring

3.2.5 Identifying IR Peaks

We do need to memorize the most commonly seen peaks in an infrared spectrum. This task may appear overwhelming at first, but you can do it. It takes a lot of practice. The following chart is your lifeline to IR spectroscopy. Use it at first to identify the major peaks in your spectrum. When you identify a peak, look at the letter above the peak, and read the extra information about it below the chart. After a while, begin to try to solve the problems without looking at the chart.

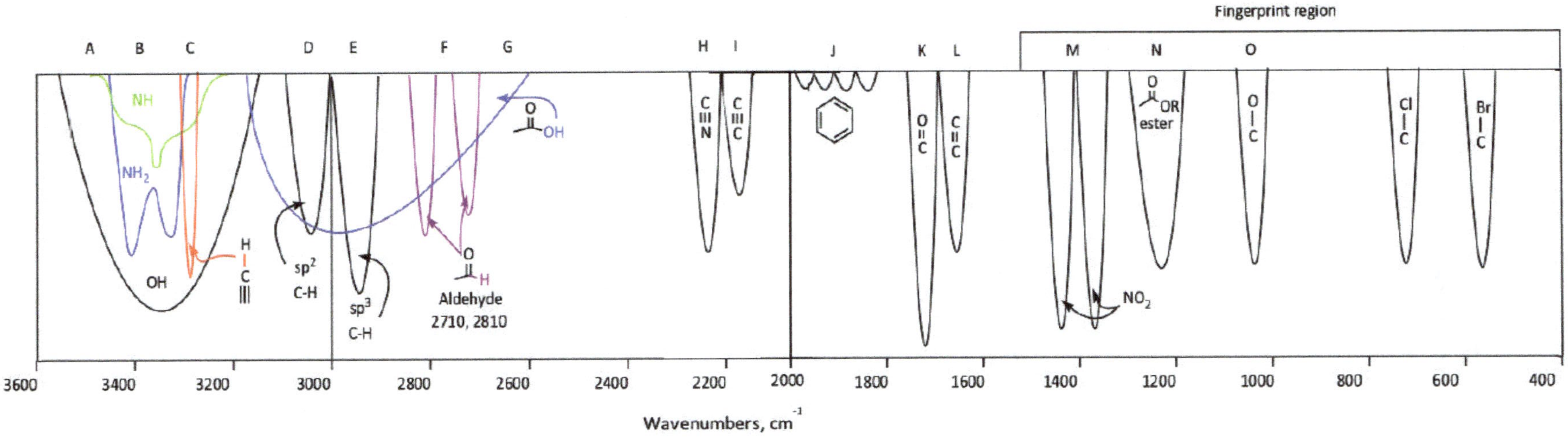

A. O-H (alcohol). Usually very broad and strong. You should also see the C-O stretch at O. BEWARE: A small amount of water in the sample will give a small OH absorption.
B. N-H (amine). Broad and with one spike for one N-H bond. Usually two humps for two N-H bonds in NH_2.
C. C-H for a terminal alkyne. It is a sharp peak and will show up only if the C≡C peak at I shows up.
D. C-H on an sp^2 carbon. This could be on an aromatic ring or regular C=C. Greater than 3000 cm^{-1}.
E. C-H on an sp^3 carbon. Also called aliphatic C-H. Less than 3000 cm^{-1}.
F. Aldehyde C-H peak. You must have both peaks around 2700, 2800 and the C=O peak at K.
G. O-H (carboxylic acid). A broad peak usually centered around 3000 and trailing off to the right. Notice how this is different than the O-H of an alcohol at peak A.
H. C≡N (nitrile). Right above 2200cm^{-1}. BEWARE: CO_2 absorbs at 2350 cm^{-1} and is a common contaminant if you do not carefully obtain your IR spectrum.
I. C≡C (alkyne). Right below 2200cm^{-1}. Weaker. Usually will not see if it is not a terminal alkyne, so also look for the peak at C.
J. Aromatic overtones. Very characteristic of aromatic compounds. Their absence does not preclude an aromatic ring, though. Usually will have peak D with it.
K. C=O (carbonyl). Very strong.
   a. 1730' s for ester (look for ester peaks at 1250 and 1050) or a ring strained ketone (like cyclobutanone)
   b. 1720' s for aldehyde (look for aldehyde peaks at 2710, 2810)
   c. 1700-1720 for regular C=O (ketones, carboxylic acids). Look for the broad peak, G, to confirm the carboxylic acid OH.
   d. 1680' s is a conjugated carbonyl, O=C-C=C. Look for peak L.
   e. 1660' s is an amide, O=C-N. If N-H bonds are on the nitrogen of the amide, you can see them at B.
L. C=C. Regular alkene. Weaker than C=O peaks. Usually will have peak D with it. Will be absent or very small for internal C=C bonds.
M. NO_2 (nitro) Two strong peaks that surround 1400 will be present for a nitro group. They must both be there.
N. Ester. If have an ester, will also have high carbonyl for K and the C-O bond around 1050.
O. C-O. Will see this in esters, alcohols, ethers, and carboxylic acids.

Alkanes, Alkenes, Alkynes

Hexane has all single bonds. The sp^3 C-H bonds show a strong absorbance *below* 3000 cm^{-1}. Go back to the chart on the previous page and locate the sp^3 CH peak below 3000 cm^{-1} and compare it to the peak in the IR spectrum for hexane below. As you begin to learn these IR peaks, keep going back to the chart on the previous page to identify the peaks and begin memorizing them. You will need to know the locations of all of these peaks!

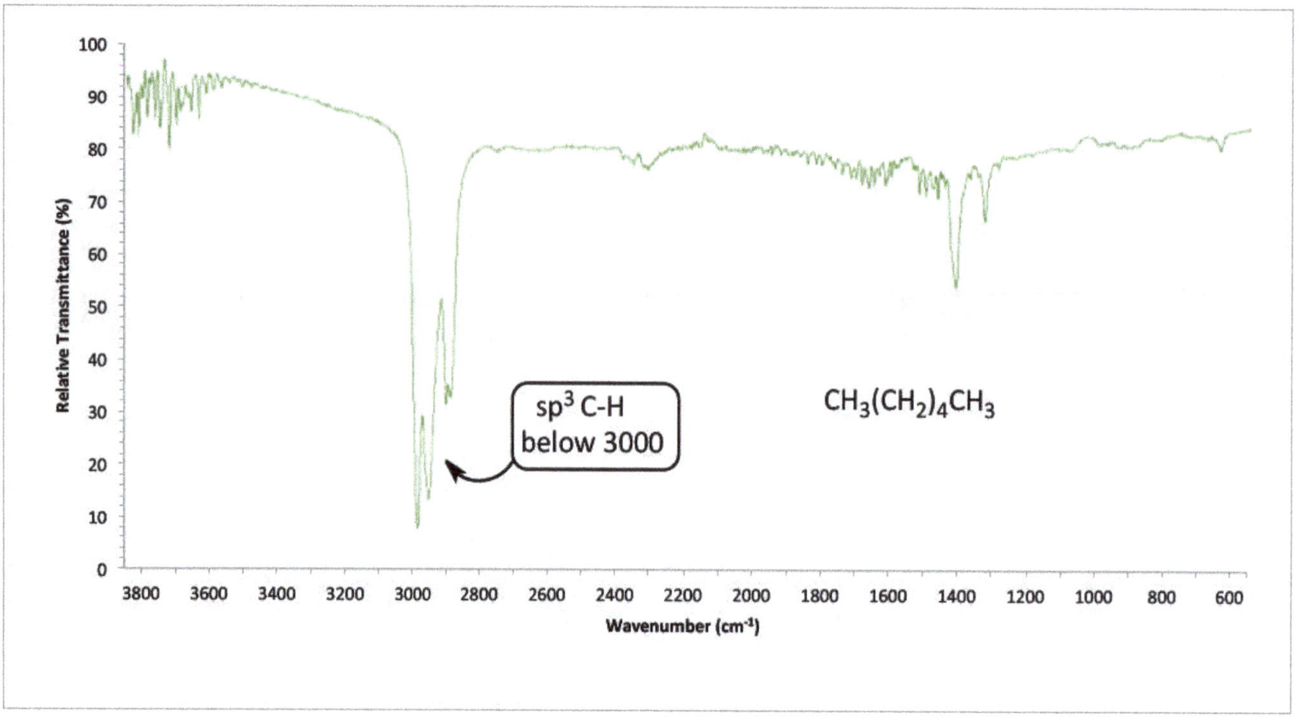

Figure 3.20 IR spectrum of hexane

For 1-hexene, we notice the sp3 C-H bonds of the alkane part below 3000. There are sp^2 C-H bonds above 3000. Since there are sp^2 C-H bonds, we know that there is some type of C=C. We see a regular alkene C=C at 1660.

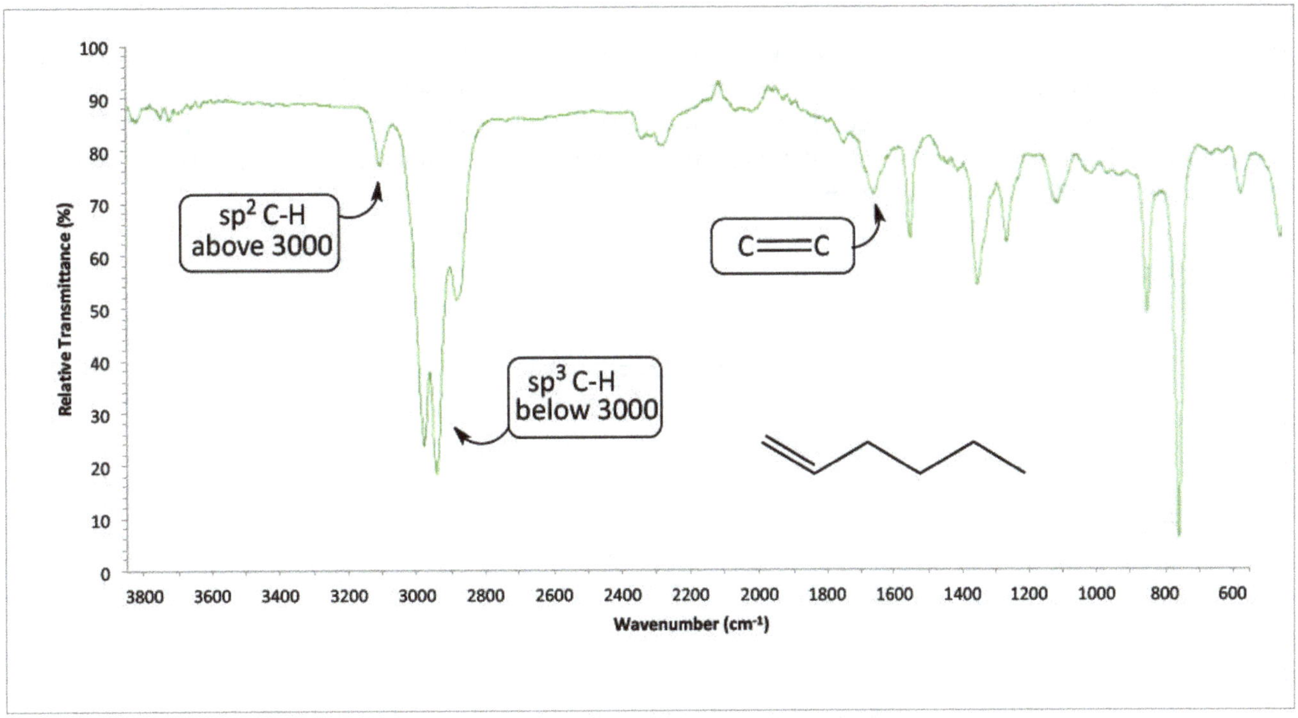

Figure 3.21 IR spectrum of 1-hexene

For ethylbenzene, we see the sp^3 and sp^2 C-H bonds. When we see the sp^2 C-H bonds right above 3000, we look to see what type of C=C we have. In this case, we see aromatic overtones below 2000 indicating an aromatic ring.

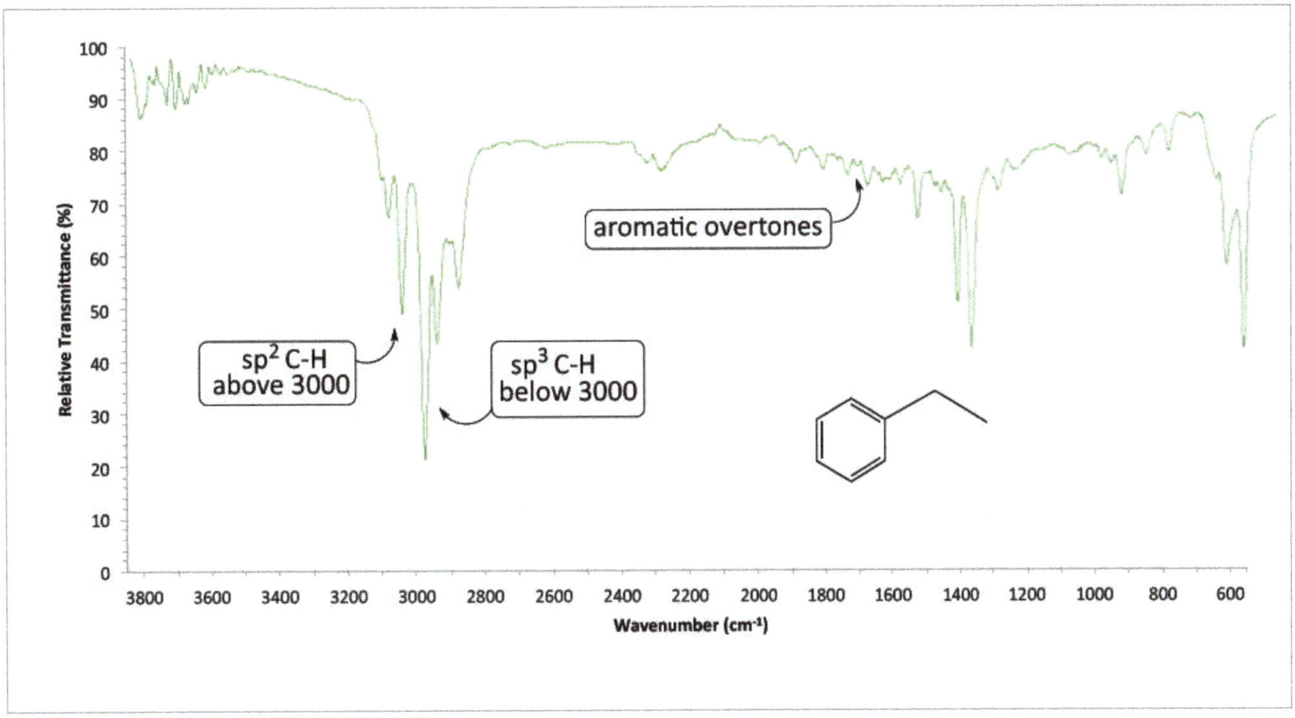

Figure 3.22 IR spectrum of ethylbenzene

IR active/IR inactive

The reason that infrared light causes a bond to vibrate is that the frequency of the IR light matches the resonance frequency of the bond. A bond must have a dipole moment in order to absorb IR light. If it does, the bond is said to be IR active. Not every molecular bond absorbs IR light and can be seen in an IR spectrum. Symmetrical bonds in the middle of symmetrical molecules do not change a dipole moment when they are stretched or compressed. These bonds are called IR inactive and do not absorb IR light.

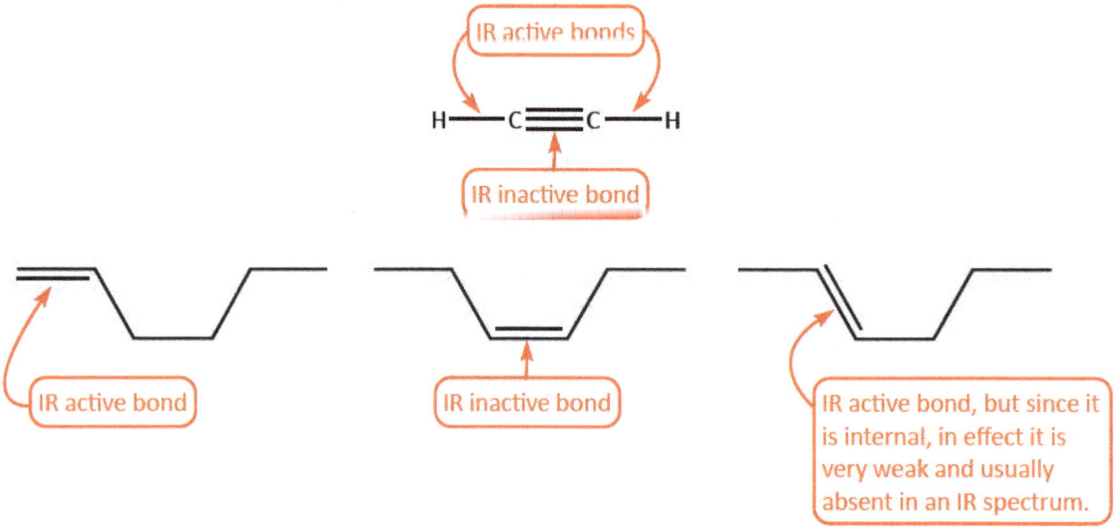

Figure 3.23 IR Active and Inactive examples

For 2-octene, we see the sp^3 and sp^2 C-H bonds. When we look for what type of C=C is present, we look for the alkene peak at 1650 or aromatic overtones. We do not see either. An alkene C=C is present in the molecule, but it is absent in the IR spectrum because it is an internal double bond.

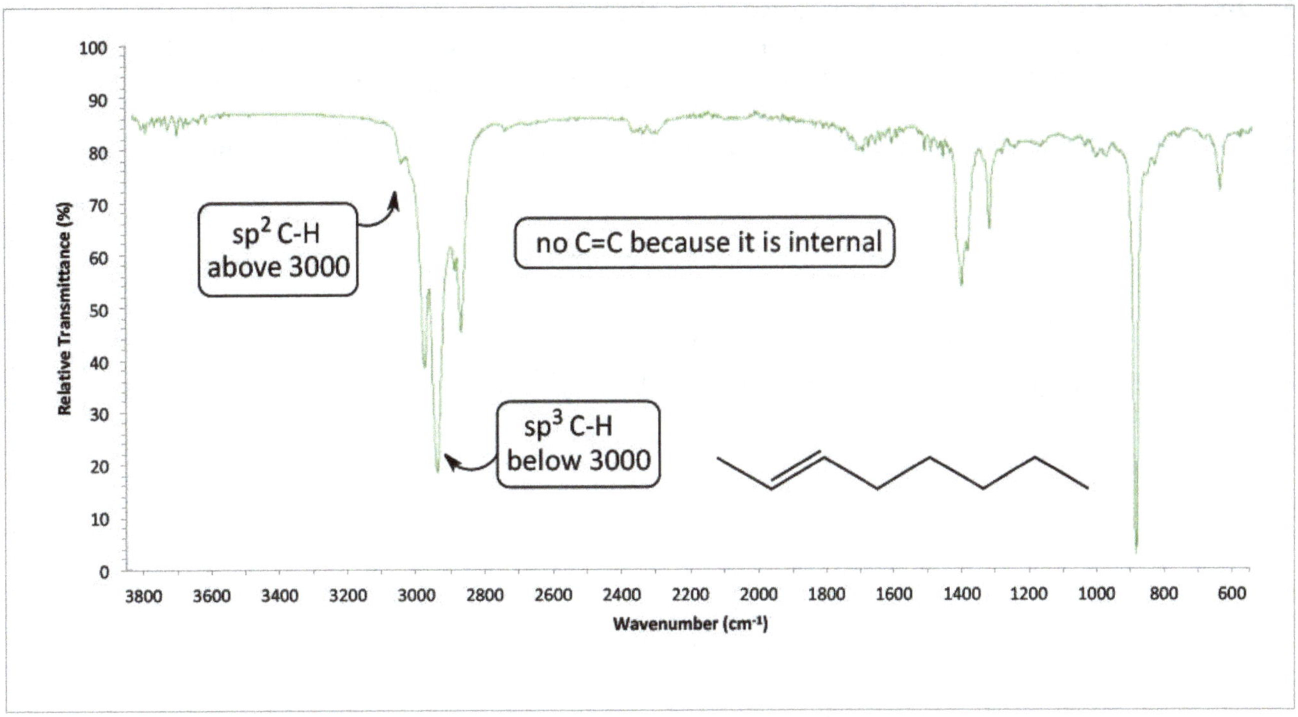

Figure 3.24 IR spectrum of 2-octene

1-Pentyne is a terminal alkyne (alkyne on the end of the molecule). The sp^3 C-H bonds of the methyl group are below 3000. The sp C-H stretch is sharp and at 3300. The C≡C is weak but visible right around 2100.

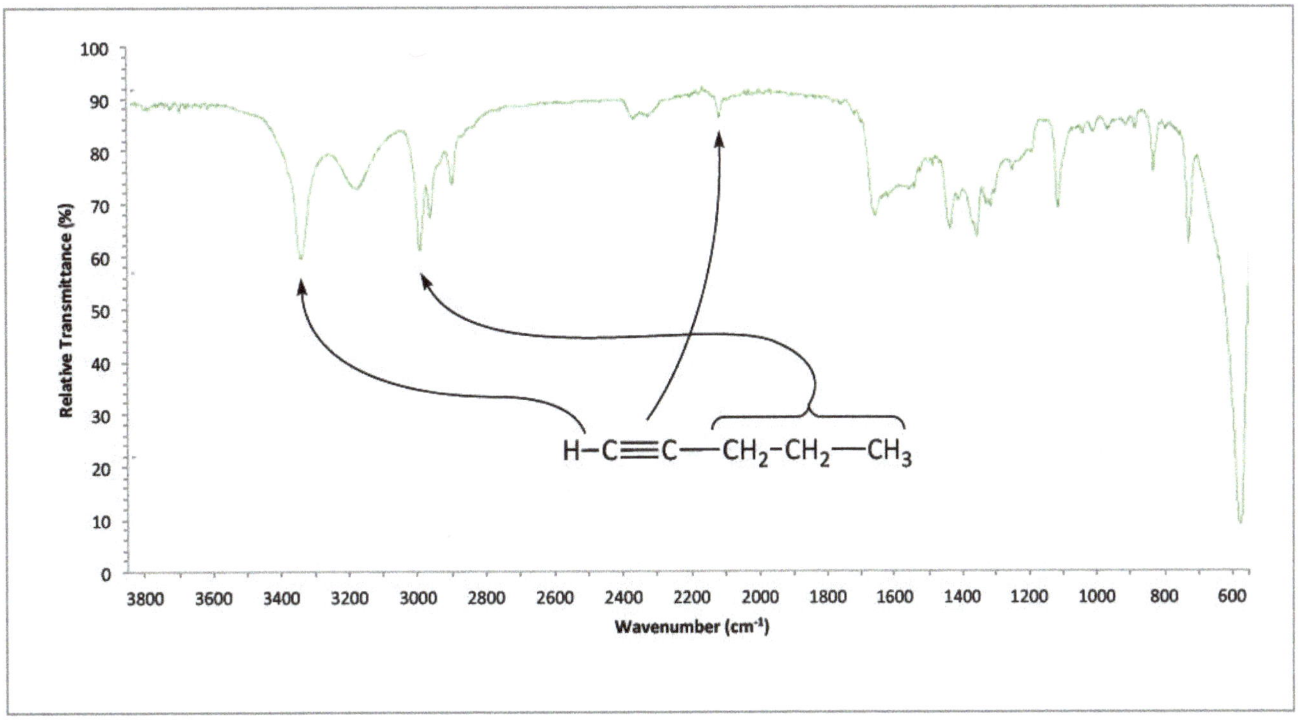

Figure 3.25 IR spectrum of 1-pentyne

For 2-pentyne, there are sp³ C-H stretches below 3000. There are no sp C-H peaks because there is not a hydrogen atom on the C≡C. Also, since the C≡C is internal, it is pretty much IR inactive and absent from the IR spectrum. The sp CH peak at 3300 and the C≡C peak below 2200 go together. We usually do not see one without the other.

We see then, that internal C=C and C≡C are difficult to determine strictly from an IR spectrum.

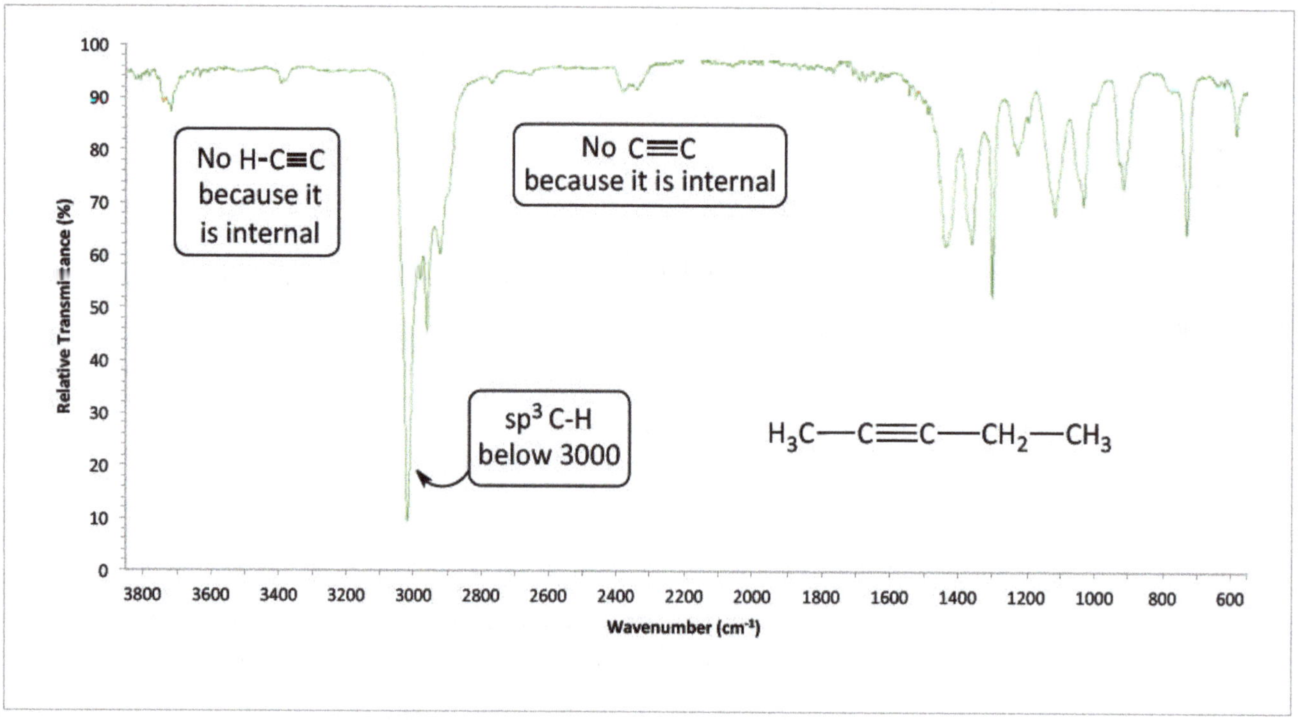

Figure 3.26 IR spectrum of 2-pentyne

3.4 Circle the molecule below if its bond that is marked red is and IR ACTIVE bond.

$CH_3–CH_3$ $CH_3CH_2–CH_3$ $HC≡CCH_3$ H–O–H

O=C=O C≡O H–C≡N H–C≡C–H

OH, NH, NH₂

OH, NH bond stretches all appear around 3300. Alcohols have an obvious, very identifiable peak. In benzyl alcohol, we see the sp³ CH peak below 3000 and the sp2 CH peak above 3000. There is a strong, broad peak at 3300 for the alcohol OH. It is much broader than the sp CH peak. We also see the aromatic overtones.

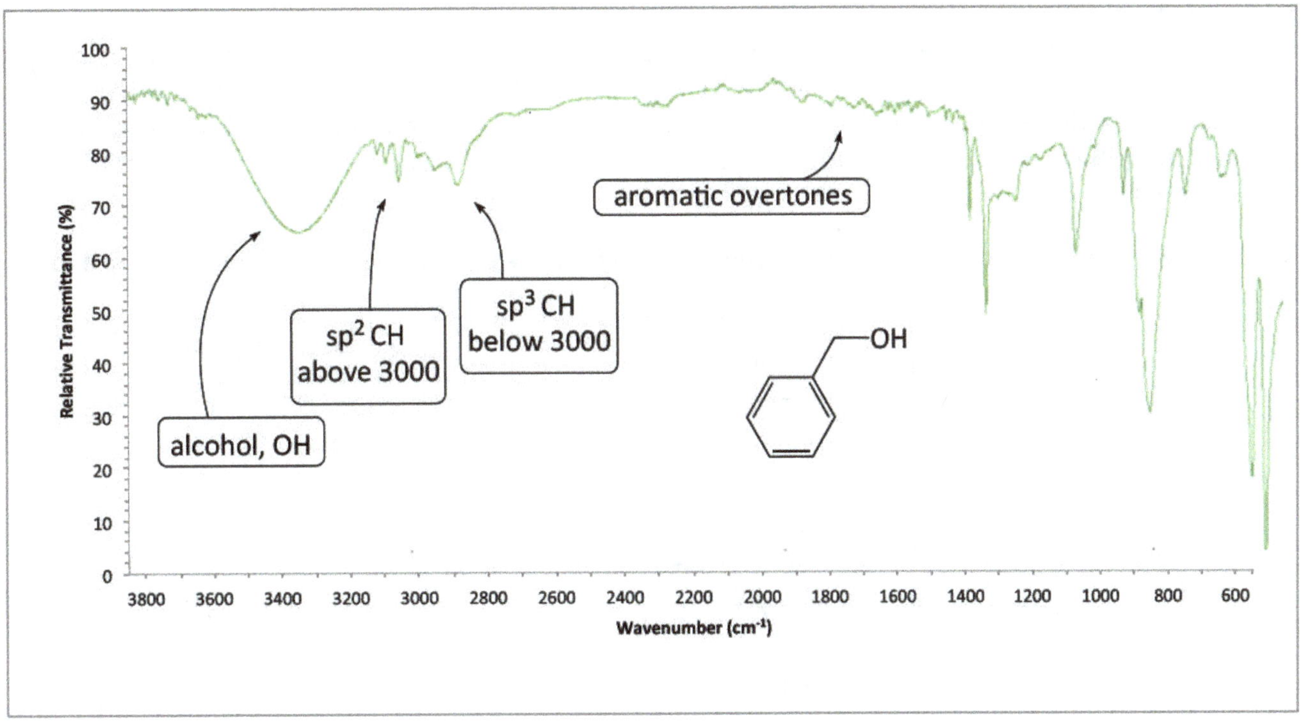

Figure 3.27 IR spectrum of benzyl alcohol

For benzylamine, there are sp^3 CH peaks below 3000 for the CH_2 and sp^2 CH peaks above 3000 for the aromatic ring. The aromatic overtones below 2000 confirm this. At 3300, we see a two-hump peak. This means there are two N-H bonds present (an NH_2).

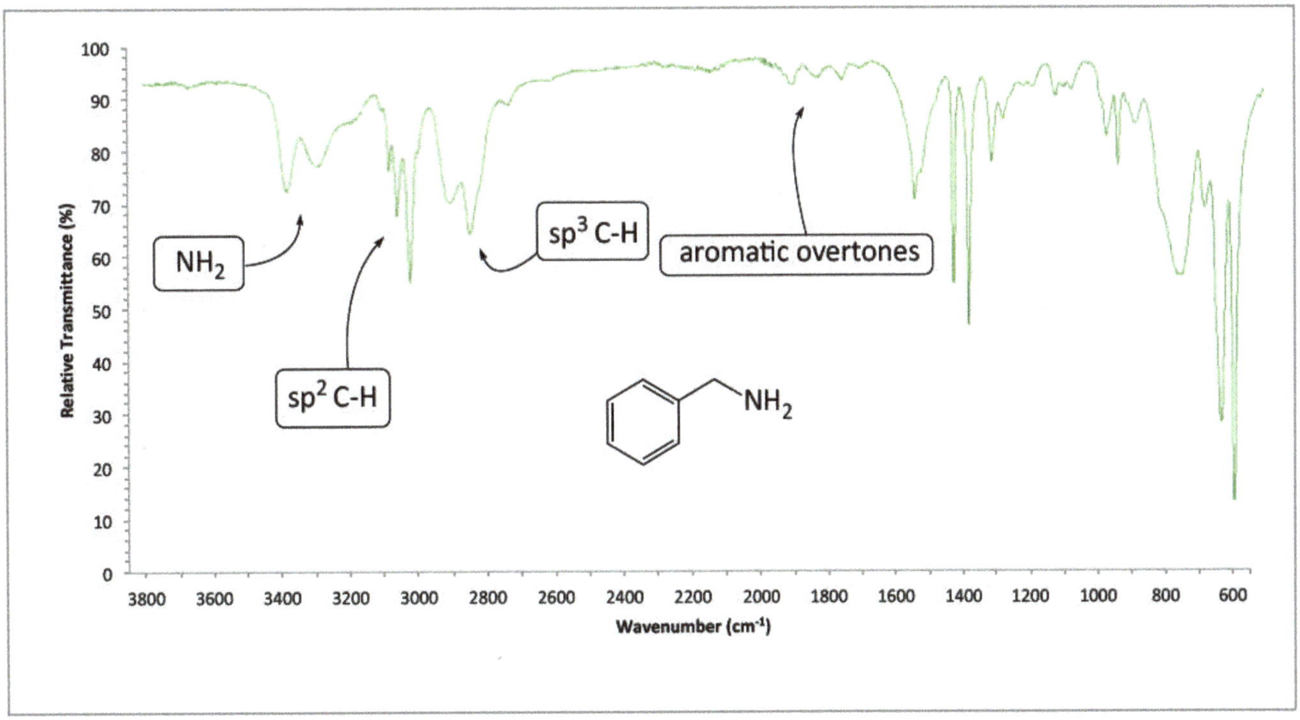

Figure 3.28 IR spectrum of benzylamine

Carbonyl compounds

The C=O, carbonyl, stretch is very easy to identify. It is usually the strongest peak in an IR spectrum. It is usually around 1700 but can range from the 1660s up to about 1800. When we see the carbonyl, we also look for secondary peaks that help us identify the functional group. Also, the frequency of the C=O stretch gives us information on the type of functional group it is. Be flexible with these frequencies. The below ranges are typical but vary slightly.

A conjugated C=O is one where a C=C is one bond removed from the C=O. It cannot be closer or farther away.

C=O Frequency	Functional Group	Example	Secondary peaks to look for. . .
1730s	Ester	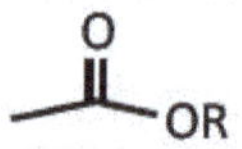	1050, 1250
1730s	Ring strained ketone	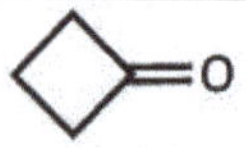	N/A
1710-20s	Aldehyde	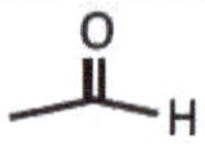	2710, 2810
1710s	Ketone	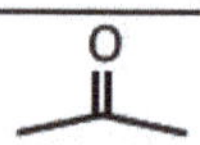	N/A
1710s	Carboxylic Acid	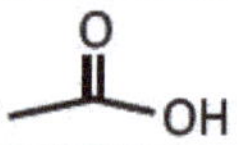	Broad OH at 3000
1680s	Conjugated C=O		sp^2 CH at >3000 C=C at 1650
1660s	Amide	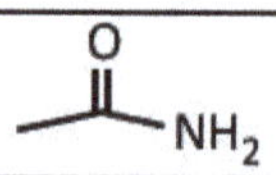	NH peaks at 3300 if present

Figure 3.29 IR carbonyl frequencies

For butanoic acid, a regular carbonyl peak at 1710 appears. The secondary peak of a broad OH that is centered at 3000 and slowly tapers off indicates a carboxylic acid. The sp^3 CH peak appears just above 3000 on top of the OH peak.

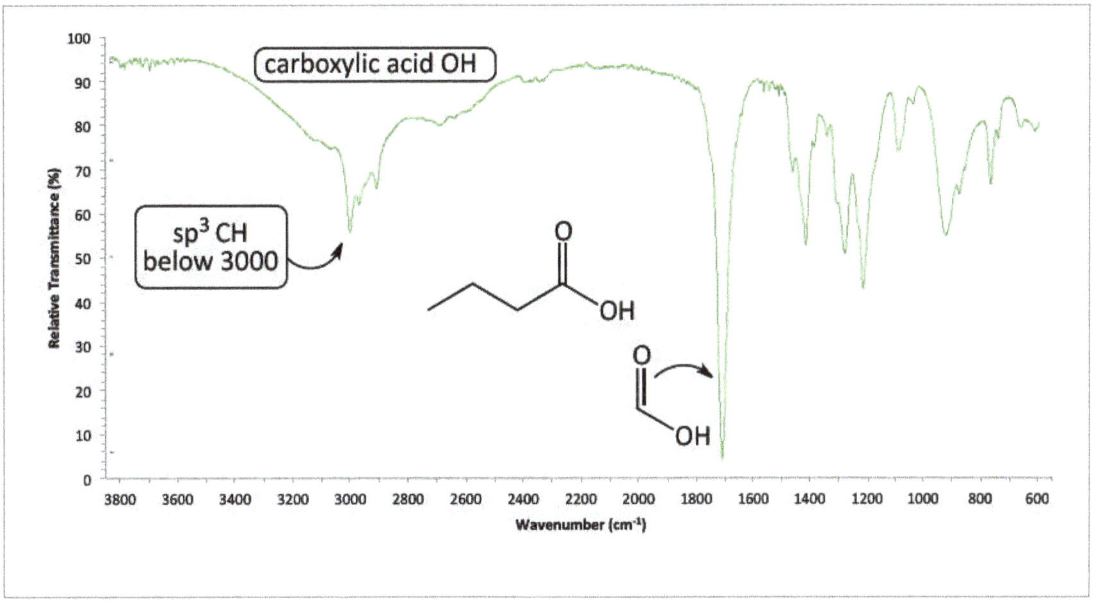

Figure 3.30 IR spectrum of butanoic acid

For pentanal, a carbonyl C=O peak appears at 1725. We also see two peaks near 2700 and 2800 to help confirm aldehyde.

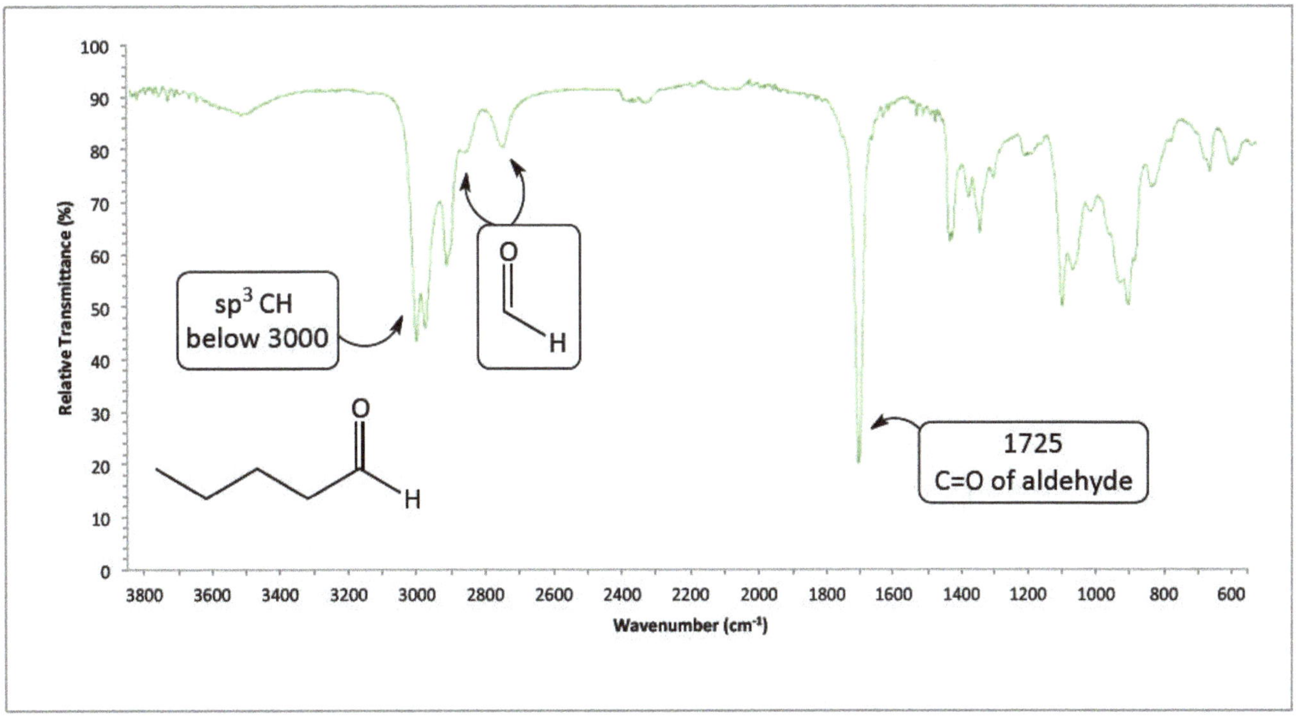

Figure 3.31 IR spectrum of pentanal

In acetone, there is a regular C=O at 1710. There are no secondary peaks that suggest carboxylic acid or aldehyde, so we conclude it is a ketone.

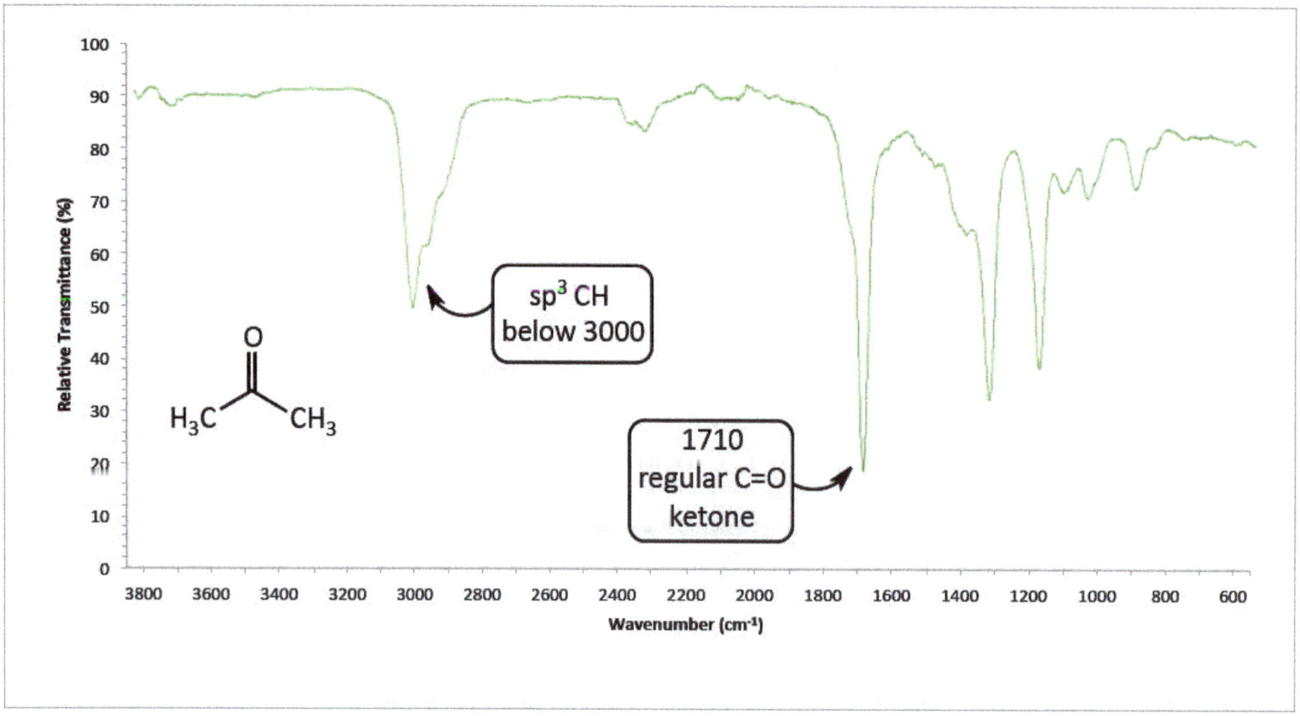

Figure 3.32 IR spectrum of acetone

For ethyl acetate, the carbonyl peak is high in the 1730s. This indicates ester. The two fairly strong secondary peaks at 1250 and 1050 confirm ester.

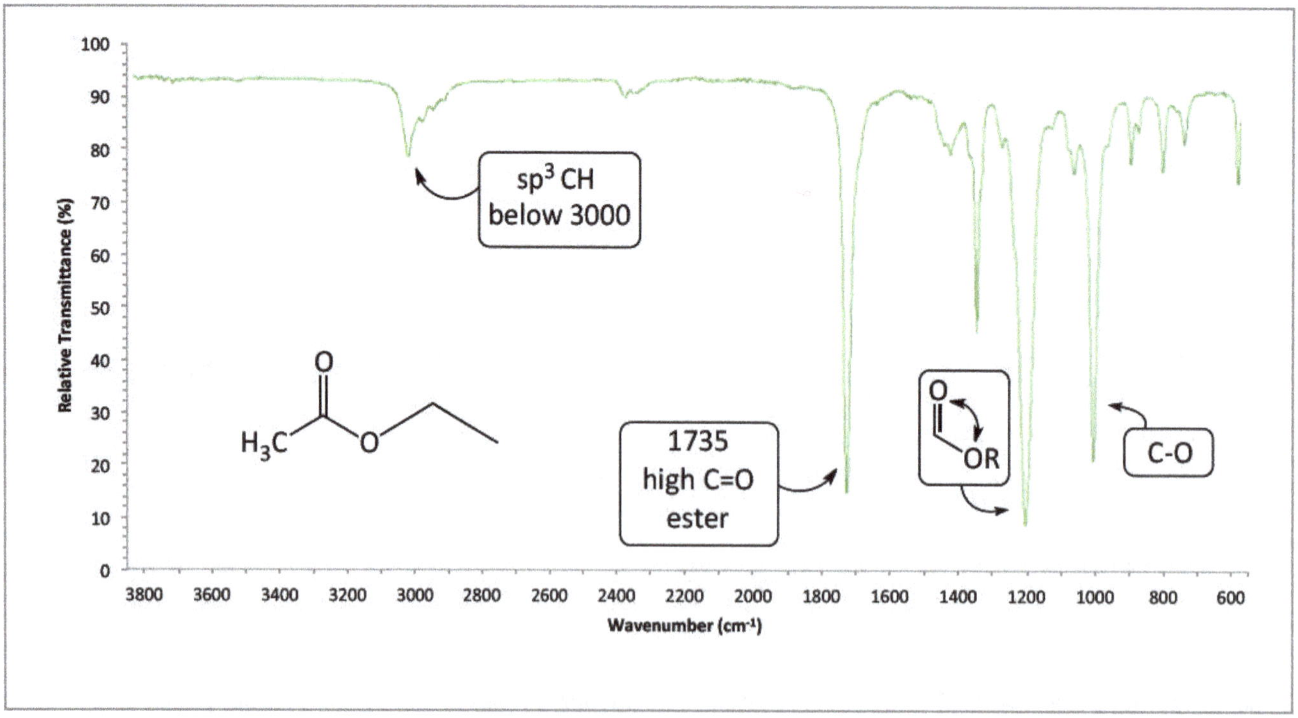

Figure 3.33 IR spectrum of ethyl acetate

For cyclopentanone, we see the sp^3 CH peak below 3000. There is also a high carbonyl peak which could indicate an ester or maybe a ring-strained ketone. Normally when we analyze an IR structure with a high carbonyl frequency, we would be trying to determine if we have an ester or a ring-strained ketone. We can look at the fingerprint region to look for secondary peaks for an ester. There are not the two strong peaks around 1250 or 1050, so we lean towards a small ring that is a ketone.

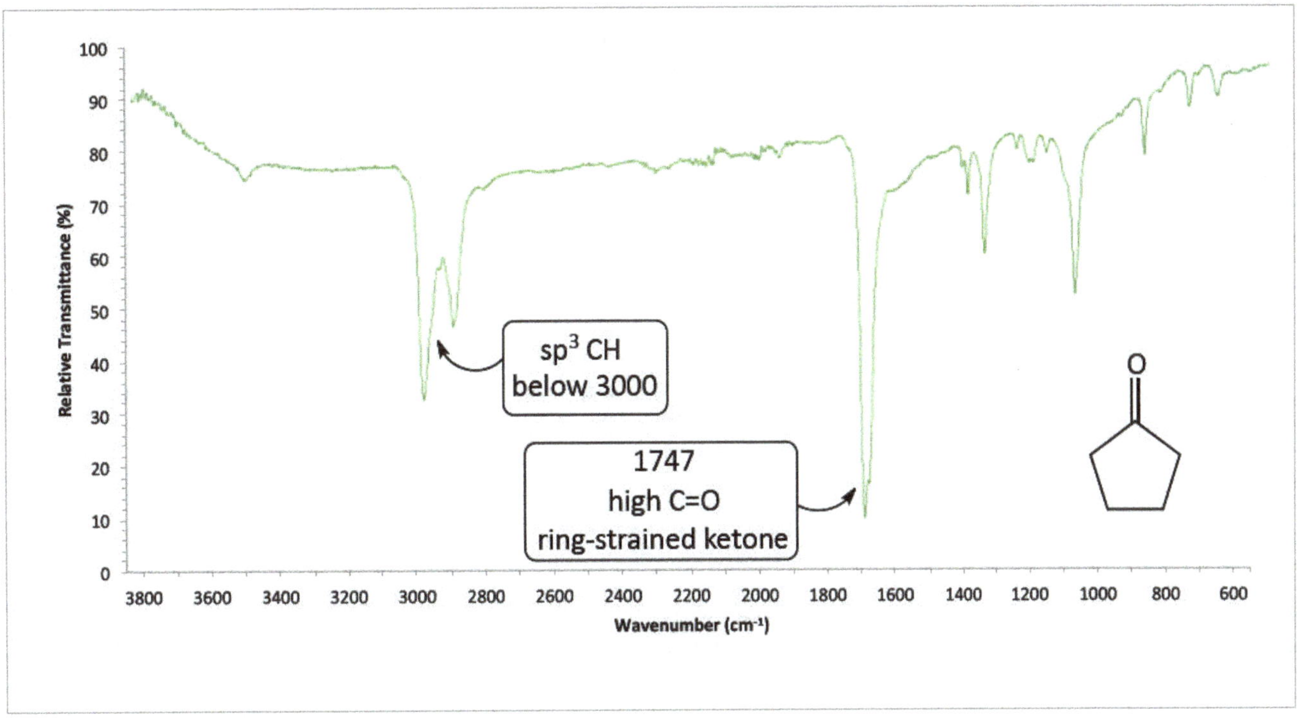

Figure 3.34 IR spectrum of cyclopentanone

For acetamide, we see that the carbonyl peak is quite low in the 1660s indicating an amide. We look for secondary peaks at 3300 and find two N-H bonds. We also see sp_3 CH peaks below 3000. It is fairly small since there are only three C-H bonds.

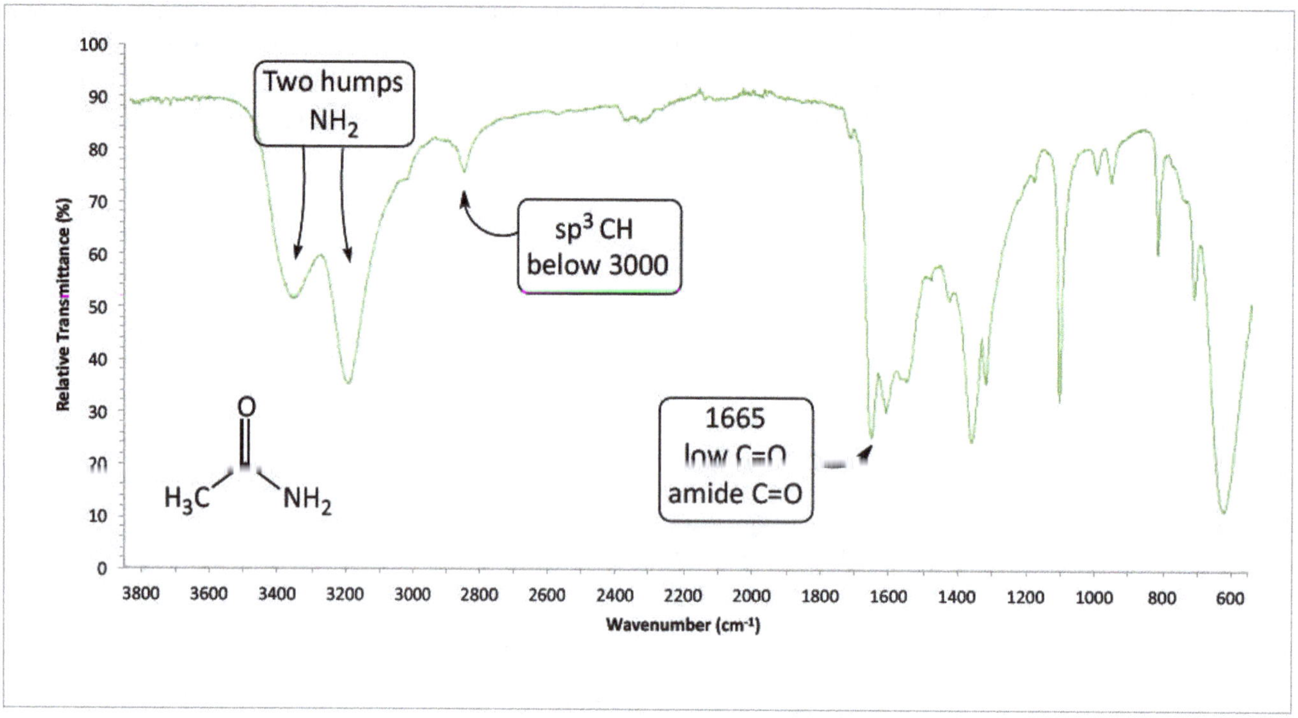

Figure 3.35 IR spectrum of acetamide

In acetophenone, we see sp^2 and sp^3 CH bonds. Since we have sp^2 CH, we look for what type of C=C is present. We notice the aromatic overtones below 2000. We also see a carbonyl peak, but it is a little low in the 1680s. This tells us that our aromatic ring is next to the carbonyl so we can get a conjugated C=O.

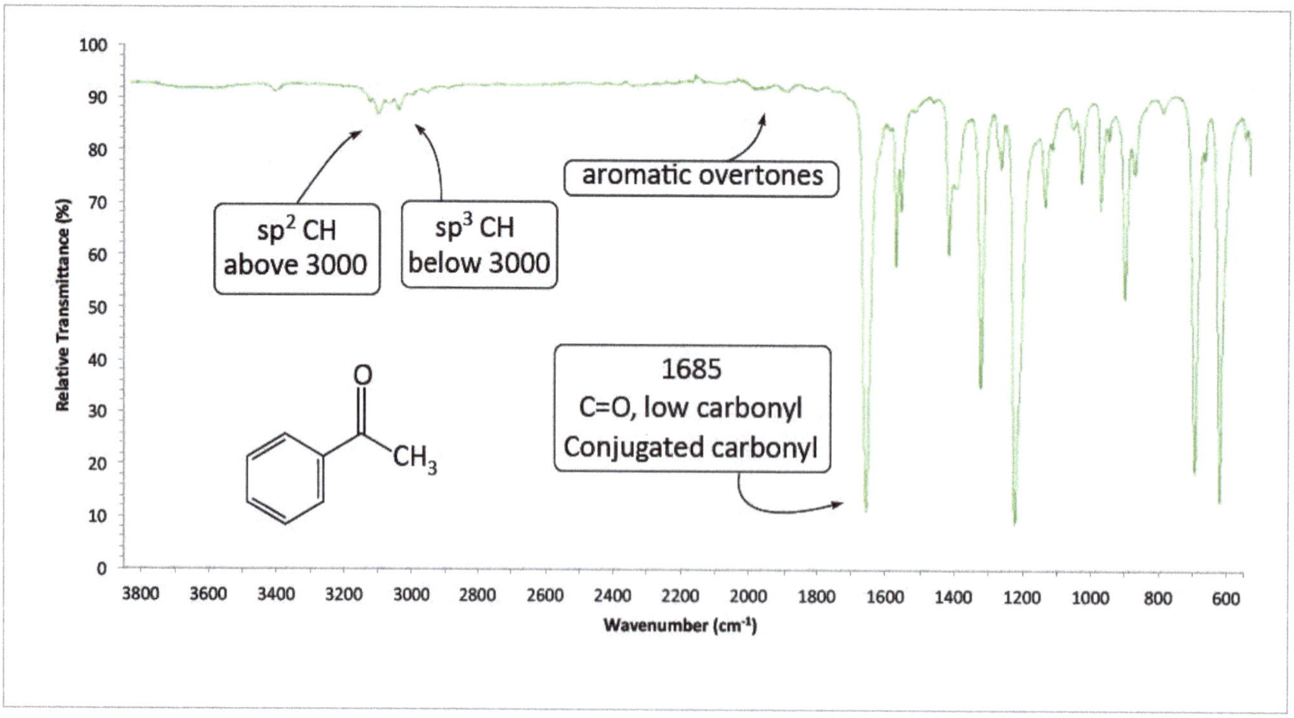

Figure 3.36 IR spectrum of acetophenone

Other functional groups

In acetonitrile, the C≡N nitrile is the sharp peak just above 2200. C≡C IR peaks also show up near C≡N. We can tell this is not C≡C because C≡N shows up just above 2200 while C≡C is usually just below 2200. Additionally, we usually only see C≡C if it is at the end of a molecule (terminal alkyne). In that case, we also see the sp CH stretch at 3300, which is absent here.

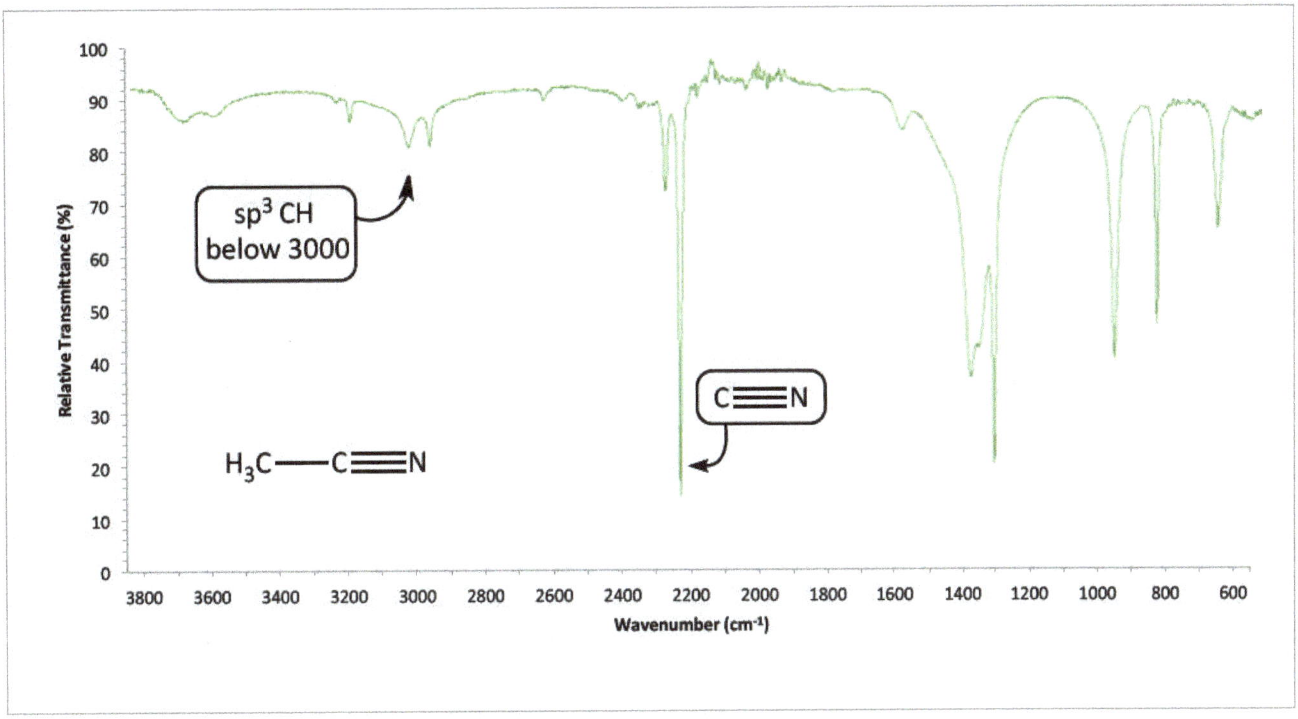

Figure 3.37 IR spectrum of acetonitrile

For ethyl ether, we see the sp^3 CH peak below 3000, but not much else. All other major peaks, like OH, C=O, etc. are missing. Since we don't have many clues yet, we venture down into the fingerprint region and see a strong C-O peak at 1100 indicating ether.

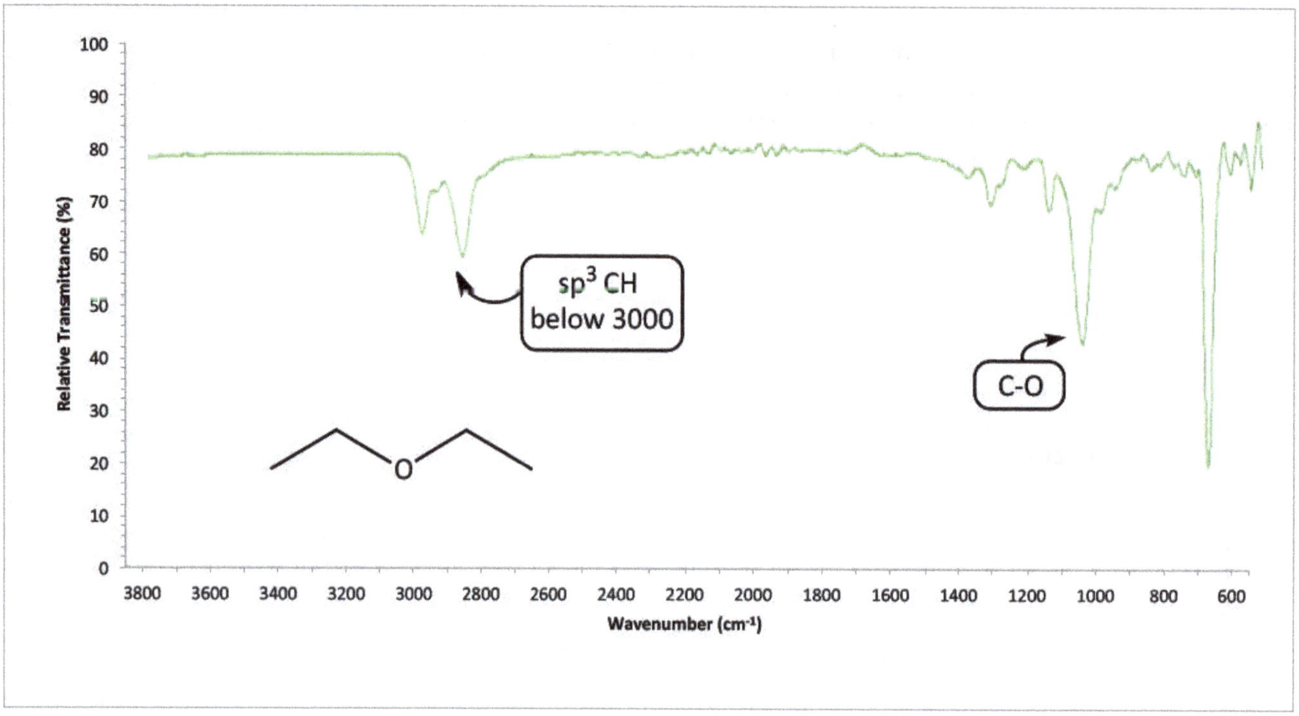

Figure 3.38 IR spectrum of ethyl ether

Nitrobenzene shows the sp^2 and the aromatic overtones of the benzene ring. With no other major peaks identified, we venture into the fingerprint region and see two strong, sharp peaks surrounding 1400. With both of them there, that indicates nitro, NO_2.

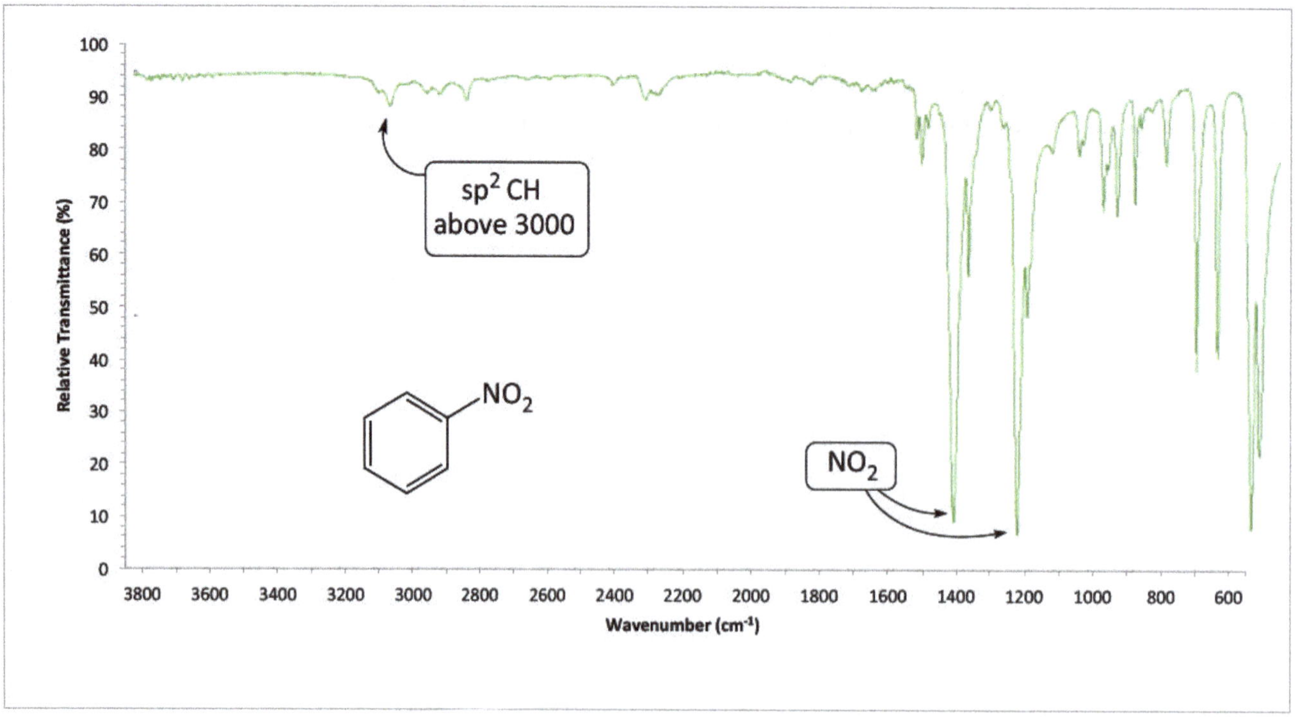

Figure 3.39 IR spectrum of nitrobenzene

For N-methylaniline, we see the sp^2 and sp^3 CH peaks surrounding 3000. We also see the aromatic overtones indicating aromatic ring. We do see a one-hump peak at 3300. We know this is an N-H stretch or an sp CH stretch. It is fairly sharp. An OH stretch is usually broader and NH_2 is two humps. If it was an sp CH stretch, we would also see the C≡C peak below 2200. Therefore, we know that we have an amine with one N-H bond.

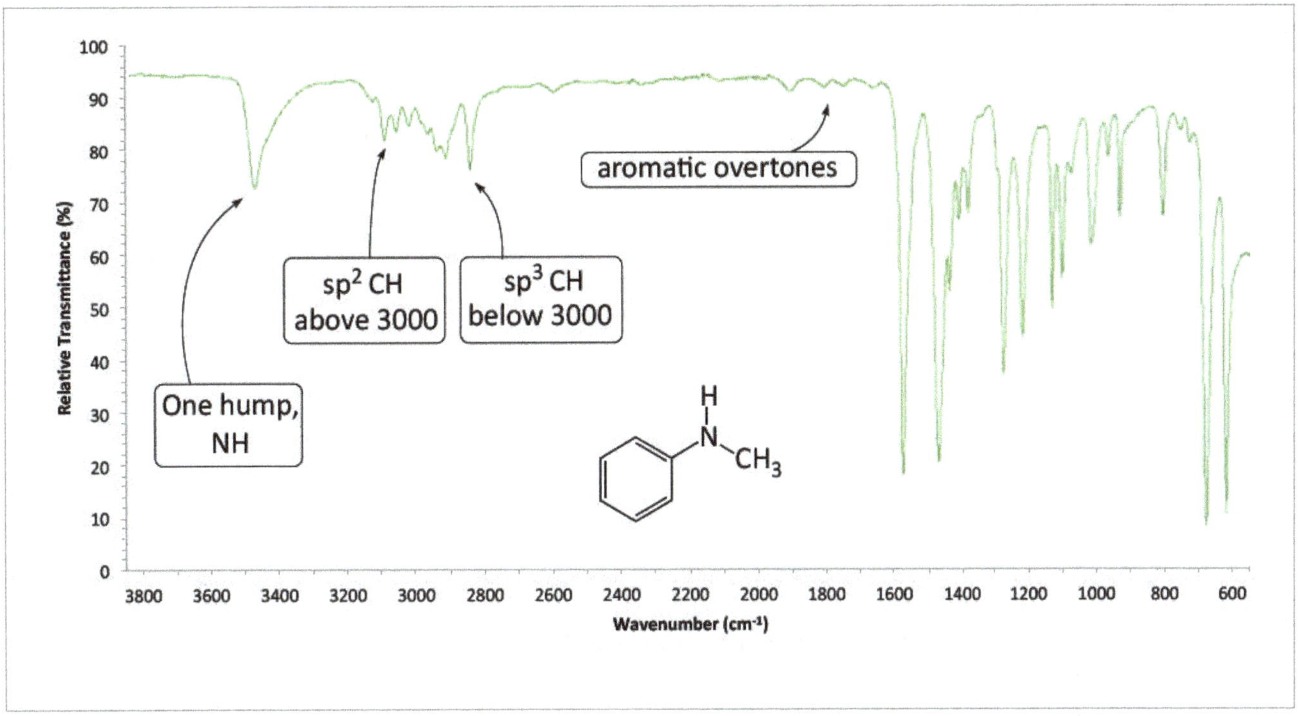

Figure 3.40 IR spectrum of N-methylaniline

In chloroform, we see no major peaks so we venture into the fingerprint region. There, we see an unusually strong peak at 700 which indicates the C-Cl stretch.

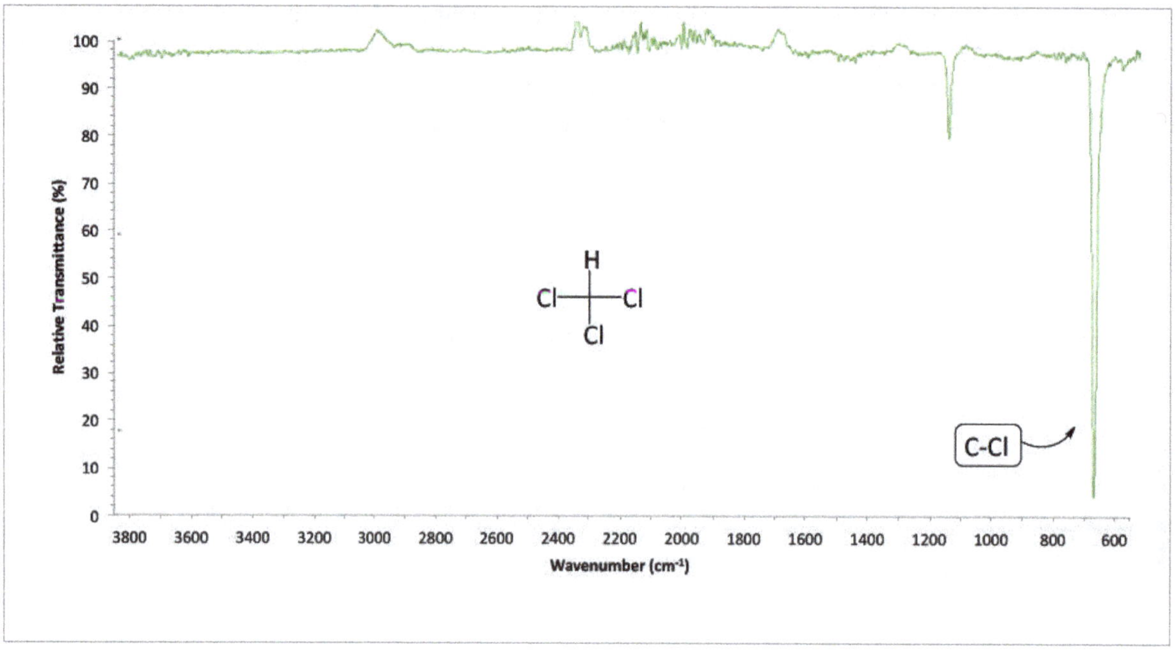

Figure 3.41 IR spectrum of chloroform

3.5 Identify the functional group(s) indicated in each generic IR spectrum.

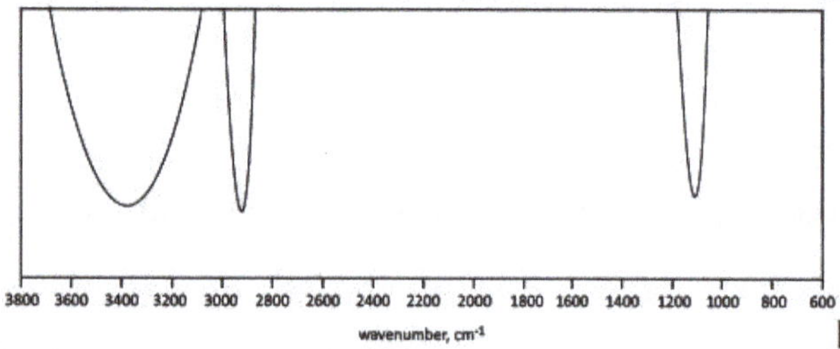

a)

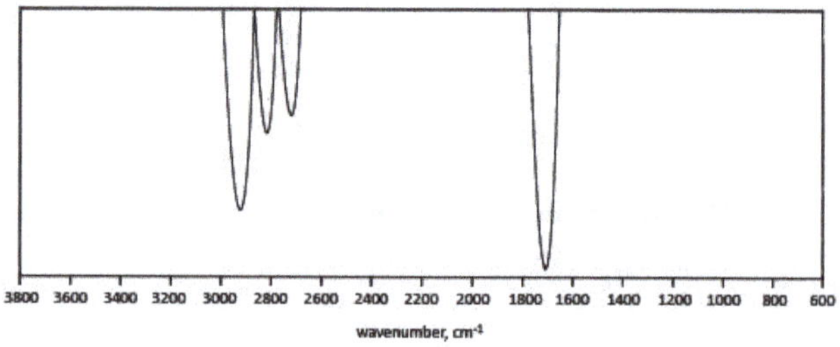

b)

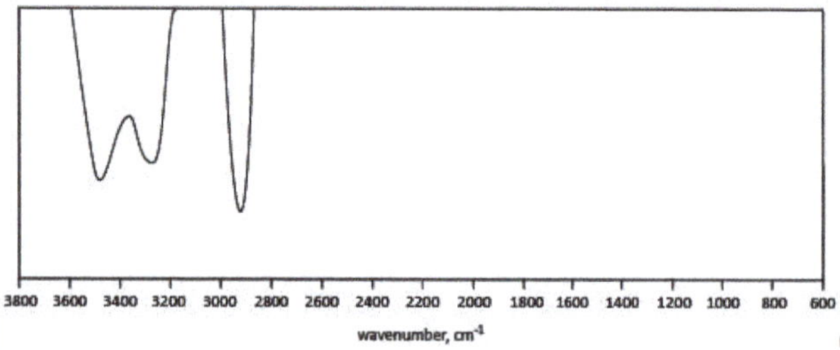

c)

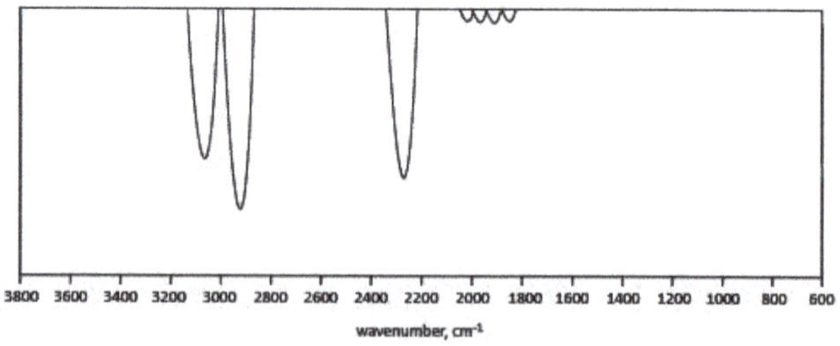

d)

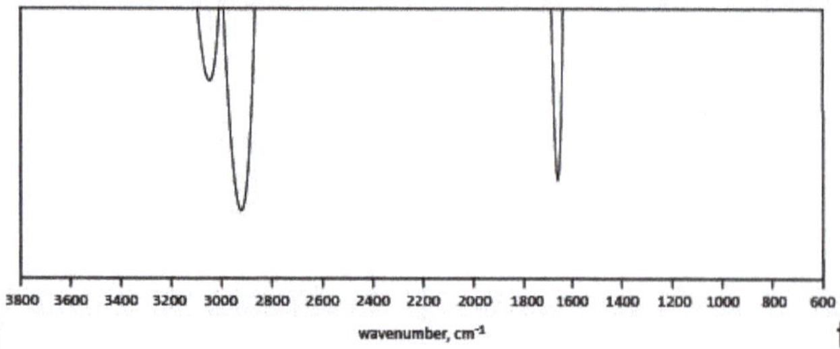

e)

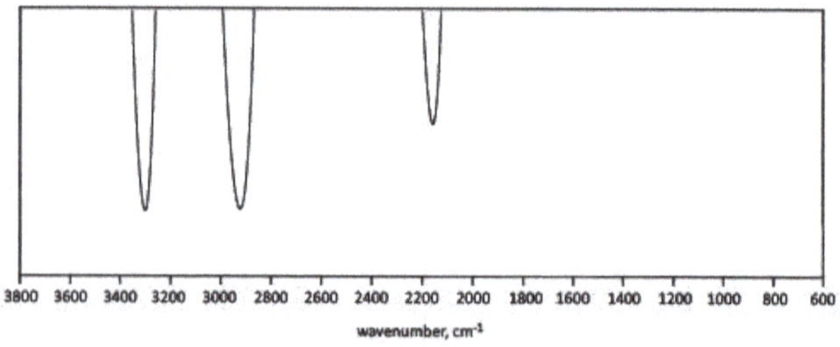

f)

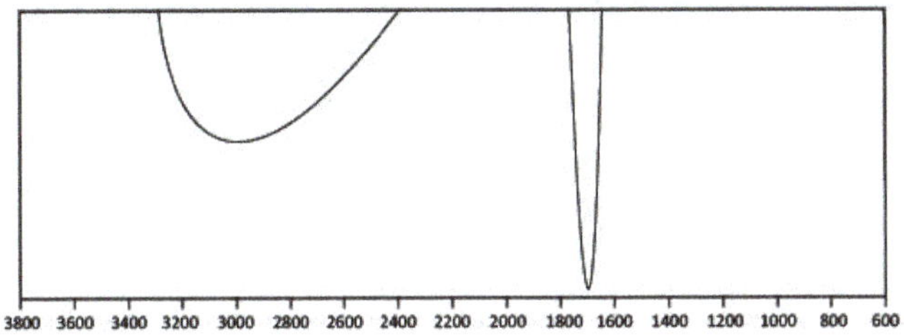

g)

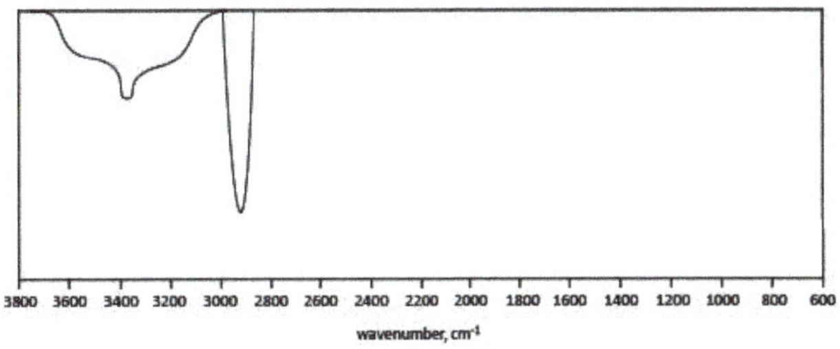

h)

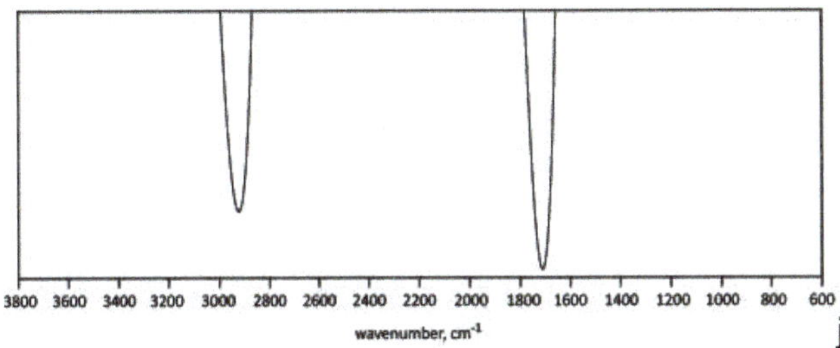

i)

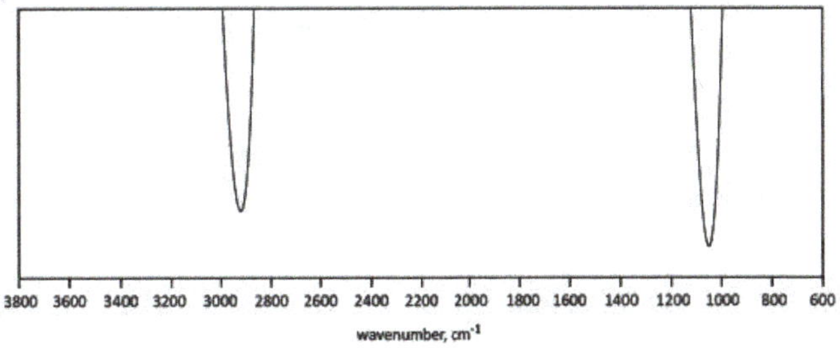

j)

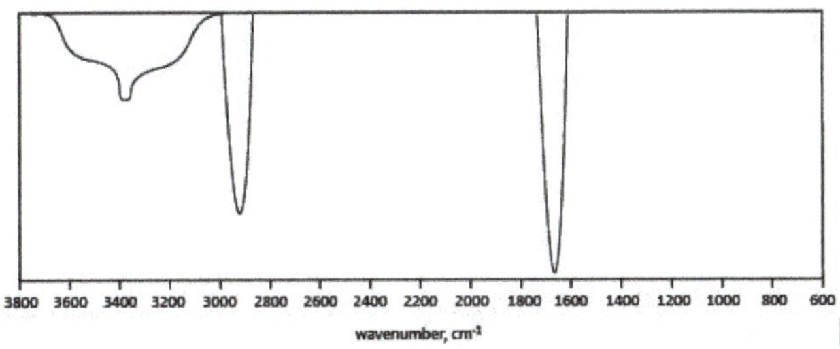

k)

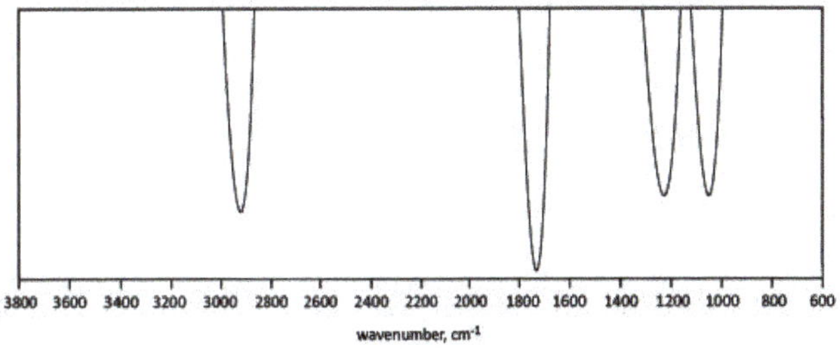

l)

3.6 Match the following four compounds with their IR spectrum.

___

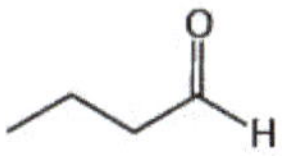

___

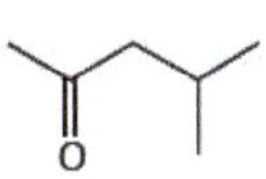

___

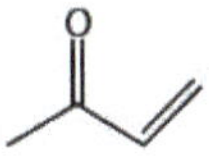

___

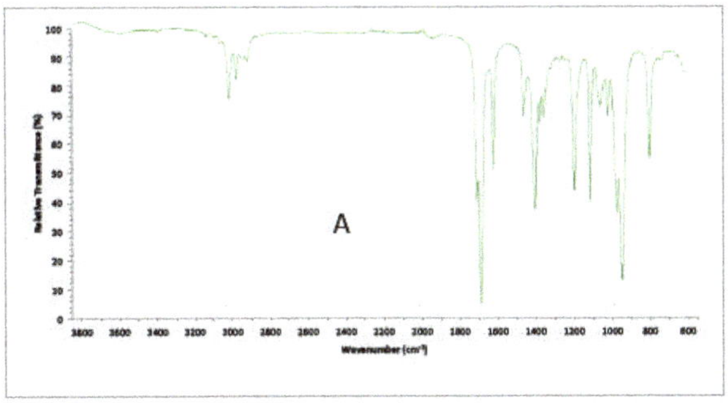

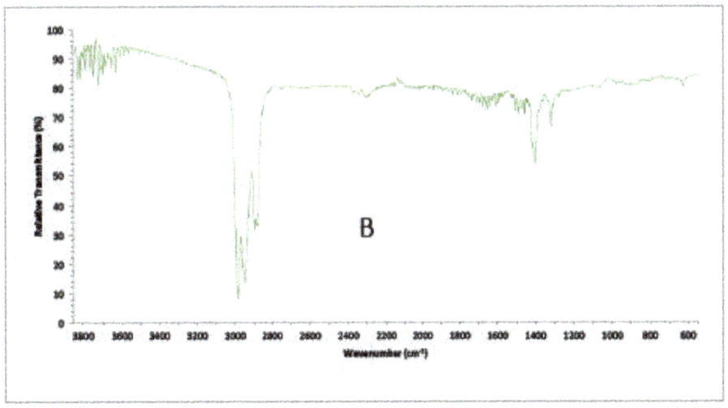

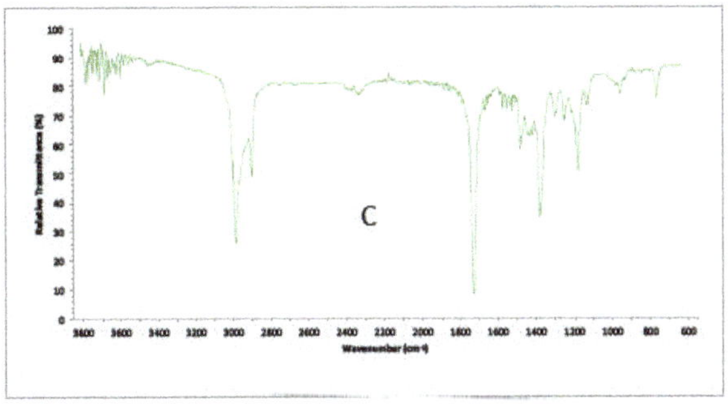

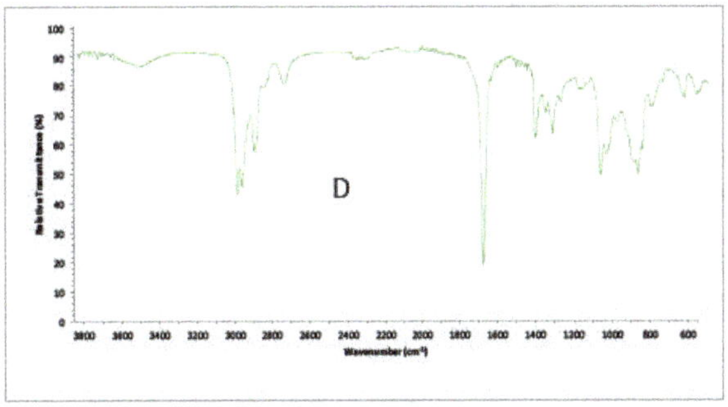

3.7 **Match the following four compounds with their IR spectrum.**

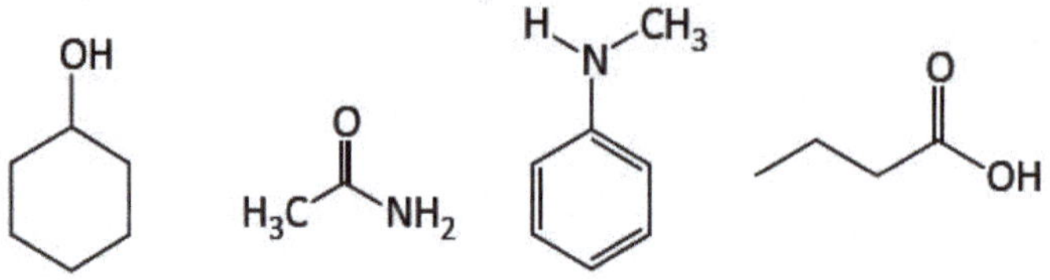

_____ _____ _____ _____

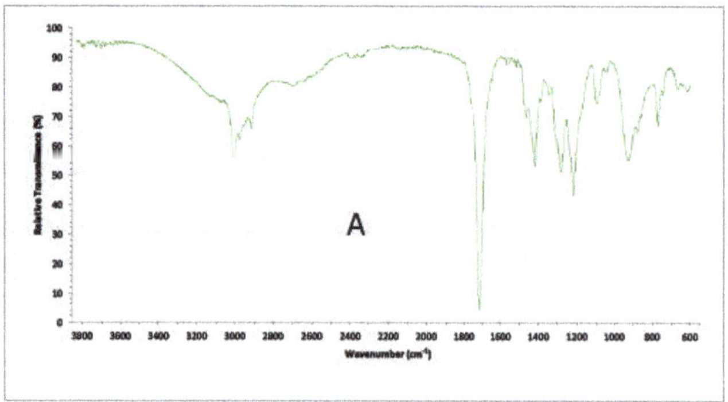

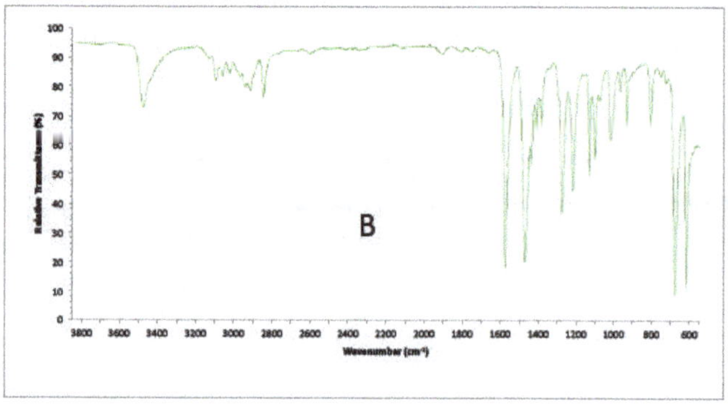

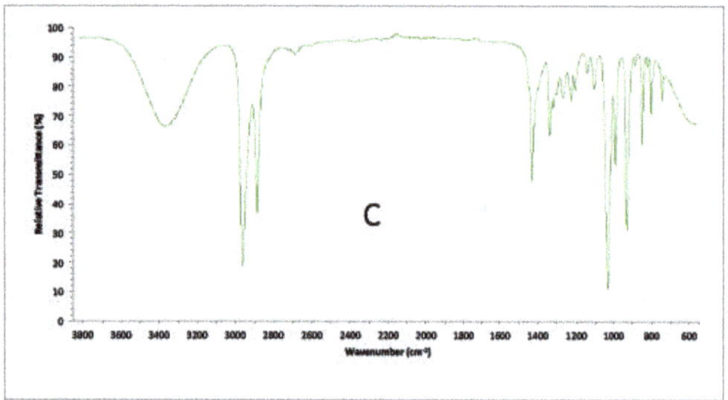

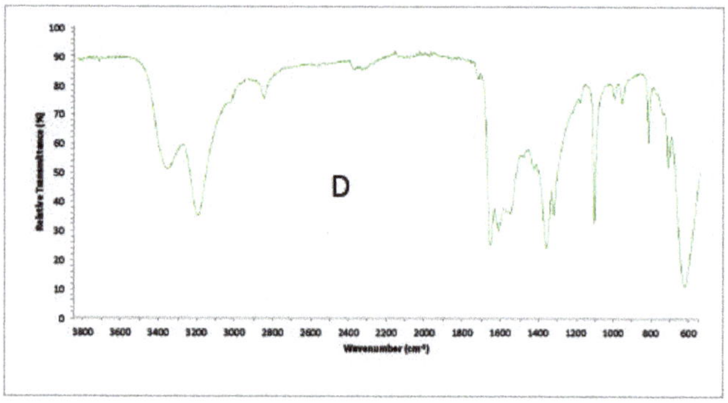

3.3 Problems

3.8 A common, organic solvent gives a mass spectrum that shows an M^+ peak at 41. The IR spectrum for this solvent shows peaks at about 2950 and 2254 cm^{-1}.

a) Propose a viable molecular formula for this solvent. Viable means it corresponds to at least one stable, neutral compound.

b) Propose a structure that is consistent with all of the spectroscopic information given above.

3.9 Based on their molecular formula, determine the number of elements of unsaturation in the following compounds.

a) $C_7H_{10}N_2O$ b) $C_6H_{10}O$ c) $C_7H_{11}OCl$ d) C_6H_7N e) C_5H_5N f) $C_{10}H_{18}$ g) $C_5H_{11}Br$

3.10 For the following MS molecular ion isotope patterns, describe which heteroatom is indicated.

a) 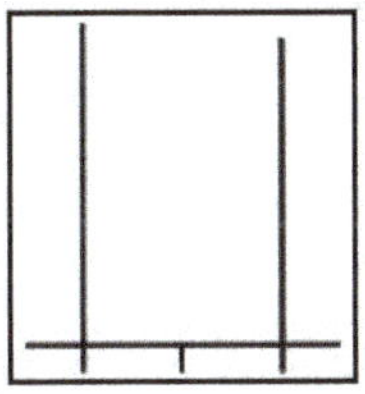

b) 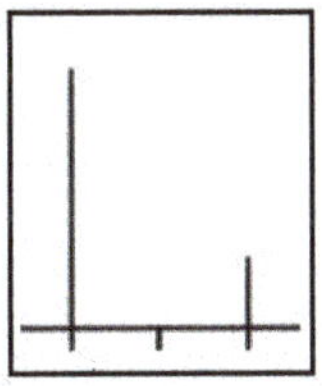

c) 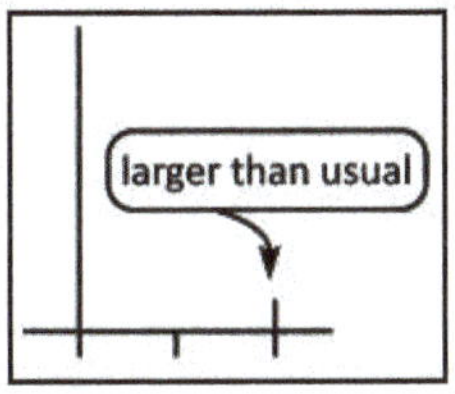

d) Odd M^+ e) M^+-127 is prominent

3.11 Two bottles were found that contained fingernail polish remover. One bottle was acetone and the other was non-acetone remover (ethyl acetate). The bottle labels were removed, so you wanted to identify which bottle contained which compound. IR spectra were obtained to identify them. Spectrum A is acetone and spectrum B is ethyl acetate. Clearly identify which IR spectrum goes with which compound structure.

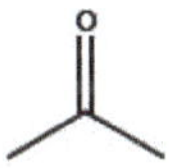
______

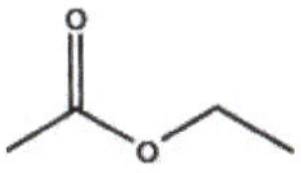
______

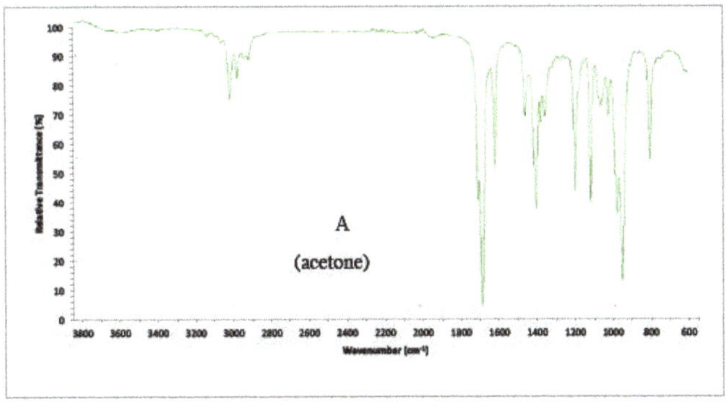

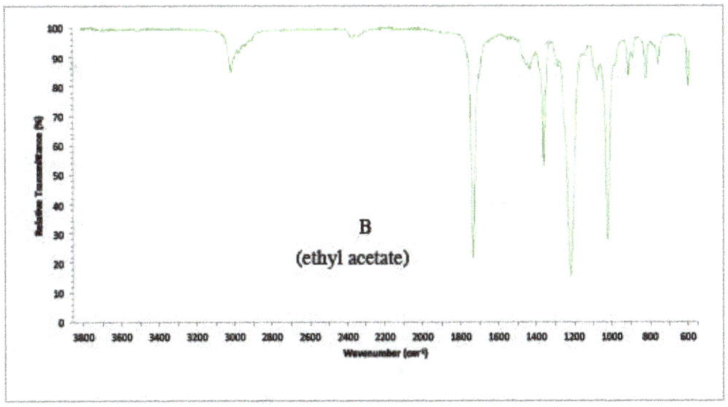

3.12 Your summer job is to clean up an organic chemistry laboratory. You must identify some bottles with illegible labels. In each problem below, describe one difference that you would expect to see in the IR spectra of the two compounds that will allow you to identify which compound it is.

a) Bottle 1 contains one of the following:

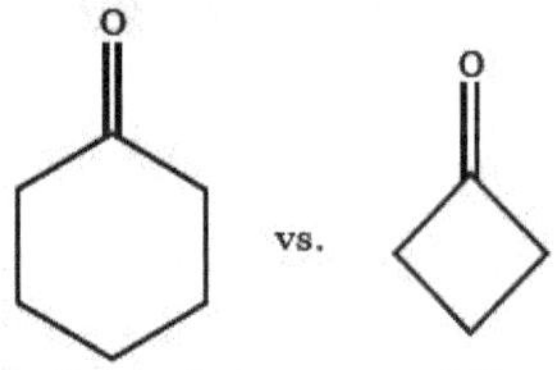

b) Bottle 2 contains one of the following:

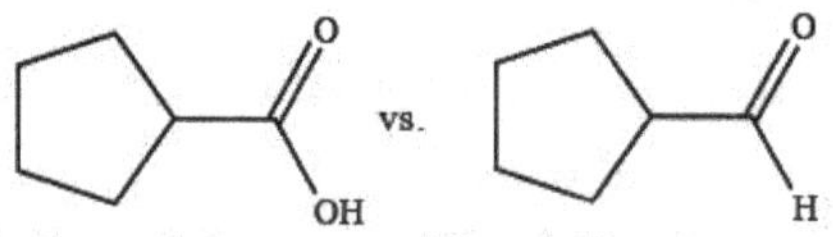

c) Bottle 3 contains one of the following:

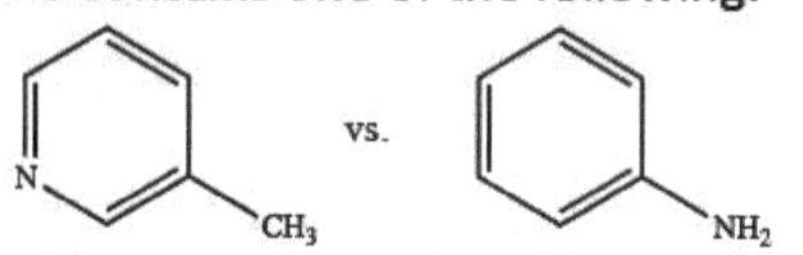

d) Bottle 4 contains one of the following:

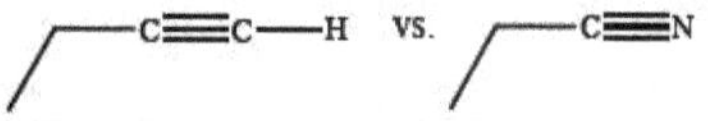

e) Bottle 5 contains one of the following:

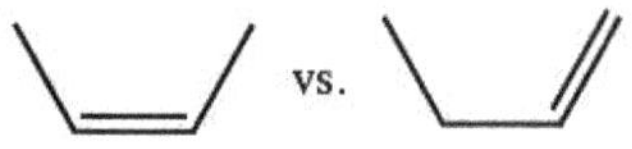

f) Bottle 6 contains one of the following:

$CH_3CH_2CO_2CH_3$ vs. $CH_3CH_2CONHCH_3$

3.13 For the following reactions, indicate the changes you would see in the IR spectra between then reactant and the product.

a)

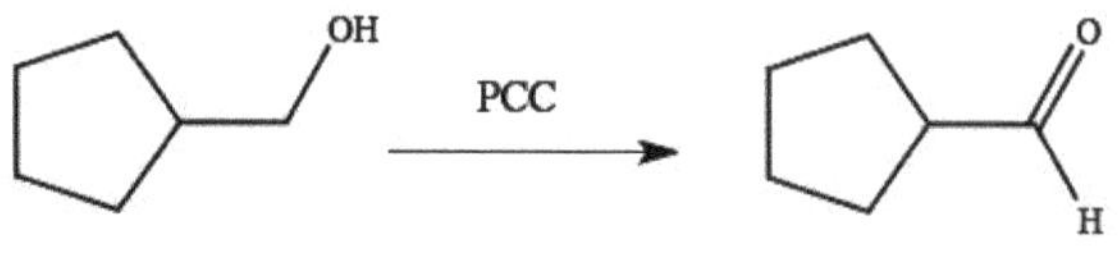

b)

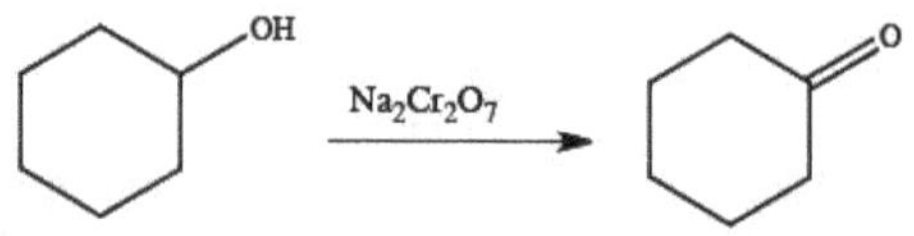

c)

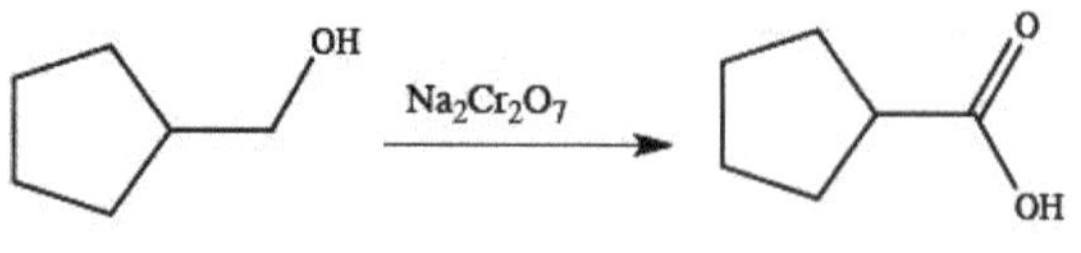

d)

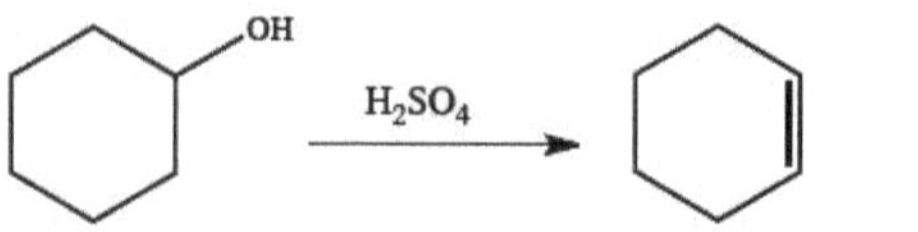

e)

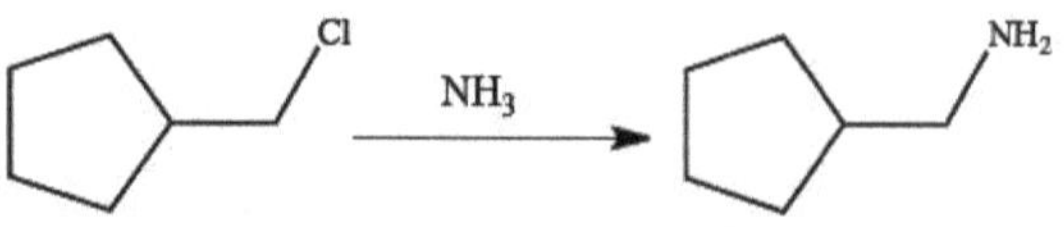

f)

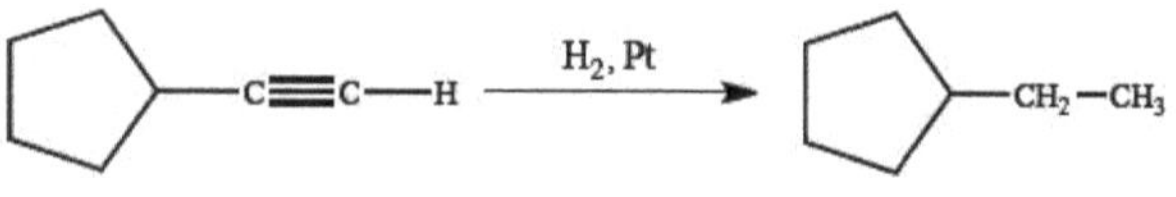

3.14 The following IR spectrum was taken of a compound. A student identified the compound as an aldehyde because of the two peaks marked. a) Why was this student incorrect? What major peak is missing from this IR spectrum if it is indeed an aldehyde?

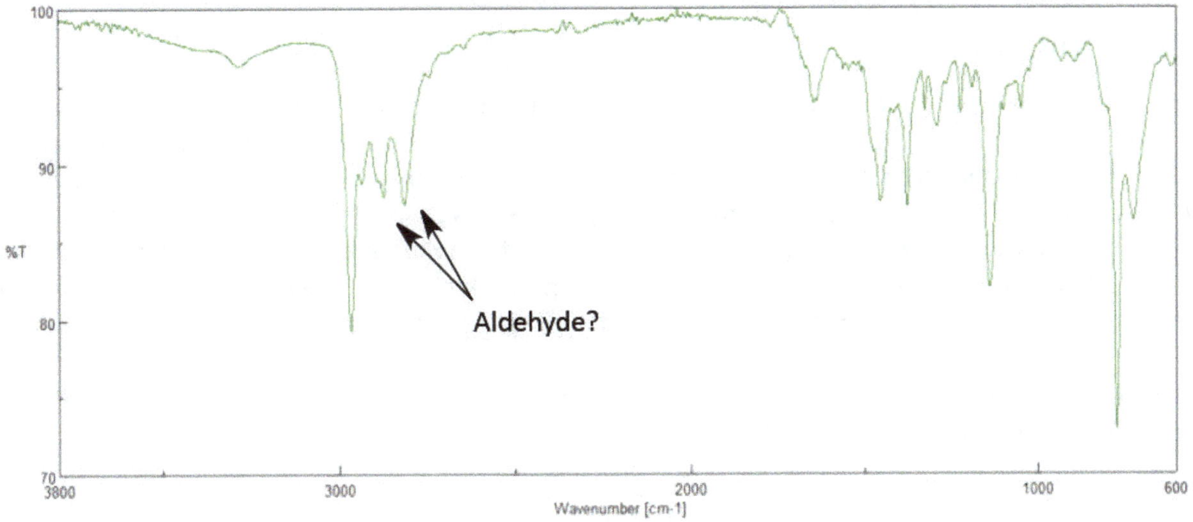

b) After realizing those were not aldehyde peaks, but sp^3 C-H peaks, the student noticed that this compound had, in its mass spectrum, an M^+ peak at 73. Then, the student noticed the following, previously overlooked peak in the IR spectrum. This was indeed a significant peak. What type of compound might this student have?

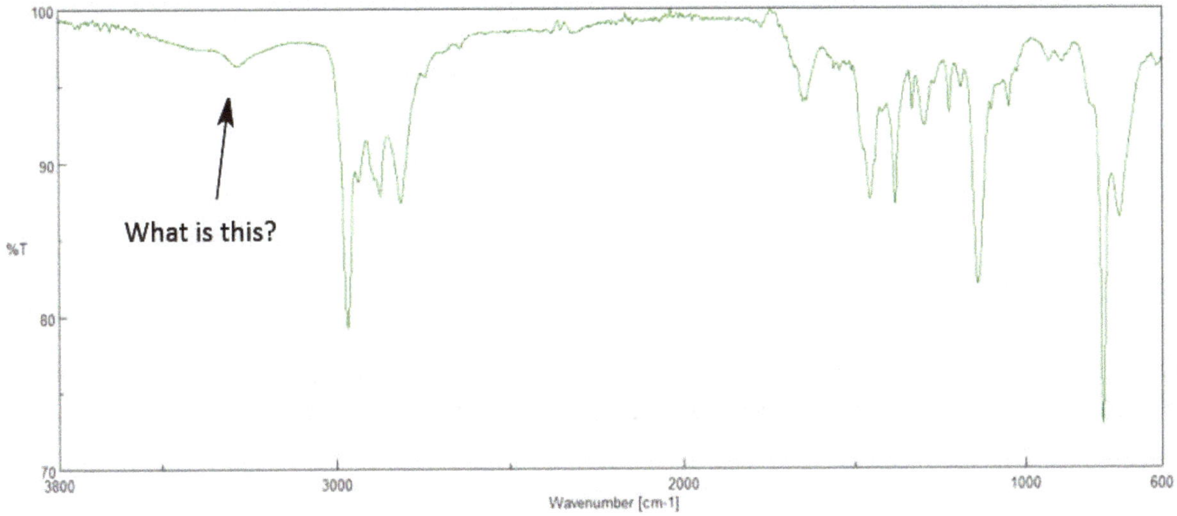

Chapter 4. Resonance

4.1 Introduction

Organic chemists have agreed to show the movement of electrons by using arrows. A regular, two-headed arrow, →, shows the movement of two negative electrons. Did you get that? That last sentence may be one of the most important statements in organic chemistry. *Arrows show the movement of negative electrons.* Arrows DO NOT show the movement of positive charges, atoms, or anything else you can think of. Arrows in organic reaction mechanisms show the movement of negative electrons. Students who fail to learn this are almost never successful in organic chemistry. If you don't learn to focus on the movement of these electrons, organic chemistry will remain a mystery. It is my belief that successful organic chemistry students learn to think in terms of arrows.

Since opposite charges attract each other, we would expect the following reaction of hydroxide with a positive cation intermediate to proceed as follows.

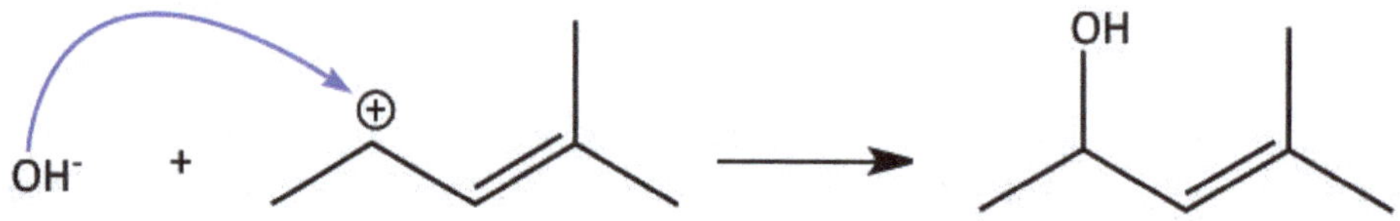

Hydroxide Cation intermediate Expected product

This is, in fact, one of the products we find. Surprisingly, we also find the following product is made.

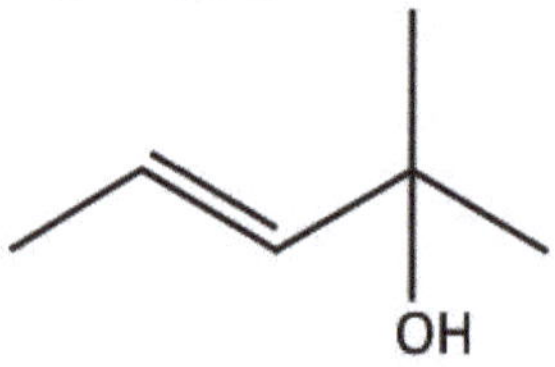

Surprise product

By analyzing this surprising product that is made, we can think backward and figure out what unknown cationic or positively charged intermediate could have been attacked by the hydroxide.

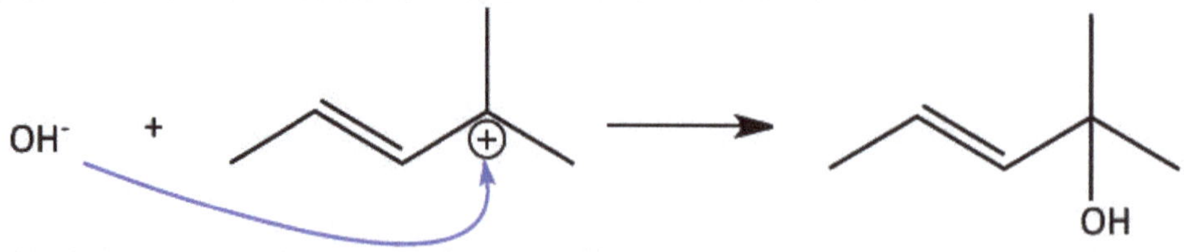

Hydroxide Surprise cation intermediate Surprise product

Let's compare these two cationic intermediates we must have had in order to give us these two products.

Given cation intermediate Surprise cation intermediate

These two cation intermediates are called *resonance structures* of each other. Neither one actually exists, but they are convenient representations of the true intermediate. They are useful to us because it makes it much easier for us to explain why a reaction proceeds the way it does.

The symbol we use to signify that two structures are resonance structures of each other is a two-headed arrow, ↔. Sometimes, brackets are put around the two structures to identify them as resonance structures.

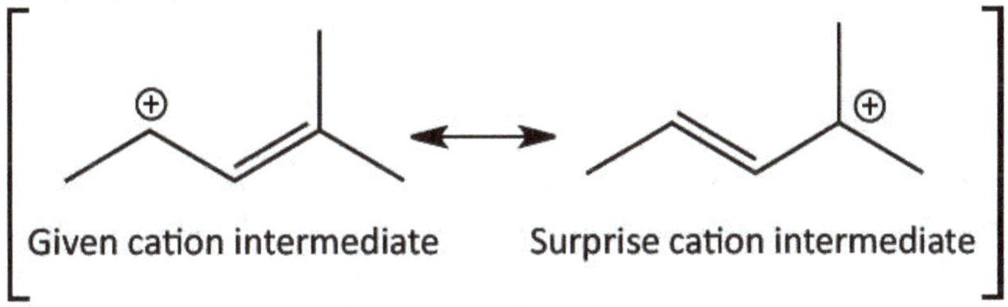

Resonance structures

Well, which is it? Is the intermediate the given one, the surprise one, both, or neither? Some people even mistakenly think that these two resonance structures are flipping back and forth from one form to the other. This is not true. The resonance hybrid, the "true" form of the intermediate, is a combination of these two resonance structures. It really looks more like the following.

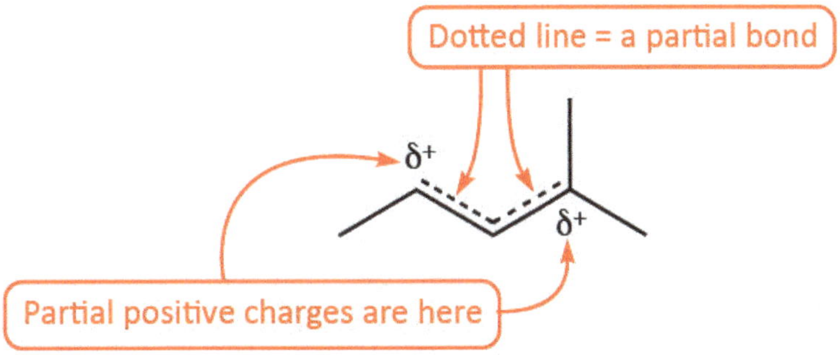

Resonance Hybrid

The two electrons of the pi bond are shared between second and third bond. A solid line drawn between two atoms represents one full, two-electron, bond. A dotted line drawn between two atoms represents a partial bond. So, instead of a double bond at each site, there is really more like 1½ bonds. The one positive charge is shared between the second and fourth carbon atoms. So, instead of having a full positive charge, each of those two atoms carries about ½ of the positive charge. It can get a little confusing. So, in order to simplify our thinking, organic chemists prefer to write the two, distinct, resonance structures.

These two resonance structures are very similar to each other. Notice that the atomic framework in both structures is exactly the same. The only difference between them is the location of the pi bond and the location of the positive charge. If we are given one structure, we can find, or predict, other resonance structures by moving pi electrons around.

Given this intermediate I can find this one.

Notice, only two pi electrons were moved. The arrow shows the movement of these two electrons, NOT the positive charge. We'll spend quite a bit of time learning how to find good, valid resonance forms of various structures.

Let's take a break from this example to discuss resonance structures in more detail. We'll return and finish up this example later.

4.2 Half steps or full steps

Whenever resonance structures are drawn, there are two ways to move electrons. Electrons can move from an atom to an adjoining bond, from a bond to an adjoining atom, or from a bond to a neighboring bond. That may sound a little confusing if you read it too fast. To put it more simply, I use the terminology that electrons can move one full step or ½ step. You won't find these terms in any other organic text, but I think it makes our discussion of resonance forms more convenient.

Electrons move ½ step if they move from a pi bond to an attached atom or if a set of lone pair electrons moves from an atom to an attached bond. Study some examples of valid ½ steps below.

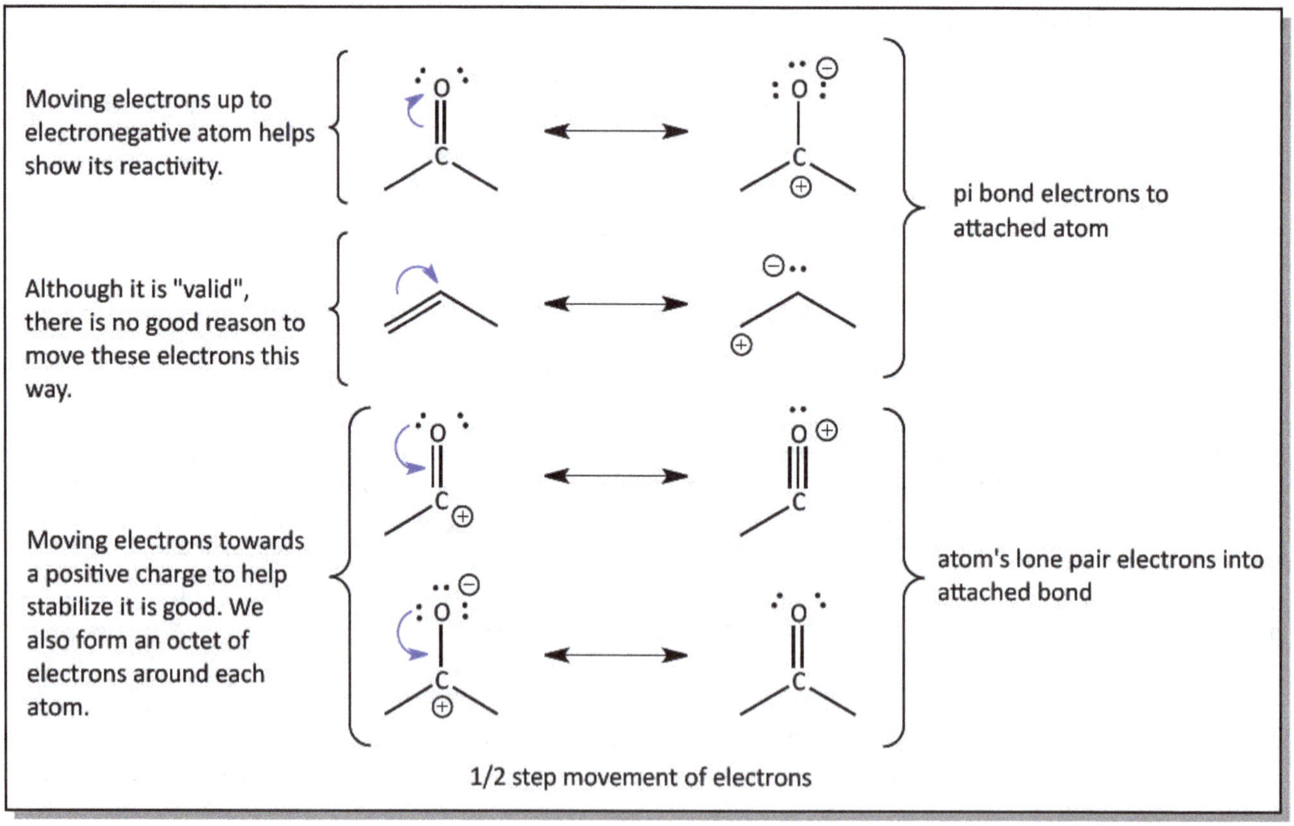

Figure 4.1 Examples of 1/2 step movements of electrons

A full step of moving the electrons is moving from a pi bond to a neighboring bond, skipping over the atom between the two bonds.

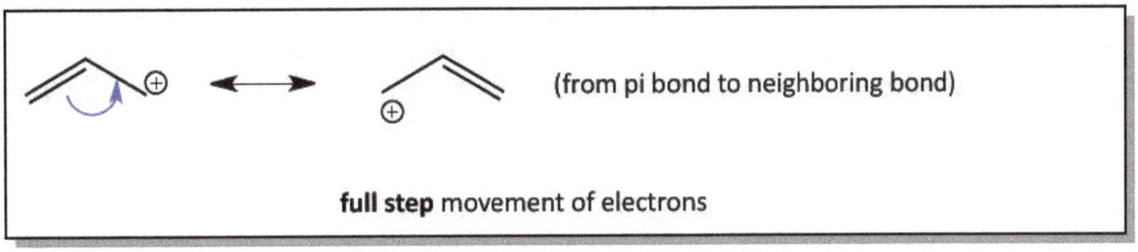

Figure 4.2 Example of full step movement of electrons

The reason I call the above pi bond to neighboring bond movement of electrons a "full step" is that it really is the sum of two ½ steps.

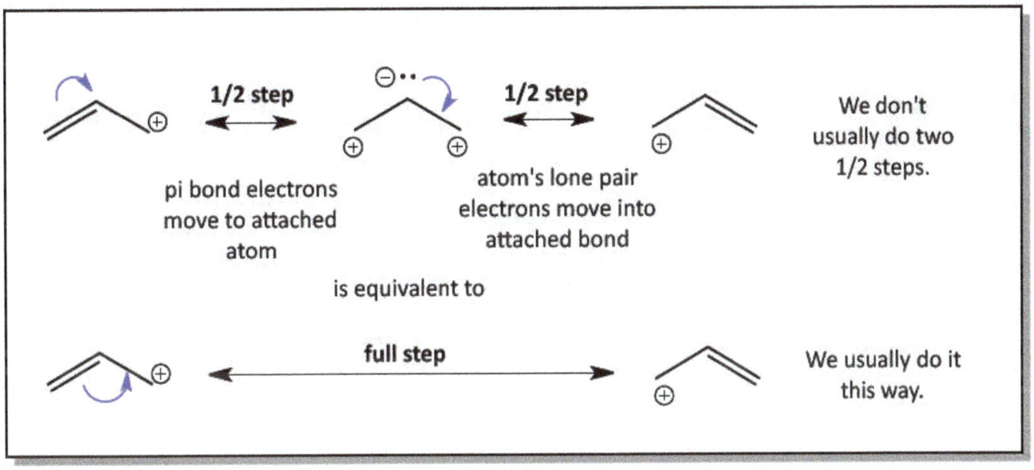

Electrons may be moved ½ step or one full step, but they may never be moved farther than one step. This means that electrons are not allowed to move 1½ steps, 2 steps, or more.

4.1 Each of these problems has an arrow indicating the movement of electrons. Write how many "steps" each movement would be. Then, indicate if the movement is allowed or not allowed by circling the appropriate word.

a) 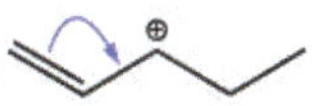

___ step(s)

allowed or not allowed

b) 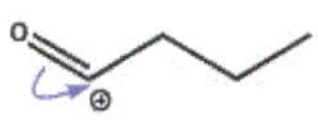

___ step(s)

allowed or not allowed

c) 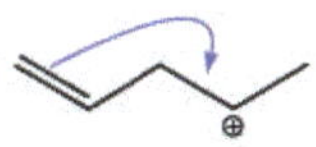

___ step(s)

allowed or not allowed

d) 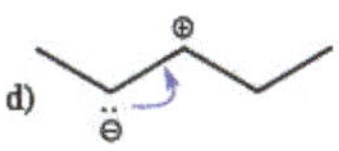

___ step(s)

allowed or not allowed

e) 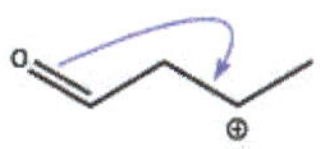

___ step(s)

allowed or not allowed

f) 

___ step(s)

allowed or not allowed

Resonance

Let's further investigate how we find good, valid resonance structures. The most important thing is to not forget our mechanism arrow training. It is important to still think in terms of arrows moving electrons. For resonance structures, our arrows will move electrons from one spot to another within the same molecule. We will do this by moving the electrons ½ step or one full step at a time. Since we are moving electrons, our arrows will always start at a pi bond (in a double or triple bond) or a lone pair of electrons. We can only move these electrons certain places.

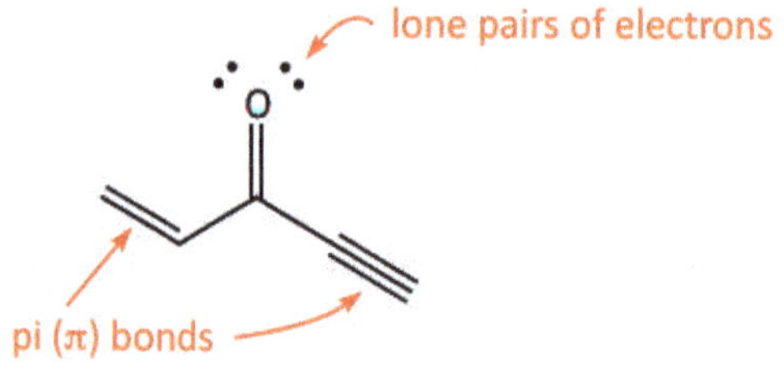

Figure 4.3 Electrons that can be moved in resonance forms

4.2 For the following structures, put a check in the box if the curved arrow STARTS at a valid location (lone pair of electrons or a pi bond). Put an X in the box if curved arrow STARTS at an incorrect location.

a) 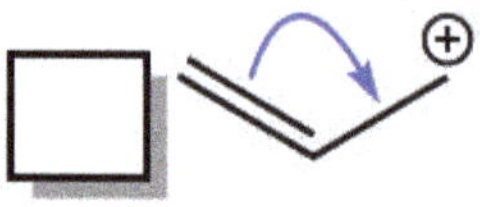

b) 

c) 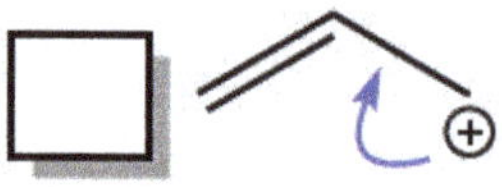

d) 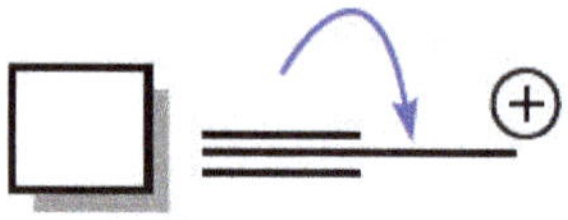

e) 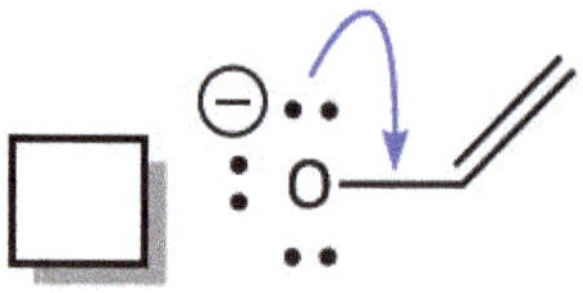

f) 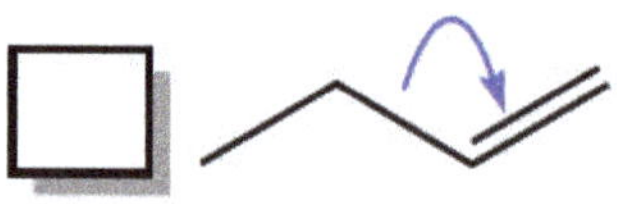

4.3 Resonance rules

There are a few rules for us to consider when we find resonance forms.

1. Don't start any arrows at single bonds. Your molecule will fall apart if you move single bond electrons. That's bad.
2. One atom cannot contain more than eight electrons. If you end up with five bonds around a carbon atom, you've made a mistake.
3. Electrons cannot move more than one full step.
4. If we have the option, it is better to move the negative electrons towards positive charges (cations) or on to electronegative elements.
5. The molecular structure must remain intact. Atoms are not allowed to be rearranged or moved. Only electrons may be moved around.
6. The sum of the charges on resonance structures must be the same.

4.3.1 Pi electrons towards cations

Let's try to find valid resonance forms of the following carbocation.

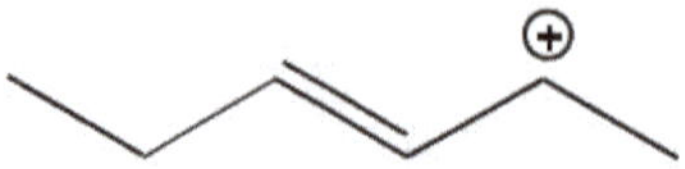

As we are learning how to do this, it is helpful to draw in all hydrogen atoms. This is sometimes called Grossman's rule. It is named after chemistry professor Bob Grossman at the University of Kentucky. It can trip us up if we forget how many hydrogen atoms are on each carbon atom. That is easy to mess up for beginning organic chemistry students. Sometimes even experienced organic chemists make this mistake! Grossman's rule can help us out.
So, let's rewrite the compound with the hydrogen atoms attached.

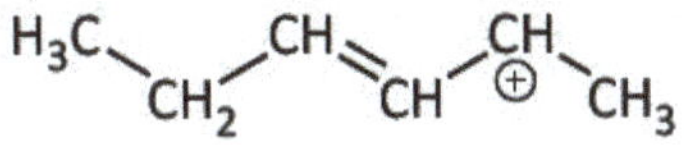

Let's begin looking for a good resonance form. We'll look at some possible ways electrons could be moved on this compound. We'll work through several bad attempts until we come to the correct one.

Attempt 1

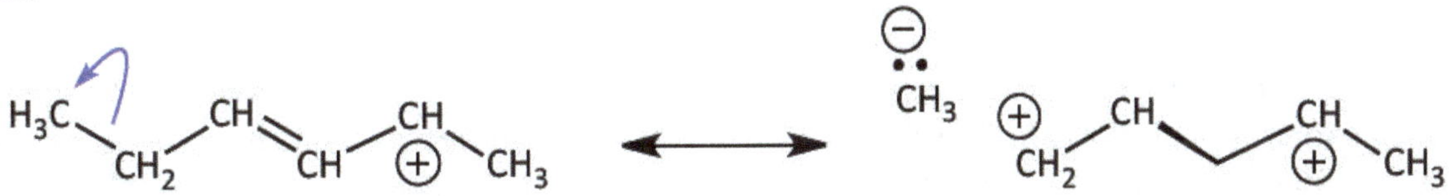

This is a violation of rule 1. If we move these electrons, we could break our molecule apart. Yikes! A good way to avoid this problem is to ALWAYS start our arrows at pi electrons (from a double or triple bond) or from lone pairs of electrons. In attempt 2, we'll start our arrow at the double bond (pi bond). We'll move two of those electrons around.

Attempt 2

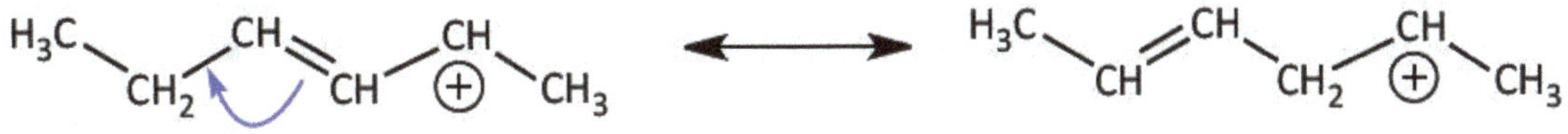

This is a violation of rule 5. We rearranged the atomic structure when we were not allowed to change the molecular structure or the arrangement of the atoms. We cannot rearrange the atoms. Notice that there were two hydrogen atoms on the second carbon, but there is only one after the electrons move. There was only one hydrogen atom on the fourth carbon, but there are two after the electrons move.

Attempt 3

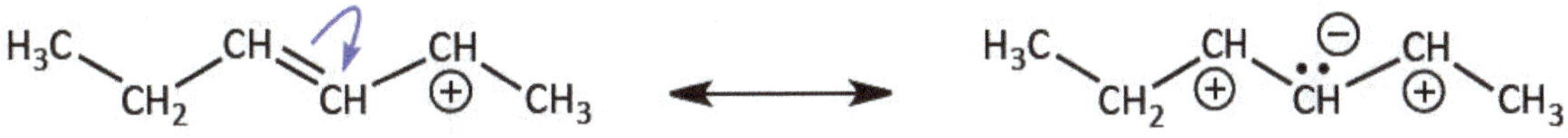

Attempt 3 is a ½ step movement of electrons. This is a legitimate way to move electrons. The problem with it is that we get a new resonance structure that has three charges drawn on it. This makes this resonance structure fairly unstable and not very useful. This attempt does follow all of the rules, but it simply takes more experience for us to learn that it is not very useful.

Attempt 4

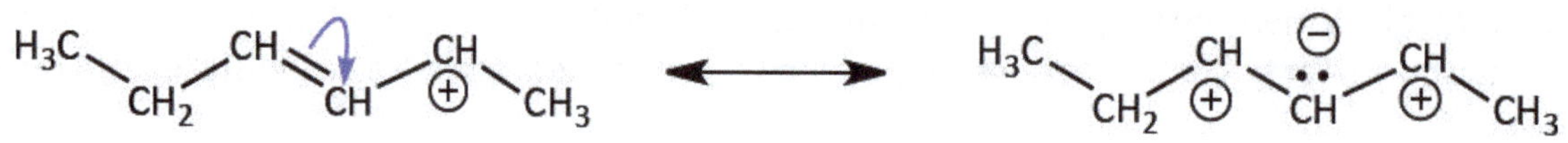

Attempt 4 is a ½ step movement of electrons, but we simply moved the electrons the other way as compared to attempt 3. Again, it is a legitimate way to move electrons, but not very useful because it makes a triply charged, unstable, resonance structure.

Attempt 5

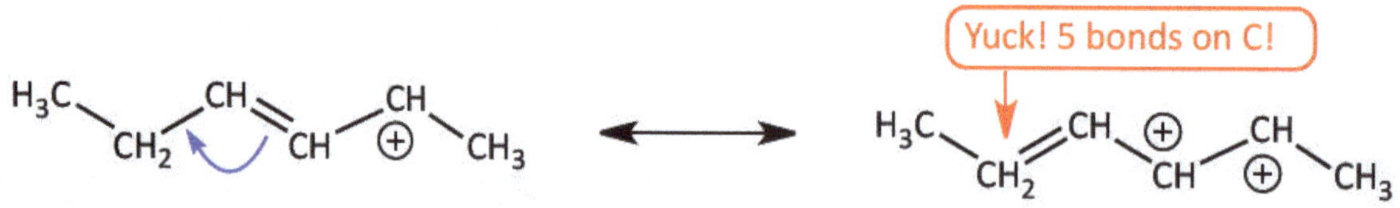

Attempt 5 is a full step movement of the pi electrons. This movement of the electrons violates rule 2. We have too many electrons around an atom. The carbon atom has the most unfortunate situation of having five bonds! It is also a violation of rule 6. The left resonance form has a +1 charge. The right resonance form has a +2 charge.

Attempt 6

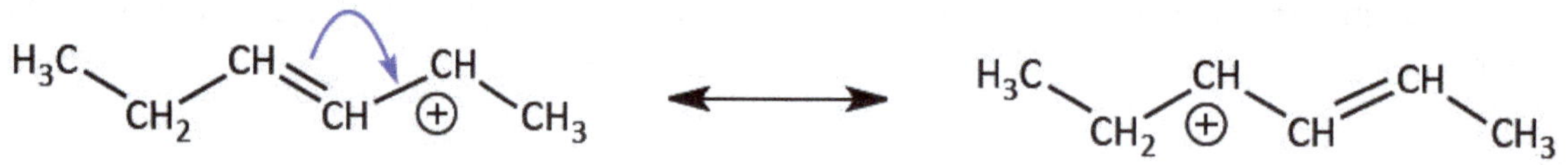

This is a full step movement of pi electrons towards a positive carbocation. A few things should be noted. It is preferred to have negative electrons flow towards a positive site. The atomic structure has not changed from one resonance form to the other. No atoms have more than an octet around them. This is because the positively charged carbon atom only started with six electrons (3 bonds) around it. When the pi electrons moved towards it, it made an octet. The bottom line is a pi bond can move one full step towards a carbocation to make a resonance structure. If the pi bond is further removed from the cation more than one full step, they *cannot* be pushed towards the cation.

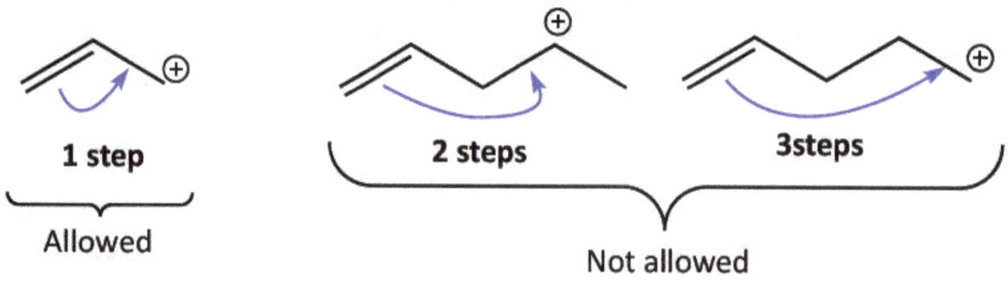

4.3 One of the following cyclopentene cations has a valid resonance structure and the other does not. For the one that has a valid resonance structure, draw the curved arrow needed and the resonance form. For the one that does not have a resonance form, explain why. It may be helpful to use Grossman's rule.

a) 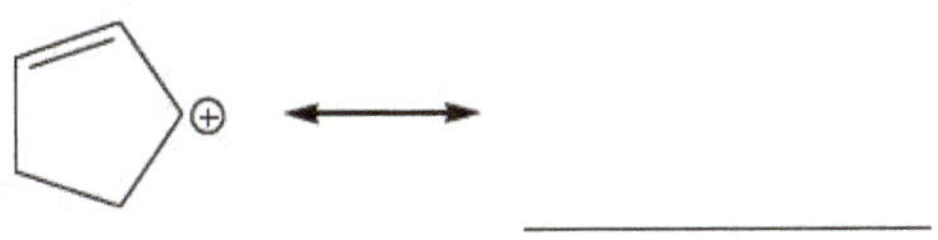

b) 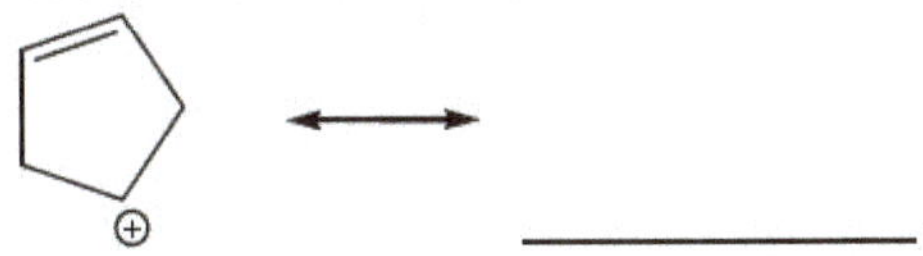

4.4 Draw in the appropriate resonance structures and curved arrow for the following. Pay careful attention to where the positive charge is located in the resonance structures. Grossman's rule can help you out.

a)

Draw resonance structure then draw in one arrow.

Draw resonance structure.

Circle the three atoms on the resonance hybrid of this molecule that have a partial positive charge. You can tell by noticing which atoms in **part a** have a positive charge drawn on them.

b) 

4.5 The cycloheptatrienyl cation has seven valid resonance structures. Draw in each resonance structure and the appropriate arrows showing how each resonance structure if made. The first arrow has been drawn for you.

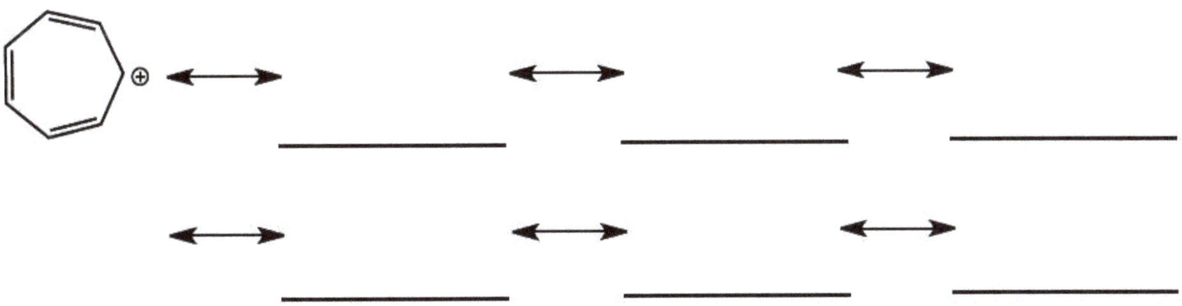

Circle the atoms on the cycloheptatrienyl cation hybrid that have a partial positive charge.

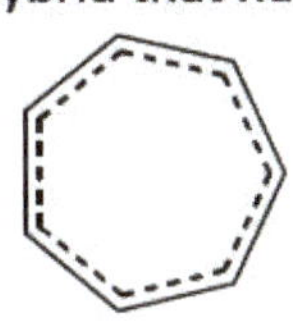

Resonance

Let's return to our original example and take a closer look at how we would find this new resonance structure. Until we become experts, and even then, it is wise to draw in hydrogen atoms near all of our reactive sites so we don't lose track of them

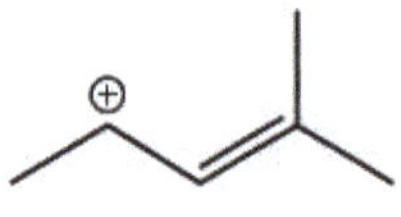

Given intermediate

becomes

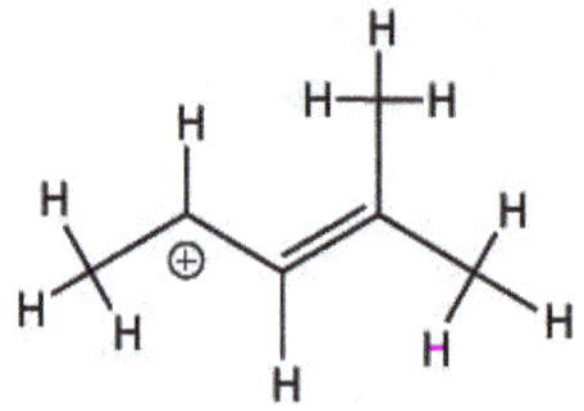

Notice that the carbocation only has one hydrogen atom attached to it. All of the other carbon atoms need four bonds.

The carbocation, being positively charged, can be stabilized if some negatively charged electrons can be "pushed" towards it. In our example, we have a pi bond located right next door with plenty of juicy, negative electrons. We can push them over one full step towards the carbocation.

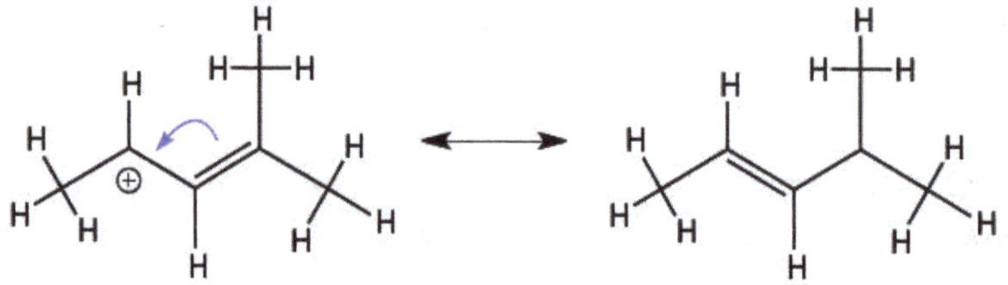

Given intermediate

New resonance structure
(not yet complete)

To finish, we need to evaluate our new resonance structure and make sure all of our formal charges are correct. Since our given intermediate has an overall +1 formal charge, we need to have a +1 formal charge in our new resonance structure (rule 6). After a quick analysis, we find that one of the carbon atoms only has three bonds; it must be our carbocation.

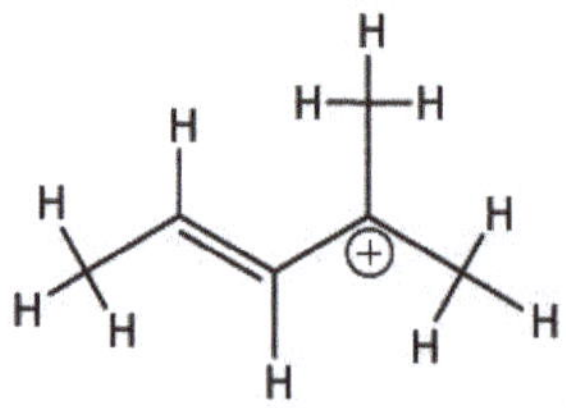

New resonance structure
(complete)

This is a correct resonance structure. Let's look at some things beginning students try to do that are *not* correct.

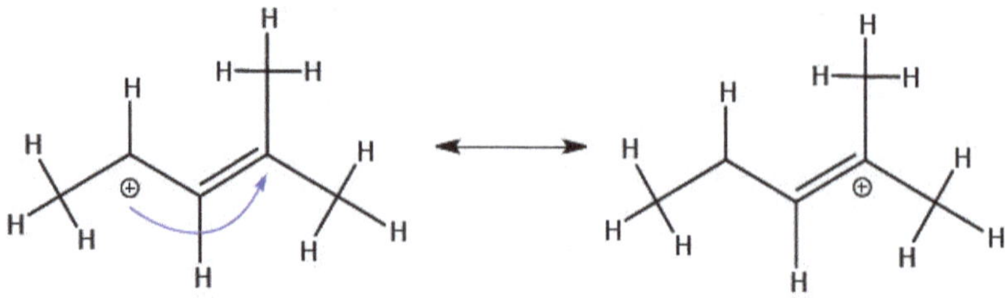

Given intermediate

Why this is incorrect. Arrows *always* show the movement of electrons, not positive charges. Never do this. Even if we broke the rules and moved the positive charge, our new resonance form does not make sense. The carbocation should be on a carbon atom with three bonds, not the one with four bonds.

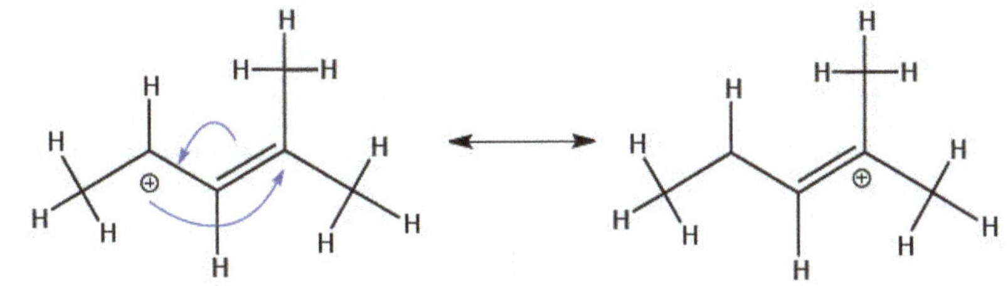

Given intermediate

Why this is incorrect. These two arrows give a resonance structure that looks right, but again, arrows *always* show the movement of electrons, not positive charges. Never do this. Many novice organic chemistry students draw too many arrows for no reason. Do not draw more than one arrow to make a resonance form unless you need to. If an atom gets more than an octet of electrons because a first arrow has moved two electrons onto it, then a second arrow should be drawn to remove electrons. For example, if an atom will have ten electrons, then some electrons need to be immediately removed in order to maintain an octet (see 4.3.3).

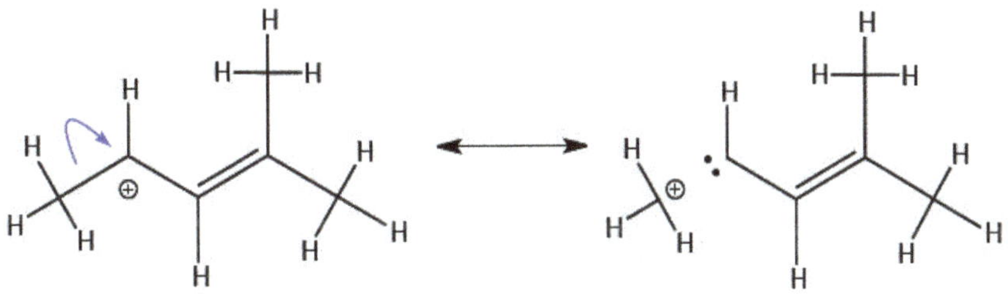

Given intermediate

Why this is incorrect. Looking at the resonance structure made from this movement of electrons makes it obvious what the problem is. Resonance structures must keep the same basic atomic structure and only differ by the location of pi and nonbonding electrons. If an arrow starts at a sigma bond, it breaks apart the structure. Never do this.

4.3.2 Pi electrons towards electronegative atoms

We've focused on pi electrons moving towards positive cations, but cations are not the only things that pull negative electrons towards them. Electrons are also drawn towards electronegative atoms. Pi electrons can move ½ step onto an electronegative atom if the pi electrons are located adjacent to an electronegative atom like oxygen or nitrogen.

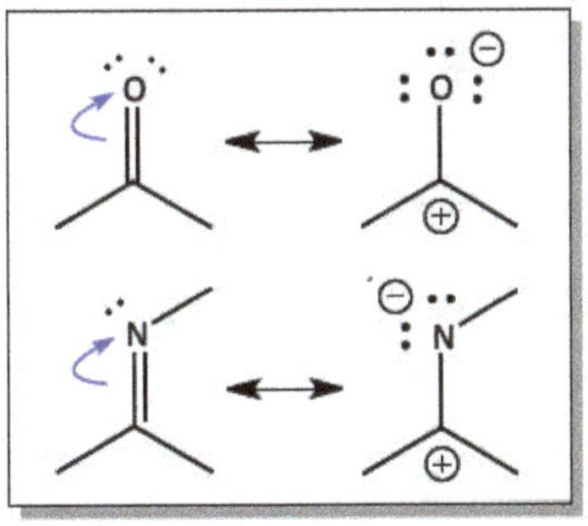

Resonance

When pi electrons move onto an electronegative atom, we get a resonance structure with a charge separation. We get a positive atom and a negative atom. The electronegative oxygen or nitrogen atom becomes the negative one.

4.6 In the following reaction, hydroxide is going to react with cyclohexenone. The question is, where? Hydroxide has a negative charge. Since opposites attract, the most positive atoms on cyclohexenone will be where hydroxide attacks. There are two atoms on cyclohexenone which carry a partial positive charge. Let's try to find them.

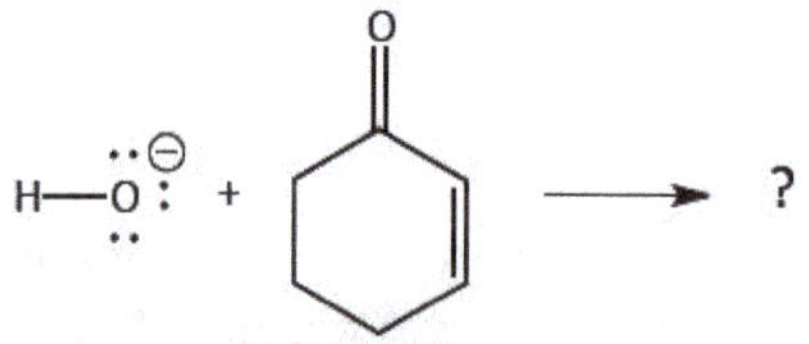

cyclohexenone

Where are the two positive sites?

Let's work through it together. First, we notice the electronegative oxygen atom with a pi bond attached. We can move the pi bond ½ step onto the oxygen atom to find one of the partially positive atoms.

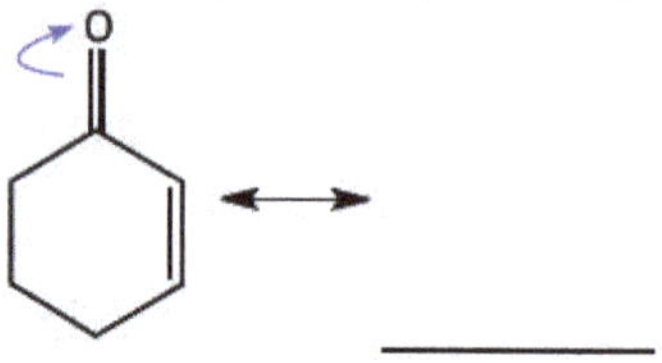

6a

a) Draw resonance structure

This resonance structure has a pi bond adjacent to a positively charged carbon atom (carbocation). Move some pi electrons one full step towards the carbocation.

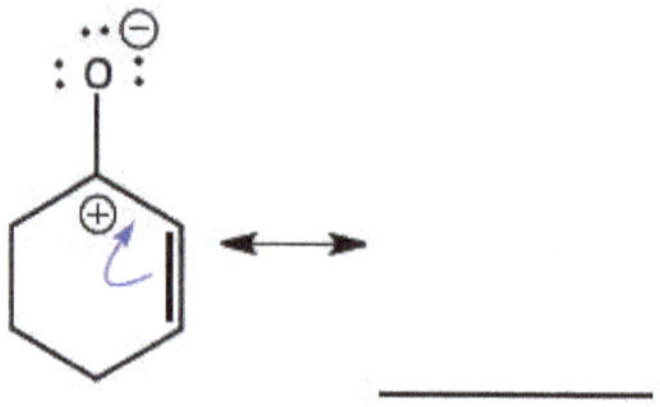

6a

6b

b) Draw resonance structure

Now, since we have two resonance structures with positively charged atoms, we have insight into the reactivity of cyclohexenone.

c) Circle the two atoms on cyclohexenone hybrid that have a partially positive charge.

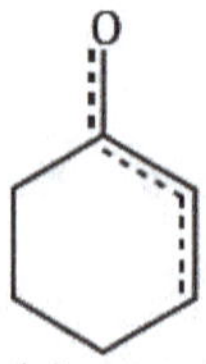

cyclohexenone

4.7 a) Find resonance forms to help find the three atoms on the following compound that have a partially positive charge. Draw in all appropriate arrows and resonance forms.

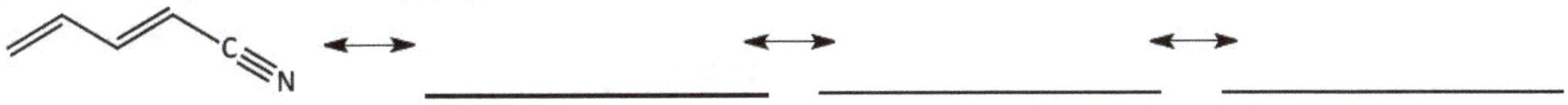

b) Circle the three atoms that are partially positive.

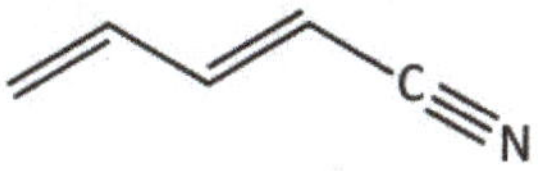

4.8 Four atoms of benzaldehyde have a partial positive charge. Draw resonance forms to find them all.

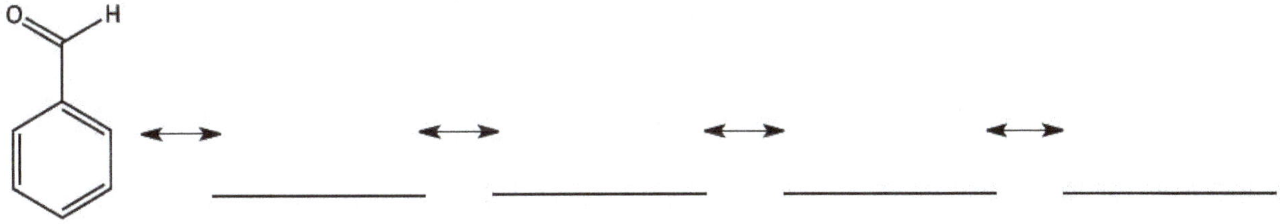

Circle the four atoms on benzaldehyde that have a partial positive charge.

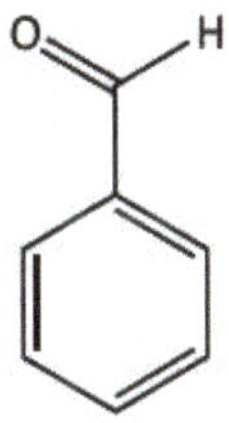

benzaldehyde

4.3.3 Lone pairs moving towards pi bonds.

Pi bonds are not the only electrons that can be moved to make resonance forms. A lone pair of electrons (especially at a position with a formal negative charge) is a great place to start pushing electrons. They can be pushed ½ step IF AND ONLY IF a situation does not occur where an atom ends up with greater than eight electrons on it (an octet). An atom can temporarily have more than eight electrons around it, IF DURING THE SAME STEP, electrons are removed from the back side of said atom to correct the problem of having too many electrons.

Let's take a closer look to see what I mean. Consider the following anion.

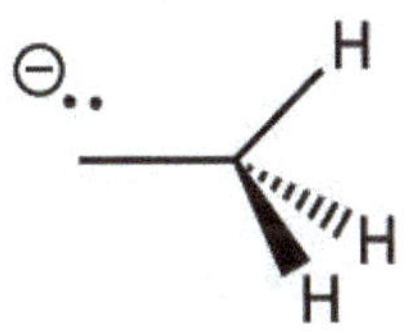

If I push the lone pair from the negative site ½ step, we run into a bad situation.

Resonance

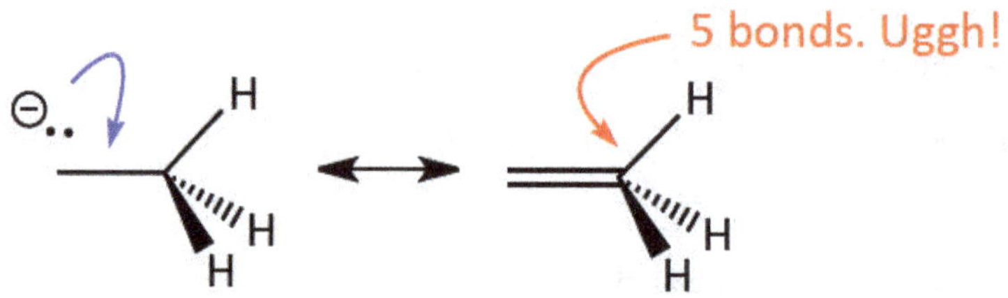

We have a carbon atom with five bonds! That's bad unless we can fix the situation by moving some electrons away from it. But, in this example, the problem cannot be fixed because all of the bonds on the backside of this carbon are all single bonds and cannot be moved away.

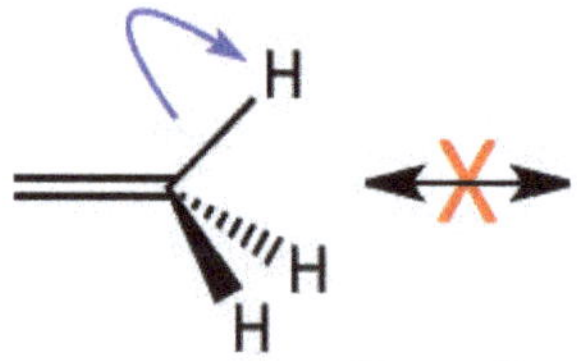

We cannot start arrows at single bonds to make resonance forms because our molecule will fall apart.

How about the following carbanion?

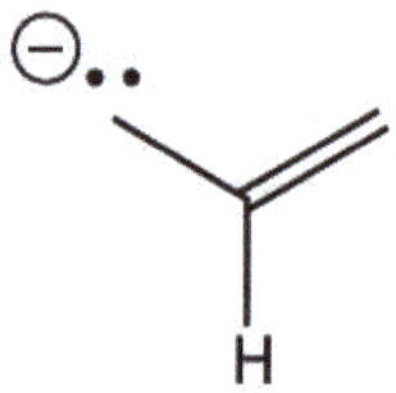

We move the lone pair of electrons from the negative site ½ step and see what we get.

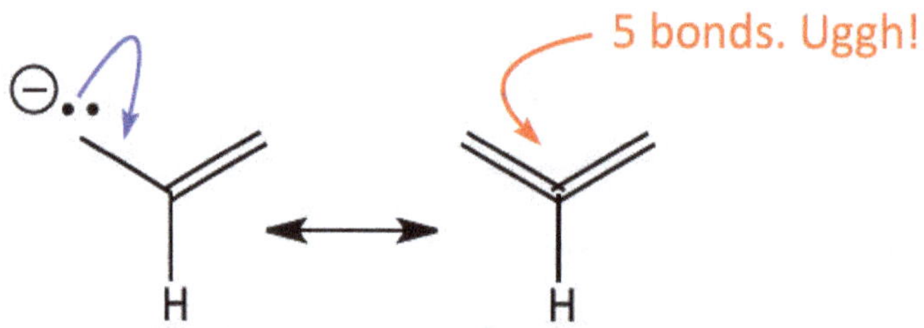

Again, we get the bad situation of having 5 bonds around a carbon. We have more than eight electrons around the carbon and have broken the octet rule. But, don't panic just yet. We'll be OK if we can move electrons away from this overloaded carbon atom. In this case, we have a pi bond that can be pushed away ½ step onto another carbon atom to remove some electrons. We're saved!

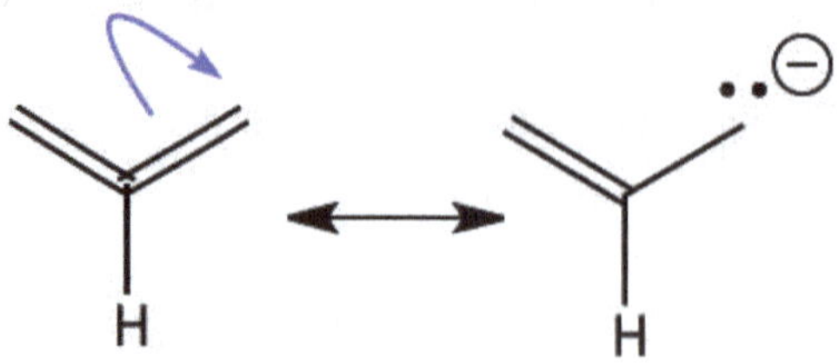

Normally, organic chemists do not draw the bad looking structure with five bonds around the carbon. We know that it is not a good structure, so we don't want to pretend that it is valid in the least. So, we typically draw both arrows on one structure like this.

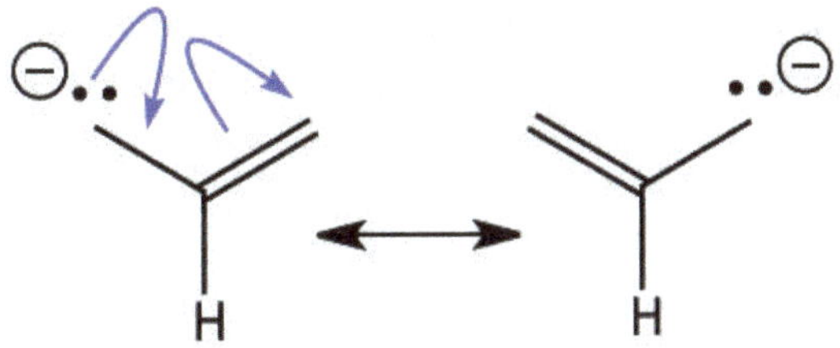

4.9 Finish the movement of the electrons on this structure and draw the resonance structure made.

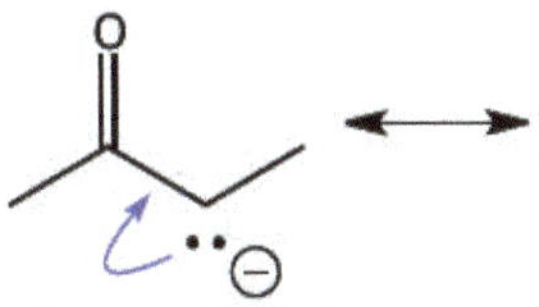

Draw one more arrow. Draw resonance structure.

This a very good resonance structure because electrons end up on an electronegative oxygen atom!

4.10 Nitropropane has one other good resonance structure. Draw it.

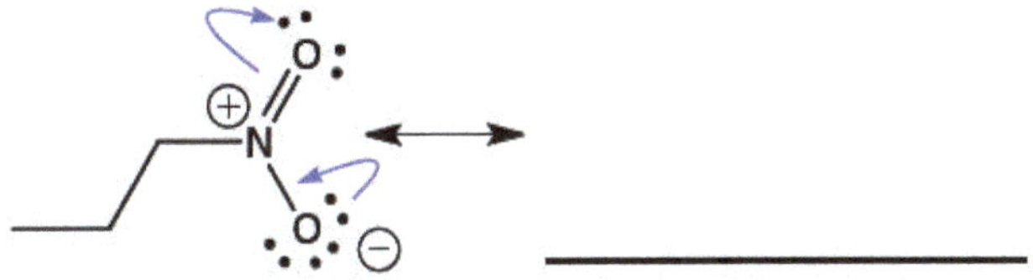

Since the two oxygen atoms share 1 negative charge, the resonance hybrid (or the true structure of the molecule) has ½ of a negative charge on each oxygen atom.

4.11 The cyclopentadienyl anion has five valid resonance structures. Draw them and the appropriate arrows.

Draw resonance structure then draw two arrows | Draw resonance structure then draw two arrows | Draw resonance structure then draw two arrows | Draw resonance structure

Circle the atoms on the cyclopentadienyl anion that have a partial negative charge.

4.12 The phenoxide anion has four atoms with a partial negative charge on them. One is the oxygen atom which is clearly delineated in the given structure. Draw three other resonance forms to find the other three negatively charged atoms.

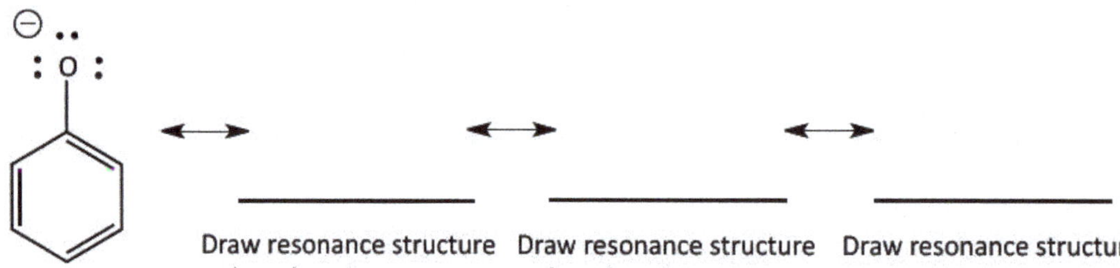

Draw resonance structure then draw two arrows | Draw resonance structure then draw two arrows | Draw resonance structure

Circle the atoms on the phenoxide anion that have a partial negative charge.

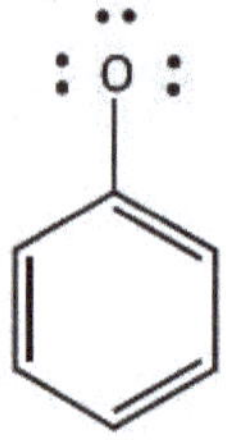

4.13 Draw in the two resonance forms as well as all appropriate arrows to show the movement of electrons in the following.

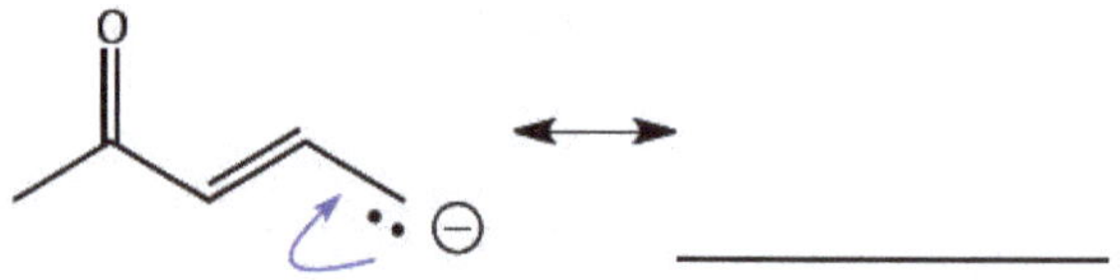

a) 13a Draw another arrow to move pi electrons 1/2 step

13b Draw resonance structure

In this new resonance structure, we can push the carbanion electrons towards the carbonyl C=O. Draw in the second arrow needed and then the resonance structure made.

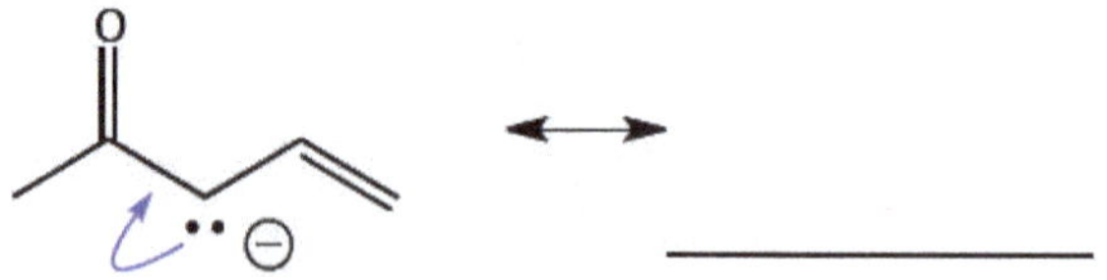

b) 13b Draw another arrow to move pi electrons 1/2 step

13c Draw resonance structure

We can put it all together and get the following.

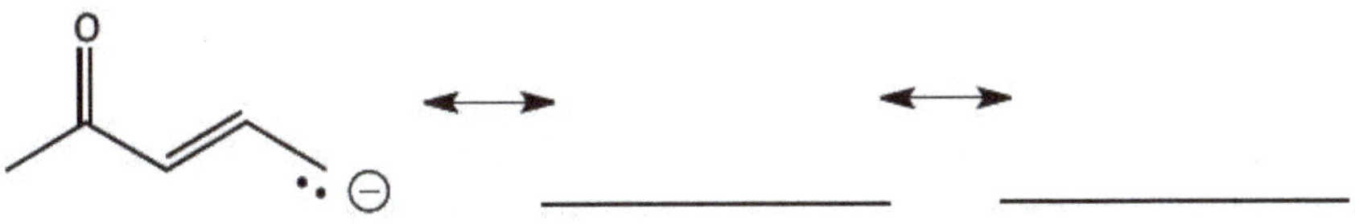

c) 13a 13b 13c

Circle the three atoms that are partially negative.

d)

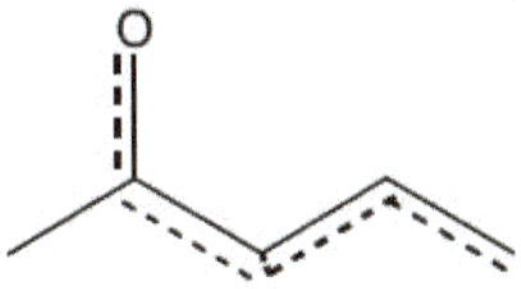

4.14 The following movement of electrons is *not* allowed in order to make a new resonance form. Why?

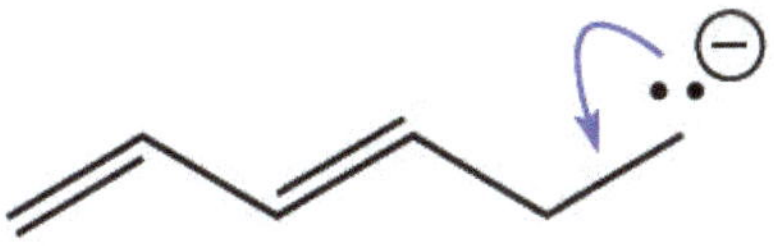

4.15 The following compound does have two valid resonance forms. Draw them and appropriate arrows.

a)

b) Circle the three atoms that are partially negative.

4.4 Problems

4.16 For the following structures, if possible, push the lone pair electrons to generate other good, valid resonance structure(s). If no other good resonance structures can be drawn, say "no resonance structures."

a) 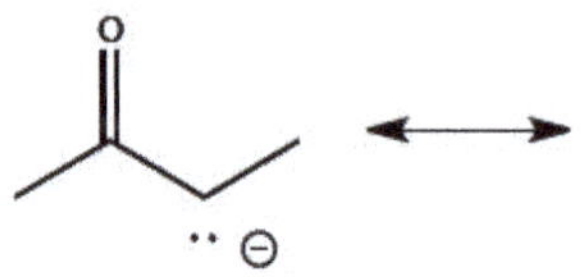

b) 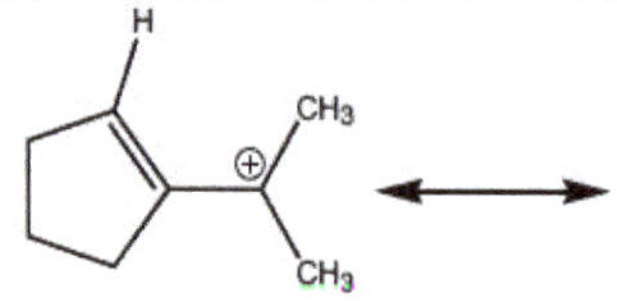

c) 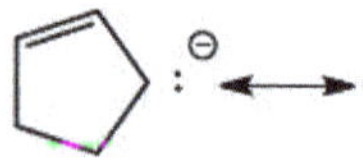

d) 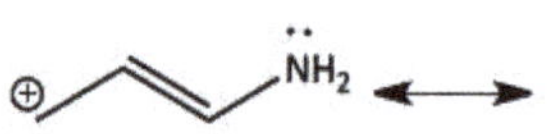

e) 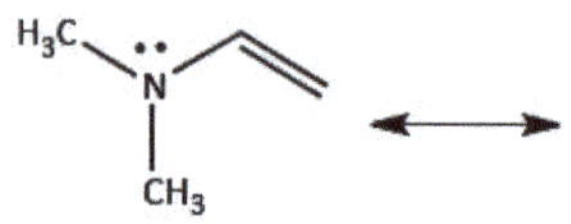

f) 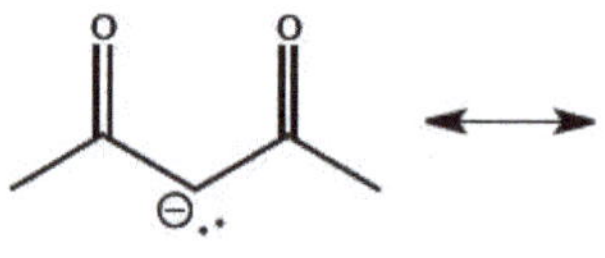

g) 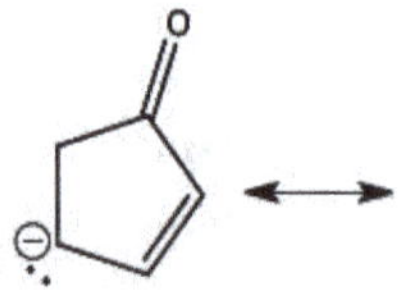

h) 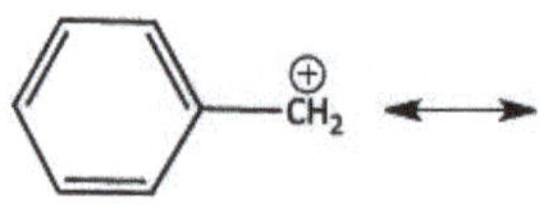

i) 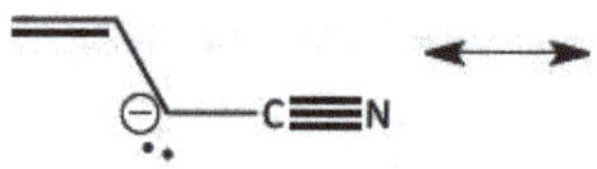

j) 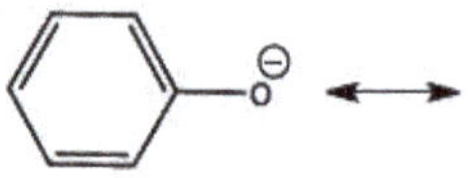

k) 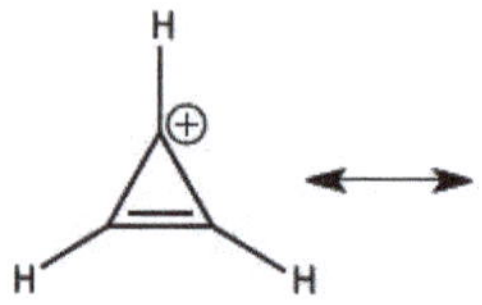

l) 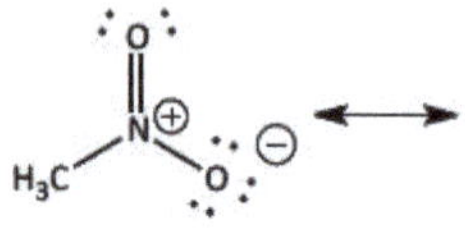

m) 

n) 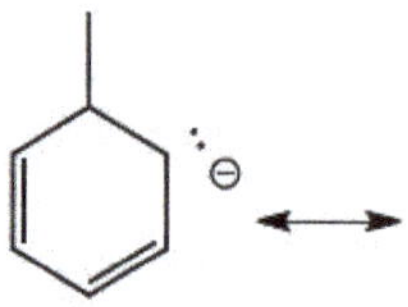

o) 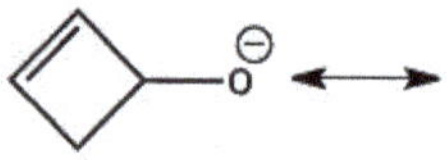

p) 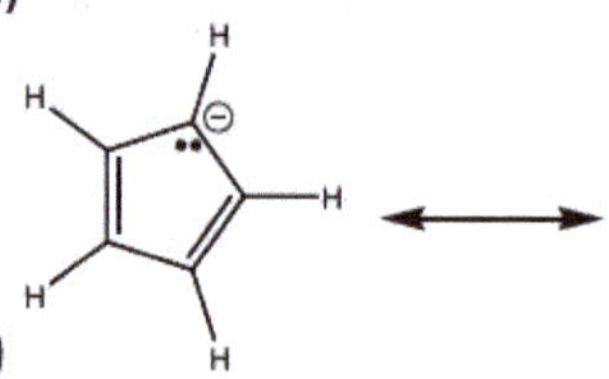

q) 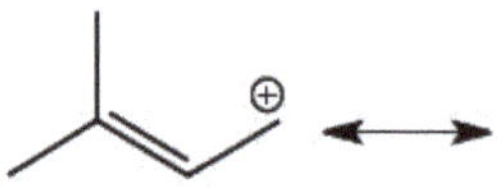

r) 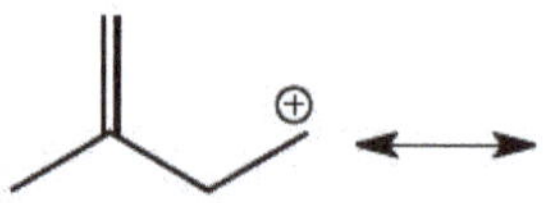

s) 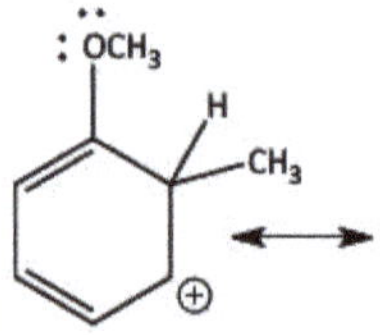

t) 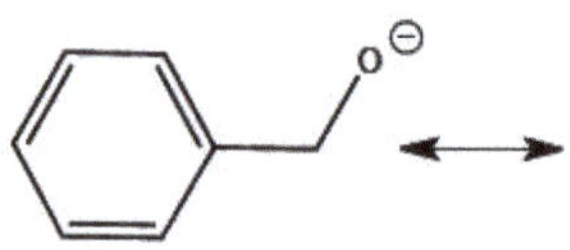

u) 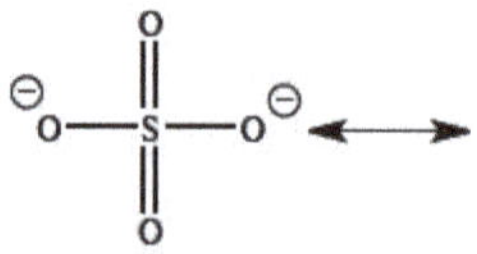

v) 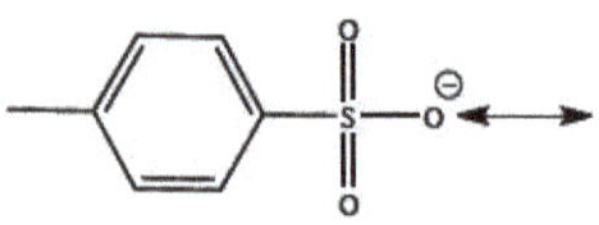

w) 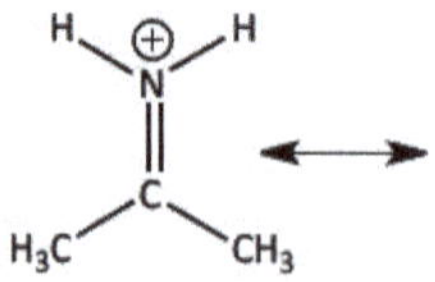

x) 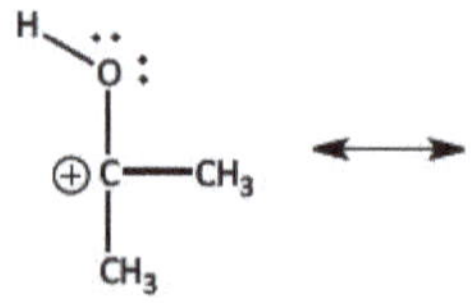

4.17 Which of the following is NOT a viable resonance structure of the others?

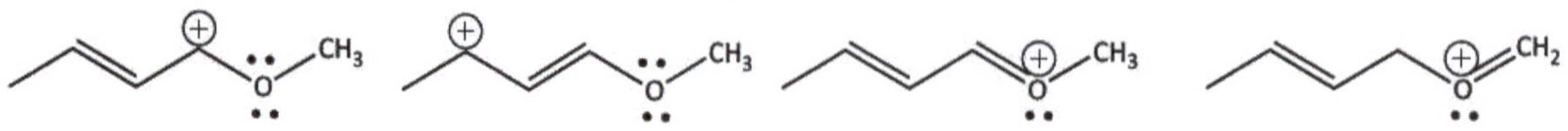

4.18 The following intermediate has four atoms that have a partial negative charge. Use resonance forms to identify these four negative atoms.

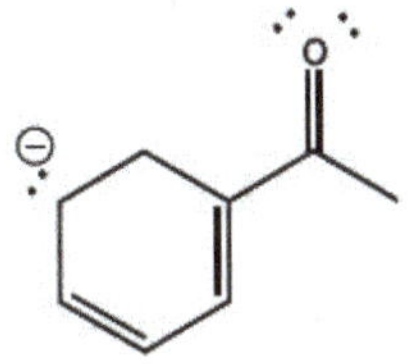

4.19 For each pair of ions, identify the one that is more stable because the charge can be spread out through resonance forms.

a) 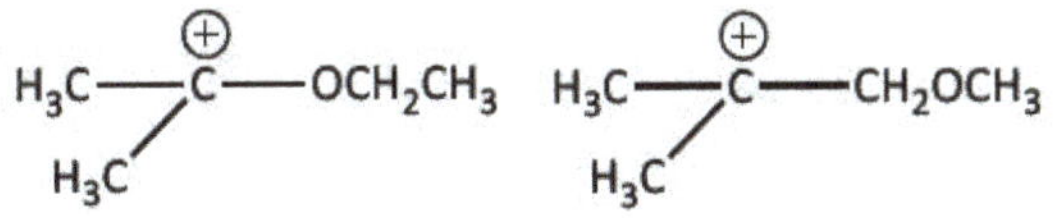

b) 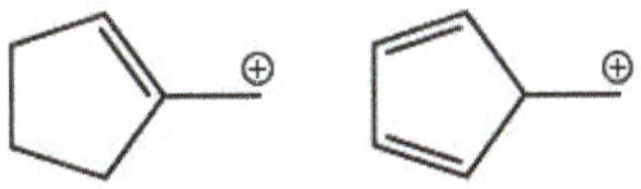

c) 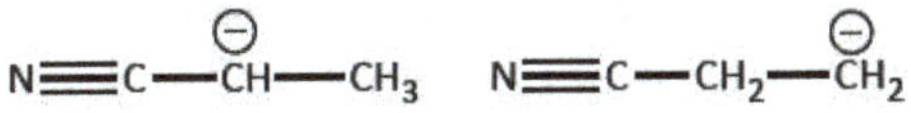

d) 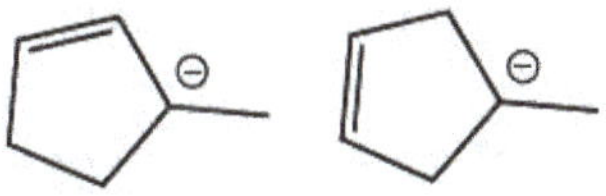

e) 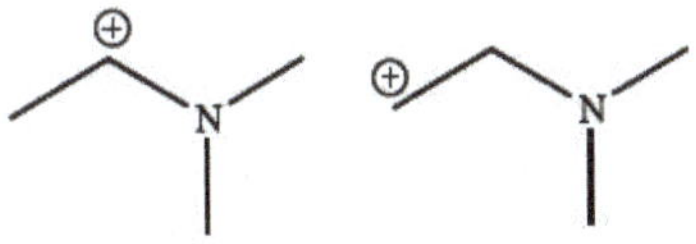

f) 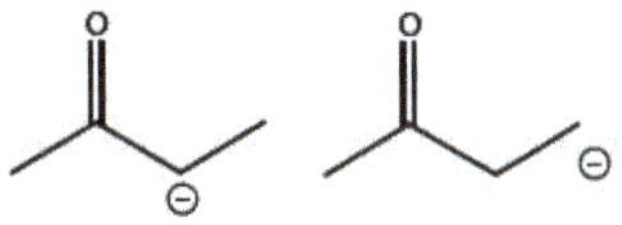

g) 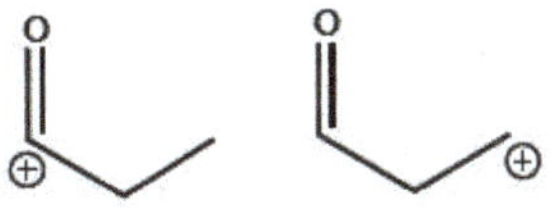

4.20 The following resonance structures have an error in them. Either the arrow(s) are incorrect, or the resonance structure drawn is incorrect. Correct the errors.

a) 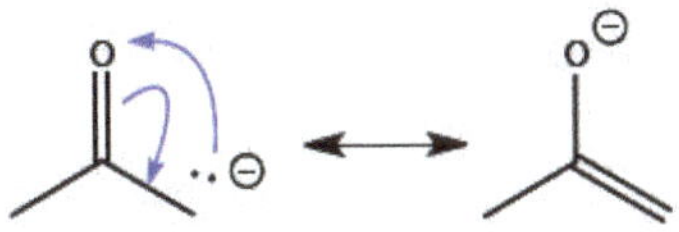

b) 

c) 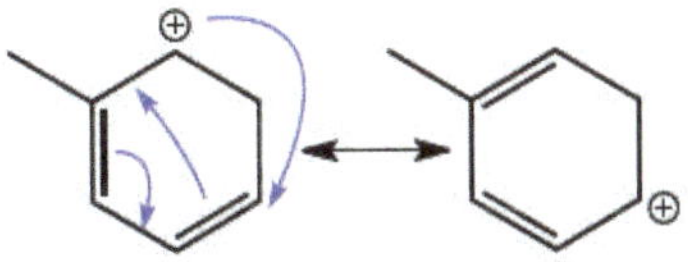

d) 

4.21 Draw all good, valid resonance forms of the following intermediates.

a) 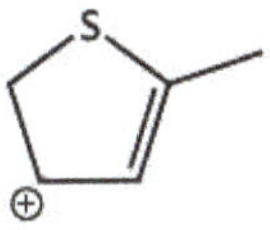

b) 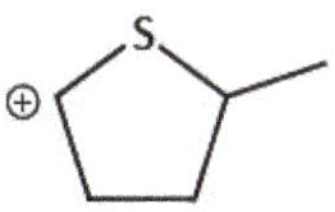

c) 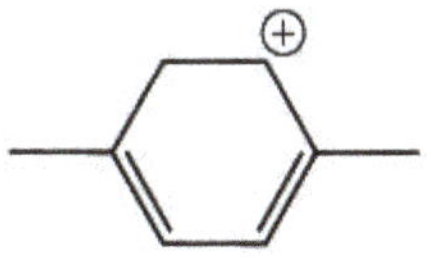

d) 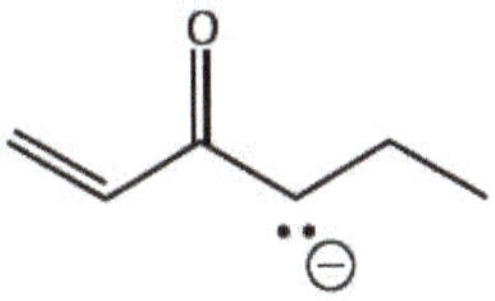

e) 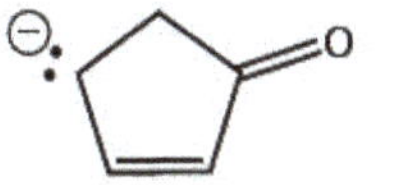

Chapter 5. Acids and Bases

5.1 Introduction

Acids and Bases. For those of you who have taken general chemistry, you might be thinking, "Yuck" or "Uh-oh". There are many chemical definitions for acids and bases. That's confusing. We use mathematical terms like negative log. That's confusing because when we first learned about logs in junior high or early high school math, we did not believe our teachers when they told us it was important. And, then there is the small "p" (pH, pK_a, pK_b) and numbers like 10^{-14}.

So, at the start of this chapter, I need to reassure you that the way organic chemists usually talk about acids and bases is much simpler than what you learned in general chemistry. Don't be intimidated. Give me a chance to explain this before you turn off your mind. Also, let me assure you that all of those terms in the previous paragraph become much less confusing and much more understandable as you continue your chemistry education. It is something you need to look at several times before it "clicks". That's like a lot of things in life that are worth knowing and doing.

The definitions of acids and bases that are typically taught in general chemistry are the Arrhenius, Brønsted-Lowry, and Lewis definitions. Believe it or not, there are even more ways to define acids and bases. In organic chemistry, when one speaks of an acid or base, they are most likely speaking of the Brønsted-Lowry definition. The Arrhenius definition is too simple. You might hear an organic chemist use the terms Lewis acid or base, instead they usually will use the terms electrophile or nucleophile, respectively.

The Brønsted-Lowry acid is a proton (H^+) donor. A Brønsted-Lowry base is a proton (H^+) acceptor.

Why do they call H^+ a proton? Consider the difference between a hydrogen atom (which has one proton and one electron) and a hydrogen ion (H^+). Once the electron is removed from the hydrogen atom, all that remains is the proton from the nucleus.

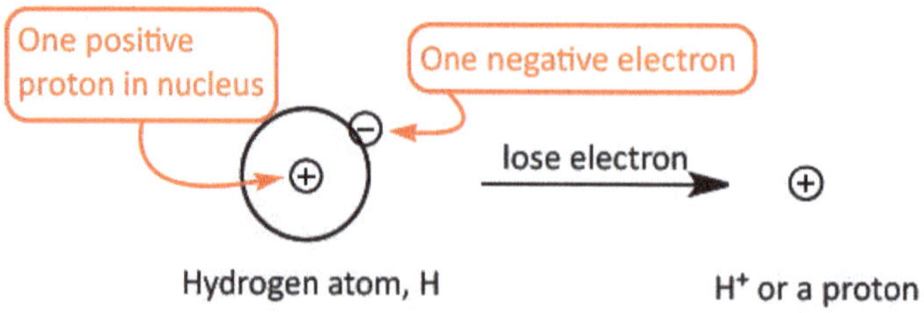

Figure 5.1 Why H^+ is called a proton

The following figure shows the movement of electrons in the acid-base reaction.

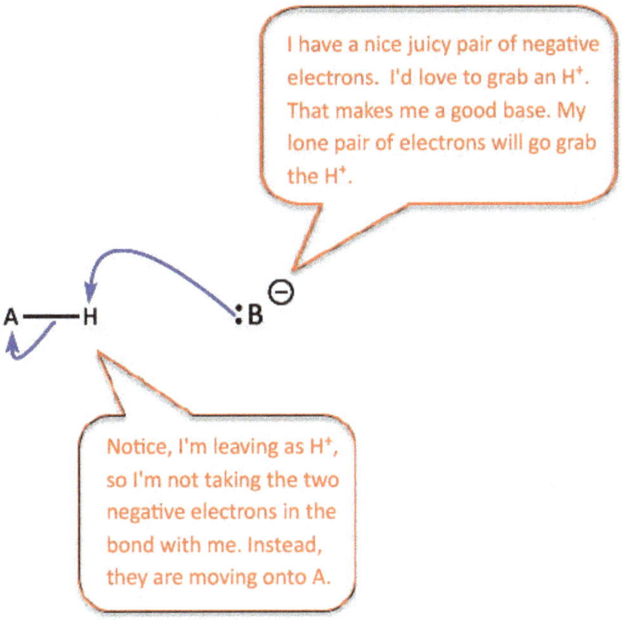

Overall, the acid-base reaction—including the products would look like this.

B-L Acid

B-L Base

Conjugate base, or what the B-L acid became after it lost its H^+.

Conjugate acid or what the B-L base became after it grabbed H^+.

When an acid gives up its proton, H^+, it becomes something we call a conjugate base of that acid.

5.1 Given the acid, draw the conjugate base. The acid proton that needs to be removed is circled.

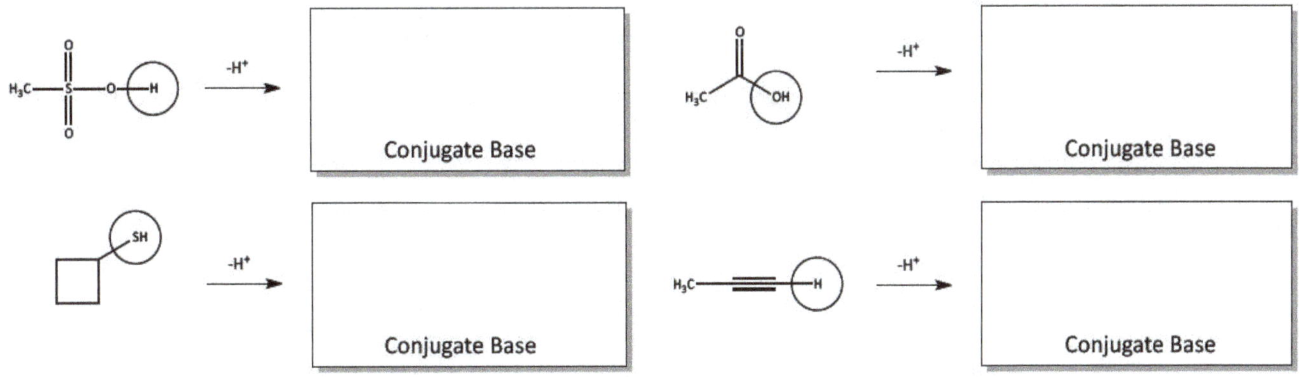

5.2 Acid Strength

In the previous exercise, I circled the protons that needed to be removed. But, how did I know which ones to remove? I needed to identify which protons were the most acidic and were most likely to be pulled off by a base.

Acids and Bases

In organic chemistry, you will be faced with the question on which proton in a molecule is the most acidic. You might even be asked to compare several compounds and decide which is the strongest acid. I hated these questions on tests when I took organic chemistry. I would usually just guess and move on. Somehow, I thought I probably had to memorize some magical list somewhere that ranked acid strengths. But, I never saw that list and probably would not have memorized it even if I did see it. The good news is that you do not need to memorize any list. The way to compare acid strengths is much simpler than that and is quite easy to understand.

Whenever we want to know how strong an acid is, we look at its conjugate base. Another way to say this is, we need to look at what the acid becomes after its proton is removed.

$$\text{H—A} \longrightarrow \text{A}^-$$

Acid — Conjugate Base

Happy or not?

Then, we look at the conjugate base, A^-, and determine how happy it is. A^- wants to be made if it is really happy. If it is happy and wants to be made, then H-A is a good acid that would gladly give away its H^+ proton so it can make A^-. A^- does not want to be made if it is not happy. Then, H-A would be a weak or poor acid because it would tightly hold onto its H^+ proton so it would not have to make A^-.

So, how do we tell if A^- is happy and stable or not? We need to remember that nature hates localized charges. Localized charges are those that are stuck in one spot. In general, the more we can spread out the negative charge, the better. Also, the more the negative charge is on atoms that like negative charges, the better.

There are four things we look for.

1. What type of atom is the negative charge on?
2. Can it be stabilized with resonance?
3. Does the inductive effect help or hurt it?
4. What type of orbital is the lone pair of electrons in?

5.2.1 Rule 1. Look at the atom that has the negative charge.

First, we need to know where several elements are in the periodic table. We learned these in Chapter 1, but let's make sure we remember them.

5.2 Fill in the chart below. Try to do it from memory, but if you can't do it, go back and refresh the mnemonics you learned.

Na	Mg											Al	Si				Ar
K	Ca	Sc	Ti	V	Cr	Mn	Fe	Co	Ni	Cu	Zn	Ga	Ge	As	Se		Kr
Rb	Sr	Y	Zr	Nb	Mo	Tc	Ru	Rh	Pd	Ag	Cd	In	Sn	Sb	Te		Xe
Cs	Ba	Lu	Hf	Ta	W	Re	Os	Ir	Pt	Au	Hg	Tl	Pb	Bi	Po		Rn
Fr	Ra																

Consider this problem. Which of the following compounds is the most acidic?

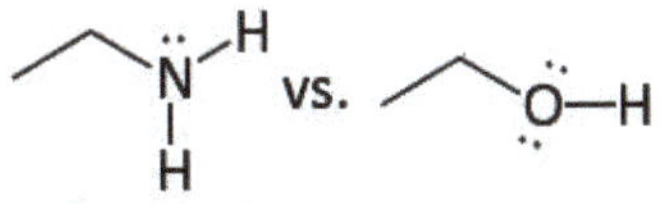

Whenever we solve a problem like this, we ALWAYS look at the conjugate bases. Let's draw those.

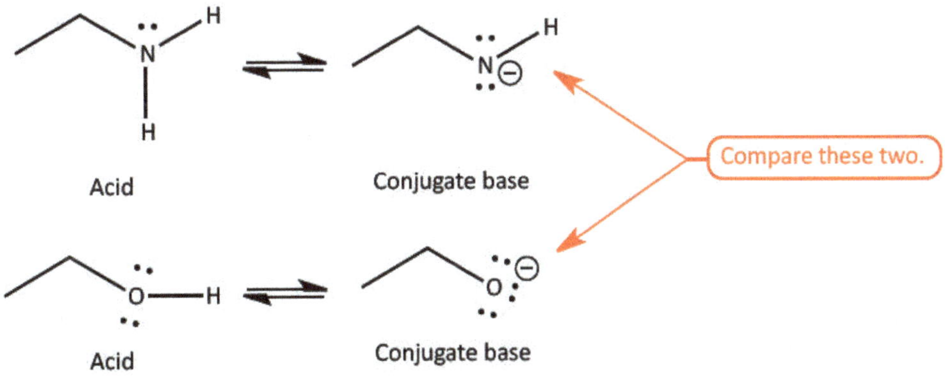

The real question is which of these negative species is happier? We have N$^-$ versus O$^-$. Remembering LiBe BCNOFNe, I see that N and O are in the same row of the periodic table. Between these two, which likes a negative charge on it better? Well, fluorine is the most electronegative element. So, the closer an element is to fluorine, the more electronegative it is and the more it wants the negative charge. With O being to the right of N, we know that O$^-$ is happier or more stable than N$^-$.

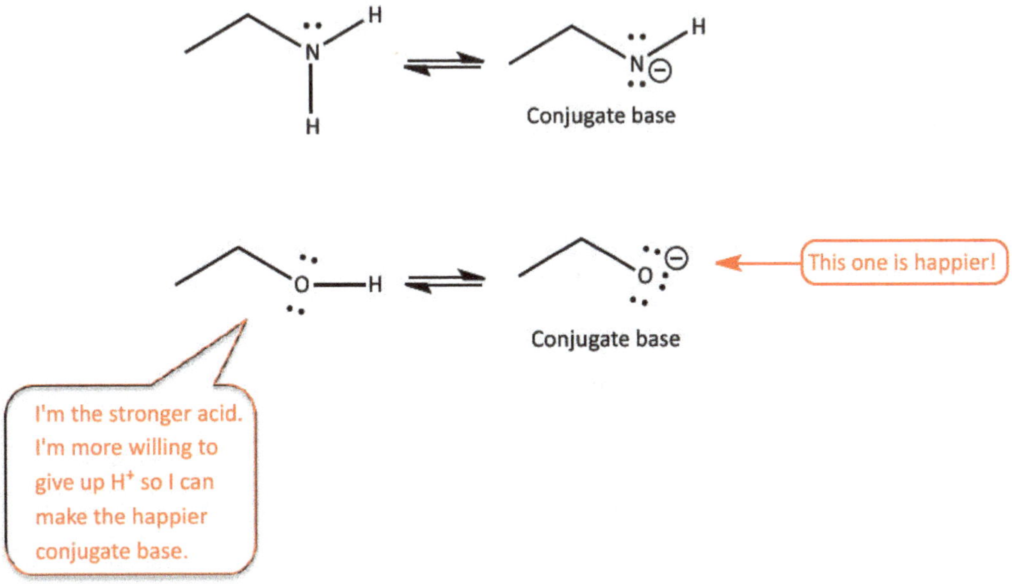

Important point

If we are comparing two atoms in the same horizontal row of the periodic table, the element farthest to the right likes the negative charge better. Most of the time in organic chemistry, you will be comparing C, N, and O, so know where those are.

5.3 In each pair below, circle the compound that is the MOST acidic.

$R\text{-}NH_2$ vs. $R\text{-}CH_3$ $\qquad$ H_2O vs. NH_3 $\qquad$ $R\text{-}CH_3$ vs. $R\text{-}OH$

Consider this problem. Which of the following compounds is the most acidic?

H—I vs. H—Cl

Once again, we ALWAYS look at the conjugate bases.

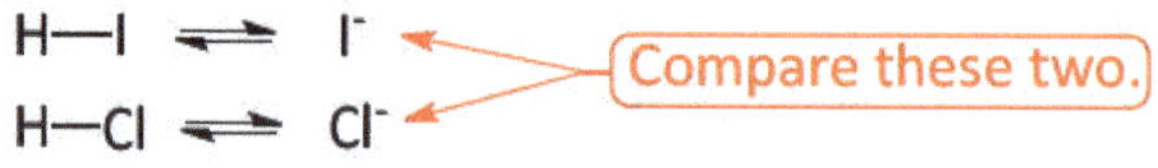

Here is where things get a little tricky. If I compare the electronegativities of I and Cl, I see that Cl is closer to fluorine, so it is more electronegative. I might then think, incorrectly, that Cl^- is happier than I^-. Why is that wrong? Another trend comes into play that ends up being more important than electronegativity. This trend is size. We learned in Chapter 2 that as we go down the periodic table, atoms get larger—because there are more electron shells for the atom. When we compare the sizes of I^- and Cl^-, we see that I^- is quite a bit larger.

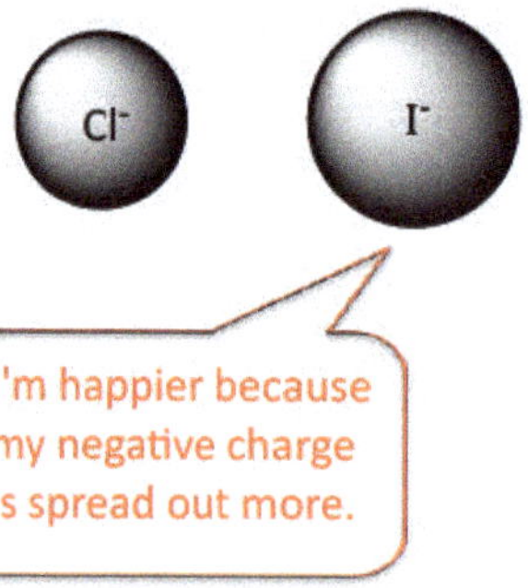

Being larger means that the negative charge can spread out and smear over a much larger area. Remember, nature hates localized charges. It is much better the more the negative charge spreads out. Cl^- is unhappier than I^- because its negative charge is confined to a smaller space.

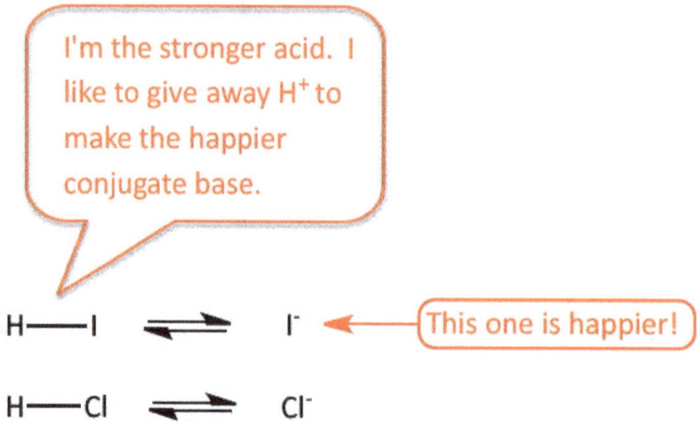

Important Point

If we are comparing two atoms in the same vertical column of the periodic table, the lower one is larger and likes the negative charge better. This may seem counterintuitive at first because the larger one is usually less electronegative, but size trumps electronegativity in this case.

5.4 In each pair below, circle the compound that is the MOST acidic.

HBr vs. HI CH_3-SH vs. CH_3-OH HF vs. HCl

5.2.2 Rule 2. Can it be stabilized with resonance?

Point 1 showed us that spreading out negative charge over a larger atom is very beneficial. Spreading out the charge trumps electronegativity arguments we can make. It would be even better if we could spread out a negative charge over two or more atoms! This is sometimes possible if we can make resonance forms.

Consider ethanol and acetic acid. Acetic acid is about 100 billion times more acidic than ethanol. Why is this?

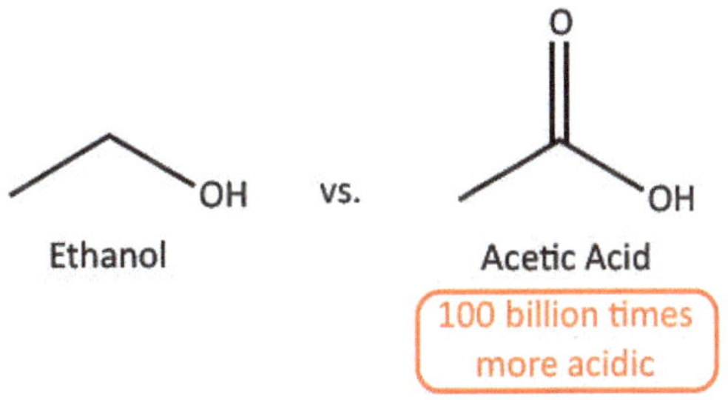

Whenever we compare acidity, we always, always, ALWAYS look at the conjugate bases and compare them.

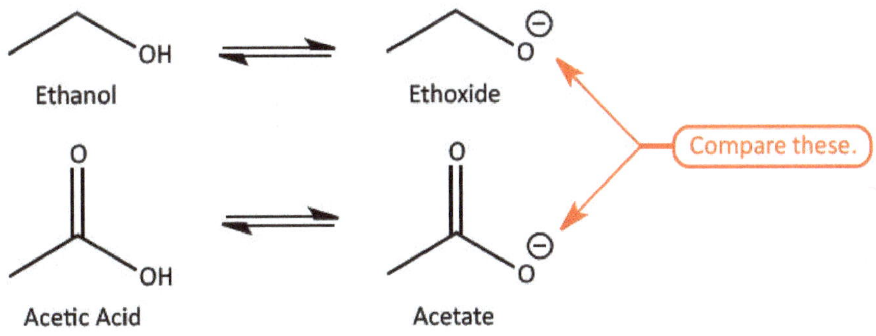

So, which of the negative charges is happier or more stable? First, we compare what atoms the negative charge is on. We see that they are both negative oxygen atoms. This is a tie and cannot explain the big difference in acidity of the two compounds. So, then what is it?

The answer is resonance. We can draw a good, valid resonance form for acetate. We cannot do this for ethoxide. The acetate negative charge can be spread over two oxygen atoms while the ethoxide negative charge must remain solely on one oxygen atom. Therefore, it is a much more localized charge. And nature hates localized charges.

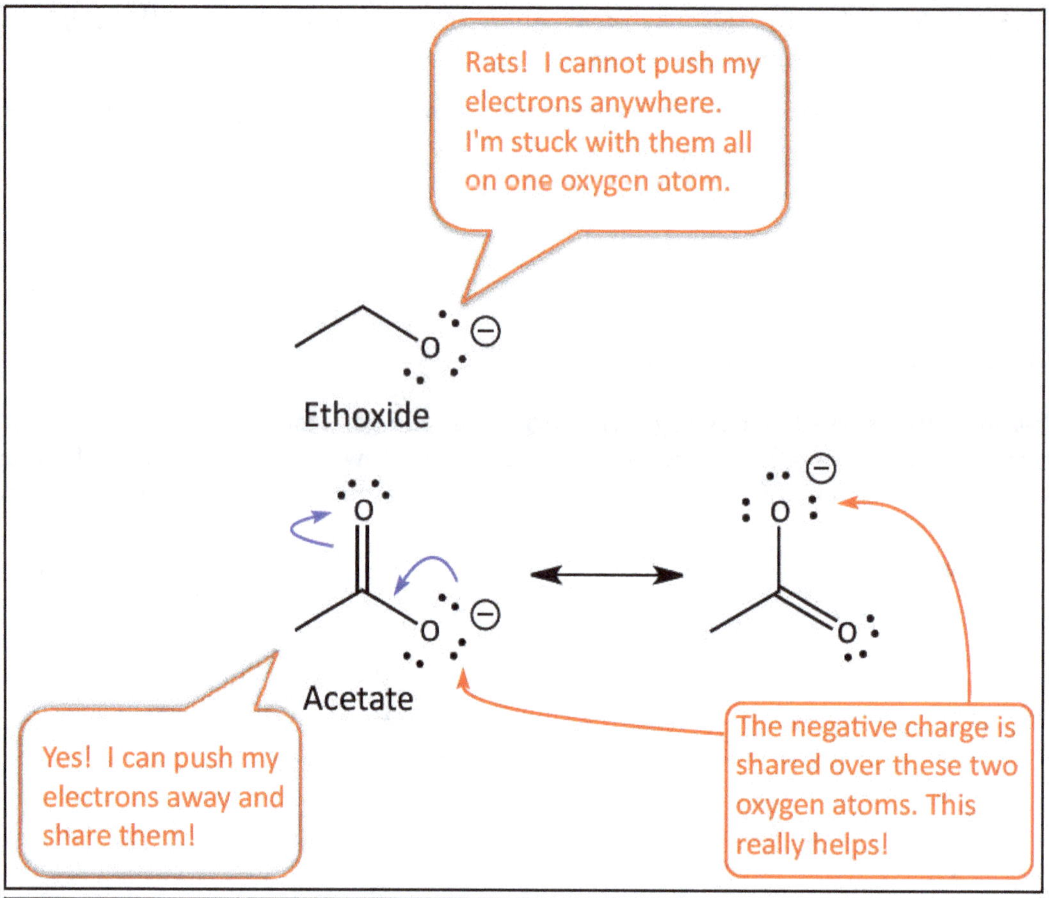

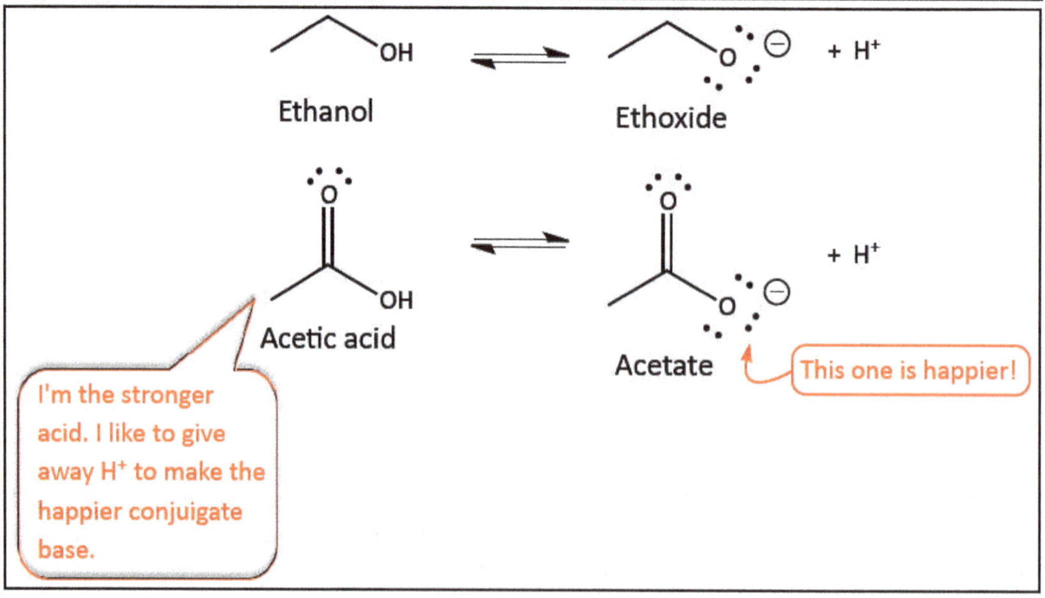

Resonance also explains why sulfuric acid is such a strong acid.

Sulfuric acid

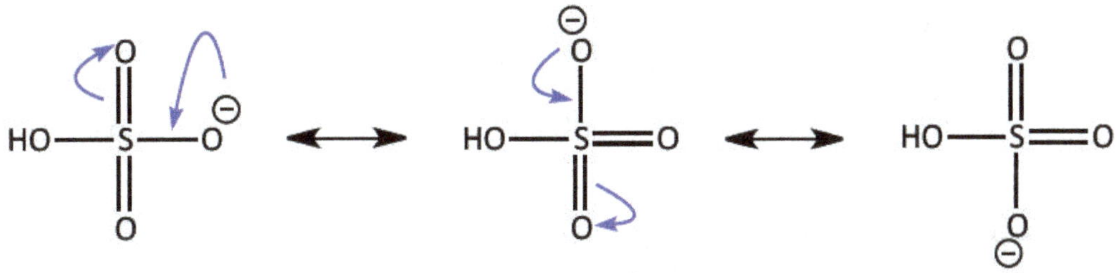

Three resonance forms can be drawn for the conjugate base of sulfuric acid. The negative charge can be spread over three electronegative oxygen atoms. This makes this a very happy conjugate base and sulfuric acid a strong acid.

Important Point

If resonance forms can be drawn to spread out a charge over several atoms, it makes it much happier.

5.5 In each pair below, circle the compound that is the MOST acidic.

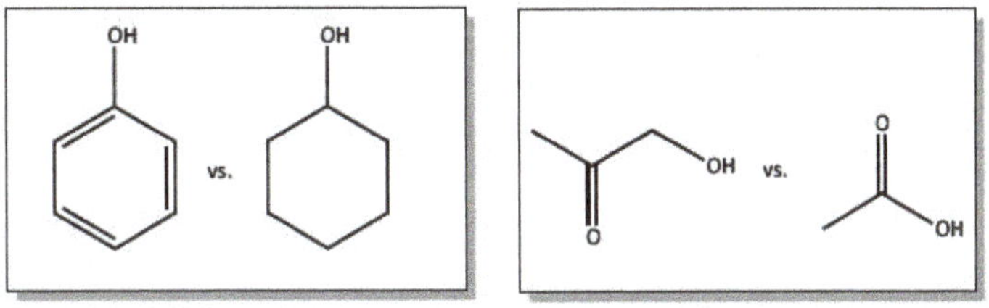

5.2.3 Rule 3. Does the inductive effect help or hurt it?

The inductive effect, or induction, is changing the electron density through the sigma bond framework.

Consider this problem. Which of the following compounds is the most acidic?

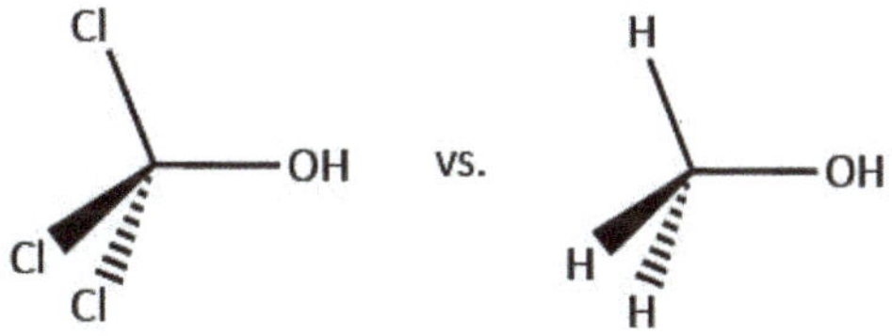

Whenever we compare acidity, we ALWAYS look at the conjugate bases and compare them.

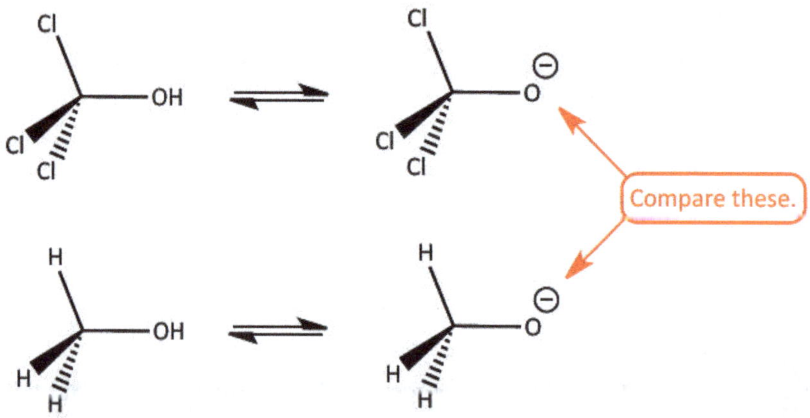

So, we begin to compare them. They both have negative oxygen atoms, so that is a tie. Neither one has a resonance form to share the negative charge, so they are still tied. So, what can make a difference? One has chlorine atoms attached to the alpha carbon and one has hydrogen atoms attached. We know that chlorine is more electronegative than hydrogen. We can think of the chlorine atoms as vacuum cleaners for negative electrons. They begin sucking the negative electrons through the sigma bond framework and help spread out the negative charge. This does not happen near as much with the hydrogen atoms since hydrogen atoms are not as electronegative. Notice, the electrons are being sucked through the sigma bonds.

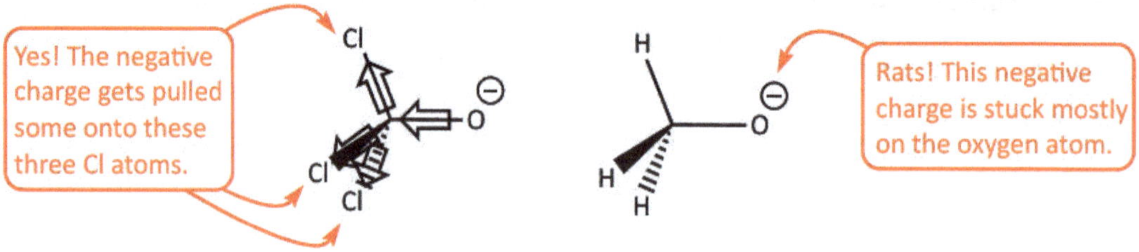

It is better the more a charge is spread over a molecule. The chlorine atoms take part of the negative charge upon themselves spreading the charge more.

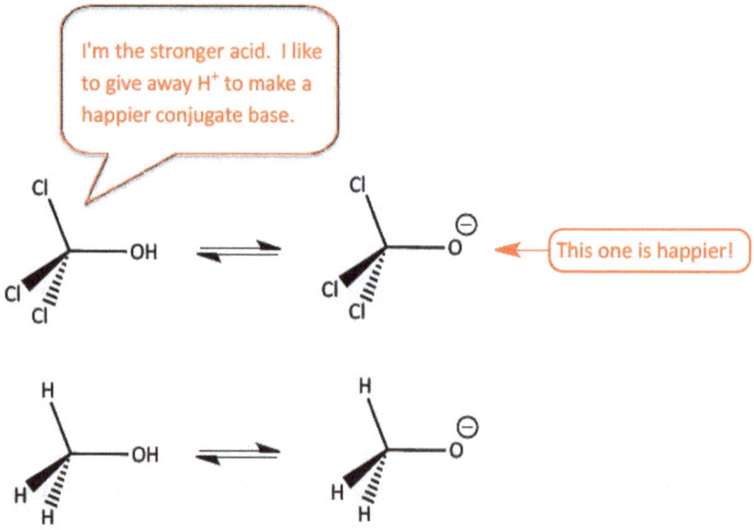

It is better the closer the electronegative atoms are to the negative charge.

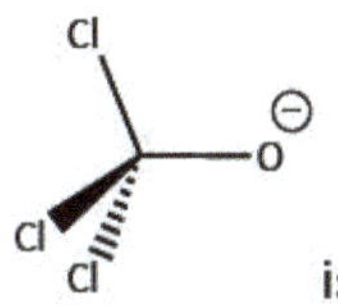
is much happier than
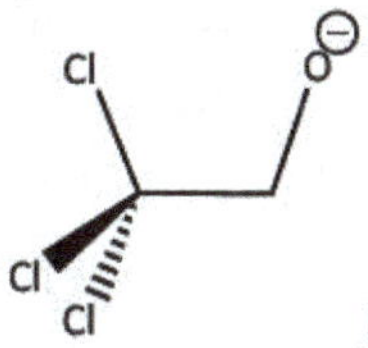
because the Cl atoms are now one bond farther away and cannot suck the electrons from the negative oxygen as well.

Electronegative atoms like F, Cl, Br help stabilize a negative charge through induction. Groups that donate electron density would hurt a negative charge. Silicon-based groups or carbon-based groups (alkyl groups) tend to donate electron density and destabilize negative charges.

Consider this problem. Which of the following compounds is the most acidic?

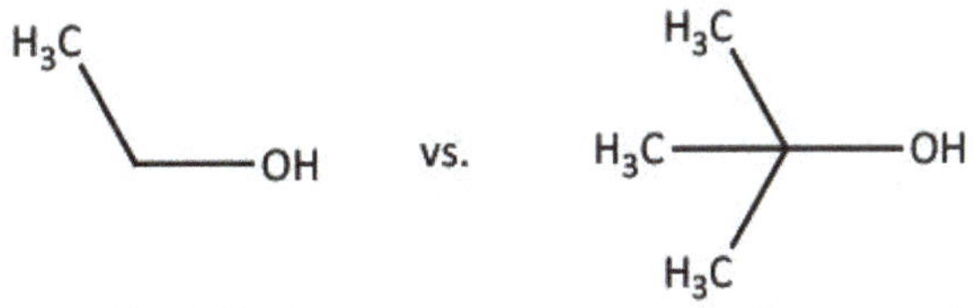

Whenever we compare acidity, we ALWAYS look at the conjugate bases and compare them.

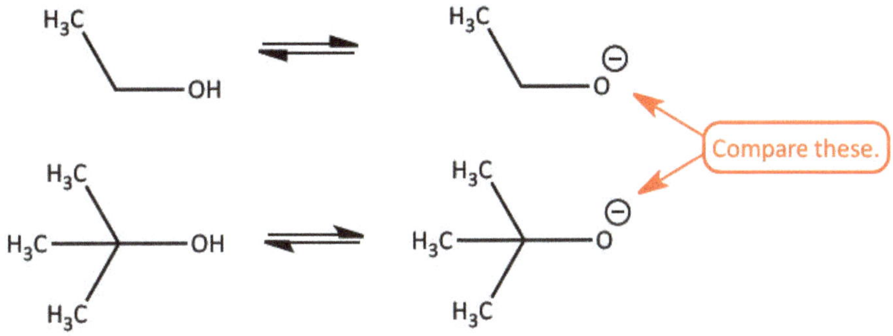

The negative charge is on the same type of atom. This is a tie. There are no resonance forms for either one.How does the inductive effect influence it?

The CH_3 groups are not electronegative. In fact, alkyl groups like CH_3 are a little electron donating. They push electron density into the molecule. You could also say that instead of sucking electron density out of the molecule like a halogen does, an alkyl group "blows" electron density into the molecule.

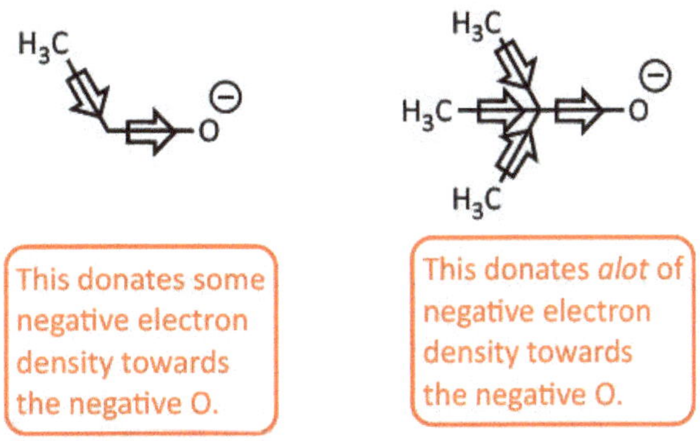

Does pushing or "blowing" more electron density towards the negative charge on the oxygen help or hurt the situation? It hurts us of course! The negative charge on the oxygen wants to be spread out over more of the molecule. But, if even more negative electron density is being heaped upon it, it makes the situation even worse.

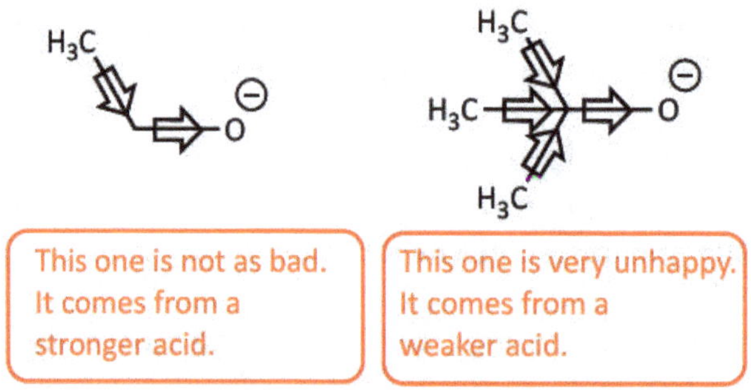

So, which of the protonated compounds is most likely to give up its H⁺ to make the conjugate base? Definitely not the one that makes something very unhappy. It will hold onto its H⁺ as much as it can in order to avoid making something so unhappy. It is the weaker acid.

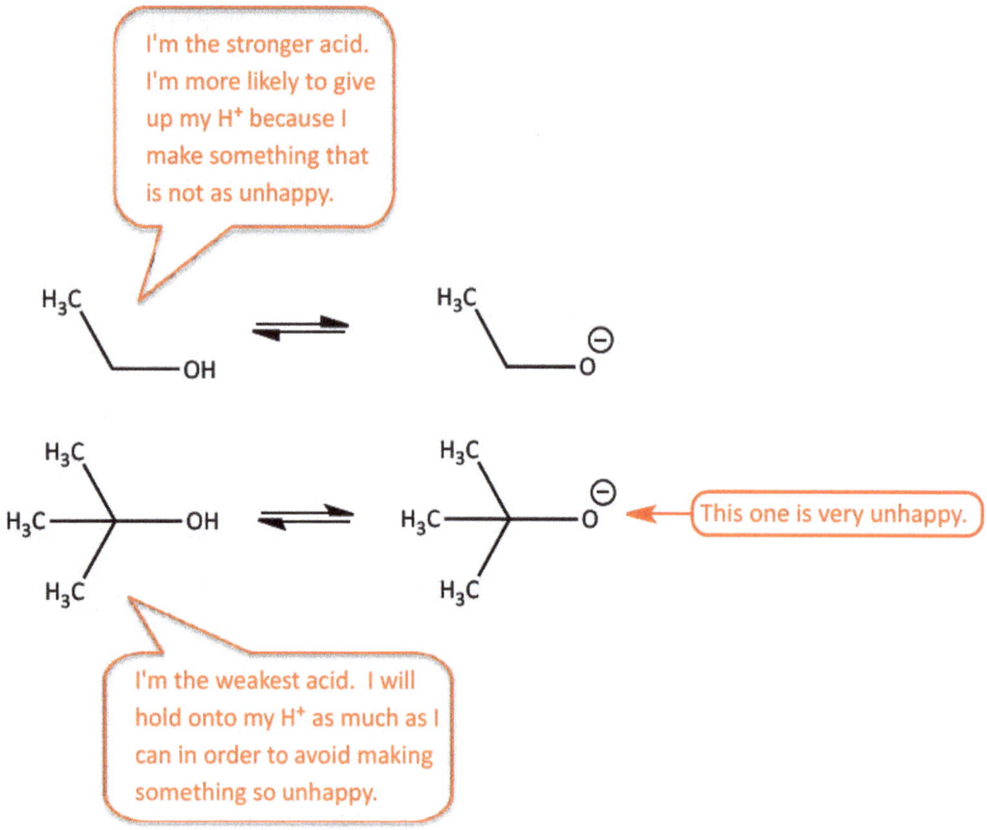

Important Point

Electronegative groups help stabilize negative charges and electron donating groups (alkyls) hurt negative charges.

5.6 In each pair below, circle the compound which would be the MOST acidic.

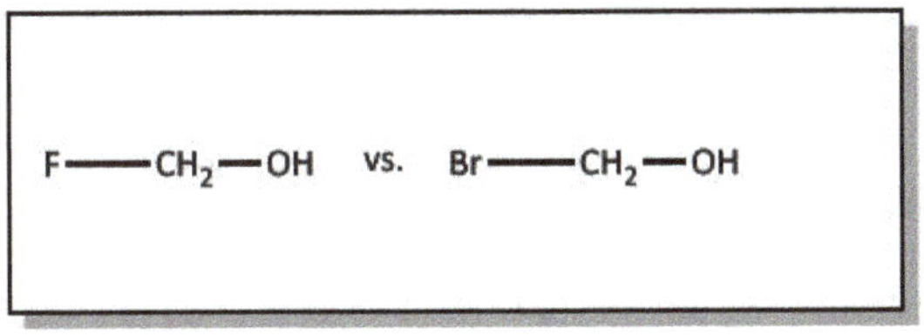

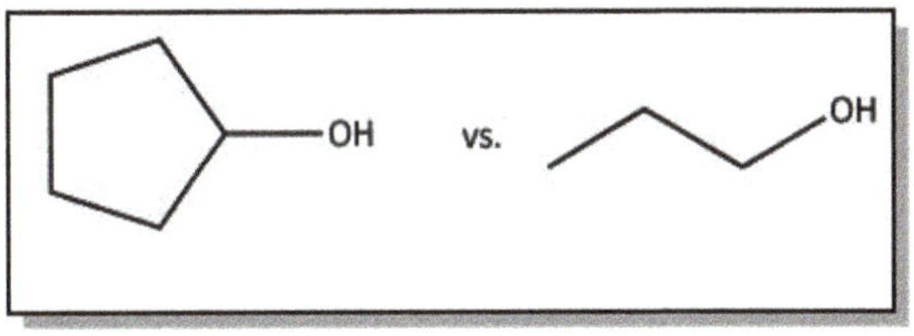

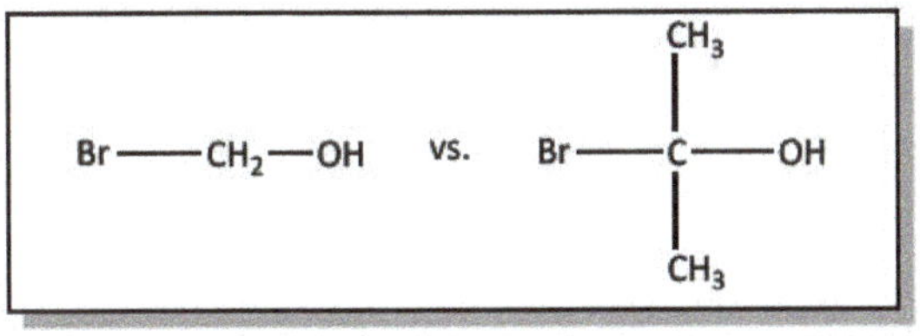

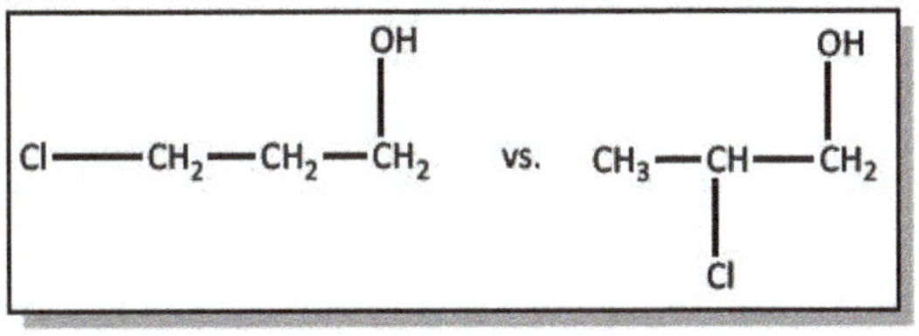

5.2.4 Rule 4. What type of orbital is the lone pair of electrons in?

Finally, if the negative charge is on carbon atoms, you need to look at what type of hybrid orbital the lone pair of electrons is in. Carbon atoms in organic chemistry are sp, sp^2, or sp^3 hybridized. Here they are with ⊕ drawn where the nucleus of the atom is located. These hybrid orbitals are not all the same shape and size.

Electrons are closer to positive nucleus.

Electrons are farther from positive nucleus.

Atomic s orbitals, ○, hold their electrons close to its nucleus. Atomic p orbitals, , hold their electrons above and below the nucleus. The sp hybrid orbitals are made from mixing one s and one p atomic orbital. They have 50% "s" character. The sp^2 hybrid orbitals are made from mixing one s and two p atomic orbitals. They have 33.3% "s" character. The sp^3 hybrid orbitals are made by mixing one s and three p atomic orbitals. They have 25% "s" character. The more "s" character a hybrid orbital has the more it holds its electrons close to the nucleus. So, sp hybrid orbitals hold their electrons closer to its positive nucleus. Sp^2 hybrid orbitals hold their electrons a little farther away from the positive nucleus and sp^3 hybrid orbitals hold their electrons the farthest away from the positive nucleus. Lone pairs of negative electrons prefer to be in sp hybrid orbitals because they are then held closer to the positive nucleus.

Consider this problem. Which of the following compounds is the most acidic?

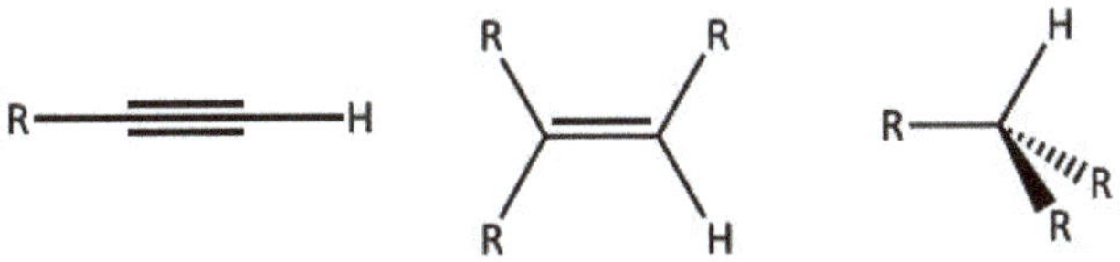

Whenever we compare acidity, we ALWAYS look at the conjugate bases and compare them.

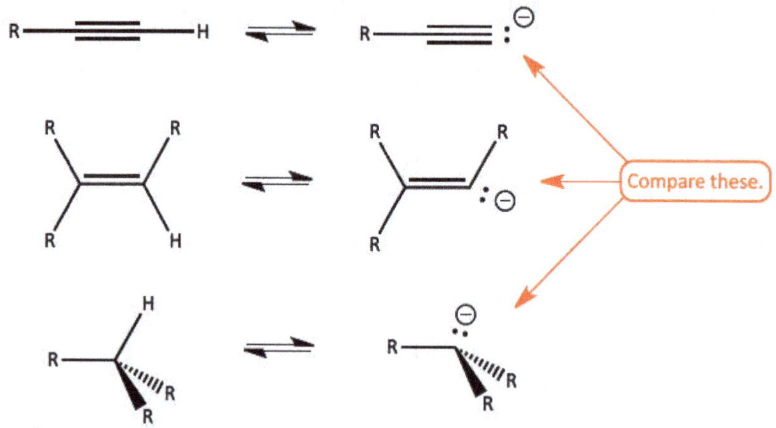

- The negative charge is on a carbon atom for each one. This is a tie.
- There are no resonance forms for any of them.
- The inductive effect might have a small effect.
- The largest effect is the hybrid orbital the lone pair of electrons is in.

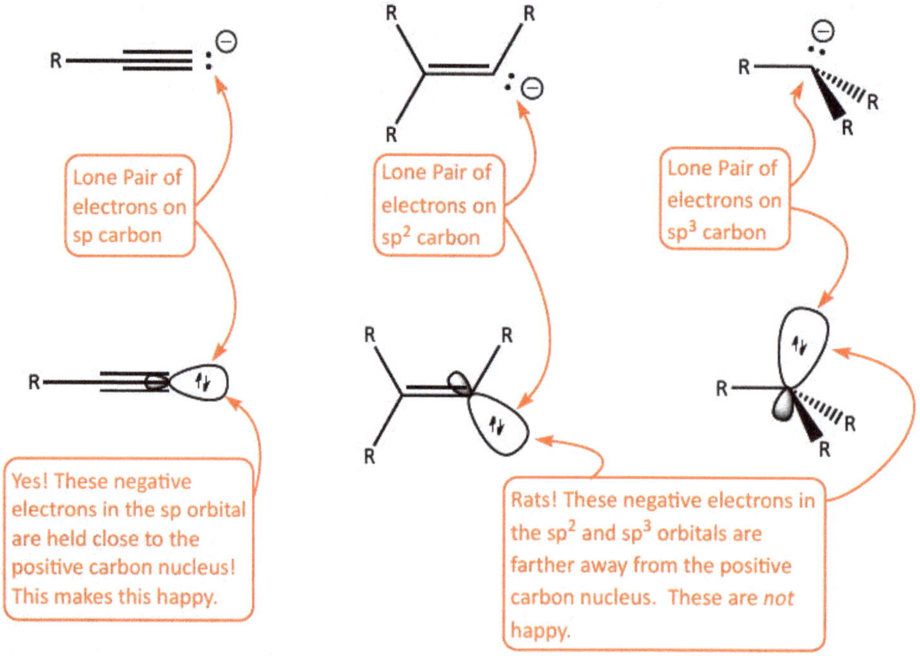

So, R—≡—H is the most acidic of these compounds because it makes the happiest conjugate base. In fact, it is 10^{19} times more acidic than a proton on an alkene!

Important Point

Protons on sp carbons are more acidic than on sp^2 or sp^3 carbon atoms. This usually means protons on triple bonds, C≡C-H, are more acidic than on double bonds, C=C-H, and single bonds C-C-H.

5.7 Label the circled protons from most to least acidic (1=most, 3=least).

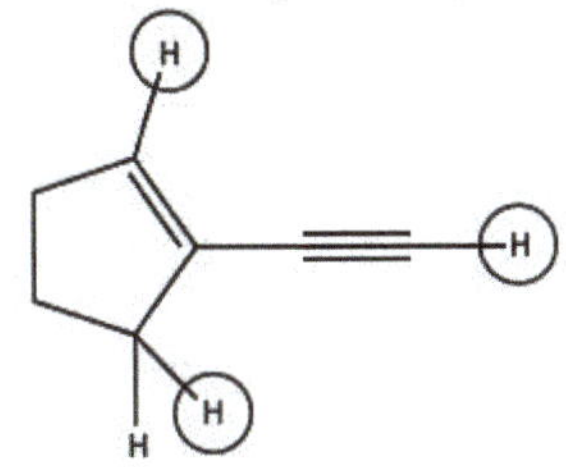

5.3 Which is the better base?

A base is a compound that grabs a proton, H^+. If we are asked the question, "Which of two compounds is the better base?" we are really being asked, "How likely is the compound to go and grab a proton, H^+? If the compound is very happy, it will stay the way it is and not grab a proton. It would be a poor base. If it is not happy, it is more likely to go grab an H^+ and be a good base. Another way to think about it is a good base has the most concentrated negative charge and is most likely to go grab an H^+.

For the following, the strongest base of each pair is circled.

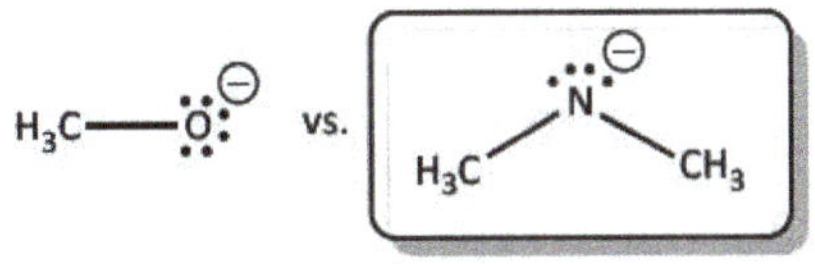

Rule 1. Oxygen and nitrogen are on the same row of the periodic table. Oxygen is more electronegative than nitrogen, so is more stable with the negative charge. Being more stable, negative oxygen is less likely to go grab H^+ compared to negative nitrogen.

I^- vs. Cl^-

Rule 1 explains this. I and Cl are in the same column of the periodic table—both halogens. I^- is a larger anion than is Cl^-. The negative charge can spread out more on I^-, so it is happier, or more stable than Cl^-. Since Cl^- has a more localized negative charge and it is less happy, it will be more likely to go grab an H^+. It is a better base.

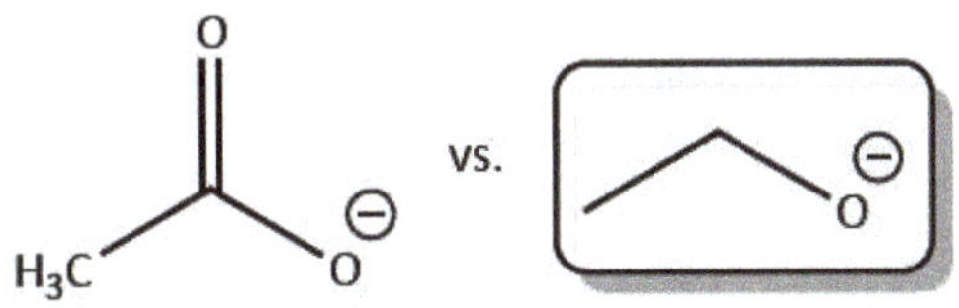

Rule 2 explains this. The negative charge of the acetate molecule on the left can be shared over two oxygen atoms through resonance. The negative charge on the right cannot be shared, so it is a localized negative charge. This is unhappy and more likely to go grab an H^+, making it a better base.

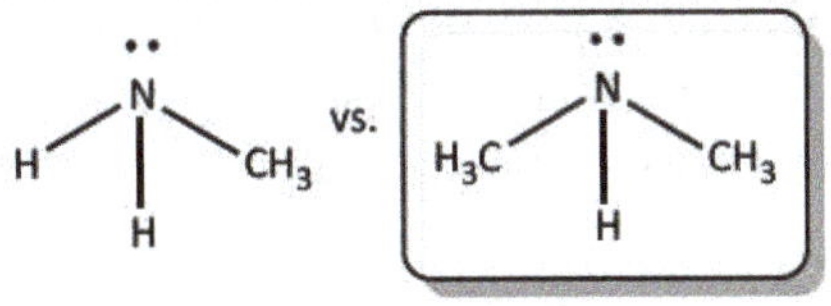

Rule 3 explains this. They both have a juicy lone pair of negative electrons on a nitrogen atom. Since they are both nitrogen atoms, it is a tie. The difference is that one nitrogen atom has one methyl group attached while the other has two methyl groups attached. The methyl groups are electron donating. Therefore, the molecule on the right has a greater negative electron density on its nitrogen atom, making it more likely to grab an H^+. It is a better base.

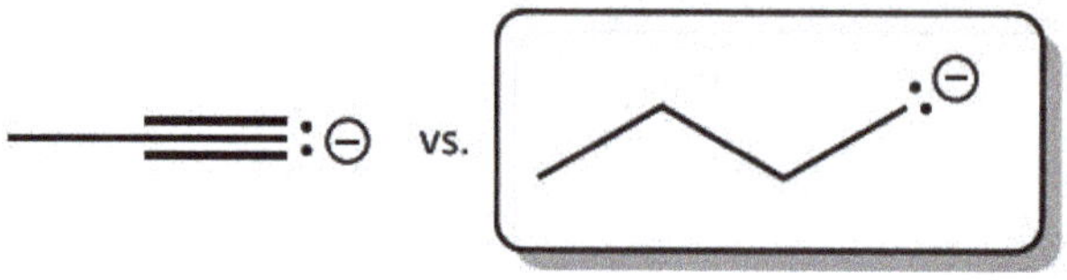

Rule 4 explains this. The negative charge on the triple bond is in an sp hybrid orbital. The negative charge on the single bond is in an sp^3 hybrid orbital. The lone pair of electrons in the sp hybrid orbital is held closer to its positive carbon nucleus and is more stable. Being more stable, it is less likely to go grab an H^+, so it is a worse base.

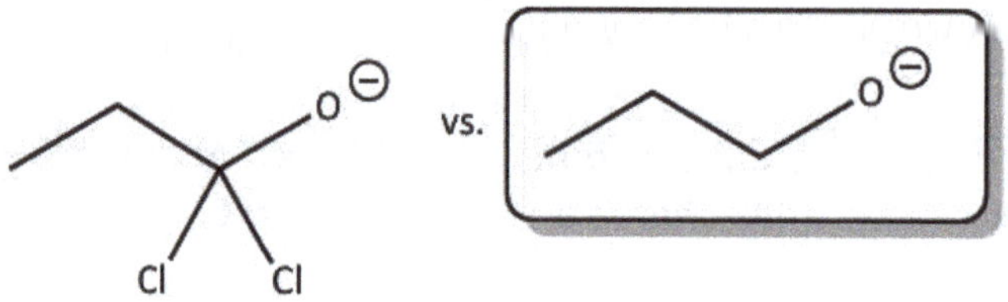

Rule 3 explains this. The molecule on the left has electronegative chlorine atoms which help pull the negative charge through the sigma bond framework by induction. This makes the negative charge more spread out and the molecule happier. The anion on the right has the negative charge stuck on one oxygen all by itself. It is unhappy. It has a localized negative charge. It is more likely to grab an H^+ and is a better base.

5.8 In each of the following pairs, circle the one that is the stronger base.
a) NH_3 and PH_3 b) F^- and Br^- c) NH_2^- and OH^- d) HS^- and Cl^- e) CH_3O^- and $(CH_3)_2CHO^-$

5.9 What is the conjugate base, if any, for the following compounds?
a) CH_3NH_2 b) CH_3OH c) HNO_3 d) Br^- e) H_2O

5.10 Methyl mercaptan (CH_3SH) is a stronger acid than methanol (CH_3OH). Explain.

5.11 The butyl anion ($CH_3CH_2CH_2CH_2{:}^-$) is a much stronger base than bromide (Br^-). Explain.

5.4 In the lab

In the lab, an unknown compound's acidity or basicity can help us begin to identify what type of compound we have. To explain how this works, let's first take a look at how four different compounds react with aqueous acid or base. Of course, aqueous acid or base means the acid or base is dissolved in water. We will use a *strong acid* like 10% aqueous hydrochloric acid (10% HCl), a *strong base* like 10% aqueous sodium hydroxide (10% NaOH), and a *weaker base* like 10% aqueous sodium bicarbonate or baking soda (10% $NaHCO_3$). In each case, when an organic compound reacts with the acid or base, it forms a salt (a positive cation with a negative anion). Salts are soluble in water. This is key for us.

Notice how a neutral compound will not react with any of the acids or bases. It will not dissolve in any of the aqueous acids or bases. A fairly strong acid, like a carboxylic acid, will not dissolve in aqueous acid (HCl), but can react with a strong base (NaOH) or will even react with a weak base ($NaHCO_3$) forming a salt that will dissolve in the aqueous solution. A weaker acid, like phenol, will not react with aqueous acid (HCl) or a weak base ($NaHCO_3$). Therefore, it will not dissolve in those aqueous solutions. But, a weak acid like phenol can react with a strong aqueous base (NaOH) to form a salt and dissolve. Finally, a basic compound like an amine will react with aqueous acid (HCl) to form a salt and dissolve, but will not react the bases. Study the following chart until it makes sense to you.

Extractions, using separatory funnels, is a separation technique in the organic chemistry lab that uses this idea of different compounds forming water-soluble salts with various acids or bases. This is a very common organic lab technique. These are crucial ideas to understand.

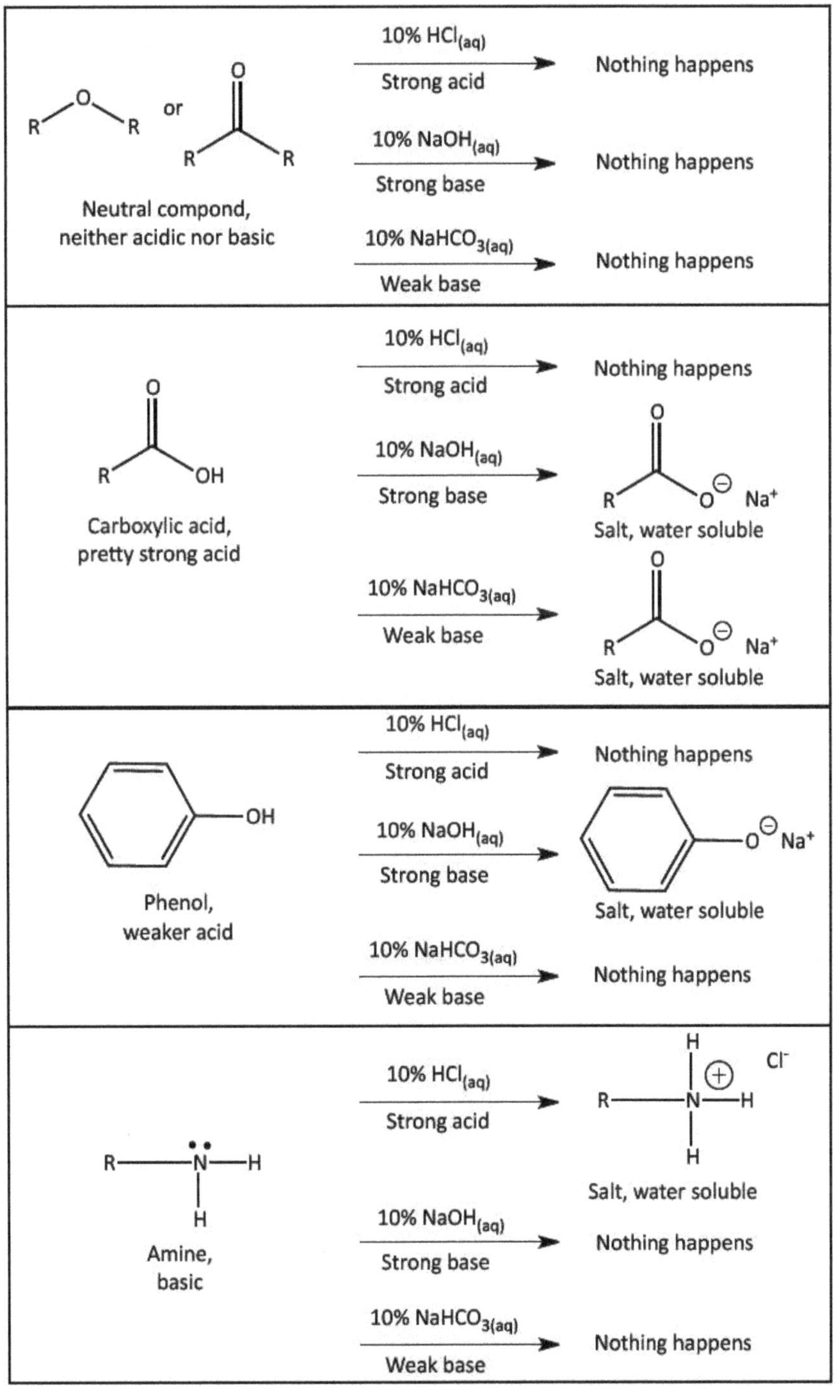

5.5 Problems

5.12 Rank the relative acidity for the following compounds. Put the most acidic on the right.

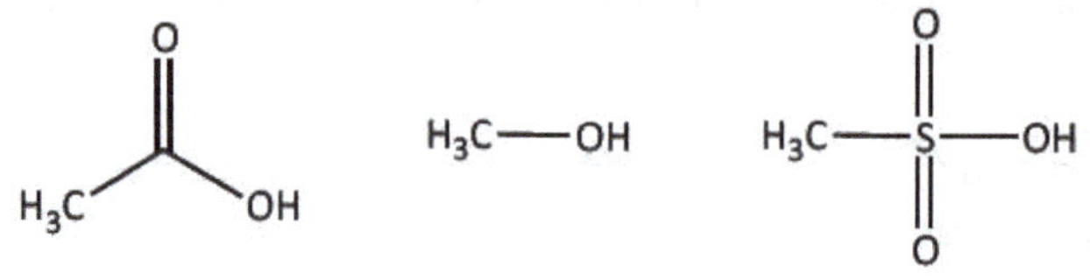

A B C

_______ < _______ < _______

5.13 Consider the set of compounds, NH_3, HF, and H_2O. Rank these compounds in order of increasing acidity.

5.14 Circle the more acidic compound out of each pair of compounds below.

a) 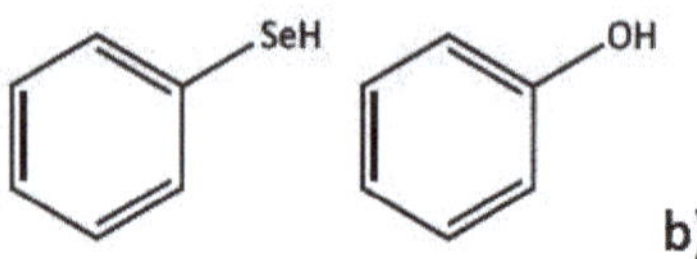 b) 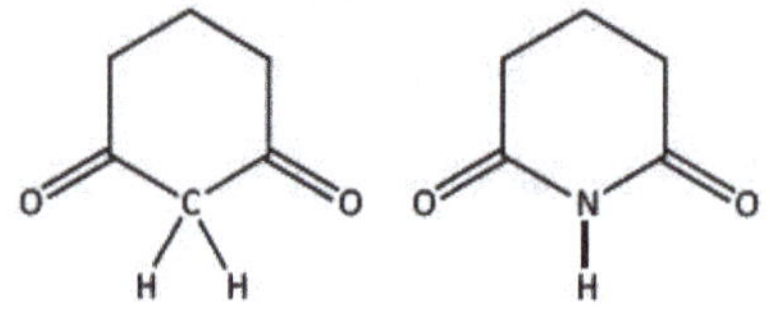 c) 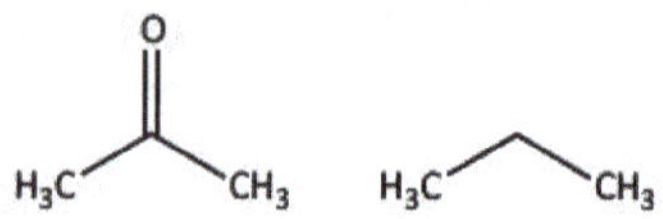

5.15 Circle the stronger base in each of the following pairs.

a) I^- vs. Cl^- b) H_2O vs. HO^- c) HO^- vs. H_2N^- d) CH_3COO^- vs. HO^- e) H_2N^- vs. CH_3COO^-

5.16 Rank the following from least to most acidic.

A B C D

5.17 Rank the following circled protons from least to most acidic.

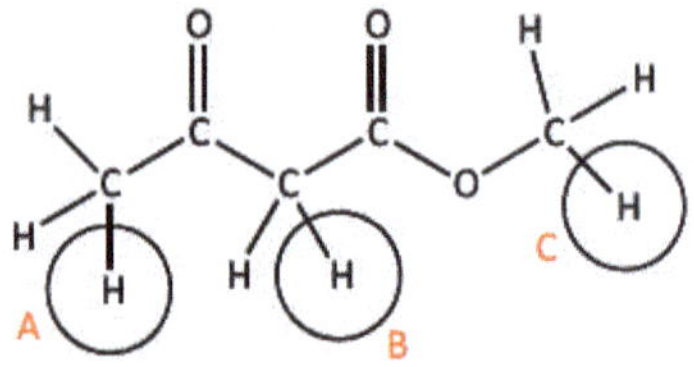

5.18 Rank the following compounds from least to most acidic.

$BrCH_2OH$ vs. CH_3NH_2 vs. CH_3OH

5.19 Rank the following protons from least to most acidic.

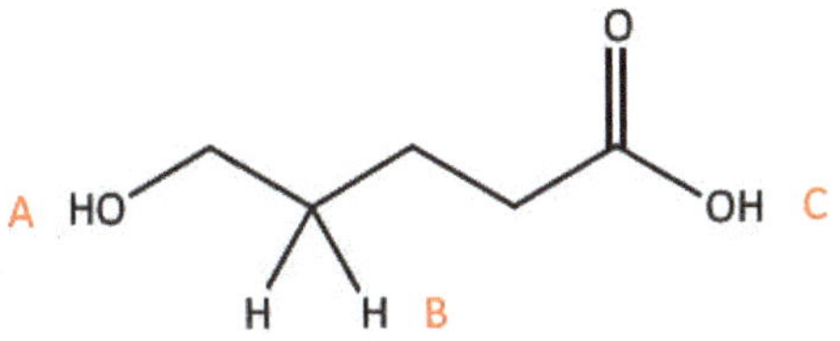

5.20 Rank the following protons from least to most acidic.

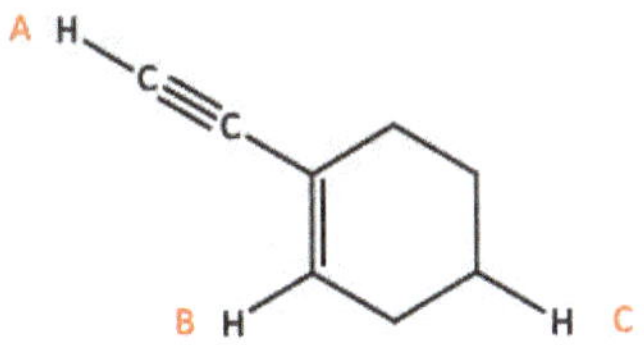

5.21 Which molecule (nitric acid or sulfuric acid) plays the role of acid in this reaction?

$$HNO_3 + H_2SO_4 \rightarrow HSO_4^- + H_2NO_3^+$$

5.22 Which is the strongest base?

LiOH vs. LiF vs. $LiCH_3$

5.23 Imidazole is a ring that contains two nitrogen atoms. One of the nitrogen atoms is much more basic than the other. There are a couple of ways to explain this, but we will focus in this problem on using resonance.

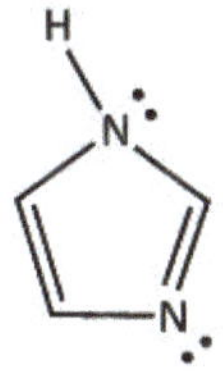

imidazole

a) Draw structures of the two conjugate acids that are formed when each of the nitrogen atoms is protonated (the lone pair of electrons on the nitrogen atom reacts with H^+).
b) If possible, draw good resonance forms for each of these conjugate acids. Remember; do not break the octet rule for the nitrogen atoms.
c) If a resonance form can be drawn, the positive charge can be spread out over more than one atom. This makes that conjugate acid more stable and its nitrogen atom more basic. Which nitrogen atom of imidazole is more basic?

5.24 Predict the products of the following acid/base reactions.

a) 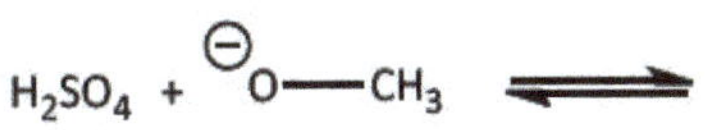

b) 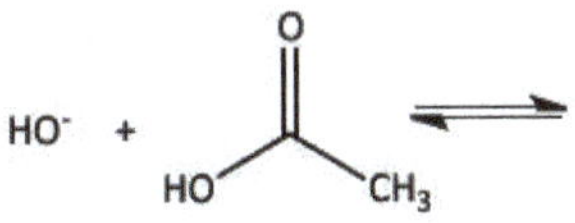

c) $HNO_3 + NH_3 \rightleftharpoons$

d) 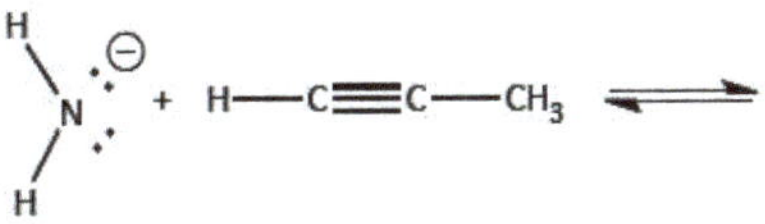

e) 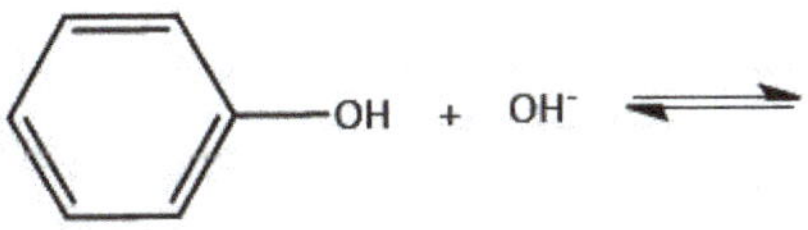

f) 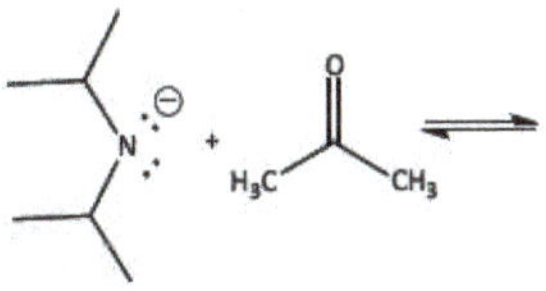

Chapter 6. Alkanes

6.1 Introduction

Alkanes are hydrocarbons with only single bonds between the carbon atoms. They do not contain a reactive functional group, so they are pretty unreactive chemically. They do burn well, though. I'm very glad the alkane, propane, burns well so I can cook my burgers and brats on my gas grill.

Since most organic compounds have a framework built out of alkanes, we need to have a good understanding of them. Let's look at some of the simplest straight-chain alkanes to start.

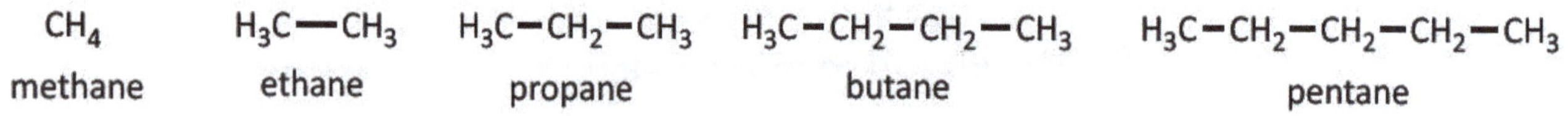

Figure 6.1 Normal alkanes

These straight-chain alkanes can be thought of as a string of $-CH_2-$ groups, called methylene groups, with two more H atoms—one on each end—to make the two $-CH_3$ groups, called methyl groups.

The general formula for an alkane with n carbon atoms is $(CH_2)_n$+2H, or maybe more simply, C_nH_{2n+2}.

Let's look at ethane to see that it works. Ethane's formula is C_2H_6 which fits the formula with n=2. Compare that to ethene and ethyne.

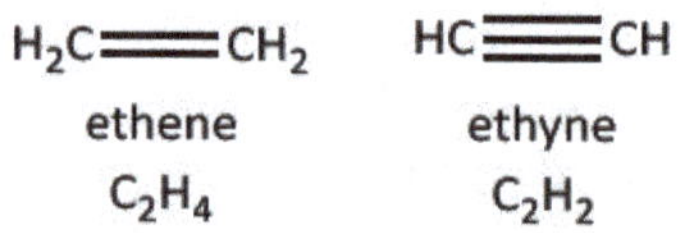

The formula is reduced by two hydrogen atoms for every π-bond a compound has. For one π-bond in a double bond as in ethene, there are two fewer hydrogen atoms than ethane. For two π-bond in a triple bond, like in ethyne, there are four fewer hydrogen atoms. Alkanes are *saturated*. They are saturated with all of the hydrogen atoms the carbon atoms can possibly take. They are all single bonds. *Unsaturated* compounds contain π-bonds, double or triple bonds.

6.1 Are the following compounds saturated or unsaturated?

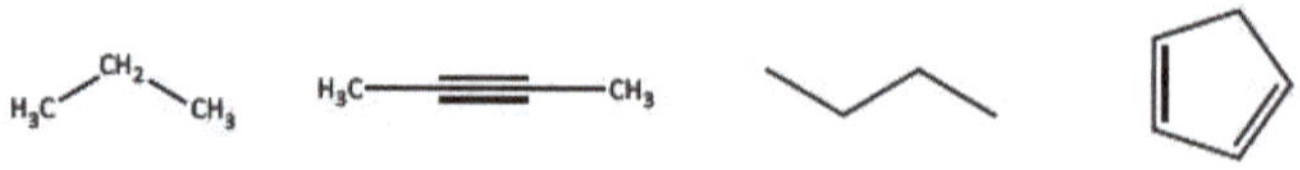

_________ _________ _________ _________

6.2 Nomenclature (Naming Compounds)

6.2.1 Common names

Some compounds have common names. These are historical names of compounds. We need to learn a few of them that are commonly used. Different compounds that have the same formula are called isomers. For alkanes, there are common names for the various isomers. We will only consider those that have four or five carbon atoms to get a feel for common names. There are two ways to arrange four carbon atoms alkanes. The straight chain, unbranched, form of butane is also called *n*-butane, which stands for *normal* butane. Four carbon atoms can be arranged in another way. So, we need another name for this new arrangement. We cannot call it butane since it is different. Therefore, the common name for this is isobutane (an isomer of butane). Isobutane is **branched** whereas *n*-butane is **unbranched.**

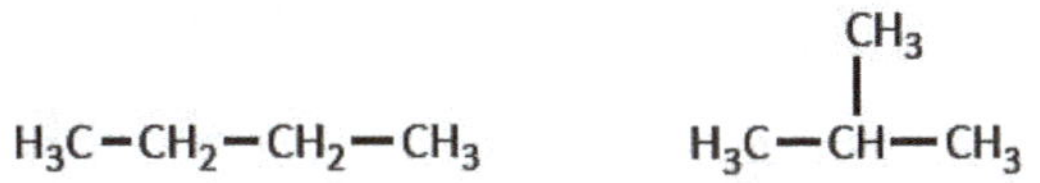

butane, *n*-butane; isobutane

Figure 6.2 Isomers of butane

Five-carbon alkanes can be arranged three different ways. The normal chain is called *n*-pentane. Then, there is an isomer of pentane called isopentane. There is one more isomer of pentane, but the name isopentane was already taken. So, this last isomer is called neopentane for "new" pentane.

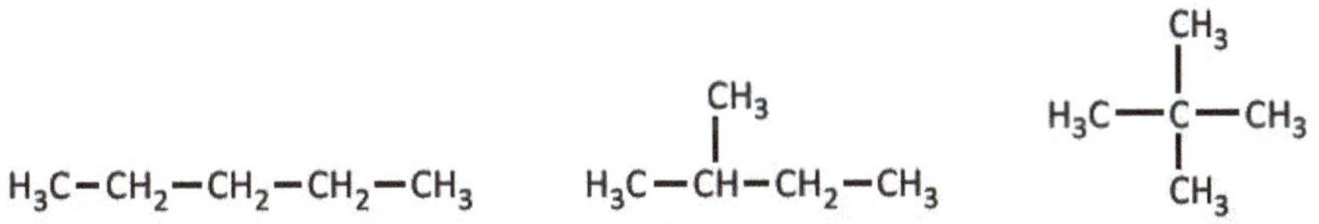

pentane; isopentane; neopentane

Figure 6.3 Isomers of pentane

So, how do we decide which isomer gets the *iso-* designation? Notice that the *iso-* compounds each have a "Y" shape at its end.

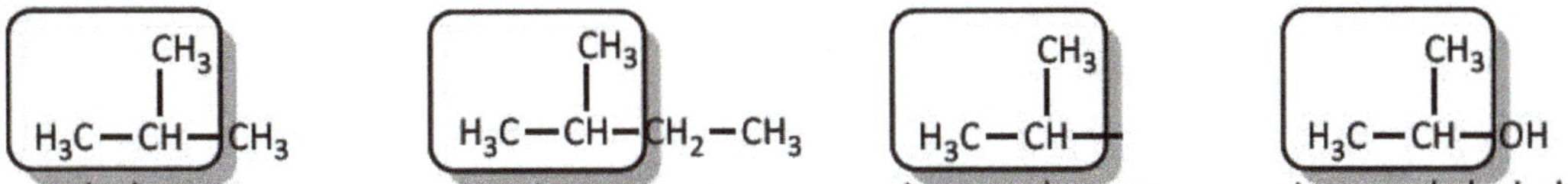

isobutane; isopentane; isopropyl group; isopropyl alcohol

Figure 6.4 The "iso" designation for isomers

For six carbons, there are five isomers. So, we'll stop looking at common names. Thankfully, a system has been devised to help us name compounds so we do not need to memorize thousands of common names. The group responsible for naming chemicals is the International Union of Pure and Applied Chemistry, IUPAC. The rules for naming compounds they have developed are called IUPAC rules.

6.2.2 IUPAC names for alkanes

We will look at the rules for naming alkanes by investigating why the following compound is named 3-ethyl-2,2-dimethylhexane.

Alkanes

Rule 1. Find the longest continuous chain of carbon atoms. This is the base chain. This name will be at the end of the name. Substituents are all of the groups attached to this main, base chain.

a. Watch out, the base chain may not be in a straight line!
b. If two pathways have the same chain length, choose the one that gives the most substituents. This keeps the substituent groups shorter making it easier to name the compound.

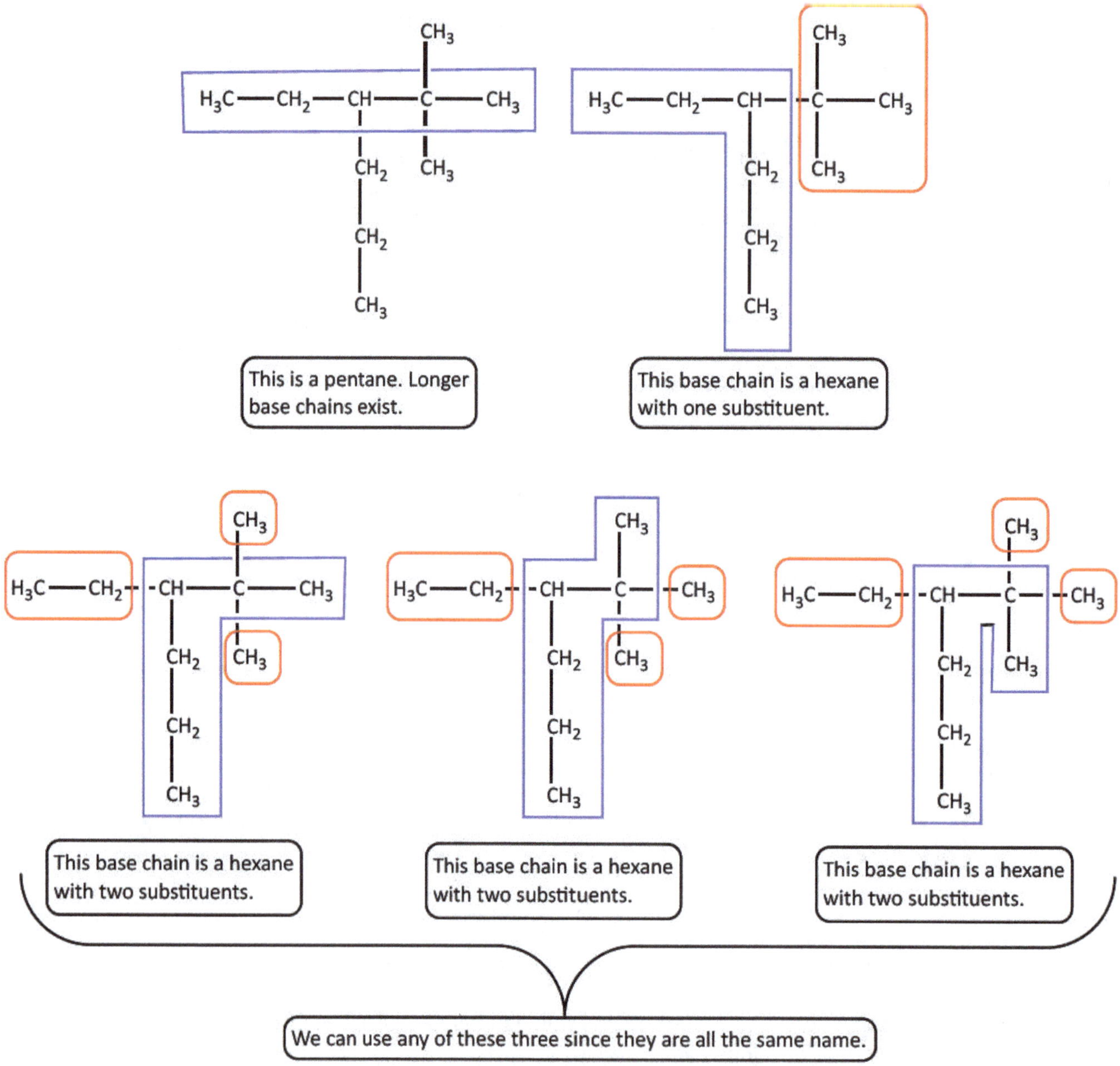

This is how we get the last part of the name, 3-ethyl-2,2-dimethyl*hexane*.

Rule 2. Number the base chain so we can tell where the substituents are located.

We want the smallest numbers we can get. So, begin numbering the chain from the end *closest to a substituent*.

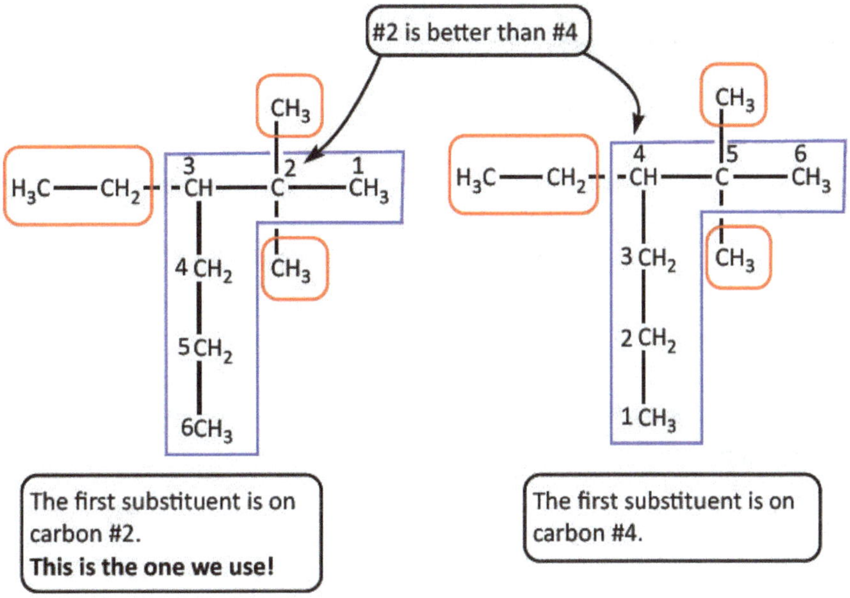

The numbers 2 and 3 are used (the first numbering system) and *not* 4 and 5 (the second numbering system) in the name 3-ethyl-2,2-dimethylhexane.

Rule 3. Name the alkyl group substituents. $-CH_3$ = methyl, $-CH_2CH_3$ = ethyl, $-CH_2CH_2CH_3$ = propyl, etc.

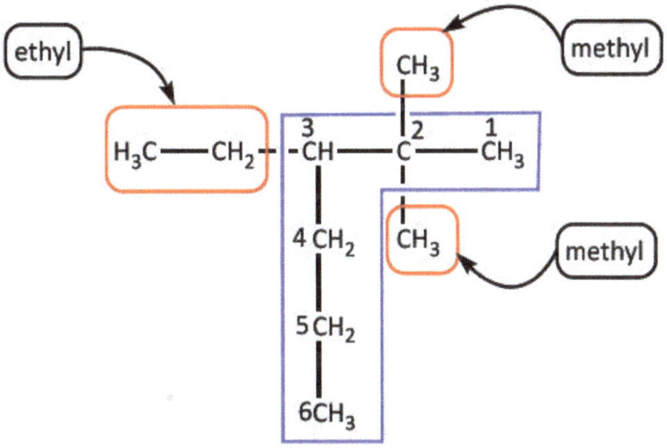

3-ethyl-2,2-dimethylhexane

Rule 4. If there is more than one of the same kind of substituent, use di-, tri-, etc. to tell how many of them there are.

di = 2, tri = 3, tetra = 4, penta = 5, hexa = 6

In our example, we have two methyl groups. The two methyls are put together as "dimethyl" in our name, 3-ethyl-2,2-dimethylhexane. Notice, every substituent gets a number. We name the two methyl groups as 2,2-dimethyl (using "2" twice) and do not simply say, 2-dimethyl.

Rule 5. Put substituents in alphabetical order, not including the di-, tri-, tetra- prefixes.

Notice, the name is NOT 2,2-dimethyl-3-ethylhexane. Even though dimethyl starts with a "d" and ethyl with an "e". We ignore the *di-* prefix and use the "m" of methyl for our alphabetical arrangement. The name we use is named 3-ethyl-2,2-dimethylhexane because "ethyl" comes before "methyl" alphabetically.

Rule 6. Use commas to separate numbers and dashes to separate numbers and letters.

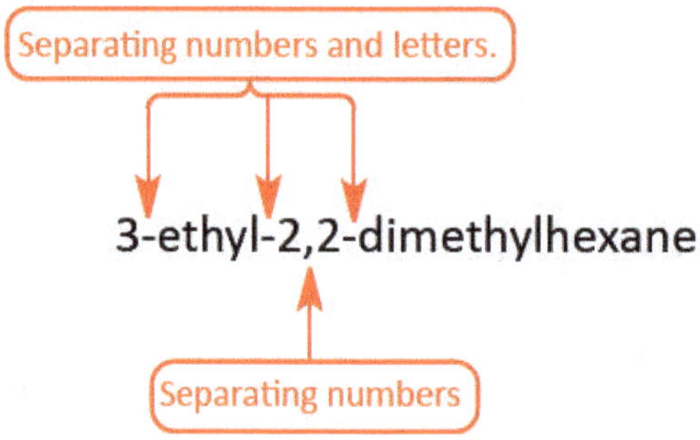

6.2 Name the following compounds.

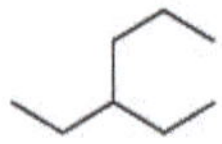
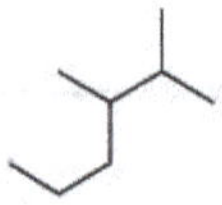
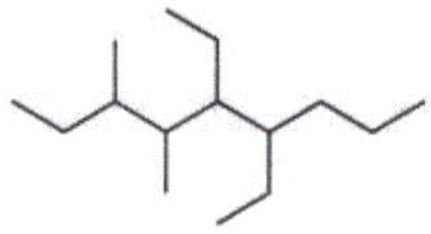
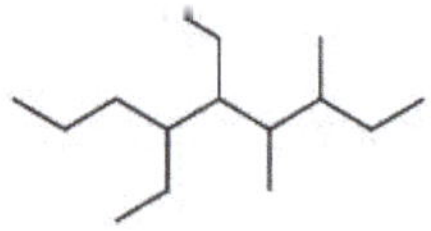
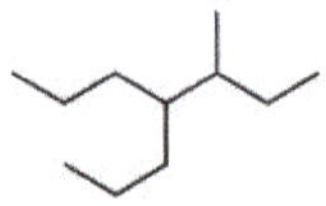
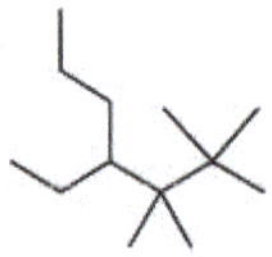
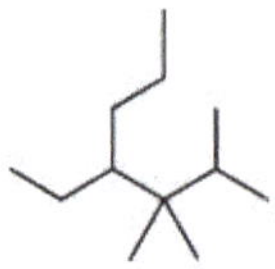
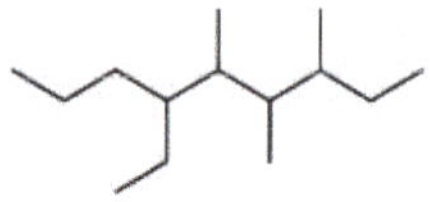

6.2.3 Nuances

If we were to name the following compound, we would run into a small issue we should look at. The longest chain is a hexane with a methyl group and an ethyl group attached.

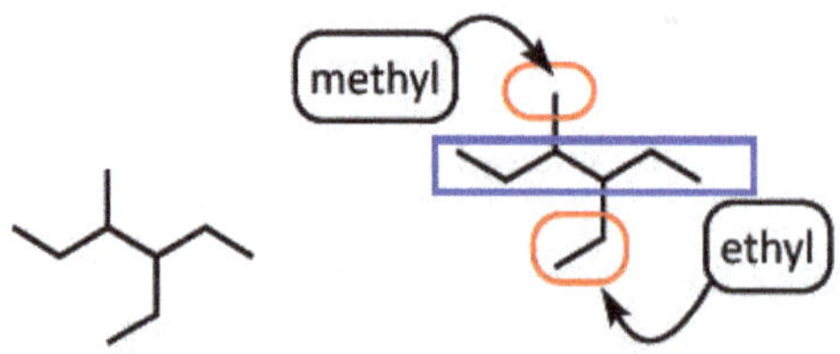

If we number the hexane chain from left to right, we get the name 4-ethyl-3-methylhexane. If we number the hexane chain from right to left, we get the name 3-ethyl-4-methylhexane. Since there is a tie with the numbers, we choose the one with the lower number written first. So, we prefer the name 3-ethyl-4-methylhexane.

6.3 Name the following.

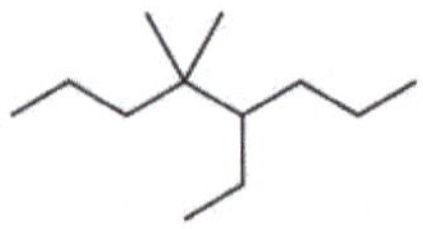
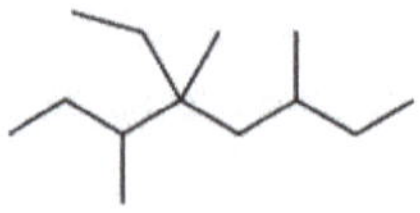
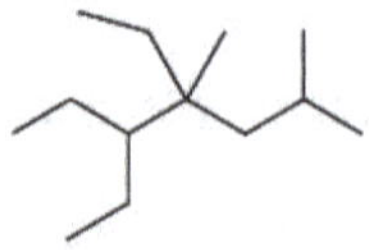

But, what if a substituent is a little more complicated than a normal, n-alkyl group like methyl, ethyl, propyl, or butyl? There are some common names we should learn for isomers of propyl and butyl. We are already familiar with propyl. The only isomer of propyl that exists is isopropyl. Notice the "Y" formation in isopropyl.

propyl (Pr) isopropyl (^{i}Pr)

Figure 6.5 Isomeric propyl groups

For butyl, there are more possible isomers. The first isomer to learn is isobutyl. This isomer again has the "Y" at the end of it. When the carbon attached to the R group is a secondary carbon, meaning it is attached to two other carbon atoms in the group, the group is called secondary-butyl or more commonly, sec-butyl. The tertiary-butyl group is so named because the carbon atom attached to the R group is a tertiary carbon atom, meaning it is attached to three other carbon atoms in the group.

butyl (Bu)	isobutyl	secondary-butyl sec-butyl	tertiary-butyl tert-butyl t-butyl (Bu^{t})

Figure 6.6 Isomeric butyl groups

Sometimes these common group names are used when naming a compound. This naming system is very common for the above groups.

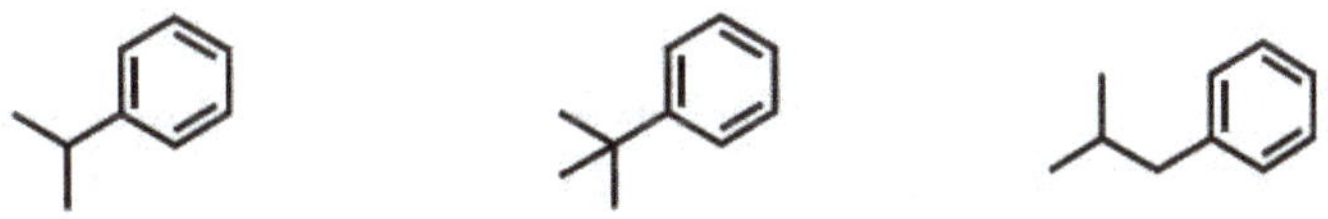

isopropylbenzene	tert-butylbenzene t-butylbenzene	isobutylbenzene

6.4 Draw the following molecules.
a) 4-isopropylheptane b) 4-tert-butyloctane c) isobutylcyclopentane d) tert-butylchloride

But, sometimes, these groups will be named using the more technical IUPAC system. When these groups are named this way, they are put in parenthesis. The longest base chain of the substituent group is found and named with the *-yl* suffix. This alkyl base chain is always numbered starting closest to the "main" part, or the longest base chain, of the molecule. The other groups are named as substituents on this base chain.

Alkanes

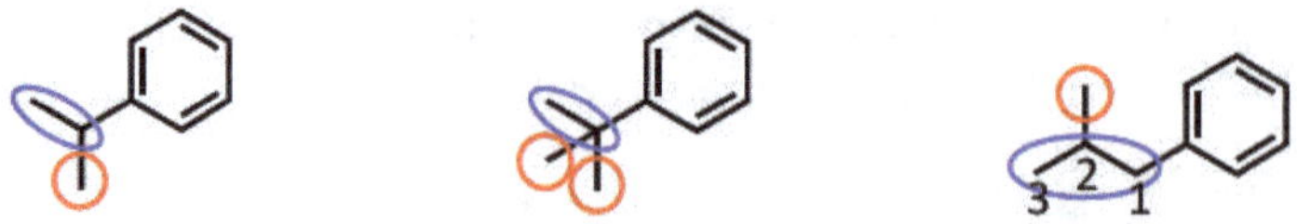

(methylethyl)benzene (dimethylethyl)benzene (2-methylpropyl)benzene

This technique is useful for more complicated alkyl groups like:

5-(ethylpropyl)decane or
5-(1-ethylpropyl)decane

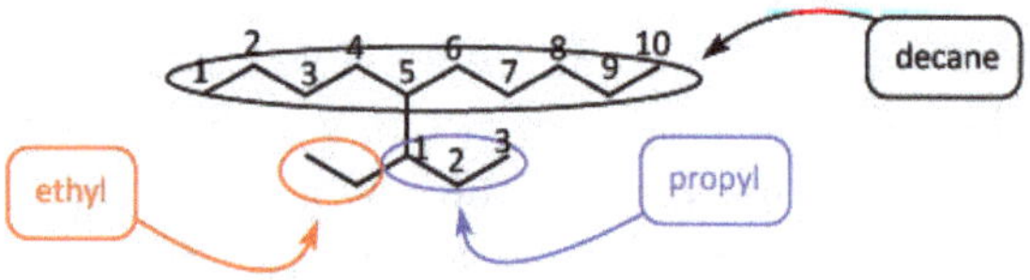

6.5 Name the following molecules (one which contains an isopropyl group and one that contains a tert-butyl group) using the IUPAC name for all substituents.

a) 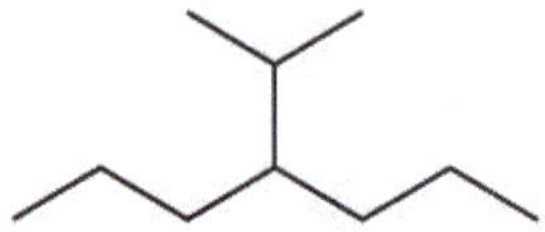 b) 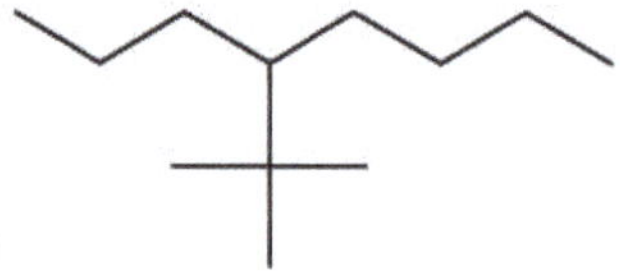

6.6 Draw the following compounds.
a) 4-(1-ethylpropyl)nonane b) 4-(1,1-dimethylbutyl)decane

6.3 Structural Isomers

Isomers are two different compounds with identical molecular formulas. We saw this previously with two compounds with the formula C_4H_{10}, n-butane and isobutane. These two compounds share the same molecular formula, yet they are different, distinct compounds. These two compounds are **structural isomers** of each other. Structural isomers are also sometimes called **constitutional isomers**. Structural isomers have different structures. They usually look like a piece was pulled off of a molecule and placed in a different location. The actual structure is changed.

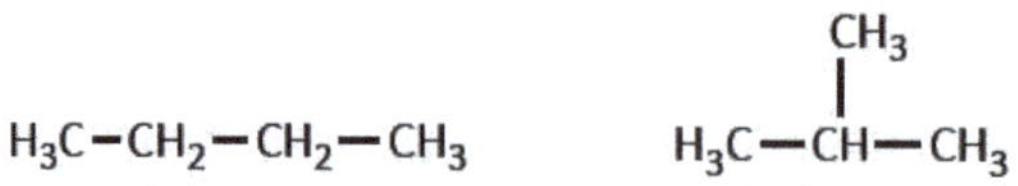

butane, *n*-butane isobutane

Figure 6.7 Isomers of butane

6.7 Identify whether each of the following pairs of compounds are structural isomers of each other or not.

a) 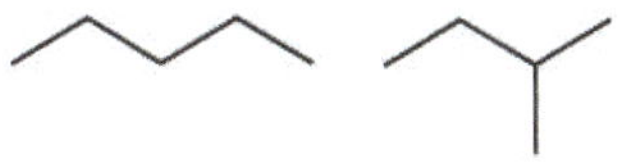

b) 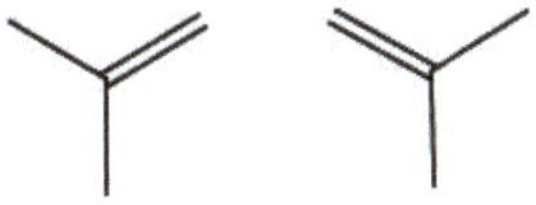

c) 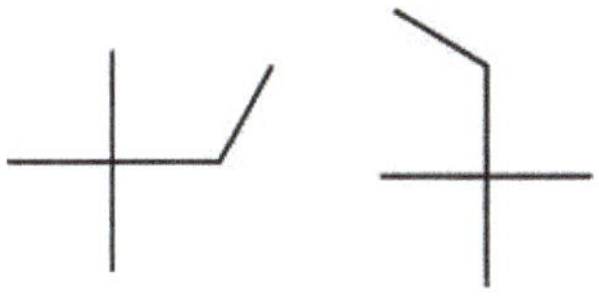

d) 

e) 

f) 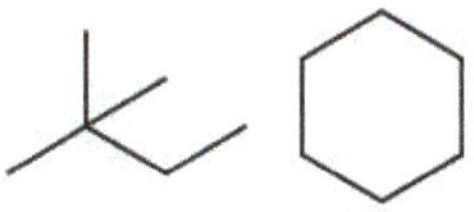

6.4 Conformations

6.4.1 Newman Projections

There are various ways to look at the conformation of a molecule. Let's look at ethane, CH_3CH_3 from different angles. The first two drawings are good (from the side and the sawhorse structure). The third (end-on) is not. This is not a good way to draw an end-on version of a molecule because two of the carbon atoms down our line of sight are on top of each other. This makes the molecule look like one carbon atom with six hydrogen atoms on it. We need a better way to draw molecules end-on.

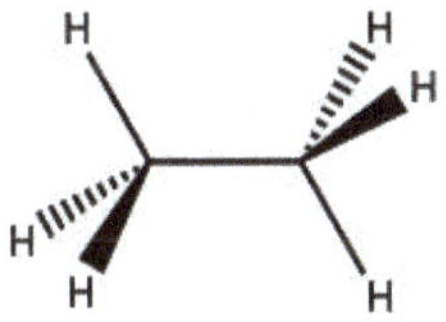

from the side

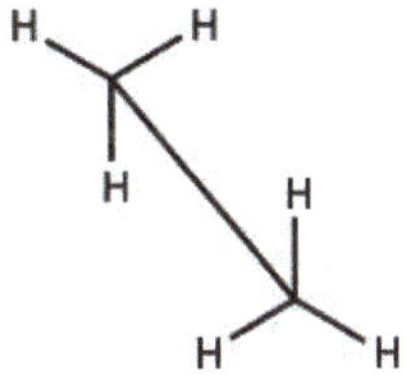

slightly askew sawhorse structure

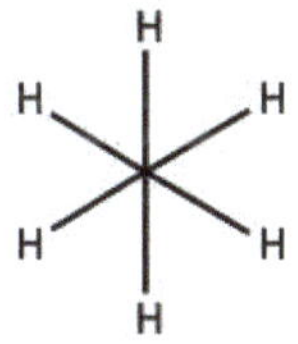

end-on

There is a better way to look at the conformations, end-on. We can use a system called Newman projections to describe what we see. Newman projections look down one carbon-carbon bond. First, let's briefly look at how Newman projections are drawn. First, we draw a circle to represent the back carbon atom. We then draw three bonds on it evenly spaced out to represent the three groups attached to it.

A carbon atom

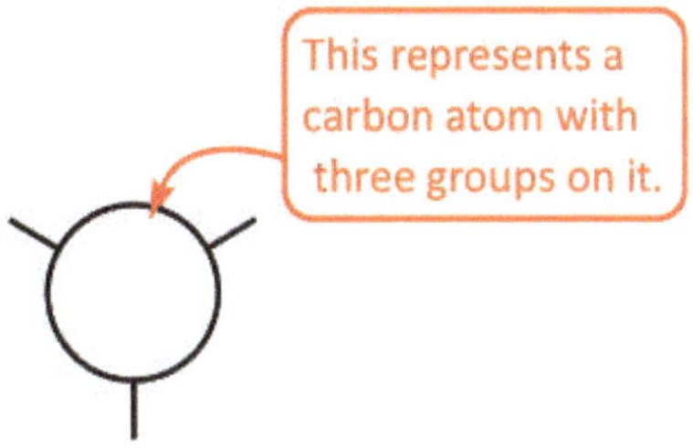

We use a point to represent the front carbon atom and draw three bonds on it to represent the three groups attached to the front carbon.

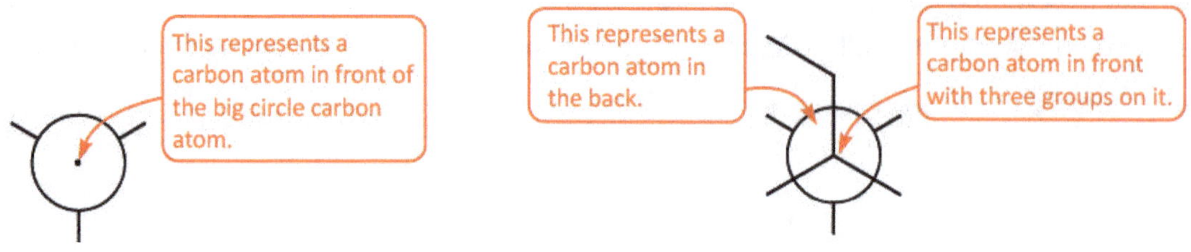

Newman Projection

So, if we are looking at ethane from his vantage point, we would make this Newman projection.

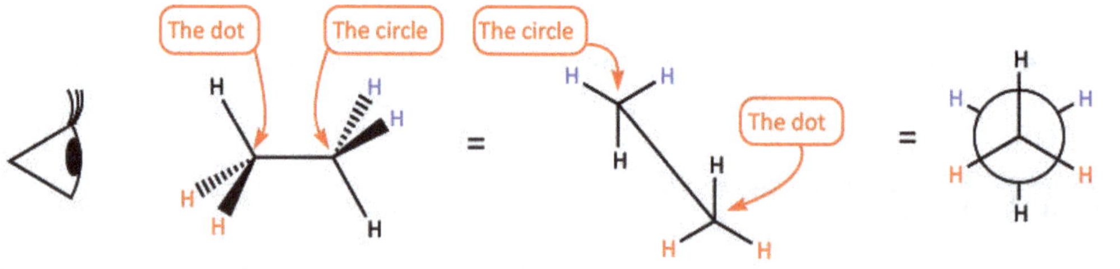

Sawhorse structure

Newman Projection

6.4.2 Rotation of sigma bonds

Sigma bonds can rotate. Because of this, alkanes are always twisting and spinning around. Let's look at two of these conformers.

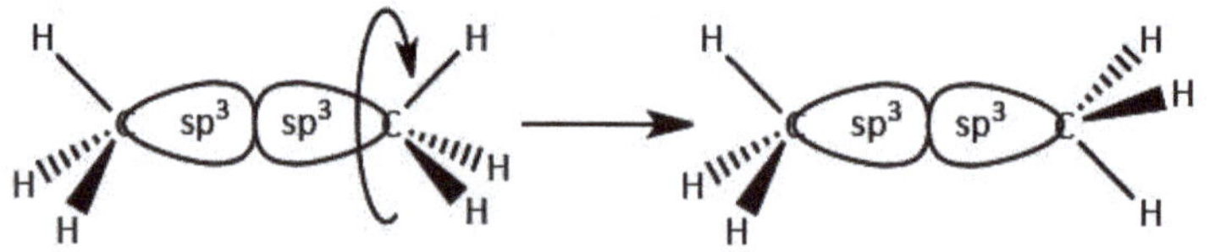

ethane rotating

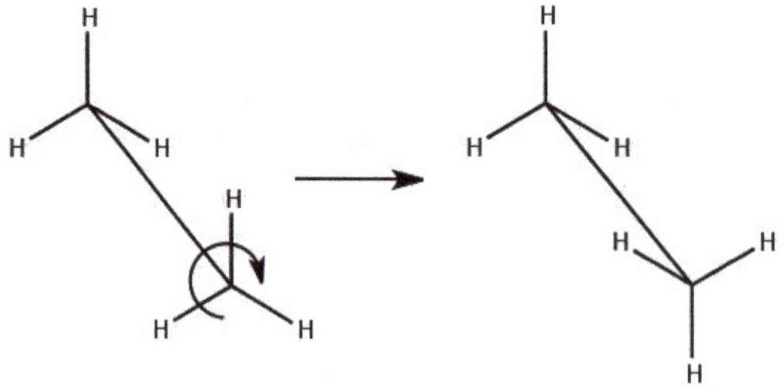

eclipsed form

staggered form

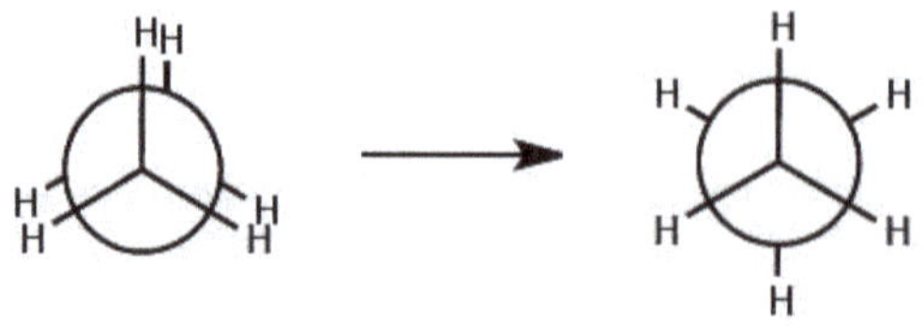

eclipsed form
higher energy

staggered form
lower energy

The **eclipsed** form with its bonds overlapping in the same plane is higher energy than the **staggered** form, which does not have overlapping bonds. The eclipsed form is not as happy. There are electrons in each C-H bond, so when they overlap, as in the eclipsed form, the negative electrons in one C-H bond repel the negative electrons of another C-H bond. In the staggered form, these negative electrons in the C-H bonds are as far away from each other as is possible. This extra energy the eclipsed form has over the staggered form is called **torsional strain.**

Let's look at the Newman projections of butane, looking down the second and third carbon atoms.

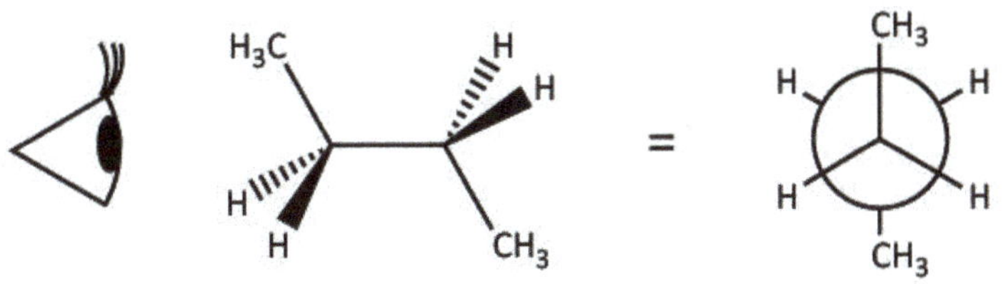

As we look at the various conformations, notice that the lowest energy, best conformation, is when the large methyl groups are as far away from each other as possible. This is called the **anti**-conformation. The highest energy, worst conformation, is when the methyl groups are closest to each other and overlapping in **the totally eclipsed** form.

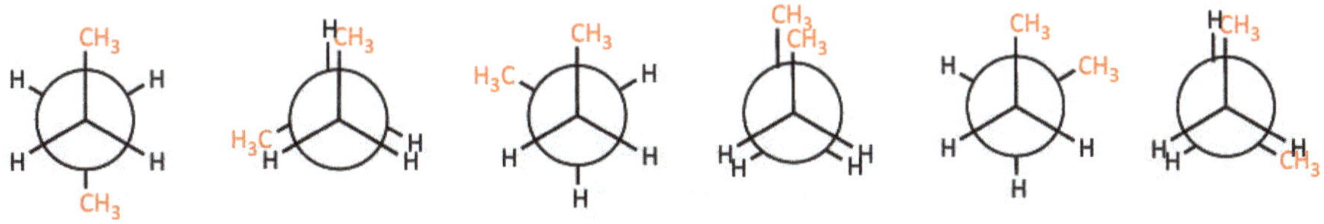

Figure 6.8 Newman projection conformers of butane

When two groups come close like this, it is a high-energy interaction. This bad interaction caused by large groups overlapping and interacting is called **steric hindrance.** Molecules will twist or rotate to avoid steric hindrance when they can.

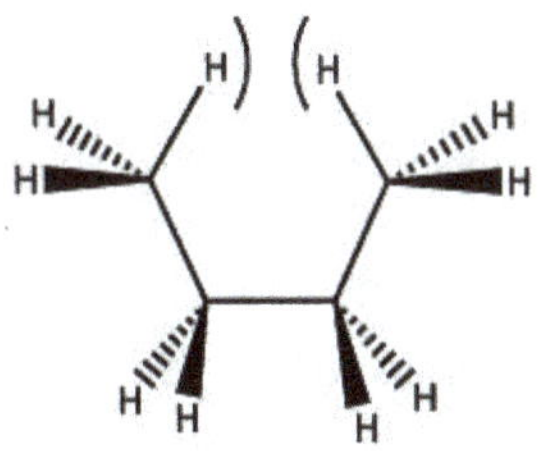

Figure 6.9 Steric Hindrance

The **eclipsed** conformer is not as bad as totally eclipsed because the methyl groups are not overlapping, but the methyl groups are overlapping with hydrogen atoms, so it is still bad. **Gauche** (French for "left") is better than eclipsed because the bonds are not overlapping, but the methyl groups are only 60° apart in the gauche conformer, while they are 180° apart in the lower energy **anti** conformer.

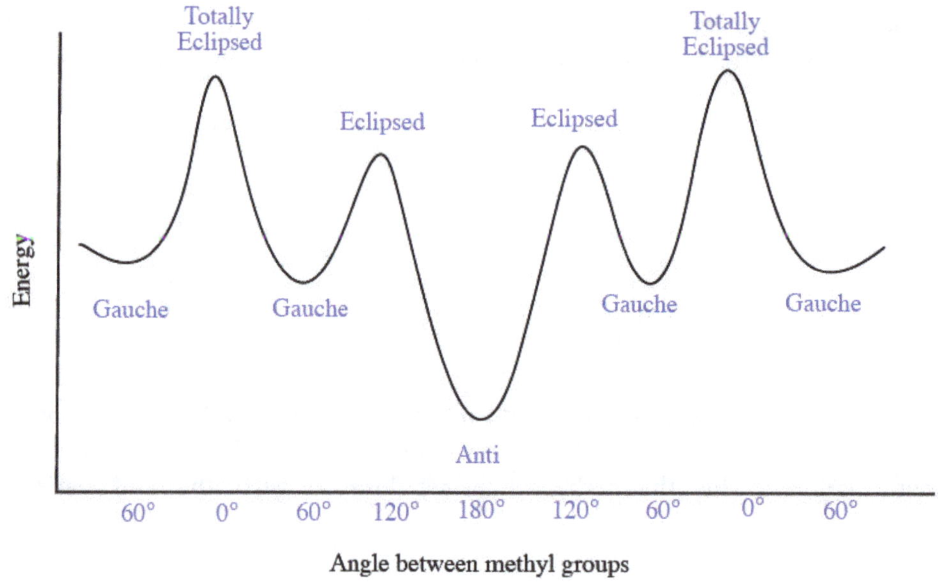

Figure 6.10 Energy Diagram for Conformers of Propane

6.8 Draw the Newman projections for pentane to represent the view from which they are seen.

a) 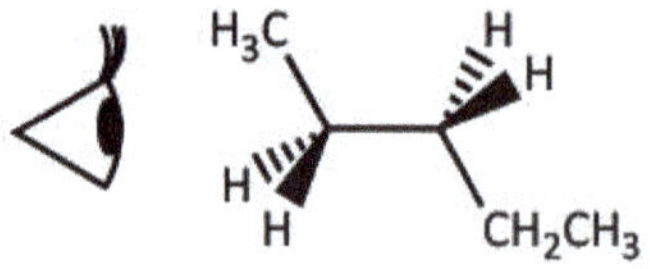

____

b) 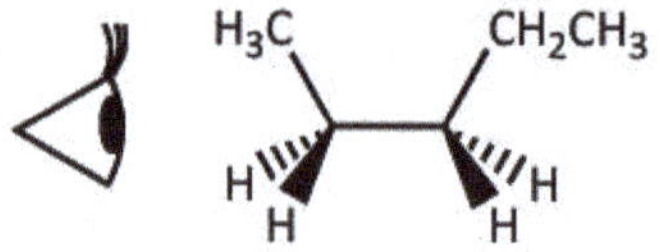

____

c) 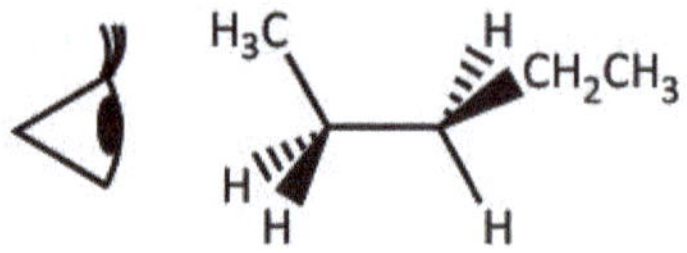

____

d) 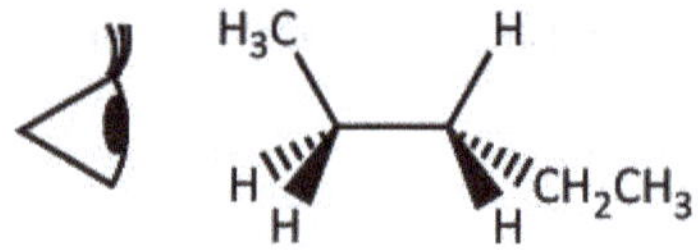

____

6.9 Label the Newman projections in the previous problem with their appropriate labels of Anti, Eclipsed, Totally Eclipsed, and Gauche.

6.10 Rank the Newman projections of pentane, as shown in 6.8, from lowest energy to highest energy. Circle the correct ranking below.

a < b < c < d   a < c < d < b   a < d < c < b   b < d < c < a   c < a < d < b

6.11 Review problem 6.8. Count the five carbons in pentane and notice that the Newman projections drawn were looking down the bond of the 2nd and 3rd carbon atoms (C2-C3). Now draw a lowest energy Newman projection looking down the C1-C2 bond.

Because of this easy rotation about sigma bonds, most molecules will often twist and turn. This twisting and turning of a molecule does not change what the molecule is. For instance, we could place a length of string down on the table, and it could look like the image on the left. We could place another piece of string, the same length, on the table, and it look like the image on the right.

In both cases, we have the exact same strings. It is only that their conformations are different. These would be called conformations or **conformers** of each other. These are not isomers of each other.

6.12 Identify each pair of compounds as *conformations* of each other or as *structural isomers*.

a) 

b) 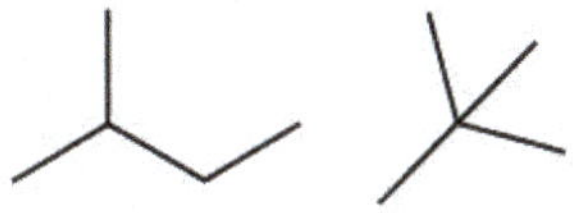

c) 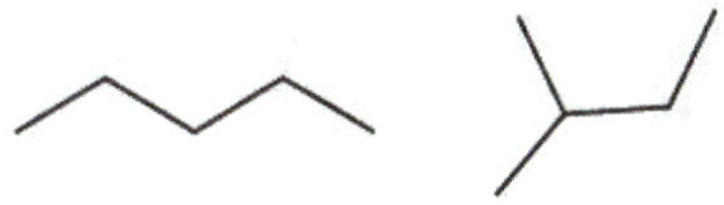

d) 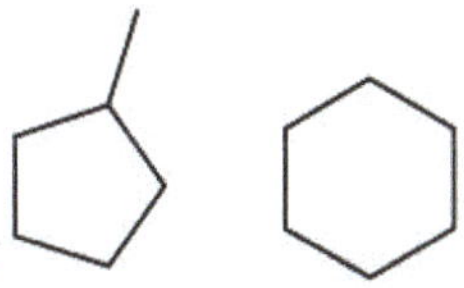

e) 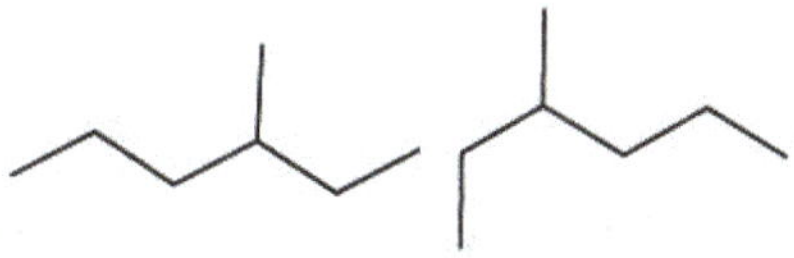

f) 

6.5 Cycloalkanes

If an alkane is in a ring, a cycloalkane, we name it by using the prefix *cyclo-*.

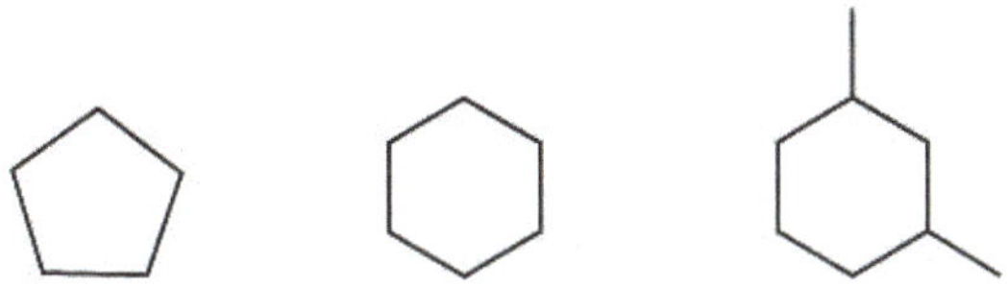

cyclopentane cyclohexane 1,3-dimethylcyclohexane

Figure 6.11 Cycloalkane examples

If there are two groups on a ring, we need to say whether the groups point towards the same face of the ring (cis) or opposite faces of the ring (trans).

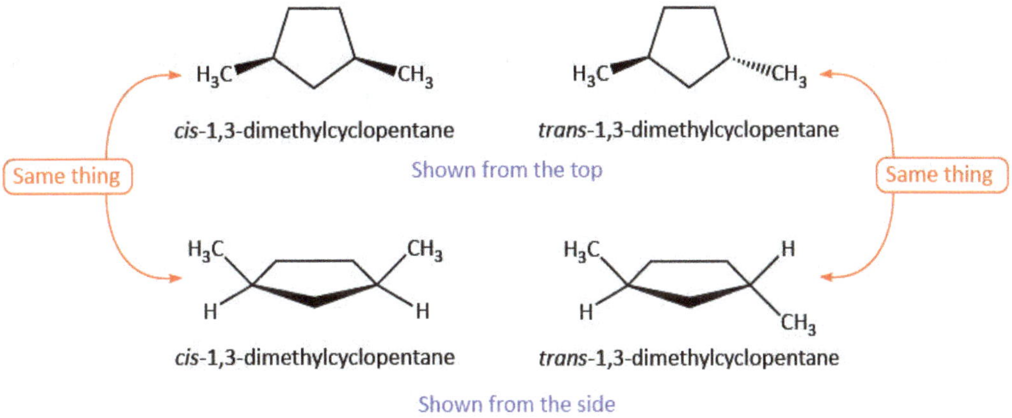

Figure 6.12 Examples of cis/trans cycloalkanes

6.13 Name the following compounds (use cis/trans when appropriate).

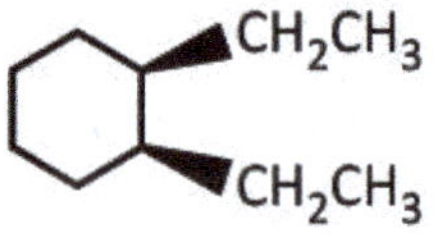

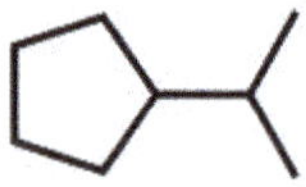

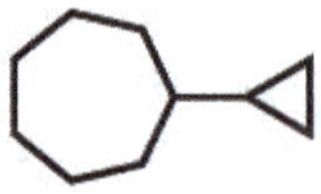

6.14 Draw the following compounds.

a) *trans*-1-ethyl-3-propylcyclohexane

b) *cis*-1-propyl-3-(2,2-dimethylpropyl)cycloheptane

c) cyclopropylcyclobutane

d) *cis*-1,3-dichlorocyclopentane

6.5.2 Bicycloalkanes

When two cycloalkanes are fused together, they are called **bicycloalkanes** or a **bicyclic compound**. These two rings always share two carbon atoms. These two shared carbon atoms are the **bridgehead carbons**. In the following bicyclic compounds, the bridgehead carbons are circled.

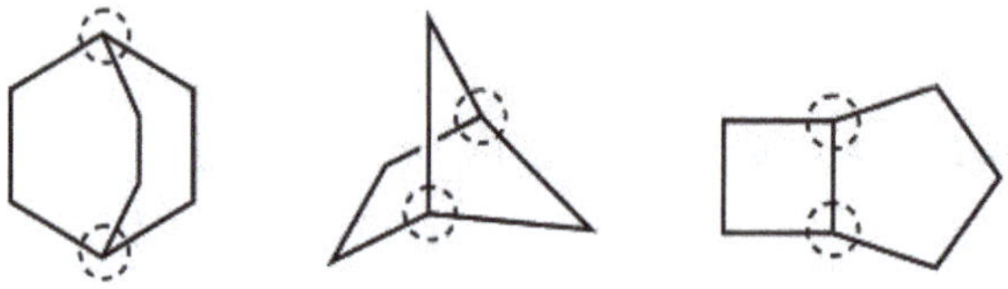

Figure 6.13 Bicycloalkane examples

Three bridges connect these bridgehead carbon atoms. The number of carbon atoms in each of these three bridges need to be counted, but do not include the bridgehead carbon atoms in this count!

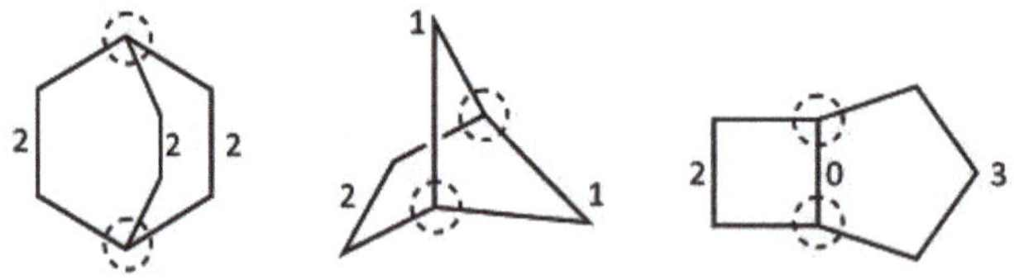

Bicycloalkanes are named by writing bicyclo, followed by brackets that contain three numbers separated by periods, followed by the alkane name for the total number of carbon atoms in the entire molecule. The three numbers are the number of carbon atoms in the three bridges arranged from highest to lowest. It takes the form bicyclo[#.#.#]alkane.

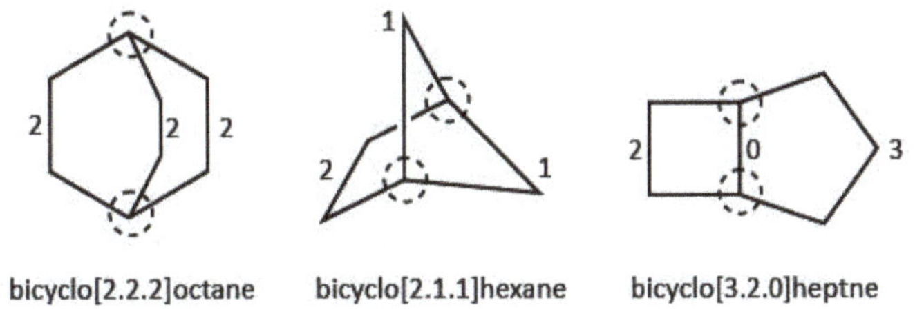

Figure 6.14 Names of some bicycloalkanes

6.15 Name the following compounds.

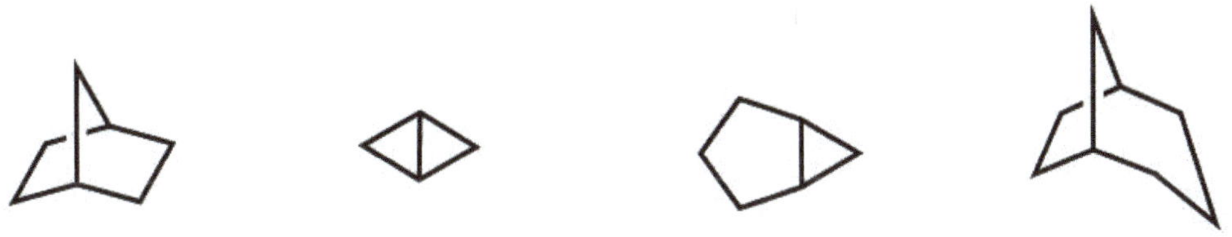

__________ __________ __________ __________

6.16 Draw the following compounds.

bicyclo[2.1.1]hexane bicyclo[2.1.0]pentane bicyclo[2.2.2]octane

6.5.3 Stability of cycloalkanes

Alkane carbon atoms are sp^3 hybridized which naturally produces tetrahedral bond angles of 109.5°. In order to form cycloalkanes, the bond angle must deviate from this desired 109.5. For instance, in cyclopropane, the bond angle inside the ring is 60°. This is 49.5° away from the ideal bond angle. **Angle strain** is the strain that comes from pulling the bonds away from the idea bond angle into this strained conformation. Also, cyclopropane has some torsional strain because it is locked into a conformation that caused the C-H bonds to overlap. The strain caused by angle and torsional strains combined is called **ring strain**.

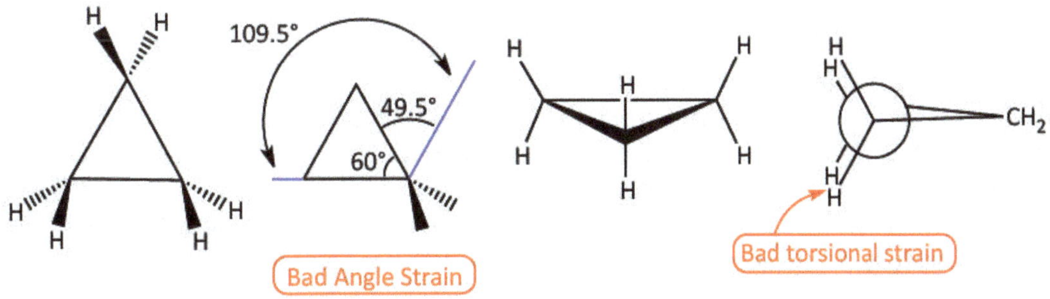

Figure 6.15 Ring strain examination of cyclopropane

Cyclobutane also has quite a bit of ring strain. If cyclobutane was a planar, regular square, it would have high angle strain with the deviation of a 90° ring being 29.5° away from the idealized 109.5°. But, if it is planar, it has terrible torsional strain. So cyclobutane twists or folds a bit so it is not planar. It hurts the angle strain because by folding, it becomes an 88° angle in the ring. But, the torsional strain is much better. We do see cyclobutane rings at times, but they are still fairly rare because of the high ring strain.

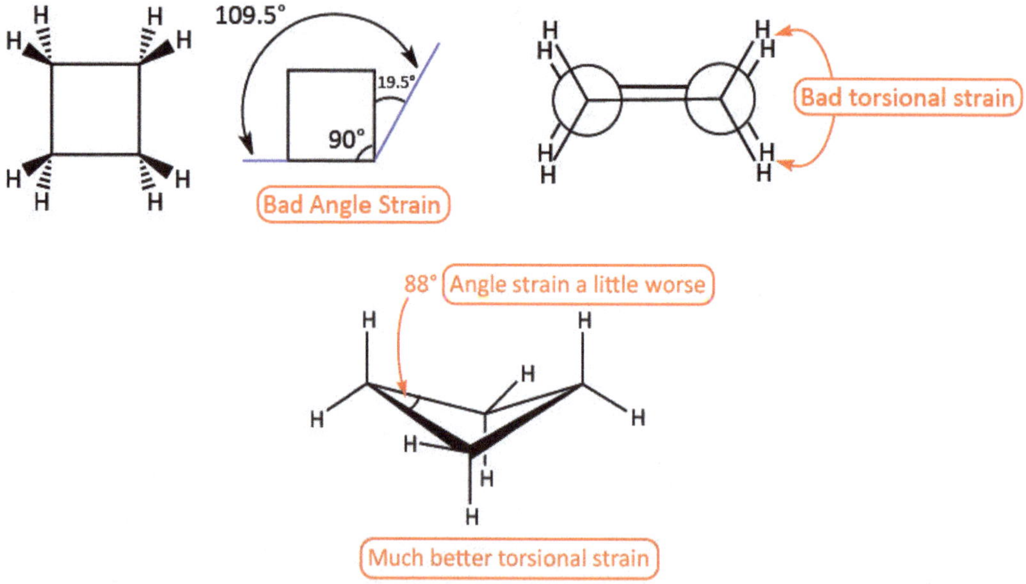

Figure 6.16 Ring strain examination of cyclobutane

If cyclopentane was a planer, regular pentagon, it would have bond angles of 108°. This does not deviate much from the ideal 109.5°. In order to reduce the torsional strain cyclopentane would have if it was planer, it folds a bit. Cyclopentane is quite stable with low ring strain. Cyclopentane is a fairly common cycloalkane.

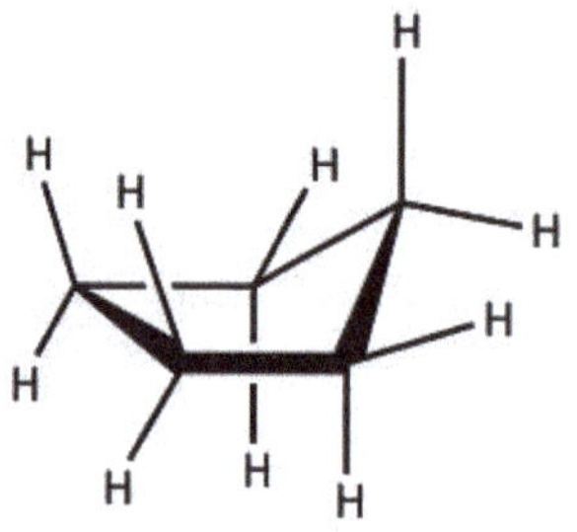

6.6 Cyclohexane Chair

Cyclohexane is the most stable of the cycloalkanes. It is a very, very common ring structure in organic chemistry. Many organic molecules have this ring structure. Therefore, it is incumbent upon us to investigate and have a very good understanding it. Cyclohexane is not flat. It has what is called a chair conformation. This chair conformation has no angle strain. Each carbon atom is a perfect 109.5°. The C-H bonds are also all in a staggered conformation. In fact, the chair form of cyclohexane rings is so stable, they are found as repeating units in the carbon structure of diamond. Look carefully at the piece of diamond structure. Can you see some chairs in it?

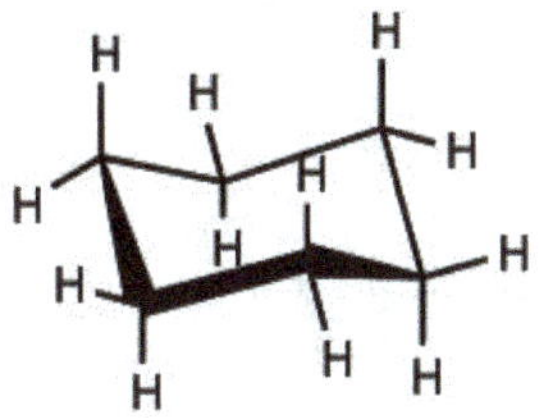

Cyclohexane in chair form

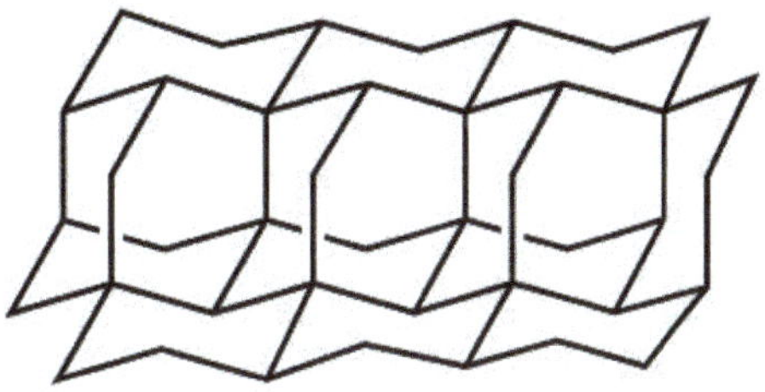

Diamond

Let's practice drawing this chair conformation. First, draw two parallel lines at an angle.

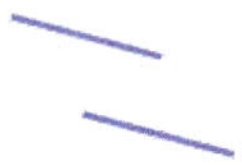

At the top of the two lines, draw a V down connecting the two lines. At the bottom of the two lines, draw a v up connecting the two lines. This makes the base form of the cyclohexane chair.

Let's begin to put hydrogen atoms on the chair. There are two types of bonds on a chair, axial and equatorial chair bonds. Axial bonds point straight up or down, much like the axis of the earth goes through the poles. Equatorial bonds point more towards the side, much like the equator of the earth.

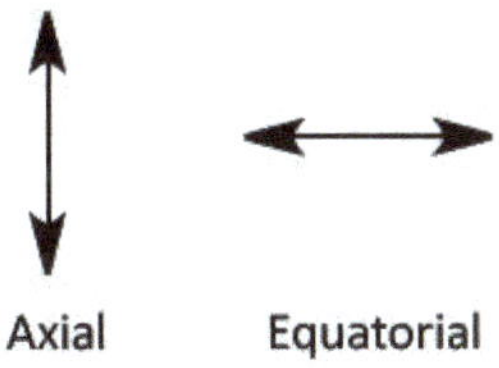

Figure 6.17 Axial and equatorial

First, on each corner, draw a line straight up on carbon atoms that point up, and draw a straight line down on carbon atoms that point down. These bonding positions that are straight up or straight down are called axial positions. Axial bonds point straight up on up carbons and point straight down on down carbons.

Figure 6.18 Axial bonds

Next, let's look at the equatorial bonds. For every up carbon, an equatorial bond is drawn down and for every down carbon, an equatorial bond is drawn up. The four equatorial bonds on the ends are easiest to draw, so draw them first.

Figure 6.19 Equatorial Bonds

The last two equatorial bonds are sometimes confusing for students at first. For the up carbon atom, we know the equatorial bonds needs to point down, but do we angle it down to the left or the right? As a beginner, both may look OK to you.

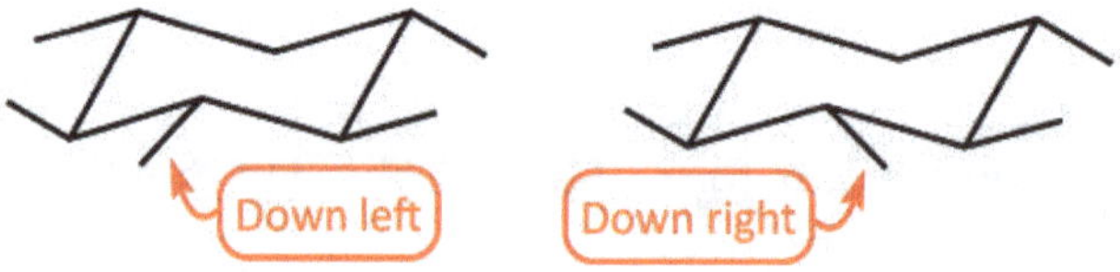

Thisis the correct way. **This is the wrong way.**

The way to remember which way to point these last two equatorial bonds is to look for the "M" and "W". Below is the correct way to draw in the equatorial bonds.

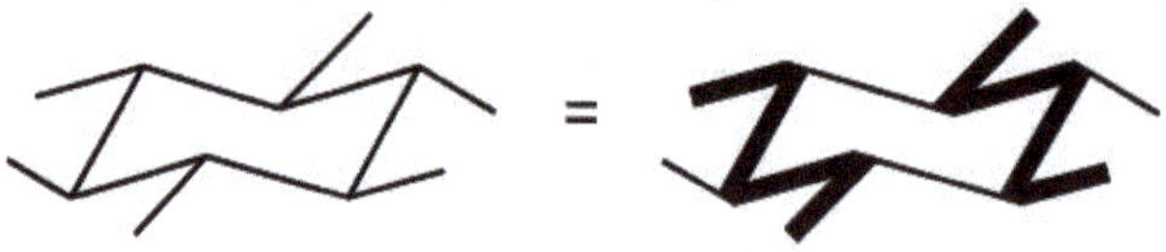

6.17 Practice drawing a chair. Draw the chair five times without the axial and equatorial bonds on it.

6.18 On this chair, label the axial position with an "A" and the equatorial positions with an "E".

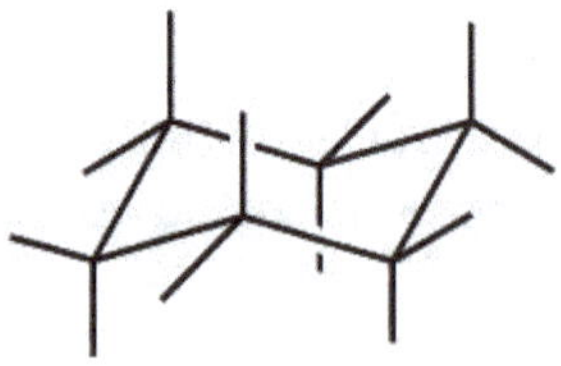

6.19 Draw a chair and put the six axial bonds on it.

6.20 Draw a chair and put the six equatorial bonds on it.

6.21 Draw a chair and put all axial and equatorial bonds on it.

There is enough energy for a cyclohexane chair to convert to another cyclohexane chair conformation. This chair interconversion, or chair flip, goes through a boat conformation. It is called the boat conformation for its obvious similarities to a boat. This boat conformation is less stable than the chair because of steric hindrance. Two groups end up getting close to each other. But, overall, this problem is easily overcome with room temperature thermal energy.

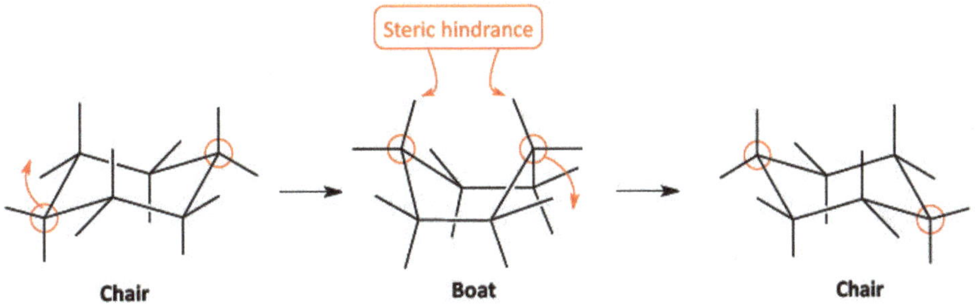

Figure 6.20 Example of a cyclohexane chair flip

6.6.2 Monosubstituted chair

When a chair flip occurs, all of the axial groups become equatorial groups, and all of the equatorial groups become axial groups. This can be seen in the example of methylcyclohexane.

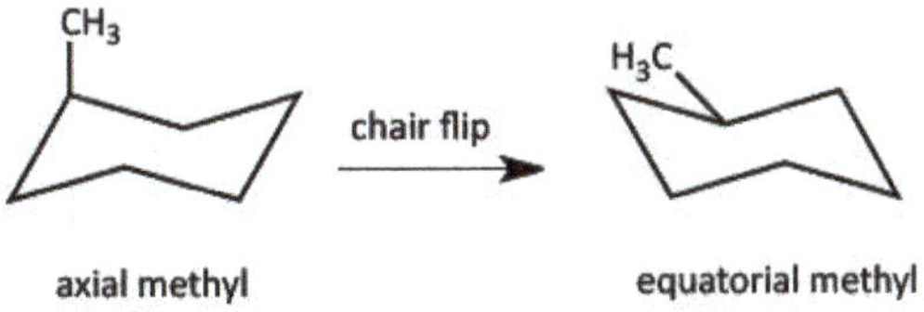

Figure 6.21 Axial group to equatorial group with chair flip

Larger groups prefer to be in the equatorial positions on a chair. Steric hindrance is a problem when a large group is in an axial position because the large axial group ends up too close to other axial groups. This particular type of steric hindrance is called 1,3-diaxial interaction because the two groups are both axial on carbon atoms 1 and 3 of a cyclohexane chair.

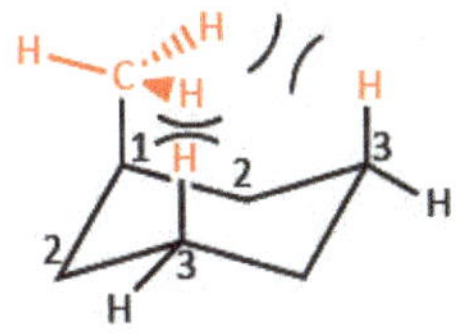

Figure 6.22 1,3-diaxial interaction

If the group is larger than a methyl, it is even more likely to be found in an equatorial position. In fact, a *t*-butyl group is so large, that you may *never* see it in an axial position.

Important point: *On a chair, a large group in an equatorial position is more stable than in an axial position.*

6.22 a) Use the following two lines to begin to draw a chair. Finish drawing an ethyl group on the chair to make the lowest energy form of ethylcyclohexane. Be very careful and particular about the direction your draw your bonds.

b) Draw the chair conformation after your ethylcyclohexane drawing from *part a* undergoes a chair flip. You may want to use these parallel lines to begin to draw your chair.

6.23 Identify the following groups on cyclohexane chairs as axial or equatorial.

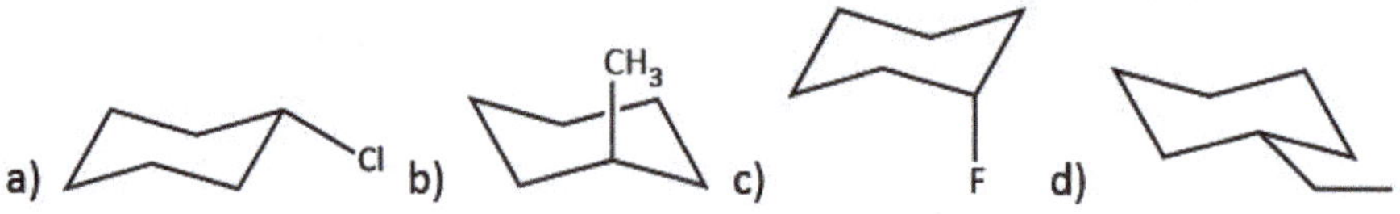

6.6.3 Disubstituted chairs

When two groups are attached to a cyclic compound, cis-trans isomers exist. This is because there are two sides to the ring. Because cyclohexane is a ring, cis-trans isomers exist when two groups are attached to it. It is easy to see if the two groups point in the same direction when the ring is drawn planar.

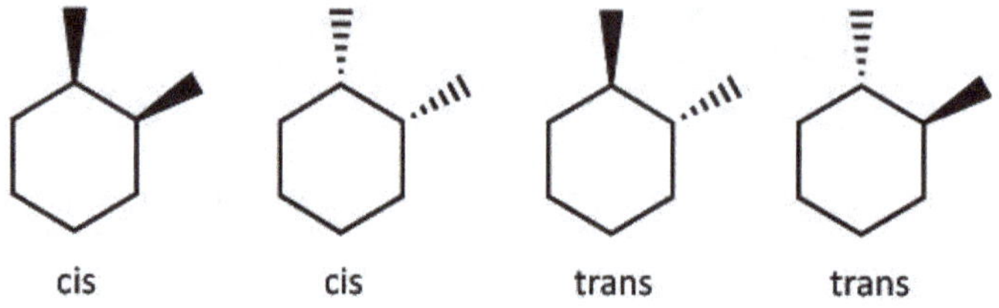

Figure 6.23 *cis* vs. *trans* disubstituted cycloalkanes

But, sometimes in a chair conformation, students have more difficulty at first when trying to determine cis or trans. It is really still pretty simple. You need to remember that if two groups both point up or both point down, it is cis. If the two groups point in opposite directions, one up and one down, then it is trans. On the following chair, all of the bonds are labeled either up (u) or down (d) based on the direction the bond points.

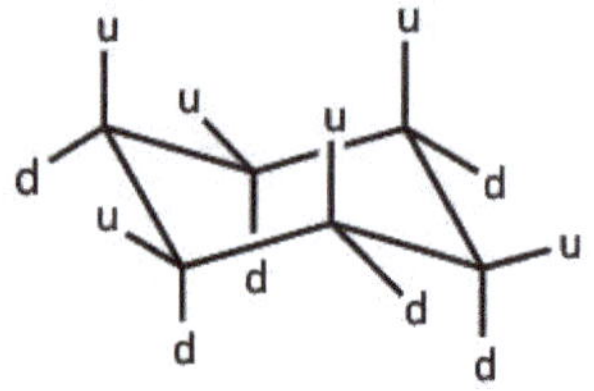

So, let's look at the various possible forms of 1,2-dimethylcyclohexane.

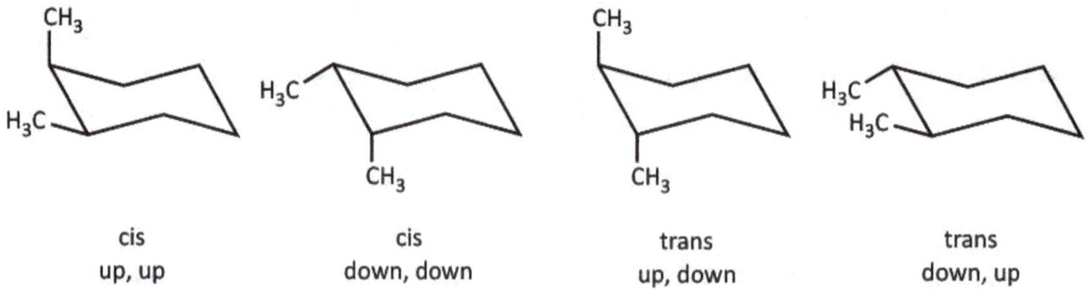

Figure 6.24 *cis* vs. *trans* on disubstituted cyclohexane chairs

Notice that the cis and trans determination of a chair is solely based on the direction (up or down) of the groups and not by axial or equatorial designations.

6.24 Name the following cyclohexanes. Remember to use *cis* or *trans* in the name.

a) 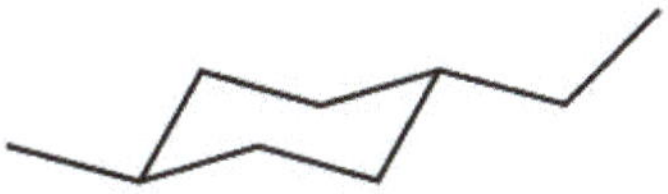 b)  c) 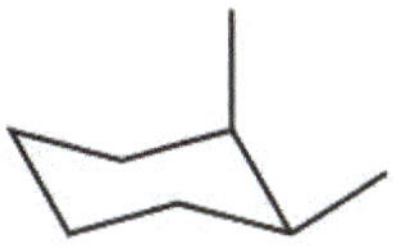

Let's draw a structure from a name. Let's draw *trans*-1,4-dimethylcyclohexane in its lowest-energy chair conformation.

First, let's draw a chair. You can draw either chair-flipped form of the chair.

Draw a methyl group on any carbon atom. We'll call this carbon atom #1. This methyl is in the "up" position.

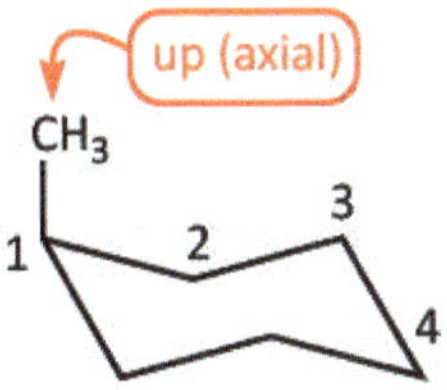

Draw in the axial and equatorial bonds on carbon atom #4. Since carbon #4 is a down carbon atom, the axial bond goes straight down and the equatorial bond must be up.

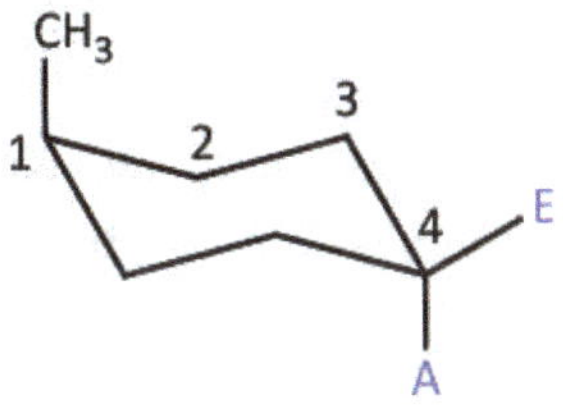

Since the methyl we have already drawn is "up" and we need *trans*, the second methyl group must point in the opposite direction. Therefore, we draw the methyl group down, which is the axial position on carbon #4. We can draw a hydrogen atom in the equatorial position, or we can simply delete it.

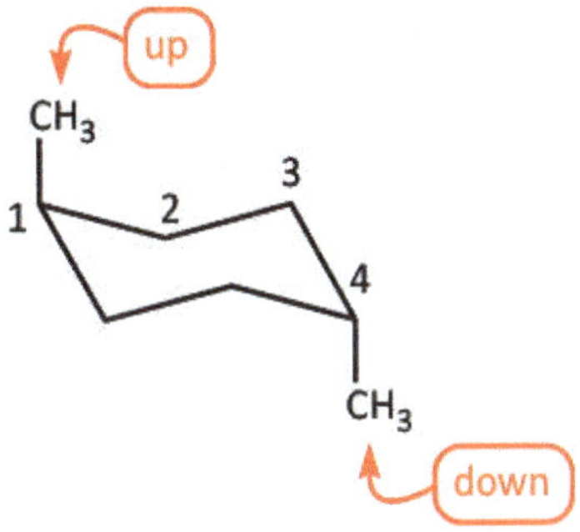

We must also consider the chair flipped version of this compound. Every axial group then moves to an equatorial position.

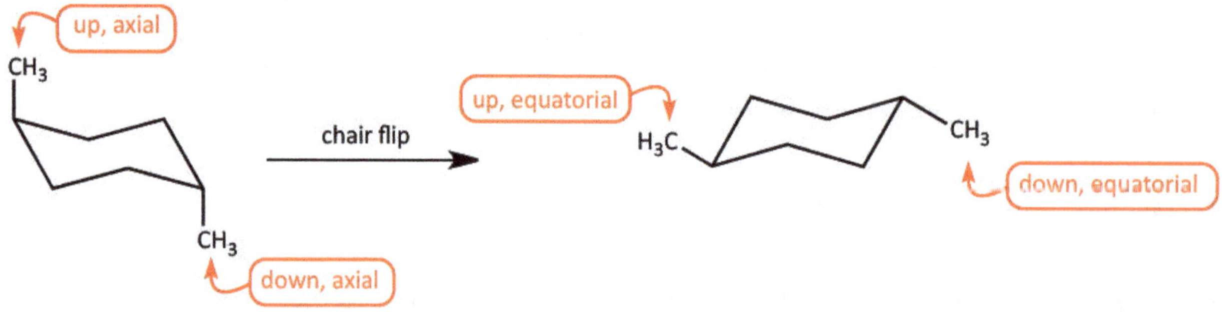

We then compare the two forms to see which one is the lowest energy form. We like the one on the right with the larger methyl groups equatorial instead of both axial like on the left. Our final answer is:

trans-1,4-dimethylcyclohexane

What if we have two different groups? Let's look at the lowest energy conformation of *trans*-1-isopropyl-3-methylcyclohexane.

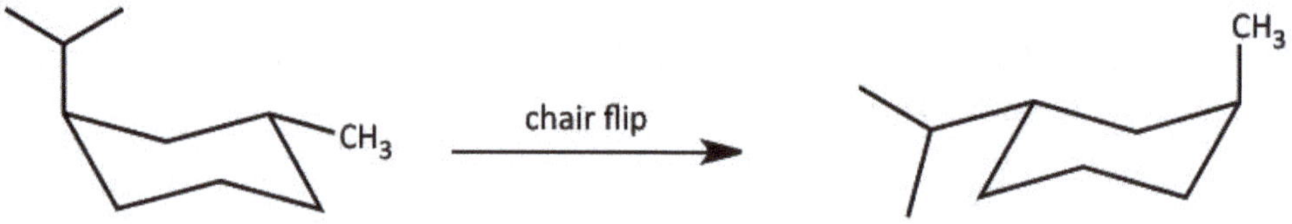

In each of these, there is one axial and one equatorial group. Which do we choose? Since larger groups prefer to be equatorial, we prefer the one with the isopropyl group in the equatorial position.

6.25 Draw the following cyclohexane compounds as chairs in the lowest energy, most stable configuration. Pay careful attention how you draw your axial and equatorial groups!

a) *trans*-1,2-diethylcyclohexane b) *cis*-1-(1,1-dimethylethyl)-4-propylcyclohexane

6.7 Problems

6.26 Draw the following compounds.
a) 1-isopropylcyclohexane b) *tert*-butyl chloride c) *sec*-butyl bromide d) ethyl chloride
e) *tert*-butylcyclopentane f) 2-phenyl-5-isopropyldecane g) 2-chloro-4-ethyl-3-methylheptane
h) *cis*-1-bromo-2-*sec*-butylcyclopentane i) 6-ethyl-2,4,5-trimethyloctane
j) 4-ethyl-2,3,3,5-tetramethylhexane k) 4-(1,1-dimethylethyl)nonane l) bicyclo[3.2.1]nonane
m) bicyclo[2.2.2]octane n) 2,2,3,3-tetramethylheptane o) 1,1-diethylcyclopentane

6.27 Name the following compounds:

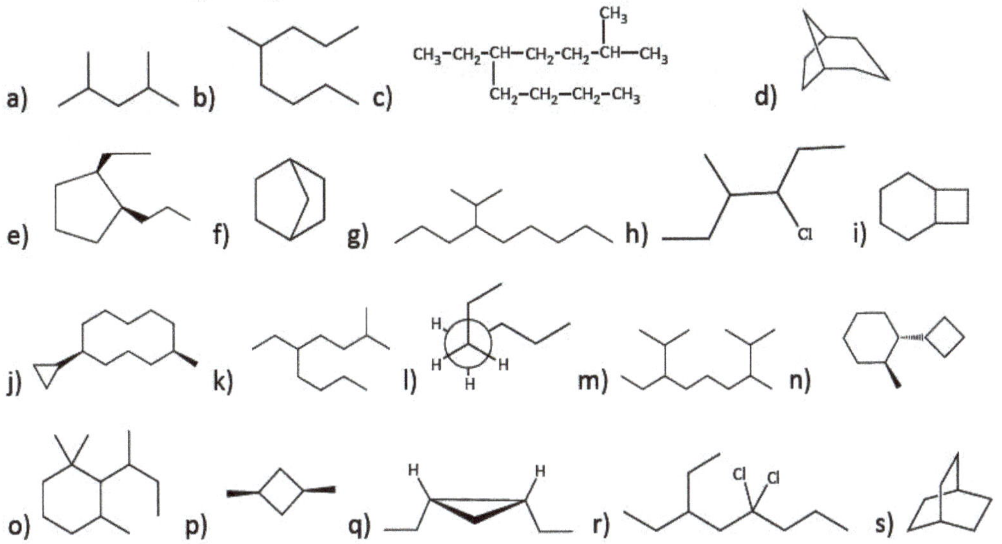

6.28 On an exam, student named some compounds incorrectly. These names are incorrect, but they represent real structures. Draw the structures of the compounds, and name them correctly.
a) 2-propylpentane b) 3-methylbutane c) 3-isopropylpentane d) 1,2-diethylbutane e) cyclohexylethane

6.29 In the following groupings, identify all of the compounds that would have the name listed.

a) *cis*-1,2-dimethylcyclohexane

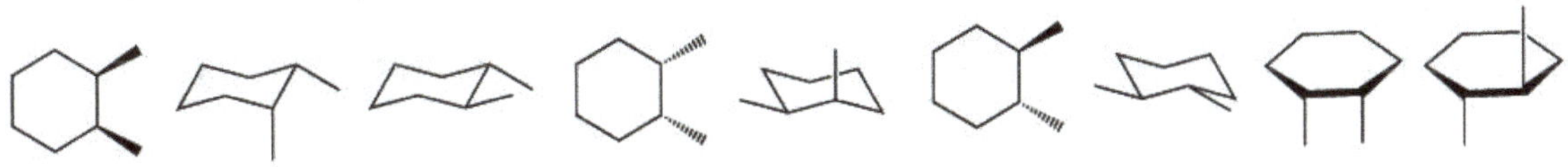

b) pentane

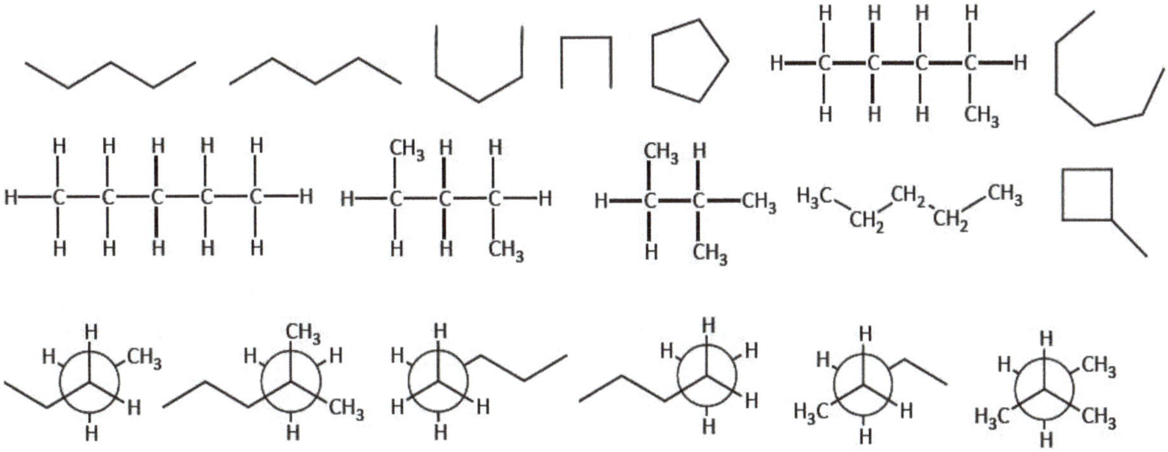

c) 2-methylbutane

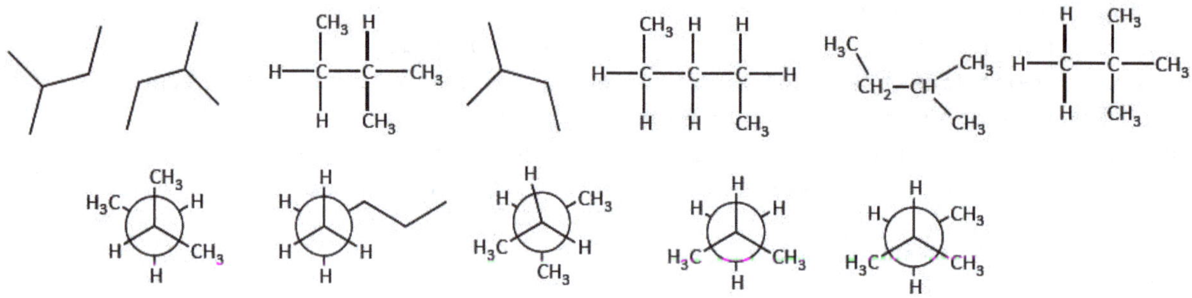

6.30 a) Draw a Newman projection of propane in a staggered conformation. b) Draw the rotational energy diagram for propane. On your diagram, clearly indicate the position of staggered conformations and eclipsed conformations. Your diagram should include rotation through any two staggered and any two eclipsed conformations.

6.31 Draw the following compounds in their most stable chair conformation. Be sure that your drawing clearly shows which groups are axial (if any) and which are equatorial (if any).
a) propylcyclohexane b) *cis*-1,2-dimethylcyclohexane c) *trans*-1-isopropyl-2-methylcyclohexane

d) 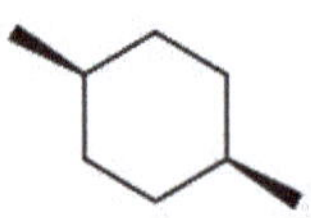 e) 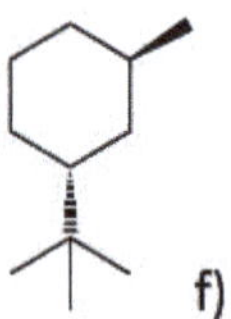 f) 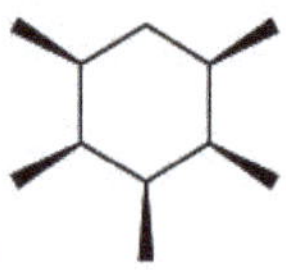 g)  h) *trans*-1,4-diethylcyclohexane

6.32 a) Draw *trans*-1-methyl-4-isopropylcyclohexane in planar form (no chair). Be sure to use wedge and dash lines to indicate stereochemistry.
b) Now, draw *trans*-1-methyl-4-isopropylcyclohexane in its lowest energy chair conformation.
c) Now, draw *cis*-1-methyl-4-isopropylcyclohexane in its lowest energy chair conformation.
d) Based on your answers to (b) and (c), is *trans*-1-methyl-4-isopropylcyclohexane higher in energy, lower in energy, or about the same energy as the *cis* isomer, or is there no way to tell? Explain your answer.

6.33 In the lowest energy conformation of the compound below (chair form), how many alkyl substituents are axial?

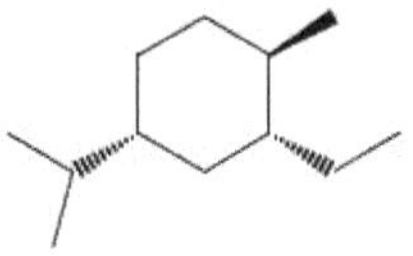

6.34 a) Draw *trans*-1-bromo-3-ethylcyclohexane in planar form (no chair). Be sure to indicate the stereochemistry with wedge and dash lines.

b) Now, draw *trans*-1-bromo-3-ethylcyclohexane in its lowest energy conformation.

c) Now, draw *cis*-1-bromo-3-ethylcyclohexane in its lowest energy conformation.
d) Draw what *cis*-1-bromo-3-ethylcyclohexane would look like if it underwent a chair flip.
e) Based on your answers for the lowest energy confirmations in (b) and (c), is *trans*-1-bromo-3-ethylcyclohexane higher in energy than, lower in energy than, or about the same energy as the *cis* isomer, or is there no way to tell? Briefly, explain your answer.

6.35 a) Draw 1,2,3,4,5,6-hexachlorocyclohexane in its chair conformation with all chloro groups in the axial positions. b) Then, draw the chair flipped version with all chloro groups in the equatorial positions.

6.36 Name the following cyclohexanes.

a) 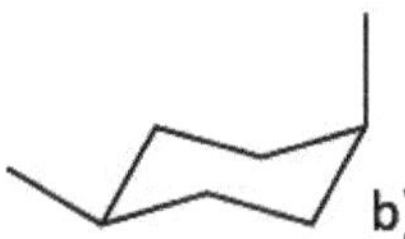 b) 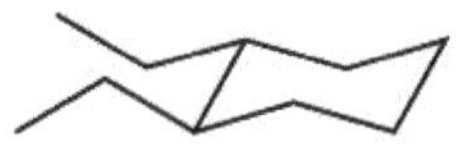 c) 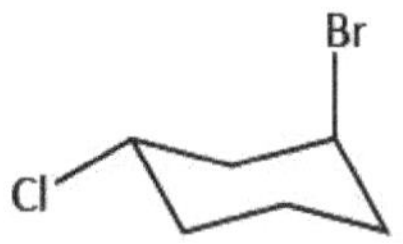

d) 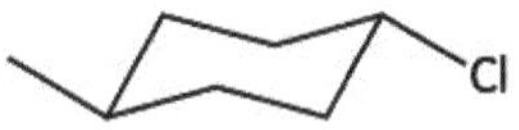

6.37 Consider 2-methylpropane. Sighting along the C2-C3 bond, draw and label the Newman projections of *both* the most stable *and* least stable conformations.

6.38 What is the relative position of the substituents A and B to each other in the following conformation? (anti, eclipsed, or gauche)

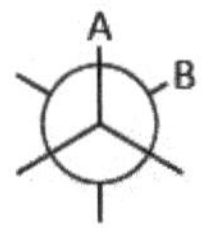

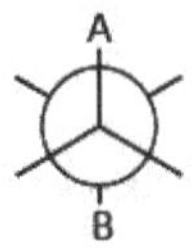

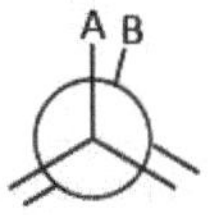

__________ __________ __________

6.39 Using Newman projections, draw all the staggered conformations of 1-chloropropane looking down carbon1-carbon2 bond. Label all anti and gauche forms.

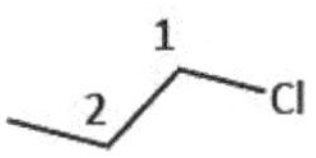

6.40 Draw the Newman projections for the following compounds in their anti-conformations.
a) pentane looking down C1-C2 bond b) butane looking down C3-C4 bond
c) 2-dimethylbutane looking down C2-C3 bond d) 2,3-dimethylhexane looking down C2-C3

6.41 a) Draw a Newman projection of the gauche conformation of 2-methyl pentane, looking down the C2-C3 bond.

b) Now, draw the Newman projection of the *most stable* conformation of 2-methyl pentane. This may or may not look like your answer for part (a).

Chapter 7. Stereochemistry

7.1 Introduction

Stereochemistry is the study of 3D structures of molecules. How are the atoms of a molecule arranged in space? This may seem boring, but it is literally a matter of life and death. In the late 1950's, thalidomide was an over-the-counter drug prescribed to pregnant women in West Germany. This drug helped with nausea associated with morning sickness. It worked well, but soon, nearly 7000 babies were born with birth defects. They had a malformation of their arms and legs. More than half of them died in infancy. Because of this, thalidomide was pulled from the market in the early 1960's.

Thalidomide is a compound that exists as two different mirror images. One of these mirror images helped fix nausea. The other mirror image caused birth defects. Many times, when a medicine is produced that has two different mirror images, one is bioactive while the other is not. After thalidomide, we now only sell the bioactive enantiomer unless it is proven that the other one does not cause harm. Now, we try to produce only one of the mirror images, or if we produce a mixture of the two, we separate them. Unfortunately, for thalidomide, it would not have mattered if only the good mirror image was sold. In the body, the good thalidomide mirror image gets converted into a mixture of the good and bad.

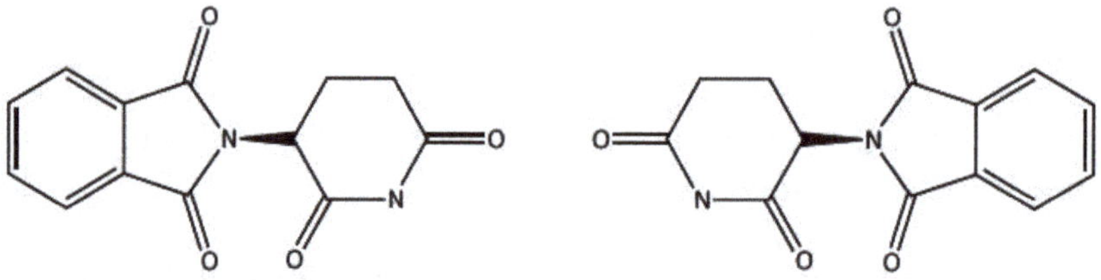

Figure 7.1 Thalidomide mirror images

7.2 Chirality - handedness

If early some morning you groggily put your right foot into your left shoe, you would be able to tell because your foot would hurt. Why? Your feet are mirror images of each other. Your shoes are mirror images of each other. But, there is a difference between your left and right foot. They are not identical. There is a difference between your left and right shoe. They are not identical. Your feet are chiral. Your shoes are chiral. They have a handedness. Hands and gloves are also chiral. You cannot put your right hand in your left glove. A chiral object has a non-superimposable mirror image. Another way to think about it is it is an object whose mirror image is different or not identical.

Not every object is chiral. Most socks are not. You are equally comfortable if you put a tube sock on either foot. It makes no difference. They do not have handedness. Socks are achiral. Achiral means "not chiral". It is an object with a mirror image that is identical. The chairs in the picture below are achiral. The two green rocking chairs are mirror images of each other. If you sit in either chair, you could tell no difference. They are identical. Not all chairs are achiral. Some desks with a writing surface are chiral. They are right or left-handed. You can tell the difference if you sit in one or the other.

Chiral

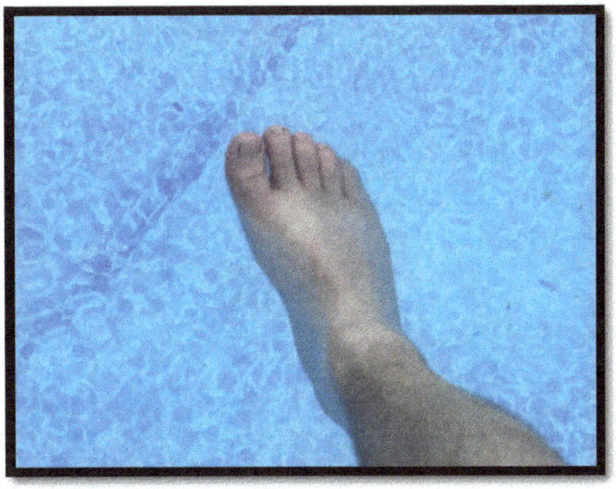

Chiral

Achiral

Chiral

Whenever you are trying to identify if an object is chiral, you compare it to its mirror image. You take the mirror image and can turn it, rotate it, or flip it any way you can. If you can make it look identical to the original object, it is NOT chiral. If you cannot make it look the same and are convinced that the mirror image is different than the original object, it is chiral.

7.1 Are the following objects chiral or achiral?

a) a pair of scissors b) a screw c) a screwdriver d) a hammer

Molecules can also be chiral or achiral. Let's look at a carbon atom with different groups on it. This is how we determine if a molecule is chiral or achiral.

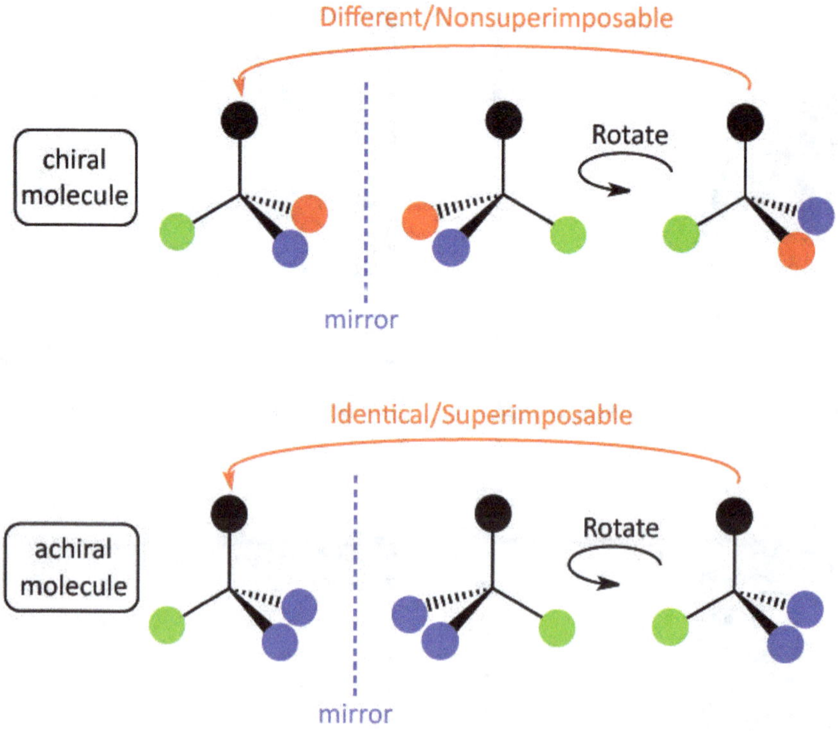

Figure 7.2 Test for chirality

The first molecule with four different groups is chiral. The second with only three different types of groups is achiral. A carbon atom with four different groups on it is called an *asymmetric carbon atom, stereocenter, or stereogenic center*. Some call it a chiral carbon or a chiral center, but the term chiral should really be reserved for entire molecules or objects.

7.2 The following molecules each have one stereocenter in them. Find the stereocenter (the one with four different groups on it) and circle it. Hint: Remember that hydrogen atoms are not always drawn.

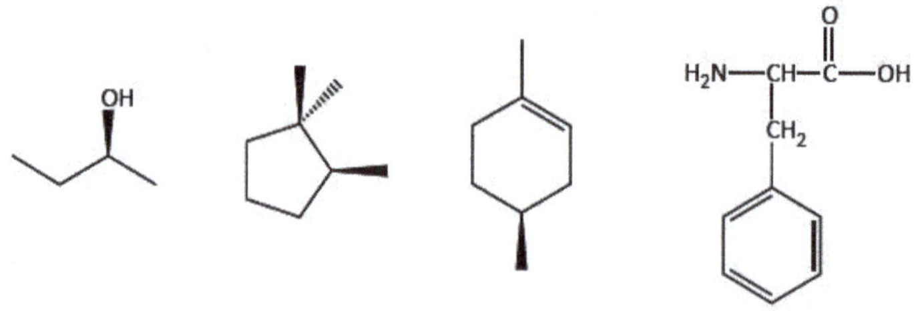

All of the previous compounds with one stereocenter are chiral. This is a big help in helping identify if a compound is chiral or not! *A compound is chiral if it has only one stereocenter*. If a compound has zero or two stereocenters, we need to look more carefully. These molecules can be either chiral or achiral. Let's look at how compounds with two stereocenters can be either. The carbon atoms labeled with * are stereocenters.

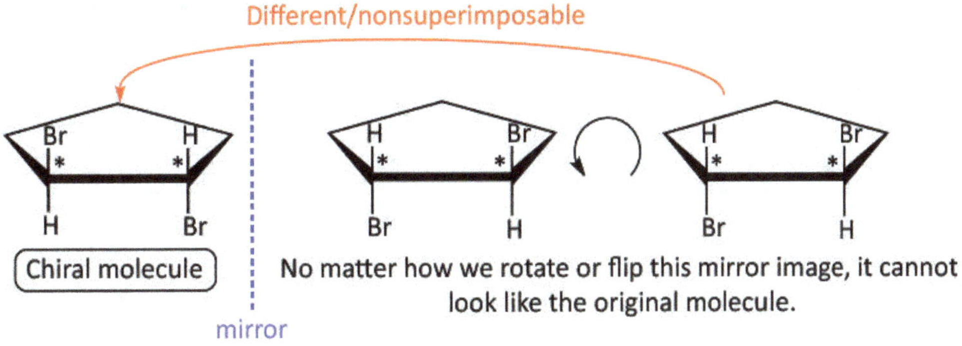

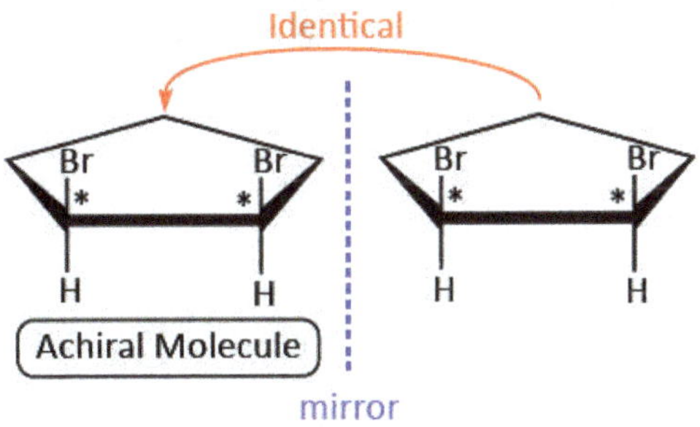

Achiral molecules with two stereocenters are usually very symmetric. If a mirror plane can be found in the middle of the molecule then it is achiral.

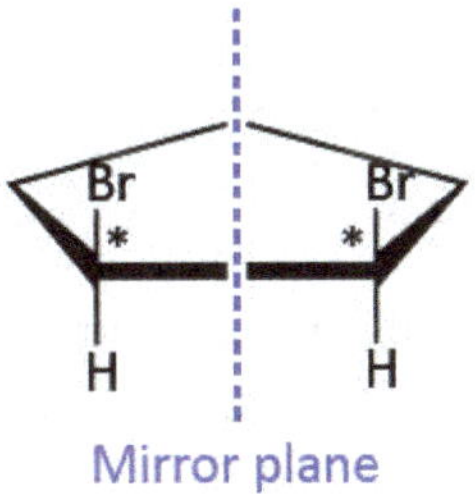

A molecule is achiral if it has a mirror plane in it.

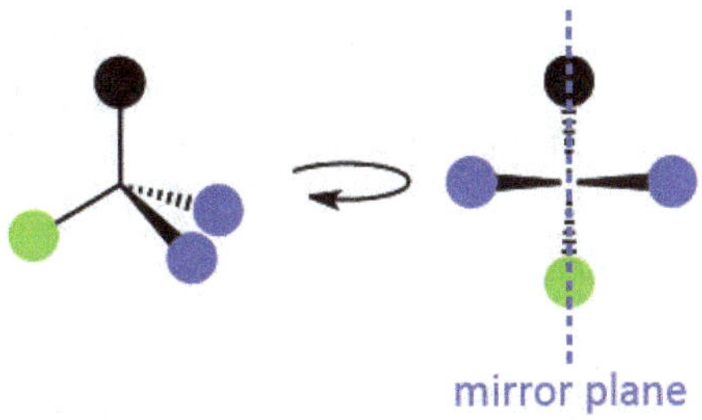

A molecule is usually achiral if it has zero stereocenters. The very few zero stereocenter molecules that are chiral have fan blade type twist in them. For example, the following molecule has zero stereocenters and is chiral.

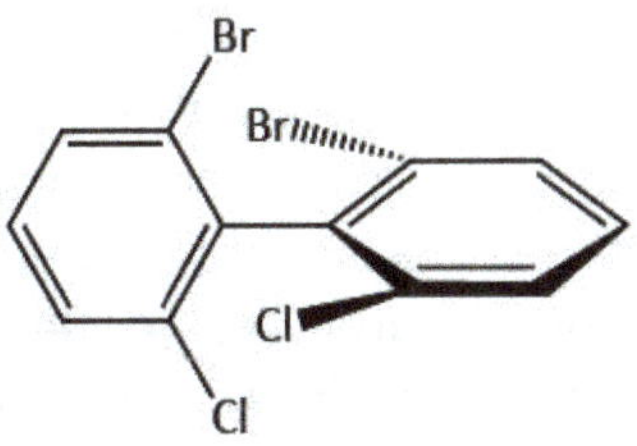

7.3 Are the following molecules chiral or achiral?

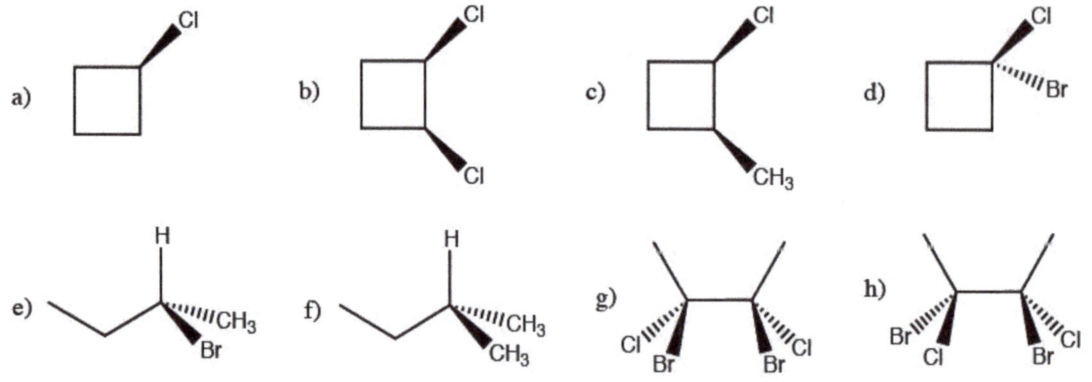

7.3 R and S Nomenclature

A method is needed to help distinguish between mirror images of a stereocenter. For instance, both of these enantiomers would be named 2-bromobutane.

2-bromobutane 2-bromobutane

But, since these two enantiomers are different compounds, they need to have different names. Robert Sidney Cahn, Sir Christopher Ingold, and Vladimir Prelog developed a system of rules to differentiate the names of compounds like these. The Cahn-Ingold-Prelog system is used to assign a letter (R) or (S) to each stereocenter of a compound based on its 3D orientation.

How is it done? Each stereocenter is considered all by itself.

1. Assign a priority to each group on the stereocenter. 1 is the highest priority while 4 is the lowest priority.
   a. We begin by looking at the first atom of the group. Higher atomic numbers are a higher priority. If the atoms are different isotopes of the same atom, the heavier isotope is a higher priority.

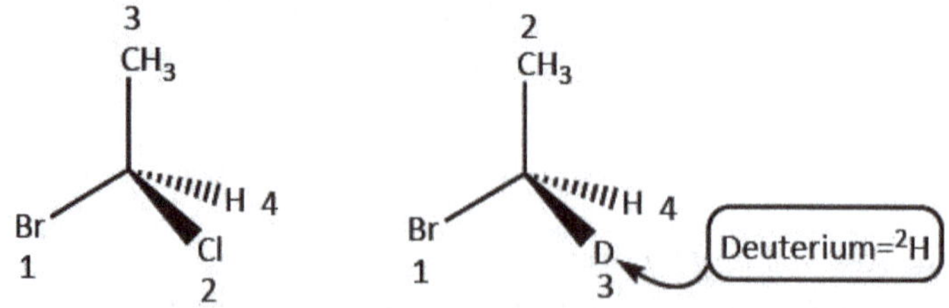

   b. If the first atoms of two groups are identical and there is a tie, we look at the next atoms along the chain in the group to break the tie. We keep going until the first place where the tie is broken. In this example, we begin by looking at the first atoms attached to the stereogenic carbon. H is obviously the lowest priority and is assigned 4. The first atoms of the other three groups are all carbon, so there is a tie. Therefore, we look at the next atoms. For the methyl, CH_3, the next atoms are three hydrogen atoms that are attached. For the ethyl, the next three atoms attached to the carbon are one carbon atom and two hydrogen atoms. For the isopropyl group, the next atoms are two carbon atoms and one hydrogen atom. This breaks the tie. (C,C,H) > (C,H,H) > (H,H,H), so the priorities of 1, 2, and 3 are assigned respectively.

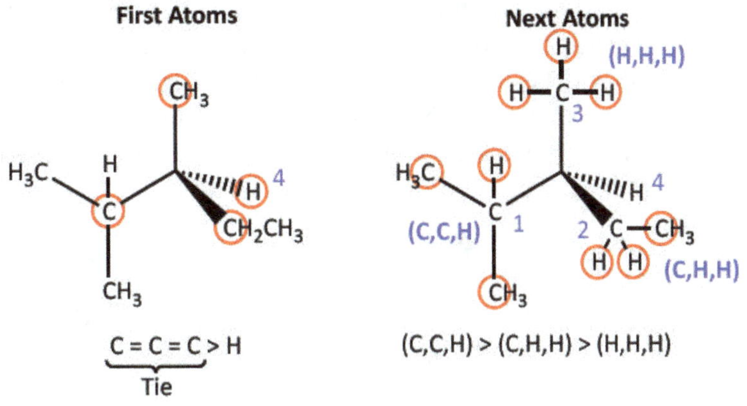

c. We count double bonds as being two atoms and triple bonds as being three atoms when assigning priority.

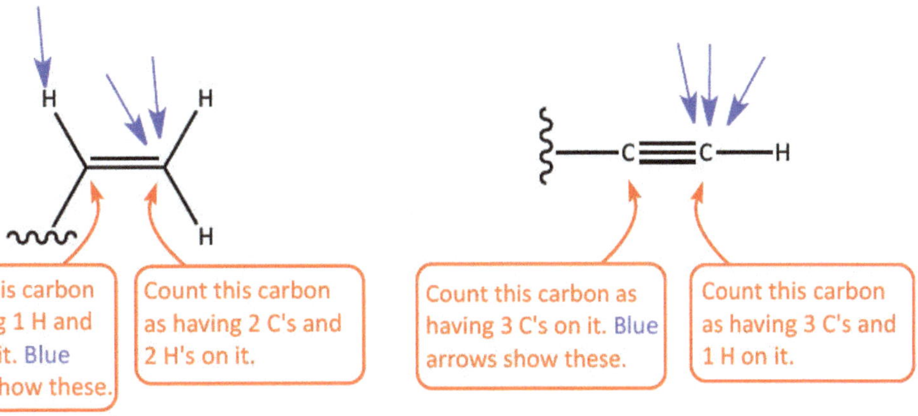

In the example, we have a tie between three carbon atoms when we look at the first atoms of the groups. When we go to the next atoms, the methyl group has three hydrogen atoms. The ethene (C=C) group has two carbon atoms (one for each bond of the double bond) and one hydrogen atom. The ethyne (C≡C) group has three carbon atoms (one for each bond of the triple bond).

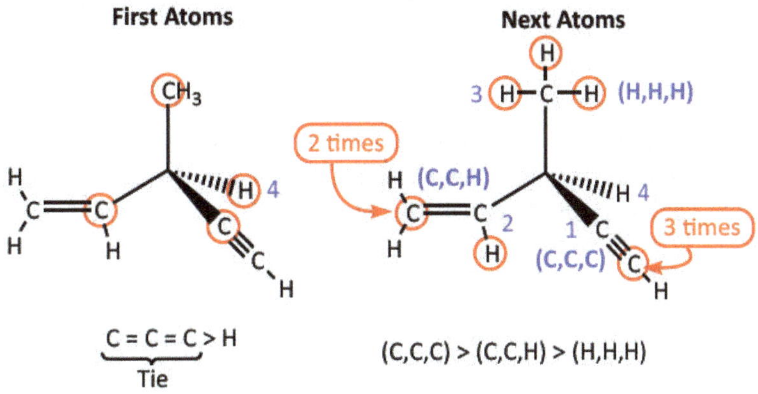

2. Take the 3D representation of the molecule and put the lowest priority group (#4) in the back. This makes groups 1-3 face you. Draw an arrow from 1 to 2 to 3. If the arrow is a clockwise arrow (to the right), then we assign an R to the stereocenter. "R" comes from the Latin word Rectus-"upright"). If the arrow is a counterclockwise arrow (to the left), we assign an S to the stereocenter. "S" comes from the Latin word Sinister-"Left"). Another way to think about it is if you turn the steering wheel of a car in the direction of an (S) arrow, the car would turn left. If you turn the steering wheel of a car in the direction of an (R) arrow, the car would turn right.

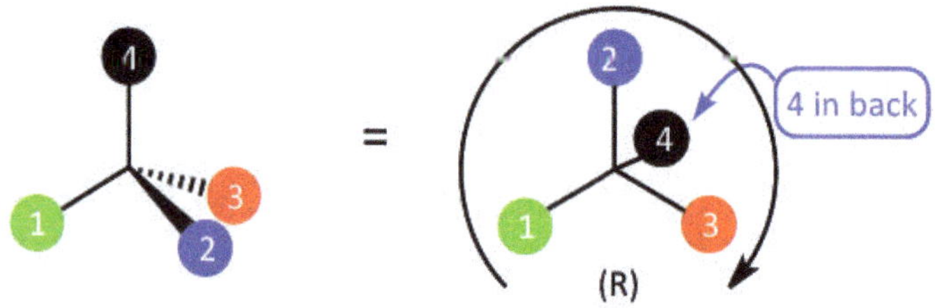

Group 4 in the back

The situation is very easy if group 4 is already in the back. We can draw an arrow from 1 to 2 to 3 and see if it is an S arrow or an R arrow.

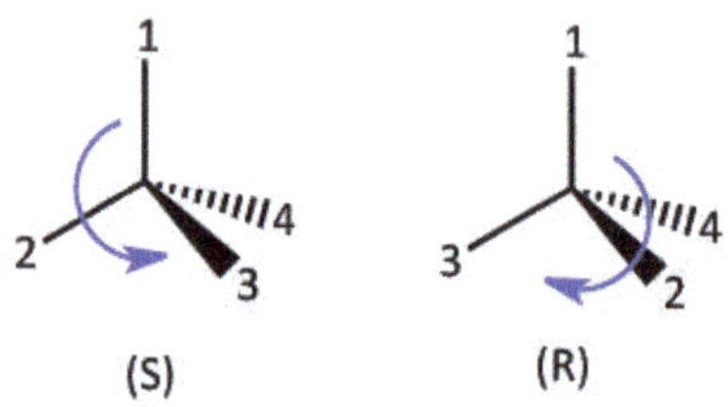

Group 4 in the front

If group 4 is in the front, it is still fairly easy. There are three ways you can do it. Pick your favorite.

1. One way to tackle this problem is to rotate the molecule so group 4 is in the back. Then the arrow is drawn to determine (R) or (S). This takes a bit of mental gymnastics, though.

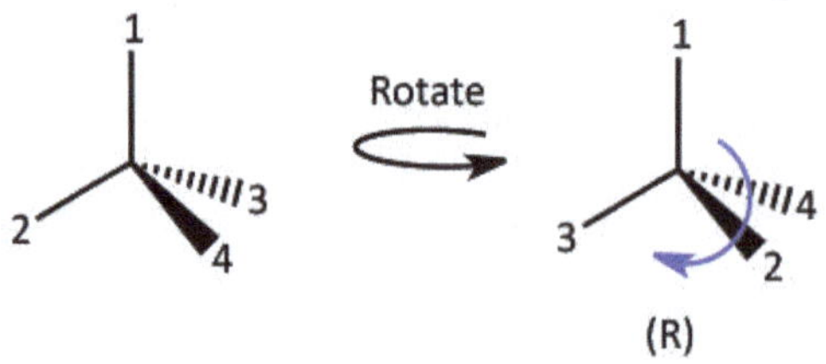

2. Another way to tackle the problem is to take yourself, in your mind's eye, for a walk behind the page. From behind the page, the molecule would appear to have group 4 in the back. You can look at the arrow from that position and it would appear as an (R) arrow. This also takes a bit of mental gymnastics.

3. Probably the easiest way to tackle this problem is to leave the molecule as it is with group 4 in the front. Draw the arrow from 1 to 2 to 3. In this case, it is an (S) arrow. Since group 4 is in the front instead of the back where it should be, the arrow needs to be flipped. We know it is really an (R) arrow.

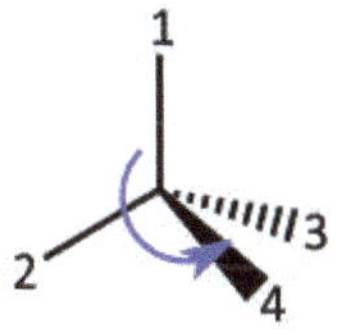

(S) arrow, but 4 is in the front. If we rotate 4 to the back, the arrow reverses and becomes an (R) arrow.

Group 4 to the side

If group 4 is to the side, it is the most difficult situation. Here are two ways to figure it out.

1. One way is to rotate the molecule. For simple molecules, this is possible, but for bigger more complicated molecules, it is more difficult.

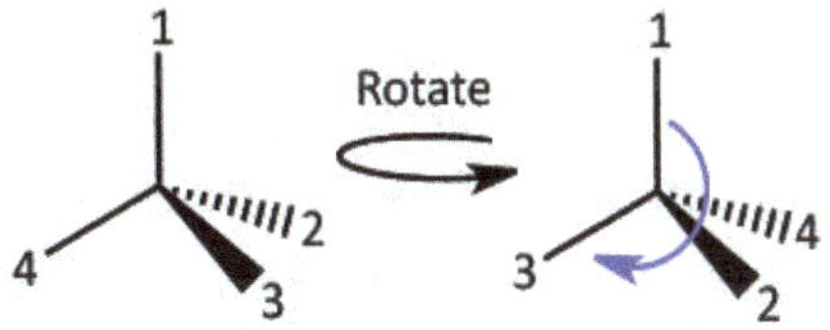

(R)

2. A second way is to take a walk in your mind's eye. Position yourself so that group 4 is in the back and take a look at the arrow. In this case, the arrow goes from 1 at the top to 2 in the back to 3 in the front. This is an (R) arrow.

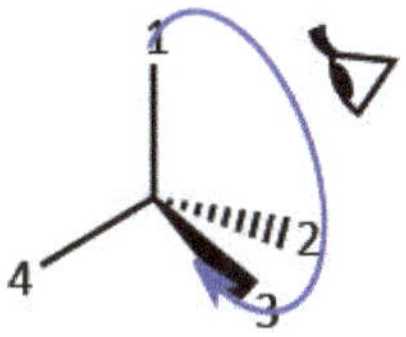

(R)

Carefully think through the following compounds and make sure you see why they are named the way they are.

Example:

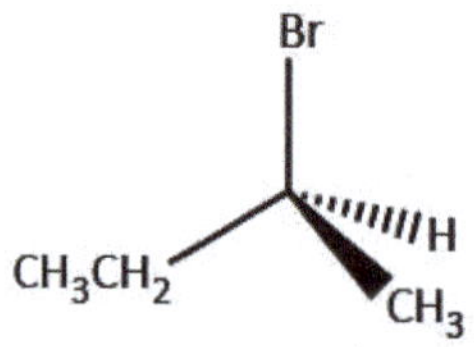

(S)-2-Bromobutane

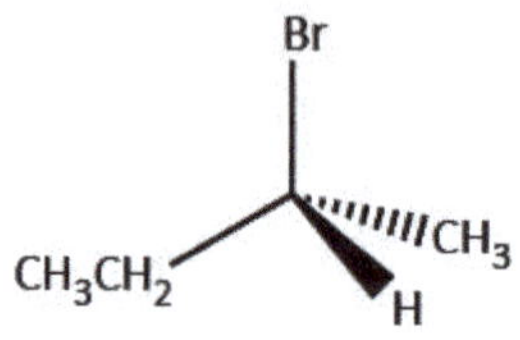

(R)-2-Bromobutane

7.4 Find the stereocenters in the following molecules and label them as R or S.

a) 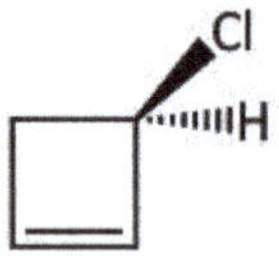

b) 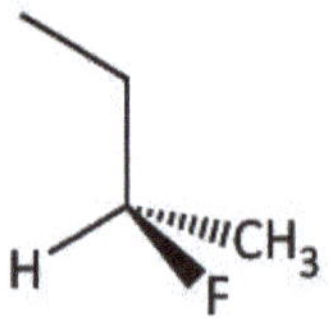

c) 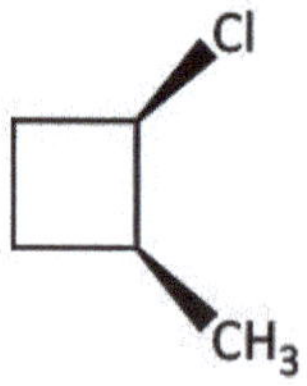

d) 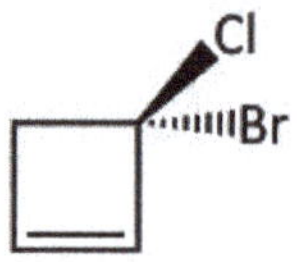

e) 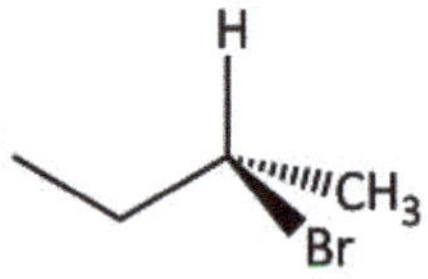

f) 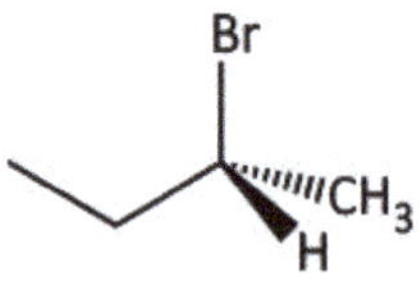

g) 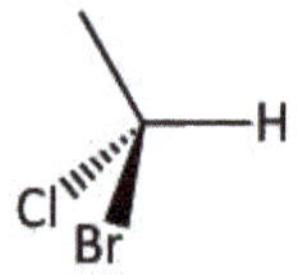

h) 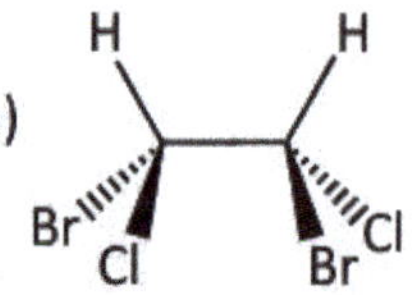

7.4 Fischer Projections

When we are paying attention to stereochemistry, we need to pay attention to which direction groups are pointing. Wedges and dashes become important to us. It also makes for messy and tedious drawings at times. There is a simpler method that can be used to draw three-dimensional drawings in a two-dimensional manner. In order to draw a Fischer projection, the molecule is twisted so all of the horizontal bonds come towards the observer. When the Fischer projection is drawn, the vertical lines are all in the plane of the page or pointing away from us. The horizontal lines represent groups that are coming towards us. One way to think about it is every set of arms on a Fischer projection are pointing towards us as if they want to hug us.

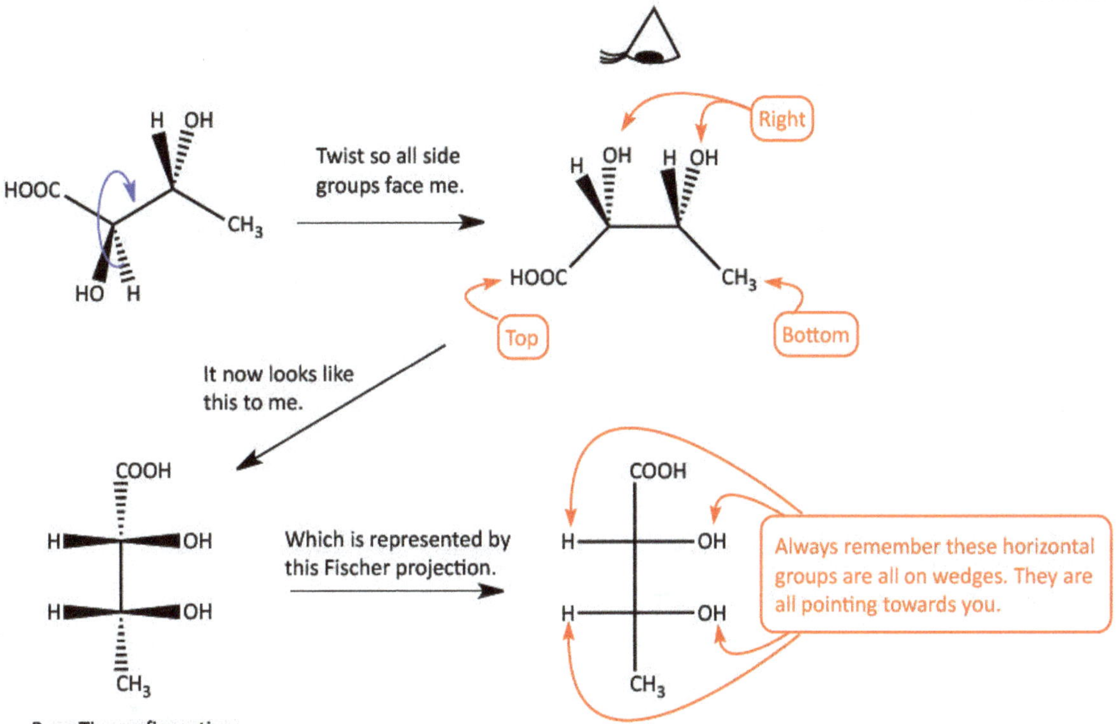

Bow-Tie configuration
Notice it is "hugging" us.

Fischer Projection

Fischer projections are commonly used in biology and biochemistry to represent sugars and carbohydrates. For example, here are the Fischer projections for two common sugars.

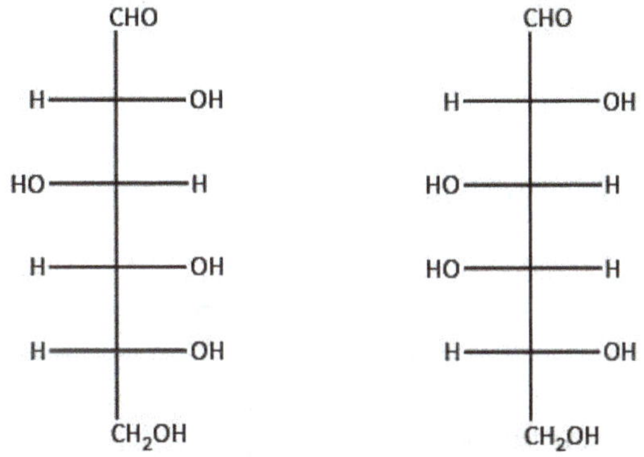

D-Glucose D-Galactose

Figure 7.3 Fischer projections of D-glucose and D-galactose

Fischer projections are helpful in determining if a compound is chiral or not. They make finding mirror planes (sigma planes) much easier. If a mirror plane is found in a molecule, it is achiral.

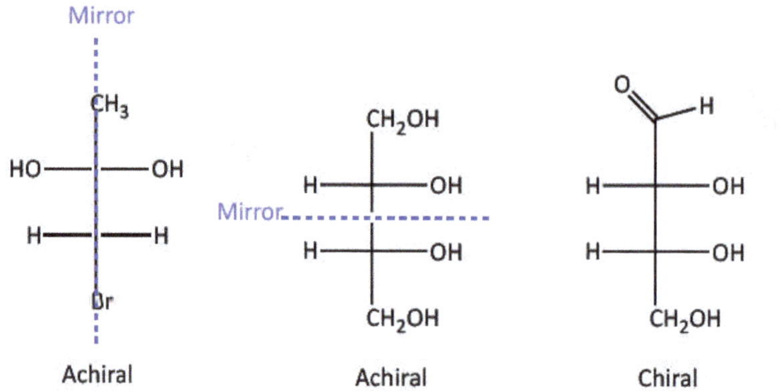

Figure 7.4 Chiral and achiral Fischer projections

7.5 Are the following molecules chiral or achiral?

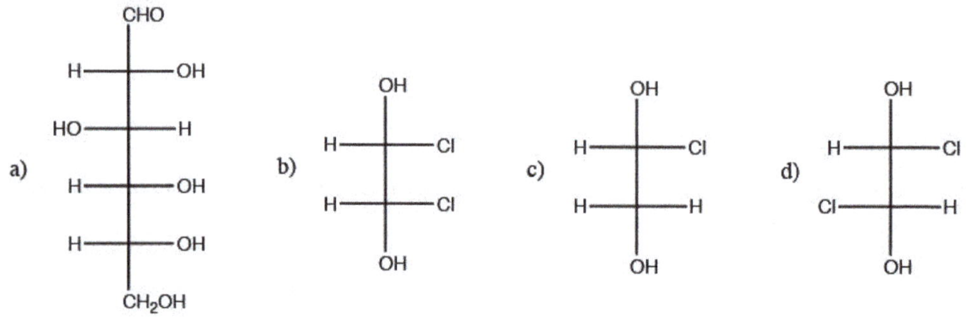

7.4.2 (R) and (S) in Fischer Projections

If a carbon atom in a Fischer projection has four different groups on it, that carbon atom is a stereocenter and can thus be given an (R) or (S) designation. When assigning an (R) or (S) designation, we must remember that if the lowest priority group (4) is on a horizontal line, that is a wedge. That group is coming towards us. For the (R) or (S) designation, 4 needs to be in the back. The simplest way to work this problem is to draw the arrow from 1 to 2 to 3, and then reverse the arrow to get our (R) or (S) designation.

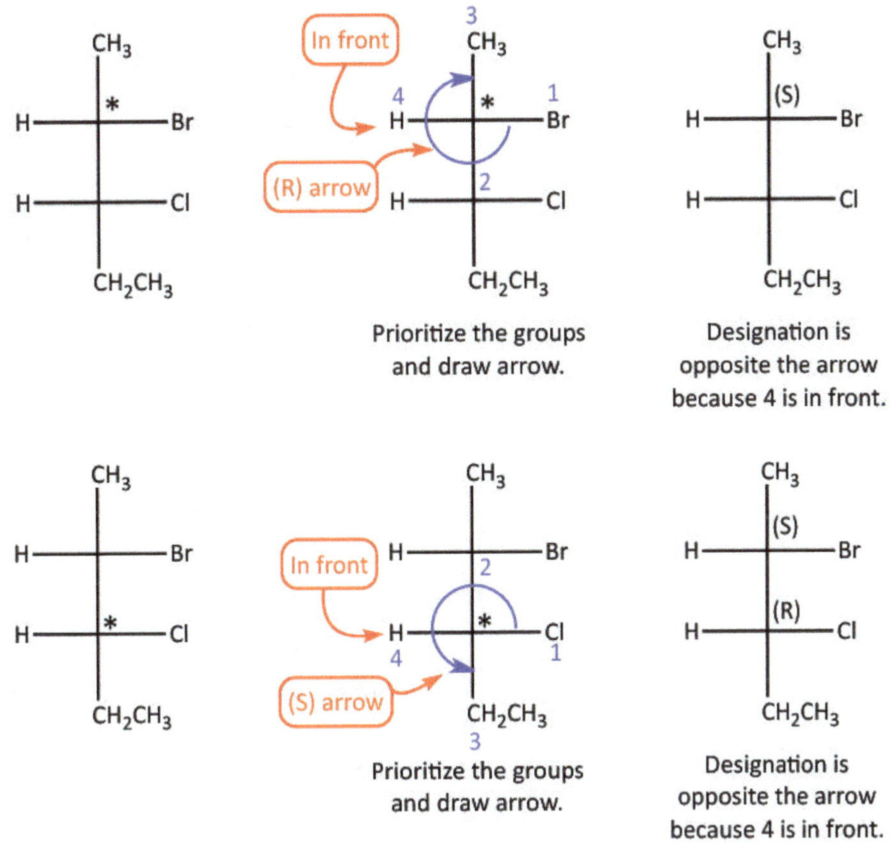

Figure 7.5 R and S of Fischer projections

This compound is called (2S,3R)-2-bromo-3-chloropentane.

7.6 Identify each of the stereocenters of D-glucose as R or S. The * indicates the stereocenters.

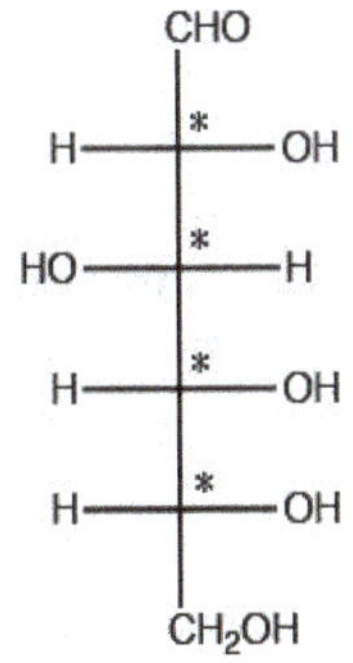

7.4.3 Compounds with 2 or more stereocenters

Let's look at all of the possible stereoisomers of 2-bromo-3-chlorobutane and how they are related to each other. Two of the stereoisomers, (2S,3R) and (2R,3S) are enantiomers. They are non-identical mirror images of each other. (2R,3R) and (2S,3S) are also enantiomers. They are non-identical mirror images of each other.

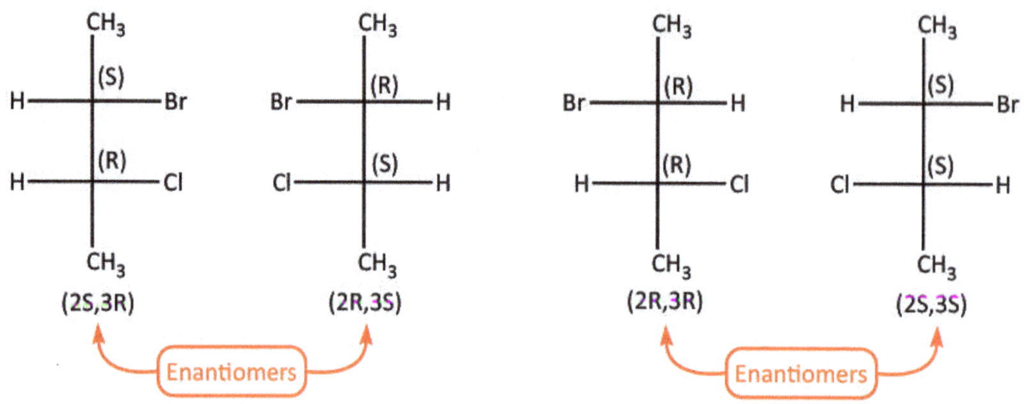

But, how are (2R,3S) and (2R,3R) related to each other? They are stereoisomers of each other, but they are not mirror images of each other. There is a name for this relationship. They are *diastereomers* of each other. *Diastereomers* are stereoisomers that are not mirror images of each other.

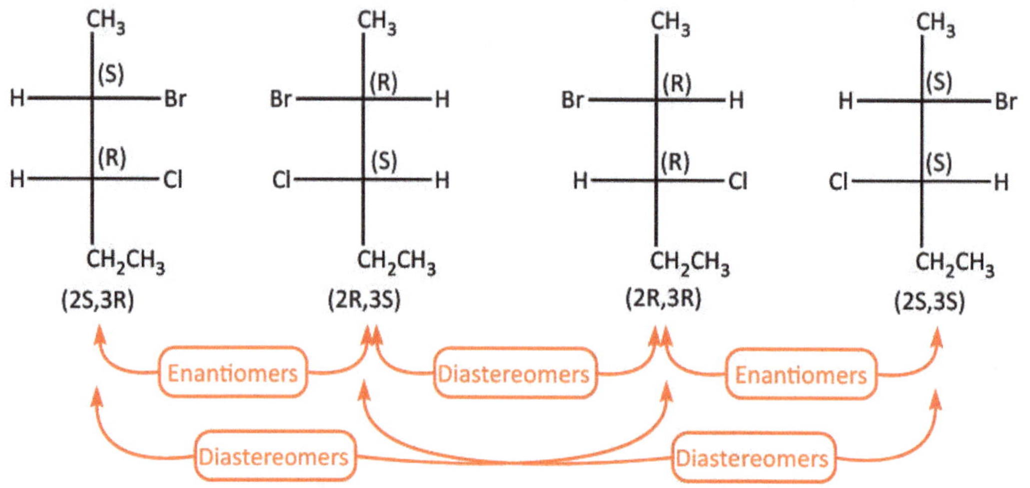

Figure 7.6 Enantiomers and Diastereomers

7.4.4 Other examples of diastereomers.

Geometric isomers, also called cis-trans isomers, are another example of diastereomers. They are stereoisomers that are not mirror images of each other.

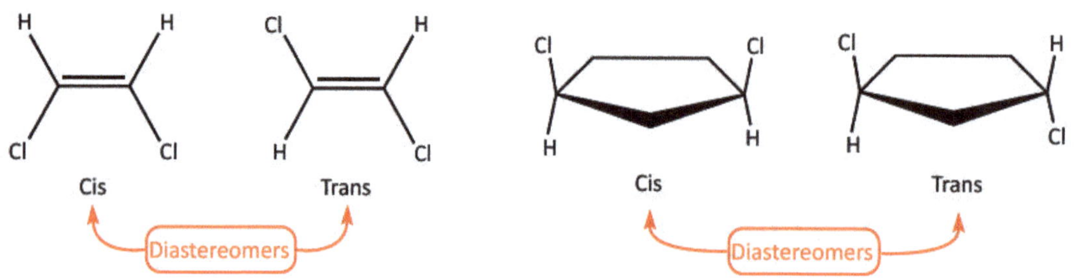

Figure 7.7 Geometric isomers are also diastereomers

7.7 Label the following compounds as enantiomers, diastereomers, the same compound, or neither.

a) 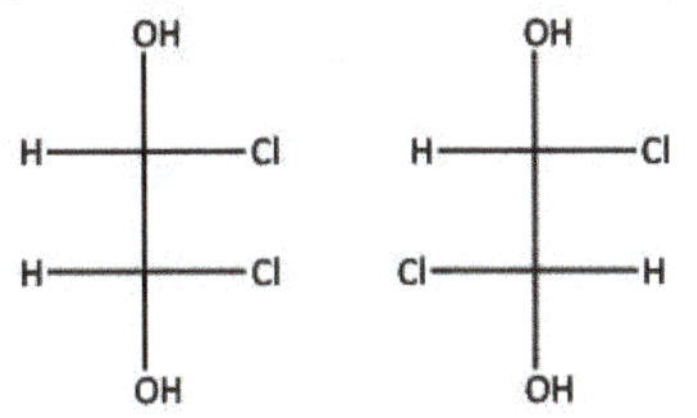

b) 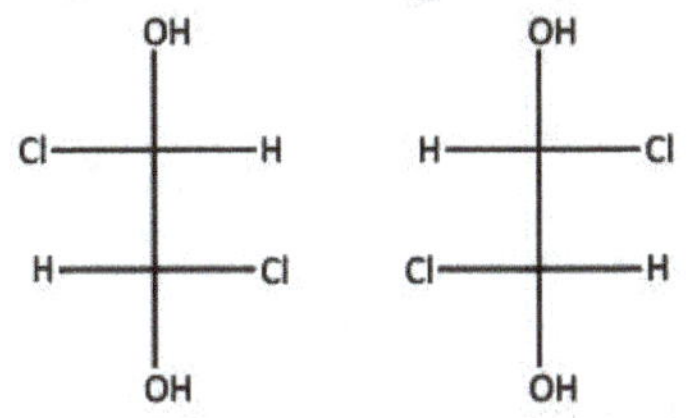

c) 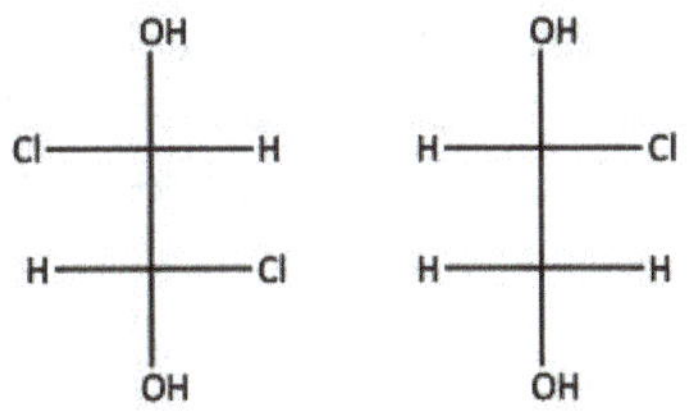

d) 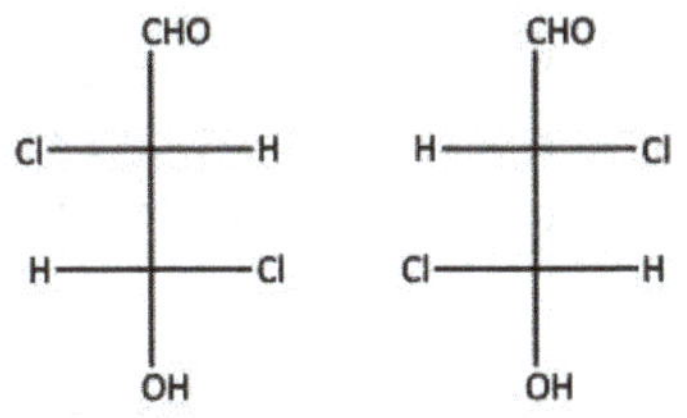

7.4.5 Meso Compounds

Compounds that have stereogenic atoms but are achiral are called *meso compounds*. These compounds typically have stereogenic atoms, but they have a mirror plane of symmetry to prevent them from becoming chiral.

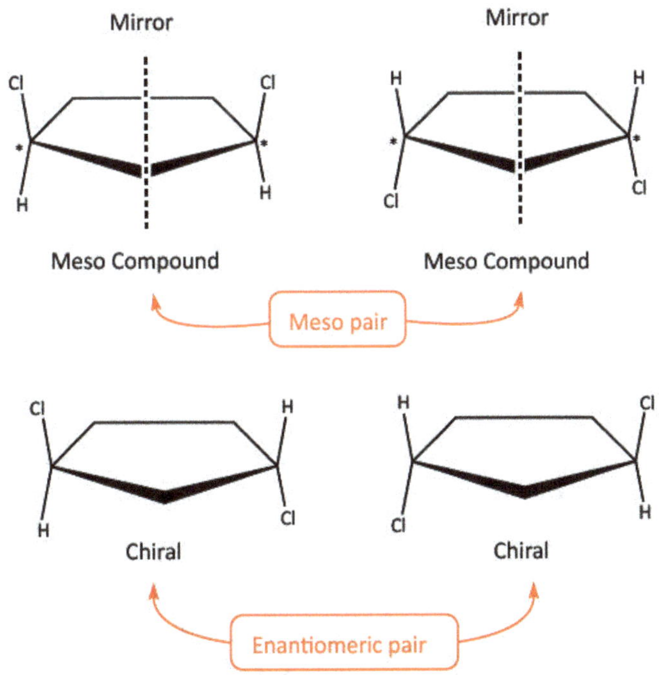

Figure 7.8 Meso vs. enantiomeric pair

7.8 Circle each of the following compounds that is a meso compound.

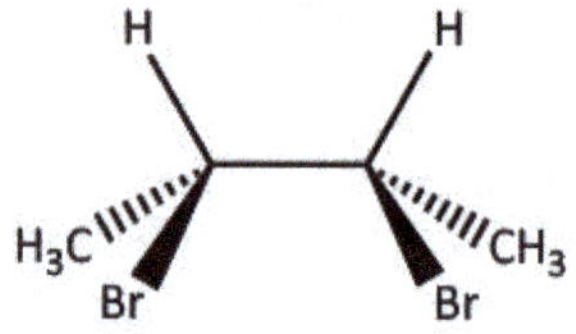

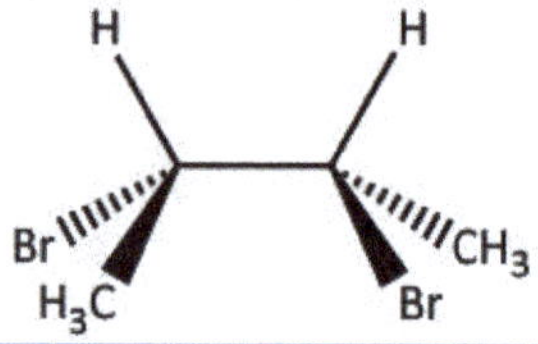

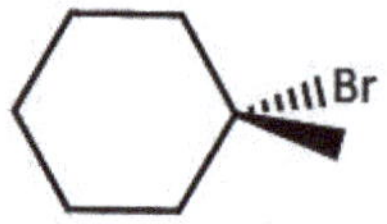

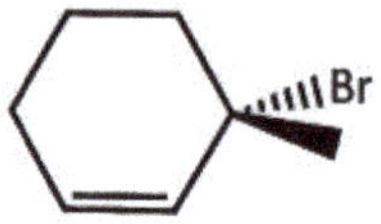

7.5 Types of isomers

Isomers are compounds that have the same molecular formula but are not the exact same compound. We have learned about several types of isomers. The following flow chart will help you understand how they fit together. This is something that takes some time to learn well. Study the flowchart. Make sure you can think of examples of each type of isomer.

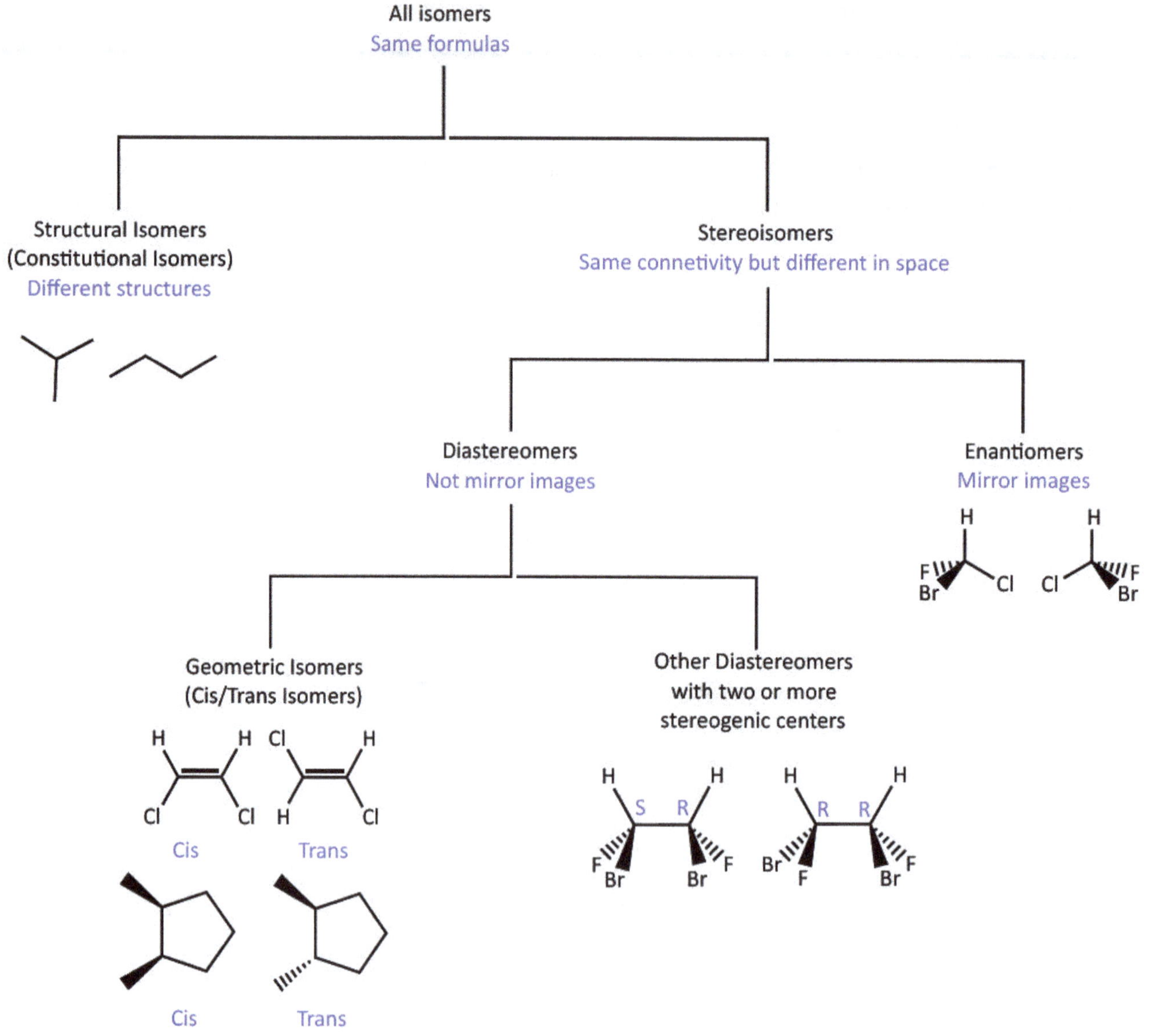

Figure 7.9 Types of isomers

7.9 Identify each of the following pairs as structural isomers, enantiomers, geometric isomers, other diastereomers, or not isomers.

a) 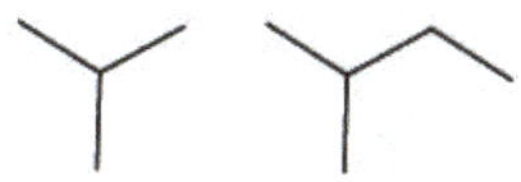

b) 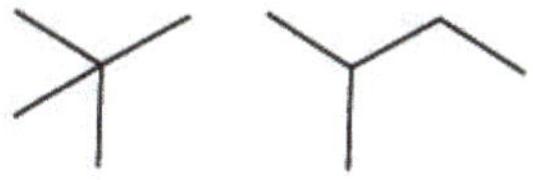

c) 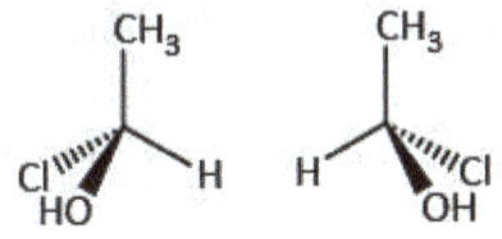

d) 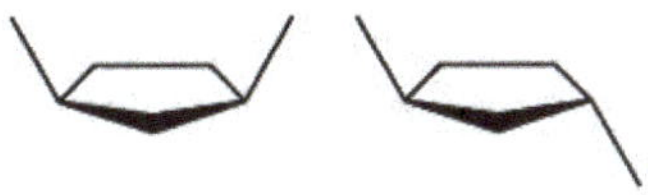

e) 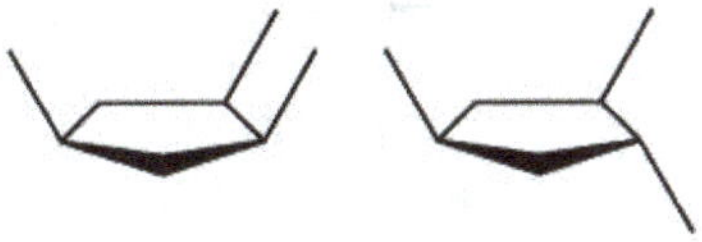

f) 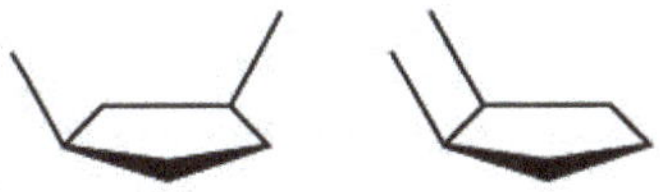

7.6 Optical Activity

There are very few differences between enantiomers of chiral compounds. They have the same melting points, the same boiling points, and the same IR spectra. One difference they do have is that they rotate plane-polarized light in different directions. This rotation of plane polarized light is called optical activity. This rotation of plane polarized light is detected in an instrument called a polarimeter.

7.6.1 Plane Polarized Light

Most light you encounter is not polarized. The electromagnetic waves vibrate in all directions. But, if this unpolarized light passes through a polarized filter, the light vibrates in only one direction. This light that vibrates in only one direction is plane-polarized light. Polarizing filters that are aligned in parallel allow light to pass through them. If polarizing filters are aligned perpendicularly, light cannot pass through them. This is useful for us to analyze how the plane-polarized light is rotated.

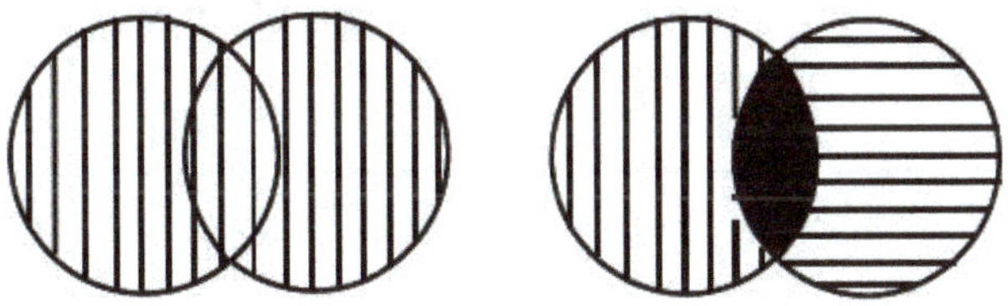

Figure 7.10 Polarized lenses aligned parallel and 90° to each other

7.6.2 Polarimeter

A polarimeter is an instrument that lets us tell how a chiral sample rotates plane-polarized light. A light source emits unpolarized light. This light passes through a polarizing filter. The plane-polarized light transmits through a sample tube containing the chemical sample. Typically, the chemical sample is an enantiomer or a mixture of enantiomers. The rotated plane-polarized light that transmits through the sample tube is invisible; we cannot see it with our eyes. The way it is analyzed is by using a polarizing filter. This analyzing filter is rotated until a maximum amount of light comes through it. This angle we turn this filter is equal to the angle of the plane-polarized light.

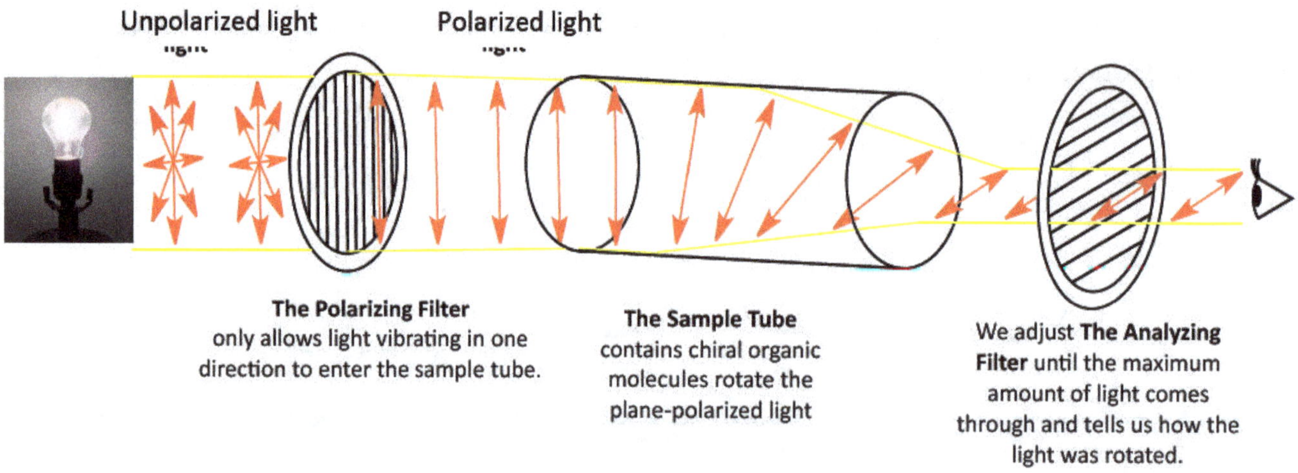

Figure 7.11 A polarimeter

If an enantiomer in the sample tube rotates the plane-polarized light to the right (clockwise), that enantiomer is called dextrorotatory (Latin: *dexter* = right). These enantiomers may also be called d or +. If an enantiomer in the sample tube rotates the plane-polarized light to the left (counterclockwise), the enantiomer is called levorotatory (Latin: *laevus* = left). These enantiomers may also be called l or -.

7.6.3 Why does an enantiomer rotate plane-polarized light?

When plane-polarized light enters an enantiomer, it encounters an unsymmetrical set of atoms with an unsymmetrical amount of electrons around the atoms. This helical electron density that is encountered rotates the electromagnetic wave of the plane-polarized light. The mirror image enantiomer has the exact opposite helical electron density and therefore rotates the plane-polarized light in the opposite direction (with the same amount of rotation). The helical electron density is what determines the rotation of light. This rotation must be determined experimentally. The (R) or (S) designation does not tell us which way the plane-polarized light will rotate. This is because the (R) or (S) designation is an accounting trick. It doesn't necessarily correspond to electron density. So, don't assume the optical rotation unless the experiment is done.

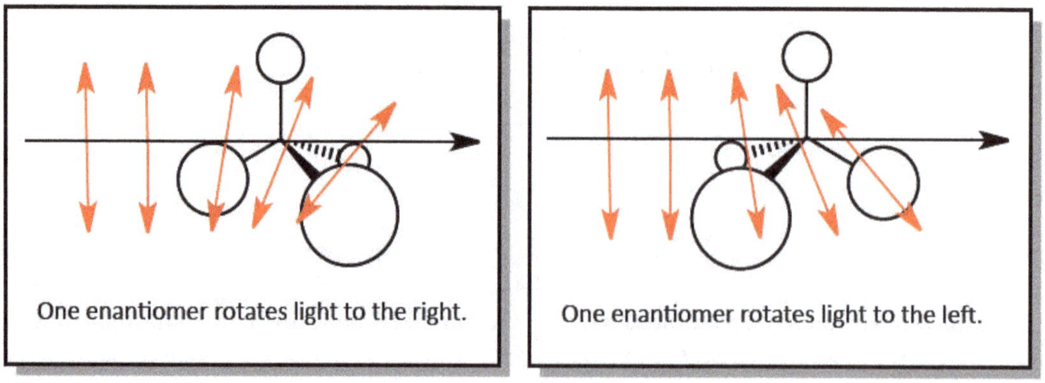

Figure 7.12 Enantiomers rotate plane polarized light in opposite directions

7.6.4 Specific Rotation

How much an enantiomer rotates plane-polarized light can be useful information about a compound. The length of the sample tube affects the amount the light rotates (the longer the tube is, the more it can rotate the light).

The concentration of the enantiomer solution in the sample tube also affects the amount the light rotates (the more concentrated it is, the more it will rotate the light). Chemists have devised a standard way to measure this rotation. This standardized rotation is called *specific rotation* [α]. If the specific rotation is being very carefully measured, the color of light used and the temperature can affect the rotation. For example, you may see the specific rotation for a dextrorotatory enantiomer written as:

$$[\alpha]_D^{25} = +15°$$

The superscript, 25, means the rotation was observed at a temperature of 25°C. The subscript, D, means the D line of a sodium lamp (λ = 589.6 nm) was used. This is a yellow color of light.

The dextrorotatory enantiomer always has a + specific rotation. The levorotatory enantiomer always has a – specific rotation.

Typically, the polarimeter contains a 1-decimeter (10-centimeter) long sample tube. The concentration of the enantiomer in the tube is 1 g/mL.

If these conditions vary, the following equation can be used to correct for it.

$$[\alpha] = \frac{\alpha}{c \cdot l}$$

[α] = specific rotation (degrees)
α = observed rotation (degrees)
c = concentration (in g/mL)
l = the length of the sample tube in decimeters (1 decimeter = 10 centimeters)

If the dextrorotatory form of an enantiomer has a specific rotation of $+15°$, the levorotatory form of the mirror image enantiomer has a specific rotation of $-15°$.

Example:

Question: The specific rotation of (S)-2-butanol is -13.5°. What is the specific rotation of (R)-2-butanol?

Answer: (R)-2-butanol = +13.5°. We would call this (R)-(+)-2-butanol.

7.10 Without looking it up, would we predict (R)-2-bromobutane to be dextrorotatory or levorotatory?

7.6.5 Racemic mixture

What happens if we have an equal amount of the dextrorotatory and levorotatory enantiomers in the polarimeter's sample tube? A solution with equal amounts of d and l enantiomers is called a *racemic mixture* or a *racemate*. A racemic mixture of 2-butanol could be called (±)-2-butanol or (*d,l*)-2-butanol. If there is a racemic mixture of 2-butanol, *d*-2-butanol would rotate the plane-polarized light to the right and *l*-2-butanol would rotate the plane-polarized light equally to the left. The net effect will be a 0° rotation of the light.

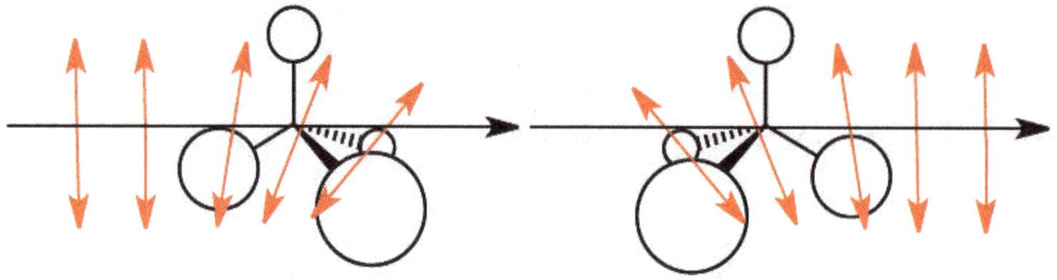

l-enantiomer rotates light to the left

d- enantiomer rotates light to the right. If there are equal amounts of *d* and *l*, it ends up at 0° rotation.

Figure 7.13 How racemic mixtures bend plane polarized light

7.6.6 Enantiomeric Excess

A racemic mixture of two enantiomers rotates plane polarized light 0°, but what happens if there is not an equal mixture of the two enantiomers? What if there is a more of one enantiomer than the other. In this case, there is an excess of one enantiomer. If there is more of the *d*-enantiomer, the light will rotate more to the right and will give an overall rotation to the right (+). If there is more of the *l*-enantiomer, the light will rotate more to the left and will give an overall rotation to the left (-). How much more of one enantiomer there is over the other is called the *enantiomeric excess, e.e.*

The enantiomeric excess (e.e.) is measured by the following equation:

$$enantiomeric\ excess\ (e.e.) = \frac{amount\ of\ one\ enantiomer - amount\ of\ other\ enantiomer}{total\ amount\ of\ both\ enantiomers}\ X\ 100$$

Example 1:

There is 98% of the *d*-enantiomer and 2% of the *l*-enantiomer in a mixture. What is the enantiomeric excess of *d* over *l*?

$$e.e. = \frac{(98\% - 2\%)}{100\%}\ X\ 100 = 96\%\ e.e.$$

Example 2:

A mixture contains 7 grams of the *l*-enantiomer and 3 grams of the *d*-enantiomer. What is the enantiomeric excess of *l* over *d*?

$$e.e. = \frac{(7g - 3g)}{10g}\ X\ 100 = 40\%\ e.e.$$

Example 3:

A pharmaceutical company developed a new drug that was produced in 98% e.e. The *d*-enantiomer was produced in excess. This drug mixture contains how much of the *d*-enantiomer and how much of the *l*-enantiomer?

Answer: There is 99% of the *d*-enantiomer and 1% of the *l*-enantiomer.

$$e.e. = \frac{(99\% - 1\%)}{100\%} X\ 100 = 98\%\ e.e.$$

Notice, the difference between the two percentages is 98%. The common wrong answer a student might make here is 98% of *d* and 2% of *l*. This is NOT the case as this mixture would give 96% e.e, not 98%.

7.6.7 How can we experimentally measure the e.e. of a mixture?

If we know the specific rotation of a compound, we can find the enantiomeric excess using a polarimeter.

For example, (R)-(+)-2-butanol has a specific rotation of +13.5°. If we have a racemic mixture of (±)-2-butanol (equal amounts), the rotation in the polarimeter would be 0°. If we have 100% of (+)-2-butanol, the rotation in the polarimeter would be +13.5°. What if the polarimeter gave a result of +5.4°? We would know that there is a little more of (+)-2-butanol than (-)-2-butanol because the rotation is positive. We can tell that the polarimeter only rotated 40% of the way to +13.5° (+5.4/+13.5 X 100% = 40%). The e.e. is therefore 40%. So, the excess of (+) over (-) is 40%. We could quickly determine that we have 70% of (+)-2-butanol and 30% of (-)-2-butanol.

$$e.e. = \frac{(70\% - 30\%)}{100\%} X\ 100 = 40\%\ e.e.$$

7.11 What is the enantiomeric excess of a mixture of 80% dextrorotatory enantiomer and 20% levorotatory?

7.12 What is the enantiomeric excess of a mixture of 1.2 grams of levorotatory enantiomer and 0.8 grams of dextrorotatory?

7.13 What percentage of the dextrorotatory enantiomer is in a sample that is 50% enantiomeric excess of the *d*-enantiomer?

7.7 Enantiomers and biochemistry

The way enantiomers interact with plane-polarized light is not the only way enantiomers act differently. As was mentioned earlier with thalidomide, enantiomers can interact with the human body in different ways. The reason enantiomers behave differently is that much of the human body is chiral. Our amino acids are chiral. We only have one enantiomer of amino acids in our body. These enantiomeric amino acids make enantiomeric proteins and enantiomeric enzymes. One enantiomer of a compound can fit into an enzyme, but the mirror image enantiomer cannot.

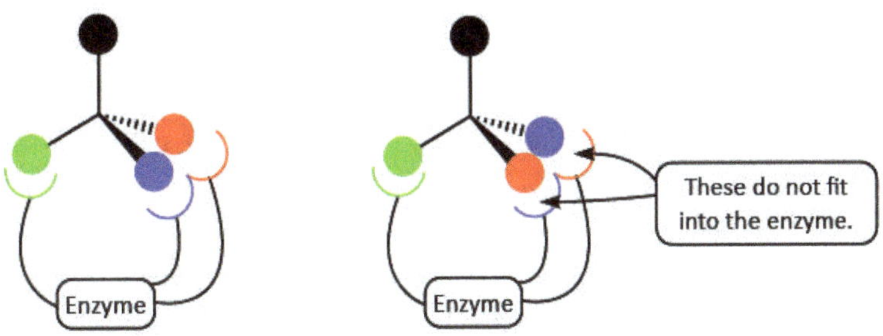

Figure 7.14 Activity of enzyme enantiomers

Enantiomers of compounds can even smell differently to us. (R)-carvone smells like spearmint while (S)-carvone smells like caraway seeds. (R)-limonene smells like oranges while (S)-limonene smells like lemons.

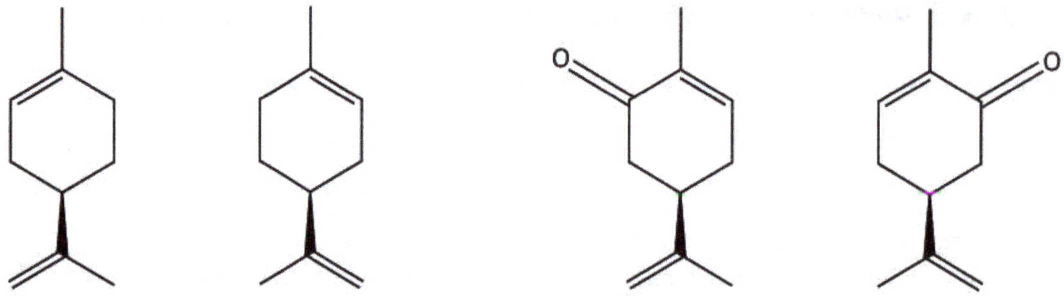

(R)-limonene orange odor | (S)-limonene lemon odor | (R)-carvone spearmint odor | (S)-carvone caraway seed odor

Figure 7.15 Enantiomers of limonene and carvone

We need to be very careful when making medicines because of this effect in the human body. If a chiral compound helps cure a disease, we need to make sure its enantiomer is removed from the sample given to people, a very expensive proposition, or does not cause harm.

7.8 Problems

7.14 Identify each set of compounds as conformational isomers (conformers), constitutional isomers (structural isomers), stereoisomers, or identical.

a) *cis*-1,3-dibromocyclohexane vs. *trans*-1,4-dibromocyclohexane b) CH_2=CH-CH_3 vs. CH_3-CH=CH_2

c) 2,3-dimethylhexane vs. 2,3,3-trimethylpentane d)

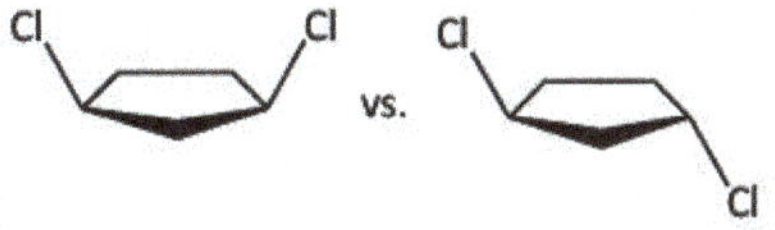

e)

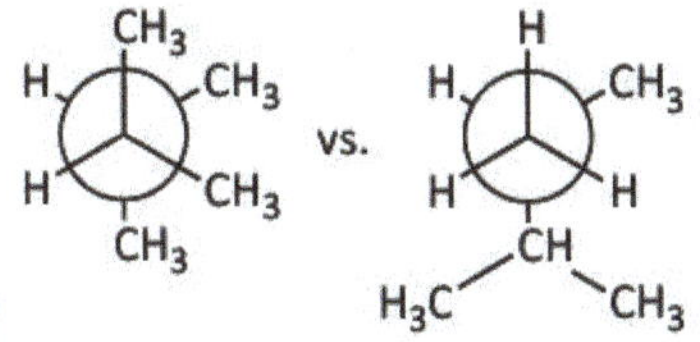

f)

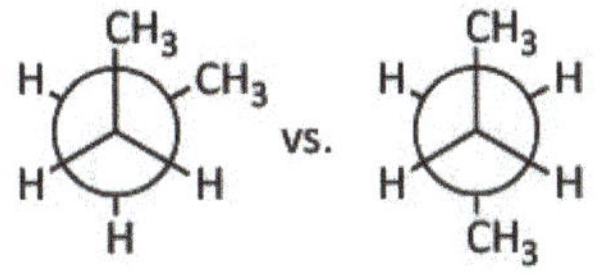

g)

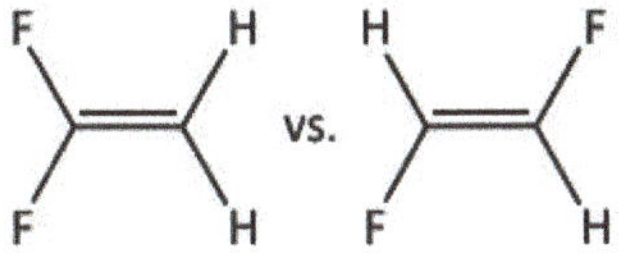

h)

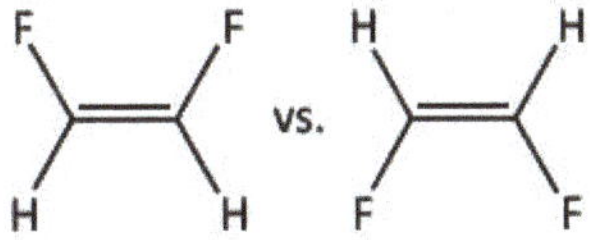

i)

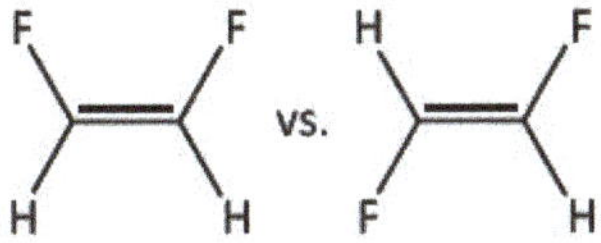

j)

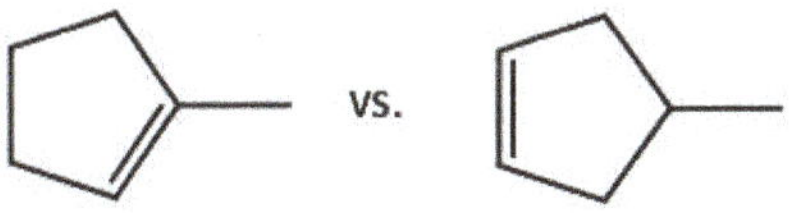

7.15 This molecule is a form of D-glucose. In each box, indicate if the arrow points to a carbon that is R, S, or NA. Use NA for not applicable for non-stereogenic carbon atoms.

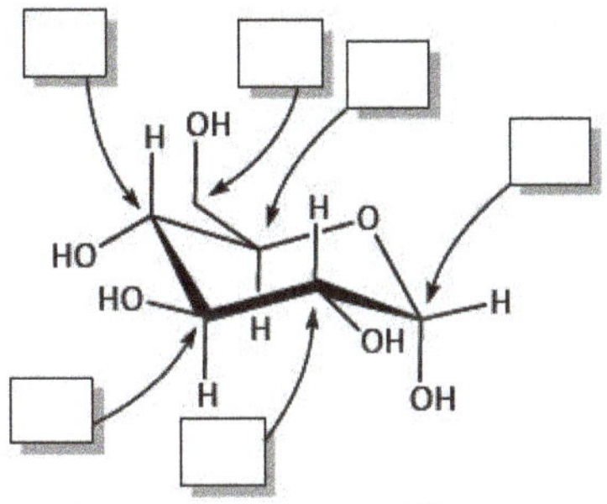

7.16 For each of the following compounds, identify whether the stereogenic atom is R or S.

a)

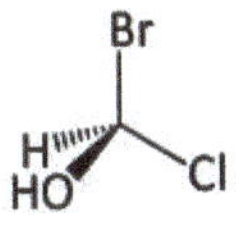

b)

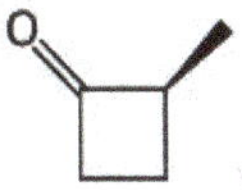

c)

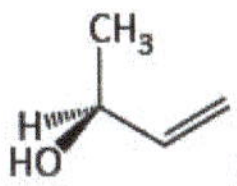

d)

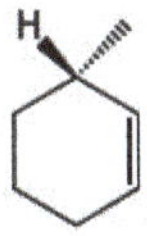

e)

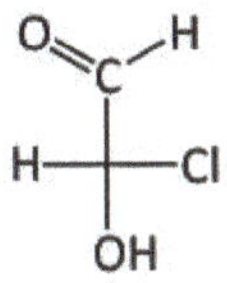

f)

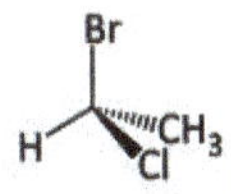

g)

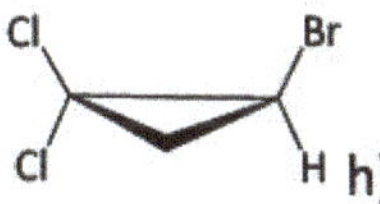

h)

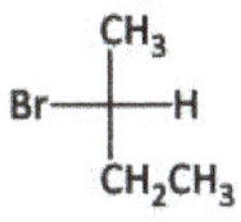

7.17 Draw the following compounds and identify the molecules as chiral or achiral.

a) (R)-2-chloropentane b) (2R,3S)-2-chloro-3-methylpentane c) (2R,3R)-2,3-dichlorobutane
d) (2R,3S)-2,3-dichlorobutane e) *meso*-1,2-dibromo-1,2-dichlorocyclobutane f) (S)-1-bromo-1-chloroethane
g) (1R,3S)-1,3-dibromo-1,3-dichloropropane h) (S)-1-bromo-1-chloro-1-fluoro-1-iodomethane

7.18 Circle each of the following compounds that is a meso compound.

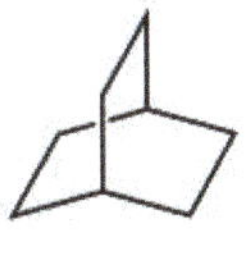

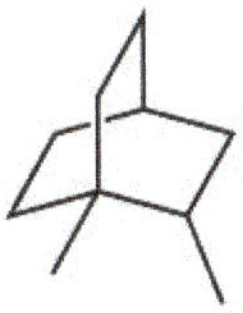

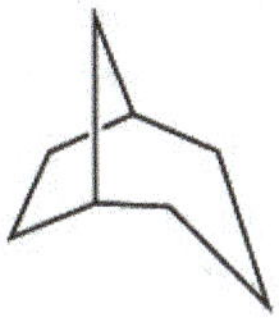

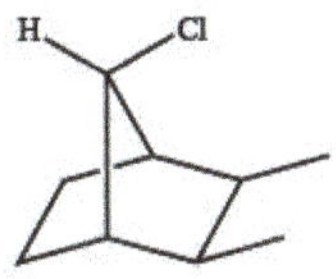

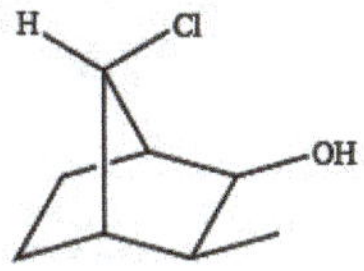

Stereochemistry

7.19 Identify each of the following molecules as chiral or achiral.

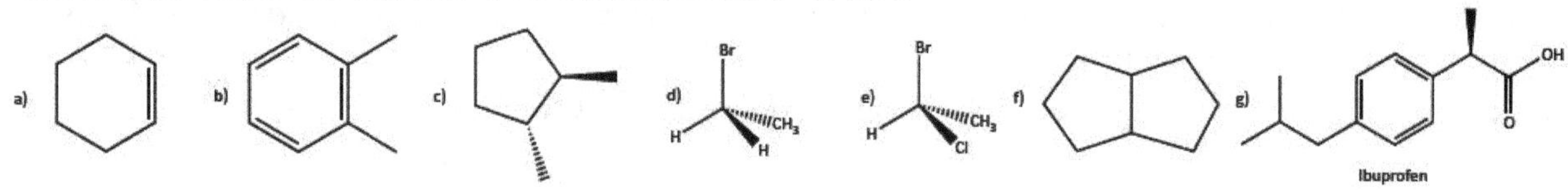

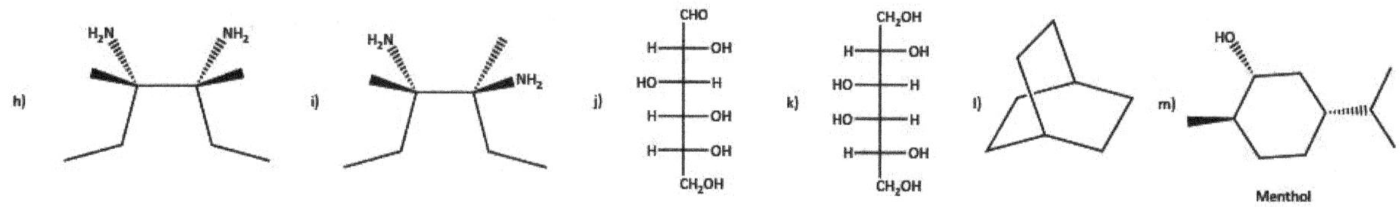

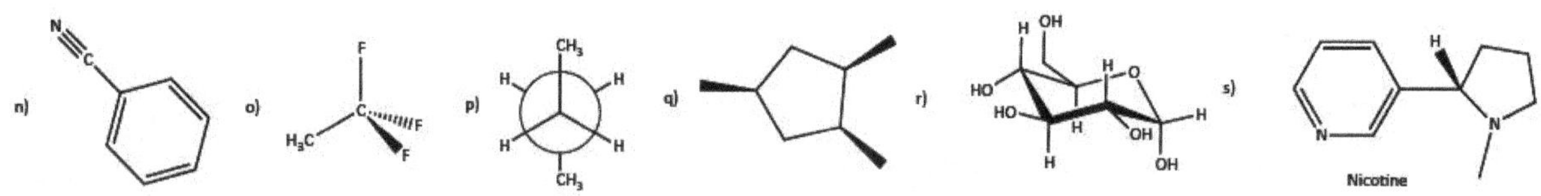

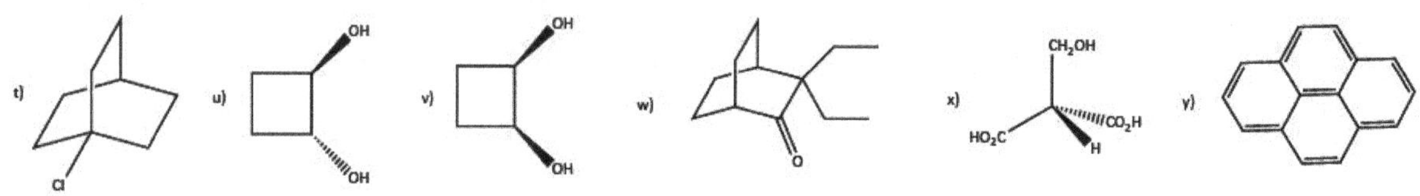

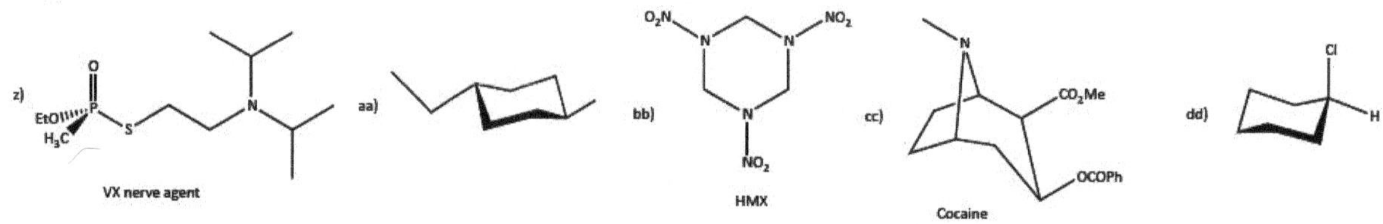

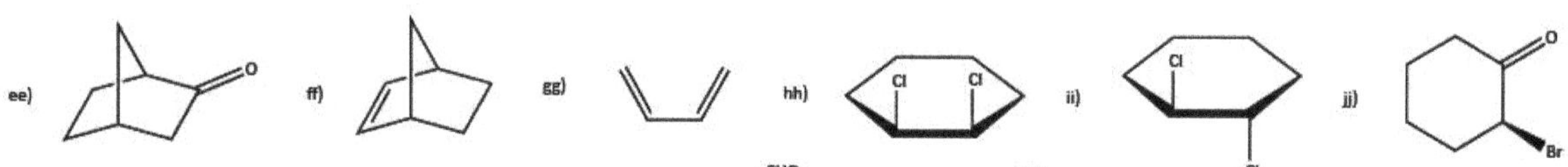

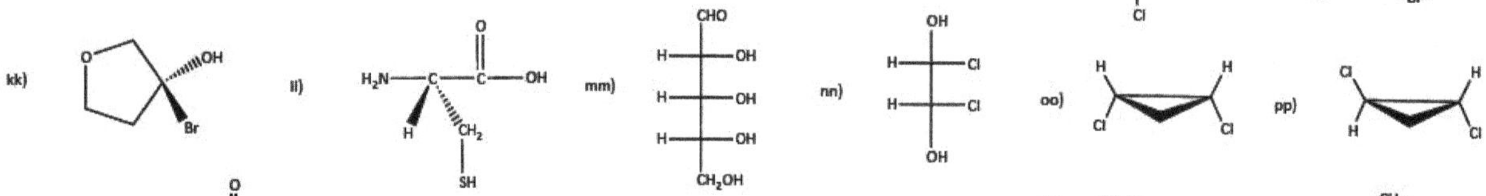

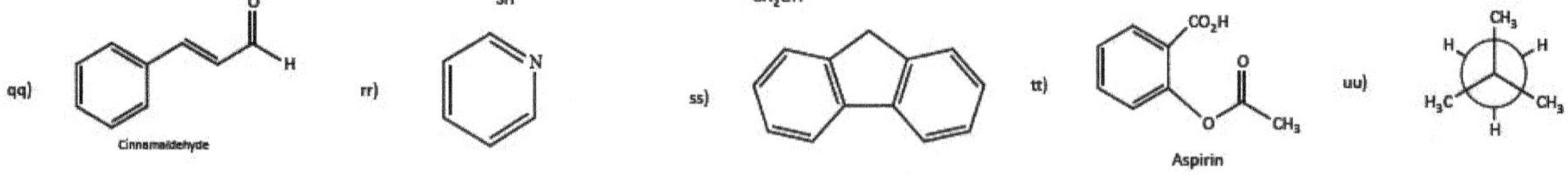

7.20 A reaction was performed giving the S-enantiomer in 97% e.e. Which of the following is true?

a) 97% of the S- enantiomer and 3% of the R- enantiomer were formed.
b) 97% of the R- enantiomer and 3% of the S- enantiomer were formed.
c) 98.5% of the S- enantiomer and 1.5% of the R- enantiomer were formed.
d) None of these is correct.

7.21 Draw the enantiomer (if one exists) for the following compounds.

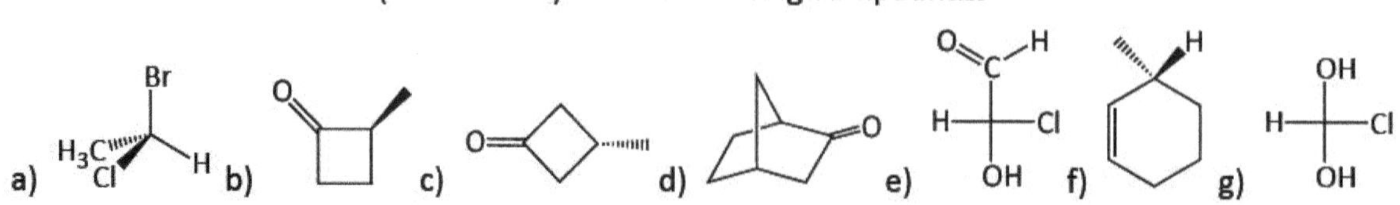

7.22 (R)-glyceraldehyde has a specific rotation of +8.7°,

a) What is the specific rotation of (S)-glyceraldehyde?
b) A mixture of 75% of (S)-glyceraldehyde and 25% of (R)-glyceraldehyde was made. What is the e.e. of this mixture?
c) What is the specific rotation of the mixture in part (b)?
d) Because (R)-glyceraldehyde rotates plane polarized light to the right, +8.7°, it can also be called what two names? (Use (d)-, (l)-, (+)-, or (-)-.)
e) A mixture of 6.0 grams of (R)-glyceraldehyde and 4.0 grams of (S) glyceraldehyde has an e.e. of _____% and a specific rotation of _____°.
f) A mixture of equal amounts of (R)-glyceraldehyde and (S)-glyceraldehyde has an e.e. of _____%. This type of mixture is called a ______________.
g) If a mixture of glyceraldehyde enantiomers has a specific rotation of -6.96°. What is its e.e.?

7.23 Give the following pairs of compounds the appropriate label from the following: enantiomers, diastereomers, structural isomers, or the same compound.

a) 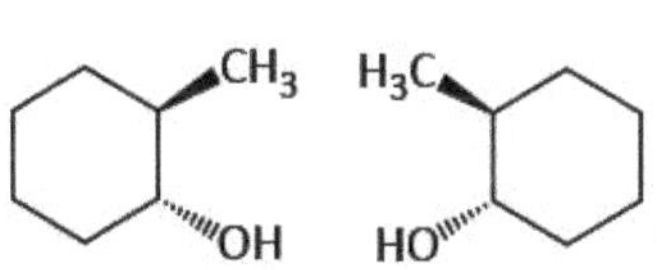
b) 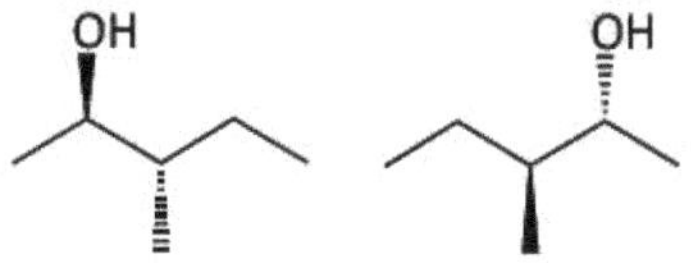
c) 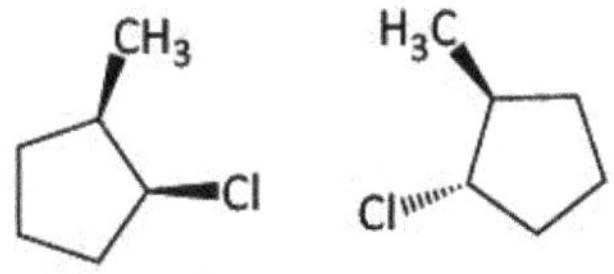
d) 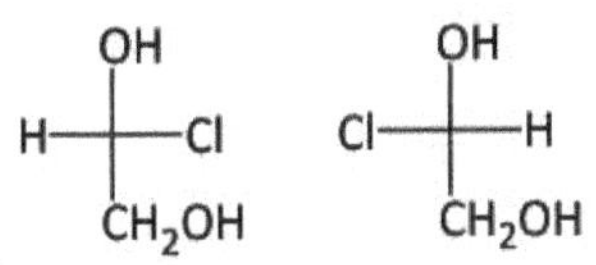
e) 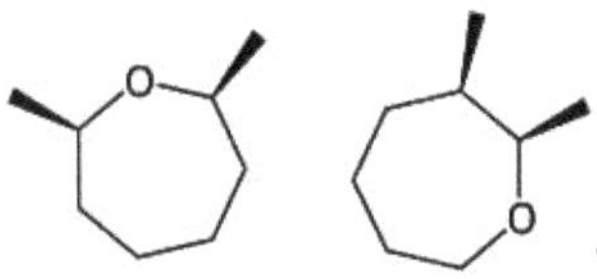
f) 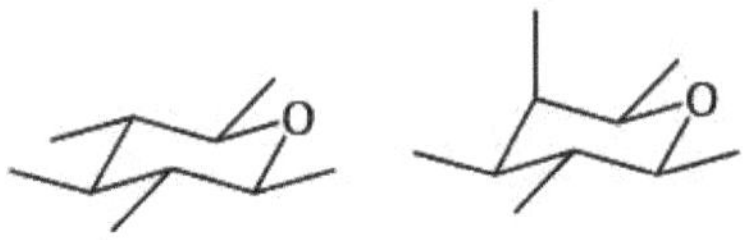
g) 
h) 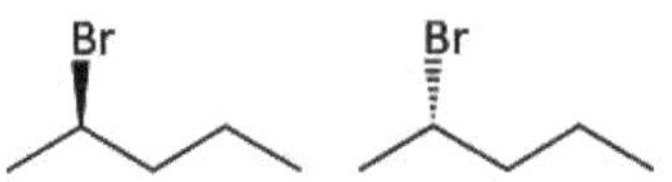
i) 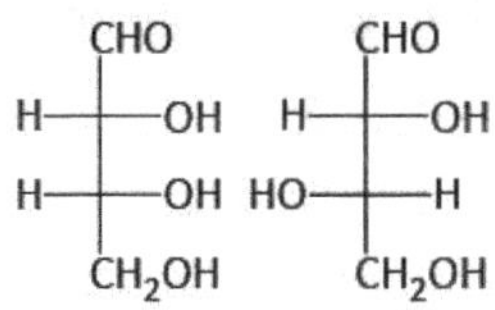
j) 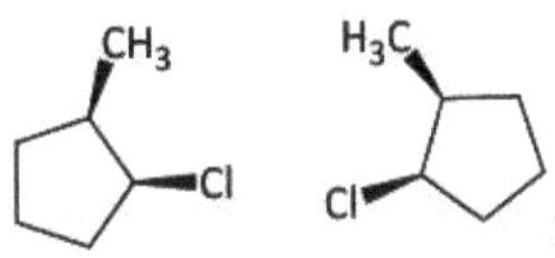
k) 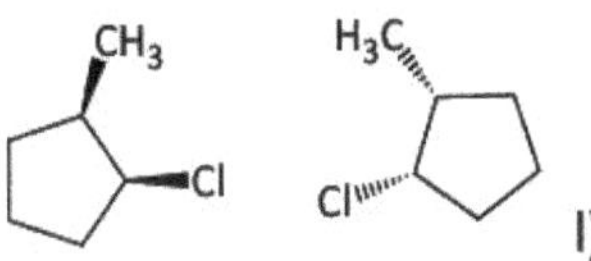
l) 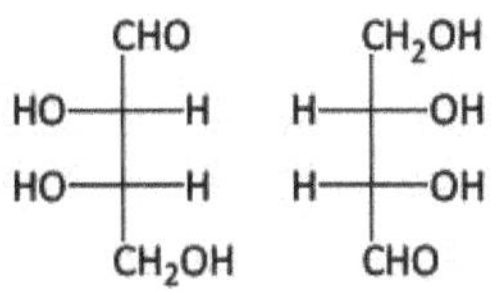
m) 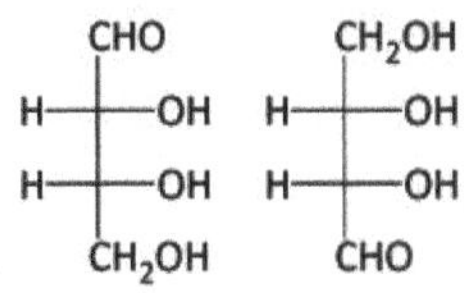
n) 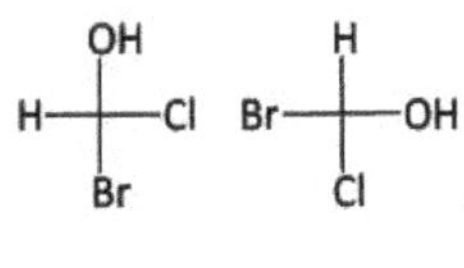
o) 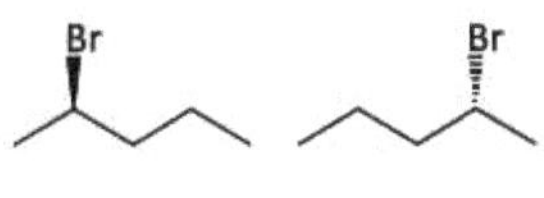
p) 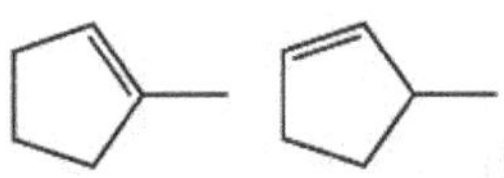
q) 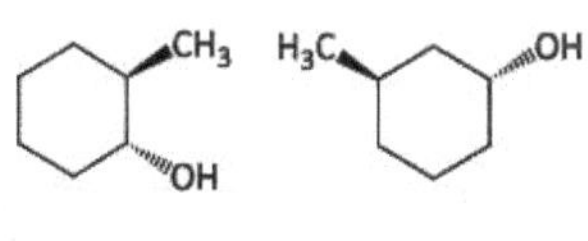

7.24 Name the following compounds. Their Fischer projections are shown. Use R and S as appropriate.

a) 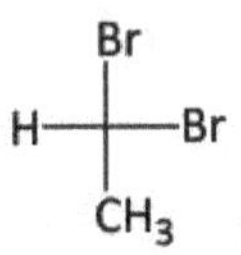
b) 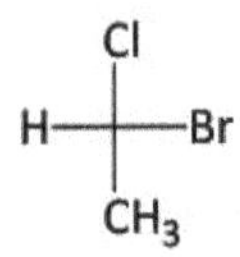
c) 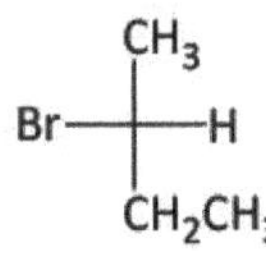
d) 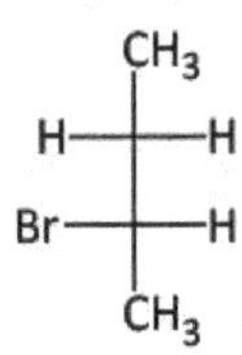
e) 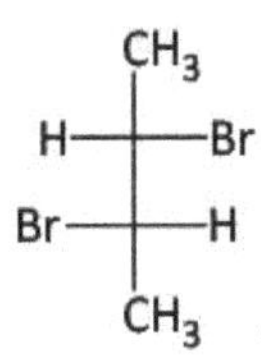
f) 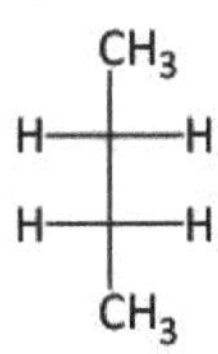
g) 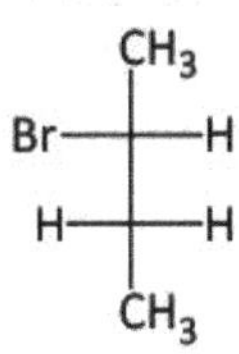

7.25 The following is a synthesis of a compound that has been proven to be highly inhibitive to HIV-1 protease. The reaction sequence involved the following:

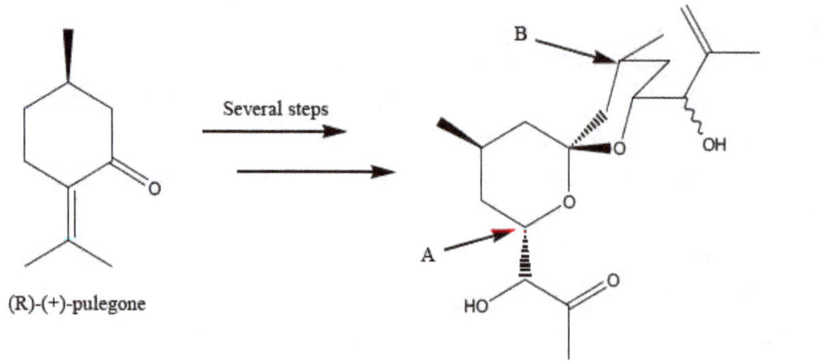

(2S,4R,6R,8S,10S,1'R,1"R)-2(acetylhydroxymethyl)-4,10-dimethyl-8(isopropenylhydroxymethyl)-1,7-dioxaspiro[5,5]-undecane

a) Why is the starting material labeled (R)?
b) What does the (+) mean for the pulegone starting material?
c) Do (R) and (+) always go together or can some compounds have (S) and (+)?
d) Draw and name the enantiomer of (R)-(+)-pulegone.
e) The carbon marked A in the product is either carbon #2 or carbon #4. Which is it? Explain your reasoning.
f) Is the methyl group on the carbon atom marked B in the product axial or equatorial?

Let's identify some other parts on the product above. . .

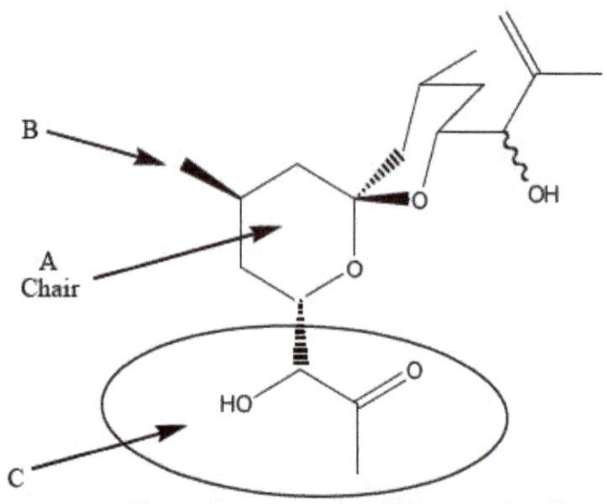

The 6-membered ring marked with an arrow is really a chair. We are looking straight down on the molecule.

g) Are the two groups (B and C) cis or trans to each other?
h) For B and C, state if you think B is axial or equatorial *and* if you think C is axial or equatorial. Explain your reasoning. (Hint: It may help to draw out a chair to answer this question.)

Chapter 8. Organic Reaction Basics

8.1 Types of Reactions

Organic reactions can be placed into several different categories.

1. **Substitutions.** Much like how a substitute teacher replaces the regular teacher, a substitution reaction is one where one atom or group is replaced, or substituted, with another atom or group. One group replaces another group. In this case, hydroxide (OH^-) replaces chloride (Cl^-).

$$H_3C{-}Cl \;+\; {}^{-}OH \longrightarrow H_3C{-}OH \;+\; Cl^-$$

2. **Additions.** In an addition, two atoms or groups are added to a molecule's π bond. In effect, two molecules are added together. Two molecules become one molecule.

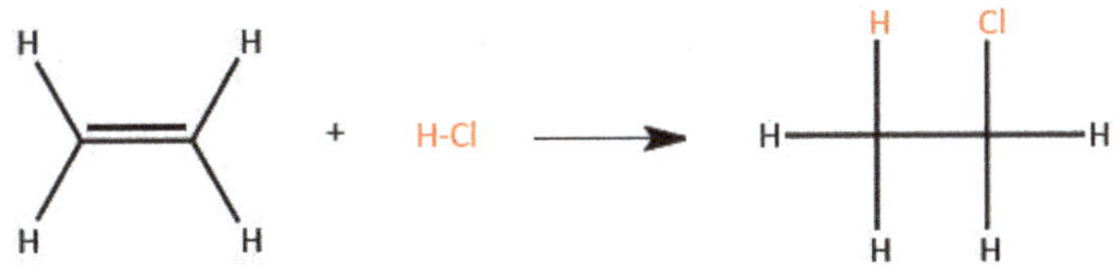

3. **Eliminations.** When something is eliminated, it is expelled or removed. In an elimination reaction, two atoms or groups are expelled from a molecule to generate pi bond and make two molecules. Elimination reactions are the opposite of addition reactions. Notice in this example, both H and Cl are removed to form a pi bond.

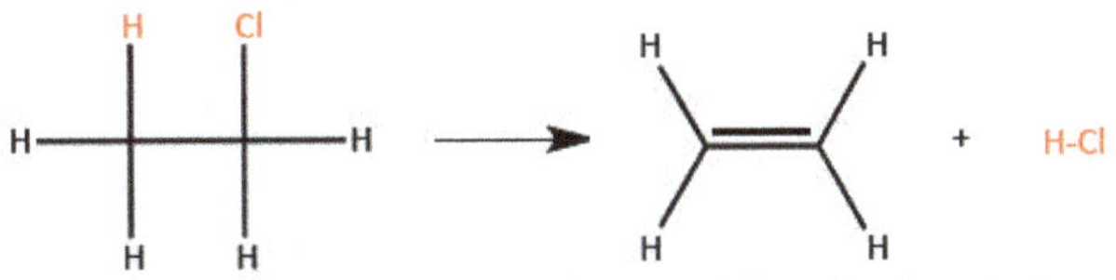

4. **Rearrangements.** In a rearrangement, a molecule reorganizes its bonds and atoms to make an isomer of itself.

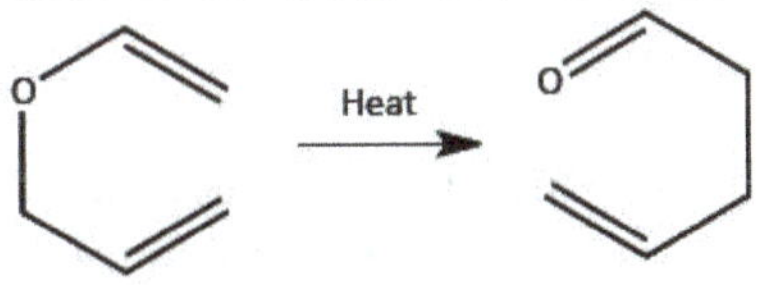

Claisen Rearrangement

5. **Free radical.** In a free radical reaction, the reactive species is one that contains a single, unpaired electron (a free radical). Ultraviolet light is often used to generate this free radical.

$$Cl_2 \xrightarrow{h\nu} 2\,Cl\cdot$$

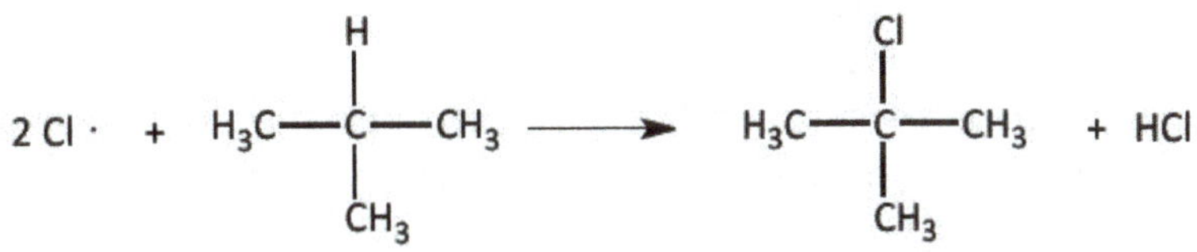

6. **Oxidations/Reductions.** In organic oxidations or organic reductions, an organic species either gains oxygen or loses hydrogen. If a molecule makes more carbon-oxygen bonds and/or loses carbon-hydrogen bonds, we say that the organic species is oxidized. If a molecule gains hydrogen bonds, we say it is reduced.

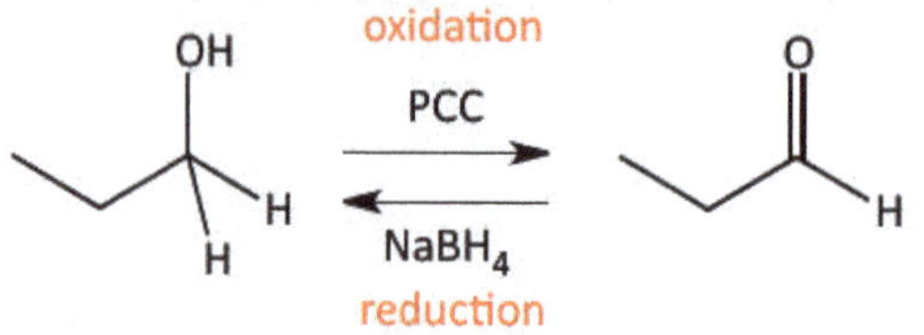

8.1 Identify each of the following reactions as a substitution, addition, elimination, rearrangement, free radical, oxidation, or reduction.

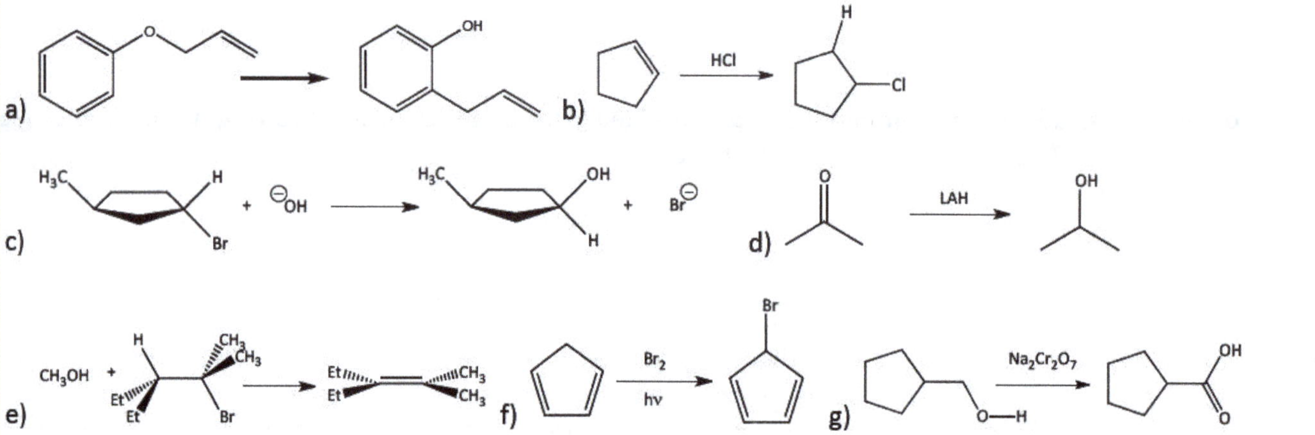

Each organic reaction has a story. Bonds are broken and bonds are formed. Reactants become products. Organic chemists call this story the mechanism of a reaction. It is how a reaction proceeds. The mechanism helps organic chemists understand why and how the reaction happened. Mechanisms are very important in organic chemistry.

A mechanism is a step-by-step story of how a reaction proceeds. Sometimes, you will be given both the reactants and the products of a reaction and be asked to give the mechanism of how the reaction went. At other times, you may be asked to predict what will happen in a reaction given two or more reactants. This is usually the more difficult task.

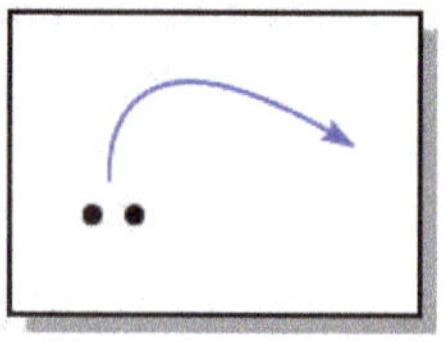

To be able to tell the mechanism story, you need to know the conventions organic chemists have agreed upon. The agreement is that we will show where electrons move in a reaction. To show this movement of electrons, we use arrows. A regular two-headed arrow shows the movement of two negative electrons. Did you get that? That last sentence may be one of the most important statements I will teach you. *Arrows show the movement of negative electrons.* Arrows DO NOT show the movement of positive charges, atoms, or anything else you can think of. Arrows in organic reaction mechanisms show the movement of negative electrons. Students who fail to learn this are almost never successful in organic chemistry. If you don't learn to focus on the movement of these electrons, organic chemistry will remain a mystery. It is my belief that successful organic chemistry students learn to think in terms of arrows.

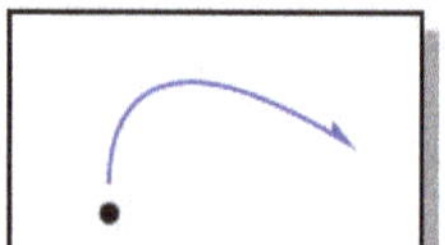
Fishhook arrow

There is a small subset of organic reactions called free radical reactions where only one electron is moving at a time. In these cases, we use an arrow with only one barb on its head, sometimes called a fishhook arrow.

For this chapter, we'll focus on regular, two-headed arrows. Arrows in mechanisms always start at negative electrons and show you where they go. Every time I look at a mechanism, I first ask myself where the electrons are moving. And since electrons have a negative charge, they almost always move towards the most positive atom on another reactant.

8.2 Draw one arrow in each of the following boxes. Your arrow should ALWAYS begin at the negative electrons and travel towards a positive spot. You can curve your arrows a little bit if you want, but it doesn't matter.

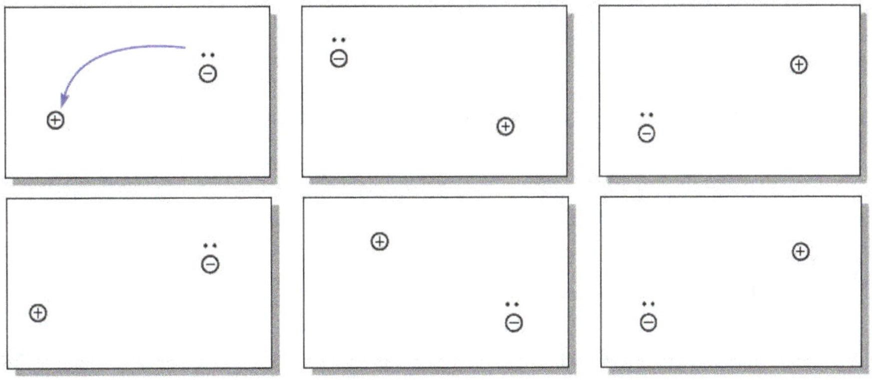

Of course, for real molecules, there are several atoms. In the following exercise, I've made several molecules. On each molecule, I've identified one most negative atom and one most positive atom.

8.3 Draw one arrow in each box showing the movement of the electrons from the negative charge on the molecule on the left to the positive charge on the molecule on the right.

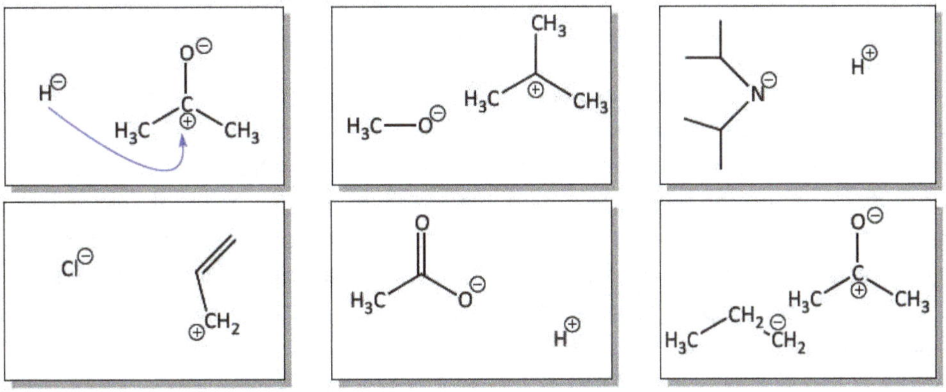

8.2 Nucleophiles/Electrophiles

To be a little more technical, the negative molecules are called **nucleophiles** and the negative spots are called **nucleophilic sites**. Nuclei are positively charged and "philic" means "lover of". Opposites attract, so negative things love positive things; they're nucleophilic.

Positive molecules are called **electrophiles**. Positive spots are called **electrophilic sites**. Electrons are negatively charged, so electrophiles are lovers of negative things. Negative things are called nucleophiles, positive things are called electrophiles.

8.4 Circle the appropriate word in each of the following sentences.

a) Negatively charged molecules are called (electrophiles/nucleophiles).

b) Positively charged molecules are called (electrophiles/nucleophiles).

c) A molecule that is attacked by a negatively charged molecule is called a(n) (electrophile/nucleophile).

8.3 Finding Nucleophilic Sites

Let's first address how we recognize the most negative spots (nucleophilic sites) in real molecules. Perhaps the most obvious situation is if an atom has a formal negative charge. Examples of this are oxides, amide ions, carbanions, and hydrides. Remember that arrows always show where the negative electrons move. In the following figure, the arrows originate at the negative, nucleophilic sites.

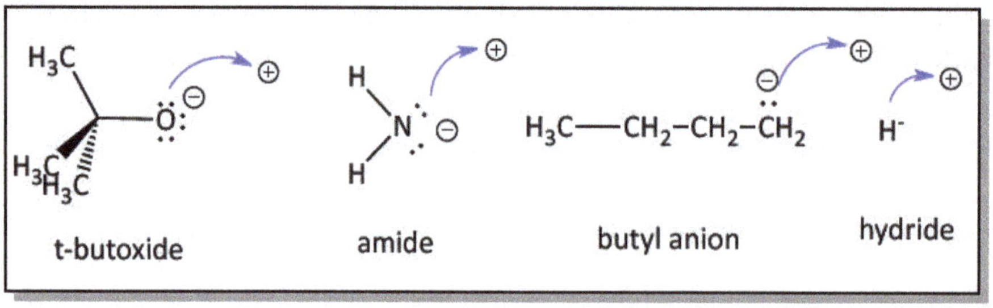

The charges are sometimes drawn on the atoms. When they are, that helps us determine where to start our arrows. But in fact, this negative charge is always balanced out by a counterion. The counterion is a positive cation, usually a metal cation. This metal cation is typically drawn attached to the most negative, nucleophilic site. This can make it a little trickier recognizing the negative atom. Remember that anytime you see a metal in a molecule, the atom attached has a negative charge. A positive metal has a negatively charged atom attached to it! It is this negative atom that will go out and attack in our mechanisms. Our arrow will start there. Here are some examples of how these compounds may be written with the metal. Remember that arrows are pointing away from the most negative, nucleophilic sites.

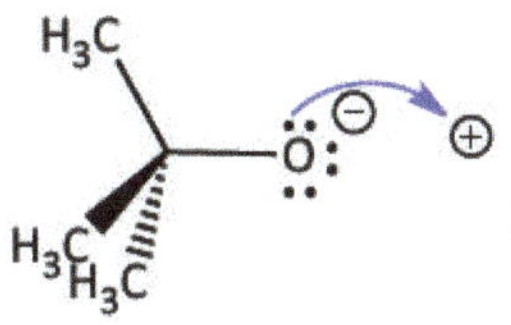

may be written as

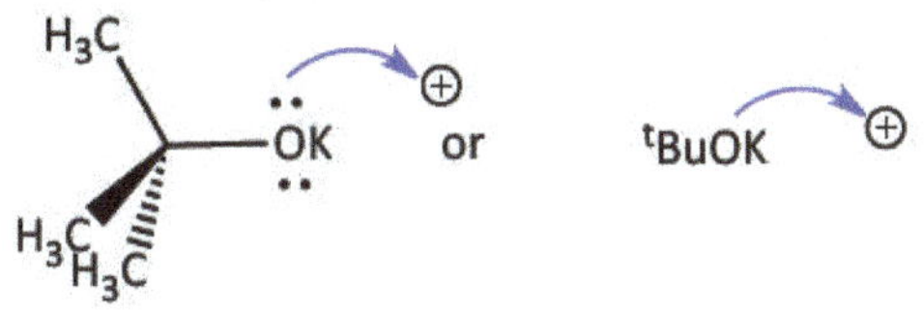

t-butoxide

potassium t-butoxide

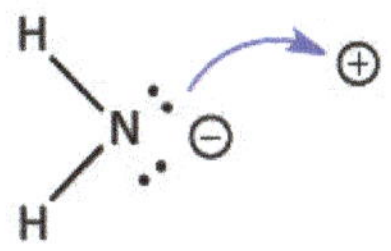

may be written as

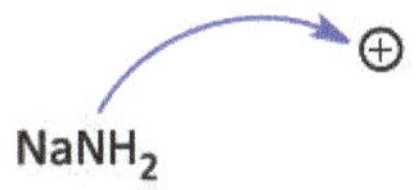

amide

sodium amide

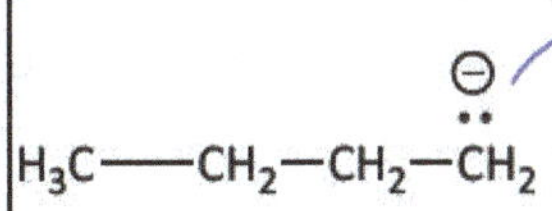

may be written as

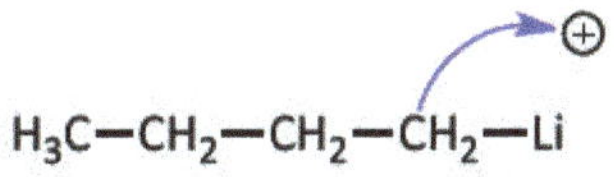

or

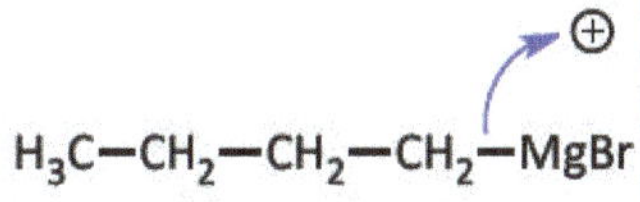

butyl anion

butyl lithium

butyl grignard or butyl magnesium bromide

Notice that when the metal cation is written, the mechanism arrow starts either at the carbon atom or at the carbon-metal bond. Either is acceptable.

may be written as

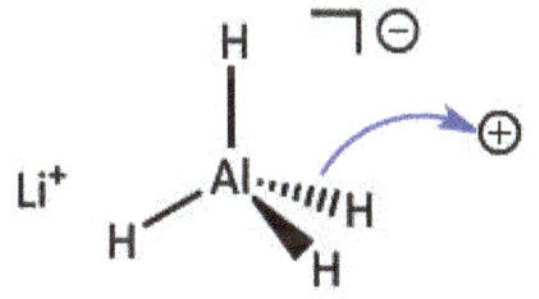

or

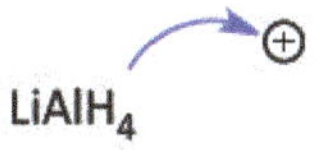

hydride

lithium aluminum hydride

Aluminum is a positively charged metal, so the hydrogen atoms attached are negatively charged "hydrides." Notice the mechanism arrow starts either at the hydrogen atom or at the hydrogen-metal bond. Either is acceptable. I prefer starting the arrow at the bond.

If you don't see a formal negative charge on an atom, or if you don't see a metal to tip you off to the location of a negative atom, look for other locations of nice, juicy negative electrons. The best place to look? Look for double or triple bonds as they contain one or two π bonds respectively. You can also look for lone pairs of electrons. This is a little tricky for beginning students because chemists are lazy when it comes to drawing structures. We quite often do not draw in all of the lone pairs of electrons. It is assumed you will know how many lone pairs there are on each atom. Remember the normal bonding patterns and you will be all right. So, if you have a nitrogen, oxygen, or sulfur atom, be sure to draw in the lone pairs of electrons.

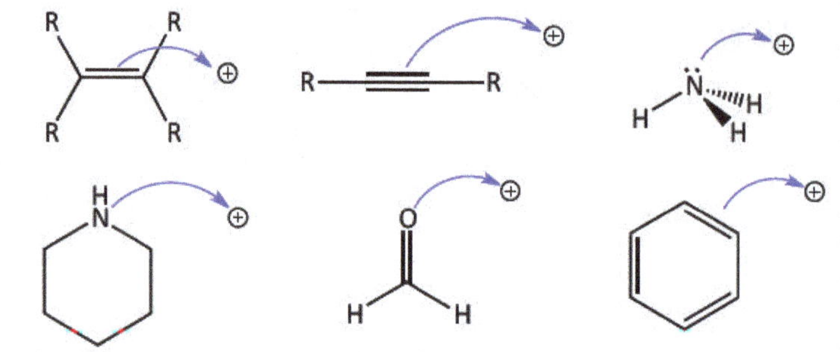

N has 1 lone pair. O has 2 lone pairs.

Sometimes, you'll have two sites to choose from. Let's consider lithium acetylide, H—C≡C—Li. In this molecule, we see the lithium metal atom, which we remember is positively charged. This means that the carbon atom attached to the lithium has a partial negative charge. We can think of it as having a lone pair of electrons on it. But, we also notice that there is a pi bond that is also available for bonding. When you see a molecule with a metal written, our mechanism arrow will almost always start at the atoms attached to it. So, typically, carbon-metal bonds are more negative (more nucleophilic) than pi bonds.

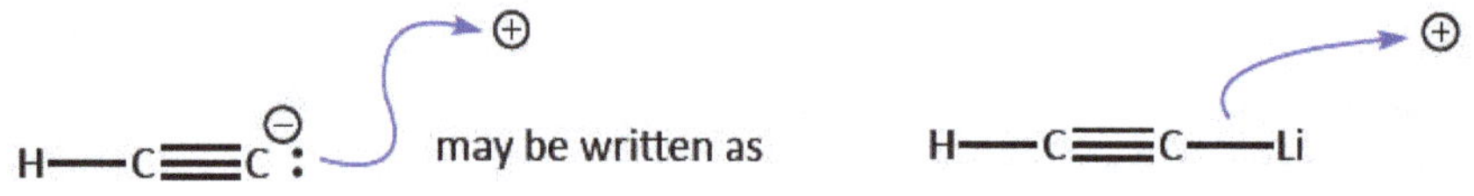

acetylide ion — lithium acetylide

Notice you have two choices in this example. The mechanism arrow could start at the pi bond of the triple bond or the negative carbon atom next to the lithium atom. The positive lithium atom makes the carbon atom especially negative, so it wins.

In summary, there are four main locations to look for negatively charged sites on molecules.

How to find negative sites

1. Formal negative charges
2. Bonds between nonmetal and metal atoms
3. Lone pairs of electrons
4. Pi bonds

8.5 Draw arrows for the following molecule starting at the most negative spots (nucleophilic sites) and pointing towards the positive charges.

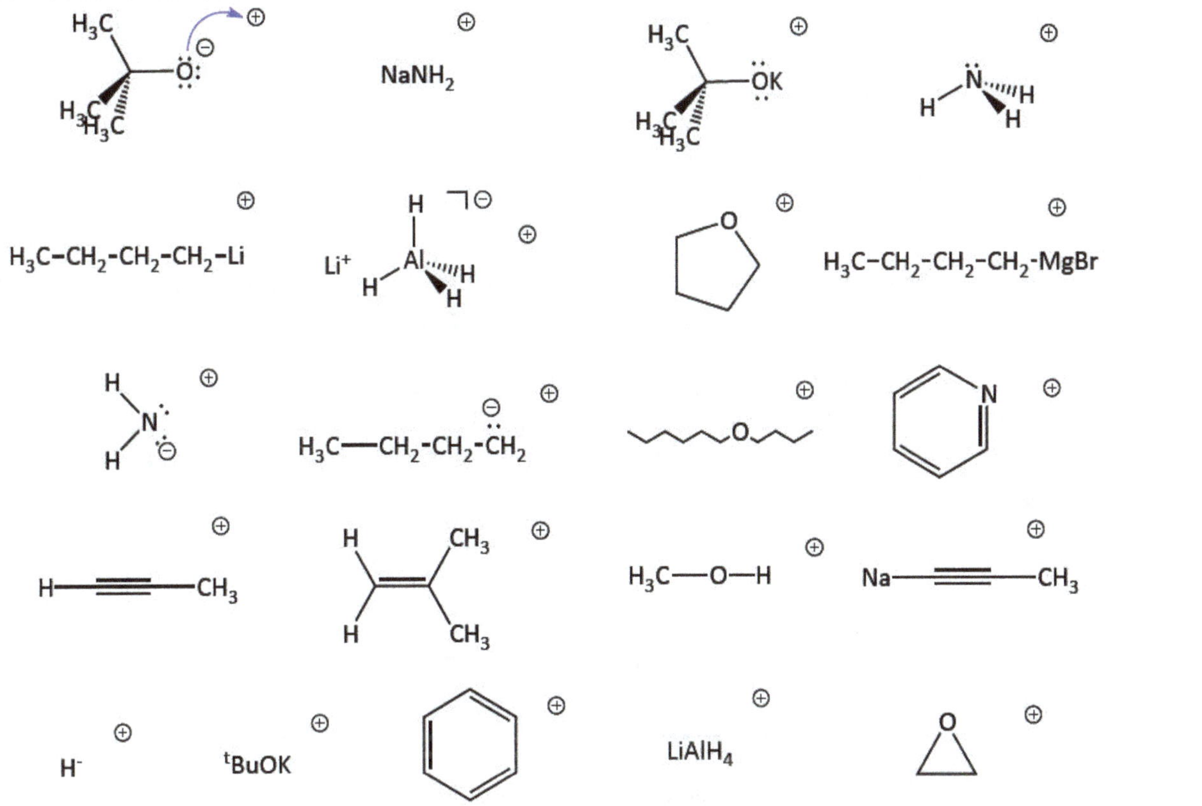

If you struggled at all or missed any of the problems in the previous exercise, go back and reread the beginning of this chapter once again. You should make sure you understand what is going on before proceeding. If you were successful in the previous exercise, great! Let's learn to recognize the positive spots (electrophilic sites) in molecules.

8.4 Finding Electrophilic Sites

The first and most obvious electrophiles are *atoms with a positive charge explicitly written on them*. A carbocation (positively charged cation on a carbon atom) is a prime example. Carbocations have a carbon atom with three bonds and an empty p orbital.

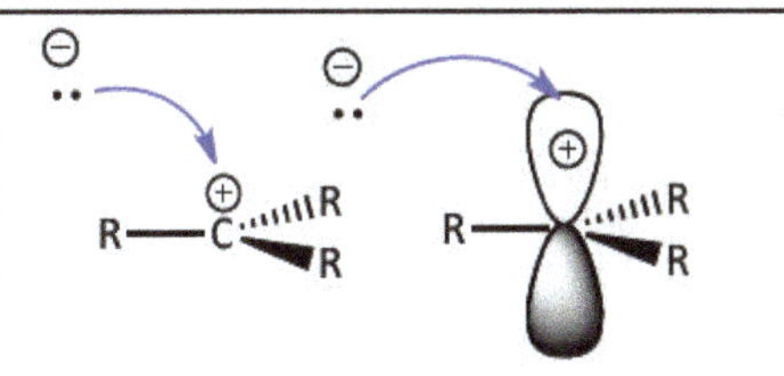

Carbocation

Other sources of atoms with a positive formal charge, which may seem "hidden", are the acids. Recall that acids are a source of protons, H^+. So, HCl, H_2SO_4, H_3PO_4 are all sources of the electrophilic H^+.

If a carbon or hydrogen is attached to a more electronegative atom, the carbon or hydrogen atom has a partially positive charge. Alkyl halides, carbonyls, and proton acids are also electrophiles.

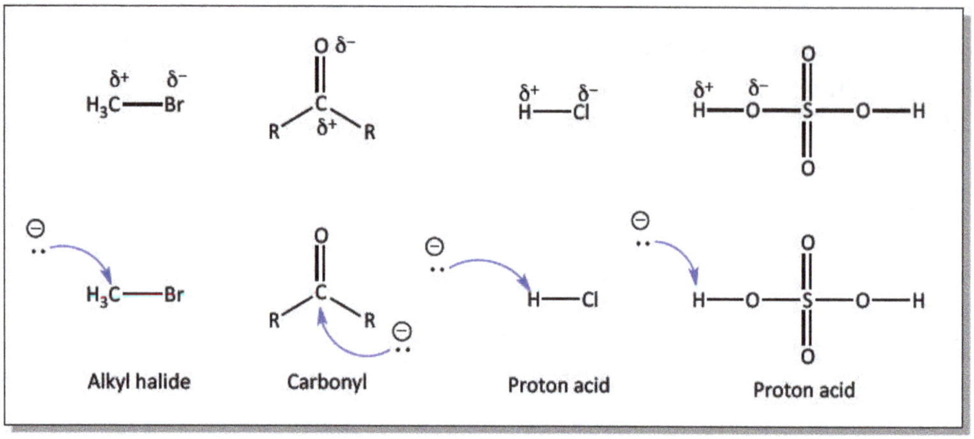

You may remember from general chemistry the strange chemical behavior of *boron (B) and aluminum (Al)*. They are atoms that are electron deficient because they normally have six electrons in their Lewis structures. They have an incomplete octet. The B and Al atoms contain a partial positive charge because they normally lack negative electron density around them.

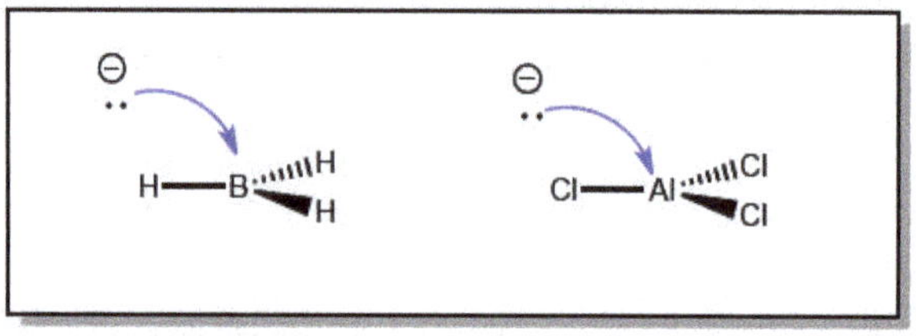

To summarize how to find positively charged electrophilic sites, look for the following.

1. Atoms with a formal positive charge written on them.
    a. Sometimes the positive formal charge is written right on the atom. Hooray for those times!
    b. Strong acids are a source of H^+.
    c. If you cannot find a positive formal charge, it might help to draw valid resonance forms until you find some good sites.
2. Atoms attached to electronegative elements.
3. Al or B atoms

8.6 Draw an arrow from the negative electrons to the most positive spot (most electrophilic site) in the following molecules.

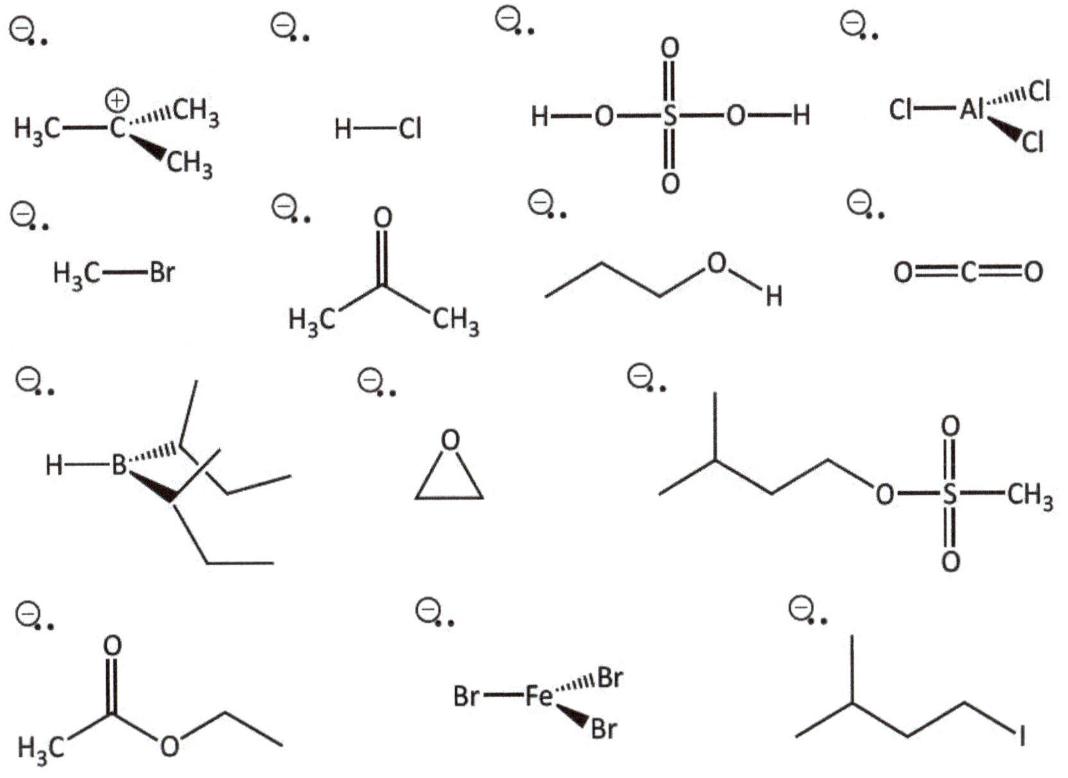

8.7 Draw arrows from the most negative, nucleophilic sites on one molecule on the left towards the most positive, electrophilic sites in the molecule on the right.

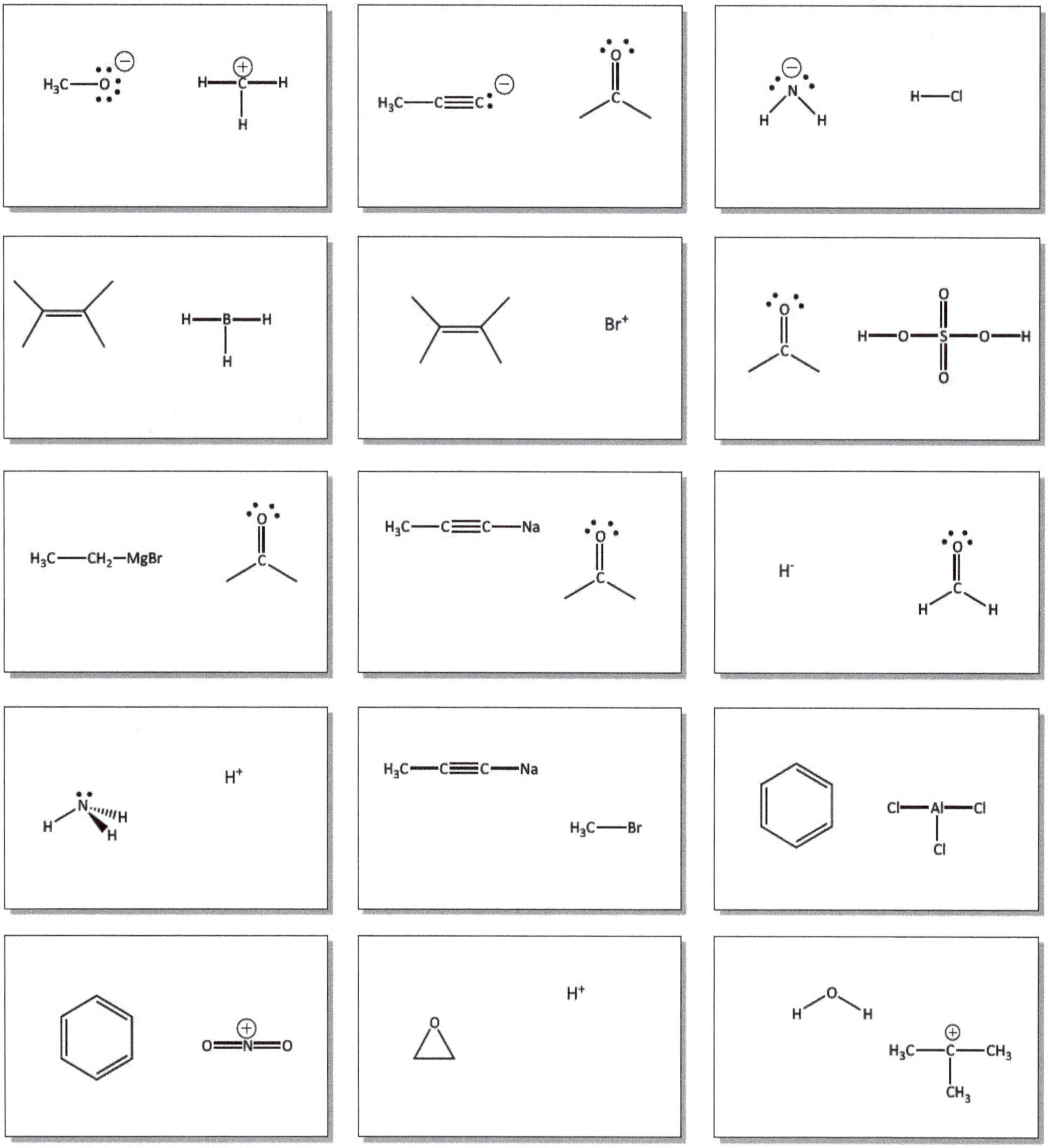

8.5 One arrow or more?

In the previous exercise, we drew one arrow from the nucleophile to the electrophile. This is an important first step for our mechanism. Only one arrow is drawn in each step unless it is necessary to draw more. If only one arrow is needed, then draw the product and move on to the next step of the mechanism. More arrows can be drawn in subsequent steps, but for goodness sakes, draw only the necessary arrows in each step!

The second step of arrow drawing is learning whether a second arrow or more is needed. One common mistake of novice organic chemistry students is drawing too many arrows for no reason. A mechanism is a step-by-step story. There are two reasons to draw more arrows. One reason to draw more than one arrow is if you know that the reaction is concerted (that two or more things happen at the same time). If two things happen in the same step, they should be drawn in the same step. If two things happen sequentially, one after the other, they should be drawn in two steps, one after the other.

Another reason to draw more arrows in one step is if after the first arrow has moved two electrons, an atom has more than an octet of electrons. For example, if an atom will have ten electrons, then some electrons need to be immediately removed in order to maintain an octet. In the following figure, after the first two electrons are moved, the atom receiving electrons has an octet, so no more arrows are needed. Notice, only one arrow is used.

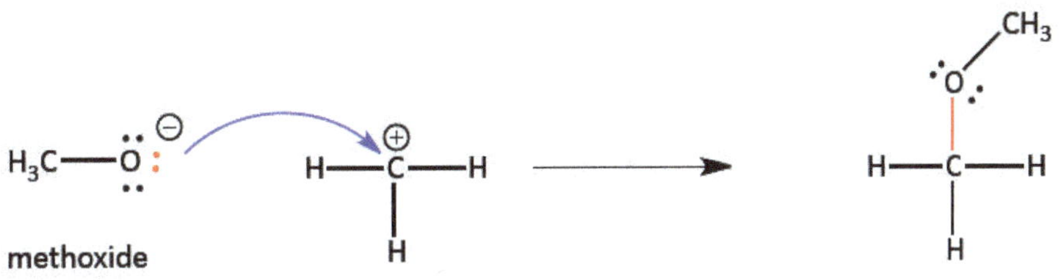

There are several things about the above mechanism that I want you to observe.

1. The arrow shows where the two red electrons moved. The electrons from the lone pair on the oxygen atom in the methoxide reactant move to become the bond between the carbon and oxygen atom in the product.
2. The carbon atom in the product has eight electrons around it (an octet). The carbon atom is happy, so only one arrow was required. If the carbon atom had more than eight electrons around it, some electrons would need to be removed with a second arrow.
3. Notice how the formal charges changed in the above reaction. The oxygen atom started with one bond and three lone pairs of electrons giving it a -1 formal charge. The carbon atom started with three bonds and no lone pairs of electrons giving it a +1 formal charge. After the oxygen atom donated two electrons to the carbon atom, we see in the product that both atoms have a normal bonding pattern. They have a zero formal charge in the product.

How about the following reaction? Observe what the product looks like after one arrow has moved two electrons.

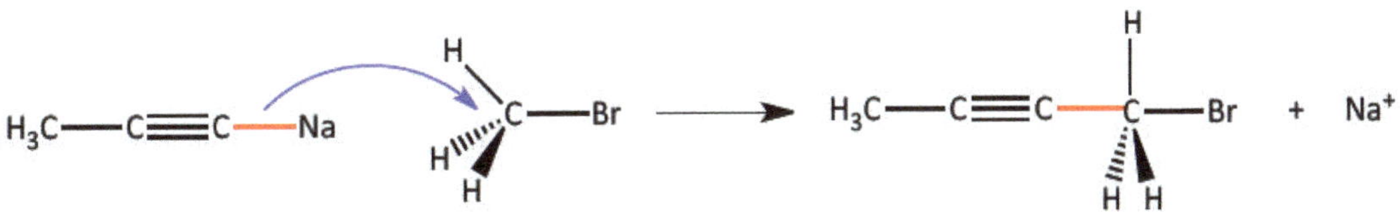

Do you see a problem with the product? Of course, there are five bonds around a carbon atom! Ten electrons around the carbon atom break the octet rule. This molecule needs to get rid of two electrons from that carbon. A better mechanism would be the following with two arrows happening at the same time, a concerted mechanism.

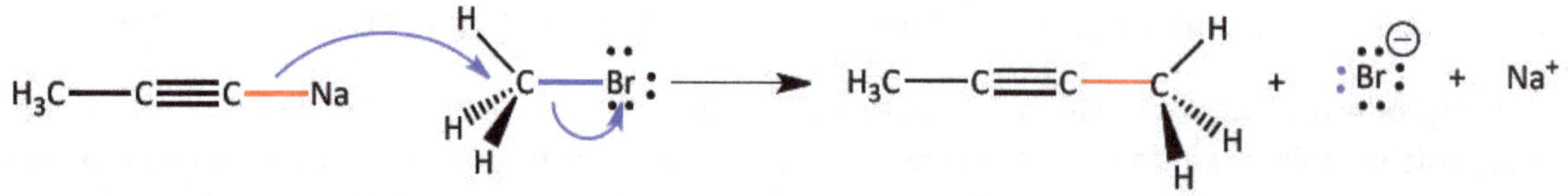

Using this two-arrow, concerted, mechanism we avoid the unsavory 10-electron species from above. Now, the carbon atom in the product only has its octet. So, a rule to follow is that if electrons are coming to an electrophile and it makes more electrons than an octet at an atom, then electrons MUST move away from this atom. If this is not the case, *use only one arrow*.

A similar situation occurs with acids. Consider what the following reaction looks like after one arrow has moved two electrons.

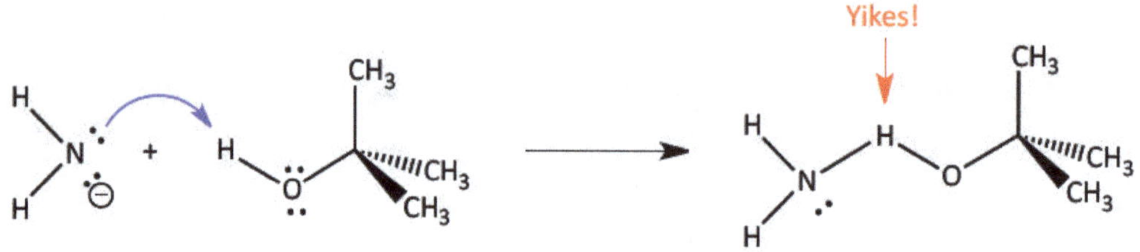

Do you see the problem? A hydrogen atom with two bonds in organic chemistry is not going to happen. Let's please have only one bond for every hydrogen atom. In order to avoid a situation like this, we again need two arrows.

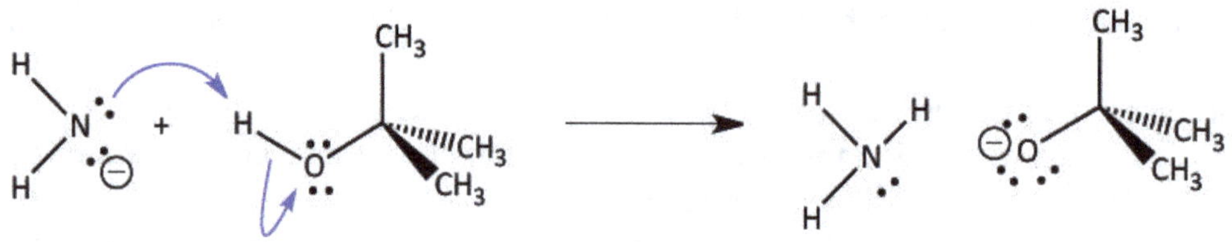

8.8 Draw in either one or two arrows, as appropriate, to complete the mechanism.

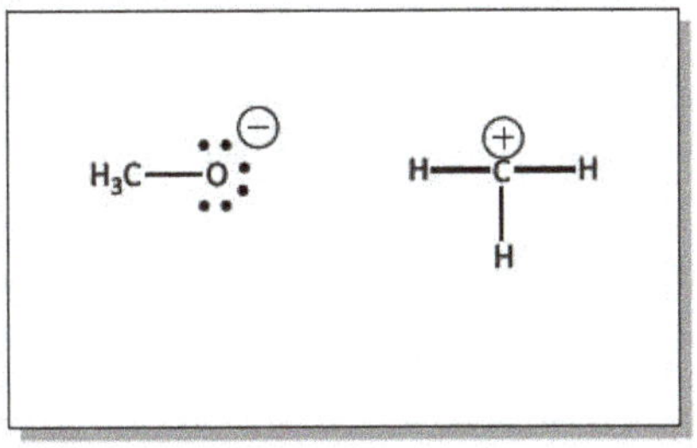

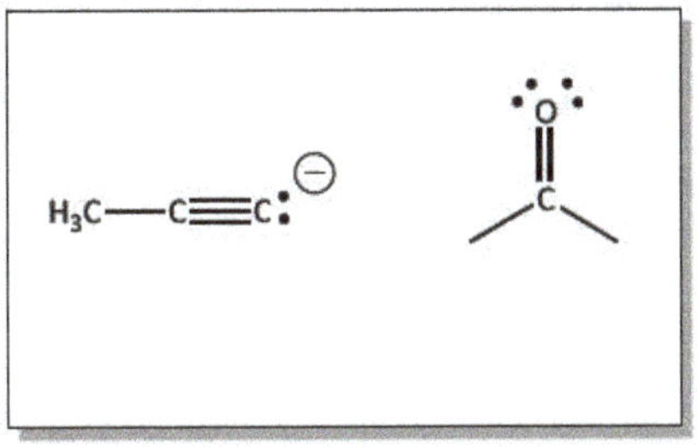

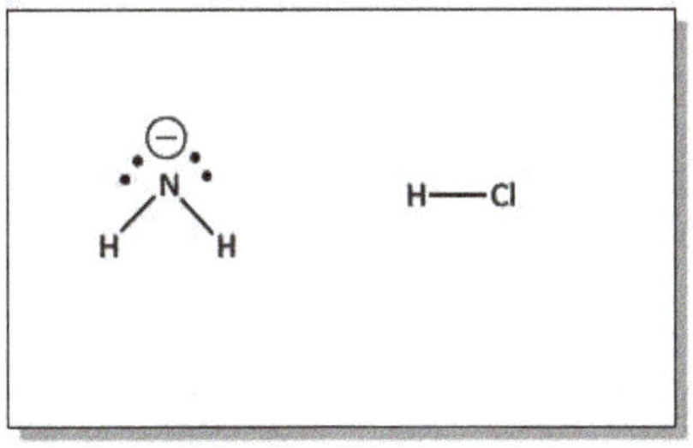

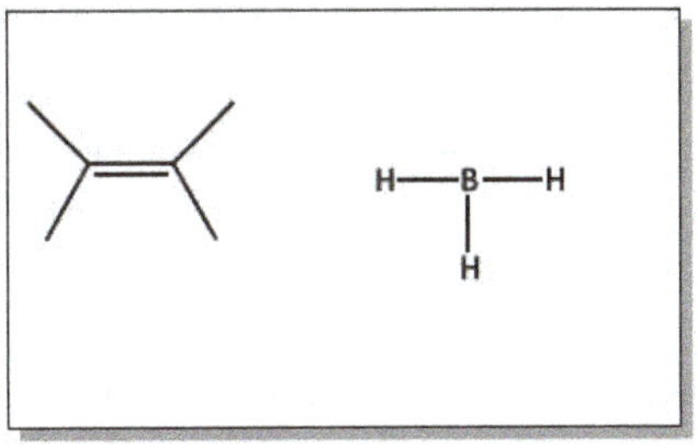

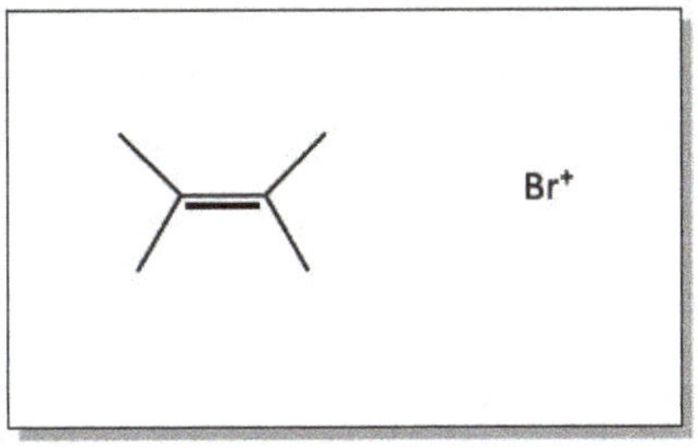

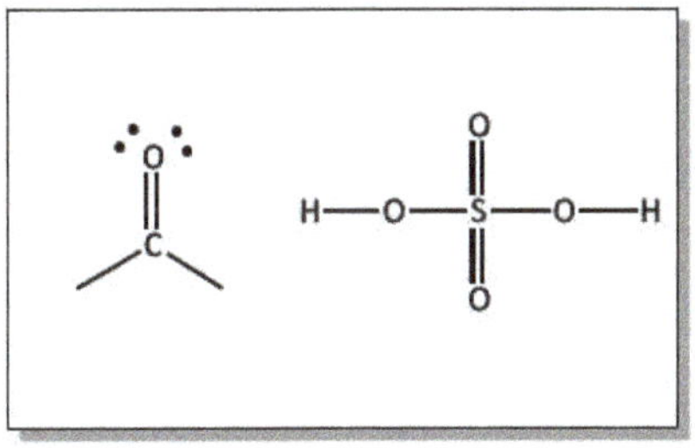

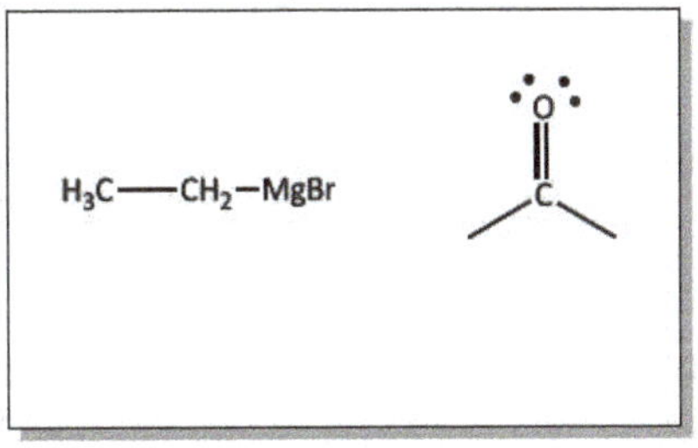

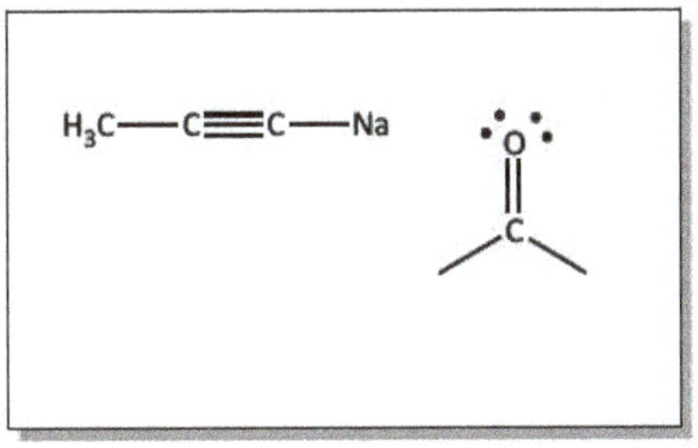

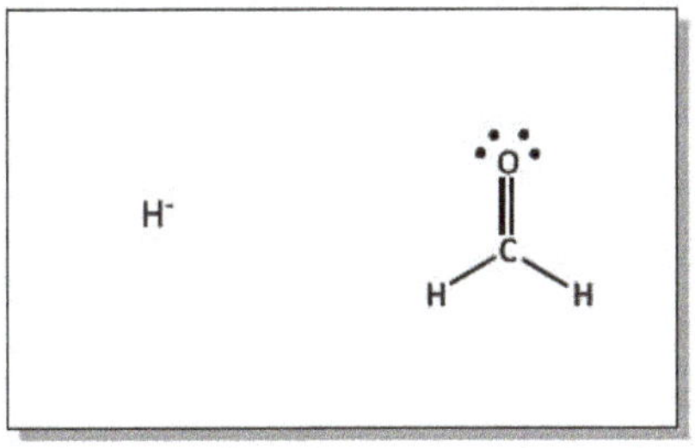

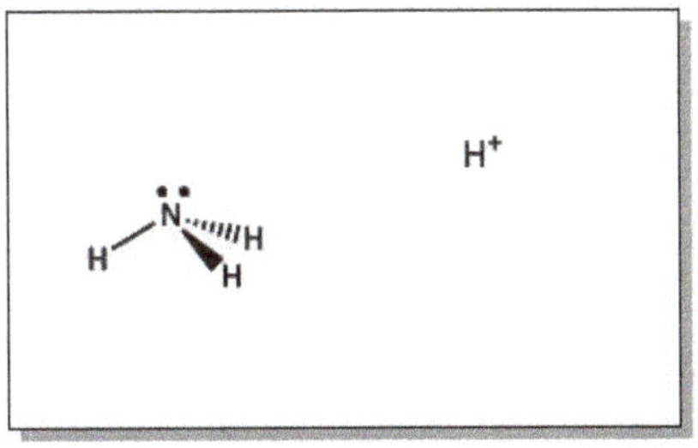

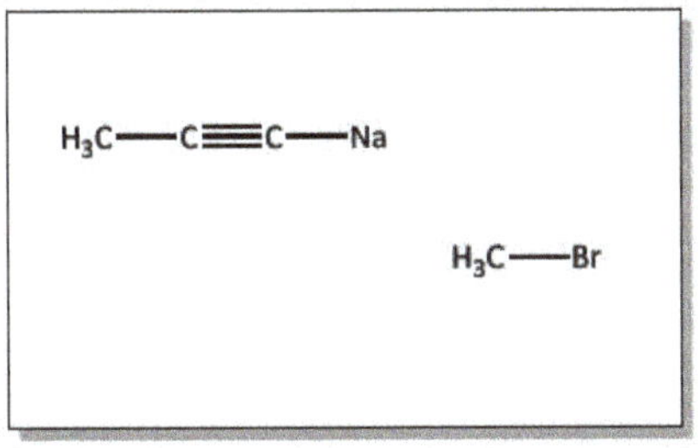

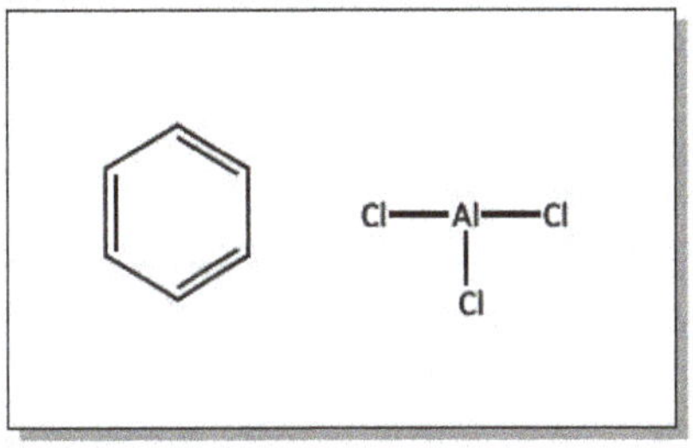

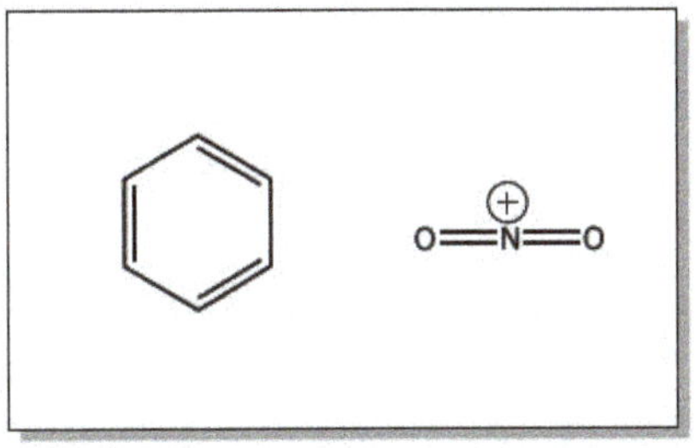

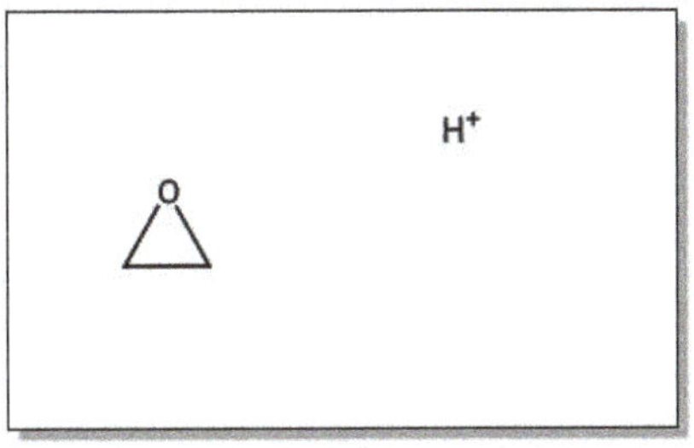

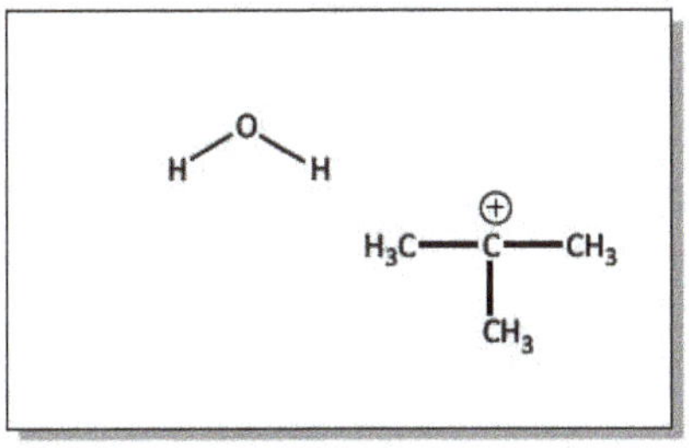

After you have completed this exercise, go and check your answers. If you missed any, make sure you understand why. If you did not miss any, you are ready for the next step.

8.9 Draw in the product or resonance form that is made after pushing the electrons that the arrows show. It is recommended that you draw in all lone pairs of electrons and hydrogen atoms near where the arrows are. This will help you identify formal charges in the products.

a) 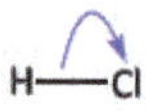 ⟶ ________ + ________

b) 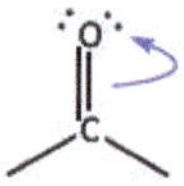 ⟷ ________

c) 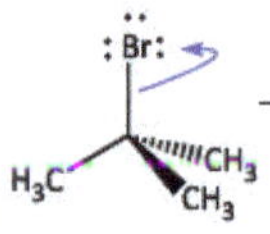 ⟶ ________ + ________

d) 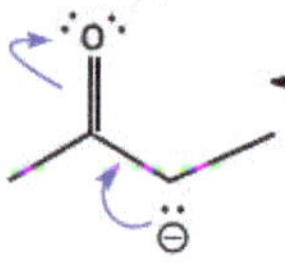 ⟷ ________

e) 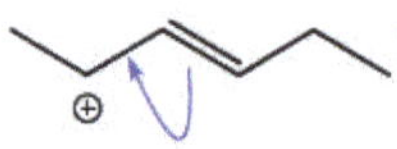 ⟷ ________

f) 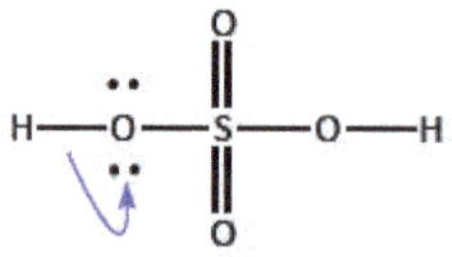 ⟶ ________ + ________

8.10 Let's put it all together. This time, draw the mechanism arrow(s) as well as the product that is formed.

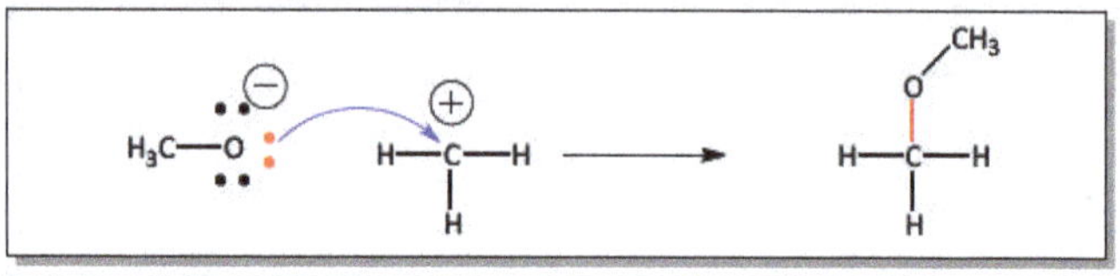

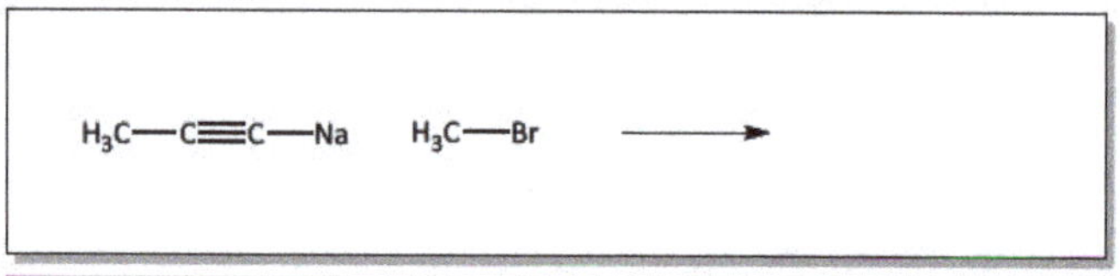

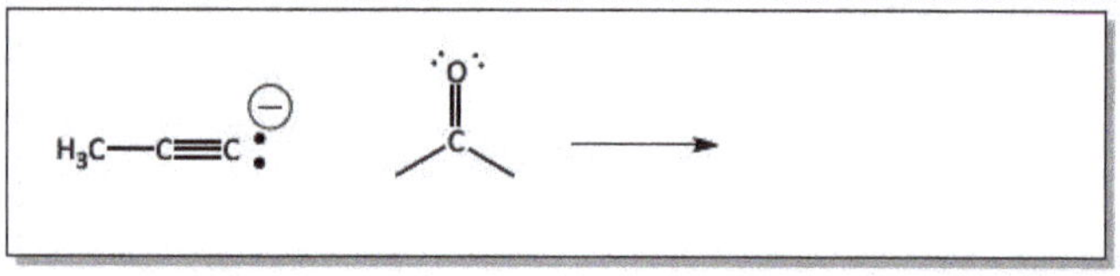

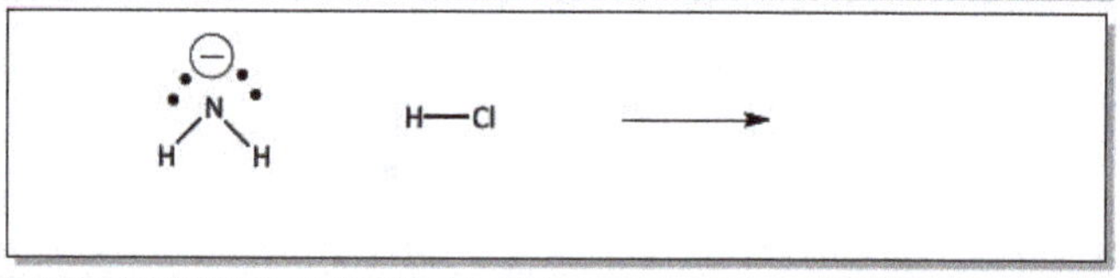

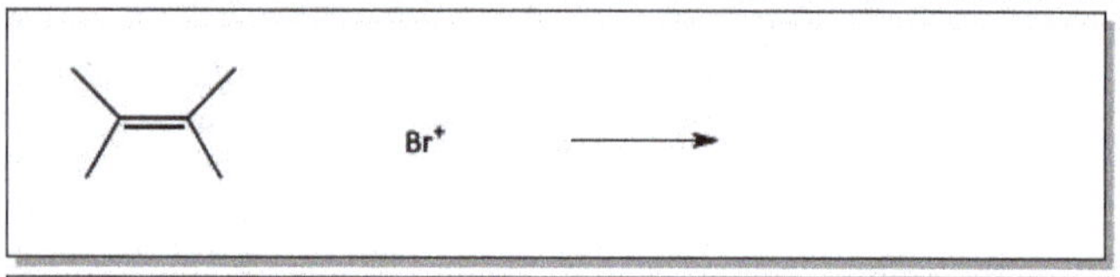

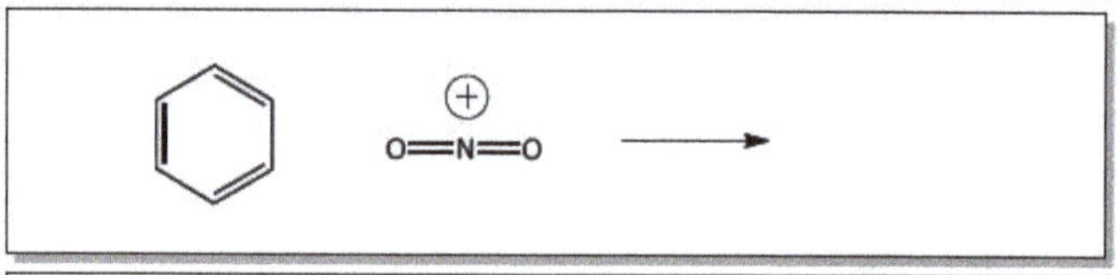

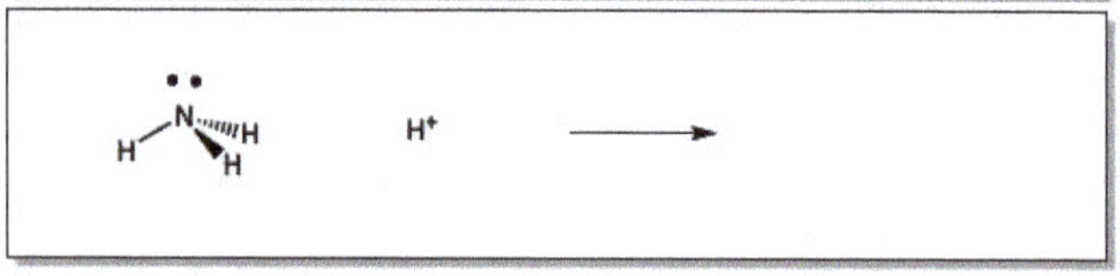

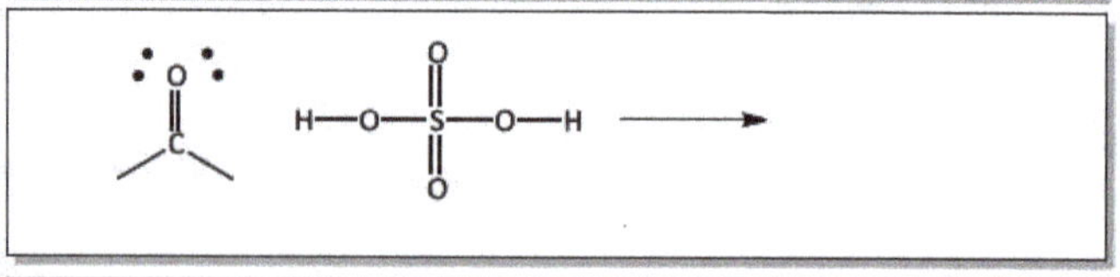

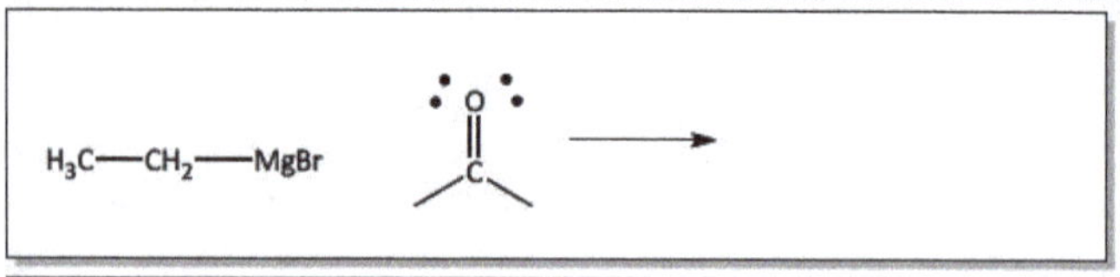

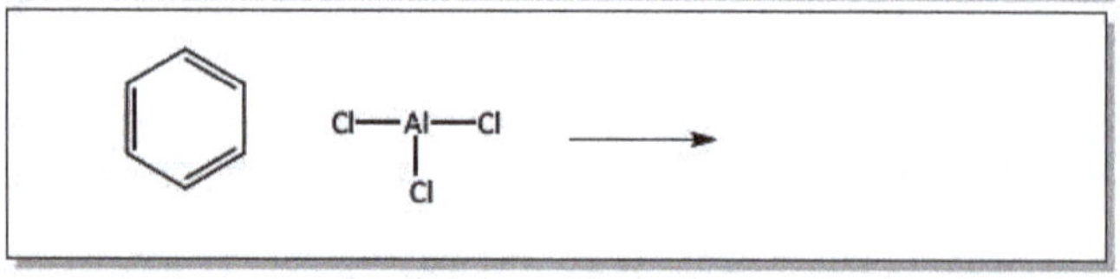

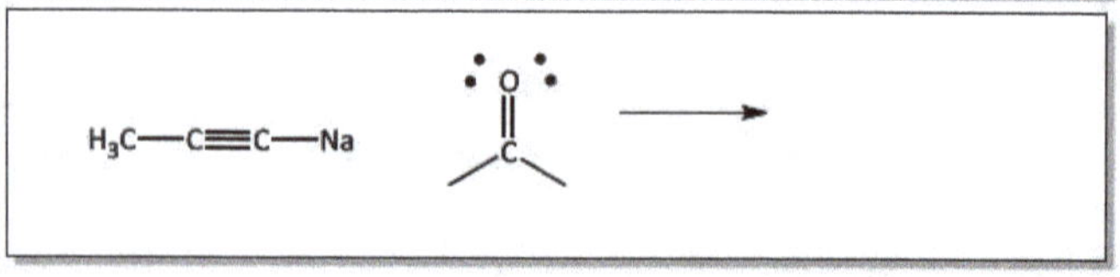

8.6 Arrow Review

We have seen several types of arrows so far. Let's review the different types of arrows found in organic chemistry. Most of these are review, but here is one new one. A retrosynthetic arrow is the opposite of a regular reaction arrow. We use it when we are thinking about how we want to synthesize a particular product. When we ask, "What reactants would we mix together to make this product?" we are doing a retrosynthetic analysis. We use a retrosynthetic arrow to show we are thinking how to synthesize something.

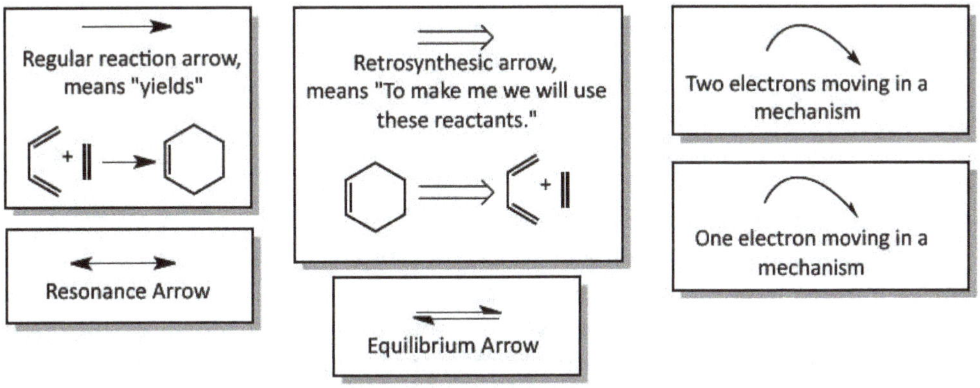

Figure 8.1 Various arrows

8.7 Reaction conditions

Symbols are sometimes written above the reaction arrows to indicate the reaction conditions. A small triangle is written above the arrow if the reaction is heated. It looks like the tip of a flame. One way to heat up an organic reaction to the boiling point of the solvent without losing the solvent is to perform a reflux. The solvent is boiled. The vapors travel up a condenser where they cool down, condense back into a liquid, and fall back into the reaction flask. An arrow pointing up and one pointing down is drawn over the reaction arrow if a reaction is heated to reflux. It represents the path the solvent travels. The symbol hν is drawn above the reaction arrow if a reaction is placed under ultraviolet light. This comes from the equation for the energy of light E=hν.

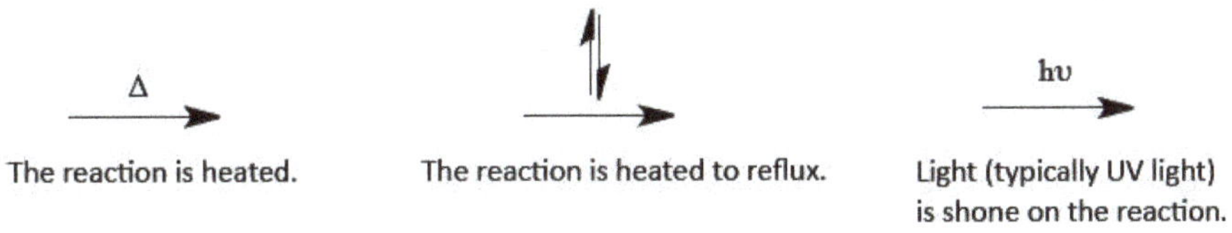

Figure 8.2 Symbols for various reaction conditions

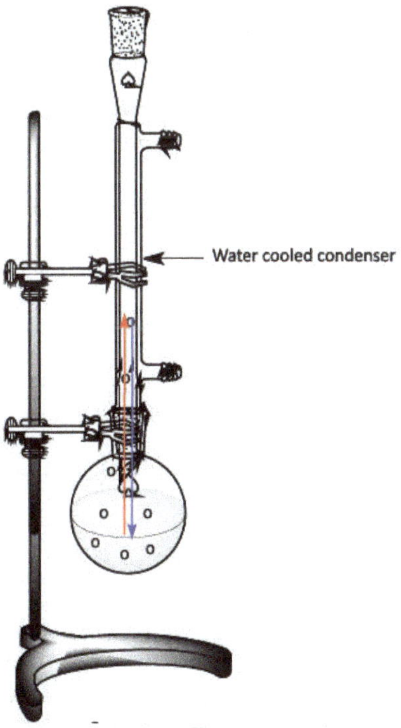

Figure 8.3 Reflux apparatus

Sometimes, instead of writing all of the reactants in a reaction to the left side of the reaction arrow, they are written above or below the arrow to signify they are added to the reaction. If numbers are placed before the reactants, then they are added in that particular order. If numbers are not written before the reactants, then all reactants are added to the reaction at the same time.

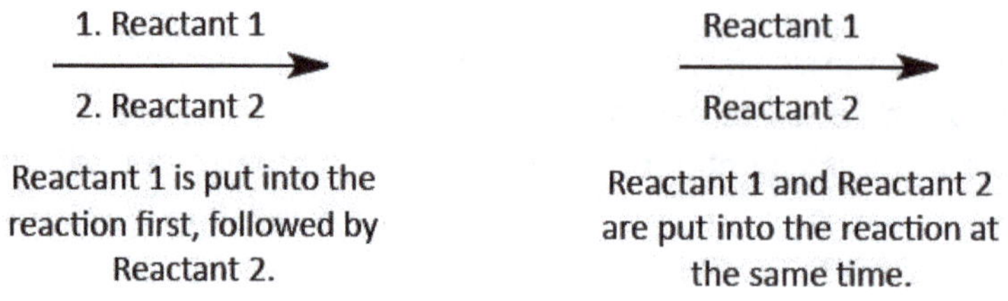

Figure 8.4 How to write reactants on reaction arrows

8.8 Thermodynamics

8.8.1 Bond Dissociation Energy

In every compound, there is a certain amount of energy in all of the bonds. This heat energy, or *enthalpy* (H), can be measured. In a reaction, bonds are broken and bonds are formed. There is an exchange of energy each time this happens. Energy must be put into the system in order to break a bond. Imagine trying to break a chain and the energy you exert trying to do that with the sweat on your brow and strained muscles. Remember that it always takes energy to break a bond. Since we are using energy to break a bond, the bond dissociation energy (BDE) is positive. If a weak bond is broken, the BDE is a small positive number, because it is easy and takes little energy to break the bond. If a strong bond is broken, the BDE is a large positive number, because it is difficult and takes a lot of energy to break the bond.

Energy is released when bonds are formed. The individual ions do not like to be by themselves with a charge on them. They would much prefer to be in a neutral compound. So, when they form a bond and become a compound, they give a sigh of relief. They can relax and have less energy. A negative BDE represents the energy that is lost when this bond is formed. If a weak bond is formed, the BDE is a small negative number. If a strong bond is formed, the BDE is a large negative number (the ion feels more secure and relaxes more). The following is an example reaction to form methanol.

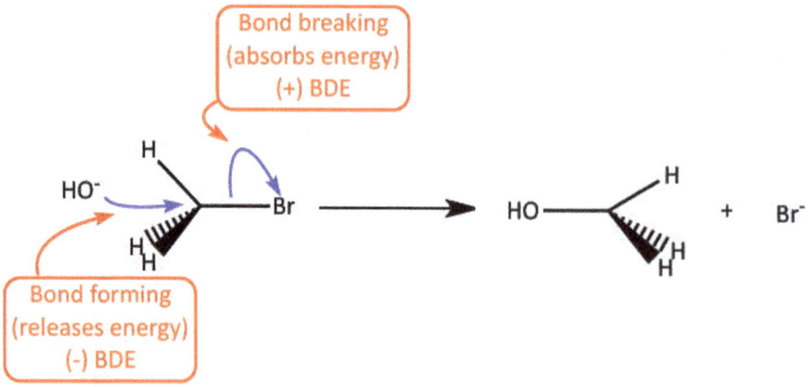

8.8.2 Enthalpy

When a reaction occurs, we look at all of the bonds that are broken (absorbing energy) and all of the bonds that are formed (releasing energy) and add up all of these BDE's. This overall measurement on how much energy (heat) is released or absorbed is called the *heat of reaction* or the *change in enthalpy* (ΔH). The symbol, Δ, means "change in". Overall, if more energy is absorbed than released, the reaction is called an endothermic reaction (+ΔH). If more energy is released than absorbed, the reaction is called an exothermic reaction (-ΔH). Reactions that are exothermic and give off energy are typically favored.

Breaking bonds	Endothermic = + ΔH
Forming bonds	Exothermic = - ΔH

8.11

a. In a reaction, one strong bond is broken and one weak bond is formed. ___ ΔH

b. In a reaction, one weak bond is broken and one strong bond is formed. ___ ΔH

There is a method chemists use to quickly communicate the changes in energies of the reactants and products throughout a reaction. A Reaction-Energy Diagram can be drawn. The *reaction coordinate* is like the timeline of the reaction, from the reactants to the products. The energies of all species along this reaction coordinate are mapped out. Below is the reaction-energy diagram for our one-step reaction to make methanol.

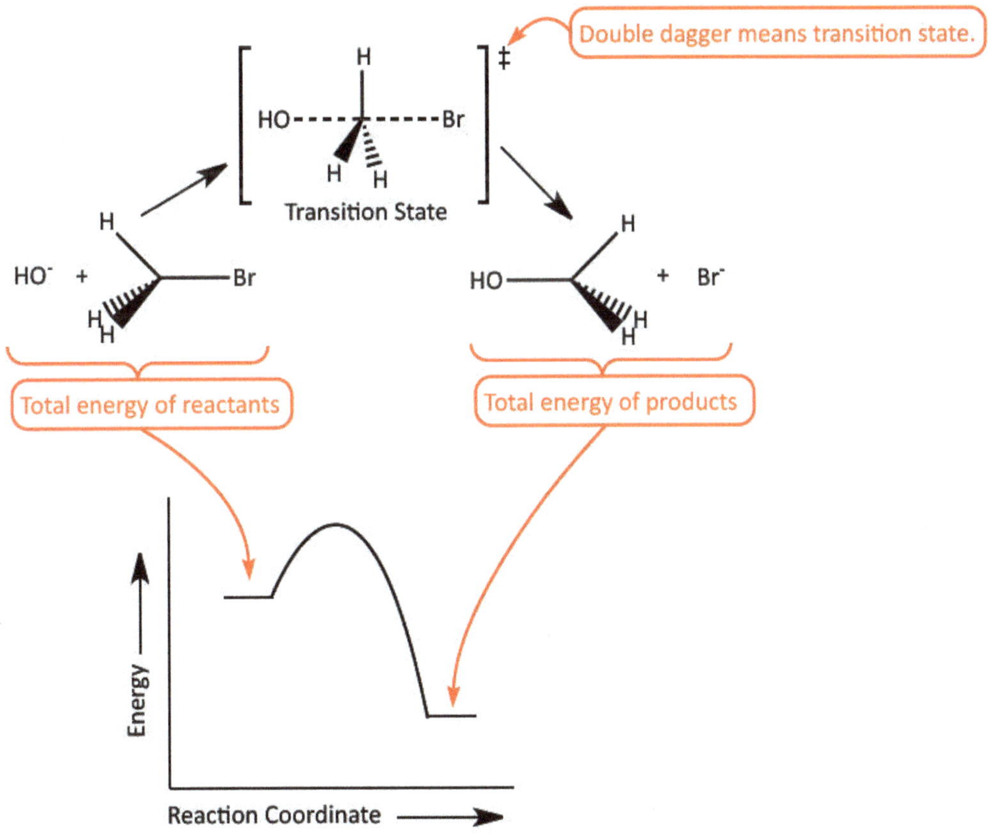

Figure 8.5 Reaction Coordinate Diagram

So, from a reaction-energy diagram, we can see the relative energies of the reactants and products. We can also determine the activation energy of the reaction, the change in enthalpy (ΔH), whether ΔH is positive (endothermic) or negative (exothermic). We know that ΔH is the difference in energy between the products and the reactants. If the products have less energy than the reactants, heat is given off and the reaction is exothermic (-ΔH). This is the case in this reaction-energy diagram.

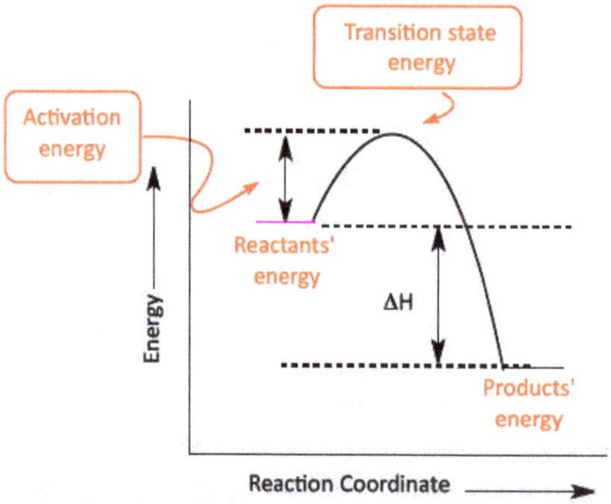

Figure 8.6 Various parts of a reaction-energy diagram

The *activation energy* (E_a) of a reaction is the amount of energy it takes to get from the energy of the products to the energy of the transition state. It is the amount of energy needed to get up the first hill to make the reaction go. For instance, if there is a high activation energy, perhaps the reaction may need to be heated in order to proceed.

8.12 Match the following statements with its reaction-energy diagram.

A. Endothermic, high activation energy
B. Endothermic, low activation energy
C. Exothermic, high activation energy
D. Exothermic, low activation energy

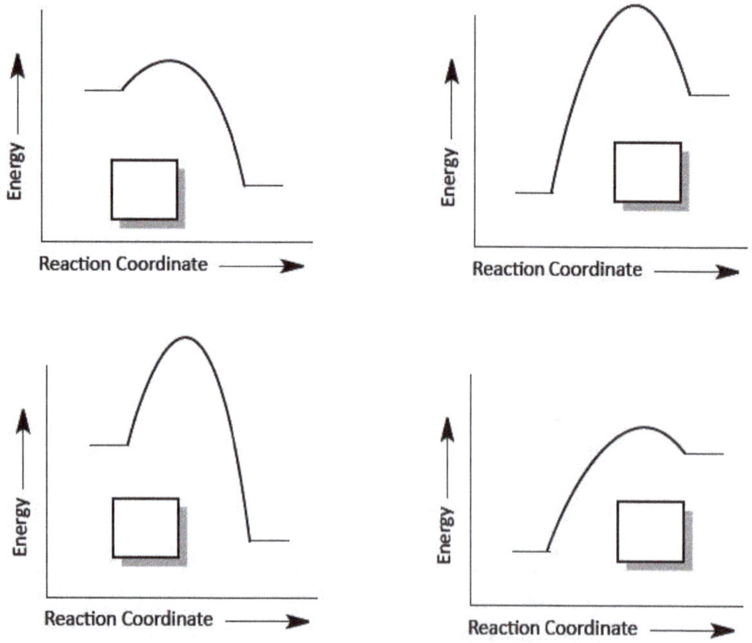

Catalysts lower the activation energy of a reaction and speed up the reaction. Catalysts are not consumed in the process. They must be regenerated so only a small amount of catalyst is needed. In biology, certain proteins called enzymes act as catalysts. Notice, catalysts only change the energy of the transition state by lowering the E_a. Catalysts do not influence the energies of the reactants or products at all.

Even though exothermic reactions (-ΔH) are more likely to proceed, more needs to be considered.

8.8.3 Entropy

Entropy (S) is the amount of disorder or randomness. The change in the entropy (ΔS) of a reaction is a consideration. If the products of a reaction are more disordered than the reactants, ΔS is positive. If the products are more ordered, ΔS is negative. How can we predict ΔS? Reactions that go to more disorder (+ΔS) are favored.

If two reactants react to form one product, there is less disorder in the system. ΔS is negative.

A + B	⟶	A—B	**(-ΔS)**
Two reactants		One product (Fewer products, less disorder)	

If one reactant forms two products, there is more disorder in the system. ΔS is positive.

A—B	⟶	A + B	**(+ΔS)**
One reactant		Two products (More products, more disorder)	

So, ΔH and ΔS both influence whether a reaction will proceed towards the products. If a reaction proceeds or not is measured by the Gibbs free energy change (ΔG). If ΔG is positive, energy must be put into the system for the reaction to proceed. If ΔG is negative, the reaction will proceed spontaneously. The relationship between ΔG, ΔH, and ΔS can be expressed by the following equation.

$$\Delta G = \Delta H - T\Delta S$$

ΔG = Change in Gibbs free energy (Negative proceeds)
ΔH = Change in enthalpy (Negative helps)
T = Temperature (High temp. helps or hurts, depending on ΔS)
ΔS = Change in entropy (Positive helps)

Usually, ΔH is a greater term than TΔS, so the change in energy of the system is a good predictor of the spontaneity of the reaction. $\Delta G \cong \Delta H$.

8.9 Kinetics

The process of diamond turning into graphite has a negative ΔG. It happens spontaneously. But, my wife's wedding ring is not turning into a worthless hunk of graphite any time soon. This process takes eons to occur. It is important for us to understand something about how fast a particular reaction will occur. We need to understand the *kinetics* of a reaction. Kinetics is the study of the rates of reactions.

The kinetics of a particular reaction is expressed using a rate equation. For a typical reaction, the rate equation takes the following form.

$$A + B \longrightarrow C + D \qquad \text{Rate} = k_r[A]^a[B]^b$$

[A] = the concentration of A
[B] = the concentration of B
a = the order of the reaction with respect to A
b = the order of the reaction with respect to B
k_r = the rate constant. If k_r is high, the reaction is fast, if k_r is low, the reaction is slow.

We can know the values for [A] and [B] as we know how much of each reactant we use. But, what about a, b, and k_r? The good news is that you will never be asked to predict these. These values **can only be discovered experimentally**! An experiment needs to be carefully performed to measure how quickly A and B are used up in the reaction. This experiment can give us the values of k_r, a, and b. Let's look at a couple of example reactions to make sense of this.

8.9.1 A one-step reaction

$$\text{Rate} = k_r[HO^-]^1[CH_3Br]^1 \quad \text{or} \quad \text{Rate} = k_r[HO^-][CH_3Br]$$

Figure 8.7 SN_2 Backside Attack. A one-step reaction.

In this reaction, hydroxide reacts with bromomethane to make methanol. The rate constant can be measured experimentally, and changes depending on the solvent and the temperature of the reaction. From the rate equation, we see that the superscript for each reactant is 1. The 1 can be written, or like superscripts of 1 in mathematics, they may not be written. We then say that the reaction is first order with respect to hydroxide and first order with respect to bromomethane. Overall, the reaction is a second-order reaction (1+1=2). Let's start with a concentration of 1 for each reactant.

$$\text{Rate} = k_r[HO^-][CH_3Br] \;\ldots\; \text{Rate} = k_r[1][1] \;\ldots\; \text{Rate} = k_r$$

Now, let's double the amount of hydroxide in the reaction.

$$\text{Rate} = k_r[HO^-][CH_3Br] \;\ldots\; \text{Rate} = k_r[2][1] \;\ldots\; \text{Rate} = 2k_r$$

The rate of the reaction doubled. This makes sense if you think about the reaction. For the reaction to proceed, a hydroxide ion must bump into a bromomethane molecule. When the amount of hydroxide in the reaction flask is doubled, there is twice the chance for this encounter to happen.

If instead of doubling the amount of hydroxide, we doubled the amount of bromomethane, we get the following.

$$\text{Rate} = k_r[HO^-][CH_3Br] \;\ldots\; \text{Rate} = k_r[1][2] \;\ldots\; \text{Rate} = 2k_r$$

The reaction rate again is doubled. Again, there is twice the chance for a hydroxide ion and a bromomethane molecule to encounter each other if the amount of bromomethane is doubled.

The reaction coordinate diagram for a one-step reaction has only one transition state.

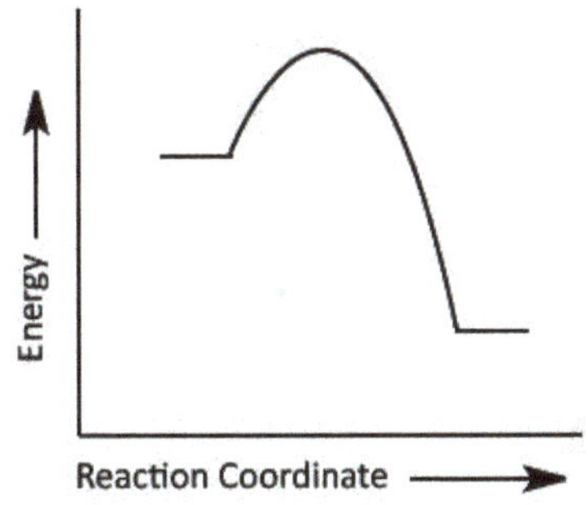

Figure 8.8 Reaction Coordinate Diagram for a one-step reaction

8.9.2 Two-step reactions

Many reactions are not first order with respect to each of its reactants. This may be a little more difficult to understand, but let's try with an everyday example first. Imagine a group of people making peanut butter and jelly sandwiches. Some people are putting peanut butter on bread. Others take this peanut butter-bread and put jelly on it. Finally, somebody cuts the sandwich in half. We could express this process as a reaction.

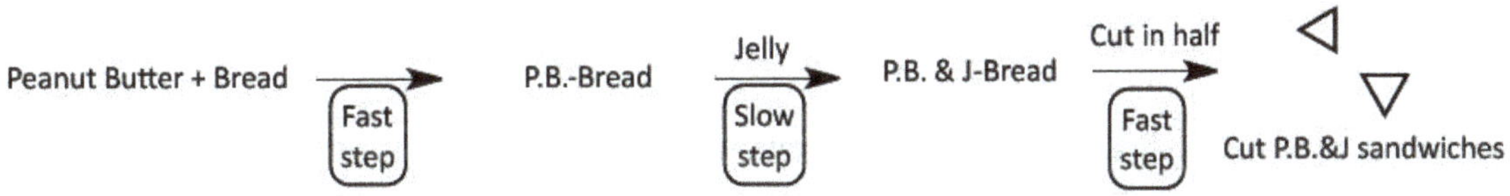

Figure 8.9 The kinetics of making peanut butter and jelly sandwiches

As the reaction proceeds, you can see that if the first step is fast, the second step is slow, and the third step is fast again, that the rate of the reaction (how fast we make peanut butter and jelly sandwiches) is determined by the slow step. The people putting jelly on the sandwiches slow us down. After a bit, you can bet that there will be a pile of peanut butter-bread stacking up at that step. What happens if we double the concentration of the folks putting peanut butter on bread? We do not make sandwiches any faster because more peanut butter-bread would stack up at the jelly station. The reaction is zeroth-order with respect to peanut butter. What happens if we double the amount of people that are cutting the sandwiches? Again, we do not make sandwiches any faster because they would be sitting around waiting for jelly to be put on the sandwiches. The reaction is zeroth-order with respect to sandwich cutting. The only way to step up the rate in which we make sandwiches is by increasing the number of jelly spreaders. If we double the number of people putting jelly on sandwiches, we double the rate at which we make sandwiches. In the end, the rate of this reaction is first-order with respect to jelly. We could write this rate equation as

$$\text{Rate} = k_r[\text{Jelly}]$$

The only way to predict the rate equation for this reaction was to do the experiment. If we tried to make sandwiches and the people spreading the peanut butter and adding jelly were working very quickly (fast steps) and the people cutting the sandwiches in half were talking and goofing off and had VERY dull knives (the slow step), then the reaction rate would be zero order with respect to peanut butter and jelly and first order with respect to cutting. The rate equation would be

$$\text{Rate} = k_r[\text{Cutting}]$$

The experiment needs to be performed to determine the rates. That is good news as the pressure is off on us to guess!

We've learned that the rate is determined by the slowest step in a process. This slowest step is then called the *rate-determining step* (r.d.s). The highest peak, the highest energy transition state, in a reaction-energy diagram is the transition state for the rate-determining step of a reaction.

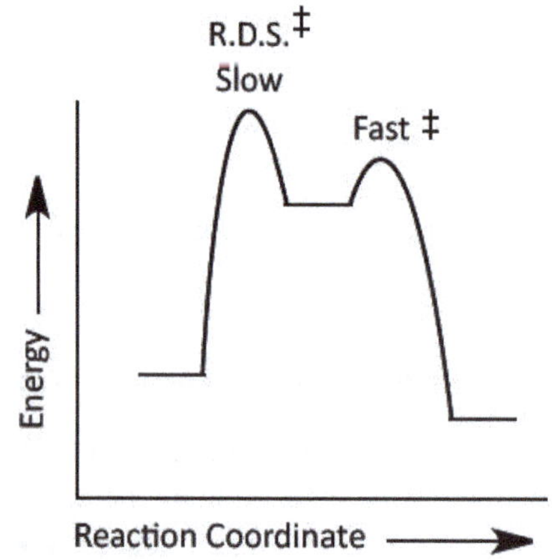

Figure 8.10 Reaction Coordinate Diagram for a two-step reaction

8.10 Ions

An *ion* is an atom or molecule that has a net charge- either positive or negative. An *anion* is a negatively charged ion. Some people remember this by using the following mnemonic.

ANION = **A Negative ION**

A *cation* is a positively charged ion. Some people remember this because the "t" in the middle of the word "cation" looks like a positive sign.

Cation = ca+ion

8.11 Reactive Intermediates

In an organic reaction, sometimes a species is made that is short-lived and reactive. They usually have an unusual number of bonds. These species are created on the way to making some product. They are intermediates in the reaction. We call these species reactive intermediates. There are four main types of reactive intermediates.

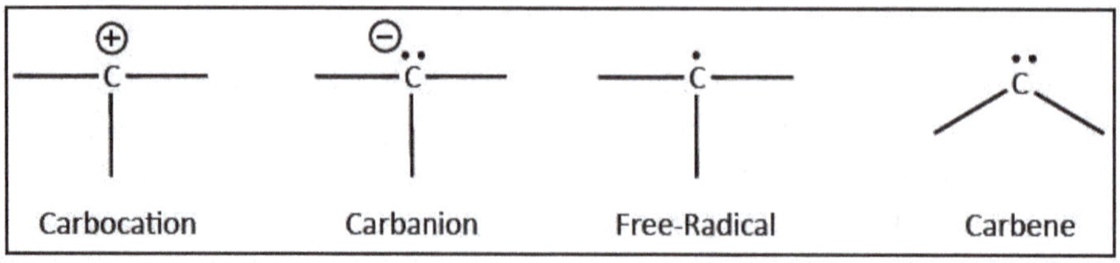

Figure 8.11 Reactive intermediates

8.11.1 Carbocations

A carbocation is a carbon atom that is a cation (a positive ion). The simplest example of a carbocation is methyl carbocation.

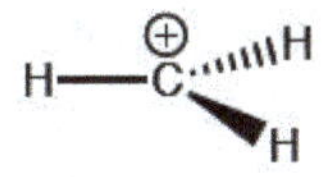

Figure 8.12 Methyl carbocation

A methyl carbocation only has hydrogen atoms attached to the carbocation. A primary carbocation has one carbon atom attached to the carbocation. A secondary carbocation has two carbon atoms attached to the carbocation. A tertiary carbocation has three carbon atoms attached to the carbocation.

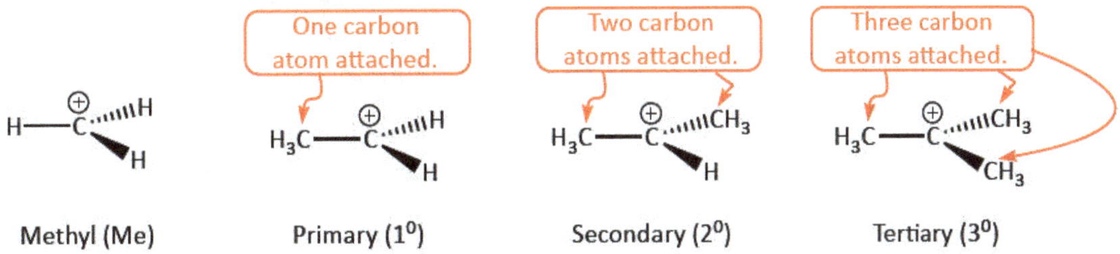

We use the terms primary, secondary, and tertiary often in organic chemistry to describe the generic forms of molecules. For example, below is how we name generic amines and generic alcohols. R = the Rest of the molecule. In this case, it means at least a carbon atom is there.

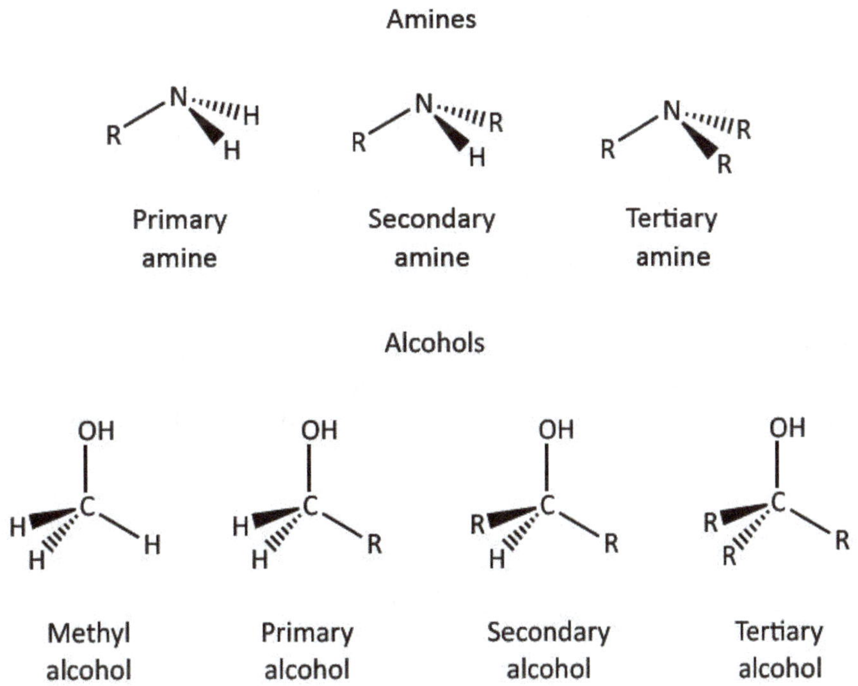

Figure 8.13 Primary vs. secondary vs. tertiary

Carbocation stability

A carbocation has three sigma bonds and an empty p-orbital on the carbon atom. The empty p-orbital has no negative electrons in it. This is what gives it its positive charge. Nature abhors a localized charge like this. So, in order to try to fix this positively charged problem, the positively charged, empty p-orbital, wants to have more negative electrons donated to it. If it could only get some negative electrons given to it, it would be relieved. It is like a dry mouth that wants to be quenched with water. Water needs to flow towards the dry mouth to help, not away from it.

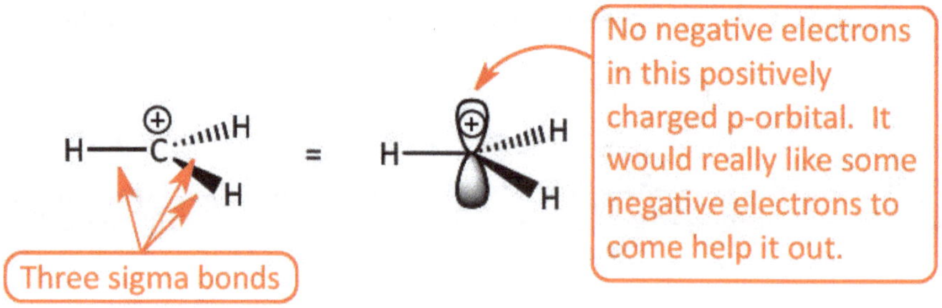

Through Hyperconjugation

So, how can water flow towards the dry mouth? How can negative electrons flow towards the empty p-orbital to help stabilize it? One very common way is through a simple process that uses a ten-dollar word, *hyperconjugation*. Don't let the big word discourage you. Hyperconjugation is simple to understand. To begin to understand it, let's look at primary carbocations.

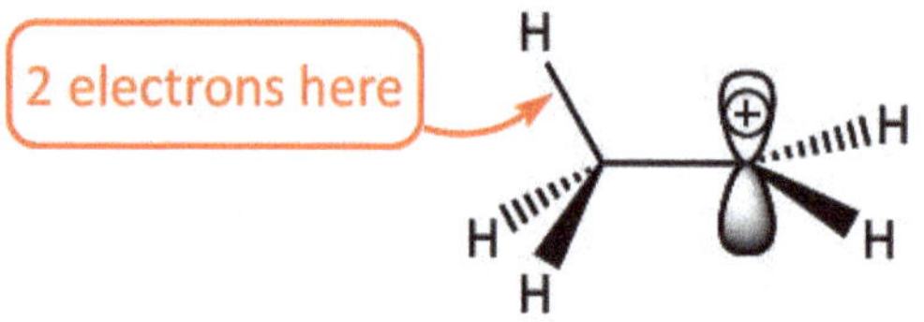

Primary (1^0) carbocation

There are two electrons in each C-H bond. This bond overlaps with the empty p-orbital. Both the C-H bond and the empty p-orbital are in the plane of the paper. When they overlap, the electrons in the bond can donate "through space" to the empty p-orbital. This helps quench the thirst of the p-orbital. It helps stabilize the carbocation.

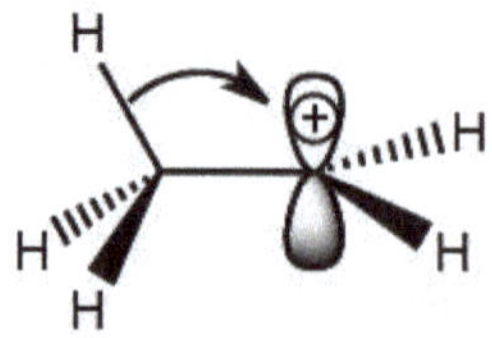

If we twist it to another angle, you can maybe better see that the C-H bond and the empty p-orbital are both pointing vertically and can overlap.

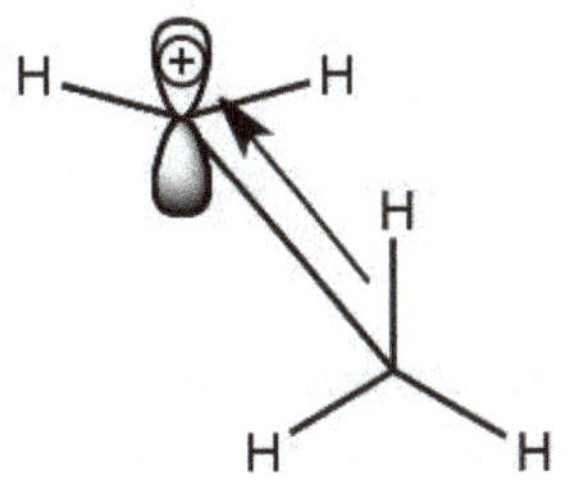

Because of the way we draw molecules on paper, it may seem strange that these negative electrons in the C-H bond are pushing or moving through such a long distance to help stabilize the positive charge. But, remember that bonds are not really thin lines like we draw. The C-H bond is an sp^3 hybridized carbon orbital overlapping with a hydrogen 1s orbital.

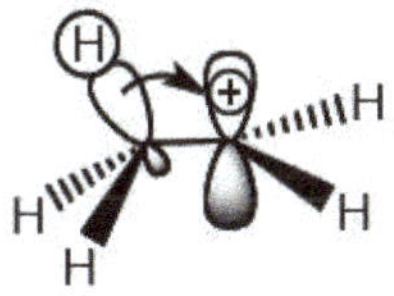

In fact, both the C-H bond and the empty p-orbital are so fat they actually overlap and touch each other. The electrons in the C-H bond can "leak" into the positive p-orbital of the carbocation to help stabilize it. They don't have to jump or travel very far at all.

Now, let's look at the hyperconjugation that can occur in all four types of carbocations we have seen.

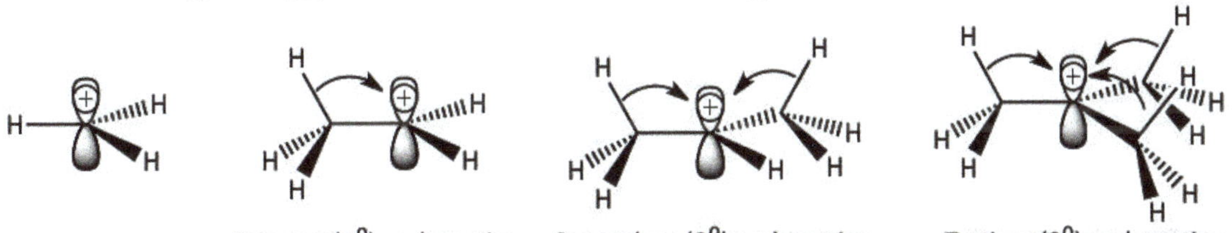

Primary (1^0) carbocation Secondary (2^0) carbocation Tertiary (3^0) carbocation

Figure 8.14 Hyperconjugation stability of carbocations

Notice that with no sp^3 C-H bond neighbors the methyl carbocation cannot hyperconjugate at all. Its thirst is not quenched. The positive charge is all alone and not stabilized in any way. This is why methyl carbocations are very rare. Primary carbocations have one attached carbon group that can help the positive charge. Its thirst is quenched a little, but it is still quite thirsty. There is a little stabilization of the positive charge. Secondary carbocations have two carbon groups that can help. Its thirst is beginning to be satisfied. Tertiary carbocations have three carbon groups that can help through hyperconjugation. This positive charge is the most stabilized with three sets of negative electron bonds donating into the positive p-orbital.

$$Me < 1^0 < 2^0 < 3^0$$

Worst Best

Figure 8.15 Order of stability of carbocations

You will find the order of stability of carbocations to be very useful when we predict reactions in the future. But, I warn you to not simply memorize this list. Think deeply about hyperconjugation until it simply makes sense that you want more carbon groups attached to the positively charged, empty p-orbital. If you can truly understand this concept, you will never again forget that tertiary carbocations are the most stable.

Through Resonance

Another way to stabilize carbocations is through resonance. If there is a pair of negative electrons in a pi bond one full step away from the carbocation, those electrons can be pushed towards the carbocation through resonance. This helps by spreading out the positive charge.

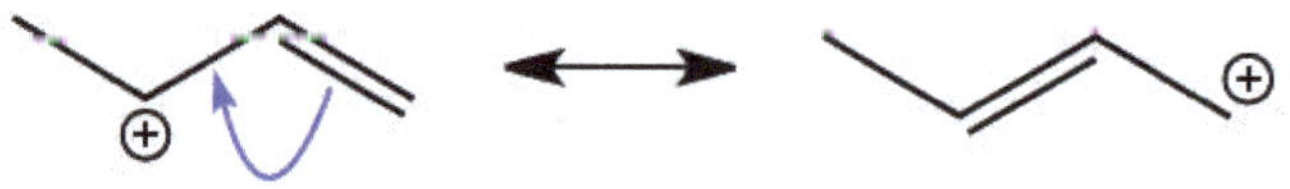

The positive carbocation can be stabilized by the negative electrons of the pi bond one full step away.

The secondary carbocation above is stabilized by two carbon groups through hyperconjugation. It is further stabilized through resonance. In the end, it has similar stability to a tertiary carbocation.

Another way resonance can help stabilize a carbocation is if a lone pair of negative electrons is nearby.

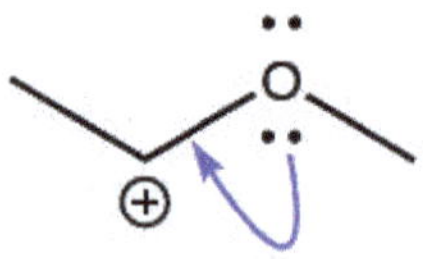

In this case, the lone pair of electrons can be pushed towards the carbocation to help stabilize it. Notice, this only works if the lone pair of electrons is on an atom one full step away from the carbocation. Below are several examples of heteroatoms that can help stabilize carbocations.

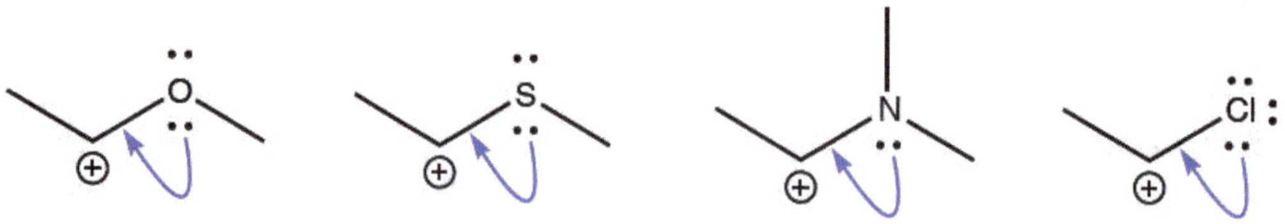

Be careful though, the lone pairs of electrons are not always drawn, so we need to be alert to when we draw them in ourselves to make resonance forms. You will often see them drawn without the lone pairs of electrons.

8.13 Draw in the appropriate number of lone pairs of electrons on each heteroatom. Then, draw an arrow to push electrons on each molecule to show how the carbocation is stabilized through resonance.

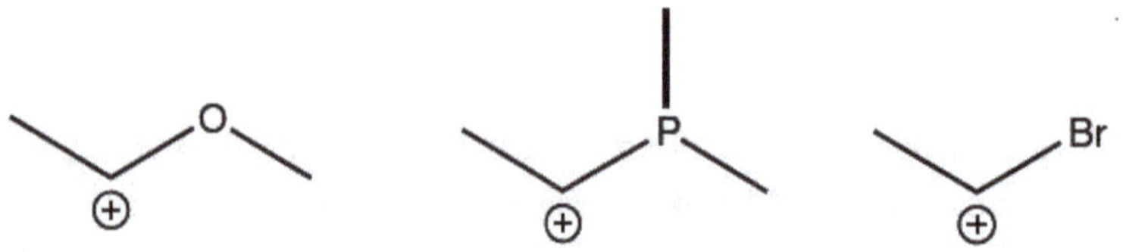

Through Induction

One final consideration for the stability of a carbocation is to look at inductive effects or induction. The inductive effect is an effect that takes place through the sigma bonds or the sigma bond framework. Notice that in hyperconjugation and resonance, the stability took place above or below the sigma bonds. Hyperconjugation can happen above or below the sigma bond and resonance can happen above or below the sigma bond. That depends on the orientation we happened to draw the molecules.

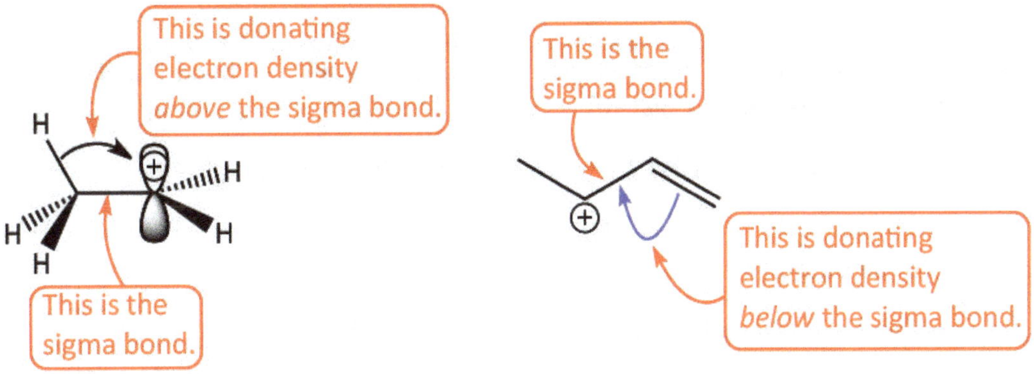

Hyperconjugation — Resonance

But, the inductive effect does not take place above or below the sigma bonds, but right through them! Usually, there is some atom or group attached that in some way affects the electron density. This group either sucks electrons towards itself (if it is electronegative or has a positive charge) or it donates electrons away from itself (if it is not very electronegative or has a negative charge). Below are four carbocations with a group (G) that either donates electron density away or attracts electron density towards itself.

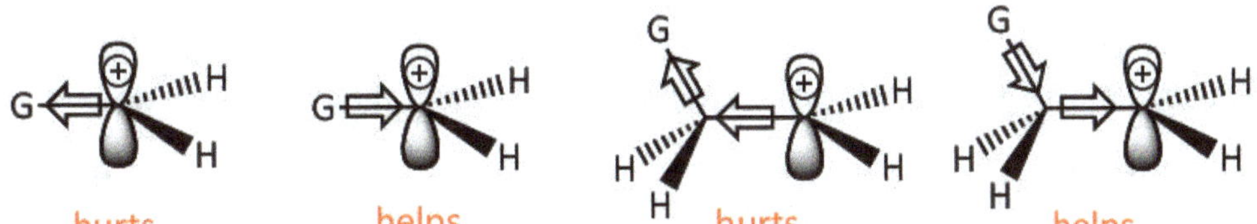

hurts	helps	hurts	helps
G is one bond away and sucks electrons away.	G is one bond away and donates electrons in.	G is two bonds away and sucks electrons away.	G is two bonds away and donates electrons in.

Figure 8.16 Induction stability of carbocations

Below are two similar carbocations. The only difference between them is one atom, a different halogen.

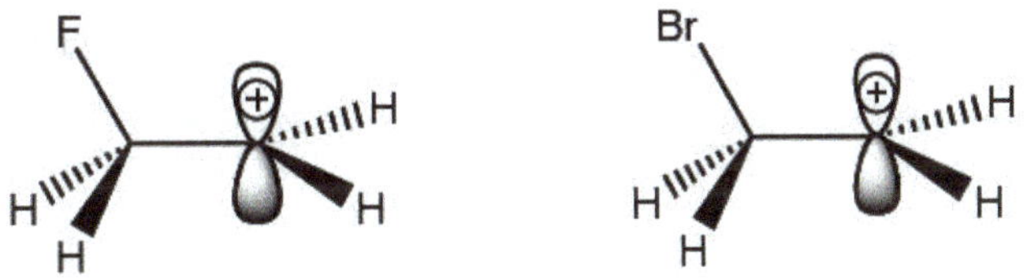

One of these carbocations is more stable, and one is less stable. Let's reason through this together.

8.14 If negative electrons are pushed towards the positive p-orbital of a carbocation, does this help or hurt the carbocation? ____________

8.15 If negative electrons are sucked away from the positive p-orbital of a carbocation, does this help or hurt the carbocation? _____________

You should be able to do the next problem in your head. If not, review chapters 1 and 2.

8.16 The strength of an atom to pull negative electrons towards itself in a bond is called electronegativity. Between F and Br, which is more electronegative? _________

8.17 Since F and Br are both fairly electronegative, do they help or hurt the carbocation they are on? _____________

8.18 Which hurts the carbocation more, F or Br? __________

8.19 Circle the structure below that is the most stable (or is hurt the least).

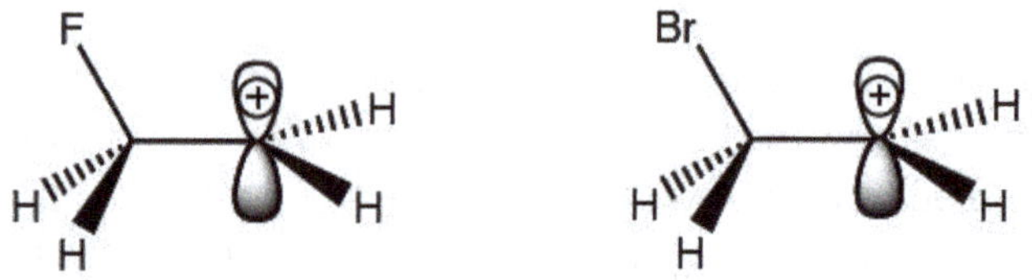

Sometimes the group that affects the electron density is close to the carbocation p-orbital, and sometimes it is farther away. Of course, when the group is close, it has a larger effect than if it is far away.

In the following two structures, having the F atom attached hurts one carbocation worse than the other.

8.20 Draw a circle around the molecule that is hurt worse by the electronegative fluorine atom.

8.21 Now, think about which is the happiest and draw a square around the most stable carbocation of the two.

8.11.2 Free Radicals

Free radicals are much like carbocations. They are sp^2 hybridized but the p-orbital has one electron in it. Free radicals are not as positively charged as carbocations, but they are slightly electron poor. Much like carbocations, free radicals want more electrons!

Stability of Free Radicals

Free Radicals are stabilized the same way as carbocations. Tertiary free radicals are the most stable because alkyl groups would donate electrons towards the electron poor free radical carbon through hyperconjugation.

Me < 1^0 < 2^0 < 3^0
Worst Best

Figure 8.17 Order of stability for free radicals

Resonance forms can help stabilize free radicals if the lone electron can be shared over more than one atom.

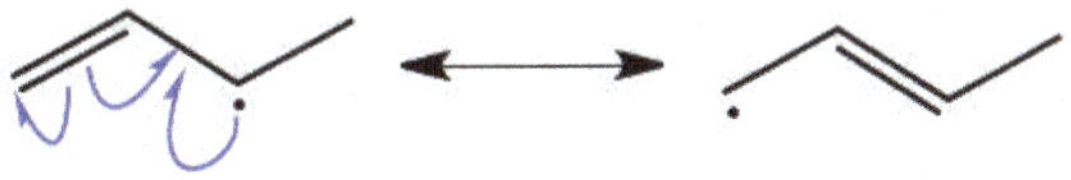

Figure 8.18 Resonance stability of free radicals

Electronegative atoms attached to the free radical would hurt the electron deficient free radical by trying to pull the negative electrons away from it. Free radicals prefer groups, like alkyl groups, that donate electrons towards them.

8.11.3 Carbanions

A carbanion is an anionic carbon atom. It is a carbon atom with three bonds and a lone-pair of electrons. Carbanions are sp^3 hybridized with the lone pair of electrons in one of the hybrid orbitals. Carbanions are nucleophilic because of their negative charge.

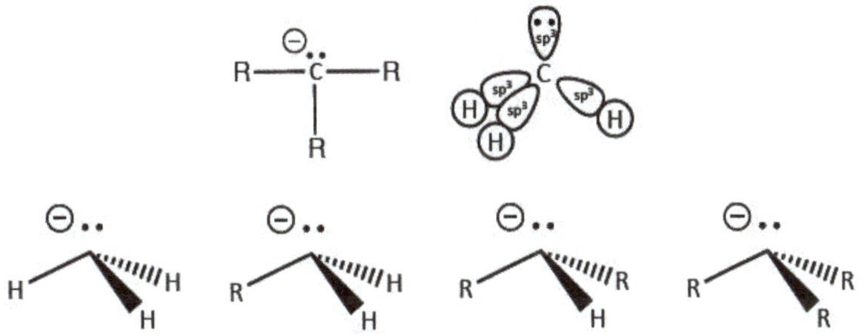

Methyl (Me) Primary (1°) Secondary (2°) Tertiary (3°)

Figure 8.19 Carbanion examples

Carbanion stability

Nature abhors a localized charge like this. So, in order to try to fix this negatively charged problem, the negatively charged carbon wants to push the negative electrons away from it. If negative electrons try to come towards the carbanion, the carbanion is destabilized.

Me > 1^0 > 2^0 > 3^0
Best Worst

Figure 8.20 Carbanion stability

How about Hyperconjugation?

What do alkyl groups do on a carbanion? Do they help or hurt the negative charge? We know that alkyl groups (R groups) donate electrons through hyperconjugation. If they donate electrons in towards the lone pair of electrons on our carbanion, this hurts the stability of the carbanion. This makes the negative charge even more negative! So, alkyl groups hurt carbanions and make them less stable. Therefore, the stability of carbanions is:

How about Resonance?

Resonance can help spread out the negative charge. If the lone pair of electrons can be pushed away using resonance, it helps. Often, if the carbanion is next to a carbonyl, a nitrile, or a nitro group, it is stabilized through resonance.

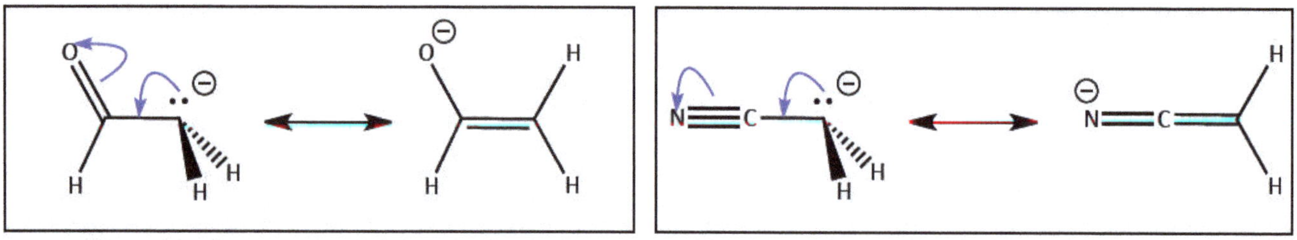

Figure 8.21 Resonance stability of carbanions

How about Induction?

Induction is through the sigma bond framework. Let's look at a generic group, G, on a carbanion.

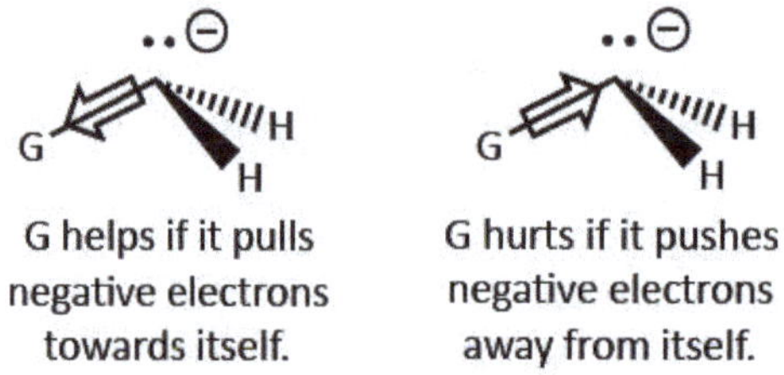

Figure 8.22 Induction stability of carbanions

So, what type of group do we want G to be to help stabilize the negative carbanion? Electronegative atoms help. Fluorine would be an excellent group to have on a carbanion.

8.22 Rank the following carbanion from most stable (1) to least stable (3).

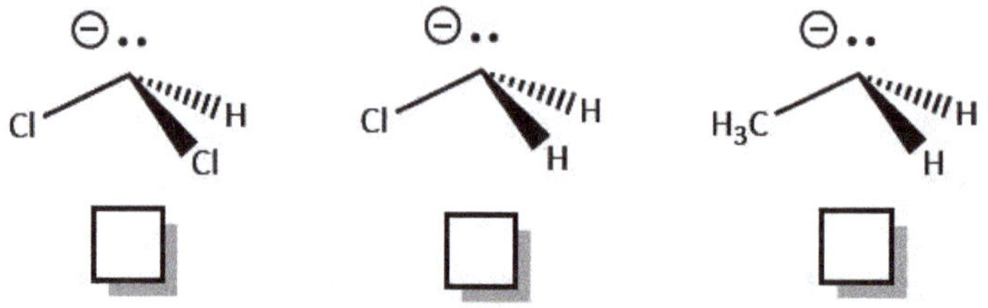

8.23 Rank the following carbanions from the most stable (1) to the least stable (4).

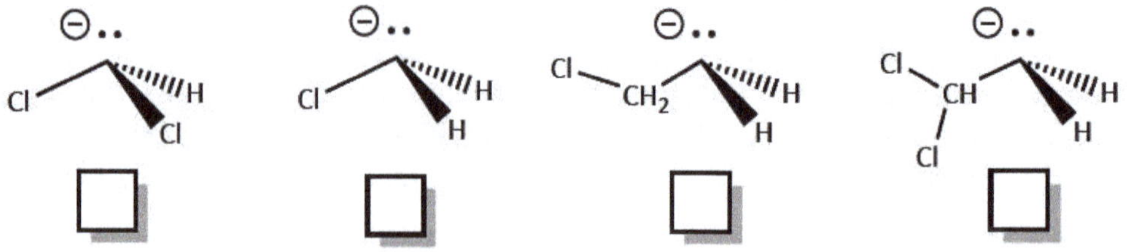

8.11.4 Carbenes

Carbenes have a formal charge of zero. They are very reactive species. Carbenes are carbon atoms with two bonds and a lone pair of electrons. Carbenes are sp^2 carbon atoms with a lone pair of electrons in an sp^2 hybrid orbital and an empty p-orbital. Carbenes are both electrophilic and nucleophilic. They are electrophilic because of the empty p-orbital and nucleophilic because of the lone pair of electrons.

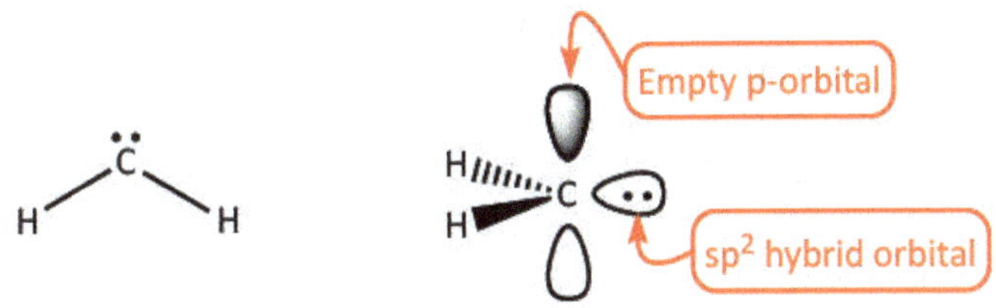

Figure 8.23 Carbenes

We will learn more about carbene chemistry when we study alkenes, but in general, they turn alkenes into cyclopropanes.

8.24 Carbenes are:

a) electrophilic
b) nucleophilic
c) both electrophilic and nucleophilic

8.12 Problems

8.25 In the reaction below, label each reactant as a nucleophile or an electrophile. Draw the appropriate arrow(s) on the reactant side of the reaction to show the movement of electrons for how this reaction proceeds.

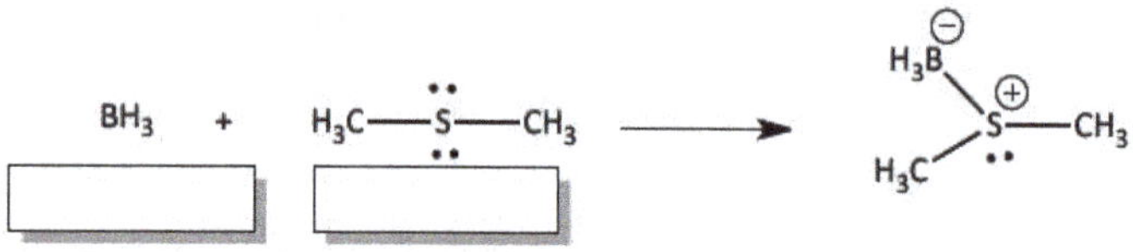

8.26 Label the electrophile and nucleophile of the following. Draw the mechanism of the reaction (two arrows) and the product it forms.

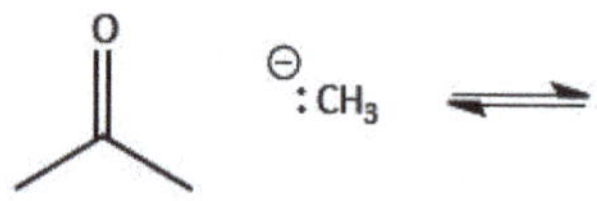

8.27 The triphenylmethyl cation is very stable.

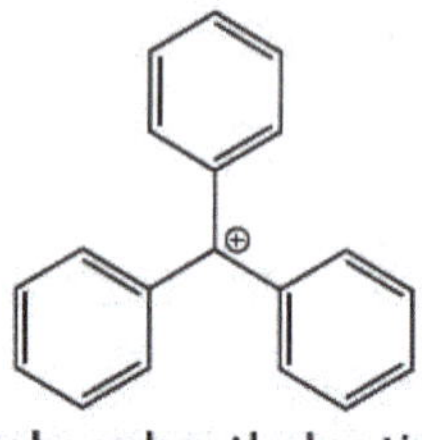

triphenylmethyl cation

a) What is the hybridization of the carbocation?
b) Specifically describe how hyperconjugation helps stabilize the triphenylmethyl cation.
c) What would happen to the stability of this cation if fluorine atoms were attached to each of the carbons of the phenyl groups? Would this cation be more or less stable? Why?

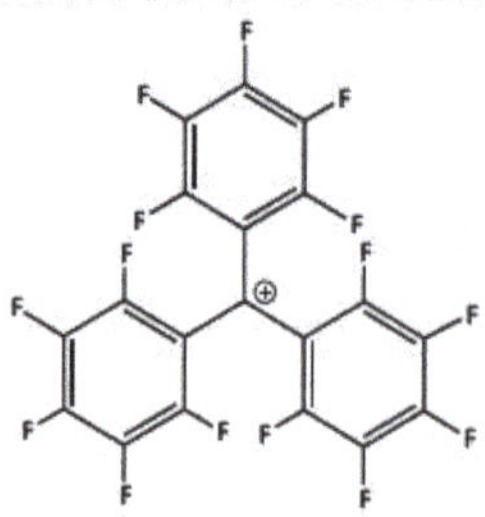

8.28 The following reaction occurs readily: $CH_3Br + I^- \rightarrow CH_3I + Br^-$
Experimentally, a chemist finds that if the concentration of I^- is doubled, the rate doubles. Also, if the concentration of CH_3Br is doubled, the rate is doubled.

a) What is the order of the reaction with respect to CH_3Br?
b) What is the order of the reaction with respect to I^-?
c) What is the overall rate order of the reaction?
d) Fill in the superscripts for the rate equation for this reaction: rate = $k[CH_3Br][\,I^-]$

8.29 The following reaction was performed: $(CH_3)_3CBr + H_2O \rightarrow (CH_3)_3COH + HCl$
Experimentally, a chemist finds that if the concentration of $(CH_3)_3CBr$ is doubled, the rate doubles. Also, the reaction rate never changed when the amount of water changes in the reaction.

a) What is the order of the reaction with respect to $(CH_3)_3CBr$?
b) What is the order of the reaction with respect to H_2O?
c) What is the overall rate order of the reaction?
d) Fill in the superscripts for the rate equation for this reaction: rate = $k[(CH_3)_3CBr][H_2O]$

8.30 Which of the following is a carbene?

a) $CH_2{=}CHCH_2O^-$
b) Br^-
c) $:CCl_2$
d) $CH_3CH_2^+$
e) NC^-

8.31 For the following carbocations, a) label them as methyl, primary, secondary, or tertiary. b) If any of the carbocations are resonance stabilized, make note of that. c) Finally, rank these carbocations in order of increasing stability.

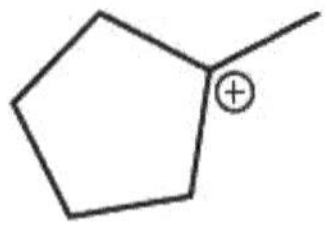
A

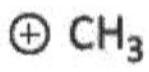
B

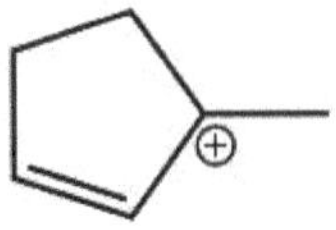
C

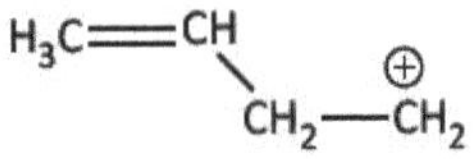
D

E

8.32 For the following free radicals, a) label them as methyl, primary, secondary, or tertiary. b) If any of the free radicals are resonance stabilized, make note of that. c) Then, rank these free radicals in order of increasing stability.

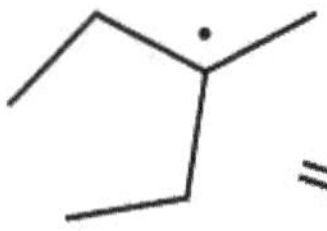
A

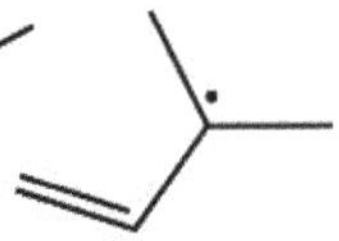
B

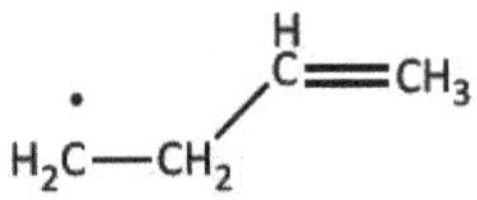
C

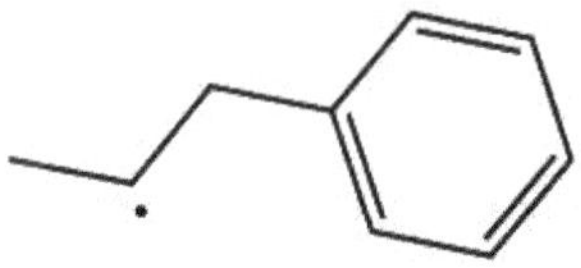
D

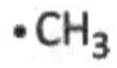
E

8.33 For the following carbanions, a) label them as methyl, primary, secondary, or tertiary. b) If any of the free radicals are resonance stabilized, make note of that. c) Then, rank these free radicals in order of increasing stability.

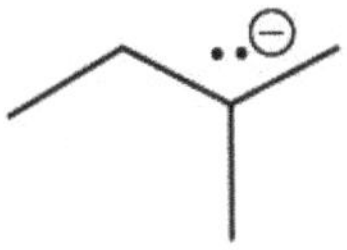
A

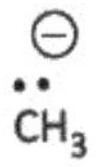
B

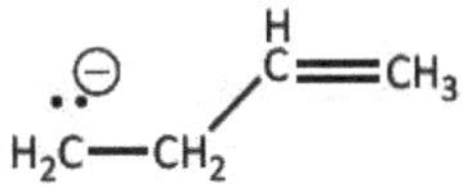
C

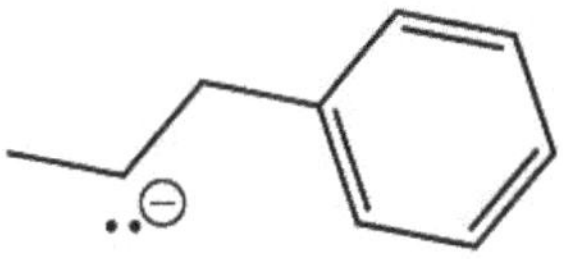
D

8.34 The protons drawn on the carbon atom in between the two carbonyl groups in the following diketone are fairly acidic. Following deprotonation, the resulting carbanion is resonance stabilized. Draw the carbanion and the other good resonance forms for it.

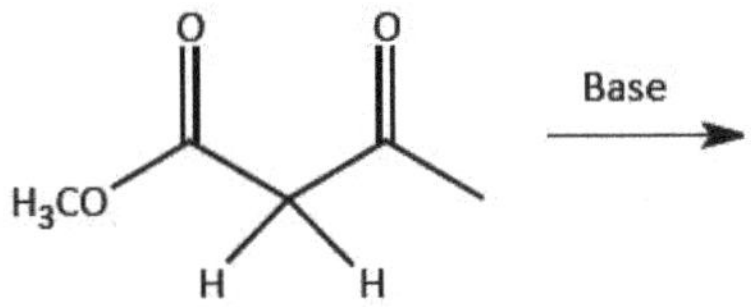

8.35 When diazomethane is heated, it gives off nitrogen gas and makes a methyl carbene. Draw the Lewis structure of the methyl carbene created.

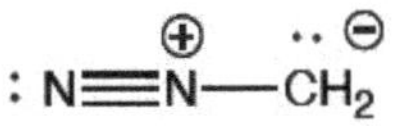
Diazomethane

8.36 From the terms supplied below, insert letters in blanks that correspond to the appropriate labels. You may use a letter more than once.

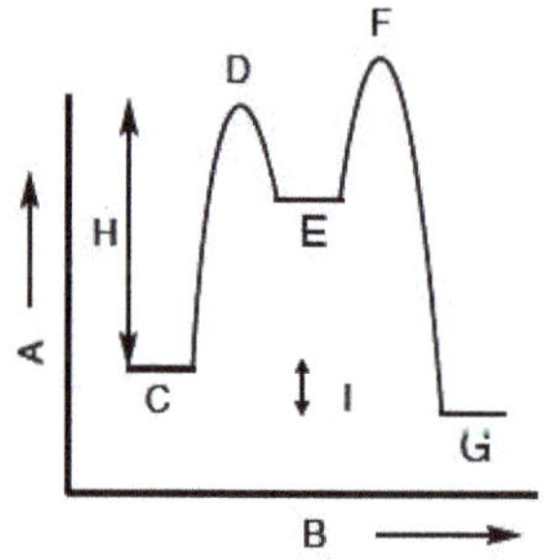

____ Energy ____ Reaction coordinate ____ intermediate

____ E_{act1} ____ products ____ 1st transition state

____ reactants ____ rate-determining transition state

8.37 Sketch reaction coordinate diagrams that have the following features.

a) a 1-step reaction with an activation energy of +10 kcal/mol and a ΔH of -30 kcal/mol

b) a 1-step reaction with an activation energy of +30 kcal/mol and a ΔH of +25 kcal/mol

c) a 1-step exothermic reaction with a high activation energy

d) a 2-step reaction where the first step is the rate-determining step. The reaction is exothermic overall.

8.38 Rank the following circled protons from least to most acidic.

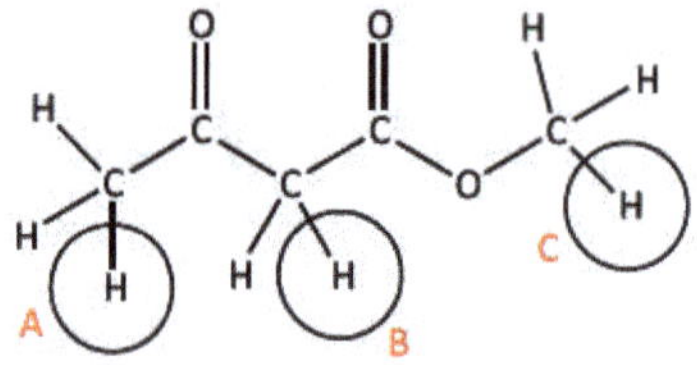

8.39 Alcohols can be protonated to make water as a good leaving group. Once the water leaving group leaves, it forms a carbocation.

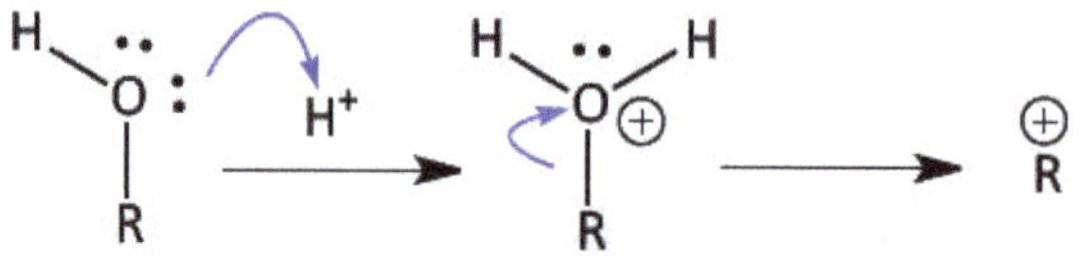

Rank the following alcohols in how quickly they will perform this reaction from the quickest (1) to the slowest (4).

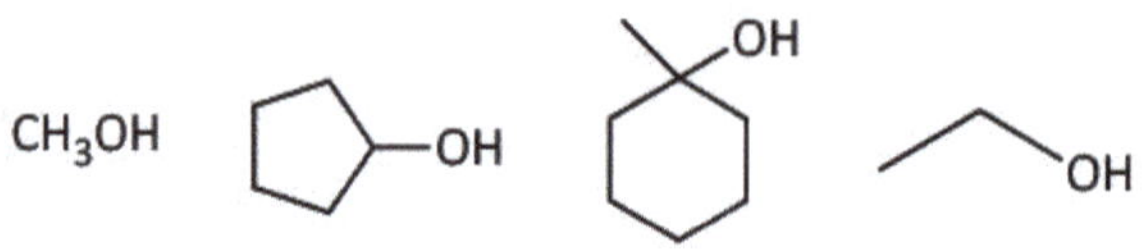

Chapter 9. Alkyl Halides

9.1 Introduction

A good starting point for learning organic chemistry reactions is to investigate organohalides. Organohalides are organic compounds that contain a halogen atom. They can be used to synthesize many different compounds. This is the first step into a wonderful world of making organic compounds. From new materials to life-saving medicines, synthetic organic chemistry is incredibly important to the human condition.

Organohalides can be categorized into several broad categories.

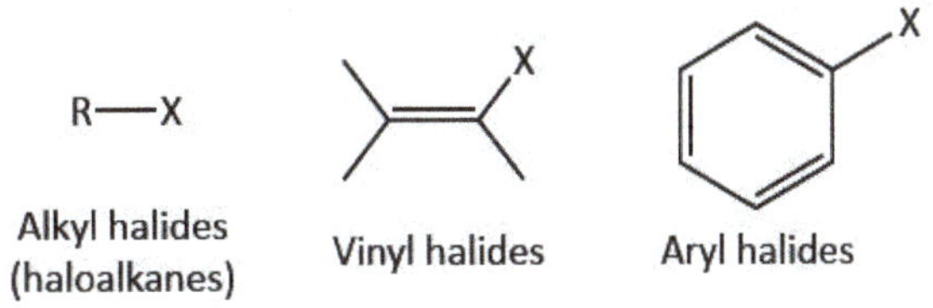

Figure 9.1 Generic organohalides

9.2 Nomenclature

9.2.1 Trivial Names

There are some trivial names for organohalides. The common name typically takes the form of "alkyl halide".

CH_3Br	CH_3CH_2I
Methyl bromide	Ethyl iodide

Some halogenated solvents are very common solvents used in the organic chemistry lab. Methylene chloride (or dichloromethane) is a very common solvent. At one time, methylene chloride was used to decaffeinate coffee. There were concerns that traces of it remaining in the coffee might cause cancer. This fear was largely unfounded as evidenced by animal studies. But, methylene chloride did extract some of the compounds that add to the wonderful taste and flavor of coffee. For these reasons, coffee is now decaffeinated using liquid carbon dioxide, so drink up.

Methylene chloride is still often used in the organic chemistry lab as a solvent. Chloroform is also often used in the organic lab and it was used industrially as a degreaser. It is more toxic than methylene chloride and a stronger carcinogen. It was also used historically as an anesthetic. Today, Halothane, $CF_3CHClBr$ is used as a general anesthetic because it is less toxic. Carbon tetrachloride is an excellent spot remover and was used as the primary dry-cleaning solvent for many years. Again, it is toxic and a carcinogen so the less harmful 1,1,1-trichloroethane has replaced it.

CH_2Cl_2	$CHCl_3$	CCl_4
Methylene chloride	Chloroform	Carbon tetrachloride (Carbon tet.)

9.2.2 IUPAC Nomenclature

The IUPAC system used the prefixes fluoro-, chloro-, bromo-, and iodo- followed by the alkane name to name alkyl halides.

2-fluorobutane dichloromethane 2-bromopropane iodomethane

9.2.3 Dihalides

If two halides are on the same carbon atom of a molecule, the molecule is a geminal dihalide. If two halides are on neighboring carbon atoms of a molecule, the molecule is a vicinal dihalide.

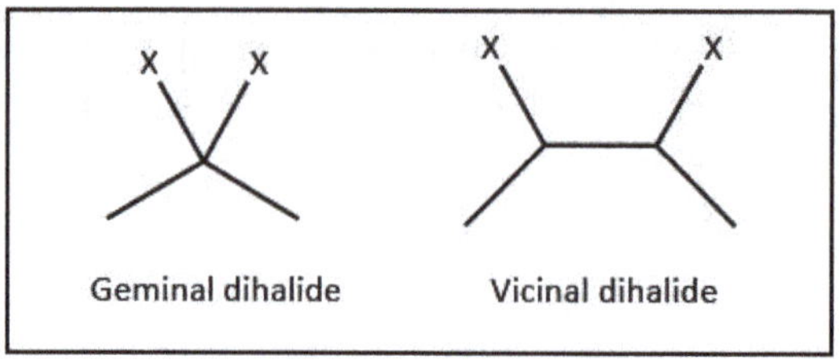

X = halide

9.1 Name the following compounds using IUPAC nomenclature.

______ ______ ______ ______

9.3 Preparation of Alkyl Halides

9.3.1 Free radical Halogenation

Industrially, alkyl halides are usually produced through a mechanism called free radical halogenation. In this reaction, a hydrogen atom of an alkane is replaced by a halogen. Free radicals are, of course, involved in the mechanism. There are three steps in this mechanism that you should memorize.

The *initiation* step of free radical halogenation generates free radicals. In free radical halogenation, this usually means generating Cl· or Br· by shining ultraviolet light on Cl_2 or Br_2. Once the reaction is initiated, a chain reaction occurs. This chain reaction is a two-step process called *propagation*. First, the chlorine radical reacts with the alkane. It pulls off a hydrogen atom generating an alkyl radical. This alkyl radical then reacts with Cl_2 to make the alkyl halide and another chlorine radical. Look closely at the propagation step. Notice that in each of the two propagation steps, a radical is produced. Radicals are always produced in propagation in order to propagate the chain reaction. For most of the reaction's life, a radical is most likely to bump into another molecule that is not a radical. The propagation continues. But, at the end of the reaction, the number of molecules that are not radicals gets low. At that point, the radicals in the reaction mixture are more likely to find another radical and combine together.

The *termination* step is when two radicals combine. A radical is not produced, so the reaction is not propagated, but it is terminated.

Step 1. Initiation (Getting it going by generating radicals.)

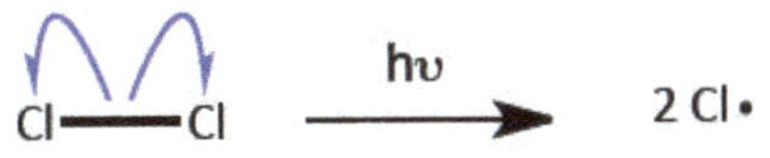

Remember that fishhook arrows show the movement of one electron.

Step 2. Propagation (A chain reaction that keeps making radicals.)

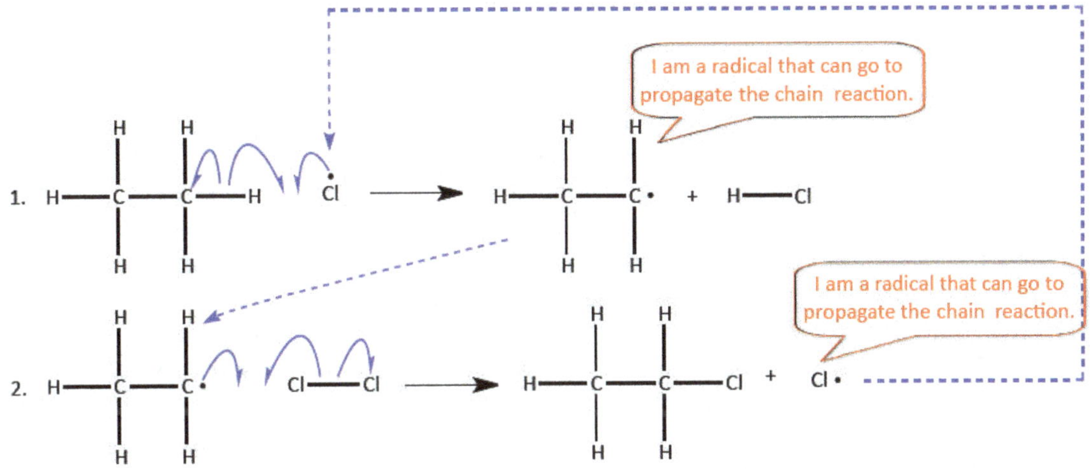

3. Termination (Putting a stop to it all because it combines two radicals and does not make a radical.)

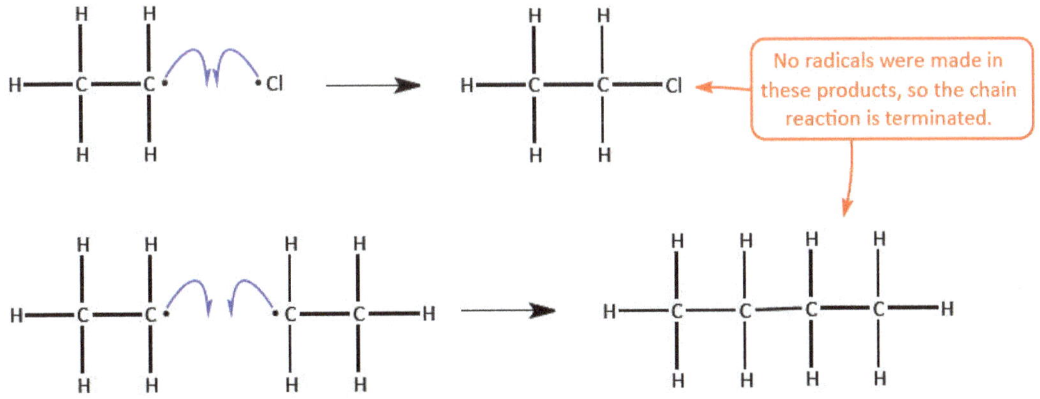

Chlorination of Propane

Sometimes, there is more than one type of hydrogen that can be replaced by the halogen. In propane, there are two secondary hydrogen atoms and six primary hydrogen atoms. If the chlorination of propane is a totally random process, we would expect more primary chloride (1-chloropropane) than secondary chloride (2-chloropropane) in a 6:2 (3:1) ratio. But, when the reaction is performed, more of the secondary chloride is produced! Why is this? When we look at the mechanism for the chlorination, we see that a secondary radical or a primary radical are formed. We've learned that secondary radicals are lower in energy and more likely to be made because they are better and more stable than primary radicals (through hyperconjugation).

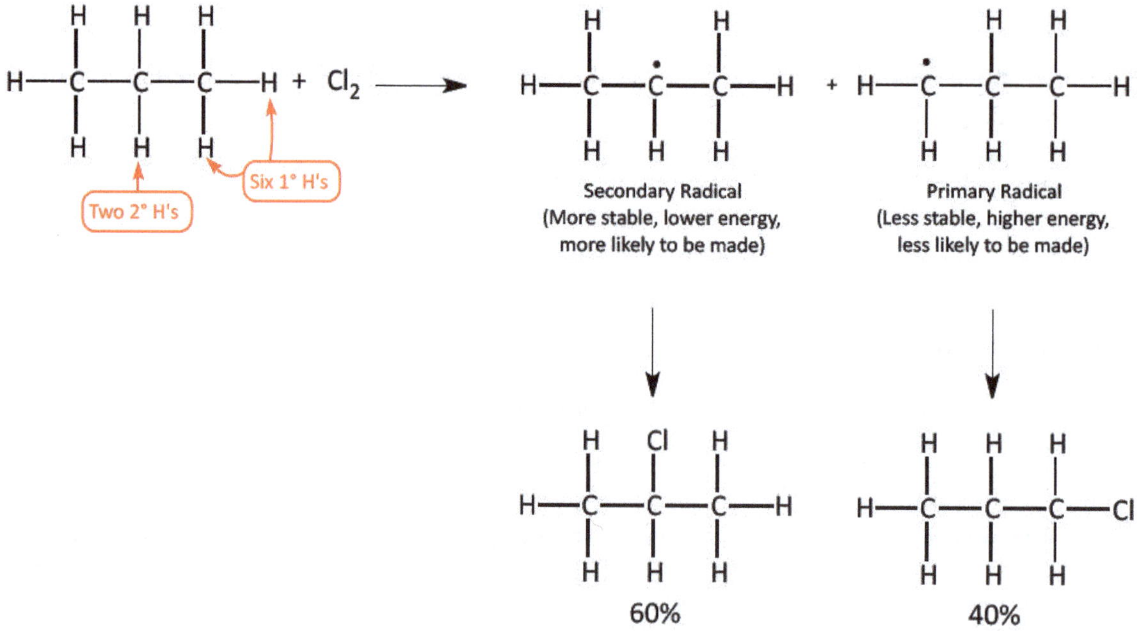

We can see why more of the secondary radical is produced by looking at the reaction-energy diagram. It is easier to slip over the secondary transition state because the activation energy is lower. It is more difficult to slip over the primary transition state because the activation energy is higher. It is more difficult, but not impossible, to slip over the primary transition state. That is why we get 40% 1-chloropropane. Knowing that if the process was random, we would expect 3:1 primary:secondary but we actually get 60:40 secondary:primary, chemists can do some math to get an understanding of how much more reactive secondary radicals are than primary radicals in a free radical chlorination reaction. Chemists have done many such reactions and have found that tertiary radicals form 5 times faster than primary radicals and secondary radicals form 3.8 times faster than primary radicals.

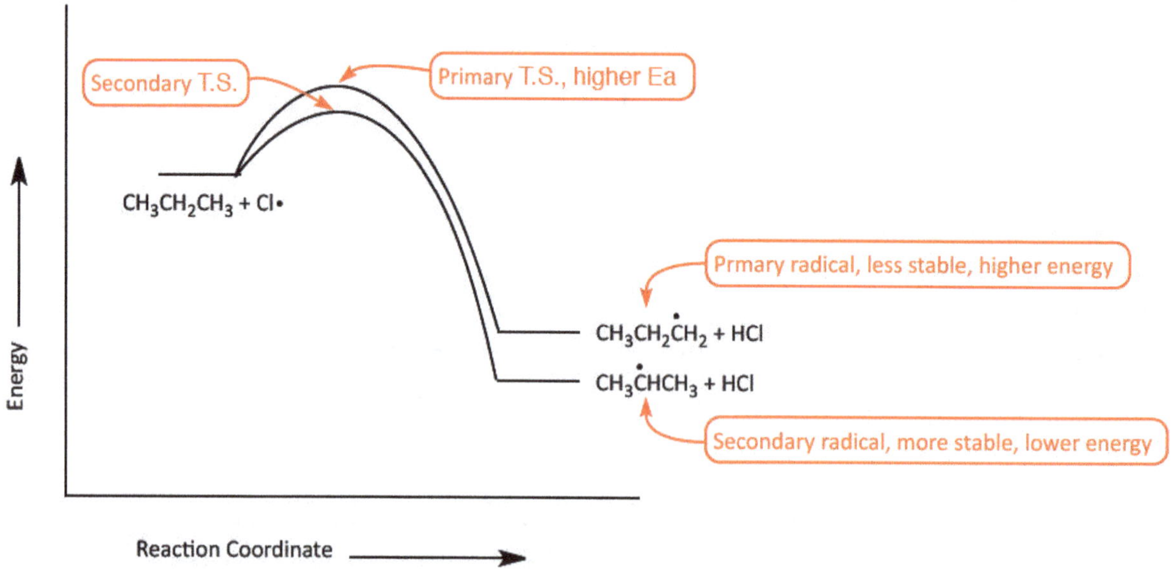

Figure 9.2 Reaction coordinate diagram for the chlorination of propane

Bromination of Propane

In the bromination of propane, we have similar considerations to think about. Once again, there are six primary hydrogen atoms and two secondary hydrogen atoms on propane. If the process was completely random, we would expect a ratio of 6:2 (3:1) 1-bromopropane:2-bromopropane. When the experiment is performed, we find the following results.

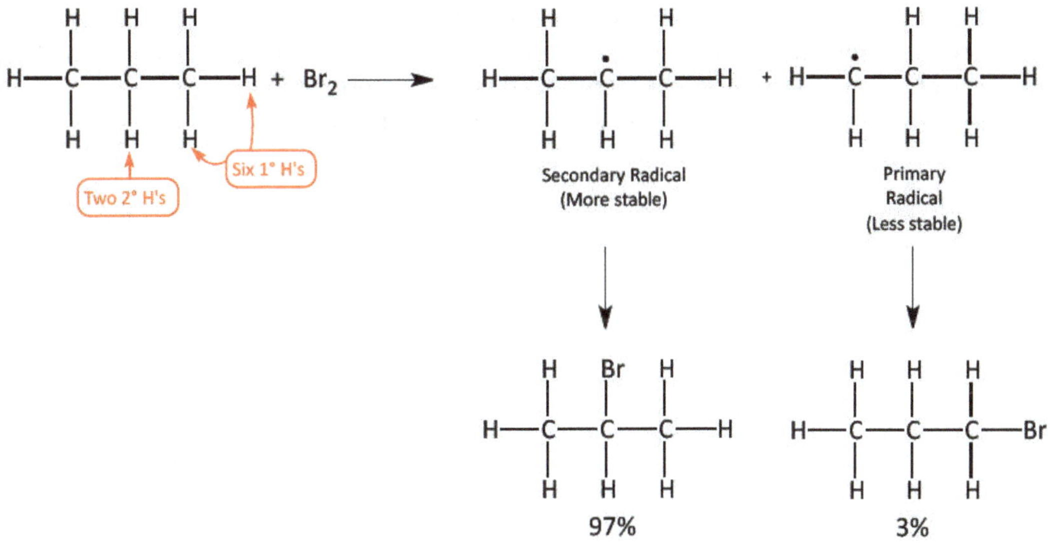

We get 97% of 2-bromopropane! This is a much higher percentage than in the chlorination reaction. We say that this reaction is more selective. The secondary radical is highly favored over the primary radical in this bromination reaction. Why is this? The reaction-energy diagram helps explain it. Unlike the chlorination reaction which is exothermic, the bromination reaction is endothermic. Because of this, there ends up being a bigger gap between the energies of the two transition states than there was in the previous chlorination reaction.

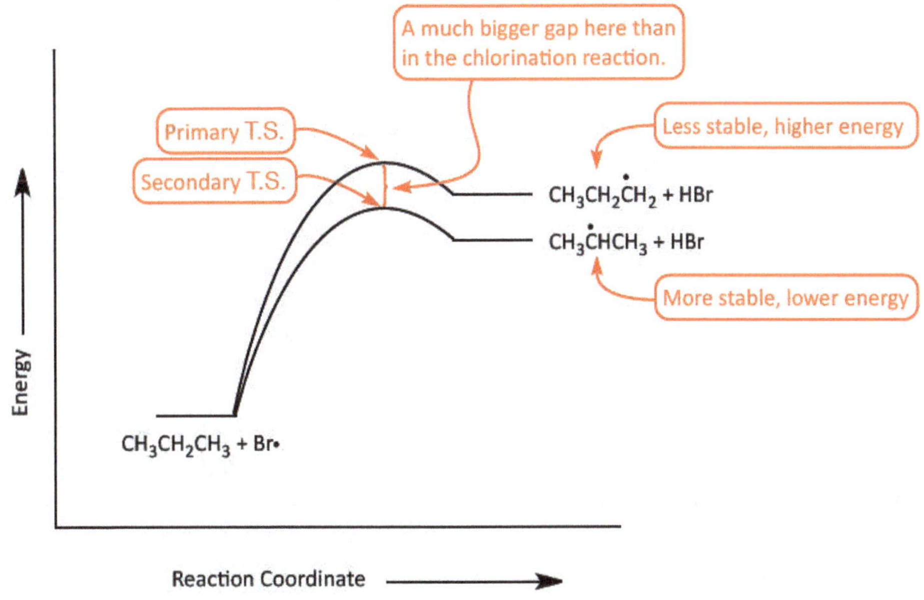

Figure 9.3 Reaction coordinate diagram for the bromination of propane

After chemists have done the experiments and math on bromination reactions, they found that tertiary radicals are 1600 times more reactive than primary radicals. Secondary radicals are 82 times more reactive than primary radicals. So, in bromination free radical reactions, the most stable free radical sites are where the bromination usually occurs.

The reason that there is a bigger difference in the energies of the transition states in an endothermic reaction over an exothermic reaction is because of *The Hammond Postulate*. The Hammond Postulate states that the transition state's structure is closest to the species of similar energy. For example, in the exothermic chlorination reaction, both the primary and secondary transition states are similar in structure to the starting propane material. In the endothermic bromination reaction, the transition states are more similar to the free radical products. This causes the difference between the two energies of the transition states to be larger.

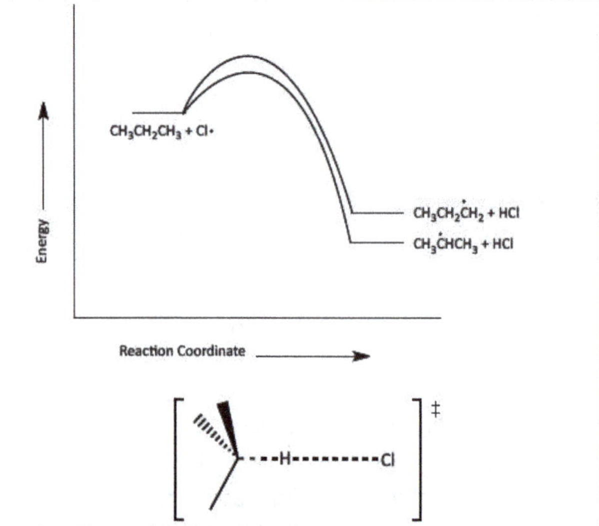

Exothermic chlorination
The T.S. is more similar to the reactants.

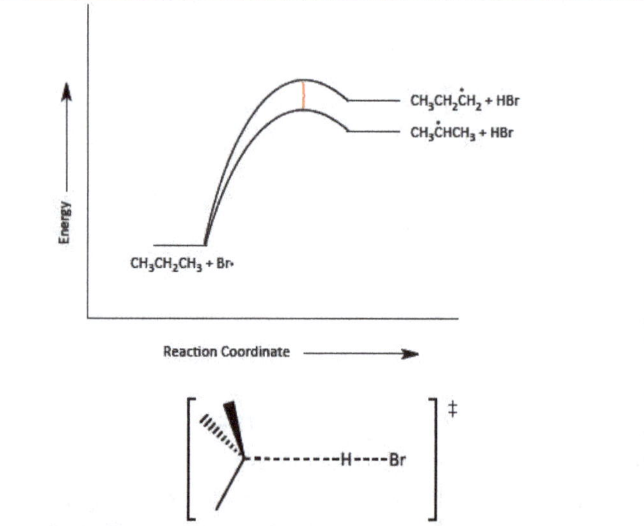

Endothermic bromination
The T.S. is more similar to the products (primary and secondary free radicals).

Allylic and Benzylic Radicals

Let's look at free radical reactions when there is a pi-bond present. When cyclohexene reacts with a bromine radical, the following product predominates.

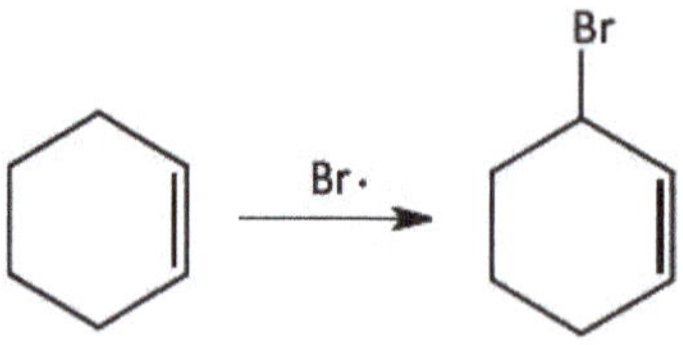

Let's analyze the hydrogen atoms that we could have replaced on cyclohexene. All of the hydrogen atoms on cyclohexene are secondary, yet their reactivity is very different. The hydrogen atoms on the double bond are called vinyl or vinylic hydrogen atoms. Hydrogen atoms on a carbon that neighbors the double bond are called allyl or allylic hydrogen atoms. This carbon is an allylic carbon atom. The other hydrogen atoms are regular alkyl hydrogen atoms. Allylic hydrogen atoms are the most reactive. Why is this since they are all secondary hydrogens? They are all secondary because each of the hydrogen atoms is on a carbon atom that is attached to two carbon atoms.

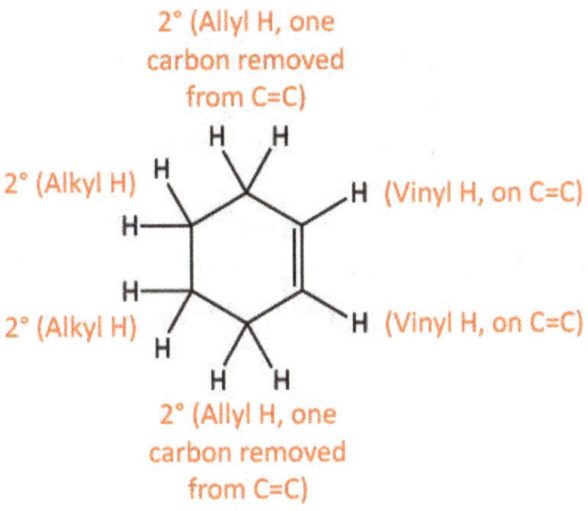

Figure 9.4 Vinyl vs. allyl vs. alkyl protons

If we look at the radical that is formed when the allylic hydrogen atom is removed, we find that resonance forms can be drawn for it. This greatly stabilizes the free radical.

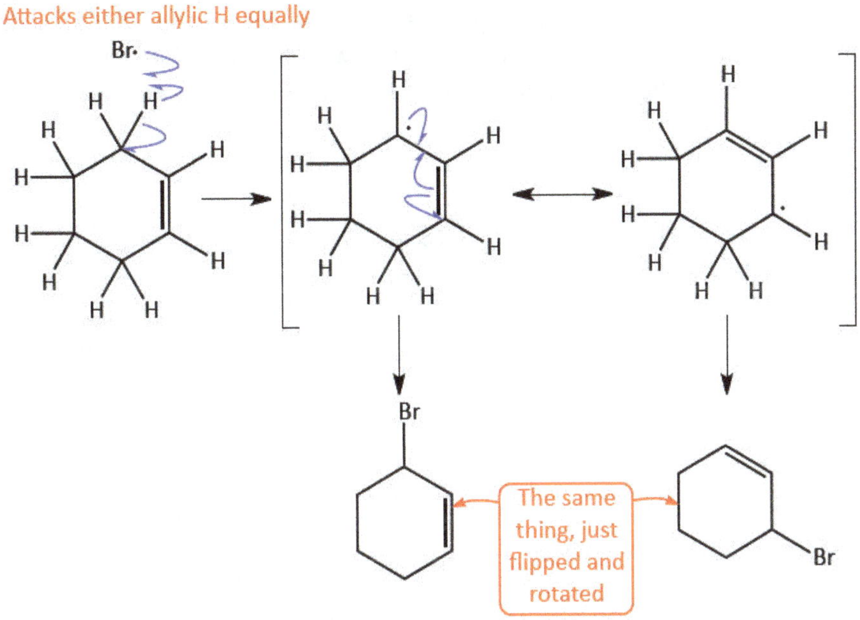

Figure 9.5 Free radical bromination at an allylic position

The 2-bromocyclohexenes that were formed above were drawn flat. In fact, they are not flat. A racemic mixture of two enantiomers is formed.

9.2 Draw a star on the stereogenic center of 2-bromocyclohexene. Then, draw the two enantiomers of 2-bromocyclohexene. Hint: you may want to draw in a missing hydrogen atom.

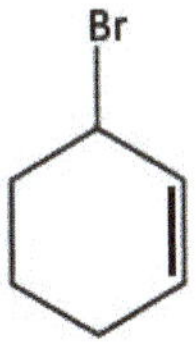

If a vinylic or alkyl hydrogen are removed, resonance forms cannot be drawn. Vinylic radicals are especially unstable. In a vinyl radical (much like in a vinyl cation), the electron deficient partially positive location is in an sp^2 hybrid orbital and held closer to the positive carbon nucleus than it is in an sp^3 hybrid orbital. This is a bad situation.

When we generate bromine radicals in the presence of a double bond, we need to be careful. If we did an initiation reaction with Br_2 and ultraviolet light like we have already learned, we would run into problems. The problem is that alkenes react with Br_2. We'll learn more about this in the alkene chapter, but a dibromination of the double bond can happen.

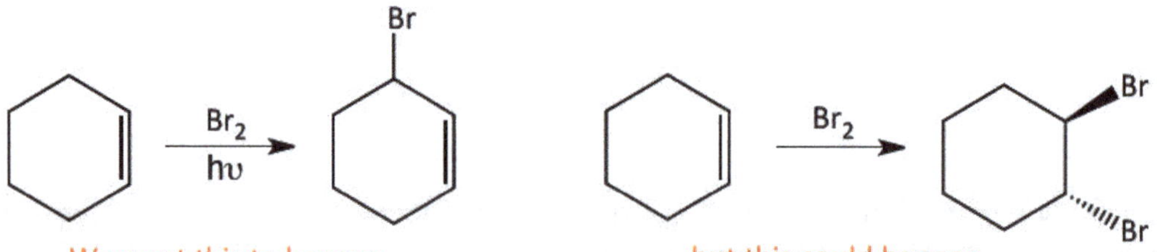

We want this to happen. . . but this could happen.

How do we avoid this problem? We need to make sure we don't have too much Br_2 present in our reaction mixture. The best way to do this is to perform this reaction with NBS (N-bromosuccinimide) and not Br_2. NBS generates Br_2 in small amounts so the dibromination of the double bond does not happen.

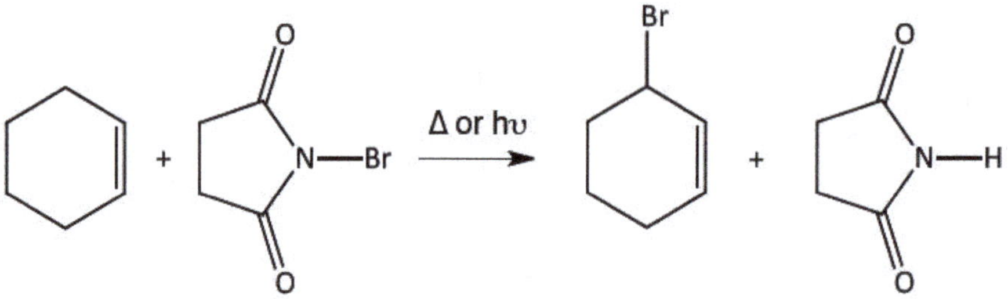

NBS Succinimide

Now that you know about allylic bromination, predict the product of the following reaction.

9.3 Predict the product of the following reaction.

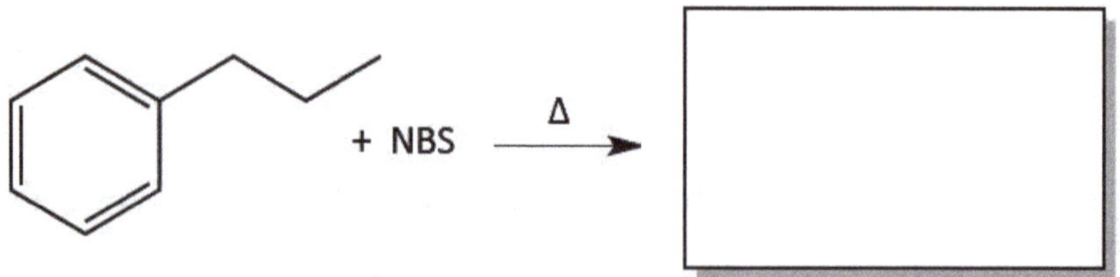

We call the carbon that neighbors a benzene ring a benzylic carbon atom. The hydrogen atoms on that carbon are benzylic hydrogen atoms.

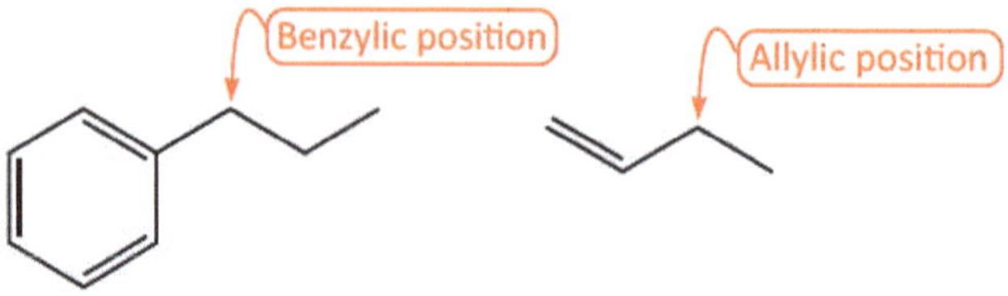

We can now include these resonance-stabilized positions in a rank of the stability of free radicals.

Least stable Most stable

vinyl ≈ methyl < 1° < 2° < 3° < allylic ≈ benzylic

Not resonance stabilized (vinyl through 3°) Resonance Stabilized (allylic, benzylic)

Figure 9.6 Stability of free radicals

Always consider the various resonance forms when a free radical reaction occurs in a compound with a pi-bond. Sometimes, if the molecule is symmetrical, the products we get are the same, so it does not matter. We saw this in the bromination of cyclohexene above and in the following bromination of propene.

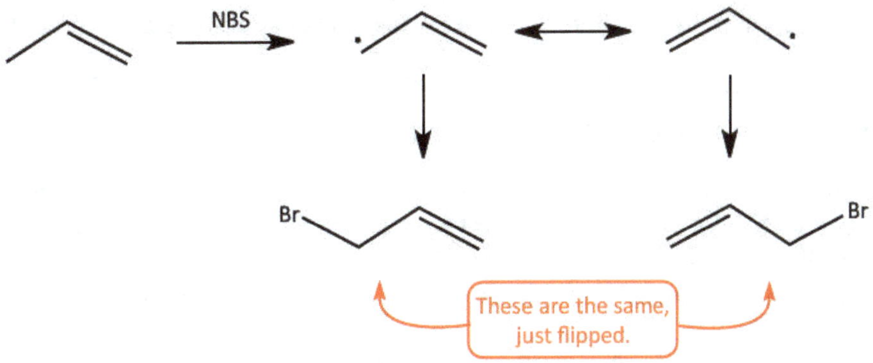

So, if we draw the overall reaction, we only need to draw one product.

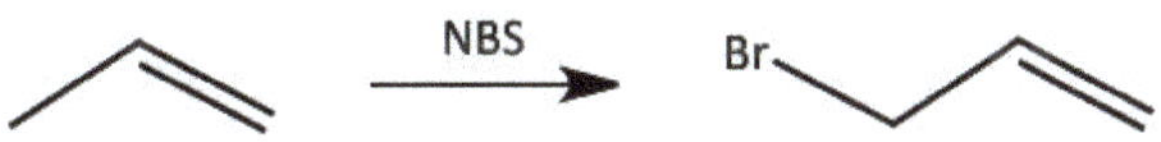

But, when the alkene is less symmetrical, it can lead to a mixture of products.

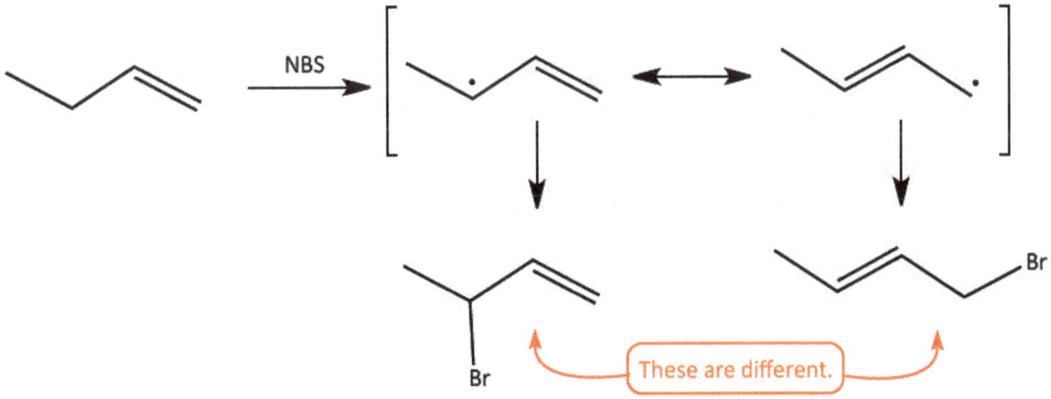

Since the products are different, we need to include both products when we draw out the overall reaction.

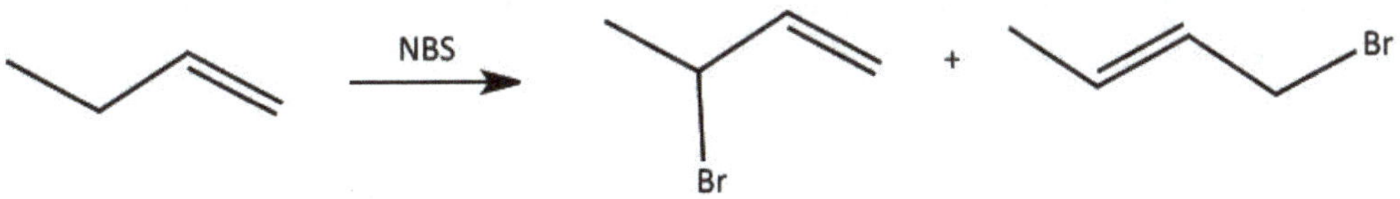

9.4 Identify the most reactive hydrogen atom(s) in the following molecules if they undergo free radical reactions.

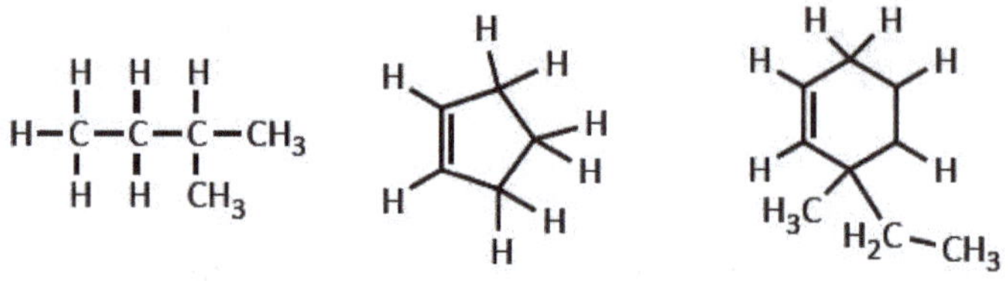

9.5 For the three molecules in the previous problem, draw the *monobrominated* product(s) of each if they undergo a free radical bromination. Monobrominated means only one bromine atom is on the molecule.

9.4 Alkyl halides from alcohols

If you need an alkyl halide for a reaction, it is possible to make it from a free radical halogenation. But, there are other ways to obtain alkyl halides other than this. If I was in the lab and needed an alkyl halide for a reaction, I usually open a chemical catalog and order it. It shows up in a day or two in the mail. But, if I cannot order it, I might make the alkyl halide from an alcohol. The common method to make them is from an alcohol.

$$R\text{—}OH \longrightarrow R\text{—}X$$

Which reagent we use to make an alkyl halide from an alcohol depends on two things, the type of halogen we desire and the type of alcohol we start with.

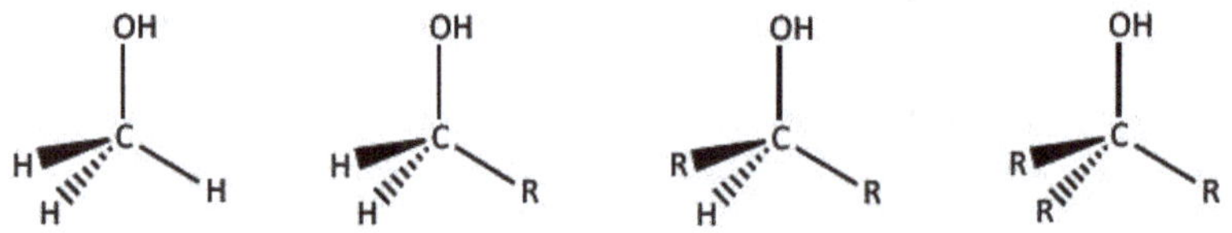

Methyl alcohol | Primary alcohol | Secondary alcohol | Tertiary alcohol

Tertiary alcohols react with HCl, HBr, or HI to make the appropriate alkyl halide. Secondary alcohols work slower this way. Primary and methyl alcohols do not work well with this method.

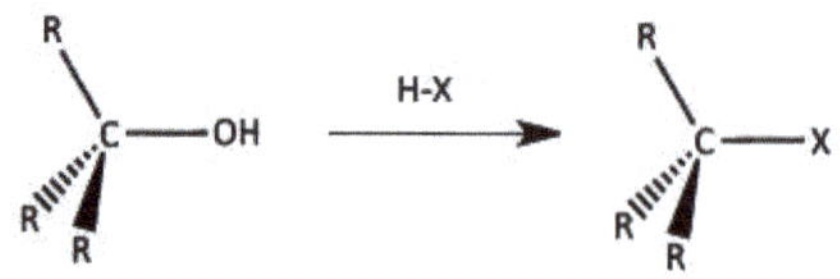

Tertiary alcohol

Sometimes a reagent called the **Lucas reagent** is used to turn a secondary or tertiary alcohol into a chloride. For the Lucas reagent, zinc chloride, $ZnCl_2$, is used with HCl. This combination of $ZnCl_2$ and HCl works better than HCl alone.

If we try to turn a primary or secondary alcohol into a halide, we need to use different reagents. We use thionyl chloride ($SOCl_2$) to make chlorides, phosphorus tribromide (PBr_3) to make bromides, and a combination of phosphorus and iodine (P/I_2) to make iodides.

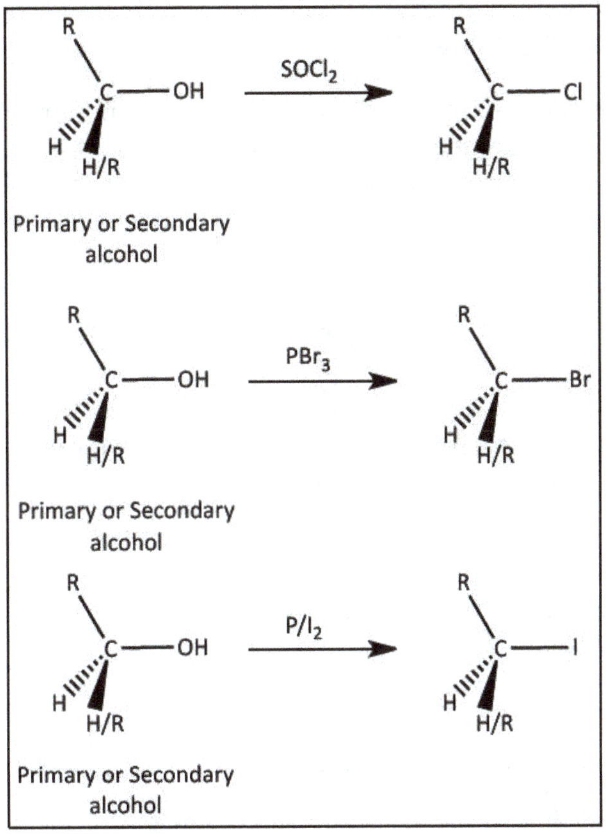

Figure 9.7 Conversion of hydroxyl groups to halides

Alcohol	Bromide	Chloride	Iodide
Tertiary	HBr	HCl or HCl/$ZnCl_2$ (Lucas Reagent)	HI
Secondary	PBr_3	$SOCl_2$	P/I_2
Primary	PBr_3	$SOCl_2$	P/I_2

Table 1 Best Reagents to Transform Alcohols to Halides

9.6

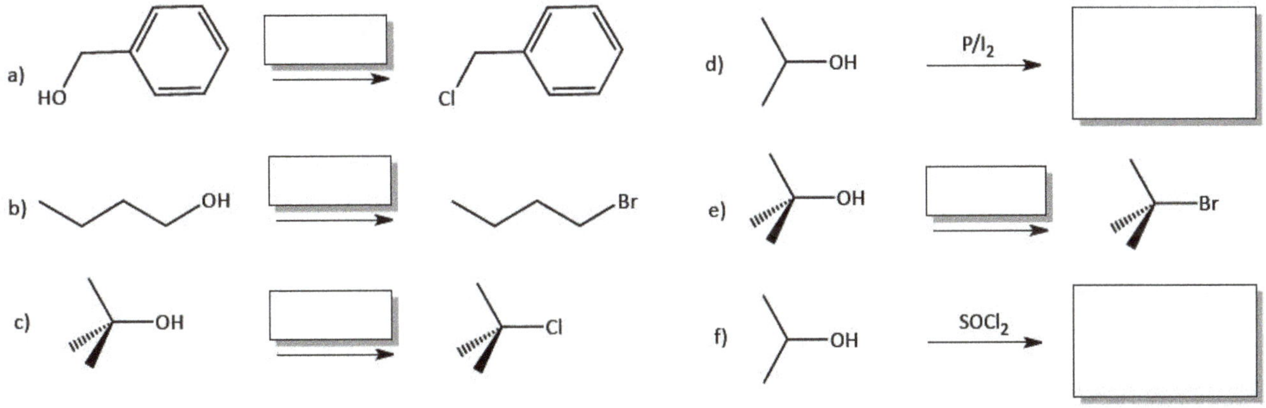

9.5 Reactions of Alkyl Halides

Consider the molecule, bromomethane. The halide of bromomethane is more electronegative than the carbon atom to which it is attached. This makes the carbon atom partially positive, so a negatively charged nucleophile will attack it. This is the basis of the reactivity of alkyl halides.

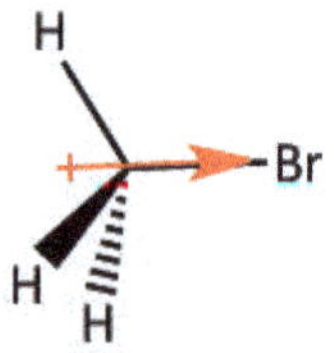

There are four reactions for alkyl halides that we will be learning about for the remainder of this chapter. They come in two main types, substitutions and eliminations.

9.5.1 Substitutions

Substitutions are when one group replaces another, much like when a substitute teacher replaces a regular teacher who is absent from school.

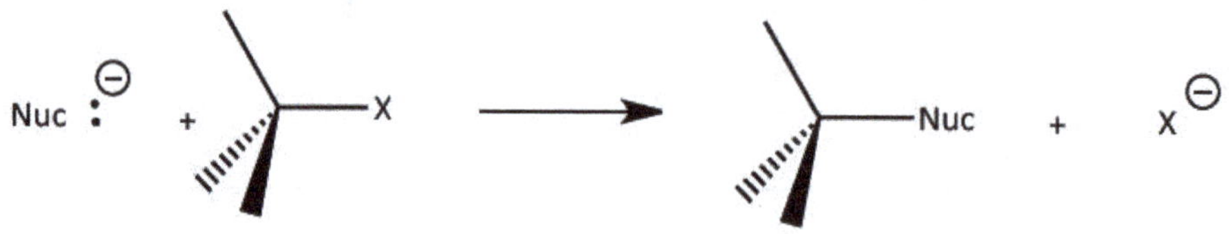

Figure 9.8 Generic substitution reaction

9.5.2 Eliminations

When something is eliminated, that means something is removed or expelled. When we eliminate waste from our bodies, we remove or expel the waste. In organic chemistry, *eliminations* are two groups being removed from a compound (H and X in this case) forming a π bond.

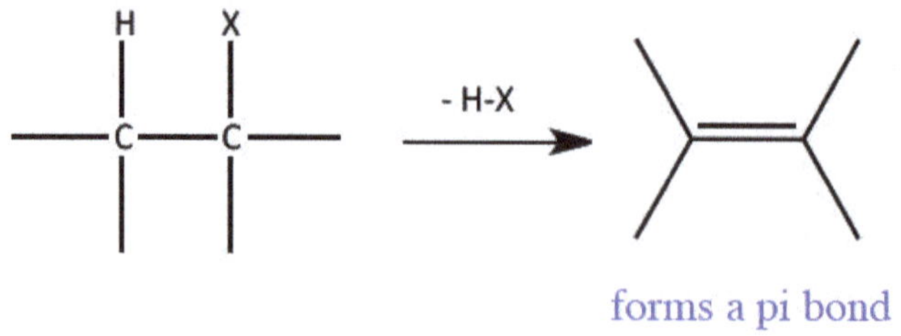

Figure 9.9 Generic elimination reaction

9.7 Label each of the following reactions as either a substitution or an elimination.

a) 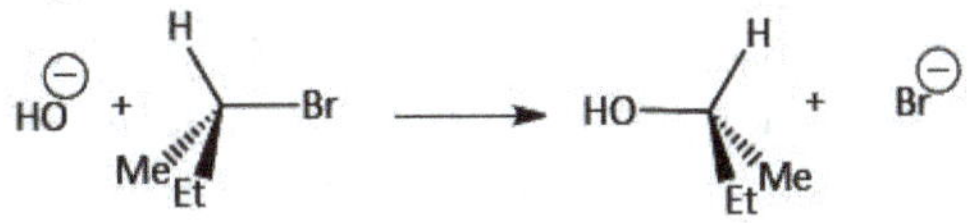

b) 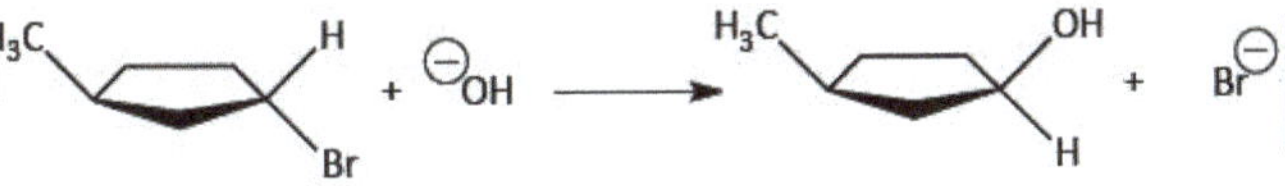

c) 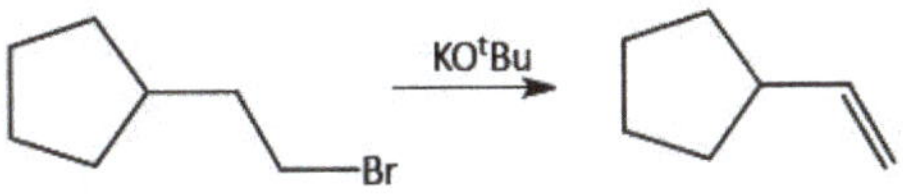

d) 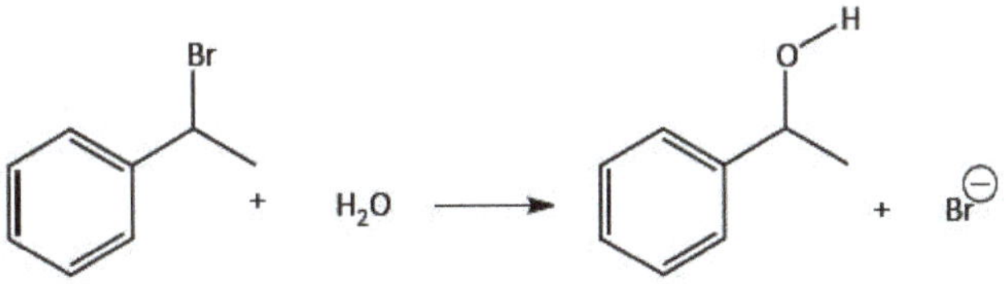

e) 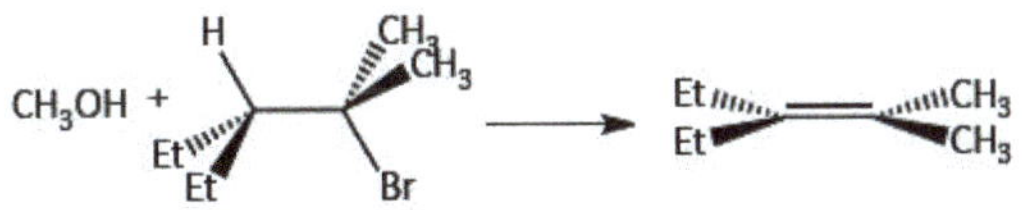

9.6 The S_N2 Reaction- The "backside attack"

Let's take a closer look at the substitution reactions first. There are two main mechanisms for the substitution of an alkyl halide. The first substitution mechanism we will look at is called the S_N2 mechanism. Let's walk through the S_N2 mechanism by analyzing the reaction of two species, a hydroxide ion and bromomethane.

In this substitution, an aggressive nucleophile attacks the carbon atom of the alkyl halide from the back side opposite the halide. If this nucleophile makes a bond with the carbon atom, that would make five bonds on the carbon atom! This cannot happen, so a group needs to leave. As the "backside attack" happens, the halide leaves. The halide is called a *leaving group*. The S_N2 reaction all happens in one step.

This is an example of a typical S_N2 reaction.

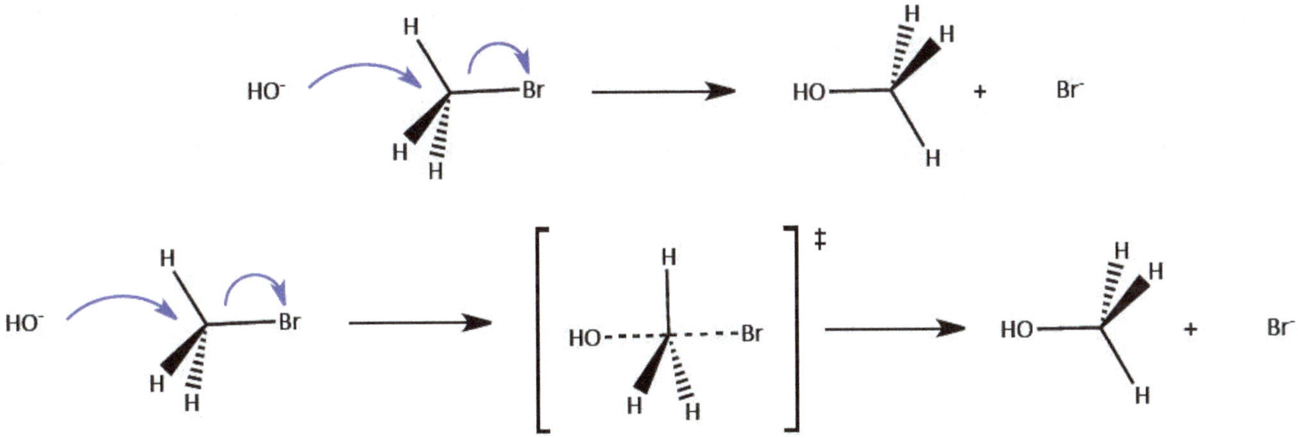

Figure 9.10 Example of S_N2 reaction

If you wanted to do a reaction with these two species, you could buy them from a chemical company. You would be able to find bromomethane in the chemical catalog, but you would not find hydroxide. That is because hydroxide does not exist by itself because of its negative charge. It exists with a positive counterion. You would need to buy sodium hydroxide (NaOH) or potassium hydroxide (KOH). But, in the reaction, the alkali metal is not involved in the reaction. It is called a *spectator ion*. A spectator ion is not involved in the reaction. It just floats around and acts like a spectator. It watches. Therefore, at times, the spectator ions are not explicitly drawn in our reactions. Spectator ions are sometimes ignored.

So, why the weird name, S_N2?

S: Substitution. This is a substitution reaction.
N: Nucleophilic. A nucleophile is attacking an alkyl halide.
2: A Bimolecular mechanism. Two species are involved in the rate-determining step.

The substitution and nucleophilic part are self-explanatory. But, what does the bimolecular part mean? The S_N2 reaction mechanism is a one-step mechanism. It is a *concerted* mechanism, meaning the steps happen at the same time. The nucleophile attacks at the same time as the halide leaving group leaves.

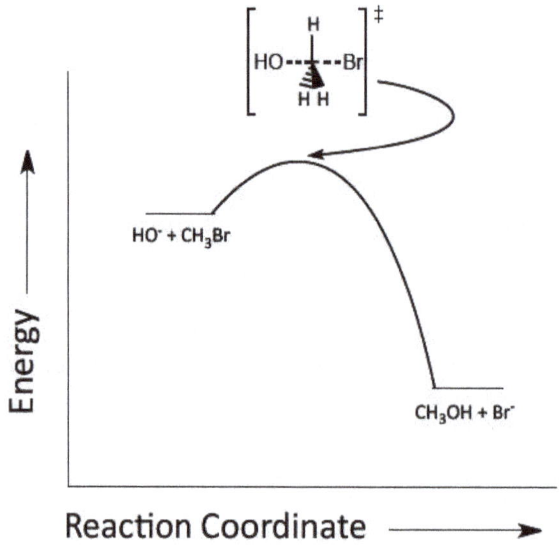

Figure 9.11 Reaction coordinate diagram of an S_N2 reaction

This one step is also the slowest step in the reaction because it is the only step! In this slowest step, the rate-determining step, two molecules are involved, the hydroxide and the bromomethane. For the S_N2 reaction to occur, these two molecules need to meet each other. If the amount of hydroxide were doubled in the reaction, what would happen to the odds of the two colliding? It would double, or it would double the rate of the reaction. If the amount of bromomethane in the reaction were doubled, what would happen to the odds of the two colliding? It would double, so it would double the rate of the reaction. The reaction is first order with respect to each reactant and is, therefore, 2nd order overall. The rate equation for the S_N2 mechanism is:

$$\text{Rate} = k_r[\text{Nuc}][\text{Substrate}]$$

or

$$\text{Rate} = k_r[OH^-][CH_3Br]$$

9.7 Factors that affect the S_N2:

9.7.1 Why a backside attack?

One consideration is sterics. A frontside attack of the nucleophile is blocked by the halide leaving group. Another consideration is electronics. The leaving group is negatively charged and the nucleophile is typically negatively charged. There is an electronic repulsion between the two preventing the nucleophile from attacking from the front side. A backside attack is therefore what happens.

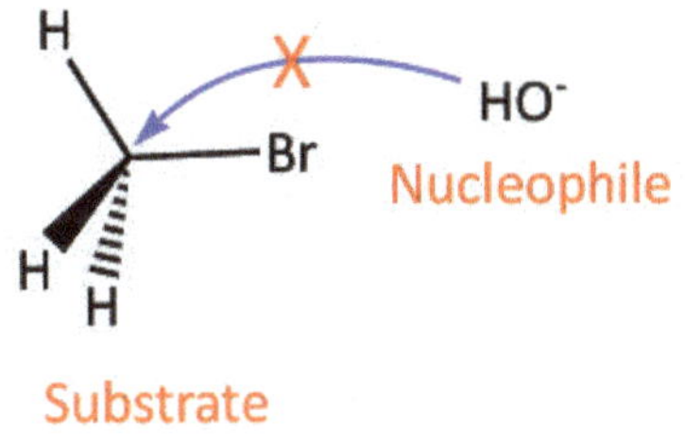

9.7.2 The Substrate

A nucleophile attacks a substrate (the alkyl halide) in an S_N2 reaction. For the S_N2 reaction, since the substrate is attacked from the backside by the nucleophile, the steric hindrance to that backside attack is important. In order for the nucleophile to be able to reach the carbon atom, it needs to navigate past three groups pointing towards the backside of the substrate. If these groups are small, like hydrogen atoms for a methyl halide, the nucleophile can more easily reach the carbon atom. If there is one alkyl group (primary halide), it still happens pretty easily. With two alkyl groups (secondary halide), it becomes more difficult. With three alkyl groups (a tertiary halide), it is nearly impossible for the nucleophile to reach the carbon atom. This is steric hindrance. The alkyl groups play defense on the backside and prevent the nucleophile from reaching the carbon atom.

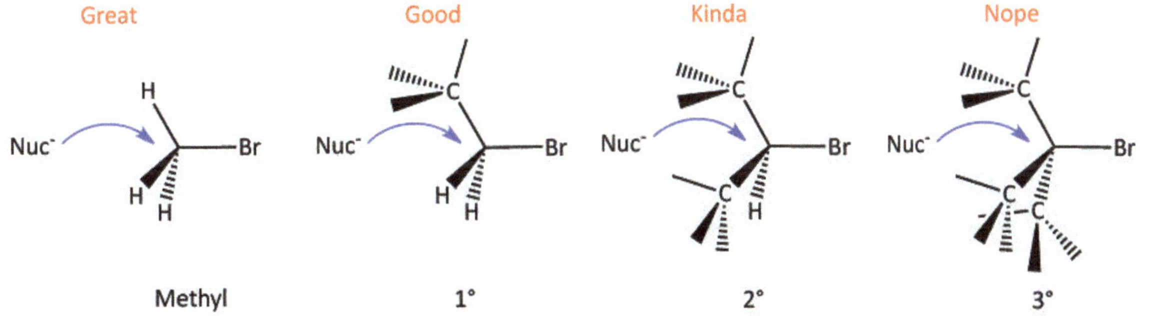

Figure 9.12 The backside attack of an S_N2 reaction

In general, the rate of reaction of the substrates in an S_N2 reaction is:

Me > 1° > 2° >>> 3°

Figure 9.13 The rate of reaction for an S_N2 reaction

The substrate does not need to be tertiary for the backside attack to be sterically hindered. For example, neopentyl bromide, a primary alkyl halide, is very slow to the S_N2. It is as bad as most tertiary alkyl halides.

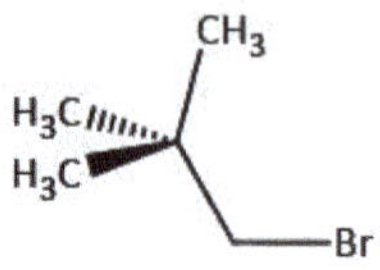

Neopentyl bromide

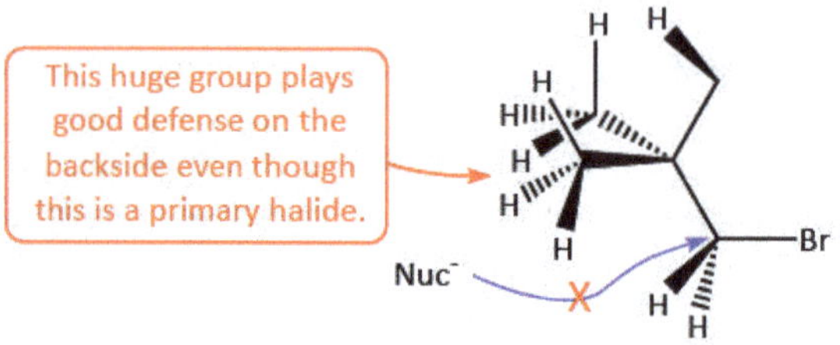

Neopentyl bromide

9.7.3 The Leaving Group

In an S_N2 reaction, the leaving group usually has a negative charge on it. Better leaving groups are those that are happier and more stable with the negative charge. Nature abhors an isolated charge, so it is better the more the negative charge can spread out. For halide leaving groups, the larger atoms (lower in the periodic table) are more stable because they can spread out the negative charge more. They make better leaving groups.

$$I^- > Br^- > Cl^- > F^-$$

F^- is a poor leaving group because it is so small. Since oxygen and nitrogen are in the same row of the periodic table, O^- and N^- are also small. They are less stable than F^- because they are less electronegative than F. Therefore, HO^-, RO^-, and H_2N^- are poor leaving groups. Spreading the negative charge through resonance is the only way to make O^- a good leaving group. Resonance is why the following are excellent leaving groups. In mesylates and tosylates, the negative charge is spread over three oxygen atoms. Mesylates and tosylates can be made from alcohols as we will learn later.

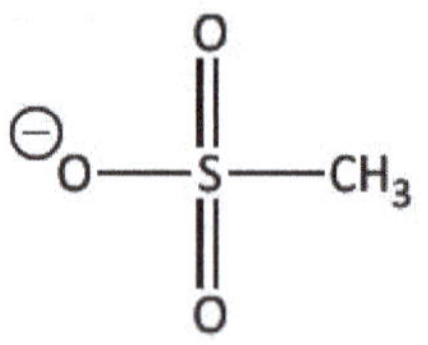

Mesylate ($^-$OMs)

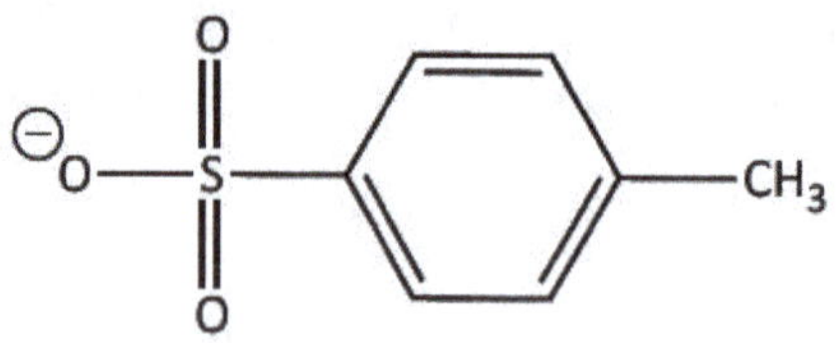

Tosylate ($^-$OTs)

The order of leaving groups stability is

Resonance		large		smaller		smallest		isolated, very small negative charge
$^-$OTs, $^-$OMs	>	I^-	>	Br^-	>	Cl^-	>	F^-, NH_2^-, HO^-
Good L.G.								Poor L.G.

Figure 9.14 Stability of leaving groups

9.7.4 The Solvent

The solvent that is chosen for an S_N2 reaction can either speed up the reaction or slow it down. Polar solvents are generally used to dissolve the reactants in an S_N2 reaction since ions are generally involved as the nucleophile and leaving groups. The polar solvent used can be either a polar protic solvent or a polar aprotic solvent.

Polar Protic Solvents

Polar protic solvents are those with protons (hydrogen atoms) attached to an oxygen or nitrogen atom. The protons are small, concentrated centers of positive charge. They, therefore, attach themselves to the negatively charged nucleophile. The nucleophile is solvated, meaning the solvent surrounds it. Therefore, the nucleophile is surrounded and has difficulty attacking the carbon atom of the substrate. The reaction proceeds very slowly.

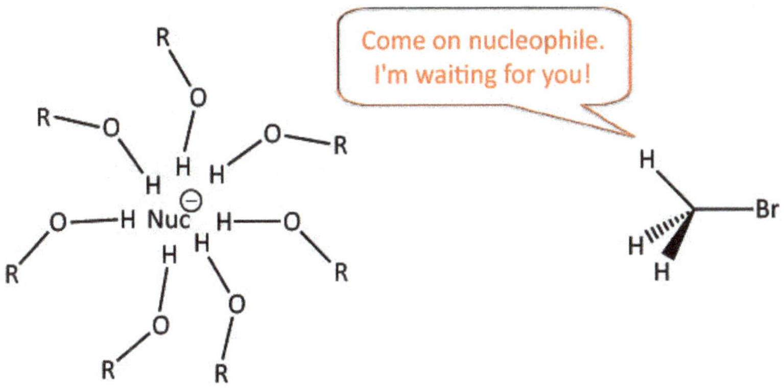

Nucleophile surrounded by polar protic solvent

Polar Aprotic Solvents

Polar aprotic solvents are those with no O-H or N-H bonds. Examples are:

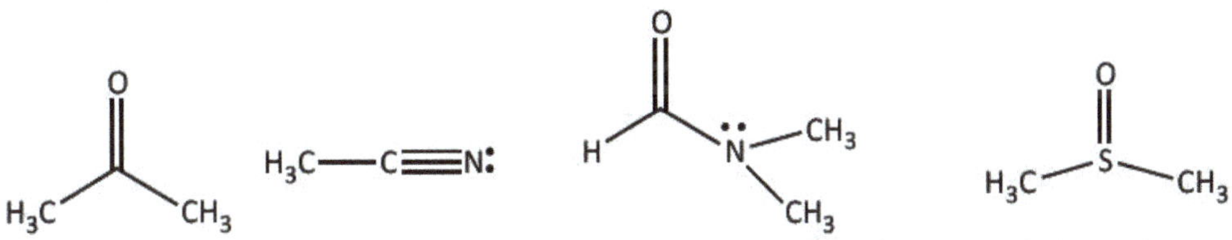

Acetone   Acetonitrile   Dimethylformamide (DMF)   Dimethylsulfoxide (DMSO)

Figure 9.15 Polar aprotic solvents

Polar aprotic solvents are polar enough to dissolve the substrate and the nucleophile, but they do not solvate the nucleophile. In fact, the solvent surrounds the positive counterion (Na^+ or K^+) instead. Since the cation is bound up by the solvent, it cannot attach to the nucleophile. This makes the nucleophile extra negative and very strong. Polar aprotic solvent speed up an S_N2 reaction. Most S_N2 reactions are therefore performed in polar aprotic solvents. For example, if KOH is added to bromomethane in acetone, the following occurs.

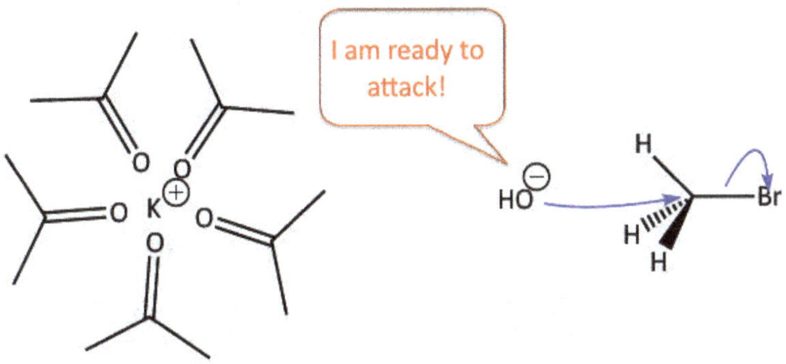

K^+ surrounded by acetone molecules

9.7.5 The Nucleophile

For an S_N2 reaction, the nucleophile is aggressive. The nucleophile must attack the substrate before anything else happens. So, what makes a strong, aggressive nucleophile? Remember, nucleophiles are lovers of nuclei (which are positively charged). Therefore, nucleophiles are species that attack positive charges. In this case, a good nucleophile will attack the carbon atom of the substrate that has a partially positive change. One thing that makes a nucleophile aggressive is if it has a negative charge on it or a great big cloud of negative electron density. The hydroxide ion is an excellent nucleophile because it is negatively charged.

We need to define a couple of terms since negatively charged species act as both nucleophiles and bases.

Nucleophilicity is a measurement of how much a species wants to attack a positively charged atom (typically an electrophilic carbon atom).

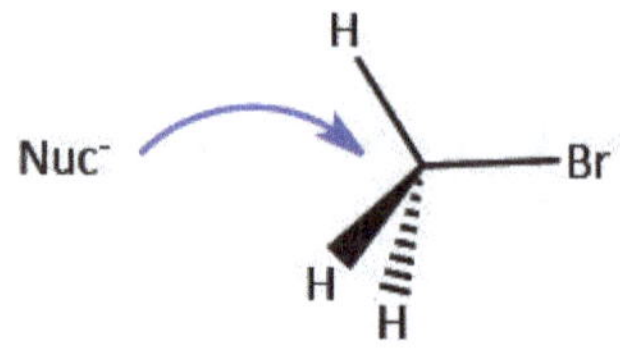

Basicity is a measurement of how well a species abstracts a proton (H^+) by donating its electrons to it.

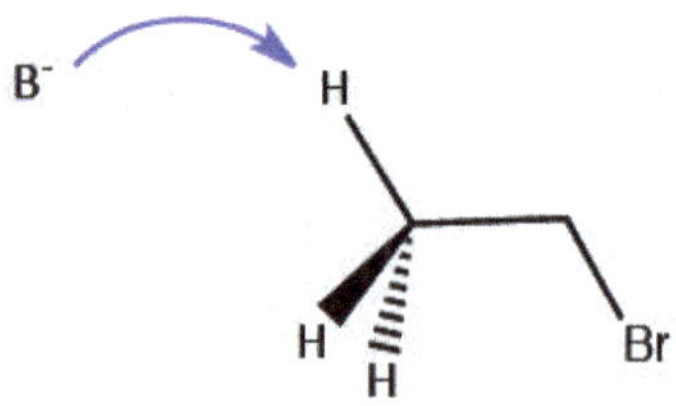

In the S_N2 reaction, we need a good, strong nucleophile. So, what makes a good nucleophile?

Negative charge helps.

Negative charges typically make stronger nucleophiles.

$CH_3CH_2O^{\ominus}$	vs.	CH_3OH
ethoxide Good nucleophile Good base, too		ethanol Poor nucleophile Poor base, too

Remember that anions exist in their salt form, so many times they are written with their positively charged cationic counterions attached. Look for metals like Na, K, or Li with them.

$LiOCH_2CH_3$ (LiOEt)
Lithium ethoxide

$NaOCH_3$ (NaOMe)
sodium methoxide

Down a column helps.

Nucleophilicity increases for atoms down a column in the periodic table. This is because these atoms are larger. We say that they are more polarizable or softer. I like to think of them as big, soft, fluffy, squishy electron clouds. These large, squishy electron clouds can reach out their electrons and squeeze around the backside alkyl groups more easily to reach the carbon atom of the substrate. So, nucleophiles are better if they are softer or more polarizable.

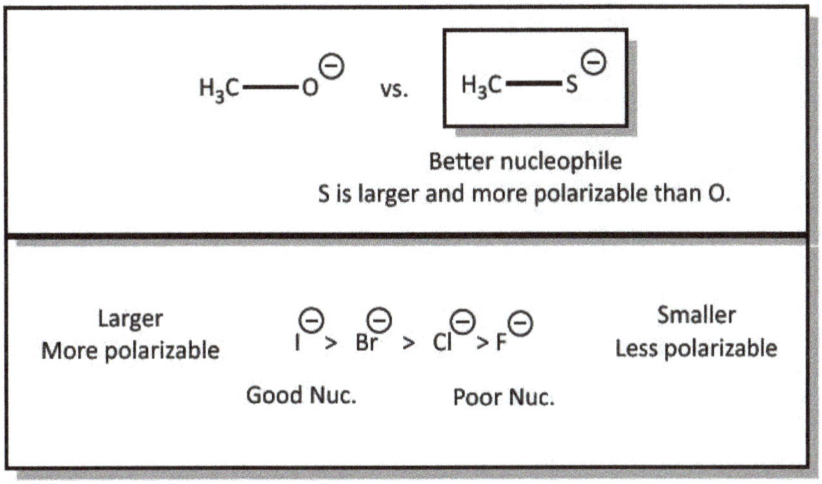

Across a row hurts.

This is not as big of an effect, but nucleophilicity decreases across the atoms in a row or period in the periodic table. This is because, across a row, the atoms become more electronegative and hold onto their electrons more. Since they hold their electrons stronger and closer to themselves, they are less polarizable, harder, or less squishy. They cannot reach out their electrons as well to the carbon atom of the substrate.

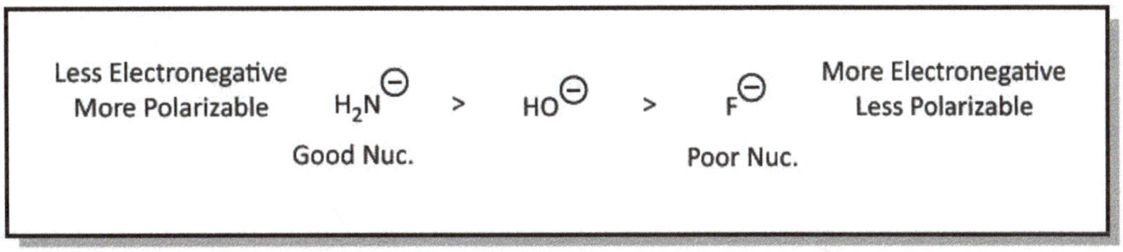

Resonance hurts.

Nucleophiles are less likely to go out and attack the substrate the more stable they are. Good nucleophiles are unhappy, negatively charged species that want to give away that negative charge. So, if resonance forms can be drawn to spread out a negative charge or a lone pair of electrons on a nucleophile, it becomes more stable. The attacking atom has less negative electron density on it and is less likely to go attack a positive carbon atom. It is then a worse nucleophile.

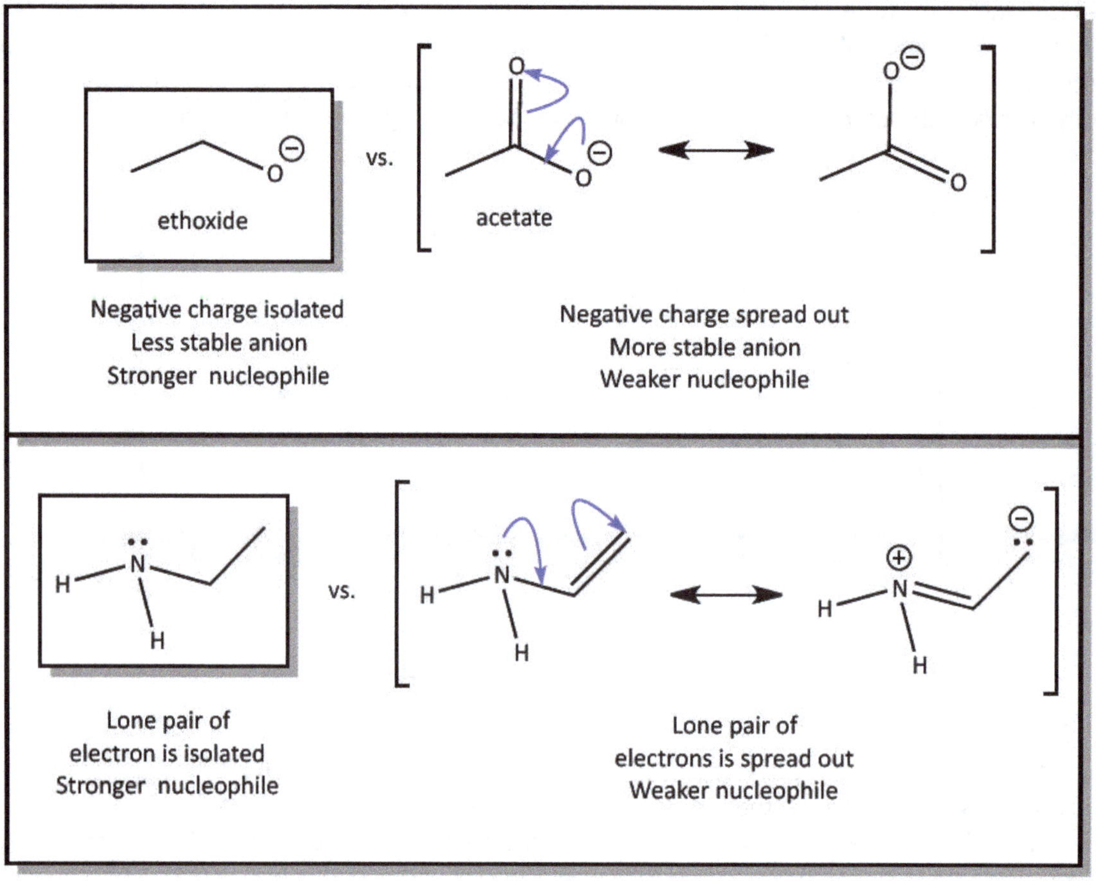

Inductive effects

Resonance is not the only way to spread out a negative charge. Negative charges can also be spread throughout a molecule through induction. If an electron-withdrawing group is on the nucleophile, it can pull electron density away from the attacking atom.

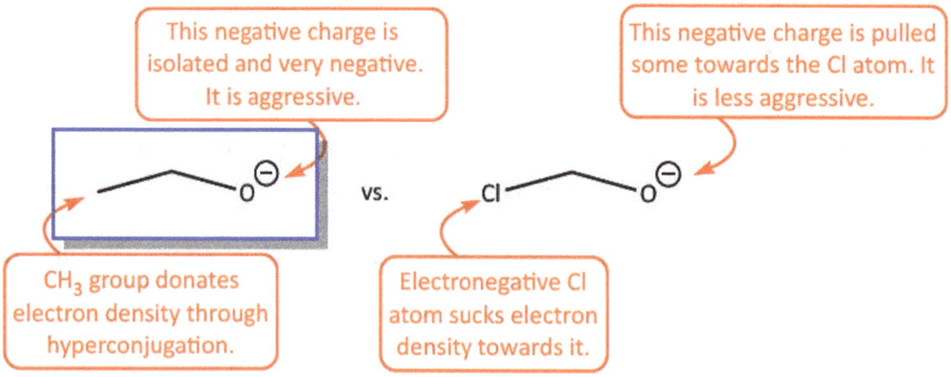

Better Nucleophile **Weaker Nucleophile**

Sterics hurts

Large, polarizable attacking atoms increase nucleophilicity because the electron cloud can squeeze and reach out to the carbon atom of the substrate. But, it is different if the nucleophile is large because of attached alkyl groups. These groups do not increase the polarizability of the attacking atom, rather they make the nucleophile big and bulky. For example, t-butoxide is so large, the negative oxygen atom cannot reach the carbon atom of the substrate.

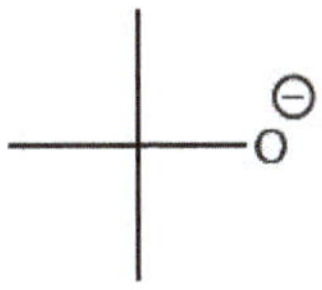

t-butoxide

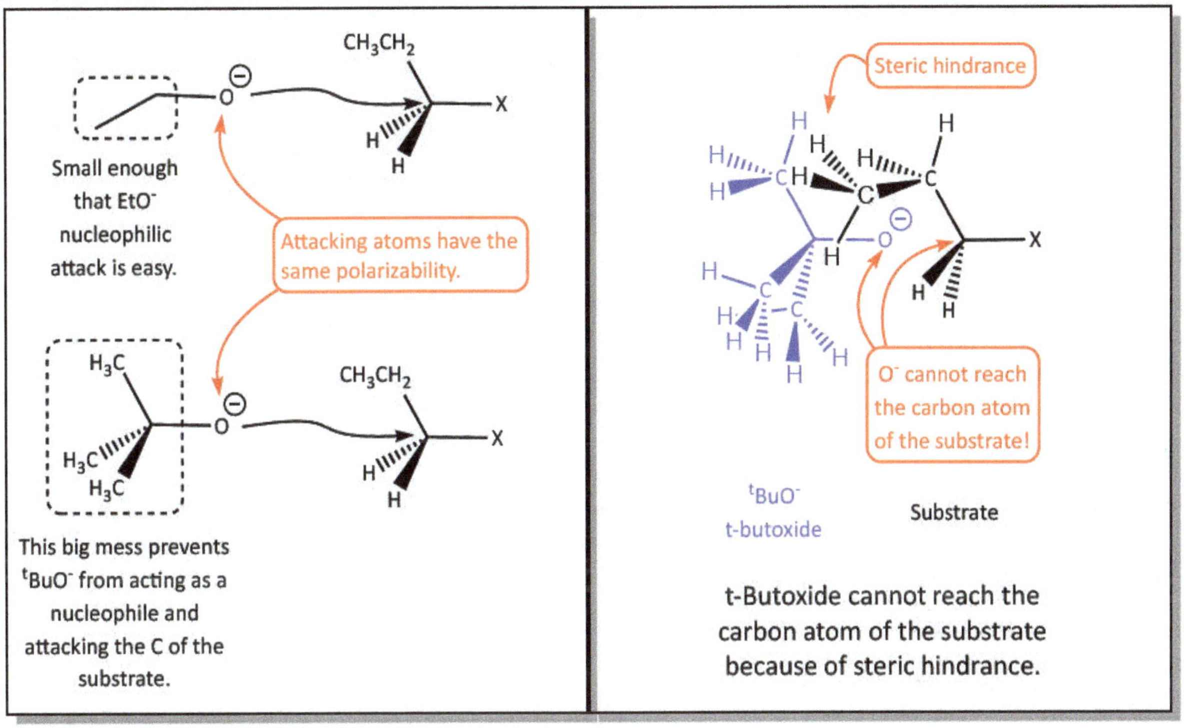

Large, bulky anions like t-butoxide are poor nucleophiles, but they do have a negative charge that will still aggressively attack positive charges. t-Butoxide will grab a proton (H^+) quickly making it a good base. It can do this because protons are typically on the outside of a substrate where they can be easily reached, unlike the carbon atom.

9.7.6 Stereochemistry - inversions

When the backside attack happens in the S_N2 mechanism, the substrate begins to open up in the transition state and then flips inside out or inverts. This is called a *Walden inversion*. This often flips the stereochemistry at the substrate's carbon atom. In this example, an (S) configuration flips to an (R) configuration. Since there is always an inversion in the S_N2 reaction, it is a *stereospecific reaction*. The same stereochemistry changes always occur. One stereoisomer reacts to give one specific stereoisomer of a product.

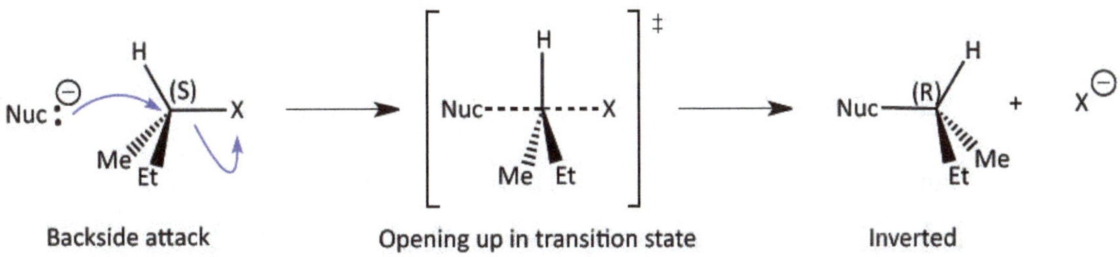

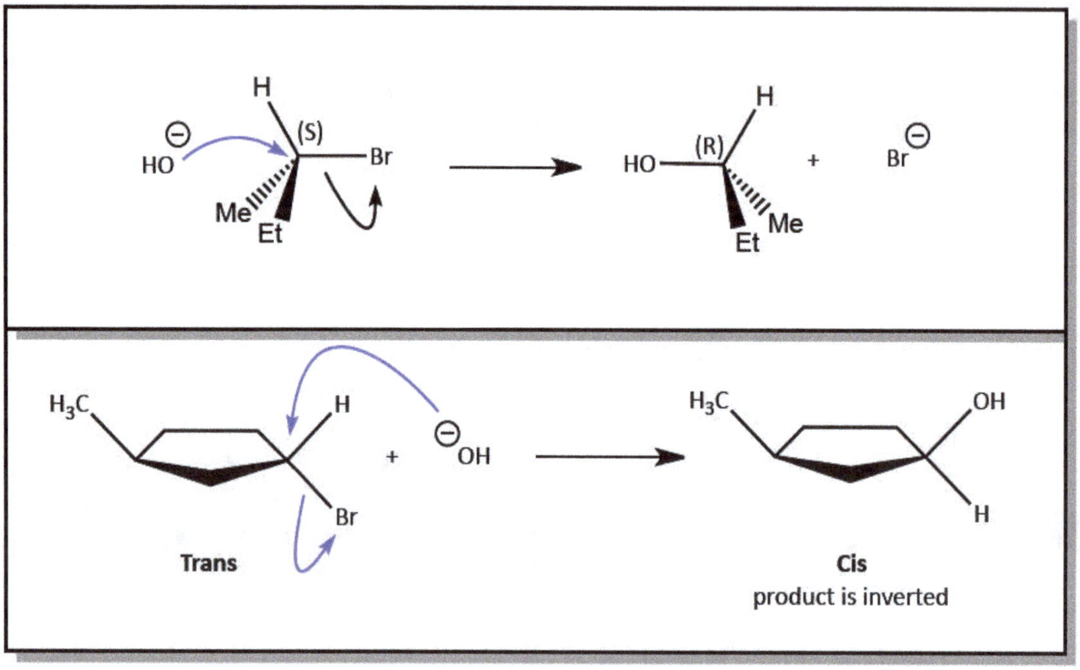

9.7.7 Anionic vs. Neutral nucleophiles

We have seen that nucleophiles can be both anionic and neutral. They have very slightly different mechanisms. If an S_N2 reaction involves a neutral nucleophile, it makes a positively charged product. Typically, there is a cleanup step at the end where a proton (H^+) is removed.

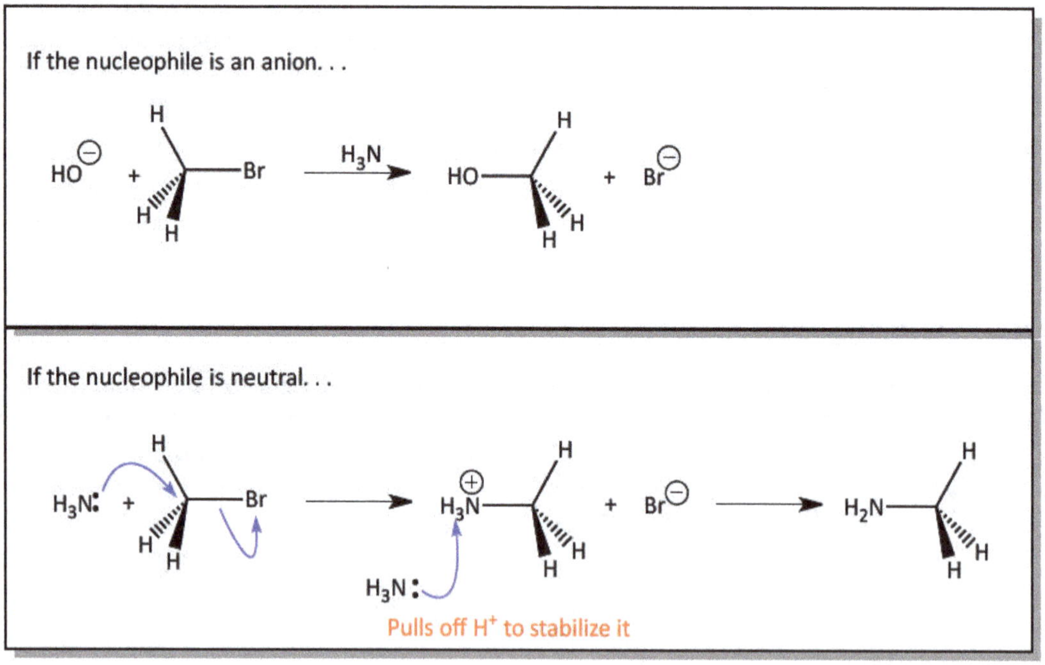

When ammonia, NH_3, was the nucleophile in the S_N2 reaction, it made a positively charged ammonium salt. Another molecule of ammonia acted as a base to deprotonate the ammonium salt making a neutral amine.

9.7.8 Some common nucleophiles

All of the strong nucleophiles have a negative charge, are fairly large, or have a lone pair of electrons on nitrogen or phosphorus. The weak nucleophiles are neutral molecules that are not nitrogen or phosphorus bases, or they have a negative charge on a very electronegative element like F, or resonance stabilized like acetate.

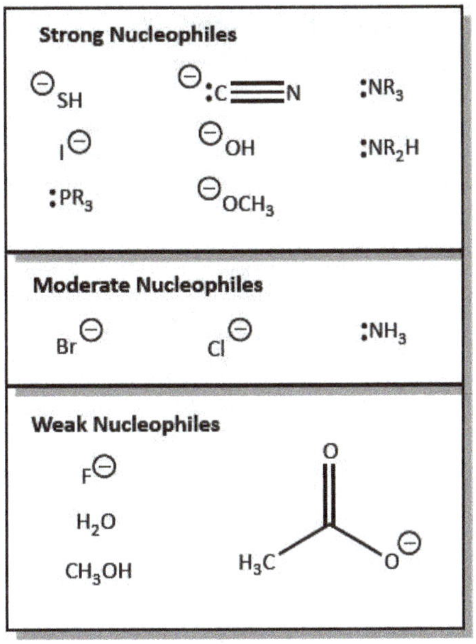

Figure 9.16 Strength of common nucleophiles

9.7.9 S_N2 Examples

The S_N2 reaction is a powerful reaction. We can use it to produce many different types of compounds from an alkyl halide.

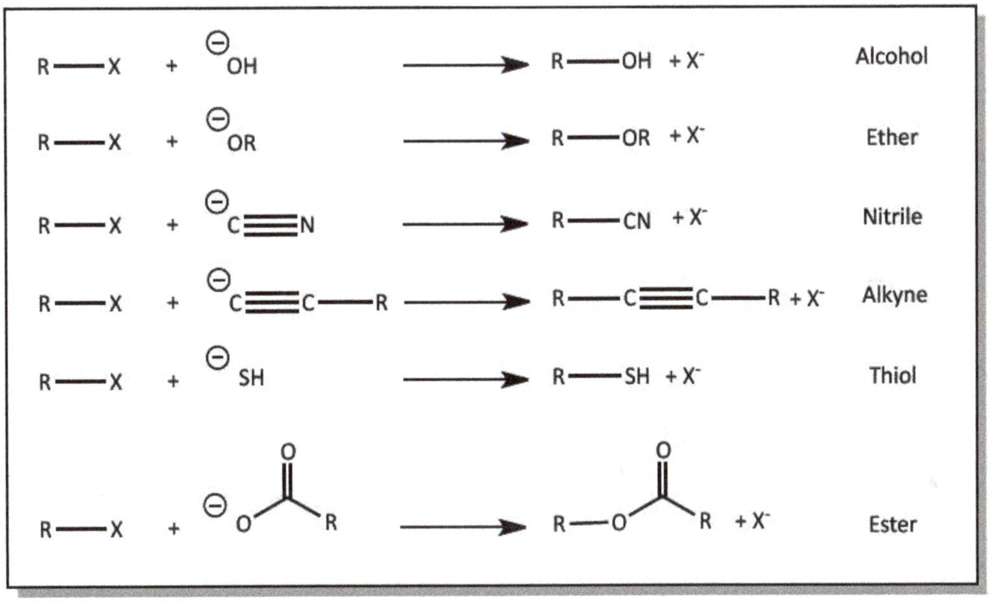

Figure 9.17 Example S_N2 reactions

9.7.10 Why S_N2 reactions do not occur on sp^2 carbon atoms

We also do not see S_N2 reactions occur on sp^2 carbon atoms because the backside attack is prevented. If the halogen is on a benzene ring (an aryl halide) or an alkene (a vinyl halide) S_N2 reactions do not occur. There are a couple of reasons for this. Two of the simpler explanations are: 1. the electron cloud of the aromatic ring or alkene prevent the negative nucleophile from reaching the carbon atom. 2. The carbon-halogen bond is stronger than usual. It is more double bond like because of resonance forms that can be drawn.

Reason 1

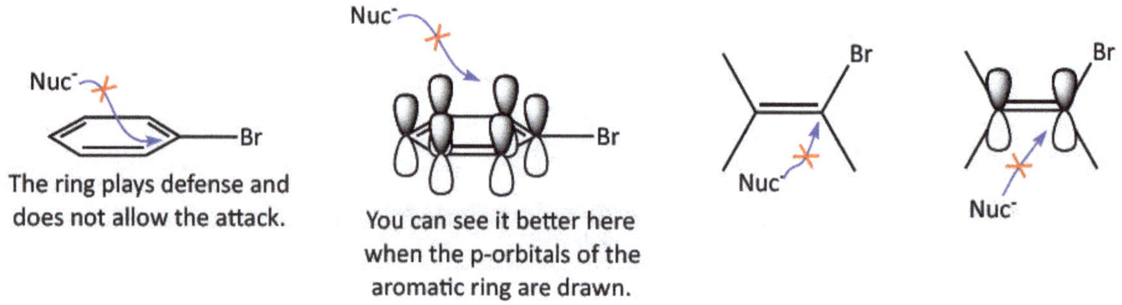

Reason 2

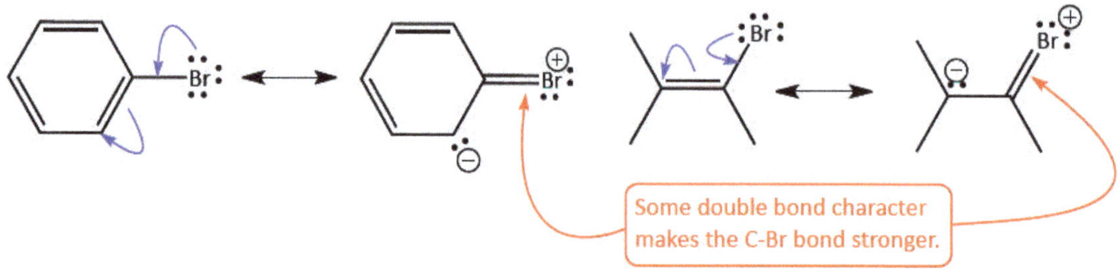

9.8 The S_N1 Reaction

Unlike the S_N2 reaction that is one-step, the S_N1 reaction occurs in several steps. The first step of the S_N1 reaction is the leaving group leaving the substrate to make a carbocation. Once the carbocation is formed, a nucleophile attacks it. If this nucleophile is neutral, it makes a protonated, positively charged species. This species is deprotonated (usually with a solvent) to make the product.

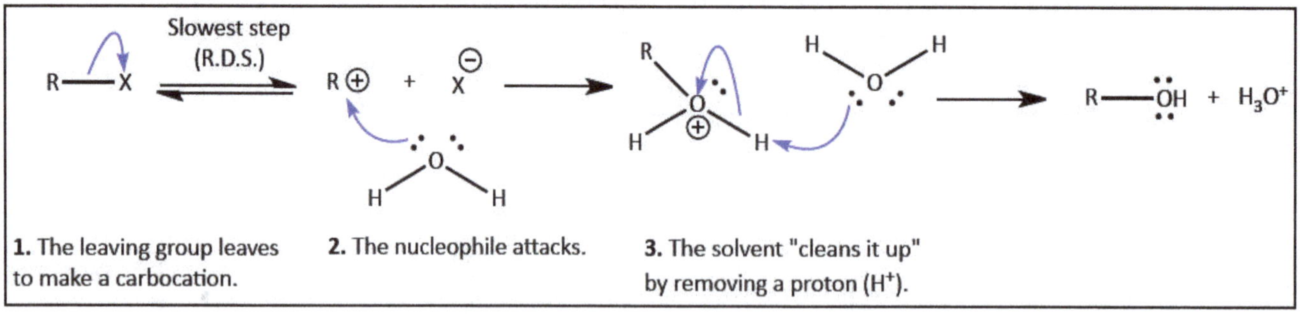

Figure 9.18 Generic S_N1 reaction mechanism

S: Substitution. This is a substitution reaction.
N: Nucleophilic. A nucleophile is attacking an alkyl halide.
1: Unimolecular. One species is involved in the rate-determining step.

The S_N1 reaction is at least a two-step reaction mechanism. The reaction energy diagram is below. The first step is the slowest, rate-determining step (R.D.S.). The second step is a fast step. In a two-step reaction energy diagram, there are two transition states and one intermediate. The intermediate, in this case, is a carbocation.

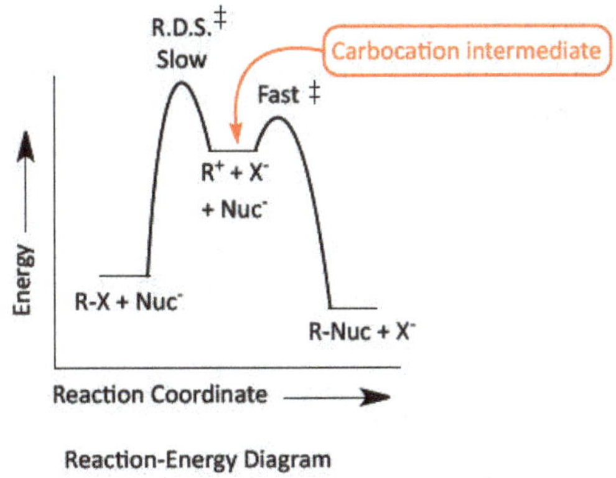

Figure 9.19 Reaction coordinate diagram for an S_N1 reaction

The S and N of this mechanism's name are because the S_N1 reaction is a nucleophilic substitution. This is obvious. The 1 in the name is because it is unimolecular (involves one molecule) in the rate-determining step. The first step of the S_N1 reaction is the slowest step. The reaction cannot proceed until the leaving group leaves. In step 1, notice that there is only one molecule involved, the alkyl halide, R-X. If more nucleophile is added to the reaction (water in our example), the reaction will not go any faster. The extra water molecules will stand around waiting for a carbocation to be formed. They cannot do anything until that happens. But, if more alkyl halide, R-X, is added to the reaction, the reaction will go faster because the carbocation will be made faster. Therefore, the reaction is first order with respect to R-X and zeroth order with respect to the nucleophile. The S_N1 reaction is first order overall.

$$\textbf{Rate} = \mathbf{k_r[Substrate]}$$

or

$$\textbf{Rate} = \mathbf{k_r[R\text{-}X]}$$

9.8.1 The Substrate

The S_N1 reaction makes a carbocation. The happier or more stable this carbocation is the easier it will be for the halide to leave. Therefore, the reaction goes faster. So, what makes a good carbocation?

At a first glance, we look for the substitution of the carbocation atom. We remember that tertiary carbocations are the most stable because of hyperconjugation. Primary and methyl carbocations very unstable. Therefore, tertiary alkyl halides make excellent substrates for the S_N1 reaction, secondary carbocations are OK but primary and methyl alkyl halides cannot undergo S_N1 reactions unless something else is involved. This is the opposite of what we saw in the S_N2 reaction.

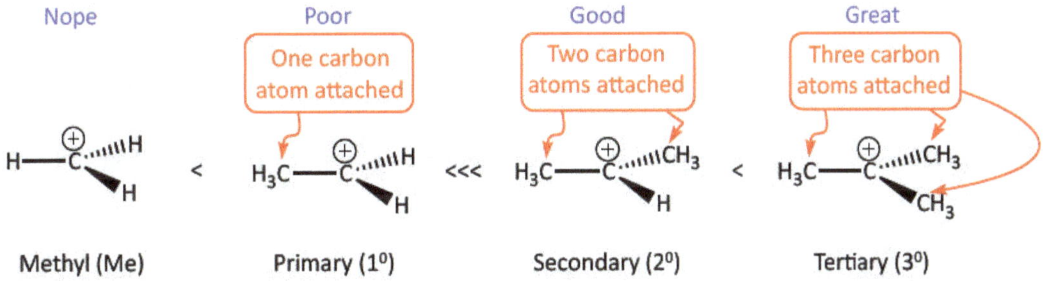

Figure 9.20 The carbocation substrate of an S_N1 reaction

The other thing to look for in regards to the stability of the carbocation is resonance. Both allylic and benzylic carbocations are quite stable because resonance forms can be drawn. Allylic carbocations are carbocations that neighbor a double bond. In allylic carbocations, the nucleophile can attack either carbocation spot to make different products. Benzylic carbocations are carbocations neighboring a benzene ring. In benzylic carbocations, the only resonance form the nucleophile attacks is the one with the carbocation next to the benzene ring, not the with the carbocation on the benzene ring.

Allylic carbocations

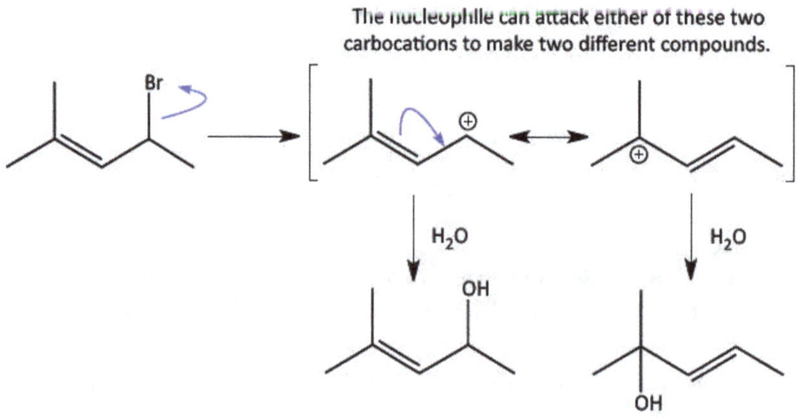

Benzylic carbocations

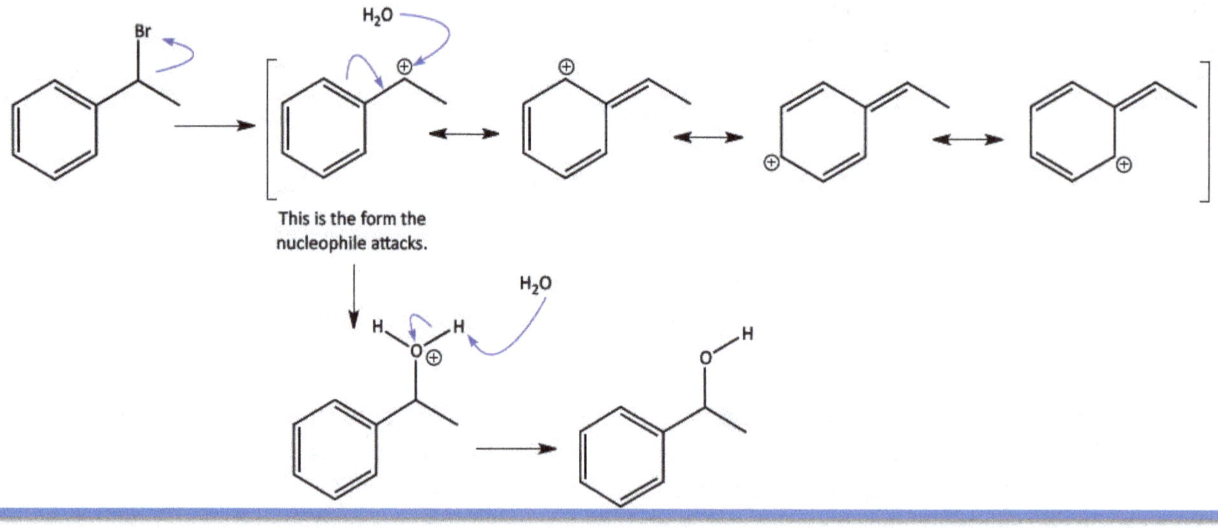

9.8 Are the following substrates more likely to undergo a substitution reaction via S_N1, S_N2, or both?

a) 

b) 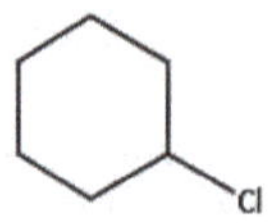

c) 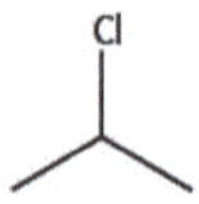

d) 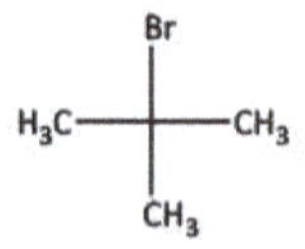

9.9 Indicate if the following substrates would form a resonance-stabilized carbocation after the leaving group leaves.

a) 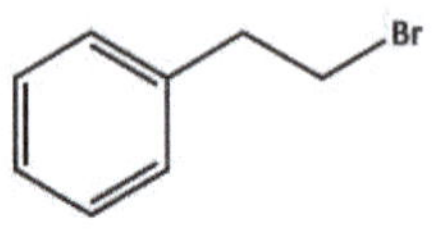

b) 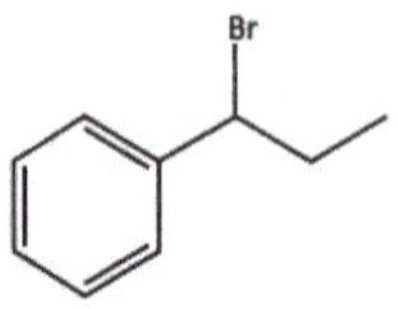

c) 

d) 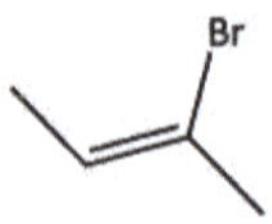

9.8.2 The Leaving Group

Like in the S_N2 reaction, good leaving groups must be stable after they leave. One small difference is that in the S_N1 reaction, sometimes the leaving group is a neutral molecule. Often, this happens when a reaction is run under acidic conditions. For example, the hydroxyl group of an alcohol is a poor leaving group. But, in acidic conditions, the hydroxyl group can be protonated and converted to water, which is a very good leaving group.

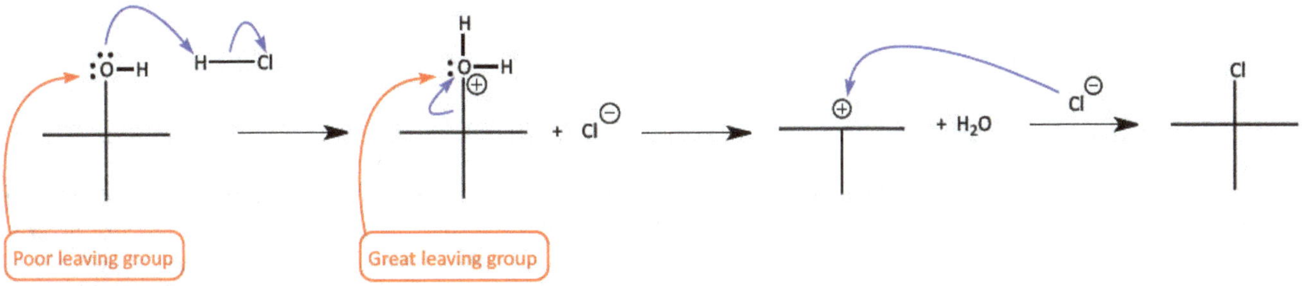

9.10 Label the leaving groups on the following compounds as good or bad.

a) $PhCH_2Cl$ b) 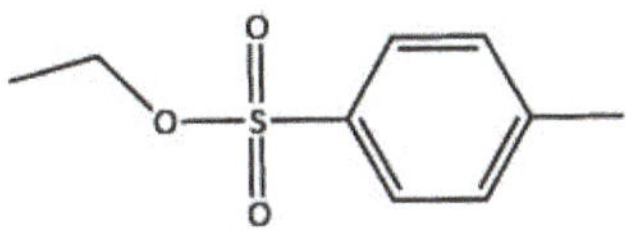 c) 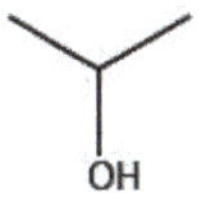 d) 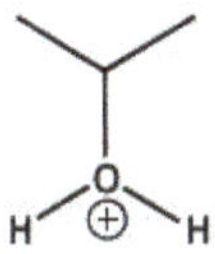

9.8.3 The Nucleophile

Since the first step of the S_N1 reaction is the leaving group leaving to make a carbocation, the nucleophile in the S_N1 is typically a weak, neutral molecule. We do not want an aggressive nucleophile that attacks right away, but we want a nucleophile that will wait around until after the leaving group leaves. Also, if the nucleophile is a negatively charged base, it will perform an elimination reaction before allowing the S_N1 to proceed. In an S_N1 reaction, the nucleophile is often the solvent. S_N1 reactions are called *solvolysis* reactions because of this.

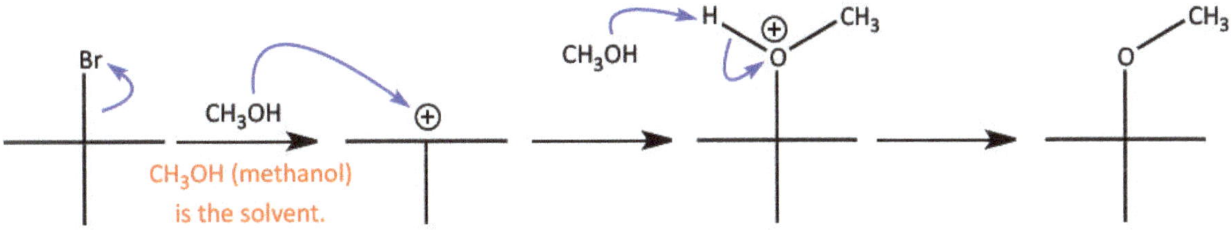

9.11 Identify whether the following nucleophiles are more likely to be involved in an S_N1 or an S_N2 reaction.

a) CH_3OH b) $(CH_3)_2N^-$ c) $CH_3CH_2CH_2CH_2^-$ d) CH_3CH_2SH e) H_2O

9.8.4 The solvent

The first step in the mechanism is the formation of a positive carbocation and a negative leaving group, which are ions. The formation of these ions is helped by the presence of a polar solvent that can surround the ions and help stabilize them. Polar solvent molecules have a positive end, which can surround, attach to, and help stabilize the negative leaving group. The polar solvent molecule also has a negative end, which can surround, attach to, and help stabilize the positive carbocation. Nonpolar solvents cannot attach to the ionic species and therefore do not help stabilize or lower their energy. S_N1 reactions prefer polar solvents.

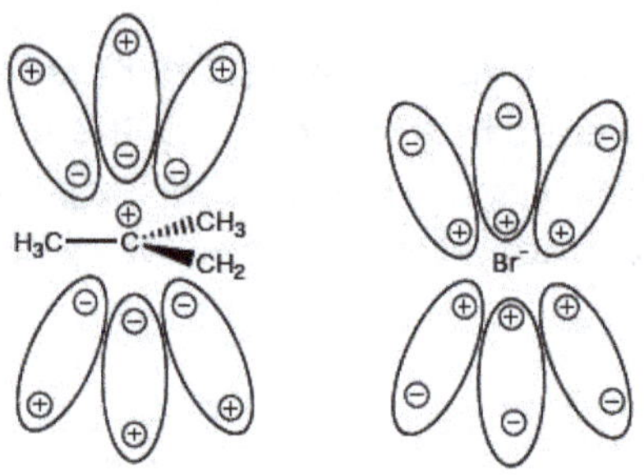

9.8.5 The Stereochemistry

The carbocation intermediate is a flat, planar species. Because it is flat, the nucleophile can attack the empty p-orbital from the topside of the intermediate or the bottom side of the intermediate. This leads to two different products, the *retention product* if the nucleophile attacks the same side as where the leaving group left and the *inversion product* if it attacks from the backside of where the leaving group left. This leads to a racemic mixture of products if there are three different groups on the carbocation leading to chiral molecules.

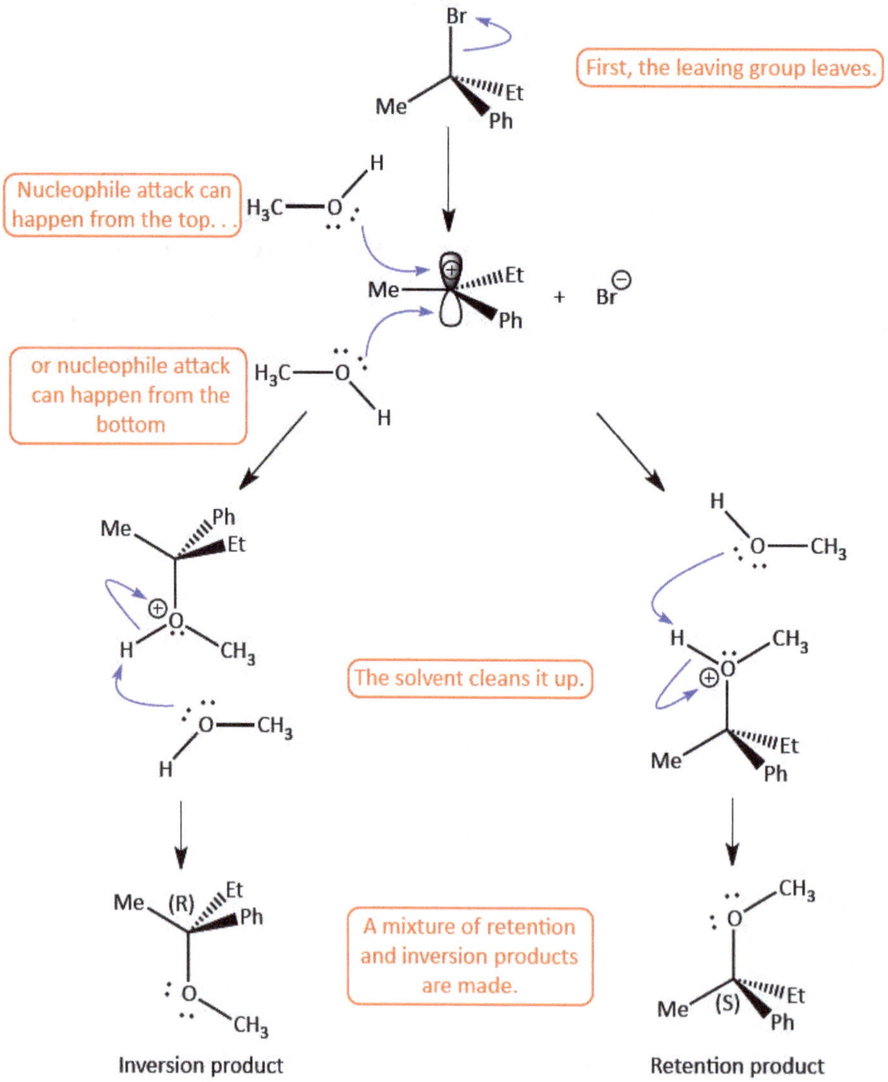

Figure 9.21 Formation of a racemate in the S_N1 reaction

9.8.6 Rearrangements

Any time a carbocation forms, like in S_N1, it is possible that a side-reaction called a *rearrangement* may occur.

1,2-Hydride shift

Let's look at the simplest of the rearrangements, the 1,2-hydride shift, and see why it happens. In this S_N1 reaction, a secondary carbocation is formed after the halide leaves. At this point, the nucleophile could come in and make the product. But, in this particular case, a rearrangement can happen. The carbocation wants electrons. A hydride, a hydrogen atom with its two electrons, moves from a neighboring carbon to the carbocation. This is called a 1,2-hydride shift. Why would this happen? What is the driving force? In this particular case, once the hydride shifts over, a secondary carbocation becomes a more stable tertiary carbocation. This tertiary carbocation is happier, lower energy, and is preferred. If more stable carbocations are formed, a rearrangement may occur.

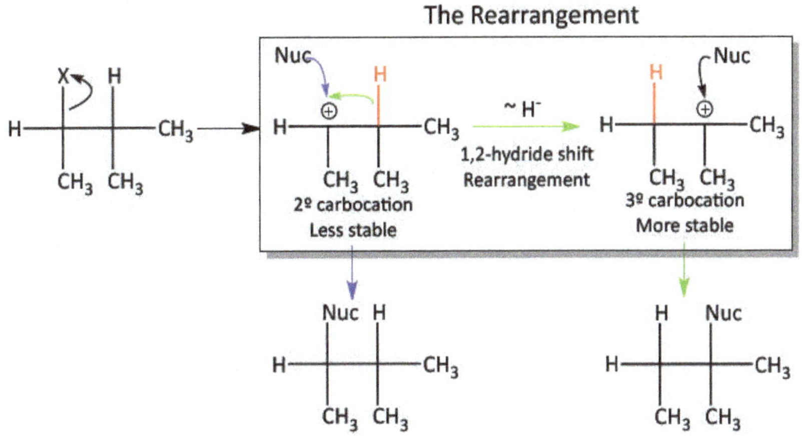

Figure 9.22 The 1,2-hydride shift rearrangement

Usually, at this point, a sense of panic starts to set in on the part of the typical student. They begin to wonder how in the world they will be able to predict when this will happen. Don't worry too much about that. Often, you will not be asked to predict when a rearrangement will happen. Instead, you will more often be asked to identify, after the fact, when a rearrangement has happened. This is much easier to do. It is often an indication of a rearrangement if you ever notice that the carbon framework has changed.

1,2-Methyl shift

Alkyl groups can also shift. In this next example, a methyl group takes two electrons from its bond and does a 1,2-methyl shift to make a more stable tertiary carbocation.

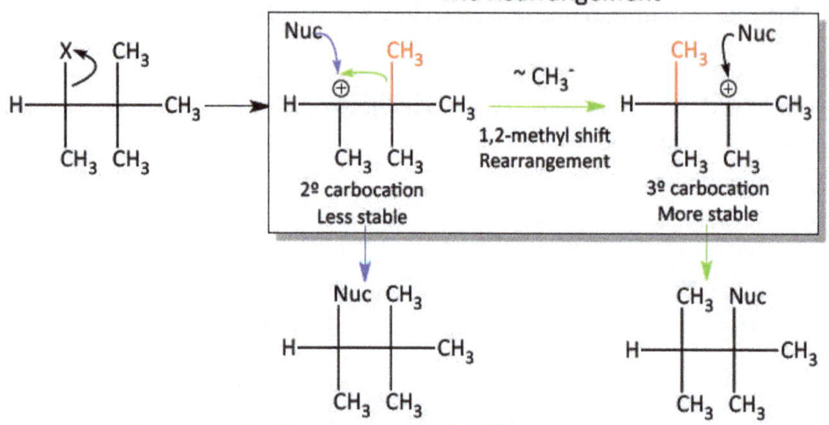

Figure 9.23 The 1,2-methyl shift rearrangement

Ring expansion

Larger alkyl groups than methyl can also shift. In this example, the alkyl group that shifts is a longer chain that wraps around and hooks onto another part of the molecule. The cyclopentyl group opens up as the CH_2 does a 1,2-shift. Therefore, the ring gets one carbon larger becoming a cyclohexyl ring. This ring expansion rearrangement is usually the trickiest rearrangement, so study it until you clearly understand it.

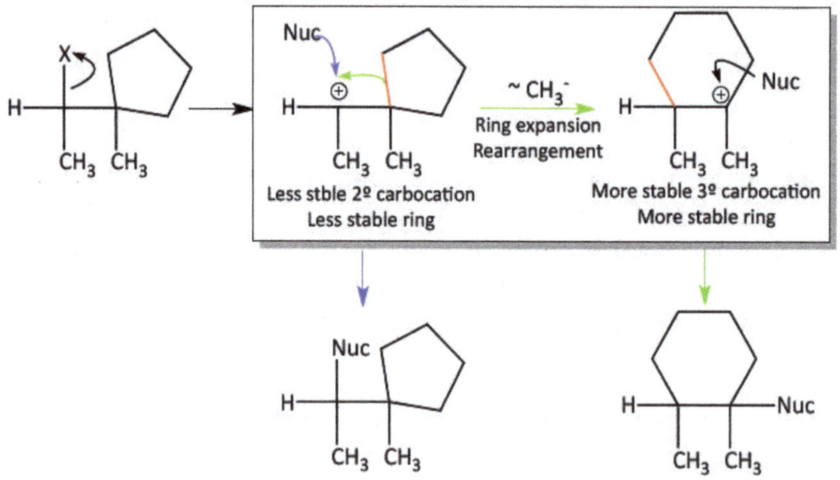

Figure 9.24 A ring expansion rearrangement

9.12 CH_3CH_2OH does not react with NaBr, but adding H_2SO_4 forms CH_3CH_2Br. Explain.

9.13 Explain why neopentyl bromide, even though it is a primary alkyl halide, does not do an S_N2 reaction.

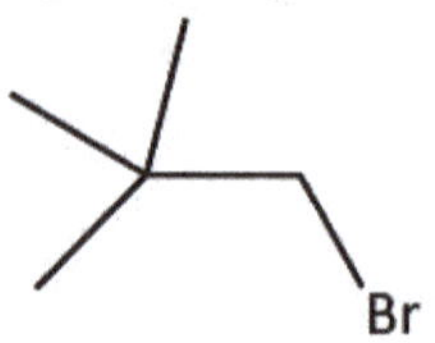

neopentyl bromide

9.14 Rank the nucleophilicity of OH^-, H_2O, CH_3O^-, and CH_3COO^-.

9.15 Rank the nucleophilicity of NH_3 and PH_3.

9.16 The primary alkyl halide, $CH_2{=}CHCH_2Cl$, undergoes an S_N1 reaction faster than the secondary alkyl halide, $(CH_3)_2CHCl$. Explain why.

9.17 Explain why $PhCH_2Cl$ is hydrolyzed (reacts with water) slowly to form $PhCH_2OH$, but forms the alcohol quickly if a small, catalytic amount of potassium iodide, KI, is added to the reaction.

9.18 Give the organic product in the following substitution reactions.

a) $HS^- + CH_3CH_2CHBrCH_3 \rightarrow$

b) $I^- + CH_3CHBrCH_3 \rightarrow$

c) CH_3NH_2 + ethyl bromide →

d) ethyl bromide + PPh_3 →

9.19 Explain why the following bicyclic alkyl halide does not undergo S_N1 or S_N2 reactions.

9.20 Label the following reactions as more likely to react via an S_N1 or an S_N2 mechanism.

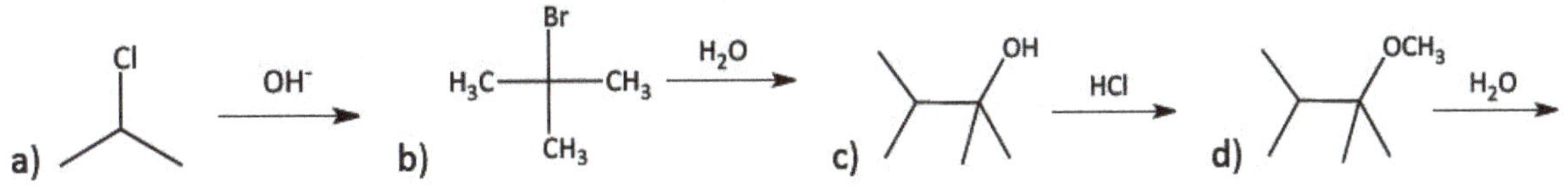

9.9 Beta-Elimination reactions (E1 and E2)

In beta-elimination reactions, a base reacts with a substrate (like an alkyl halide) to make a π-bond. Two groups are removed or eliminated from the substrate. One is a hydrogen atom and one is a leaving group. The carbon atom attached to the leaving group is the alpha carbon. There must be a hydrogen atom on an adjacent carbon atom (a beta carbon atom). The molecule cannot eliminate if there is no beta hydrogen atom. There are two types of elimination reactions, the E1 and the E2.

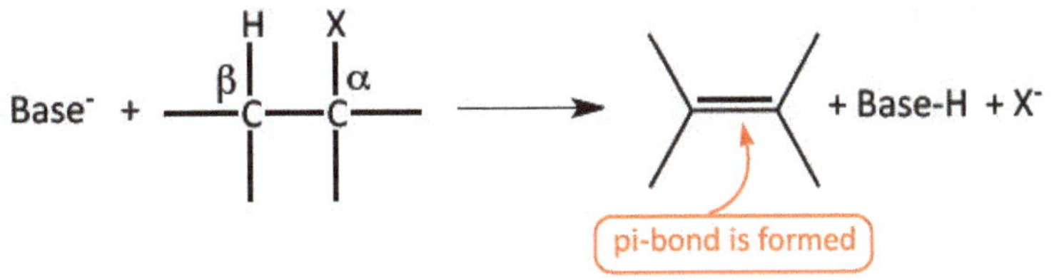

Figure 9.25 A generic elimination reaction

9.10 The E1 reaction

The E1 reaction is an elimination that is unimolecular. It is unimolecular because there is only one species, the substrate, which is involved in the rate-determining step. The rate of the reaction depends on the rate of the leaving group leaving the substrate. The amount of base in the reaction has nothing to do with that.

E: Elimination. Two groups are eliminated to form a π-bond.
1: Unimolecular. One species is involved in the rate-determining transition state.
The rate of the E1 reaction only depends on the concentration of the alkyl halide substrate. It does not matter how much of the base is in the reaction.

$$\textbf{Rate} = \mathbf{k_r[Substrate]}$$

or

$$\textbf{Rate} = \mathbf{k_r[R\text{-}X]}$$

The mechanism for the E1 reaction is as follows.

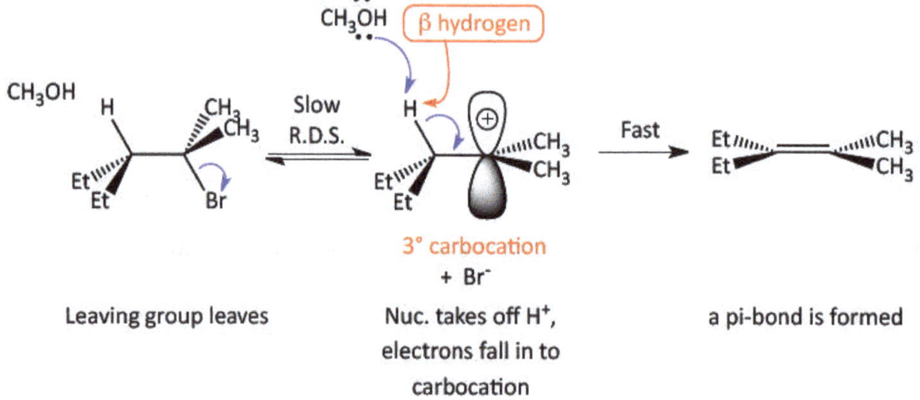

Figure 9.26 An E1 reaction mechanism

The first step of the E1 reaction is the same as the S_N1 reaction. First, the leaving group leaves to form a carbocation. Once a carbocation is formed, the base removed a beta-hydrogen atom to form the π-bond.

The E1 mechanism competes with the S_N1. Once a carbocation forms, the weak nucleophile can attack the alpha carbon to do S_N1 or it can act as a base and remove a beta-hydrogen to do E1.

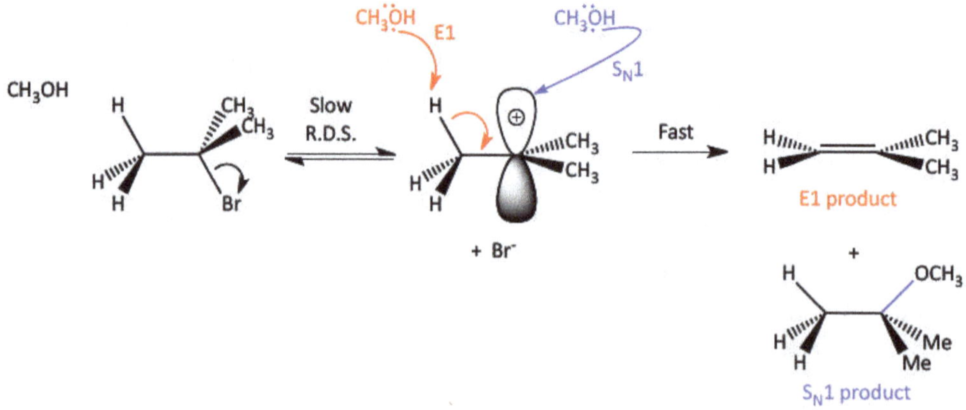

Figure 9.27 The E1 mechanism competing with the S_N1 mechanism

Since the first step of an E1 reaction is to form a carbocation, do you think the E1 reaction could have rearrangements? Of course it could. This is something for us to watch for.

9.11 The E2 reaction

The E2 reaction is a concerted reaction. All steps happen at the same time. An aggressive base pulls a proton (H^+) off of a beta carbon. At the same time, the electrons between the proton and beta carbon atom form a π-bond at the same time as a leaving group leaves the alpha carbon. Since this is a concerted, one-step mechanism, there is only one transition state. In this transition state, which must be the rate-determining step, two species are involved, the base and the substrate. This is why there is a "2" in the E2 name.

E: Elimination. Two groups are eliminated to form a π-bond.
2: Bimolecular. Two species are involved in the rate-determining transition state.

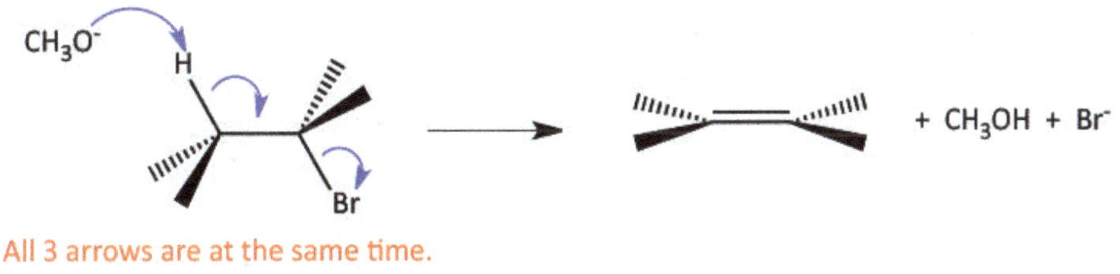

Figure 9.28 A generic E2 mechanism

Since two species are involved in the rate-determining step, the rate equation for the E2 reaction is:

$$\textbf{Rate} = k_r[\text{Base}][\text{Substrate}]$$

9.11.1 Nucleophiles and Bases

A base grabs an H^+ (think E2). A nucleophile attacks a carbon atom (think S_N2).

The E2 reaction typically requires a strong, aggressive base. The deprotonation needs to occur immediately. If the leaving group leaves first, it is an E1 or S_N1 mechanism.

Strong Nucleophile/Strong Base

What makes a strong, aggressive base? Strong bases often have negative charges like HO^-, RO^-, NH_2^-, etc. These are both good bases and good nucleophiles.

Some strong bases won't eliminate but will substitute instead. A good, small base like HO^- or CH_3O^- may S_N2 backside attack an alkyl halide that is not crowded instead of performing an E2 reaction. In fact, the S_N2 reaction is favored under these conditions.

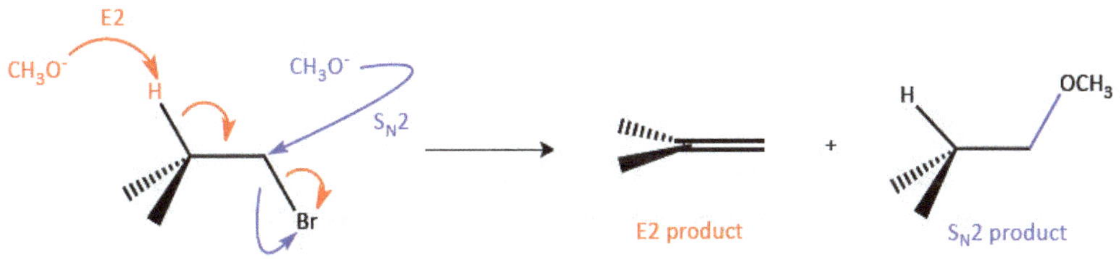

Figure 9.29 E2 competes with S_N2

Strong Nucleophile/Weak Base

Sometimes, though, a negatively charged species is a strong nucleophile while being a weak base. It is a weak base if the negatively charged species is pretty stable. Examples of strong nucleophiles that are weak bases are negatively charged halides or when a negative charge can be spread out on the species using resonance forms. These are poor bases because they don't want to grab an H^+. But, because they are large and polarizable, they are good nucleophiles and can attack a carbon atom in an S_N2 reaction.

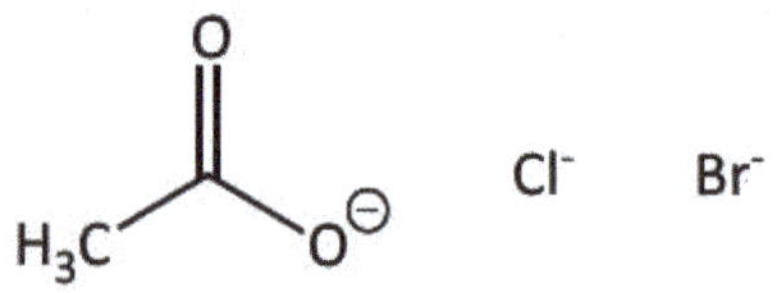

Strong Base/Weak Nucleophile

Using a bigger, bulkier base is one way to get more of the E2 reaction to occur over S_N2. For eliminations, it is better to use the larger t-butoxide ($^-O^tBu$) instead of using the smaller CH_3O^-, methoxide. Big, bulky bases like t-butoxide are much worse nucleophiles. They have difficulty reaching the alpha carbon as a nucleophile in an S_N2 reaction and therefore can only remove protons (H^+) that are on the outside of the molecule.

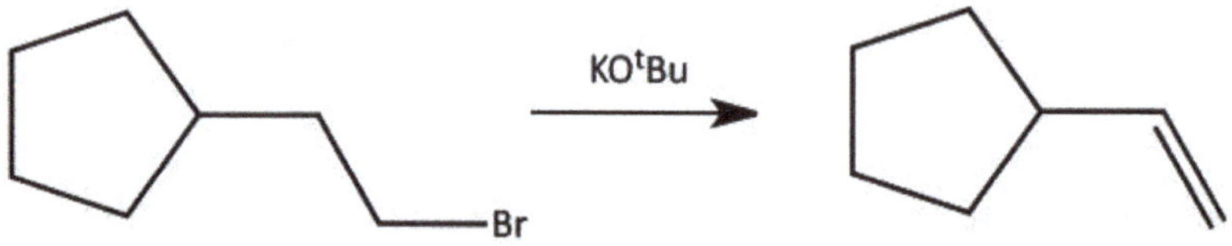

The S_N2 reaction is prevented if the alpha carbon is crowded on the backside of the halide (like a tertiary alkyl halide). A backside attack could not occur. In this case, it doesn't really matter if the base is large or small. As long as it is strong and aggressive, an E2 reaction will occur because the crowded substrate prevents the S_N2 reaction.

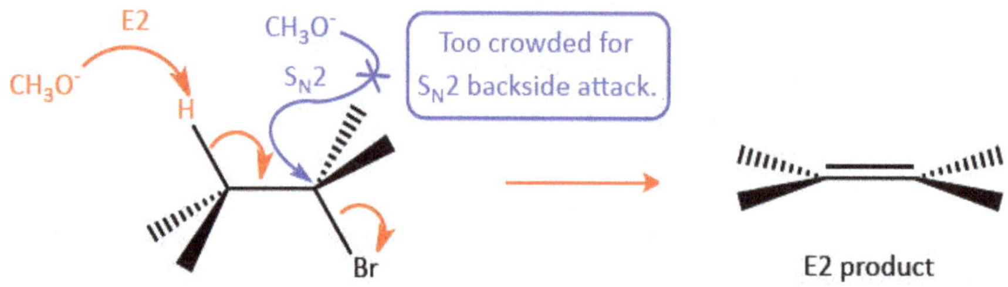

> 9.21 Are the following bases more likely to be involved in E1 or E2 reactions?
> a) water b) CH_3OH c) OH^- d) $(CH_3)_2N^-$

9.11.2 E2 Regiochemistry

Sometimes more than one beta carbon atom has a hydrogen atom that can be removed in the E2 reaction. Notice that we always remove the proton from a beta carbon, never the alpha carbon or farther away than the beta carbon, like gamma, etc. The reason for this is that the electrons need to fall between the beta and alpha carbons to make a π-bond.

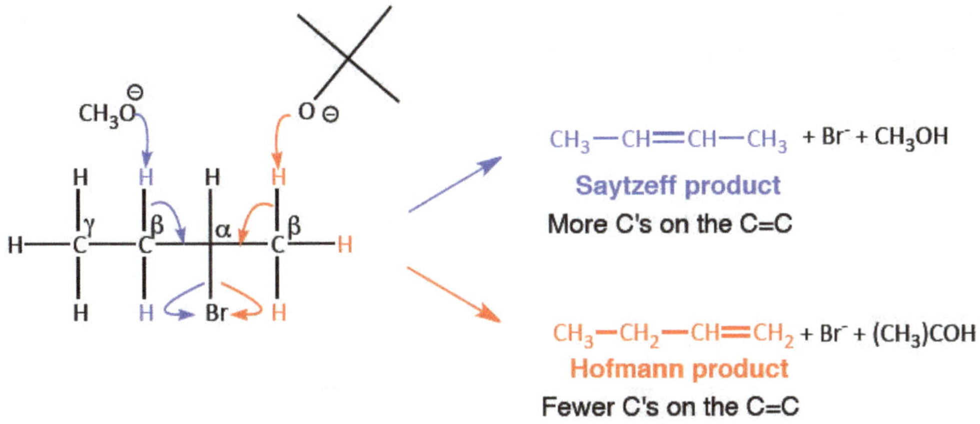

Figure 9.30 Hofmann vs. Saytzeff product

The Saytzeff product is more stable than the Hofmann product. The Saytzeff product is the one with the most carbon groups attached to the alkene carbons. The Hofmann product has fewer carbon groups attached to the alkene carbons. All alkenes have four groups on them. Look at those four groups to see how many of them are carbon groups.

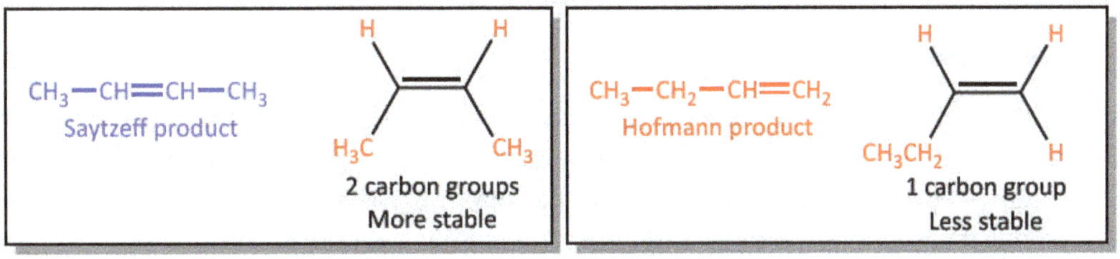

If the Saytzeff product is more stable, why would we ever get more of the Hofmann product? If a smaller base, like HO^-, CH_3O^-, EtO^- is used, the more stable Saytzeff product is made. But, if the Hofmann product is desired, a very large base like t-butoxide, ${}^tBuO^-$ is used. A very big, bulky base like t-butoxide cannot reach the blue, interior protons. It can only reach the outside red beta protons. This leads to the Hofmann product.

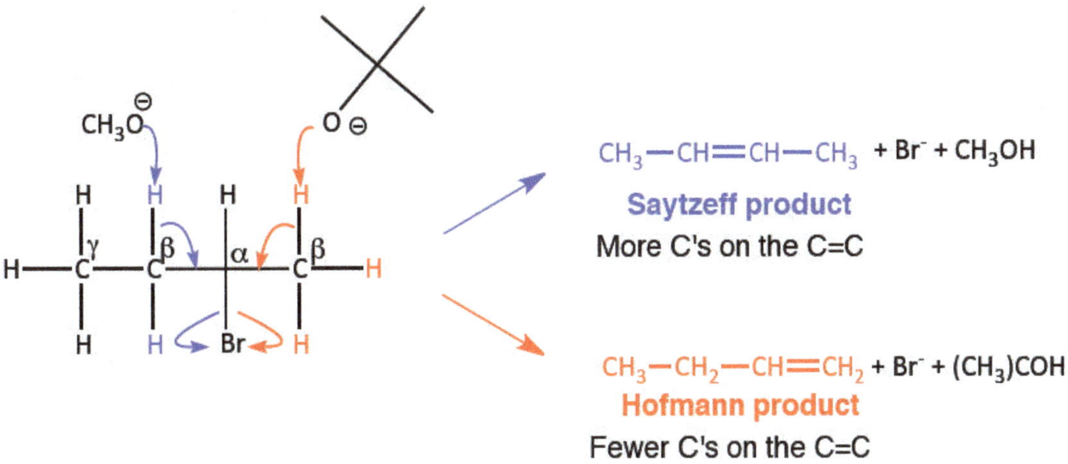

9.22 For each of the following compounds, draw the Saytzeff and Hofmann products made following an E2 reaction.

a) 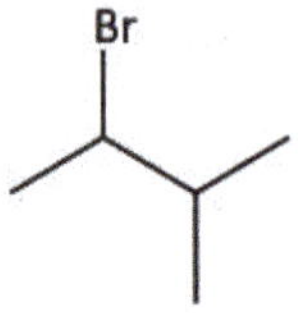

Saytzeff ______ Hofmann ______

b) 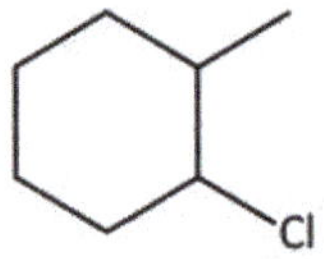

Saytzeff ______ Hofmann ______

c) 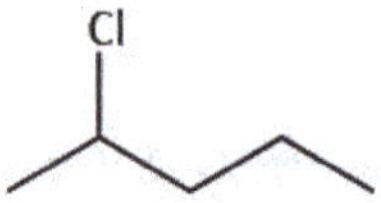

Saytzeff ______ Hofmann ______

d) Circle the product in part (a) of this problem if KO^{t}Bu was used as the base for the elimination.

e) Circle the product in part (b) of this problem if $NaOCH_3$ was used as the base for the elimination.

9.11.3 E2 Stereochemistry

In order for an E2 reaction to occur, the beta proton that is extracted by the base and the leaving group must be in a coplanar arrangement. The sp^3 hybrid orbitals turn into p-orbitals and need to overlap to form the π-bond. The overlap can be syn-coplanar or anti-coplanar (anti-periplanar). Syn-coplanar is when the beta hydrogen and the leaving group have a dihedral angle of 0° and anti-coplanar has a dihedral angle of 180°.

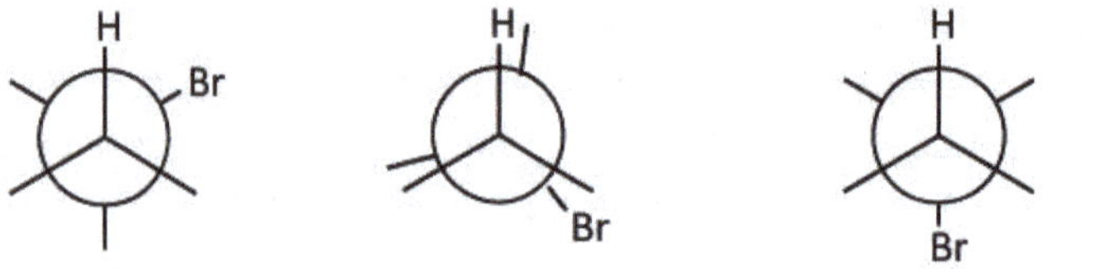

60° dihedral angle 120° dihedral angle 180° dihedral angle

Anti-coplanar (anti-periplanar) is preferred because the negatively charged incoming base is farther away from the negatively charged leaving group. You can also see in the Newman projection that anti-coplanar is in a more stable staggered conformation while *syn-coplanar* is in an unstable eclipsed conformation.

Anti-coplanar (anti-periplanar) is preferred.

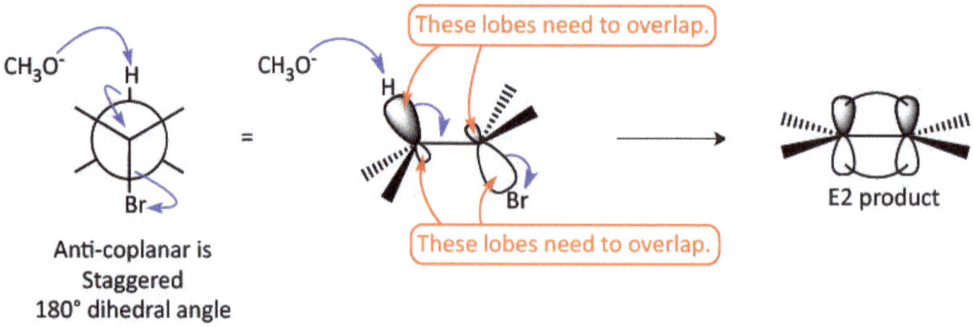

Figure 9.31 Anti-periplanar E2 mechanism

Syn-coplanar can also happen, but it is not as preferred because the molecule must be in the higher energy eclipsed form.

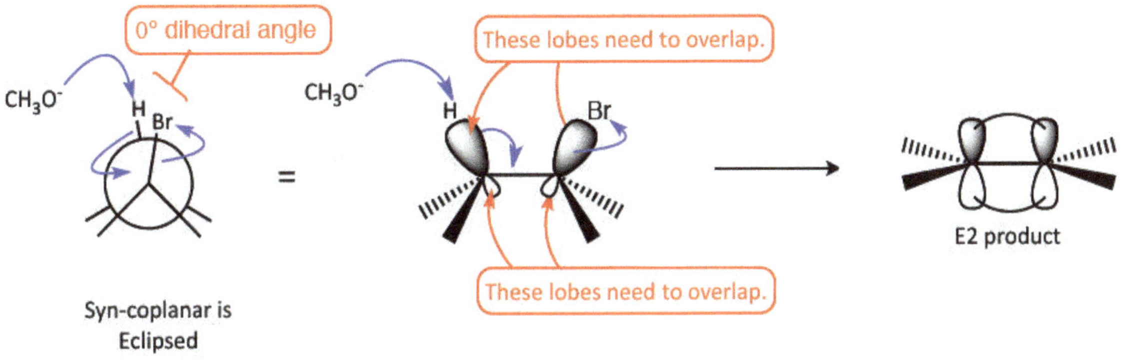

Figure 9.32 Syn-coplanar E2 mechanism

Typically, the syn-coplanar arrangement does not occur unless it is forced and the anti-coplanar arrangement cannot be obtained. In the following example, because of the bicyclic ring structure, only the syn-coplanar arrangement can be obtained with a dihedral angle of 0°. The red hydrogen atom and the red bromine atom are syn-coplanar to each other. The molecule cannot rotate to become anti-coplanar.

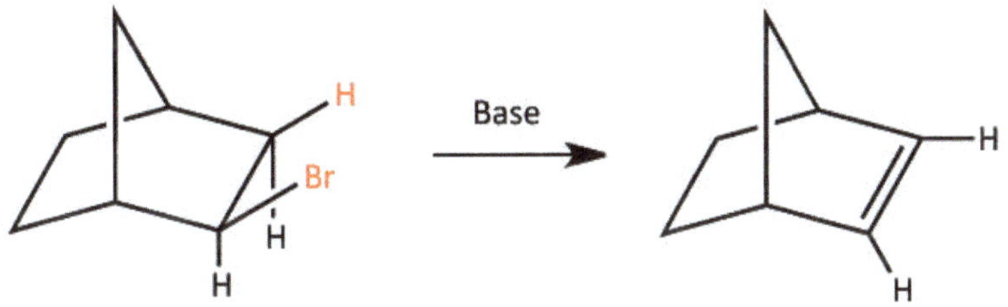

On cyclohexane chairs, the anti-coplanar formation can often be made. When 1-bromo-3-methylcyclohexane is attacked by a base, it can do an E2 reaction. In the lowest energy chair conformation on the left, there are no blue beta hydrogen atoms that are coplanar to the red bromine leaving group. Look closely until you convince yourself that they are not coplanar. But, if a chair flip occurs, now two hydrogen atoms (in red) are anti-coplanar to the bromine. The red hydrogen atoms and the bromine are in a 1,2-diaxial configuration. This is the only way to get a coplanar relationship in a cyclohexane chair.

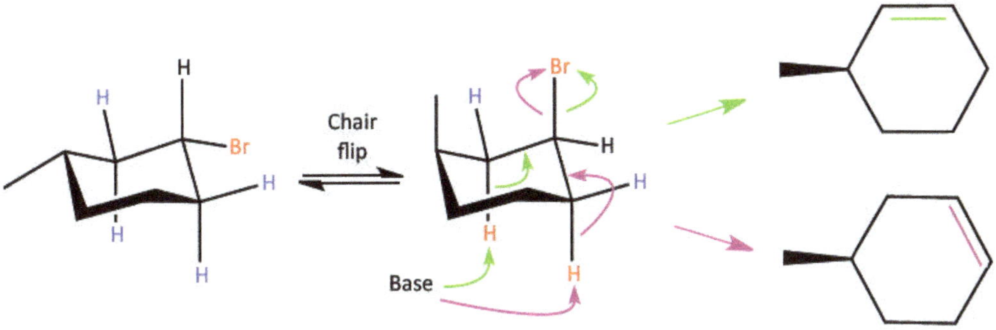

Stereospecific

Because the anti-coplanar arrangement is preferred, that arrangement is most often formed. The E2 mechanism is also concerted (all happening at once), so there is no time for the molecule to rotate. The E2 mechanism can be considered a *stereospecific* reaction. One stereoisomer is made preferentially over another. In this example, we end up with the methyl groups on the same side as each other because that is how they started in the anti-coplanar arrangement.

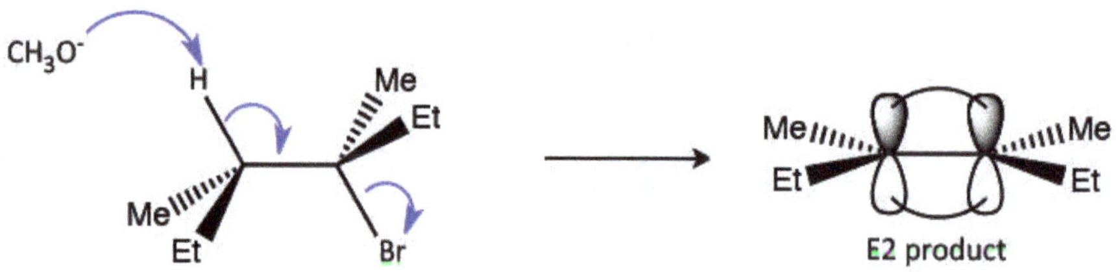

Anti-coplanar

9.23 Are the following reactions more likely to occur via the E1 or E2 mechanism (or neither)?

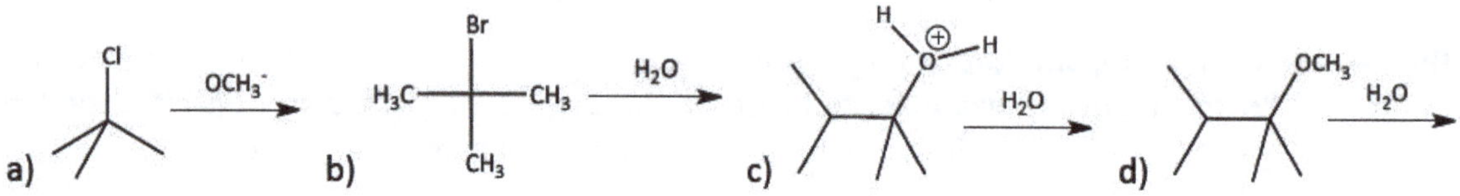

9.12 Comparison Summary

Do not memorize this chart. But, you can self-test yourself by making sure you understand why each of these is the way they are. Again, do not memorize, but develop a deep understanding of the underlying principles.

	S_N1	S_N2	E1	E2
Nucleophile/ Base	Not important Is usually weak	Strong nucleophile	Not important Is usually weak	Strong base
Substrate	3° > 2° >> 1° > Me Stabilizes Carbocation	Me > 1° > 2° >> 3° Not crowded backside	3° > 2° >> 1° > Me Stabilizes Carbocation	3° > 2° > 1° > Me More substituted alkene preferred
Solvent	Polar Helps ionize. Is often the base.	Polar aprotic preferred	Polar Helps ionize. Is often the base.	Unimportant
Kinetics	Rate = k[R-X]	Rate = k[R-X][Nuc]	Rate = k[R-X]	Rate = k[R-X][Base]
Stereochemistry	Racemic mixture Flat carbocation	Walden inversion	Flat carbocation	Coplanar H and X Usually anti-coplanar
Rearrangements	Yes Forms carbocations	No	Yes Forms carbocations	No
Regiochemistry	N/A	N/A	Saytzeff with smaller bases. Hofmann with very large bases.	Saytzeff with smaller bases. Hofmann with very large bases.

9.24 Explain the following observation. $(CH_3)_3CBr + H_2O \rightarrow (CH_3)_3COH + HBr$, but $(CH_3)_3CBr + OH^- \rightarrow (CH_3)_2C{=}CH_2 + H_2O + Br^-$.

9.25 Indicate the products of the following reactions and indicate whether the mechanism is S_N1, S_N2, E1, or E2.

a) propyl bromide + $LiAlH_4$

b) tert-butyl bromide + ethanol (CH_3CH_2OH) heated to 60°C.

c) $CH_3CHBrCH_3 + NaOCH_3$ in CH_3OH

d) $CH_3CH{=}CHCH_2Cl + NaNH_2$

e) $(CH_3)_3CBr$ + KOtBu

9.13 Reaction Summary

Free-radical halogenation

a) *Initiation*

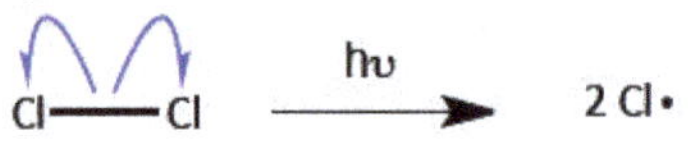

b) *1st propagation step*

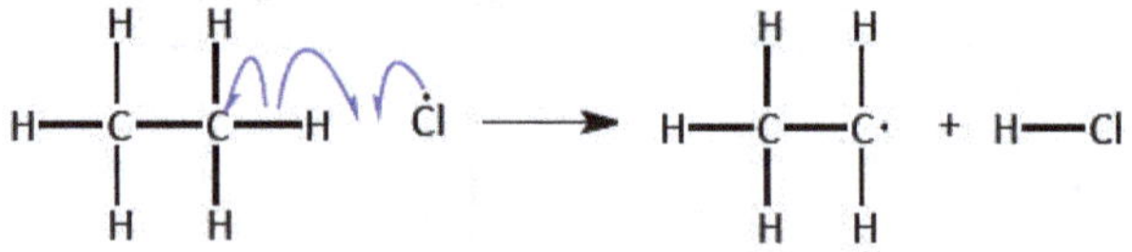

c) *2nd propagation step*

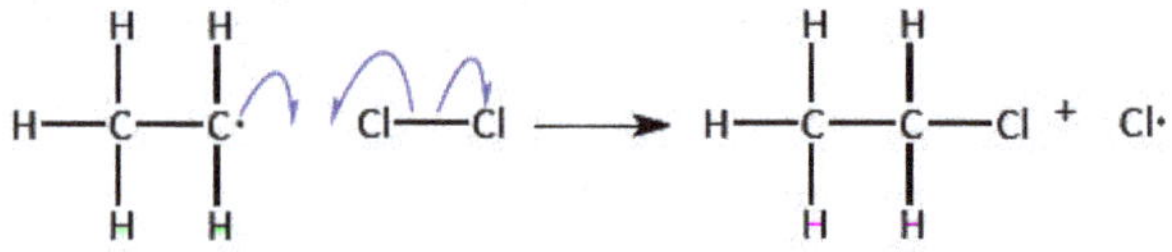

d) *Termination*

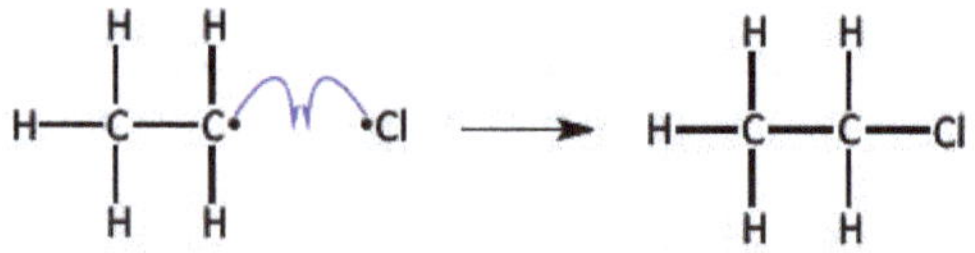

Allylic bromination

e)

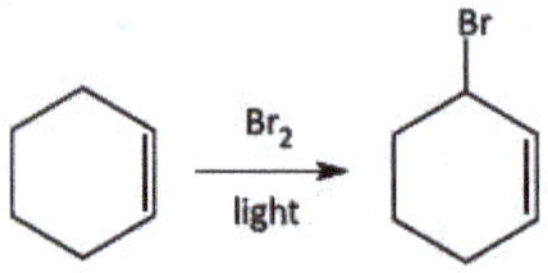

f)

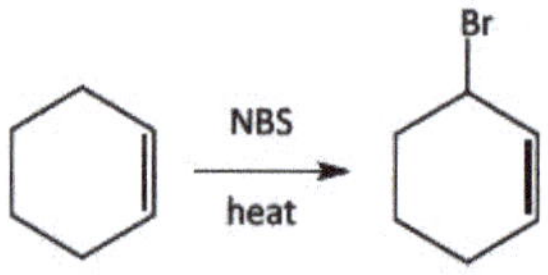

Alcohols to good leaving groups

g) 1° alcohol

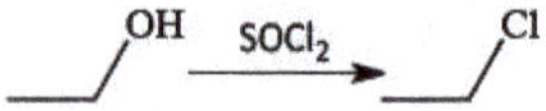

h) 2° alcohol

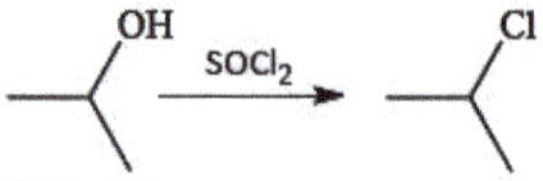

i) 3° alcohol

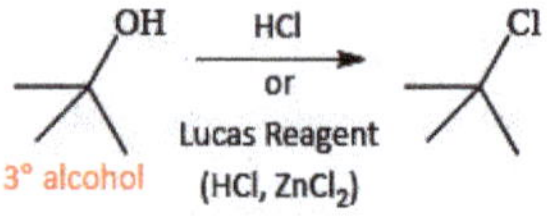

j) 1° alcohol

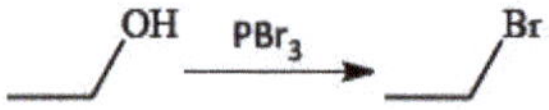

k) 2° alcohol

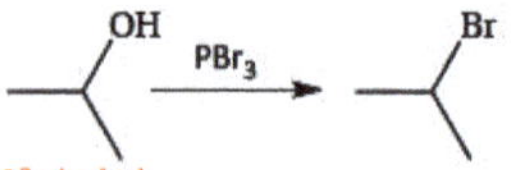

l) 3° alcohol

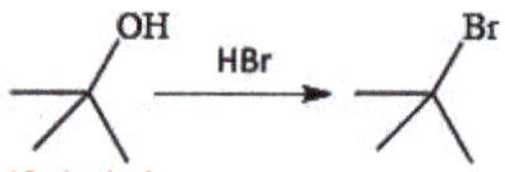

m) 1° alcohol

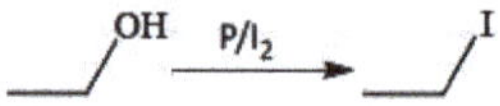

n) 2° alcohol

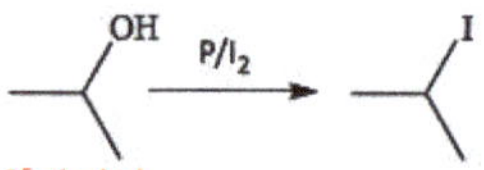

o) 3° alcohol

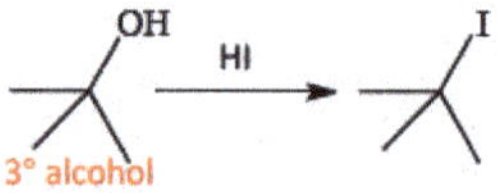

S_N2

p)

S_N1

q) 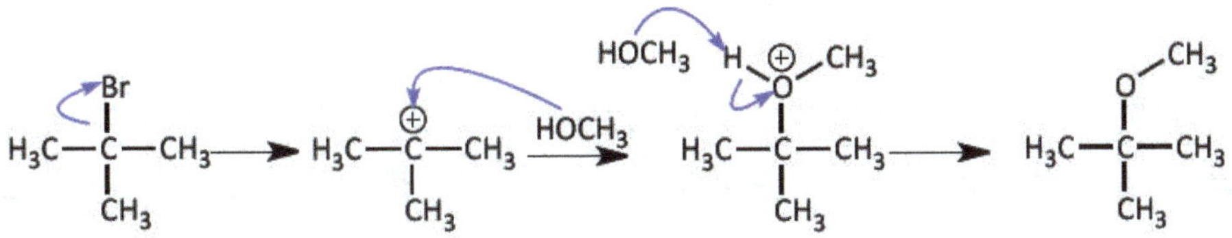

Rearrangement

r) 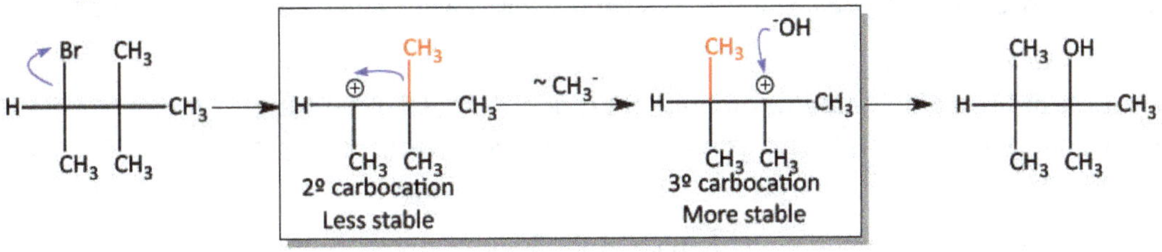

E1

s) 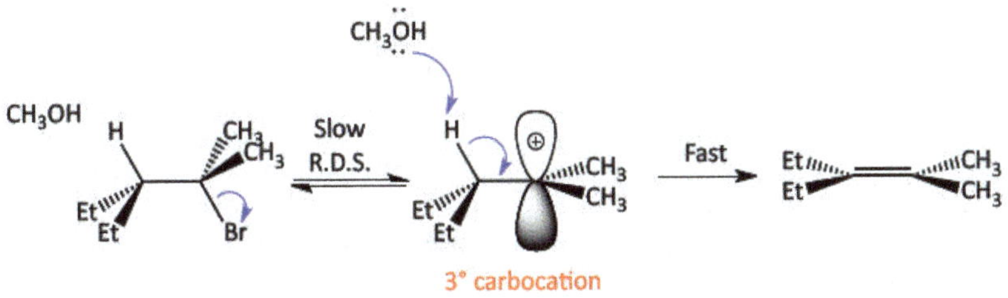

E2

t) 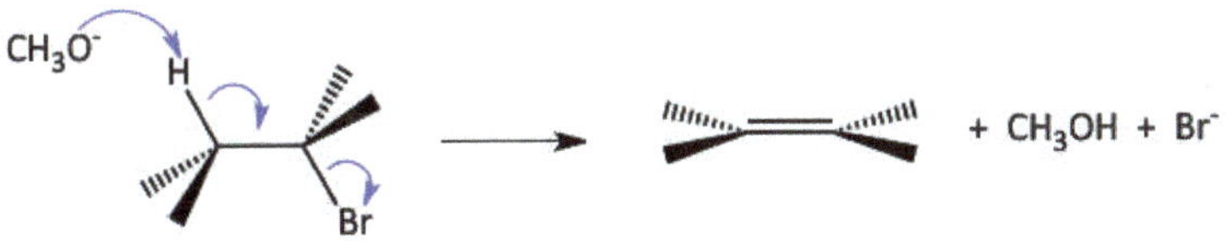

E2 regiochemistry

u) 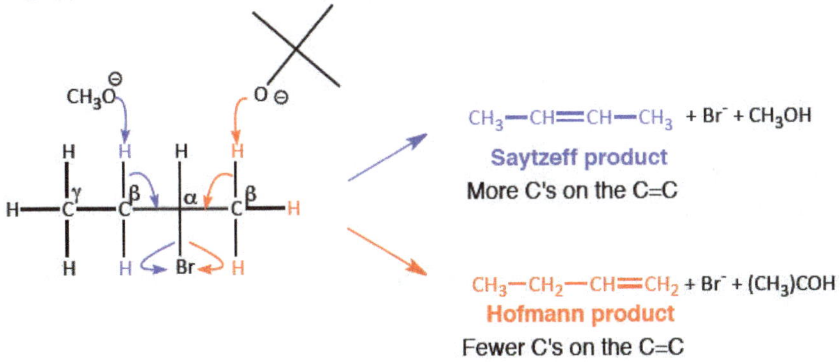

9.14 Problems

9.26 Draw the following:

a) isopropyl iodide b) *tert*-butyl chloride c) 1,2-dichlorobutane d) chloroform e) carbon tetrachloride

9.27 Write out the *initiation*, two *propagation* steps, and one *termination* step for the free-radical chlorination of methane.

9.28 One major monobrominated product is made when methylcyclopentane is brominated with Br_2 under ultraviolet light. a) Draw that product. b) Draw the mechanism for this reaction (initiation, propagation, and termination steps).

9.29 a) Write the structures of both of the monobromination products of 1,1,2,2-tetramethylcyclopropane. b) Which of your monobromination products would you get more of when 1,1,2,2-tetramethylcyclopropane reacts with bromine under ultraviolet light?

9.30 Upon bromination with Br_2 under ultraviolet light, each of the following makes one primary monobrominated product. Draw that product.

a) cyclopentane b) 2-methylbutane c) propylbenzene d) methylcyclobutane e) cyclopentene

9.31 Show how free radical halogenation can be used to synthesize the following molecules. For each compound, explain why it is the major product formed.

a) 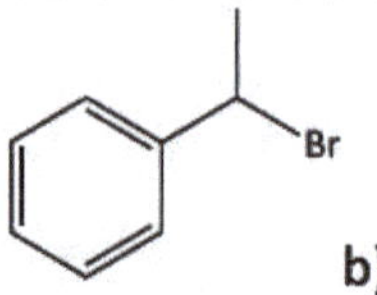

b) 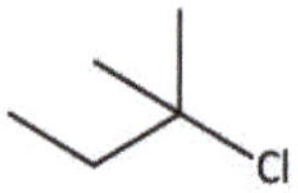

c) 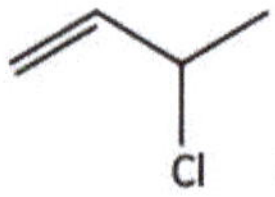

d) 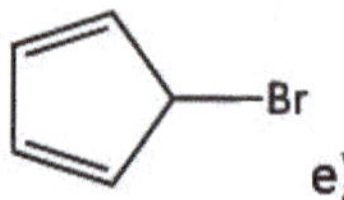

e) 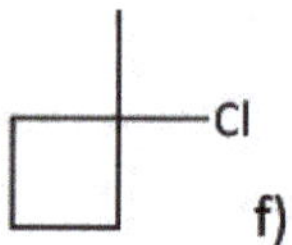

f) 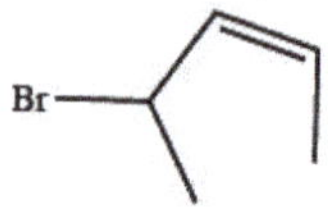

9.32 Draw good resonance structures for the following free radicals.

a) 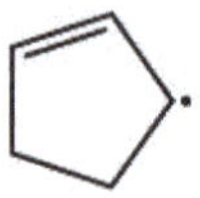
b) 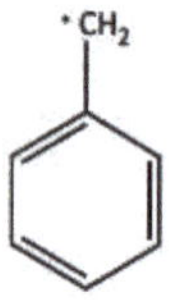

c) 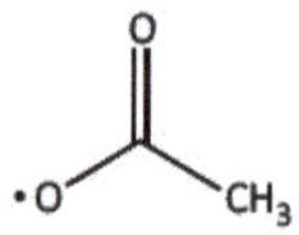

d) 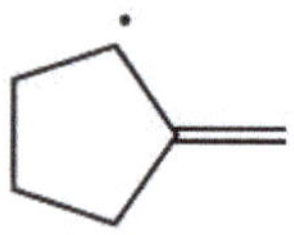
e) 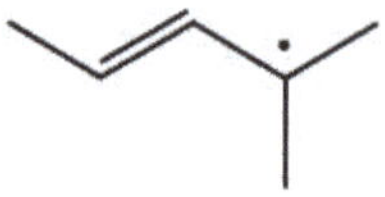

9.33 When 4-methylpent-2-ene reacts with bromine under ultraviolet light, the following brominated products are made.

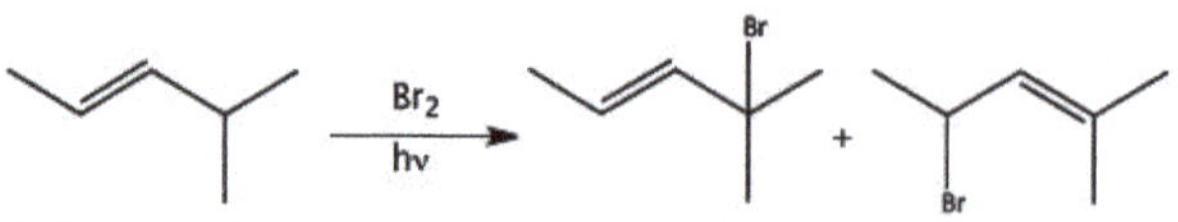

a) Draw the initiation step of the mechanism.
b) Write a mechanism for the propagation steps. Your mechanism should explain how both products are formed.
c) Why are these hydrogen atoms replaced with bromine compared to any of the other hydrogen atoms in the molecule?
d) Show any termination step.

9.34 Draw the mechanism for the S_N1 substitution reaction of 2-methyl-2-bromobutane to 2-methyl-2-butanol in water.

9.35 Provide a detailed mechanism for the S_N1 reaction of ethanol (CH_3CH_2OH) with 1-bromo-1-methylcyclohexane.

9.36 Which of the following compounds will undergo an S_N2 reaction most readily?

$(CH_3)_3CCH_2I$ $(CH_3)_3CCl$ $(CH_3)_2CHBr$ $(CH_3)_2CHCH_2CH_2CH_2I$ $(CH_3)_2CHCH_2CH_2CH_2Cl$

9.37 For each of the following molecules, predict which of the halogens would be more reactive in the S_N2 reaction.

a) 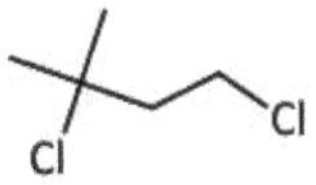
b) 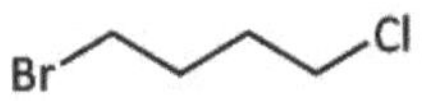
c) 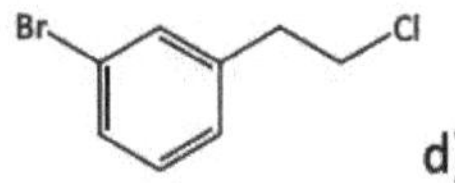
d) 

9.38 For each pair, circle the compound that will react the fastest as the electrophile (substrate) in an S_N2 reaction.

a) 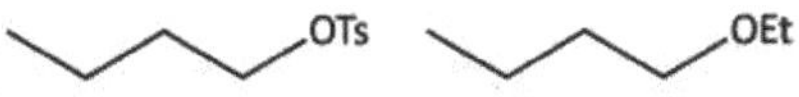
b) 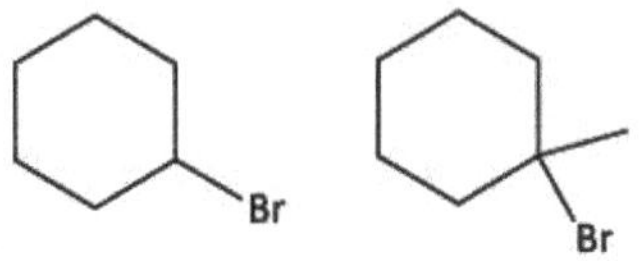
c) 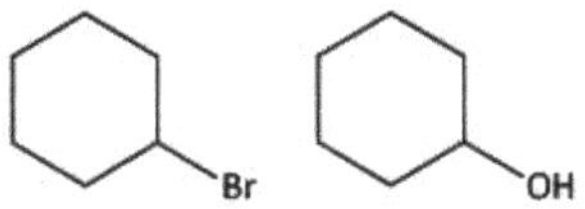
d) 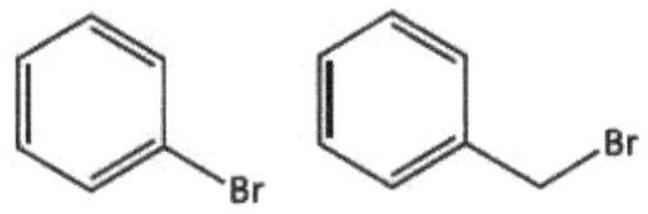
e) 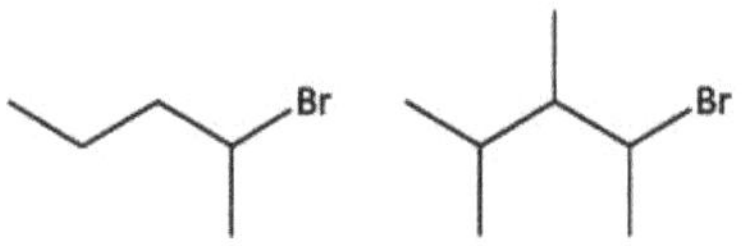

9.39 Show the major product for each of the following S_N2 reactions.

a) $CH_3—Cl + Ph—C\equiv C:^{\ominus} \longrightarrow$

b) $CH_3—CH_2—Br + KCN \longrightarrow$

c) $H_3C—O^{\ominus} Na^{+} + Ph—CH_2—Br \longrightarrow$

d) 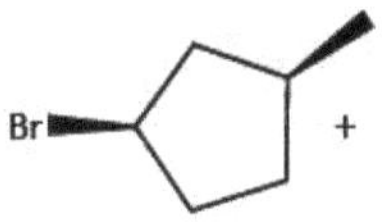 + NaOH ⟶

e)  + $Ph—NH_2$ (excess) ⟶

f) 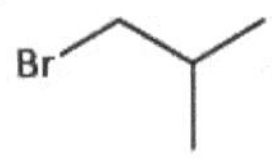 + NaI ⟶

9.40 Show the reagent you would use to make each of the following compounds from an S_N2 reaction using benzyl bromide ($Ph-CH_2-Br$) as your substrate.

a) $Ph—CH_2—CN$ b) $Ph—CH_2—OH$ c) $Ph—CH_2—O—CH_2—CH_3$

d) $Ph—CH_2—I$ e) $Ph—CH_2—NH_2$ f) $Ph—CH_2—F$ g) $Ph—CH_2—SH$

9.41 For each pair, circle the compound that will react the fastest as the electrophile (substrate) in an S_N1 or E1 reaction.

a) 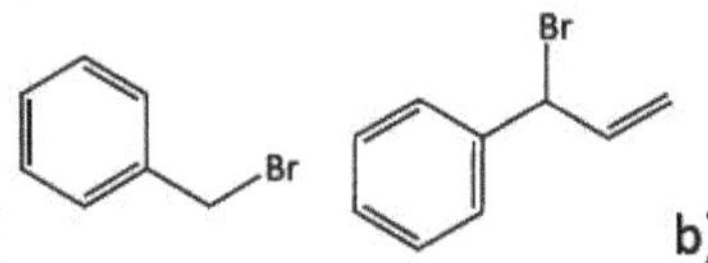
b) 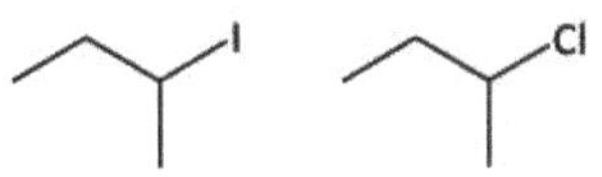
c) 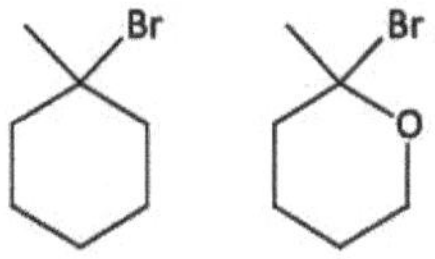
d) 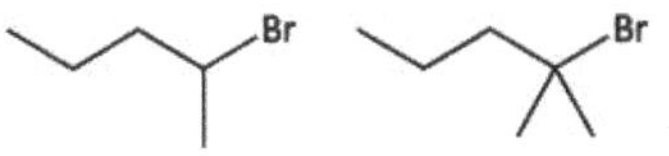
e) 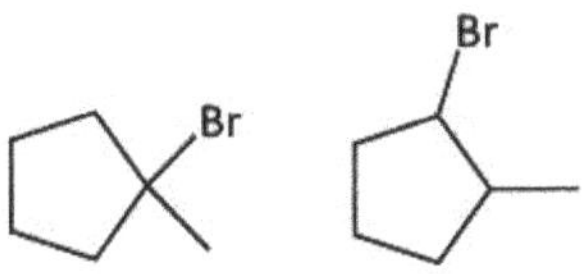
f) 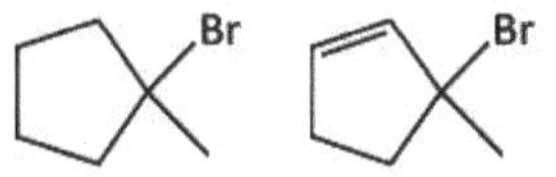

Alkyl Halides

9.42 List the following compounds in order of increasing reactivity in an S_N1 reaction.

$(CH_3)_3CBr$ $CH_3CHBrCH_3$ CH_3CHICH_3, CH_3Br $CH_3CH_2CH_2I$

9.43 Pick the stronger nucleophile from each of the following pairs.

a) HO^- vs. H_2O b) I^- vs. Cl^- c) CH_3OH vs. CH_3O^- d) Et_2NH vs. Et_3N e) HO^- vs. HS^- f) $tBuO^-$ vs. MeO^-

9.44 Rank the following species in order of **INCREASING** nucleophilicity.

I^- Cl^- CH_3OCH_3 $CH_3CH_2CH_3$

9.45 Draw the missing products of the following reactions:

a) 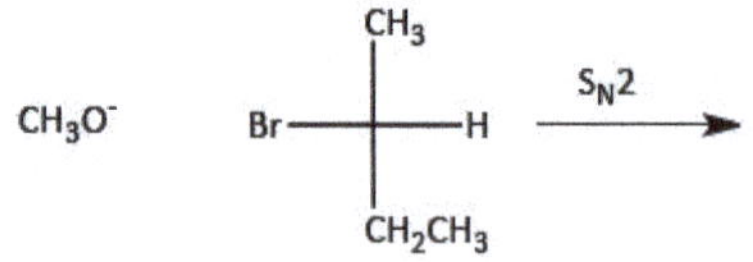

b) 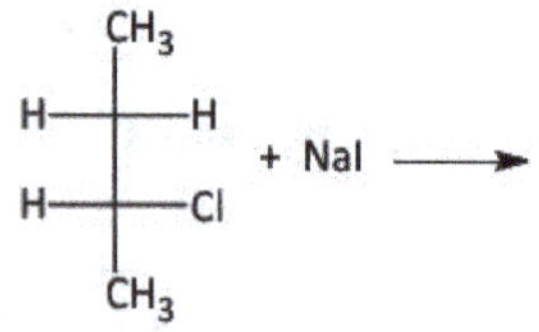

c) 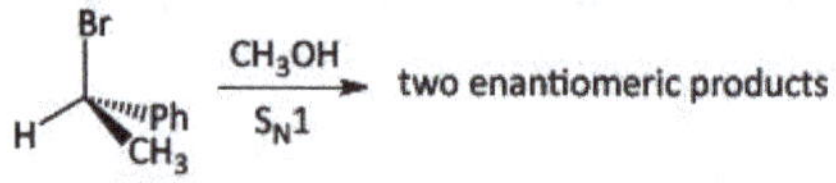

d) 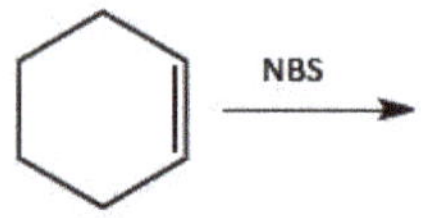

e) 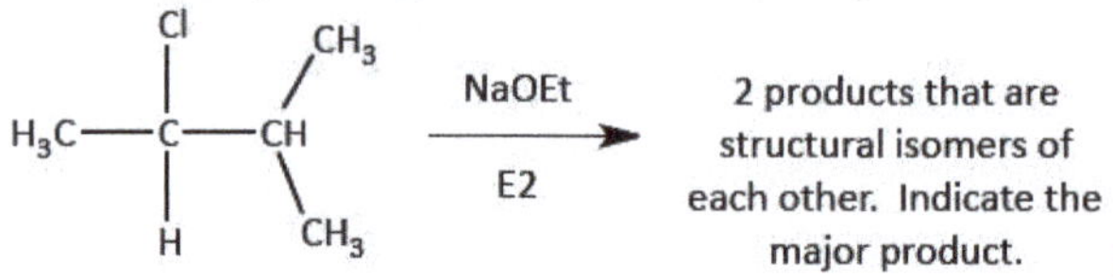

f) 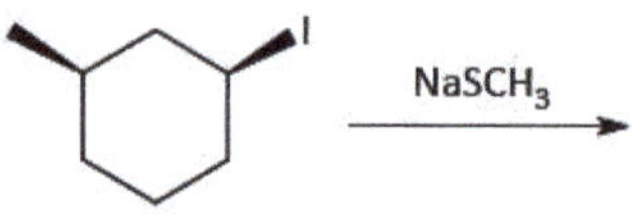

g) 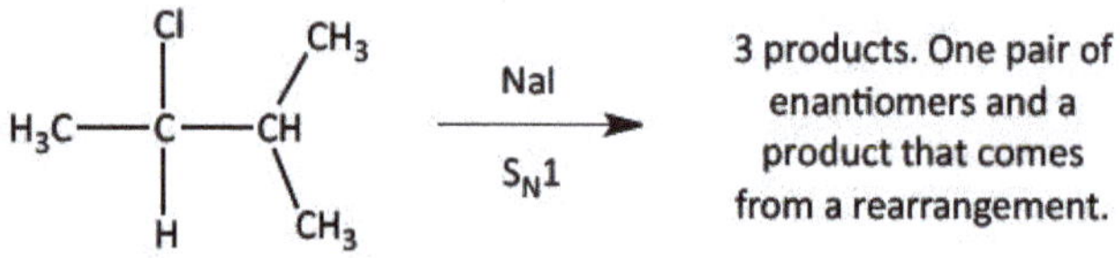

h) 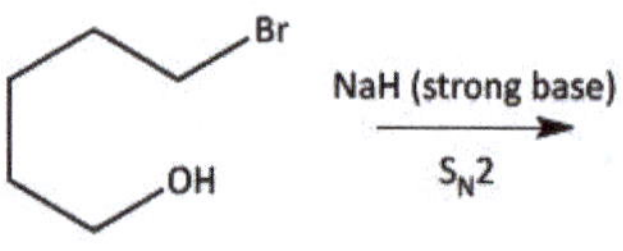

i) 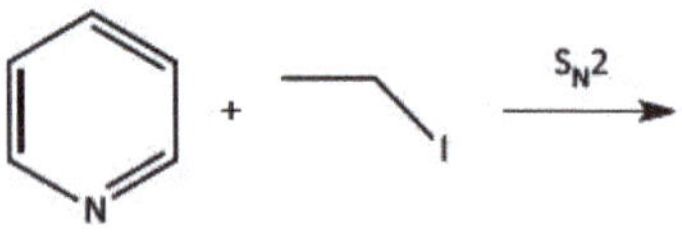

j) 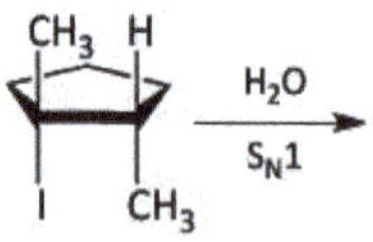

k) 1-bromobutane + $NHMe_2$ ⟶

l) 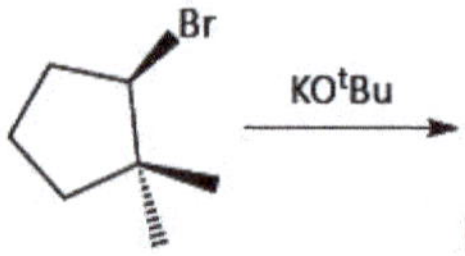

m) 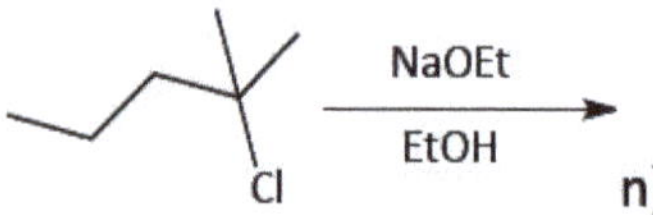

n) 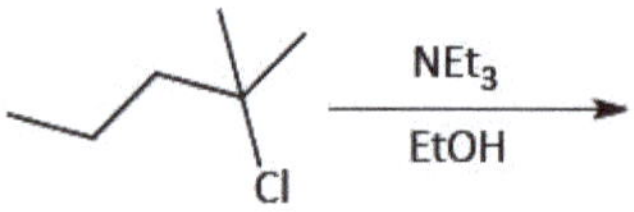

o) 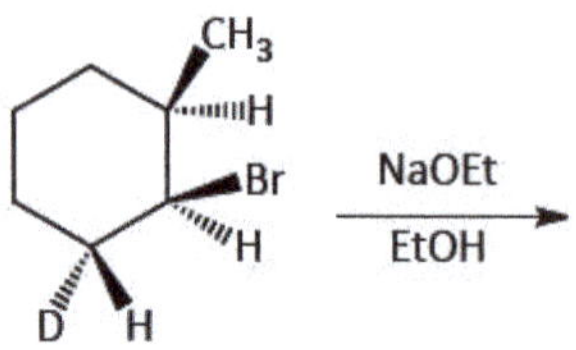

p) 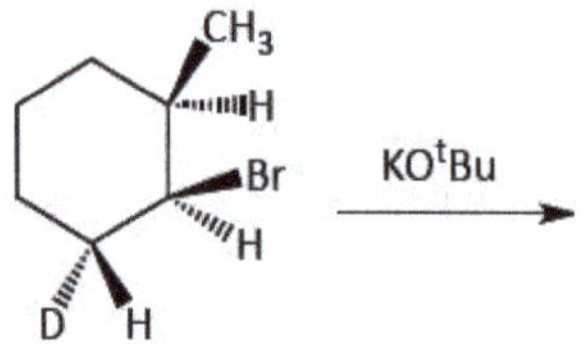

q) (S)-2-butyl tosylate + NaBr ⟶

9.46 Give suitable reactants/reagents for the following reactions.

a) 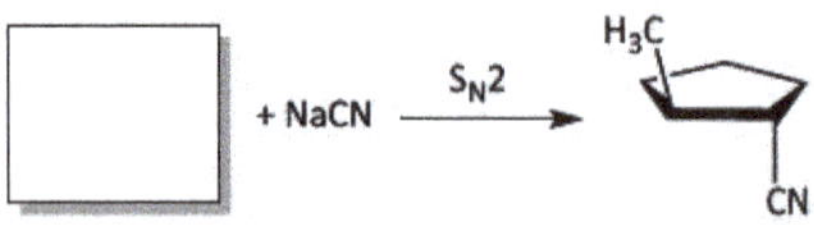

b) 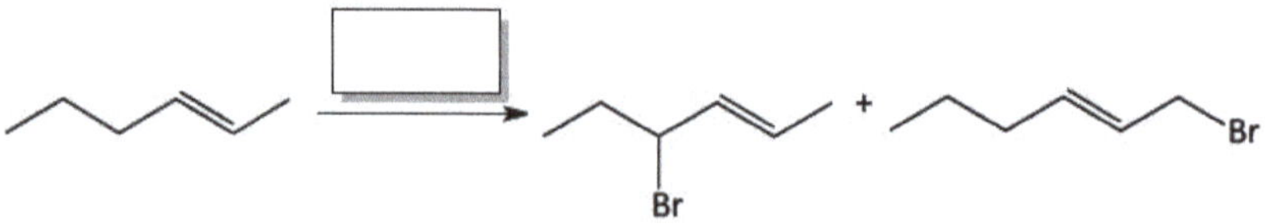

9.47 Draw good resonance forms for the following carbocations.

a) 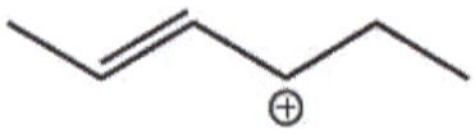

b) 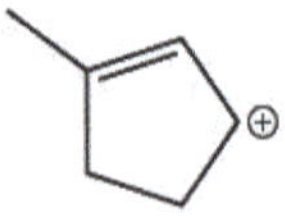

c) 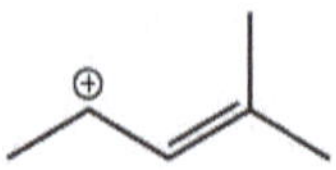

d) 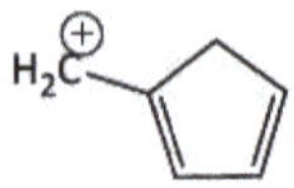

9.48 Show the mechanism for the solvolysis (S_N1) reaction of *tert*-butyl chloride in water.

9.49 Predict the E1 products of the following substrates. If more than one alkene can be formed, predict the major product.
a) 1-bromo-2-methylcyclopentane b) 2-bromohexane c) chlorocyclohexane d) 3-iodo-3-ethylpentane

9.50 When 1-bromomethylcyclohexene undergoes a reaction in ethanol, the compound shown below is one of the products formed. Give a detailed mechanism for its formation.

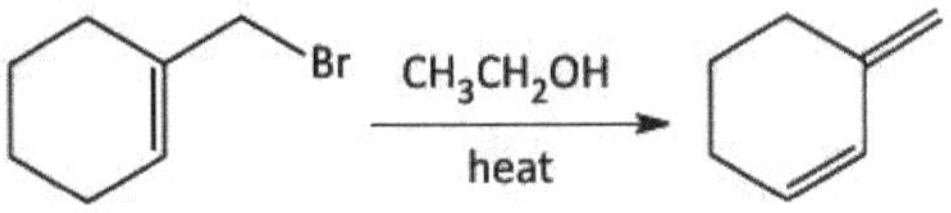

9.51 When 1-bromo-2,2-dimethylcyclopentane is heated in methanol (CH_3OH), one of the products which results is 1,2-dimethylcyclopenene. Provide a detailed, stepwise mechanism for the production of this compound, and give the name of the mechanism by which it is produced.

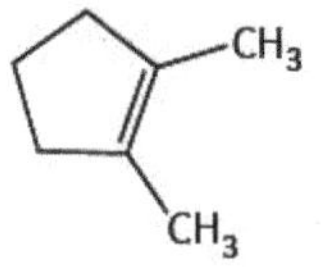

9.52 a) What two elimination products could be formed from the reaction of a base with 2-bromo-3-methylbutane? b) Label these two elimination products with the appropriate term of Saytzeff product or Hofmann product. c) What base would you use in the experiment to make more of the Saytzeff product? d) What base would you use in the experiment to make more of the Hoffman product?

9.53 Determine whether the following substitution reactions would be first order (S_N1) or second order (S_N2).

a) 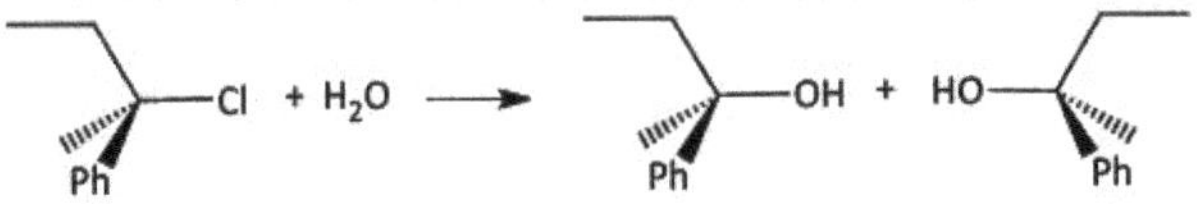

b) 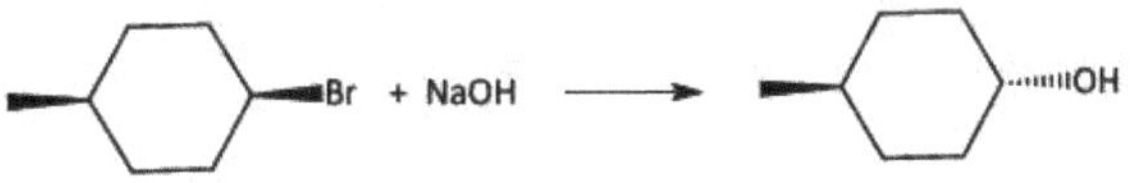

c) 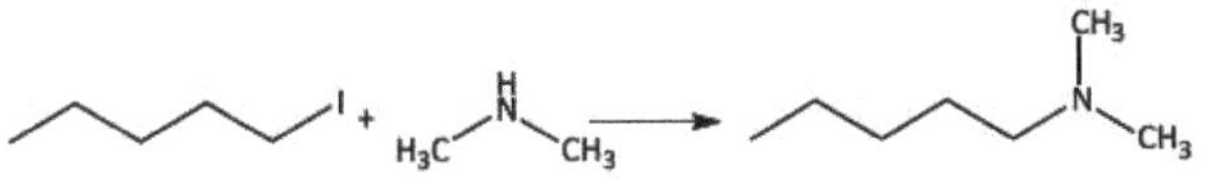

d) 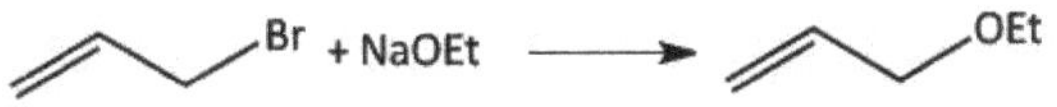

e) 

f) 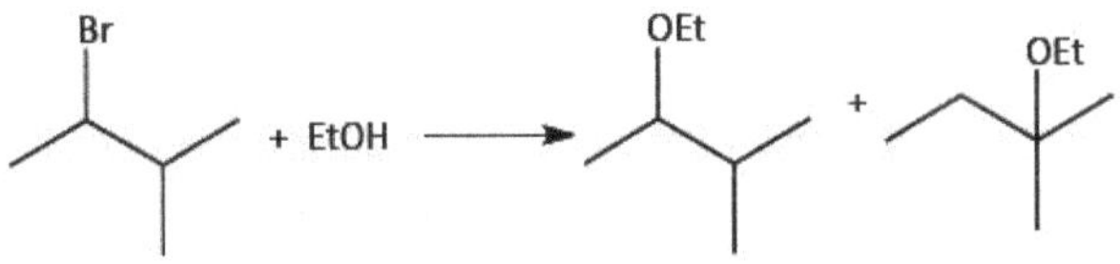

9.54 Looking only at the reagents, predict whether the following substitution reactions would be first order (S_N1) or second order (S_N2).
a) 2-iodo-2-methylbutane + methanol b) bromopropane + sodium hydroxide c) iodopentane + ethylamine
d) 2-chloro-2-methylpentane + water e) 1-bromo-1-methylcyclopentane + ethanol

9.55 The following structure is formed when 1,4-dibromobutane reacts with sodium sulfide, Na_2S. Show the mechanism for this reaction.

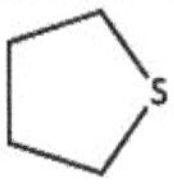

Alkyl Halides

9.56 For the following reactions, determine the substitution and elimination products that would be formed. For the elimination products, if more than one alkene is formed, predict which one is formed in a greater amount.

a) iodocyclohexane + $AgNO_3$ in H_2O

b) 1-bromo-1-methylcyclobutane heated in EtOH

c) 3-chloro-3-methylpentane heated in CH_3OH

d) 2-chloro-2-ethylbutane heated in methanol

9.57 We have learned in this chapter that tertiary alcohols can be converted to tertiary chlorides by reacting an alcohol with Lucas Reagent or HCl. a) Show the mechanism for the reaction of *t*-butanol with HCl. b) Is the mechanism for this reaction S_N1, S_N2, E1, or E1?

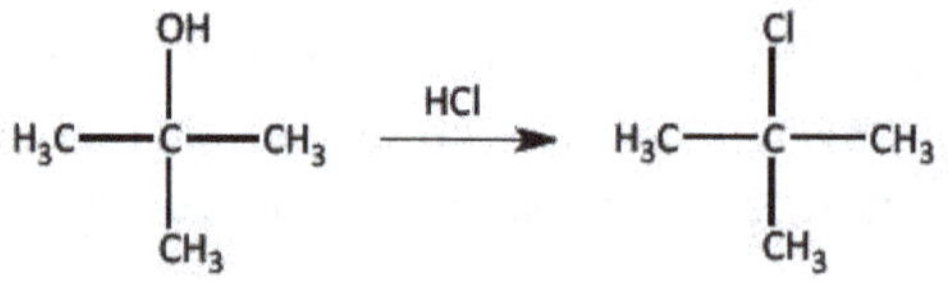

9.58 When the following alkyl bromide reacts with a base, only the *trans* isomer is made, not the *cis* isomer. Explain why.

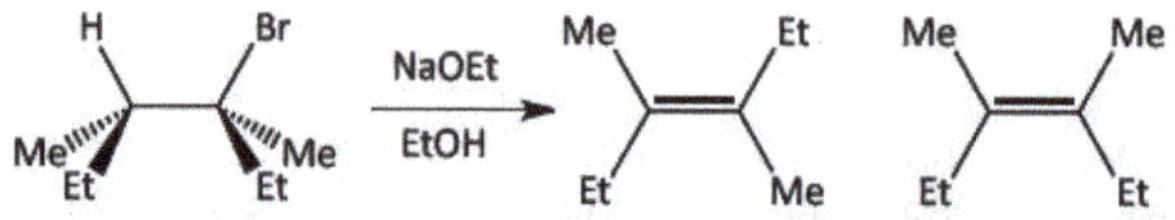

Mostly this and not this.

9.59 A student performed the following E2 elimination reaction of an alkyl halide with potassium t-butoxide expecting two cyclohexenes to form. But, only one of them formed.

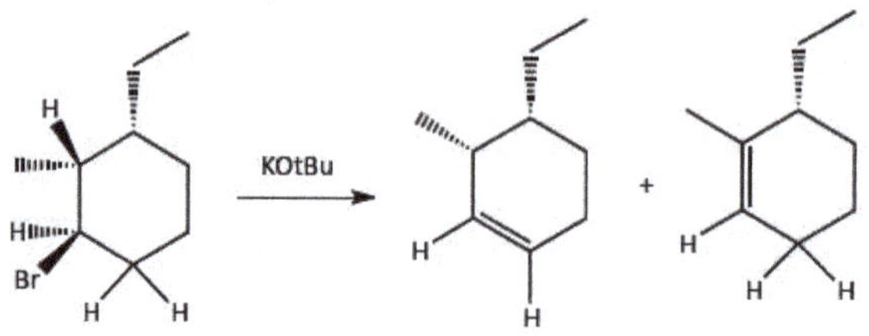

This one made. This one not made.

To try to understand this why, the student drew out the two chair forms of the reactant. Only one chair form can do an E2 elimination. Draw in the hydrogen atom(s) on that conformer that can be removed and then show the mechanism of the elimination forming the only alkene that is made.

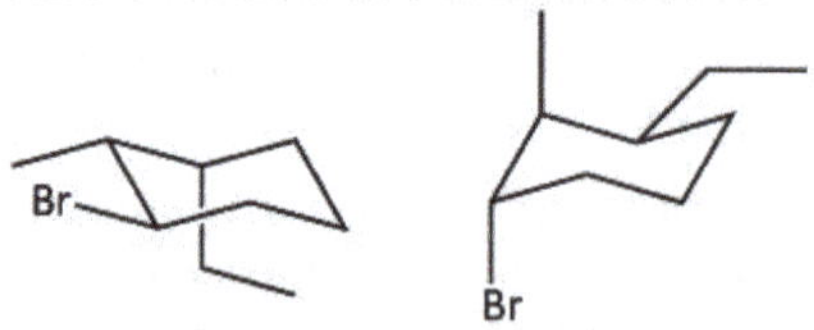

9.60 When 1-chloro-2-methylcyclohexane undergoes solvolysis in methanol, two substitution products are formed. One is the "normal" product and the other is a product that follows a rearrangement. a) Draw out the mechanism for the formation of both products. One of the hydrogen atoms (the one that shifts) is labeled in red.

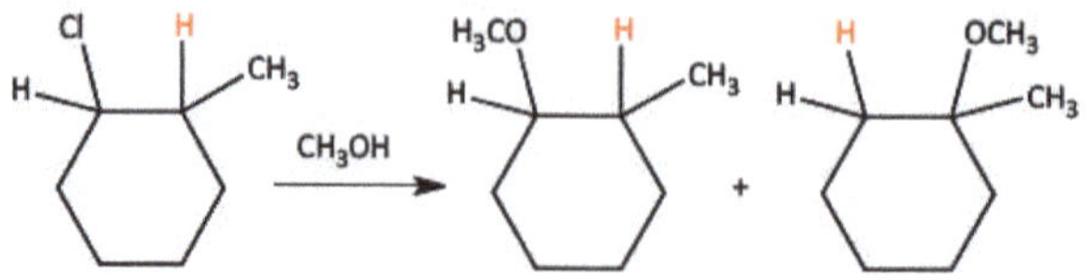

Rearranged product

b) Why does the rearrangement happen? What is its driving force?

9.61 One trick to help pull off a halide for form a carbocation is to use $AgNO_3$. Ag^+ pulls I^- off of this alkyl iodide to form silver iodide precipitate and a carbocation. a) Show the mechanism for the formation of cyclohexanol and 1-methylcyclopentanol. (Hint: look for rearrangements). b) Is this reaction first or second order?

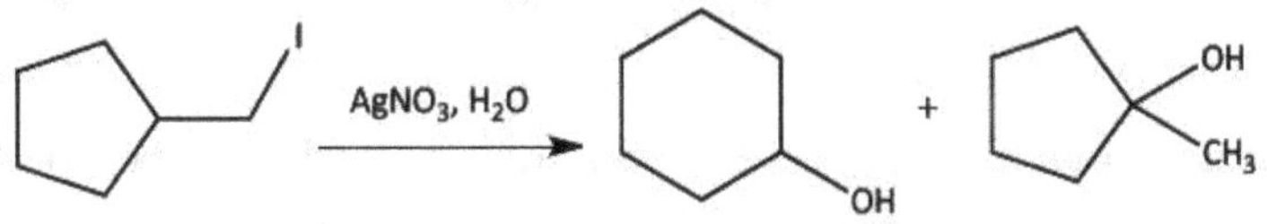

9.62 Why does $CH_2{=}CHCHBrCH_3$ undergo an S_N1 reaction much more rapidly than 2-bromobutane?

9.63 Which diastereomer of 1-bromo-4-tert-butylcyclohexane, the cis or trans, undergoes elimination more rapidly when treated with sodium ethoxide? Explain your answer. (Hint: pull up a chair and think about this one.)

9.64 Show the best way to prepare $CH_3OCH(CH_3)_2$ by an S_N2 reaction.

9.65 Predict the two most likely mechanisms for the reaction of 2-bromoheptane with sodium methoxide from S_N1, S_N2, E1 and E2.

9.66 For each of the following pairs of alkyl halides, choose the one that is more likely to undergo a rearrangement during an S_N1 or E1 reaction?

a) 3-bromohexane vs. 1-chloro-3,3-dimethylpentane
b) 1-bromo-1-methylcyclopentane vs. 1-bromo-2-methylcyclopentane
c) 1-chloropropane vs. 2-chloropropane

9.67 One ether and one alkene are the major product that results when *t*-butyl chloride is heated in ethanol.

a) Draw the ether.
b) This reaction proceeds via which mechanism, S_N1, S_N2, E1, or E2?
c) Draw the alkene.
d) This reaction proceeds via which mechanism, S_N1, S_N2, E1, or E2?
e) When *t*-butyl chloride is treated with sodium methoxide in methanol, the major organic product is:
f) The reaction in part e proceeds via which mechanism, S_N1, S_N2, E1, or E2?

9.68 Using Saytzeff's rule, choose the most stable alkene among the following.

1-methylcycloheptene vs. 3-methylcycloheptene vs. 3,4-dimethylcycloheptene

9.69 Draw all likely alkene products in the following reaction and circle the product you expect to predominate.

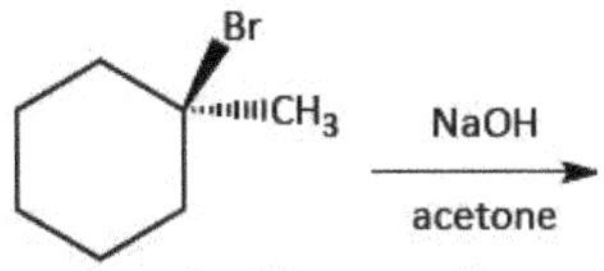

9.70 Propose a detailed, step-by-step mechanism for the reaction pathway shown below.

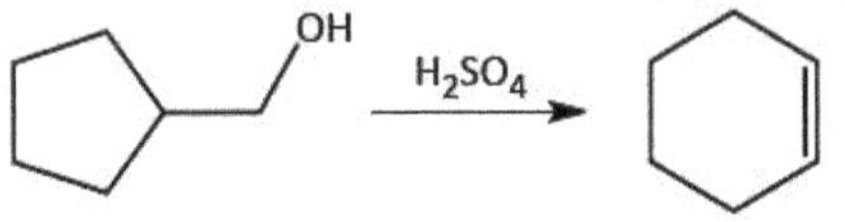

Alkyl Halides

9.71 What alkyl bromide undergoes dehydrohalogenation in the presence of a strong base to give 2-pentene as the only alkene product?

9.72 Provide the reagents you would use to carry out the transformation of cyclohexane to cyclohexene.

9.73 Provide the reagents you would use to carry out the transformation of 2-methylpentane to 2-methylpent-1-ene.

9.74 a) Explain why the molecule shown below does not react with an S_N2 mechanism.

b) Explain why this same molecule does not react by an E2 mechanism.

9.75 Label the following reactions by their mechanism, E1, S_N1, E2, or S_N2.

a) 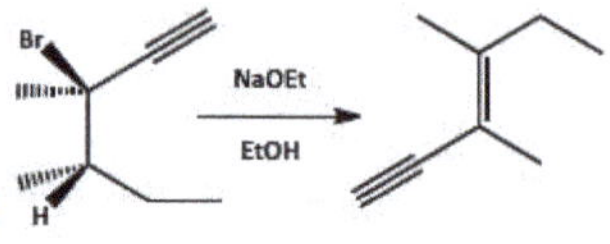

b) 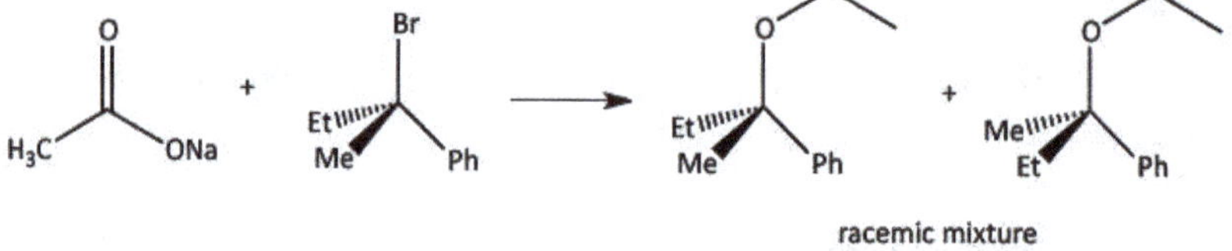

c) 

d) 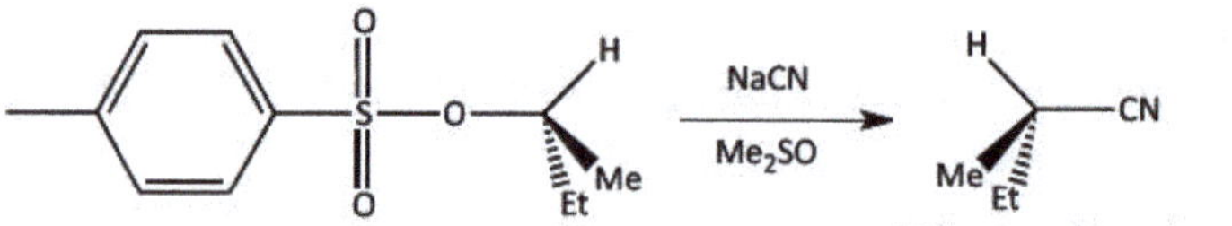

Major organic product

e) 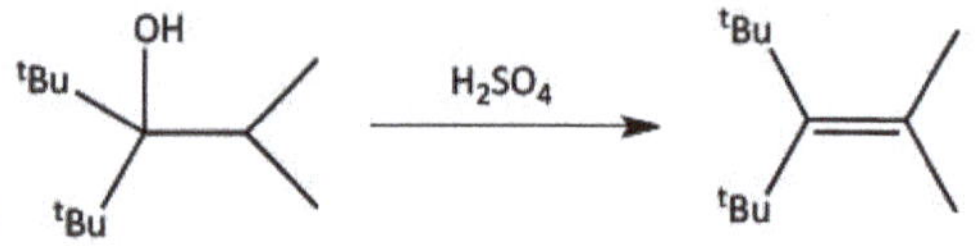

f) 

g) 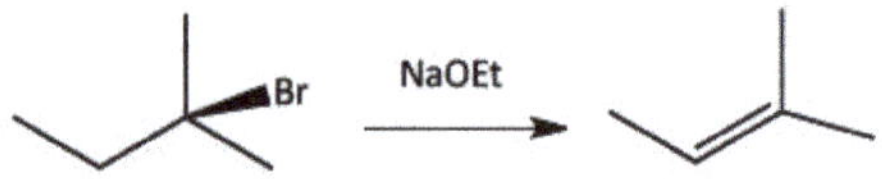

h) 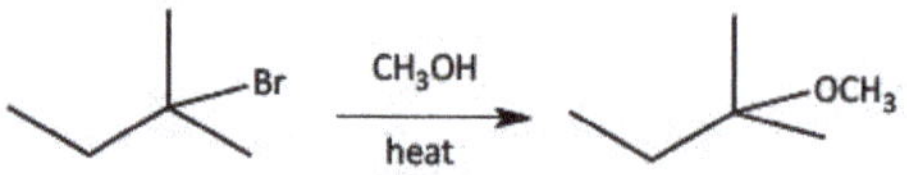

i) 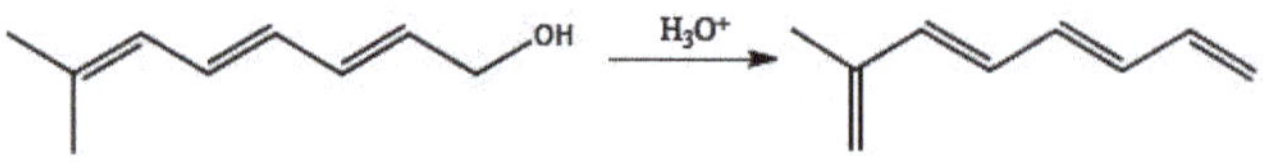

j) 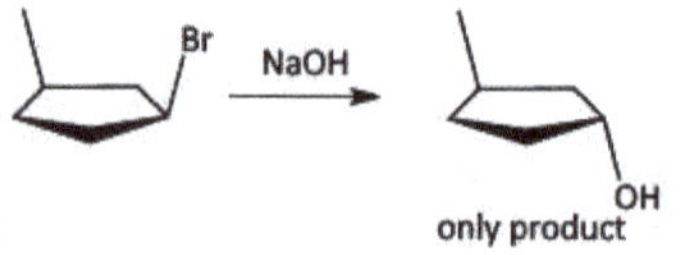

k) 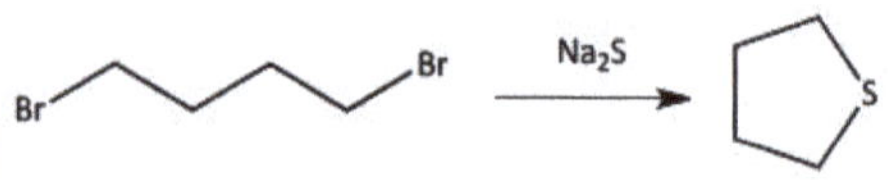

l) 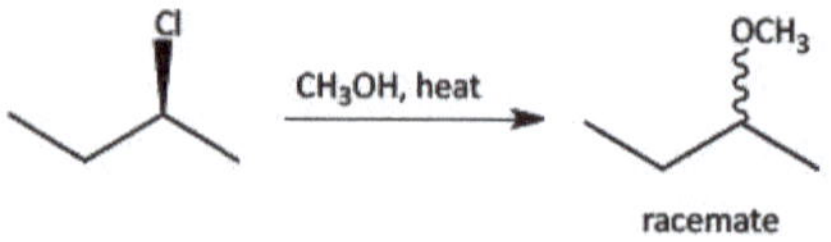

9.76 This alcohol undergoes an S_N1 reaction with methanol and acid. It produces three different products. Show the mechanism for the formation of these three compounds.

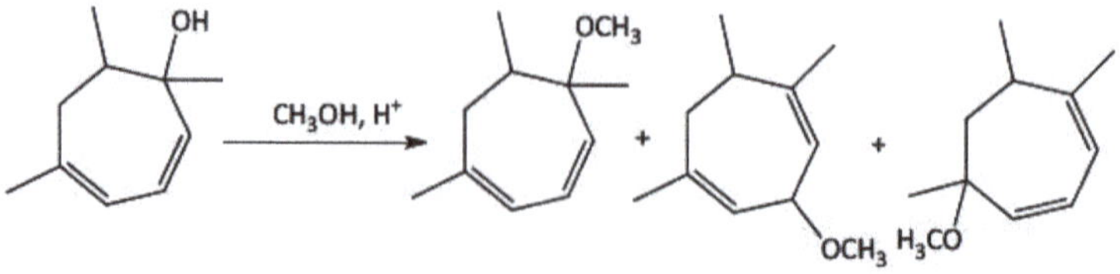

9.77 The following elimination does not work. Pull up a chair, think about it, and explain why.

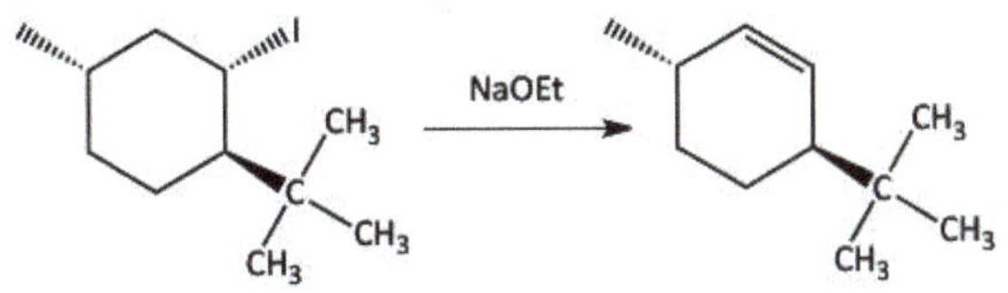

9.78 Show the mechanism for the following reactions. Be sure to show how each product is made.

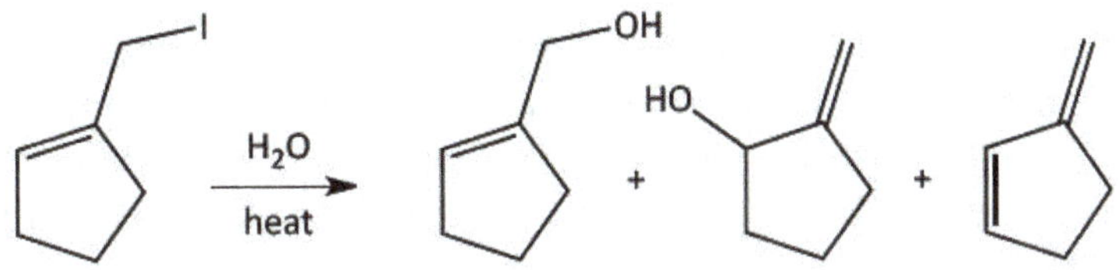

9.79 Show the mechanism for the following reactions. Be sure to show how each product is made.

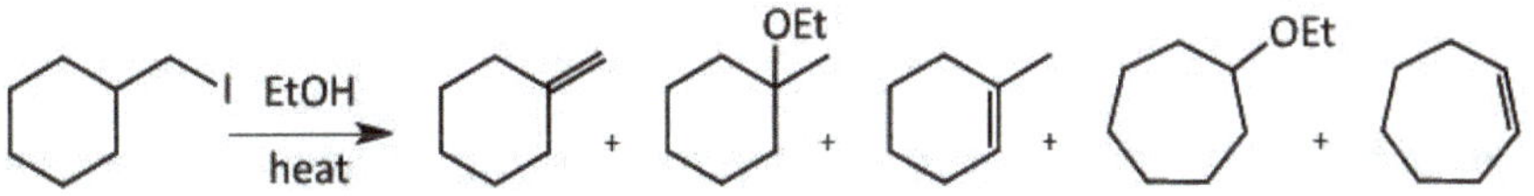

9.80 Show the mechanism for the following reactions. Be sure to show how each product is made.

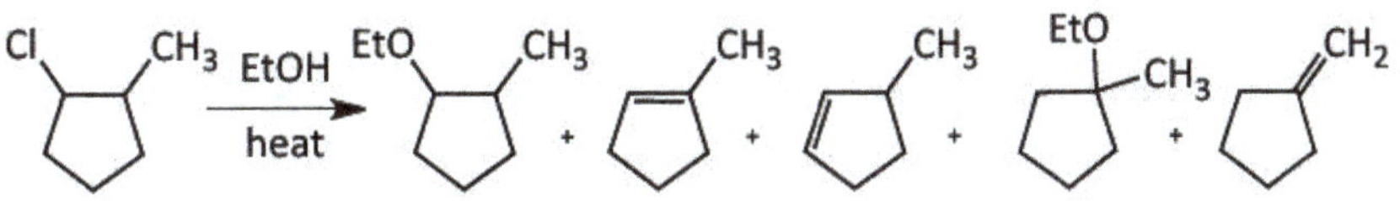

9.81 The pinacol rearrangement is a reaction that we will study later in this text. The mechanism is, mostly, shown below. a) Complete the mechanism by drawing an arrow on the intermediate in the box to show the methyl shift. b) Explain the driving force for this rearrangement. Why does it happen?

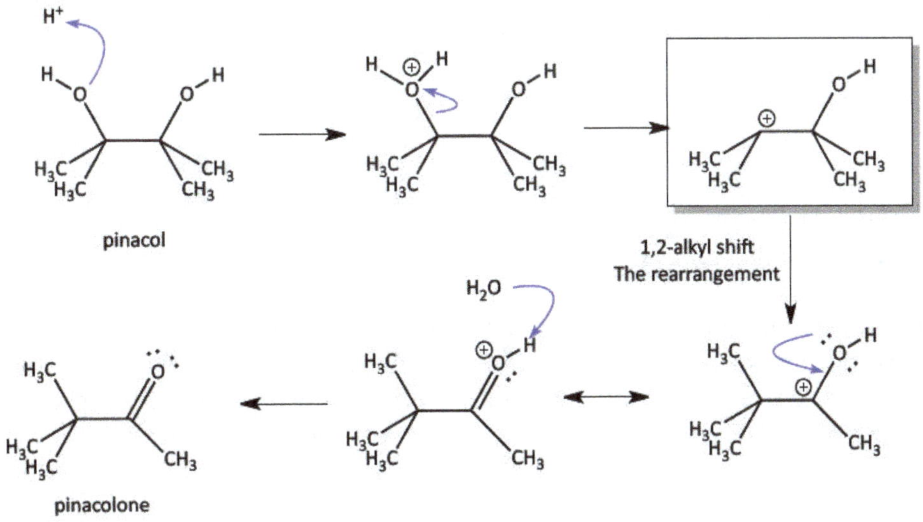

Chapter 10. Alkenes & Alkynes

10.1 Introduction

Alkenes are compounds that contain a carbon-carbon double bond. Years ago, it was common to call alkenes olefins. This term is still heard at times today. Alkenes are found in many natural and biological compounds. Alkenes contain one sigma bond and one pi bond and are sp^2 hybridized. The pi bond is more reactive than sigma bonds because it is a little weaker. The ubiquitous nature of alkenes and their reactivity make them important to study.

Alkynes contain carbon-carbon triple bonds. They contain one sigma bond and two pi-bonds. They are sp hybridized.

The more bonds that exist between two carbon atoms, the shorter the distance is between the two carbon atoms. Another way to think about it is the greater the "s-character" of the hybridization, the closer an atom will hold its electrons to its nucleus because s-orbitals have the electron density closer to the nucleus than p-orbitals.

sp (50% s-character) > sp^2 (33% s-character) > sp^3 (25% s-character)

Figure 10.1 Percent s-character for hybrid orbitals

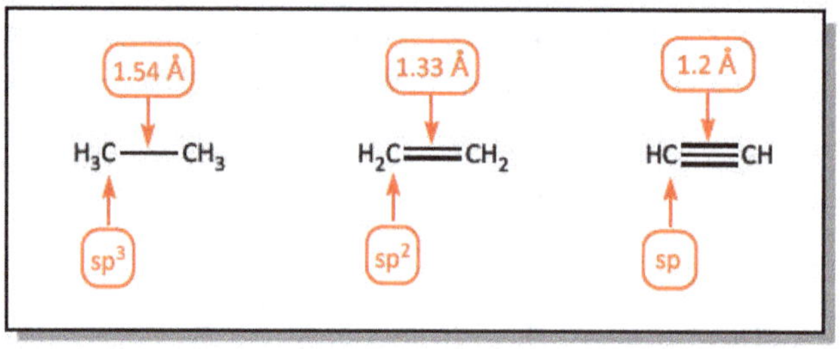

Figure 10.2 Hybridization and lengths of various bond types

10.2 Degrees of Unsaturation

Compared to the chemical formula for a saturated alkane, each pi bond in a compound reduces the chemical formula by two hydrogen atoms. We learned in the Mass Spectrometry chapter a formula to determine the degrees of unsaturation of a molecule from its chemical formula. Each degree of unsaturation represents a pi bond or a ring in the molecule. The formula we use to determine degrees of unsaturation is

$$\textbf{\textit{Degrees of Unsaturation}} = \frac{(2C + 2 - H)}{2}$$

N counts a ½ a carbon atom.
O does not count in the equation.
Halogens count as hydrogen atoms.

Figure 10.3 Formula to calculate degrees of unsaturation

For alkenes, the C=C contains one pi-bond and accounts for one degree of unsaturation. For alkynes, the C≡C contains two pi-bonds and accounts for two degrees of unsaturation.

10.1 How many degrees of unsaturation are in the following compounds?

a) C_7H_{12} b) C_5H_8O c) C_2H_7N d) C_4H_5Cl e) 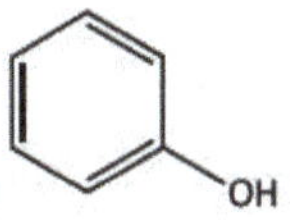

f) 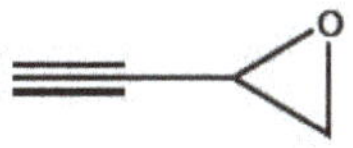

10.3 Nomenclature

10.3.1 Alkenes and Alkynes

In IUPAC nomenclature, alkene names have the *–ene* suffix.

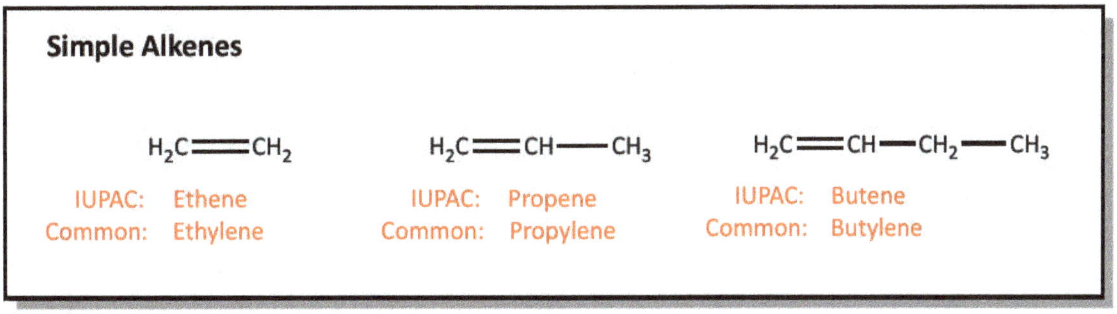

Figure 10.4 Examples of simple alkenes

For the IUPAC system, alkyne names have the *–yne* suffix. The simplest alkyne, ethyne, is more commonly called acetylene. Acetylene is a gas that is commonly used in torches and welding. Nearly everybody calls this compound acetylene. Very few people would call it "ethyne". Common names for alkynes are named as substituted acetylenes. The groups on the alkyne (the acetylene) are named followed by the word "acetylene".

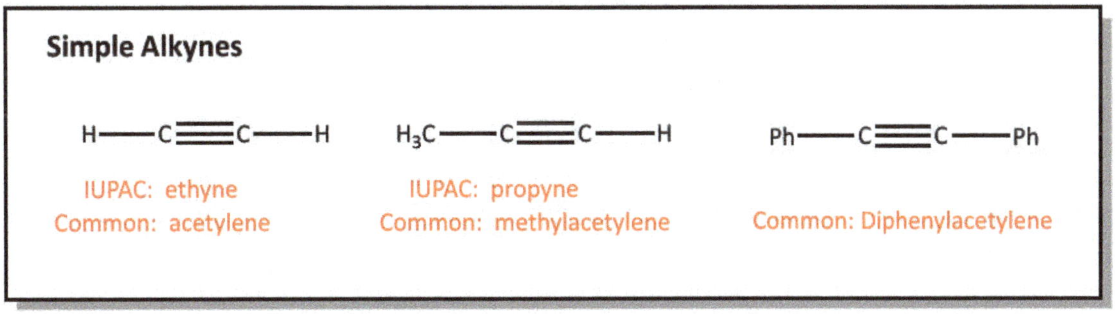

Figure 10.5 Examples of simple alkynes

When alkynes are on the end of a molecule, with a hydrogen atom attached to the acetylene, they are called terminal alkynes. When two carbon groups are attached to the acetylene, the alkyne is called an internal alkyne.

When naming alkenes, the longest carbon chain that contains the double bond is identified. The chain is numbered starting from the side nearest the C=C. The location of the double bond is identified with the lowest carbon number of the two carbon atoms that make up the double bond. The older IUPAC system put the numbers before the name. In 1993, IUPAC made a change to put the number location of the C=C directly before the *–ene* suffix. Both are very commonly used, so you should learn both of them.

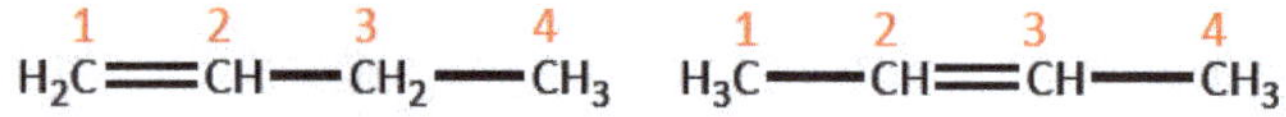

Older: 1-butene 2-butene
Newer: but-1-ene but-2-ene

Alkynes are identified by the *–yne* suffix. The location of the triple bond is also identified with a number.

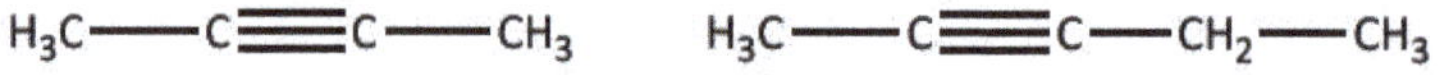

2-butyne 2-pentyne

The longest carbon chain, the base chain, must include the alkene. In molecule below, a seven-carbon chain can be found, but the six-carbon base chain on the right is the one we use.

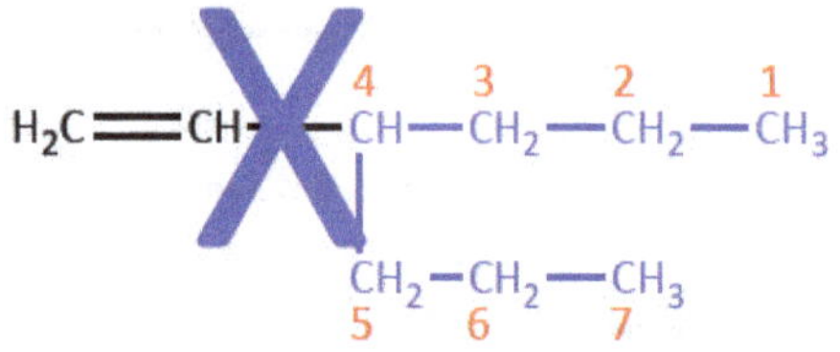

We don't use this numbering system because the C=C is not included.

vs.

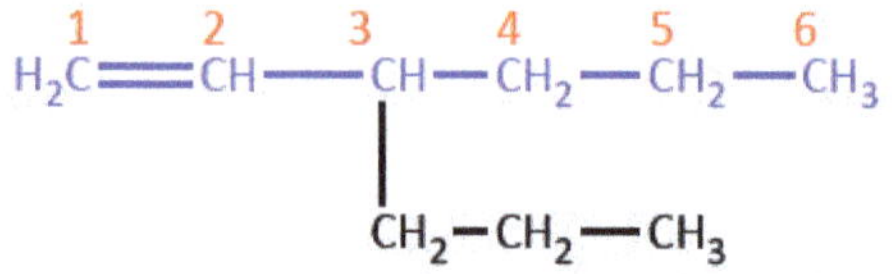

3-propyl-1-hexene
or
3-propylhex-1-ene

If a compound contains both a C=C and C≡C, the base chain should include both functional groups.

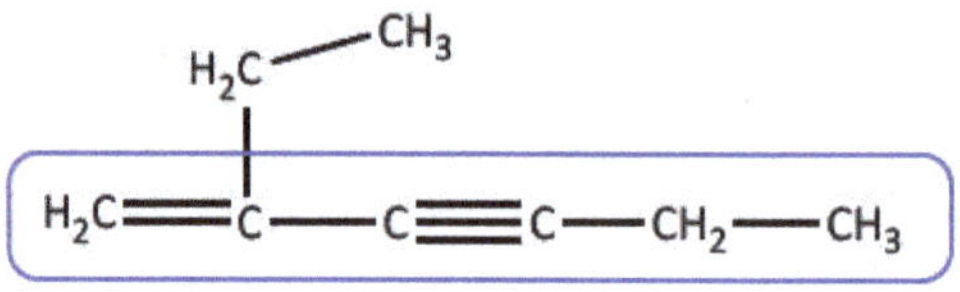

2-ethylhex-1-ene-3-yne

We identify two carbon-carbon double bonds in a molecule by using the suffix *diene*. Three double bonds is *triene*, and four is *tetraene*, etc.

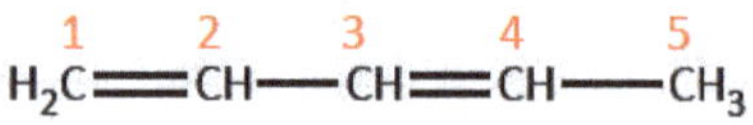

1,3-pentadiene
or
penta-1,3-diene

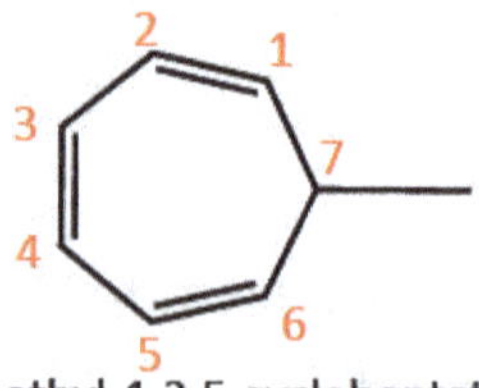

7-methyl-1,3,5-cycloheptatriene
or
7-methylcyclohepta-1,3,5-triene

Like alkanes, alkenes can be named as substituents of a chain with names like methyl, ethyl, etc., alkenes can also be named as substituents. The following substituent common names should be memorized.

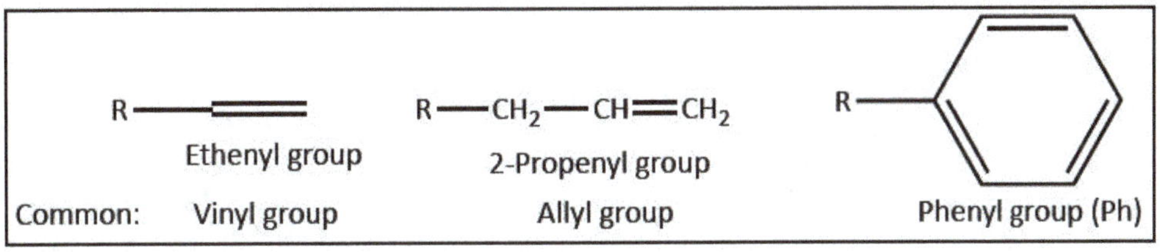

Figure 10.6 Common names for alkenyl groups

10.2 Name the following compounds.

a) 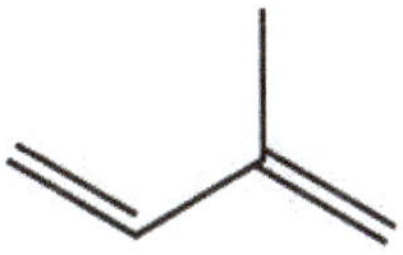 b) 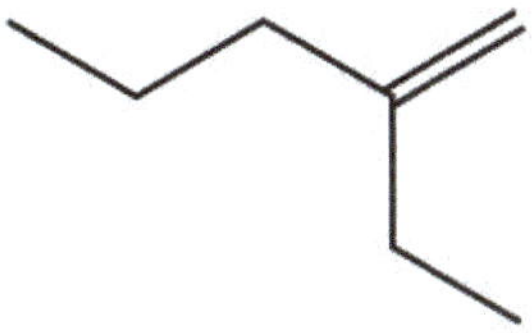 c) 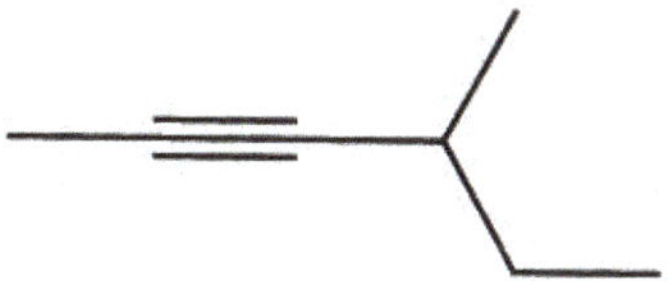

10.3 Draw the following compounds.

a) 2-methyl-1-butene b) 3-propylheptene c) 1,5-hexadiene d) 2-pentyne

10.3.2 Alkenes - Cis/Trans and E/Z

Sigma bonds can easily rotate around.

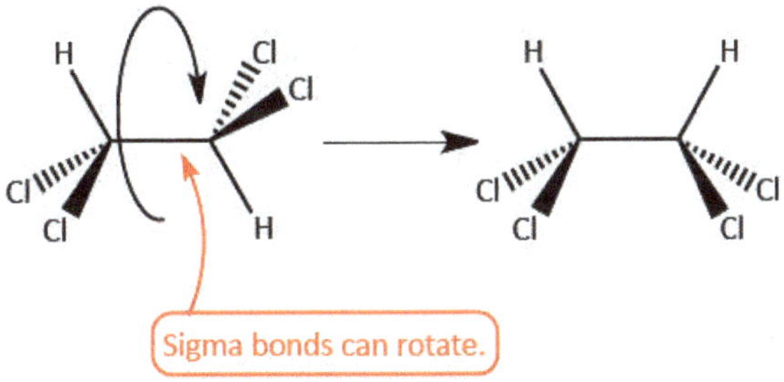

But pi-bonds cannot rotate. Since part of the pi-bond is on the top of the molecule and part is on the bottom, the molecule cannot spin or rotate.

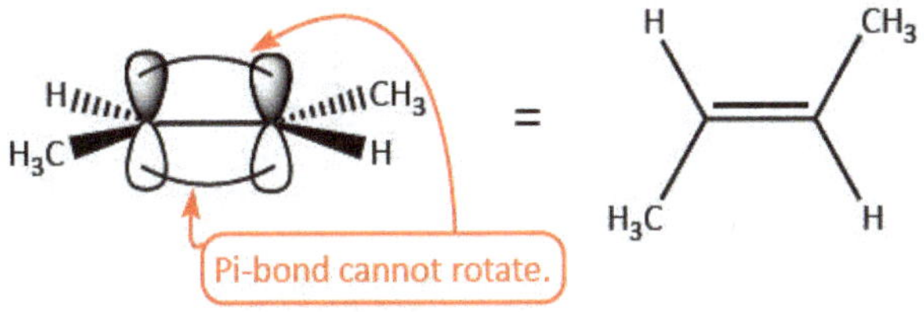

Because of the pi bond's inability to rotate, molecules like these 2-butenes below are different. The one with the methyl groups on the same side cannot rotate to become the one with the methyl groups on opposite sides. These molecules are different from each other. These 2-butenes must therefore have different names. We use the prefixes cis or trans.

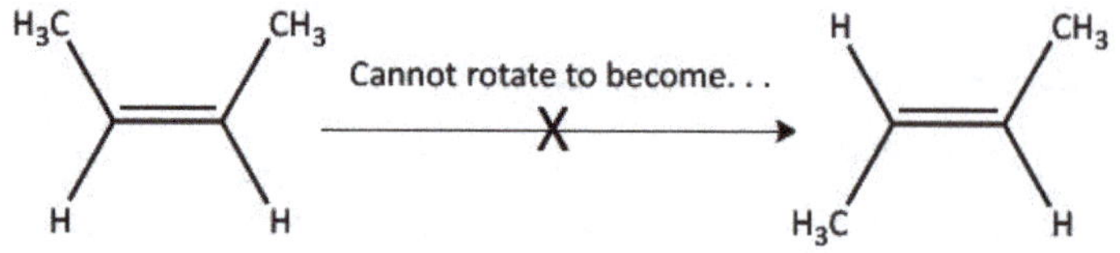

2-Butene can exist as a cis or trans isomer. We use cis if identical groups are on the same side and trans if they are on opposite sides.

cis-2-butene *trans*-2-butene

But, how would we name the following alkene isomers? Which is *cis* and which is *trans*?

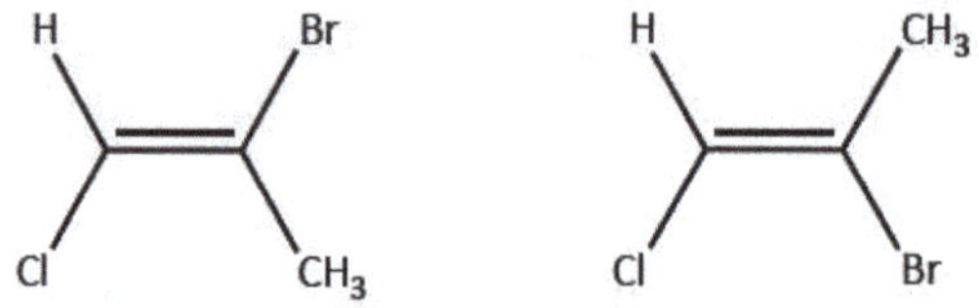

With no rotation of the pi bonds, these are obviously different compounds. But, sometimes, it is not easy to see if a molecule is cis or trans like it was for the 2-butenes. In compounds like this, we use the E-Z system. How easy!

10.3.3 The E-Z system

Each side of the double bond has two groups attached to it. These two groups are compared and assigned a priority (1-highest or 2-lowest) using the Cahn-Ingold-Prelog system like for R & S designations. If the two highest priority groups (1) are on the same side of the double bond (like cis), the prefix "Z" (from the German: Zusammen "together") is assigned to the name. If the groups are on opposite sides of the double bond, like trans, the prefix "E" is used (from the German: Entgegen "opposite").

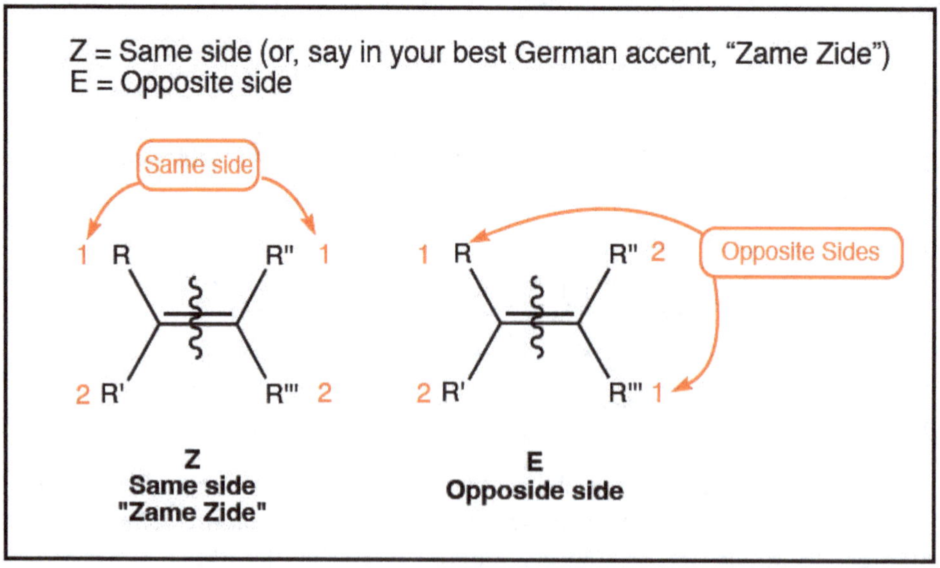

Figure 10.7 E and Z examples

For our compounds above, we would get the following names.

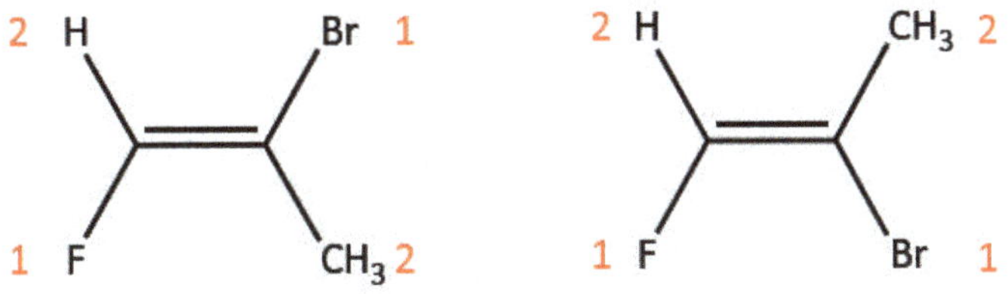

(E)-2-bromo-1-fluoropropene *(Z)*-2-bromo-1-fluoropropene

Let's name the following compound.

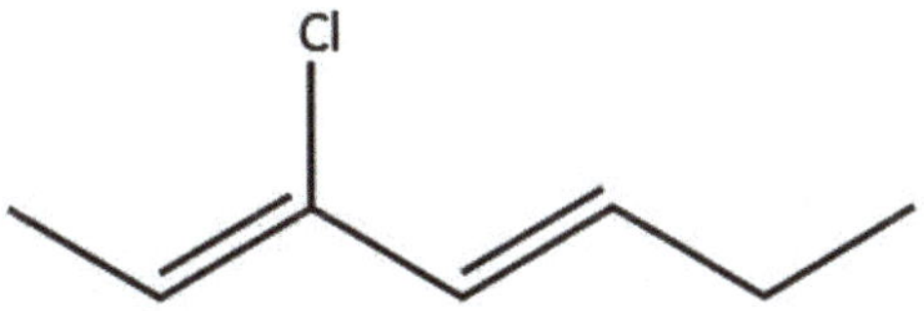

The longest chain is 7-carbons long and contains two double bonds. We label the chain from the left to the right so the double bonds have lower numbers. With the double bonds between carbons 2-3 and 4-5, the double bonds are at locations 2 and 4. The chlorine atom is on carbon 3. We therefore have the name 3-chloro-hepta-2,4-diene.

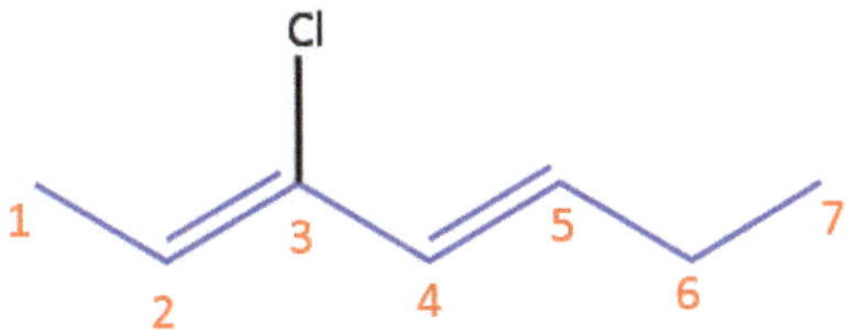

But, we haven't described the orientation of the double bonds. We need to describe them. Let's look at bond-2 first. We need to remember that hydrogen atoms are not always drawn. It helps if we draw them in. We see on the left side that CH_3 has higher priority than H and on the right side, Cl has higher priority than carbon. With the top priority groups on the same side, this is a "Z" double bond.

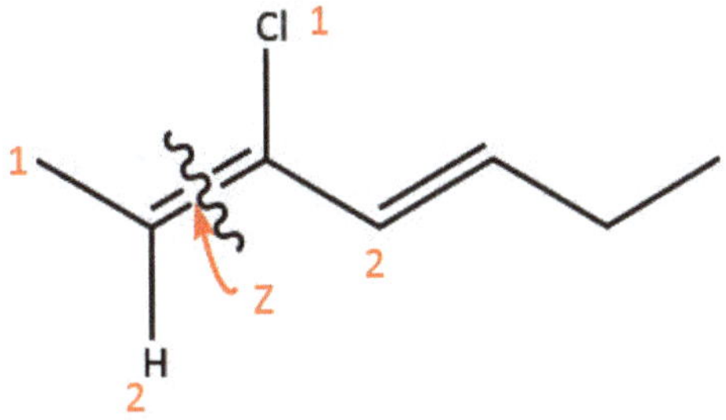

Now, let's analyze double bond-4. On the left side, carbon has higher priority than H. On the right side, CH_2 has higher priority than H. Since the top priority groups are opposite each other, this double bond is labeled "E".

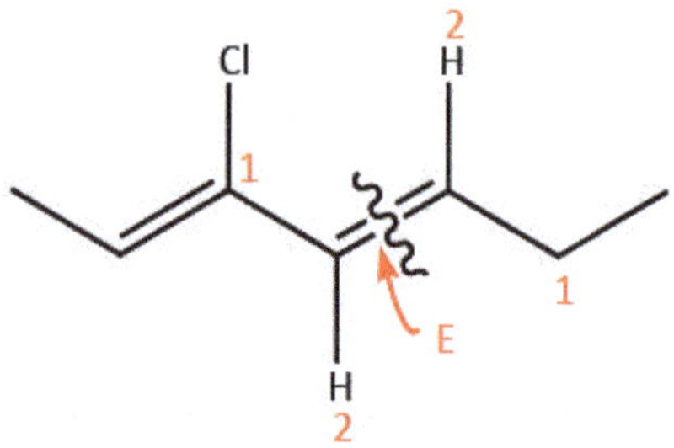

We can now give a complete name to the compound.

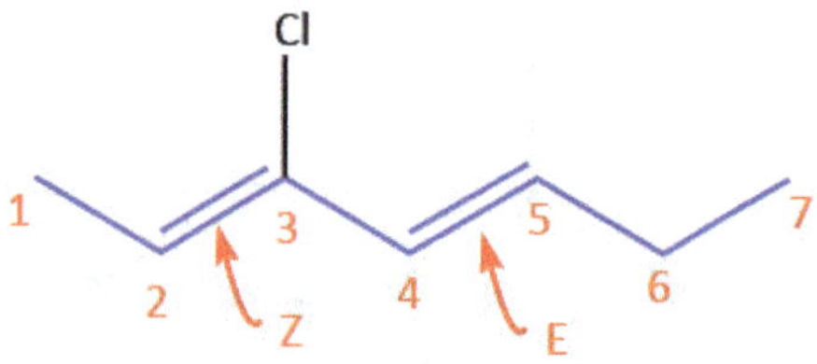

3-chloro-(2Z,4E)-heptadiene
or
3-chloro-(2Z,4E)-hepta-(2,4)-diene

10.4 Label each of the following alkenes as cis or trans.

a)  b) 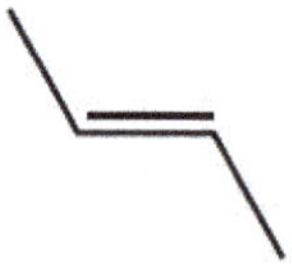 c) 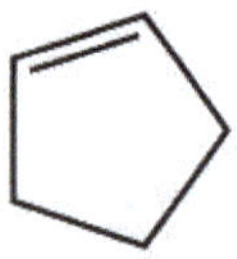 d) 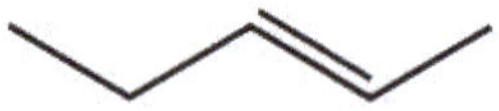

10.5 Label each of the following alkenes as E or Z.

a) 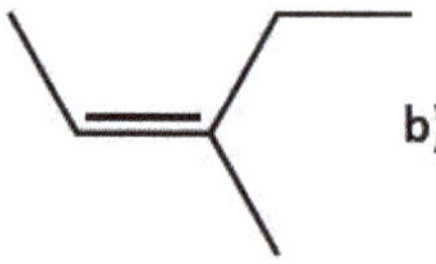 b) 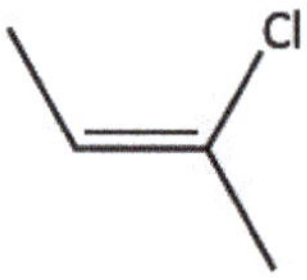

c) 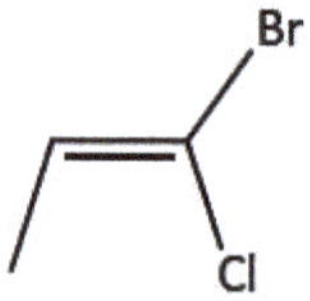

d) 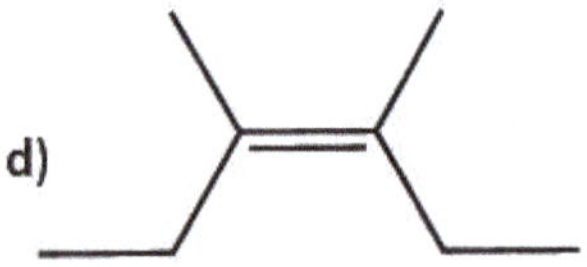

10.6 Draw the following alkenes.

a) 2-chloro-3-methyl-2Z-pentene
b) *cis*-2-pentene
c) *trans*-1,2-dibromoethene
d) 1-chloro-1-fluoro-(1E,3Z)-pentadiene

10.7 Name the following alkenes. Remember to use cis/trans or E/Z as appropriate.

a) 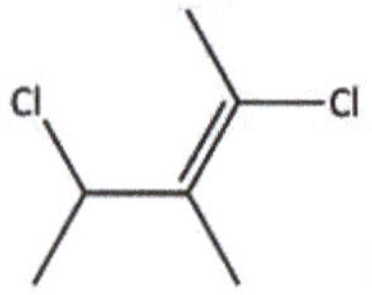

b) 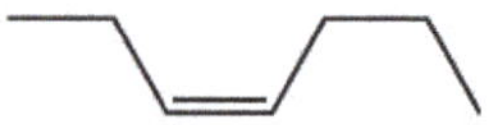 c) 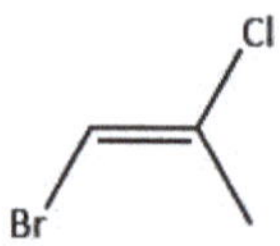

d) 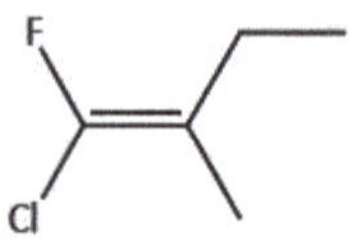

10.3.4 Cycloalkenes

Most cycloalkenes contain cis-double bonds. The cis prefix for cycloalkenes does not need to be written. It is assumed. The prefix only needs written if it is trans because the only way it is possible to have a trans-double bond is to have at least eight carbon atoms in the ring.

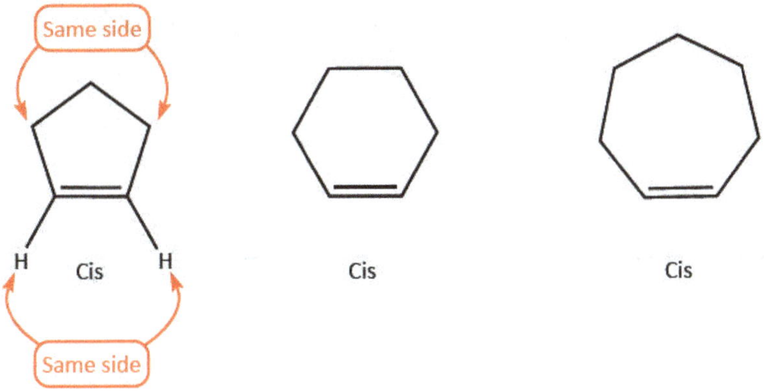

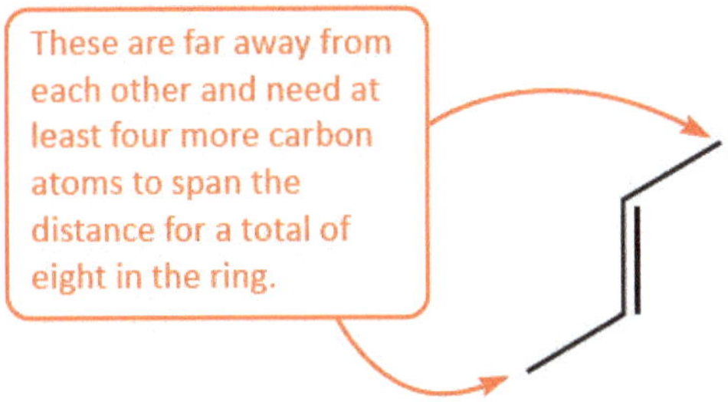

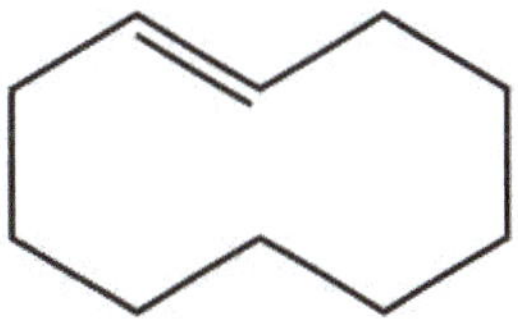

trans-cyclodecane

10.3.5 Bredt's Rule

Let's look at bicyclic molecules that are alkenes. A double bond cannot be attached to a bridgehead carbon of a bicyclic compound if the ring it is in does not have at least eight carbon atoms in it. This is **Bredt's Rule.**

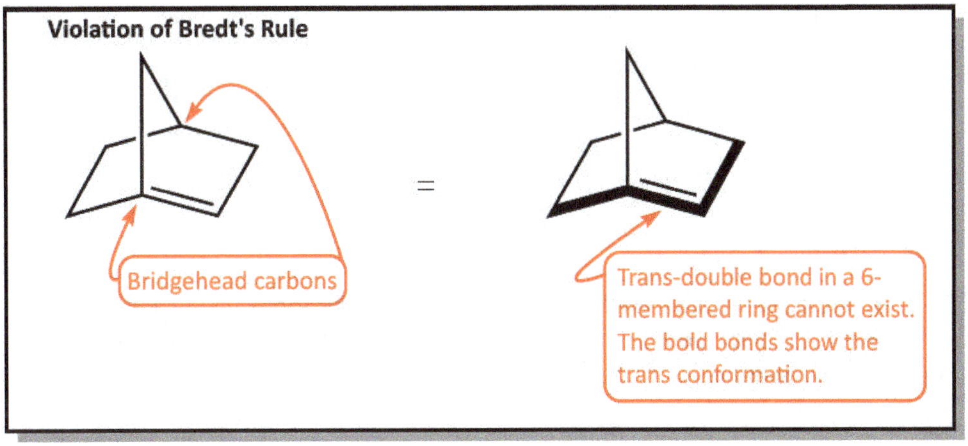

Figure 10.7 Violations of Bredt's Rule

Not a violation of Bredt's Rule

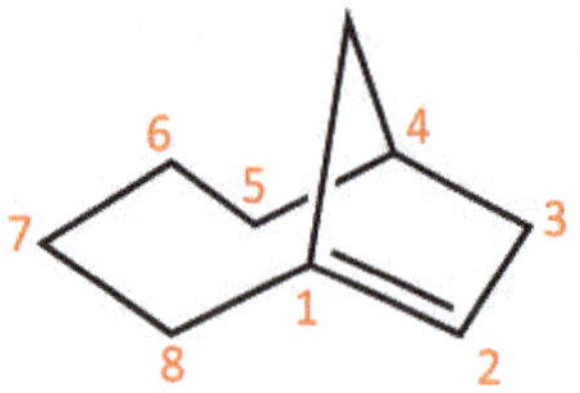

Bridgehead alkene in 8-membered ring. This can exist.

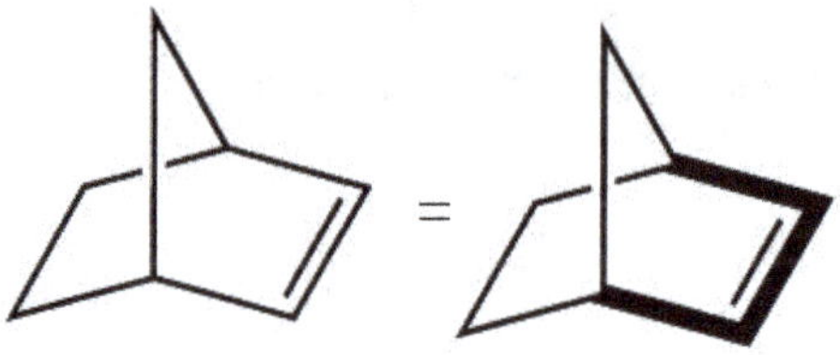

Double bond is NOT on a bridgehead carbon. It is a cis alkene. Bold bonds show the cis conformation. This can exist.

10.8 State whether the following compounds *violate* Bredt's rule.

a)  b)  c)  d)  e) 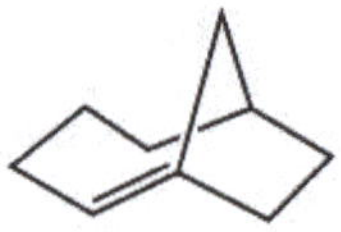

10.4 Stability of Alkenes

10.4.1 Cis vs. Trans

Trans alkenes are usually more stable than *cis*. In *trans*-alkenes, the bulkier alkyl groups are farther away from each other and do not bump into each other.

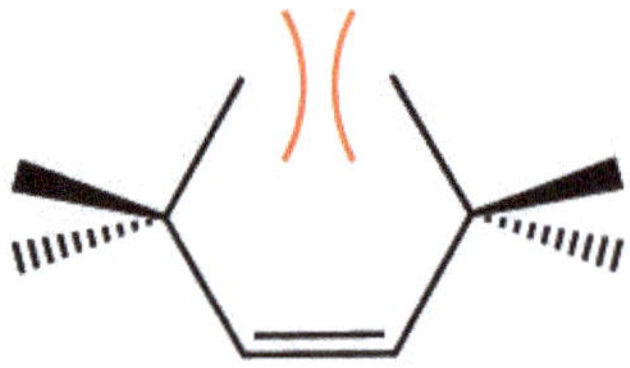

Cis

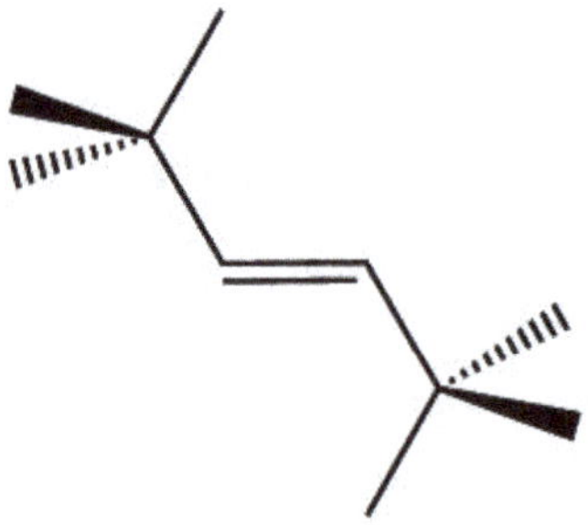

Trans

10.4.2 Saytzeff's Rule

We've learned Saytzeff's rule previously. Saytzeff's Rule states that the more highly substituted double bonds are usually more stable than less substituted double bonds. So, more stable alkenes are those that have more carbon groups of the four groups on the C=C.

More stable — Less stable

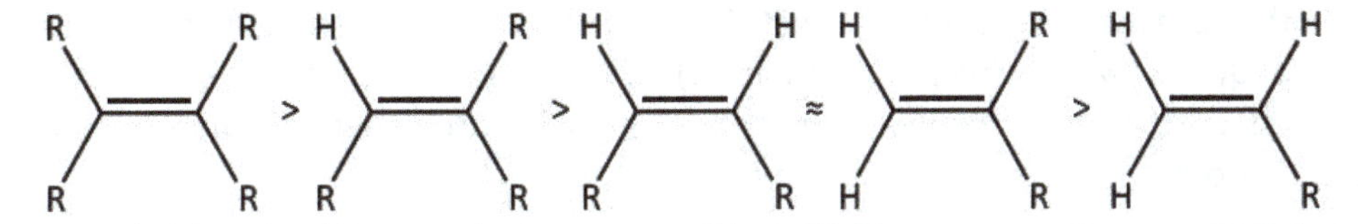

Tetrasubstituted 4 carbon groups on C=C | **Trisubstituted** 3 carbon groups on C=C | **Disubstituted** 2 carbon groups on C=C | **Monosubstituted** 1 carbon group on C=C

10.9 For each of the following pairs of compounds, circle the one that is more stable.

a) 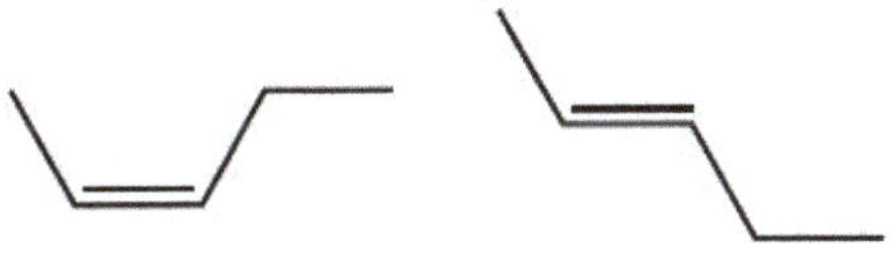 b) 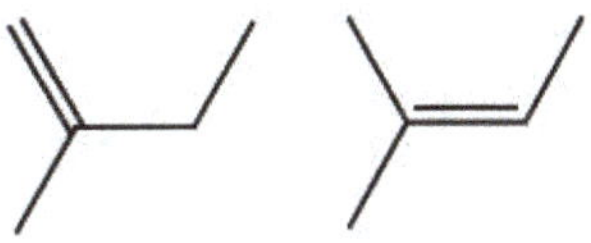

10.5 Alkene Synthesis

10.5.1 Review

We have seen that alkenes can be synthesized by the E2 and E1 reactions.

E2-Dehydrohalogenation to form alkenes

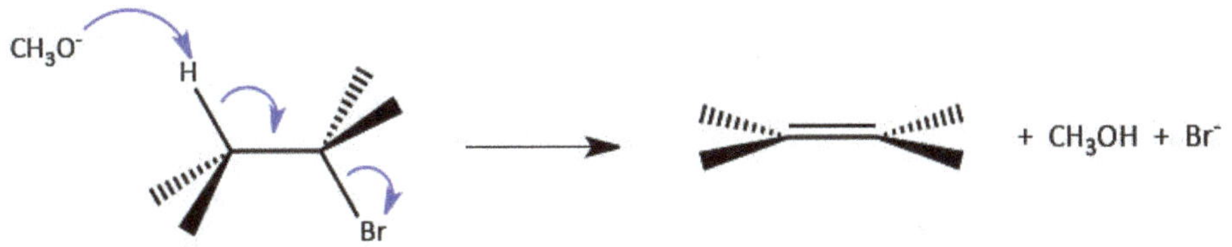

- E2 needs a strong base to attack right away.
- It happens in a 1-step, coplanar arrangement.
- Small, aggressive bases like $^-$OEt make more of the Saytzeff product (if they don't S_N2). Large, bulky bases like ^{t}BuO$^-$ make more Hofmann product.

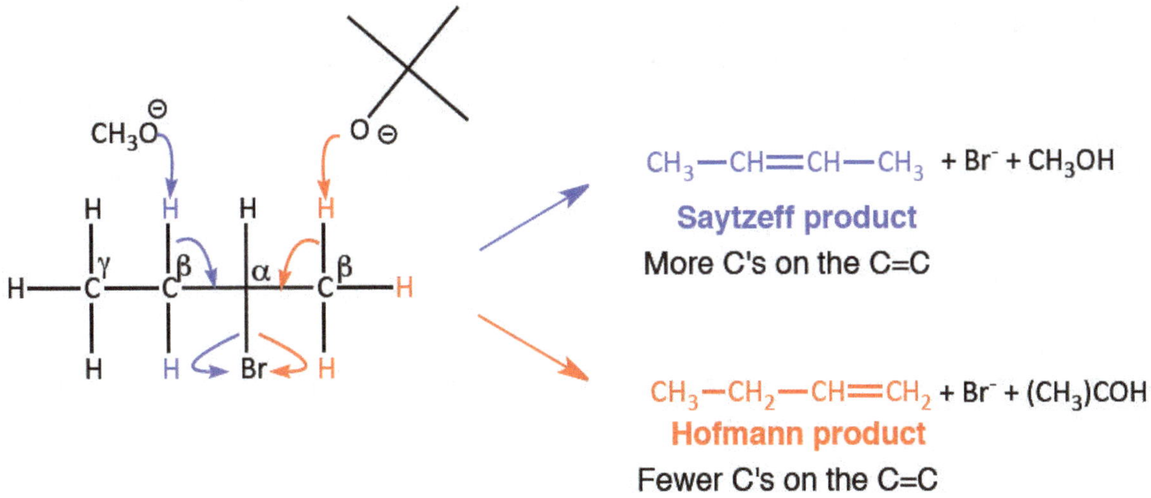

E2-Dehydrohalogenation to form alkynes

Alkynes can be synthesized by an E2 reaction as well. The elimination must happen twice to form the two pi-bonds of the alkyne. This requires two leaving groups (halides) and two hydrogen atoms on neighboring carbon atoms. This could be a vicinal dihalide, with halides on neighboring carbon atoms, or a geminal dihalide with two halides on the same carbon atom.

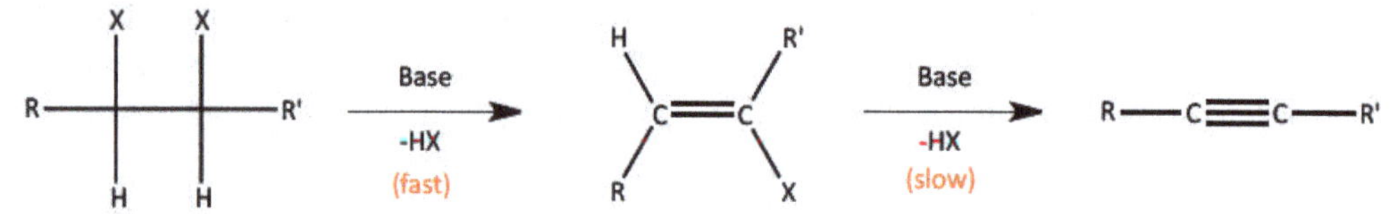

vicinal dihalide

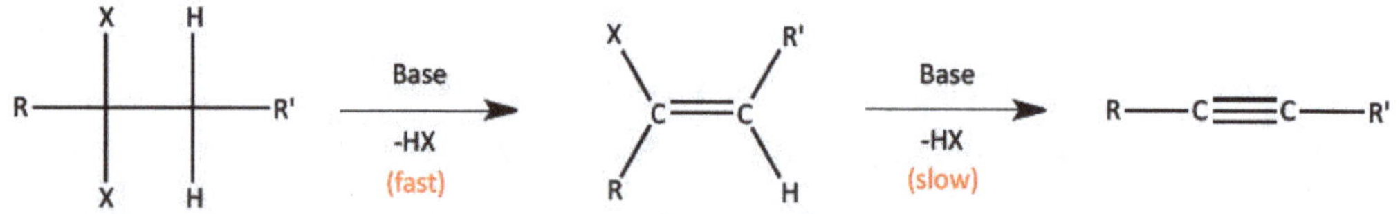

geminal dihalide

The second, slow, step of each of these dehydrohalogenations is a difficult reaction to force. It only happens under the most basic and harsh conditions. This reaction can be performed with molten KOH or KOH in alcohol heated to 200°C. Under these harsh conditions, if any other base sensitive functional groups are present, they could be destroyed.

E1 Formation of Alkenes

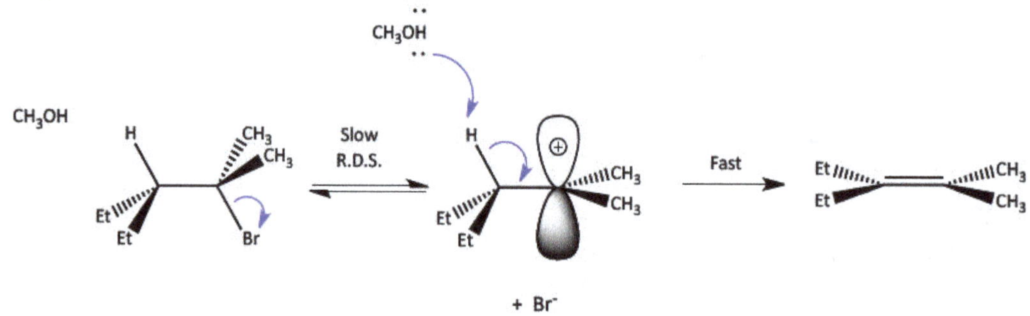

- E1 works better with a polar solvent to help with the ionization.
- E1 typically involves a weak base. Stronger bases may deprotonate too quickly causing an E2 reaction.
- S_N1 competes with E1. The base/nucleophile can attack either the carbocation (S_N1) or deprotonate a beta carbon (E1).
- Because carbocations are formed in the mechanism, rearrangements are common. Rearrangements can happen in E1 and S_N1 reactions.

10.5.2 New Alkene Syntheses

Dehalogenation of Vicinal Dibromides

If two bromides are on neighboring carbon atoms (vicinal dibromides), they can react with I^- (from a salt like NaI or KI) in an E2 style reaction to form an alkene.

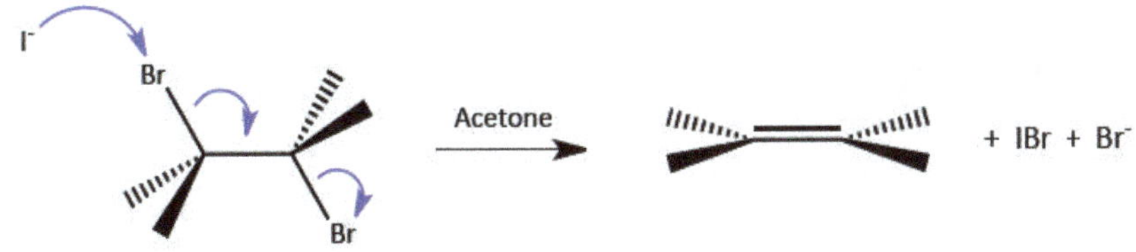

The two bromides are usually in an anti-coplanar arrangement when they react with I⁻.

Another way to accomplish a dehalogenation of a vicinal dibromide is to use zinc and acetic acid.

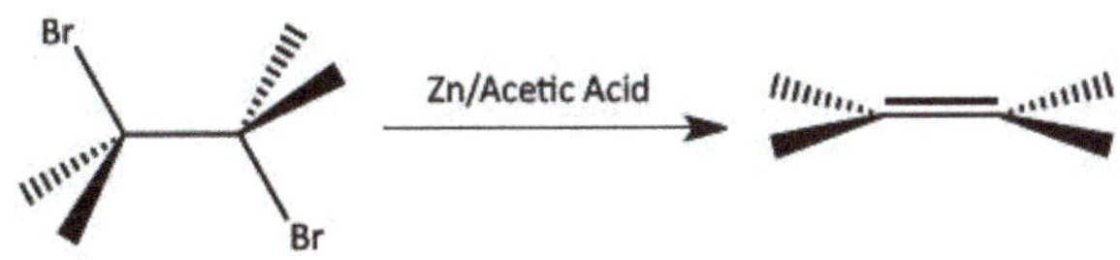

Dehydration

The hydroxyl group of alcohols is a poor leaving group. But, in the presence of a strong acid like sulfuric acid, H_2SO_4, phosphoric acid (H_3PO_4) or toluenesulfonic acid (TsOH) it can be turned into a good leaving group (H_2O). Once this good leaving group is formed, an E1 reaction can occur to form an alkene.

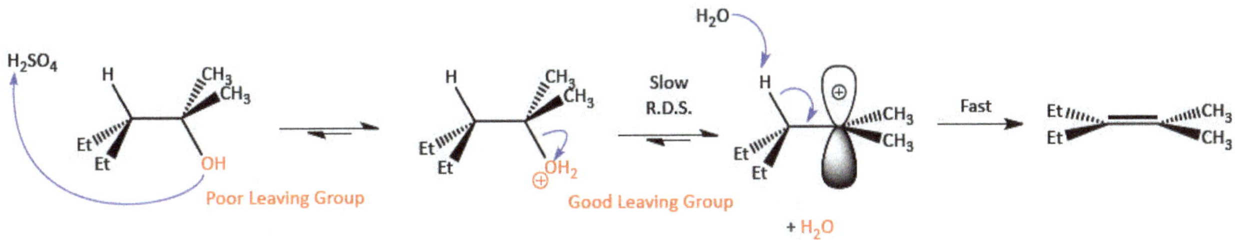

Because a carbocation is formed in this E1 reaction, rearrangements are common.

10.10 Draw the products of the following reactions.

a) 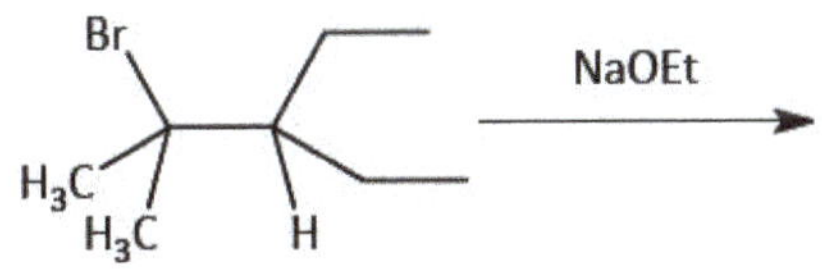

b) 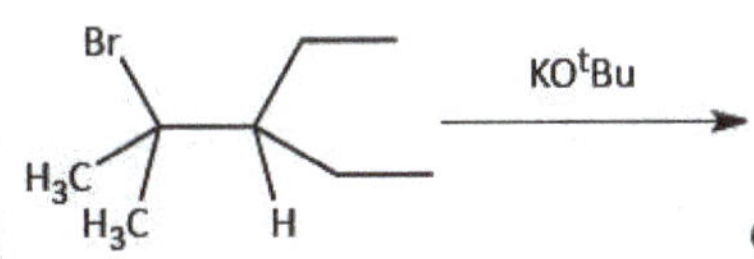

c) 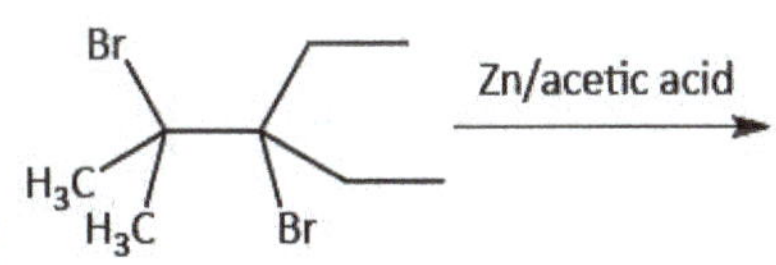

d) 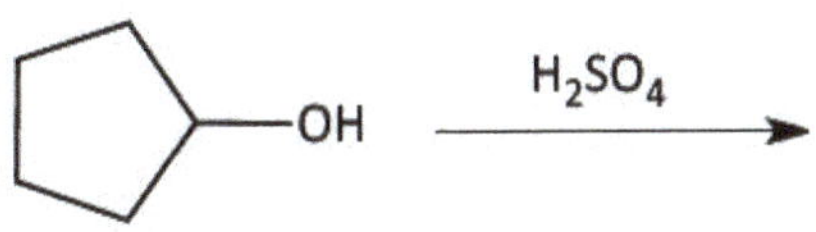

e) 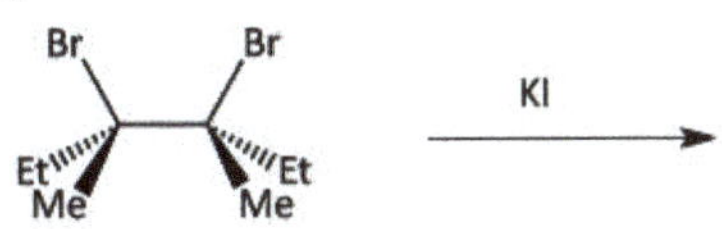

10.6 Reaction Summary

E2 Dehydrohalogenation to form alkene

a) 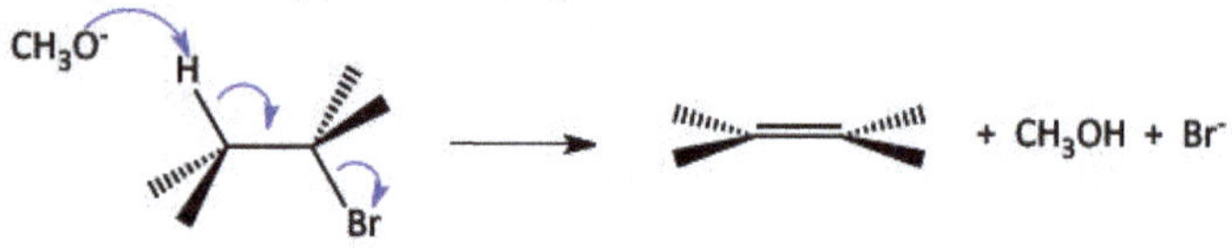

E2 regiochemistry

b) 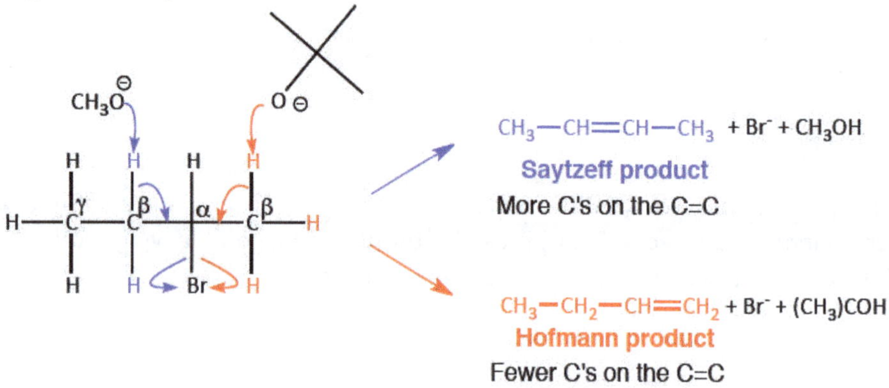

E2 Dehydrohalogenation to form alkyne

c) 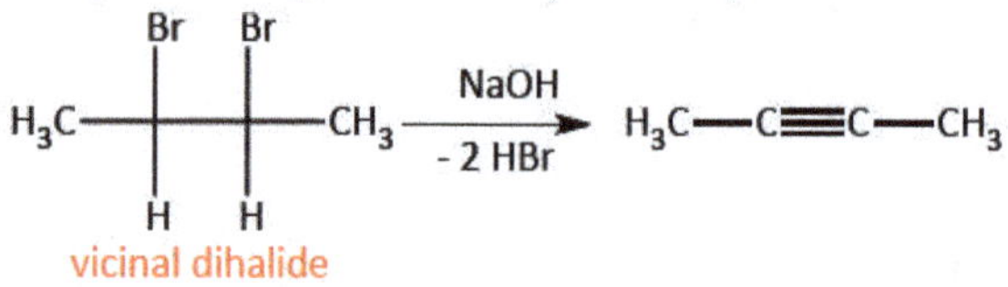

d) 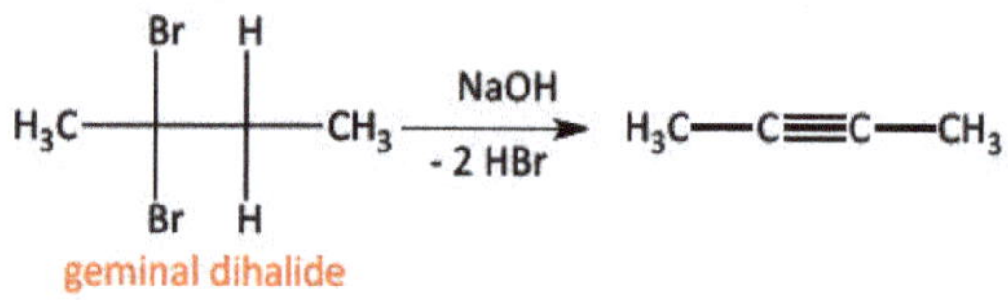

E1

e) 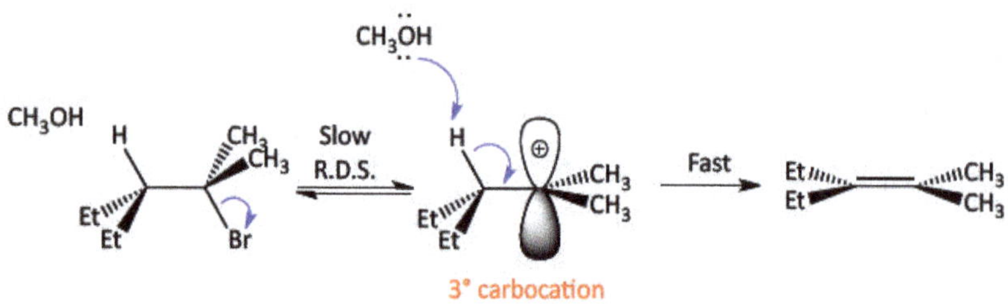

Dehalogenation

f) 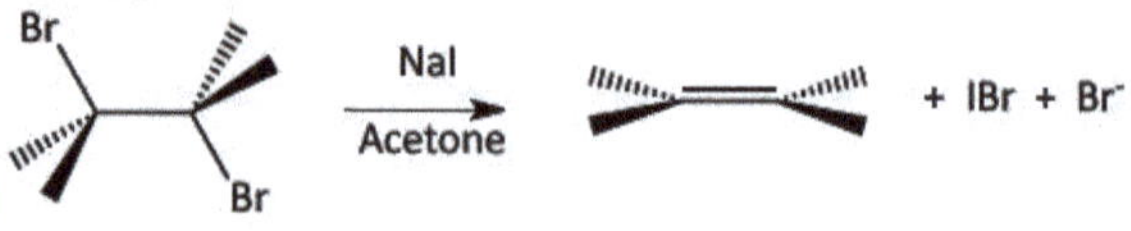

g) 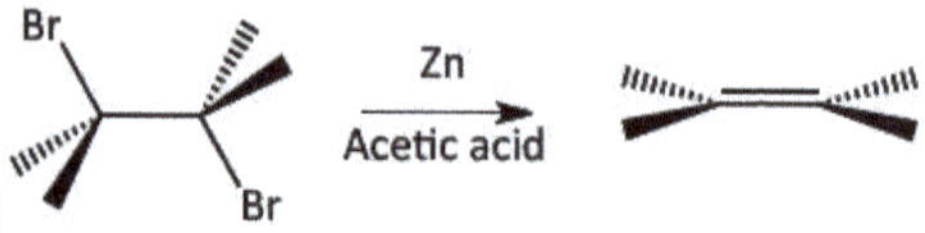

Dehydration

h) 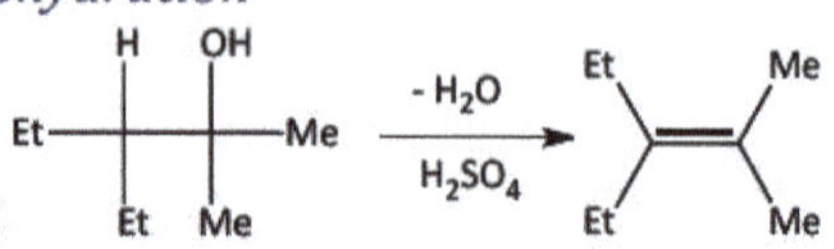

10.7 Problems

10.11 Determine the degrees of unsaturation for each of the following formulas. Then, draw two possible structures for each formula.

a) C_5H_8O b) C_4H_9N c) C_6H_{10} d) C_5H_5Cl e) C_4H_8

10.12 Label the following compounds as *E*, *Z*, or neither.

a) 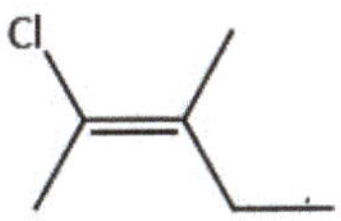

b) 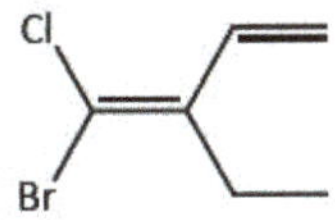

c) 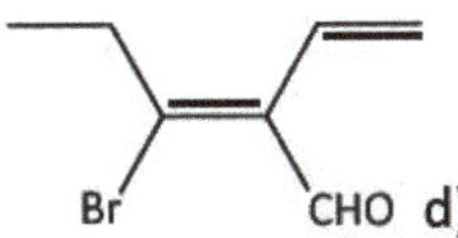

d) 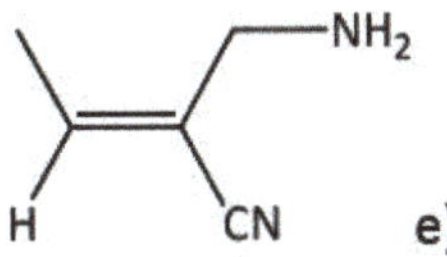

e) 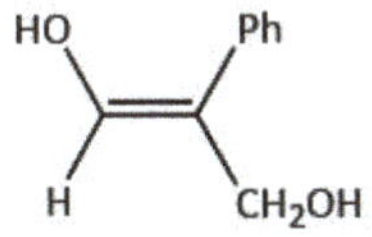

10.13 Name the following compounds. Do not forget cis-, trans-, E-, or Z- when appropriate.

a) 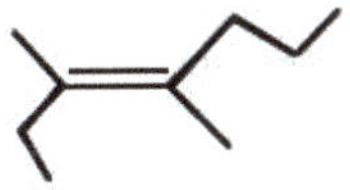
b) 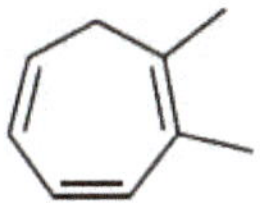
c) 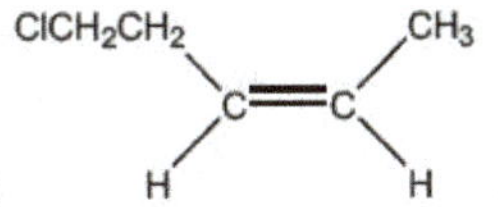

d) 
e) 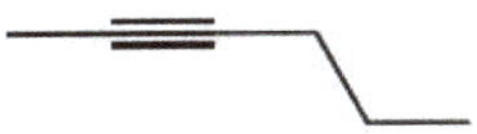
f) 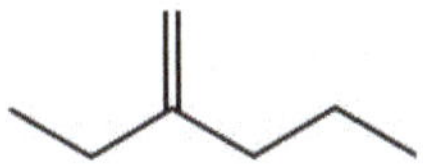
g) 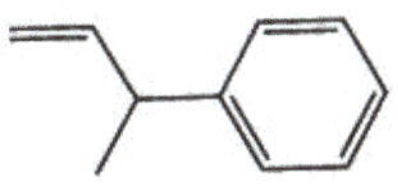
h) 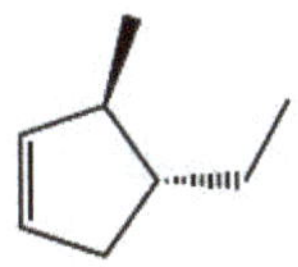
i) 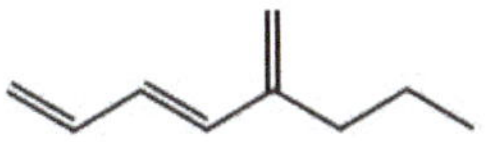
j) 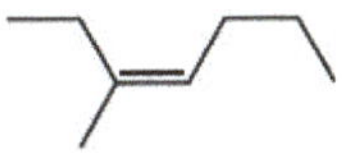
k) 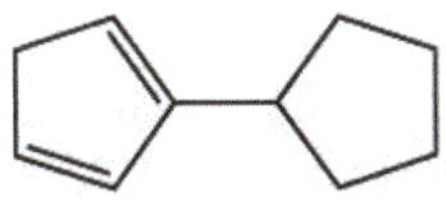
l) 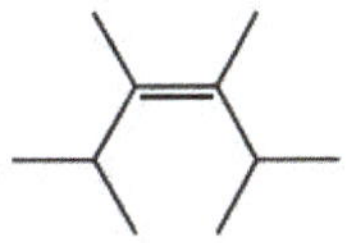

10.14 Draw the following compounds.

a) 4-phenyl-1-butene
b) 3-heptyne
c) acetylene
d) polyethylene
e) (*E*)-4- methyl-2-pentene
f) 2-phenyl-1,3-cyclopentadiene
g) ethylene
h) polypropylene
i) (3*E*, 5*Z*)-2-methyl-3,5-nonadiene
j) *cis*-3,4-dibromo-3-hexene
k) (2*E*,4*E*)-octadien-6-yne
l) methylacetylene
m) cyclohepta-1,3-diene
n) *cis*-3,4-dimethyl-1-cyclopentene
o) 2,4-dimethylcyclopentadiene
p) 1-cyclohexyl-1,3,5-cycloheptatriene

10.15 Compounds don't necessarily need a double bond to have E and Z, you just need geometric isomers (like if they contain rings). Label the following as E or Z.

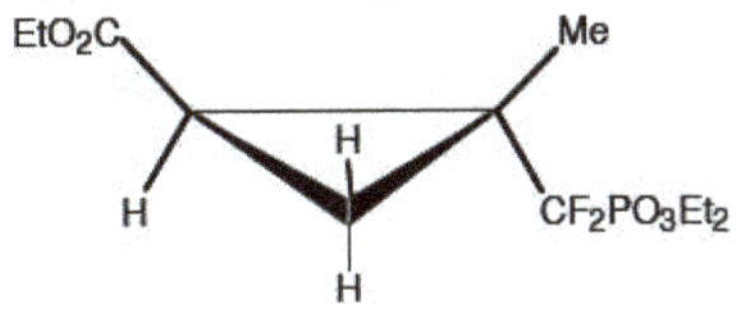

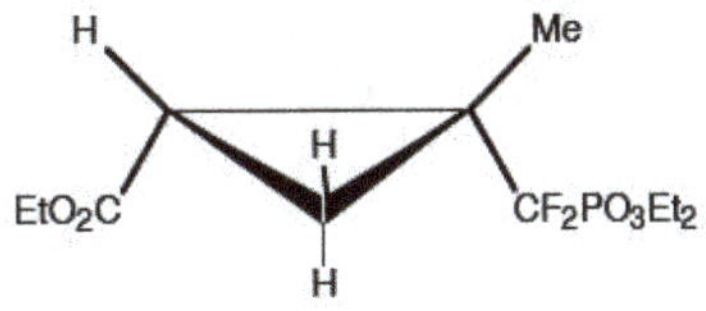

______ E or Z?

______ E or Z?

10.16 Components of the sex pheromone of the leopard moth have been synthesized. They are **A** and **B**, shown below.

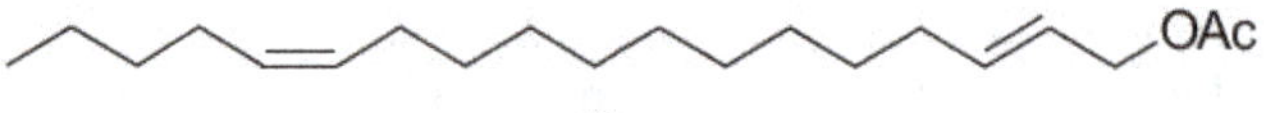

A

B

Complete the names for **A** and **B**.
A is (,)-2,13-octadecadienyl acetate.
B is (,)-3,13-ocdecadienyl acetate.

10.17 For the following compounds, identify if they have cis-trans isomerism. If they do, draw all of the cis/trans or *E/Z* isomers. Label them cis, trans, *E*, or *Z*.
a) 2-methyl-2-butene b) 2-pentene c) 1,3-pentadiene d) 2,3-dichloro-2-butene e) 1-butene
f) 1-chloro-1-fluoropropene g) 1-bromo-1-chloro-2-methylpropene h) 2-bromo-1-chloro-1-iodobutene

10.18 a) Draw the two elimination products formed in the following reaction and indicate which of the two products would predominate.

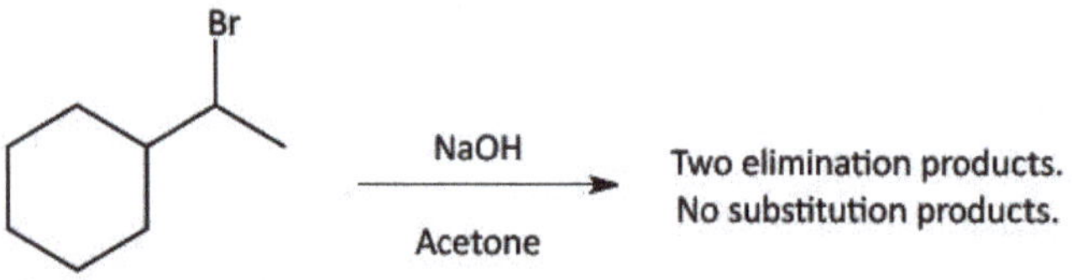

b) If KOtBu was used instead of NaOH, which of the two elimination products would predominate?

10.19 Draw the products of the following reactions.

a) 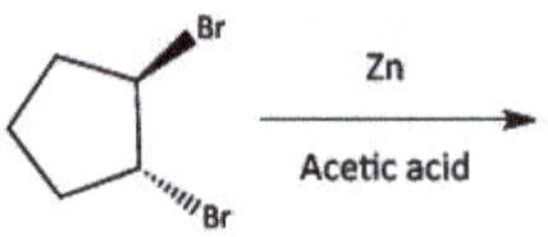

b) 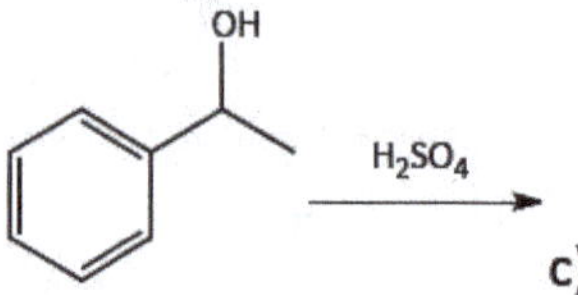

c) 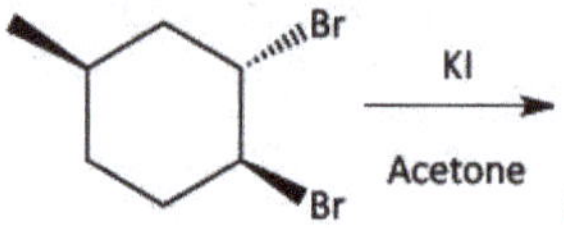

d) 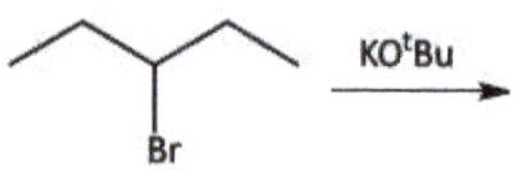

10.20 Indicate whether or not the following violate Bredt's rule.

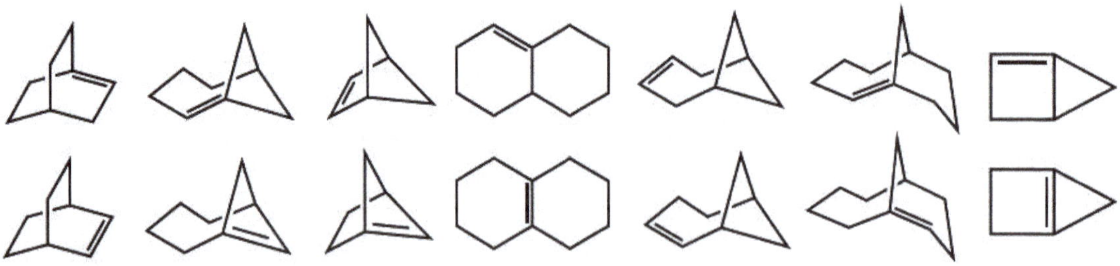

10.21 Fill in appropriate reagents that would perform the following reactions.
a) 1,2-dibromopentane to pentene b) 1-butene to 1,3-butadiene (2 reactions)

c) 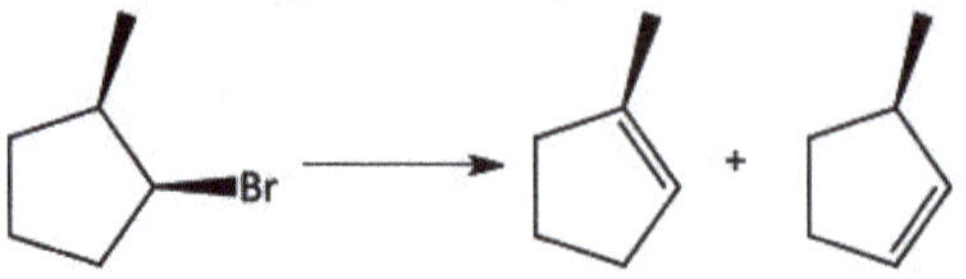

d) 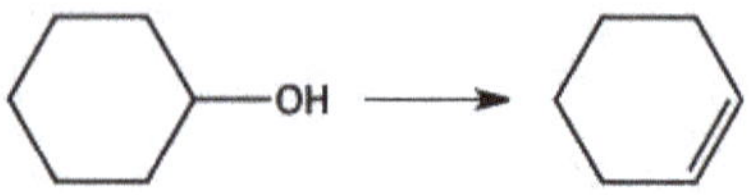

10.22 Propose a mechanism for the following reactions.

a) 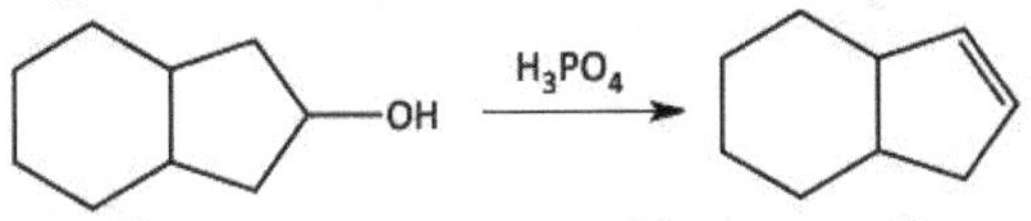

b) 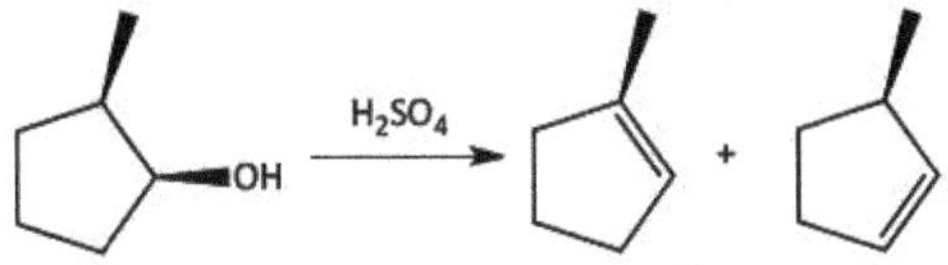

c) 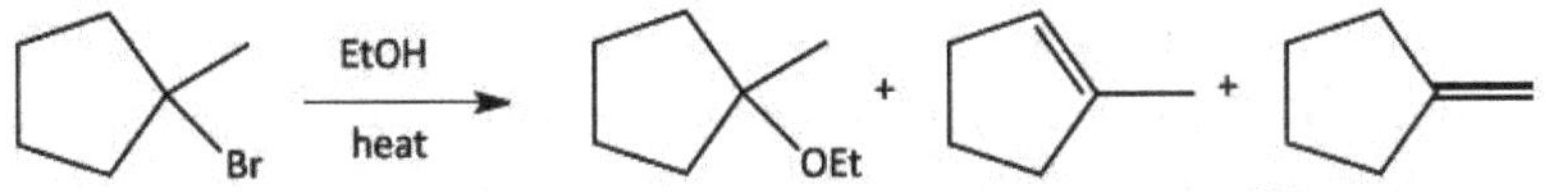

d) 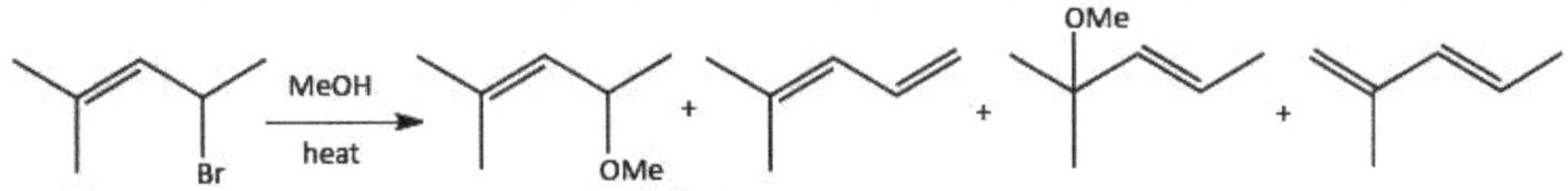

e) 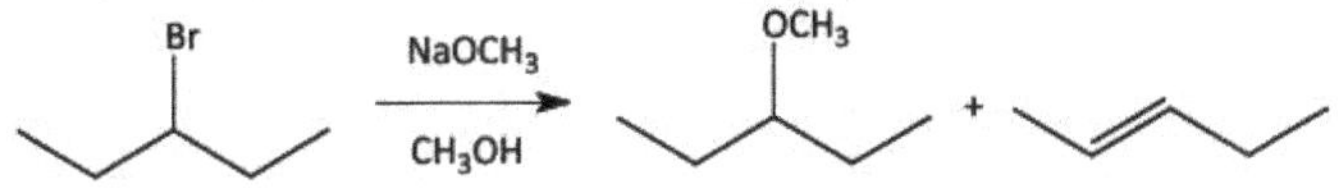

f) 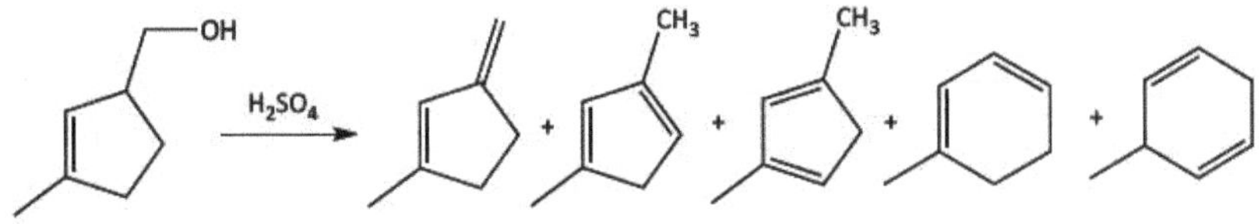

g) 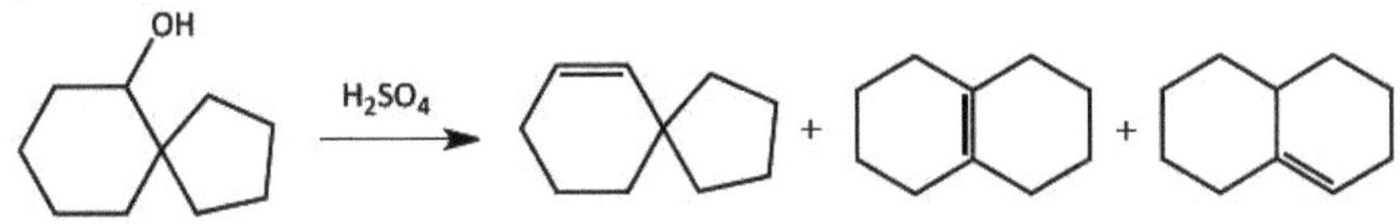

10.23 The following dehydration does not occur. Explain why.

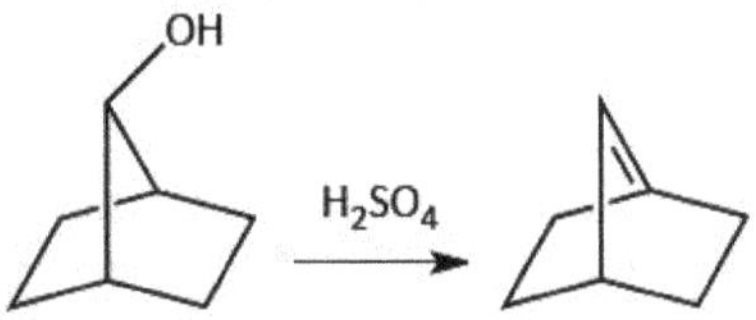

10.24 Predict the products of the following reactions. If more than one alkene is formed, identify the one that would be made in the greatest amount.

a) 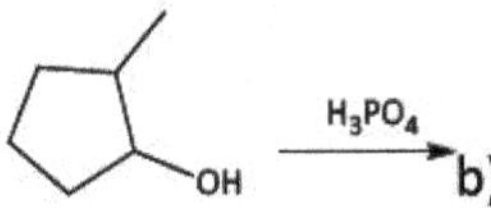

b) 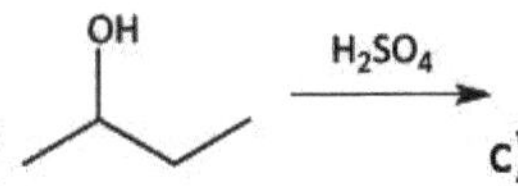

c) 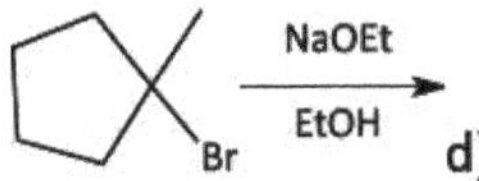

d) 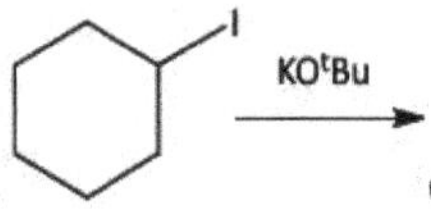

e) 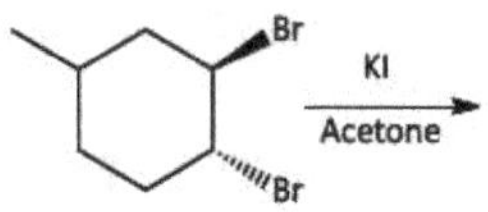

Chapter 11. Alkene & Alkyne Reactions

11.1 Electrophilic attack

Electrophiles are species that love negative electrons. They are normally electron deficient or have a slight positive charge. The pi-bond of an alkene can be thought of as a large, juicy, pair of electrons that can go out and attack these electrophiles. When the pi-bond grabs hold of an electrophile, it leaves a positive charge behind on the other carbon atom.

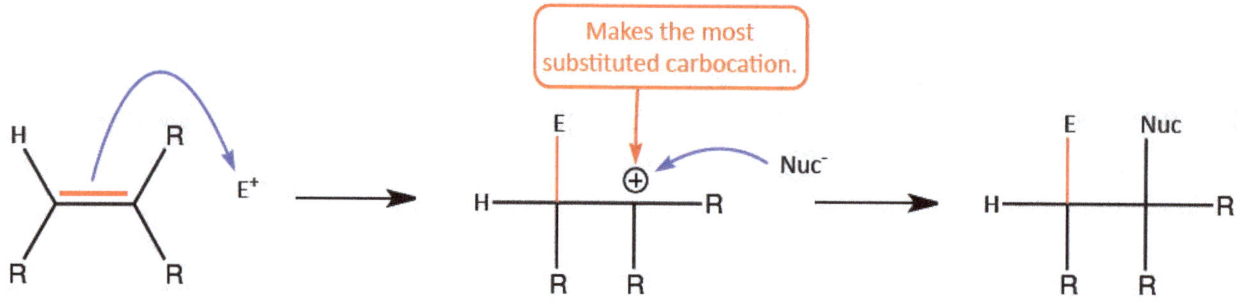

Figure 11.1 A Generic Electrophilic Attack

11.2 Markovnikov hydrohalogenation

11.2.1 Of an alkene

The addition of hydrogen halide, HBr, HCl, HI, is a regiospecific reaction. One of the two possible orientations is favored. Vladimir Markovnikov, a Russian chemist, noticed this preference for one regioisomer over the other. In 1869, Markovnikov wrote how to predict which regioisomer is favored. He wrote, when H-X is added to an alkene, the hydrogen atom adds to the carbon atom of the double bond that starts with more hydrogen atoms on it. Another way to think about it is the "special group", the non-hydrogen group, is added to the more substituted carbon atom of the double bond (the carbon atom with the most R groups). Reactions that follow this rule are called Markovnikov reactions. The products of these reactions are Markovnikov products.

Although Markovnikov wrote his rule for hydrogen halide reactions of alkenes, chemists today use this rule for any reactions that add a hydrogen atom to one side of an alkene and a different group to the other carbon atom of the alkene.

Why does Markovnikov's rule work? For the reaction of H-X, the more substituted carbocation is lower in energy. Its pathway is easier to follow, resulting in more of the Markovnikov product, where the special group is on the more substituted carbon atom (where the carbocation was).

Markovnikov's rule is not always followed. Sometimes what appears to be the wrong product, the anti-Markovnikov product, is formed. But, in these instances, the more stable intermediate is still the one whose path is followed. This idea, where the more stable intermediate is the path that is followed more, is ALWAYS obeyed.

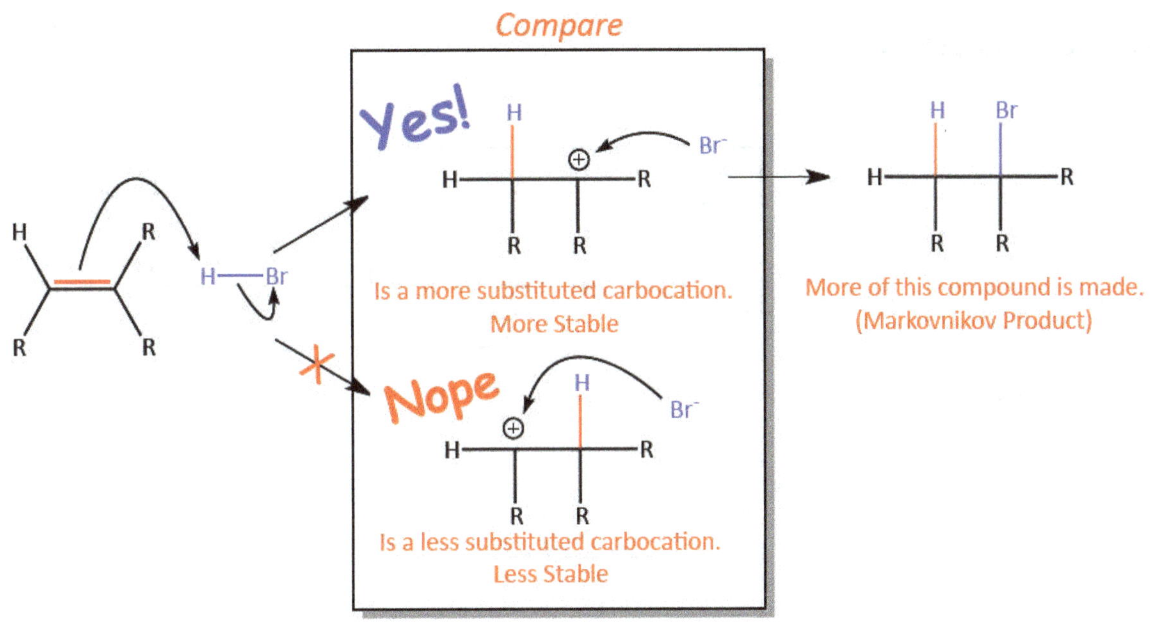

Figure 11.2 Markovnikov hydrohalogenation of an alkene

11.2.2 Of an alkyne

The same reaction can be performed on an alkyne. But, since an alkyne contains two pi-bonds, the reaction with H-X (HCl, HBr, or HI) can occur twice. If a terminal alkyne is a reactant, a Markovnikov product is made.

Mechanism

1. Addition to first pi-bond.

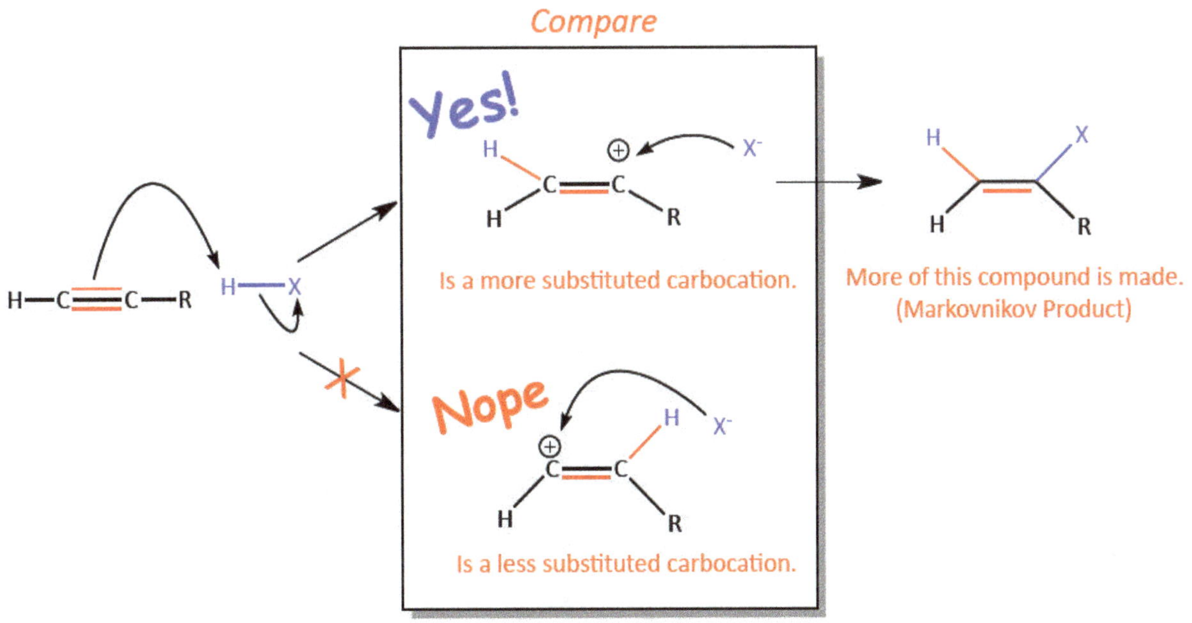

2. Addition to second pi-bond.

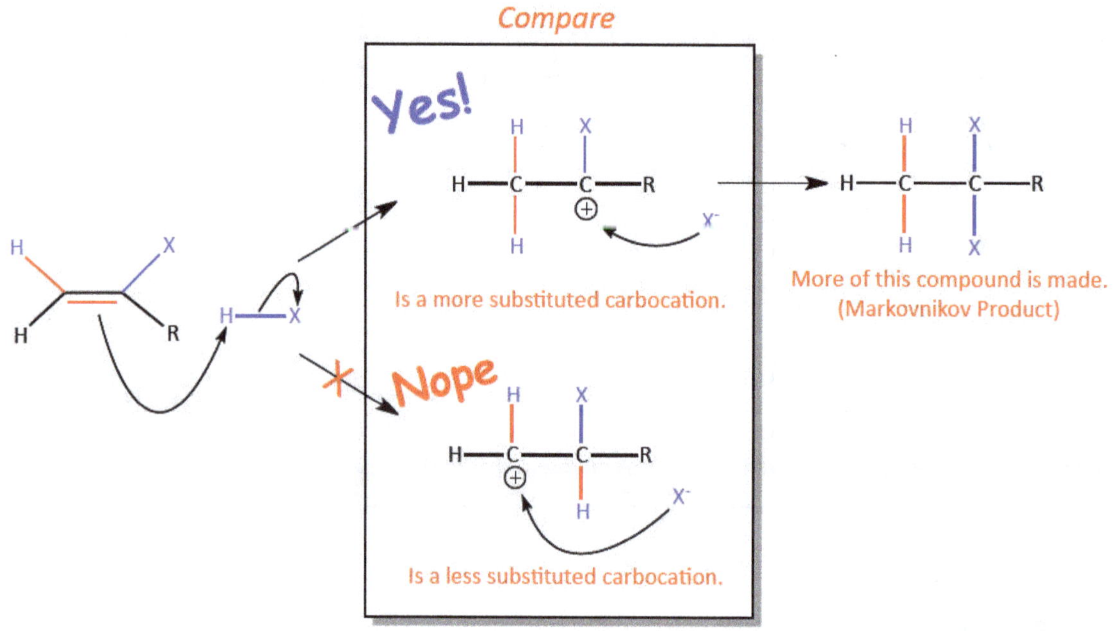

If an internal alkyne is a reactant, it doesn't matter which side the hydrogen adds. Both are equally stable. Therefore, a mixture of products is formed.

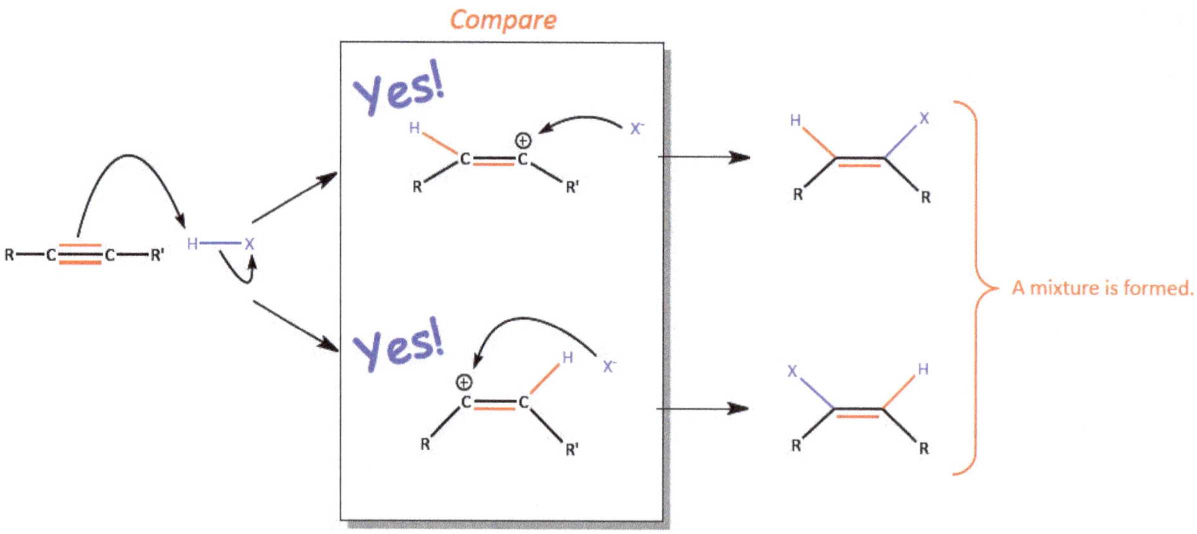

11.1 Draw the product of the following reactions.

a) 

b) 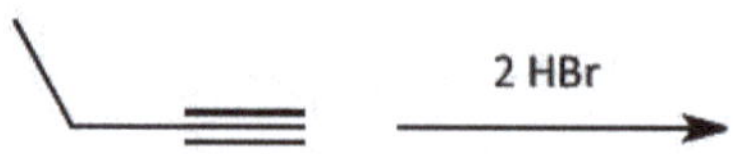

c) 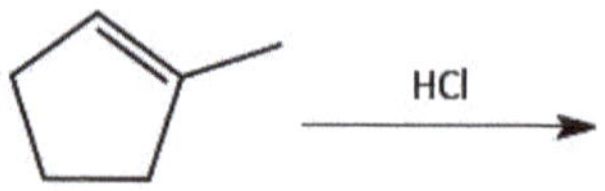

11.3 Anti-Markovnikov hydrobromination

11.3.1 Of an alkene

Anti-Markovnikov halides can also be synthesized. When peroxide is added to the reaction of HBr and an alkene, a different mechanism occurs than the one we saw previously. When peroxide is added, a free radical mechanism occurs.

The first step of this mechanism is that the peroxide breaks apart to form free radicals.

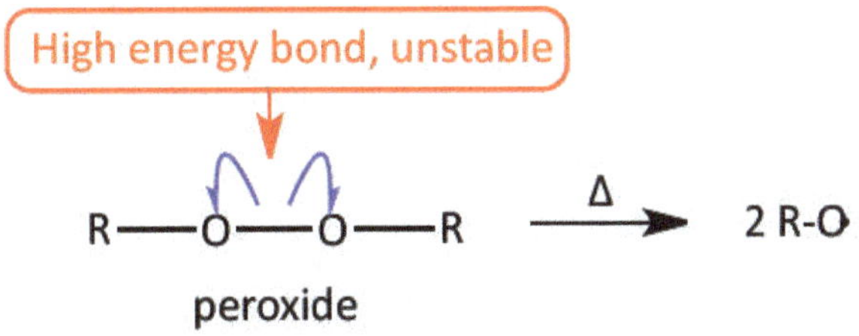

The oxygen free radicals react with H-Br to make bromine radicals.

The bromine radical can attack one of two ways. We'll look at both possibilities and choose the one with the more stable intermediate. Remember, we ALWAYS do this. But, in the end, we will make more of the anti-Markovnikov product.

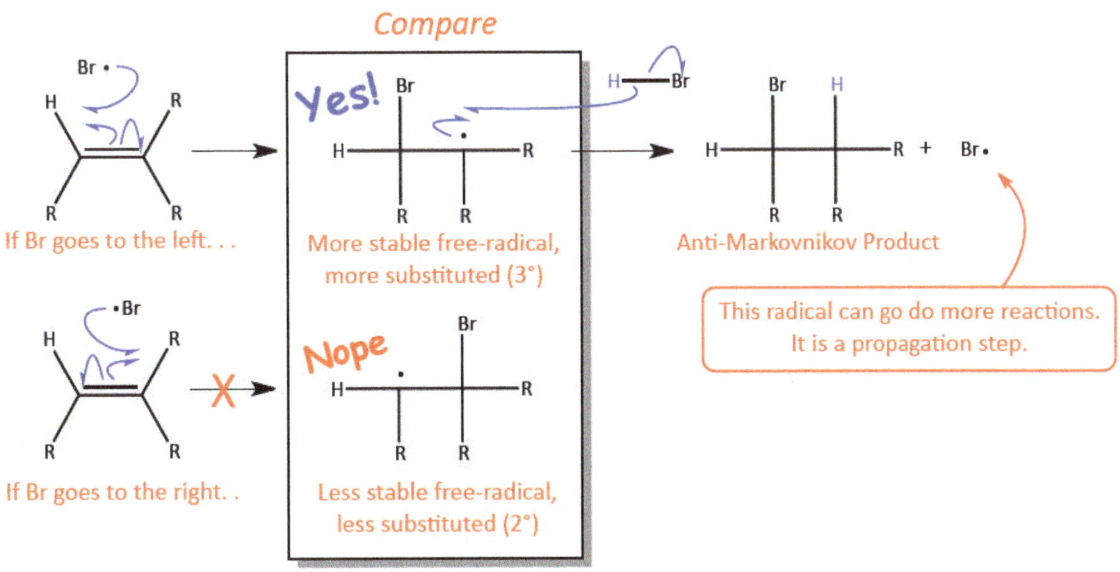

11.3.2 Of an alkyne

If HBr and peroxide are reacted with an alkyne, a free radical reaction occurs giving anti-Markovnikov addition of HBr.

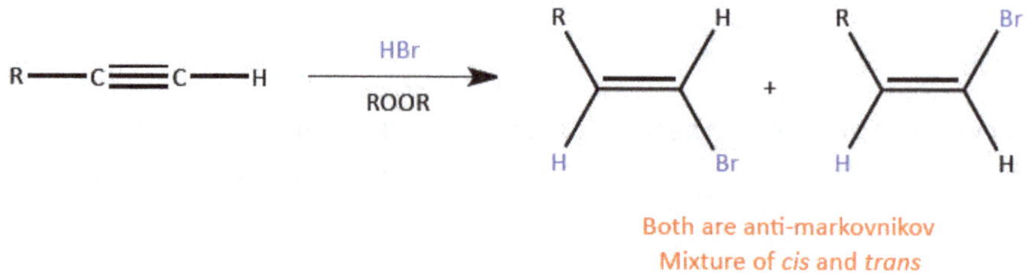

The reaction usually does not occur again. It tends to stop at the double bond.

11.2 Draw the products of the following reactions.

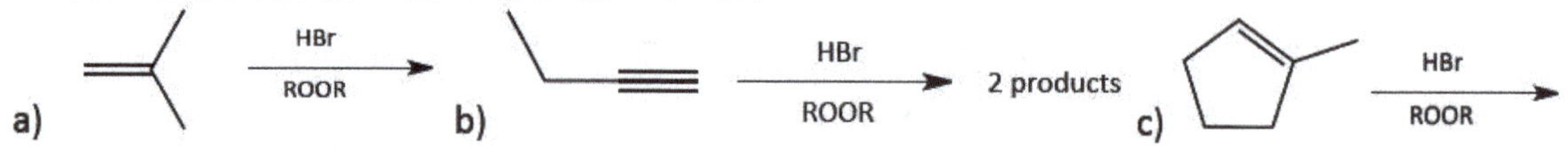

11.4 Hydration of alkene

In the hydration of an alkene, acid and water are added to an alkene. In the mechanism, the pi bond of the alkene attacks H^+ of the hydronium (H_3O^+). The two possible carbocations that can be formed are compared. The more stable one, which is usually the more substituted one, is lower energy and its path followed. This leads to more of the Markovnikov alcohol.

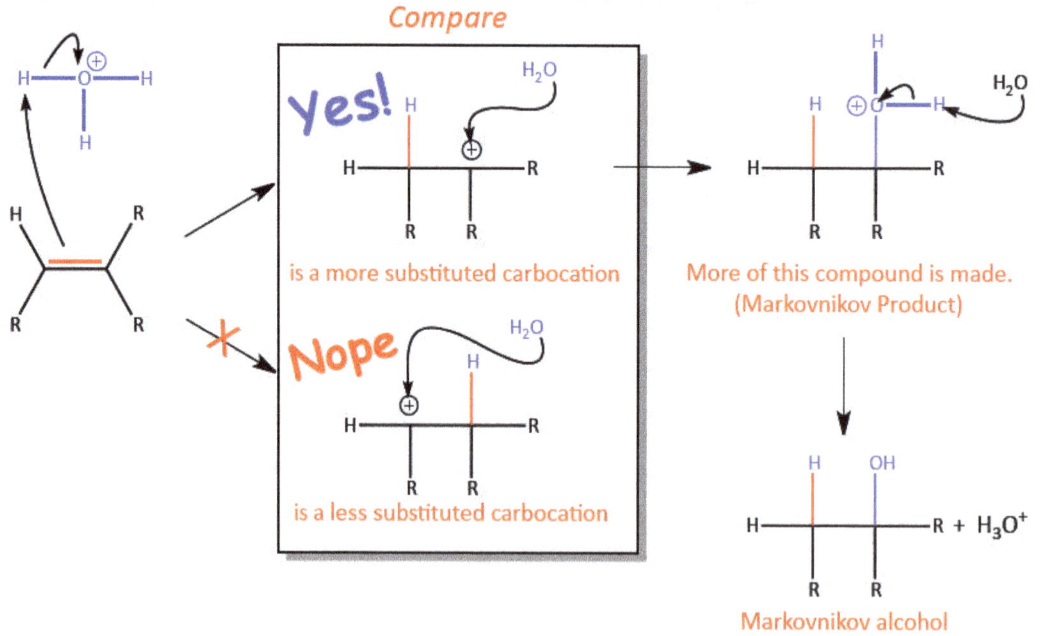

If the reaction is performed with concentrated H_2SO_4 or H_3PO_4, the reaction equilibrium is driven to the alkene starting material because H_2SO_4 or H_3PO_4 are dehydrating acids and absorb the water. If the acid is diluted with plenty of water, the reaction equilibrium is driven to the alcohol. One difficulty with this reaction is that alkenes are not usually very water soluble, so it is difficult to run this reaction. Also, rearrangements are possible since a carbocation is formed. For these reasons, there are better ways to make Markovnikov alcohols.

11.5 Hydrosulfonation

Hydrosulfonation could be a better way to make Markovnikov alcohols. Since alkenes do not dissolve very well in water, concentrated sulfuric acid (H_2SO_4) can be used. Normally, sulfuric acid drives the equilibrium to the alkene, but sulfuric acid does dissolve the alkene. If the reaction is run cold, it leads to the Markovnikov hydrogen sulfonate. At this point in the reaction mechanism, if plenty of water is added and the reaction is heated, the hydrogen sulfonate becomes a Markovnikov alcohol.

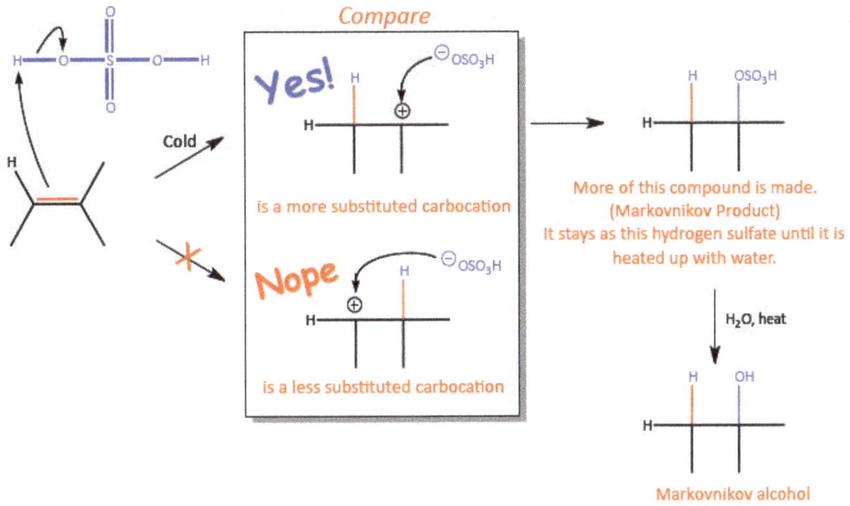

The problem with this reaction is concentrated sulfuric acid is used for the first part of the reaction. Sulfuric acid reacts with many functional groups. If there are any sensitive functional groups on the alkene, the sulfuric acid will chew them up. If a chemistry student spent six months of research making a large, beautiful, complicated molecule that contains an alkene and they want to make a Markovnikov alcohol out of it, they would not want to add concentrated sulfuric acid to it! Also, since a carbocation is formed, rearrangements are possible.

11.6 Triangle reactions

There are two types of reactions that I call "triangle reactions". It is important to have a general understanding of them before we encounter them in the remainder of this chapter. An analogy helps explain how triangle reactions work. This analogy is the story of a victim and two criminals. This analogy is a little bizarre, but I think it helps us remember how these reactions happen, so bear with me. The victim of the crime is somebody who is totally honest. He is so honest; he cannot help but announce to the world the amount of money he has by carrying a sign above his head. The victim has two options available to him. One is to hold a sign that reads "O" indicating that he has zero money on him. The other option is a sign that reads "+" indicating that he is carrying a positive amount of money on him. The victim sometimes has bags slung over his shoulders. In order to protect any money he holds, the victim puts his money in the pocket that is hidden under the most bags.

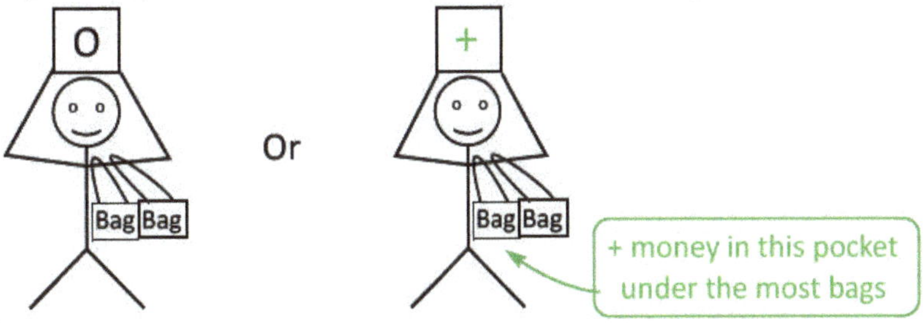

Figure 11.3 The victim of a Triangle Reaction

One of the two criminals is a weak, scared, but greedy pickpocket. This weak pickpocket will not attack a victim unless he knows the victim is *definitely* carrying money. If he doesn't know this, he will leave the victim alone. So, if the victim is carrying the "O" sign, the weak criminal leaves him alone. If the victim is carrying a "+" sign, the pickpocket will reach into his pocket under the bags, where the money is kept.

The second criminal is a strong, aggressive, violent criminal. The strong criminal doesn't really care about money. He is so violent; he wants to punch the victim in the gut immediately. This aggressive criminal will attack any type of victim. If the strong criminal encounters the victim carrying the "O" sign, he punches the victim immediately in the gut. The strong criminal wants to do as much damage as *quickly* as he can to the victim, so he punches him in the side that has the fewest bags on it.

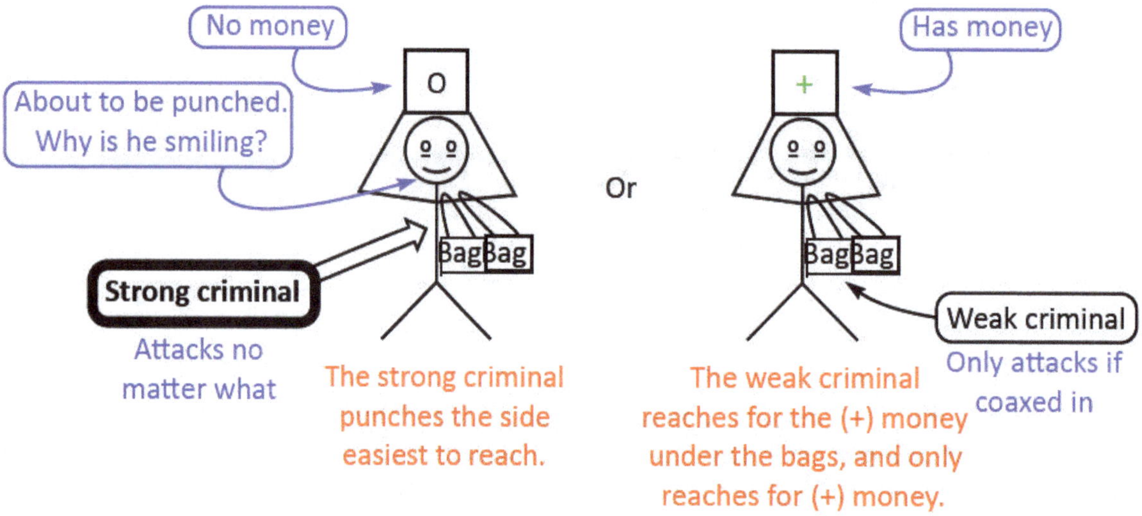

Figure 11.4 How criminals attack in a triangle reaction

If we have a "triangle reaction" the top atom is usually either neutral or it has a positive charge. If the top atom is neutral, it is like a victim that is holding an "O" sign. If the top atom has a positive charge, it is like a victim that is holding a "+" sign. The criminals in the analogy are either strong nucleophiles (the strong criminal) or weak nucleophiles (the weak criminal). Strong nucleophiles tend to attack the least crowded side of the triangle because it is easiest to get to. Weak nucleophiles tend to attack the side of the triangle with the most alkyl groups on it. The bags the victim is carrying are "R" groups or alkyl groups. If the top atom has a positive charge, the positive charge is shared somewhat on the other two atoms of the triangle. The atom of the triangle with the most R groups attached holds more of the partially positive charge. This is because more substituted carbon atoms are more stable carbocations. That more substituted carbon atom of the triangle has enough positive charge (+ money) that even a weak nucleophile (weak criminal) cannot resist and will attack it. Below is an example of a neutral epoxide on the left and a positive, protonated epoxide on the right.

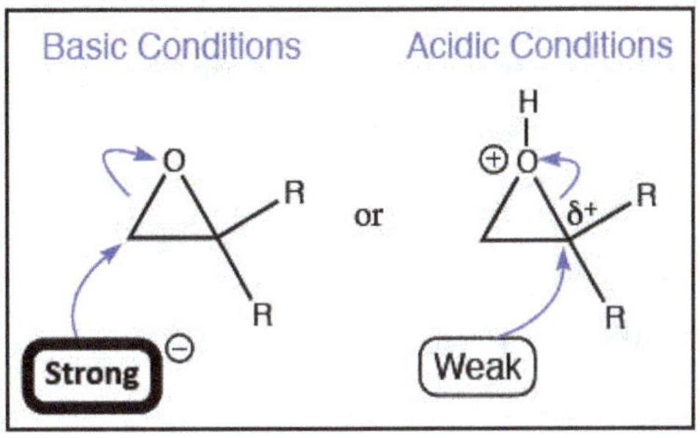

Figure 11.5 How strong or weak nucleophiles attack epoxides

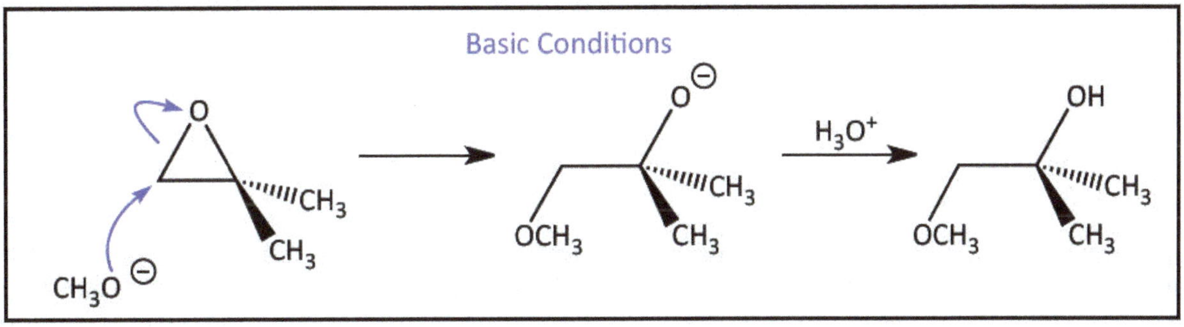

Figure 11.6 Example of a strong nucleophile attacking a neutral epoxide

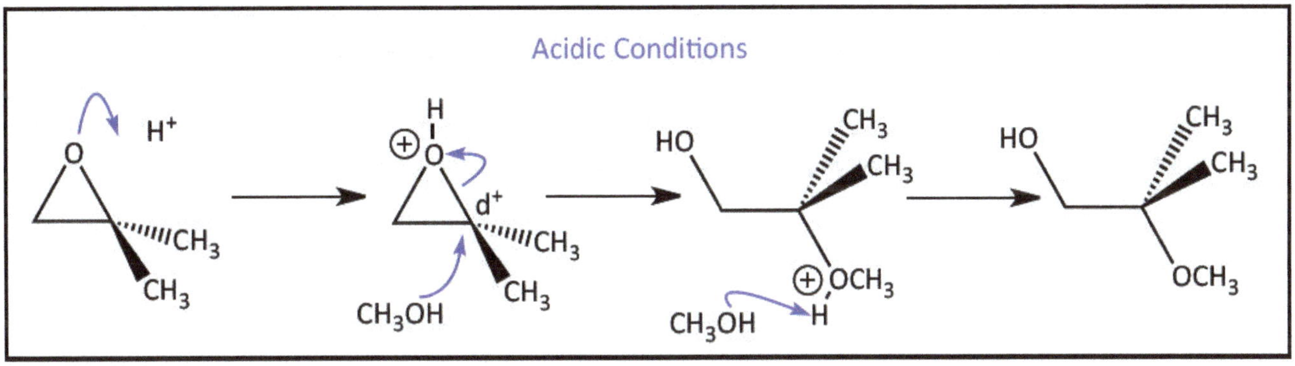

Figure 11.7 Example of a weak nucleophile attacking a protonated epoxide

11.7 Oxymercuration/Demercuration-with water- A triangle reaction

A good, gentle way to make Markovnikov alcohols is oxymercuration in water followed by demercuration.

11.7.1 Oxymercuration in water

To perform the oxymercuration, mercuric acetate, $Hg(OAc)_2$, is used. There are two common oxidation states of mercury, 1^+ and 2^+. Mercuric refers to the higher oxidation state, Hg^{2+}. Acetate (OAc) is $CH_3CO_2^-$. There needs to be two of these 1^- acetates to balance the 2^+ charge of the mercury. The key first step in the oxymercuration of an alkene is the breaking apart of the mercuric acetate to form the electrophilic ^+HgOAc.

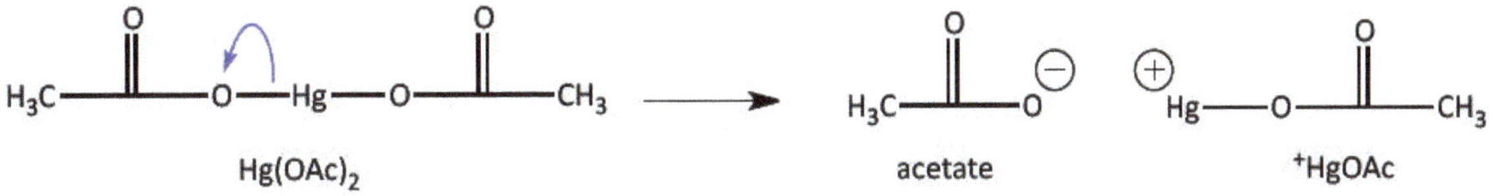

The negative pi-electrons of the alkene eagerly attack the positively charged, electrophilic $^{+}HgOAc$. The pi-electrons attack to form the triangular mercurium ion. The two electrons from the pi bond go towards the Hg^+, and two electrons return towards the pi bond. This causes the oxidation state of the mercury not to change so it remains Hg^+. These triangles are formed when very large, soft atoms like mercury are used in the electrophile. The mercury atom is so large, it takes both carbon atoms to hold up. The victim in our triangle analogy is now holding a "+" over his head indicating he has money.

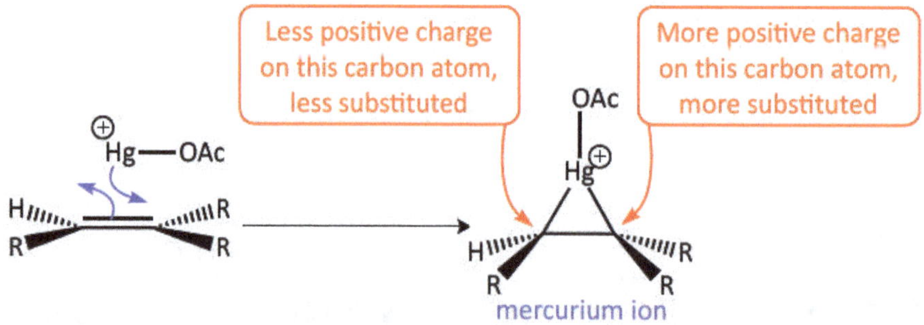

When the reaction is run in water, a weak nucleophile is present. Water is a stable, happy molecule. It will not attack a triangle compound unless it is coaxed in by money in some bags. In this case, the positive charge is shared over the two carbon atoms of the triangle. The more substituted carbon atom (on the right) carries a greater partial positive charge. Water then attacks the carbon atom with the greatest partial positive charge. In our criminal/victim analogy, the right carbon atom of the triangle is holding more bags and the + money is kept on the right carbon atom of the triangle under the bags. From what direction will the water attack the triangle? The large HgOAc blocks the topside of the mercurium ion triangle from attack, so the water must attack from the backside. It attacks, and the electrons go up onto the positive mercury. The neutral mercury acetate that is formed is a good leaving group—even though it is still attached to the left triangle carbon of our molecule. Finally, another water molecule cleans up the protonated alcohol to give us a mercury-alcohol compound.

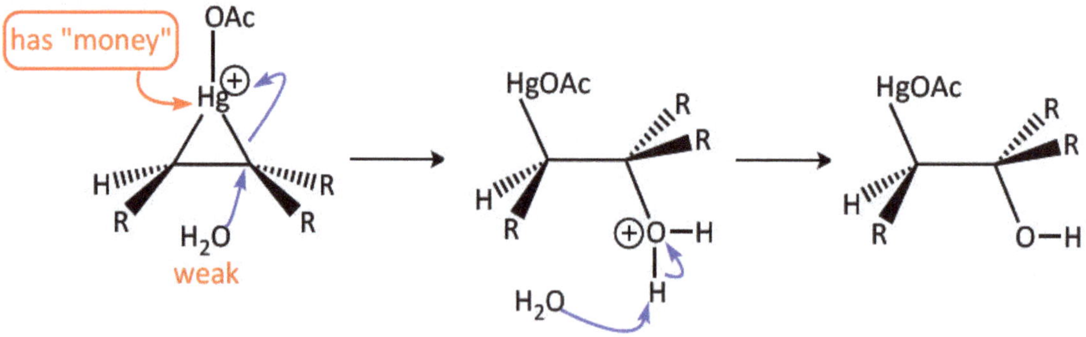

11.7.2 Demercuration

The final step is demercuration, where HgOAc is replaced with H. Sodium borohydride ($NaBH_4$) is a source of hydride (H^-). This H^- directly replaces the HgOAc. Don't worry about the mechanism of this step. The mechanism will only confuse you at this state, so this is one of those instances where you should simply memorize this.

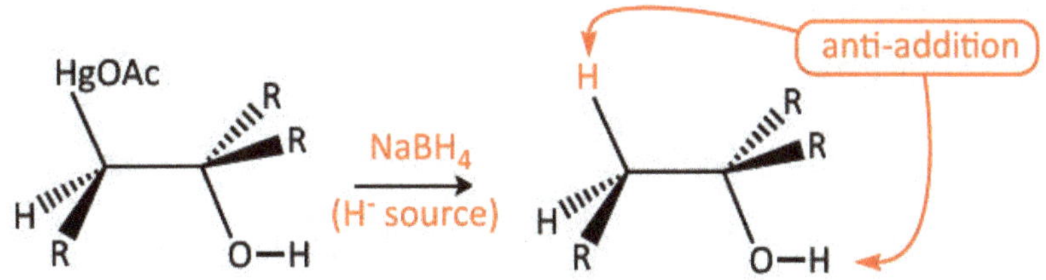

The benefits of oxymercuration/demercuration are: 1) a regular carbocation is not formed, so there is no chance for rearrangements. 2) the reaction is gentler than using sulfuric acid. 3) the reaction tends to go in high yield.

Are there any drawbacks to the oxymercuration/demercuration reaction? You may have already guessed it. Any time a reaction uses mercury, we need to be careful. Mercuric acetate and our mercury-containing products are toxic. We need to be very careful using them and be careful that we properly dispose of our waste.

11.8 Oxymercuration/Demercuration-with alcohol- A triangle reaction

11.8.1 Oxymercuration in alcohol

Once again, mercuric acetate is used to make the triangular mercurium ion. If water is not used as our solvent, it cannot react as the nucleophile. If alcohol is used as the solvent, the lone pair of electrons on the alcohol can act as a weak nucleophile and attack the mercurium. Again, it attacks the carbon atom of the mercurium triangle with the greatest partially positive charge. The electron bond goes up onto the mercury cation to make a good, neutral mercury leaving group. The protonated –OR group is deprotonated with another molecule of solvent to make a mercury-ether containing compound. Many different Markovnikov ethers can be made depending on the alcohol used as the solvent. R = CH_3 for methanol, R = CH_2CH_3 for ethanol, etc.

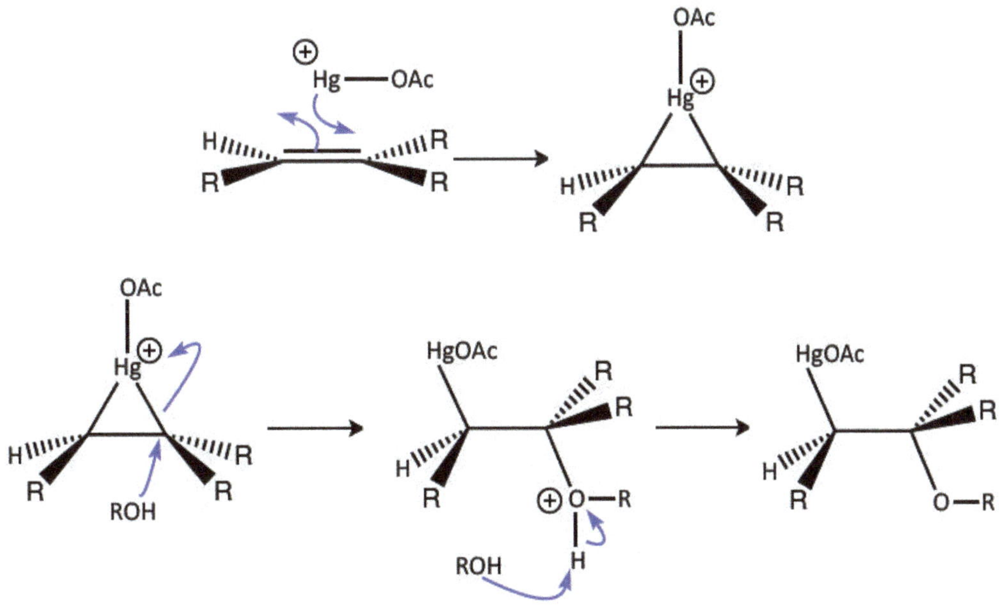

11.8.2 Demercuration

Sodium borohydride ($NaBH_4$) is once again used to replace HgOAc with H⁻ making a Markovnikov ether.

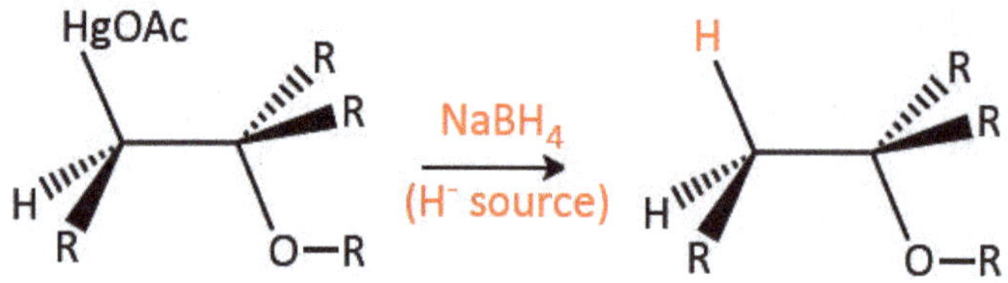

11.9 Markovnikov hydration of alkyne

Alkynes, with their two pi-bonds, can also react in an oxymercuration-demercuration reaction. The reaction is very similar to that for an alkene with a twist at the end. Instead of mercuric acetate, the oxymercuration of alkynes usually involves mercuric sulfate and aqueous sulfuric acid. Like with alkenes, this results in the addition of water across a pi-bond resulting in a Markovnikov alcohol. Only one pi-bond of the triple bond reacted, so alkene remains. This vinyl alcohol is called an enol (pronounced "een-awl"). The "enol" name comes from this compound being an alkene (*ene*) and an alcohol (*ol*).

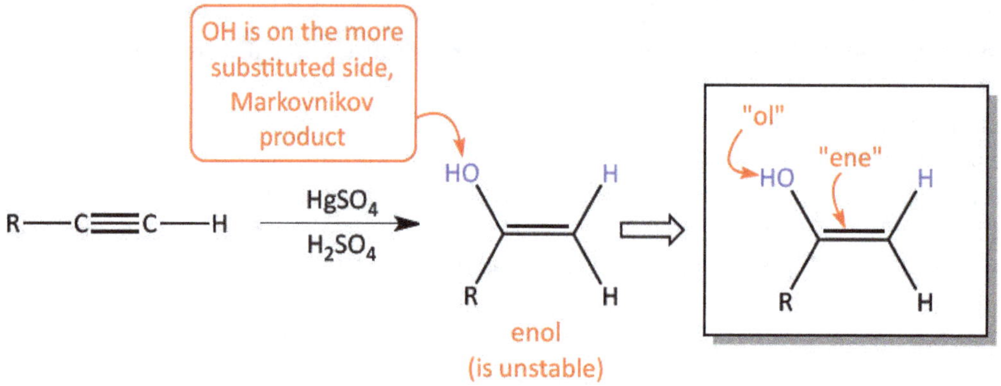

Enols tend to be unstable. Enols are in equilibrium with their isomer, the ketone. These isomers differ only in the location of one proton and a double bond. Such isomers are called tautomers (pronounced "taw-toe-mers"). The ketone and enol are called keto-enol tautomers. The transformation of one into another is called a keto-enol tautomerization. The process itself is called keto-enol tautomerism. In the equilibrium of the two, the keto form predominates. Take careful notice of the transformation. The hydrogen atom moves from the alcohol of the enol to the alpha carbon. The pi-bond of the alkene of the enol moves to make a carbonyl (a ketone).

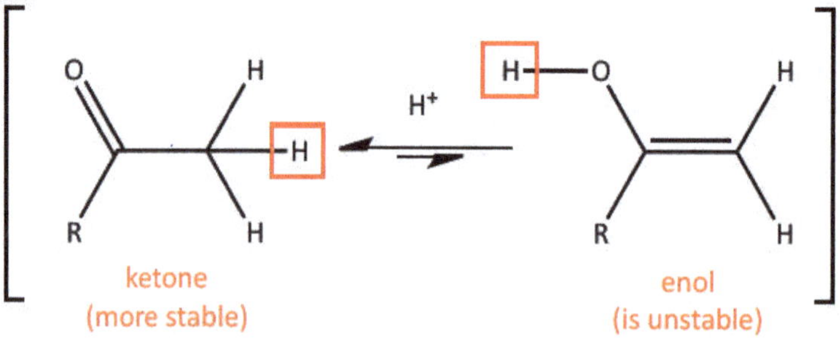

Figure 11.8 Keto-enol tautomerization

Overall Reaction

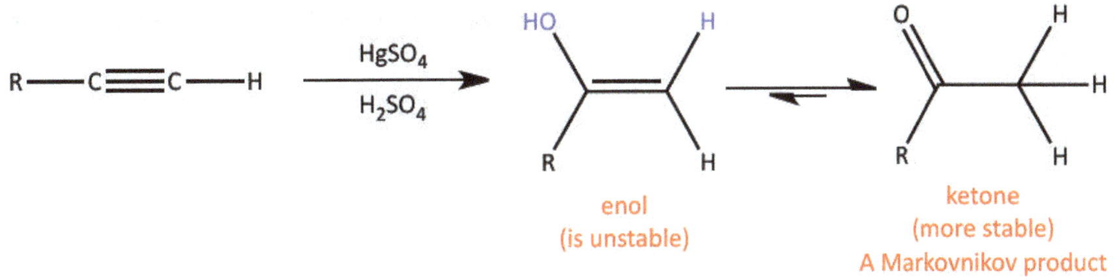

The tautomerization is acid-catalyzed since sulfuric acid is used in the reaction. H^+ is available because of the acid. To write the mechanism of the acid-catalyzed keto-enol tautomerization, protonate the enol at the alpha-carbon, draw resonance forms, and clean it up to make a ketone.

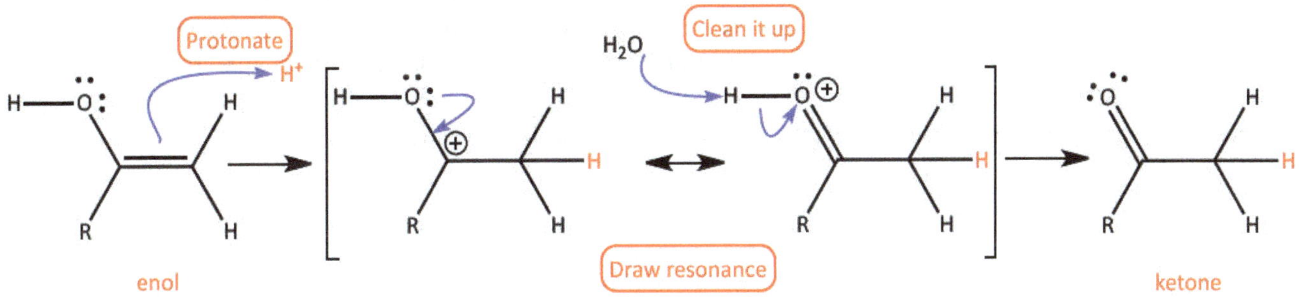

Figure 11.9 Acid-catalyzed keto-enol tautomerization

11.3 Draw the products of the following reactions.

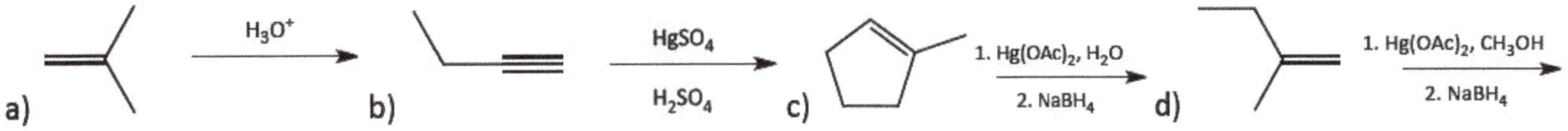

11.10 Hydroboration – Anti-Markovnikov hydration

11.10.1 Hydroboration of alkene

In 1979, Purdue University professor Herbert C. (H.C.) Brown won the Nobel Prize in chemistry for his work on the hydroboration of alkenes. Brown discovered that hydroboration is a way to make anti-Markovnikov alcohols. The simplest hydroborane, BH_3, has six valence electrons on the boron atom. Remembering the octet rule, we should realize this is electron deficient and therefore an electrophile. It just so happens that there is a nice, juicy pair of negative electrons in an alkene which is more than happy to attack the BH_3!

BH_3 is available in a couple of forms. One possibility is to buy diborane, B_2H_6. Diborane breaks apart to make BH_3.

$$B_2H_6 \rightleftharpoons BH_3$$

But, diborane is a toxic, flammable, explosive gas. So, this is not the preferred method. More commonly, BH_3 is purchased dissolved in a solvent like THF. This is a much safer way to use BH_3. Since it is an adduct with the solvent, it is a liquid that can be measured out.

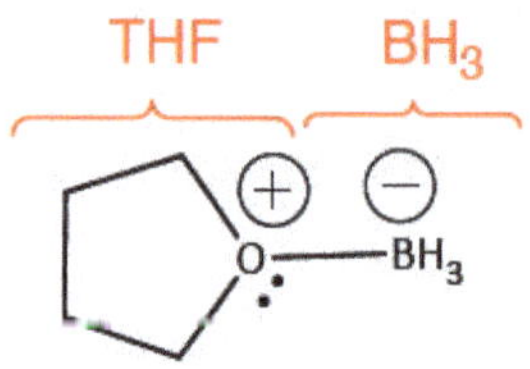

The mechanism for the hydroboration reaction begins when a B-H bond of BH_3 adds across the C=C in one step to make a square transition state. The boron atom gains a little negative electron density and therefore becomes partially negative. The carbon atom opposite the boron in the square loses some negative electron density and becomes partially positive. When the two transition states are compared, the partially positive carbon atom that is more substituted is more stable. It is the pathway that is followed more. When this carborane is oxidized with H_2O_2 and NaOH, the BH_2 is replaced with OH making an anti-Markovnikov alcohol. This is an extremely important reaction. Now we can make either Markovnikov or anti-Markovnikov alcohols. And, we'll learn later in organic chemistry, that we can turn alcohols into many different types of functional groups.

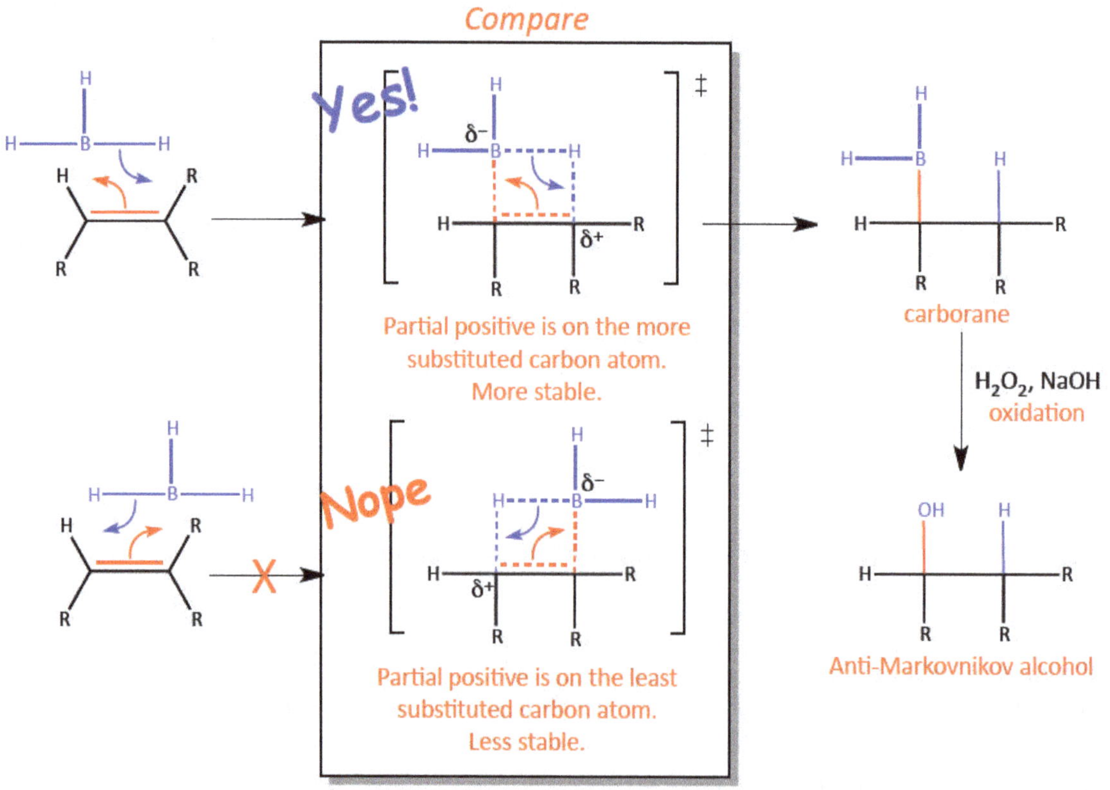

11.10.2 Hydroboration of alkyne

The hydroboration of an alkyne can lead to an anti-Markovnikov enol. Since alkynes contain two pi-bonds, two molecules of borane could add to it. A big, bulky, sterically hindered borane must be used in order to prevent this addition of two boranes. Usually, the borane used is di(*sec*-isoamyl)borane, commonly called disiamylborane. Instead of the small BH_3 molecule, disiamylborane has two big, large, deer antler-like alkyl groups attached to it. There are five carbon atoms in each alkyl branch. Amyl is an old, common name for pentyl.

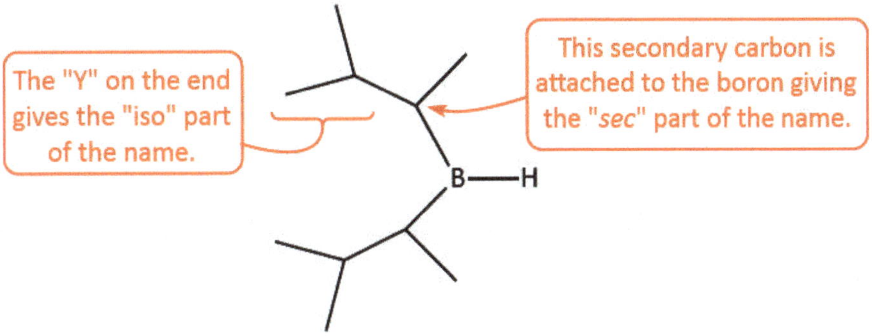

Figure 11.10 Di(sec-isoamyl)borane or disiamylborane

Disiamylborane reacts with an alkyne to make an anti-Markovnikov borane. This borane can be oxidized with peroxide and sodium hydroxide to make an anti-Markovnikov enol. This enol can tautomerize to the keto-form. In the case of a terminal alkyne, this keto-form is an aldehyde with a hydrogen attached to the carbonyl group.

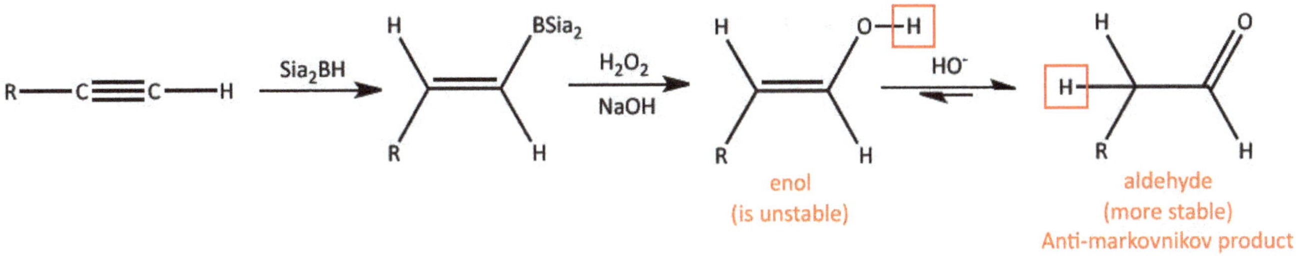

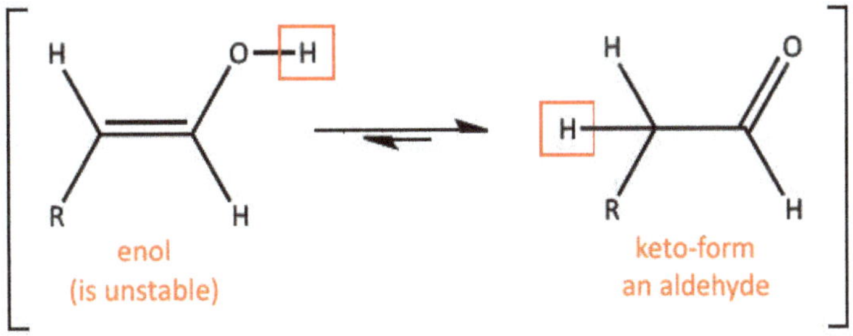

The mechanism of the base catalyzed keto-enol tautomerization is different than the mechanism was in acid. Bases deprotonate. So, the first step of this mechanism is the deprotonation of the enol at the alcohol position. This is pretty easy to remember because we are turning the alcohol portion of the enol into a carbonyl group. Once deprotonated, we draw resonance forms to get us closer to the keto product, then clean it up with some water to make the final aldehyde.

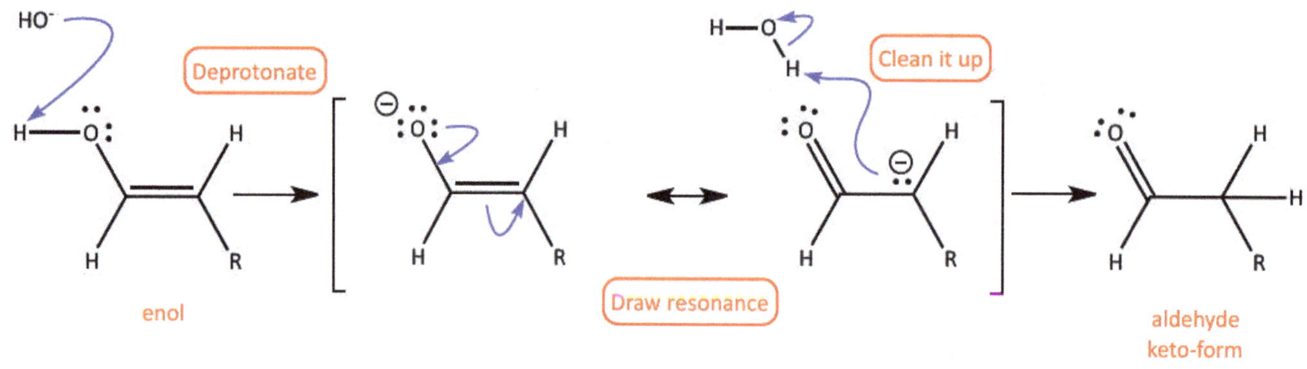

Figure 11.11 Base-catalyzed keto-enol tautomerization

11.4 Draw the products of the following reactions.

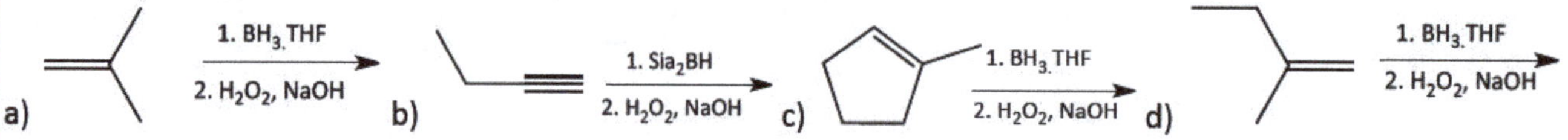

11.11 Halogenation-A triangle reaction

11.11.1 Halogenation of alkenes

Alkenes can undergo a halogenation reaction. This is a reaction with Cl_2 or Br_2 to make vicinal dichlorides or dibromides. This reaction must be run in an inert, non-reactive solvent like CCl_4.

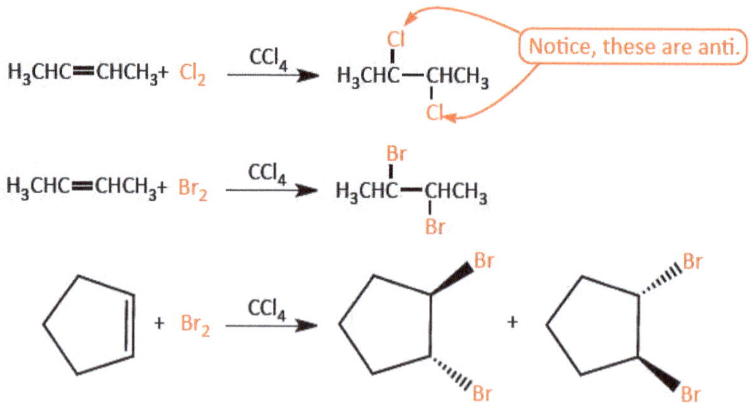

If one looks closely, the two halogens end up on neighboring carbon atoms which makes the product a vicinal dihalide. The third reaction, the bromination of cyclopentene, clearly shows the two bromides of the product are on opposite sides of the ring to each other in a *trans* configuration. This addition where the two groups end up on opposite sides is called *anti-addition*. This anti-addition occurs because a halonium intermediate is formed in the course of the reaction.

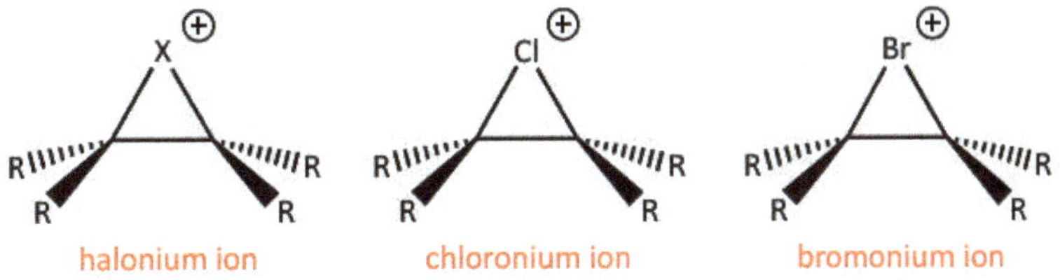

Figure 11.12 Halonium ions

The mechanisms for the chlorination or bromination of alkenes are identical. Bromination is shown here. When the alkene reacts with Br_2, a triangular bromonium ion is produced with Br^- as the leaving group. This Br^- does a backside attack at either carbon atom of the bromonium ion. In this triangle reaction, the Br^- would attack using pathway (b) more often than (a) since this carbon is more substituted and would have a greater partial positive charge.

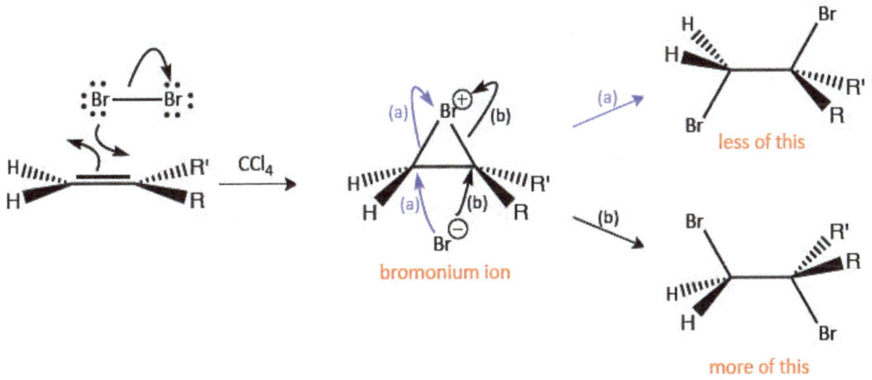

11.11.2 Halohydrin formation - A triangle reaction

If the solvent used in a bromination or chlorination reaction is not inert but instead is slightly nucleophilic, different products are formed. As an example, a bromohydrin is formed if water is present in the bromination reaction. In this mechanism, a bromonium ion is formed as before. But, if water is the solvent, there are numerous water molecules for every one Br- nucleophile. Water ends up being the nucleophile that attacks. In this triangle reaction, a "+" sign is being held up. So, the weak H_2O nucleophile attacks the more substituted carbon atom where more of the + money is held in the pocket under the bags. Once another water molecule cleans it up by deprotonation, it leads to the bromohydrin.

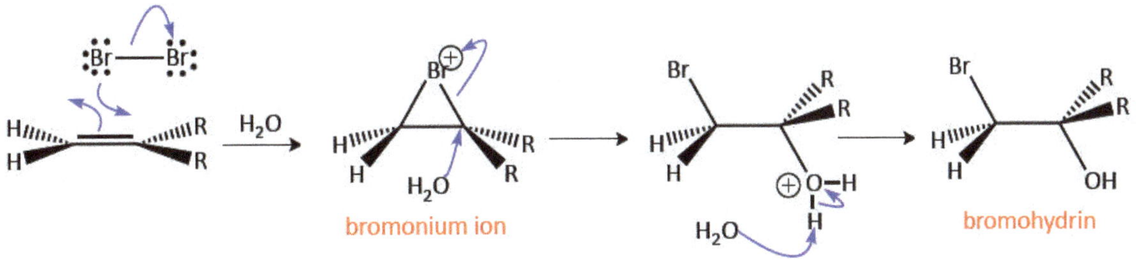

If Cl_2 and water are used, a chlorohydrin is formed. These *halohydrins* are molecules with neighboring halogens and hydroxyl groups.

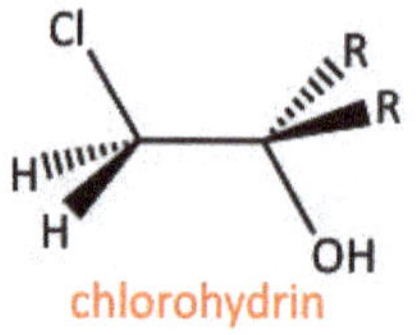

chlorohydrin

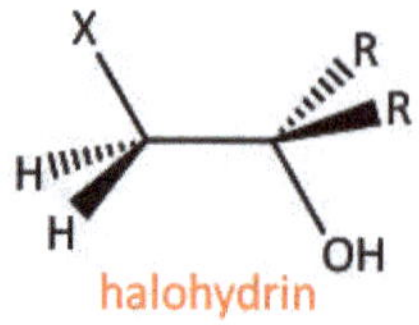

halohydrin

If instead of water, an alcohol is used as the solvent, it leads to a haloether as the product. The alcohol acts as a weak nucleophile.

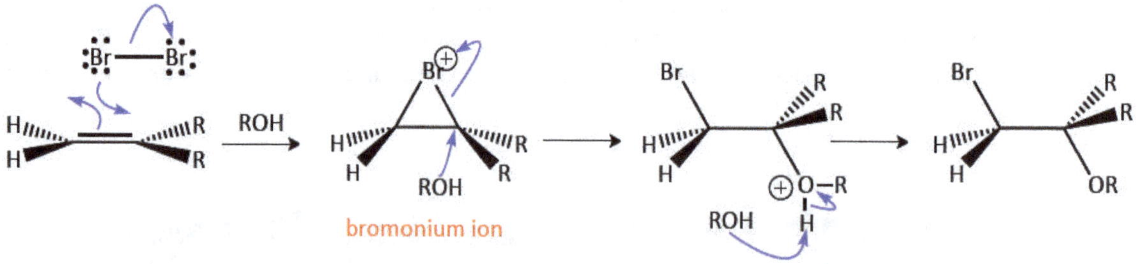

11.11.3 Halogenation of alkynes

If an alkyne with two pi-bonds is halogenated, the addition can happen twice.

The first addition of X_2 to the alkyne results in a dihaloalkene in a mixture of E and Z isomers.

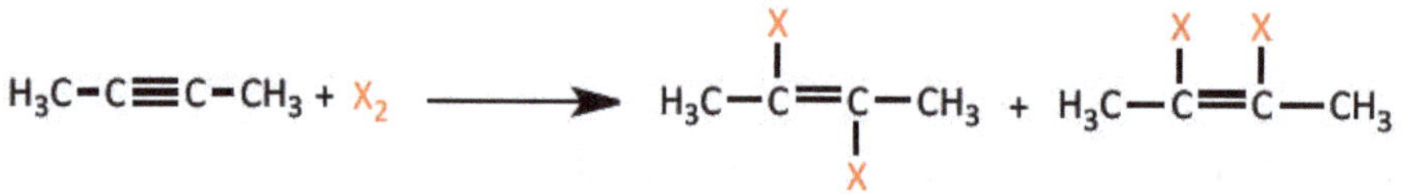

But, the alkene pi-bond further reacts to form tetrahaloalkanes if there is an excess of halogen.

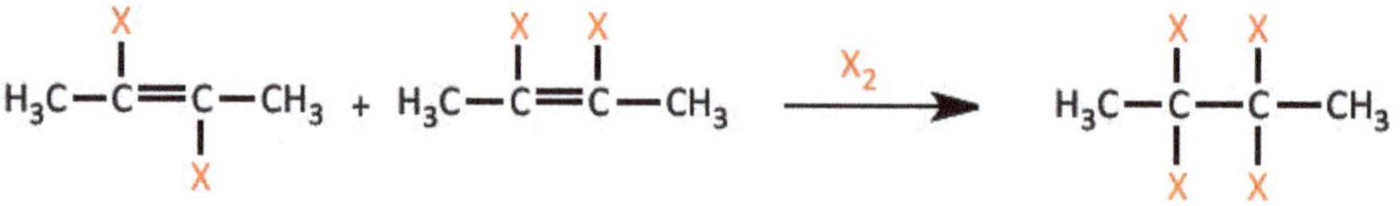

11.5 Draw the products of the following reactions.

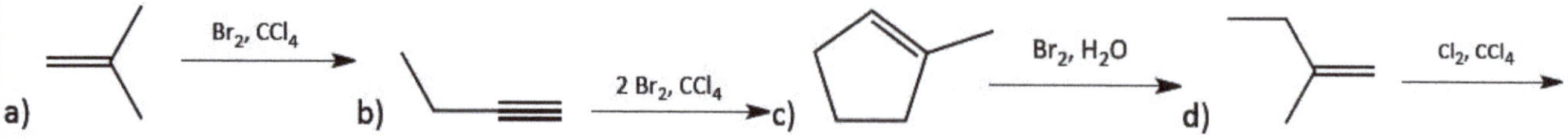

11.12 Epoxidation

An epoxide is a three-membered ring of two carbon atoms and one oxygen atom.

New: epoxide
Old: oxirane

Epoxides can be synthesized from alkenes by the addition of a peroxyacid, also called a peracid. A peroxyacid looks much like a carboxylic acid, but it contains three oxygen atoms instead of two. Any time a compound is encountered with "extra" oxygen atoms in it, that compound is probably an oxidizer.

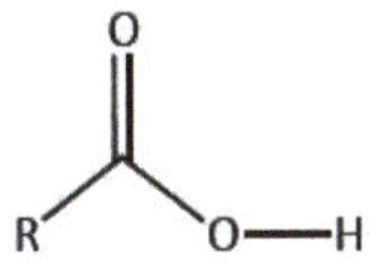

Carboxylic acid (RCO_2H)

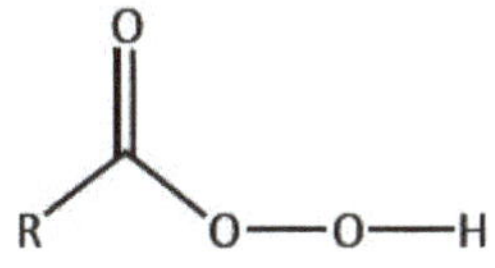

Peroxyacid (RCO_3H)

The epoxidation reaction is usually run in an inert, non-nucleophilic solvent in order to avoid an unwanted side reaction. Chlorinated solvents like carbon tetrachloride, CCl_4, or methylene chloride, CH_2Cl_2 are good solvents for this reaction. Often, MCPBA is the peroxyacid used in an epoxidation reaction. With its chloro group, it is quite soluble in the needed chlorinated solvents.

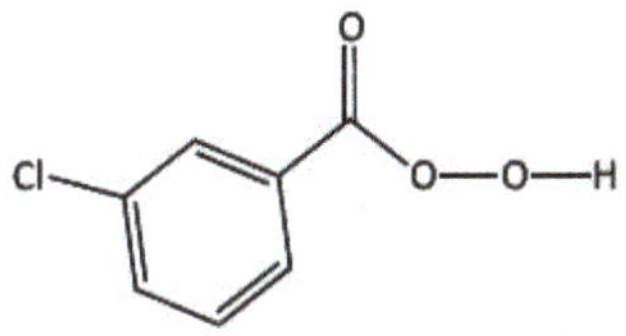

meta-chloroperoxybenzoic acid
(MCPBA)

Epoxidation Example:

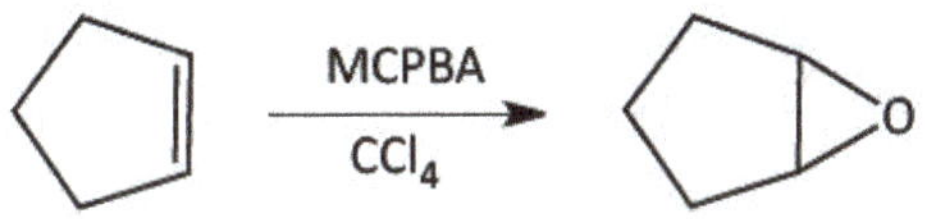

11.13 Anti-hydroxylation-A triangle reaction

If instead of using MCPBA in a chlorinated solvent, a peroxyacid in water is used, the epoxide formed can further react. This is a triangle reaction. In this reaction an acid (the peroxyacid) is present. Once the epoxide is formed, the peroxyacid can protonate the epoxide. The epoxide is holding a sign overhead announcing it has positive money. Water, a weak nucleophile, can backside attack the more substituted carbon atom resulting in a glycol with anti-stereochemistry.

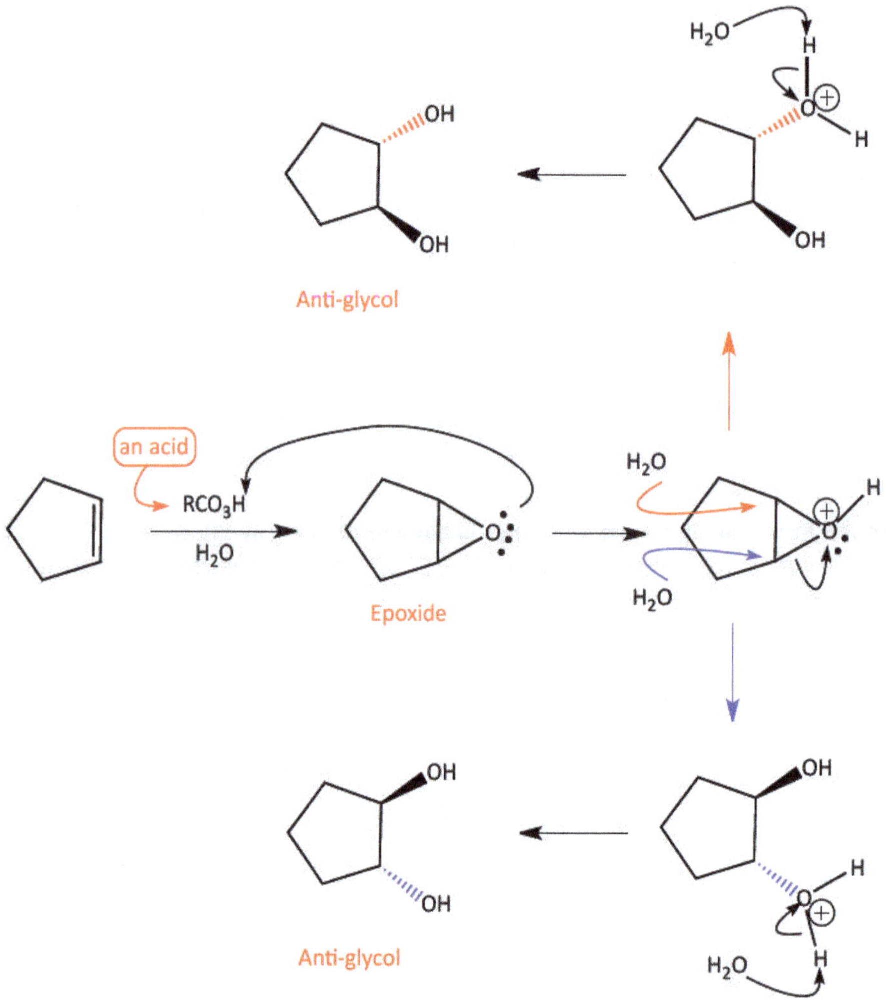

11.6 Draw the products of the following reactions.

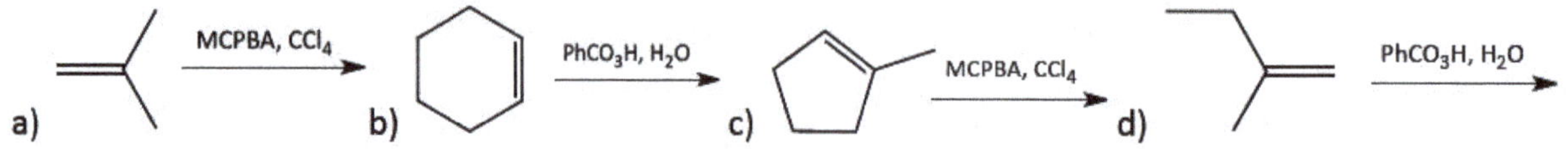

11.14 Cyclopropanation

A carbene is a neutral reactive intermediate of a divalent carbon atom with a lone pair of electrons. The lone pair of electrons makes a carbene a nucleophile. Carbenes are also electrophilic because they contain an empty p-orbital. Carbenes are used to make cyclopropane rings. There are a couple of ways to generate carbenes.

11.14.1 Alpha elimination making dihalocarbenes

If a proton (hydrogen atom) and a good leaving group are on the same carbon atom, alpha elimination can occur. *Alpha elimination* is when two groups are removed from the same carbon atom. In this case, a good base like hydroxide is used to deprotonate the carbon while simultaneously, the leaving group leaves. This generates a carbene.

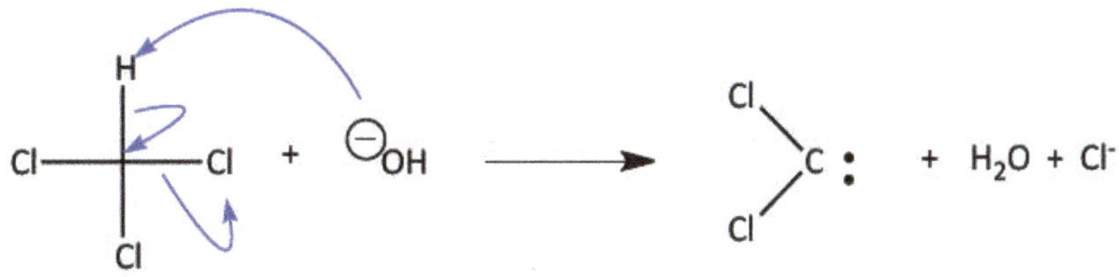

The carbene can react with an alkene to form a cyclopropane ring.

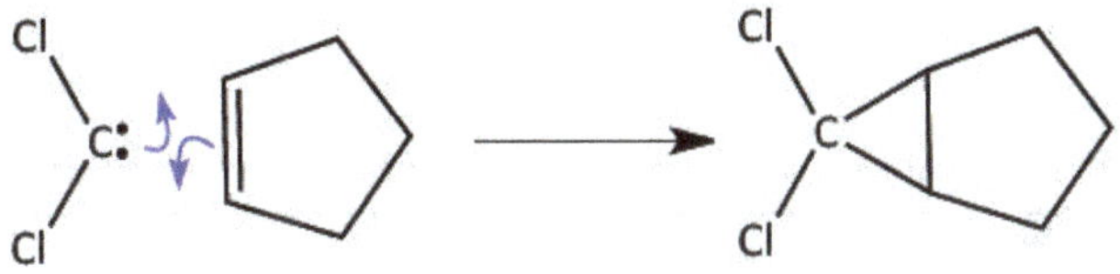

11.14.2 Diazomethane

To make the simplest of the carbenes, methylene (:CH_2), diazomethane can be used. Azo means "containing nitrogen. Diazomethane has two nitrogen atoms on one carbon. These two nitrogen atoms make up the world's greatest leaving group, nitrogen gas, N_2. The nitrogen gas leaving group is very stable and simply bubbles away as air.

Diazomethane
Contains the world's greatest leaving group, N_2!

Methylene (a carbene)

Very stable N_2. This is air!

The methylene carbene can then react with an alkene to form a cyclopropane.

carbene alkene cyclopropane

Although very effective, diazomethane is not necessarily the best way to make cyclopropane rings. Diazomethane is quite reactive. It can explode if heated above 100°C. A blast shield should be used when it is used in reactions. Diazomethane also has an acute toxicity and should be handled with care. Also, diazomethane is so reactive that an unwanted side reaction can also occur. The methylene carbene might insert between carbon-carbon bonds or between carbon-hydrogen bonds.

Unwanted side reactions

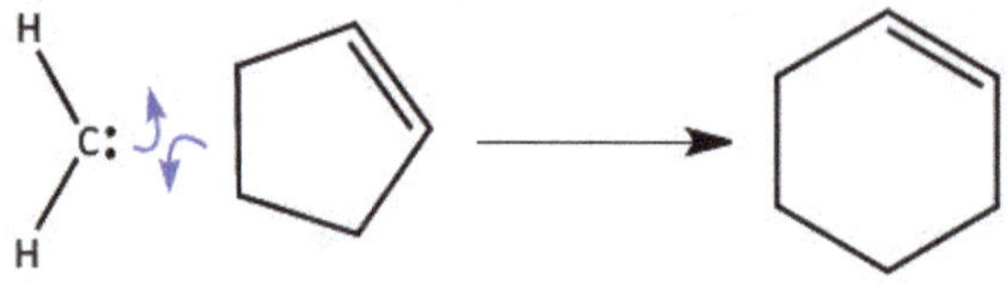

Insertion into C-C bond.

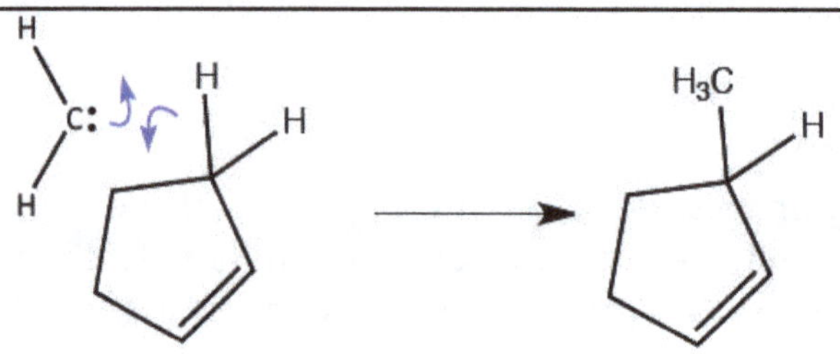

Insertion into C-H bond.

Because of these unwanted side-reactions when diazomethane was used, a gentler method was developed to synthesize cyclopropane rings. This method is the Simmons-Smith reaction. It uses what is called the Simmons-Smith reagent, which is a carbenoid. A carbenoid is like a carbene, but not exactly the same thing.

11.14.3 Simmons-Smith Reaction

In this reaction, diiodomethane is reacted with zinc-copper couple. This makes a carbene-like species called a carbenoid. This carbenoid can be thought of as $:CH_2$, methylene, but it is not as reactive as the methylene carbene made from diazomethane. The advantage is the unwanted side reactions do not occur, only the cyclopropanation.

$$CH_2I_2 + Zn(Cu) \longrightarrow ICH_2ZnI$$

Carbenoid — Can think of this as "$H_2C:$"

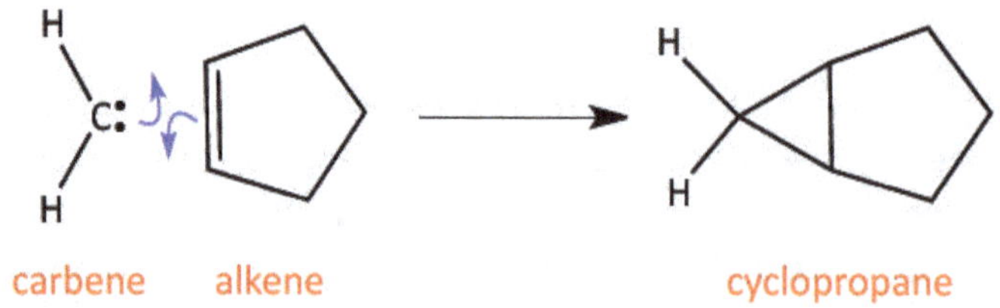

11.7 Draw the products of the following reactions.

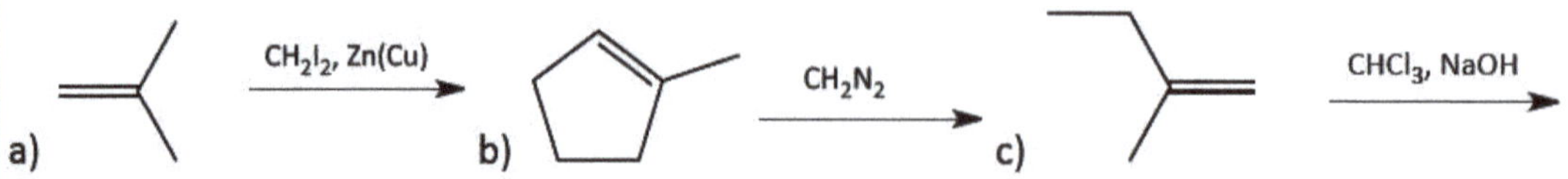

11.15 Syn-Addition: Giving it the horns

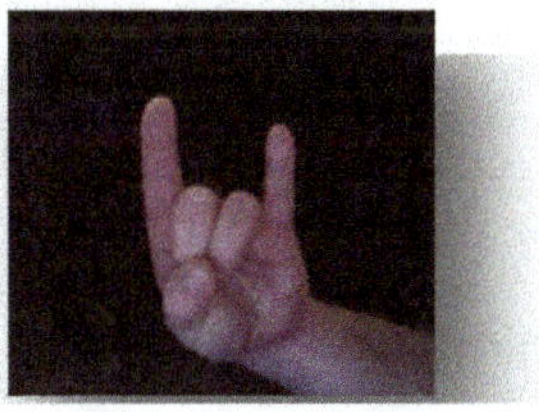

In some alkene addition reactions, the product is the result of syn-addition. This means the two groups added across the pi-bond both end up on the same side. I call such reactions "Giving it the horns" reactions. I do so because I use a mental image of a bull with its horns goring the alkene. Its two horns go into the alkene and break off leaving the horns sticking out on the same side of the alkene.

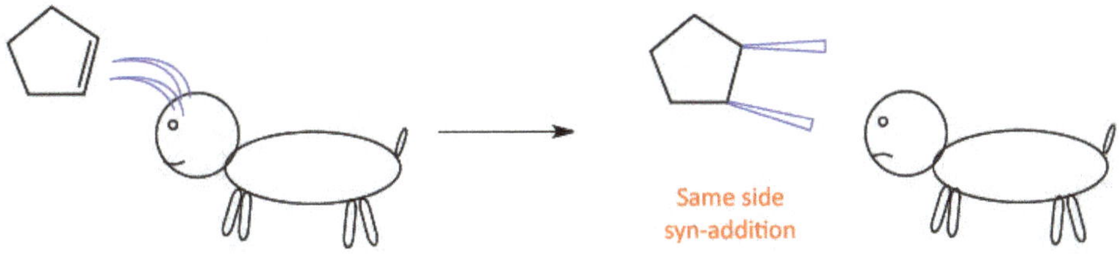

Figure 11.13 Syn-addition. Giving it the horns.

11.16 Hydrogenation-Giving it the horns.

11.16.1 Of an alkene

Hydrogenation is the reaction of an alkene with hydrogen gas in the presence of a metal catalyst. The metal catalyst used is often platinum (Pt), palladium (Pd), or Nickel (Ni). The hydrogen adds across the pi-bond turning the alkene into an alkane. The two hydrogen atoms add to the same side of the alkene in a syn-addition fashion.

Syn-addition. Giving it the horns

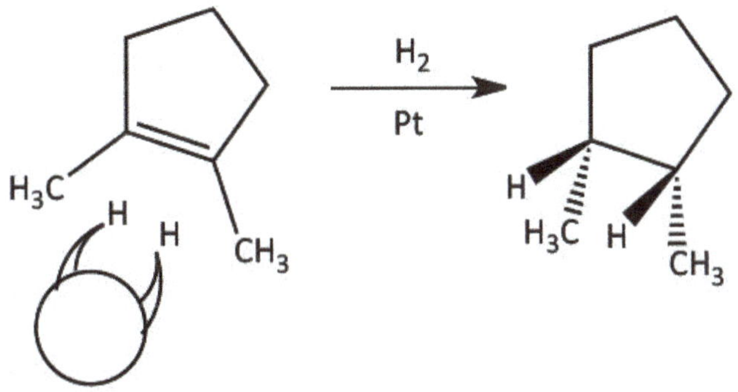

For the mechanism of hydrogenation, the hydrogen first adds to the metal catalyst. The alkene then adsorbs onto the metal catalyst. One carbon atom of the alkene inserts into the metal-hydrogen bond and then the other. Since both hydrogen atoms come from the same metal catalyst surface, both hydrogen atoms end up on the same side of the alkene. This is syn-addition. The hydrogen atoms on the metal catalyst are "Giving it the horns."

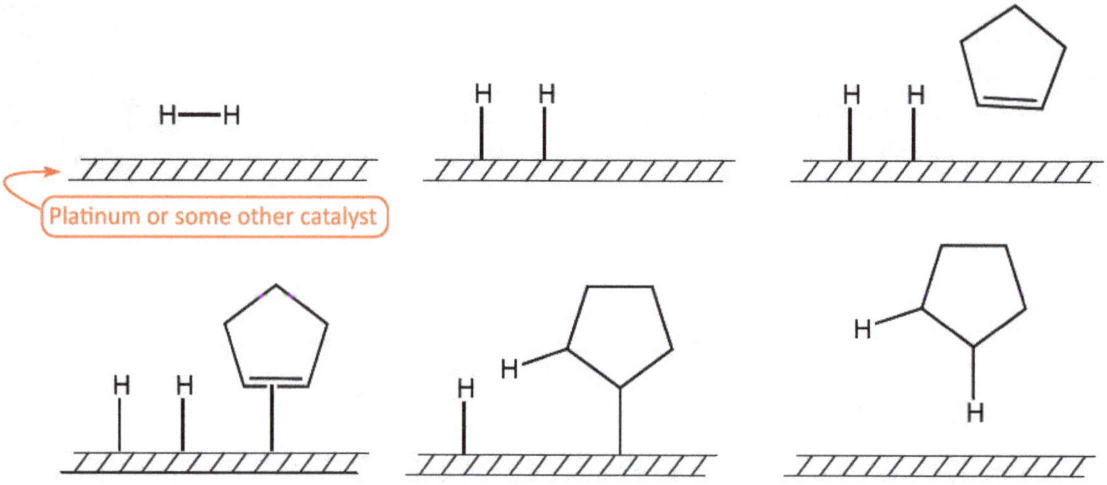

Figure 11.14 Syn-hydrogenation of alkene on a metal catalyst

11.16.2 Of an alkyne

Hydrogenation to an alkane

Alkynes have two pi-bonds compared to the one in an alkene. Therefore, the hydrogenation of an alkyne can occur twice to form a saturated alkane. Usually, good metal catalysts like Pt, Pd, or Ni are used. The hydrogen atoms on the metal catalyst are "Giving it the horns."

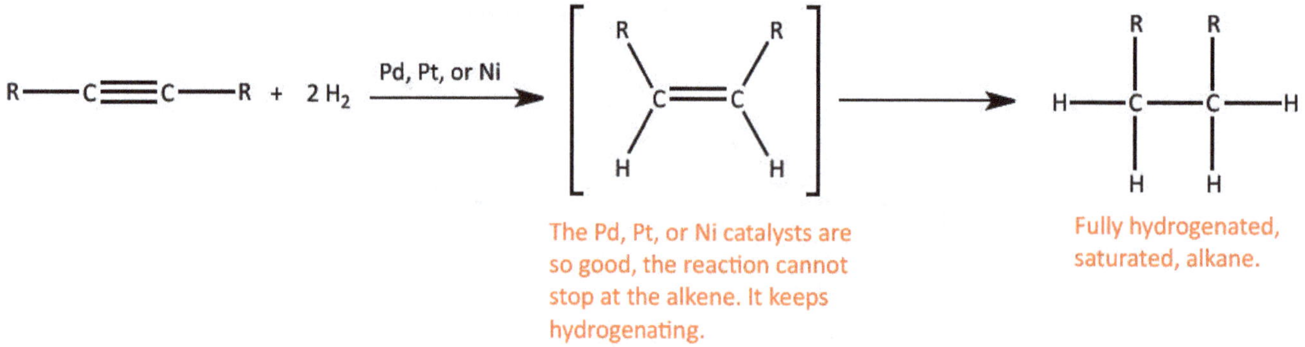

Hydrogenation to cis-alkene – Giving it the horns.

But, what if you don't want to hydrogenate an alkyne all the way down to an alkane, but you want to stop halfway at the alkene? Is there a way to only hydrogenate halfway? There is. The key is to use a less effective, partially deactivated catalyst, a poisoned catalyst. Lindlar catalyst is a poisoned catalyst. It is palladium (Pd) coated with barium sulfate powder ($BaSO_4$) in the presence of quinoline. When the hydrogenation of an alkyne takes place in the presence of Lindlar catalyst, an alkene with syn-stereochemistry is formed. It is syn-addition because both hydrogen atoms again come from the same Pd catalyst surface. The hydrogen atoms on the metal catalyst are "Giving it the horns." If an internal alkyne is hydrogenated in this way, it leads to *cis*-alkenes.

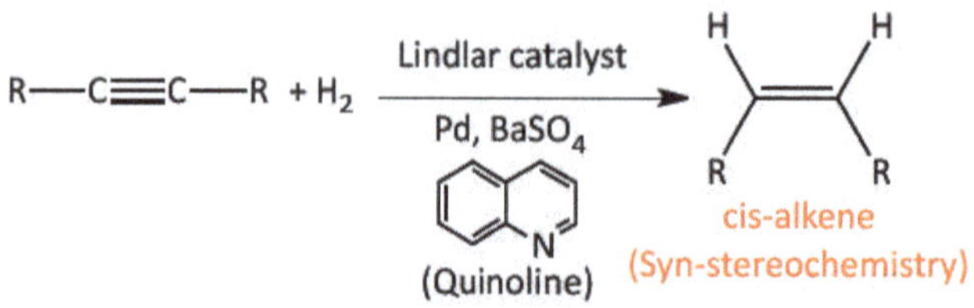

A newer reagent, nickel boride, Ni_2B, is becoming more popular to make cis-alkenes from alkynes. It is easier to make and works a little better than Lindlar catalyst.

Hydrogenation to trans-alkene

So, this begs the question, can we start with an alkyne and end up with a *trans*-alkene? We can. But, we need to change the mechanism of how the hydrogenation occurs. We can no longer use a poisoned metal catalyst because, if we do, both hydrogen atoms will add by "Giving it the horns" making *cis*-alkenes. One hydrogen atom needs to add at a time to be able to get *trans*-alkenes.

Sodium metal in liquid ammonia can turn an alkyne into a *trans*-alkene.

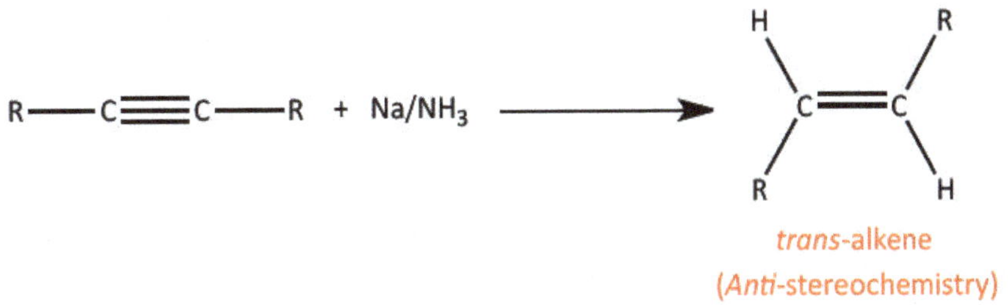

Let's walk through the mechanism to see why this works.

1. Overall, this reaction is a reduction. Reductions need electrons. It would be nice if we could simply buy a bottle of electrons. Since we cannot, we need to make some. Reacting sodium and ammonia is a good way to make a beaker full of electrons.

   For this reaction, sodium metal is dissolved in liquid ammonia. Ammonia has a boiling point of -33°C. It is a gas at room temperature. In order to turn the gaseous ammonia into a liquid solvent, it is cooled in a special chamber using dry ice. Once the ammonia is a liquid, pieces of solid sodium metal are added. Once enough sodium is added, it turns a beautiful blue color. This blue color means we have solvated electrons.

$$NH_3 + Na \longrightarrow Na^+ + NH_3 \cdot e^-$$

blue color, solvated electron

   How does this work? Sodium metal is an alkali metal. It wants to give up one electron to become Na^+. This is the source of the electrons. Ammonia has the perfect molecular shape to hold these electrons. Ammonia's trigonal pyramidal shape forms a pocket with the positively charged hydrogen atoms that can hold a negatively charged electron. These freed electrons can be thought of as "electrons in a bottle".

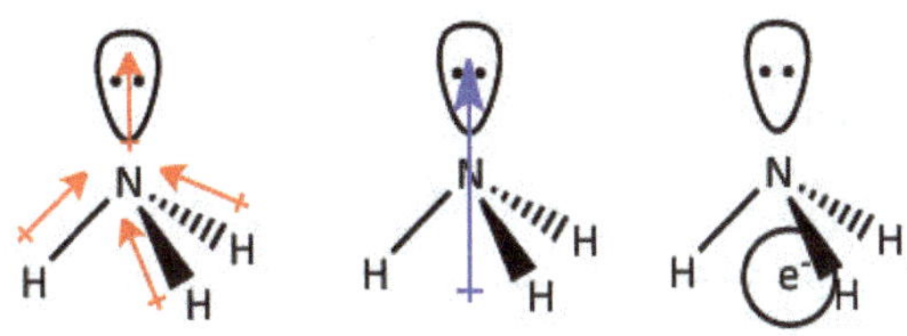

2. The free electron attacks a carbon atom of the alkyne. This would put too many electrons on the carbon atom, so the pi-bond leaves to form a lone pair of electrons making an anionic intermediate. The species formed is a radical anion because it has both a single electron (the radical) and a lone pair of electrons (the anion). Since the free electron attacking and the lone pair of electrons leaving are both negatively charged, they get as far away from each other as possible, in a trans configuration.

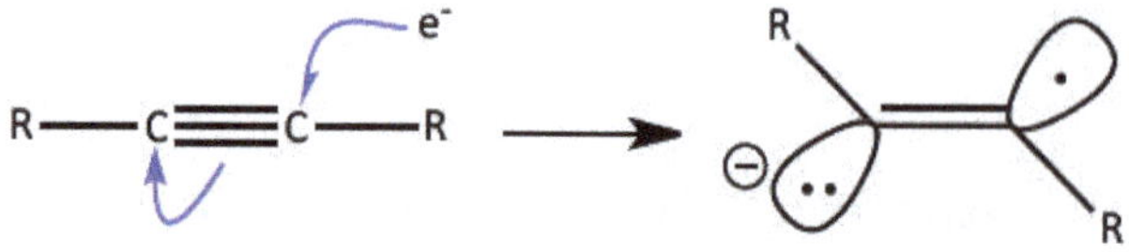

radical anion

3. In this mechanism, whenever an anionic lone pair of electrons is formed, the lone pair goes out and grabs a proton. In this mechanism, they get the proton from ammonia and form a vinyl radical. This is a radical (a single electron) on a double bond.

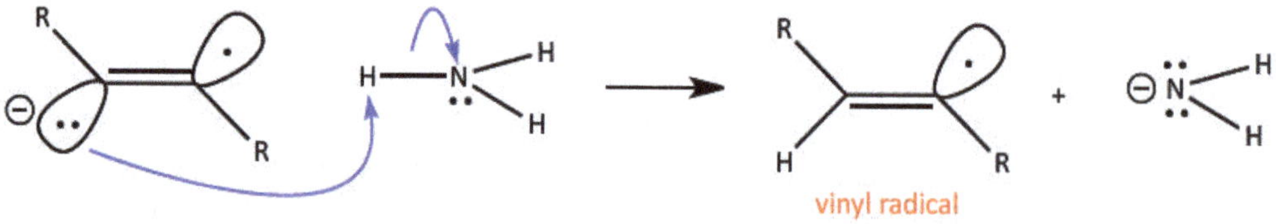

vinyl radical

4. Once the neutral vinyl radical is formed, another electron from the solvated electrons attacks. It joins the single electron to make a lone pair anion. Since this anion is directly on a double bond, it is called a vinyl anion.

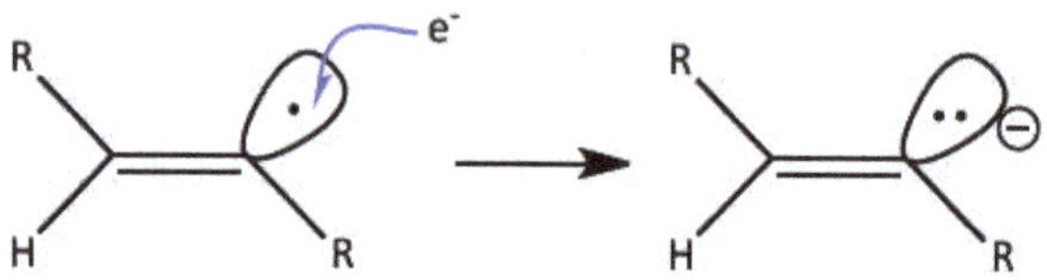

vinyl anion

5. Again, whenever the lone pair anion is formed, it grabs a proton. The ammonia once again acts as a proton source. This results in the desired *trans*-alkene.

trans-alkene

Instead of using sodium and ammonia, another reagent that can form a *trans*-alkene from an alkyne is lithium aluminum hydride, $LiAlH_4$ or LAH.

11.8 Draw the product of the following reactions.

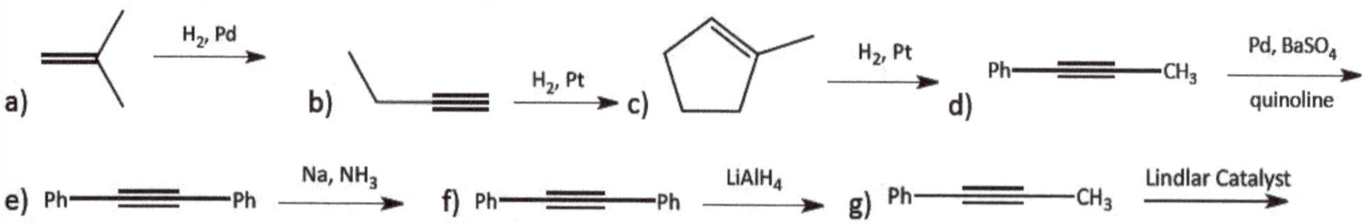

11.17 Syn-hydroxylation

11.17.1 With OsO_4

In syn-hydroxylation, an alkene is turned into a 1,2-diol, a glycol. This is another example of "Giving it the horns." In this case, osmium tetroxide (OsO_4) acts like the bull with two oxides as the horns. Osmium tetroxide reacts with the pi-bond of the alkene to make an osmate ester. The osmate ester can be oxidized with hydrogen peroxide and hydroxide to make a glycol with syn-stereochemistry. The bull has given it the horns. OsO_4 is a catalyst in this reaction. The reaction starts with OsO_4 and ends with OsO_4. Since it is regenerated, only a small amount is needed in this reaction.

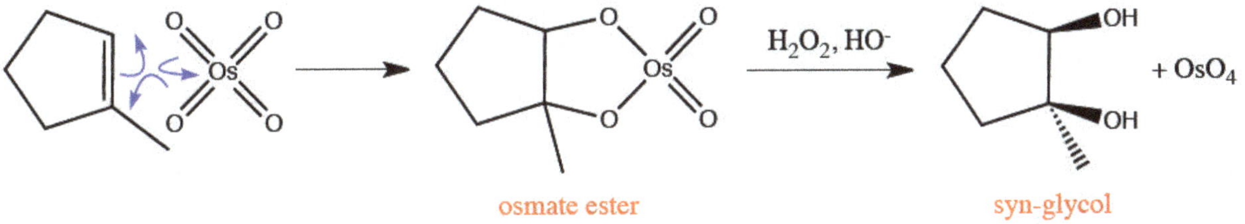

One drawback of this reaction is that OsO_4 is expensive, toxic, and volatile. Volatile means it easily evaporates. This is especially problematic if that compound is toxic!

11.17.2 With $KMnO_4$

A safer, cheaper, and easier method to perform a syn-hydroxylation is to a cold, dilute, sometimes neutral aqueous solution of potassium permanganate ($KMnO_4$). The active species is the permanganate (MnO_4^-). The permanganate acts like the bull and uses its oxides as horns. The mechanism is similar to the reaction with OsO_4. In this case, a manganate ester is formed. Following the addition of water and hydroxide, a glycol is formed with syn-stereochemistry.

Potassium permanganate is not a catalyst because it is not regenerated in this reaction. It is reduced. Notice, in the reaction with OsO_4, the oxidizer H_2O_2 is used to add more oxygen atoms to regenerate the OsO_4 catalyst. In the reaction with $KMnO_4$, no oxidizer is used. That is why the $KMnO_4$ is not a catalyst. This is OK since $KMnO_4$ is so cheap.

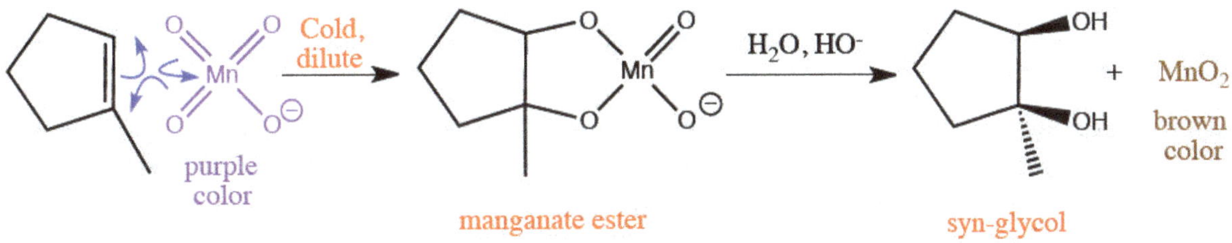

Potassium permanganate is a quick and easy chemical test for alkenes. Aqueous potassium permanganate is a purple solution. We know that it reacts with alkenes. If an unknown organic compound is added to the purple potassium permanganate and nothing happens, we could reason that the unknown organic compound is NOT an alkene. On the other hand, if the purple potassium permanganate solution turns brown, we would know a reaction occurred because the brown MnO_2 was formed and the unknown organic compound that was added must be an alkene.

11.18 $KMnO_4$ oxidation of alkynes

The same reaction can happen with the two pi-bonds of an alkyne. Since there are two pi-bonds, the dihydroxylation can happen twice, resulting in a species with four hydroxyl groups on it. This is unstable. Two water molecules are eliminated resulting in an α-diketone (alpha-diketone). Under neutral $KMnO_4$ conditions, the α-diketone is the product.

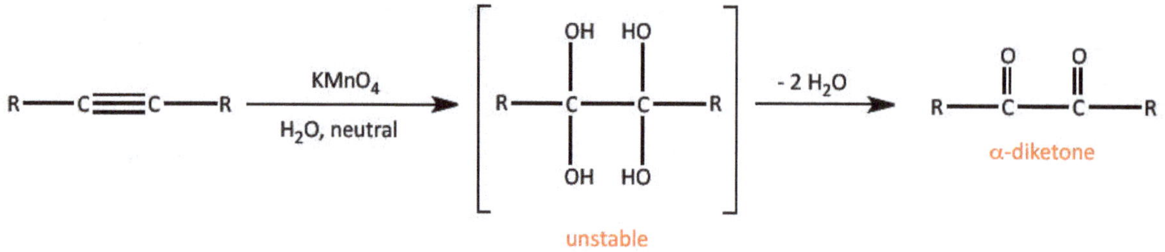

The double dihydroxylation of the alkyne makes sense. Look closely how the two molecules of water are lost leaving only two oxygen atoms behind on neighboring carbon atoms.

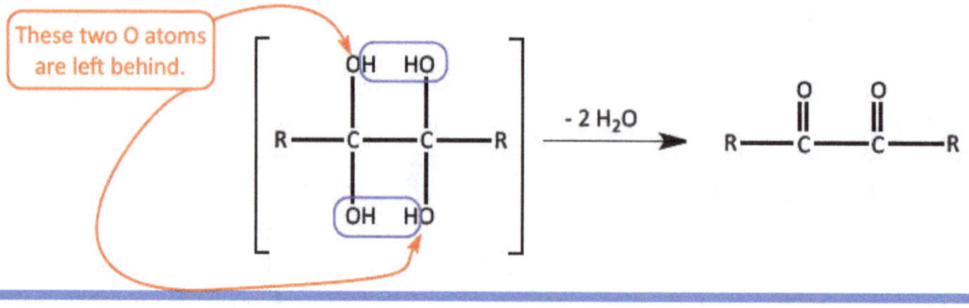

11.9 Draw the products of the following reactions.

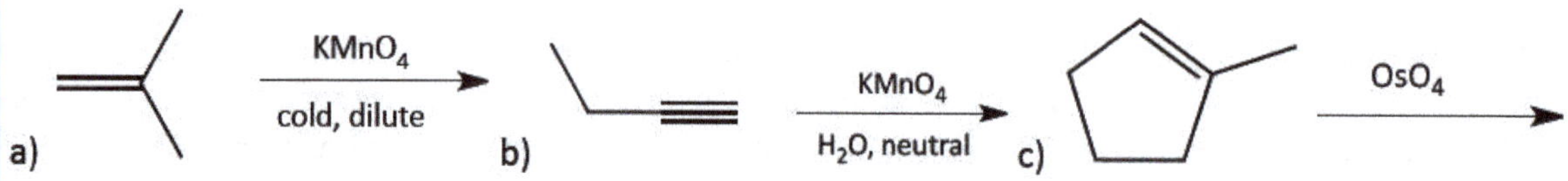

11.19 Further oxidation with $KMnO_4$

11.19.1 Of alkenes

If the alkene reaction with potassium permanganate is not conducted under dilute and cold conditions but is instead run under ***more basic*** conditions and ***warm and concentrated*** conditions, further oxidation can occur. When the glycol is further oxidized, the carbon-carbon sigma bond of the glycol is broken and the single carbon-oxygen bonds of the glycol turn into carbon-oxygen double bonds (an oxidation). $KMnO_4$ is a pretty strong oxidizer. If an aldehyde is first made by this oxidation, it is further oxidized to a carboxylic acid.

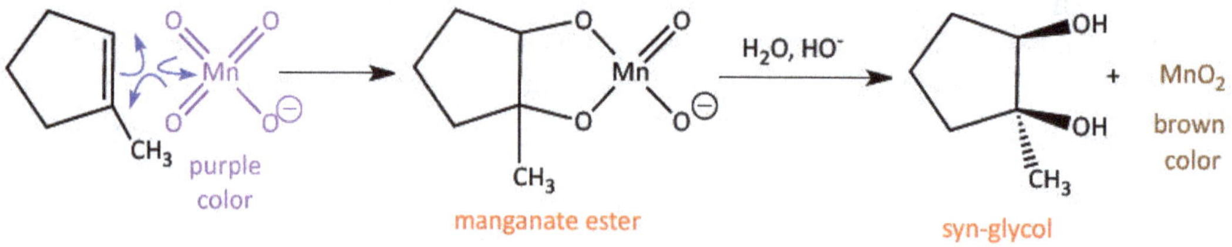

and the further oxidation of the glycol with hot, concentrated $KMnO_4$

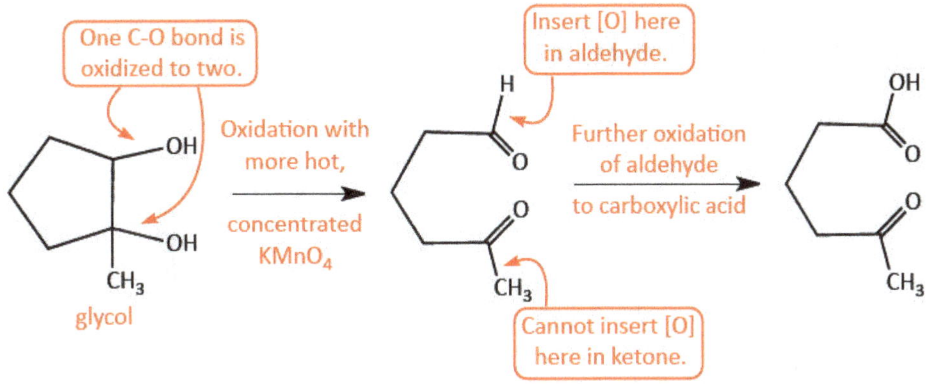

Overall, this reaction can be thought of as chopping the alkene in half and inserting two oxygen atoms. If the result is a ketone, nothing more happens to that carbon atom. If the result is an aldehyde, further oxidation to a carboxylic acid occurs.

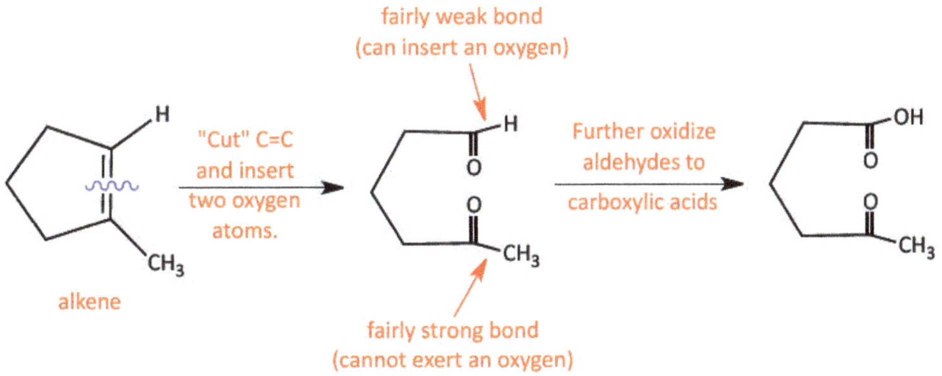

11.19.2 Of alkynes

When $KMnO_4$ reacts with an alkyne under more aggressive conditions, in the presence of hydroxide base and heat, the α-diketone is produced like under less aggressive neutral conditions, but it is then further, fully oxidized. The C-C bond between the two carbonyl groups is cleaved followed by further oxidation resulting in two carboxylic acids. With three C-O bonds, the carboxylic acid is fully oxidized and cannot be oxidized further.

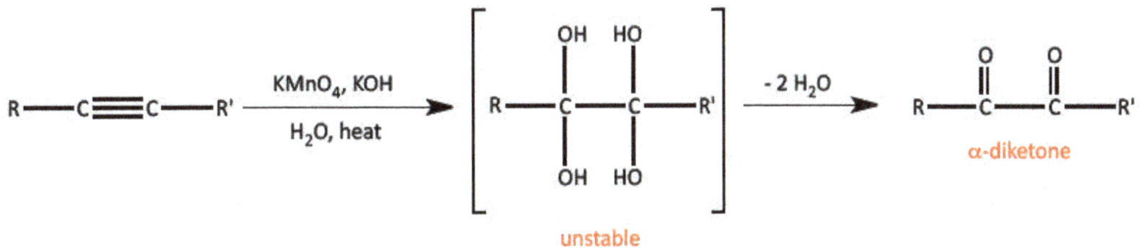

and the further oxidation of the γ-diketone with hot, concentrated $KMnO_4$

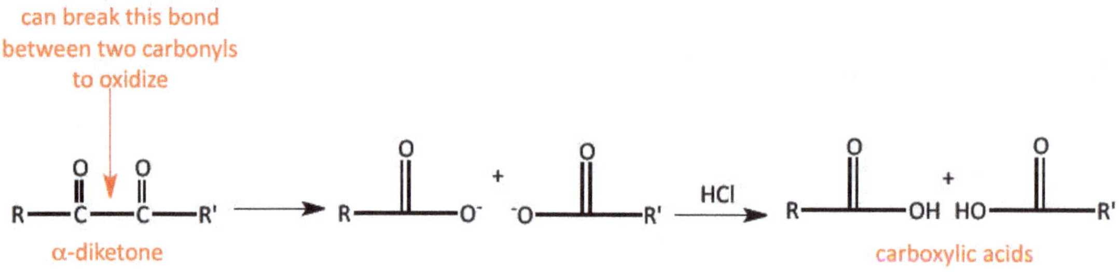

Overall, the cleavage can be thought of as…

or

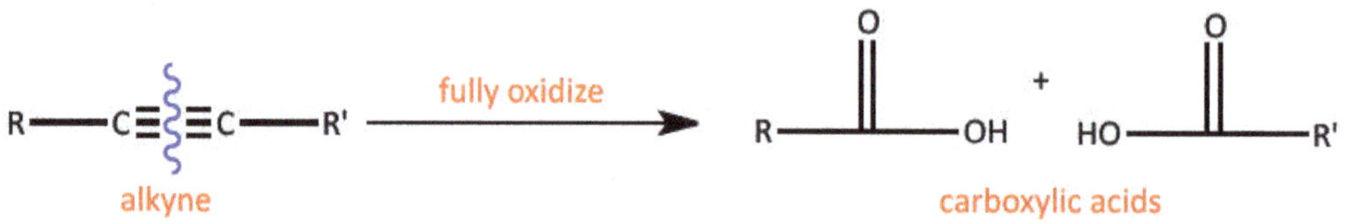

Notice that in this oxidation, three carbon-carbon bonds of the alkyne turn into three carbon-oxygen bonds of the carboxylic acid.

11.20 Ozonolysis

11.20.1 Of alkenes

Ozonolysis is very similar to oxidation by $KMnO_4$ in warm, concentrated solution. The difference is that if an aldehyde is formed, it is *not* further oxidized to a carboxylic acid. Notice that in this oxidation, two carbon-carbon bonds of the alkene turn into two carbon-oxygen bonds of the carbonyl.

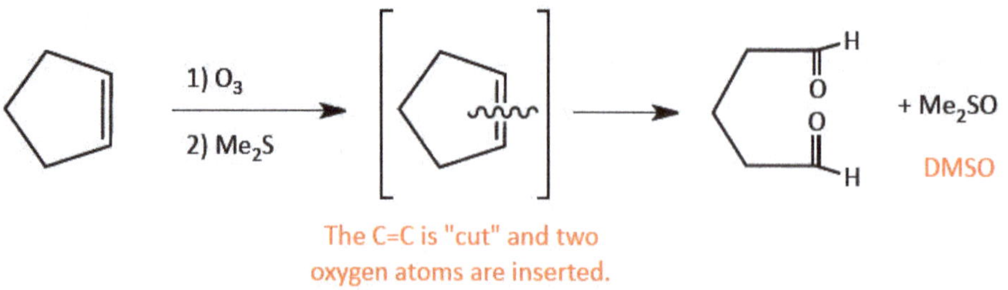

The mechanism for this reaction does not really help us learn the reaction. If you are interested in it, you can look it up elsewhere. Ozone is a compound with three oxygen atoms (O_3). Two of the three oxygen atoms are in the final product. The third oxygen atom ends up on the Me_2S to make Me_2SO.

11.20.2 Of alkynes

The reaction of an alkyne with ozone results in the same products as the full oxidation with $KMnO_4$ under basic, hot conditions. The carbon-carbon triple bond is cleaved and each carbon atom is fully oxidized resulting in two carboxylic acids. Again, notice that in this oxidation, three carbon-carbon bonds of the alkyne turn into three carbon-oxygen bonds of the carboxylic acid.

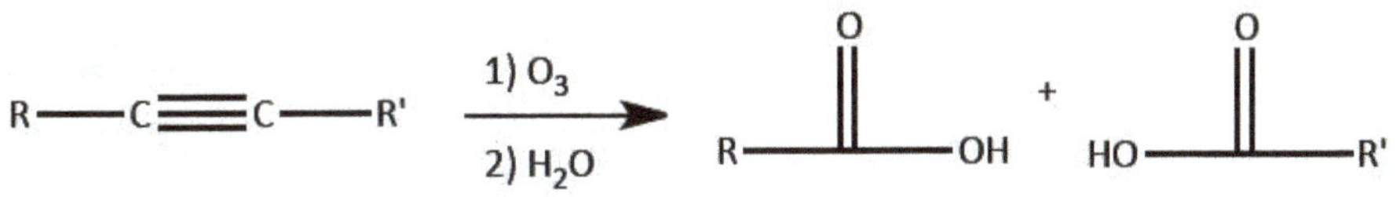

11.10 Draw the products of the following reactions.

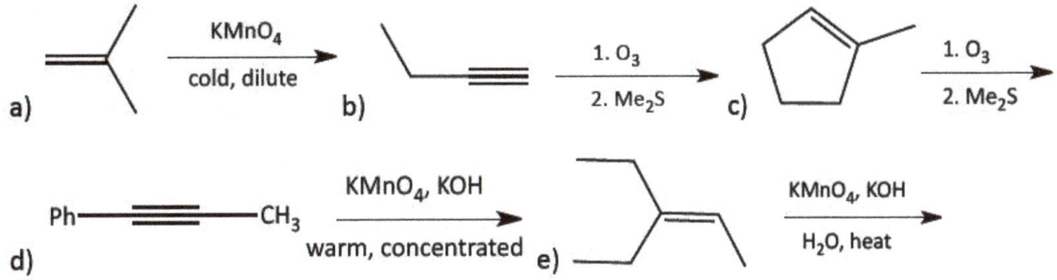

11.21 Polymers

Polymers (*poly* from Greek "many", *mer* from Greek, *méros* "part") are molecules that are long chains made of repeating units. The molecules that are the repeating units are called monomers. Many of our synthetic materials are polymers. Plastic materials are polymers. Alkenes can be monomers.

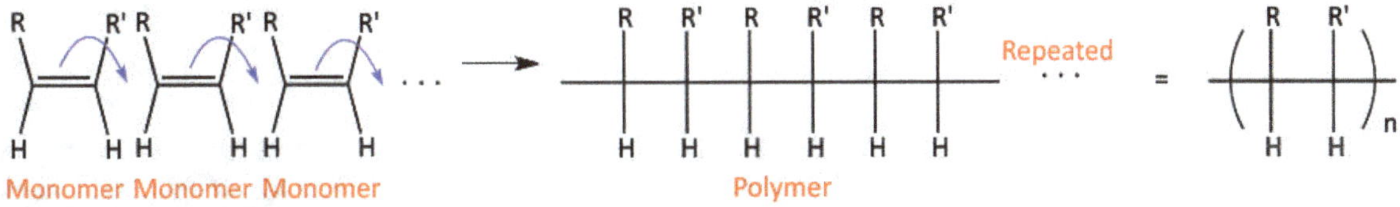

Polymers are usually represented with a simplified form. The value for "n" is usually some very large number. It represents the number of repeating monomers in the polymer.

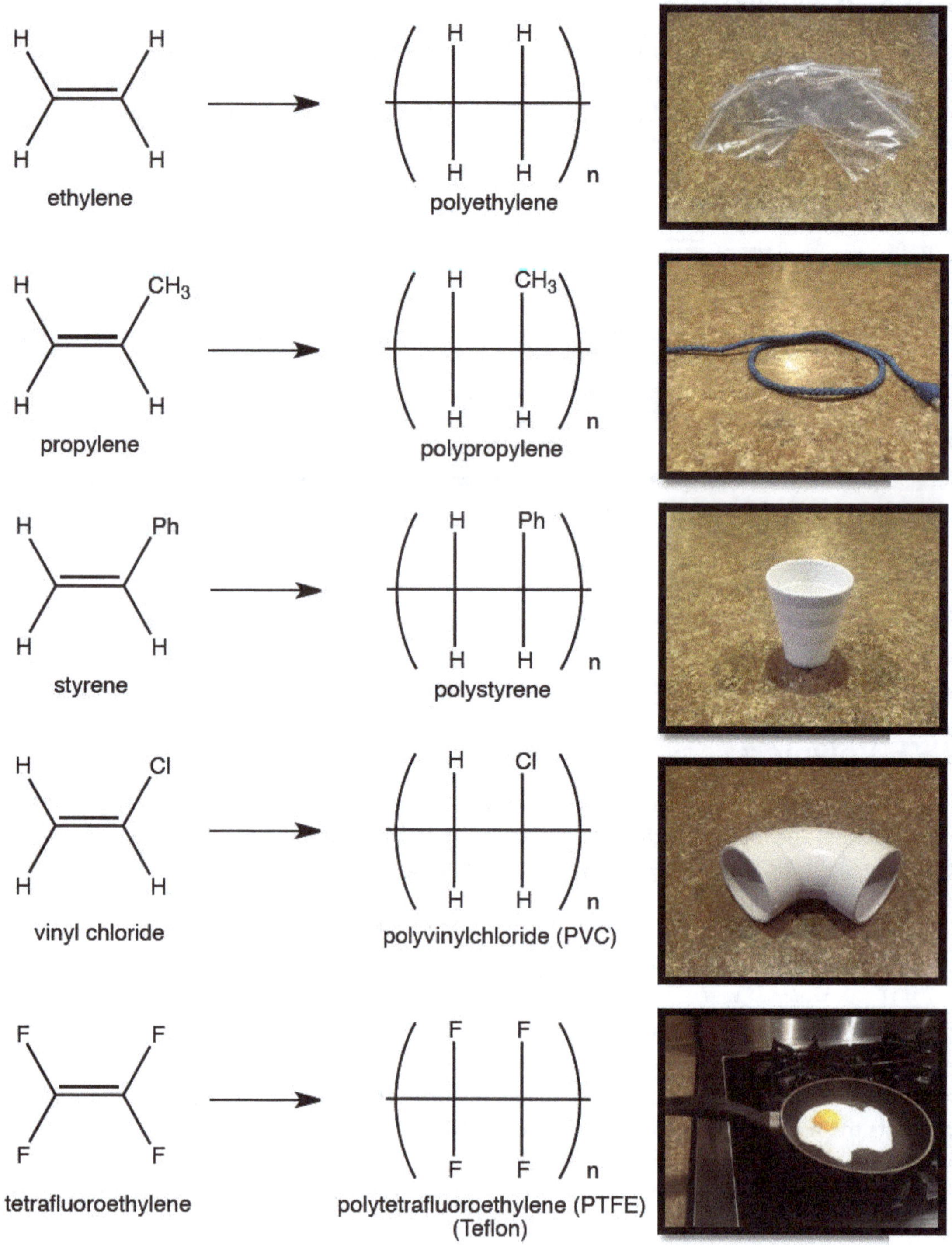

Figure 11.15 Various polymers

Alkenes undergo addition polymerization using three different mechanisms. These mechanisms are described by their reactive intermediates.

11.21.1 Anionic Polymerization

If an anion is present, like hydroxide, anionic polymerization of the alkene can occur. For an anionic polymerization to occur, the alkene must contain anion-stabilizing groups, also called electron-withdrawing groups. Carbonyl, nitro, and cyano groups are capable of stabilizing carbanions by resonance. Probably the most familiar anionic polymerization is using the monomer methyl α-cyanoacrylate. This compound can readily be purchased in a tube. Once the monomer touches something, like your fingers, a small amount of base that is usually present can attack the alkene and polymerize. Methyl α-cyanoacrylate is commonly called Super Glue. In the anionic mechanism for Super Glue, the anionic hydroxide attacks the alkene with the pi-electrons becoming a lone pair of electrons on a carbanion. This negative pair of electrons can be stabilized through resonance on both the carbonyl and the cyano (nitrile) group. Having two resonance stabilizing groups makes this polymerization occur rapidly.

First step

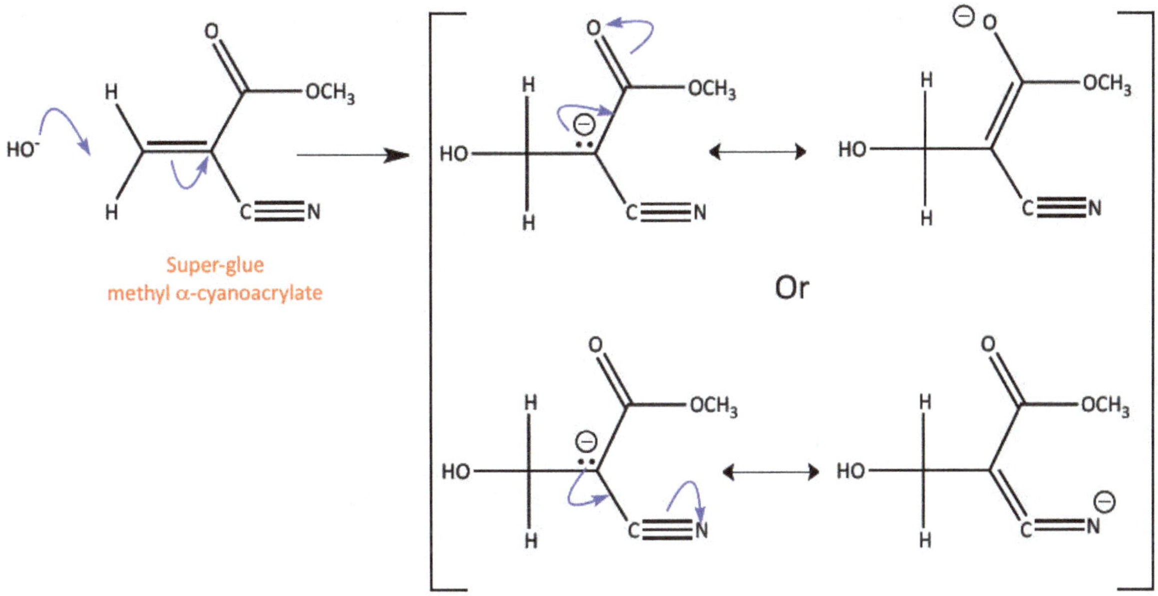

Subsequent steps

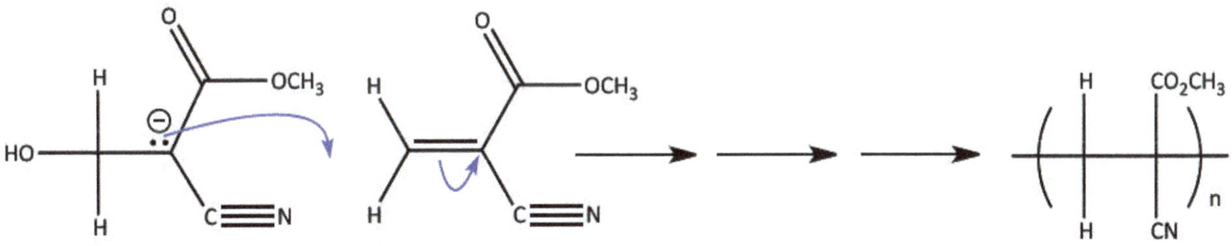

11.21.2 Cationic Polymerization

In order to help stabilize the carbocation intermediate, cationic polymerizations usually occur on alkenes with electron donating groups attached. A good electrophile is the initiator of this polymerization. A good Brønsted-Lowry acid (H^+) or a good Lewis acid ($AlCl_3$, BF_3) works well.

For the mechanism, the negative electrons of the alkene's pi bond attack the acid to form a carbocation. In the example, the carbocation is tertiary and very stable. Another alkene monomer attacks this electrophilic carbocation to form yet another. This process continues over and over.

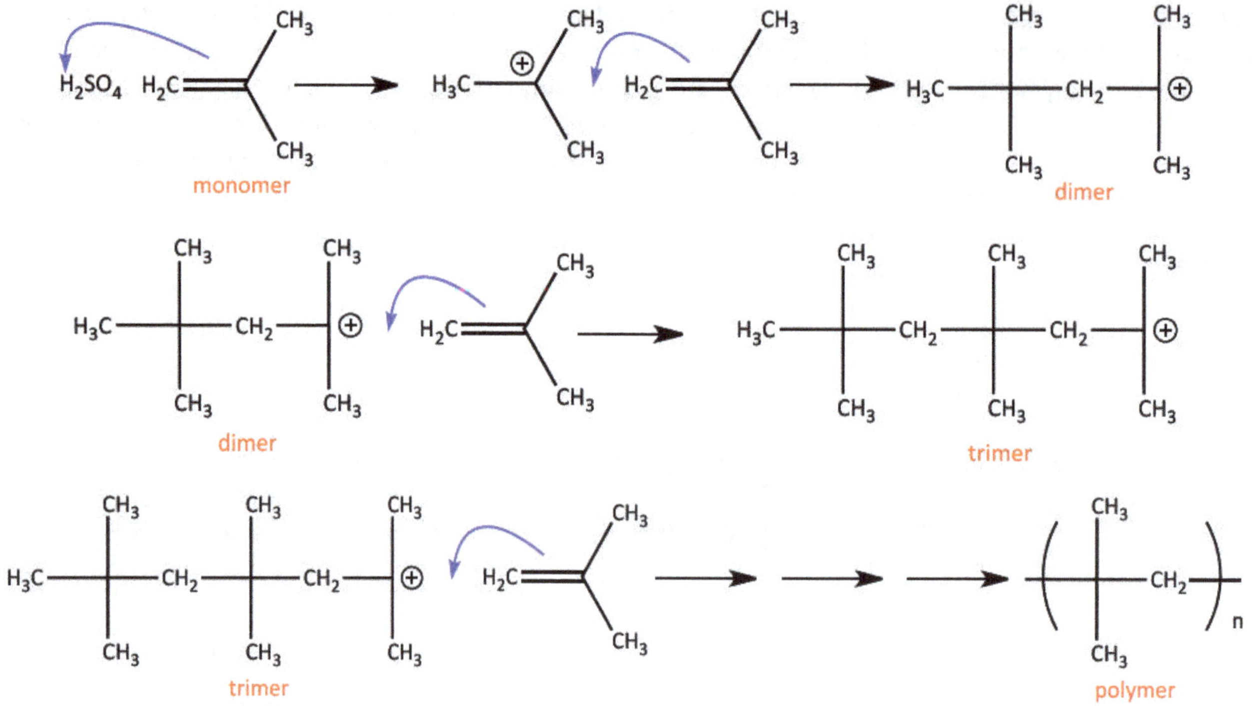

11.21.3 Free radical Polymerization

The third polymerization mechanism is a free radical polymerization. An initiation step is the first step of the free radical mechanism. Here, a peroxide is heated to form two oxygen free radicals.

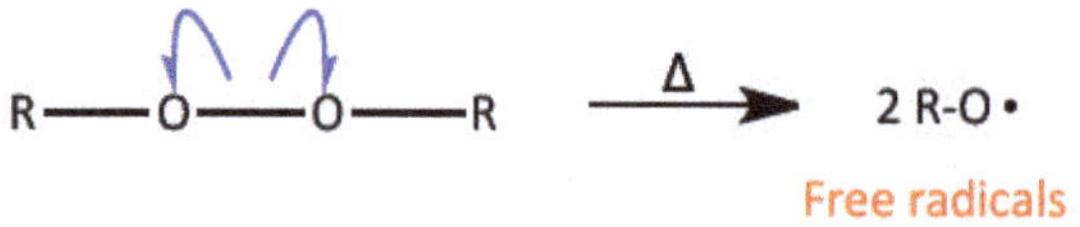

For the free radical mechanism to be successful, an alkene that can stabilize the free radical is needed. In this case, the radical ends up adjacent to a phenyl group making a benzylic free radical. This is quite stable because it can be stabilized by resonance. The reaction keeps repeating until the polymer is formed.

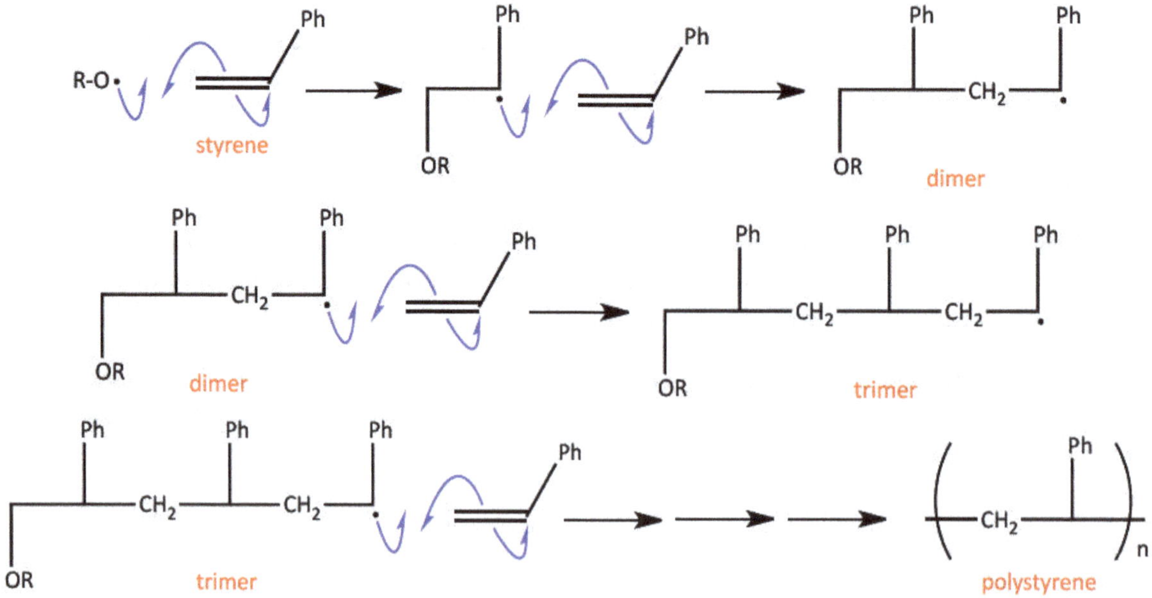

11.11 Draw the polymers that are made from the following alkene monomers?

a) 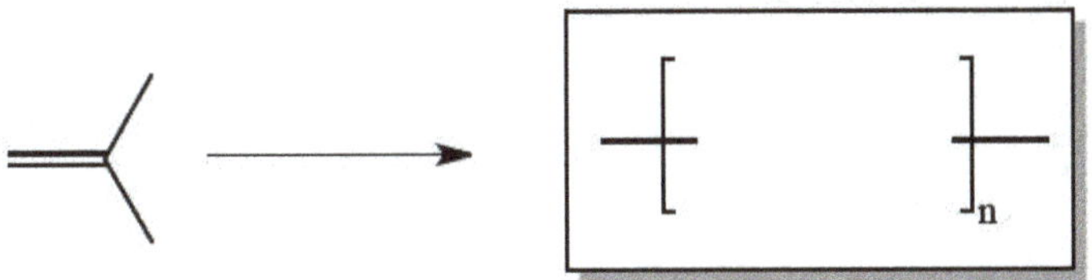

b) 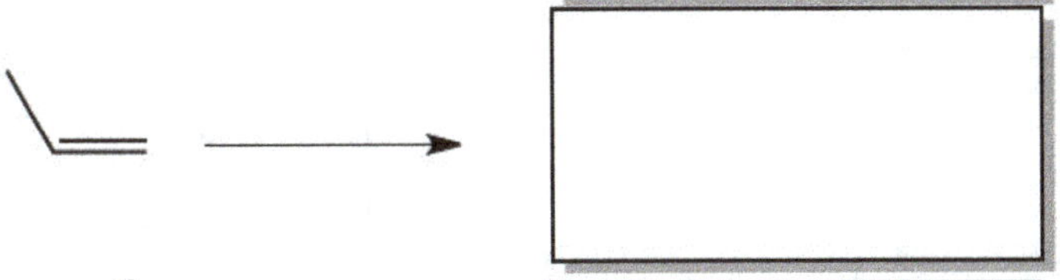

c) 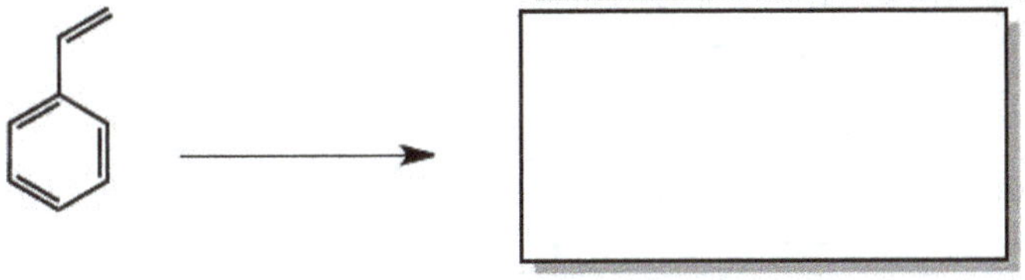

11.22 Olefin metathesis

Olefin is an old name for alkenes. Olefin metathesis is a reaction where a catalyst cuts the double bonds of two alkenes in half and reassembles them. Yves Chauvin, Robert Grubbs, and Richard Schrock developed the organometallic catalysts that are used in this reaction. They were awarded the 2005 Nobel Prize in Chemistry for this work. Probably the easiest of these catalysts to use was developed by Grubbs. Grubbs' catalyst uses ruthenium metal in it. The catalysts developed by Chauvin and Schrock are very moisture sensitive. We need to much less careful with Grubbs catalyst.

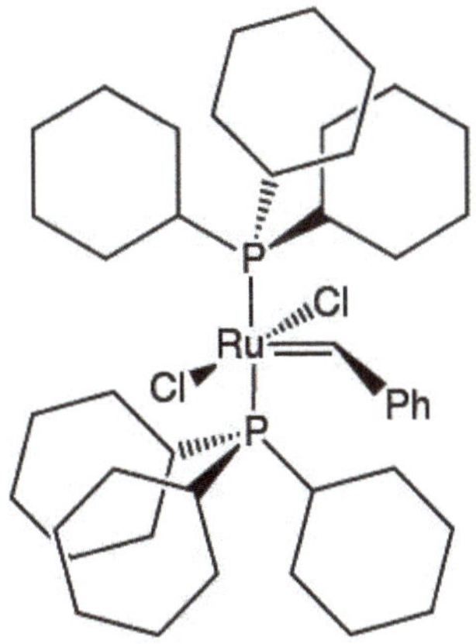

Figure 11.16 Grubbs Catalyst

Overall, you can think of the metathesis catalyst as acting like a pair of scissors. It cuts each double bond in half and makes a "sticky" part of a double bond. These "sticky" parts of double bonds can all randomly attach to each other. This sometimes remakes one of the reactants, or they can make new products. As long as the catalyst is around, cutting and reattaching keeps happening. This causes the metathesis reaction to be in equilibrium. To get more of the products, we need a driving force.

Of course, this is not really the mechanism of this reaction, but it is a good way to think about it. The mechanism is really much more elegant, but can be left for another time.

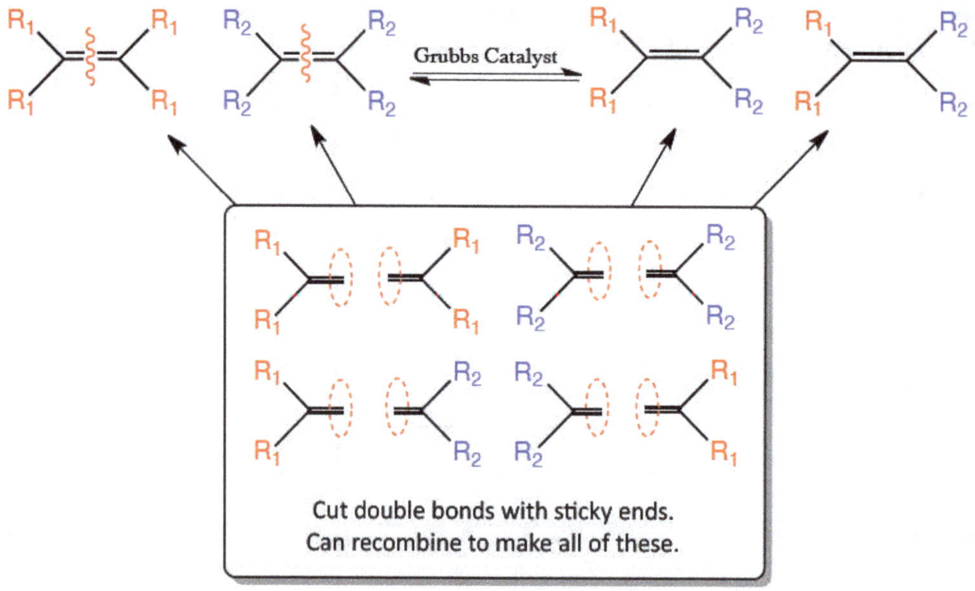

Figure 11.17 Olefin metathesis

11.22.1 Ring Closing Metathesis

If there are two alkenes on a molecule, it is possible to form a ring in a ring closing metathesis. When the alkenes are "cut" and become "sticky", the sticky ends can find each other and form a ring. In the example below, the CH_2= sticky parts can find each other, but they would form ethene which is a volatile gas and can bubble away from the reaction. Since the ethene can bubble away, one of the products is removed from the reaction driving the equilibrium to the right making more of the cyclic product.

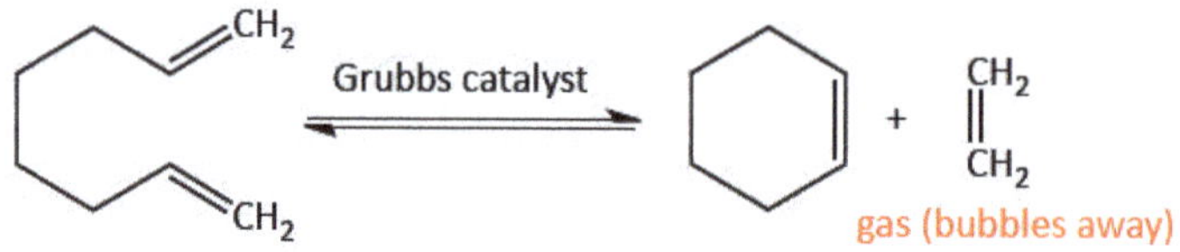

Here's how to think about it.

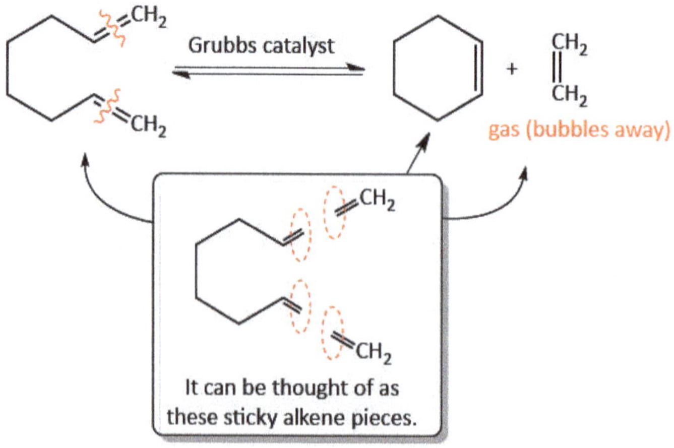

11.22.2 Ring Opening Metathesis Polymerization (ROMP)

When the alkene of a cyclic alkene is "cut", the "sticky" ends can find each other in another molecule and form a polymer. This ring opening metathesis polymerization, which is known by the acronym ROMP, is a powerful way to make polymers, especially those with double bonds in the chain. The following shows what happens when cyclobutene undergoes ROMP. The driving force for this reaction is the relief of ring strain in the reactants.

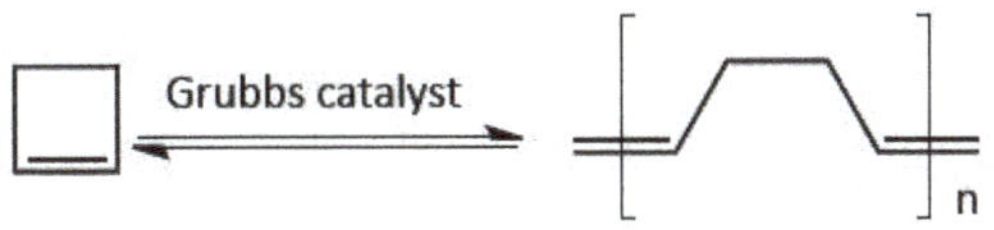

Figure 11.18 Overall ROMP of cyclobutene

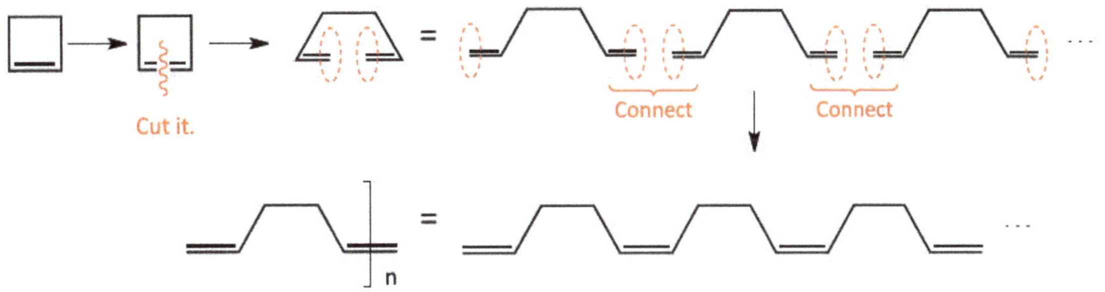

Figure 11.19 How to think about ROMP of cyclobutene

Here is another example of ROMP.

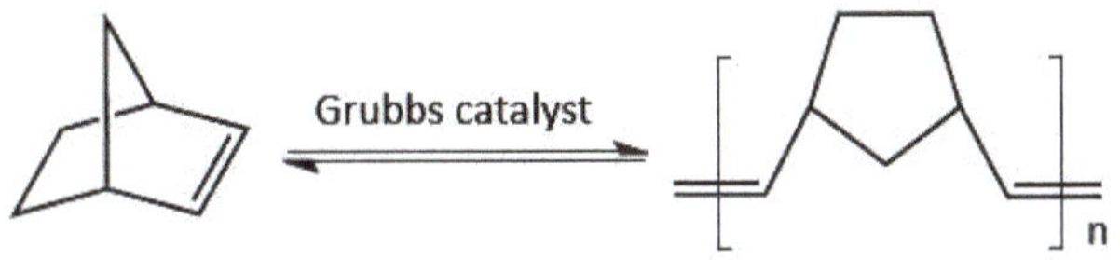

Figure 11.20 Overall ROMP of bicyclo[2.2.1]hept-2-ene

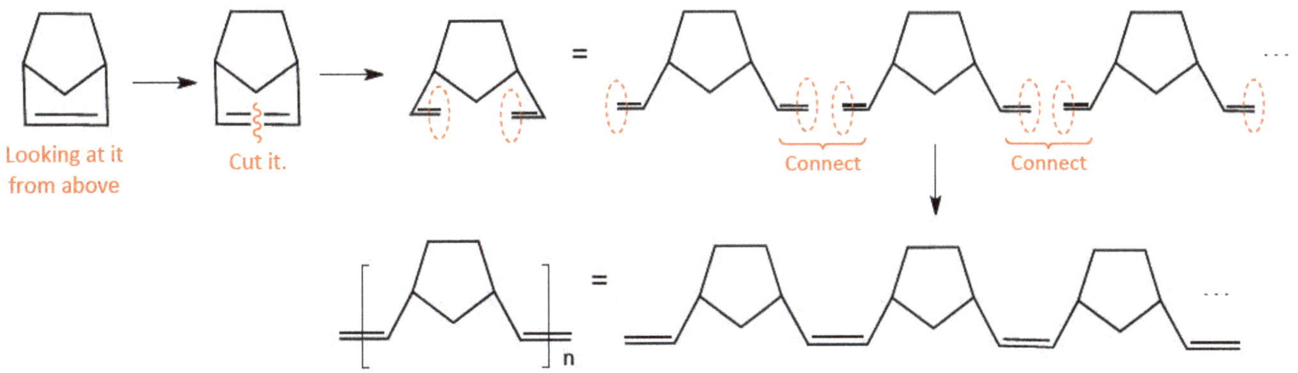

Figure 11.21 How to think about ROMP of bicyclo[2.2.1]hept-2-ene

11.12 Draw the products of the following reactions.

a) 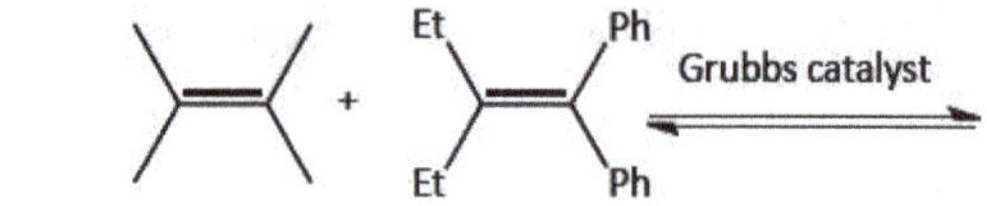

b) 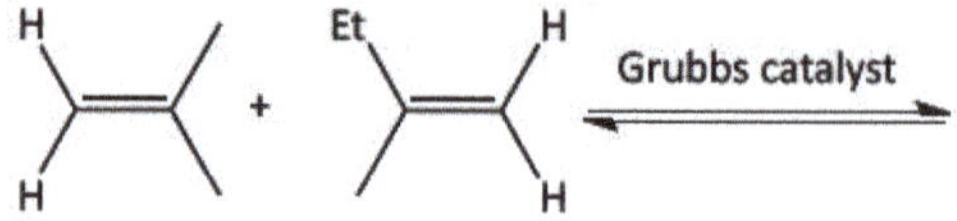

c) 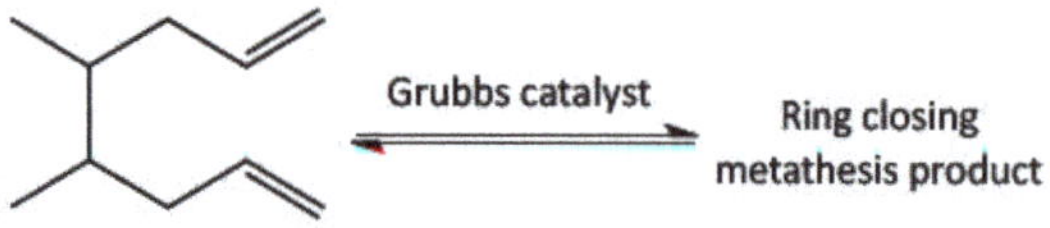

d) 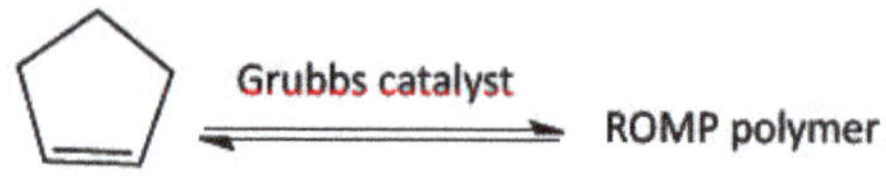

11.23 Acetylide formation

Terminal alkynes are slightly acidic. This is because the proton is on an sp hybridized carbon atom. Once deprotonated, this carbon atom will hold its negative lone pair of electrons in an sp hybrid orbital that is 50% s-character. These negative electrons will be held close to the carbon atom's positive nucleus. See section 5.2.4. The resulting anionic alkyne is called an acetylide.

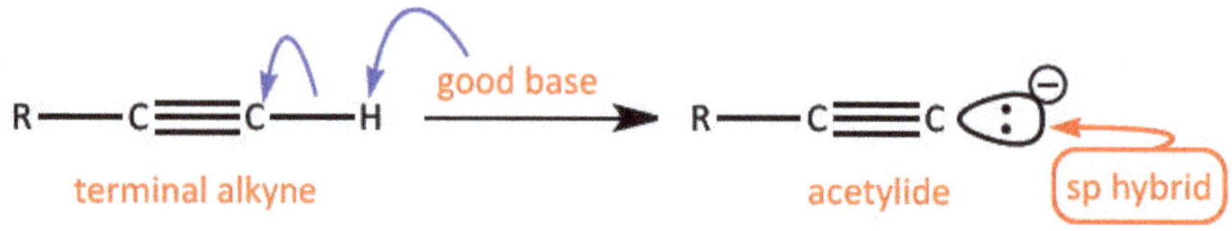

Terminal alkynes are only slightly acidic, though. A fairly strong base is required to deprotonate it. Negative oxygen bases like hydroxide, HO^-, or alkoxide, RO^-, simply do not work. Stronger bases are required. Reactive bases with negative nitrogen atoms or negative carbon atoms are needed to deprotonate terminal alkynes.

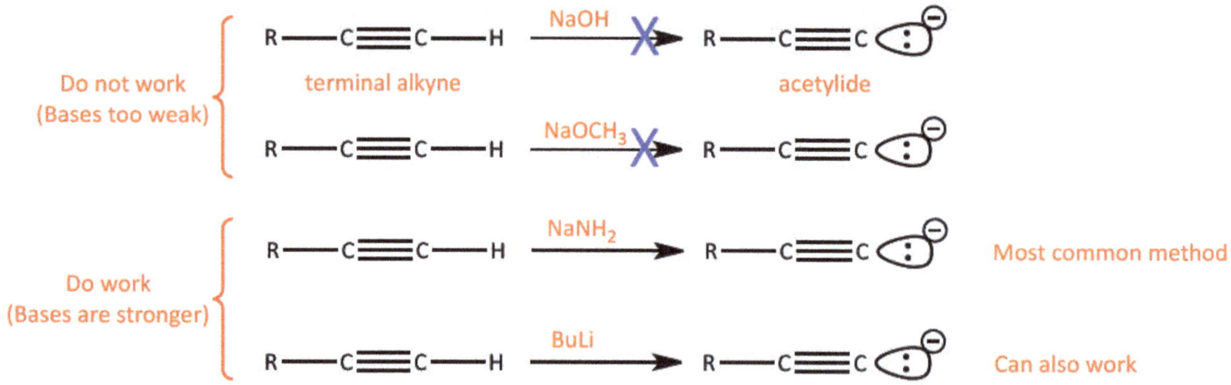

Sometimes, the acetylide ion will be drawn with its counterion. It may be drawn in the following ways.

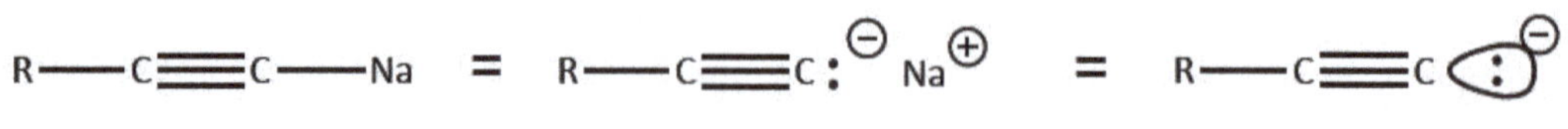

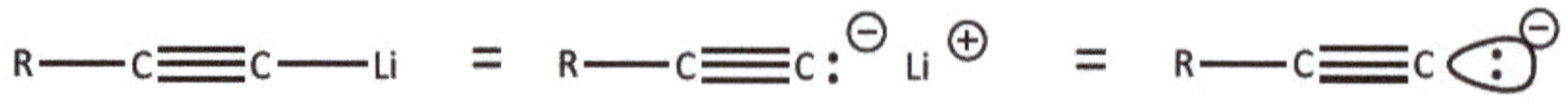

11.24 Reactions of acetylides

Acetylides are good nucleophiles and good bases. For example, they can act as the nucleophile in an S_N2 reaction or a base in an E2 reaction.

11.24.1 S_N2 Reaction of Acetylides

The negative acetylide is a very good nucleophile. It can attack the backside of a substrate with a good leaving group. Like in any S_N2 reaction, this only works if the backside attack is not sterically hindered. It is best if R' is methyl, primary, or possibly secondary, but not tertiary.

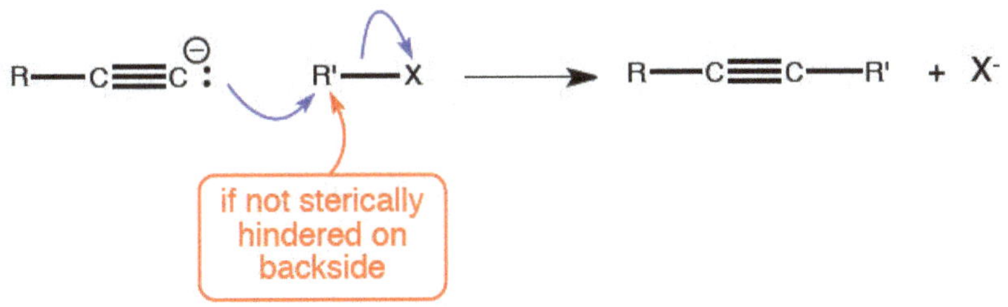

The S_N2 reactions of acetylides can be an effective way to make whatever alkynes or alkenes you might desire. The simplest alkyne, acetylene, can be deprotonated twice. The acetylides formed can undergo S_N2 reactions with various alkyl halides to make disubstituted alkynes. The disubstituted alkyne can be reduced to either cis or trans alkenes.

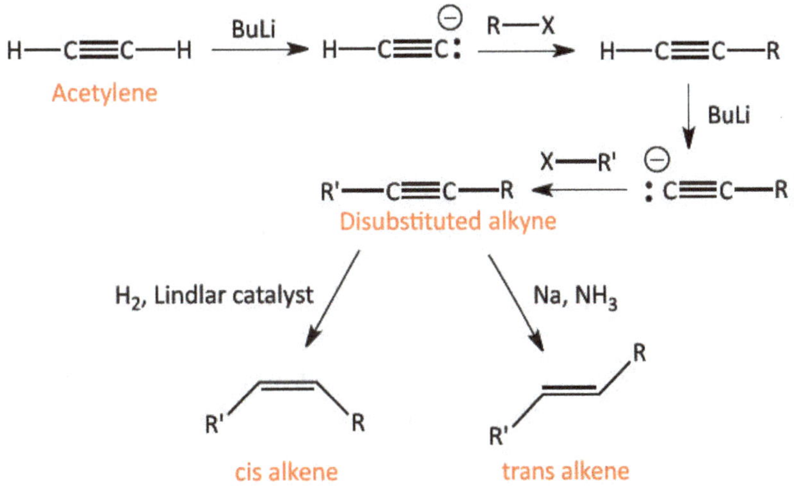

11.24.2 E2 Reaction of Acetylides

If the backside attack is hindered, an E2 reaction can occur instead. In this case, the acetylide acts as a base and deprotonates.

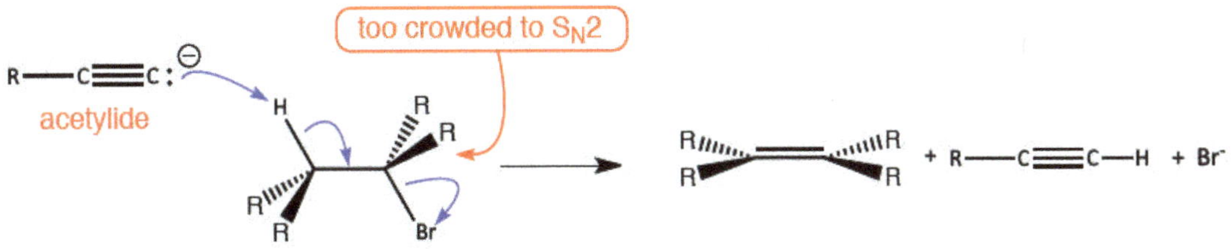

11.24.3 Acetylide Attacking a Carbonyl

Acetylides can attack carbonyl compounds. We'll learn more about this in a future chapter, but let's look at the attack of a ketone for now. The electronegative oxygen atom in a carbonyl attracts negative electron density towards itself making the carbon atom of a carbonyl partially positive.

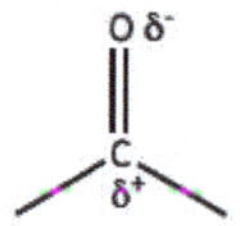

The nucleophilic, negative acetylide can attack the positively charged carbon atom of a carbonyl. Forming this bond would make five bonds on the carbon atom, so one bond must leave. The pi-bond of the carbonyl can go up onto the oxygen atom, giving it a negative charge. The compound remains this oxide until the reaction is quenched with acid and water, giving it a proton turning it into a neutral alcohol.

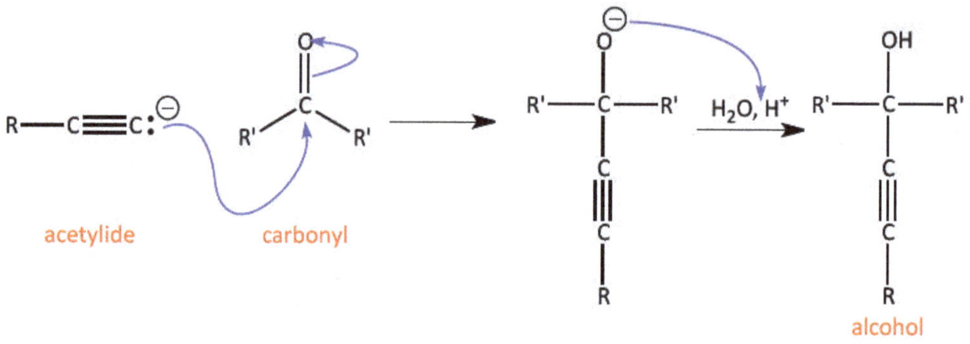

11.24.4 Acetylide Attacking an Epoxide

The nucleophilic attack of an acetylide with an epoxide is a triangle reaction. Analyzing the reaction, we see that the oxygen atom of the epoxide is neutral. The acetylide, with its negative charge, is an aggressive nucleophile. This aggressiveness makes it want to attack the less crowded carbon atom of the epoxide. In our triangle reaction analogy, the victim is holding up a sign with an 0 on it representing he is carrying 0 money. Notice the oxygen atom is neutral and not positively charged. This thief is aggressive. Even though the victim has no money, the thief punches the victim in the side. The thief punches the victim in the side that is not protected with bags. It won't punch in the side with the bags because it is more difficult to do so. There is also no money there to steal.

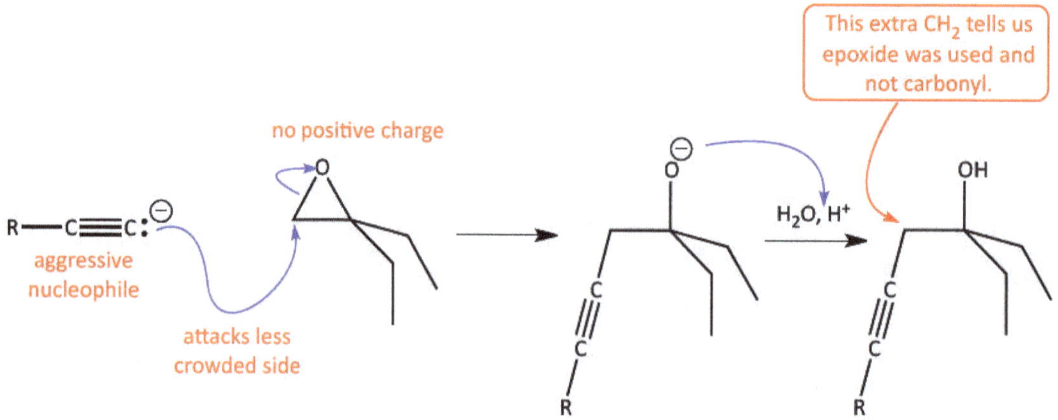

Once the acetylide attacks the epoxide, it would make five bonds on the carbon atom. Therefore, a bond must leave. The one that leaves is the sigma bond between the attacked carbon atom and the oxygen atom forming a negative oxygen oxide. The compound remains this negative oxide until the reaction is quenched with water and acid forming an alcohol.

11.24.5 Heavy metal acetylides

Terminal alkynes react with silver or copper monovalent cations to form heavy metal acetylides. These silver and copper acetylides are more covalently bonded than what we've seen with sodium or lithium acetylides. This makes them less basic and less nucleophilic. Silver and copper acetylides also precipitate out of water. Heavy metal acetylides should be handled with caution. They can explode when they are dry. So, they need to be kept wet.

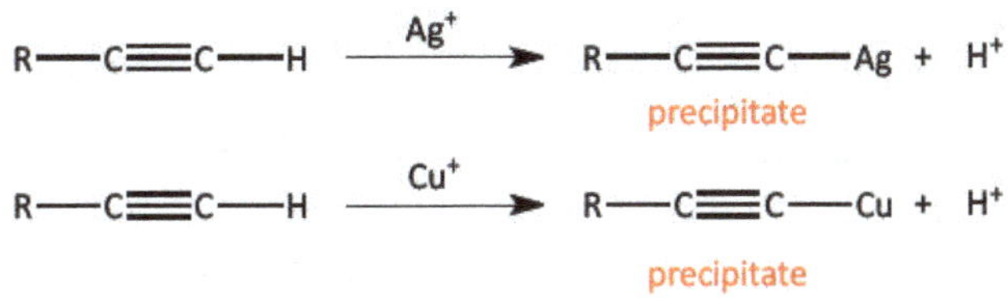

The separation of terminal alkynes from internal alkynes is one use of heavy metal acetylides. Cu^+ can be added to a mixture of both terminal and internal alkynes. Only the terminal alkyne will react with the Cu^+ forming a copper acetylide precipitate. This precipitate could be filtered out of the solution. This pulls the terminal alkyne out of the solution and leaves the internal alkyne behind.

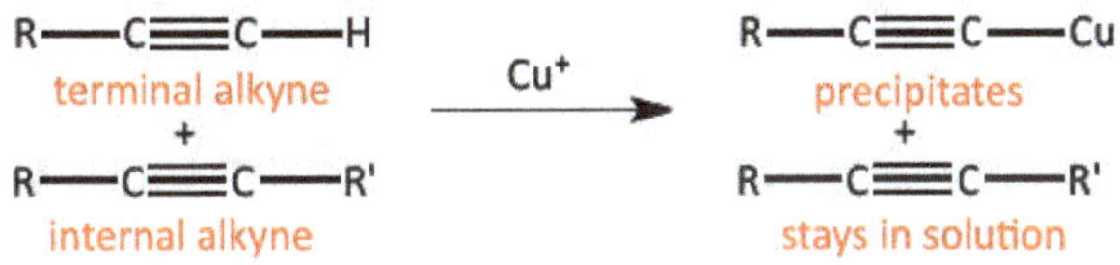

While the copper acetylide is still wet (remember dry copper acetylide is explosive), acid is added to reform the terminal alkyne.

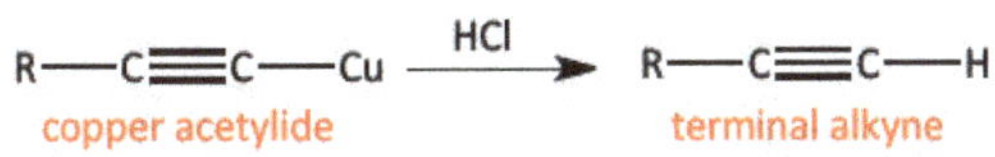

11.13 Draw the products of the following reaction.

a) 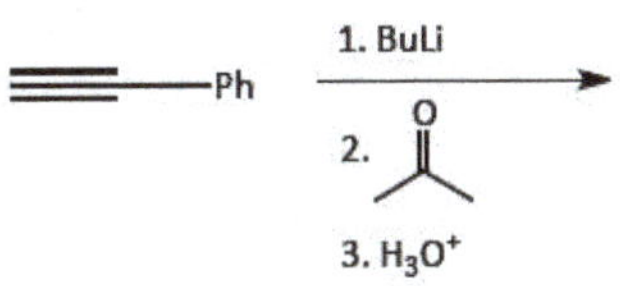

b) 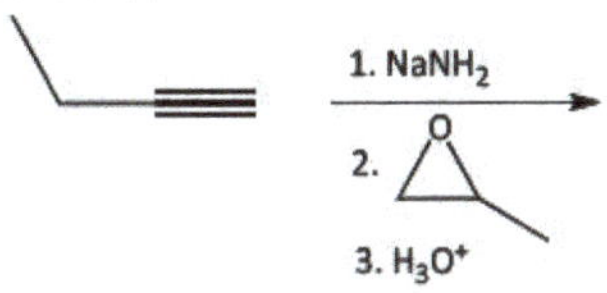

c) 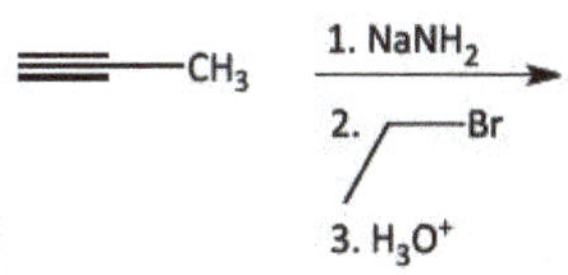

d) 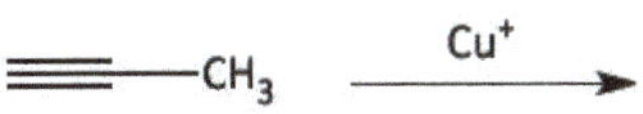

e) 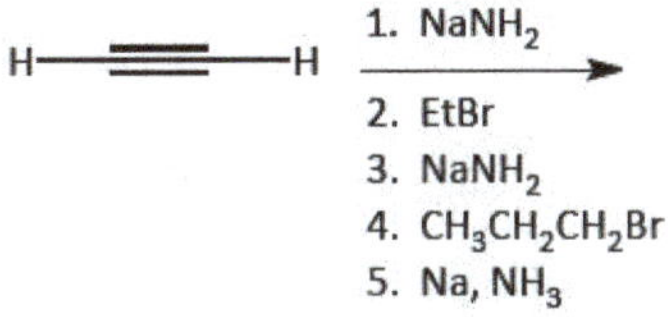

f) 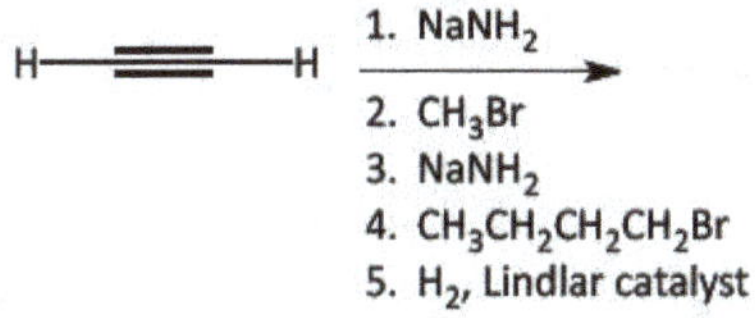

11.25 Reaction Summary

11.25.1 Alkene Reactions

a) 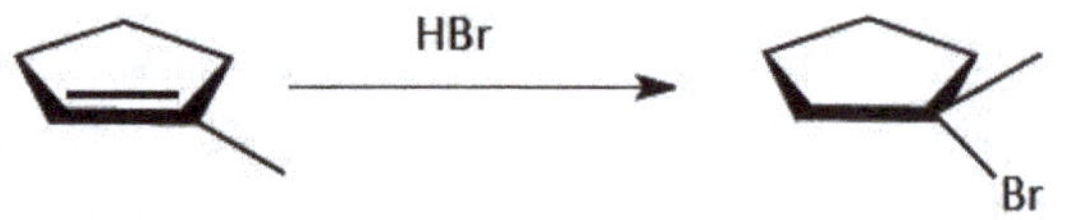

b) 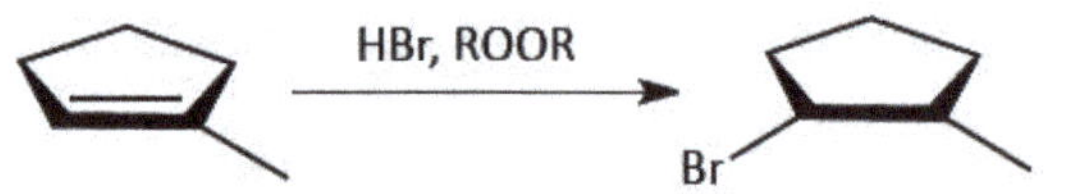

c) 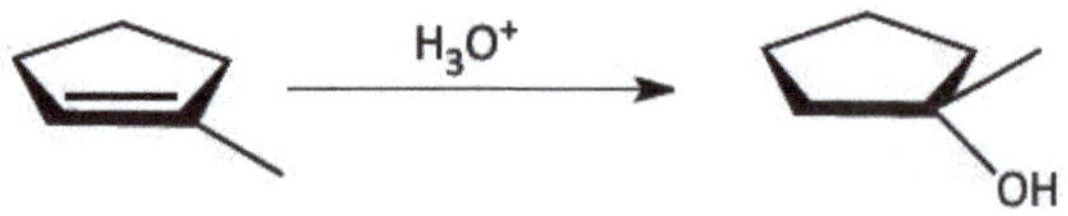

d) 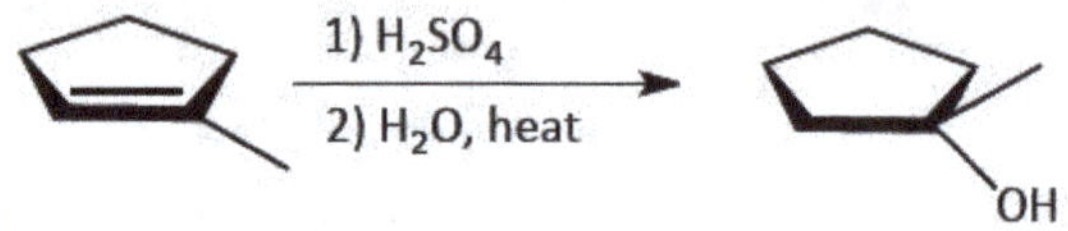

e) 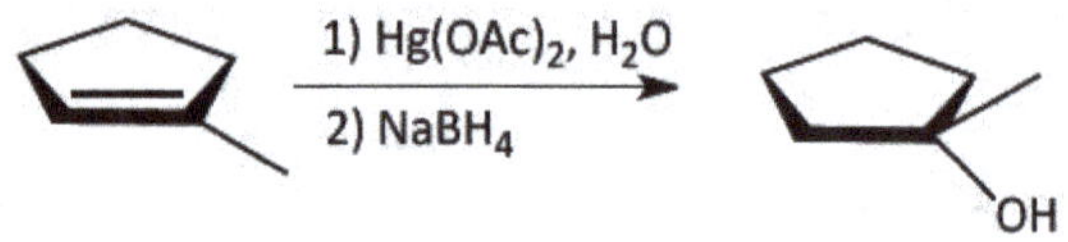

f) 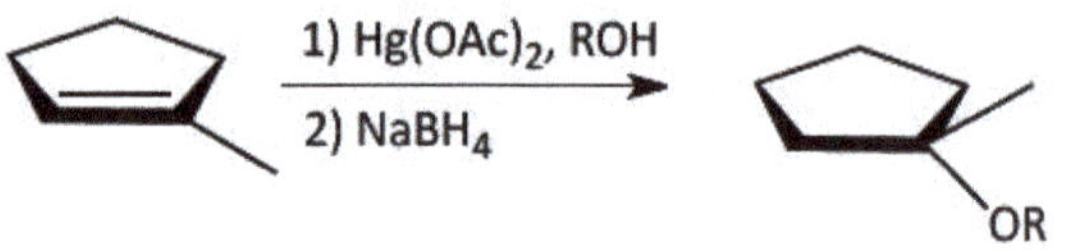

g) 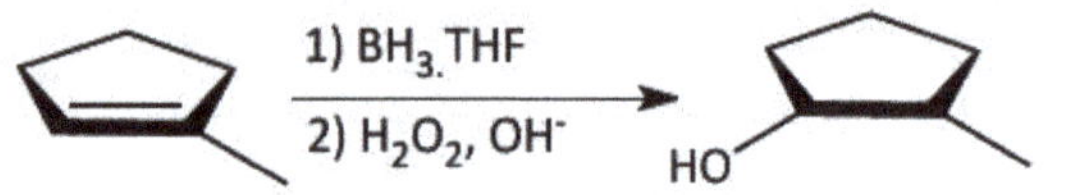

h) 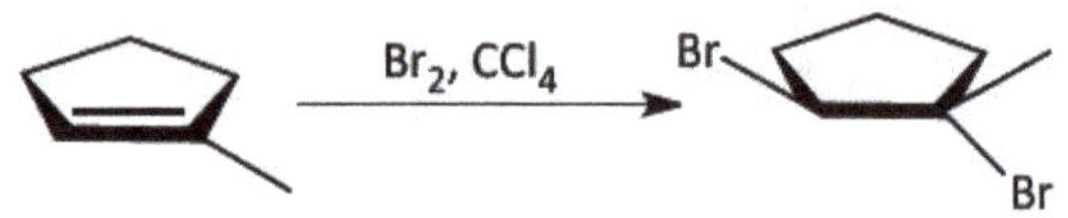

i) 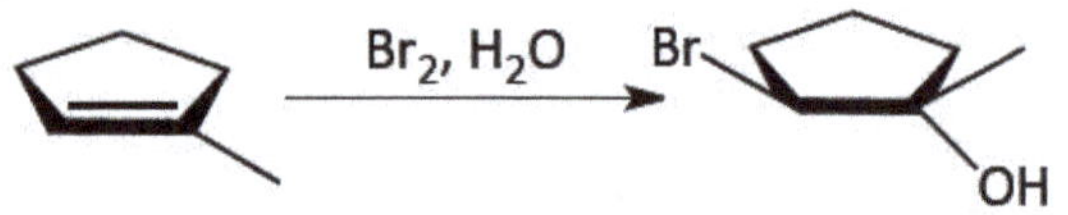

j) 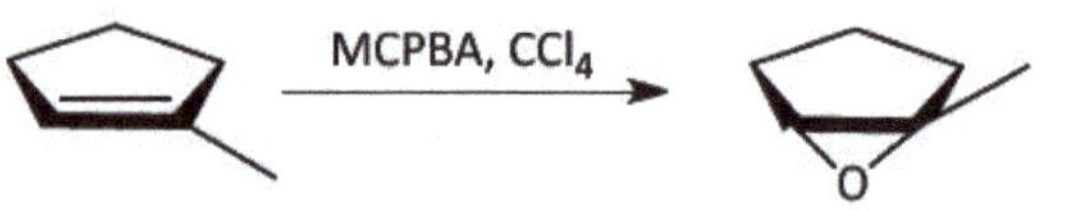

k) 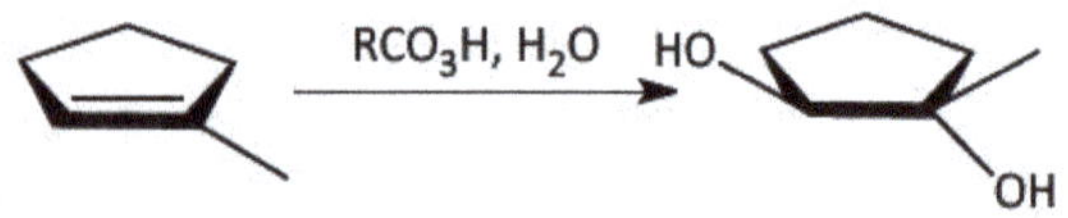

l) 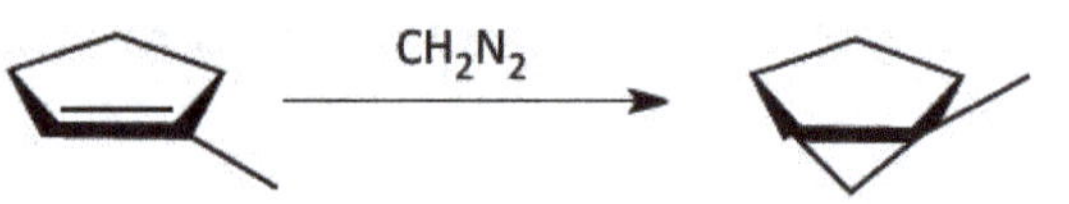

m) 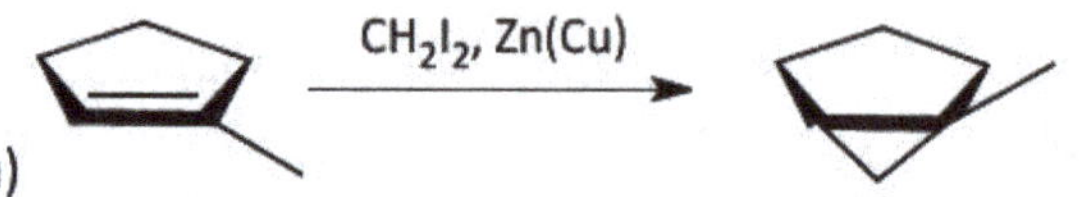

n) 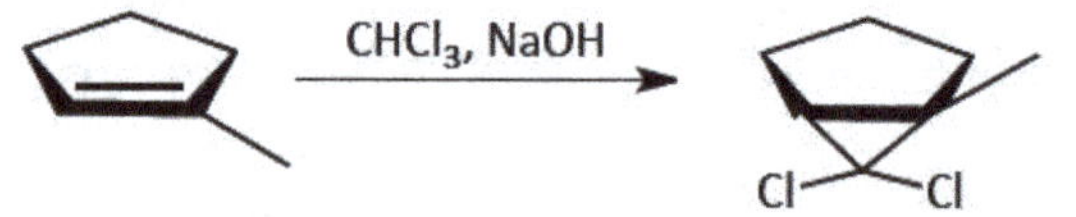

o) 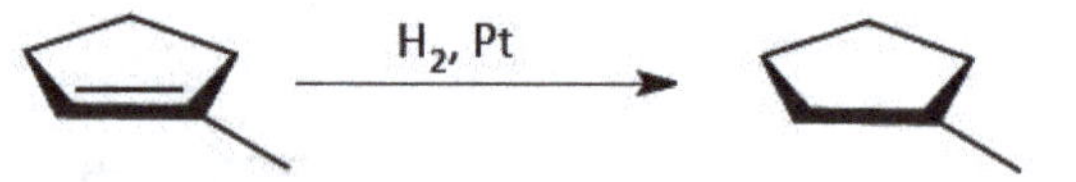

p) 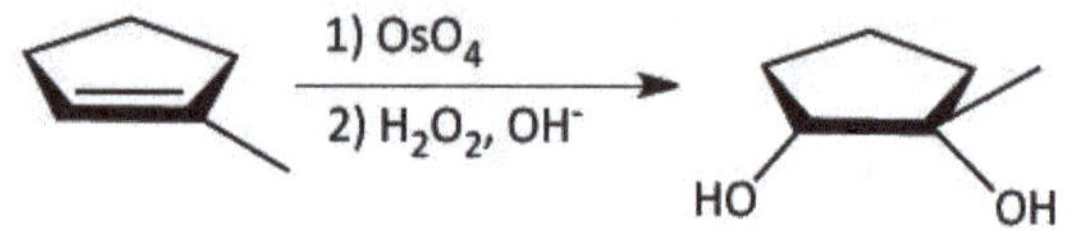

q) 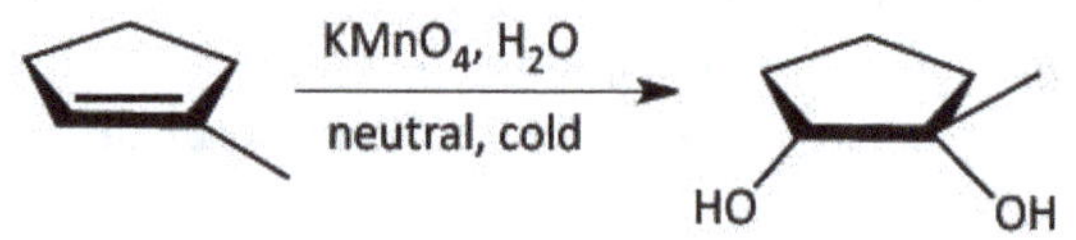

r) 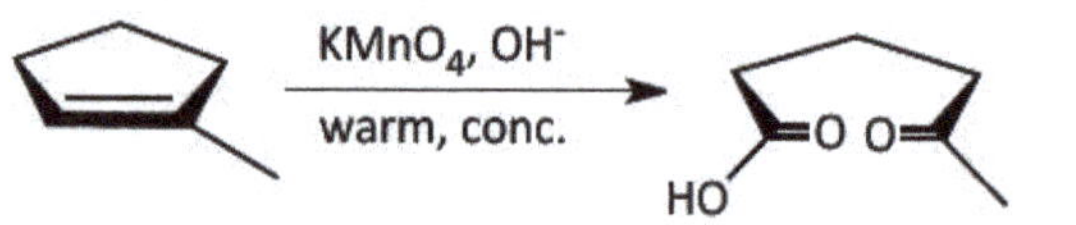

s) 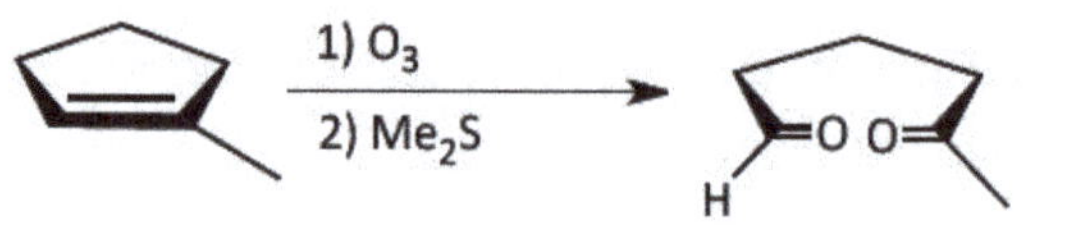

t) 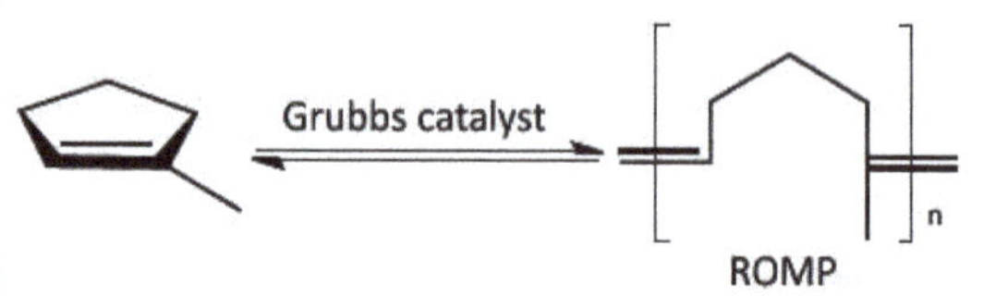

u) 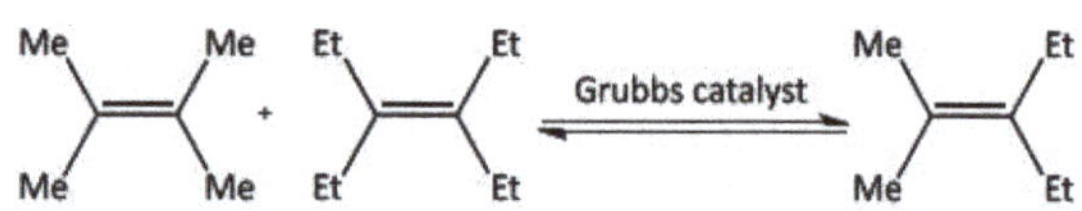

v) 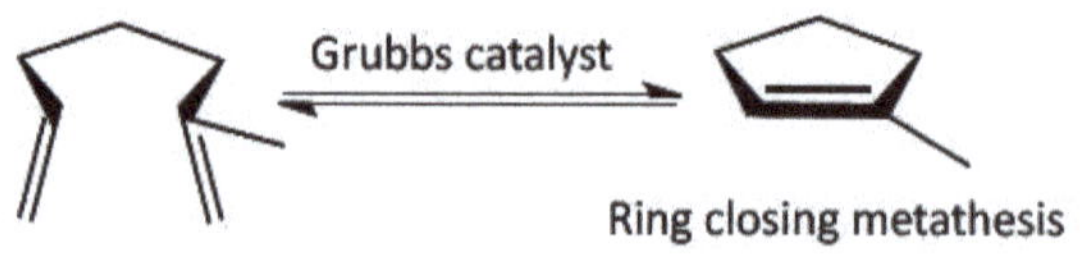

w) 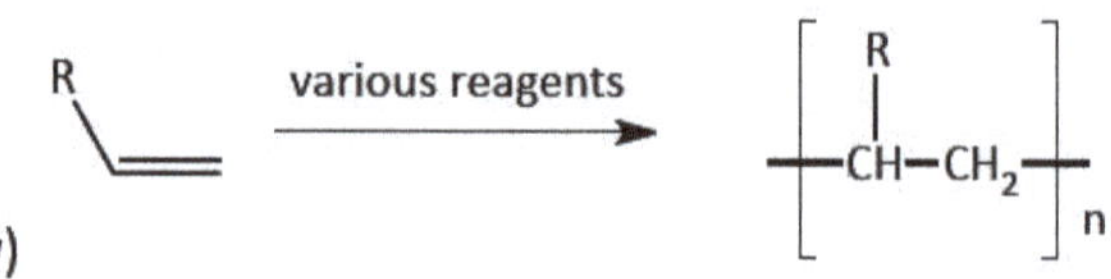

11.25.2 Alkyne Reactions

a) 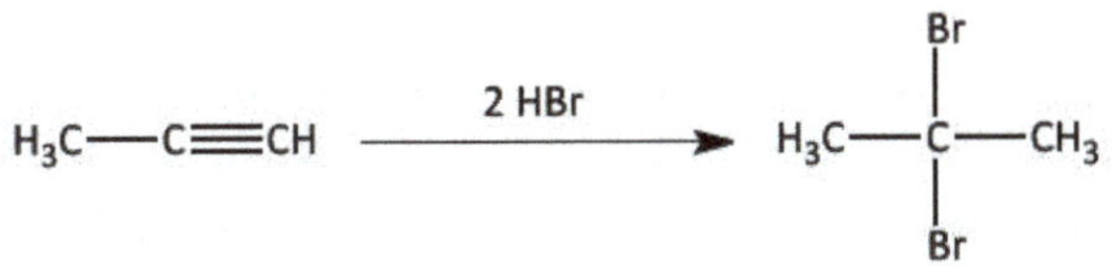

b) 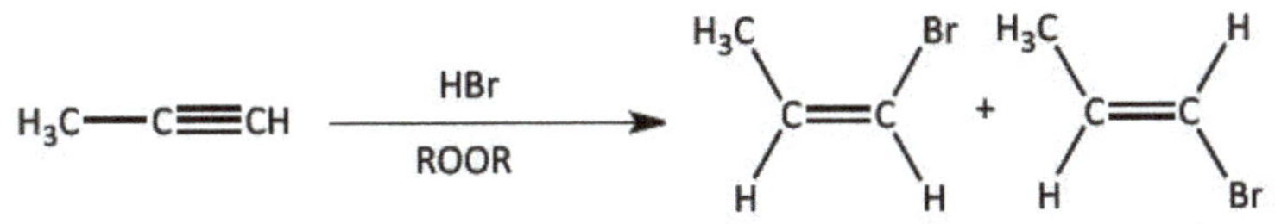

c) 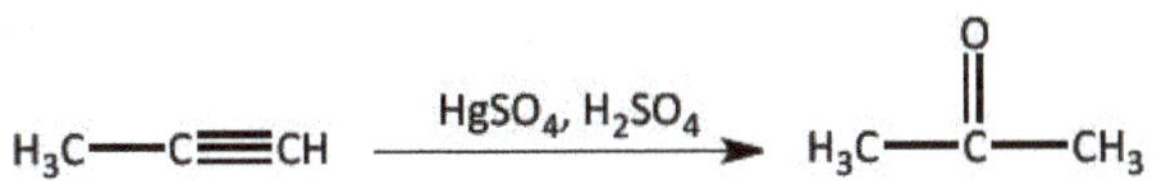

d) 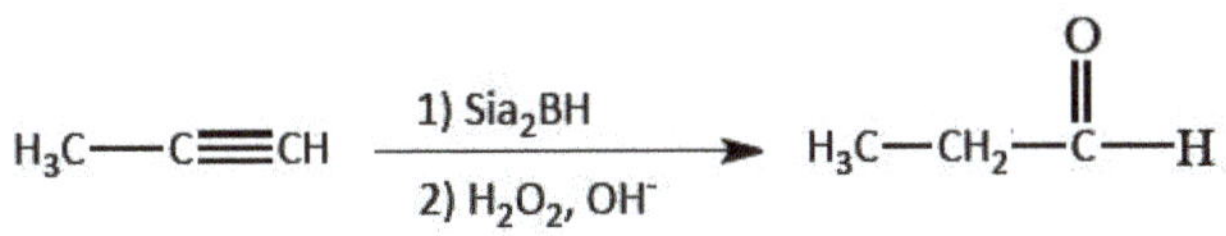

e) 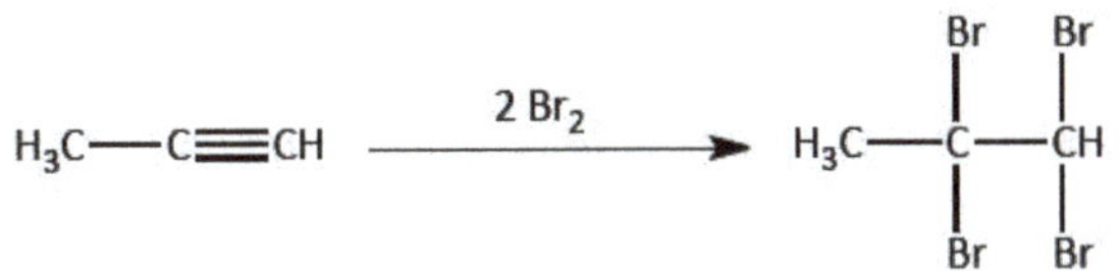

f) 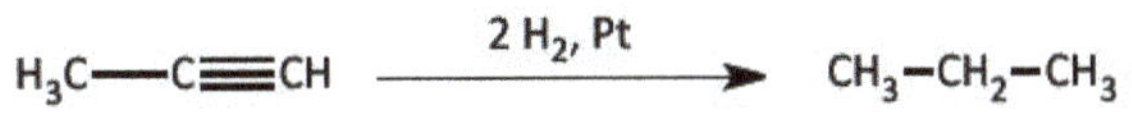

g) 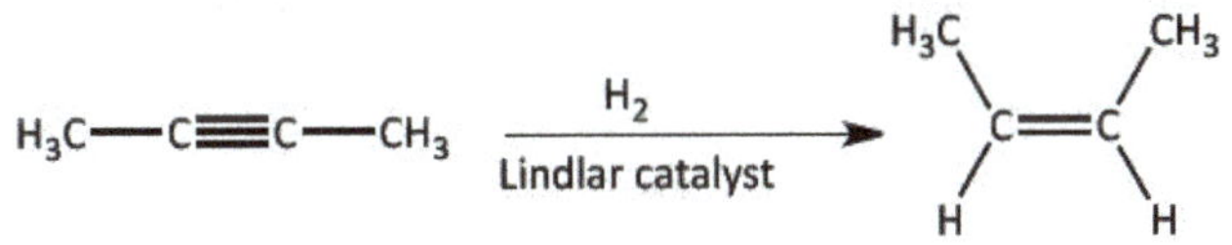

h) 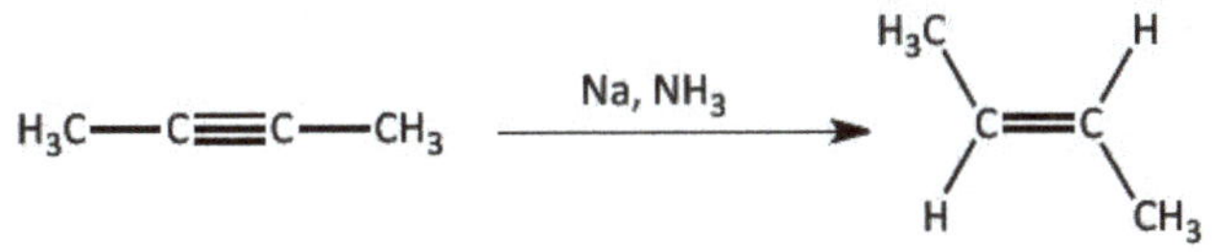

i) 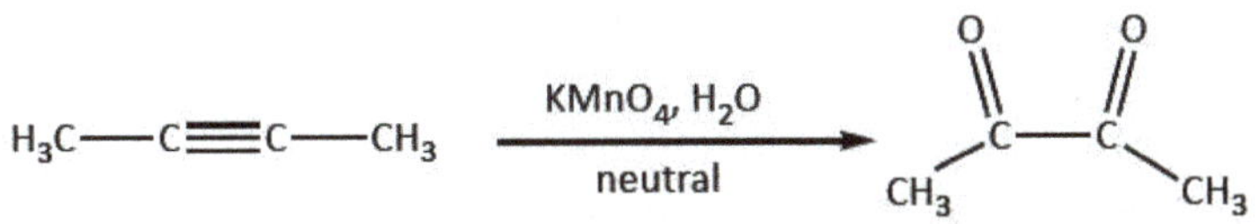

j) 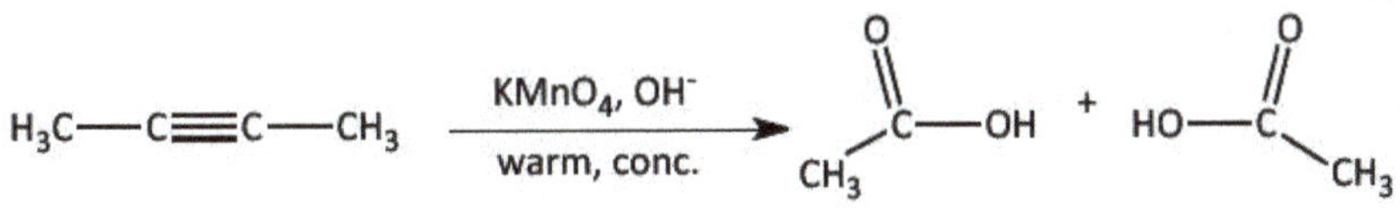

k) 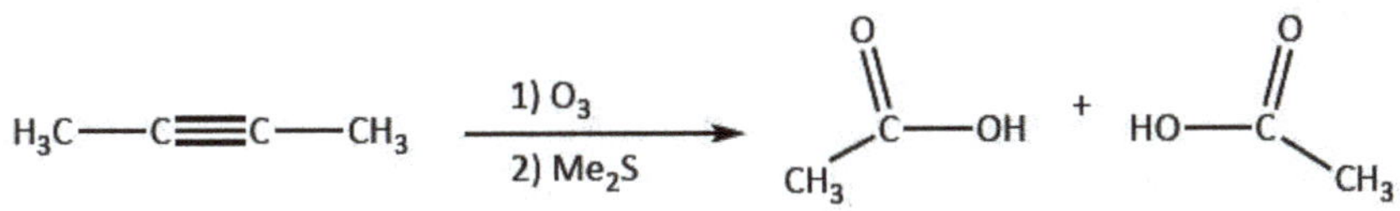

l) 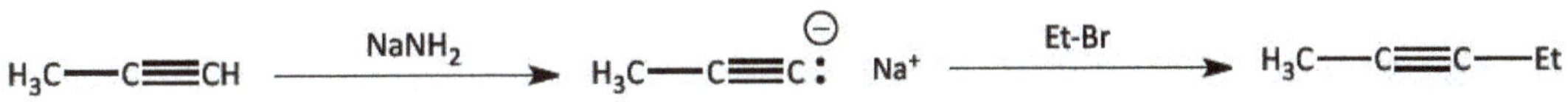

m) 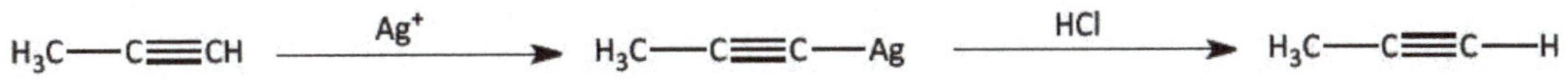

n) 

11.26 Problems

11.14 Draw the products of the following reactions.

a) 1-methylcyclopentene + HBr
b) 1-methylcyclohexene + HBr + PhOOPh
c) 1-butene + HCl
d) 2-methylbutene + HI
e) 1-ethylcyclopentene + HBr + BuOOBu
f) propene + HBr
g) propyne + 2 HBr
h) butyne + HBr + PhOOPh
i) pentyne + 2 HI

11.15 Identify the appropriate reagents needed to perform the following reactions.

a) 

b) 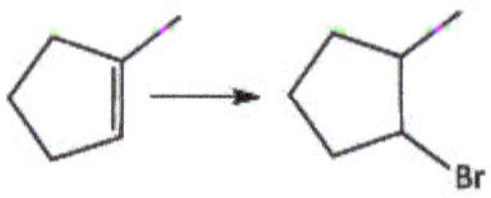

c) hexene to 1-bromohexane d) butene to 2-bromobutane

11.16 Show the mechanism for the following reactions.

a) 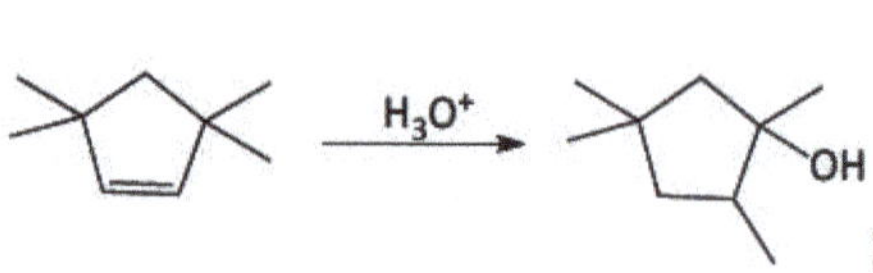

b) 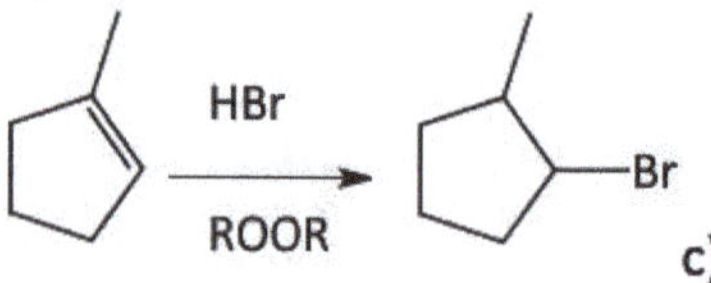

c) 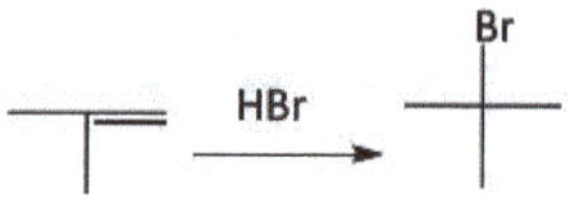

11.17 When 1-pentyne reacts with $HgSO_4$ in sulfuric acid, a Markovnikov enol is first formed. a) Draw the enol. b) Then, draw the mechanism for the acid catalyzed keto-enol tautomerization to turn the enol into a ketone. Remember to protonate and go to town.

11.18 The alkene below was reacted with BH_3 followed by H_2O_2/OH^-. Give the product of this reaction.

a) 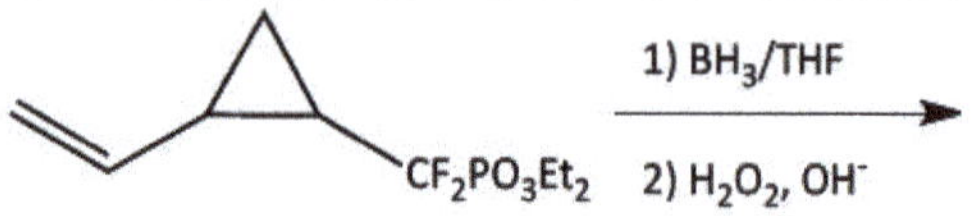

b) What would have been the product if the alkene in part a had been reacted with $Hg(OAc)_2$ in H_2O followed by $NaBH_4$?

11.19 What reagent would perform the following reactions?

a) 3,3-dimethylbutene to 1-bromo-3,3-dimethylbutane, $(CH_3)_3CCH_2CH_2Br$ b) 1-butene to 1-butanol
c) 3,3-dimethylbutene to 2-bromo-3,3-dimethylbutane, $(CH_3)_3CCH(Br)CH_3$ d) 1-pentene to 2-pentanol
e) 3,3-dimethylbutene to 3,3-dimethylbutanol, $(CH_3)_3CCH_2CH_2OH$ f) 1-pentene to 1-pentanol
g) 3,3-dimethylbutene to 3,3-dimethylbutan-2-ol, $(CH_3)_3CCH(OH)CH_3$
h) For part g), it is a bad idea to use dilute aqueous H_2SO_4 to do that reaction. Why?
i) 3,3-dimethylbutene to 3,3-dimethyl-2-methoxybutane, $(CH_3)_3CCH(OCH_3)CH_3$
j) 1-butyne to butyraldehyde, $CH_3CH_2CH_2CHO$ k) 1-butyne to 2-butanone, $CH_3COCH_2CH_3$

11.20 Draw the products of the following reactions.

a) 1-methylcyclopentene + Br_2 in $CHCl_3$
b) 1-methylcyclopentene + Cl_2 in water
c) 2-methylpropene + Br_2 in CH_2Cl_2
d) 2-methylpropene + Br_2 in water
e) 1-butyne + 2 Br_2
f) 3-hexyne + 2 Cl_2

11.21 What reagent would perform the following reactions?

a) 3,3-dimethylbutene to *trans*-1,2-dibromo-3,3-dimethylbutane b) 2-octyne to 2,2,3,3-tetrachlorooctane
c) propyne to 1,1,2,2-tetrabromopropane d) 1-methylcyclohexene to 2-bromo-1-methylcyclohexanol
e) 3,3-dimethylbutene to 2-chloro-3,3-dimethylbutan-1-ol, $(CH_3)_3CCH(Cl)CH_2OH$

11.22 Draw the products of the following reactions.

a) 1-methylcyclopentene + dilute aqueous H_2SO_4
b) 1-methylcyclopentene + aqueous $Hg(OAc)_2$ followed by $NaBH_4$
c) 1-methylcyclopentene + $Hg(OAc)_2$ in CH_3CH_2OH followed by $NaBH_4$
d) 1-methylcyclopentene + BH_3.THF followed by H_2O_2 and NaOH
e) 2-methylpropene + dilute aqueous H_2SO_4
f) 2-methylpropene + aqueous $Hg(OAc)_2$ followed by $NaBH_4$
g) 2-methylpropene + $Hg(OAc)_2$ in CH_3CH_2OH followed by $NaBH_4$
h) 2-methylpropene + BH_3.THF followed by H_2O_2 and NaOH
i) 1-pentyne + $HgSO_4$ in sulfuric acid
j) 1-pentyne + disiamylborane followed by H_2O_2 and NaOH

11.23 Draw the products of the following carbene reactions.

a) 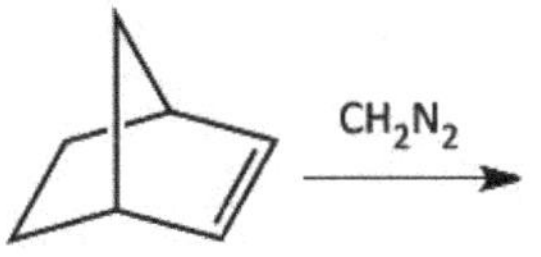

b) 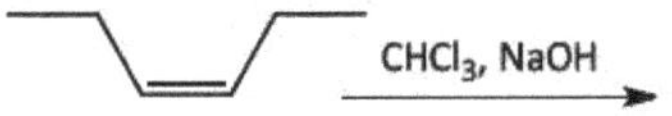

c) 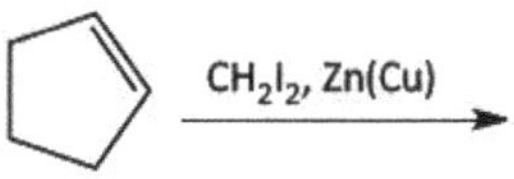

d) cyclohexene + Simmons-Smith reagent

11.24 Draw the alkene and other appropriate, carbene-generating reagents, needed to perform make the following compounds.

a) 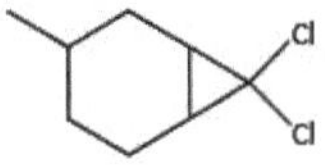

b) 

c) 

d) 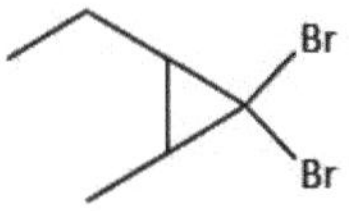

11.25 Regarding the following reaction:

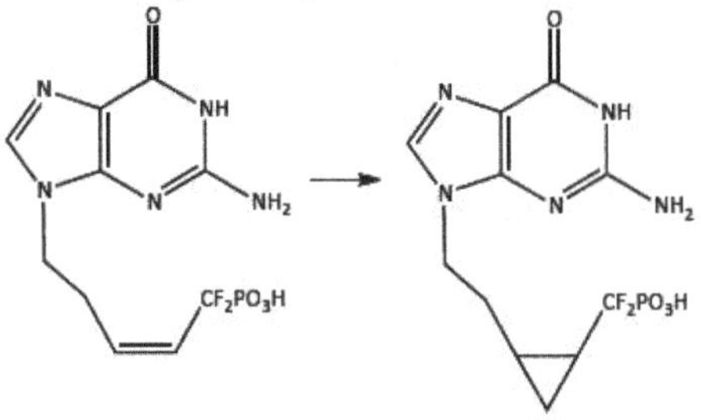

a) List the reagent(s) that could accomplish this reaction.
b) What is the reactive species, the intermediate, in this reaction called?

11.26 Draw the mechanism for the following reaction.

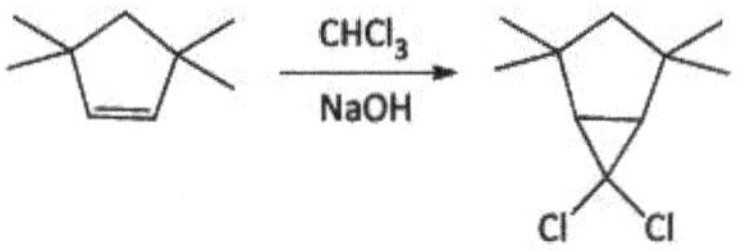

11.27 Draw the products of the following reactions.

a) 1,3-cyclohexadiene + H_2/Pt
b) trans-2-pentene + H_2/Ni
c) 1,4-pentadiene + excess H_2/Pt
d) 2-pentyne + H_2 + Lindlar catalyst
e) 2-pentyne + Na in liquid NH_3
f) 2-pentyne + excess H_2/Pt
g) 3-hexyne + H_2 + $Pd/BaSO_4$ + quinoline

11.28 The hydrogenation of the alkene below could theoretically produce two possible diastereomers (A and B). Only one, A, is formed. Explain why.

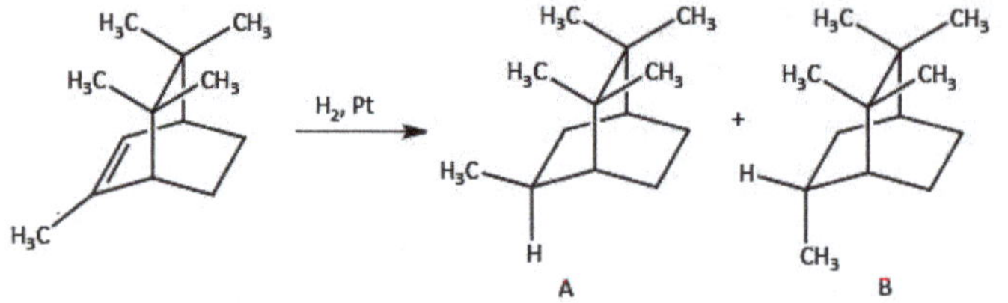

11.29 Draw the products of the following epoxidation reactions.

a)  b) 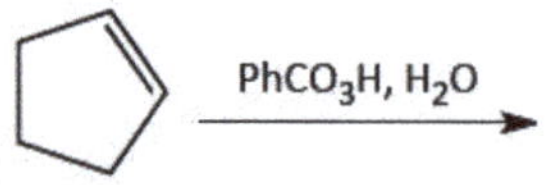 c) 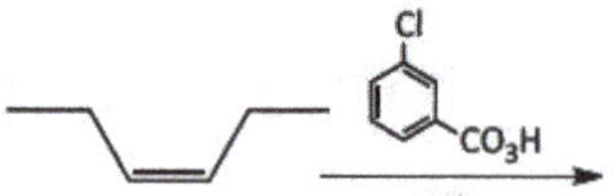 d) 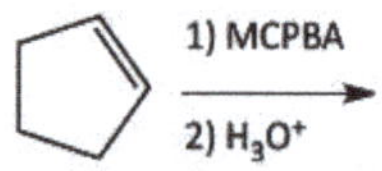

e) cyclohexene + peroxyacetic acid in water f) cycloheptene + *meta*-chloroperoxybenzoic acid in chloroform

11.30 Identify the alkene and appropriate reagents to make the following compounds.

a)  b) 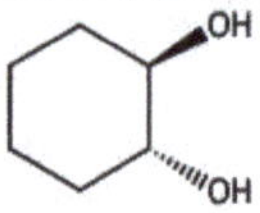 c) 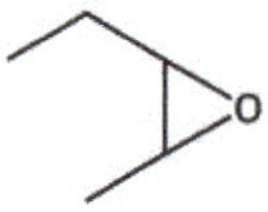 d) 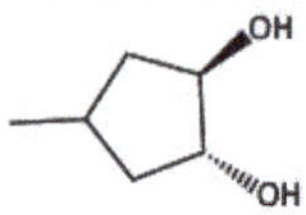

11.31 What are the products of the following alkene oxidations?

a) 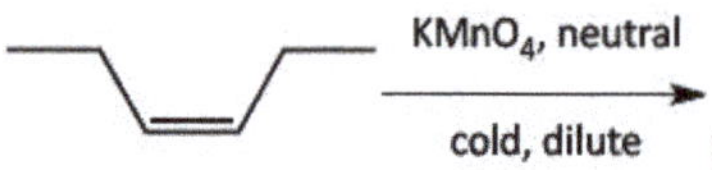 b) 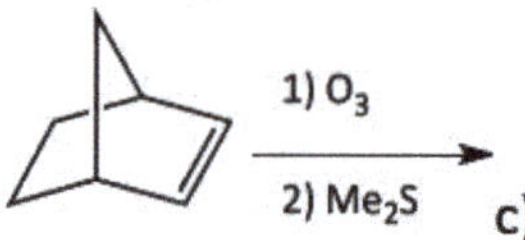 c) 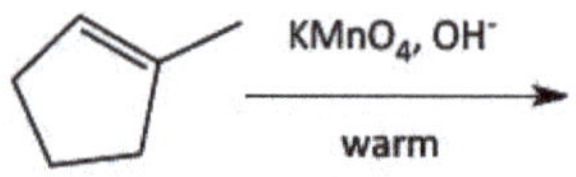

d) cyclopentene + peroxybenzoic acid in water e) 1-methylcyclohexene + OsO_4 followed by H_2O_2
f) 1-ethylcyclopentene + a cold, dilute, neutral solution of potassium permanganate
g) 2-hexyne + a cold, dilute, neutral solution of potassium permanganate
h) 3-hexyne + ozone followed by dimethylsulfide i) 3-hexyne + warm, concentrated, basic $KMnO_4$

11.32 Identify the alkene or alkyne and appropriate oxidation reagents to make the following compounds.

a) $HO_2CCH_2CH_2CH_2CH_2CO_2H$ b) 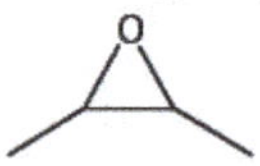 c) $CH_3CH_2CH_2CO_2H + CH_3CO_2H$ d) 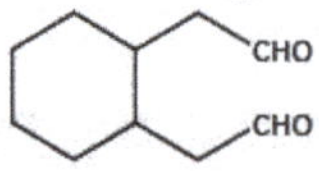

e) 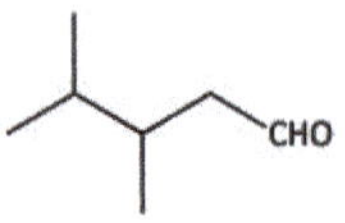 f) 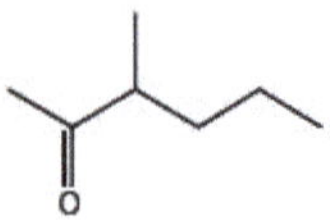 g) 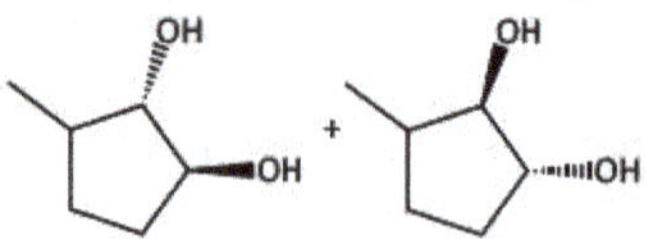 h) 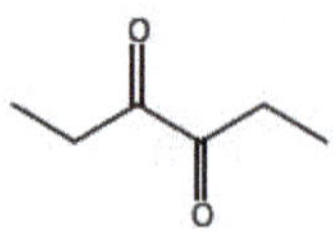

i)  Show two ways. j) 2 CH_3CH_2COOH, Show two ways, one from an alkene and one from an alkyne.

k) 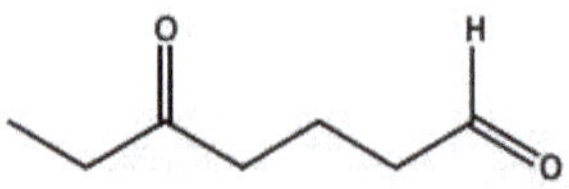 l) 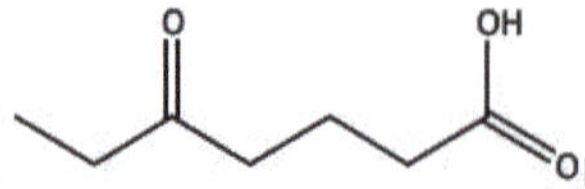 m) 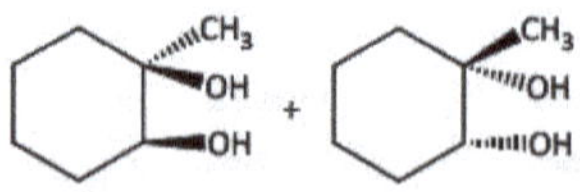

11.33 What would be the product of reacting the 2-methyl-2-pentene with:
a) OsO_4 followed by H_2O_2? b) Ozone followed by dimethylsulfide (Me_2S)? c) MCPBA in $CHCl_3$
d) peroxyacetic acid in H_2O e) $KMnO_4$, neutral, dil. f) $KMnO_4$, warm, basic g) H_2, Pt h) Simmons-Smith reagent
i) chloroform and sodium hydroxide j) Br_2 in CCl_4 k) Br_2 in water l) HBr m) HBr and benzoyl peroxide
n) BH_3.THF followed by H_2O_2/OH^- o) $Hg(OAc)_2$ in water p) $Hg(OAc)_2$ in CH_3OH

11.34 **Orlon yarn is fabric that can be knitted into sweaters. Orlon is the trade name for polyacrylonitrile. It is synthesized through a free-radical polymerization of acrylonitrile. Draw the mechanism for this polymerization, and draw the structure of the polymer.**

RO-OR + 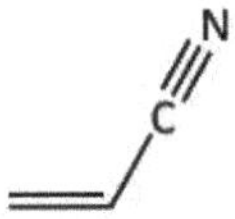 → Orlon (polyacrylonitrile)

acrylonitrile

11.35 **Give the alkene monomer that can be used to make the following polymers.**

a) 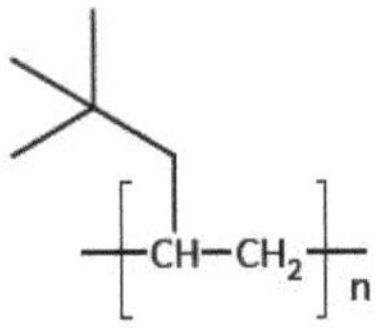
b) 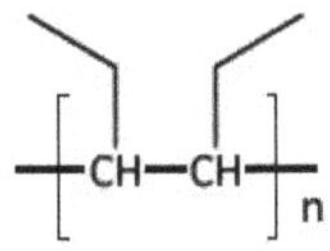
c) 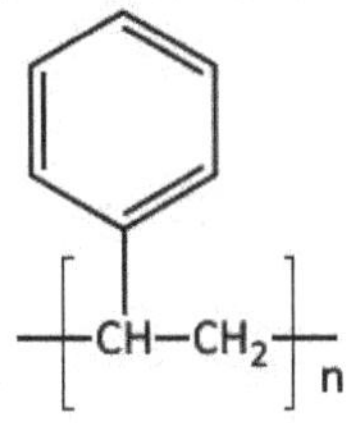
d) 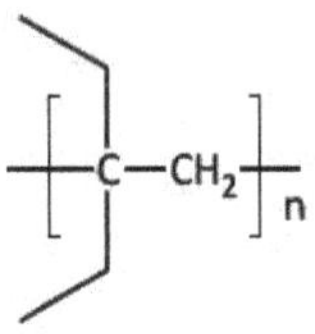

11.36 **Draw the polymer created from the addition reactions of each of the following alkene monomers.**

a) 
b) 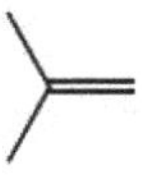
c) 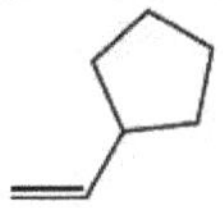
d) 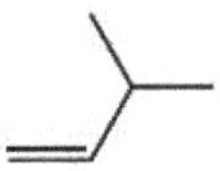
e) 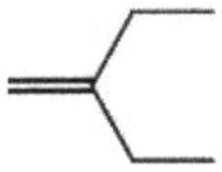

11.37 **Give the products formed in the following olefin metathesis reactions.**

a) 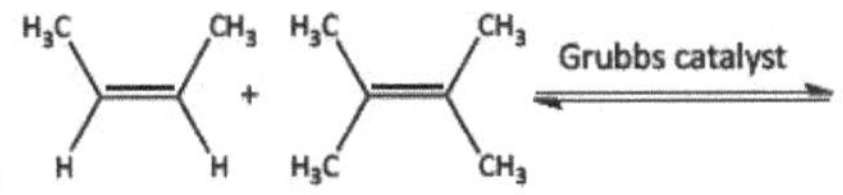

b) 

c) 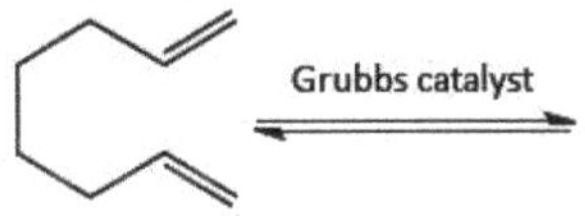

(This is a ring closing metathesis.)

d) 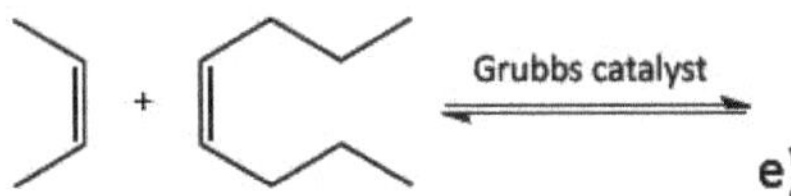

e) 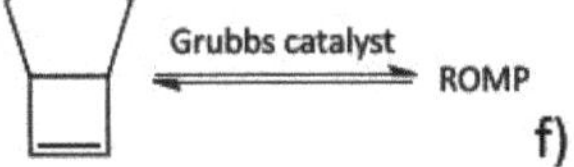

f) 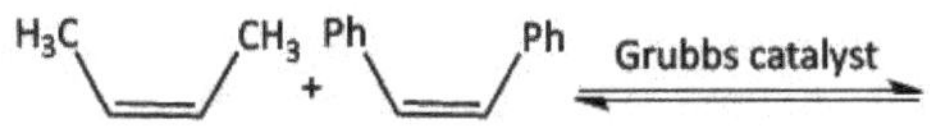

11.38 **A laboratory wants to make a shorter version of the following fatty acid ester. They react it with a smaller alkene in the presence of Grubbs catalyst. a) Draw the structure of the shorter ester they made.**

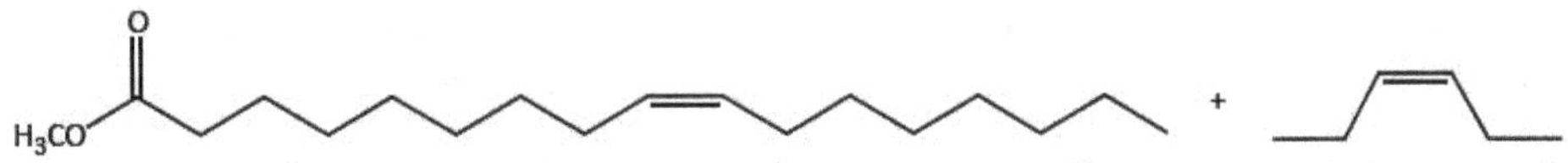

b) **Olefin metathesis can also be used to extend (or make longer) a fatty acid ester. The same fatty acid ester is reacted with a longer alkene in the presence of Grubbs catalyst. Draw this longer fatty acid ester.**

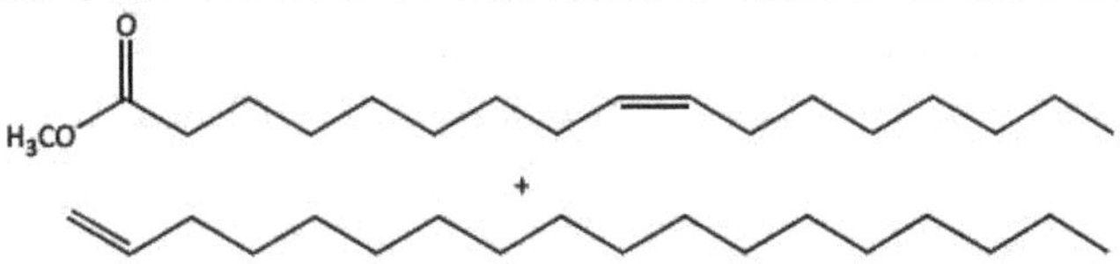

Alkene & Alkyne Reactions

11.39 An unknown cyclic alkene, A, underwent a series of reactions shown below. Identify compounds A-I.

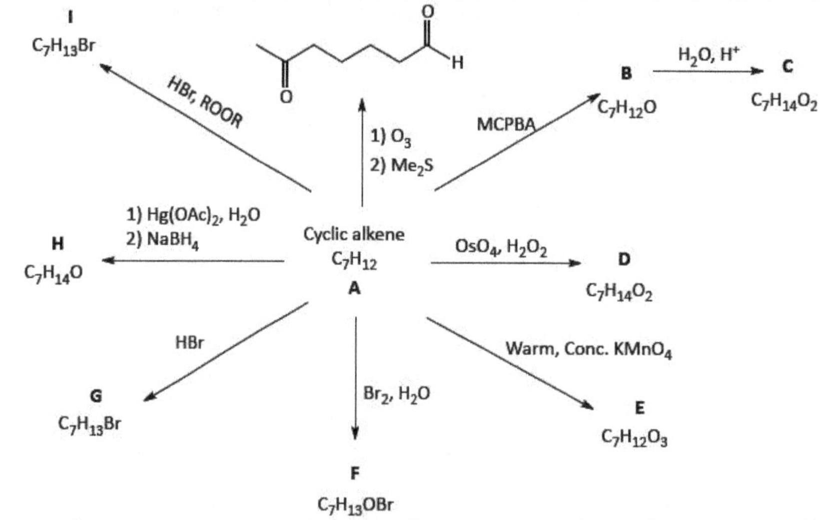

11.40 Provide the reagents necessary to complete the following transformations. (More than one step will be required for each.)

a) 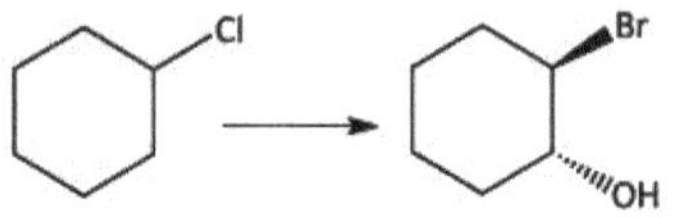

b) 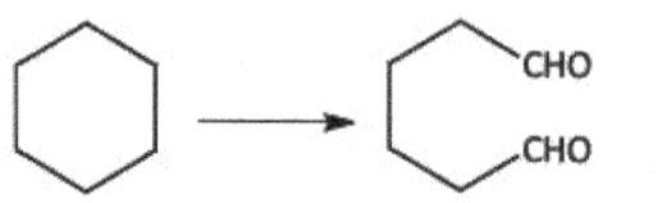

c) 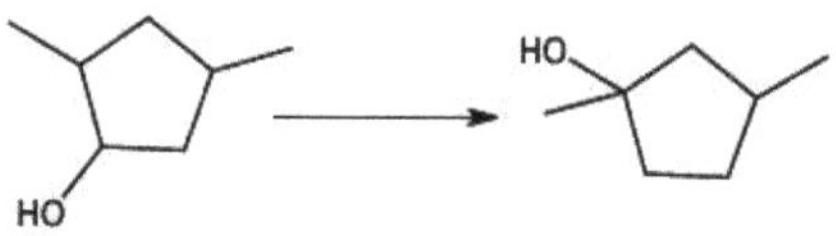

d) 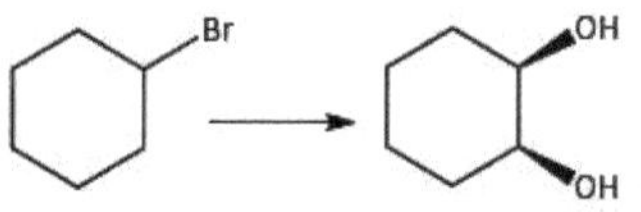

e) 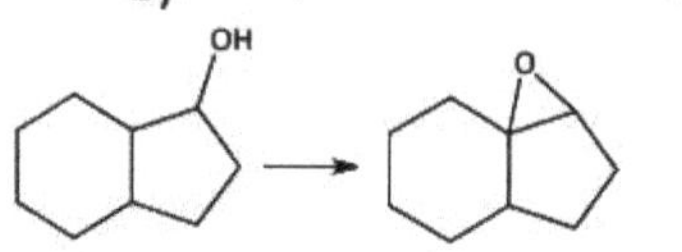

f) 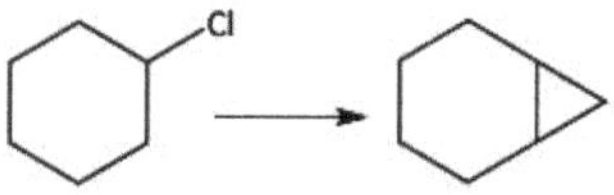

11.41 Show how you would synthesize the following alkynes starting with acetylene.
a) 1-butyne b) 2-pentyne c) 3-hexyne d) 2-methyl-3-heptyne

11.42 What reagents would perform the following reactions?
a) 2-pentyne to *cis*-2-pentene b) 3-hexyne to *trans*-3-hexene c) 1-butyne to butyraldehyde ($CH_3CH_2CH_2CHO$) d) 1-pentyne to 2-pentanone ($CH_3COCH_2CH_2CH_3$) e) 2-hexyne to hexane f) 1-butyne to 2,2-dichlorobutane g) 1-propyne to 2-hexyne h) 3-hexyne to *cis*-3-hexene i) 2-pentyne to 2,3-pentandione ($CH_3COCOCH_2CH_3$) j) 1-pentyne to a mixture of *cis*- and *trans*-1-bromopentene k) 2-butyne to CH_3CO_2H (two ways) l) 2-pentyne to 2,2,3,3-tetrabromopentane m) 2-butyne to *trans*-2-butene n) 3-octyne to octane

11.43 Show the products of the following alkynes following reactions with 1) $HgSO_4$ (mercuric sulfate) and sulfuric acid, 2) disiamylborane followed by hydrogen peroxide and sodium hydroxide, 3) H_2 and platinum, 4) H_2 and Lindlar catalyst 5) Na and NH_3, 6) dilute, neutral $KMnO_4$, and 7) warm, concentrated, basic $KMnO_4$.
a) 1-butyne b) 2-butyne c) 1-hexyne d) 2-hexyne e) 3-hexyne

11.44 What sequence of three reactions can you use to turn *trans*-3-hexee into *cis*-3-hexene? Making an alkyne is helpful.

11.45 Draw the product of the following reaction.

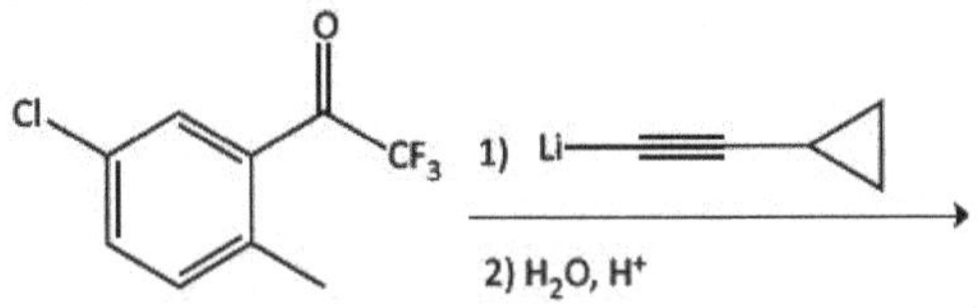

11.46 What is **A**, **B**, and **C** in the following reactions?

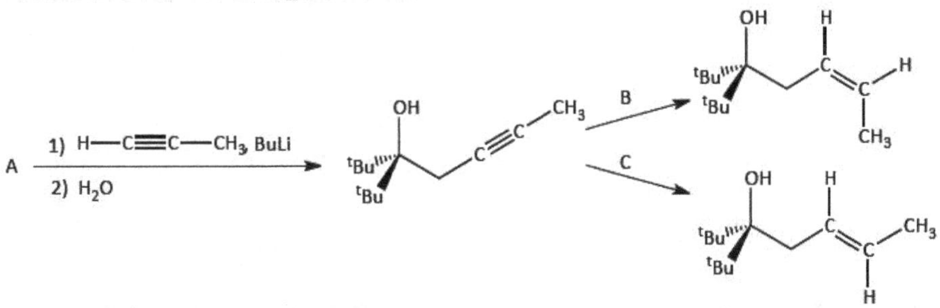

11.47 Show how you would synthesize the following compounds starting with acetylene and any necessary additional reagents.

a) 1-pentyne

b) 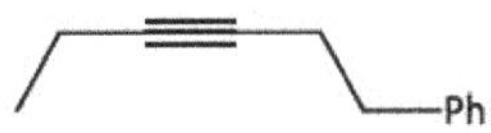

c) 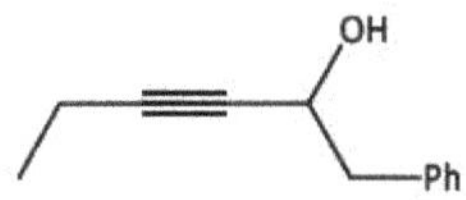

d) 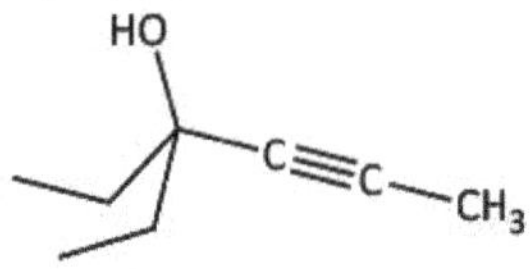

e) 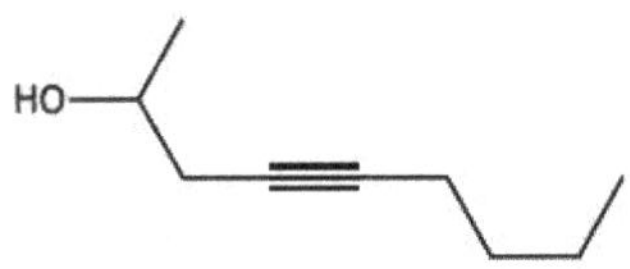

f) $CH_3CH_2CH_2CH_2CH_2CHO$

g) 1,1,2,2-tetrabromopentane

h) *cis*-3-heptene

i) *trans*-2-hexene

j) 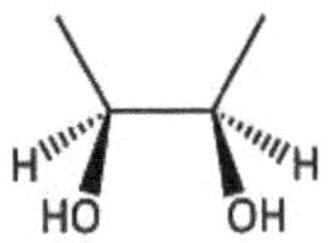

k) 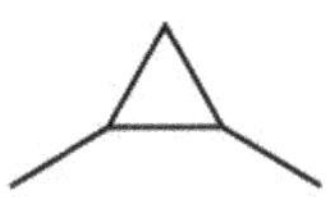

l) 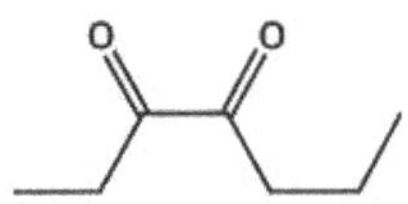

m) 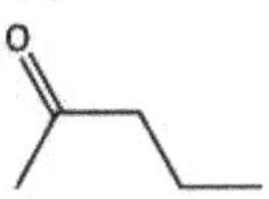

n) 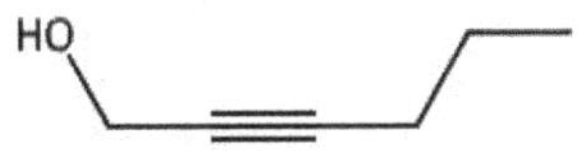

o) 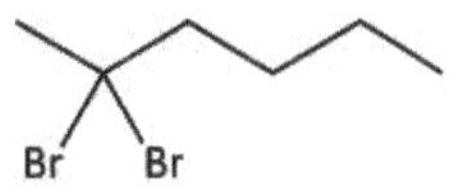

11.48 Show what alkyne, when it undergoes ozonolysis, will make the following carboxylic acids.

a) alkyne that makes $CH_3CH_2COOH + CH_3COOH$

b) alkyne that makes $HOOCCH_2CH_2CH_2COOH + 2\ CH_3COOH$

11.49 a) Show how you would convert 2-bromopropane to 1,2-dibromopropane in two steps. b) Show you would convert 1-methylcyclohexanol to 1-bromo-2-methylcyclohexane in two steps.

11.50 Draw the products for each of the following reactions.

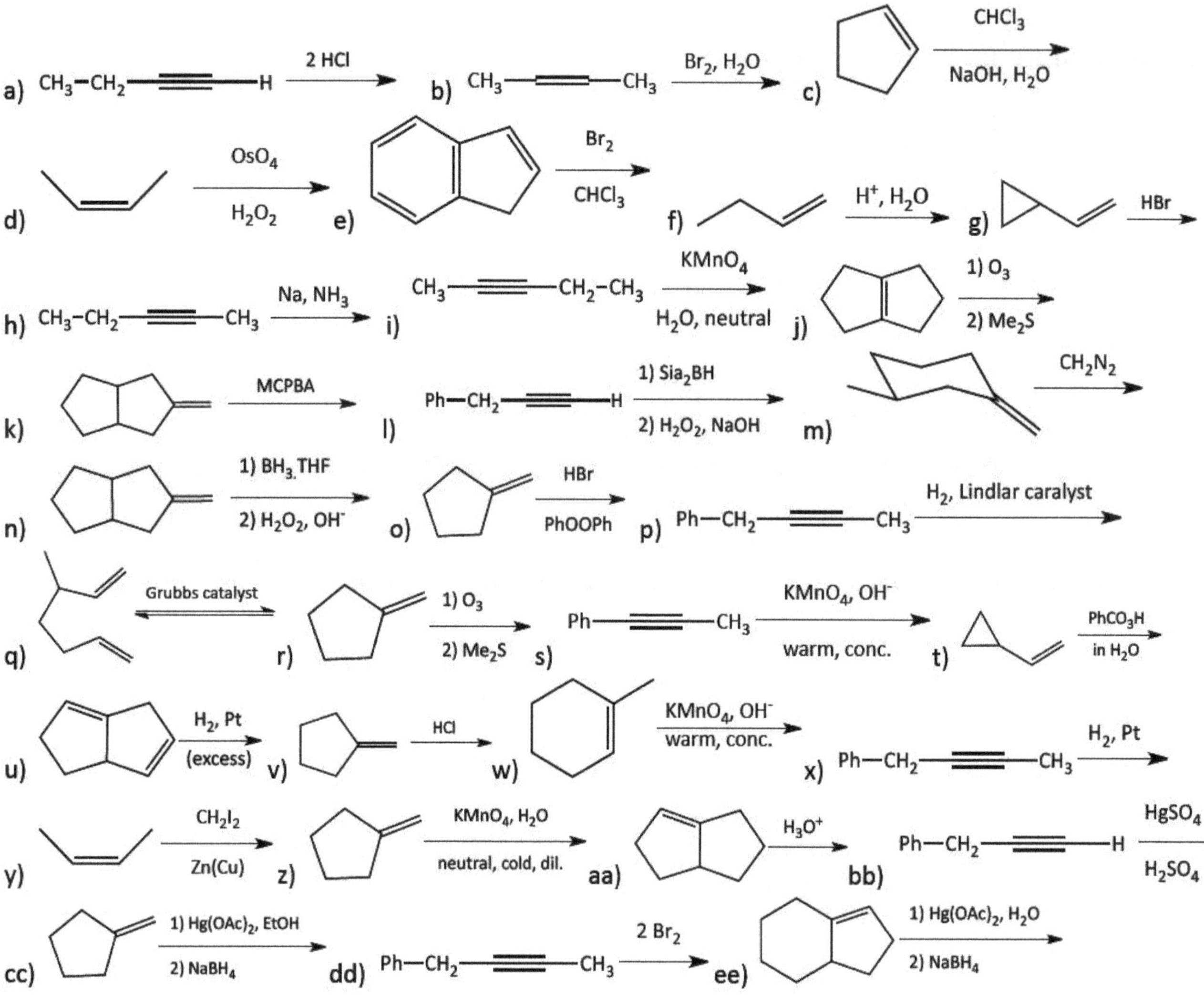

11.51 Draw in the reagents/reactants for each of the following reactions.

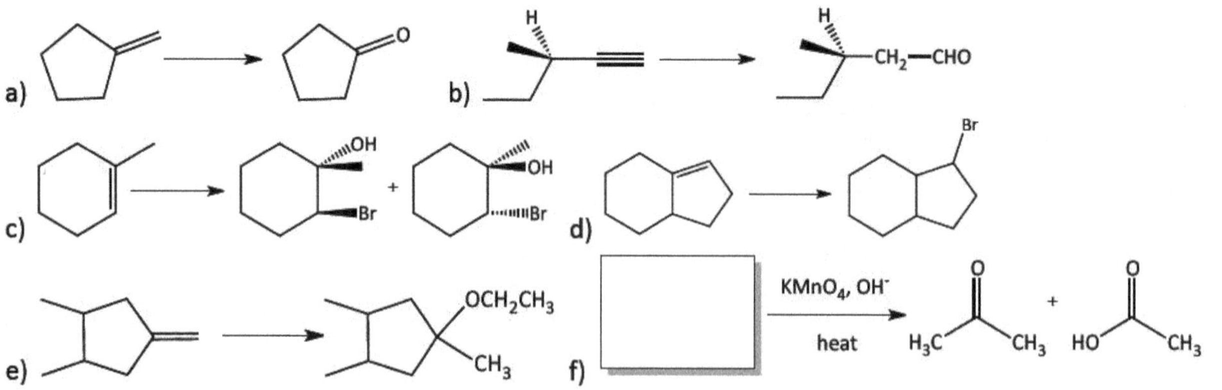

11.52 **Starting from molecules containing only five carbon atoms or less, draw the chemical reactions needed to show how you would synthesize the following compound.**

11.53 **A student was given an unknown compound (A) for an organic chemistry lab. The student was told the formula for unknown A was $C_6H_{11}Cl$.**

a) The student determined that unknown **A** contained how many degrees of unsaturation?

The student then took unknown **A** and reacted it with a bromine solution. Nothing happened. Then, the student attempted to react **A** with a $KMnO_4$ solution. Nothing happened.

b) Think about what you've learned in this chapter that reacts with bromine and/or KMnO4. What does this lack of reactivity tell the student about unknown **A**? Is the degree of unsaturation for unknown **A** from a pi bond or something else? If not a pi bond, then what?

Upon treatment of unknown **A** with ^{t}BuOK, a product, **B**, with the formula C_6H_{10} was formed.

c) How many degrees of unsaturation does compound **B** have?

Compound **B** reacts with bromine (turning the brown solution to colorless) and reacts with cold and dilute $KMnO_4$ (turning the purple solution to brown).

d) What do these reactions tell us about compound B?

Compound **B** reacts with H_2 and Pt to form a compound, **C**, with the formula C_6H_{12}. Compound **C** shows *only one peak* in its ^{13}C NMR spectrum.

e) How many types of carbon atoms are in compound **C**?

f) How many degrees of unsaturation does **C** have?

The student reacts **B** with a more concentrated solution of $KMnO_4$ and warms it up to make compound **D**, $C_6H_{10}O_4$.

g) Propose structures for **A, B, C,** and **D.**

11.54 **An unknown compound, E, decolorizes a solution of bromine in CCl_4.**

a) What does this tell us about compound **E**?

Unknown **E** undergoes hydrogenation (H_2, Pt) to make 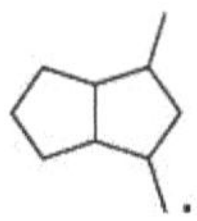. Additionally, **E** undergoes ozonolysis to make 1,2-cyclopentanedione, 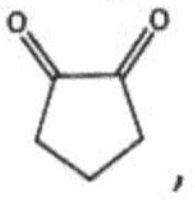, and 2,4-pentanedione, 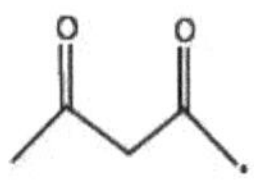.

b) What is the structure of E?

11.55 a) What is the reagent, A, in the following reaction?

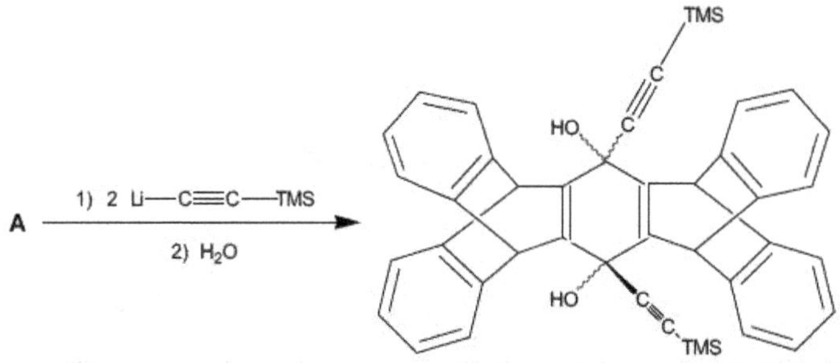

b) What two reagents would you use (one is a terminal alkyne) to make Li-C≡C-TMS.

11.56 An unknown alkyne, A, underwent a series of reactions shown below. Identify compounds A-G.

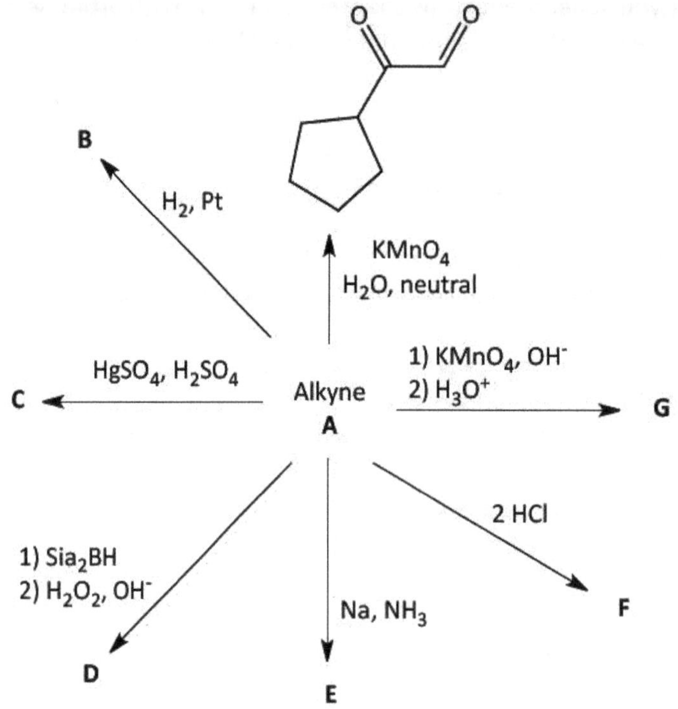

Chapter 12. Nuclear Magnetic Resonance (NMR)

12.1 Introduction

If somebody thinks they may have broken a bone, a doctor may request an X-ray image be taken. An X-ray image, or its 3-D counterpart the CAT scan, is useful for looking at bones. The calcium in the bones absorbs the X-rays well and does not let them pass through the bones and reach the detector. But, if a doctor wants to get an image of organic tissue in the patient, X-rays do not work very well. Instead, the doctor may request an MRI (magnetic resonance imaging) for the patient.

In an MRI, the patient is inserted into a tube that is surrounded by a very powerful magnet. At the same time, invisible, silent, and harmless radio waves are sent towards the patient. This combination of a powerful magnet and radio waves allows the doctor to look at the concentration of nuclei of hydrogen atoms in the organic tissue. Thankfully, our organic tissue is full of hydrogen atoms.

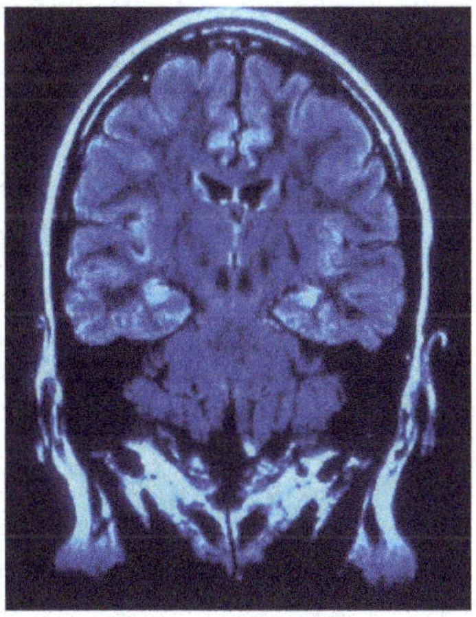

Figure 12.1 MRI of a human head

Photo by jkt_de at Morguefile.com

Using a technique that is nearly identical to an MRI, chemists can insert a chemical compound into a small tube surrounded by a powerful magnet. Radio waves are directed at the sample and information about the location of the hydrogen atoms on the molecule is obtained. Chemists call this technique nuclear magnetic resonance, NMR. Unfortunately for chemists, our pictures are not as nice to look at as what doctors get with an MRI. A chemist's NMR spectrum has information, but it ends up being a little puzzle that needs solved to determine how many and where the hydrogen atoms are on a molecule.

It is called "nuclear" magnetic resonance because the magnet affects the atomic nuclei and not the electrons of the atoms. In common everyday language, the word "nuclear" invokes thoughts of nuclear bombs, radioactivity, and danger. At one time, doctors would use nuclear magnetic resonance (NMR) on patients. But, since patients were unnecessarily avoiding the procedure because they were frightened about radioactivity, the medical establishment changed the name of this diagnostic tool to MRI to avoid the "nuclear" fear. By the way, even though X-rays and CAT scans do use high-energy light and do slightly increase the chances of causing mutations in cells and thus cancer, there is no chance that an MRI can do so. Radio waves are very low energy and *cannot* damage DNA.

12.2 The Theory

A negative electron has a spin state of either $+\frac{1}{2}$ or $-\frac{1}{2}$. Some positively charged nuclei of atoms (^{1}H, ^{13}C, ^{15}N, ^{19}F, and ^{31}P) have a $+\frac{1}{2}$ or $-\frac{1}{2}$ spin state associated with them as well. When these positively charged nuclei spin, they create a magnetic moment and become very similar to tiny bar magnets. Presently, we will focus on the nuclei of the hydrogen atoms (^{1}H or protons) in organic molecules. The NMR analysis of ^{1}H in organic molecules is called ^{1}H NMR or proton NMR.

In the absence of an external magnetic field, H_0, these spinning nuclear magnets point in random directions. All of these spinning nuclear magnets have the same energy. In the presence of a magnetic field, the spinning nuclear magnets align with or against the external magnetic field, H_0. The nuclei magnets that align with H_0 are called alpha spin and those that align against it are called beta spin. It takes a small amount of energy to flip an alpha spin nuclear magnet aligned with the external magnetic field to a beta spin magnet aligned against the external magnetic field. This low amount of energy matches the energy of low-energy, long wavelength radio frequency (Rf) waves.

In a weak external magnetic field, it is fairly easy to flip a nuclear magnet from an alpha to a beta spin. Low-energy radio waves can do this. When the Rf energy matches the "flip" energy, the nuclei are said to be in resonance. In a stronger magnetic field, the energy difference between the alpha and beta spins is greater because it is more difficult to flip the small alpha-spin magnet into a beta spin that would oppose the strong external magnetic field.

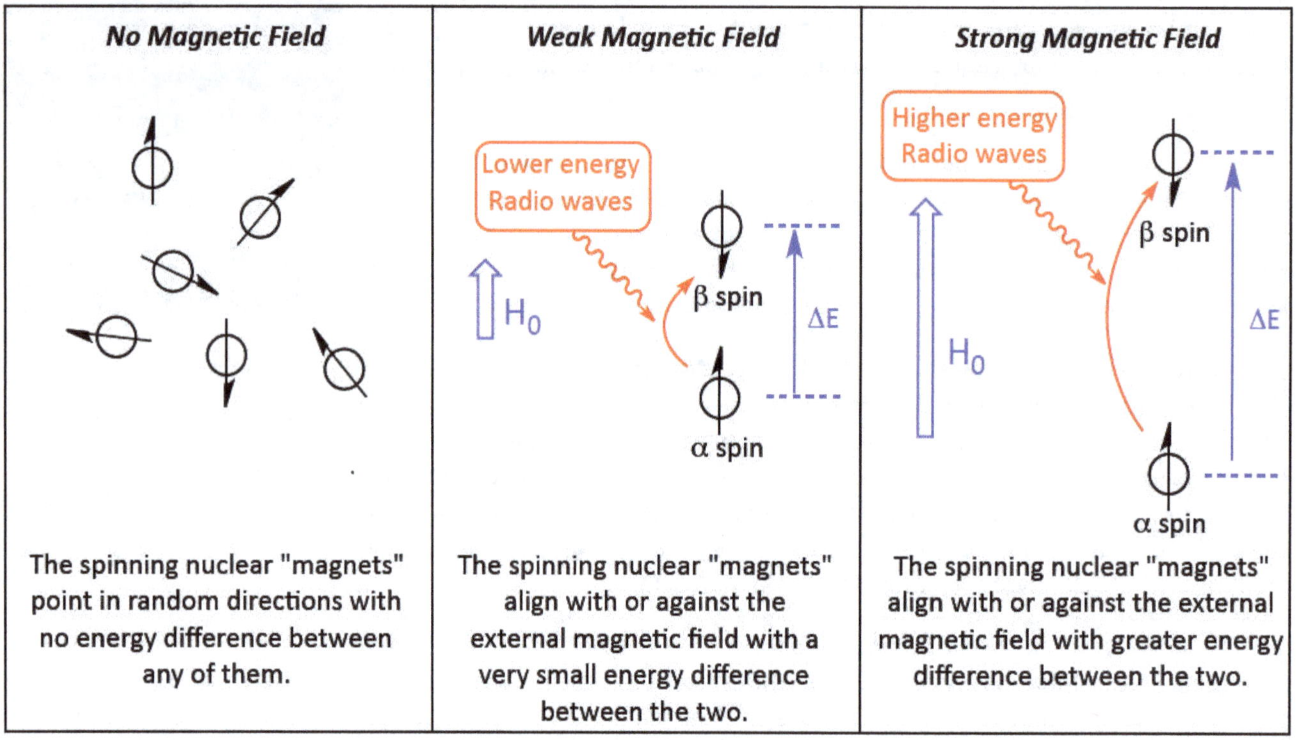

Figure 12.2 Nuclear spins separated by a magnetic field

12.2.1 Shielding

When a loop of wire enters a magnetic field, a current is induced in the wire. This induced current of electrons then induces a magnetic field that opposes the external magnetic field, H_0. A similar effect happens with the electrons that surround the positive nuclear magnet. When an atom enters a magnetic field, H_0, the negative electrons begin moving in a circular pattern and induces a current. This induces a magnetic field that opposes the external magnetic field, H_0. This makes sense, right? If the *positive* nucleus is a magnet that *aligns with* H_0, the *negative* electrons create a magnetic moment that *opposes* H_0.

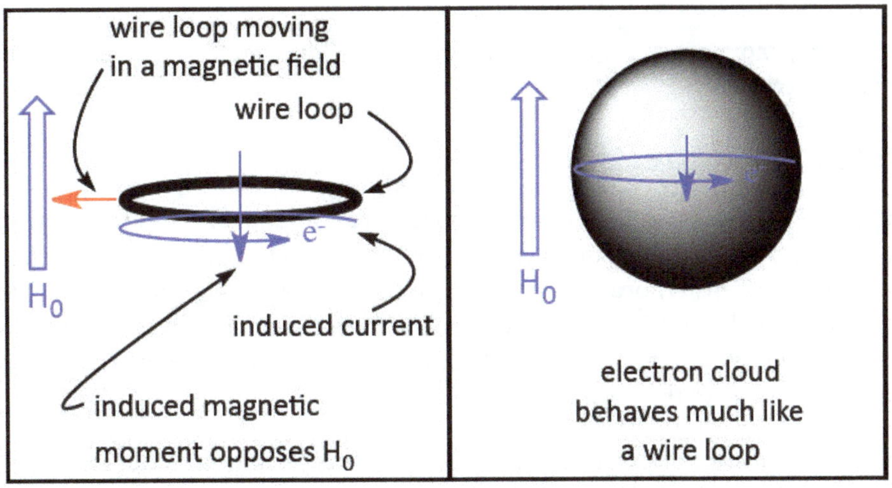

Figure 12.3 Induced magnetic field from electron cloud

In the first pane of the figure below, a naked nucleus, without its electrons, is in its alpha spin aligned with H_0. It absorbs a certain radio frequency and flips to beta spin. In pane two, negative electrons surround the positive nucleus. We say that the nucleus is shielded. With the opposing negative electron magnetic moment subtracted from the positive nucleus' magnetic moment, the overall nuclear magnetic moment is smaller than it was with the naked nucleus. The nucleus no longer absorbs the same radio frequency. In pane three, the external magnetic field is increased by H_0'. This causes the nuclear magnetic moment to grow to the same size as in pane one and the same radio frequency is once again absorbed. Different hydrogen atoms in a molecule have different electron densities around them. Therefore, they have different nuclear magnetic moments. The more electron density around a nucleus, the greater the electron induced opposing magnetic moment, the smaller the net magnetic moment will be at the nucleus. This is great news. Because the proton magnets are different, we can get information about them to help us determine the structure of the organic compound.

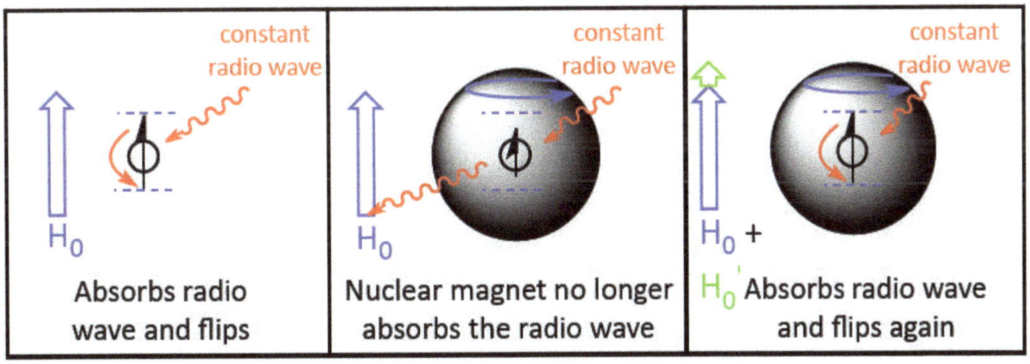

Figure 12.4 Effects of electron shielding of a nucleus

Early, continuous wave, CW, NMR experiments worked this way. A constant radio frequency was directed at the organic sample while the magnetic field strength was changed. When the correct magnetic field strength was reached to cause the net molecular moment at the nucleus to absorb a certain radio frequency, the radio frequency was absorbed. The detector noticed that less radio wave light was reaching it, and a peak was made on the spectrum.

The modern method of performing NMR uses Fourier transform. This is called FT-NMR. It keeps the magnetic field strength constant while the radio frequency is varied. Small nuclear magnetic moments are easier to flip and require a lower frequency radio wave. Larger nuclear magnetic moments are more difficult to flip and require a higher frequency radio wave.

12.3 The Proton NMR spectrum

The right side of the NMR spectrum, is where more shielded protons appear. These protons correspond to lower frequency radio waves. Higher frequency radio waves are located on the left side of the spectrum where peaks for deshielded (less electron density) protons appear. In older, CW NMR, the right side corresponded to higher magnetic fields and the left side with lower magnetic fields. Even though most NMRs are now FT-NMR, some of the older CW-NMR language still exists. The right side of the NMR spectrum is called the *upfield* side and the left side of the NMR spectrum is called the *downfield* side.

All peaks in an NMR spectrum are compared to those made by the protons of a compound called tetramethylsilane (TMS). TMS is the standard. The peak on the NMR spectrum generated by TMS is given a value of zero (0 ppm).

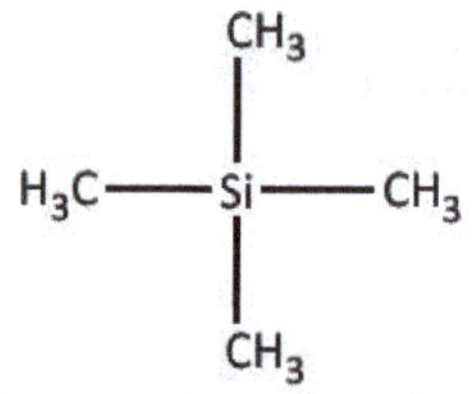

Tetramethylsilane, TMS

TMS was chosen as the standard because the protons in it are very shielded because silicon is less electronegative than carbon. Therefore, the TMS peak is to the right of most organic compounds.

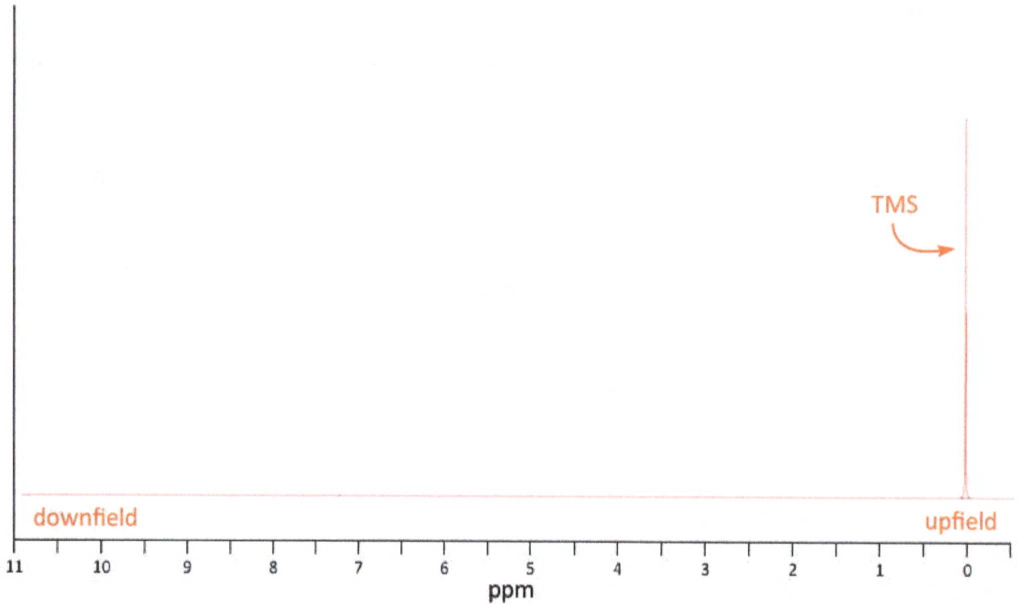

Figure 12.5 ^{1}H NMR spectrum of TMS

NMR instruments come in different magnet strengths. In a very weak magnet instrument, it is easier to flip the nuclear magnet from alpha to beta. In a very strong magnet instrument, it is more difficult to flip the same nuclear magnet from alpha to beta. If the NMR spectrum scale reported absolute values of radio frequency or magnetic strength, all reported values for a compound would be instrument dependent. In order to keep reported values consistent between any strength of instrument, the scale is reported in ppm (parts per million).

The chemical shift is determined by looking at the difference in the chemical shift in Hz compared to TMS divided by the frequency of the magnet used. Let's look at how this works for a peak that would appear at 2 ppm. On a weak, 60 MHz (60 million hertz) magnet, a peak that shows up at 2 ppm resonates at a frequency that is 120 Hz greater than that for TMS. 120 hertz divided by 60 million hertz is 2 parts per million (ppm).

$$\frac{120\ Hz}{60\ MHz} = 2\ ppm$$

On a stronger 200 MHz magnet, that same proton has a resonance frequency of 400 Hz.

$$\frac{400\ Hz}{200\ MHz} = 2\ ppm$$

This way, no matter what strength of NMR instrument is used, the same chemical shift can be reported for a particular proton. This makes things much simpler.

12.4 How many peaks?

Each hydrogen atom that has the same electron density around it shows up in the same location in an NMR spectrum. We say that these hydrogen atoms are "chemically equivalent". In the NMR spectrum for methanol, the three methyl hydrogen atoms appear in the same location on the spectrum because they are chemically equivalent. They appear more upfield. They are more shielded because they have more electron density around them. The hydroxyl (alcohol) hydrogen atom, since it is on the oxygen atom shows up more downfield. It is deshielded. It has less electron density around the hydrogen atom because it is attached to a very electronegative oxygen atom. Because the methyl hydrogen atoms and the hydroxyl hydrogen atom are in a different chemical environment, they show up in different locations.

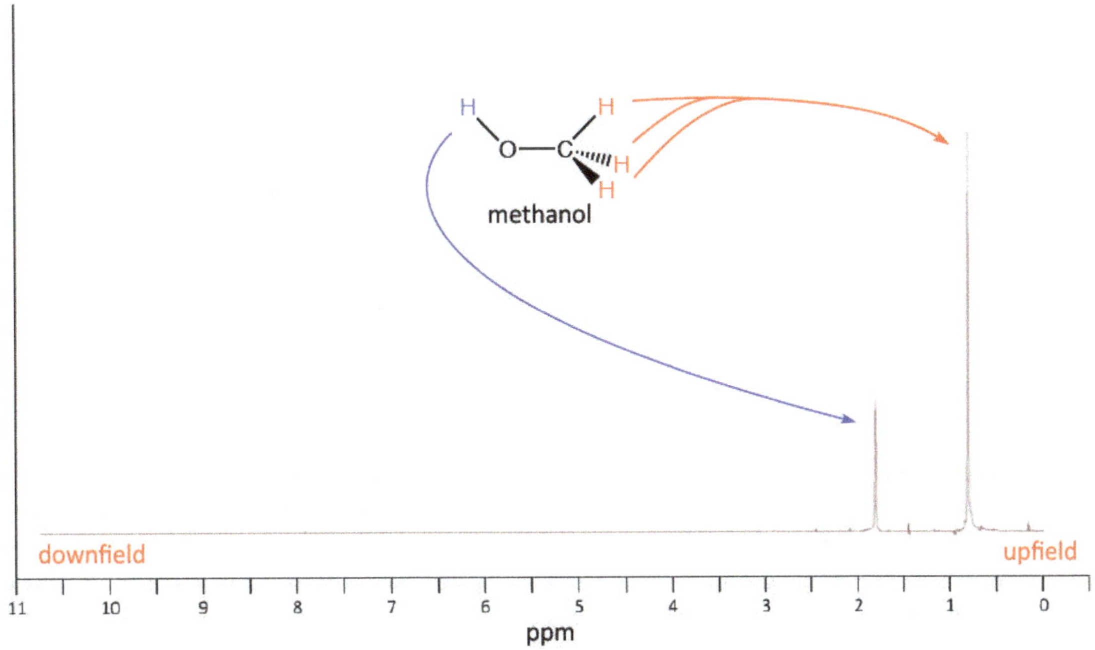

Figure 12.6 ^{1}H NMR spectrum of methanol

12.1 How many peaks would appear in the ^1H NMR spectra for the following compounds?

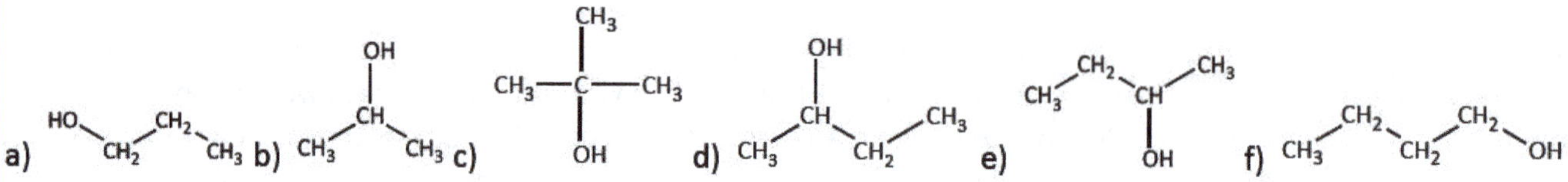

12.5 Integration of peaks

In the ^1H NMR spectrum for methanol, the NMR peak for the three methyl protons has three times the area compared to the size of the peak for the one hydroxyl proton. When protons have the same chemical shift, they add to increase the size of the peak made. The size of the peaks is called the "integration" of the peaks. We can gain information from the integration of the peaks.

Below is the ^1H NMR spectrum for methyl t-butyl ketone. The NMR spectrum shows two peaks, one for all nine of the t-butyl protons which are all chemically equivalent and one peak for the three hydrogen atoms of the methyl group which are all chemically equivalent. The relative integrations (area under the peaks) for the two peaks are 1H:3H.

12.2 Put "A" and "B" in the appropriate squares next to the NMR peak for the appropriate t-butyl and methyl hydrogen atoms for methyl t-butyl ketone.

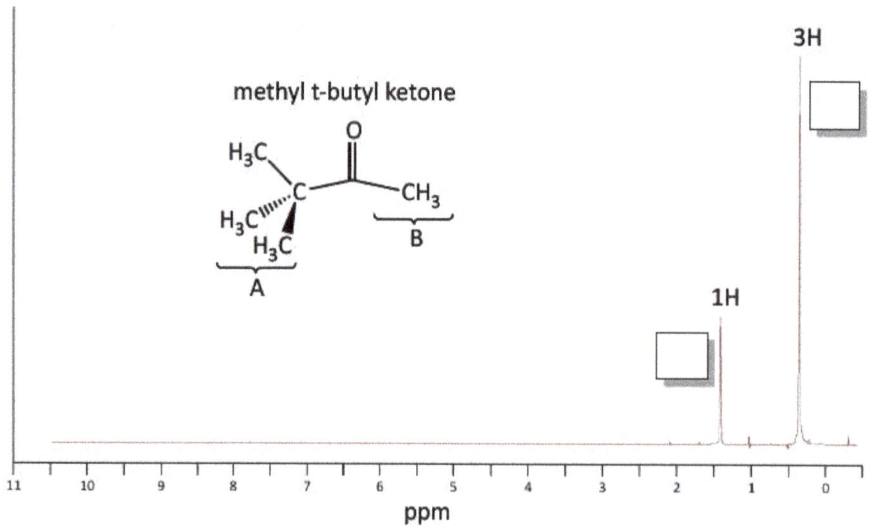

Did you put "A" next to the bigger peak at 1.2 ppm and "B" next to the smaller peak at 3.2 ppm? Of course, we can tell this because the 9 hydrogen atoms of the t-butyl group should make a peak three times larger than the three hydrogen atoms of the methyl group. But, notice that when the NMR instrument computer computes the integrations, it does not always know exactly how many hydrogen atoms each peak is worth. It tells us relative integrations. The spectrum told us one of the peaks was three times the area of the other, 3H:1H. In reality, the integrations were 9H:3H.

12.6 Chemical Shifts

The chemical shift is the location of the NMR peak in ppm. In the NMR spectrum for methanol, the chemical shift for the methyl protons is around 3.4 ppm and the chemical shift for the hydroxyl proton is around 3.9 ppm. The chemical shift for a proton depends on the electron density around the proton. We've learned the more shielded it is (the more electron density around the proton), the more upfield the chemical shift. The more deshielded it is (the less electron density around the proton), the more downfield the chemical shift. Sometimes, the Greek symbol δ is used to describe a chemical shift in ppm. For instance, you might see the chemical shift for the hydroxyl proton of methanol written as δ3.9 or δ3.9 ppm.

The following chart should be memorized to have a general idea of the chemical shifts for various types of protons.

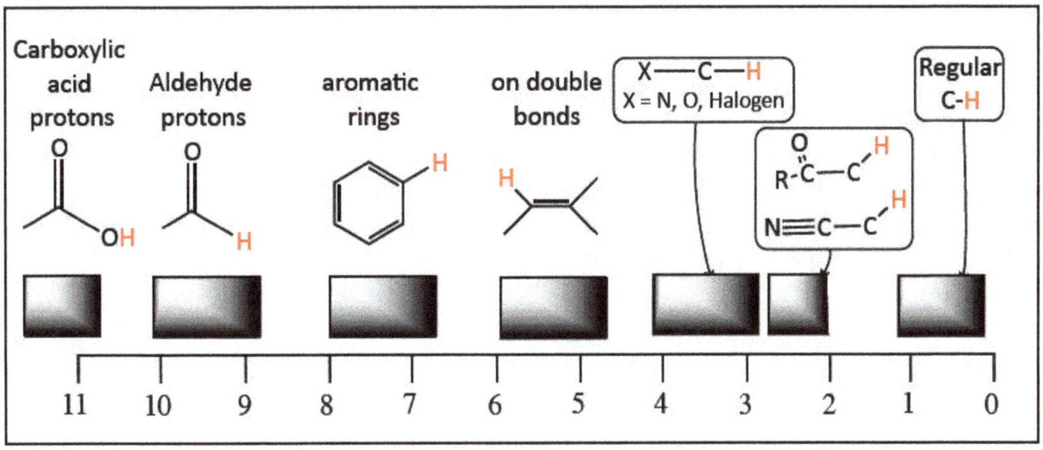

Figure 12.7 General location of peaks in ^{1}H NMR spectrum

Regular, aliphatic protons show up around 0.5 to 1 ppm. If there is a special group next door to a proton, it begins to be pulled downfield. If a carbonyl group is next door, a proton's resonance is around 2-3 ppm because an oxygen atom is three bonds away and begins to pull electrons away from the proton. If a special group like an oxygen or a halogen atom is next door, it is two bonds away from the proton and deshields the proton even more. These protons show up around 3-4 ppm. Protons on double bonds show up around 5-6 ppm and protons on aromatic rings show up around 7 ppm. This may seem strange at first because there are no electronegative atoms on these compounds to pull electron density away from protons. An interesting phenomenon happens which causes these protons to show up in this area. Let's look at the aromatic ring situation first.

12.6.1 Deshielding of benzene and alkene protons

A benzene ring contains three pi bonds that overlap around the entire ring. This overlapping of pi electrons can be thought of as a loop of wire. Inside an external magnetic field (H_0), a current of electrons flows around this ring- a ring current. When a current of electrons flows around a wire loop, which induces a magnetic field, just like in an electromagnet. This induced magnetic field reinforces the magnetic field of H_0. In effect, the little magnets of the protons on the benzene ring get a little bigger because of this induced magnetic field. They become harder to flip and end up going more downfield (as if they are deshielded) to around 7 ppm. In fact, since they go downfield, these protons are called deshielded protons.

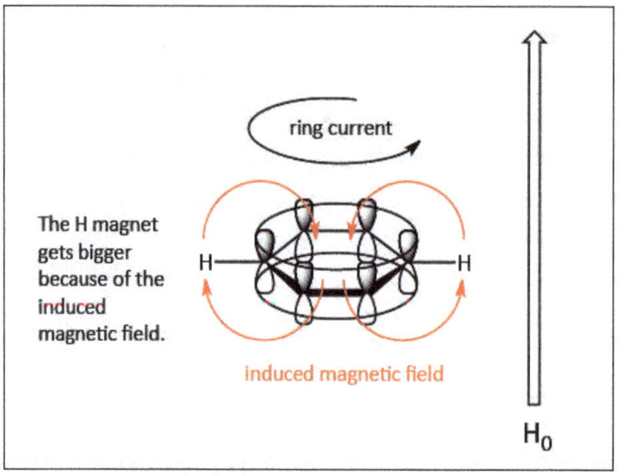

Figure 12.8 Why benzene protons are deshielded

For a double bond, a similar phenomenon is occurring. Once again, there is a pi bond because of the double bond. The external magnetic field, H_0, makes a ring current in the pi bond. This is smaller than in the benzene ring because there are fewer electrons in the double bond system. This induces a small magnetic field which causes the protons on the double bond to become slightly bigger magnets. They resonate around 5-6 ppm, not as far downfield as benzene protons.

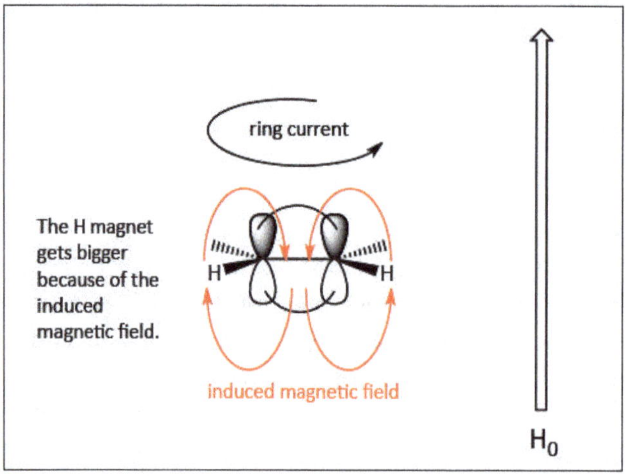

Figure 12.9 Why alkene protons are deshielded

12.6.2 Deshielding of aldehyde and carboxylic acid protons

Aldehyde protons appear between 9-10 ppm because the proton is near a double bond, pi bond so there is an induced magnetic field AND there is an electronegative oxygen atom nearby. The proton of a carboxylic acid is very far downfield around 11-12 ppm for the same reasons as the aldehyde, but it is even further downfield because there is another oxygen atom directly attached to the proton.

12.3 Give an approximate ^1H NMR chemical shift for each marked hydrogen atom in these compounds.

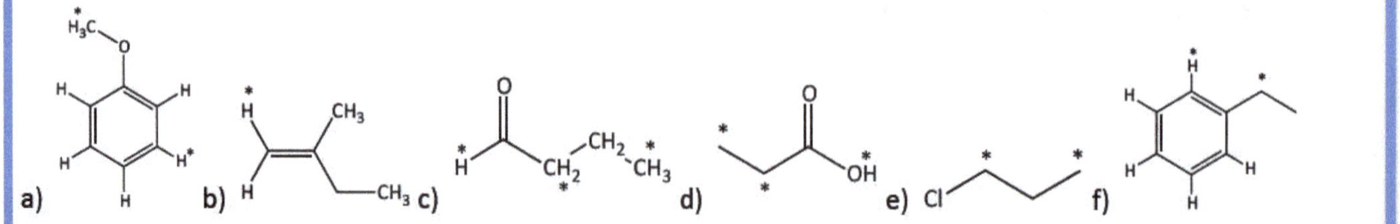

12.7 Spin-Spin splitting

The NMR peaks we have encountered so far have been single peaks. I have some good news and some bad news for you. The bad news first. . . The bad news is that not all peaks in NMR spectra are single peaks. They get more complicated. The good news is that more complicated looking peaks give us much more information about a molecule! Let's take a look at a ^{1}H NMR spectrum that has some peaks that are not single looking peaks, they are "split" into more complicated peaks. This is called spin-spin splitting. The number of peaks it splits into is called the multiplicity of the peak.

12.7.1 N+1 rule

A ^{1}H NMR resonance is split into peaks because of its hydrogen atom neighbors. We can tell how many peaks a resonance will split into, its multiplicity, by counting how many neighboring hydrogen atoms there are and adding one. This is the N+1 rule.

N+1 rule: The multiplicity of a ^{1}H NMR resonance = the number of neighboring protons + 1.

Splitting patterns can be more complex than this. But, for now, this definition will suffice. Just realize that in a real-life scenario, your peaks may be a little more complex and more advanced training will be needed to deal with those situations.

Let's walk through the ^{1}H NMR spectrum for 2-butanone to become more familiar with this. The protons of 2-butanone are labeled, A, B, and C as are the corresponding peaks in the ^{1}H NMR spectrum for these protons. Peak A is made from the three H atoms of the far left methyl group, peak B is made from the two H atoms of the CH_2 methylene, and peak C is made from the three H atoms of the far right methyl group.

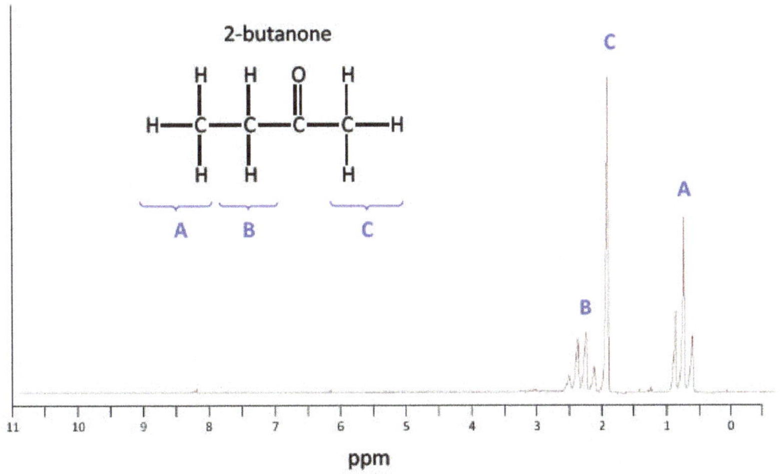

Figure 12.10 ^{1}H NMR spectrum of 2-butanone

Peak A is split into three peaks. This is called a triplet. It is a triplet because the A protons are next to the CH_2 group, B. Since there are two neighboring H atoms, the N+1 rule is 2+1 = 3 peaks, a triplet.

Peak B is split into four peaks. This is called a quartet. It is a quartet because the B protons have three H atoms, A, as neighbors to its left. It has no neighboring H atoms to its right, the C=O, carbonyl. Therefore, with three proton neighbors, the N+1 rule is 3+1 = 4 peaks, a quartet.

Peak C is not split into extra peaks. It is called a singlet. It is a singlet because its neighboring carbon atom does not contain any H atoms, it is a C=O, carbonyl. With 0 neighboring protons, the N+1 rule is 0+1=1 peak, a singlet.

The N+1 rule gives us valuable information about which proton groups are neighboring which proton groups in a molecule.

12.7.2 Why does spin-spin splitting happen?

Let's analyze the ^{1}H NMR spectrum of 1,1,2-trichloroethane.

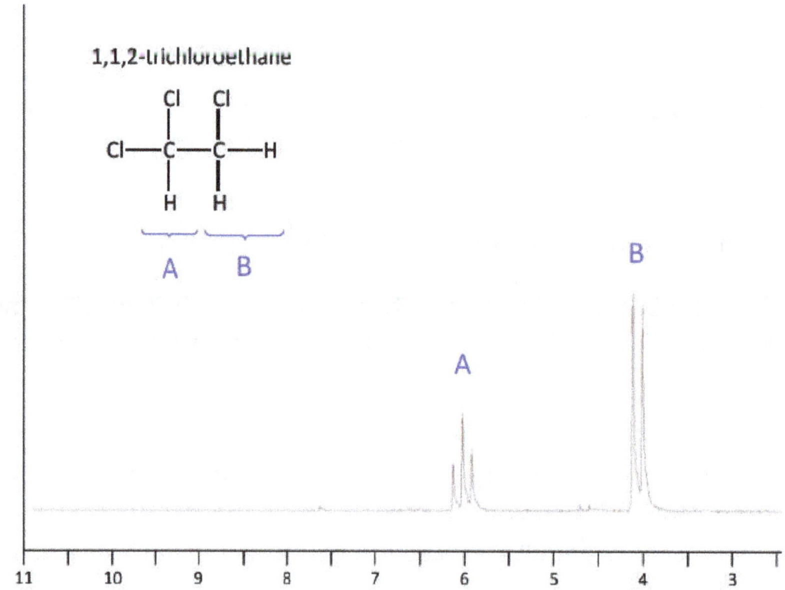

Figure 12.11 ^{1}H NMR spectrum of 1,1,2-trichloroethane

The B protons have one neighboring proton, A. The N+1 rule makes peak B two peaks, a doublet. Proton A is a tiny magnet. It is a magnet that points up half with the H_0 external magnetic field half the time and it points against H_0 half the time. When it points up, it reinforces the magnet made by protons B making it bigger. When it points down, it makes the magnet made by protons B smaller. This makes the peak for protons B show up with two different chemical shifts, a doublet. The odds of each of these happening are 1:1, so the two peaks are equal size.

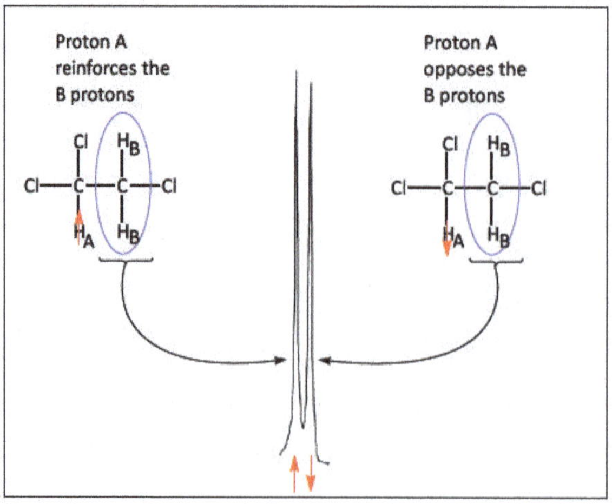

Figure 12.12 Why a doublet is formed from one neighboring proton

The A protons have two neighboring protons, the B protons. The N+1 rule makes peak A three peaks, a triplet. The B protons are both tiny magnets. The B protons can both point up, they can both point down, or one can point up while one points down (there are two ways to do that). This is how the three peaks of the triplet are formed. Because there are two ways to make the center peak of the triplet, up-down and down-up, the pattern for the triplet is a 1:2:1 pattern with the center peak twice as tall as the side peaks.

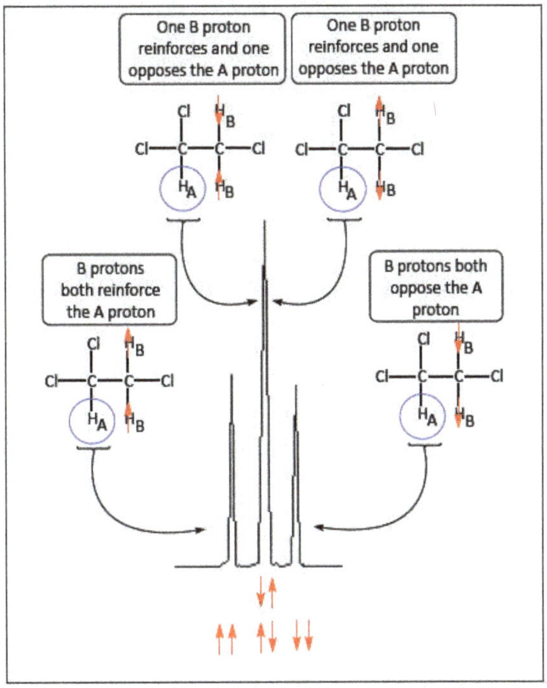

Figure 12.13 Why a triplet is formed from two neighboring protons

12.7.3 The pattern for multiplets

The shape or pattern for a normal multiplet is one where the peaks on the sides are smaller and the peaks get larger and larger towards the center peak of the multiplet. We saw this with the triplet above. The pattern can be predicted using Pascal's triangle.

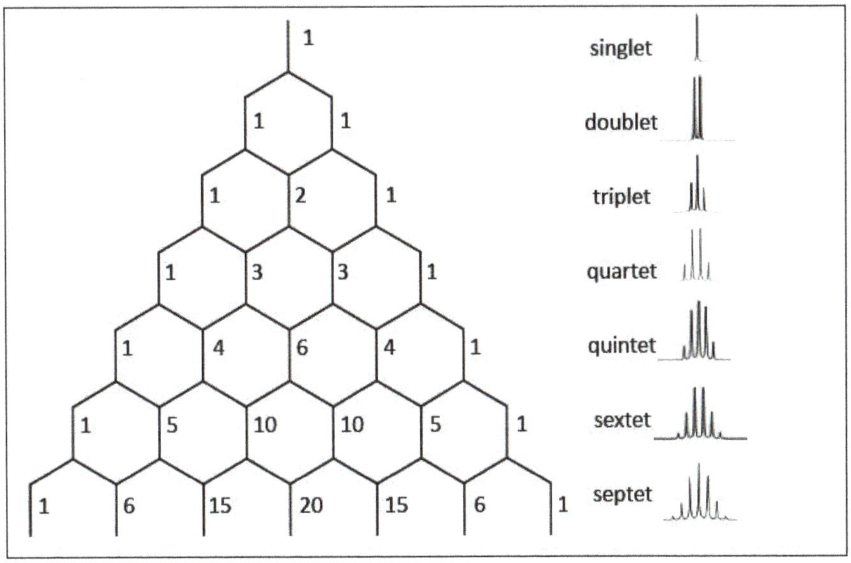

Figure 12.14 Pascal's triangle indicating the shape of multiplets

12.4 Predict the multiplicity for each set of hydrogen atoms in the following molecules. You may want to draw in hydrogen atoms if they are not drawn.

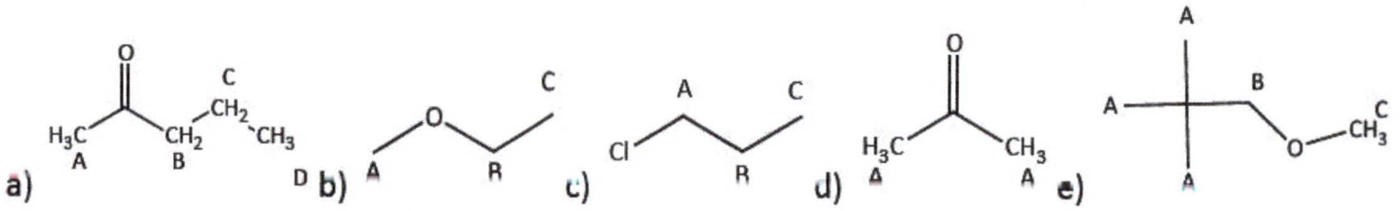

12.5 Is the following peak a quartet or two doublets? How can you tell?

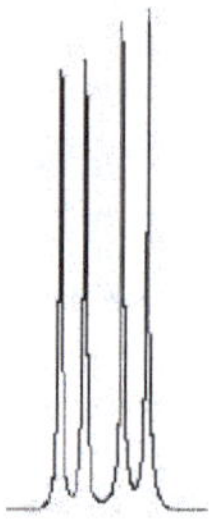

12.7.4 If they couple, they "point" towards each other

Usually, our multiplet shapes are not perfect. They give us a little hint. If a set of neighboring protons couple each other, their corresponding peaks point towards each other. If an arrow is drawn that points up, the peaks look like they point towards each other.

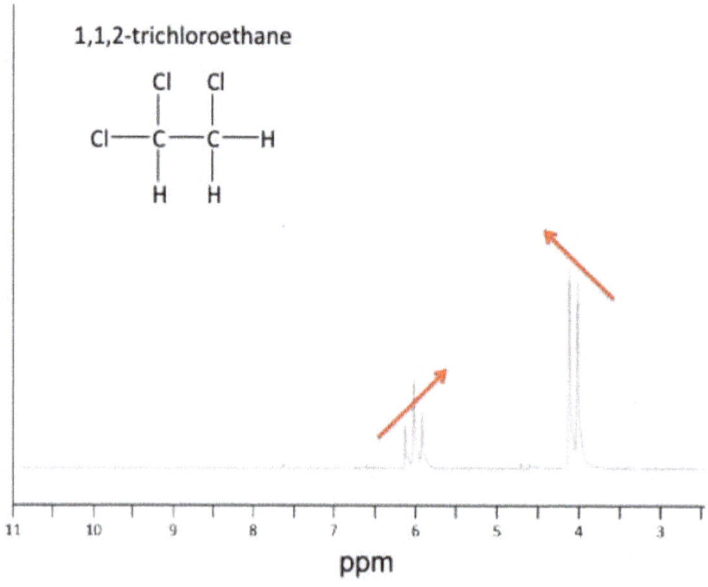

Figure 12.15 How peaks made from protons that couple each other "point" towards each other

12.6 In each box, put a letter, A, B, or C to indicate which protons of 3-chloro-2-butanone that ^{1}H NMR peak represents. Then, draw in the "pointer" arrows to show that two of the peaks hint that protons A and B couple each other.

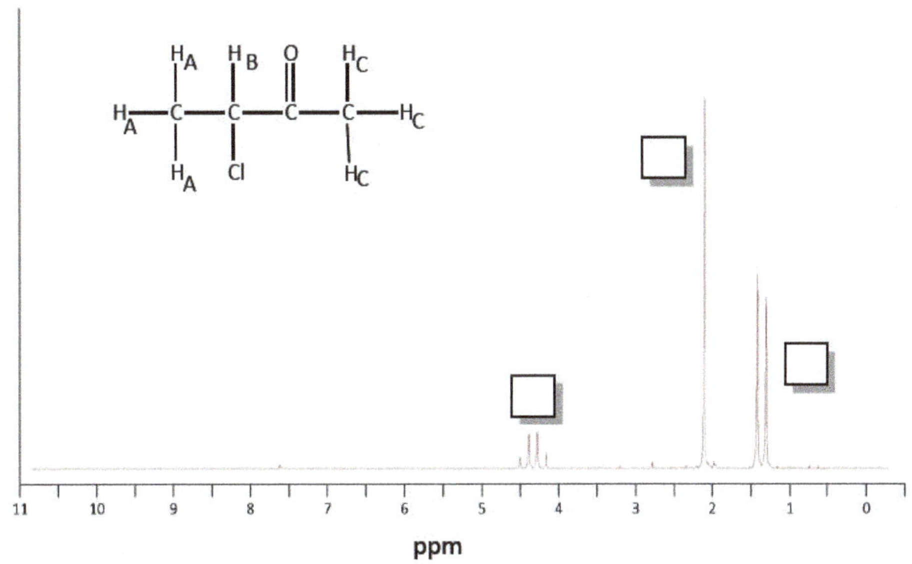

12.7 In each box, put a letter, A or B, to indicate which protons of bromoethane that ^{1}H NMR peak represents. Notice the typical triplet-quartet pattern for an ethyl group. This is a common pattern, so it helps to learn it.

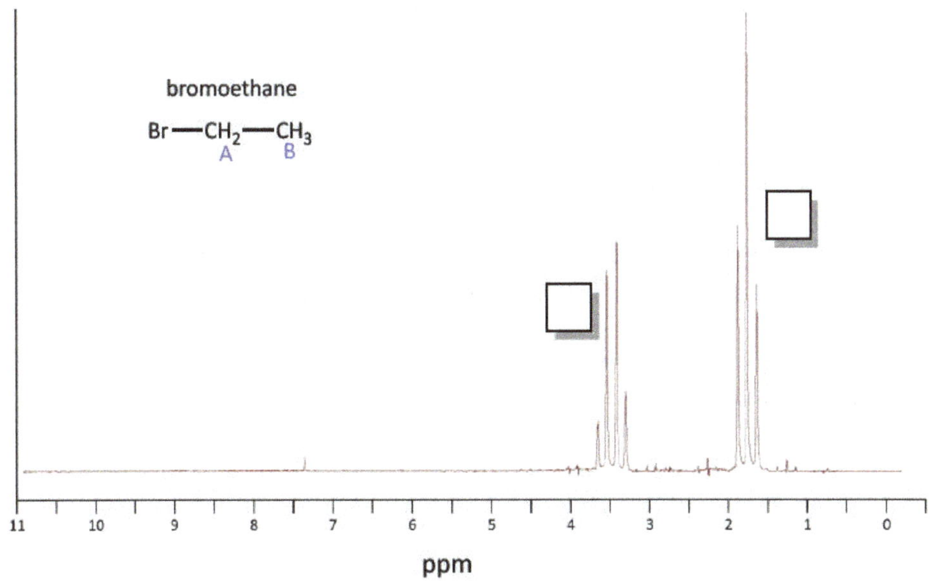

12.7.5 Coupling constants

When spin-spin splitting occurs, the distance between the peaks of the multiplet is called the coupling constant, J. The coupling constant is usually measured in Hz. The triplet and quartet of bromomethane are shown here. I've zoomed in and put them next to each other to make the coupling constants easier to see.

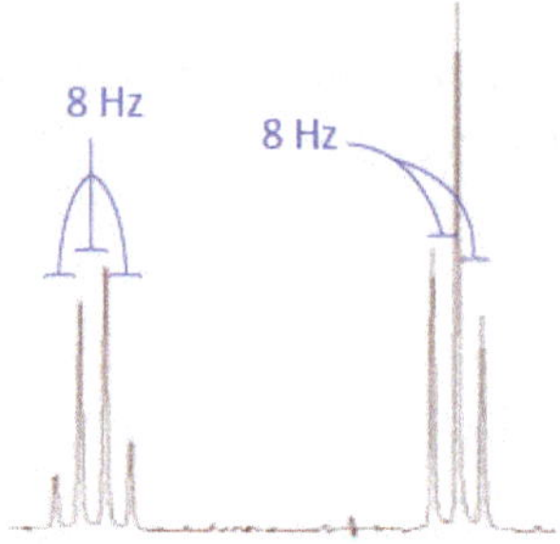

Figure 12.16 Equal coupling constants for two peaks from protons that couple each other

Notice three things.

1. The coupling constant remains the same, 8 Hz, between each of the peaks of the quartet and the triplet. The coupling constant does not change inside of a normal multiplet, it must remain constant throughout.

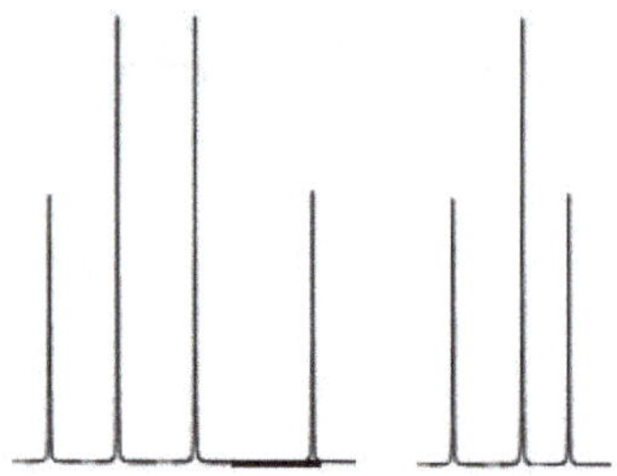

Figure 12.17 If two peaks are from coupling protons, J must be equal in both.

2. If I were to shake your hand, you shake my hand right back. It is impossible for me to shake your hand without you shaking my hand back. It is the same with neighboring protons. If two proton groups neighbor each other, they must split each other, or couple each other, in equal amounts. In the triplet-quartet pattern in our bromoethane example, notice that both the triplet and the quartet have a coupling constant of 8 Hz. This is because they are neighbors and couple each other.

In the spectrum below, there are four sets of doublets, labeled A, B, C, and D. One pair of the doublets is made by protons that couple each other and another pair is made by protons that couple each other.

12.8 Which of the following pairings makes sense?

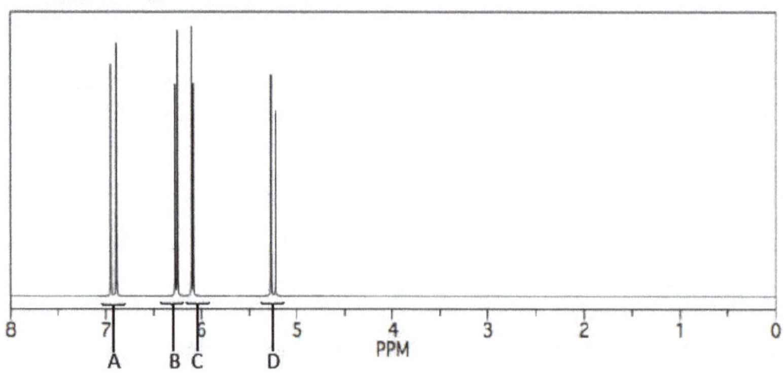

a) A-B and C-D   b) A-C and B-D   c) A-D and B-C

3. The size of a coupling constant tells us something about what types of protons are neighboring or coupling each other. We won't get too deep into this. But, a regular coupling of normal, aliphatic hydrogen atoms is 7-8 Hz. Protons on a double bond might have a coupling constant of 10-15 Hz.

12.8 OH and NH protons

Protons on oxygen or nitrogen atoms are a little strange. They can show up anywhere between 2-6 ppm. Because protons on OH and NH are involved in hydrogen bonding, their chemical shift can change depending on the temperature and concentration of the solution. They are not always broad peaks. Sometimes they stay on the molecule enough to couple.

Usually, these peaks are broad, hump-like peaks. This is because the hydrogen atom disassociates from the OH or NH bond. It falls off and returns to the molecule, so there usually is not time for it to couple with other protons.

One laboratory trick to help determine if a ^{1}H NMR peak is from an OH or NH proton is to add D_2O, heavy water, to the NMR sample. D_2O interacts with the OH or NH proton and the hydrogen atom and deuterium atoms switch to form O-D or N-D. The ^{1}H NMR peak then disappears confirming OH or NH. Protons on carbon atoms will not trade with deuterium of D_2O. A new peak will grow in the spectrum, though. HOD is formed in the reaction.

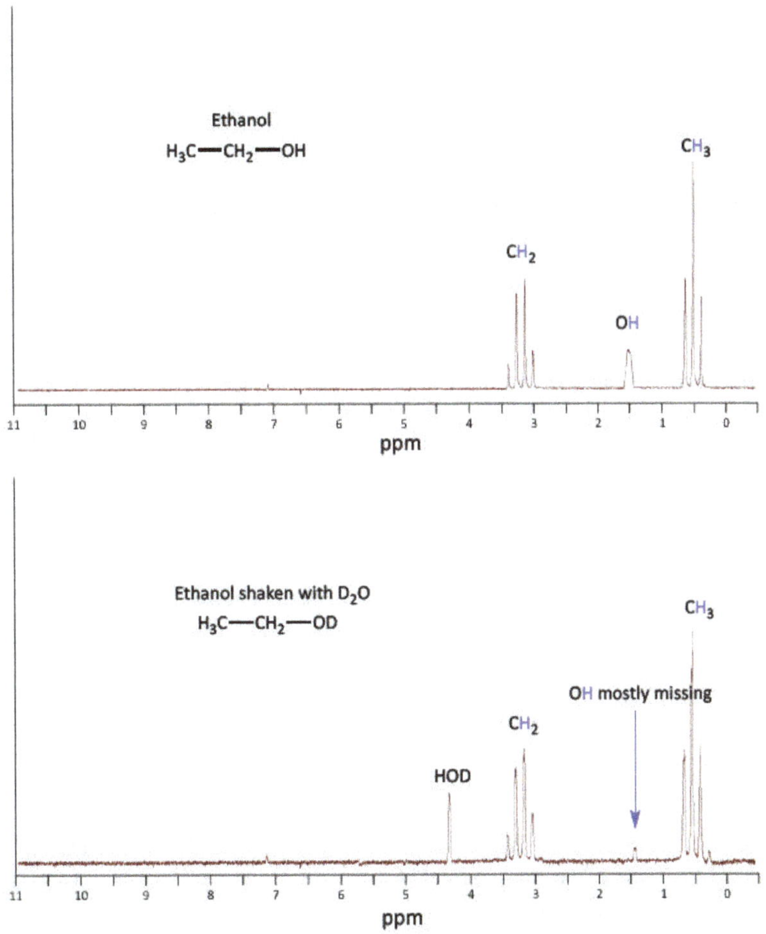

12.9 The ^1H NMR spectrum of isopropanol is below without its scale shown. Based solely on peak splittings and peak shapes, label each peak with the appropriate letter to match the protons associated with each peak.

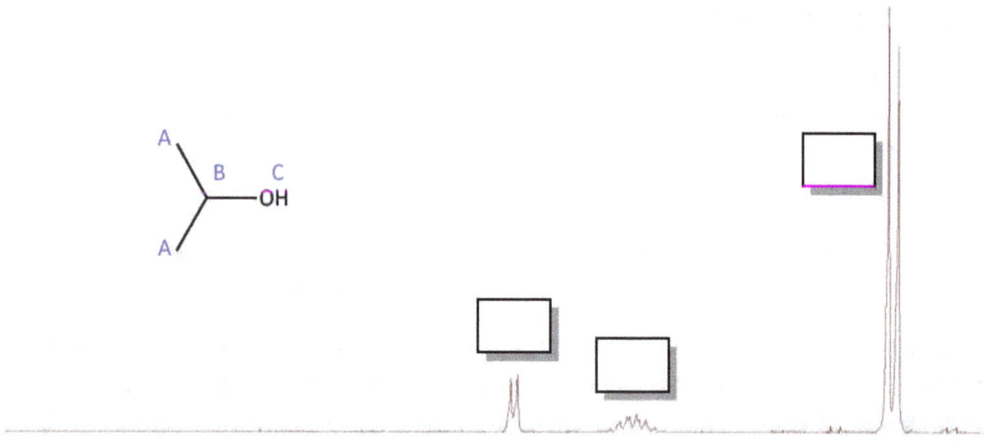

12.9 Carbon NMR

Hydrogen atoms are not the only atoms that can be observed using an NMR. Carbon atoms can also be seen. But, not all isotopes of carbon can be seen. Sadly, the nucleus of the most common isotope of carbon, carbon-12, does not have a magnetic moment. But, carbon-13, because of its odd number of nucleons, does have a nucleus with a magnetic moment. We call a carbon NMR a ^{13}C NMR (pronounced C-13 NMR).

Only about 1.1% of the carbon in nature is carbon-13. So, for a given organic sample, each carbon atom only has a 1.1% chance of being observed in an NMR. Because of this, it is a little more difficult to obtain a carbon NMR. Therefore, we usually need a more concentrated sample to scan and we usually scan the sample for a longer period of time than for a proton NMR.

12.9.1 Carbon NMR chemical shifts

The chemical shifts for carbon atom peaks are usually 15-20 times greater than their proton counterparts. An aldehyde proton shows up around 9-10 ppm in a ^1H NMR spectrum, whereas the carbon atom of an aldehyde carbonyl shows up around 190 ppm in a ^{13}C NMR spectrum. This is because the carbon atoms are one bond closer to deshielding groups than are the protons.

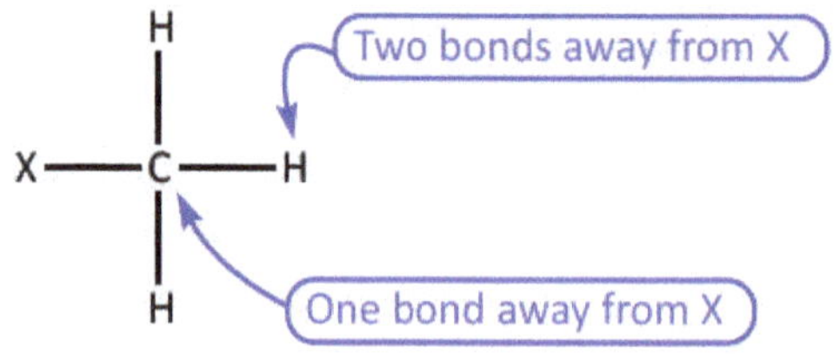

The typical ^{13}C NMR spectrum goes from 0-200 ppm compared to 0-10 ppm for ^1H NMR. In general, it is best to view a ^{13}C NMR spectrum as being broken into four different portions, 0-50 ppm, 50-100 ppm, 100-150 ppm, and 150-200 ppm. Become familiar with what types of carbon atoms appear in these different portions of the spectrum. These divisions are not perfect, so you need to have a little flexibility when interpreting a ^{13}C NMR spectrum.

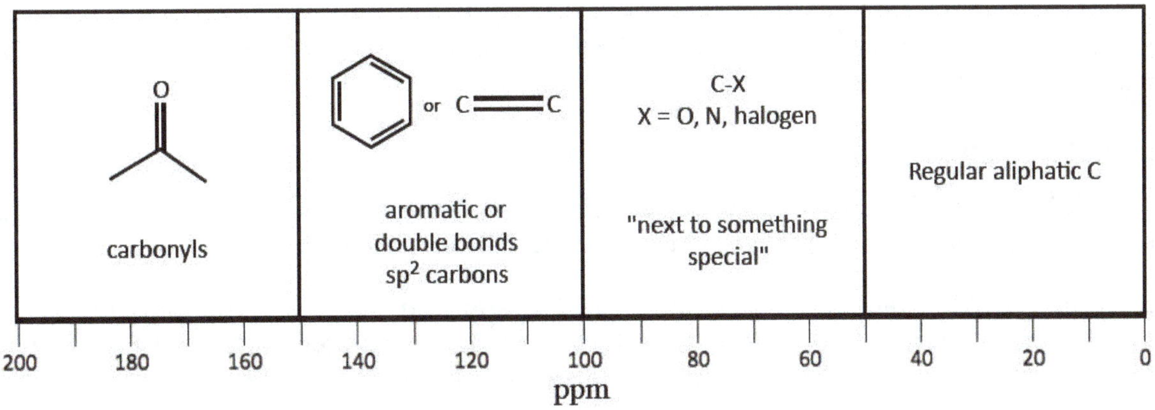

In the ^{13}C NMR spectrum for 1-penten-3-one, we have five different carbon atoms. The carbonyl, C, appears near 200 ppm. The two C=C carbon atoms, A and B, appear between 100-150 ppm, and the two aliphatic ethyl carbon atoms, D and E, appear under 50 ppm.

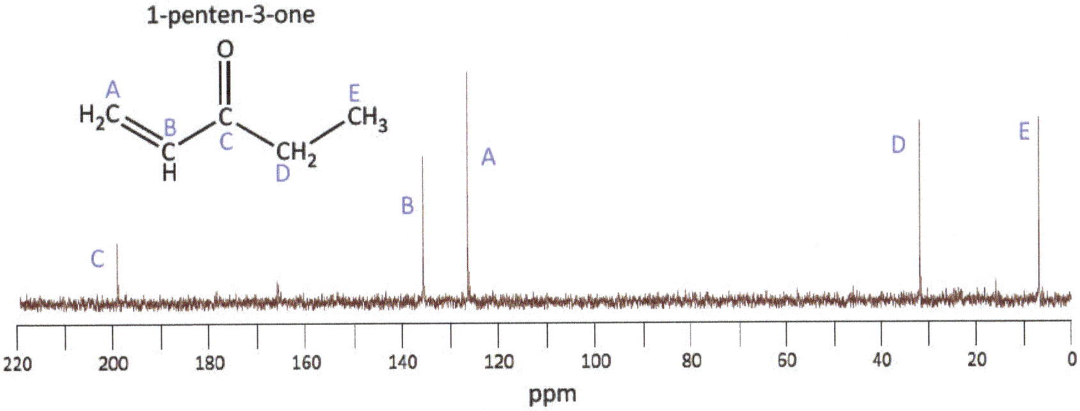

12.10 Which region in the ^{13}C NMR spectrum would you find the peak for each carbon atom marked with "*": 0-50 ppm, 50-100 ppm, 100-150 ppm, or >150 ppm?

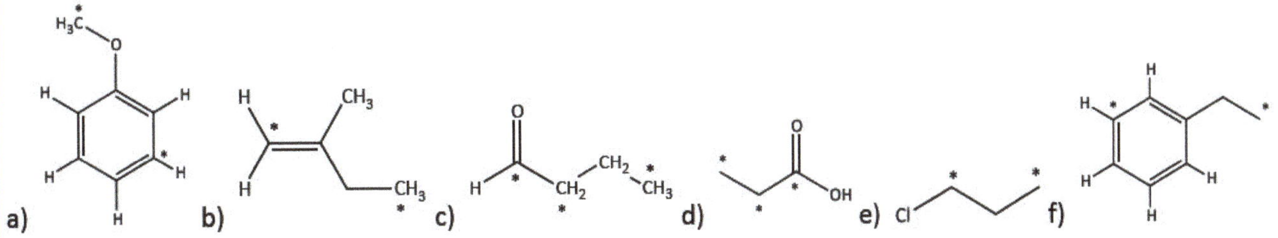

12.9.2 Integrations

Carbon NMR spectra are not usually integrated to tell how many carbon atoms are involved to make each peak. This is because not all carbon atoms are created equally. Some types of carbon atoms make larger peaks than other types of carbon atoms. The more hydrogen atoms attached to a carbon atom, the larger the carbon peak tends to be. CH_3 and CH_2 carbon peaks are the tallest, CH peaks are sometimes a little smaller, and quaternary carbon peaks are usually noticeably smaller. In the above spectrum for 1-penten-3-one, notice that carbon B, a CH, is a slightly smaller peak than carbon A, a CH_2. The quaternary carbonyl carbon, C, is the smallest peak in the spectrum. Also, there is only one C carbon and one B carbon, but carbon B is twice as tall as carbon C. That is why we cannot integrate this spectrum.

12.9.3 Off-Resonance Decoupled ^{13}C NMR spectra

So far, the ^{13}C NMR peaks have always been single peaks. This is a common way to do ^{13}C NMR. But, it is possible to get more information out of the peaks by letting them become multiplets.

Since the signals in ^{13}C NMR are so weak, an NMR method is performed to put the entire signal from a ^{13}C nucleus in one spot, to make the peaks as large as possible. This method is called proton-spin decoupling. In this method, broadband radio noise is transmitted to the proton magnets causing them to be in resonance and flip like crazy. The carbon nuclear magnet then sees an average of these proton magnetic spin states. The carbon-hydrogen splitting is eliminated.

To get more information from our peak, this broadband radio noise can be turned off and the proton magnets can then be in either a spin up or a spin down state. Allowing the carbon nucleus to couple with the protons is called *off-resonance decoupling*.

If a ^{13}C nuclear magnet is next to one proton, and that proton can be either spin up or spin down, the N+1 rule would tell us that ^{13}C nucleus would show two peaks on the NMR, a doublet. So, the peak for a ^{13}C nucleus with zero protons attached is a singlet, with one proton attached is a doublet, two protons attached is a triplet, and three protons attached is a quartet.

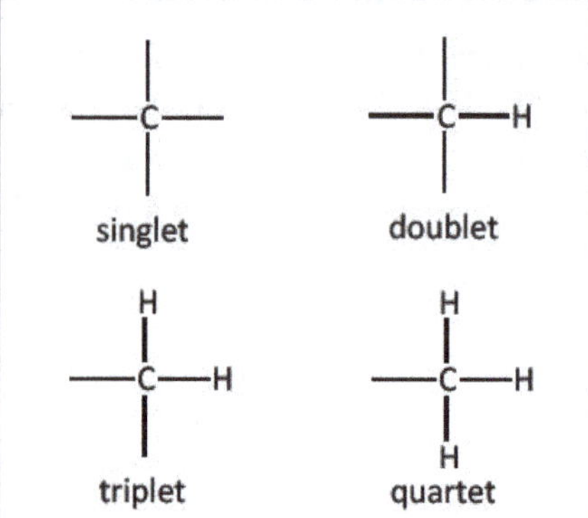

Off-resonance decoupled ^{13}C NMR

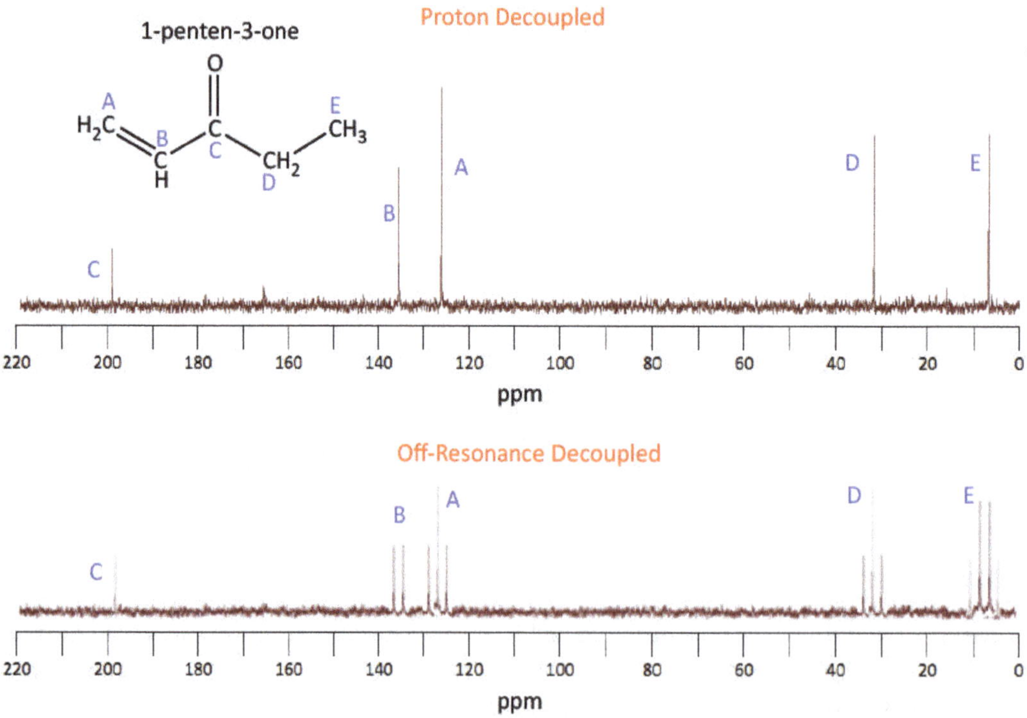

Wow! This is wonderful. If we have the off-resonance decoupled ^{13}C NMR spectrum of an organic compound, we can tell how many types of carbon atoms there are in the molecule as well as how many hydrogen atoms are on each carbon atom. With a ^{1}H NMR spectrum, we can do a little thinking and head scratching and figure out which of these carbon atoms are next to each other.

With so much information, why is a proton decoupled ^{13}C NMR spectrum the most common type to run? Notice that when the peaks of the proton decoupled spectrum are allowed to be split into doublets, triplets, or quartets, the size of those peaks gets smaller because the area for the one peak is divided into two, three, or four smaller peaks. This can be a problem because split ^{13}C NMR peaks become very difficult to see. Sometimes, a spectrum would need to be run for hours to get the carbon signals to come out of the baseline noise of a spectrum. If an off-resonance decoupled spectrum is run, it could simply take too long to make it useful. Also, if the multiplets are too close together and they overlap, they can make things more confusing.

12.10 DEPT

A DEPT spectrum is a more recent and more common way to get the same information that an off-resonance decoupled spectrum gives. DEPT is a Distortionless Enhanced Polarization Transfer experiment. DEPT keeps the peaks as single peaks so we do not lose as much signal and multiplets do not overlap.

DEPT-45: Carbon atoms with hydrogen atoms appear
DEPT-90: CH only appears
DEPT-135: CH and CH_3 point up, CH_2 points down.

Usually, a DEPT is run alongside a regular proton decoupled ^{13}C NMR spectrum. It shows how many types of carbon atoms are in the molecule. In the DEPT, three different ^{13}C NMR are shown. The DEPT-45, the bottom one in our example, shows a peak for all carbon atoms that have a hydrogen atom attached. Notice, the carbonyl, C, is not present in the DEPT-45. The DEPT-90 shows a peak for every C-H (one hydrogen atom attached) in our molecule. Notice, only the B carbon appears. Finally, the top spectrum, the DEPT-135, shows all carbon atoms with a hydrogen atom attached, but the carbon atoms with an odd number of hydrogen atoms, CH (B) and CH_3 (E), point up while the CH_2 peaks (A and D) point down. Quaternary carbon atoms, the carbonyl carbon C, do not appear.

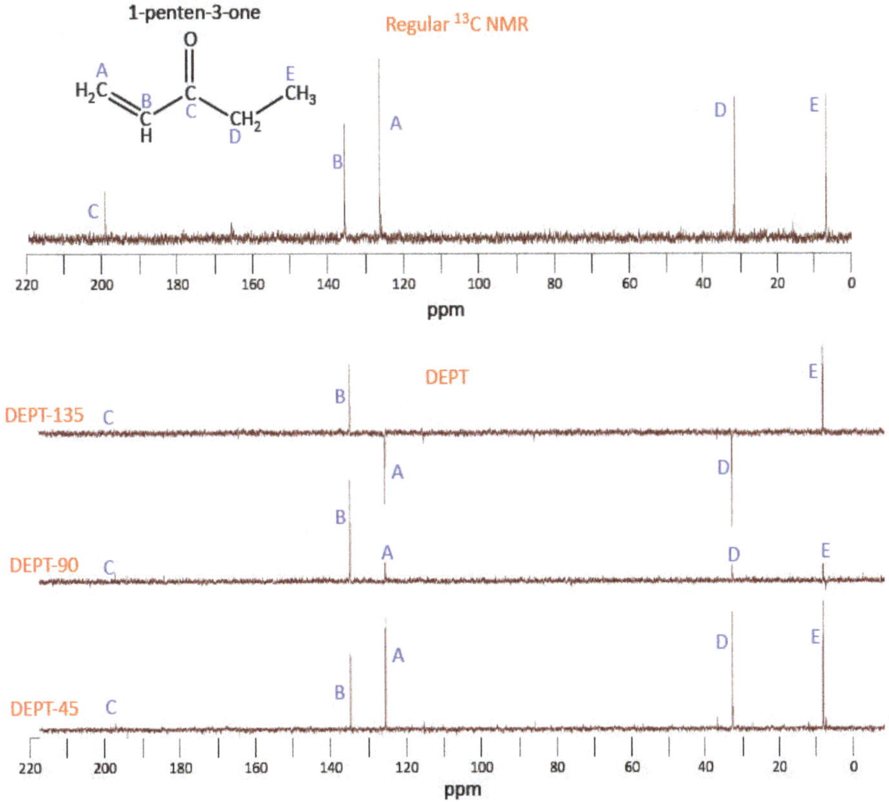

12.11 Below is a DEPT spectrum for an organic compound. Which of the following compounds made this DEPT spectrum?

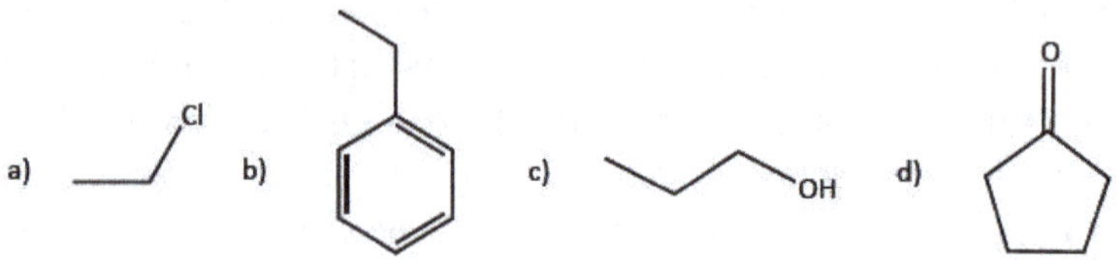

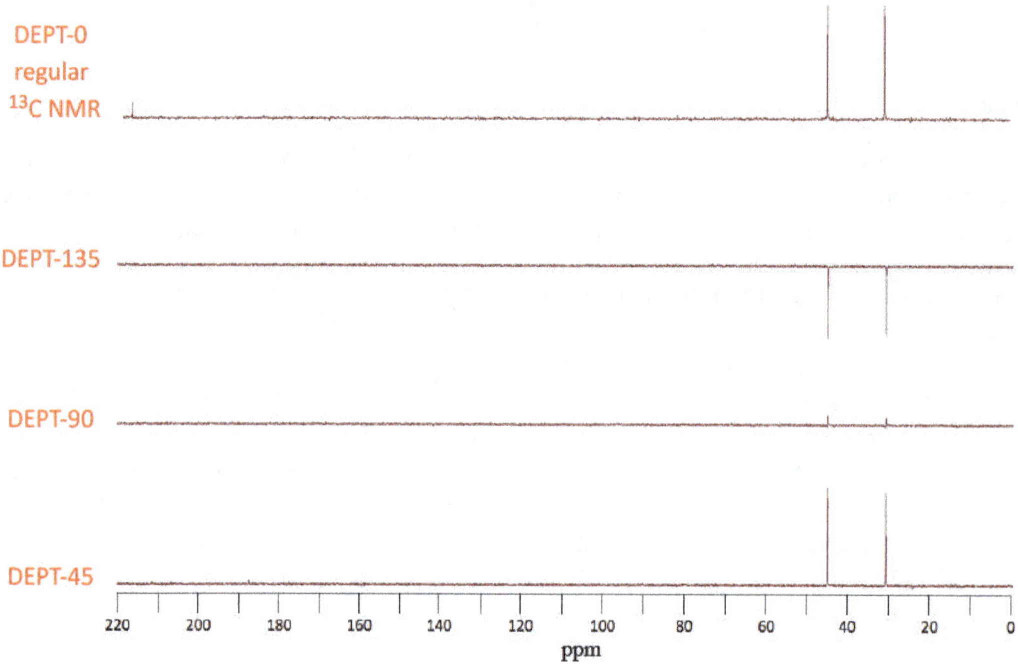

12.11 Solved Problem

Let's work through identifying the structure of an unknown compound by using spectroscopic data. The mass spectrum had an M^+ of 88 with no significant M+2 peak. This even M^+ would indicate our unknown does not have an odd number of nitrogen atoms. No M+2 peak means we do not have Cl, Br, or S. The IR spectrum is then taken.

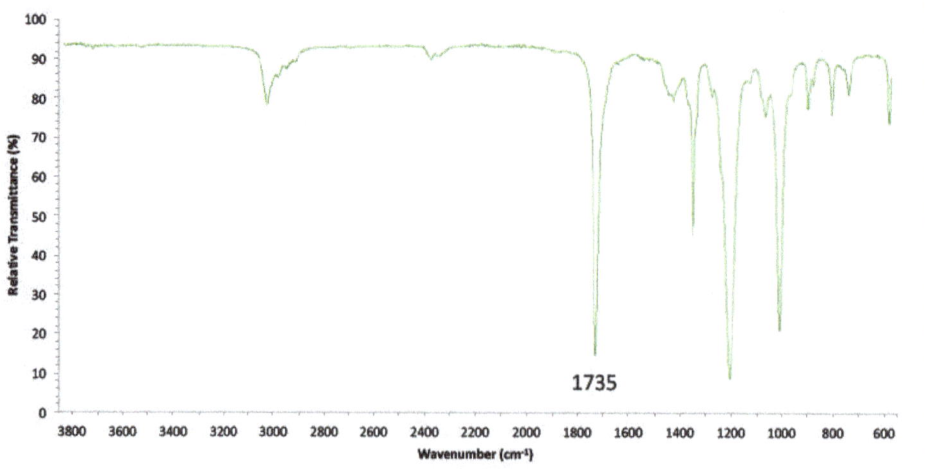

The peak below 3000 indicated sp3 C-H. The strong 1735 peak indicates a carbonyl. A normal carbonyl is around 1710. Since our number is high, in the 1730's, it indicates the carbonyl of an ester or a ring-strained ketone. The other two strong peaks at 1100 and 1200 in the fingerprint region bolster the ester guess. If we have an ester, we have at least one carbon atom and two oxygen atoms in our molecule. CO_2 weighs 44. This leaves us with 44 more in our mass to get to 88. This remaining 44 of the weight could be three carbon atoms (36) and eight hydrogen atoms (8). This would give a molecular formula for an ester of $C_4H_8O_2$. We can calculate degrees of unsaturation:

$$\#\, Degrees\ of\ Unsaturation = \frac{(2C + 2 - H)}{2}$$

This formula gives us one degree of unsaturation, which makes sense for a carbonyl of an ester. If we did not get a whole number with this calculation, we would know that our formula was incorrect.

So far, we know we have. . .

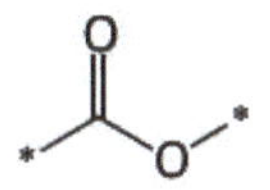

This is in the middle of the molecule (we haven't drawn the ends yet). We still have three more C atoms and eight H atoms still to place.

Now, let's turn to the NMR spectra. It is usually best to begin with the ^{13}C NMR spectrum and DEPT to begin to understand the carbon framework of our molecule.

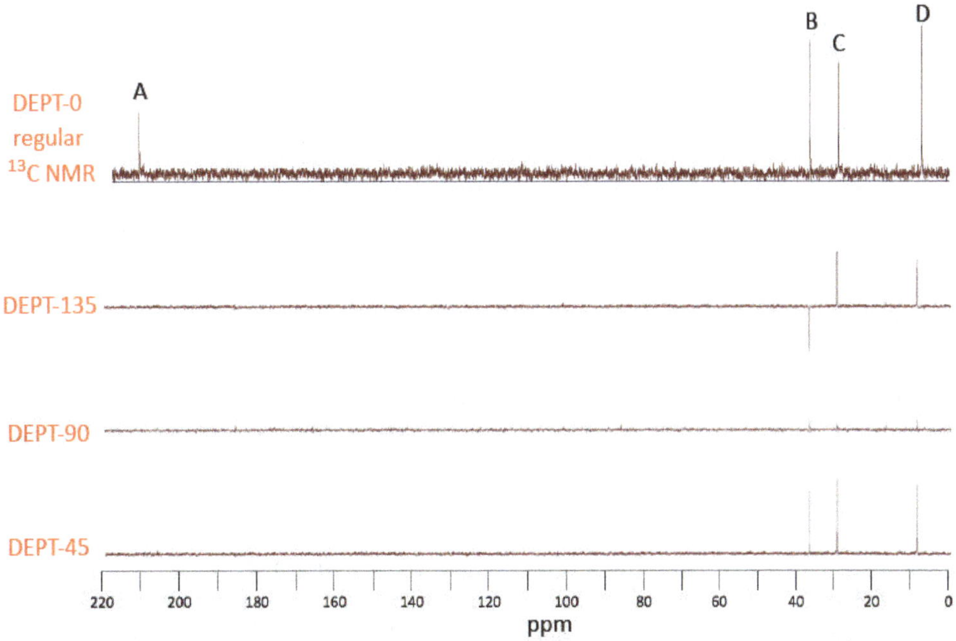

The regular ^{13}C NMR (DEPT-0) shows we have four different types of carbon atoms (labeled A-D). Peak A is above 100 ppm, so it indicates a carbonyl carbon, which is good if we have an ester. It does not show up in the DEPT-45, so it does not have a hydrogen atom on it (like an aldehyde would). This is again consistent with our ester prediction. Carbon peaks B-D represent three different carbon atoms. They are all below 50 ppm, so they are "regular" carbon atoms. B-D all show up in the DEPT-45, so they all have hydrogen atoms on them. None of them show up in the DEPT-90, so none of them are CH carbon atoms. In the DEPT-135, we see that B points down (CH_2),

NMR

C and D point up (CH or CH_3) but we know they are not CH from the DEPT-90. Therefore, we know that carbons C and D are CH_3, methyl, carbon atoms.

Let's pause and consider if the groups are on the inside of the molecule or on the outside ends. Carbons C and D are on the outside, ends, of the molecule because methyl groups, $-CH_3$, only have one more bond going to them. Carbon B is a methylene, $-CH_2-$, group. It is on the inside of our molecule as it has two more bonds going to it. If we put the pieces together, we have two possibilities.

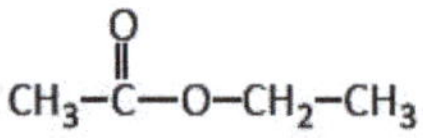
ethyl acetate

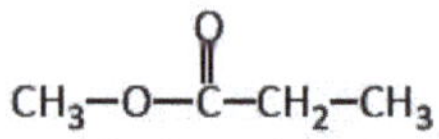
methyl propionate

Now, let's turn to the ^{1}H NMR spectrum to finish up our analysis. As we do so, let us remember that we are looking for three proton groups we have identified so far, two methyl $-CH_3$ groups and one methylene $-CH_2-$ group. Let's be careful to look for these groups and not unnecessarily complicate things. For example, if we cannot tell if a group integrates to 1H or 2H, we will guess it is 2H since we have a CH_2 and not a CH.

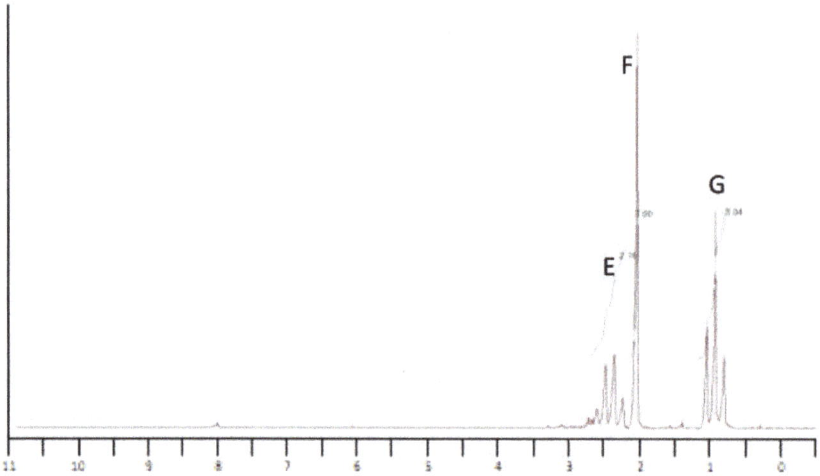

For the 1H NMR spectrum, we see that the quartet, E, integrates to 2H (CH_2) and since it is a quartet, is next to one of the CH_3 groups we identified earlier. The singlet, F, integrates to 3H (CH_3) and since it is a singlet, is next to a carbon atom with no hydrogen atoms on it. The triplet, G, integrates to 3H (CH_3) and since it is a triplet, is next to the CH_2 group (E). This shows us we were on the right track, since we have a triplet-quartet ethyl pattern ($-CH_2CH_3$) and a lone methyl group ($-CH_3$). But, which of our two possibilities is correct? Is it ethyl acetate or methyl propionate?

We turn to the chemical shifts to determine which structure is correct. Notice, E (CH_2) is farthest to the left or downfield. These protons are probably next to something special, like the oxygen atom. The singlet F, the lone CH_3, is around 2 ppm, which indicates it is next door to the carbonyl. The methyl, G, is farthest upfield and is far from any special group. This shows us that ethyl acetate is the correct structure. If the unknown was instead methyl propionate, groups E and F would be swapped in our ^{1}H NMR spectrum.

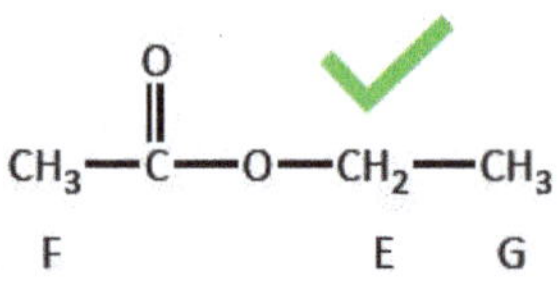

ethyl acetate

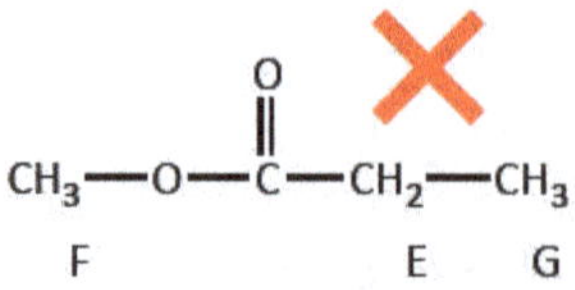

methyl propionate

12.12 Problems

12.12 For each molecule, indicate the number of unique proton groups *and* carbon atoms on the lines provided (i.e., how many you would see in the NMR).

a) 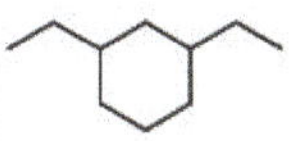 b) 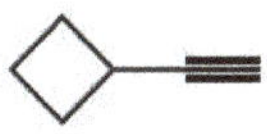 c) 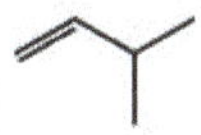 d)  e) 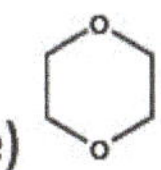 f) 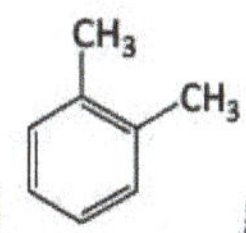 g) 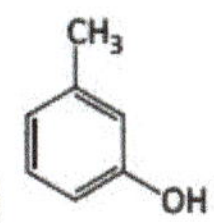

h) 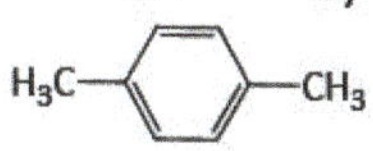 i) 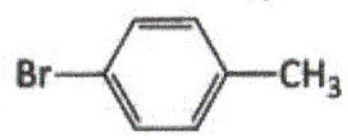 j) bromopropane k) 1-bromo-1,1-dimethylethane

12.13 Identify the approximate chemical shifts (in ppm) of the protons in the following compounds.
a) CH_3CH_2CHO b) Ph-CH_3 c) $(CH_3)_3N$ d) $H_2C{=}CH$-Ph e) $CH_3CH_2CO_2H$ f) CH_3OCH_3 g) CH_3CH_2Br

12.14 Identify the chemical shift range (0-50, 50-100, 100-150, 150-200 ppm) of the carbons in the following compounds.

a) CH_3CHO b) 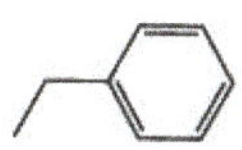 c) 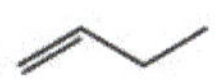 d) 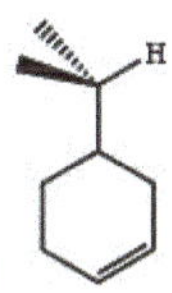 e) 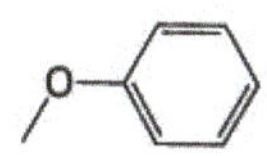 f) 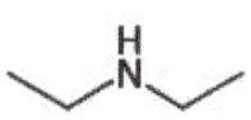 g)  h) 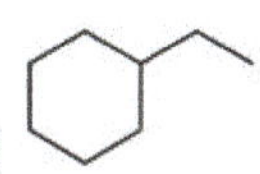

12.15 Identify the multiplicity (singlet, doublet, triplet, etc.) for each proton group in the following compounds.
a) CH_3CHO b) CH_3CH_2Br c) CH_3-O-$CH_2CH_2CH_3$ d) $(CH_3)_3C$-CH_3 e) $(CH_3)_2CHCl$ f) Br-CH_2CH_2-CN g) CH_3-Br

12.16 Sketch the ^{1}H NMR and ^{13}C NMR spectra for each compound.
a) CH_3OH b) chloroethane c) 1-bromopropane d) $CH_3CH_2OCH_3$ e) $CH_3CH_2CH_2CO_2H$

f) PhCHO g) $CH_3OCH_2CH_2Br$ h)  i) 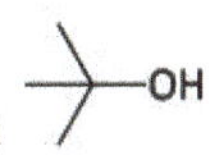 j) 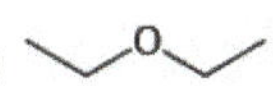

k) 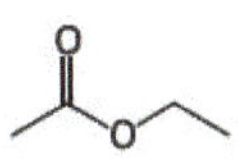 l) 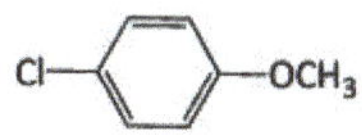 m)  n) 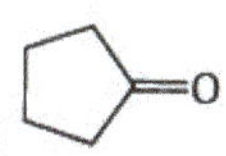 o) 2-chloropropane

12.17 Listed here are ^{1}H NMR absorption peaks for three compounds, along with empirical formula and a key IR absorption (if applicable) as well. The following, abbreviations are typically employed: **s = singlet; d = doublet; t = triplet; q = quartet; m = multiplet**. Please provide structures consistent with these data.

$C_4H_{10}O$ **1H NMR spectrum** **IR spectrum**
t δ 1.2 (3H) peak at 1126 cm^{-1}
q δ 3.5 (2H)

____
Structure

C_4H_8O **1H NMR spectrum** **IR spectrum**
t δ 1.8 (2H) peak at 1070 cm^{-1}
t δ 3.7 (2H)

____
Structure

C_7H_7OCl **1H NMR spectrum** **IR spectrum**
Broad s δ 5.23 (1H) strong, broad peak
s δ 4.83 (2H) at 3200-3400 cm^{-1}
d δ 6.76 (2H)
d δ 7.17 (2H)

____
Structure

NMR

$C_5H_{10}O$	**1H NMR spectrum** s δ 1.08 (9H) s δ 9.48 (1H)	**IR spectrum** strong peak at 1720 cm^{-1} 2698, 2791 cm^{-1}	________________ **Structure**
C_3H_8O	**1H NMR spectrum** d δ 1.2 (6H) Broad s δ 2.2 (1H) sept δ 4.0 (1H)	**IR spectrum** strong, broad peak at 3200-3400 cm^{-1}	________________ **Structure**

12.18 A ^{1}H NMR spectrum was taken of an alcohol. The sample was then shaken with D_2O. Another spectrum was acquired. What does this tell you about the peak labeled with the blue arrow in the original spectrum? What is this new peak labeled with the red arrow in the second spectrum?

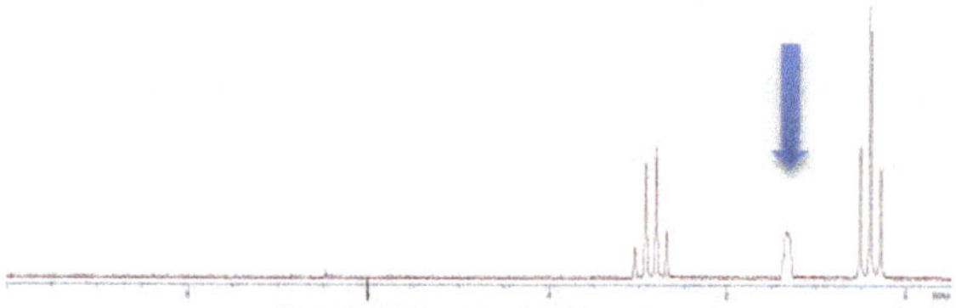

^{1}H NMR of alcohol

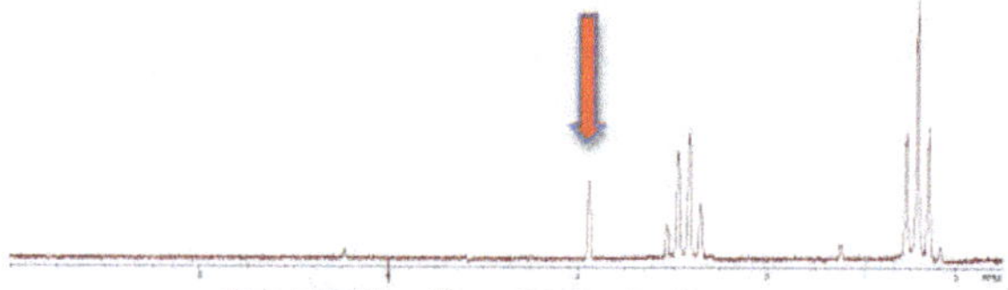

^{1}H NMR after D_2O shake

12.19 Match each of the following compounds with its ^{1}H NMR spectrum.

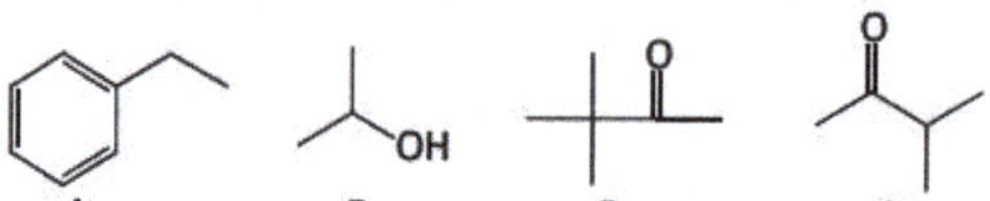

A B C D

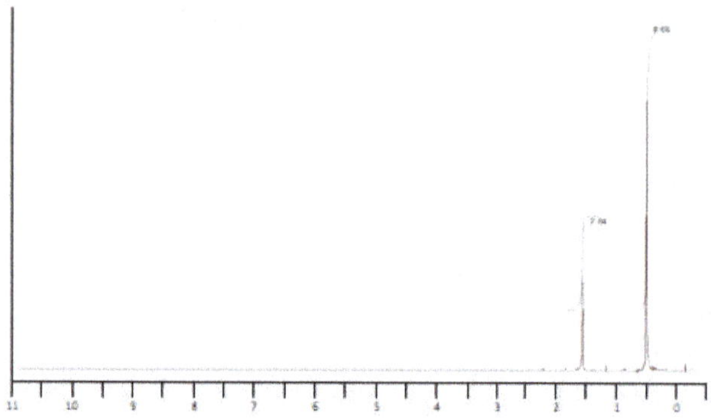

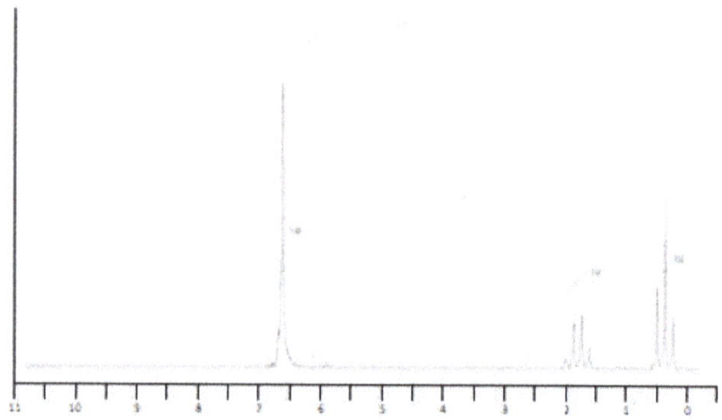

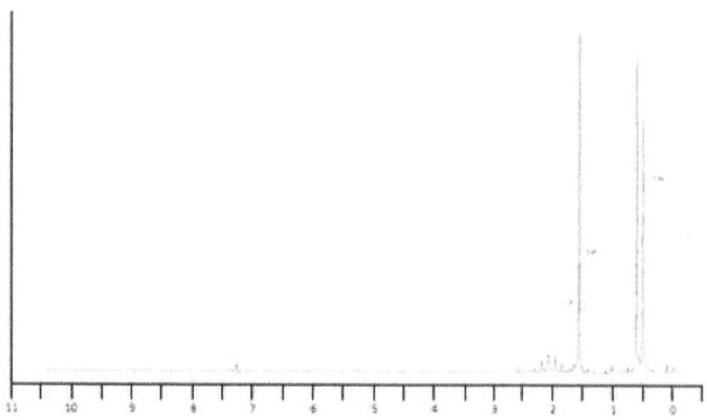

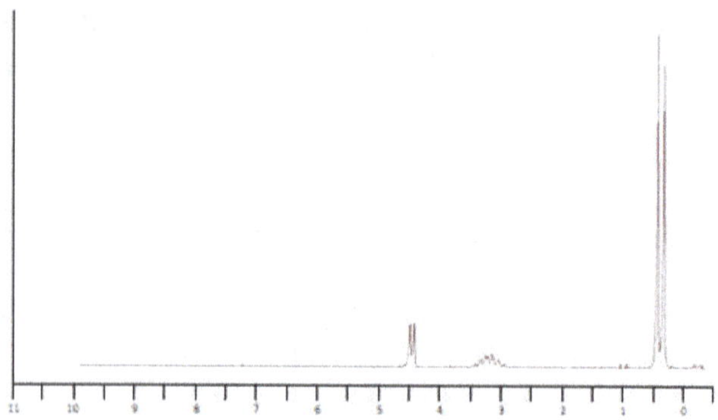

12.20 a) Draw the structure of the following ester (formula $C_5H_{10}O_2$) given its ^1H and ^{13}C NMR spectra.

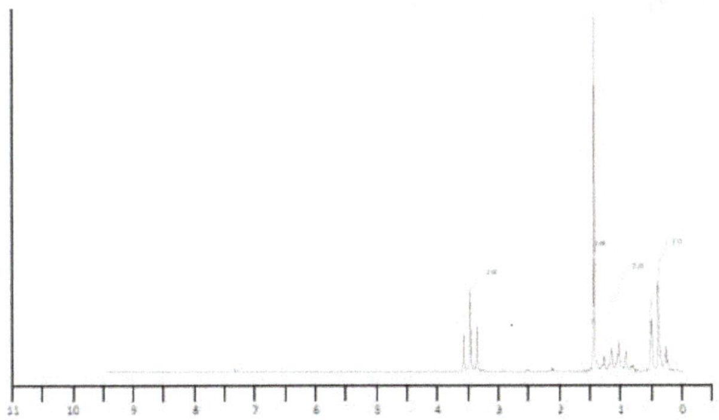

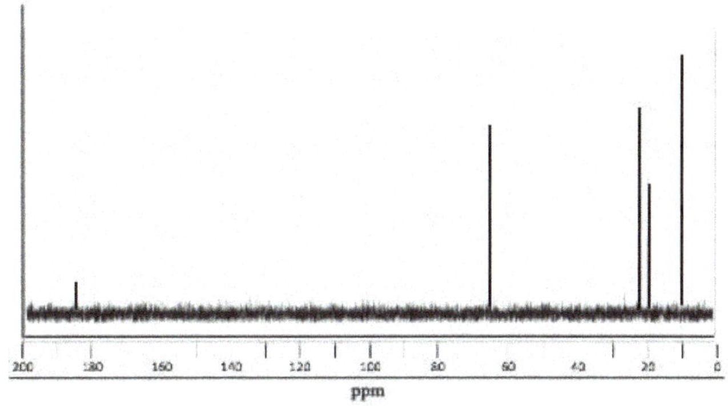

b) Once you have identified the complete structure of this compound, fill in the lines below to complete the DEPT-45, DEPT-90, and DEPT-135 spectra. Be sure to consider all five carbon peaks.

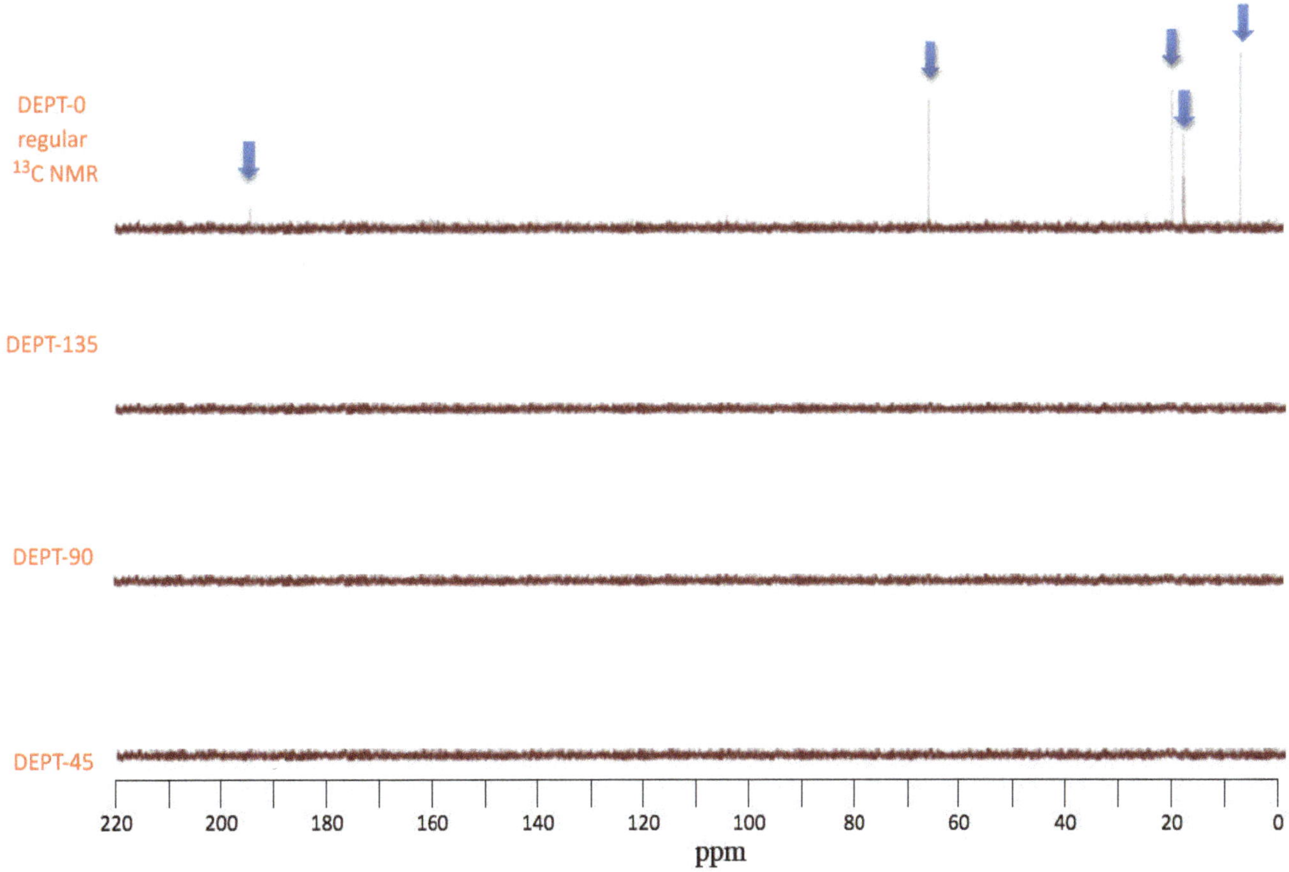

NMR

12.21 **Draw the structure of the following compound with the formula C_3H_6O given its ^{13}C NMR and DEPT spectra.**

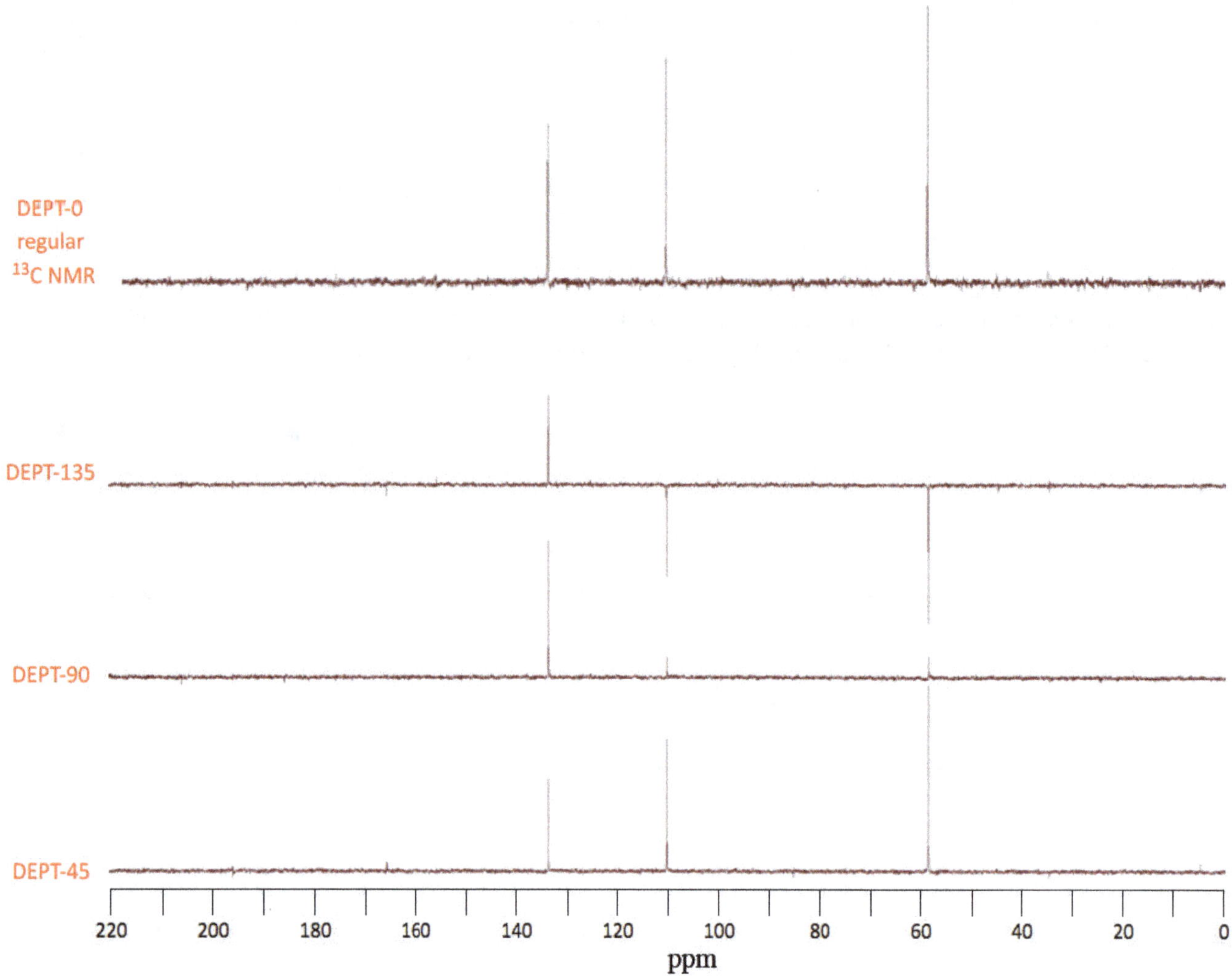

12.22 An ester with the formula $C_{10}H_{12}O_2$ has the following NMR spectra. Draw its structure.

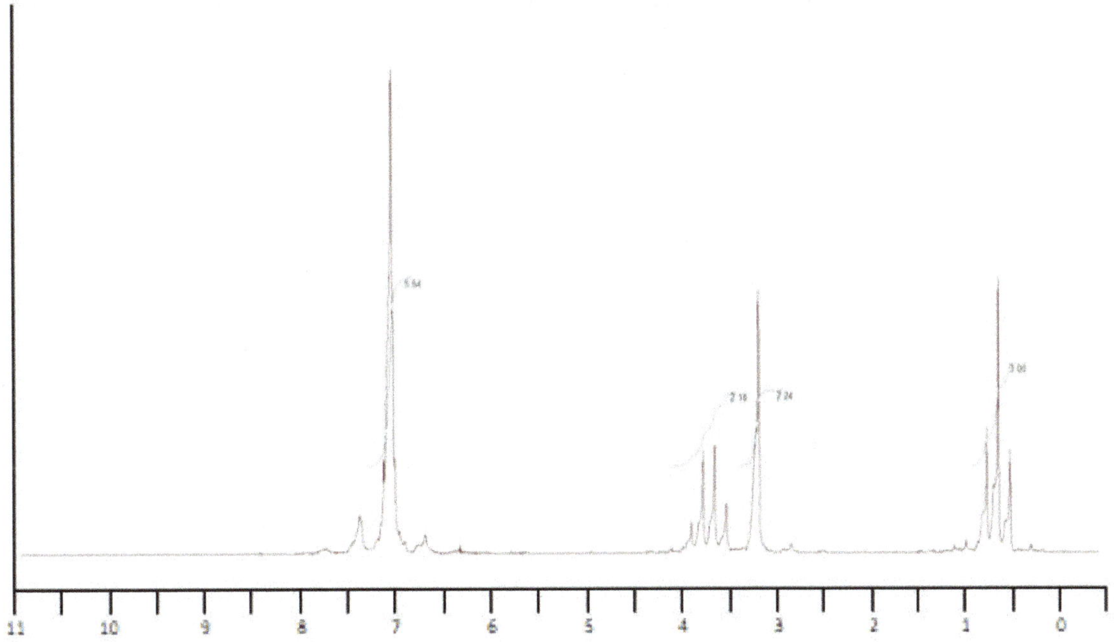

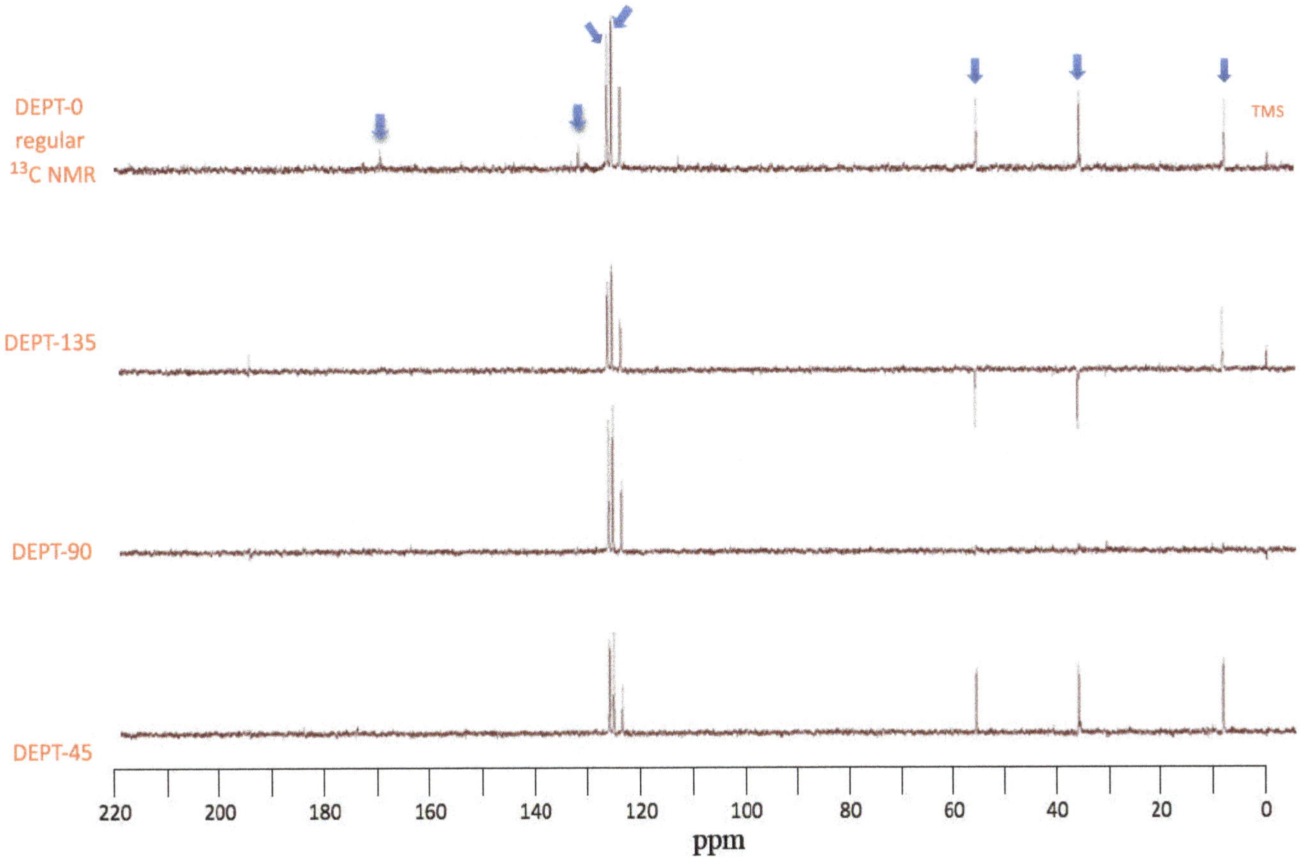

NMR

12.23 **A compound with the formula $C_5H_{10}O$ has the following ^1H and ^{13}C NMR spectra. Draw its structure.**

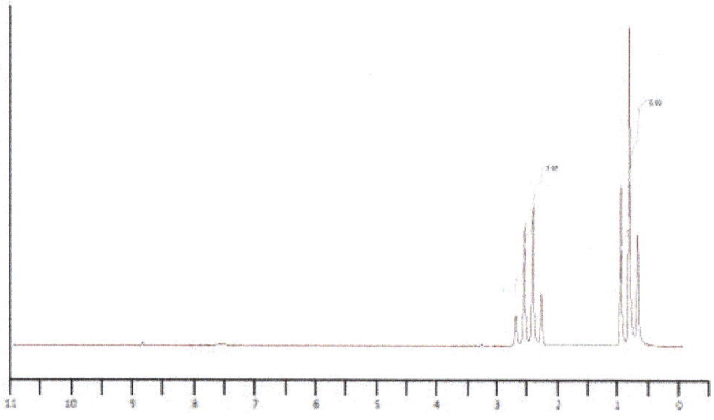

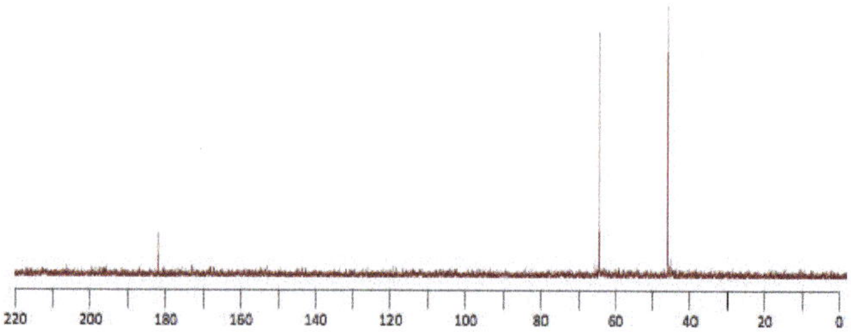

12.24 **A compound with an M^+ of 86 in its mass spectrum has the following spectra.**

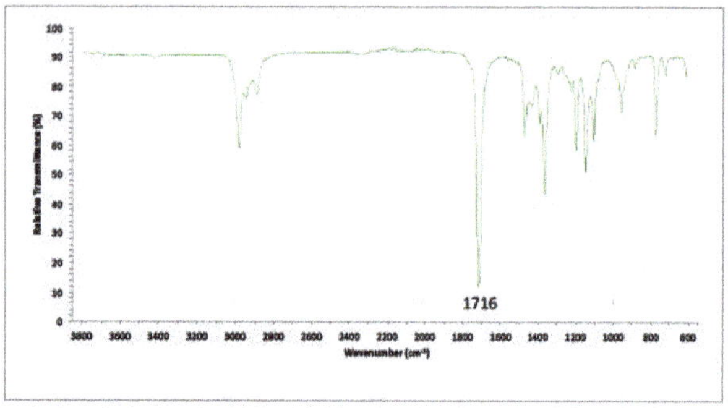

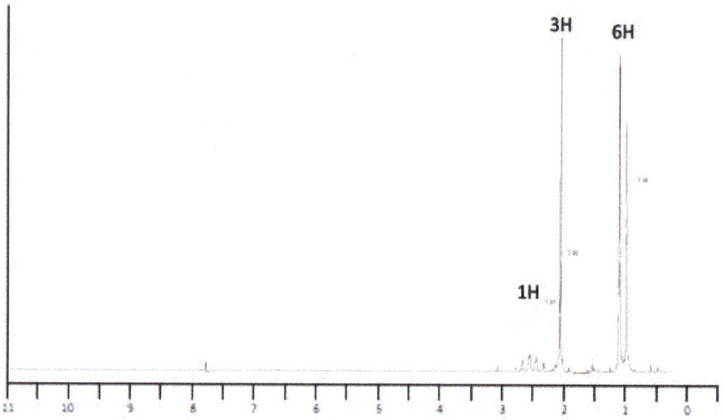

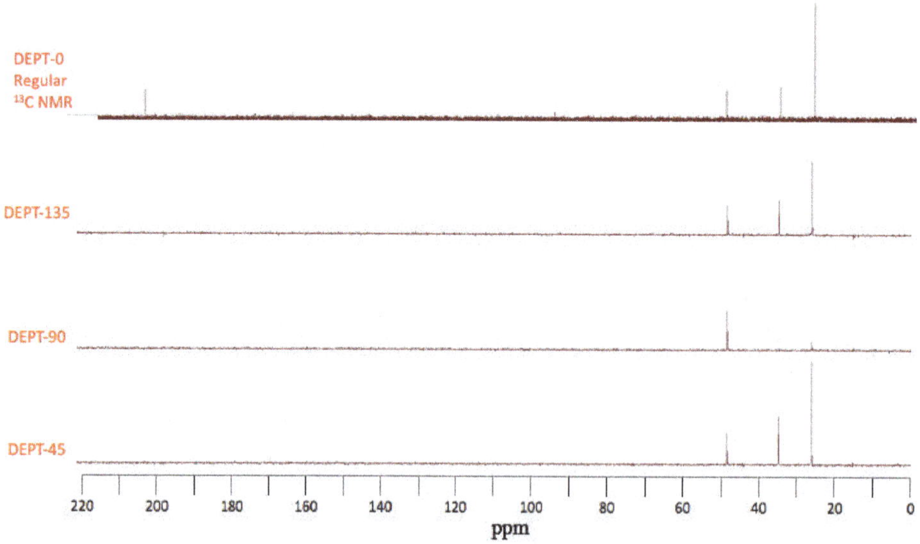

Chapter 13. Alcohols

13.1 Introduction

Alcohols are compounds that contain the hydroxyl (-OH) functional group. Alcohols can be classified as primary, secondary, or tertiary. Their classification is determined by how many carbon atoms are attached to the carbinol carbon atom (the one attached to the hydroxyl group). If only hydrogen atoms are attached to the carbinol carbon atom, it is methanol (methyl alcohol). If one carbon atom is attached, it is a primary alcohol. If two carbon atoms are attached, it is a secondary alcohol. If three carbon atoms are attached, it is a tertiary alcohol. If a phenyl group is attached to the hydroxyl group, it is classified as a phenol.

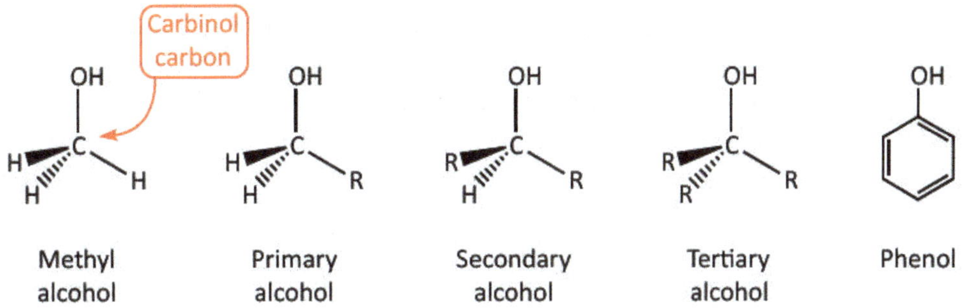

Figure 13.1 Classification of alcohols

13.2 Nomenclature

13.2.1 IUPAC Names

To name an alcohol, the longest alkane chain is identified. The *–e* suffix is removed and is replaced with an *–ol* suffix.

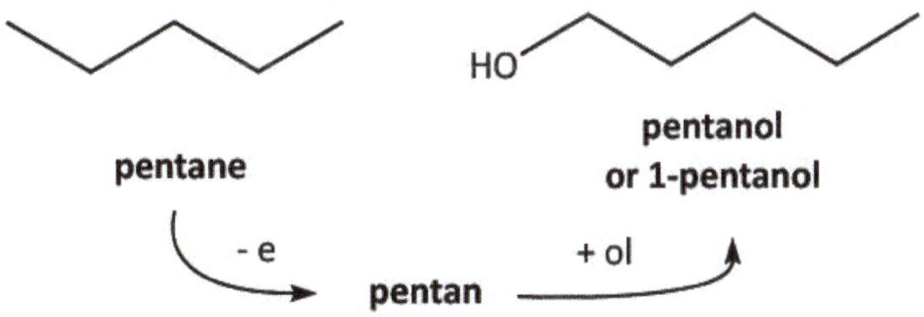

If the hydroxyl group is not on the end, #1, carbon atom, a number needs to be used to identify the location of the hydroxyl group. The carbon chain should be numbered starting at the end closest to the hydroxyl group.

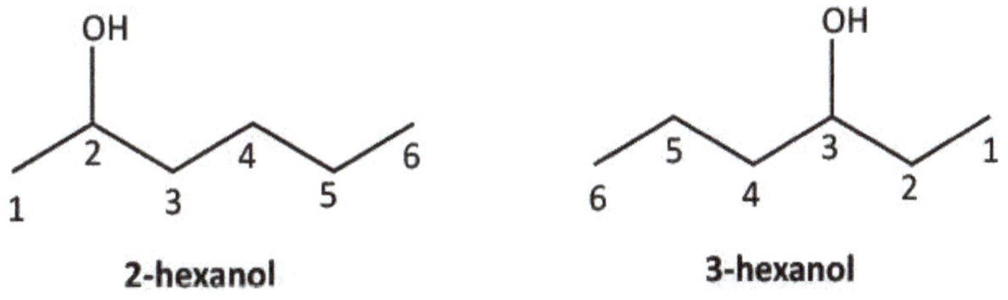

Alcohols

If we need to treat an alcohol, or hydroxyl group, as a substituent because it is attached to a more important part of the molecule, it is named "hydroxy". Let's look at the following molecule as an example. Cyclopentanol is the main part of the molecule. The ring is numbered starting at the carbinol carbon atom (in black). The substituent is on carbon #3 of the ring and is itself numbered (in blue) starting closest to the main part of the molecule, the ring. This puts the hydroxyl group on the 2nd carbon atom of the substituent.

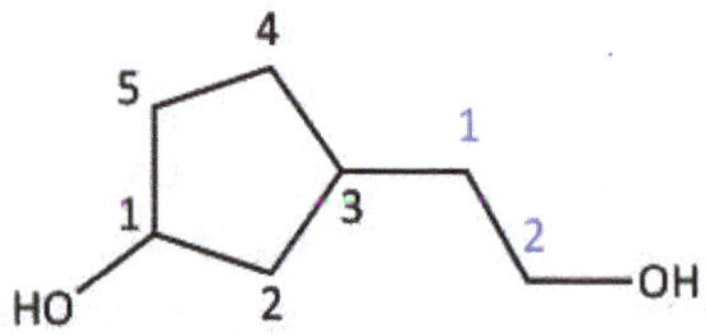

3-(2-hydroxyethyl)cyclopentanol

Sometimes, it is difficult to know which way to number a chain because two different functional groups are equally distant from the two different ends. Take for example this alkene/alcohol. On the left molecule, we see that the hydroxyl group is two carbons away from an end. On the molecule on the right, we see that the alkene is between carbons 2 and 3 and would get the number 2 associated with it. So, which is preferred?

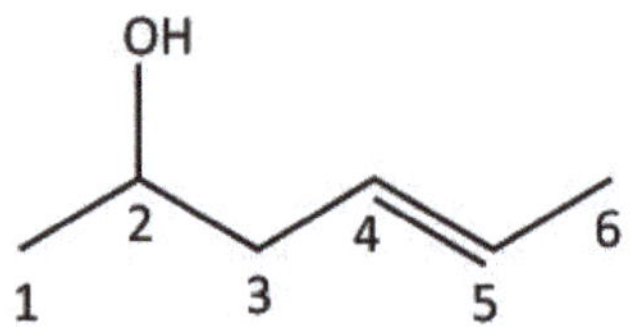

(*E*)-hex-4-en-2-ol

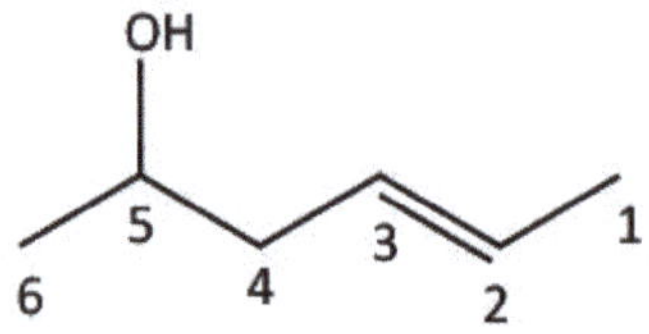

(*E*)-hex-2-en-5-ol

IUPAC has assigned a priority for various functional groups when a situation like this occurs. I wouldn't waste my time memorizing every detail of this list, but you should know it exists if you need to make nomenclature priority decisions.

High Priority
Carboxylic acids
Esters
Aldehydes
Ketones
Alcohols
Amines
Alkenes
Alkynes
Alkanes
Ethers
Halides
Low Priority

Figure 13.2 Priority of functional groups in IUPAC naming

But, you should recognize that in general, the more oxidized a functional group is (more carbon-oxygen bonds), the higher its priority. So, the hydroxyl group (with one C-O bond) is higher priority than the alkene (zero C-O bonds). We therefore assign the hydroxyl group the lowest number and find that the correct name for this molecule is:

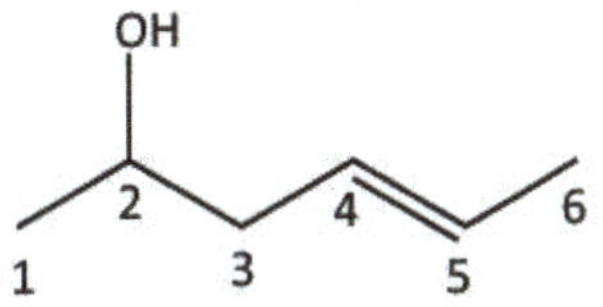

(*E*)-hex-4-en-2-ol

A hydroxyl group that is being named as a substituent attached to the main part of the molecule is called "hydroxy". In the following molecule, the cyclopentanol is the main group. When numbering the ring, the carbinol carbon atom is numbered 1. The ethyl/alcohol substituent at carbon 3 of the ring is numbered starting closest to the ring. There are two carbon atoms, so it is called "ethyl". The hydroxyl group is on carbon 2 of the ethyl substituent giving the 2-hydroxyethyl substituent name.

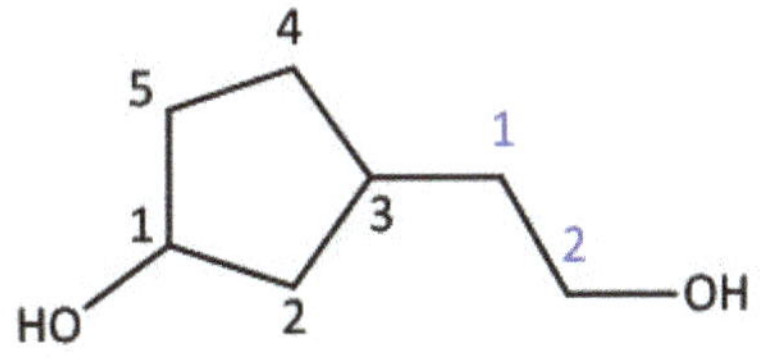

3-(2-hydroxyethyl)cyclopentanol

Diols, compounds with two hydroxyl groups, are named by finding the longest carbon chain that includes the two hydroxyl groups. The chain is numbered to give one of the alcohols the lowest number possible. If there are two hydroxyl groups, the suffix diol is used. If there are three hydroxyl groups, triol is used, etc.

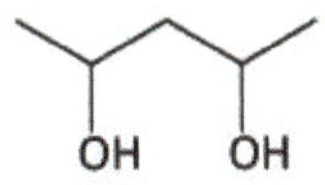

pentane-2,4-diol

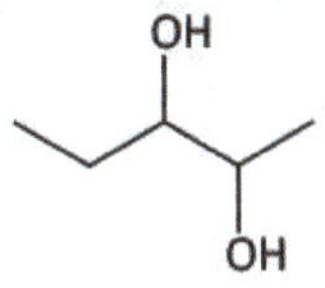

pentane-2,3-diol

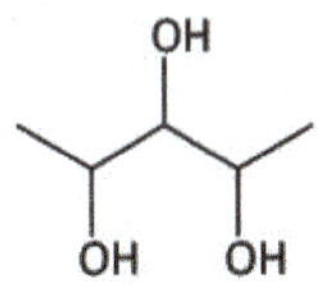

pentane-2,3,4-triol

13.2.2 Common names

Alkyl alcohols

Alcohols are commonly named with the name of the alkyl chain followed by the word "alcohol".

	$HO—CH_3$	$HO—CH_2—CH_3$	$HO—CH_2—CH_2—CH_3$	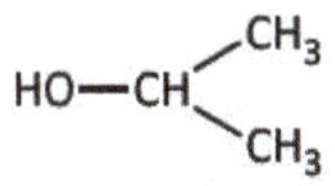
IUPAC	methanol	ethanol	1-propanol	2-propanol
Common	methyl alcohol	ethyl alcohol	propyl alcohol	isopropyl alcohol

Vicinal Diols

Vicinal diols, two hydroxyl groups on neighboring carbon atoms, are called glycols. You'll recall that glycols can be formed from the reaction of an alkene with osmium tetroxide (OsO_4), potassium permanganate ($KMnO_4$), or epoxidation followed by acid/water. Glycols are commonly named by first giving the common name of the alkene from which the glycol is made followed by the term "glycol".

Alcohols

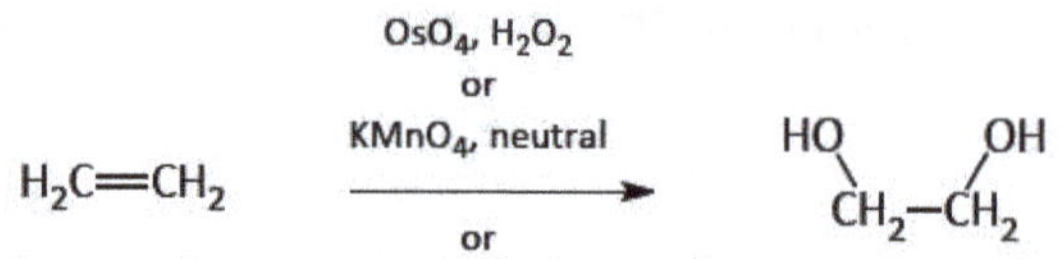

IUPAC	ethene		**IUPAC**	ethane-1,2-diol
Common	ethylene		**Common**	ethylene glycol

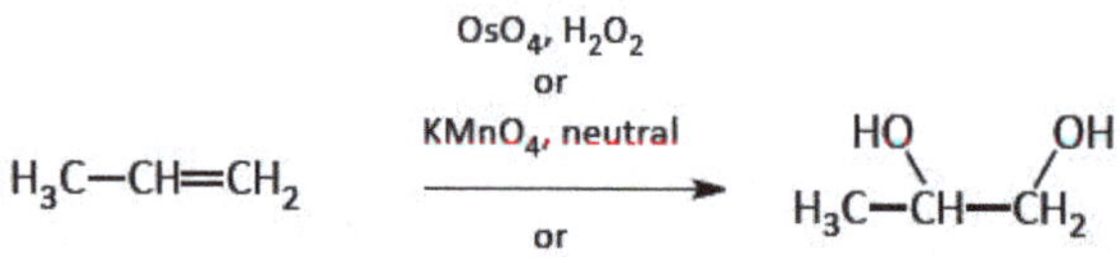

IUPAC	propene		**IUPAC**	propane-1,2-diol
Common	propylene		**Common**	propylene glycol

Phenols

A phenol is a hydroxyl group (alcohol) attached to a phenyl ring.

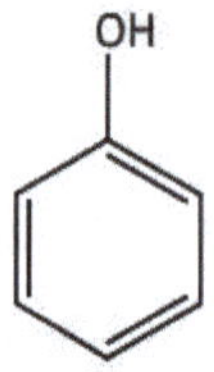

phenol

The following phenols have a chlorine group attached and are called chlorophenols. When numbering them, the alcohol is higher priority than the halide, so the carbinol carbon atom is numbered 1. The chloro groups are then assigned the appropriate number.

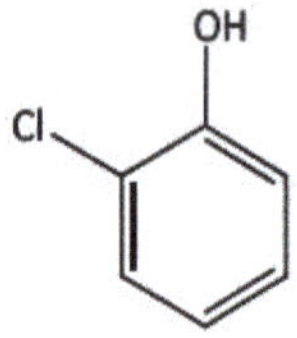

2-chlorophenol

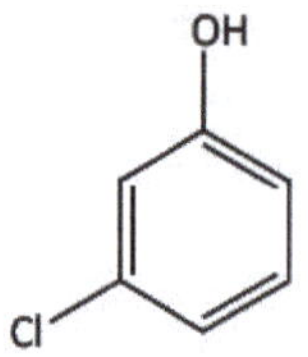

3-chlorophenol

4-chlorophenol

When two groups are attached to a benzene ring, there is a common way to name them using the prefixes ortho, meta, and para. The way to remember the order of the prefixes is to remember "ROMP". Let's look at a generic R group on a phenyl ring. The carbon atom attached to the R group is called the ipso carbon atom. The carbon atom next door to the R group, the 2nd carbon atom, is the ortho carbon atom. The 3rd carbon atom is meta, and the 4th carbon atom is para. This spells "ROMP" around the ring.

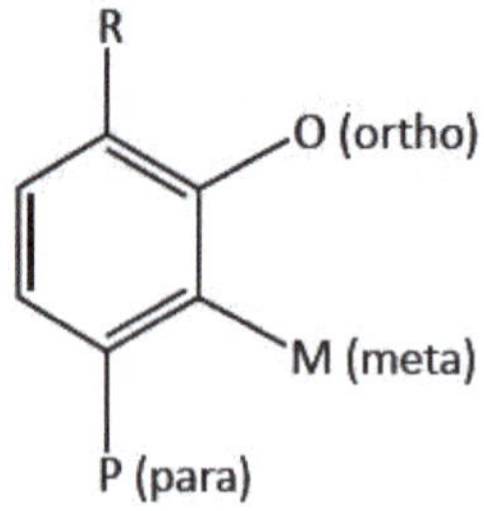

Figure 13.3 *otho, meta, para* positions in disubstituted benzenes

The chlorophenols from above have the following common names as well:

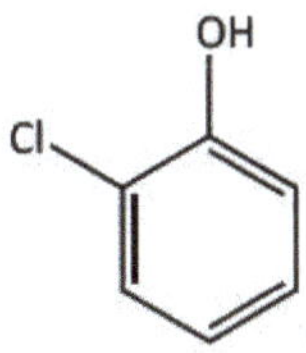
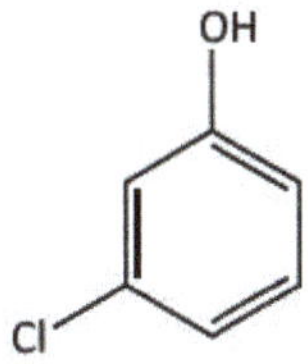

IUPAC	2-chlorophenol	3-chlorophenol	4-chlorophenol
Common	*ortho*-chlorophenol *o*-chlorophenol	*meta*-chlorophenol *m*-chlorophenol	*para*-chlorophenol *p*-chlorophenol

Methylphenols are commonly called cresols. So, a benzene ring that contains both a methyl group and a hydroxyl group can be called ortho, meta, or para cresol depending on the location of the two groups.

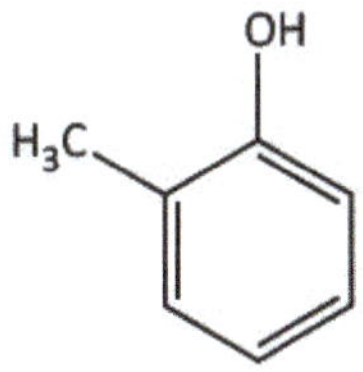
ortho-cresol
o-cresol

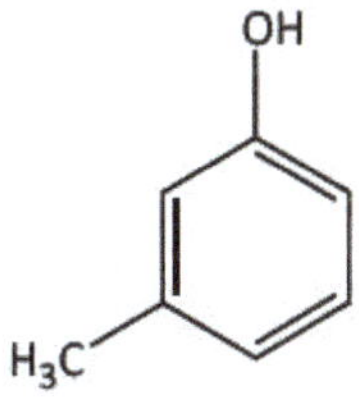
meta-cresol
m-cresol

para-cresol
p-cresol

13.1 Name the following compounds

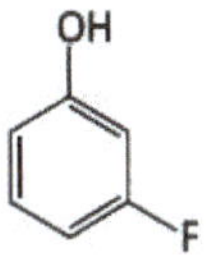
____

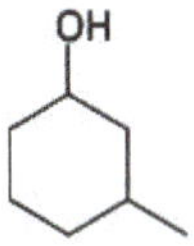
____

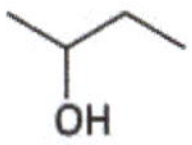
____

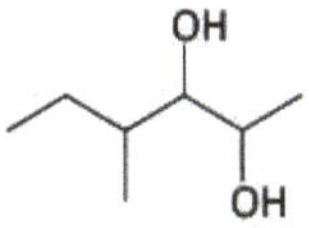
____

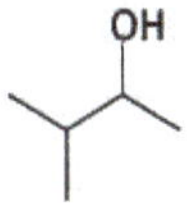
____

____

13.3 Thiols

Sulfur is directly below oxygen in the periodic table. Being in the same column, two have the same number of valence electrons and share some chemical properties. Sulfur is a little larger and less electronegative than oxygen. Thiols are the sulfur analogues of alcohols.

R—SH

Figure 13.4 Thiol

13.3.1 Nomenclature

IUPAC

Thiols are named with the alkane plus the suffix "thiol".

Alcohols

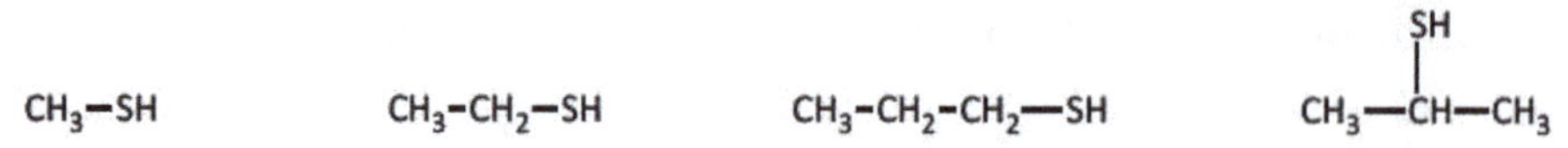

methanethiol — ethanethiol — propanethiol — 2-propanethiol, propane-2-thiol

Common

Thiols are sometimes called mercaptans. For instance, CH_3-SH has the common name of methyl mercaptan.

13.3.2 Examples

Many thiols stink. Thiols are responsible for the stink of skunks, oysters, garlic, and onions. This is a list of some common thiols.

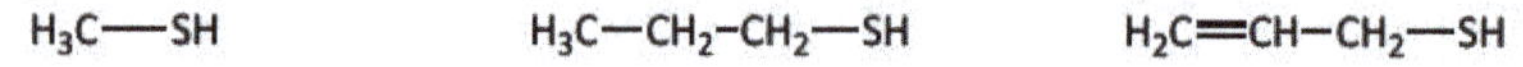

in cheddar cheese and oysters — in onions — in garlic and skunks

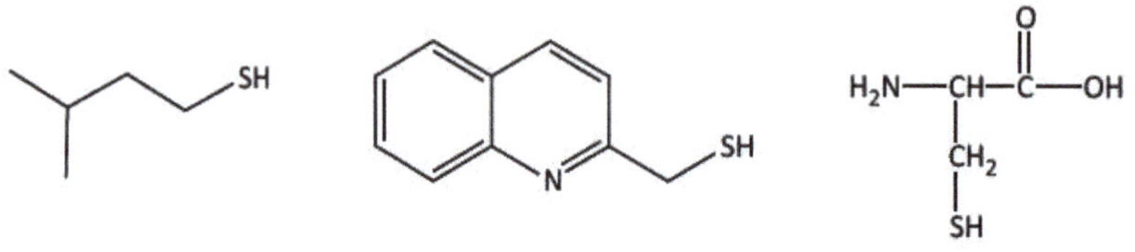

in skunks — in skunks — cysteine (amino acid)

13.3.3 Preparation

Thiols can be prepared from the reaction of sodium hydrosulfide with an alkyl halide via an S_N2 reaction.

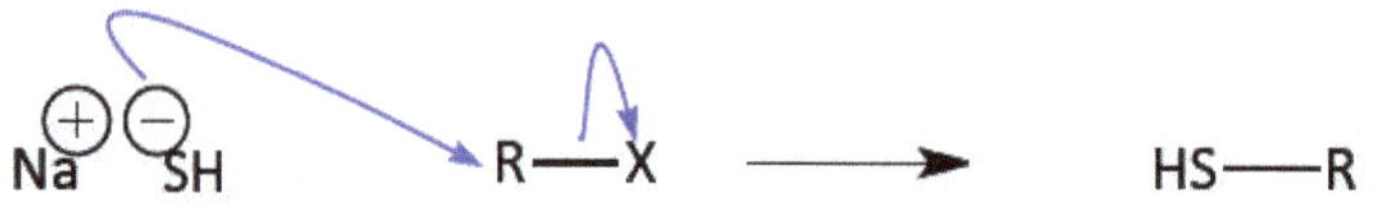

sodium hydrosulfide — Thiol

13.4 Properties of simple alcohols

13.4.1 Methanol

Methanol, or methyl alcohol, is a colorless liquid. It is a good fuel. Since 1965, it has been used in the Indianapolis racing cars. The reason it is used is it is easier to put out a fire with water if methanol is the fuel instead of gasoline. Methanol can also be made by heating wood in the absence of air. This is why methanol is sometimes called "wood alcohol".

Methanol is a good solvent. It is used in paints and varnishes. It is dangerous to drink because the liver would oxidize it into formaldehyde and formic acid. Two tablespoons of methanol can kill somebody. We will discuss this more in the next chapter.

$$H_3C-OH$$

methanol

13.4.2 Ethanol

Ethanol, or ethyl alcohol, is the alcohol in booze. It is sometimes called "drinking alcohol" or "grain alcohol". It is called grain alcohol because it is produced by the yeast fermentation of sugars from plants or grains. Ethanol usually has a little bit of water in it. To try to make ethanol have less water, to be purer and more potent, it can be distilled. But, the best a distillation can produce is 95% ethanol. In order to get 100% ethanol, or absolute ethanol, chemical drying agents must be used. Since people can drink alcohol, the government taxes it. So, in the chemistry lab, these extra taxes are costly if we simply want to use ethanol as a solvent. Therefore, we can buy "denatured" alcohol. Denatured alcohol means the alcohol purchased is poisoned. Typically, a little methanol is added to the ethanol to make it unfit to drink. So, DO NOT ever drink the ethanol you find in a chemistry lab. It is probably poisonous! The liver oxidizes the ethanol people drink. It is turned into acetic acid (vinegar) and acetaldehyde. Together, these make the scent associated with a drunk.

$$CH_3{-}CH_2{-}OH$$

ethanol

13.4.3 Isopropanol

2-Propanol, or isopropanol, is commonly found in the home. A mixture of 70% isopropanol and 30% water is called rubbing alcohol. A little more than half a cup of isopropanol can kill somebody. When it is oxidized in the liver, it turns into acetone (fingernail polish remover).

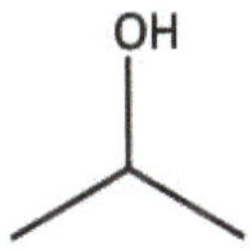

isopropanol

13.4.4 Ethylene Glycol & Propylene glycol

Ethylene glycol and propylene glycol are vicinal diols, or glycols. They are found in antifreeze. They are poisonous. Because antifreeze tastes sweet, we need to be careful if it is spilled on the garage floor. Our pets may enjoy licking it off of the floor, but it can easily kill them. So, be sure to always clean up antifreeze spills!

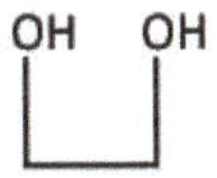

ethylene glycol

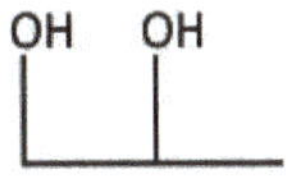

propylene glycol

13.4.5 Glycerol

Glycerol is a clear, thick liquid. Glycerol is normally found in the body. It is made as a product of fat metabolism. It is normally found in animal fats, lard, vegetable oils, etc. Glycerol is water-soluble. It is a lubricant and is used in skin lotions and soaps, and shaving creams. It is found in cake frosting to help prevent the sugar from crystallizing.

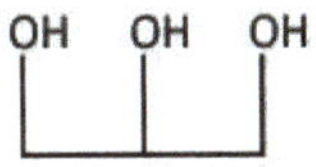

glycerol

13.5 Alcohol Spectroscopy

13.5.1 IR Spectroscopy

The OH group of an alcohol appears as a large, broad peak around 3300 cm^{-1} in an IR spectrum. This is different than the OH of a carboxylic acid that appears at 3000 cm^{-1} and tapers off.

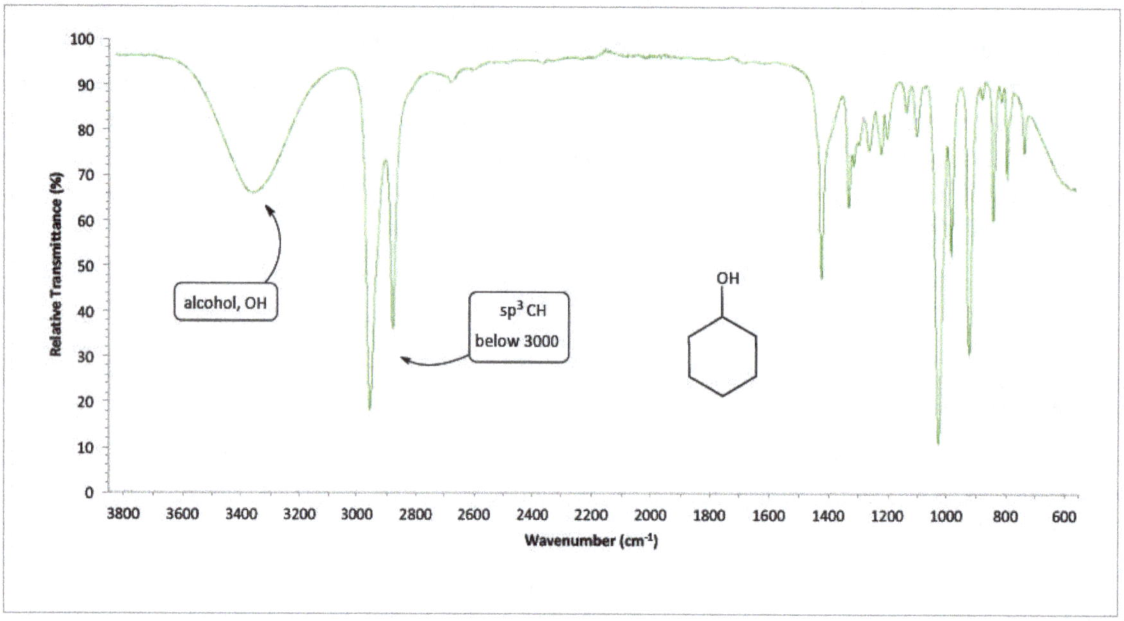

13.5.2 Mass Spectrometry

The molecular ion, M^+, peak for an alcohol usually appears, but sometimes it is small or missing. A fragment peak that is often seen is M^+-18. This is because a water molecule (which weighs 18) can easily be eliminated from the M^+.

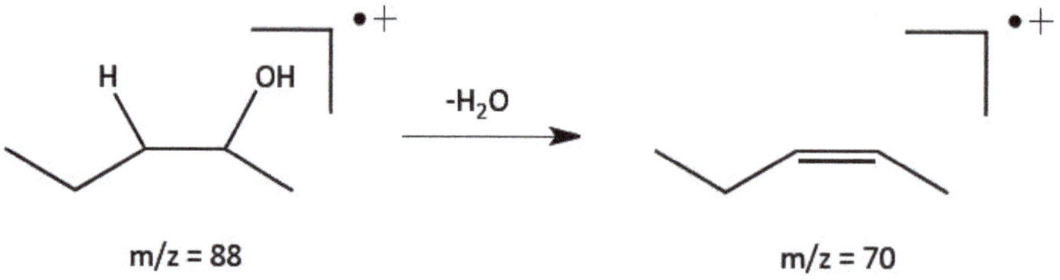

13.5.3 ^{1}H NMR

Hydroxyl protons are a little strange in the NMR. They can show up anywhere between 2-6 ppm. Because protons on OH are involved in hydrogen bonding, their chemical shift can change depending on the temperature and concentration of the solution. They are not always broad peaks. Sometimes they stay on the molecule enough to couple.

Usually, these peaks are broad, hump-like peaks. This is because the hydrogen atom disassociates from the OH bond. It falls off and returns to the molecule, so there usually is not time for it to couple with other protons.

One laboratory trick to help determine if a ^{1}H NMR peak is from an OH proton is to add D_2O, heavy water, to the NMR sample. D_2O interacts with the OH proton and the hydrogen atom and deuterium atoms switch to form O-

D. The ^{1}H NMR peak then disappears confirming OH. Protons on carbon atoms will not trade with deuterium of D_2O. A new peak will grow in the spectrum, though. HOD is formed in the reaction.

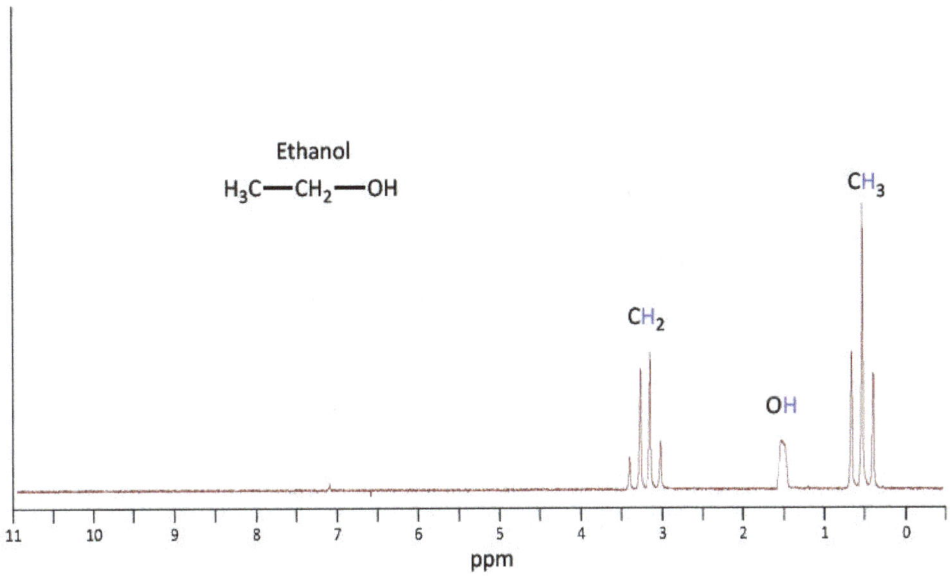

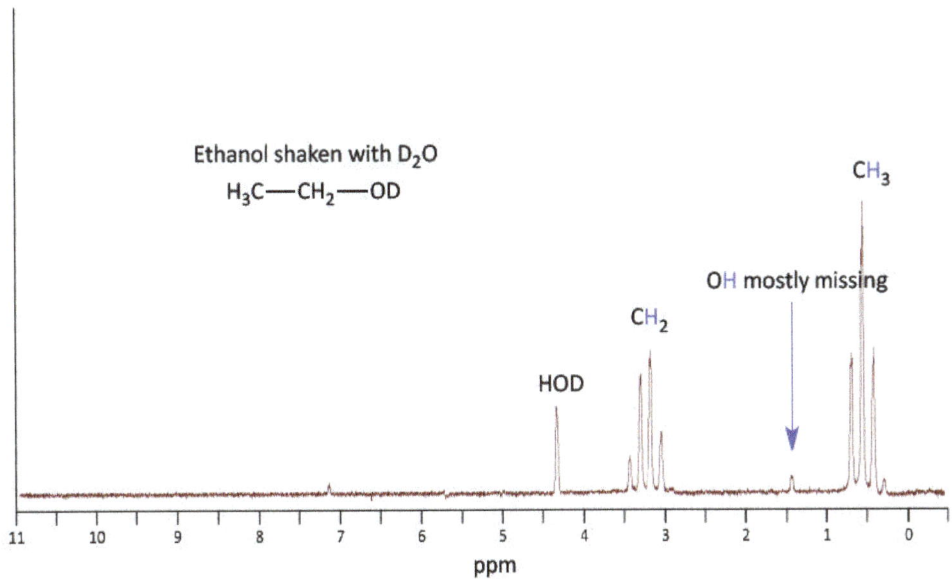

13.5.4 ^{13}C NMR

Of course, there is no carbon atom in a hydroxyl, OH group, so it is invisible in the ^{13}C NMR. But, the carbon atom attached to the hydroxyl group usually appears between 50-100 ppm.

13.6 Synthesis of alcohols

13.6.1 Review

We have already learned several ways to make alcohols. We will review them here.

Alcohols

The S_N2 reaction of hydroxide with an alkyl halide

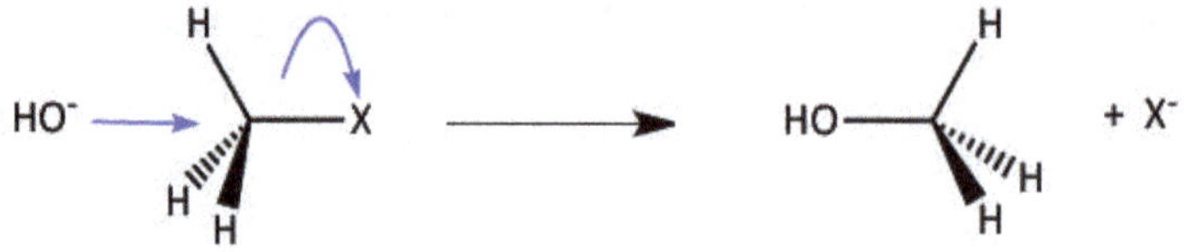

Acid hydration of an alkene

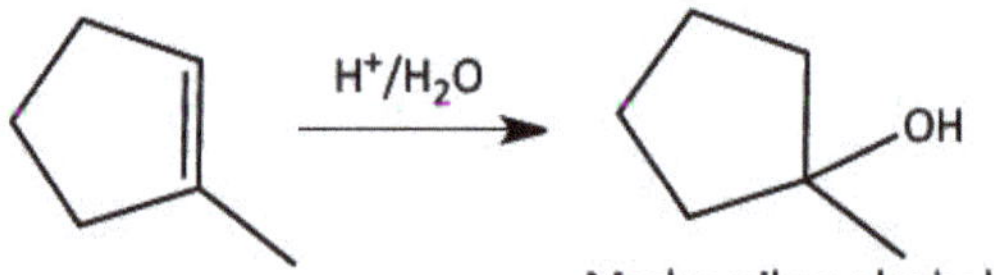

Markovnikov alcohol

Markovnikov Oxymercuration/Demercuration of an alkene in water

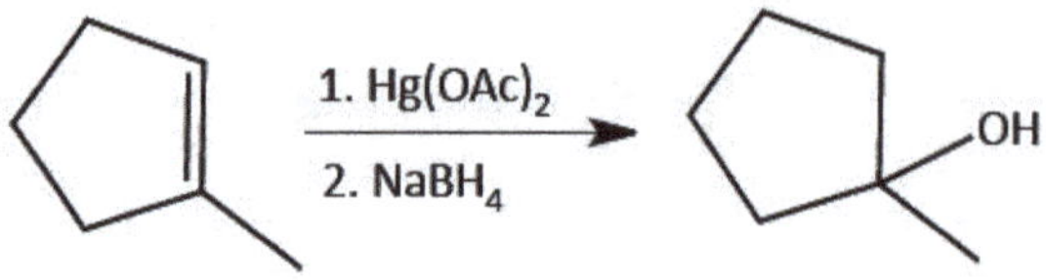

Markovnikov alcohol

Anti-Markovnikov hydroboration of an alkene

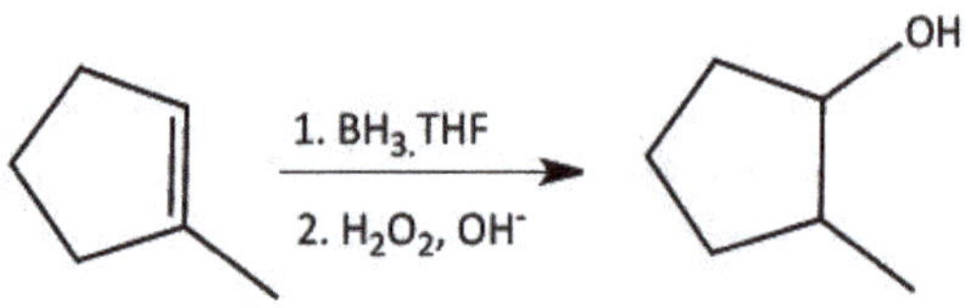

Anti-Markovnikov alcohol

Syn-dihydroxylation of an alkene using osmium tetroxide

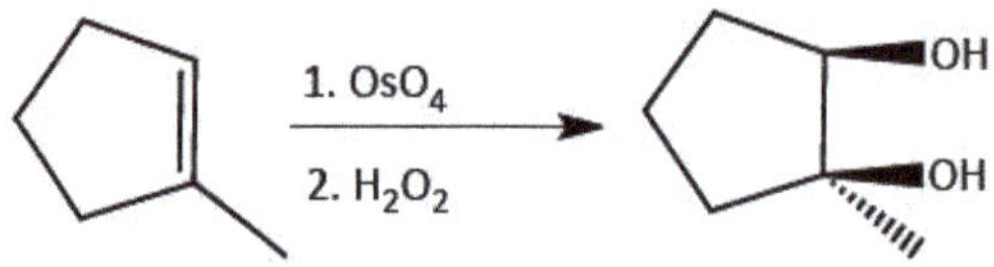

cis-glycol

Syn-dihydroxylation of an alkene using potassium permanganate

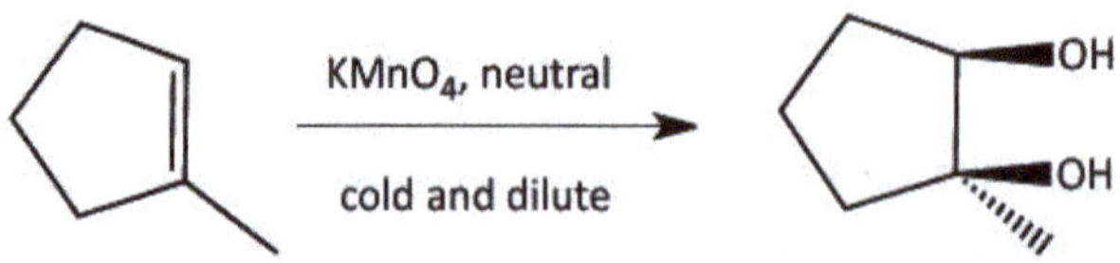

cis-glycol

Epoxidation of an alkene followed by acid hydrolysis

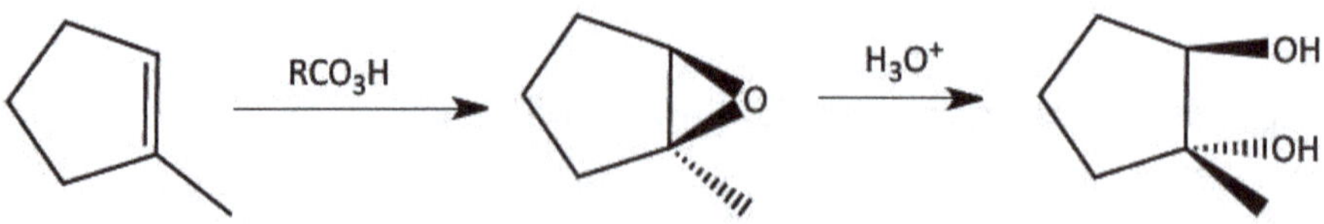

trans-glycol

Acetylide attack of an epoxide

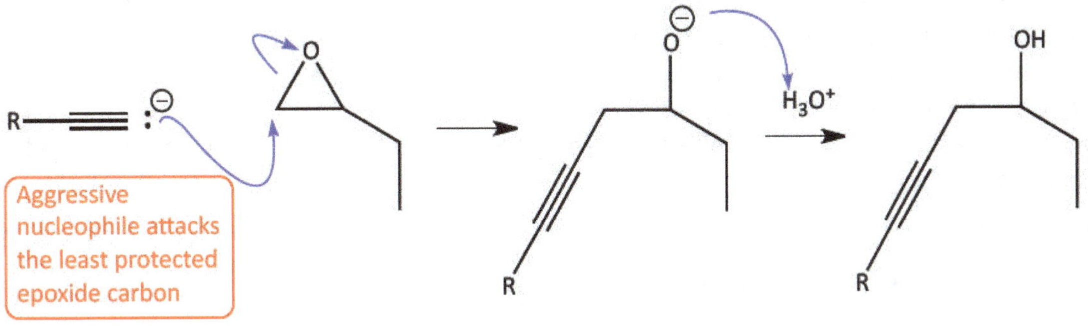

Acetylide attack of a carbonyl

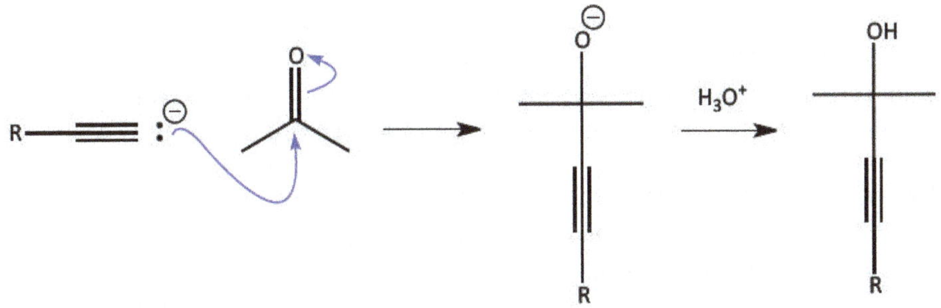

13.2 Fill in the blank for the missing reagents in the following reactions.

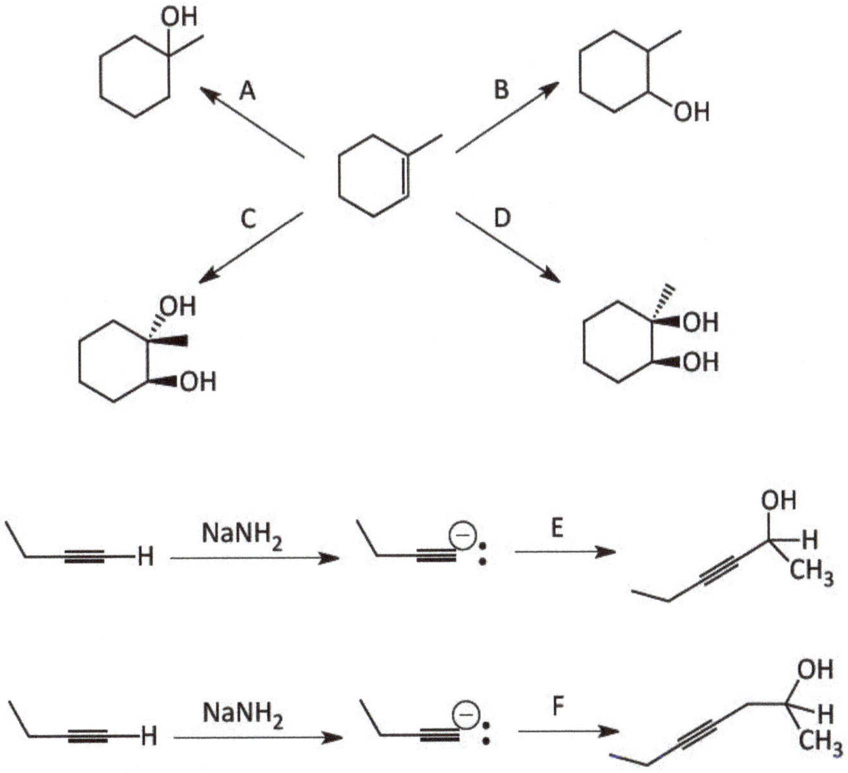

13.3 Draw the products of the following reactions.

a) 1-methyl cyclopentene with $Hg(OAc)_2$ followed by $NaBH_4$
b) 1-ethyl cyclobutene with BH_3.THF followed by hydrogen peroxide and hydroxide
c) Cyclohexene with cold, dilute $KMnO_4$
d) *cis*-2-butene with a peroxyacid in water
e) pentyne reacted with sodium amide followed by benzaldehyde (PhCHO) and then quenched with acid and water

13.7 New Syntheses of alcohols

13.7.1 More on the nucleophilic attack of carbonyls

Let's pause for a moment to more fully think about nucleophiles attacking carbonyl compounds. When a nucleophile, like acetylide or some other negatively charged species, attacks the partially positive carbon atom of the carbonyl. The electrons of the carbonyl pi bond go up on to the oxygen atom, giving it a formal negative charge. Every time this happens, a question needs to be asked. "Can the electrons come back down?"

What we are really asking with this question is whether the groups attached to the carbonyl are good leaving groups or not. If one of the groups is a good leaving group, like Cl or OR, the electrons come back down to reform the carbonyl pi bond. This would make five bonds to the carbonyl carbon, so the leaving group must leave. Acid chlorides and esters have good leaving groups on the carbonyl because Cl and O like the negative charge when they leave. Whenever the electrons come back down, a carbonyl compound is reformed. When a carbonyl is reformed, another nucleophile can attack, and we ask the same question again. Usually, at this point, the electrons cannot come back down because the remaining groups are all poor leaving groups.

Aldehydes and ketones do not have good leaving groups on the carbonyl carbon because H and alkyl groups are terrible leaving groups because H and C are not very electronegative and are too small to want the negative charge. In the case where there is no good leaving group on the carbonyl, the electrons stay up on the carbonyl oxygen as a negative charge.

Finally, a source of H^+, protons, must be added to the reaction. This is called quenching the reaction. Typically, a dilute aqueous acid is used as the proton source.

Closely study the following scheme to make sure you understand how the nucleophilic attack of a carbonyl works. You'll find knowing this process to be very helpful in understanding the reactions of carbonyls with acetylides, Grignard reagents, and reducing agents.

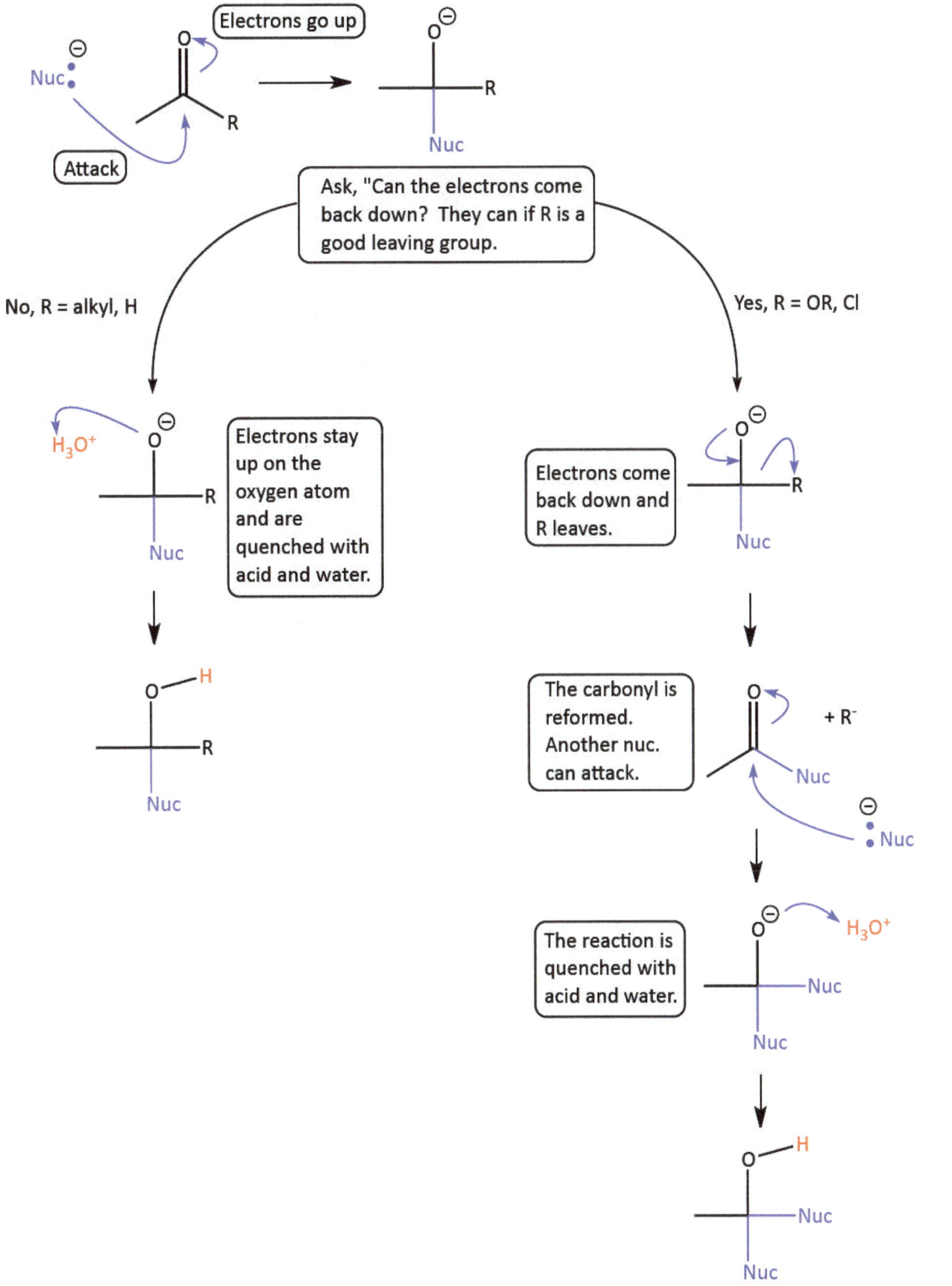

Figure 13.5 General mechanism for a nucleophilic attack of carbonyl compounds

13.7.2 Grignard and Organolithium Reagents

Organic chemistry is the chemistry of carbon-based compounds. But, surprisingly, so far, we have seen very few reactions that form carbon-carbon bonds. But, one great way to do so is to use Grignard (pronounced grin-yard) reagents. Grignard reagents, and the Grignard reaction, were discovered by Victor Grignard. Grignard was awarded the 1912 Nobel Prize in Chemistry for this discovery.

Much like acetylides, Grignard reagents have a negatively charged carbon atom. This negatively charged carbon atom makes it a good nucleophile. Grignard reagents are made by reacting alkyl halides with magnesium. Usually, a solvent like ether is used.

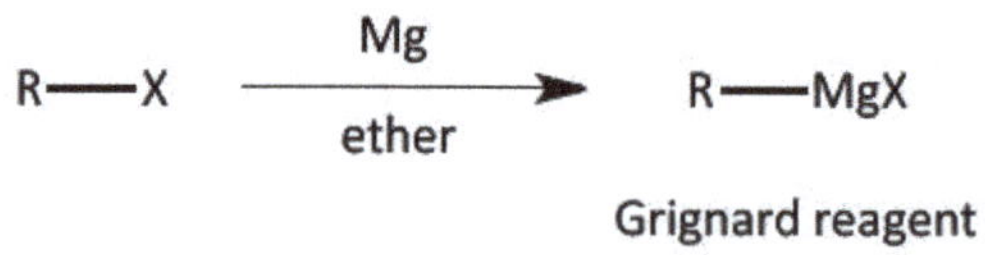

Figure 13.6 Formation of a Grignard reagent

The magnesium metal in a Grignard reagent, like other metals, carries with it a positive charge. Therefore, the carbon atom attached to the Grignard reagent has a negative charge. Grignard reagents can be thought of as an R group with a lone pair of electrons and a negative formal charge. Because of the localized, small negative charge, Grignard reagents are aggressive nucleophiles.

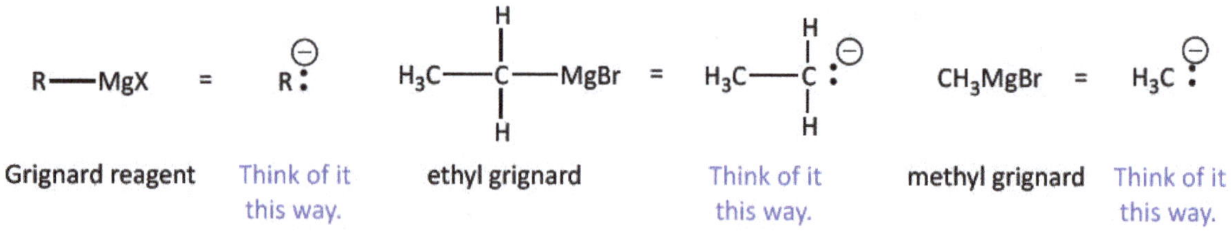

Figure 13.7 Grignard reagents

A reagent that is very similar to a Grignard reagent is an alkyl lithium or an organolithium reagent. These also behave as carbon species that contain a negative charge. They are prepared by the reaction of an alkyl halide with lithium.

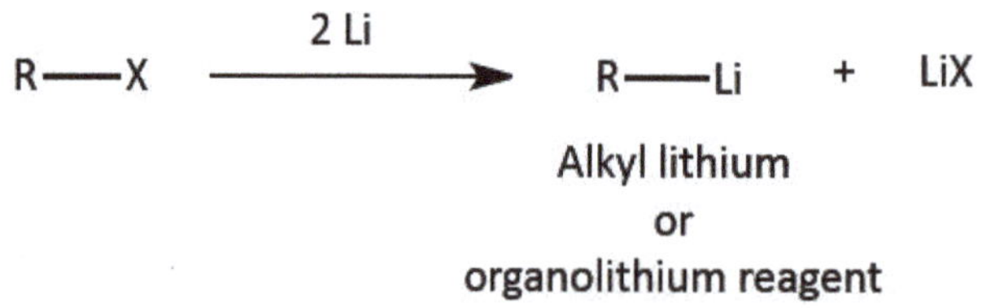

Organolithium reagents, with their positive lithium metal, also have a negative carbon atom. They react like Grignard reagents, like negative carbon species. This is how you should think about them.

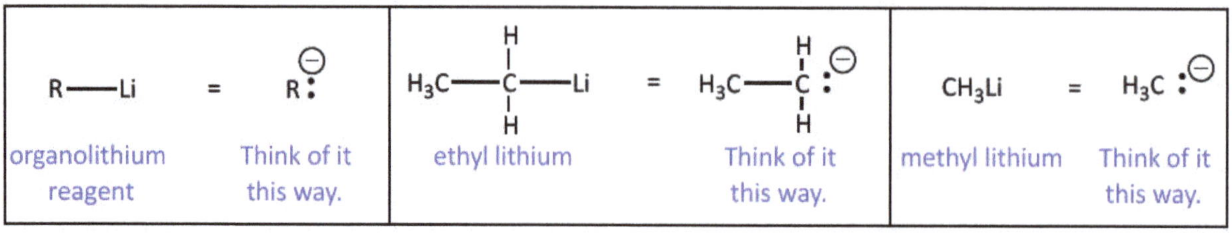

Let's look at the attack of methyl Grignard on various carbonyl compounds. As you study these, make sure each step of the mechanism makes sense to you. Learn what questions you should ask yourself as you do these mechanisms.

Grignard attack of aldehydes and ketones

The carbanion of the Grignard reagent attacks the carbon atom of the carbonyl because the carbon atom has a slight positive charge compared to the more electronegative oxygen atom. The electrons go up onto the oxygen atom. When this happens, we ask ourselves, "Can the electrons come back down?" The answer to that question for aldehydes and ketones is "No" since alkyl groups and hydrogen atoms are poor leaving groups. The negative charge stays on the oxygen atom until it is quenched with acid and water to make an alcohol. For ketones, you can notice that a tertiary alcohol is formed. For aldehydes, a secondary alcohol is formed.

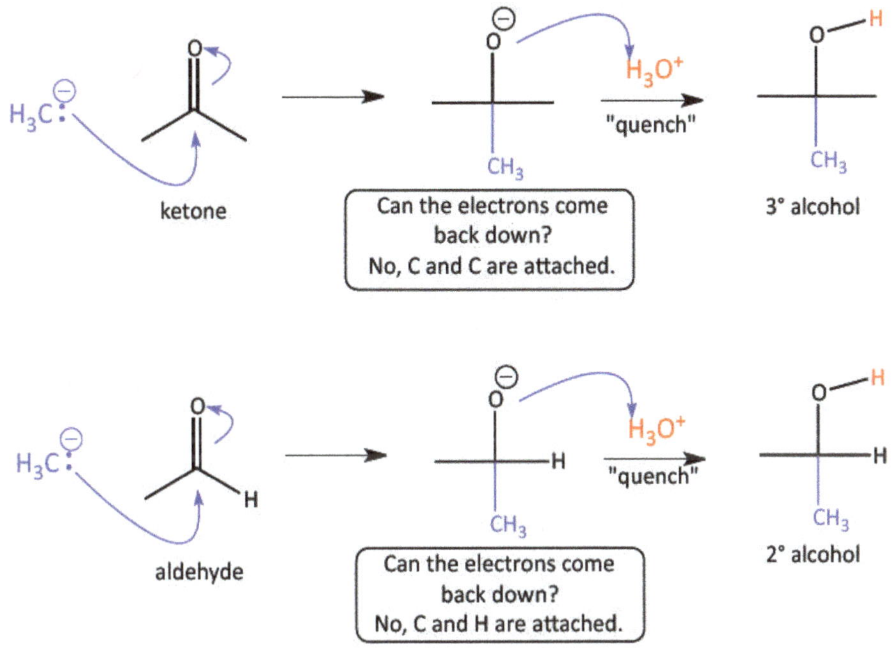

Figure 13.8 Grignard reagents attacking an aldehyde and ketone

Grignard attack of esters and acid chlorides

For esters and acid chlorides, the carbanion of the Grignard reagent again attacks the carbon atom of the carbonyl because the carbon atom has a slight positive charge compared to the more electronegative oxygen atom. The electrons go up onto the oxygen atom. When this happens, we ask ourselves, "Can the electrons come back down?" The answer to that question for aldehydes and ketones is "Yes" since chloride and $^{-}$O-R are pretty decent leaving groups. Therefore, the carbonyl can be reformed. The carbonyl can be attacked again by the Grignard reagent. The electrons go back up on the oxygen atom. Can they come back down? The answer is "No" because only alkyl groups remain, and they are poor leaving groups. The negative charge stays on the oxygen atom until it is quenched with acid and water to make an alcohol. Tertiary alcohols are then formed.

Alcohols

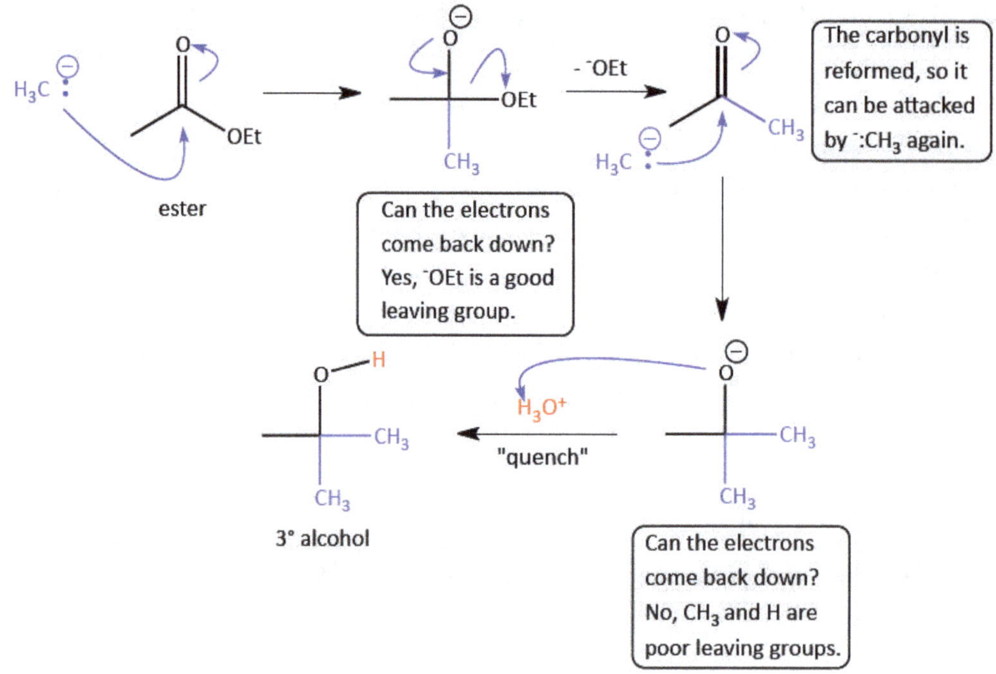

Figure 13.9 Grignard reagents reacting with an ester

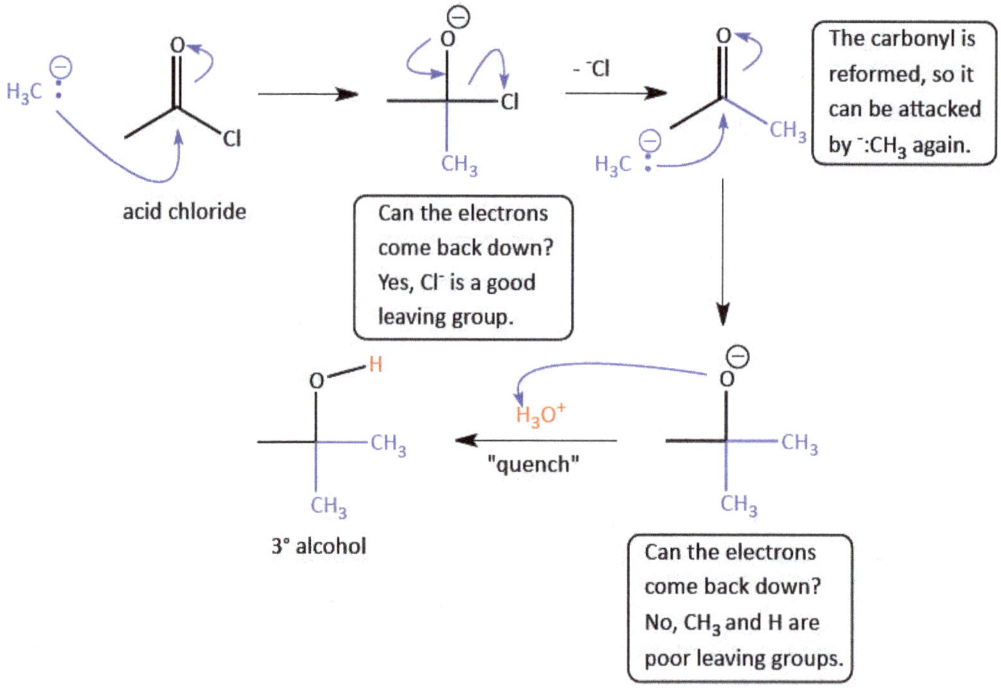

Figure 13.10 Grignard reagents reacting with an acid chloride

Grignard reagents not only react with the carbonyl functional group, but they react with other pi bonds between two atoms of differing electronegativities. Alkenes, with a pi bond between two carbon atoms, is not attacked because there is no difference in electronegativities. Neither carbon atom has a partial positive charge. The following can also be attacked by Grignard reagents.

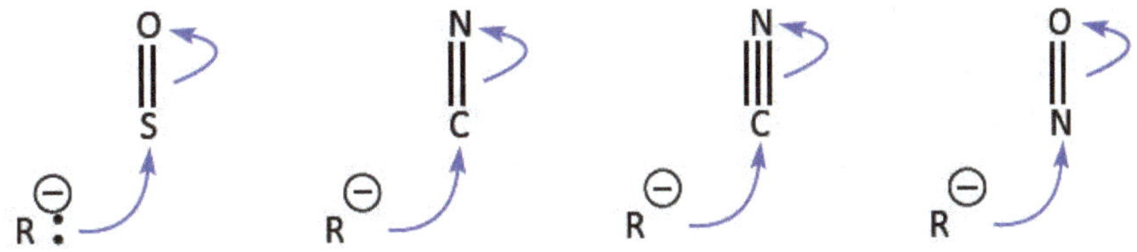

Figure 13.11 Non-carbonyl functional groups that react with Grignard reagents

Solvent for the Grignard reagent

One major drawback to Grignard reagents is that they are so negative and such reactive nucleophiles, if any type of acidic hydrogen is around, the Grignard reagent will immediately grab the H+ proton and "quench" itself. It will become an alkane and not attack a carbonyl. These protons are sometimes hidden, so we need to be on the watch for them. Anytime you see a hydrogen atom on an oxygen atom, a nitrogen atom, a sulfur atom, or a terminal acetylene, it is bad news for Grignard reagents. Hydrogen atoms on carbon atoms are no problem. They are attached to carbon atoms too well.

N-H, O-H, S-H, H-C≡C

Figure 13.12 Hydrogen atoms that protonate Grignard Reagents

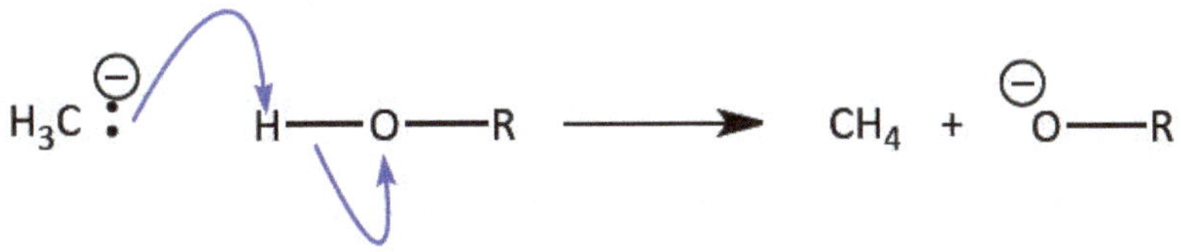

Figure 13.13 What happens when a Grignard reagent encounters an acid proton

This means that we need to be careful what types of solvents we can use for a Grignard reaction. Alcohols and water are no good as they contain acidic hydrogen atoms on oxygen atoms. Grignard reactions are usually run in diethyl ether. The ether needs to be specially prepared. It must be a dry solvent. That means that all traces of water are removed from the ether solvent.

Reaction of a Grignard with a carboxylic acid.

Look at this theoretical reaction of a Grignard reagent with a carboxylic acid. Do you foresee any problem with this reaction?

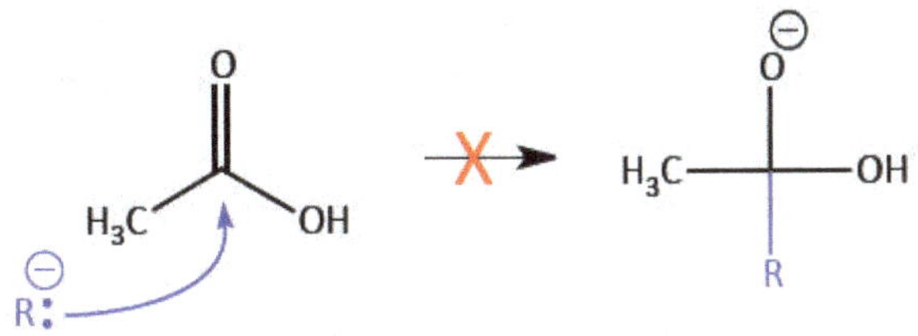

From what we just learned; the Grignard reagent will attack the proton on the oxygen making an alkane. This will effectively kill what we were trying to do in this reaction. The carbonyl will not be attacked.

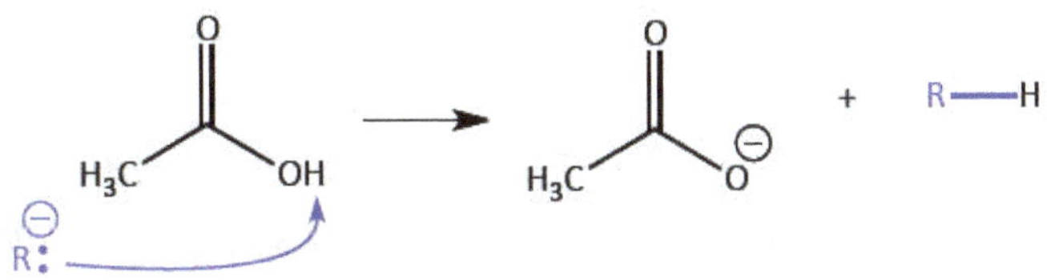

Alcohols

Grignard reaction with epoxides

The reaction of a Grignard reagent with an epoxide is a triangle reaction. The aggressive, nucleophilic, Grignard reagents act like a strong criminal in the triangle reaction and the epoxide is a neutral victim. The strong criminal, the aggressive Grignard reagent, will attack the least crowded, easiest to get to side of the epoxide. The ring opens up to make an alkoxide (alkyl group with a negative oxygen). The alkoxide can then be quenched with acid and water to form an alcohol.

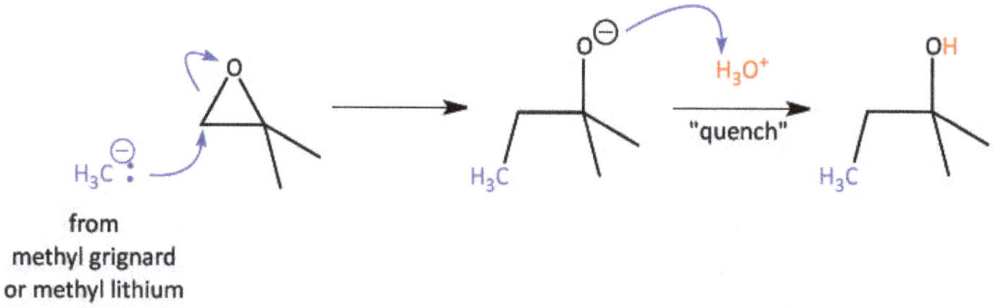

Figure 13.14 Grignard reagent reacting with an epoxide

Notice, two carbon atoms are between the CH_3 from the group added from the Grignard reagent and the hydroxyl group of the alcohol. This is because there are two carbon atoms in an epoxide. In all of the reactions of a Grignard reagent with a carbonyl, there is only one carbon atom between the group added from the Grignard reagent and the hydroxyl group.

Grignards with alkyl halides

Although Grignard reagents are excellent nucleophiles when attacking carbonyls and epoxides, we find that they are lousy nucleophiles when it comes to attacking alkyl halides in an S_N2 reaction. Grignard reagents are a little too hard to react with the alkyl halide. The electron pair on the Grignard reagent is not polarizable enough to do the reaction well. This is too bad, because you would think this would be a great way to make carbon-carbon bonds and larger molecules.

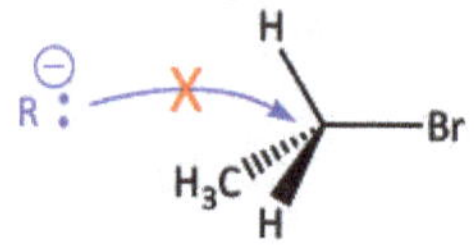

If you think about it, it is actually a good thing that Grignard reagents do not attack alkyl halides. This is because Grignard reagents are made from alkyl halides. If Grignard reagents reacted with alkyl halides, when the first Grignard reagent was made, it would immediately begin reacting with the alkyl halide starting material, and we would never be able to make a flask of Grignard reagent to react with carbonyls or epoxides!

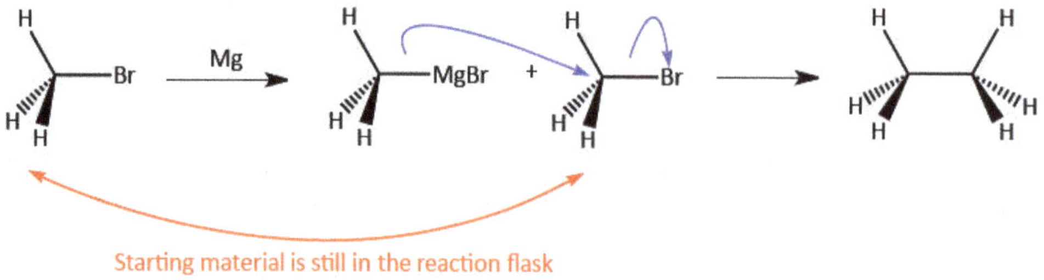

Gilman Reagents with alkyl halides

The good news is there is another way to do a similar reaction. Instead of using a Grignard reagent, another reagent called the Gilman reagent will attack an alkyl halide in an S_N2 reaction. The Gilman reagent is a lithium dialkyl cuprate. Gilman reagents are made from alkyl lithium reagents reacting with cuprous iodide (CuI). Alkyl lithium reagents can be made from the reaction of alkyl halides with lithium.

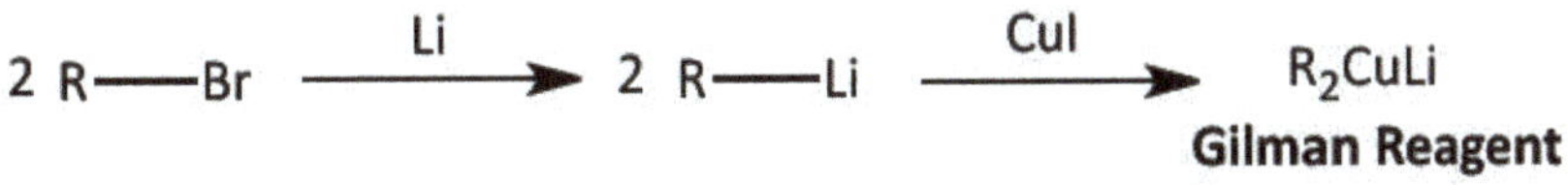

Then alkyl groups of the Gilman reagent are softer and can react with alkyl halides substituting an alkyl group for a halide.

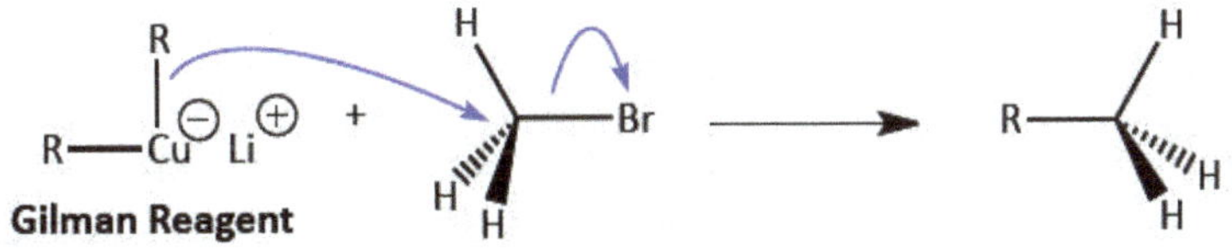

The Gilman reagent also reacts with vinyl and phenyl halides. This is not an S_N2 reaction, because S_N2 reactions do not work on sp[2] carbon atoms, but this works!

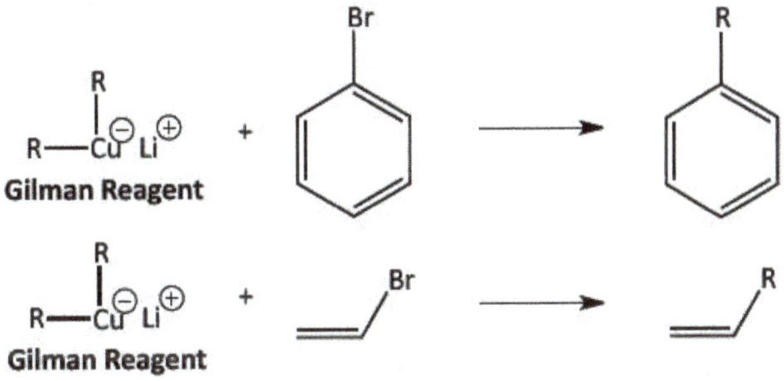

13.4 Draw the products of the following reactions.

a) 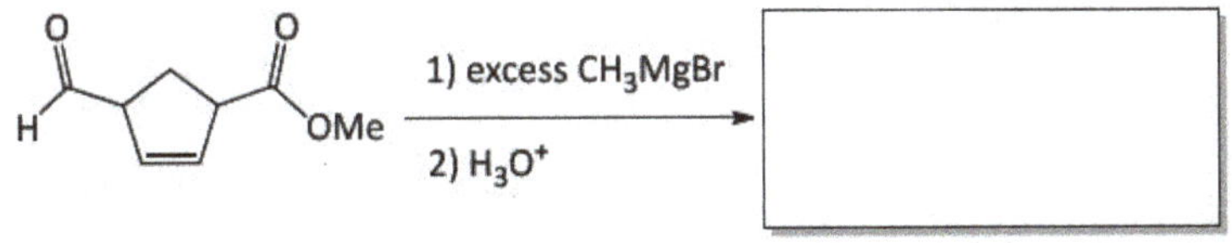

b) 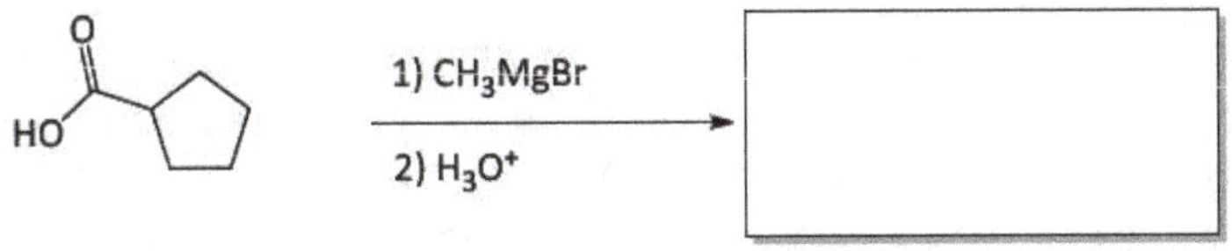

c) 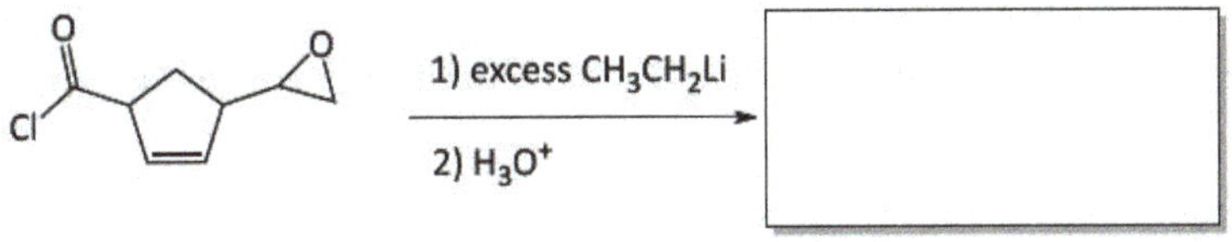

d) 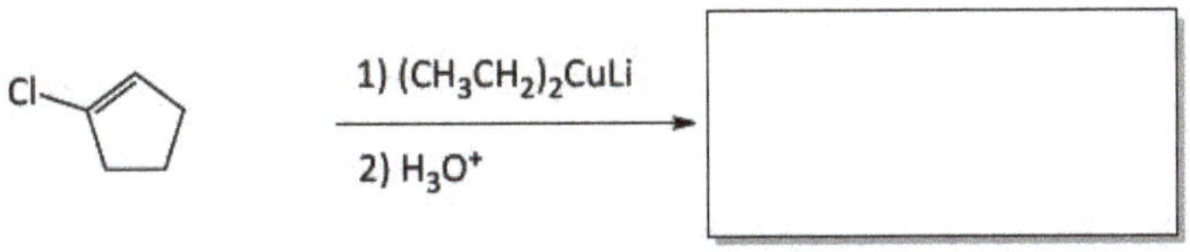

13.5 Show how to make the following organolithium, Grignard reagent, or Gilman reagent from the appropriate halide and any other necessary reagents.

a) 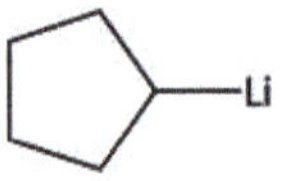

b) 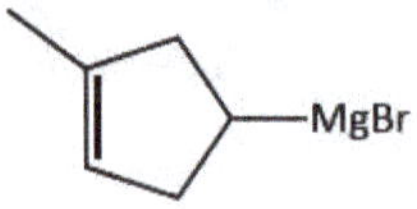

c) Ph_2CuLi

13.7.3 Reductions of carbonyl compounds

A carbonyl functional group has two C-O bonds. The two C-O bonds of a carbonyl can be taken down, or reduced, to an alcohol, one C-O bond. This process is called a reduction. Notice that in the process of reducing the carbonyl, two hydrogen atoms are attached. The hydrogen in a reduction is a hydride source like sodium borohydride, $NaBH_4$, or lithium aluminum hydride, $LiAlH_4$ or LAH. Hydride is a hydrogen atom with two electrons. Therefore, it has a negative formal charge. The weaker $NaBH_4$ works in several solvents. It can work in alcohols, ethers, and water. We must be more careful with the much stronger $LiAlH_4$ because it can react explosively with alcohols and water.

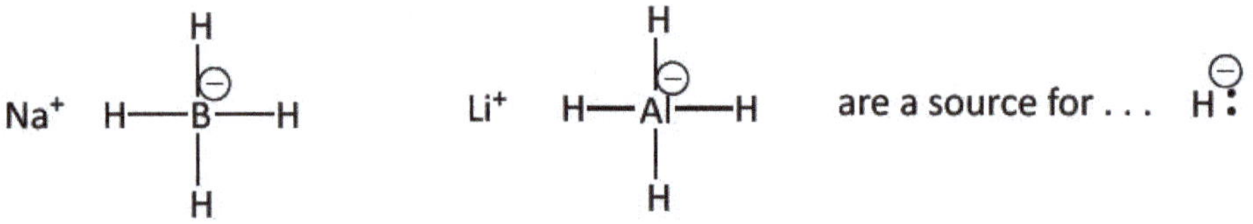

Sodium borohydride
$NaBH_4$
Weak hydride source

Lithium aluminum hydride
$LiAlH_4$
Strong hydride source

Hydride

Figure 13.15 Hydride, reducing agents

Reduction of aldehyde or ketone

In the overall reaction mechanism for the reduction of a ketone or an aldehyde, a hydride, attacks the partially positive carbon atom of a carbonyl.

The hydride attack of a ketone yields a secondary alcohol while the hydride attack of an aldehyde yields a primary ketone. But, there is no reason to try to memorize such a thing. Simply draw out the mechanism anytime you are asked such a thing and see what your product is. In the attack of the ketone, both R groups on the carbonyl are carbons and therefore terrible leaving groups. In the attack of the aldehyde, the R groups are carbon and hydrogen. These are bad leaving groups. So, after the attack of the nucleophile, the electrons stay up on the oxygen atom and do not come back down. Both the weaker $NaBH_4$ and the stronger $LiAlH_4$ can reduce aldehydes and ketones.

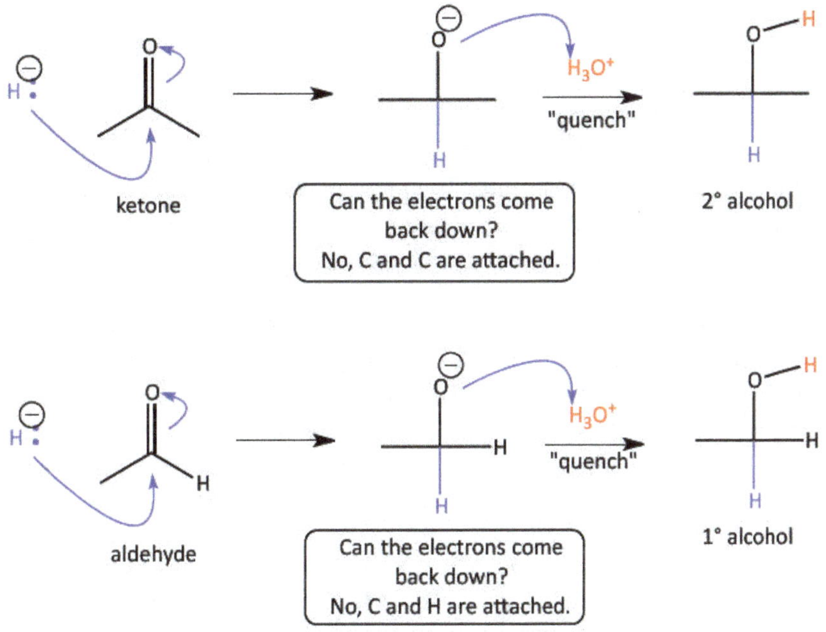

Figure 13.16 Reduction of aldehyde and ketone

Reduction of acid chloride

An acid chloride can be reduced by a strong reducer like $LiAlH_4$ or even a weak reducing agent like $NaBH_4$. The hydride attacks the carbonyl carbon atom of the acid chloride. The electrons go up onto the oxygen atom of the carbonyl. There is a good leaving group, the chloride group. The electrons come back down and kick off the chloride group. A carbonyl is re-formed as an aldehyde. Aldehydes also react with $NaBH_4$ or $LiAlH_4$. The hydride attacks the carbonyl. The electrons go up onto the oxygen atom giving it a formal negative charge. We ask, "Can the electrons come back down?" In this case, they cannot because the aldehyde only has poor leaving groups, an H and an alkyl group.

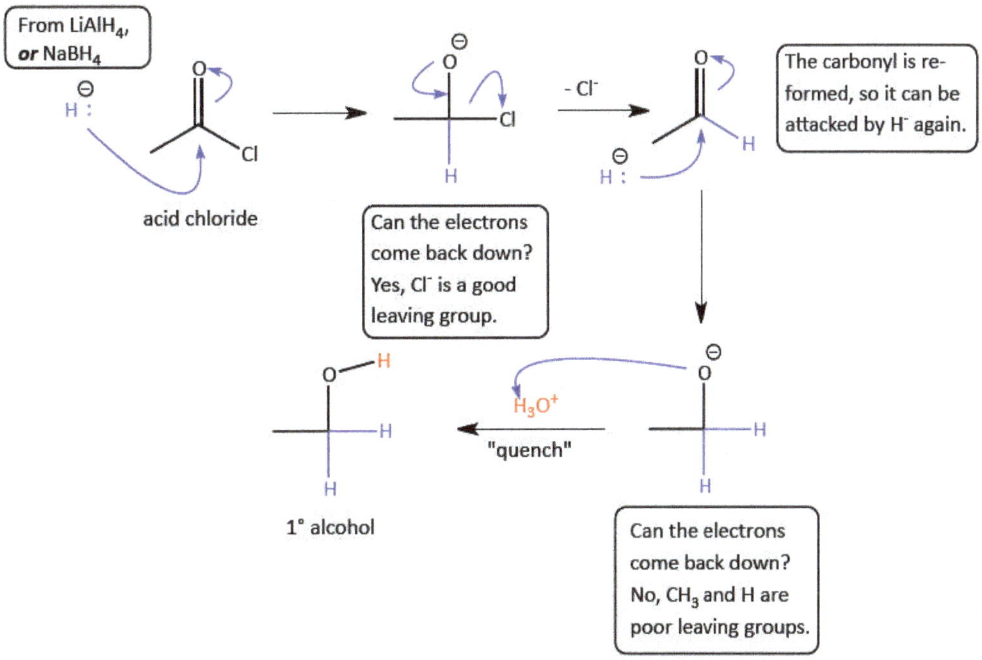

Figure 13.17 Reduction of acid chloride to alcohol

Reduction of ester or carboxylic acid

Are other carbonyl compounds like esters or carboxylic acids any different? They are. We find that the lone pair of electrons on the extra oxygen atom donated electron density to the carbonyl carbon atom. Therefore, it is not as positively charged as it is in aldehydes or ketones. The end result of this is that a weak hydride source like $NaBH_4$ is not strong enough to attack the carbonyl carbon of an ester or carboxylic acid. But, a very strong hydride source like $LiAlH_4$ *is* strong enough to attack the carbonyl carbon atom.

The hydride from the $LiAlH_4$ attacks the carbonyl carbon atom of the ester. The electrons go up onto the oxygen atom of the carbonyl. There is a good leaving group, the OR group. The electrons come back down and kick off the OR group. A carbonyl is reformed as an aldehyde. Aldehydes also react with $LiAlH_4$. The hydride attacks the carbonyl. The electrons go up onto the oxygen atom giving it a formal negative charge. We ask, "Can the electrons come back down?" In this case, they cannot because the aldehyde R groups are poor leaving groups, H and an alkyl group.

This alkoxide intermediate can be quenched with acid and water. This needs to be done carefully! $LiAlH_4$ is quite reactive with water and acid. The reaction mixture should be cooled down and the water added one drop at a time.

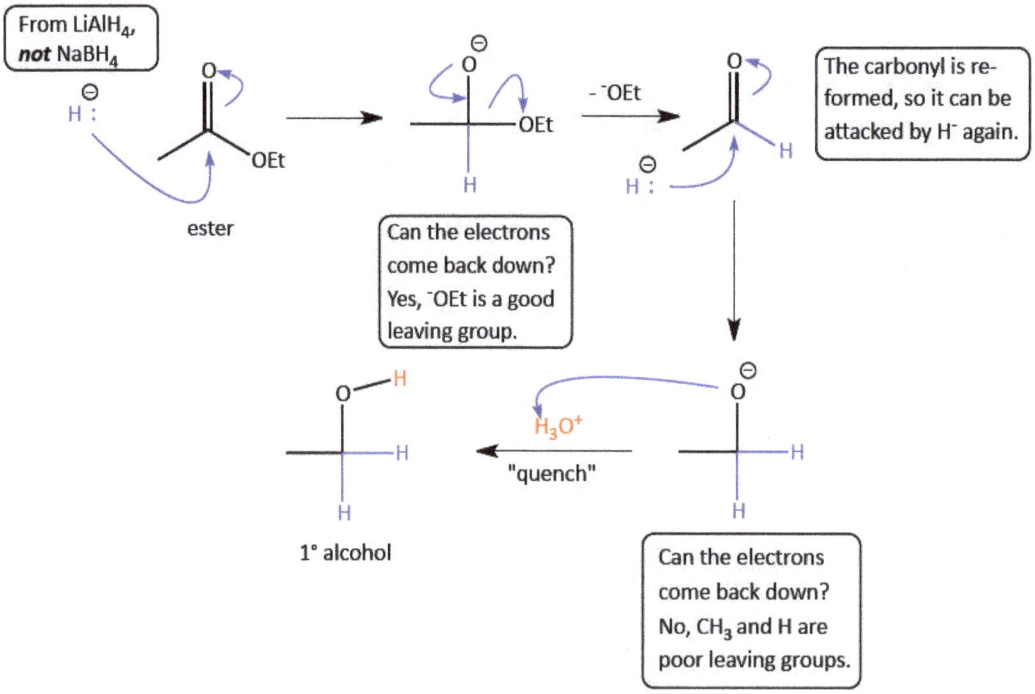

Figure 13.18 $LiAlH_4$ reduction of an ester to alcohol

Like esters, carboxylic acids are fully reduced to primary alcohols using $LiAlH_4$.

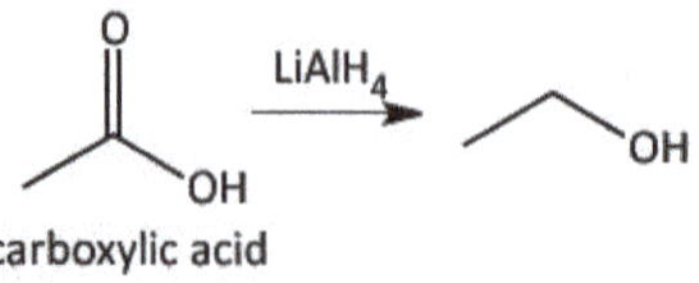

Figure 13.19 $LiAlH_4$ reduction of carboxylic acid to alcohol

13.6 Draw the product for each of the following reductions.

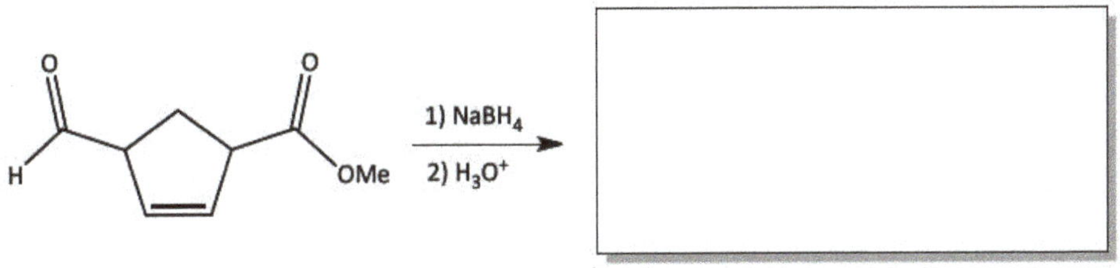

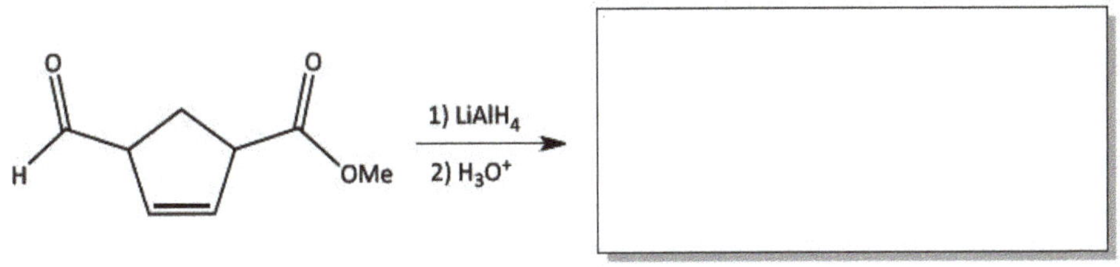

Raney nickel, H_2 reductions

While the weaker $NaBH_4$ can reduce aldehydes and ketones, the much stronger $LiAlH_4$ can of course also reduce aldehydes, ketone as well as the more difficult to reduce acid chlorides, esters, and carboxylic acids. But, neither of these reducing agents can reduce alkenes to alkanes. But, H_2 and Raney nickel can. Raney nickel/H_2 performs a hydrogenation across double bonds by adding two hydrogen atoms across any pi bond, from a carbonyl (of an aldehyde or ketone) or an alkene, with syn stereochemistry.

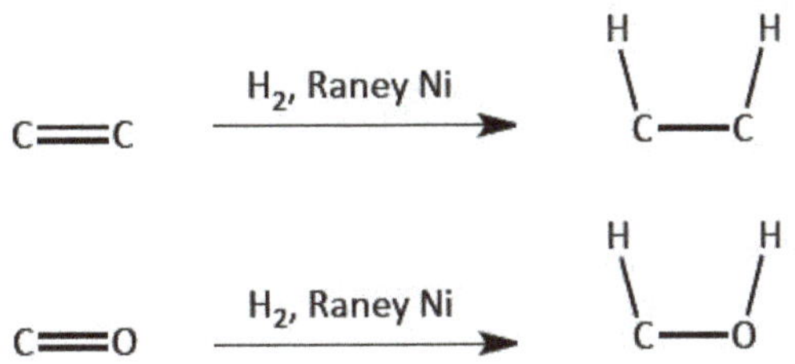

Figure 13.20 H_2 and Raney Nickel reaction with alkene and carbonyl

Notice that both the carbonyl and the alkene of cyclopentenone are hydrogenated when reacted with hydrogen and Raney nickel.

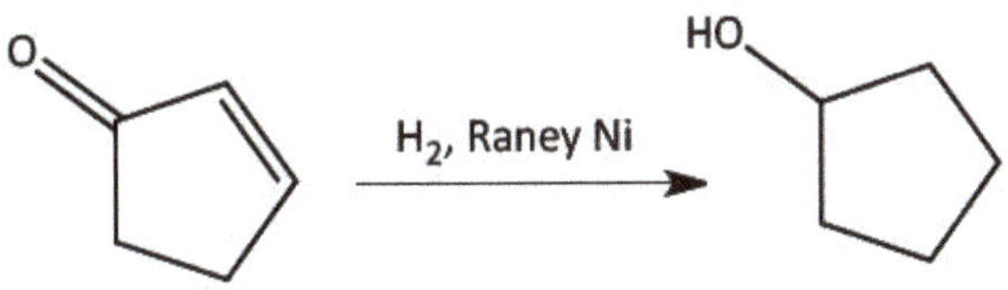

13.8 Reaction Summary

Thiol synthesis via S_N2 with sodium hydrosulfide

a) 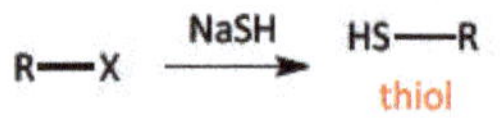

thiol

Alcohol synthesis via S_N2 with hydroxide

b) 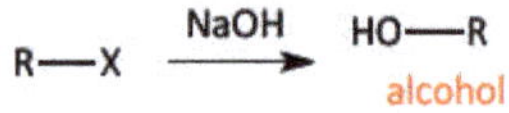

alcohol

Acid hydrolysis of an alkene

c) 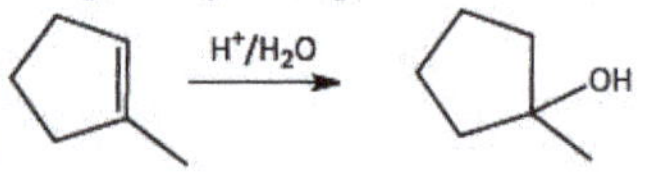

Markovnikov alcohol

Markovnikov Oxymercuration/Demercuration of an alkene in water

d) 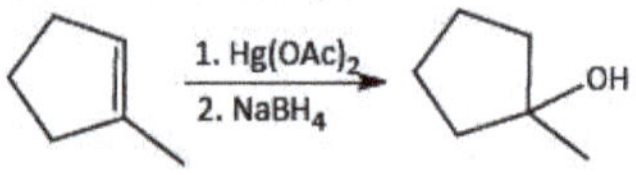

Markovnikov alcohol

Anti-Markovnikov hydroboration of an alkene

e) 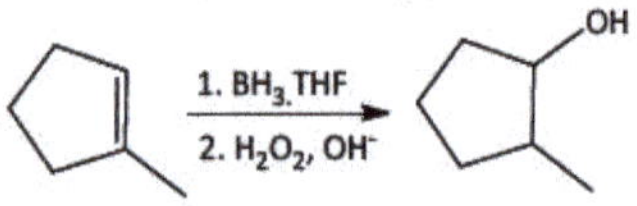

Anti-Markovnikov alcohol

Syn-dihydroxylation of an alkene using osmium tetroxide

f) 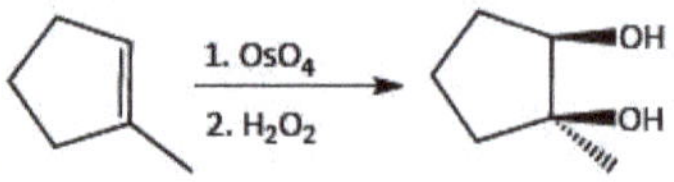

cis glycol

Syn-dihydroxylation of an alkene using potassium permanganate

g) 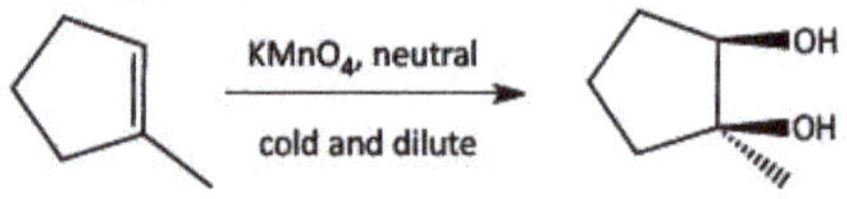

cis glycol

Epoxidation of an alkene followed by acid hydrolysis

h)

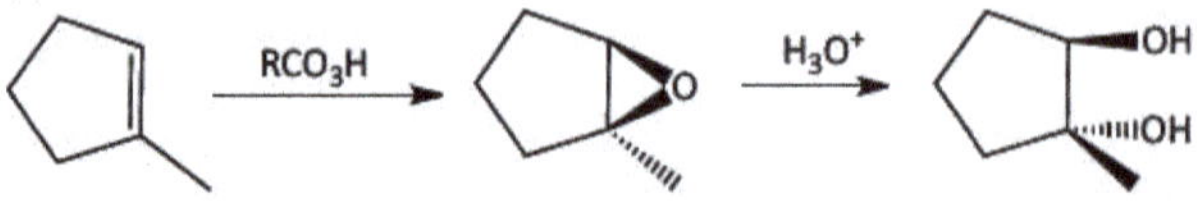

trans glycol

Acetylide attack of an epoxide

i) 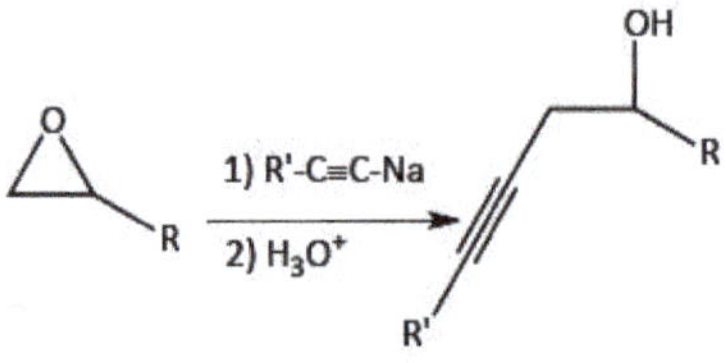

Acetylide attack of a carbonyl

j) 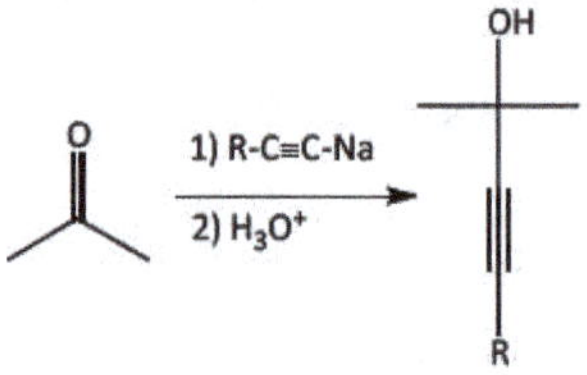

Synthesis of a Grignard reagent

k) 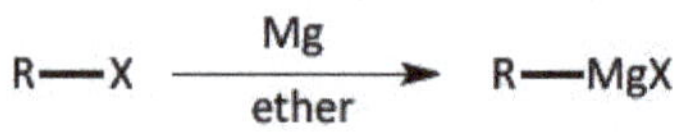

Grignard reagent

Synthesis of organolithium reagent

l) 

Alkyl lithium or organolithium reagent

Grignard reagent attacks of carbonyls

m) 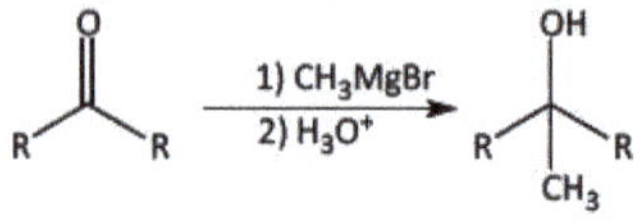

ketone — 3° alcohol

n) 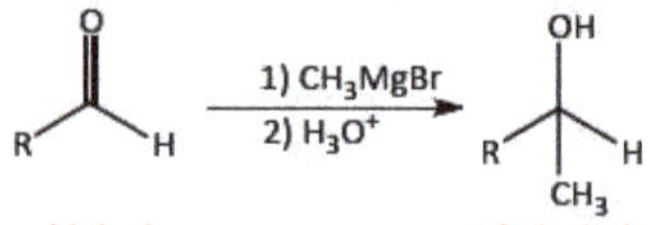

aldehyde — 2° alcohol

o) 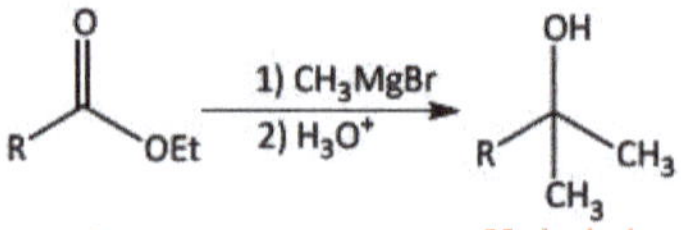

ester — 3° alcohol

p) 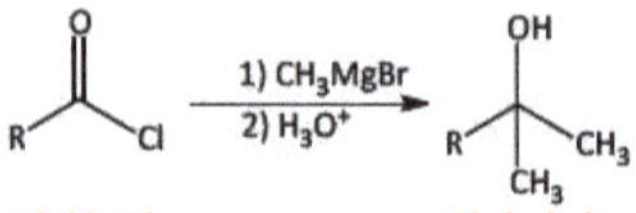

acid chloride — 3° alcohol

q) 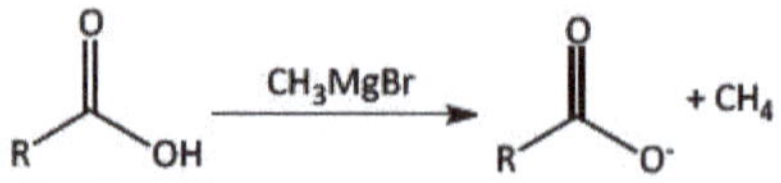

carboxylic acid

Grignard attack of epoxides

r) 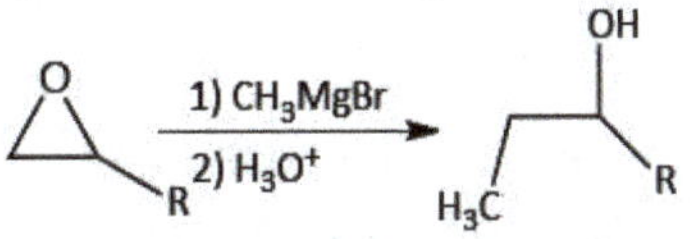

Synthesis of a Gilman reagent

s) $2\ R{-}Br \xrightarrow{Li} 2\ R{-}Li \xrightarrow{CuI} R_2CuLi$ Gilman Reagent

Gilman reagent with alkyl halide

t) $R{-}Br \xrightarrow{R'_2CuLi} R{-}R'$

Reductions of carbonyls

u) 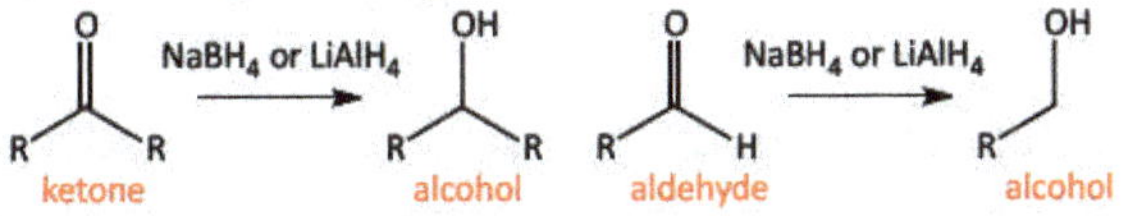

v) 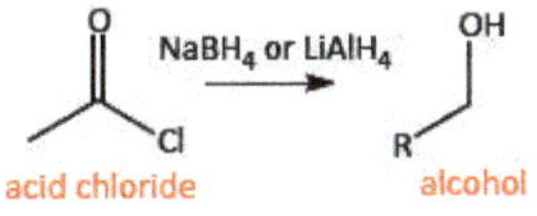

w) 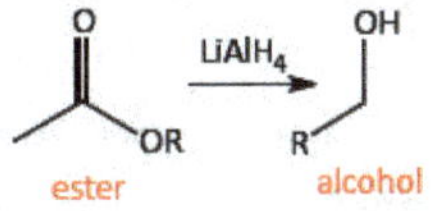

x) 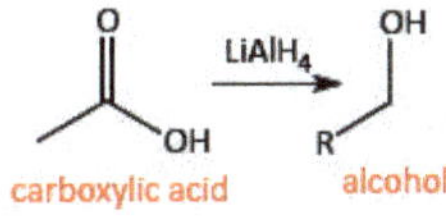

y) 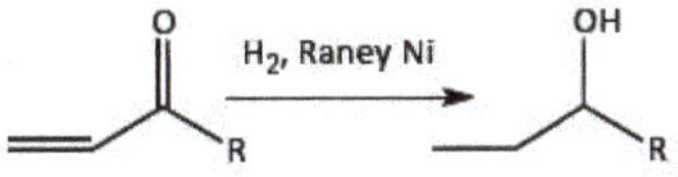

13.9 Problems

13.7 Draw the following compounds:
a) isopropyl magnesium bromide b) 3-heptanol c) cyclohexanol d) 4-iodophenol e) 3-cyclopenten-1-ol f) *p*-chlorophenol g) (*R*)-pentane-1,4-diol h) cyclohexane-1,2,3-triol i) *o*-bromophenol j) propylene glycol k) (*Z*)-4-chloro-3-buten-2-ol l) 2-methylbut-3-yn-1-ol m) *m*-cresol n) cyclohexylmethanol o) 4-bromobutan-1-ol p) propyl mercaptan q) pentanethiol r) butyl mercaptan s) butane-2-thiol

13.8 Name the following compounds:

a) 

b) 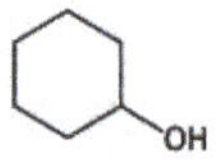

c) 

d) 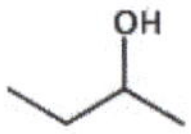

e) 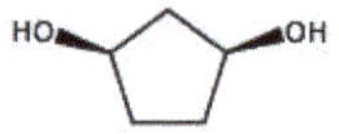

f) 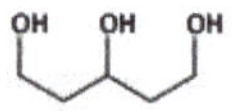

g) 

h) 

i) 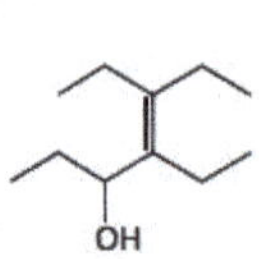

j) 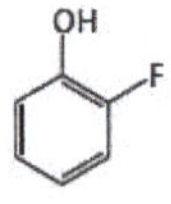

k) 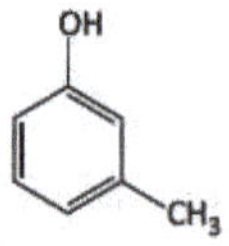

l) 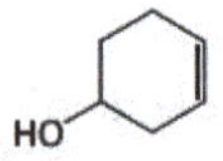

13.9 A little chemistry joke. . . What words do the following structures represent? Think *ortho*, *meta*, and *para*.

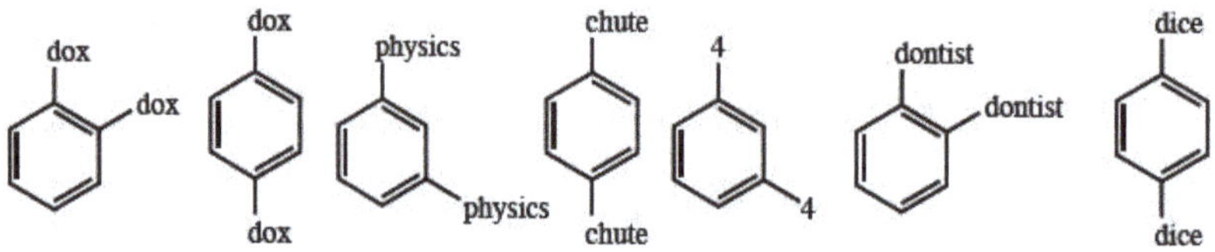

13.10 Which of the following IR spectra is that of an alcohol?

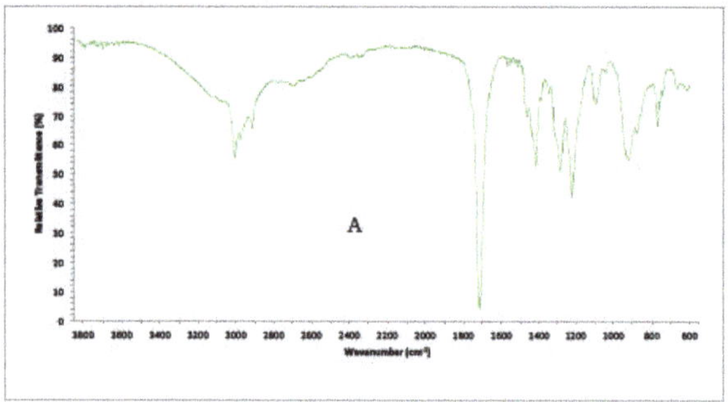

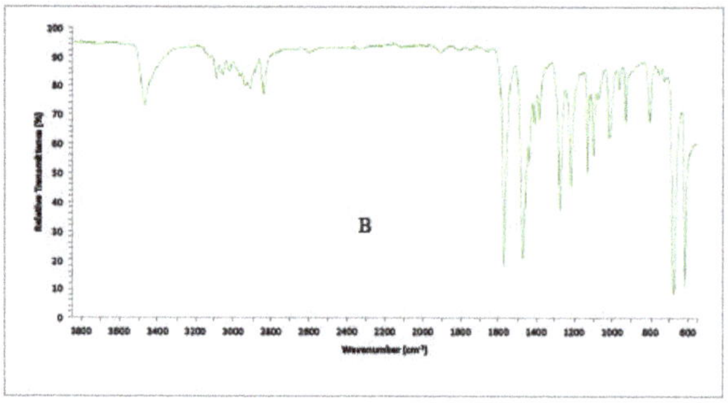

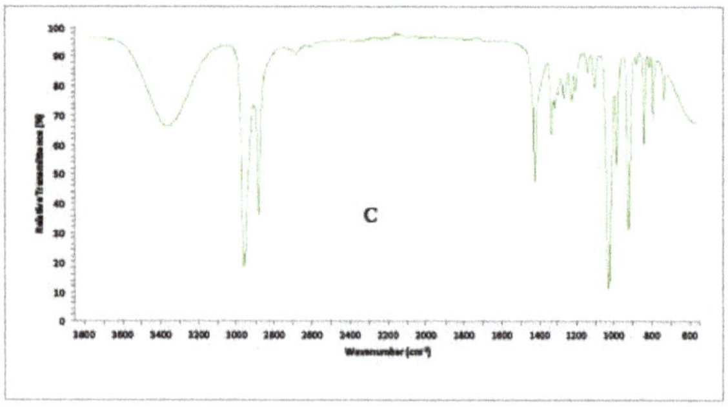

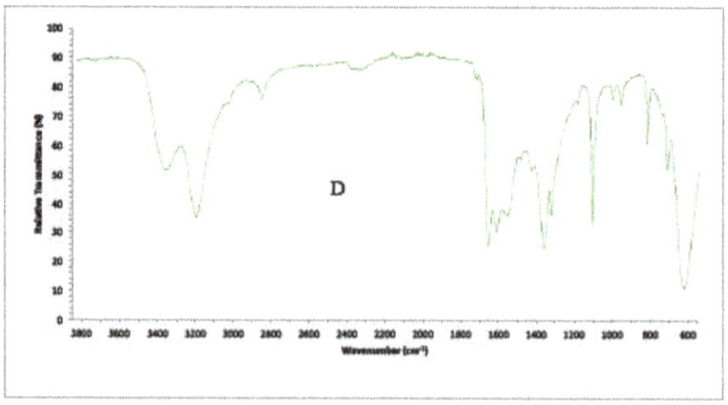

13.11 The following alcohol can be synthesized from two different alkenes. Draw these two alkenes and then give the reagents you would use to synthesize the alcohol from each of these alkenes.

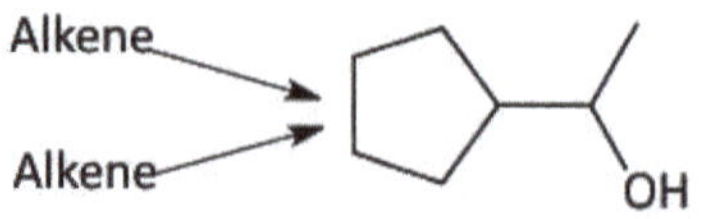

13.12 Draw the products of the following reactions.

a) 1-methylcyclopentene + dilute aqueous acid
b) 1-methylcyclopentene + aqueous $Hg(OAc)_2$ followed by $NaBH_4$
c) 1-methylcyclopentene + a cold, dilute, neutral solution of potassium permanganate
d) 1-methylcyclopentene + BH_3.THF followed by H_2O_2 and NaOH
e) 1-methylcyclopentene + OsO_4 followed by H_2O_2
f) 1-methylcyclopentene + peroxybenzoic acid in water
g) 2-methylpropene + aqueous $Hg(OAc)_2$ followed by $NaBH_4$
h) 2-methylpropene + dilute aqueous H_2SO_4
i) 2-methylpropene + BH_3.THF followed by H_2O_2 and NaOH
j) 2-methylpropene + OsO_4 followed by H_2O_2
k) 2-methylpropene + MCPBA followed by H_3O^+
l) cyclopentene + peroxybenzoic acid in water
m) cyclopentene + a cold, dilute, neutral solution of potassium permanganate
n) cyclopentene + OsO_4 followed by H_2O_2
o) cyclopentene + MCPBA followed by H_3O^+

13.13 Show how you would prepare the following compounds starting with an alkene and any other reagents you need.

a) 1-heptanol b) 2-heptanol c) 1-ethyl-1-cyclopentanol d) 2-ethyl-1-cyclopentanol

13.14 How could you make 2-methylcyclopentanol from 1-bromo-1-methylcyclopentane? More than one reaction is required.

13.15 Determine what aldehyde, ketone, or epoxide would react with ethyl magnesium bromide such that, upon quenching with aqueous acid, would make the following compounds.

a) 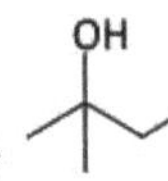 b) 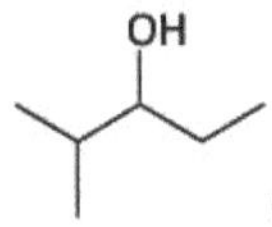 c) 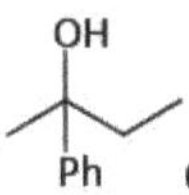 d) 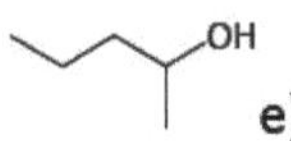 e) 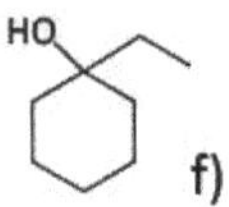 f) 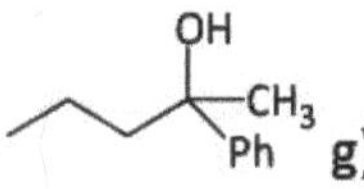 g) 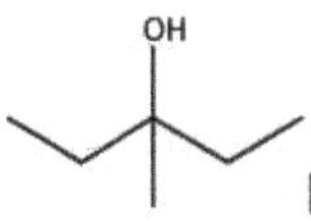 h) 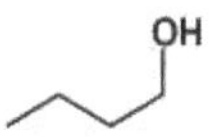

13.16 Determine an appropriate Grignard reagent that would react with acetaldehyde (CH_3CHO) such that, upon quenching with aqueous acid, would make the following compounds.

a) 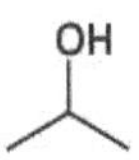 b) 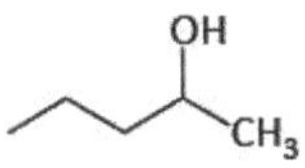 c) 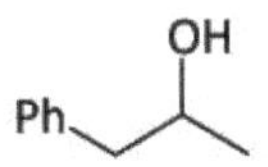 d) 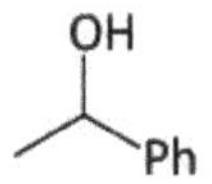 e) 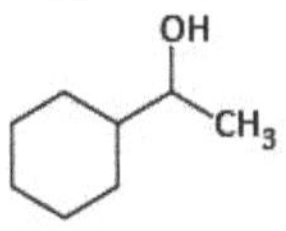

13.17 Determine an appropriate Grignard reagent that would react with 1,2-epoxypropane, , such that, upon quenching with aqueous acid, would make the following compounds.

a) 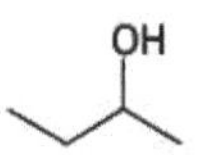 b) 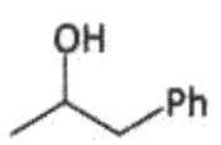 c) 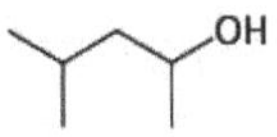 d) 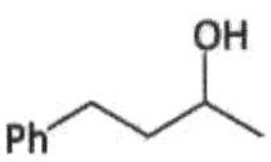 e) 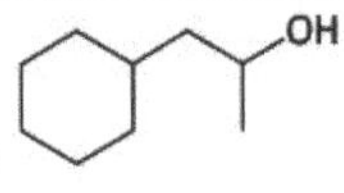

Alcohols

13.18 Draw the structure of an acid chloride that would react with an excess of methyl magnesium bromide such that, upon quenching with aqueous acid, would make the following compounds. Then, draw an appropriate ester for each problem that you could use instead of the acid chloride.

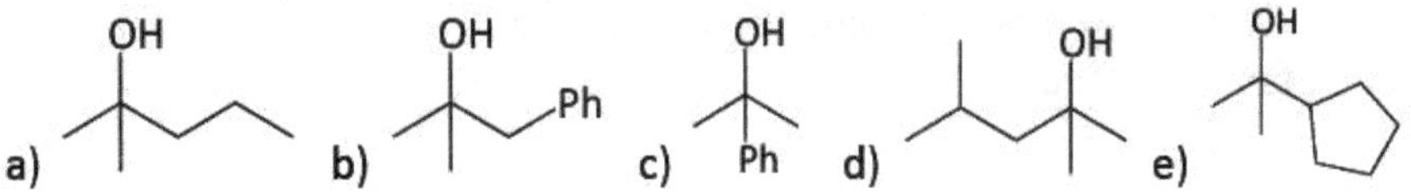

13.19 Give two ways you could synthesize the following compounds using the reaction of a Grignard reagent and an aldehyde.

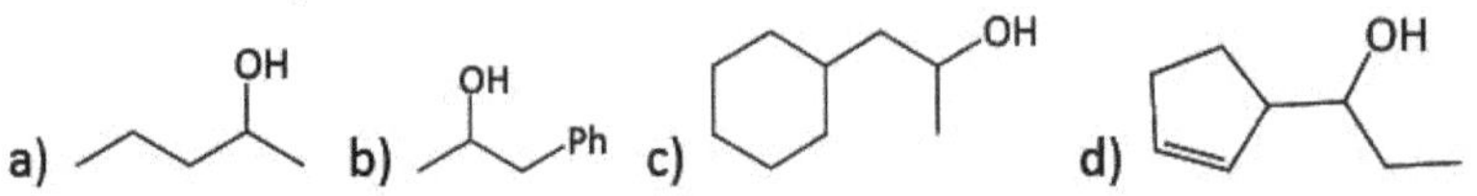

13.20 Give three ways you could synthesize the following compounds using the reaction of a Grignard reagent and a ketone.

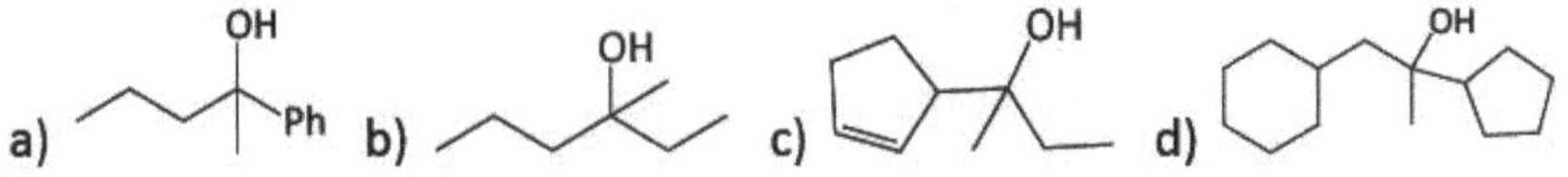

13.21 For each of the following compounds, show how it could be synthesized from an acid chloride or ester and an excess of a Grignard reagent.

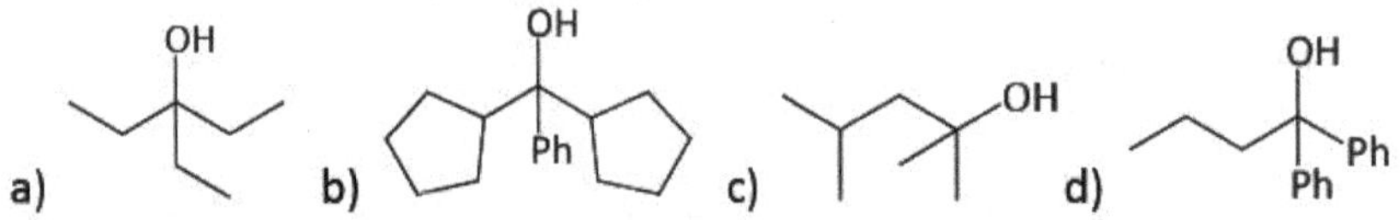

13.22 Reducing a carbonyl compound can make the following compounds. Show what carbonyl compound and appropriate reducing agent can be used to make each of the following.

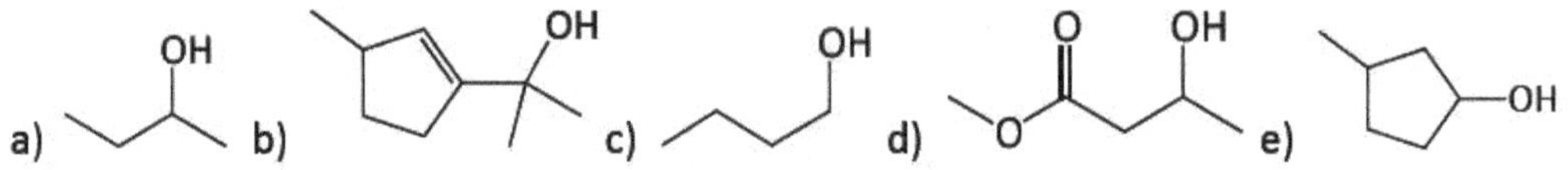

13.23 Draw the products for each of the following reactions.

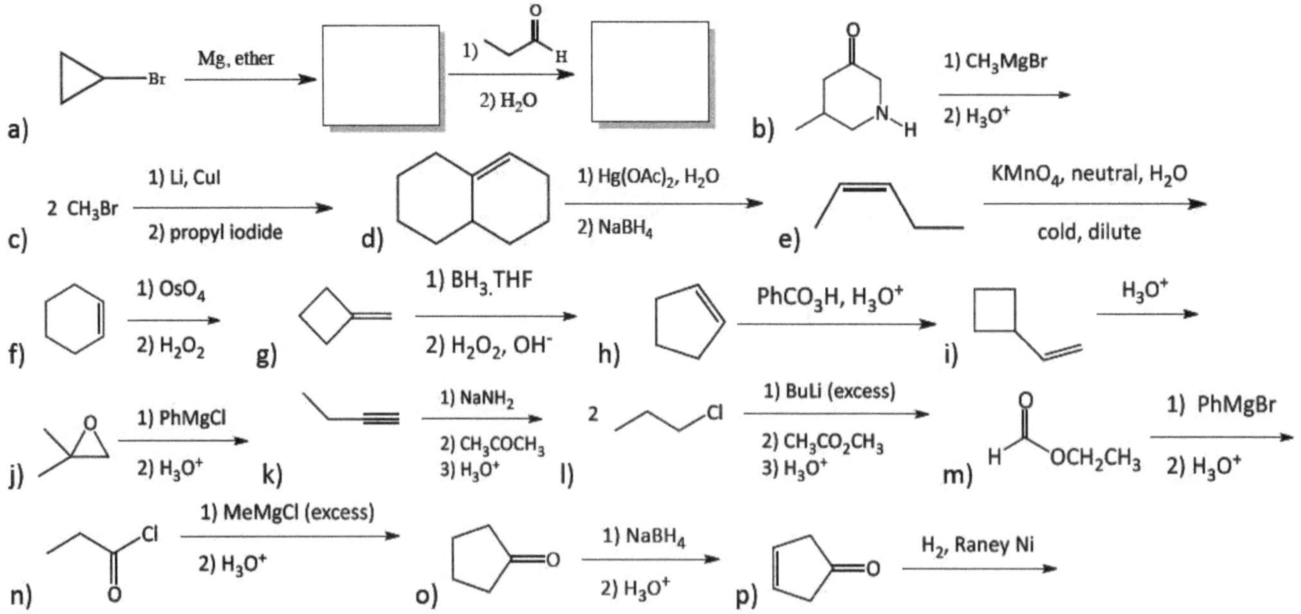

q) $H_2C{=}CHCO_2H$ + 1) $LiAlH_4$, 2) H_3O^+ r) CH_3CH_2COCl + 1) $LiAlH_4$, 2) H_3O^+ s) $CH_3CO_2CH_2CH_2CHO$ + 1) $NaBH_4$, 2) H_3O^+
t) $CH_3COCH_2CH{=}CHCH_2CO_2CH_3$ + 1) $LiAlH_4$, 2) H_3O^+ u) $PhCH_2CHO$ + 1) LAH, 2) H_3O^+
v) iodohexane + sodium hydrosulfide

13.24 What reagents are needed to perform the following transformations?
a) $CH_3CO_2CH_2CH_2CHO$ to $CH_3CO_2CH_2CH_2CH_2OH$ b) $CH_3CH_2CO_2H$ to $CH_3CH_2CH_2OH$
c) $CH_3CH_2CH{=}CHCH_2CHO$ to $CH_3CH_2CH_2CH_2CH_2CH_2OH$ d) $CH_3CO_2CH_3$ to CH_3CH_2OH

e) 
f) 
g) 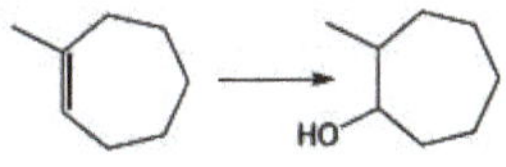

h) 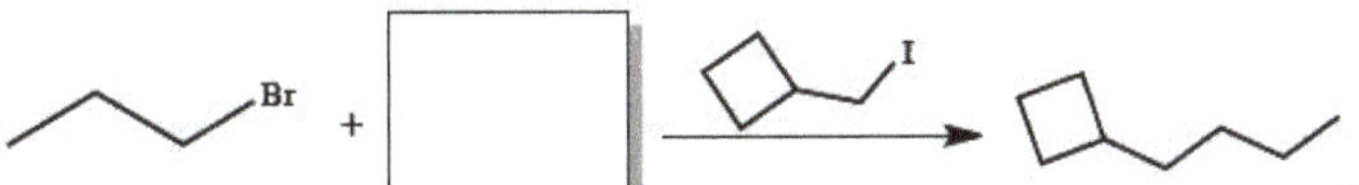
i) 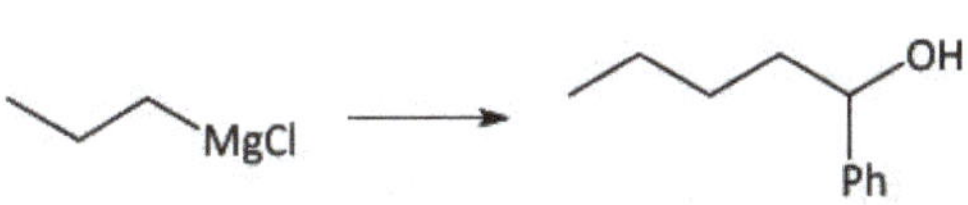

j) 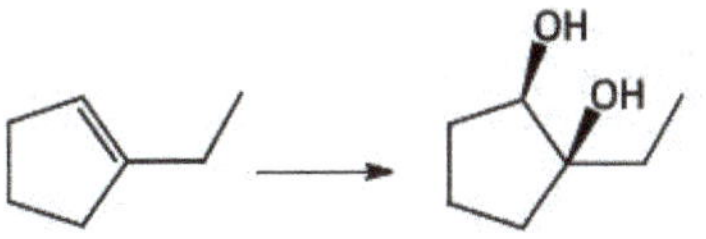
k) 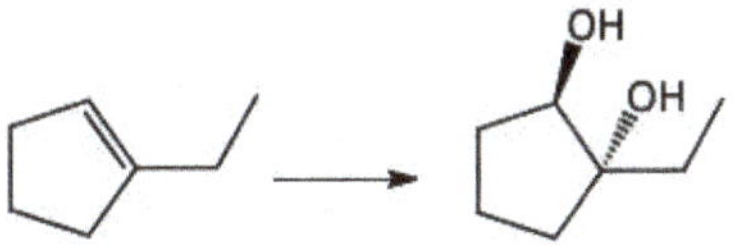
l) 

m) 

n) propyl magnesium bromide to 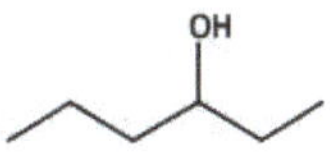

o) propyne to  p) propyne to 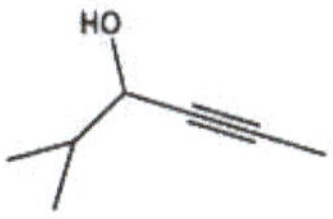 q) $CH_3CH_2CH_2CO_2H$ to $CH_3CH_2CH_2CH_2OH$
r) $PhCO_2Cl$ to Ph_3COH s) $PhCO_2Cl$ to $PhCH_2OH$ t) $HCO_2CH_2CH_3$ to $(Ph)_2CHOH$ u) 1,2-epoxybutane to 3-hexanol
v) $PhCH_2Cl$ to $PhCH_2CH_2OH$ w) 1-bromobutane to 1-pentanol x) $EtCO_2CH_3$ to $(Et)_3COH$
y) 1-bromocyclopentane to 1-cyclopentyloctane z) propene to propylene glycol
aa) bromobenzene to 2-phenylethanol bb) bromopentane to pentanethiol

13.25 A mixture of ortho, meta, and para aromatic compounds is below.

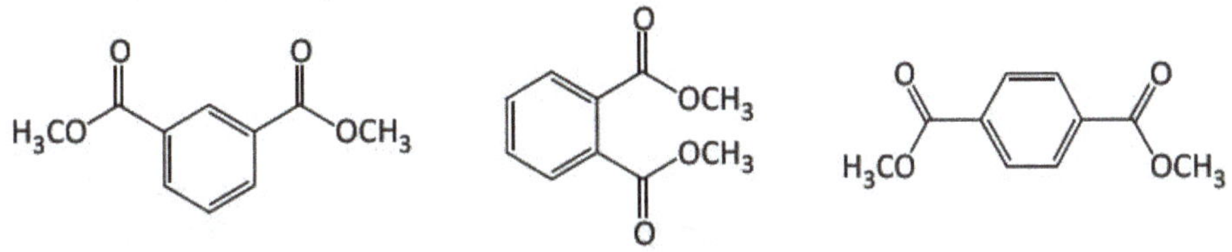

A B C

a) Label A, B, and C as *ortho-*, *meta-*, or *para-*.
b) What would be the product of the reaction of A with an excess of $NaBH_4$ followed by hydrolysis?
c) What would be the product of the reaction of C with an excess of $LiAlH_4$ followed by hydrolysis?
d) Draw the product *and* the mechanism of the reaction of B with an excess of ethyl grignard (CH_3CH_2MgBr) followed by hydrolysis.

13.26 The following reactions did not work as planned. What is wrong with each of these reactions?

a) 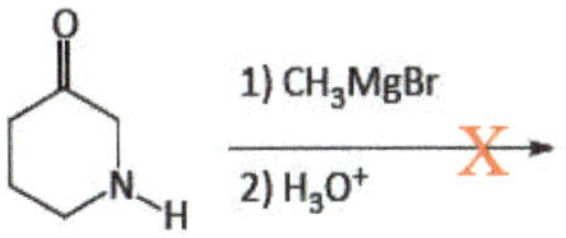

b) 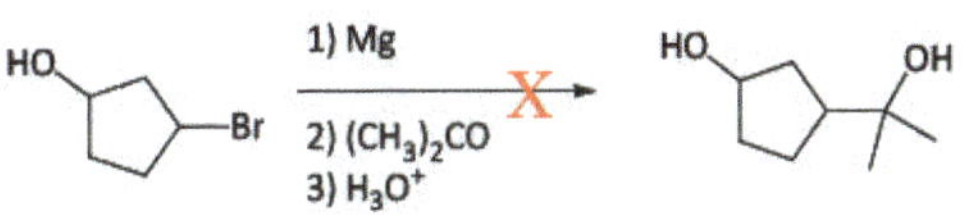

c) 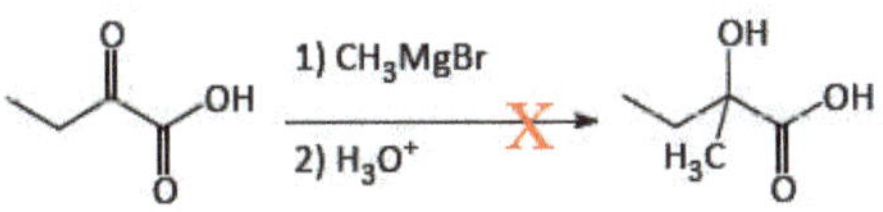

d) 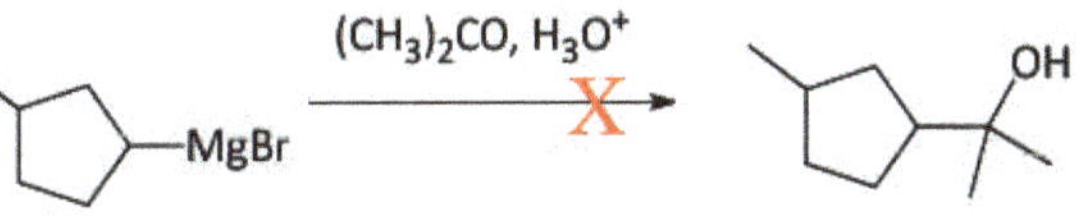

Chapter 14. Alcohol Reactions

14.1 Introduction

We've previously learned about the synthesis of alcohols. We learned about this early because alcohols are incredibly useful compounds that can react to form many different kinds of compounds. In this chapter, we will investigate many of these reactions of alcohols.

14.2 Deprotonation of alcohols

14.2.1 Synthesis of an alkoxide

The hydroxyl group of an alcohol compound contains a hydrogen atom on an oxygen atom. Because oxygen atoms are fairly electronegative and can carry a negative charge, it is possible to deprotonate the hydroxyl group to form alkoxides (groups that contain a negative oxygen atom) if we use a strong enough base. Usually, for alcohols, sodium metal, sodium hydride, potassium metal, or potassium hydride is used. The potassium compounds are a little more aggressive and stronger than the sodium compounds. Also, in the reaction, hydrogen gas is produced. One should be very careful adding these bases to a flask containing alcohol. The alcohol should be dry (without water around). The base should also be added in slow increments and never put into the reaction mixture all at once. It might also be a good idea to cool the reaction down using an ice bath. It is quite easy to start a fire if you are not careful performing this reaction.

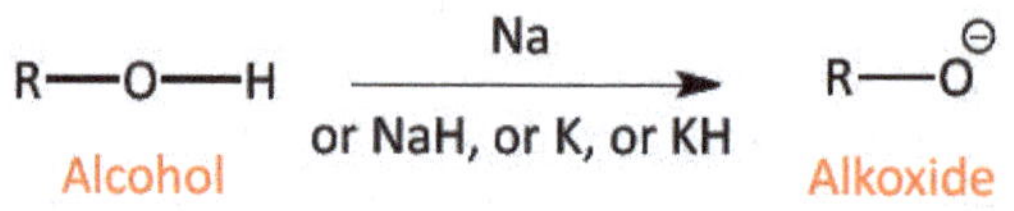

Primary alcohols react quickly. Secondary alcohols are a little slower, and tertiary alcohols need a strong base like potassium metal to deprotonate.

14.2.2 Why would we want to make an alkoxide?

An alkoxide, with its negatively charged oxygen atom is a good nucleophile. It can go attack another compound. For instance, an alkoxide can perform an S_N2 reaction on an alkyl halide to make an ether.

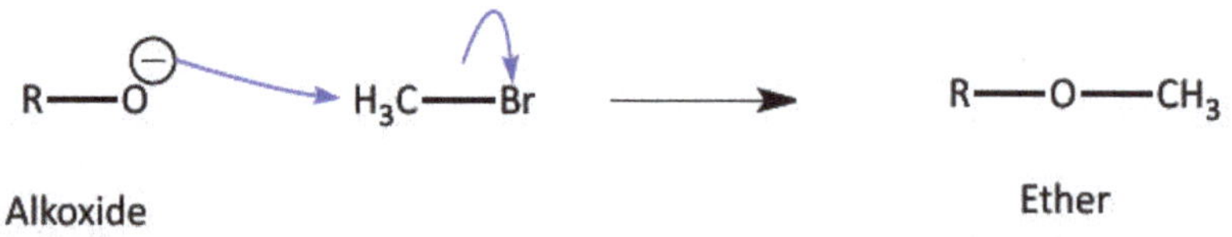

Which of the following compounds is the most acidic?

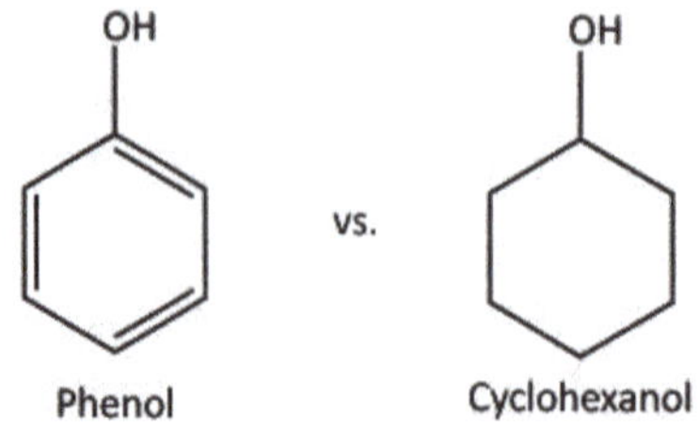

We remember that any time we try to determine relative acidities, we always, Always, ALWAYS look at the conjugate bases. In other words, what type of negative charged molecule does it become after it loses its H^+?

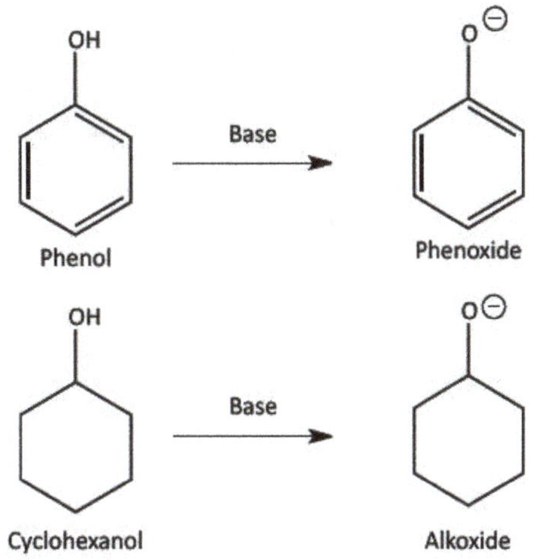

Both compounds have a negative oxygen atom. This is pretty good. The only difference we can see is the double bonds in the phenoxide where the cyclohexyl alkoxide does not. Therefore, the negative charge on the phenoxide can be spread out through resonance while the negative charge on the cyclohexyl alkoxide is stuck on the oxygen atom alone. Since the negative charge can be spread out over four atoms in phenoxide, it is much more stable than the cyclohexyl alkoxide. That is why phenol is more acidic than cyclohexanol. In fact, it is 100 million times more acidic! In fact, phenol is so acidic, it can be deprotonated with even weaker bases. Potassium hydroxide (KOH) and sodium hydroxide (NaOH) can deprotonate phenol, whereas, the stronger alkali metals (Na, K) and hydrides (NaH, KH) must be used to deprotonate regular aliphatic alcohols.

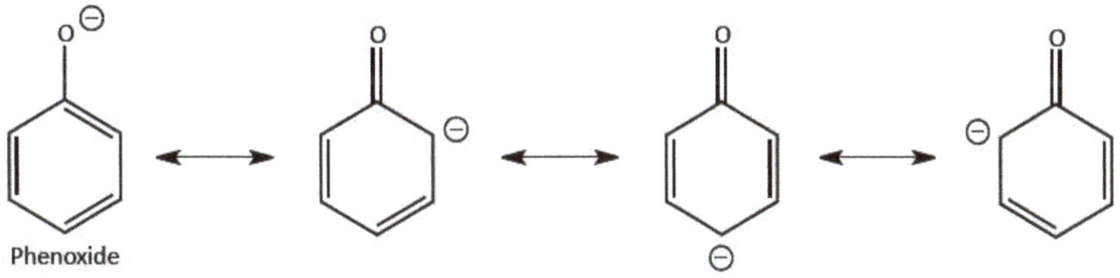

14.1 Draw the ether products of the following reactions that involve alkoxides.

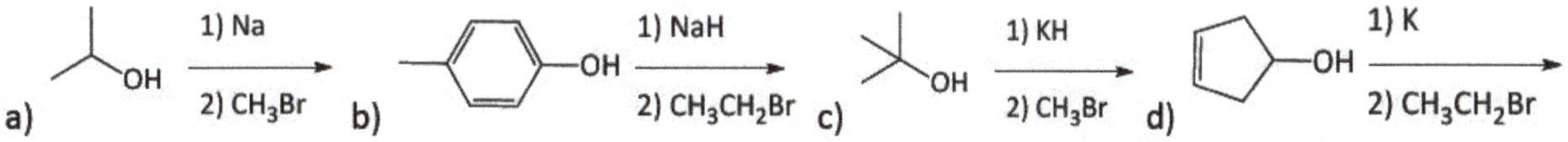

14.3 Hydroxyl, OH, is a poor leaving group

The hydroxyl group, OH is a poor leaving group. This makes alcohols poor electrophiles. Nucleophiles will not attack the R group of an alcohol because there is not a good leaving group.

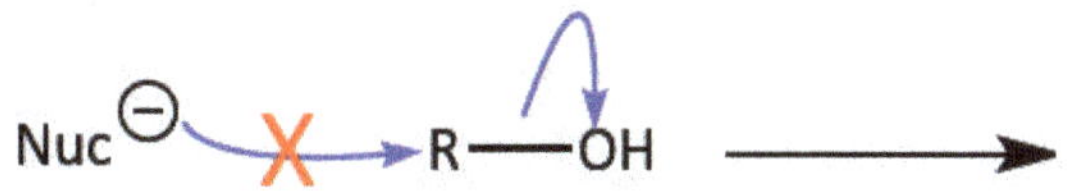

But, the good news is there are several ways to turn the poor hydroxyl leaving group into a good leaving group.

14.3.1 Turn OH into H_2O

In the presence of a strong acid, the hydroxyl group can be protonated. Once protonated, it turns into a water leaving group. Water is a wonderful leaving group because it is a stable molecule. In the example below, an alcohol reacts with HBr, a good acid. The very negative lone pair of electrons of the hydroxyl group attack the proton of HBr. This makes the positively charged water leaving group. This leaving group can leave, making a carbocation. Br- attacks the carbocation giving an alkyl bromide.

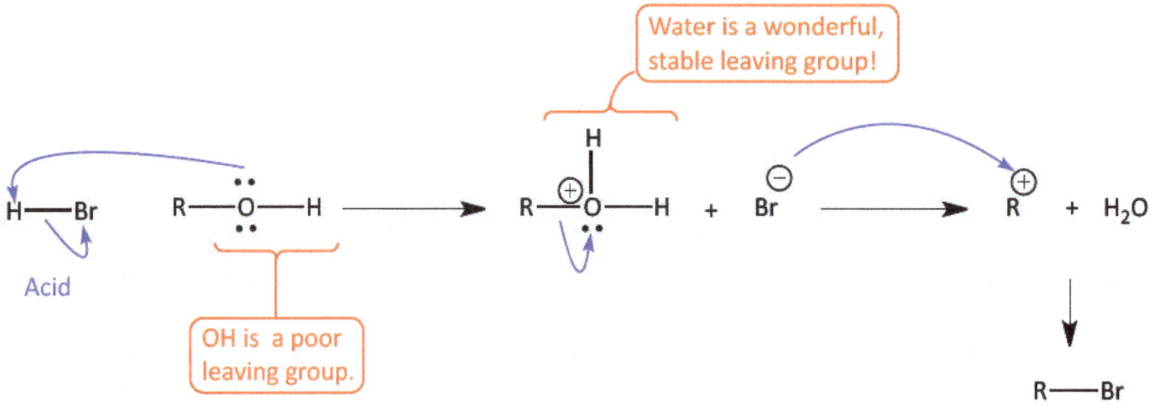

> 14.2 Which of the following alcohols would react the fastest with HBr to make an alkyl bromide? Hint: remember what type of intermediate is made in the reaction.
>
> 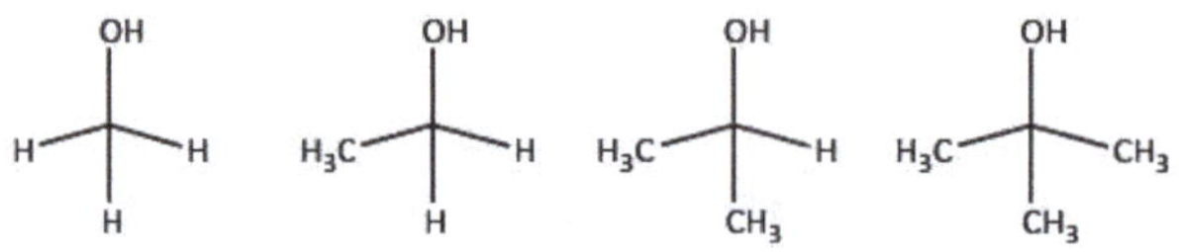
>
> A B C D

14.3.2 Dehydration

To form an ether

The hydroxyl group of alcohols is a poor leaving group. But, in the presence of a strong, dehydrating, acid like sulfuric acid, H_2SO_4, or phosphoric acid (H_3PO_4) it can be turned into a good leaving group (H_2O). The bimolecular (two-molecule) dehydration (loss of H_2O) that occurs between two alcohols produces an ether.

$$R-O-H + H-O-R \xrightarrow[H_2SO_4]{-H_2O} R-O-R$$

This is only an effective synthesis for symmetrical ethers; two identical alcohols should be used. If two different alcohols are used, it leads to a mixture of products. Because of this difficulty, this is a good way to make ethers industrially, but this method is not usually used in the lab.

$$R-OH + HO-R' \xrightarrow{-H_2O} R-O-R + R-O-R' + R'-O-R'$$

The mechanism for this ether-forming dehydration involved protonating a hydroxyl group of an alcohol to make the great, stable, H_2O leaving group. The other alcohol attacks via an S_N2, backside-attack mechanism. After cleaning it up, an ether is formed. Below, see the mechanism for the dehydration of two molecules of ethanol to form diethyl ether. This dehydration to form an ether tends to happen between 130-140°C. Above this temperature, a different, competing reaction tends to happen. This sensitivity to temperature is one of the reasons this reaction is not usually performed in the laboratory. There are better ways to make ethers.

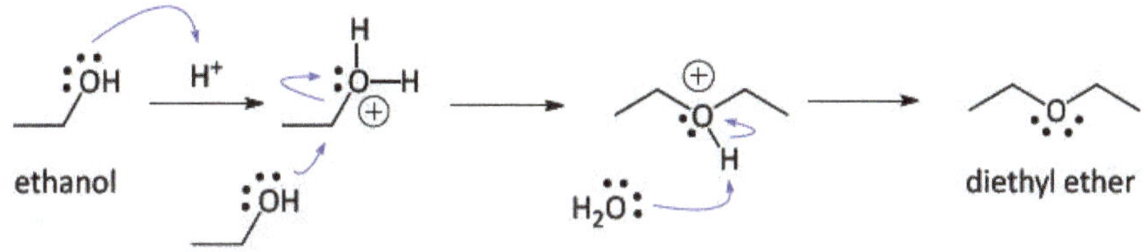

Figure 14.1 Bimolecular dehydration of ethanol to form diethyl ether

To form an alkene

The competing dehydration of alcohols is the unimolecular dehydration to form an alkene. This dehydration tends to occur at temperatures over 140°C. It occurs via an E1 reaction mechanism. The acid used is usually a good dehydrating acid like sulfuric acid, H_2SO_4, phosphoric acid, H_3PO_4, or *p*-toluenesulfonic acid (tosic acid, *p*-TsOH).

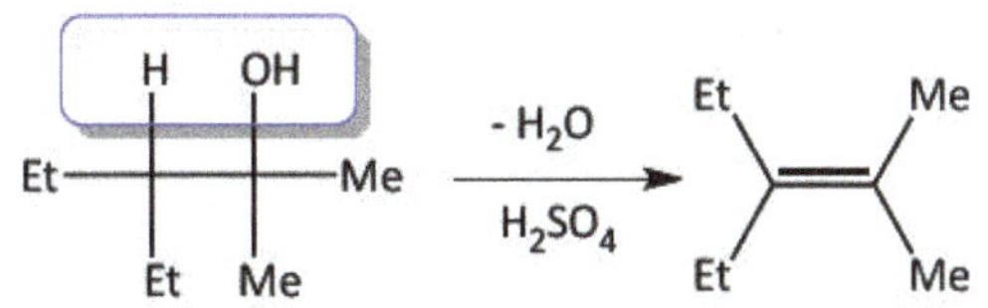

Because a carbocation is formed in this E1 reaction, rearrangements are common. Also, because a carbocation is formed in this reaction, tertiary alcohols react faster than secondary alcohols because more stable carbocations are formed. Primary and methyl alcohols react slowly because an undesirable primary or methyl carbocation would be formed.

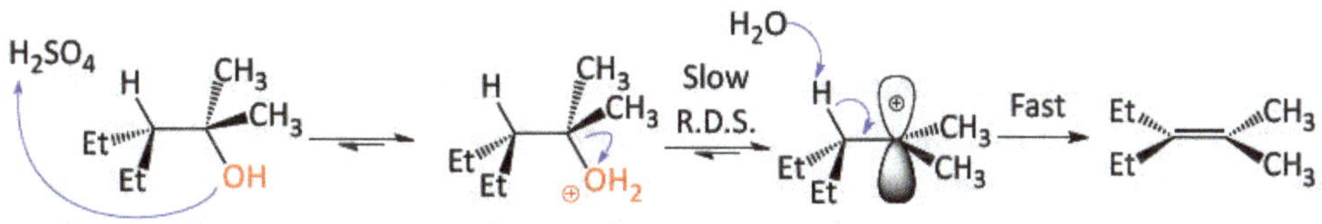

When an alkene is formed from dehydration, the Saytzeff product (more substituted double bond) is the one that is usually formed.

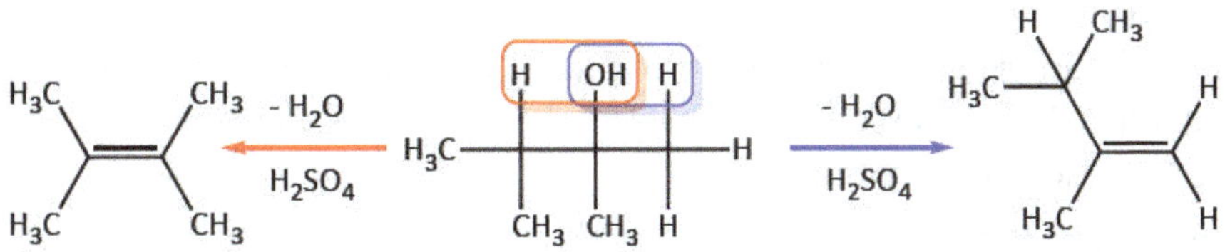

14.3 Draw the major alkene or ether products of the following dehydration reactions.

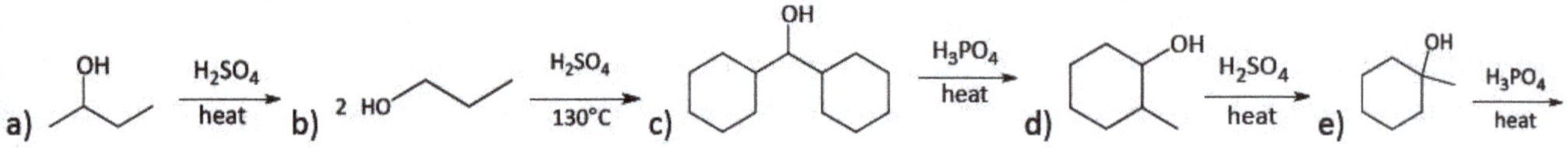

14.3.3 Turning OH into a halide

We have seen previously how an OH group can be turned into a halide. Halides, of course, are good leaving groups.

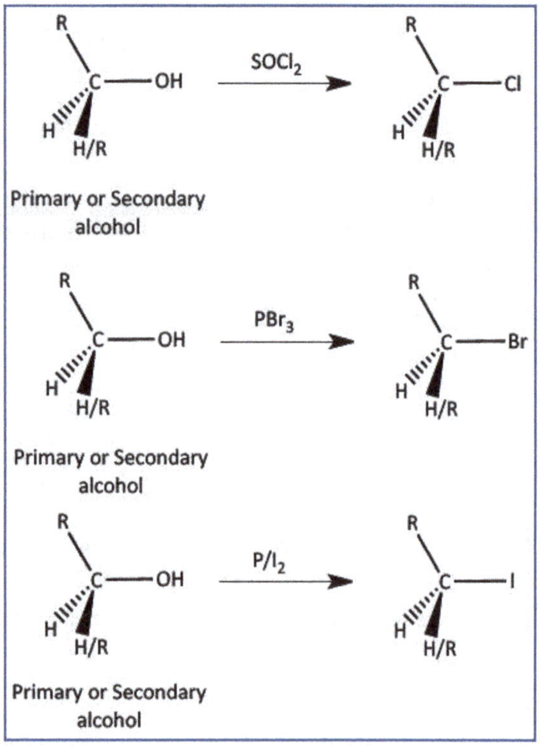

Tertiary alcohols react with HCl, HBr, or HI to make the appropriate alkyl halide. Secondary alcohols work slower this way. Primary and methyl alcohols do not work well with this method.

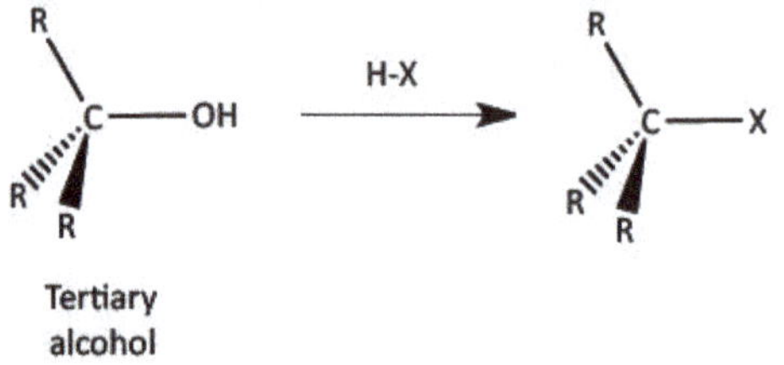

An easy way to generate HBr is to use NaBr and H_2SO_4.

Sometimes zinc chloride, $ZnCl_2$, is used with HCl to turn a secondary or tertiary alcohol into a chloride. This combination of $ZnCl_2$ and HCl is called the **Lucas reagent**.

In fact, the Lucas reagent can be a chemical test to help determine if an alcohol is a primary, secondary, or tertiary alcohol. This is called a **Lucas test**. In this test, an alcohol reacts with the Lucas reagent. If the reaction proceeds rapidly, less than 1 minute, it indicates a tertiary alcohol was used. If the reaction takes 1-5 minutes, it indicates a secondary alcohol was used. If the reaction is slower than that, it indicates an unreactive primary alcohol was used.

One drawback to using a strong acid like HCl or HBr to make an alkyl halide is a carbocation intermediate is formed in the process of the reaction. This can sometimes lead to undesired rearrangement products.

If we try to turn a primary or secondary alcohol into a halide, we need to use different reagents. We use thionyl chloride ($SOCl_2$) to make chlorides, phosphorus tribromide (PBr_3) to make bromides, and a combination of phosphorus and iodine (P/I_2) to make iodides. When $SOCl_2$ or PBr_3 are used, rearrangements are uncommon because a carbocation intermediate is not formed.

Alcohol	Bromide	Chloride	Iodide
Tertiary	HBr	HCl or $HCl/ZnCl_2$ (Lucas Reagent)	HI
Secondary	PBr_3	$SOCl_2$	P/I_2
Primary	PBr_3	$SOCl_2$	P/I_2

Table 2 Best Reagents to Transform Alcohols to Halides

Of course, there are many other ways to turn hydroxyl groups into halides. The above are a first approximation on what to try, but if they do not work for you in a real-life lab situation, do not fret. There are plenty more ways to do this.

14.4 Draw the product of the following reactions.

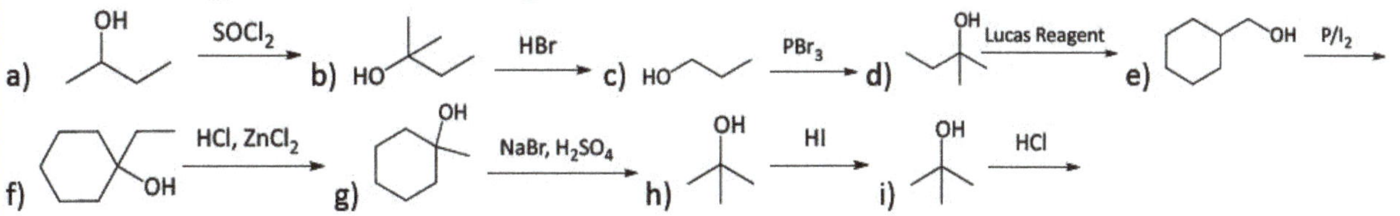

14.3.4 Turning OH into a sulfonate

Sulfonates are also very good leaving groups. Two common sulfonate leaving groups are toluenesulfonate and methanesulfonate. First, we'll take a closer look at what these sulfonates are and why they are good leaving groups, then we will look at how they are made from alcohols.

Sulfonate names

Methyl benzene is called toluene.

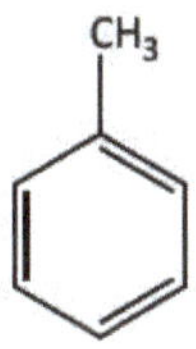

Toluene

A toluene sulfonate often called a tosylate is a toluene and a sulfonate together. A toluene sulfonate group is often called a tosyl group. When the tosylate is on an alkyl group, it is calls a tosylate ester (ROTs).

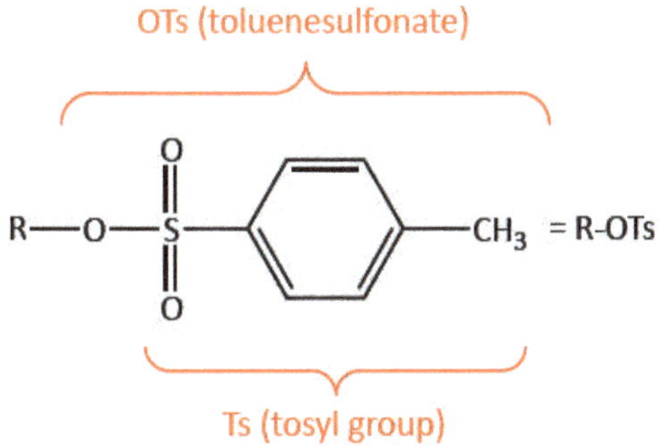

Figure 14.2 Tosylate ester

A methanesulfonate is often called a mesylate.

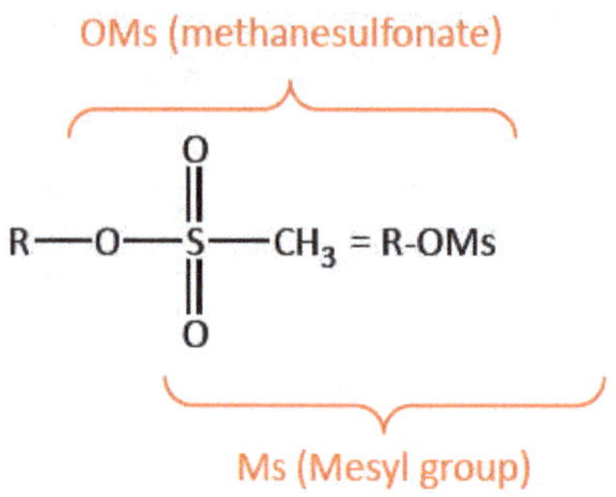

Figure 14.3 Mesylate ester

Tosylates and mesylates are great leaving groups. When they leave the alkyl group, they have a negative charge on the oxygen atom. This negative charge can be spread out over several atoms through resonance. In fact, these resonance forms make tosylates and mesylates even better leaving groups than halides like Cl^- or Br^-.

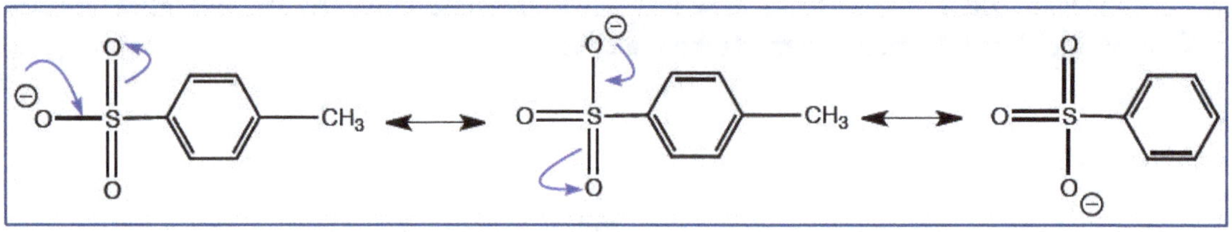

Figure 14.4 Tosylate resonance forms

14.3.5 Synthesis of tosylate esters

The most efficient way to synthesize a tosylate ester is to react an alcohol with toluenesulfonyl chloride, tosyl chloride (TsCl) and a base like pyridine.

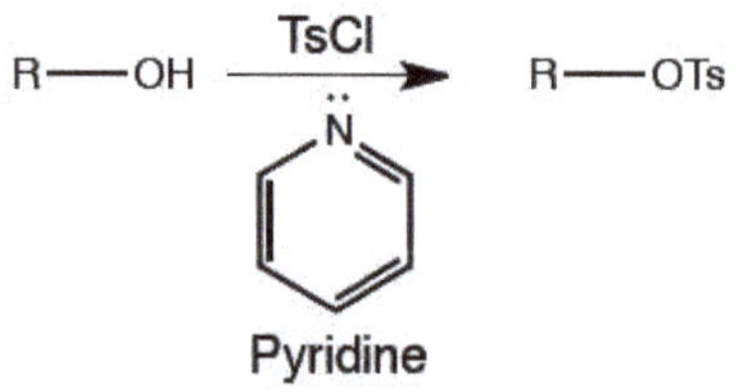

The mechanism for this reaction is as follows:

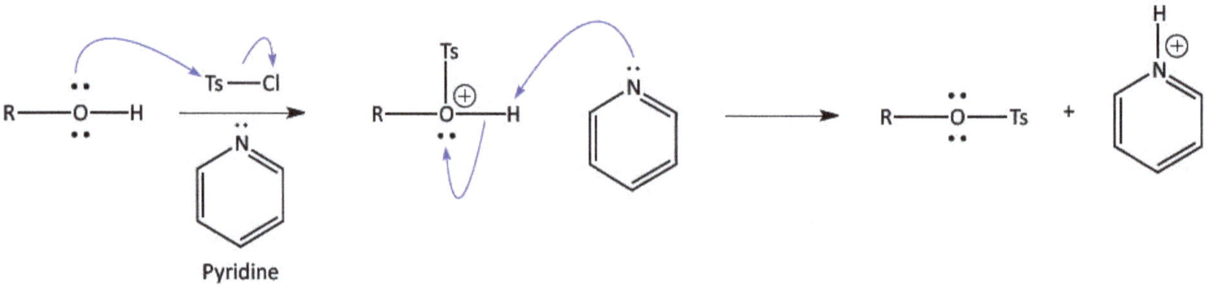

Figure 14.5 Mechanism for the tosylation of an alcohol

14.5 What reagent should be used to perform the following reactions?

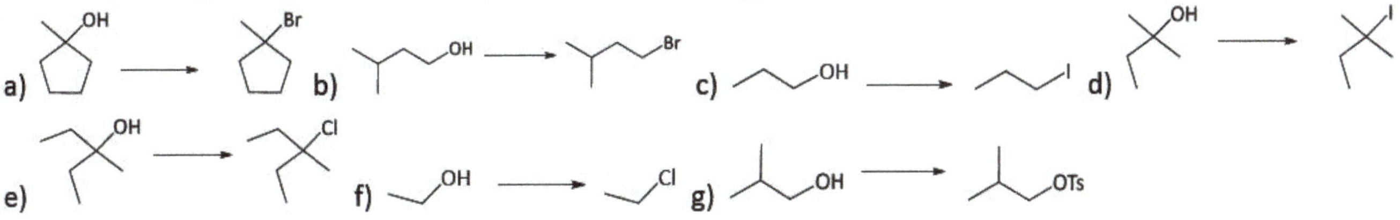

14.3.6 Why turn an OH into a good leaving group?

If a hydroxyl group is turned into a good leaving group, our new alkyl halide or tosylate ester can be a good electrophile. A nucleophile can attack it with the new leaving group leaving. This opened up the possibility for several types of reaction from S_N2 to E2 type reactions.

S_N2

Ethers are often synthesized from alcohols. One alcohol is turned into an alkyl halide or a tosylate ester. With this good leaving group, it can be attacked by an alkoxide made from an alcohol and sodium. This S_N2 reaction produces an ether. This is called the Williamson Ether Synthesis.

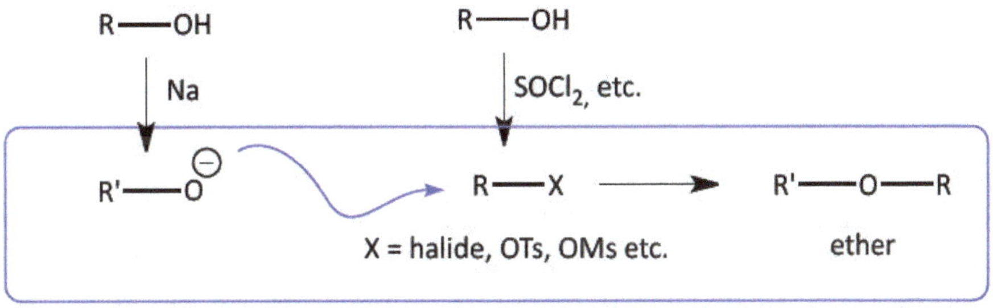

Figure 14.6 Williamson Ether Synthesis from alcohols

14.6 Synthesize the following compound, isopropyl methyl ether, using a Williamson Ether Synthesis starting from methanol and isopropanol. Hint: transform one alcohol into a halide and the other into an alkoxide.

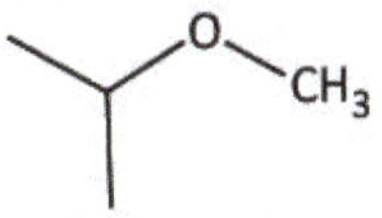

E2

Also, alkyl halides and tosylate esters with their good leaving groups can do an elimination. For example, dehydrohalogenation with a good base makes an alkene via the E2 mechanism.

Alcohol Reactions

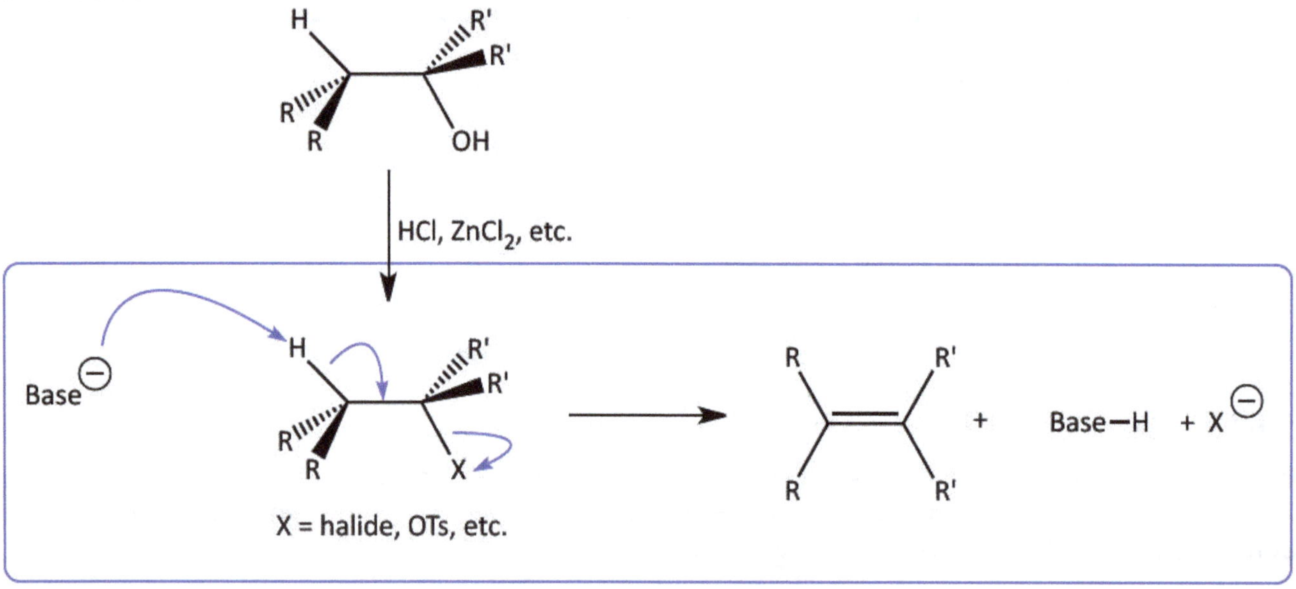

Figure 14.7 Dehydrohalogenation E2 reaction

14.7 Show the reaction of propanol with an appropriate reagent to turn it into a halide. Then add an appropriate base to make propene.

14.4 Oxidations

Previously, we saw how carbonyl compounds, with two carbon-oxygen bonds, can be reduced to make alcohols, with one carbon-oxygen bond. The opposite reaction of a reduction is an oxidation. In organic chemistry, an oxidation is essentially making a compound with more carbon-oxygen bonds. The oxidation from an alcohol to a carbonyl is accompanied by the elimination of two hydrogen atoms. An alcohol, with one carbon oxygen bond, can be oxidized to a carbonyl with two carbon-oxygen bonds. If the carbonyl compound is an aldehyde, with two carbon-oxygen bonds, it can be further oxidized to a carboxylic acid, with three carbon-oxygen bonds if a strong oxidizing agent is used. If a weak oxidizing agent is used, it can be oxidized to the aldehyde and stop there.

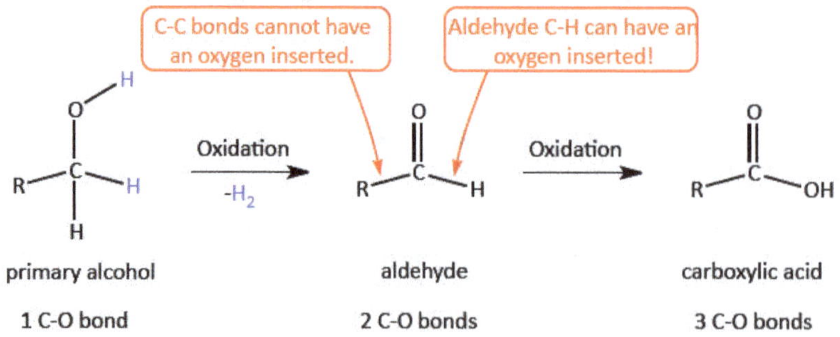

Figure 14.8 Oxidation pathway of a primary alcohol

A secondary alcohol can be oxidized to a ketone. Notice the singe carbon-oxygen bond of the alcohol loses two hydrogen atoms and is oxidized to the ketone carbonyl with two carbon-oxygen bonds. There is no need to memorize that secondary alcohols give ketones and primary alcohols give aldehydes, simply remove the two hydrogen atoms from the alcohol and see what you get. Since ketones only have carbon atoms attached to the carbonyl carbon, no further oxidation can occur. Only aldehydes with a hydrogen atom attached to the carbonyl can be further oxidized.

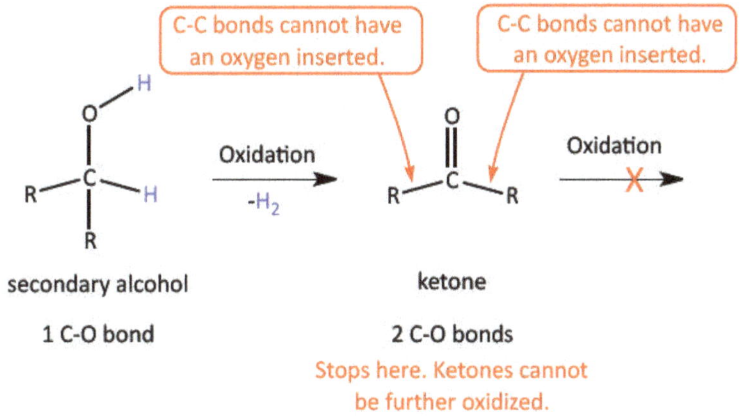

Figure 14.9 Oxidation pathway of a secondary alcohol

A tertiary alcohol cannot be oxidized. Notice that there are no hydrogen atoms on the carbinol carbon atom. Therefore, two hydrogen atoms cannot be removed from the alcohol, so it cannot be oxidized.

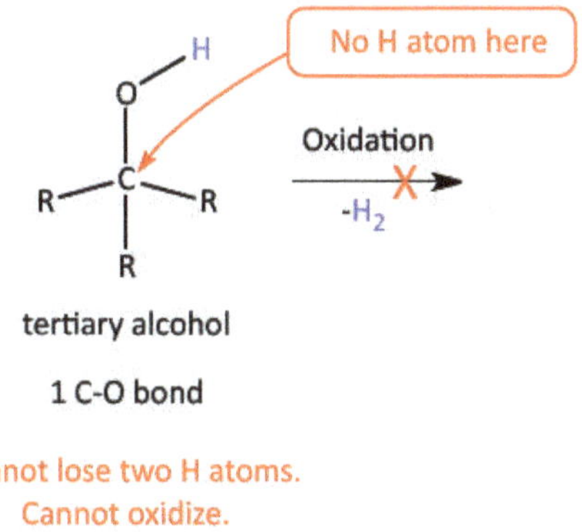

Figure 14.10 Tertiary alcohols do not oxidize

14.4.1 Oxidizing agents

There are many different oxidizing agents. One strong oxidizing agent is chromic acid. Chromic acid is made from sodium dichromate ($Na_2Cr_2O_7$) and sulfuric acid (H_2SO_4).

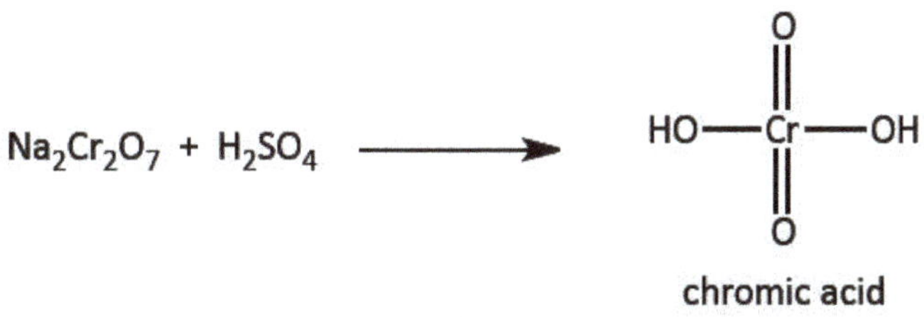

Chromic acid is such a strong oxidizing agent, that it fully oxidizes primary alcohols all the way to carboxylic acids. It proceeds from primary alcohol to aldehyde to carboxylic acid. Chromic acid oxidizes secondary alcohols to ketones.

Alcohol Reactions

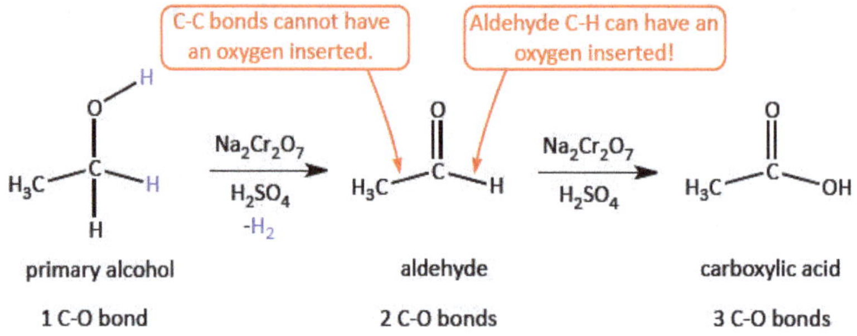

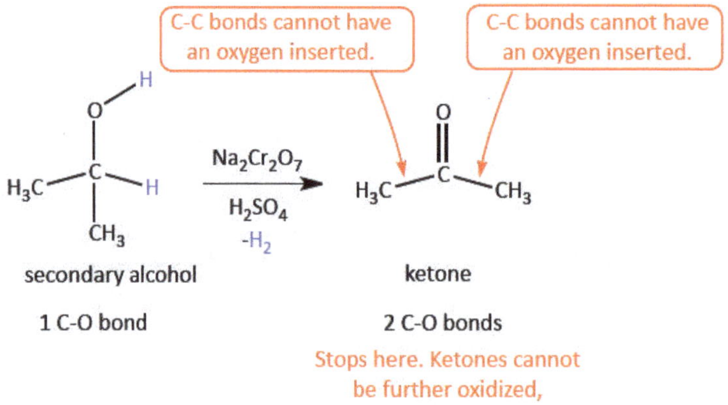

But, what if we want to oxidize a primary alcohol to an aldehyde and stop there without going further oxidizing to a carboxylic acid? We need a weaker oxidizing agent. There are several ways to do this, but one is to use a weaker oxidizing agent like PCC, pyridinium chlorochromate. PCC is a complex made from chromium trioxide, pyridine, and hydrochloric acid.

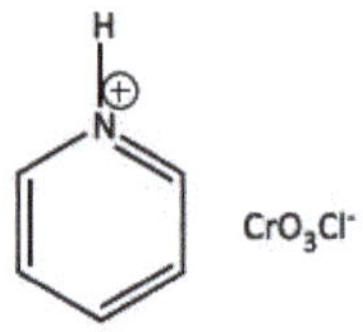

Figure 14.11 PCC

The weaker oxidizer PCC is strong enough to do the first oxidation of a primary alcohol to aldehyde, but it is too weak to completely oxidize to a carboxylic acid. The oxidation stops at the aldehyde. This is a great way to make aldehydes.

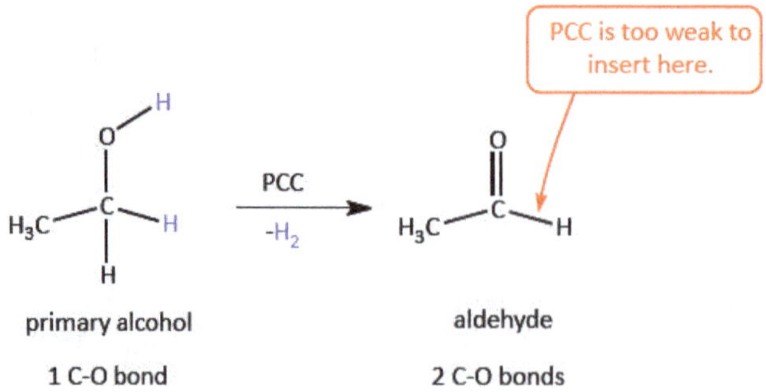

Since PCC is strong enough to do the first oxidation, it is strong enough to oxidize secondary alcohols to ketones as well.

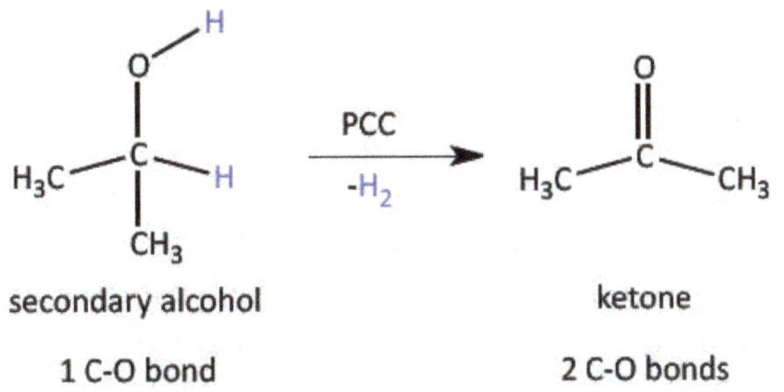

Chromium oxidizers should be handled with care. These chromium salts are hazardous to your health. They should never go down the sink drain in the lab and should be collected as hazardous waste. There are greener oxidizers.

The Swern oxidation or DMP react like PCC does. Bleach is a strong oxidizer much like chromic acid.

14.8 Show the products of the following reactions.

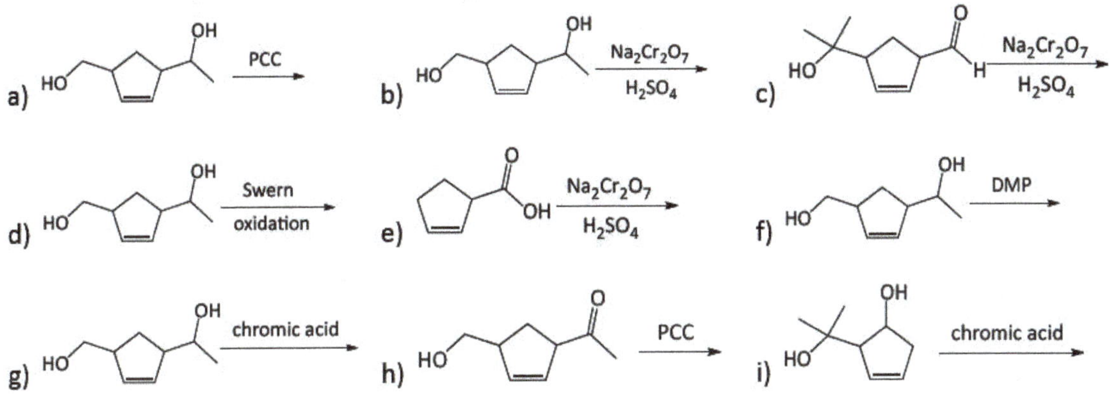

14.5 Reduction of alcohols

We have seen how alcohols can be synthesized from the reductions of carbonyl compounds. But, alcohols can also be reduced. Reductions are the opposite of oxidations we have been studying in the previous sections. So, if an alcohol, with one C-O bond, is reduced, it makes a compound with zero C-O bonds, an alkane. One simple way to reduce an alcohol is to turn the hydroxyl group into a good leaving group like a tosylate. Reacting the alcohol with tosyl chloride, TsCl, and pyridine can do this. Once the tosylate is formed, it can be reduced by the reaction with lithium aluminum hydride, $LiAlH_4$.

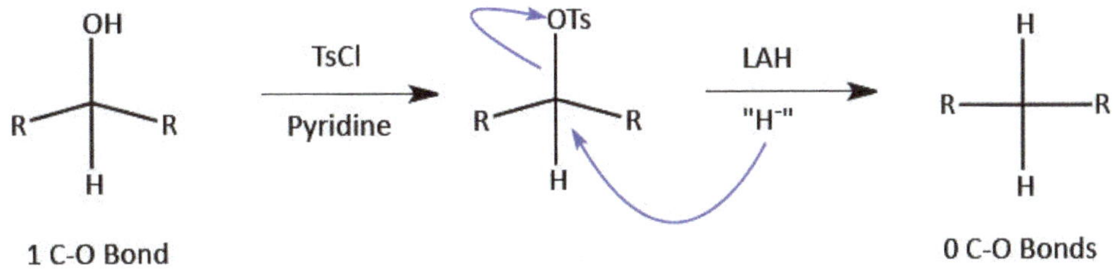

14.6 Diol reactions

14.6.1 Pinacol rearrangement

The pinacol rearrangement is a way to turn a diol, a glycol, into a carbonyl compound. Pinacol is a glycol. Under acidic conditions, one of the hydroxyl groups of the glycol is protonated forming a good leaving group, water. Water leaves forming a carbocation. A 1,2-alkyl shift occurs which is the rearrangement. Resonance forms are drawn moving the lone pair of electrons on an oxygen atom to form a carbonyl. A proton is removed to make the carbonyl of the pinacolone. The pinacol rearrangement gets its name from the reaction turning pinacol into pinacolone, but the pinacol rearrangement can happen with many glycols under acidic conditions. The rearrangement happens whenever the 1,2-alkyl shift can make a more stable carbocation. In the reaction of pinacol, the new carbocation formed after resonance forms stabilize the 1,2-alkyl shift with the lone pair of electrons on the oxygen atom.

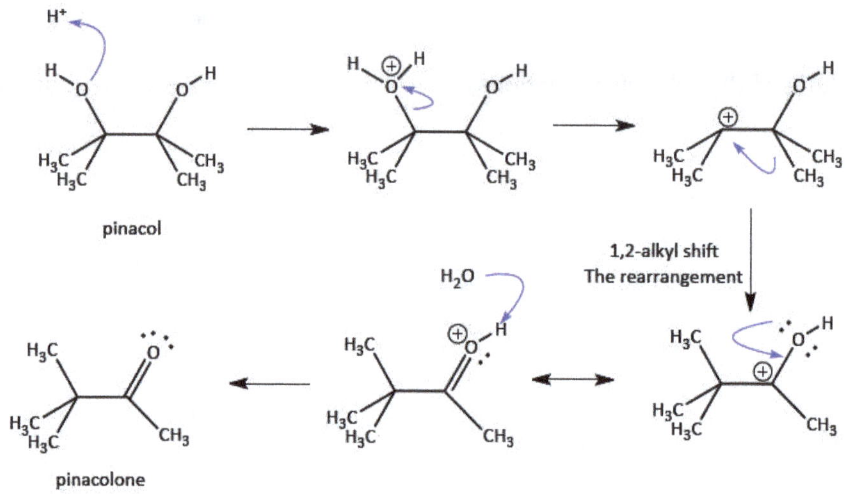

Figure 14.12 The Pinacol Rearrangement

14.6.2 Periodic Acid

The C-C bond of a vicinal diol, a glycol, can be oxidatively cleaved with periodic acid (pronounced purr-eye-oh-dik-acid), HIO_4 to make two carbonyl compounds. This is very similar to how alkenes are oxidatively cleaved in an ozonolysis reaction.

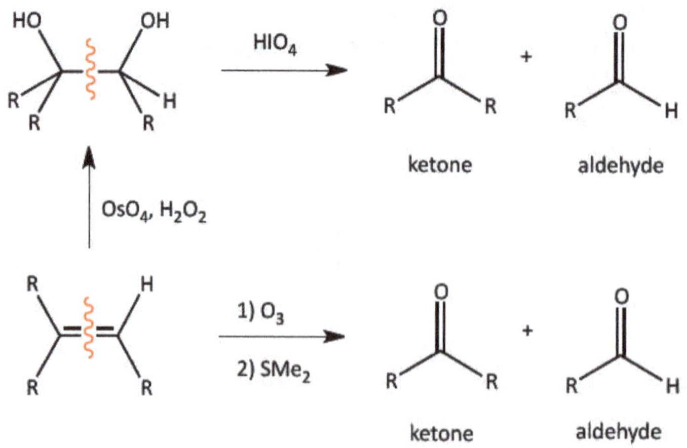

14.7 Esterifications

14.7.1 Fischer Esterification

The Fischer Esterification is the acid catalyzed reaction of an alcohol and a carboxylic acid to form an ester.

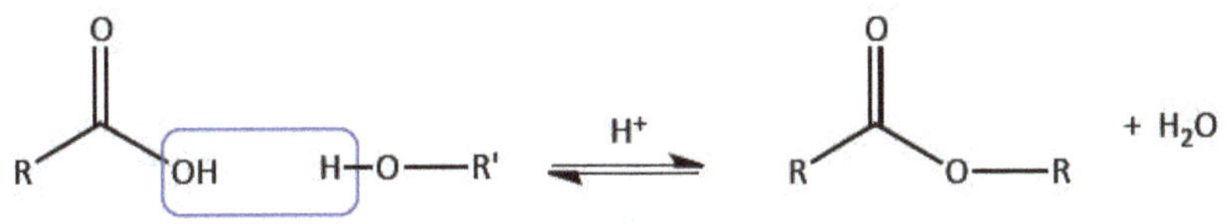

In order to drive the equilibrium of the Fischer Esterification to the ester product side, two things are commonly done. One trick is to run the reaction with an excess of starting materials. A large excess of either the alcohol or the carboxylic acid will work. The other trick is to remove the water product from the reaction using a dehydrating agent in the reaction flask that can absorb water, like molecular sieves. Also, water can be removed from a reaction using a Dean-Stark trap. This is a special attachment to a reflux apparatus that helps remove water from a reaction.

14.7.2 From an acid chloride

Another effective way to make an ester is to react an alcohol with an acid chloride. This is an exothermic reaction that is not an equilibrium reaction. The reaction usually proceeds well. Sometimes a base is put into the reaction flask to soak up the HCl that is produced. Acid chlorides can be made from reacting carboxylic acids with $SOCl_2$.

One drawback for this reaction is that it is more water sensitive than a Fischer Esterification. Water will quench the acid chloride and turn it into a carboxylic acid.

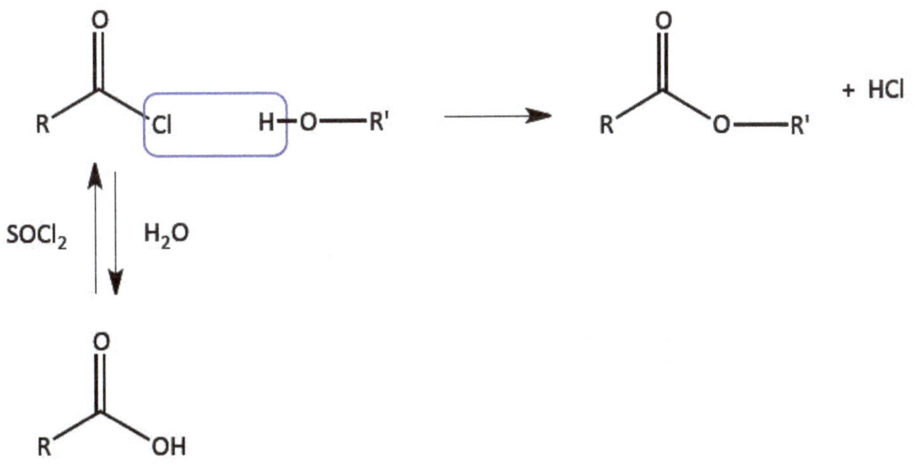

14.9 Draw the reactant or products in the following reactions, as appropriate.

a) 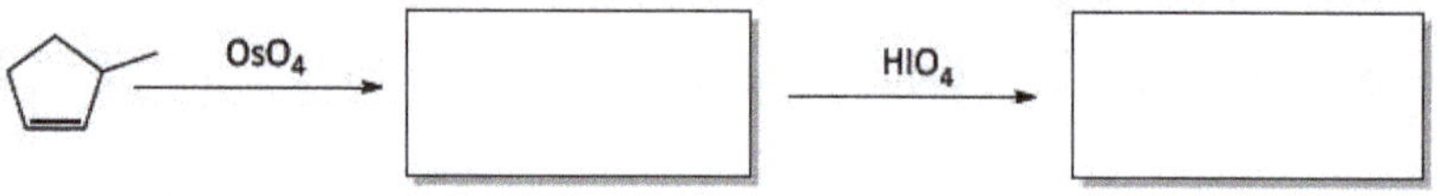

b) 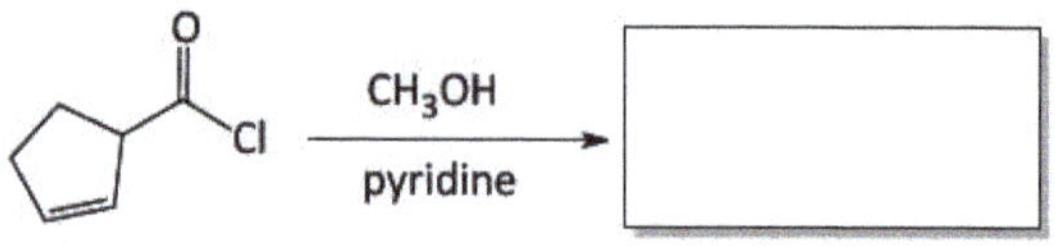

c) 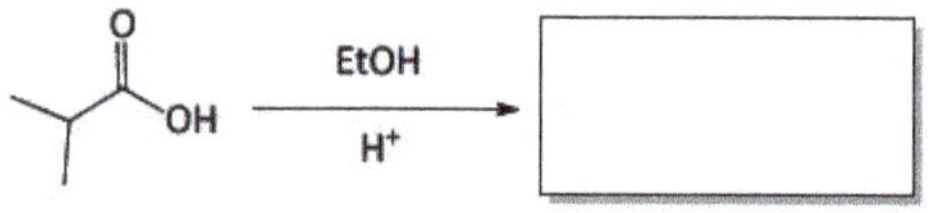

d) 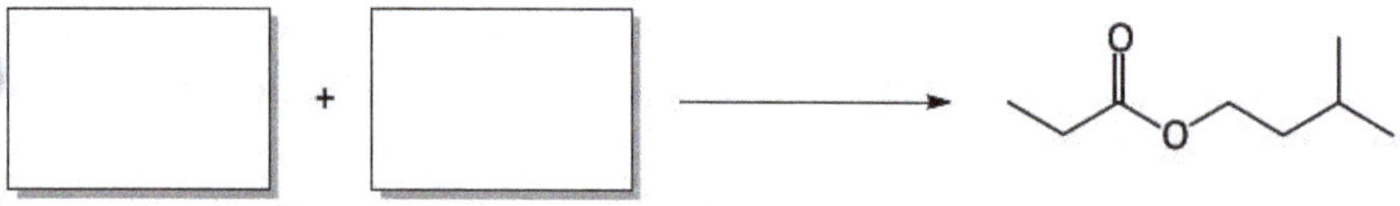

e) 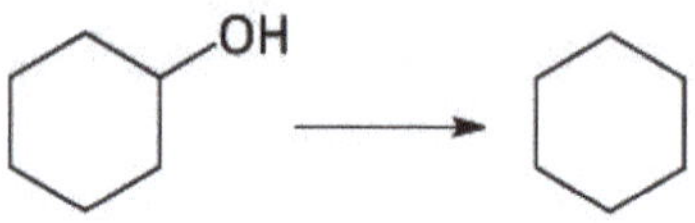

f) 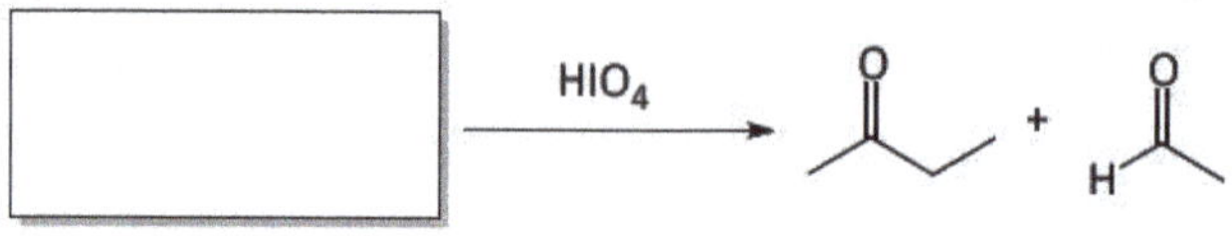

g) 

14.8 Inorganic esters

14.8.1 Sulfates

We have seen previously that sulfate esters, like tosylate or mesylate, are very good leaving groups. These sulfate esters can be synthesized from alcohols.

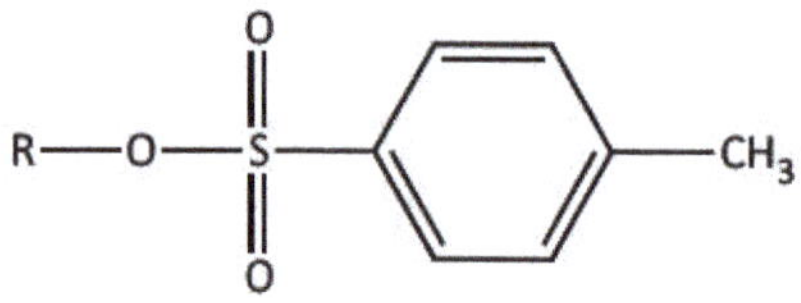

Tosylate ester

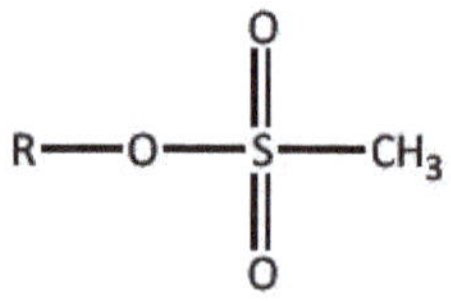

Mesylate ester

14.8.2 Nitrates

Nitrates can also be synthesized from alcohols using nitric acid.

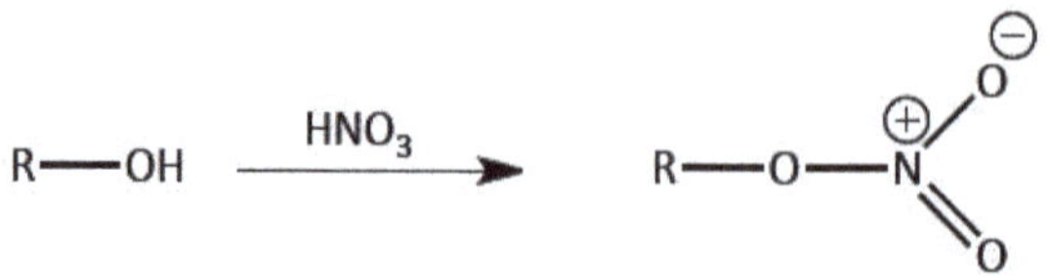

One famous example of this is the nitration of glycerine to make nitroglycerine. In 1866, Alfred Nobel discovered that nitroglycerine can be mixed into diatomaceous earth to make dynamite.

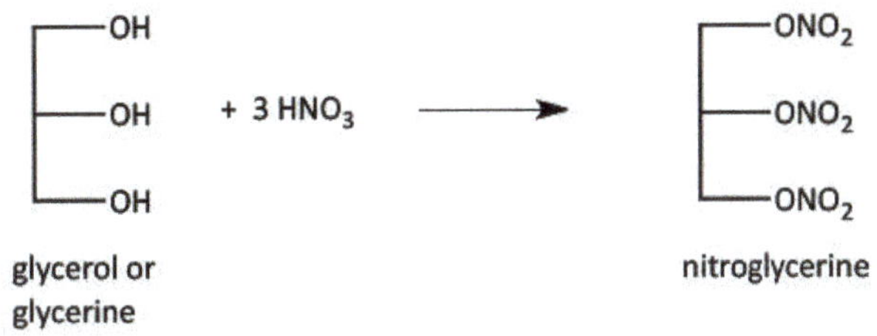

These nitrated compounds have quite a bit of energy in them. Cellulose can also be nitrated. Cellulose is a polymer of glucose, and glucose is full of hydroxyl groups on it. This is how a magician's flash paper is made. TNT, trinitrotoluene, is also a highly nitrated compound. Picric acid is also very explosive. Chemists should know about picric acid as it used to be commonly used to make picrate salts in the chemistry lab. If you ever come across a bottle of picric acid in a lab, do not touch it and call your local bomb squad. It is extremely dangerous.

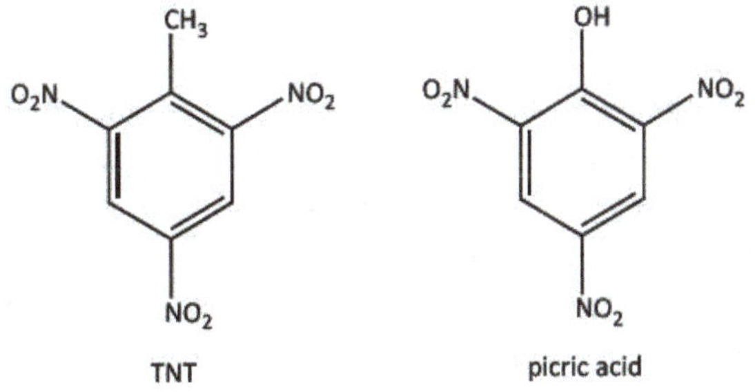

14.8.3 Phosphates

Phosphates are not as common in organic chemistry, but they play a big role in biochemistry. Phosphates are a part of the nucleic acids found in RNA and DNA. They also play a major role in ATP and ADP.

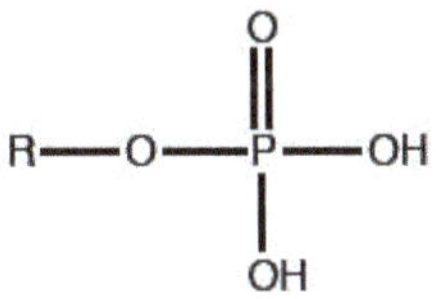

14.9 Thiols

14.9.1 Thiol/Disulfide conversion

Two thiols can react together under a mild oxidation, lose two hydrogen atoms, and form a disulfide linkage. If a reaction occurs and two hydrogen atoms are added to the disulfide in a reduction, the two thiols can once again be made.

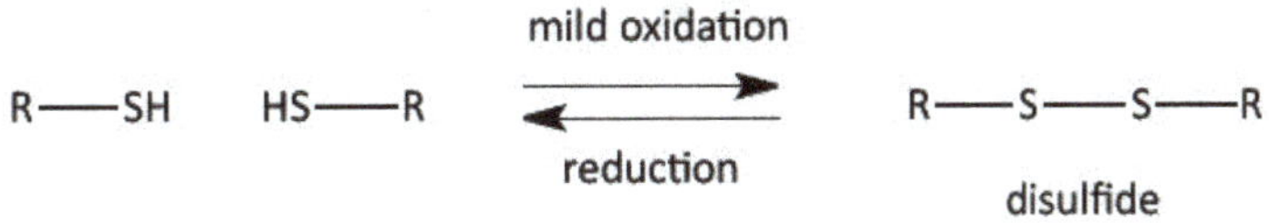

This is an important biochemical reaction. Proteins are made up of many amino acids. Cysteine is one of these amino acids. It contains a thiol group.

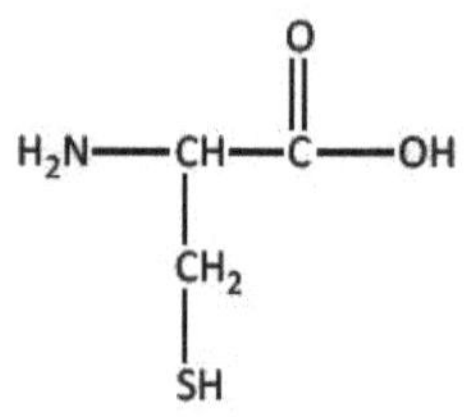

cysteine

To hold proteins in their folded shapes, the cysteine thiols often link up to form disulfide bridges. If a chemist wants to reduce the disulfides of a protein, to have it relax and lose its folded shape, a reducing agent must be used. The reducing agents we have used in the past ($NaBH_4$ and LAH) are usually *not* very effective at performing the reduction of disulfides. Instead, biochemists usually reduce disulfides in proteins by adding DTT, dithiothreitol, to their buffer solutions. A biochemist does this to keep proteins unfolded – perhaps to measure their size in gel electrophoresis.

Human hair is a protein. The cysteines make disulfide linkages in the hair protein, which helps give your hair its structure- whether it is curly or straight. A process to make hair curly is called a hair permanent, or perm. In a perm, a mild reducing agent, usually sodium thioglycolate, is places on the hair, causing the disulfide linkages to break and become thiols. The hair protein is then wrapped around curlers followed by the application of a mild oxidizer, usually hydrogen peroxide, to reform the disulfide linkages. The disulfide linkages help hold the hair in its new configuration, making it curlier. The same process can be used to straighten hair.

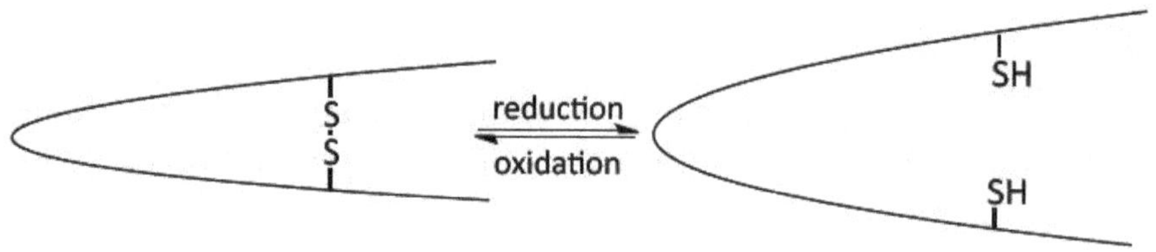

A mild oxidizer forms disulfide linkages from thiols, but a strong oxidizer like nitric acid or $KMnO_4$ forms sulfonic acids.

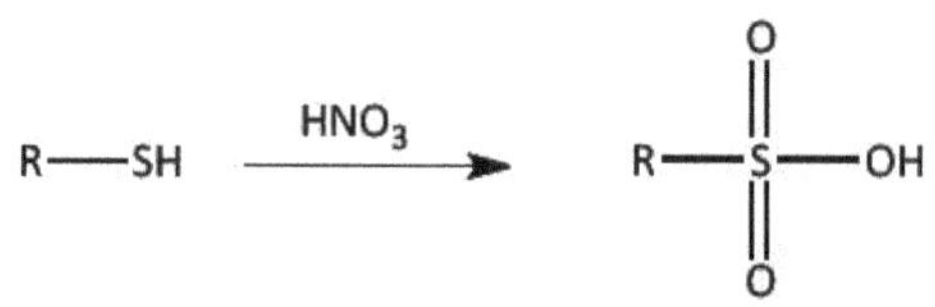

sulfonic acid

Sulfonic acids do not stink like thiols. One home remedy for being sprayed by a skunk is to use hydrogen peroxide in a mixture of soap and baking soda to try to oxidize the thiols to sulfonic acids so there is less of an odor. Let's be clear, it would *not* be a good idea to spray nitric acid on a skunk sprayed victim. It might reduce the smell but would do terrible damage to their skin.

14.10 Reaction Summary

Alkoxide formation

a) 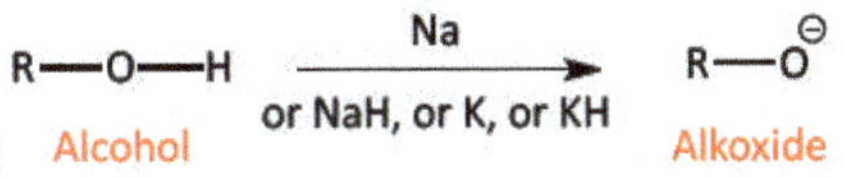

Dehydration to form ether

b) 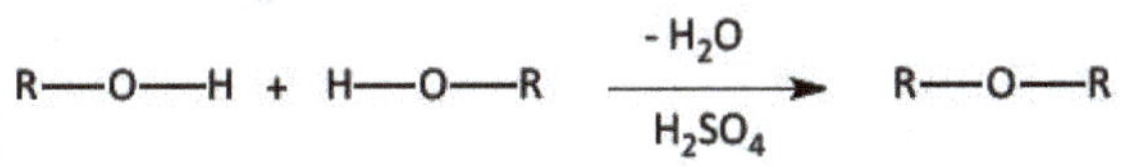

Dehydration to form an alkene

c) 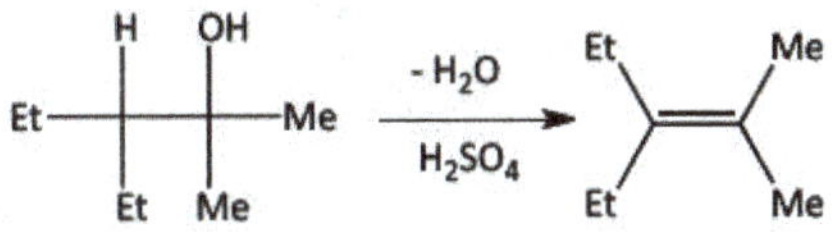

Alcohols to good leaving groups

d) 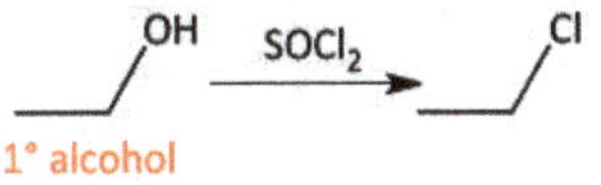

e) 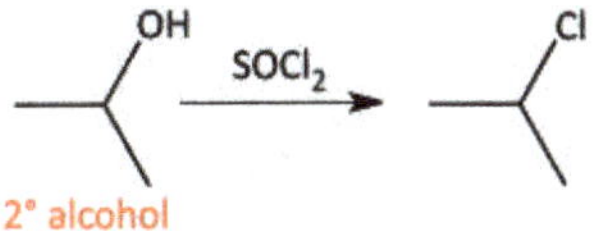

f) 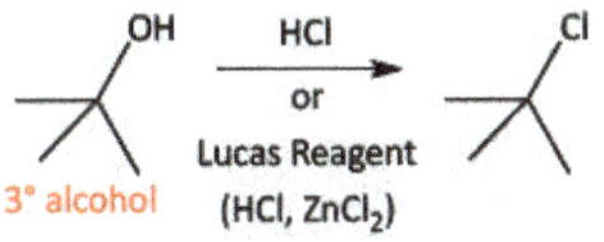

g) 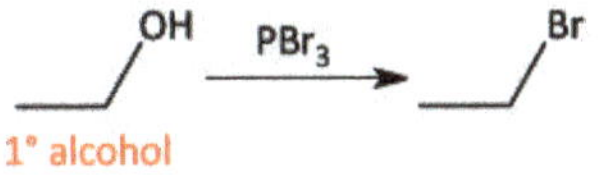

h) 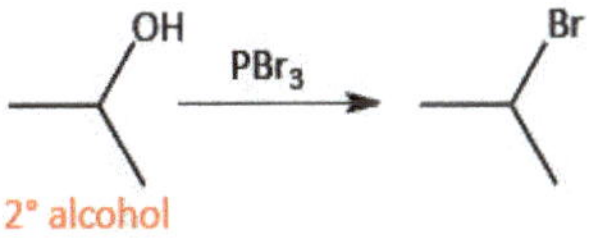

i) 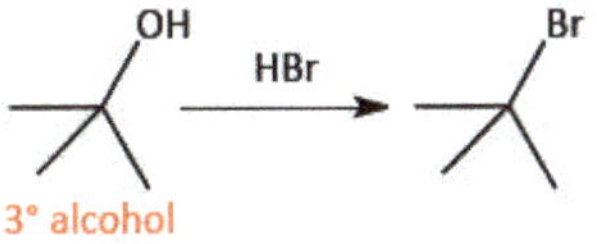

j) 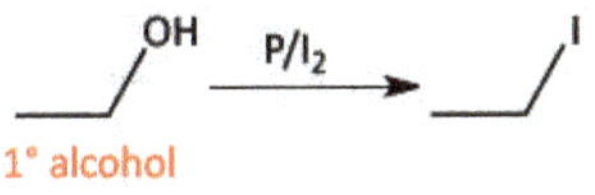

k) 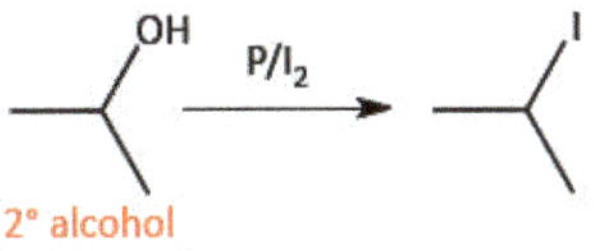

l) 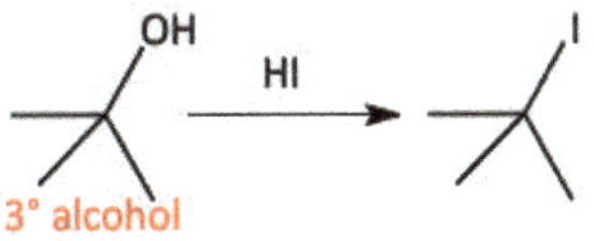

m) 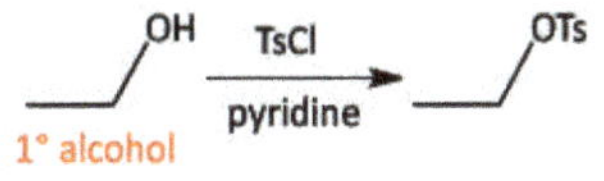

n) 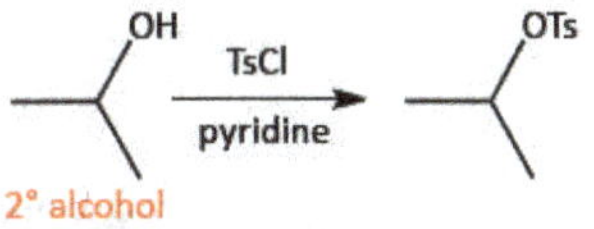

o) 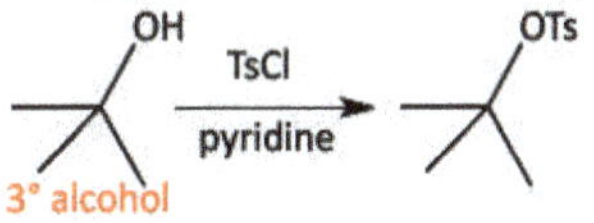

p) 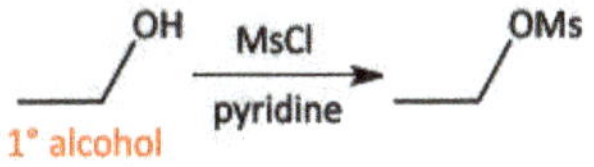

q) 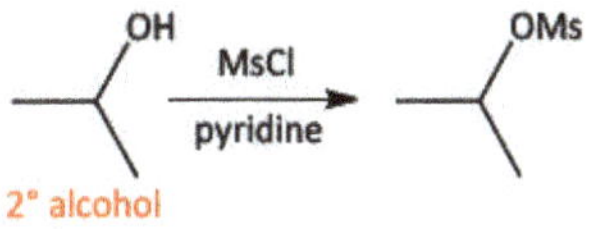

r) 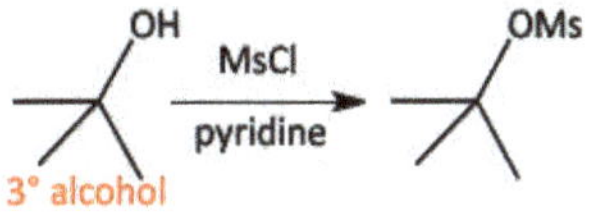

Williamson Ether Synthesis

s) 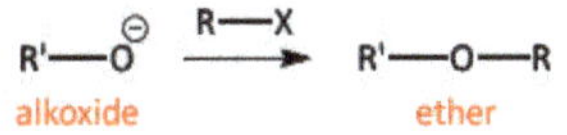

E2 Elimination

t) 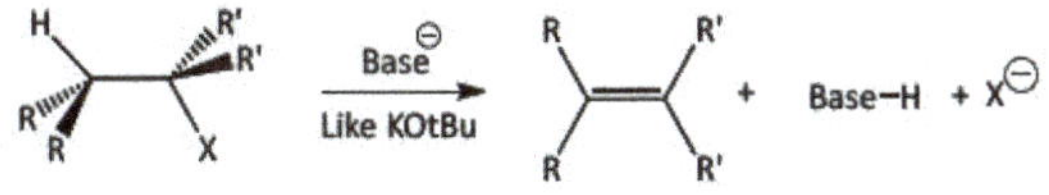

Oxidation of alcohols with weak oxidizer

u) 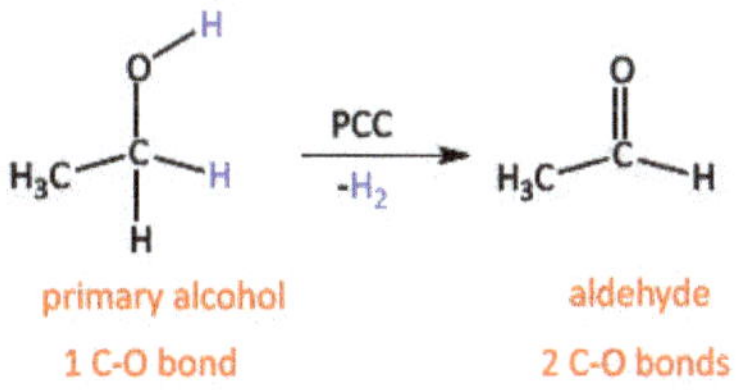

v) 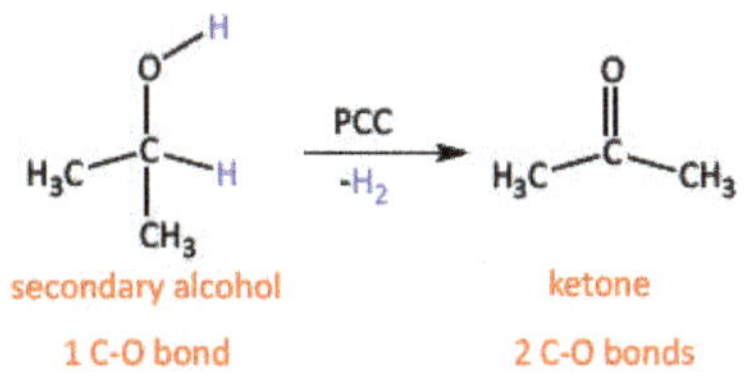

Alcohol Reactions

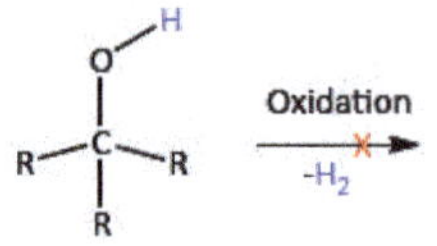

tertiary alcohol

1 C-O bond

w) Cannot lose two H atoms. Cannot oxidize.

Swern oxidation and *DMP* are weak oxidizers like PCC.

Oxidation of alcohols with strong oxidizer

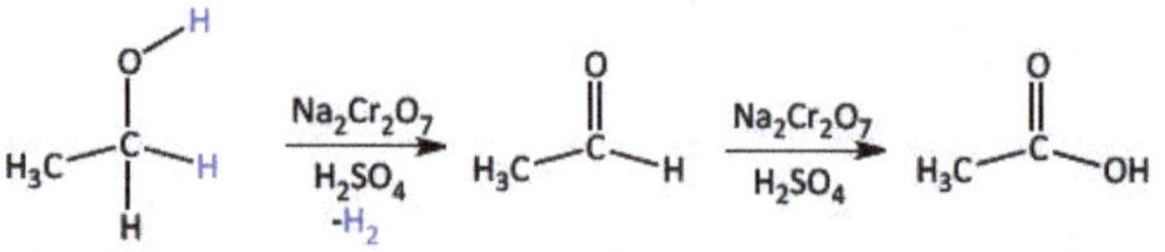

x) primary alcohol, 1 C-O bond → aldehyde, 2 C-O bonds → carboxylic acid, 3 C-O bonds

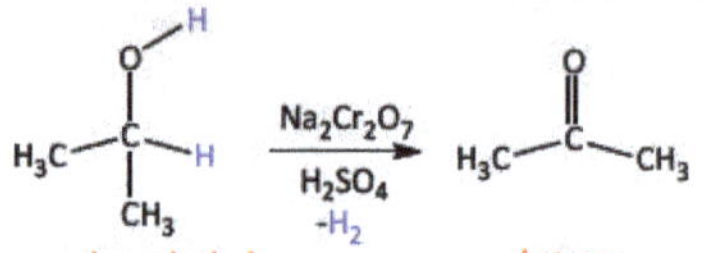

y) secondary alcohol, 1 C-O bond → ketone, 2 C-O bonds

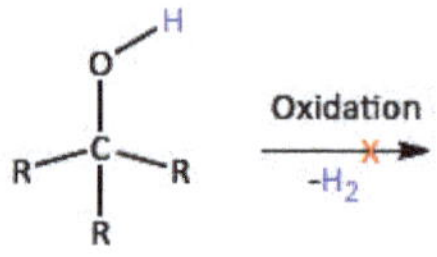

tertiary alcohol

1 C-O bond

z) Cannot lose two H atoms. Cannot oxidize.

Reduction of Alcohols

aa) 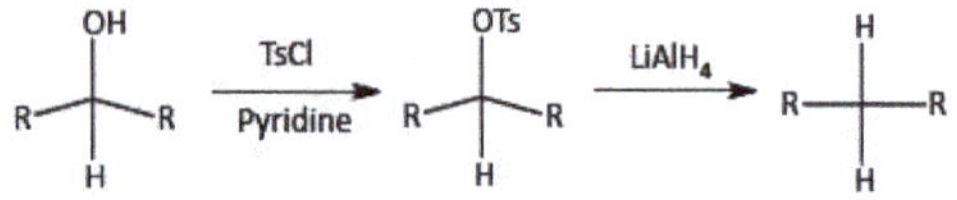

Pinacol Rearrangement

bb) 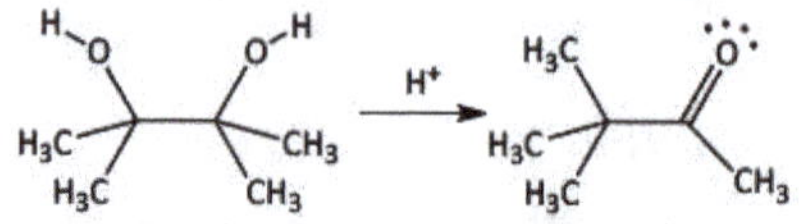

pinacol → pinacolone

Periodic acid cleavage of diol

cc) 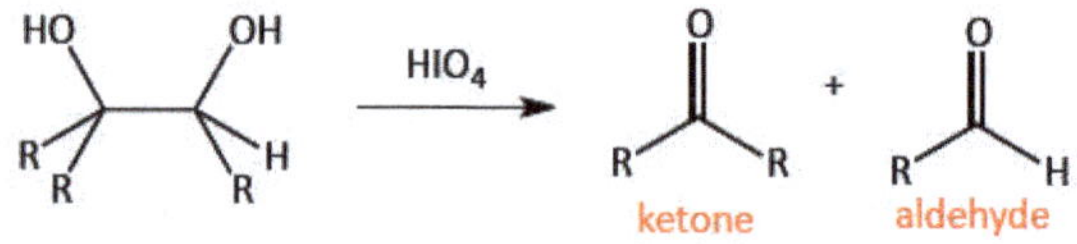

ketone + aldehyde

Fischer Esterification – carboxylic acid to ester

dd) 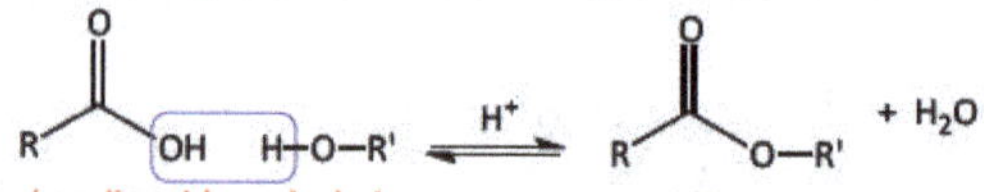

carboxylic acid + alcohol ⇌ ester

Carboxylic acid to acid chloride to ester

ee) 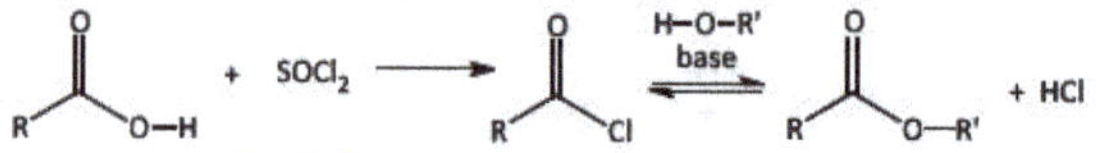

carboxylic acid, thionyl chloride, acid chloride, ester

Thiol reactions

ff) 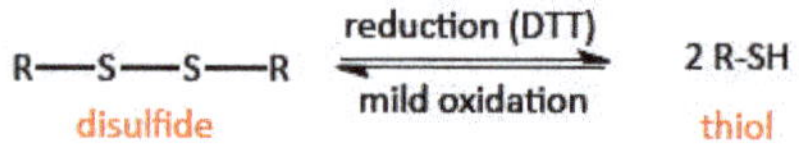

disulfide ⇌ thiol

gg) 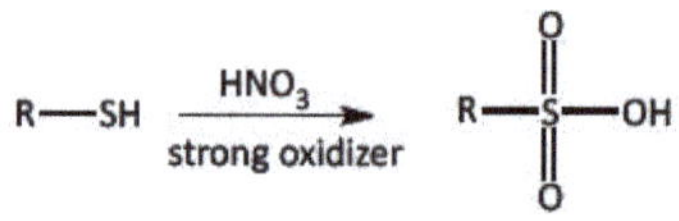

thiol → sulfonic acid

14.11 Problems

14.10 Rank phenol, 4-nitrophenol, and cyclohexanol from weakest to strongest acid? Explain why.

14.11 Draw the ethers produced in the following reactions involving alkoxides.

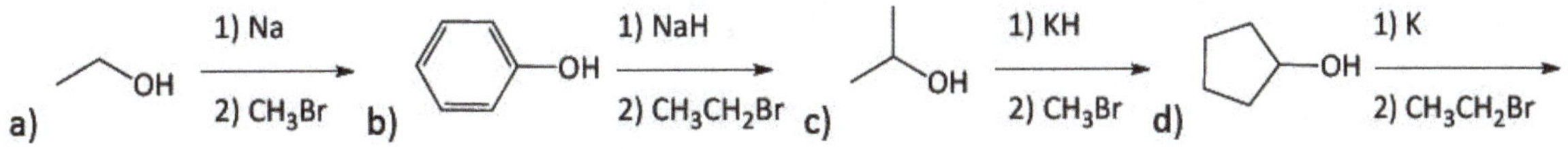

14.12 Determine appropriate reagents to make the following ethers from these alcohols.

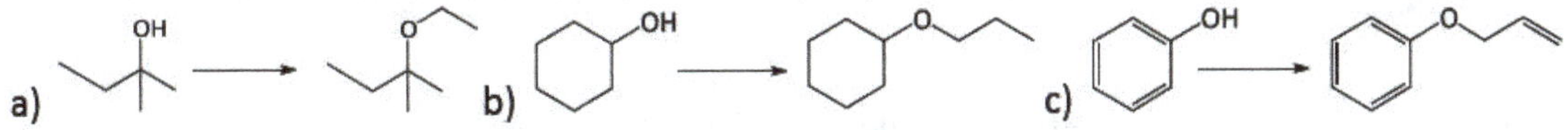

14.13 Draw the major organic product for each of the following dehydration reactions.

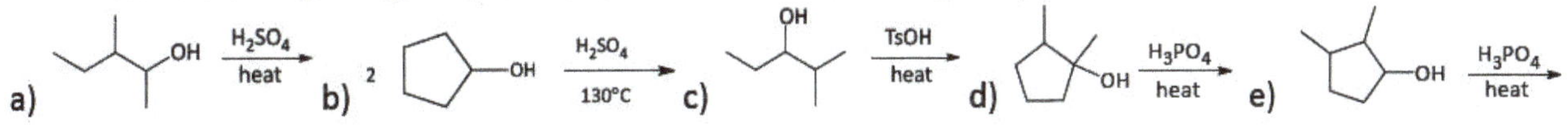

14.14 Draw the products of the following reactions.

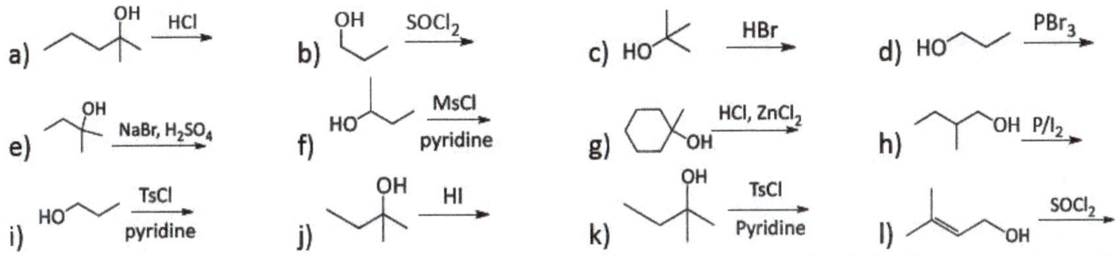

14.15 Determine appropriate reagents to make the following alkyl halides/tosylates from these alcohols.

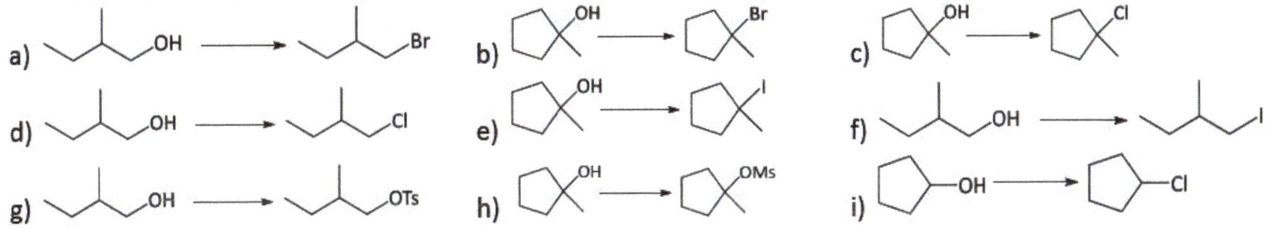

14.16 A few years ago, a graduate student was synthesizing some thiophene derivatives (**A-D**).

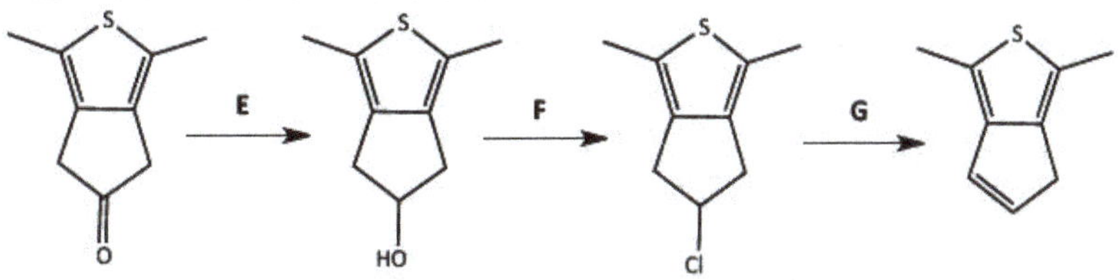

A B C D

a) What reagents (E-G) would you use to perform these reactions?
b) If the chloride leaving group of compound **C** was not good enough, what reagent would you use to turn the hydroxyl group of **B** into a better leaving group like a bromide? How about an iodide?
c) It is also possible to synthesize **D** from the dehydration of the alcohol, **B**. What reagent could do this?

Alcohol Reactions

14.17 **Draw the products of the following oxidation reactions.**

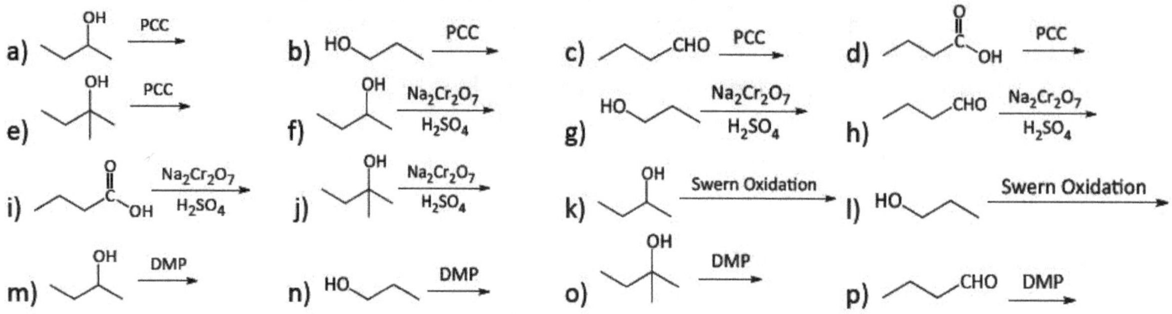

14.18 **Identify an appropriate oxidizing agent to perform the following transformations.**

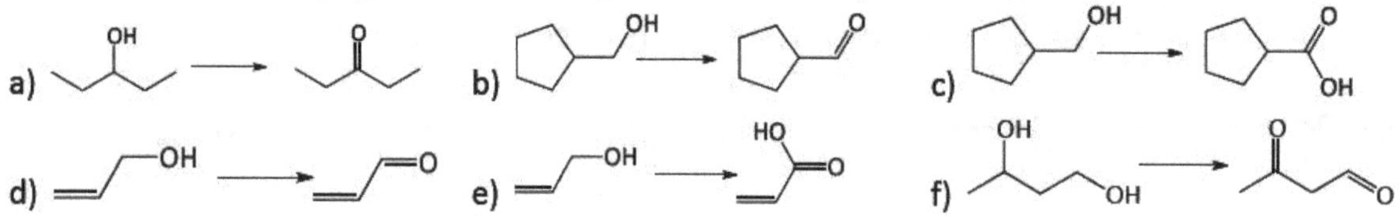

14.19 **What are the products (A-C) of the following reaction sequence?**

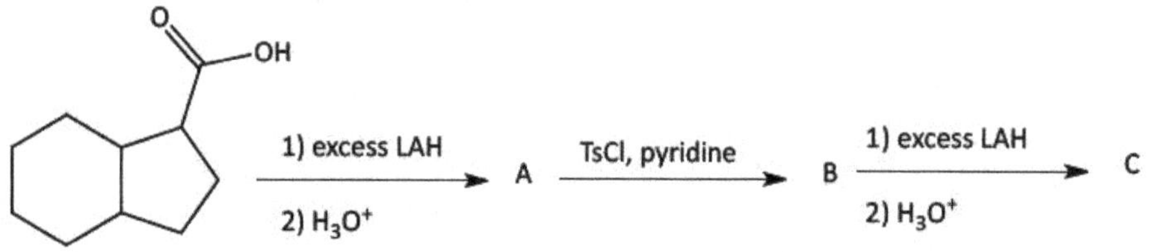

14.20 **Draw the products of the reaction of periodic acid with the following diols.**

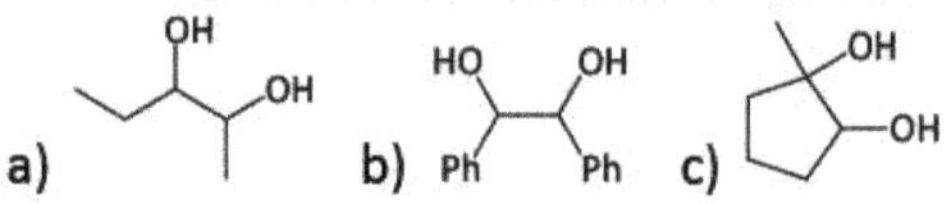

d) $CH_3CH_2CH(OH)CH(OH)CH_3$ e) propylene glycol

14.21 **Identify the alcohol and carboxylic acid that, when catalyzed with H_2SO_4, can be used to make the following esters.**

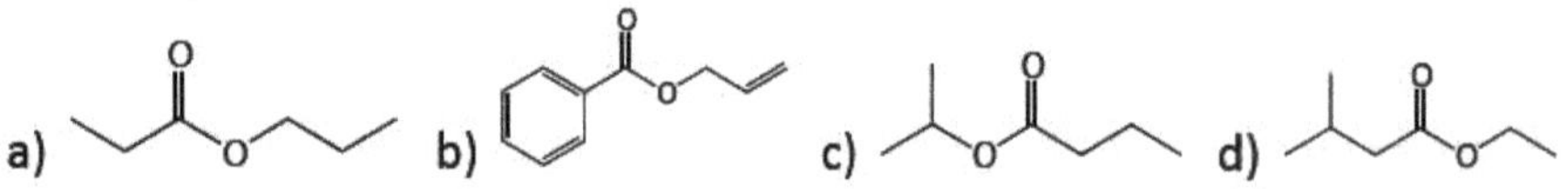

14.22 **Draw the esters formed in the following reactions.**

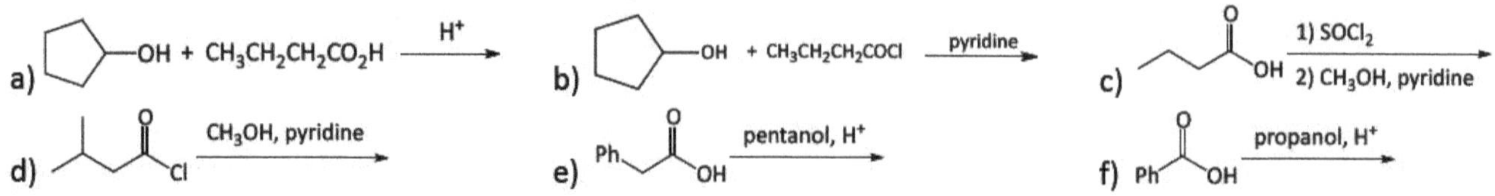

14.23 Draw the products of the following reactions.

a) 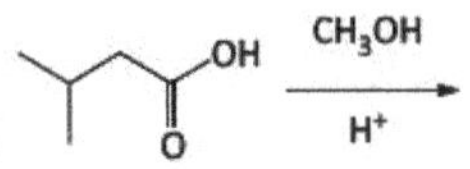

b) 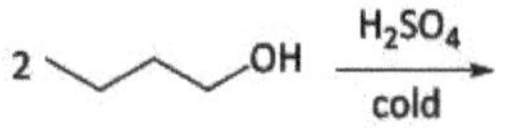

c) 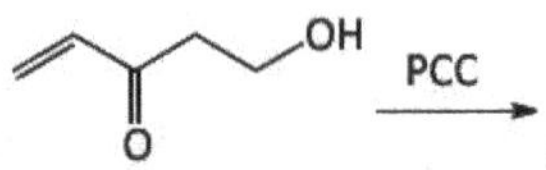

d) 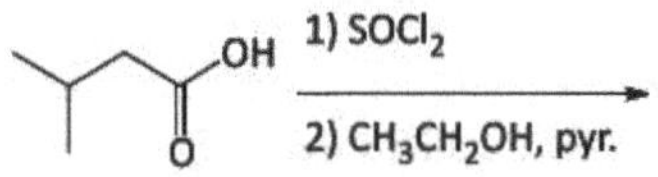

e) 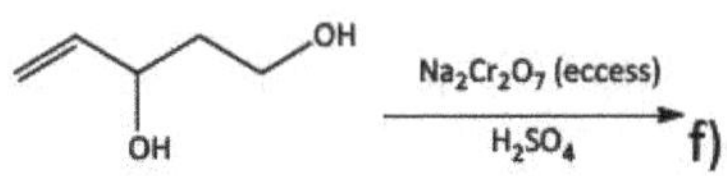

f) 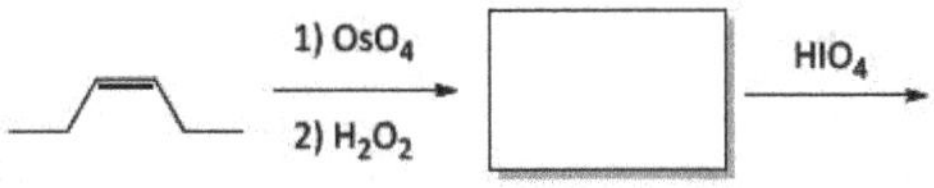

g) 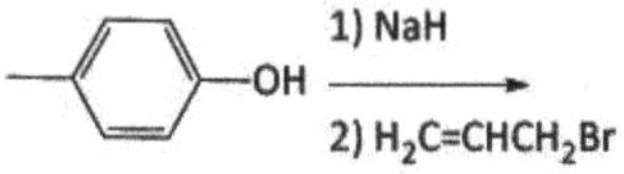

h) 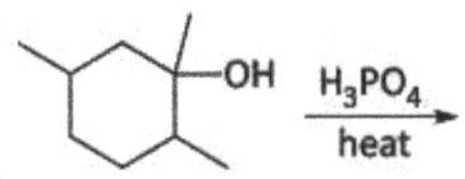

i) 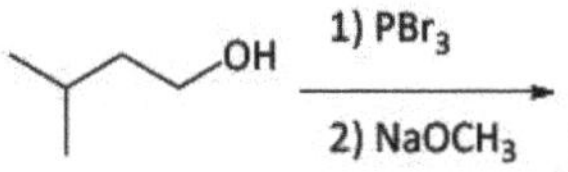

j) 1-pentanol + thionyl chloride, then sodium cyanide
k) t-butyl alcohol + Lucas Reagent l) Swern oxidation of 1-butanol m) isopropanol + HCl
n) ethanol + phosphorus and iodine then excess methylamine o) propanol + DMP p) (*S*)-2-butyltosylate + NaBr
q) hexanol + P/I$_2$ followed by sodium hydrosulfide r) heptanol + PCC s) cyclolheptanol + TsOH/heat
t) cyclopentanol + tosyl chloride/pyridine, then lithium aluminum hydride

14.24 Draw the appropriate reactants of the following reactions.

a) 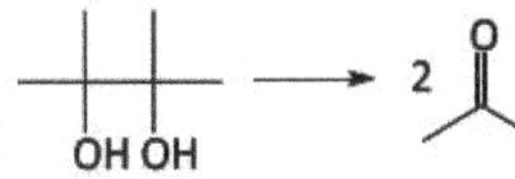

b) 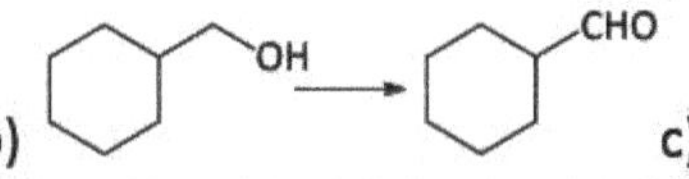

c) ethanol to $CH_3CH_2CO_2CH_3CH_2CH_3$
d) cyclohexanol to cyclohexene e) hexanol to hexanoic acid f) phenol to Ph-O-CH_3
g) propanol to propyl chloride h) pentanol to pentyl bromide i) butanol to butyl iodide
j) *tert*-butyl alcohol to *tert*-butyl chloride k) 2-methyl-2-pentanol to 2-bromo-2-methylpentane

14.25 How would you convert 1-butanol to a tosylate and then to the following compounds?
a) 1-iodobutane b) dibutylamine c) $CH_3CH_2CH_2CH_2OPh$ d) $CH_3CH_2CH_2CH_2CN$ e) butylamine

14.26 Draw the mechanism of 2-methyl-2-butanol with HBr. Hint: It is an S_N1 mechanism.

14.27 Draw the mechanism of the following reaction.

14.28 a) What reagents would turn phthalyl alcohol, **A**, to a Grignard reagent, **B**?

A B

Esters, **C**, can react with **B**, quenched with acid and water, then dehydrated with p-toluenesulfonic acid to make **D**.

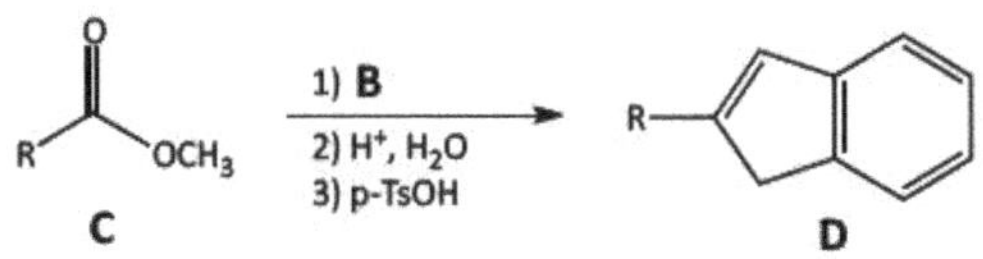

C D

b) Draw the structure of p-TsOH.
c) What does p stand for in p-TsOH?
d) Show the step by step mechanism for the formation of D from C.

14.29 In an organic chemistry lab, a student tries to identify which of the following alcohols they have. They have narrowed the list to propanol, isopropanol, *t*-butyl alcohol. They run a Lucas test where they add Lucas reagent to the alcohol and time how long it would take to react. The solution in the test tube became cloudy almost immediately, indicating the reaction proceeded very quickly. Think about the mechanism of an alcohol with HCl or Lucas reagent. Now, determine which alcohol the student most likely has.

14.30 Draw the mechanism for the following reaction. Clearly show how each product is made.

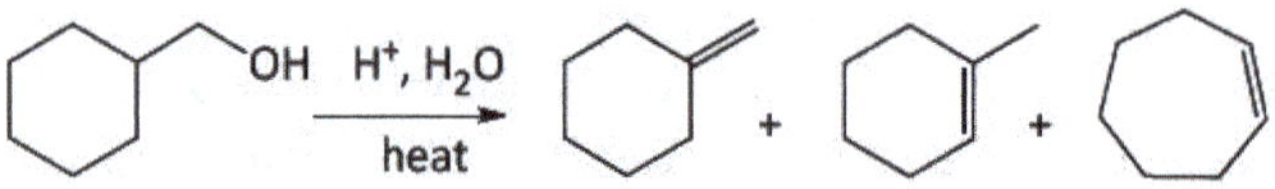

14.31 Explain how you would make the following compounds from an alcohol.

a) 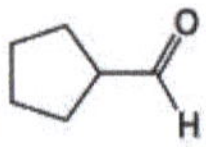 b) 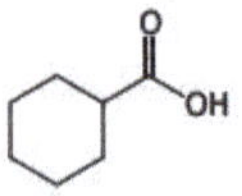 c) 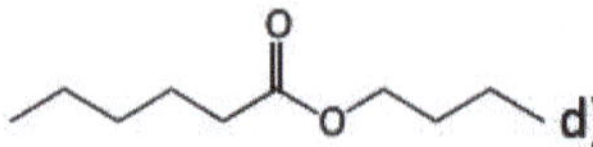 d) 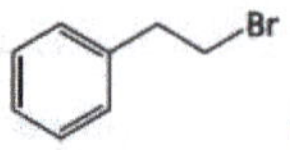 e) 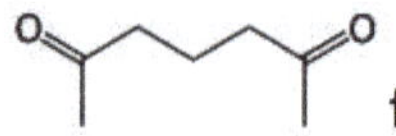 f) 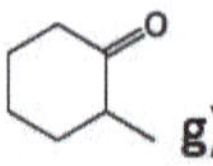 g) 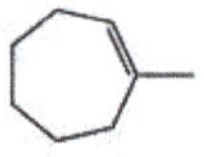

14.32 Show how you would make the following compounds starting with cyclohexanol. More than one reaction may be required.

a) 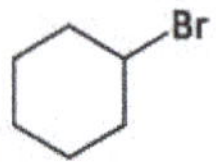 b) 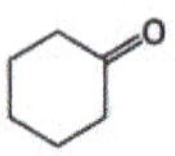 c)  d) 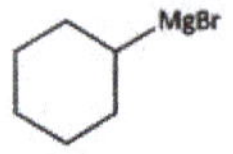 e) 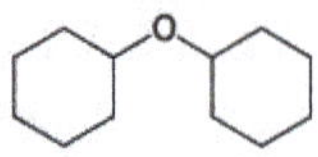 f) 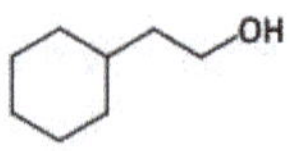

g) 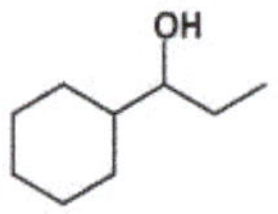 h) 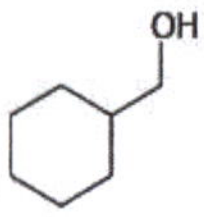 i) 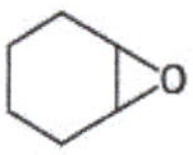 j) 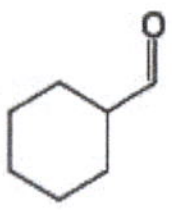 k) 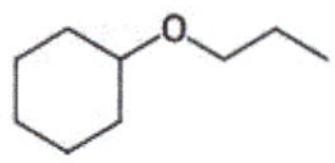 l) 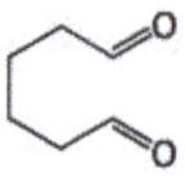

m) 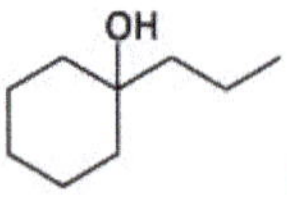 n) 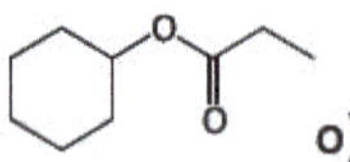 o) 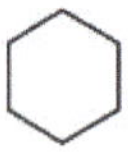 p) 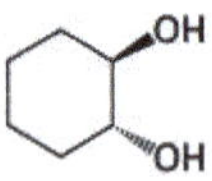 q) 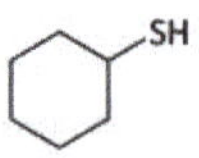 r) 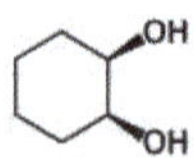

14.33 Draw the disulfide that is synthesized from the oxidation of two cysteine amino acids.

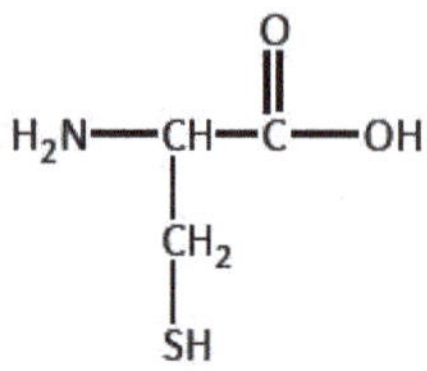

cysteine

14.34 Identify compounds/reagents B through R in the following reaction scheme.

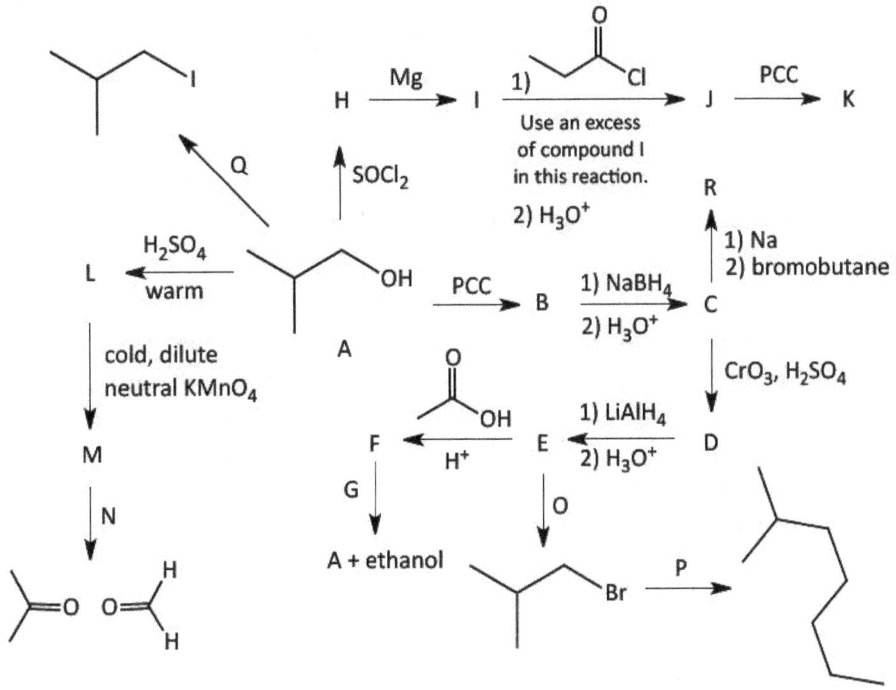

14.35 Show how you would make the following compounds starting with 1-butanol as your only organic reagent. You can use more than one equivalent of 1-butanol if necessary.

a) 1-bromobutane b) $CH_3CH_2CH_2CH_2CN$ c) 1-butoxybutane d) 1-butene e) 1,2-epoxybutane f) 3-octanol g) 3-octanone ($CH_3CH_2COCH_2CH_2CH_2CH_2CH_3$) h) butyraldehyde ($CH_3CH_2CH_2CHO$) i) 4-octanol j) 4-octanone ($CH_3CH_2CH_2COCH_2CH_2CH_2CH_3$) k) 2-butoxybutane l) propionaldehyde (CH_3CH_2CHO) m) 1,2-butanediol n) 2-butanol o) 2-bromobutane

Chapter 15. Ethers, Epoxides, Sulfides

15.1 Ethers

The ether functional group is singly bonded carbon-to-oxygen-to-carbon. The oxygen atom is sp^3 hybridized because it contains two single bonds and two lone pairs of electrons.

R—O—R

Ethers are common solvents in the chemistry lab. Two of the more common ether solvents are ethyl ether (also called diethyl ether or ether) and THF (tetrahydrofuran). Ethers are good at dissolving both polar and nonpolar compounds. Because there is no hydrogen atom on the oxygen atom, ether molecules do not hydrogen bond to each other. Therefore, they have fairly low boiling points. This makes a good solvent because when the reaction is complete, the ether solvent can be easily removed by distillation. DME and dioxane are water soluble. They are good solvents because they can be removed from an organic product at the conclusion of a reaction by simply washing with water.

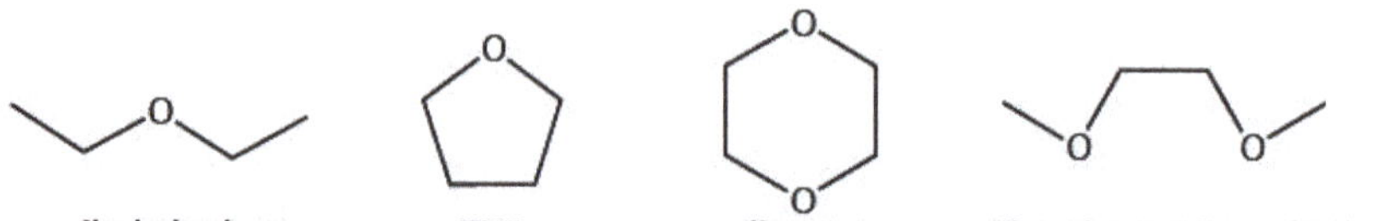

diethyl ether THF dioxane dimethoxyethane (DME)

Figure 15.1 Ether solvents

Ethers are also used as anesthetics. In 1846, Boston dentist William Morton demonstrated that ethyl ether can be used to painlessly extract the teeth of patients. Ethyl ether is also very flammable, which caused quite a problem in the days before electric lights! Ethyl ether also made patients quite nauseous. Today, halogenated ethers are commonly used as general anesthetics.

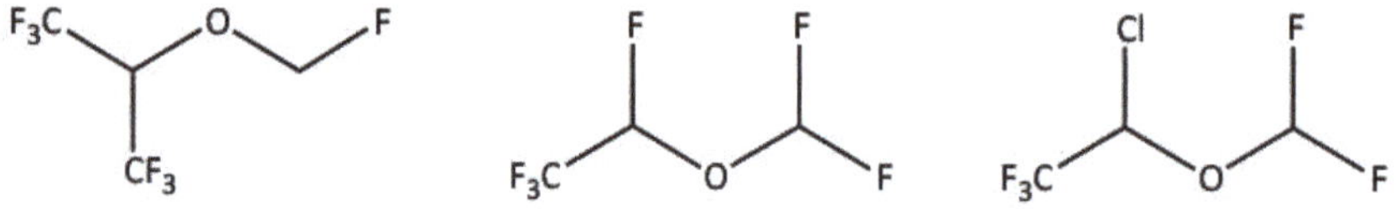

Sevoflurane Desflurane Isoflurane

Figure 15.2 Ethers as general anesthetics

Methyl *tert*-butyl ether, MTBE, is used as an oxygenate in gasoline fuel. It increases the octane rating of the fuel and works as an anti-knocking agent. MTBE is now banned in gasoline by many states because it has found its way into drinking water.

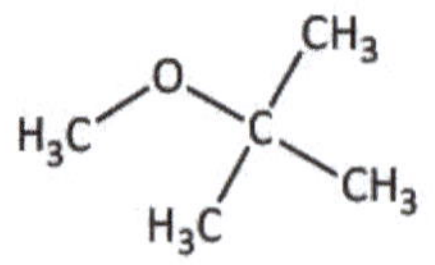

Figure 15.3 Methyl tert-butyl ether, MTBE

15.1.1 Crown ethers

Crown ethers are cyclic ethers. They are a special subset of ethers. The many oxygen atoms with their lone pairs of electrons make crown ethers especially good at binding to positive cations.

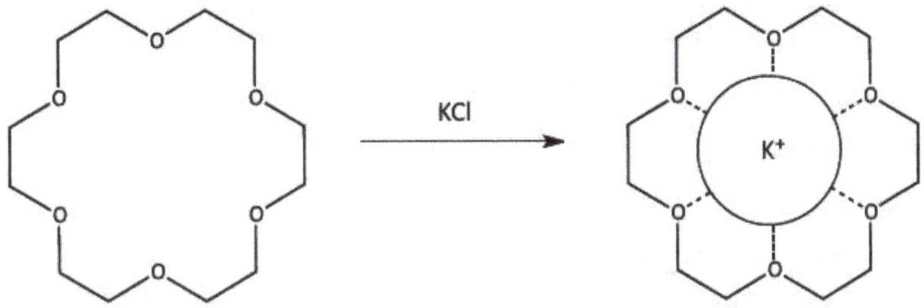

Figure 15.4 18-crown-6 crown ether binding to potassium cation

15.2 Nomenclature

15.2.1 Ether Nomenclature

IUPAC Nomenclature

Ethers are named by naming ether substituents as alkoxy groups attached to the base chain. Their names take the form *alkoxy alkane*.

1-methoxypropane 2-methoxy-2-methylpropane ethoxyethane methoxyethane

Figure 15.5 IUPAC named ethers

Common names

Ethers are commonly named by naming the two alkyl groups, usually in alphabetical order, followed by the word "ether" in the form of alkyl alkyl ether. The common name takes the form *alkyl alkyl ether*.

methyl propyl ether methyl *tert*-butyl ether diethyl ether ethyl methyl ether

Figure 15.6 Commonly named ethers

Ether rings have some common names as well. Furan is a five-membered ether ring with two double bonds in it. If four hydrogen atoms are added to furan turning the double bonds into single bonds, the solvent tetrahydrofuran, THF, is made. To locate the substituents on these rings, the numbering of the ring starts at the oxygen atom.

Ethers, Epoxides, Sulfides

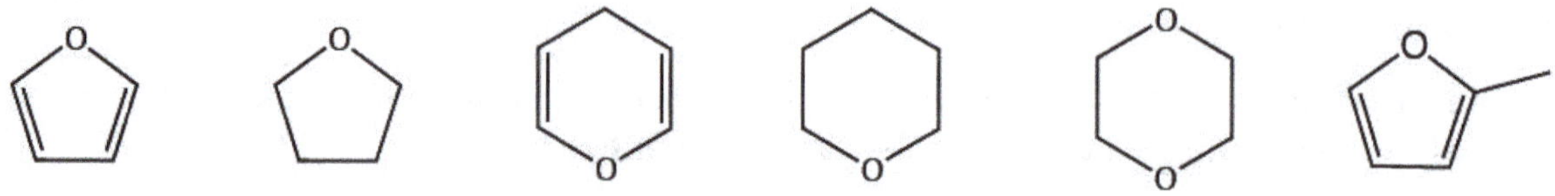

furan tetrahydrofuran (THF) pyran tetrahydropyran (THP) 1,4-dioxane 2-methylfuran

Figure 15.7 Ether rings

15.2.2 Epoxide nomenclature

Epoxides, or oxiranes in old school language, are three-membered ether rings. Epoxides are commonly made from the oxidation of an alkene using a peroxyacid. Epoxides are named by stating the alkene from which the epoxide was formed followed by the suffix "oxide".

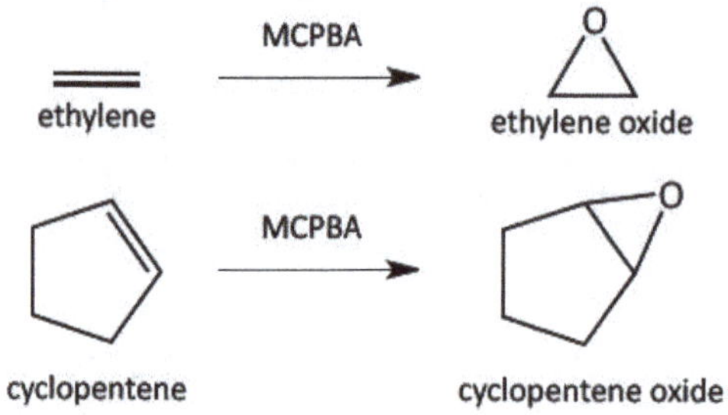

Another way to name epoxides is to name them as a substituent on a more important part of a molecule. In that case, they are named using the prefix "epoxy" and given a location using the numbers of the two carbons to which the oxygen atom is attached.

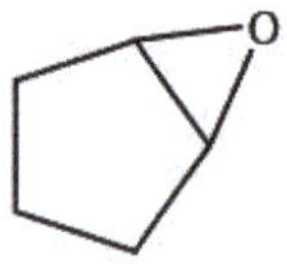

1,2-epoxy-cyclopentane

An older, common method to name epoxides is to use the old-school name oxirane. The simplest epoxide, ethylene oxide, is also called oxirane. One common naming method is to name the substituents on the oxirane followed by the suffix, oxirane.

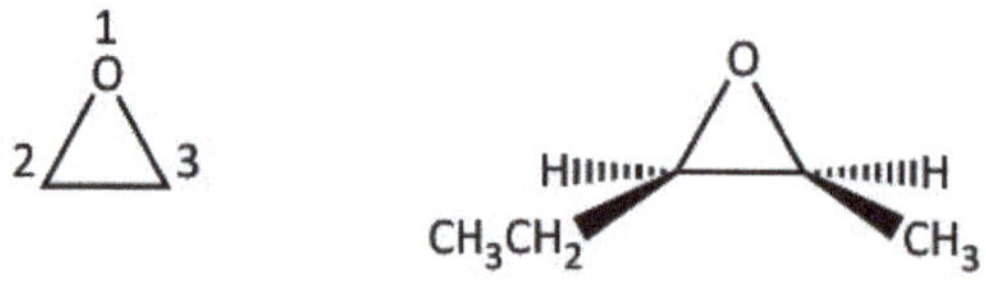

oxirane *cis*-1-ethyl-2-methyloxirane

ethylene oxide 2,3-epoxypentane

15.2.3 Sulfide nomenclature

Since oxygen and sulfur are in the same column of the periodic table, there is a sulfur analogue of ethers. Sulfides, also called thioethers, are the sulfur analogue of ethers.

IUPAC

Sulfides are named by naming an alkylthio group as an attachment on a base chain. It takes the form as alkylthio alkane.

methylthiopropane

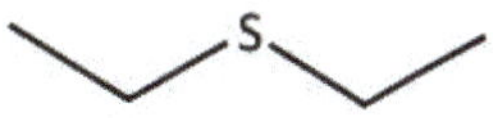

ethylthioethane

Common

The common name for sulfides is much like the common name for ethers, but the suffix "sulfide" is used instead of "ether".

methyl propyl sulfide

ethyl methyl sulfide

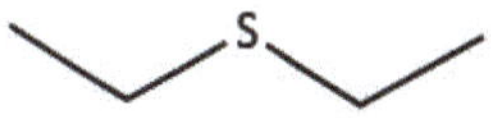

diethyl sulfide

Sulfide ring

One common sulfide ring that you should know is thiophene. Notice that thiophene is analogous to furan. To identify where substituents are, the numbering of the thiophene ring begins at the sulfur atom and works its way around the ring. The nitrogen ring analog of furan and thiophene is called pyrrole.

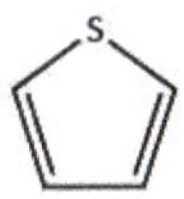

thiophene

furan

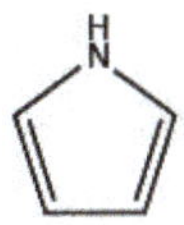

pyrrole

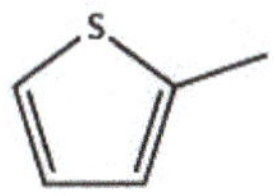

2-methylthiophene

15.1 Name the following compounds.

a) 

b) 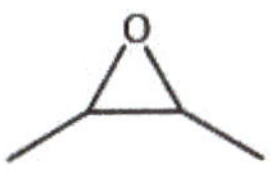

c) 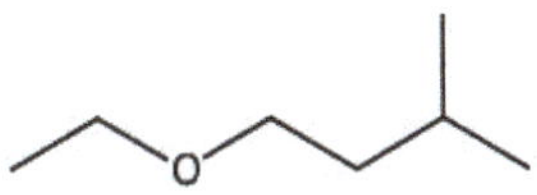

d) 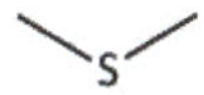

e) 

f) 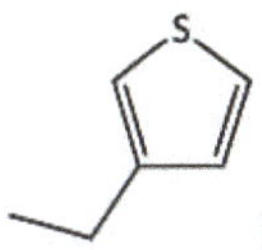

g) 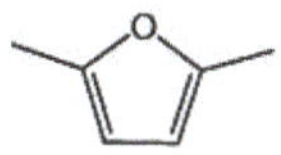

15.3 Ether spectroscopy

15.3.1 Infrared spectroscopy

The C-O stretch of ethers is found around 1000-1200 cm^{-1} and is usually a fairly strong peak. It is sometimes difficult to see this peak because it is in the fingerprint region and may be overlooked. Also, remember that alcohols also have a C-O bond, but alcohols also have the very strong, broad peak for the O-H at 3300 cm^{-1} where ethers do not.

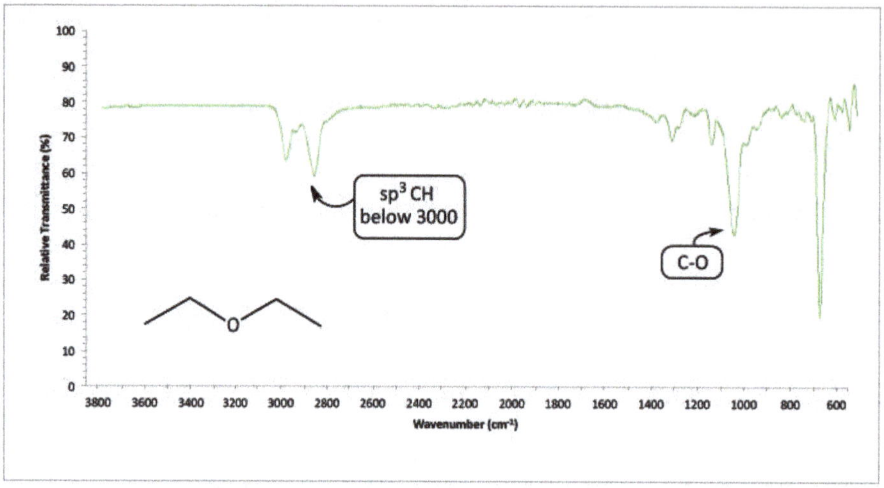

Figure 15.8 IR spectrum of ethyl ether

15.3.2 ^{1}H and ^{13}C NMR

Protons on the alpha carbon appear around 3-4ppm in the ^{1}H NMR. They are slightly deshielded by the electronegative oxygen atom. The alpha carbon atoms appear around 60-90 ppm because they too are deshielded by the oxygen.

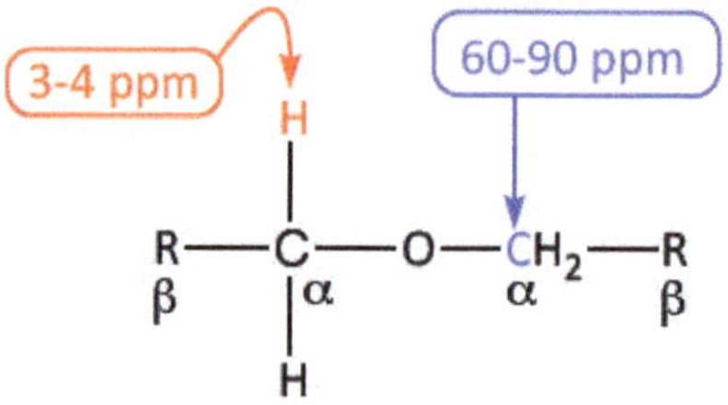

Figure 15.9 NMR of ethers

In the ^{1}H NMR of ethyl ether, the protons next door to the oxygen atom show up at 3.5 ppm in the spectrum. Those protons on the methyl group appear further upfield around 2 ppm.

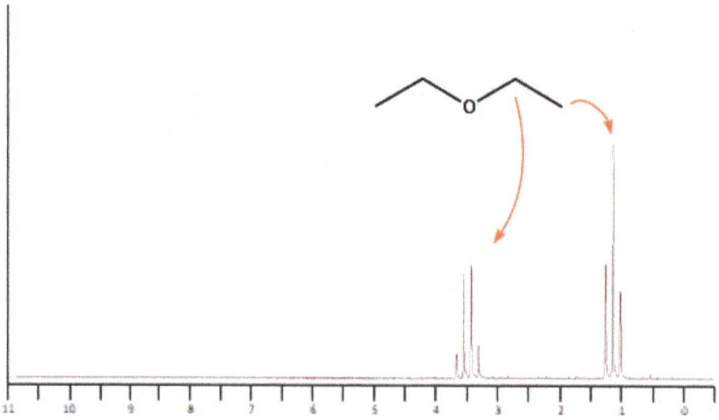

Figure 15.10 ^{1}H NMR of ethyl ether

In the ^{13}C NMR spectrum for ethyl ether, the carbon atom next to the oxygen atom appears around 85 ppm in the "next to something special (50-100)" range.

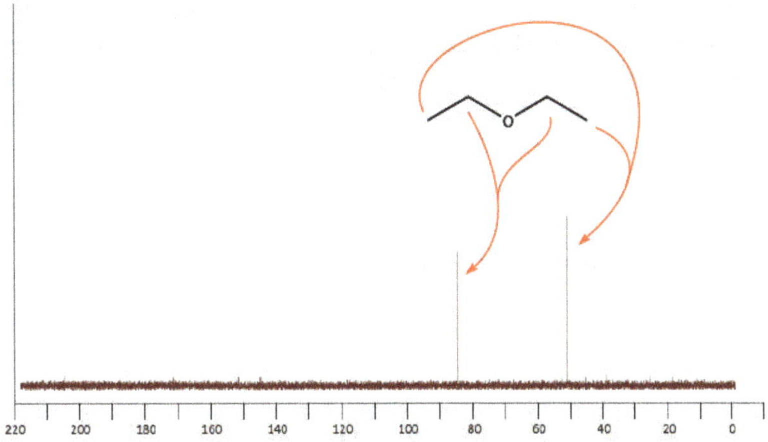

Figure 15.11 ^{13}C NMR of ethyl ether

Mass Spectrometry

A common bond cleavage in mass spectrometry is between the alpha and beta carbon atoms. This fragmentation makes the alpha carbon atom a carbocation. This carbocation can be stabilized through resonance with the lone pair of electrons of the oxygen atom.

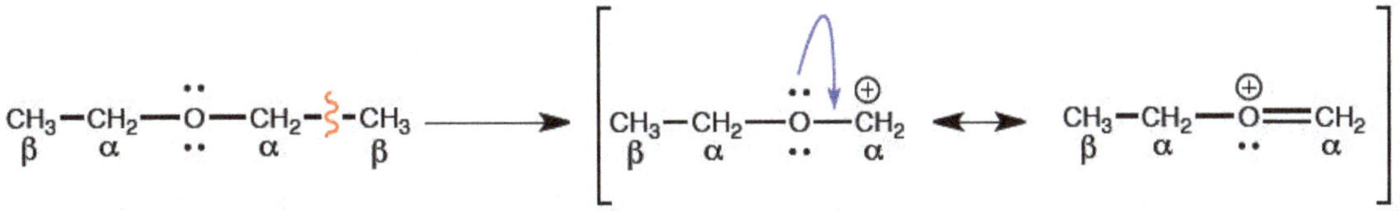

Figure 15.12 Common MS fragmentation for ethers

15.4 Ether synthesis

15.4.1 Williamson Ether Synthesis

Ethers are often synthesized from alcohols. One alcohol is turned into an alkyl halide or a tosylate ester. With this good leaving group, it can be attacked by an alkoxide made from an alcohol and sodium. This S_N2 reaction produces an ether. This is called the Williamson Ether Synthesis.

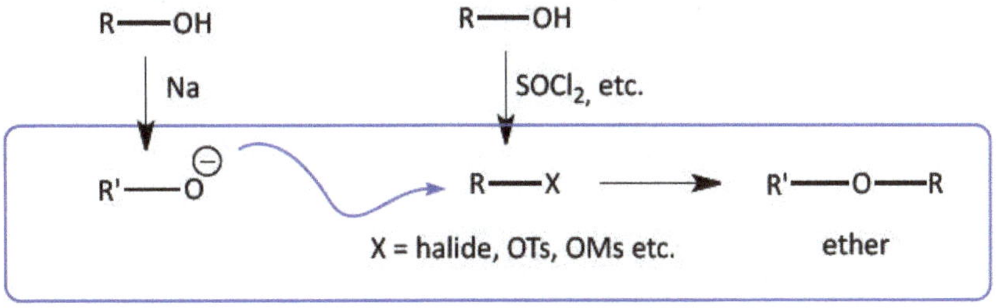

Figure 15.13 Williamson Ether Synthesis from alcohols

15.4.2 Alkoxymercuration/Demercuration in alcohol

Mercuric acetate can react with an alkene to make the triangular mercurium ion. If water is used as the solvent, a Markovnikov alcohol is formed. If water is not used as the solvent, it cannot react as the nucleophile. If alcohol is used as the solvent, the lone pair of electrons on the alcohol can attack the mercurium ion as a weak nucleophile. Again, it attacks carbon atom of the mercurium triangle with the greatest partially positive charge. The electron bond goes up onto the mercury cation to make a good, neutral mercury leaving group. The protonated –OR group is deprotonated with another molecule of solvent to make a mercury-ether containing compound. Many different Markovnikov ethers can be made depending on the alcohol used as the solvent. R = CH_3 for methanol, R = CH_2CH_3 for ethanol, etc.

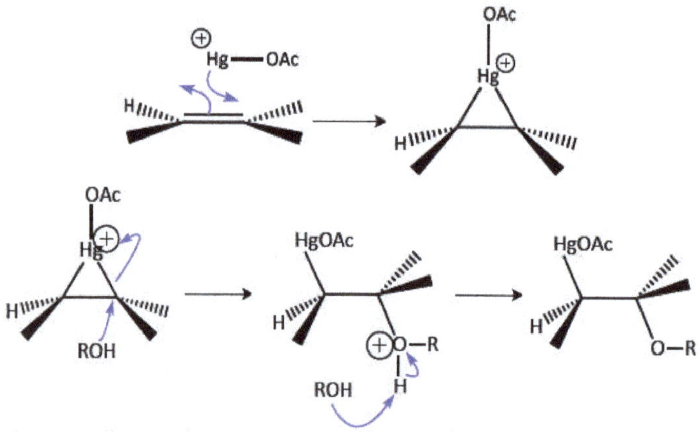

Sodium borohydride ($NaBH_4$) is used to replace HgOAc with H^- making a Markovnikov ether.

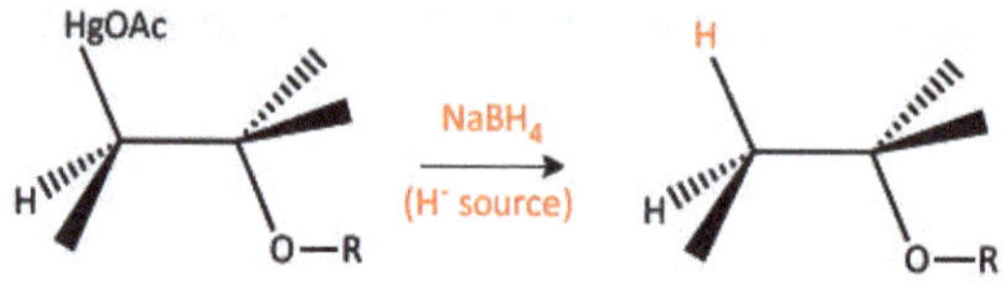

15.4.3 Bimolecular Dehydration

We saw in the last chapter that an industrial way to synthesize ethers is through the bimolecular dehydration of alcohols. In the presence of a dehydrating acid, like sulfuric (H_2SO_4), phosphoric (H_3PO_4), the hydroxyl groups can be protonated making good water leaving groups. This way to make ethers is much less common in an organic chemistry lab because the reaction conditions are fairly harsh and since carbocations are formed, rearrangements often happen.

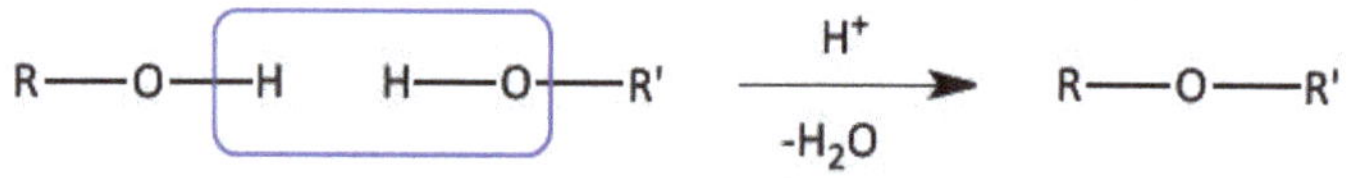

Figure 15.14 General reaction for bimolecular dehydration

It is also necessary to use two of the same alcohols to make symmetrical ethers. Otherwise, a mixture of several ethers is formed.

$$R{-}OH + HO{-}R' \xrightarrow{-H_2O} R{-}O{-}R + R{-}O{-}R' + R'{-}O{-}R'$$

The mechanism for this ether-forming dehydration involved protonating a hydroxyl group of an alcohol to make the great, stable, H_2O leaving group. The other alcohol attacks via an S_N2, backside-attack, mechanism. After cleaning it up, an ether is formed. Below, see the mechanism for the dehydration of two molecules of ethanol to form diethyl ether. This dehydration to form an ether tends to happen around 130-140°C.

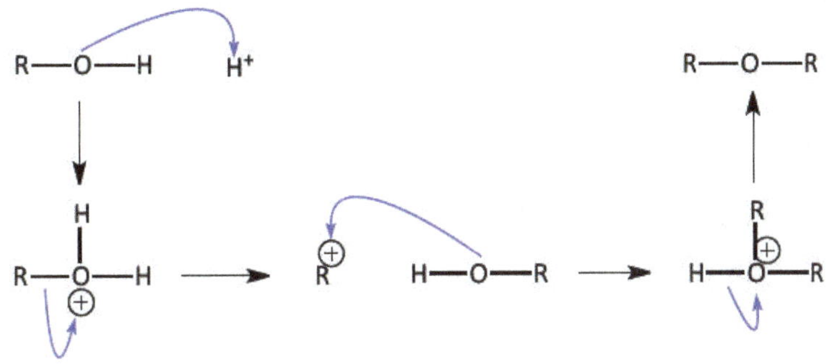

Figure 15.15 Mechanism for acid catalyzed bimolecular dehydration to make ethers

As seen in the previous chapter, a competing reaction to the bimolecular dehydration of two alcohols to form an ether is the unimolecular dehydration of an alcohol to form an alkene. This tends to happen when the temperature gets over 140°C. This dehydration can occur with sulfuric acid, H_2SO_4, phosphoric acid, H_3PO_4, and *p*-toluenesulfonic acid (tosic acid, p-TsOH).

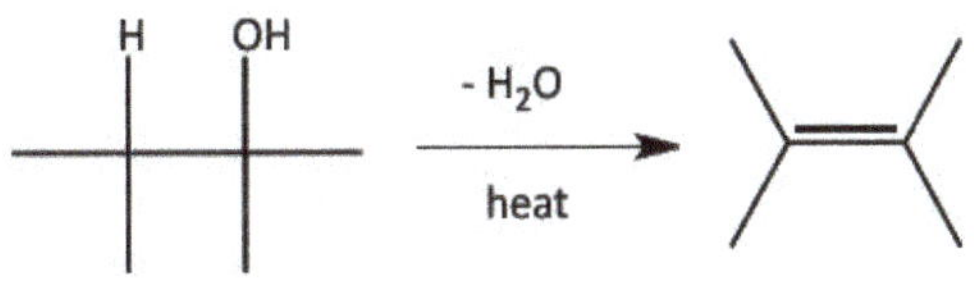

15.2 Give two distinct syntheses of ethyl isopropyl ether. Both syntheses must use ethanol as one of the starting materials.

15.5 Ether reactions

Ethers are pretty unreactive functional groups. Although they are fairly unreactive, they do undergo reactions in the presence of strong acids. What part of an ether would react with H+ from an acid? What is the most negative part of an ether? Of course, the most negative site is the juicy lone pairs of electrons on the oxygen atom. Once the oxygen atom of an ether is protonated, half of the molecule becomes a great leaving group, an alcohol. The protonated ether can then be attacked by a nucleophile in an S_N2 type reaction.

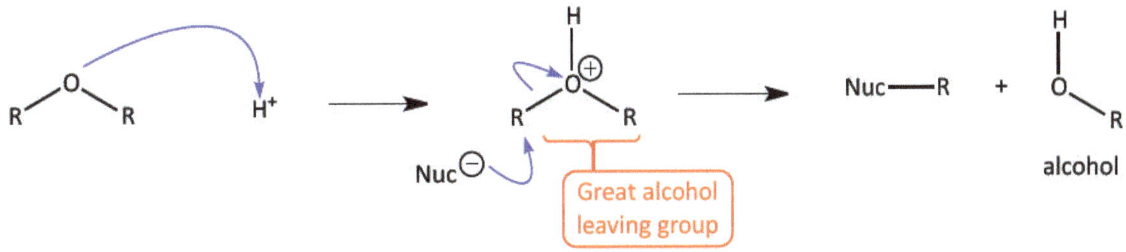

15.5.1 Ether cleavage with HBr, HI

HBr and HI can both protonate the ether and makes Br^- and I^-. Br^- and I^- are excellent nucleophiles, and they are weak bases. Think about that for a moment. If HBr and HI are strong acids, that means they want to give away H^+. They do this because Br^- and I^- are quite stable anions because they are so large. That means Br^- and I^- are terrible bases because they do not want to grab an H^+. So, in this reaction, Br^- and I^- will not deprotonate the ether, but will attack it as a nucleophile.

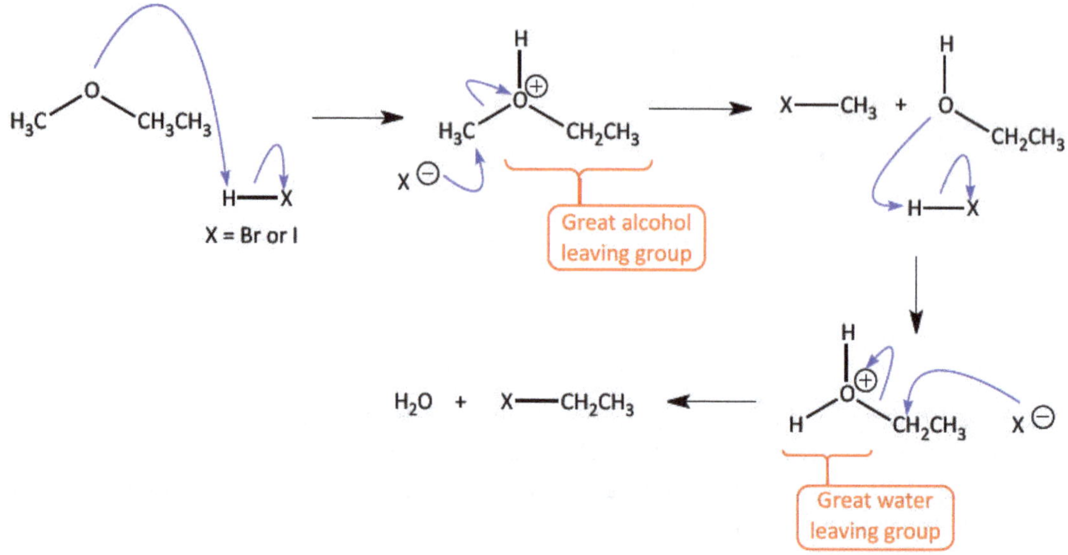

Overall:

$$R{-}O{-}R + 2\,HX \longrightarrow 2\,R{-}X + H_2O$$

15.5.2 HBr and HI with phenyl ethers

A phenyl group, because it is sp^2 hybridized, cannot undergo an S_N2 reaction with the halide. Therefore, phenyl ethers make phenol as a product and not a phenyl halide!

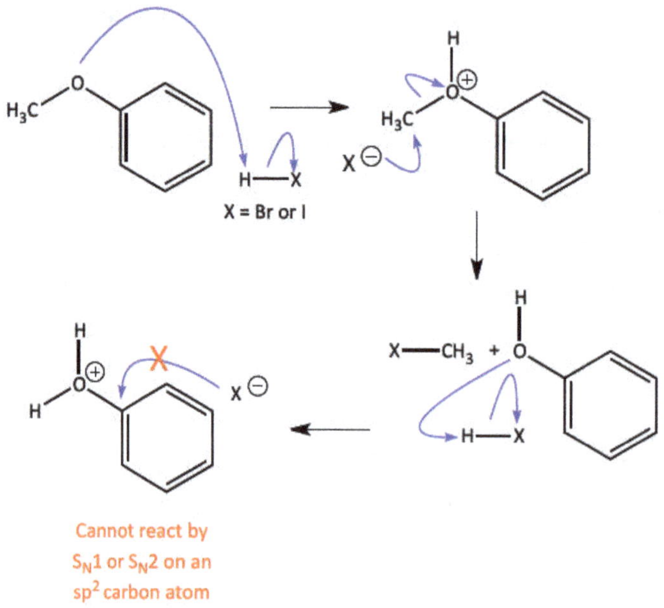

Overall:

$$R{-}O{-}Ph + HX \longrightarrow R{-}X + HO{-}Ph$$

15.3 What are the major organic products of the following reaction?

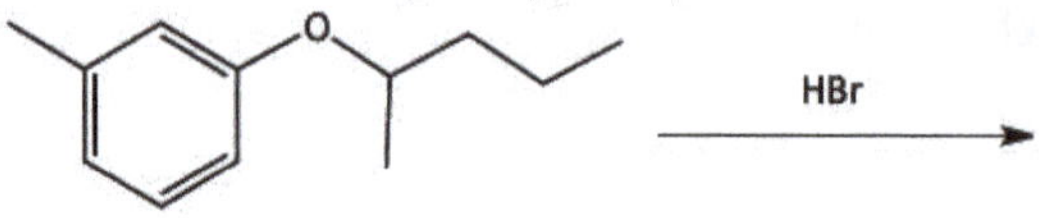

15.5.3 Beware of Autooxidation!

Ethers, over time, slowly react with oxygen. They can oxidize and form explosive peroxides. This process takes months. Solvents like diethyl ether, THF, and dioxane should not be stores for long periods of time. One should only purchase an amount that can be used in about six months.

If you ever come across an old bottle of an ether, be very careful. Look carefully at the cap of the bottle to see if there is any white, crusty powder. This is an indication the ether has been peroxidized. If the lid to the bottle is twisted, it could generate a spark right next to the fuel, the ether, which is in the bottle. This can be very dangerous. Get professional help from your local fire department/bomb squad instead of taking this matter into your own hands.

One must also be careful when distilling ethers. Most ethers that have been peroxidized are not explosive until the peroxides have been concentrated. Imagine what could happen if you are in a lab and think you have an ether compound that is not pure. Perhaps you would think it is a good idea to distill this ether. You begin heating the ether using distillation glassware. The purified ether boils and condenses into your receiving flask giving you pure ether. But, there is less ether in the impure flask. The peroxidized compounds get more and more concentrated. Eventually, the flask with peroxidizable compounds boils uncontrollably, explodes, and sends glass shrapnel everywhere. Think twice before distilling ethers.

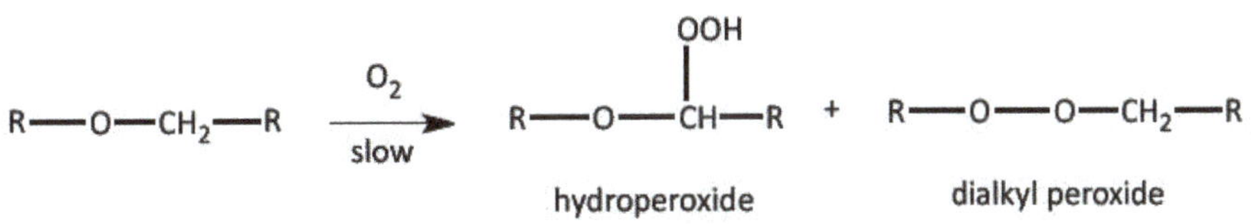

15.6 Sulfides

15.6.1 Synthesis

Sulfides, or thioethers, are the sulfur analogues of ethers. They can be synthesized by using a reaction similar to the Williamson Ether synthesis. Thiols, the sulfur analogue of alcohols, can be deprotonated making nucleophilic thiolate anions. The thiolate anions can then perform an S_N2 reaction on an alkyl halide resulting in a sulfide or thioether.

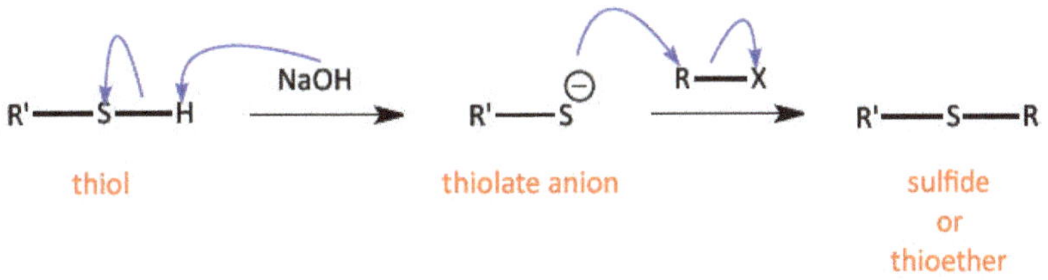

Remember that sulfur is directly below oxygen in the periodic table. Therefore, sulfur atoms are larger than oxygen atoms. Thiols are easier to deprotonate than alcohols, because the negative charge produced in a thiolate anion is spread out over a larger sulfur atom compared to the negative charge on an oxygen atom of an oxide. A base like sodium hydroxide can be used to deprotonate the thiol. The deprotonation of alcohols took a stronger base, like sodium or sodium hydride.

Thiolate anions are also very good nucleophiles because the negative sulfur atom is large and polarizable. The fluffy electron cloud of the thiolate anion can easily reach out its electrons to start the S_N2 attack of the alky halide. They are such good nucleophiles; they will even attack secondary alkyl halides.

15.6.2 Reactivity

Ethers are quite unreactive unless they react with a strong acid. Thioethers, on the other hand, are a little more reactive. Since sulfur is below oxygen in the periodic table, we know that sulfur is larger and less electronegative than oxygen. The sulfur atom of a thioether holds its electrons less tightly than the oxygen atom of an ether. There is a larger cloud of negative electrons around the sulfur atom of a thioether. These fluffy, lone pairs of negative electrons can act as nucleophiles and perform S_N2 reactions by attacking alkyl halides. The thioethers, or sulfides, are alkylated to make a sulfonium salt.

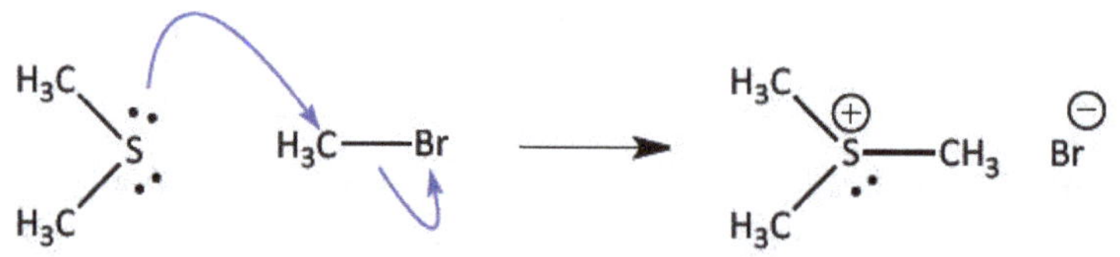

dimethyl sulfide   trimethylsulfonium bromide

Sulfonium salts are good alkylating agents. They are great electrophiles because they contain a very stable leaving group, a neutral thioether compound. Any weak nucleophile can attack a sulfonium's alkyl group.

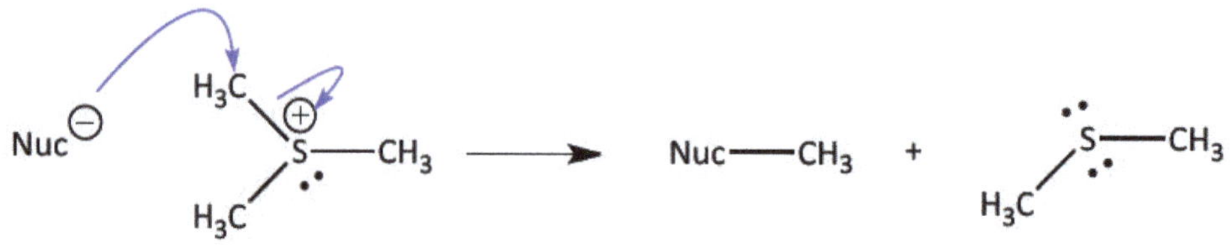

SAM

Sulfonium salts are biochemical agents that perform alkylating reactions in living things. For instance, in human adrenal glands, a molecule called SAM, S-adenosenylmethionine, is used as a methylating agent to transform norepinephrine into epinephrine (adrenaline).

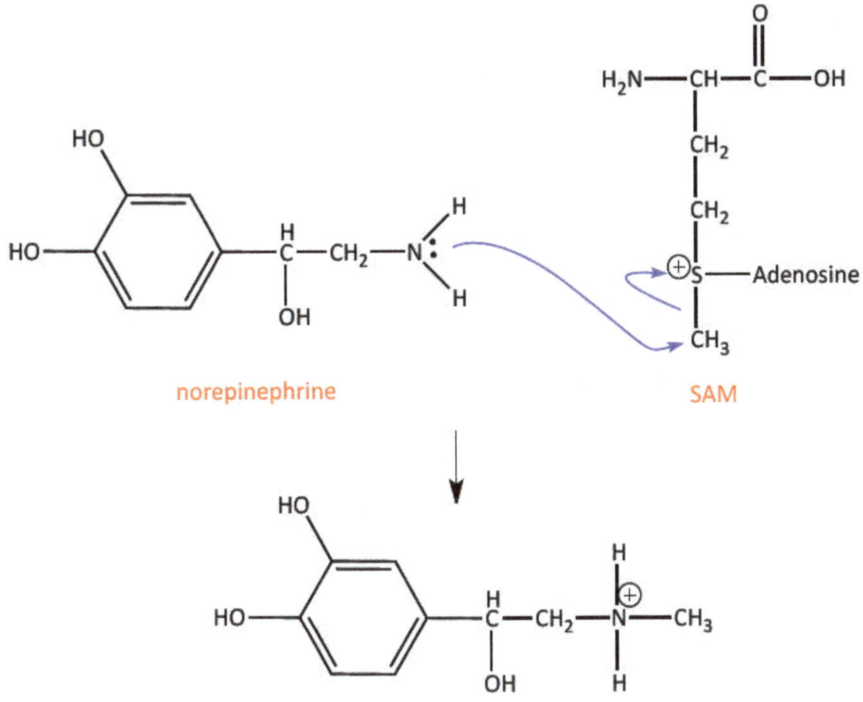

epinephrine

Sulfide oxidation

Sulfur can expand its valence beyond the normal octet rule. Sulfides are therefore also easily oxidized. Sulfides can oxidize to sulfoxides which can oxidize to sulfones. The oxidation of dimethyl sulfide with hydrogen peroxide results in dimethyl sulfoxide (DMSO), which is a common solvent and used in drug delivery. DMSO can carry products through skin into the blood stream. Dimethyl sulfoxide can be further oxidized to a sulfone.

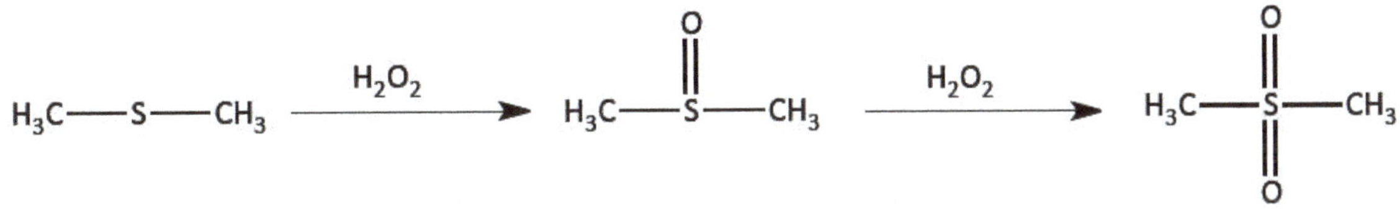

dimethyl sulfide, DMS　　dimethyl sulfoxide, DMSO　　dimethyl sulfone

We have seen the oxidation of DMS to DMSO previously in the ozonolysis reaction of an alkene. Three oxygen atoms make up ozone. Dimethyl sulfide is oxidized to make DMSO using up one of the oxygen atoms. The other two oxygen atoms are used to make the two carbonyl groups.

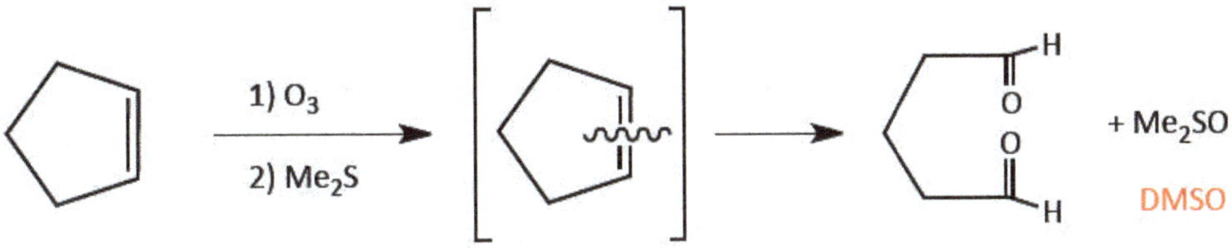

The C=C is "cut" and two oxygen atoms are inserted.

15.7 Epoxides

Epoxides are three membered rings consisting of two carbon atoms and one oxygen atom.

New: epoxide
Old: oxirane

15.8 Epoxide Synthesis

We have previously learned one method to synthesize an epoxide.

15.8.1 Epoxidation of an alkene using a peroxyacid

Epoxides can be synthesized from alkenes by the addition of a peroxyacid, also called a peracid. A peroxyacid looks much like a carboxylic acid, but it contains three oxygen atoms instead of two. Any time a compound is encountered with "extra" oxygen atoms in it, that compound is probably an oxidizer.

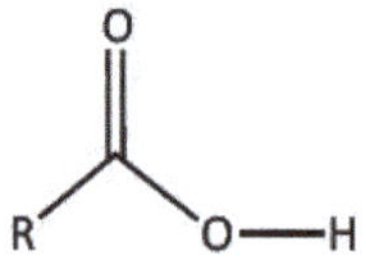

Carboxylic acid (RCO_2H)

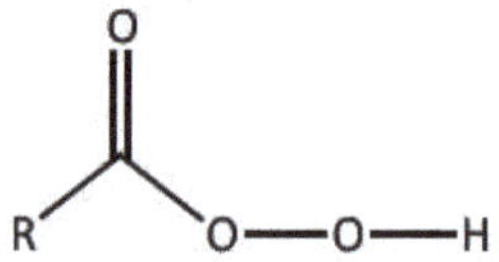

Peroxyacid (RCO_3H)

The epoxidation reaction is usually run in a non-nucleophilic solvent in order avoid unwanted side reaction. Chlorinated solvents like carbon tetrachloride, CCl_4, or methylene chloride, CH_2Cl_2 are good solvents for this reaction. Often, MCPBA is the peroxyacid used in an epoxidation reaction. With its chloro group, it is quite soluble in the needed chlorinated solvents.

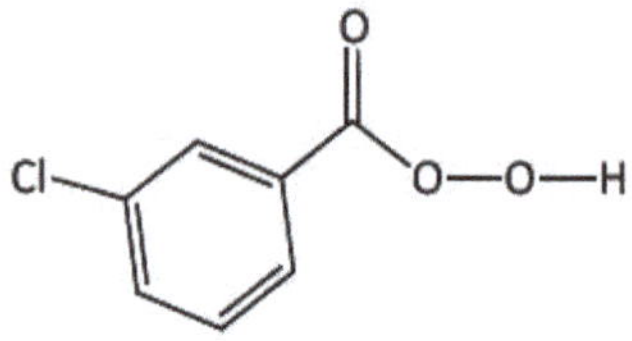

meta-chloroperoxybenzoic acid (MCPBA)

Epoxidation example:

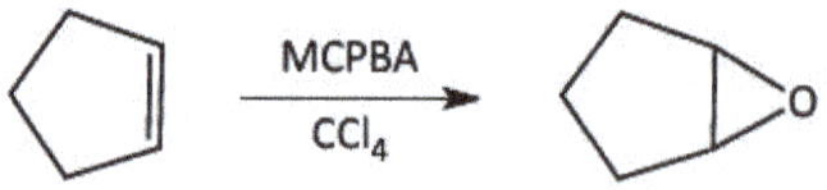

15.8.2 Halohydrin reaction with a base

When water is present in the bromination of an alkene, a bromohydrin is formed. In this mechanism, a triangular bromonium ion is formed. When water is the solvent, there are numerous water molecules for every one Br^- nucleophile. Water ends up being the nucleophile that attacks. In this triangle reaction, a "+" sign is being held up. So, the weak H_2O nucleophile attacks the more substituted carbon atom where more of the + money is held in the pocket under the bags. Once another water molecule cleans it up by deprotonation, it leads to the bromohydrin.

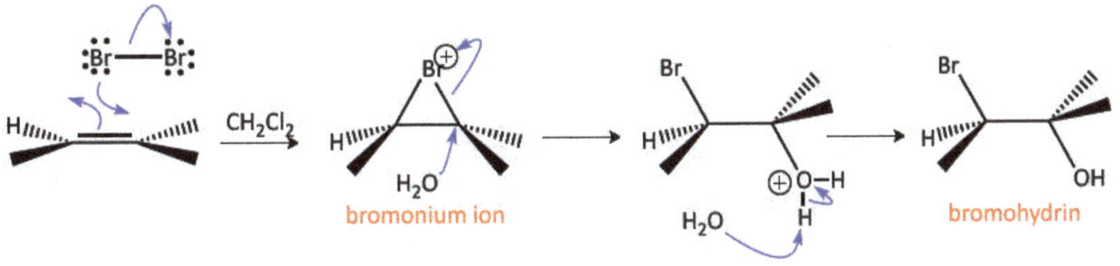

If Cl_2 and water are used, a chlorohydrin is formed. These molecules with neighboring halogens and hydroxyl groups are generally referred to as halohydrins.

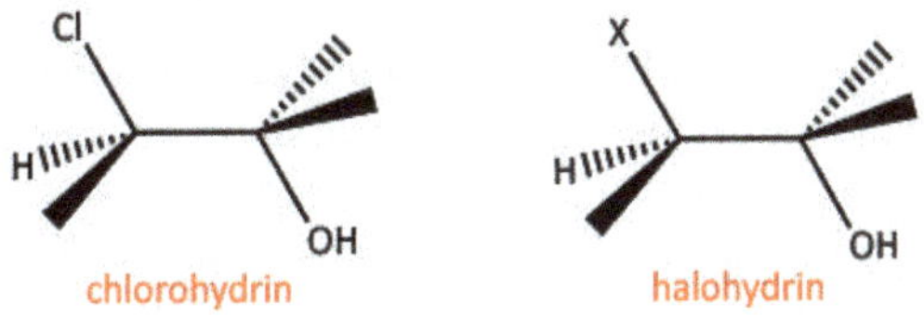

A halohydrin, like a bromohydrin or a chlorohydrin can be deprotonated with a base, like hydroxide. Once deprotonated, the oxide can S_N2, backside attack the alkyl halide part of the halohydrin. Halides like bromide or chloride are excellent leaving groups.

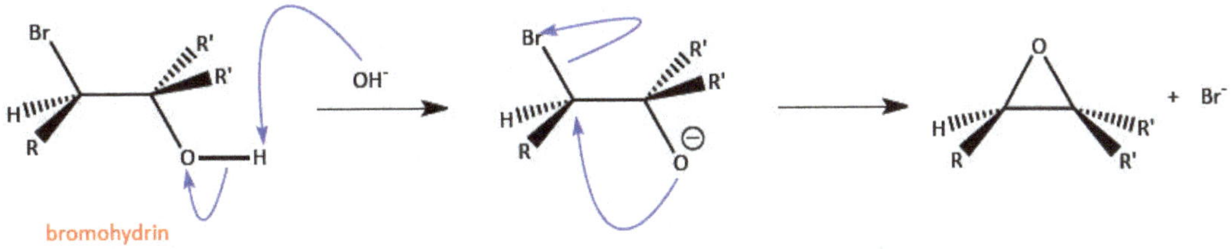

15.4 Give two syntheses for the following compound, one using a peroxyacid and the other must include a halohydrin in the reaction.

15.9 Triangle reactions

The next few reactions we will look at are triangle reactions. It is prudent to review them now.

There are two types of reactions that I call "triangle reactions". An analogy is the story of a victim and two criminals. This analogy is a little bizarre, but I think it helps remember how these reactions happen, so bear with me. The victim of the crime is somebody who is totally honest. He is so honest; he cannot help but announce to the world the amount of money he has by carrying a sign above his head. This option is a sign that reads "0" indicating that he has zero money on him. The other option is a sign that reads "+" indicating that he is carrying a positive amount of money on him. The victim sometimes has bags slung over his shoulders. In order to protect his money, he puts it in the pocket that is hidden under the most bags.

Ethers, Epoxides, Sulfides

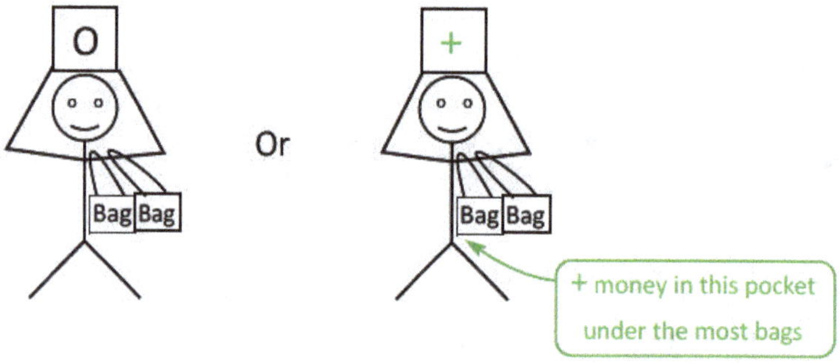

Figure 15.16 The victim of a Triangle Reaction

One of the two criminals is a weak, scared, but greedy pickpocket. This weak pickpocket will not attack a victim unless he knows the victim is *definitely* carrying money. If he doesn't know this, he will leave the victim alone. So, if the victim is carrying a "O" sign, the weak criminal leaves him alone. If the victim is carrying a "+" sign, the pickpocket will reach into his pocket under the bags, where the money is kept. The other criminal is a strong, aggressive, violent criminal. The strong criminal doesn't really care about money. He is so violent; he wants to punch the victim in the gut immediately. This aggressive criminal will attack any type of victim. If the strong criminal encounters the victim carrying a "O" sign, he punches the victim immediately in the gut. The strong criminal wants to do as much damage as quickly as he can to the victim, so he punches him in the side that has the fewest bags on it.

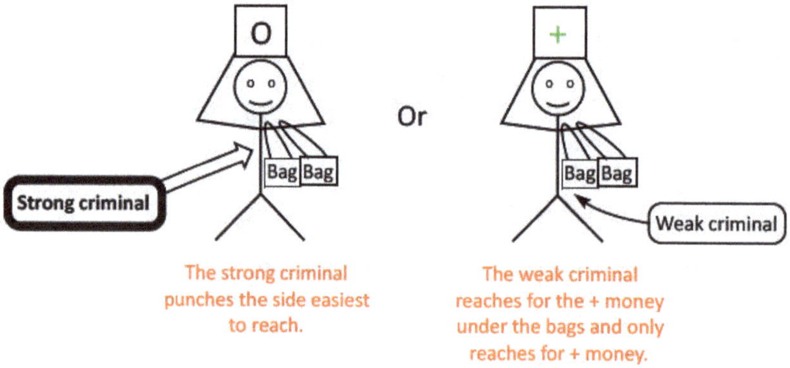

Figure 15.17 How criminals attack in a triangle reaction

If we have a "triangle reaction" the top atom is usually either neutral or has a positive charge. If the top atom is neutral, it is like a victim that is holding a "O" sign. If the top atom has a positive charge, it is like a victim that is holding a "+" sign. The criminals in the analogy are either strong nucleophiles (the strong criminal) or weak nucleophiles (the weak criminal). Strong nucleophiles tend to attack the least crowded side of the triangle. Weak nucleophiles tend to attack the side of the triangle with the most alkyl groups on it. The bags of the victim are "R" groups or alkyl groups. If the top atom has a positive charge, that positive charge is shared somewhat on the other two atoms of the triangle. The atom of the triangle with the most R groups attached holds more of the partially positive charge. This is because more substituted carbon atoms are more stable carbocations. That more substituted carbon atom of the triangle has enough positive charge (+ money) that even a weak nucleophile (weak criminal) cannot resist and will attack it. Below is an example of a neutral epoxide on the left and a positive, protonated epoxide on the right.

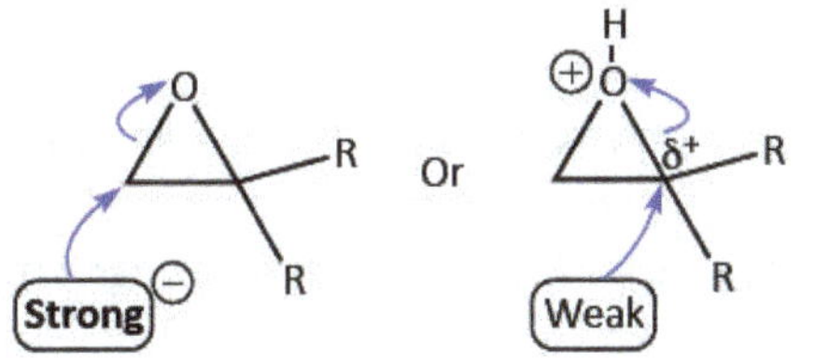

15.10 Epoxide Reactions

15.10.1 With HX

An epoxide reacting with a strong acid like HBr or HCl is a triangle reaction. In the presence of the acid, the oxygen atom of the epoxide is protonated. Once it is protonated and has a positive charge, the epoxide triangle is holding a sign overhead announcing it has positive money. The halide can backside attack the more substituted carbon atom forming a halohydrin with anti-stereochemistry.

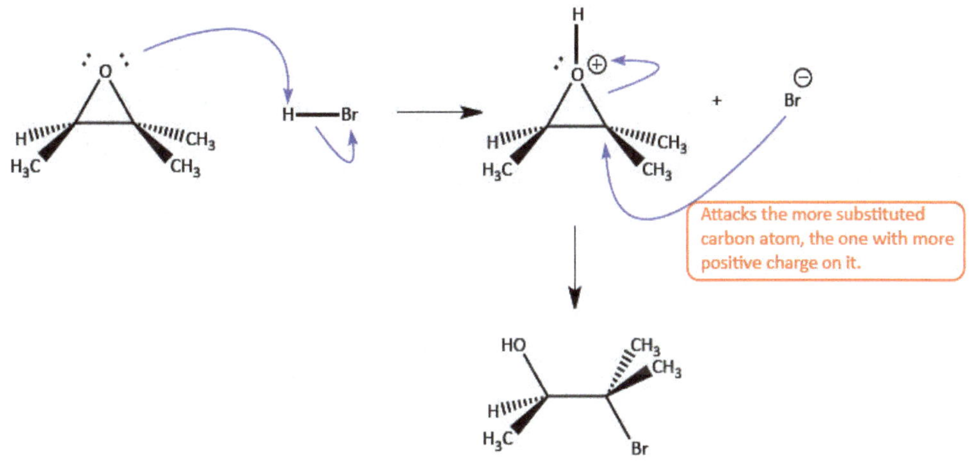

15.10.2 In acid with water to make glycols

If instead of using MCPBA in a chlorinated solvent, a peroxyacid in water is used, the epoxide formed can further react. This is a triangle reaction. In this reaction an acid (the peroxyacid) is present. Once the epoxide is formed, the peroxyacid can protonate the epoxide. The epoxide is holding a sign overhead announcing it has positive money. Water, a weak nucleophile, can backside attack the more substituted carbon atom resulting in a glycol with anti-stereochemistry.

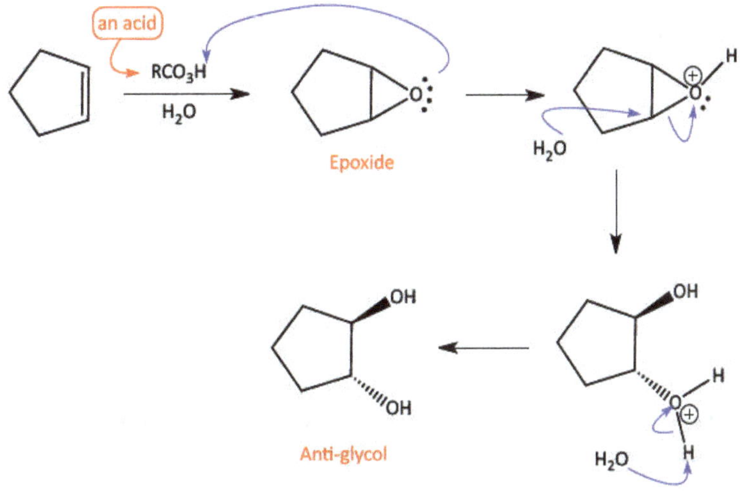

15.10.3 In acid with alcohols

A very similar reaction occurs when an epoxide reacts with acid when an alcohol is the solvent. The same protonation occurs. But, in this instance, alcohol attacks the protonated epoxide at the more substituted, more positive carbon atom. This makes an ether that is anti to a hydroxyl group.

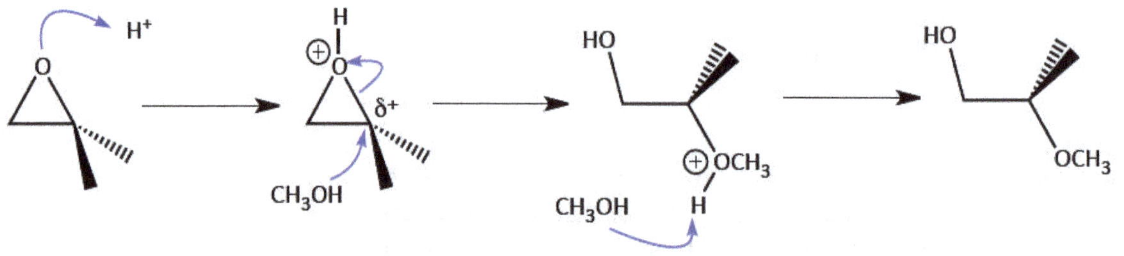

15.10.4 With base/Grignards

The previous reactions were triangle reactions in the presence of acid. They all made a positively charged epoxide that was attacked by weak nucleophiles. But, if the nucleophile is a strong, negatively charged nucleophile, epoxides react differently. In the following reaction, an epoxide is attacked by an aggressive alkoxide. In this triangle reaction, the epoxide triangle victim is holding up a sign with an "O" on it. The aggressive thief attacks the less protected side of the epoxide victim. This leads to an anti ether/hydroxyl system, but the ether is on the less substituted carbon atom. This is opposite to what happened in the acid catalyzed reaction.

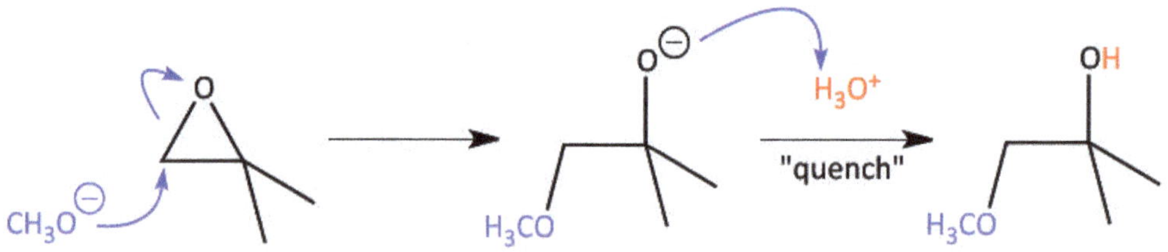

Grignard reagents attack epoxides in the same way. In the case of a Grignard reagent, the nucleophile is an aggressive negative carbon atom.

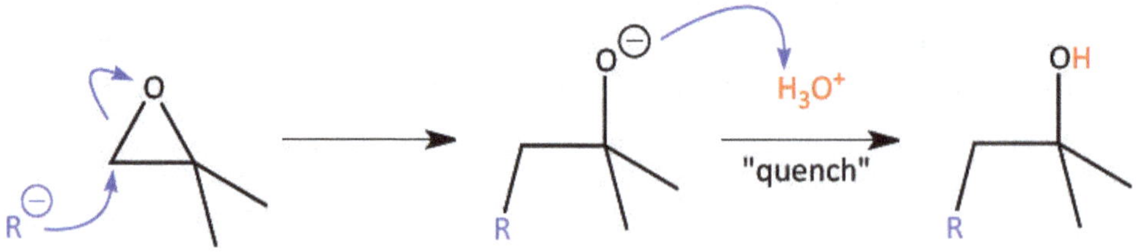

15.5 Draw the products of the following reactions.

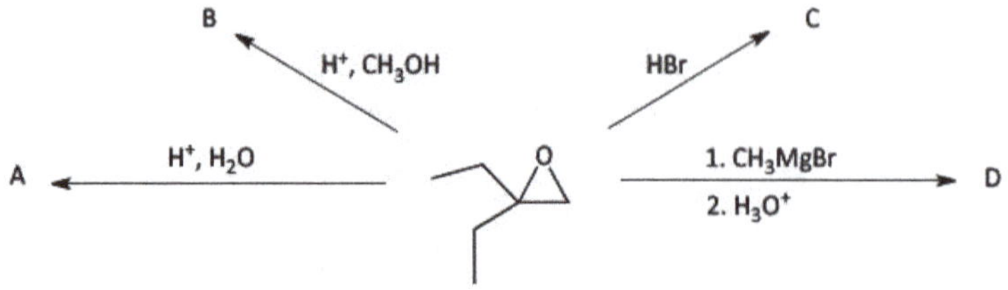

15.11 Reaction summary

Williamson Ether Synthesis

a) 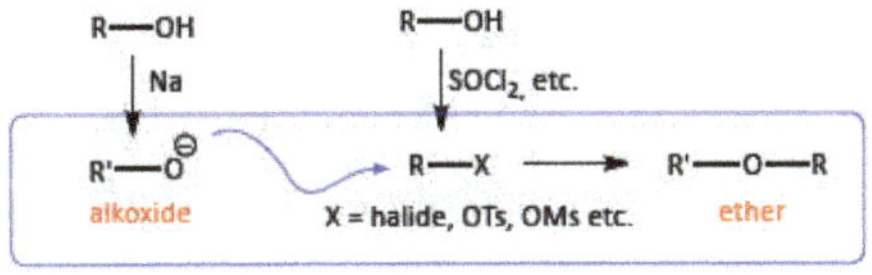

Alkoxymercuration/Demercuration

b) 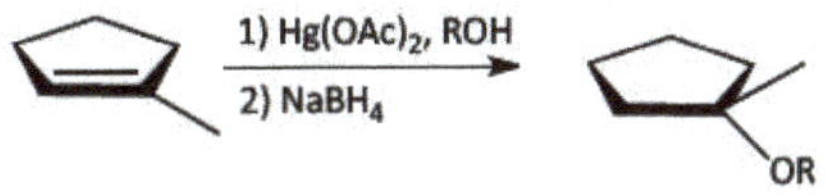

Dehydration to form ether

c) $$R{-}O{-}H + H{-}O{-}R \xrightarrow[H_2SO_4]{-\,H_2O} R{-}O{-}R$$

Dehydration to form an alkene

d) 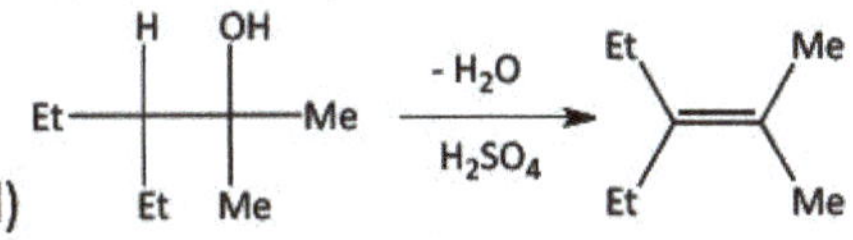

Ether cleavage with HBr or HI

f) $$R{-}O{-}R + 2\,HX \longrightarrow 2\,R\text{-}X + H_2O$$
X = Br, I

Ether cleavage of phenyl ethers with HBr or HI

g) $$R{-}O{-}Ph + HX \longrightarrow R{-}X + HO{-}Ph$$
X = Br, I

Autooxidation

h) 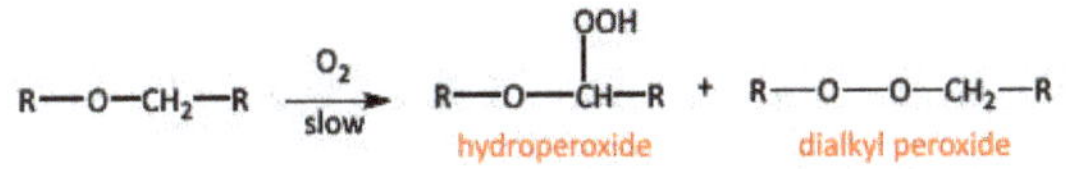

Sulfide synthesis from thiols

i) 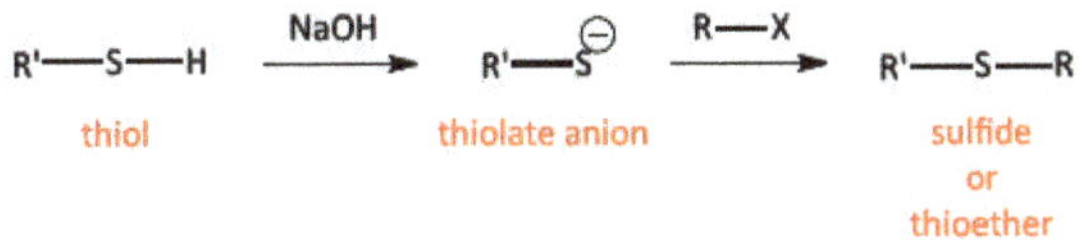

Sulfide with alkyl halide

j) 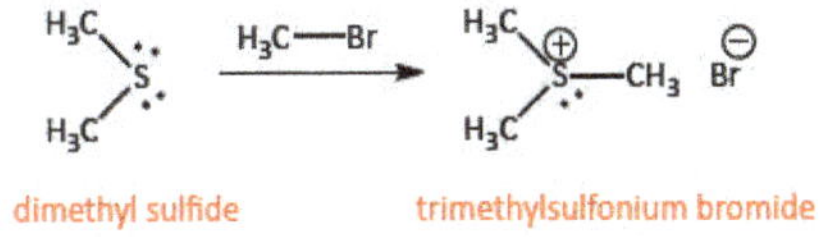

Sulfonium reacting with nucleophiles

k) 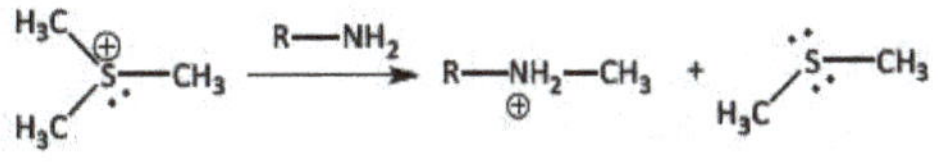

Sulfide oxidation

l) 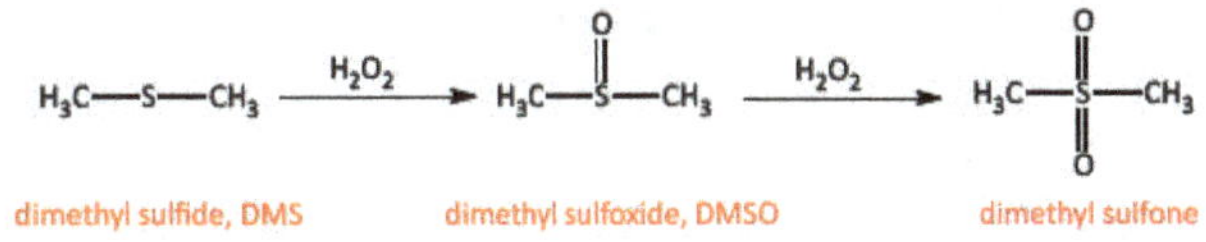

Epoxide synthesis

m) 

Halohydrin formation

n) 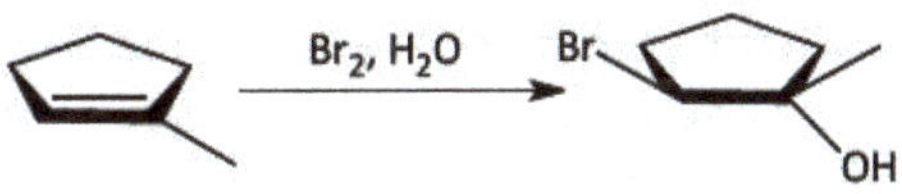

Halohydrin formation of epoxide

o) 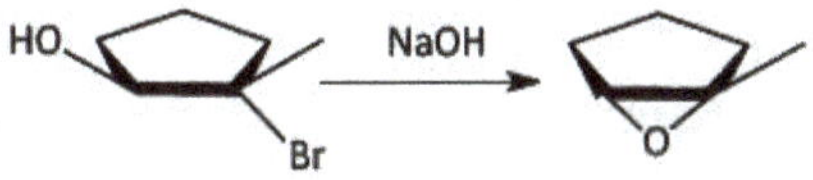

Epoxide reaction with HX

p) 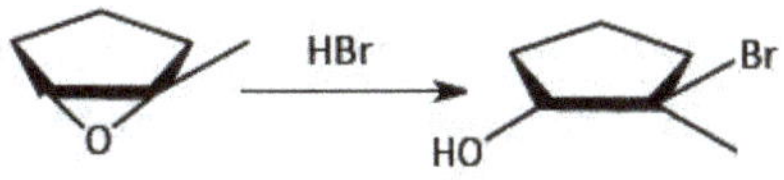

Formation of epoxide in acid and water

q) 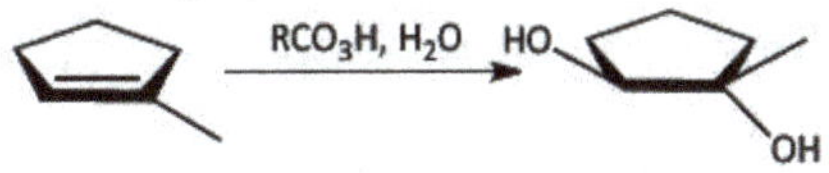

Epoxide reaction in acid and alcohol

r) 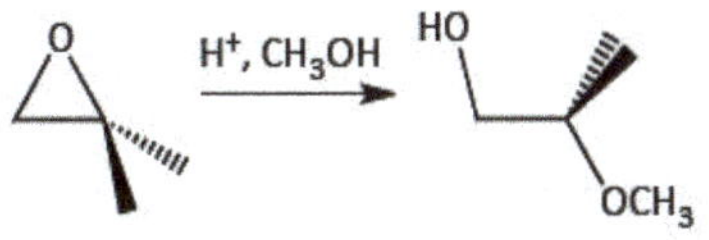

Epoxide reaction with Grignard reagent

s) 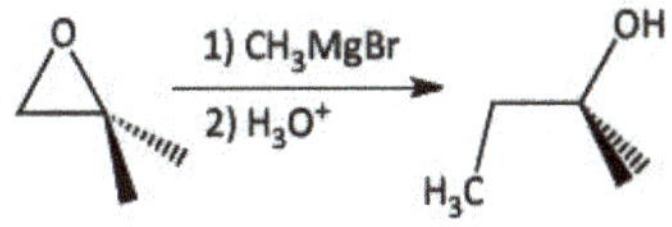

15.12 Problems

15.6 Draw the following compounds.
a) 2-ethoxypentane b) 1,2-epoxybutane c) 3-ethylfuran d) THF (tetrahydrofuran)
e) *cis*-3-ethyl-1,2-epoxycyclopentane f) *m*-propoxyphenol g) ethylthiohexane h) ethyl hexyl sulfide
i) cyclohexene oxide j) 2-butyl-1-propylsulfide k) 3-methoxyhexane l) 2,3-dimethylthiophene m) ethyl ether

15.7 Name the following compounds.

a) 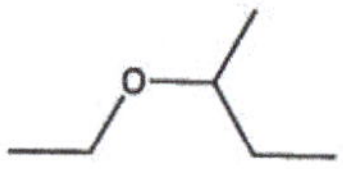
b) 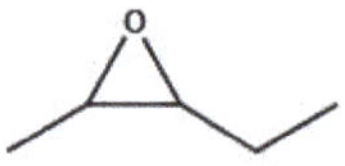
c) 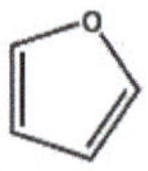
d) 
e) 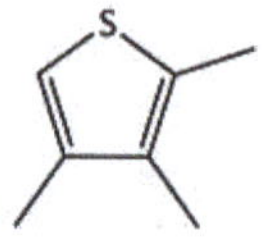
f) 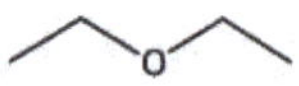
g) 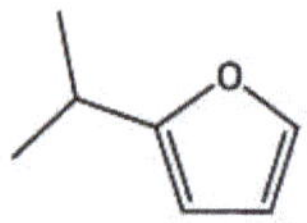
h) 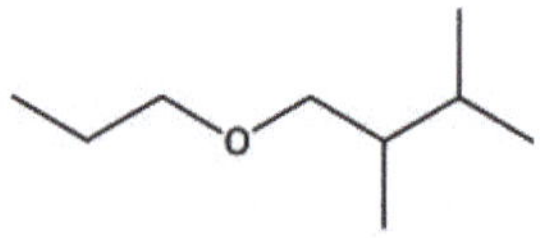
i) 
j) 

k) 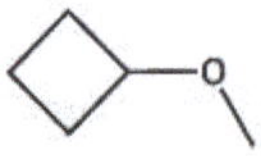

15.8 Show the best method for preparing the following ethers via the Williamson ether synthesis. You must start with alcohols or phenols as the organic reactants.
a) ethoxycyclohexane b) *tert*-butyl methyl ether c) propoxybenzene d) ethoxypropane

15.9 Use alkoxymercuration-demercuration to make the following ethers from an appropriate alkene and alcohol.
a) ethoxycyclopentane b) 2-methoxyhexane c) 1-methyl-1-propoxycyclohexane

15.10 Explain why alkoxymercuration-demercuration is not a good method to make 1-ethoxyheptane.

15.11 Determine the necessary reagents needed to perform the following reactions.
a) 1-pentene to 1,2-epoxypentane b) ethyl ether to ethyl bromide
c) cyclopentene to *trans*-cyclopentane-1,2-diol d) propyl bromide to ethyl propyl sulfide
e) butane to 1-chlorobutan-2-ol f) phenol to ethyl phenyl ether g) pentene to 2-ethoxypentane

15.12 Show the best method for preparing propoxybenzene via the Williamson ether synthesis.

15.13 Show the best method for preparing *tert*-butyl propyl sulfide from a thiol and alkyl bromide.

15.14 Show the mechanism for the following reaction.

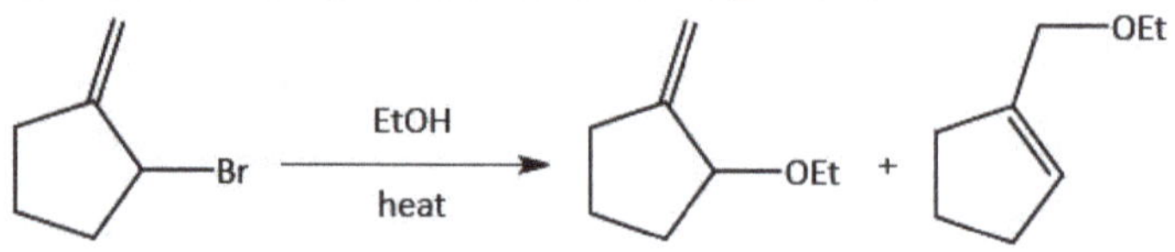

15.15 Show the mechanism for the following reaction.

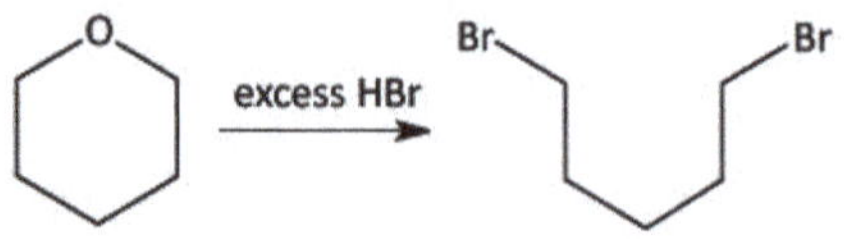

15.16 The following spectra are of a compound with the formula $C_4H_{10}O$.

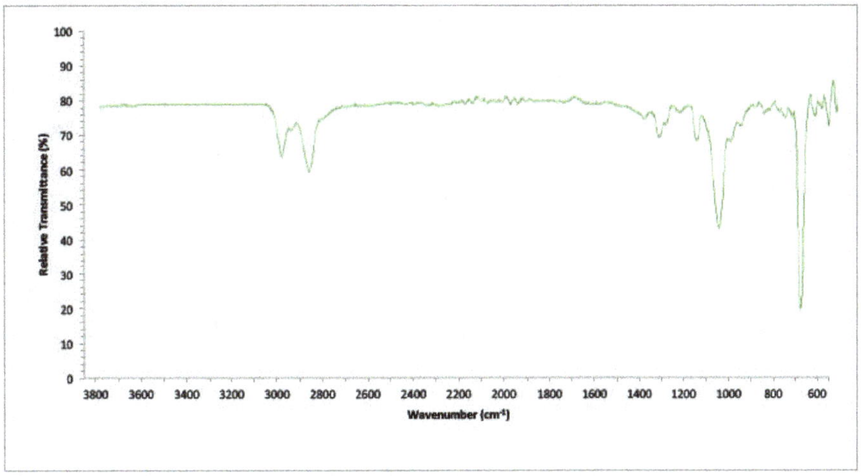

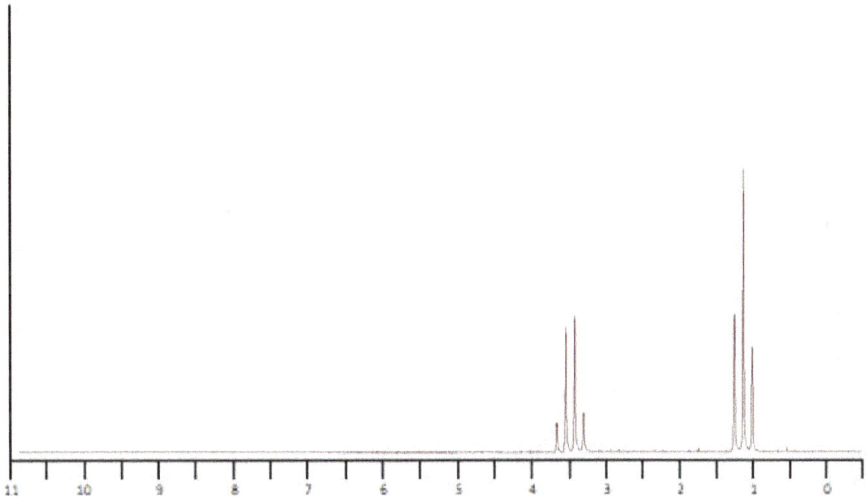

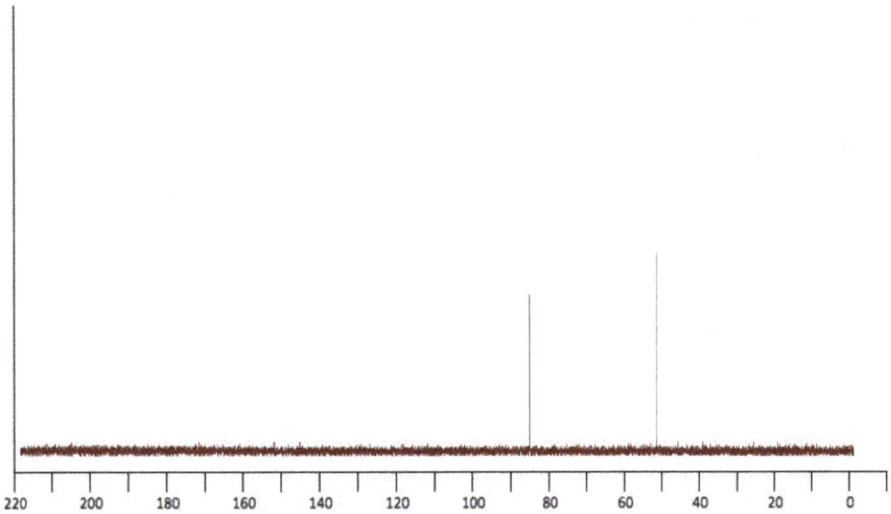

Ethers, Epoxides, Sulfides

15.17 Provide the major organic product in the following reactions.

a) 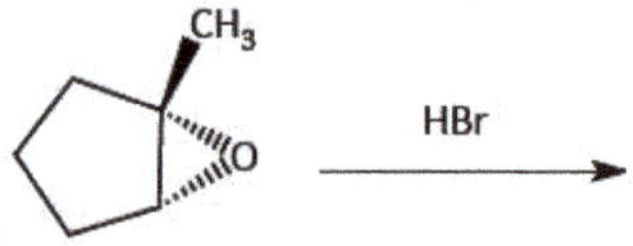

b) 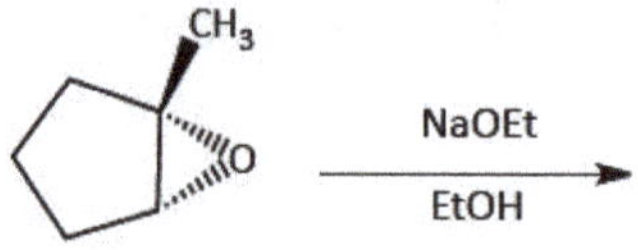

c) 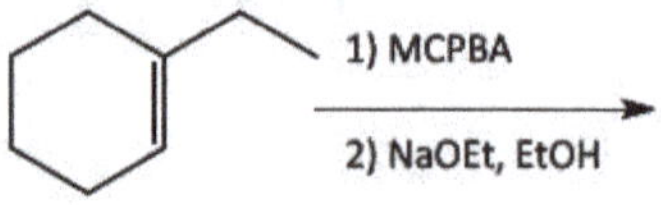

d) 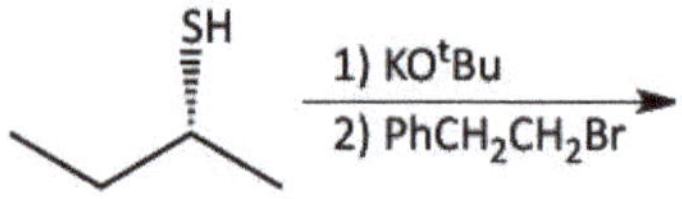

e) 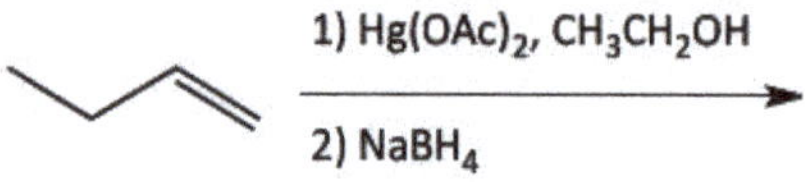

f) 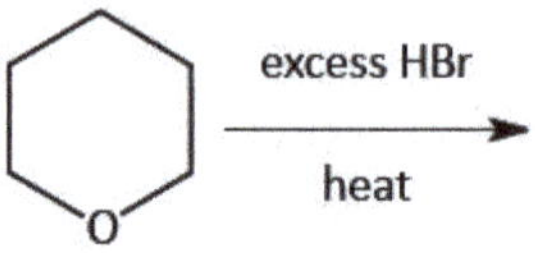

g) 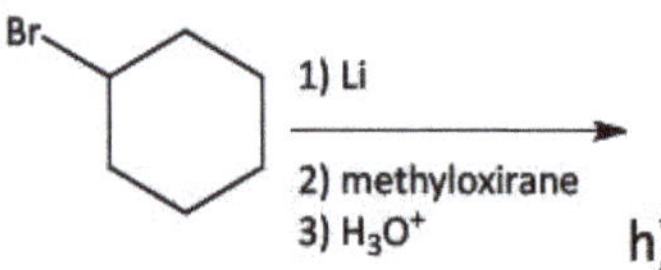

h) 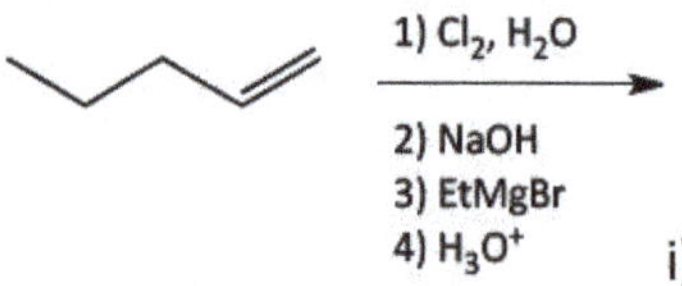

i) 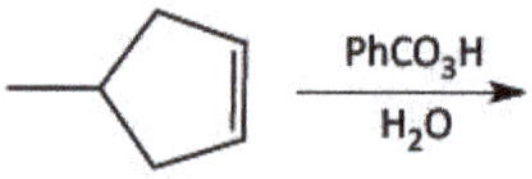

j) 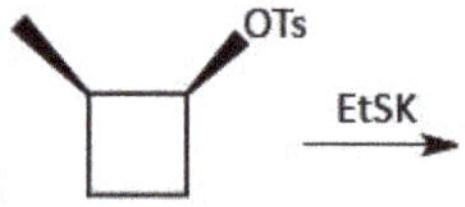

k) 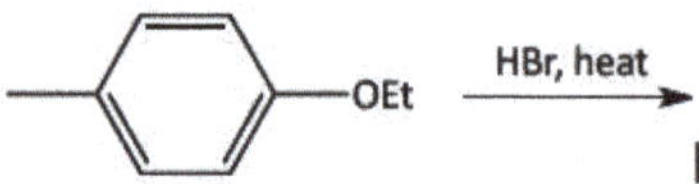

l) 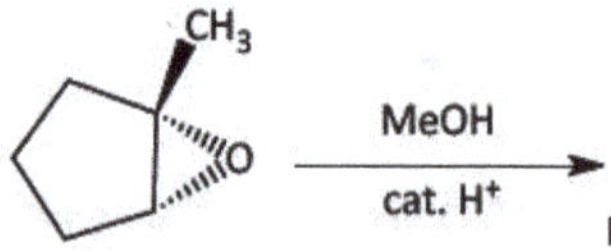

m) 2 CH_3CH_2OH + H_2SO_4/130°C n) THF + excess HBr

o) 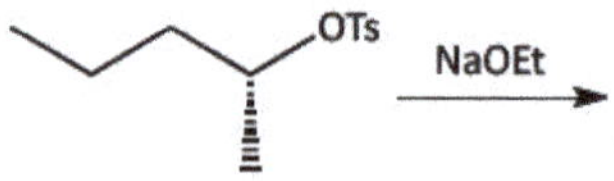

p) 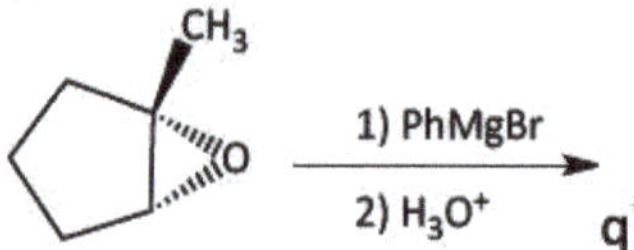

q) 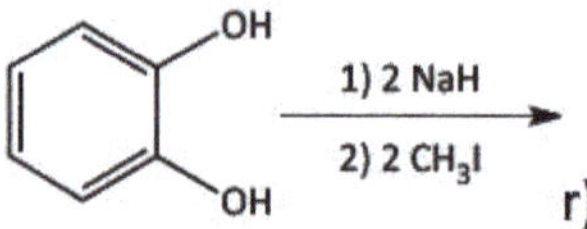

r) 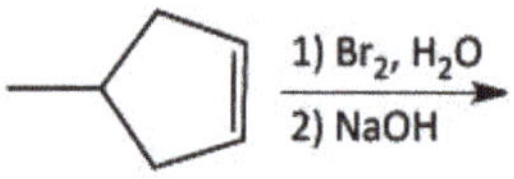

s) 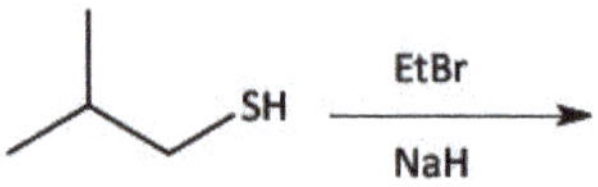

t) 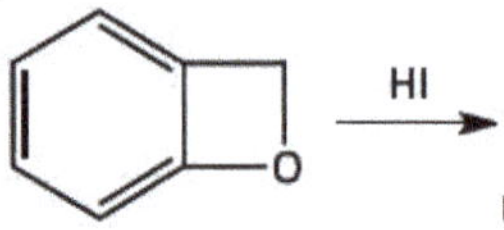

u) 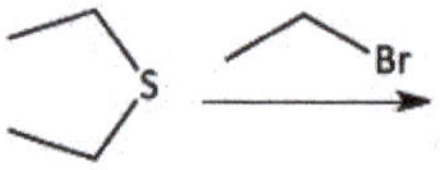

v) 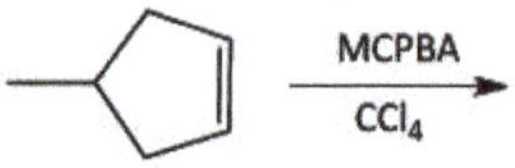

15.18 Draw the missing reactants for the following reactions.

a) 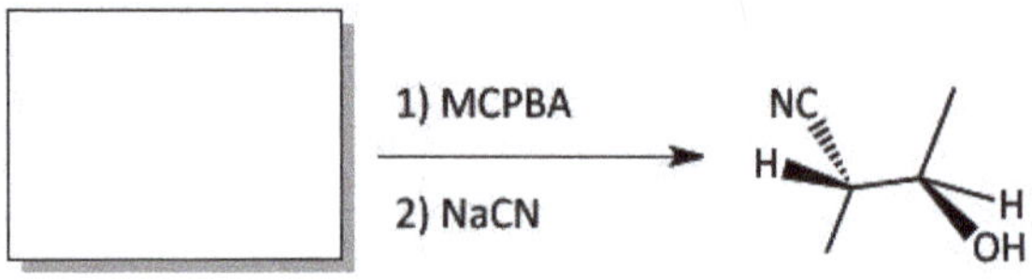

b) 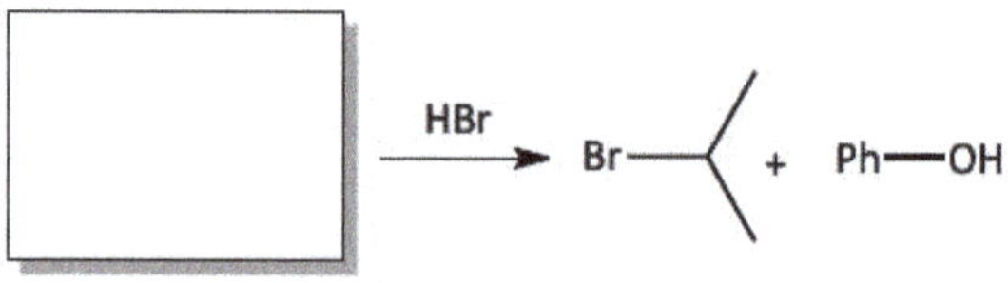

c) 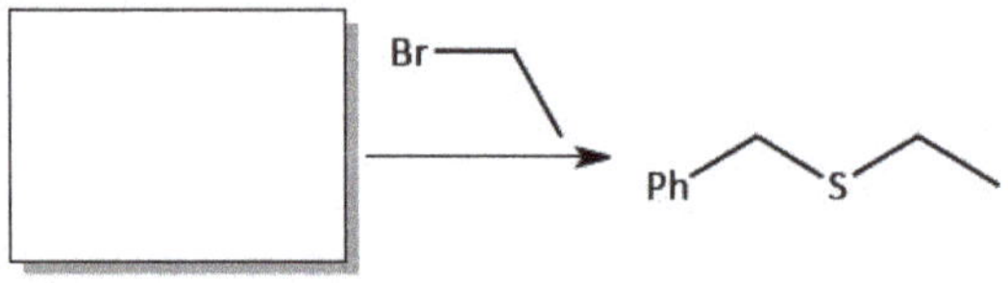

d) 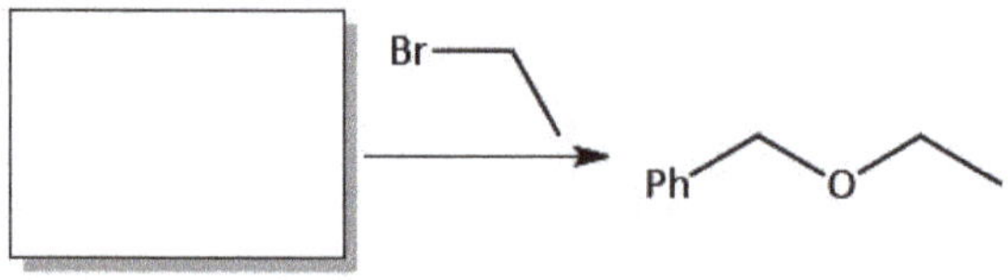

e) 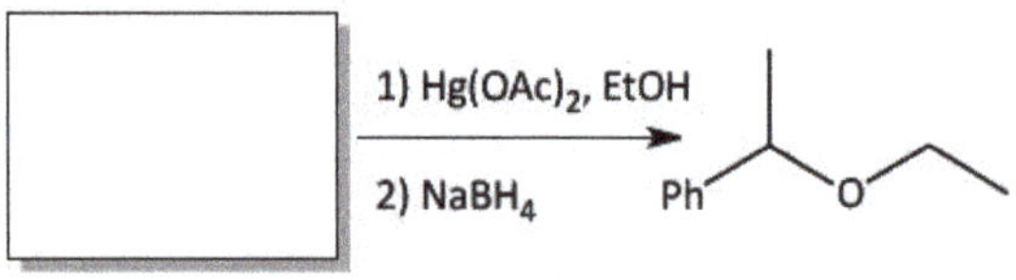

f) 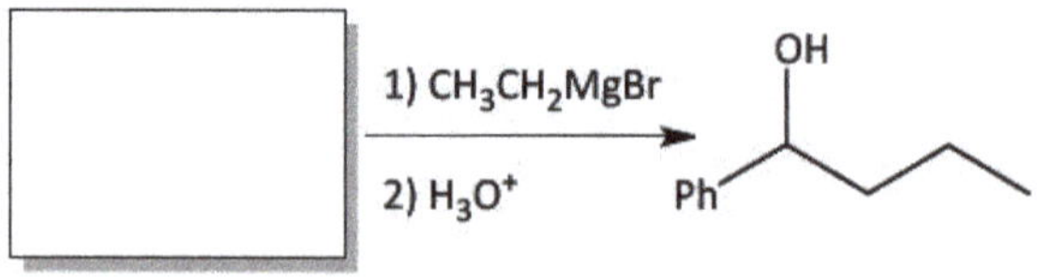

15.19 Show the mechanism for the following reaction.

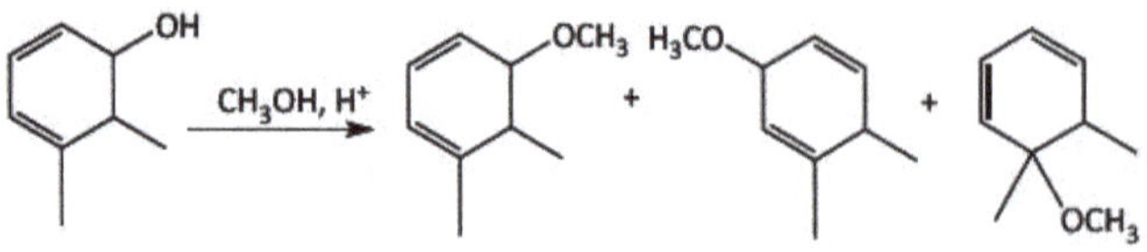

15.20 **Show the mechanism for the following reaction.**

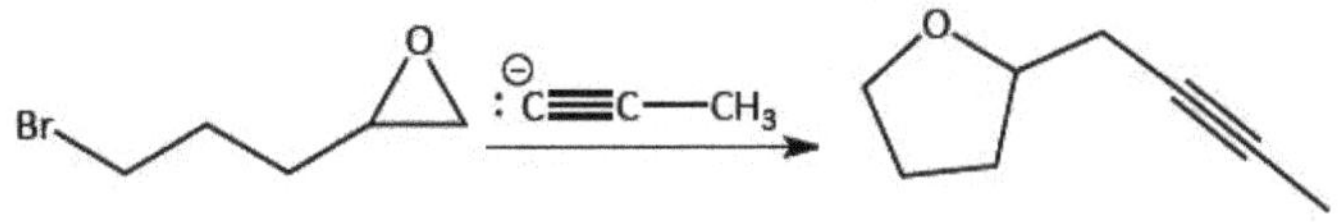

15.21 **Show the mechanism for the following reaction.**

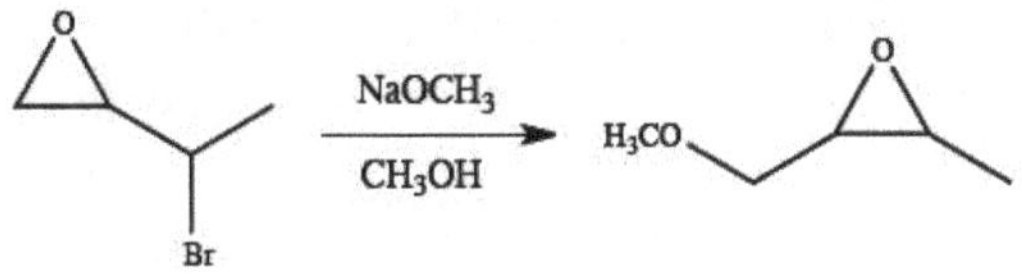

15.22 **Identify A-J in the following reaction sequence.**

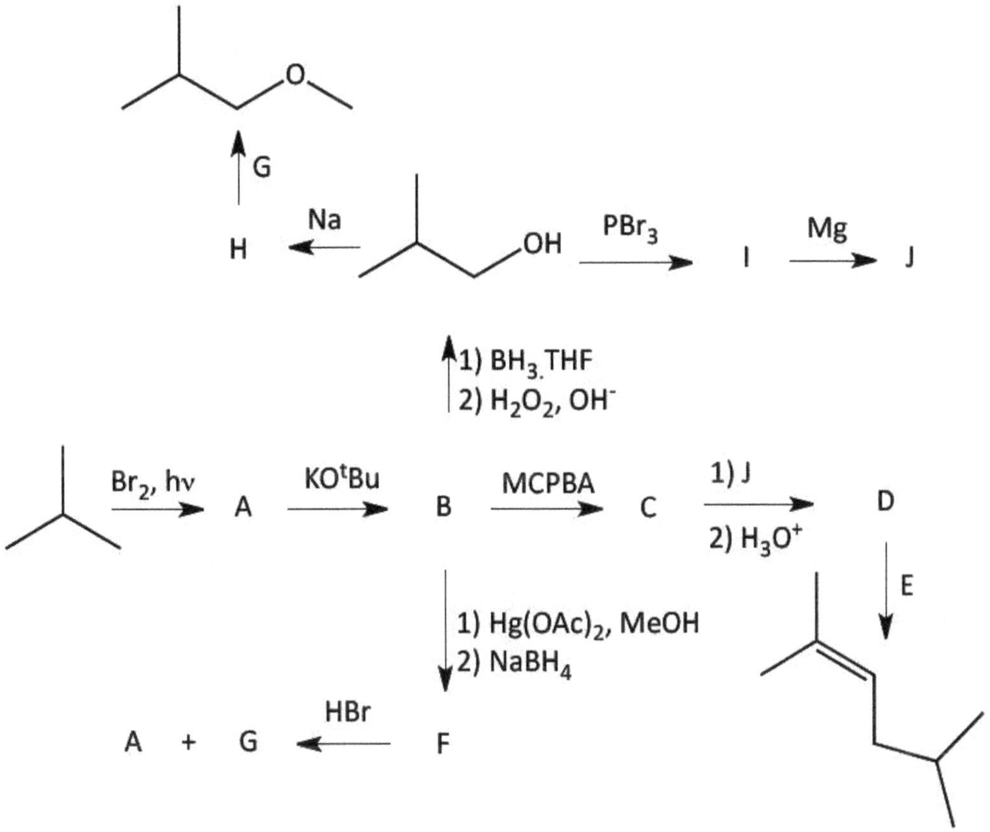

Chapter 16. Carbonyl Compounds

16.1 Introduction

The following are common carbonyl compounds. A ketone has two carbon groups attached to the carbonyl while an aldehyde has a carbon group and a hydrogen atom attached to the carbonyl group. A carboxylic acid has an OH group attached to the carbonyl. Esters, amides, and acid chlorides are carboxylic acid derivatives. As we will see later in this chapter, they can all be made from carboxylic acids. Esters have alkoxy groups (–OR) attached to the carbonyl, amides have nitrogen atoms attached to the carbonyl, and acid chlorides have chlorine atoms attached to the carbonyl.

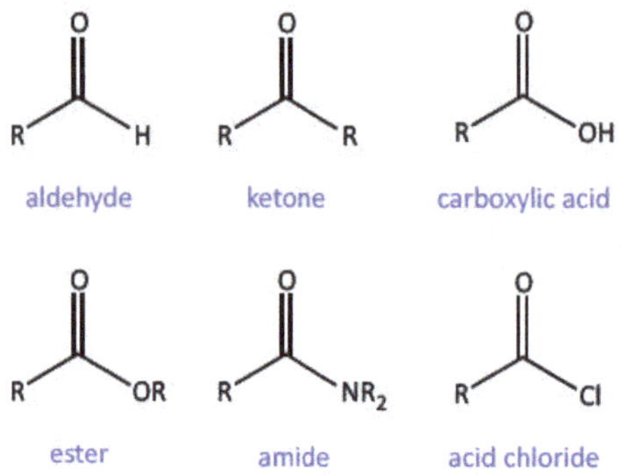

16.2 Spectroscopy

16.2.1 Mass Spectrometry

In mass spectrometry, carbonyl compounds like ketones and aldehydes often cleave at the carbonyl carbon atom. This cleavage produces a carbocation on the carbonyl carbon atom. This positive charge can be stabilized through resonance using the lone pair of the carbonyl oxygen atom.

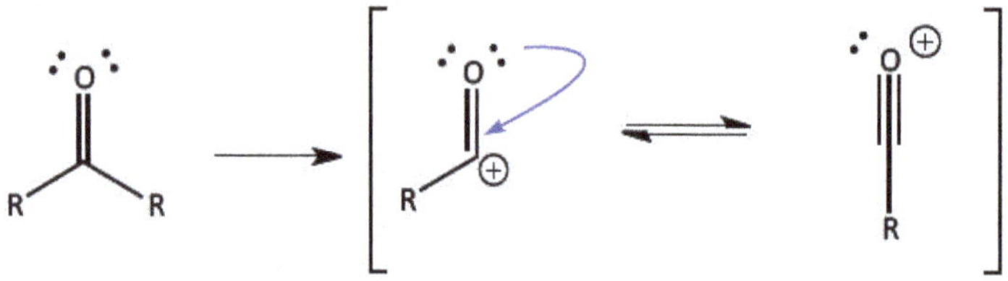

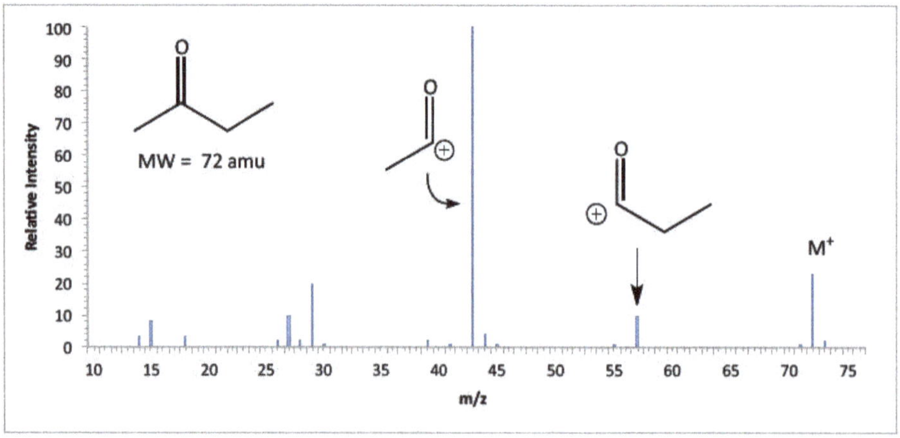

16.2.2 Infrared Spectroscopy

Carbonyl compounds

The C=O, carbonyl, stretch is very easy to identify. It is usually the strongest peak in an IR spectrum. It is usually around 1700 but can range from the 1660s up to about 1800. When we see the carbonyl, we also look for secondary peaks that help us identify the functional group. Also, the frequency of the C=O stretch gives us information on the type of functional group it is. Be flexible with these frequencies. The below ranges are typical, but vary slightly.

C=O Frequency	Functional Group	Example	Secondary peaks to look for. . .
1730s	Ester	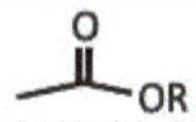	1050, 1250
1730s	Ring strained ketone	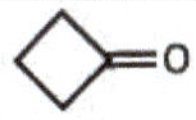	N/A
1710-20s	Aldehyde	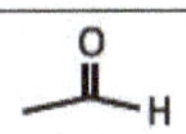	2710, 2810
1710s	Ketone	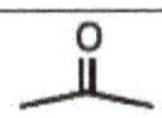	N/A
1710s	Carboxylic Acid	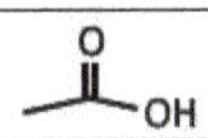	Broad OH at 3000
1680s	Conjugated C=O	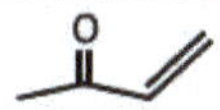	sp^2 CH at >3000 C=C at 1650
1660s	Amide	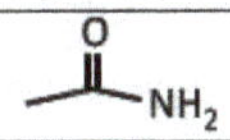	NH peaks at 3300 if present

Figure 16.1 IR carbonyl frequencies

For pentanal, a carbonyl C=O peak appears at 1725. We also see two peaks near 2700 and 2800 to help confirm aldehyde.

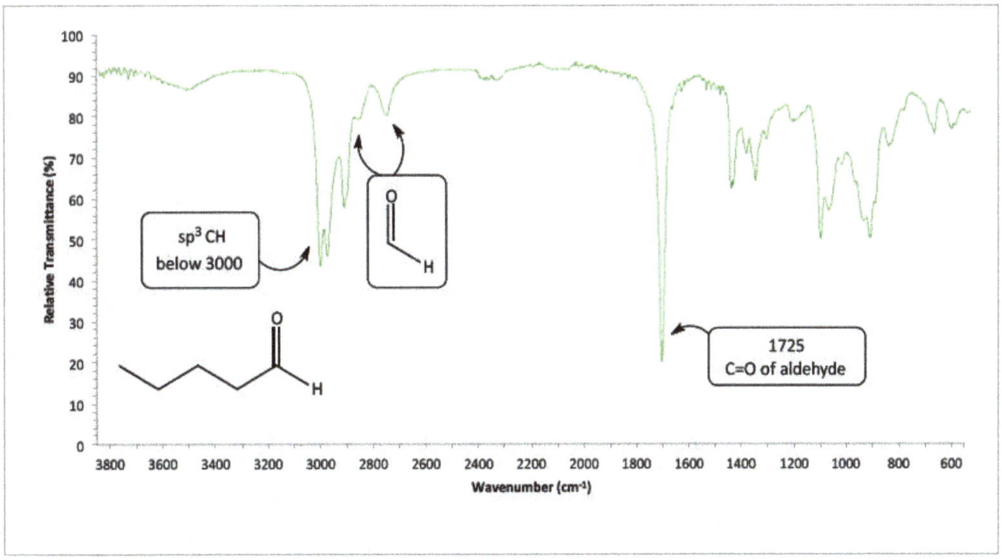

Figure 16.2 IR spectrum of pentanal

In the IR spectrum of butanoic acid, like that for most carboxylic acids, there is a carbonyl (C=O) peak at a normal 1710 cm^{-1} and a large, broad OH peak centered around 3000 cm^{-1} that slowly tapers off.

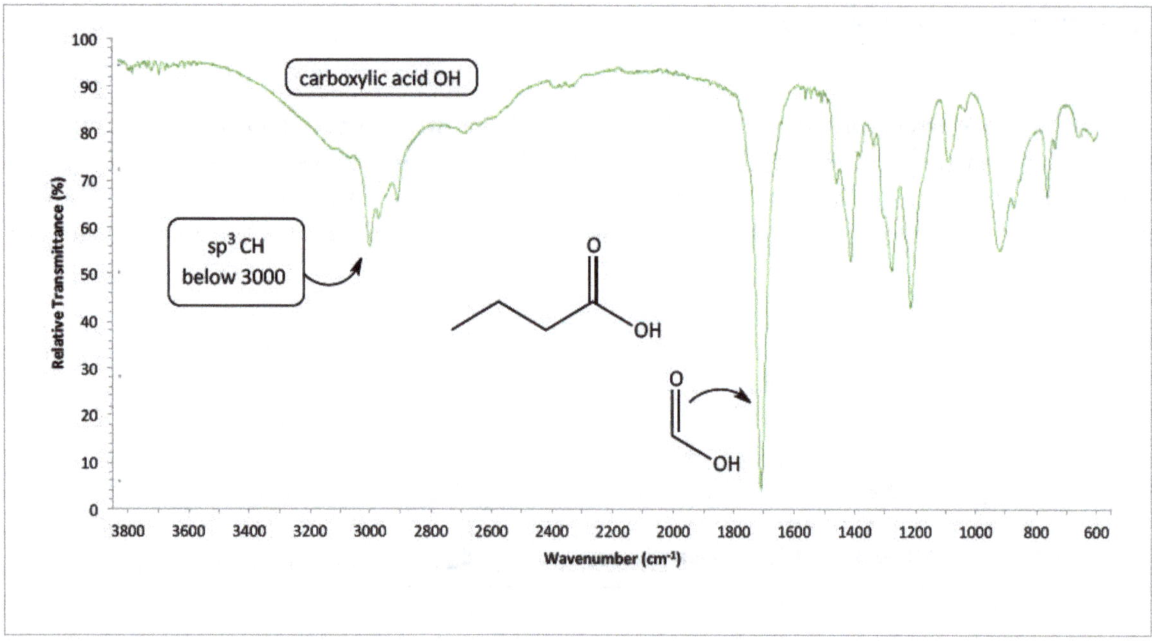

Figure 16.3 IR spectrum of butanoic acid

In acetone, like other ketones, there is a regular C=O at 1710. There are no secondary peaks that suggest carboxylic acid or aldehyde, so we conclude it is a ketone.

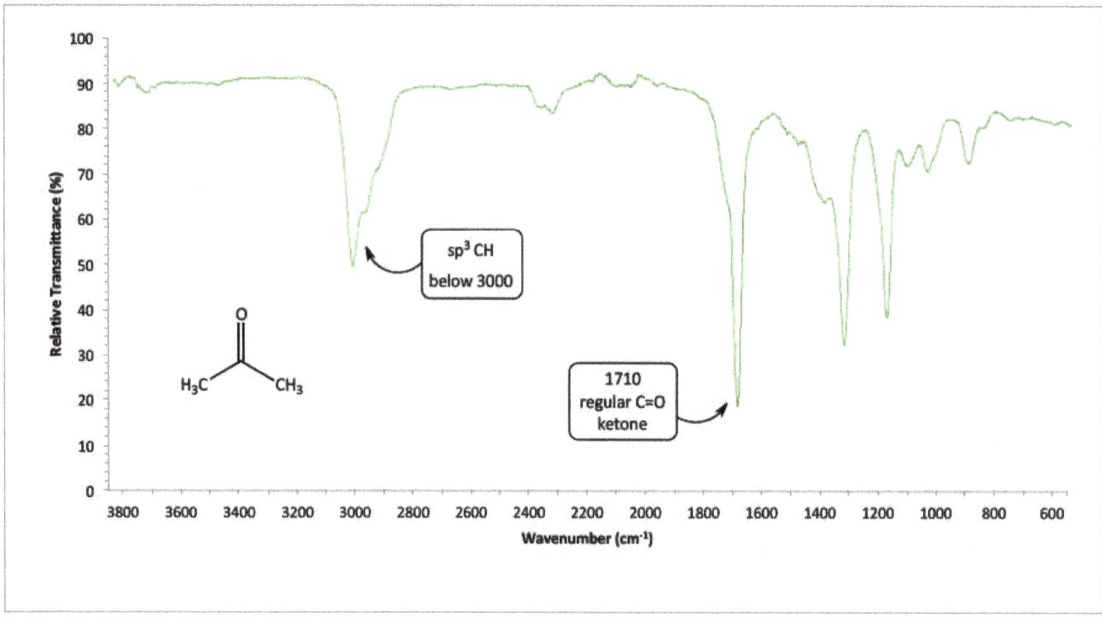

Figure 16.4 IR spectrum of acetone

For cyclopentanone, we see the sp^3 CH peak below 3000. There is also a high carbonyl peak which could indicate an ester or maybe a ring strained ketone. We venture into the fingerprint region to look for secondary peaks for ester. There are not the two strong peaks around 1250 or 1050, so we lean towards a small ring that is a ketone.

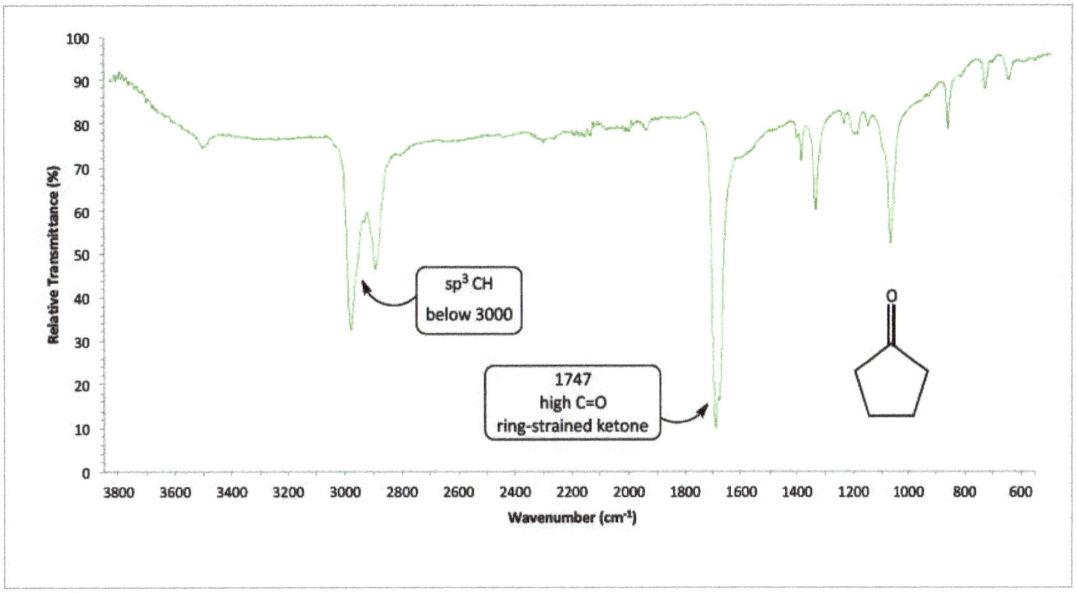

Figure 16.5 IR spectrum of cyclopentanone

In acetophenone, we see sp^2 and sp^3 CH bonds. Since we have sp^2 CH, we look for what type of C=C is present. We notice the aromatic overtones below 2000. We also see a carbonyl peak, but it is a little low in the 1680s. This tells us that our aromatic ring is next to the carbonyl so we can get a conjugated C=O.

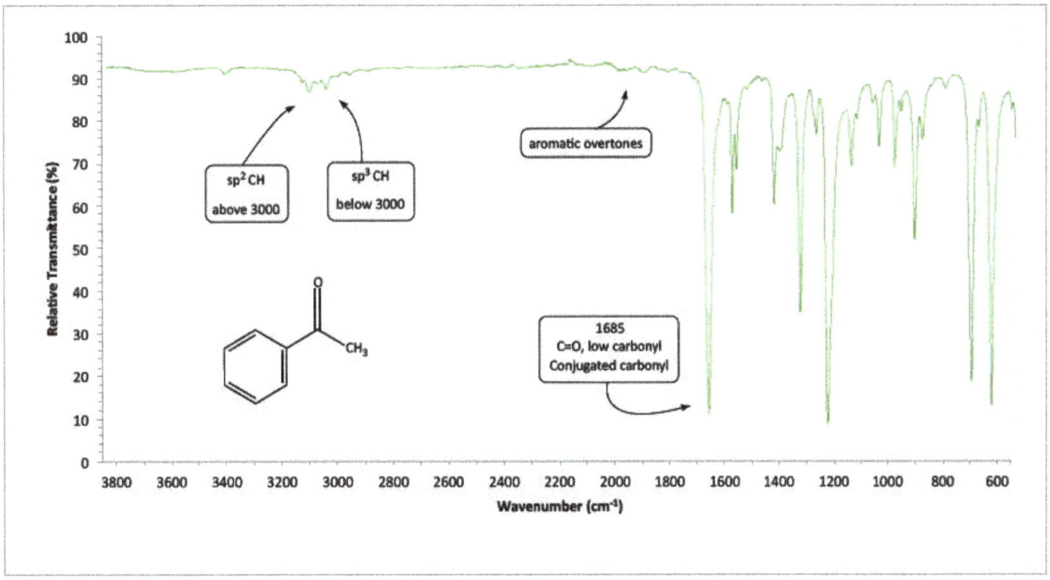

Figure 16.6 IR spectrum of acetophenone

For the ester ethyl butyrate, notice the carbonyl stretch is at a higher frequency than normal. It is in the 1730's. Also, when you have an ester, you can usually see a strong C-O stretch in the fingerprint region. You also sometimes see a strong O-C=O wag in the fingerprint region for esters.

Carbonyl Compounds

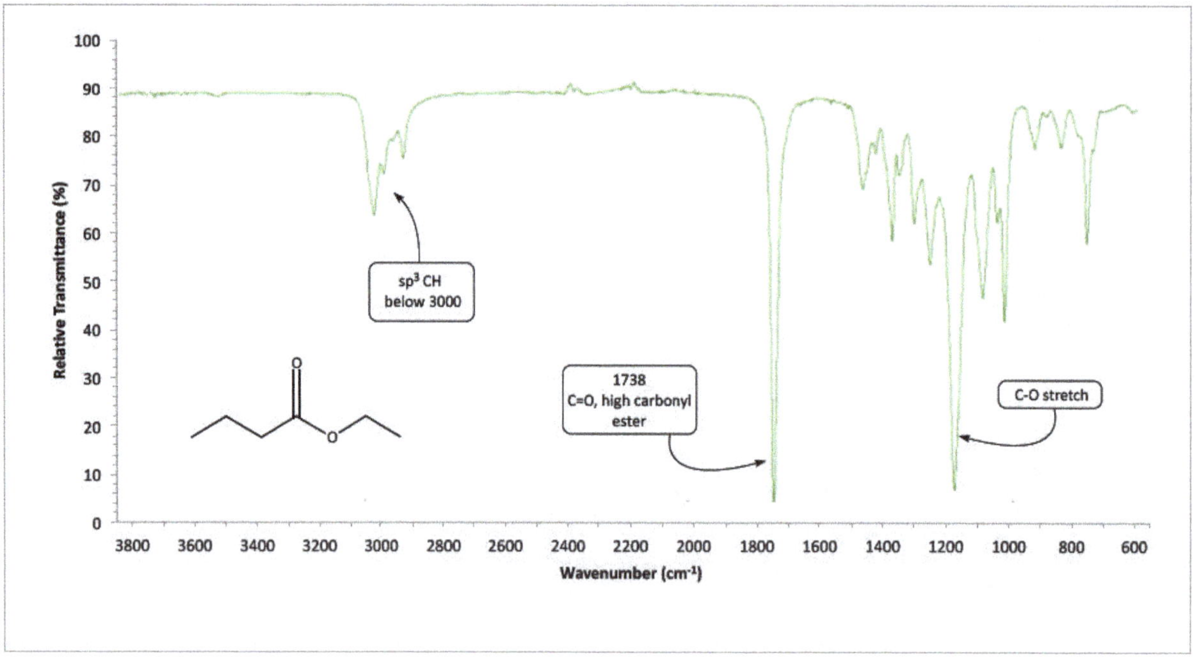

Figure 16.7 IR spectrum of ethyl butyrate

For acetamide, we see that the carbonyl peak is quite low in the 1660s indicating an amide. We look for secondary peaks at 3300 and find two N-H bonds. We also see sp_3 CH peaks below 3000. It is fairly small since there are only three C-H bonds.

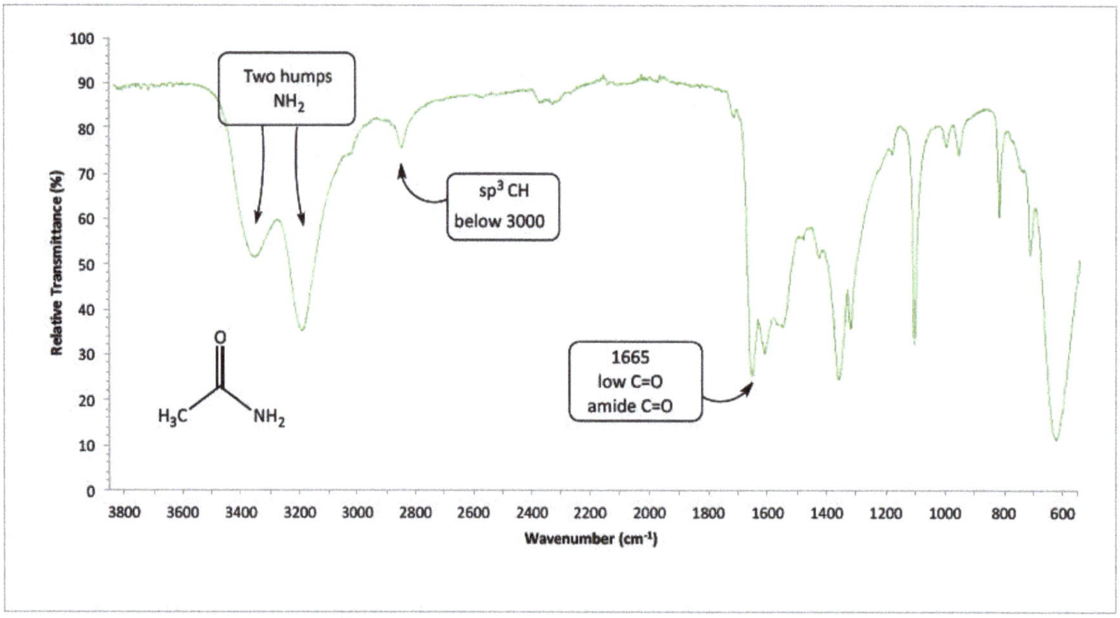

Figure 16.8 IR spectrum of acetamide

16.2.3 ^{1}H and ^{13}C NMR

Carbonyl carbon atoms appear between 150 and 200 ppm in a ^{13}C NMR spectrum. A proton on an alpha carbon atom of a ketone or aldehyde appears around 2-2.5 ppm in a ^{1}H NMR spectrum. The peak for an aldehyde proton appears between 9-10 ppm, and the peak for a carboxylic acid proton appears between 10-12 ppm.

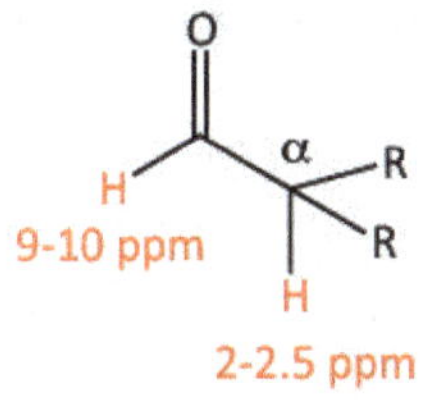

aldehyde

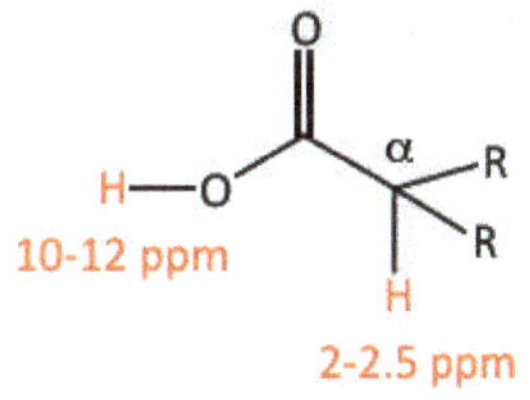

carboxylic acid

16.3 Nomenclature

16.3.1 Ketones

IUPAC

Ketone are named by using the suffix *–one*. The longest alkane chain is identified. The *–e* suffix is dropped from the alkane name and *–one* is added.

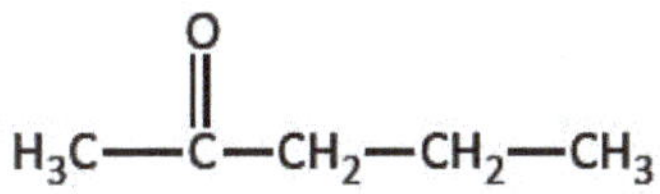

2-pentanone
or
pentan-2-one

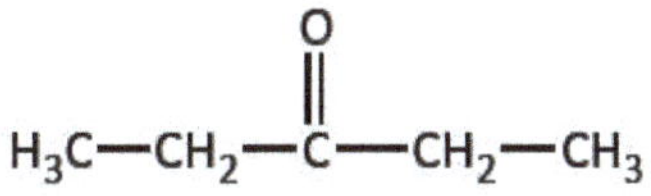

3-pentanone
or
pentan-3-one

Common

The common way to name ketones is to name the two alkyl groups of the ketone in alphabetical order followed by the suffix *–ketone*. Methyl ethyl ketone (MEK) is an exception to the alphabetical rule but it is still commonly used for historical reasons. A ketone that is a very common solvent is 2-propanone. The name 2-propanone might be correct for an IUPAC name, but it is never used. This solvent is universally called *acetone*. It is used as a fingernail polish remover as well as many other things.

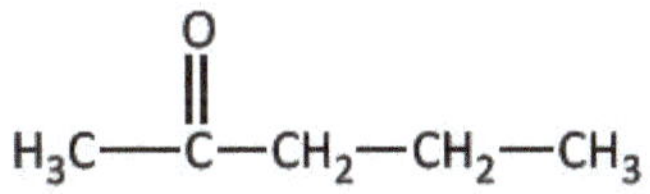

methyl propyl ketone

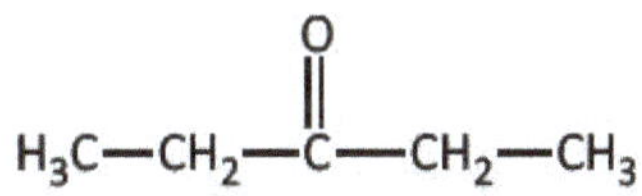

diethyl ketone

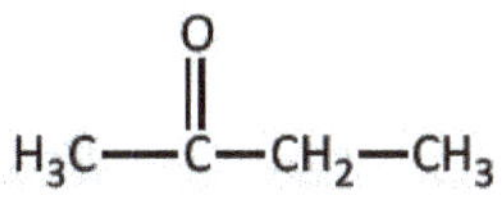

ethyl methyl ketone
or
methyl ethyl ketone (MEK)

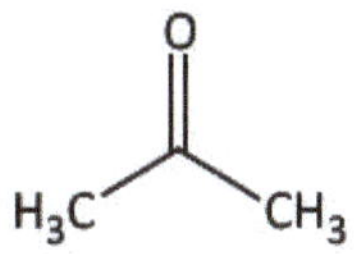

acetone

16.3.2 Aldehydes/Carboxylic acids

IUPAC

Aldehydes are named by using the suffix –al. The longest alkane chain is identified. The *–e* suffix is dropped from the alkane name and *–al* is added. Numbering for aldehydes begins at the carbonyl. The carbonyl carbon atom is

Carbonyl Compounds

#1. Carboxylic acids are named much the same way, except instead of having the *–al* suffix, they have the *–ic acid* suffix or *–oic acid* suffix.

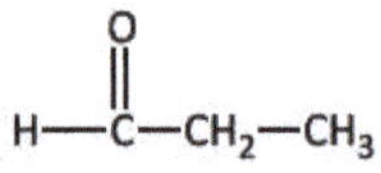

propanal

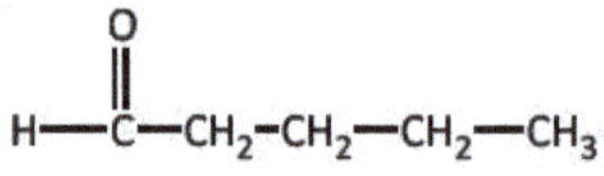

pentanal

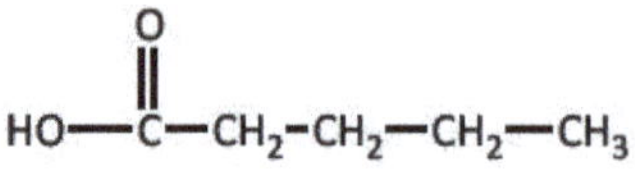

pentanoic acid

Common

Aldehydes are sometimes named using the common name for the parent carbon chain followed by the suffix *-aldehyde*. Carboxylic acids use the common parent name followed by *–ic acid*. The common name for the parent carbon chain is used in the common names for both aldehydes as well as carboxylic acids.

No. of C atoms	**Common parent name**	Common adehyde name IUPAC aldehyde name	Common carb. acid name IUPAC carb. acid name
1	form	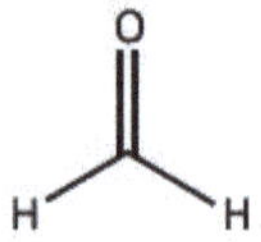 formaldehyde methanal	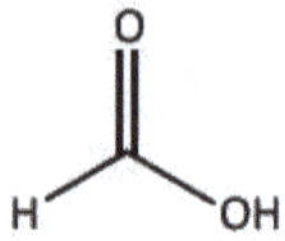 formic acid methanoic acid
2	acet	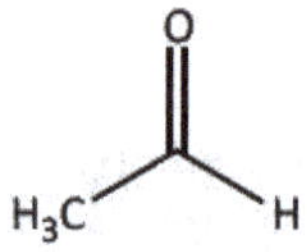 acetaldehyde ethanal	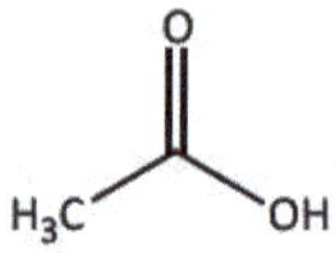 acetic acid ethanoic acid
3	propion	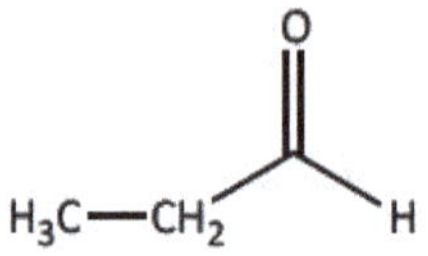 propionaldehyde propanal	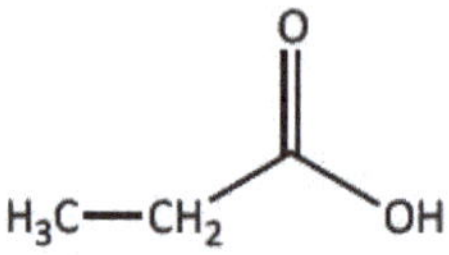 propionic acid propanoic acid
4	butyr	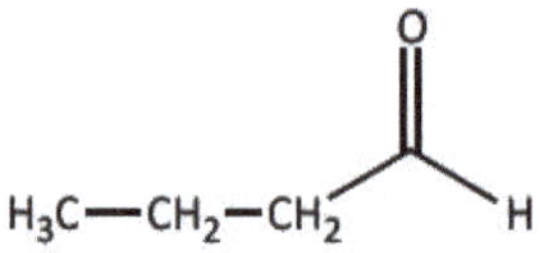 butyraldehyde butanal	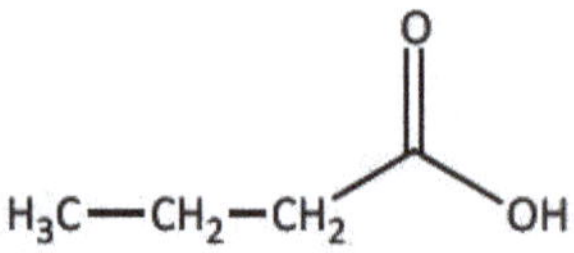 butyric acid butanoic acid

Figure 16.9 Common names for carboxylic acids and aldehydes

16.3.3 Amides

Amides are named by looking at the carboxylic acid name and replacing the *–ic acid* or *–oic acid* suffix with *-amide*.

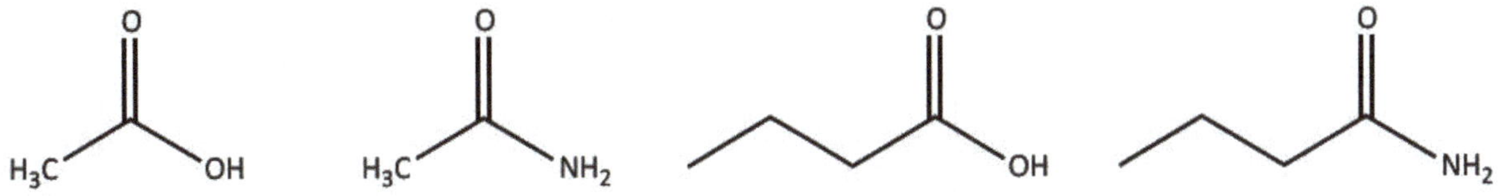

IUPAC	ethanoic acid	ethanamide	butanoic acid	butanamide
common	acetic acid	acetamide	butyric acid	butyramide

If there are substituents on the nitrogen atom of the amide, we can put a capital "N" before the substituent to indicate the substituent is attached to the nitrogen atom.

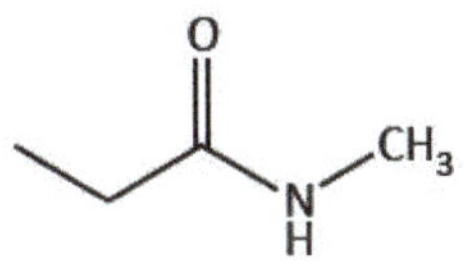

N-methylpropanamide

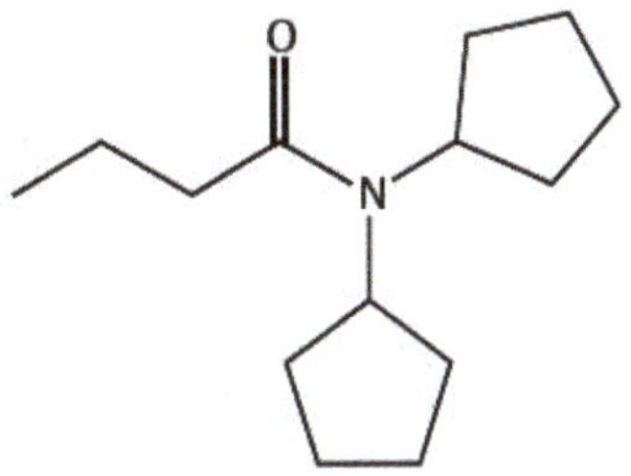

N,N-dicyclopentylbutyramide

Cyclic amides are called lactams. Penicillin is a famous lactam.

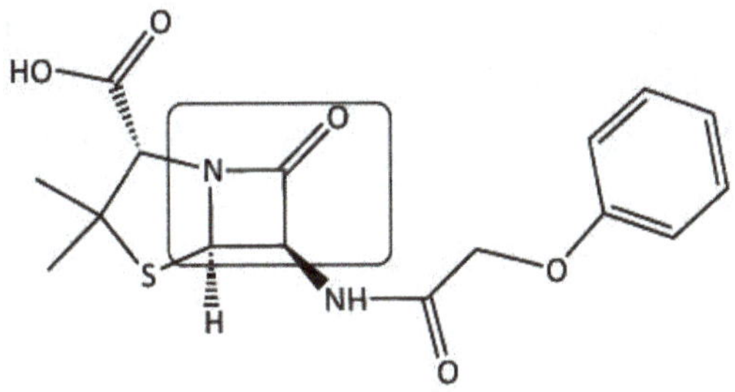

penicillin

16.3.4 Esters

We saw that one-way esters can be made is from the reaction of an alcohol with a carboxylic acid under acid catalyzed conditions (a Fischer esterification).

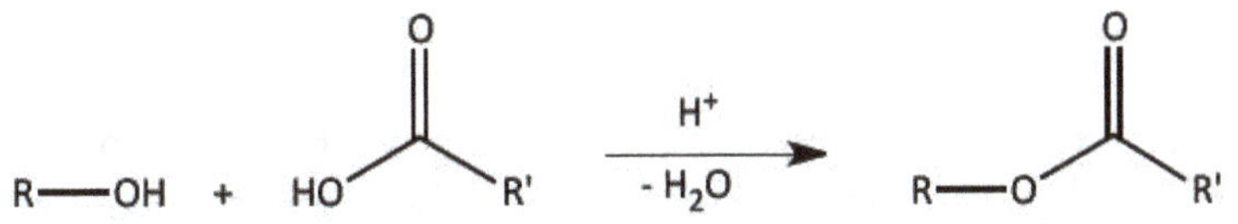

alcohol carboxylic acid ester

To name esters, we first name the alkyl group from the alcohol part of the ester first. We follow that with the carboxylate part of the ester (from the carboxylic acid) followed by the suffix *–ate*. Remember, name the alcohol part first and the carboxylic acid part last.

Carbonyl Compounds

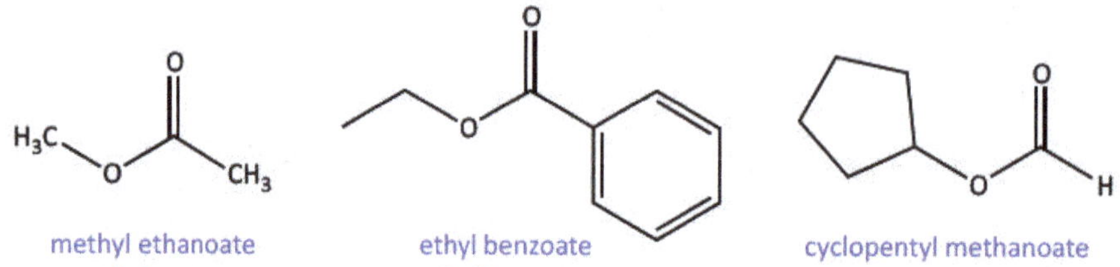

IUPAC	methyl ethanoate	ethyl benzoate	cyclopentyl methanoate
common	methyl acetate	ethyl benzoate	cyclopentyl formate

Cyclic esters are called lactones.

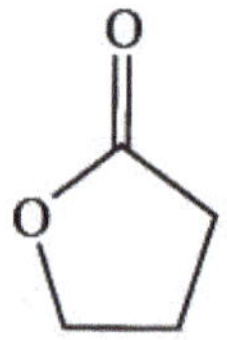

γ-butyrolactone

16.4 Synthesis of Ketones and Aldehydes - Review

16.4.1 Oxidation of alcohols

To review, in organic chemistry, an oxidation is essentially making a compound with more carbon-oxygen bonds. The oxidation from an alcohol to a carbonyl is accompanied by the elimination of two hydrogen atoms. An alcohol, with one carbon oxygen bond, can be oxidized to a carbonyl with two carbon-oxygen bonds. If the carbonyl compound is an aldehyde, with two carbon-oxygen bonds, it can be further oxidized to a carboxylic acid, with three carbon-oxygen bonds if a strong oxidizing agent is used. If a weak oxidizing agent is used, it can be oxidized to the aldehyde and stop there.

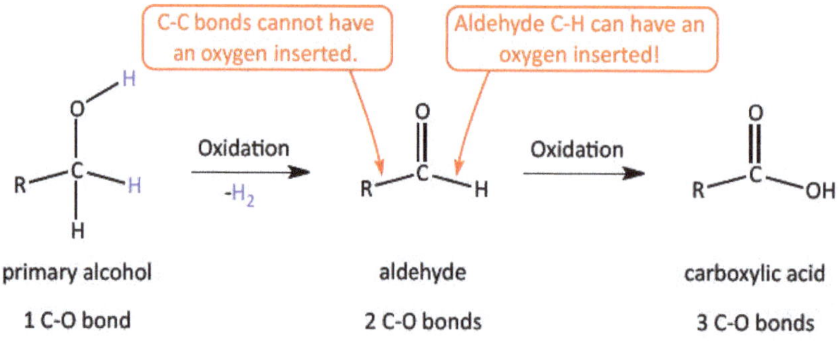

Figure 16.10 Oxidation pathway of a primary alcohol

A secondary alcohol can be oxidized to a ketone. Notice the singe carbon-oxygen bond of the alcohol loses two hydrogen atoms and is oxidized to the ketone carbonyl with two carbon-oxygen bonds. There is no need to memorize that secondary alcohols give ketones and primary alcohols give aldehydes, simply remove the two hydrogen atoms from the alcohol and see what you get. Since ketones only have carbon atoms attached to the carbonyl carbon, no further oxidation can occur. Only aldehydes with a hydrogen atom attached to the carbonyl can be further oxidized.

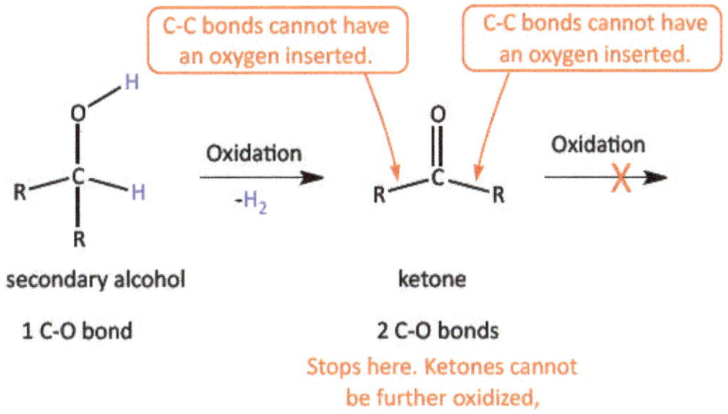

Figure 16.11 Oxidation pathway of a secondary alcohol

A tertiary alcohol cannot be oxidized. Notice that there are no hydrogen atoms on the carbinol carbon atom. Therefore, two hydrogen atoms cannot be removed from the alcohol, so it cannot be oxidized.

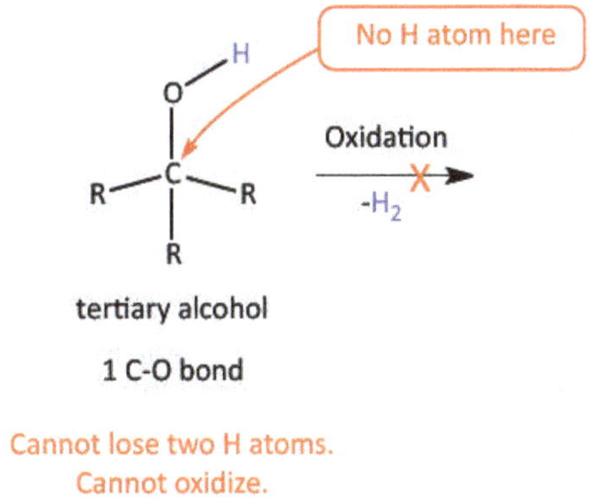

Figure 16.12 Tertiary alcohols do not oxidize

16.4.2 Oxidizing agents

There are many different oxidizing agents. One strong oxidizing agent is chromic acid. Chromic acid is made from sodium dichromate ($Na_2Cr_2O_7$) and sulfuric acid (H_2SO_4).

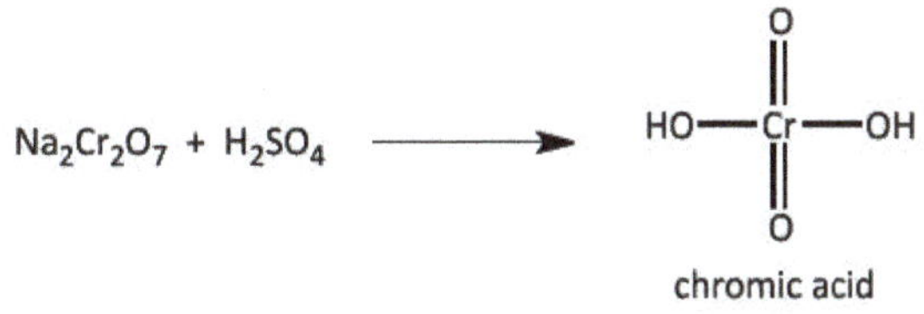

Chromic acid is such a strong oxidizing agent, that it fully oxidizes primary alcohol all the way to carboxylic acids. It proceeds from primary alcohol to aldehyde to carboxylic acid. Chromic acid oxidizes secondary alcohols to ketones.

Carbonyl Compounds

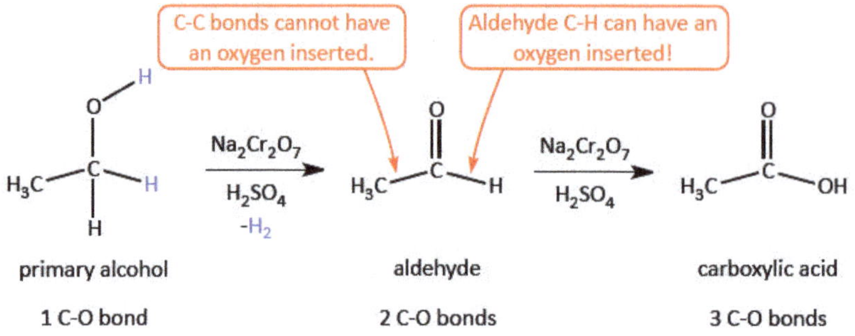

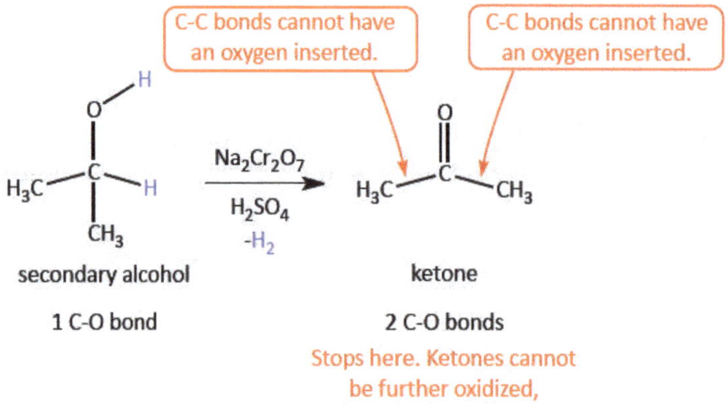

But, what if we want to oxidize a primary alcohol to an aldehyde and stop there without going further oxidizing to a carboxylic acid? We need a weaker oxidizing agent. There are several ways to do this, but one is to use a weaker oxidizing agent like PCC, pyridinium chlorochromate. PCC is a complex made from chromium trioxide, pyridine, and hydrochloric acid.

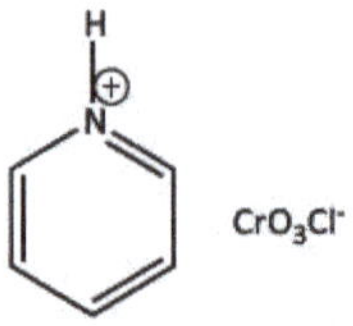

Figure 16.13 PCC

The weaker oxidizer PCC is strong enough to do the first oxidation of a primary alcohol to aldehyde, but it is too weak to completely oxidize to a carboxylic acid. The oxidation stops at the aldehyde. This is a great way to make aldehydes.

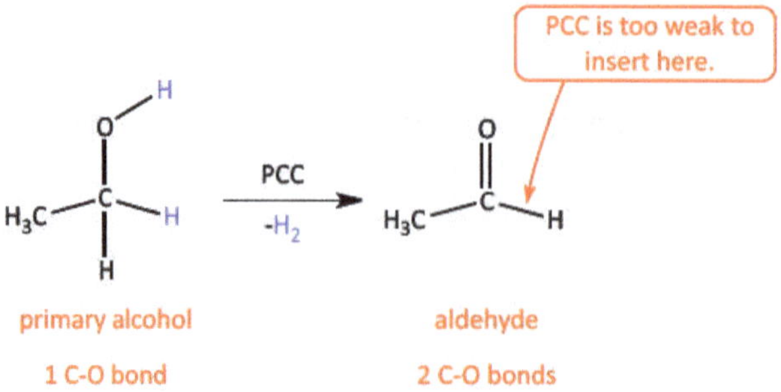

Since PCC is strong enough to do the first oxidation, it is strong enough to oxidize secondary alcohols to ketones as well.

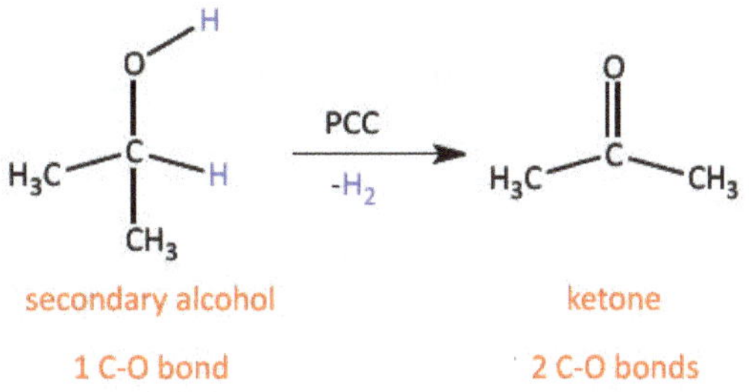

Chromium oxidizers should be handled with care. These chromium salts are hazardous to your health. They should never go down the sink drain in the lab and should be collected as hazardous waste. There are greener oxidizers.

The Swern oxidation or DMP are weak oxidizers like PPC and stop the oxidation of primary alcohols at aldehydes. Instead of chromic acid, bleach can be used as a strong oxidizer.

In fact, aldehydes can be turned into ketones using a little trick. If a Grignard reagent attacks an aldehyde, a secondary alcohol is formed. Upon oxidation, the alcohol turns into a ketone.

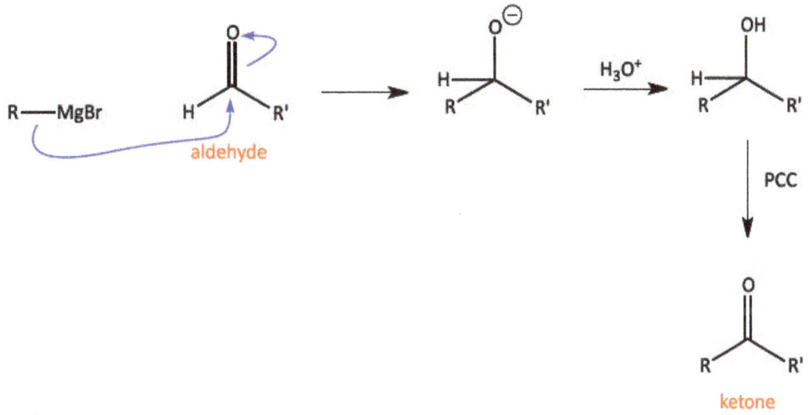

16.1 Fill in the products of the following reactions.

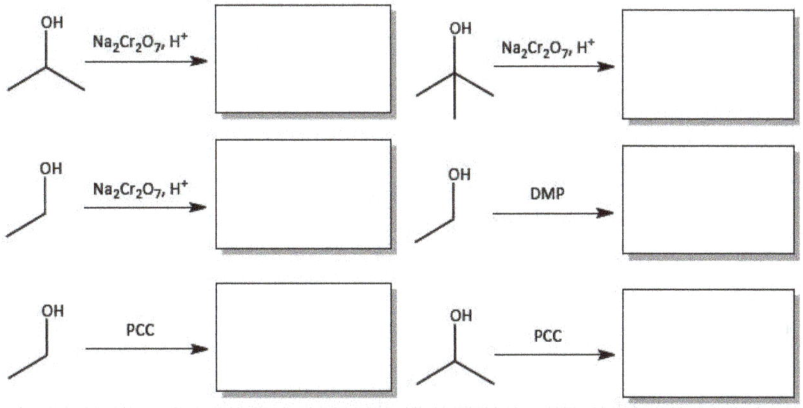

16.4.3 Reduction of carboxylic acid

The opposite of an oxidation is a reduction. The highly oxidized carboxylic acid functional group, with three a carbon-oxygen bonds, can be reduced to an aldehyde.

Carbonyl Compounds

With a strong reducing agent to alcohol

A strong reducing agent, like $LiAlH_4$, will reduce a carboxylic acid all the way down to an alcohol.

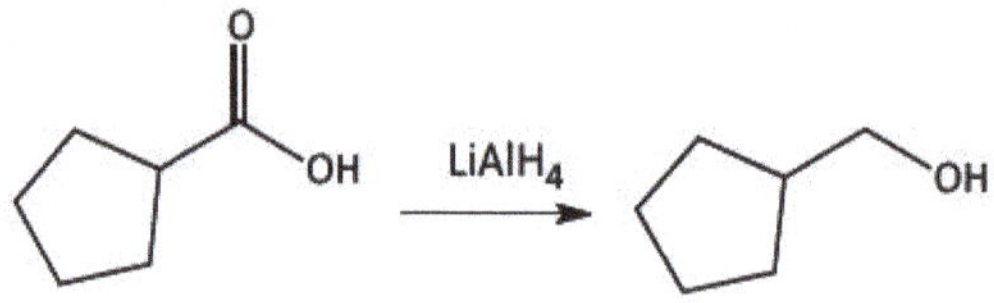

carboxylic acid   alcohol

Stopping at aldehyde

If we want to stop at the aldehyde, we need something a little less aggressive. Usually, $NaBH_4$, sodium borohydride, comes to mind as a weaker, gentler reducing agent. But, unfortunately, $NaBH_4$ is too weak to react with carboxylic acids. In fact, we don't yet have a good way to turn a carboxylic acid into an aldehyde in one step. We must use two steps. There are two common ways to do this.

1. The carboxylic acid can be turned into an acid chloride followed by reduction with tris-tert-butoxylithium aluminum hydride, $LiAlH(OtBu)_3$.

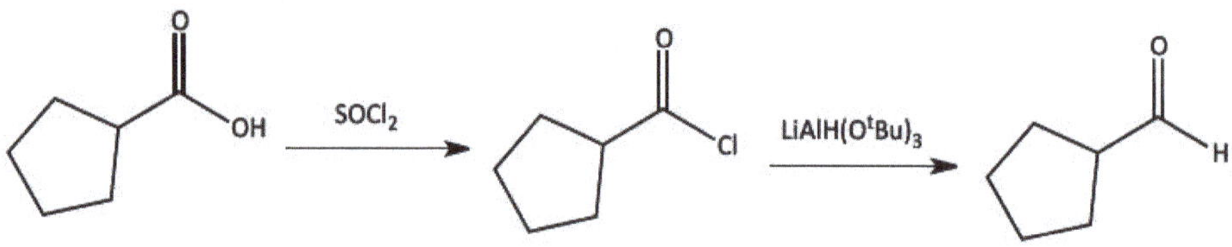

carboxylic acid   acid chloride   aldehyde

2. A carboxylic acid can be turned into an ester followed by reducing the ester with DIBAL-H to make an aldehyde.

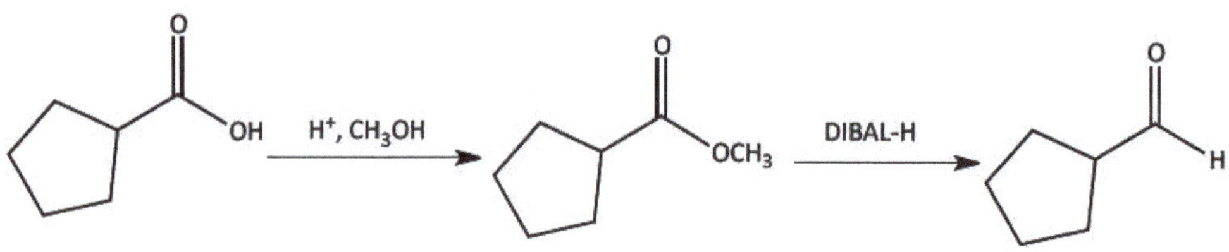

carboxylic acid   ester   aldehyde

16.2 Fill in the blank boxes.

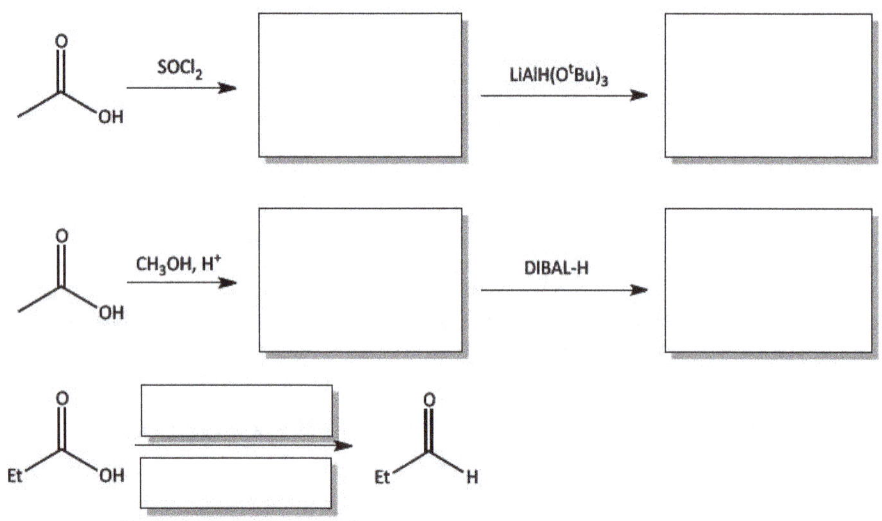

Use the acid chloride method

16.4.4 Markovnikov Hydration of alkyne

Alkynes, with their two pi-bonds can also react in an oxymercuration-demercuration reaction. The reaction is very similar to that for an alkene with a twist at the end. Instead of mercuric acetate, the oxymercuration of alkynes usually involved mercuric sulfate and aqueous sulfuric acid. Like with alkenes, this results in the addition of water across a pi-bond resulting in a Markovnikov alcohol. Only one pi-bond of the triple bond reacted, so alkene remains. This vinyl alcohol is called an enol (pronounced "een-awl"). The "enol" name comes from this compound being an alkene (*ene*) and an alcohol (*ol*).

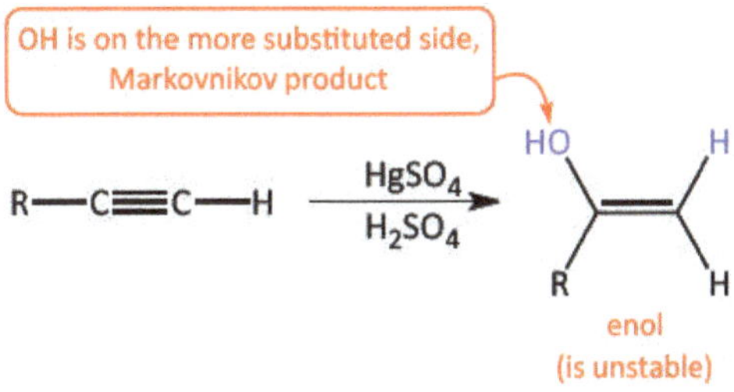

Enols tend to be unstable. Enols are in equilibrium with their isomer, the ketone. These isomers differ only in the location of one proton and a double bond. Such isomers are called tautomers (pronounced "taw-toe-mers"). The ketone and enol are called keto-enol tautomers. The transformation of one into another is called a keto-enol tautomerization. The process itself is called keto-enol tautomerism. In the equilibrium of the two, the keto form predominates. Take careful notice of the transformation. The hydrogen atom moves from the alcohol of the enol to the alpha carbon. The pi-bond of the alkene of the enol moves to make a carbonyl (a ketone).

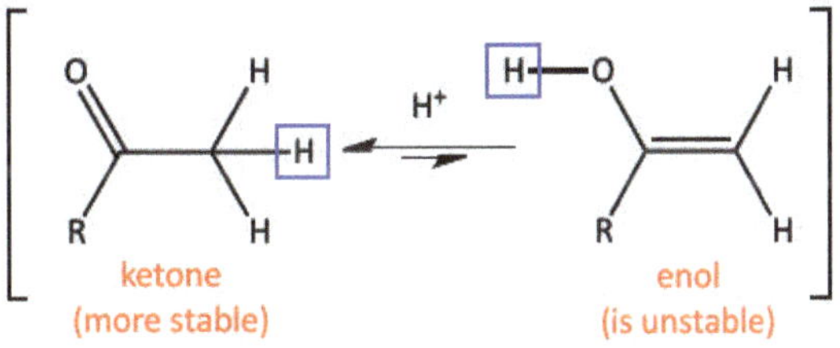

Figure 16.14 Keto-enol tautomerism

Overall Reaction:

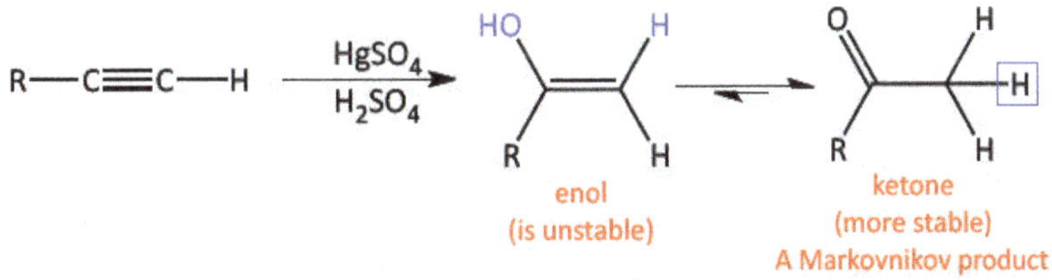

Since sulfuric acid is used in this reaction, the tautomerization is acid catalyzed. When asked to transform the enol to the ketone form, protonate the enol at the alpha-carbon, draw resonance forms, and clean it up to make a ketone.

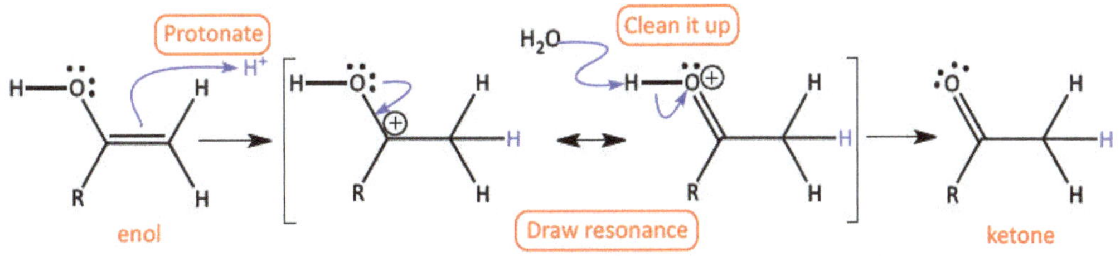

Figure 16.15 Acid catalyzed keto-enol tautomerization

16.4.5 Hydroboration of alkyne

The hydroboration of an alkyne can lead to an anti-Markovnikov enol. Since alkynes contain two pi-bonds, two molecules of borane could add to it. In order to prevent this addition of two boranes, a big, bulky, sterically hindered borane must be used. Usually, the borane used is di(*sec*-isoamyl)borane, commonly called disiamylborane. Instead of the small BH_3,disiamylborane has two big, large, deer antler-like alkyl groups are attached to it. There are five carbon atoms in each alkyl branch. Amyl is an old, common name for pentyl.

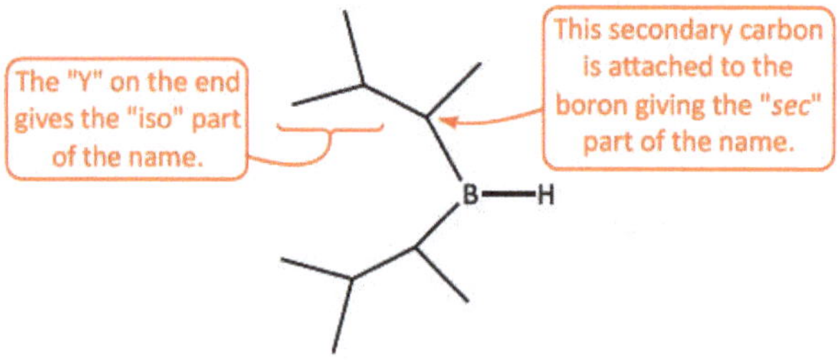

Figure 16.16 Di(sec-isoamyl)borane or disiamylborane

Disiamylborane reacts with an alkyne to make an anti-Markovnikov borane. This borane can be oxidized with peroxide and sodium hydroxide to make an anti-Markovnikov enol. This enol can tautomerize to the keto-form. In the case of a terminal alkyne, this keto-form is an aldehyde with a hydrogen attached to the carbonyl group.

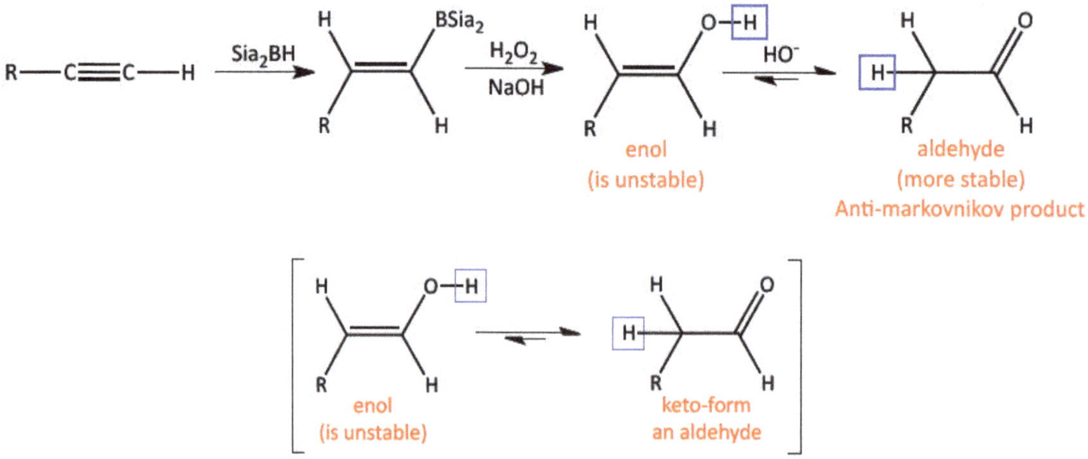

The keto-enol tautomerization in base is different than it was in acid. Bases deprotonate. So, the first step is to deprotonate the enol at the alcohol position. This is pretty easy to remember if we remember that we are turning the alcohol portion of the enol into a carbonyl group. Once deprotonation happens, we draw resonance forms to get us closer to the keto product, and then clean it up with some water to make the final aldehyde.

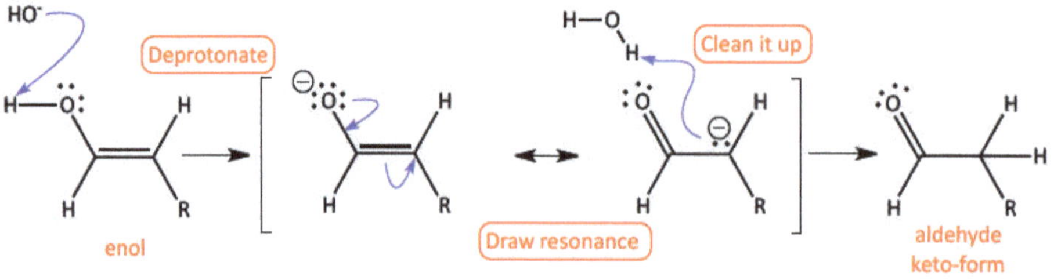

Figure 16.17 Base-catalyzed keto-enol tautomerization

16.3 Fill in the blank boxes.

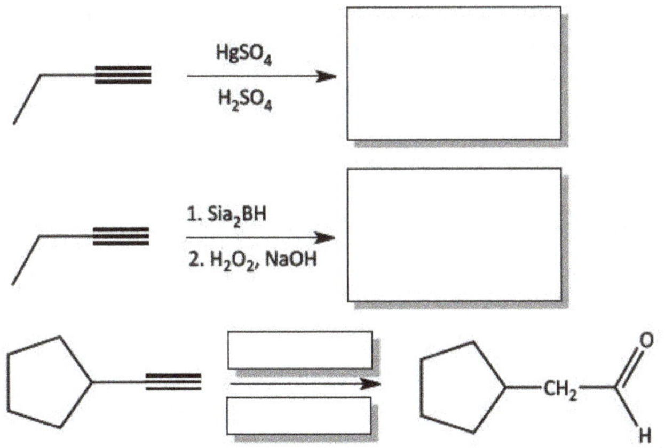

16.4.6 Ozonolysis of alkenes

Ozonolysis if very similar to oxidation by $KMnO_4$ in warm, concentrated solution. The difference is that if an aldehyde is formed, it is *not* further oxidized to a carboxylic acid.

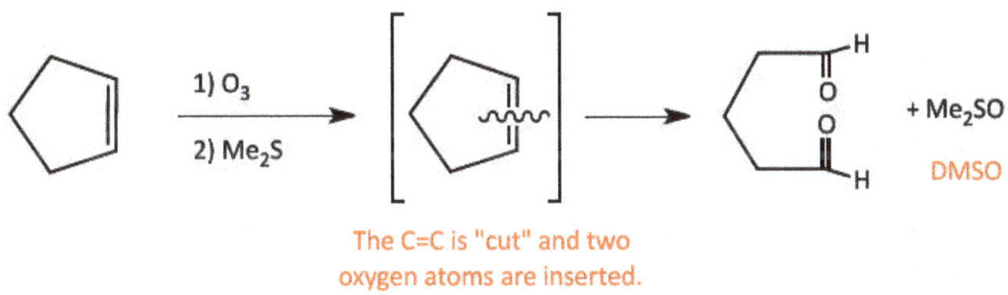

The mechanism for this reaction does not really help us learn the reaction. If you are interested in it, you can look it up elsewhere. Ozone is a compound with three oxygen atoms (O_3). Two of the three oxygen atoms are in the final product. The third oxygen atom ends up on the Me_2S to make Me_2SO.

16.4.7 Periodic Acid cleavage of glycols

The C-C bond of a vicinal diol, a glycol, can be oxidatively cleaved with periodic acid (pronounced purr-eye-oh-dik-acid), HIO_4 to make two carbonyl compounds. This is very similar to how alkenes are oxidatively cleaved in an ozonolysis reaction.

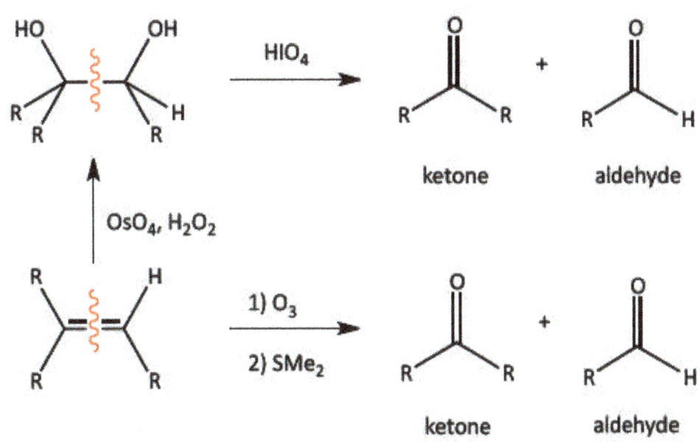

16.4.8 $KMnO_4$ oxidation of alkynes

The same reaction can happen with the two pi-bonds of an alkyne. Since there are two pi-bonds, the dihydroxylation can happen twice, resulting in a species with four hydroxyl groups on it. This is unstable. Two water molecules are eliminated resulting in an α-diketone. Under neutral KMn04 conditions, the α-diketone is the product.

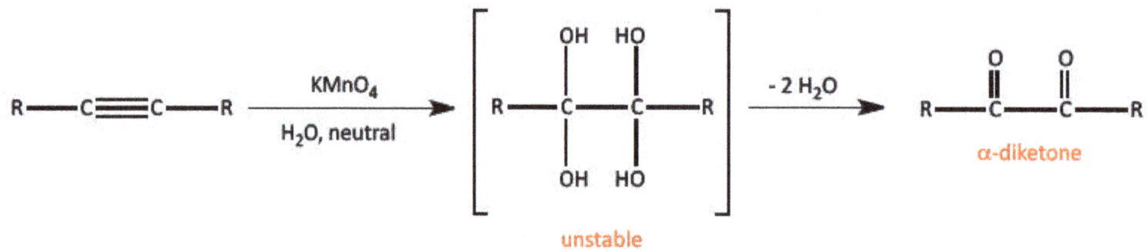

The double dihydroxylation of the alkyne makes sense. Look closely how the two molecules of water are lost leaving only two oxygen atoms behind on neighboring carbon atoms.

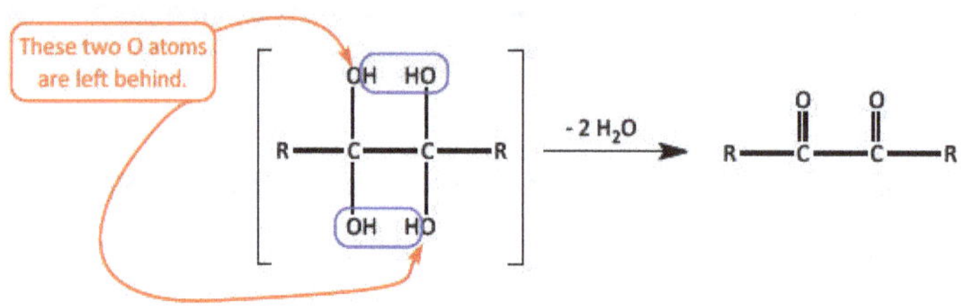

16.5 Synthesis of Carboxylic Acids- Review

16.5.1 Ozonolysis of alkynes

The reaction of an alkyne with ozone results in the same products as the full oxidation with $KMnO_4$ under basic, hot conditions. The carbon-carbon triple bond is cleaved and each carbon atom is fully oxidized resulting in two carboxylic acids.

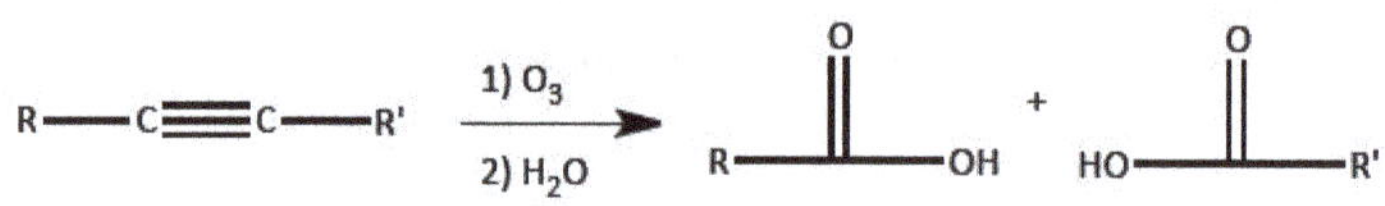

16.5.2 Further oxidation of alkenes with $KMnO_4$

If the alkene reaction with potassium permanganate is not conducted under dilute and cold conditions, but is instead run under ***warm and concentrated*** conditions, further oxidation can occur. When the glycol if further oxidized, the carbon-carbon sigma bond of the glycol is broken and the single carbon-oxygen bonds of the glycol turn into carbon-oxygen double bonds (an oxidation). $KMnO_4$ is a pretty strong oxidizer. If an aldehyde is first made by this oxidation, it is further oxidized to a carboxylic acid.

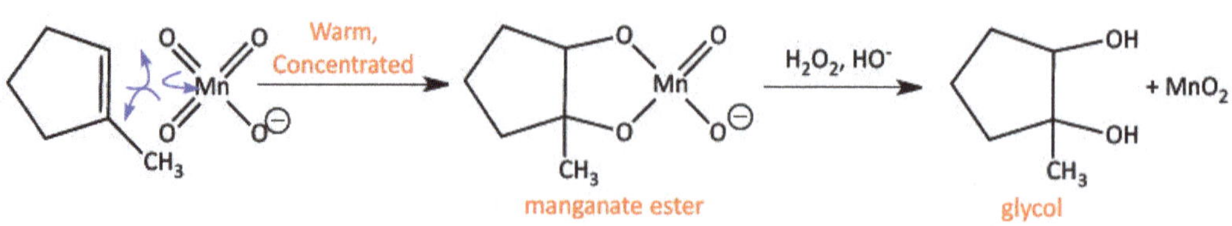

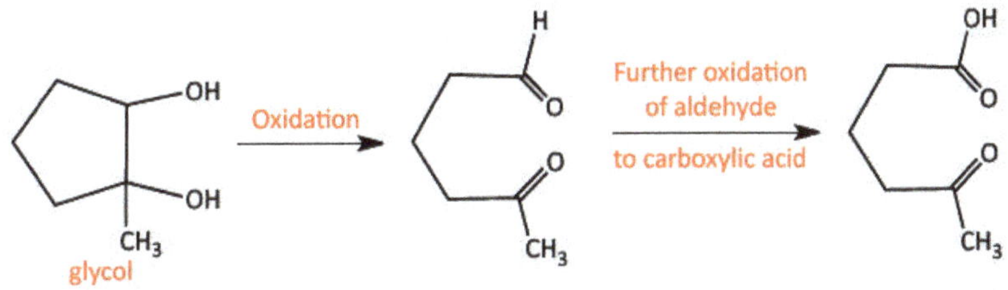

Overall, can be thought of as chopping the alkene in half and inserting two oxygen atoms. If the result is a ketone, nothing more happens to that carbon atom. If the result is an aldehyde, further oxidation to a carboxylic acid occurs.

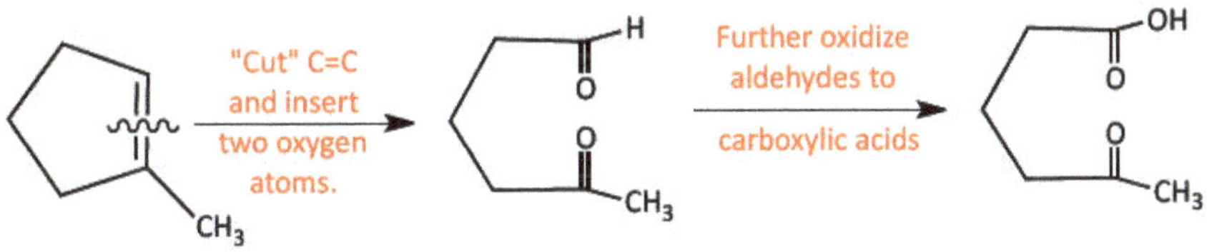

16.5.3 Further oxidation of alkynes with $KMnO_4$

When KMnO4 reacts with an alkyne under more aggressive conditions, in the presence of hydroxide base and heat, the α-diketone is produced like under less aggressive neutral conditions, but it is then further, fully oxidized. The C-C bond between the two carbonyl groups is cleaved followed by further oxidation resulting in two carboxylic acids. The three C-O bonds of a carboxylic acid is the most it can be oxidized. The three carbon-carbon bonds of the alkyne become three carbon-oxygen bonds of the carboxylic acid.

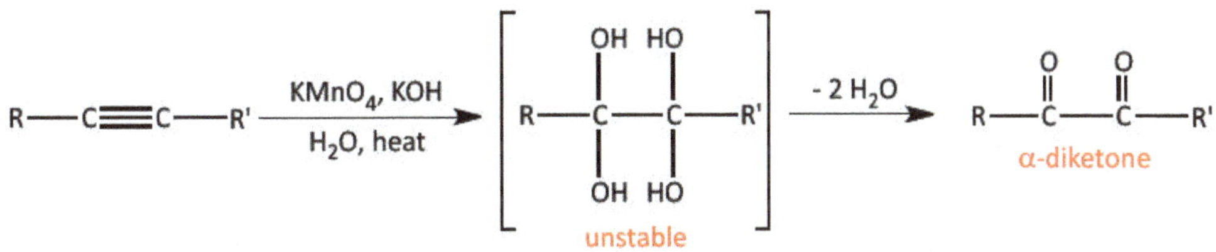

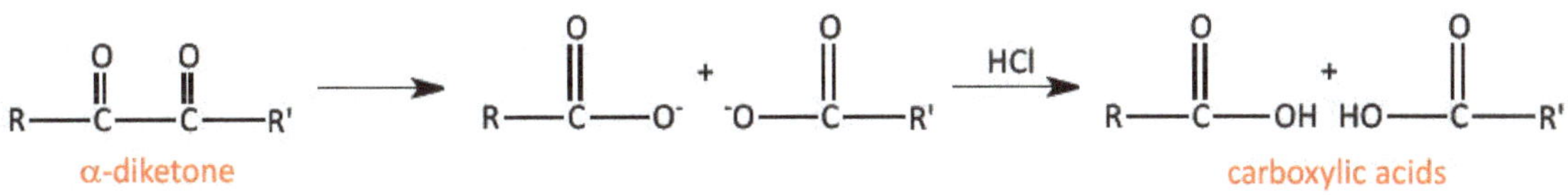

Overall, the cleavage can be thought of as. . .

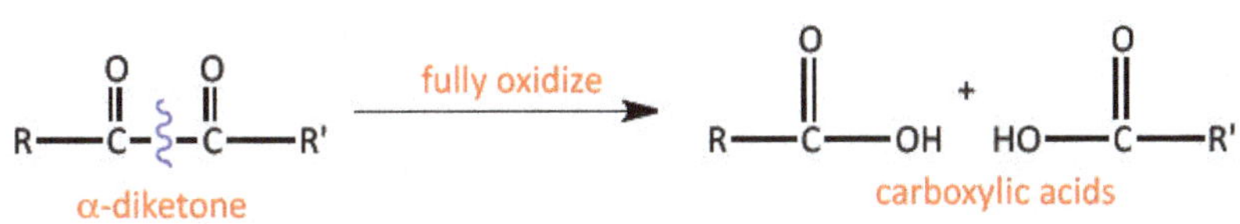

or

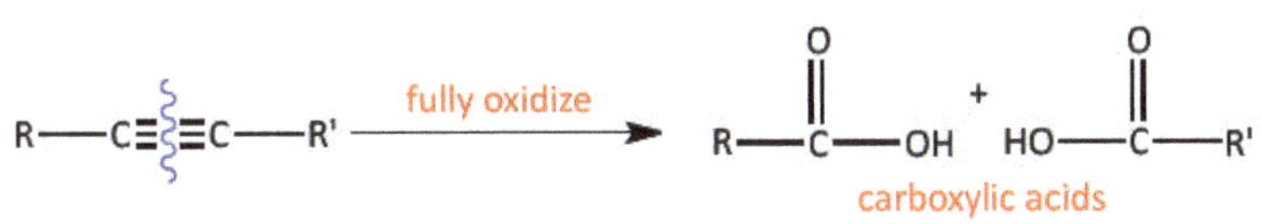

16.4 Fill in the blank boxes.

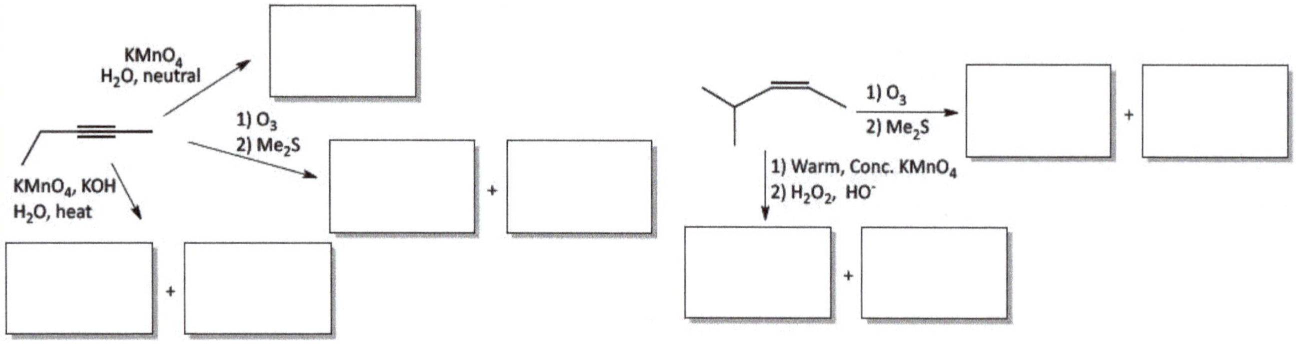

16.6 New Syntheses of Ketones and Aldehydes

16.6.1 Ketones and Aldehydes from 1,3-Dithiane

Ketones and aldehydes can be synthesized in a nice, controlled fashion from 1,3-dithiane.

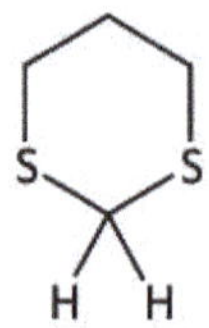

Figure 16.18 1,3-dithiane

The two protons shown on 1,3-dithiane are acidic. The dithiane can be easily deprotonated using butyl lithium (BuLi) as the base. Once deprotonated, the negatively charged carbon atom of the dithiane can act as the nucleophile in an S_N2 reaction to alkylate the dithiane. This attaches one alkyl group to the dithiane making a monoalkylated dithiane (a thioacetal). If mercuric chloride ($HgCl_2$) is added in acidic water, this alkylated dithiane would become an aldehyde. Alternatively, this monoalkylated dithiane can be deprotonated once again, followed by another S_N2 reaction. This puts two alkyl groups on the dithiane. If this dialkylated dithiane is reacted with $HgCl_2$, a ketone is produced.

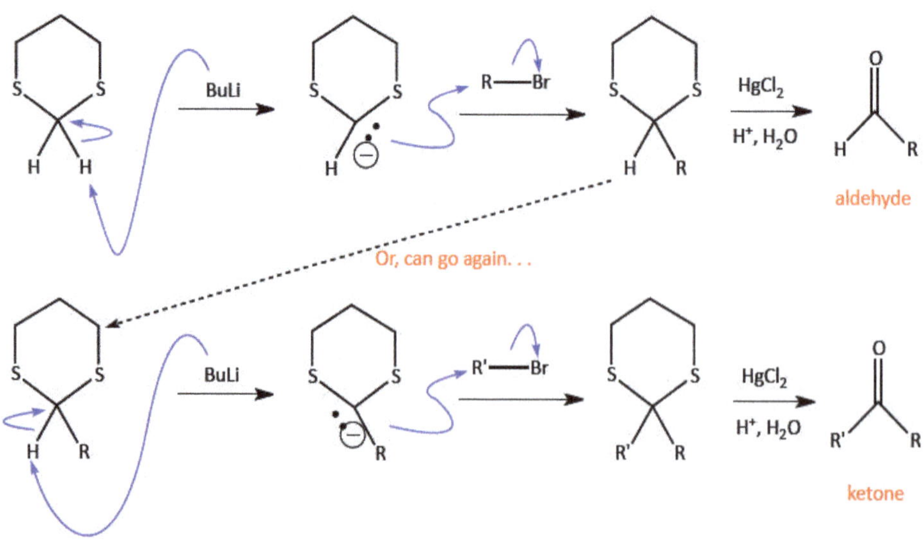

16.5 Fill in the blank boxes.

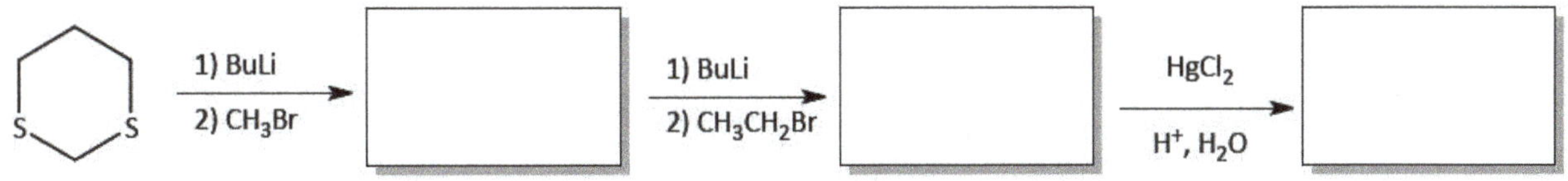

16.6.2 Ketones from Carboxylic Acids

Ketones can be synthesized from carboxylic acids. To do this, the proton of the carboxylic proton must first be removed using a base. This can be done by using lithium hydroxide (LiOH) as the base. Any alkyl lithium compound could also work. This deprotonation forms a carboxylate ion. A nucleophilic, negatively charged alkyl group can attack the carbonyl of this carboxylate. The electrons go up onto the oxygen atom. This forms a species with two negatively charged oxygen atoms. Following protonation using acid and water, a geminal diol is formed. Geminal diols like this are unstable, and easily dehydrate forming a carbonyl. In this case, the dehydration forms a ketone.

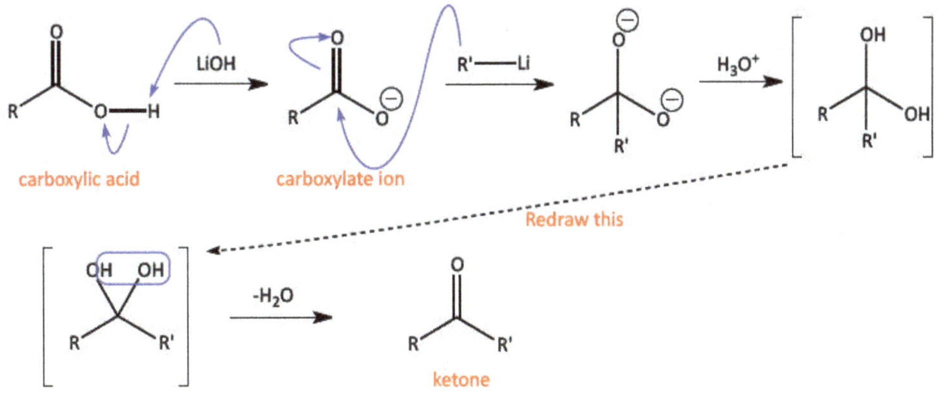

16.6 Fill in the blank boxes.

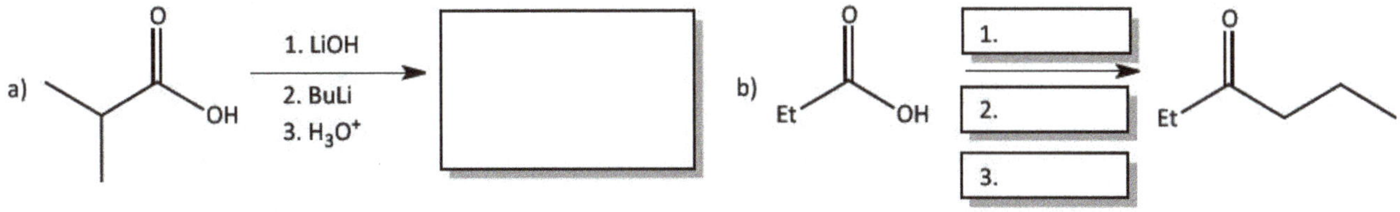

16.6.3 Ketones From Nitriles

Grignard reagents can attack nitriles. When they do, imines are formed. Upon protonation, imines form ketones.

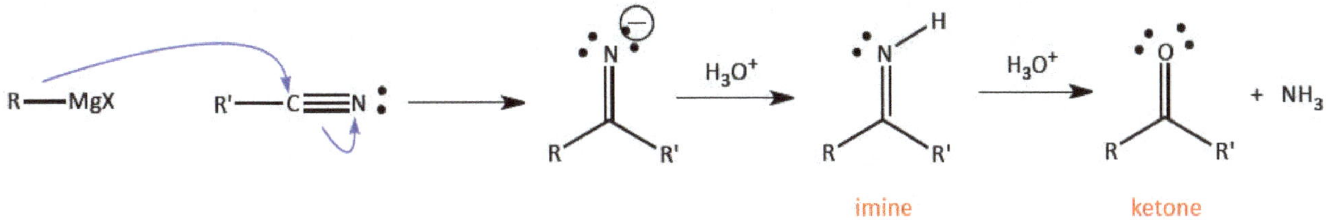

16.7 Fill in the blank boxes.

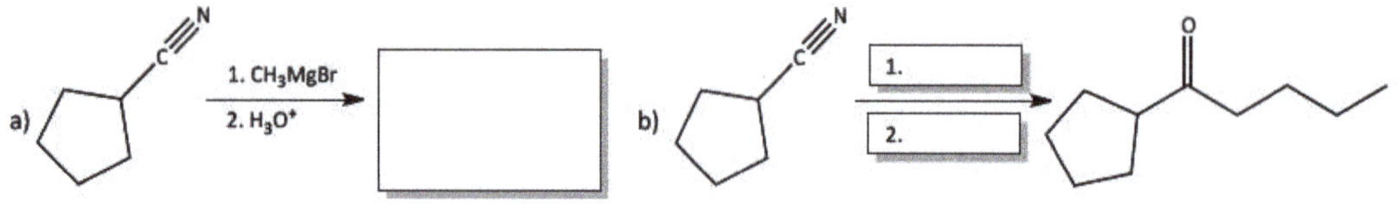

16.6.4 Hydrolysis of an Imine (Protonate and "Go to town.")

The mechanism for the hydrolysis of an imine is several steps. When asked to recall a mechanism like this, sometimes a sense of dread comes over a student. It seems like too much to memorize. But, notice that the first step is that the nitrogen atom of the imine is protonated. I think it's best to remember than in acidic conditions, this is the first step of the mechanism. When asked to recall this mechanism, I do not panic. I simply remind myself of a simple phrase, "Protonate, and go to town." This calms me down and I just write down the first, protonation step of the mechanism and let the water attack. After that, I "go to town". That means, I figure out the remaining parts of the mechanism as I go along. I keep protonating the nitrogen until it comes off and deprotonate the water until all of the protons are removed from the oxygen atom. Also, notice that never, in any of the mechanism's steps, is there ever a situation where there are two positive charges. There is one positive charge, it becomes neutral, then positive again, then neutral, but never doubly positive.

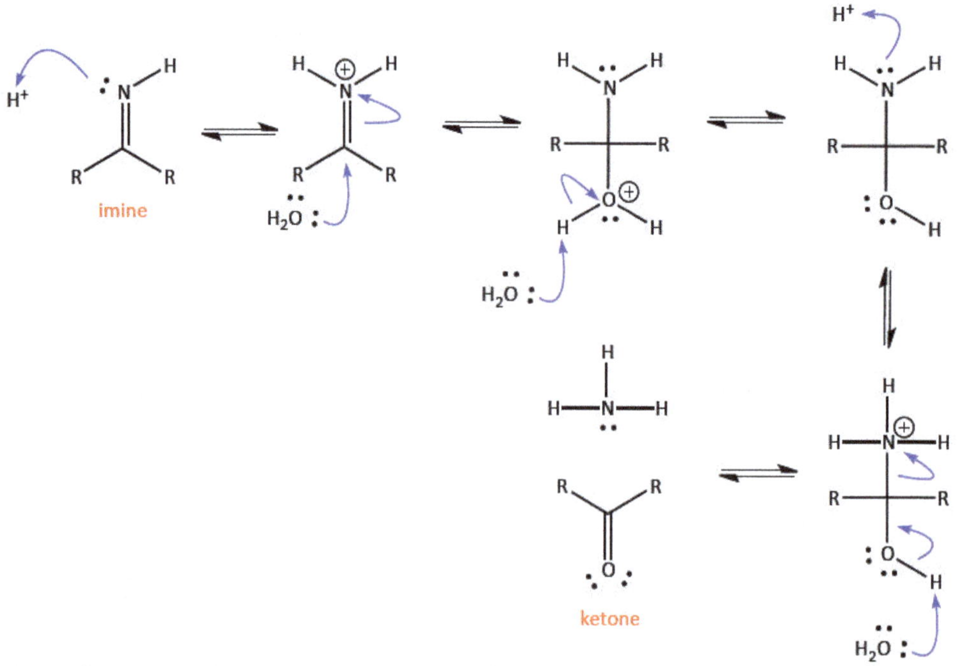

16.7 New Syntheses of Carboxylic Acids

16.7.1 Grignard attack of carbon dioxide

A Grignard reagent can attack carbon dioxide to make a carboxylate, which upon protonation, forms a carboxylic acid. This is a simple, sleek way to make carboxylic acids. The negative carbon of the Grignard attacks the partially positive charge of the carbon atom of carbon dioxide. The pi electrons go onto oxygen, making the carboxylate.

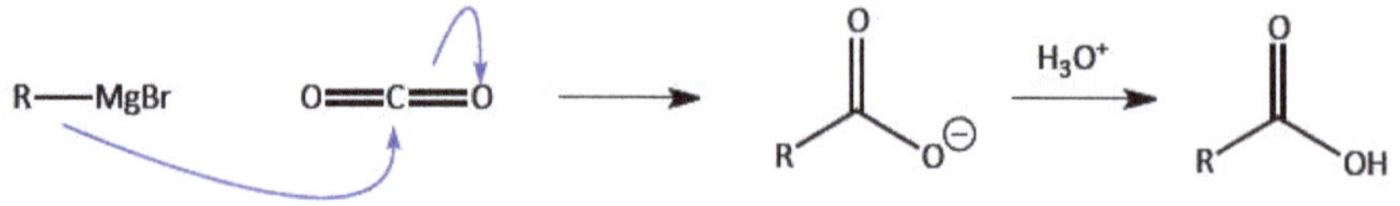

16.7.2 From a nitrile

Nitriles can be hydrolyzed to carboxylic acids. The write out the mechanism of this reaction, just "protonate and go to town." The mechanism is quite similar to the hydrolysis of an imine shown earlier.

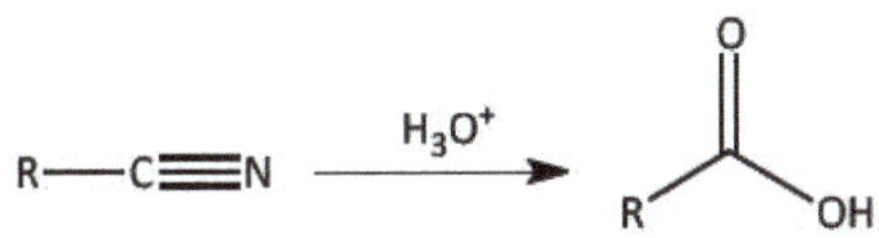

This is a convenient mechanism, because nitriles can be made from alkyl halides in an S_N2 reaction. Below, an alkyl bromide reacts with cyanide ion (from potassium cyanide salt, KCN) to form a nitrile. Hydrolysis in acid water forms the carboxylic acid.

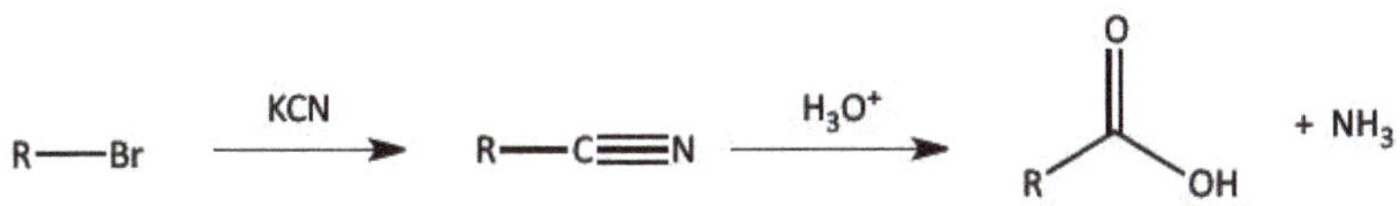

16.8 Fill in the blank boxes

a)

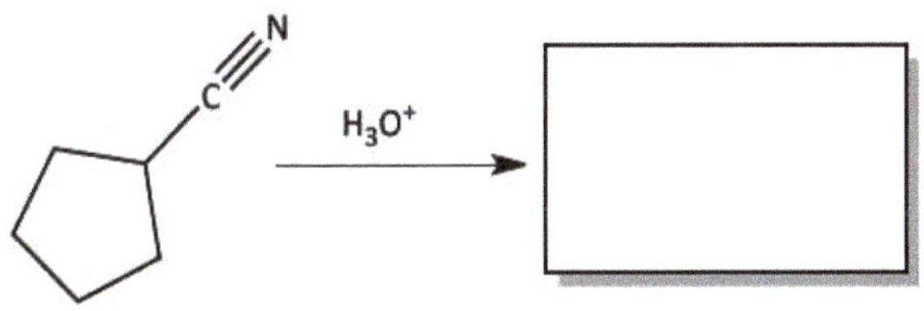

b)

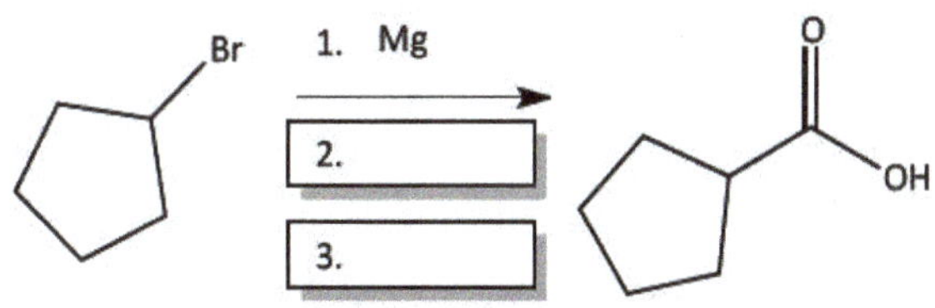

16.7.3 Permanganate oxidation of alkylbenzene

Potassium permanganate ($KMnO_4$) can oxidize any alkyl group that is attached to a benzene ring to a one-carbon atom carboxylic acid.

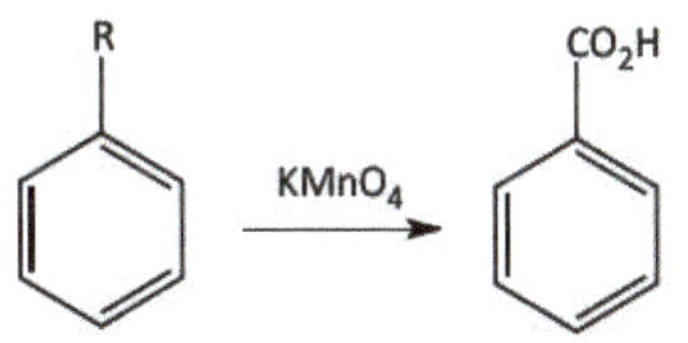

Benzoic acid

Notice that no matter how long the alkyl group is, a one-carbon carboxylic acid is all that is formed.

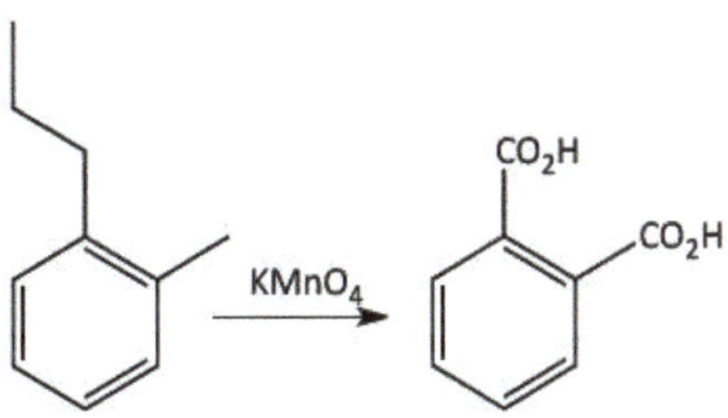

16.8 Reaction Summary

Oxidation of alcohols with weak oxidizer

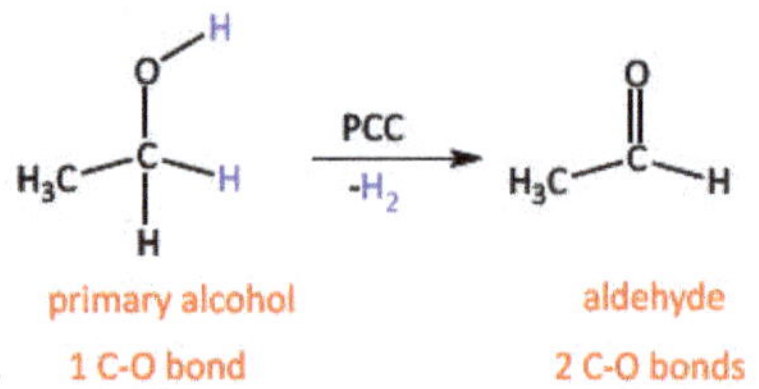

a)

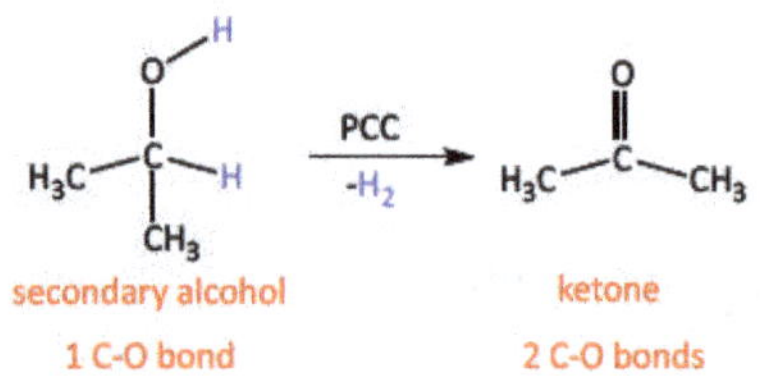

b)

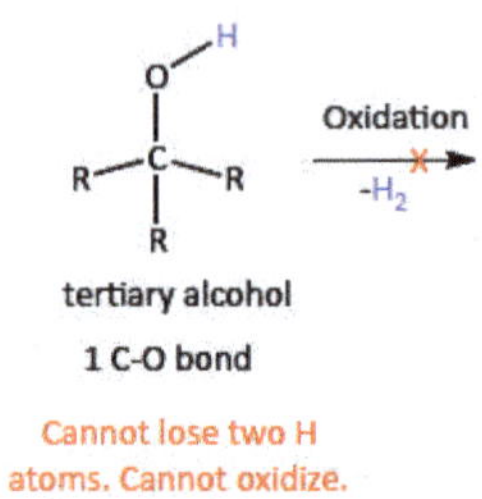

c)

Swern oxidation and *DMP* are weak oxidizers like PCC.

Oxidation of alcohols with strong oxidizer

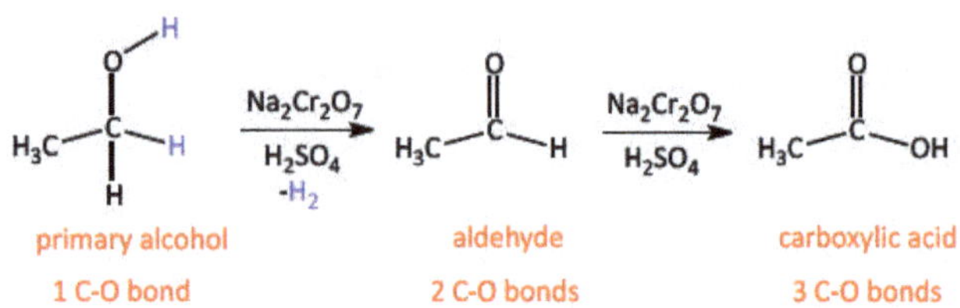

d)

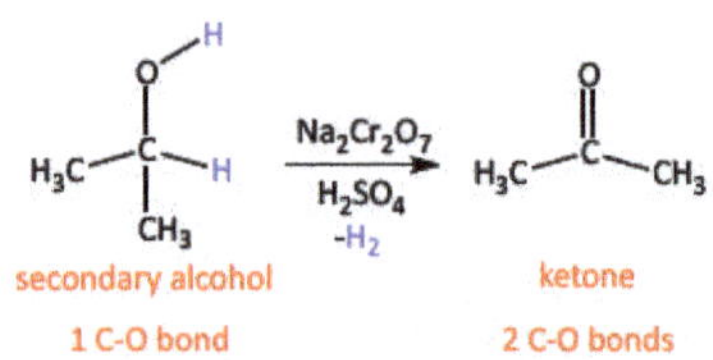

e)

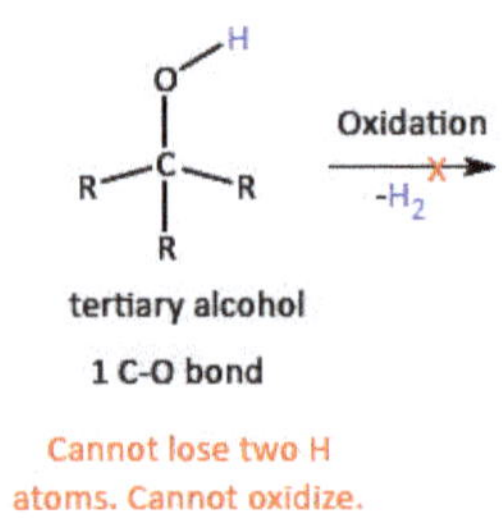

f)

Reduction of carboxylic acid

g) 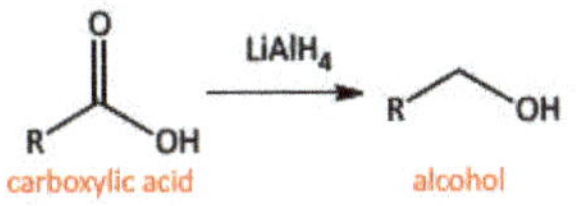

h) 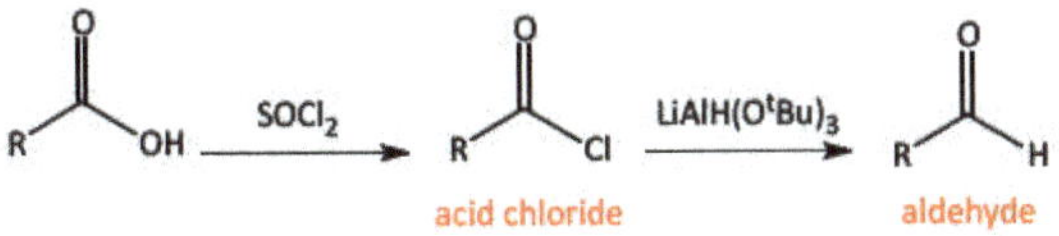

i) 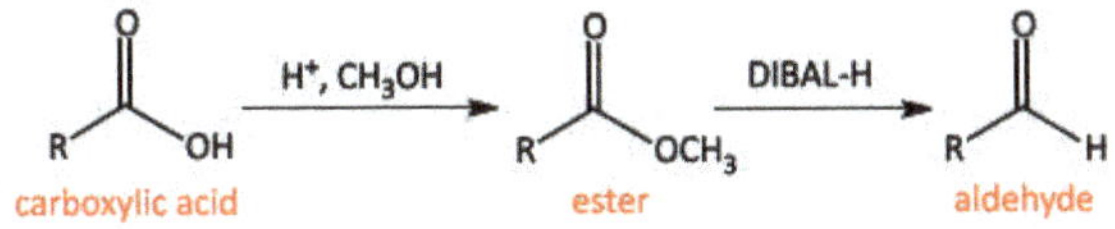

Alkene/Alkyne oxidations

j) 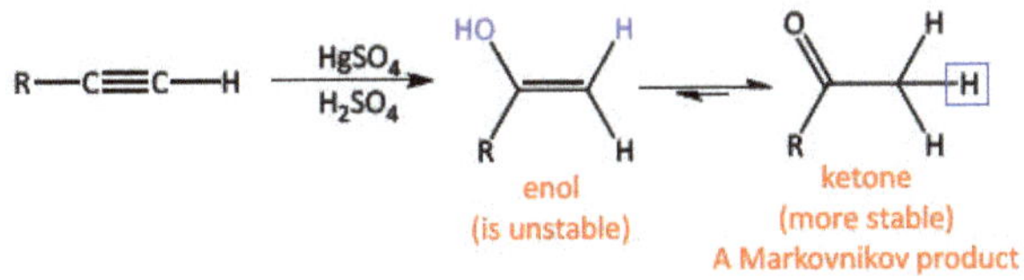

k) 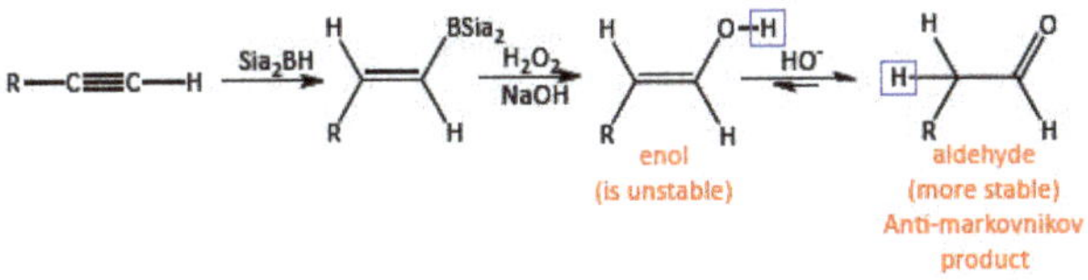

l) 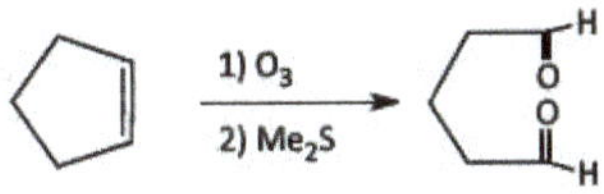

m) 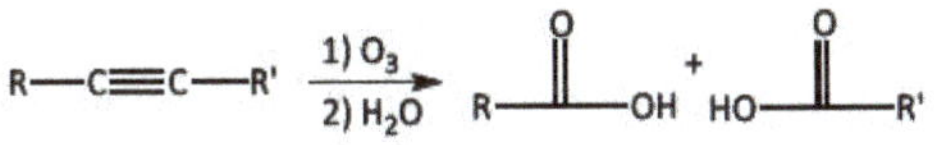

n) 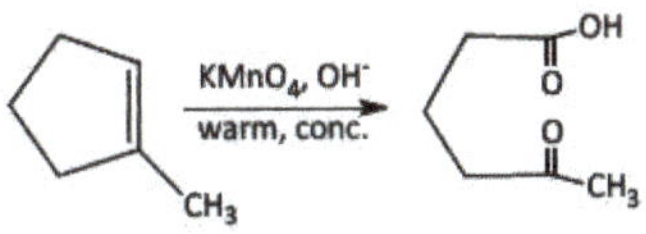

o) 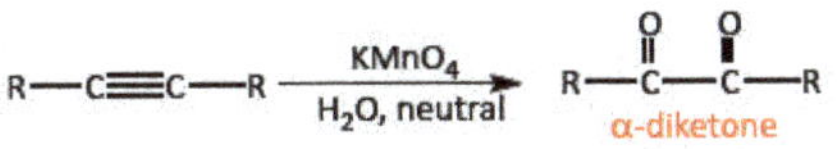

p) 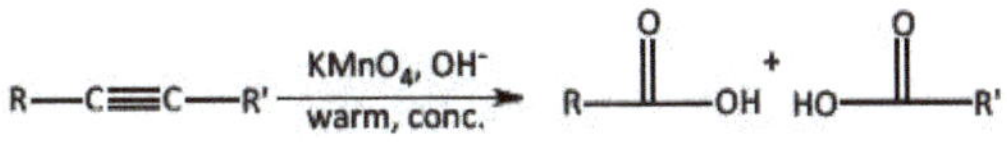

q) 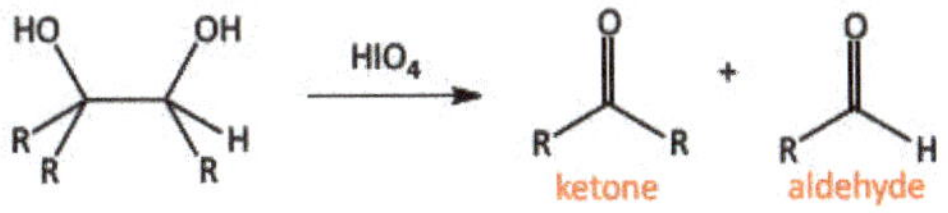

Carbonyl Compounds

1,3-dithiane aldehyde/ketone synthesis

r) 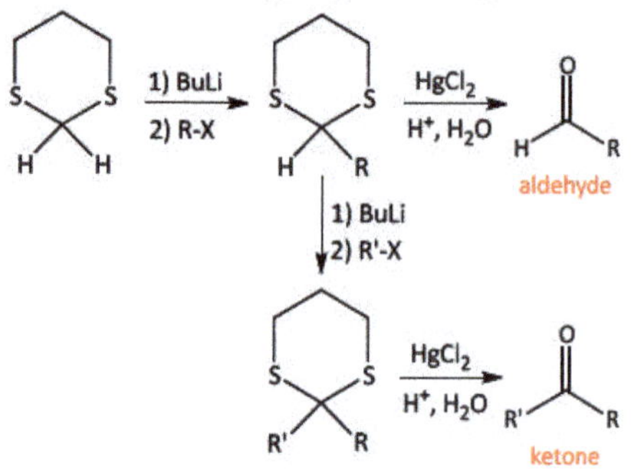

Ketones from carboxylic acids

s) 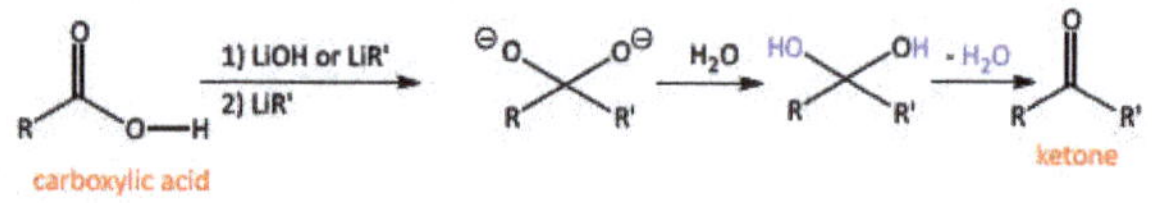

Ketones from nitriles

t) 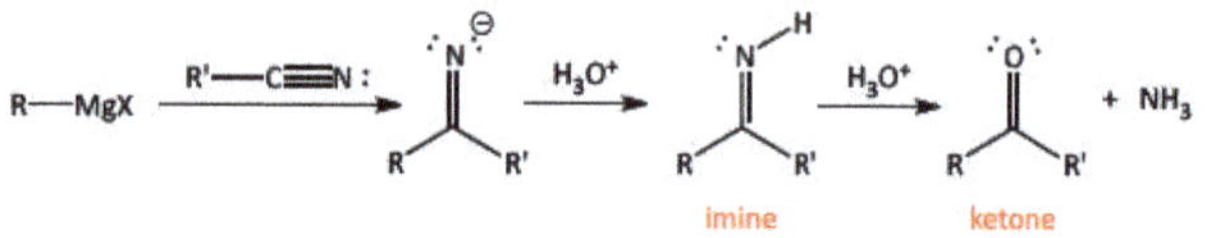

Carboxylic acids from Grignard reagent and CO_2

u) 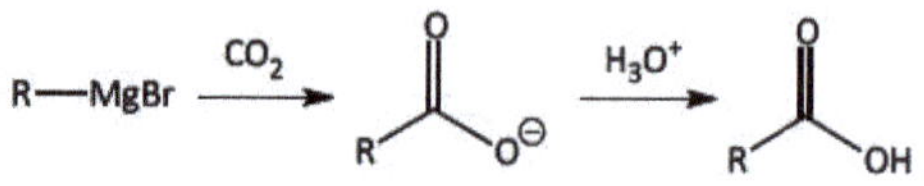

Carboxylic acids from the hydrolysis of nitriles

v) 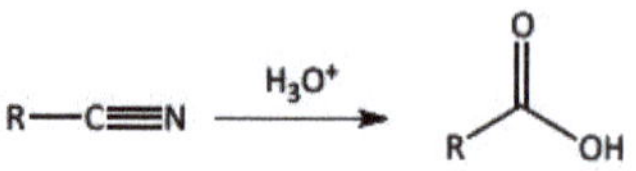

Carboxylic acids from the permanganate oxidation of alkylbenzene

w) 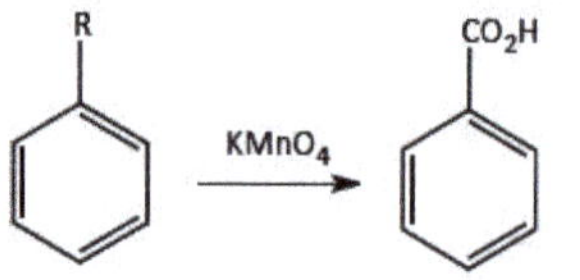

Benzoic acid

16.9 Problems

16.9 Draw the following compounds.

a) propionaldehyde b) 2-pentanone c) 2-methylhexanoic acid d) methyl propyl ketone e) 2-methylpentanal f) acetic acid g) MEK (methyl ethyl ketone) h) acetone i) pentanamide j) butyramide k) ethyl acetate l) acetamide m) butyl formate n) ethyl ethanoate o) phenyl benzoate p) N-methylbutanamide q) methyl methanoate

16.10 Name the following compounds.

a) 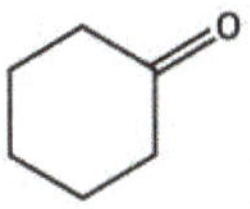
b) 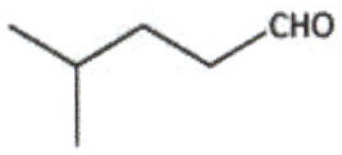
c) 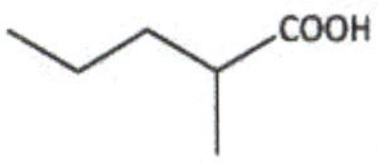
d) 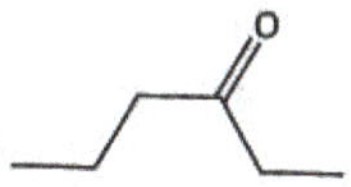
e) 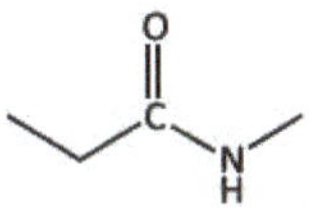
f) 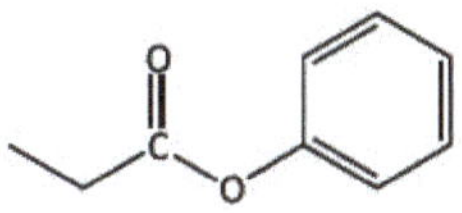

16.11 Identify the functional group(s) indicated in each generic IR spectrum.

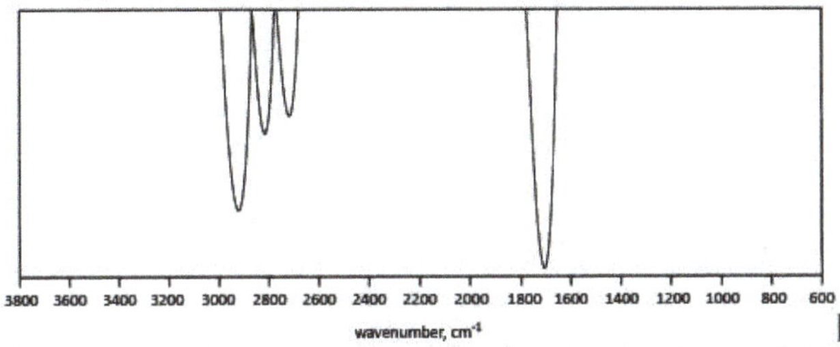

a)

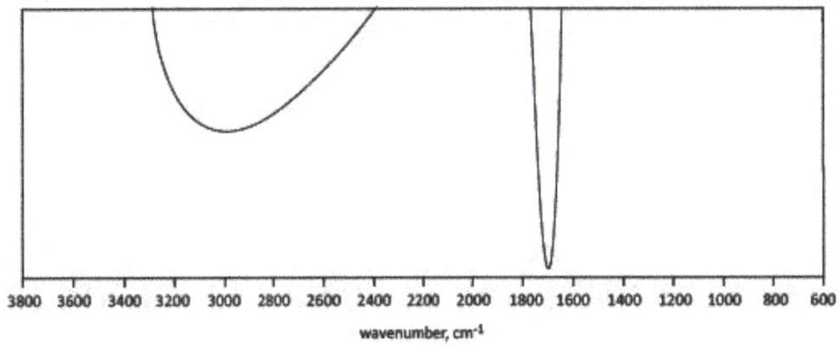

b)

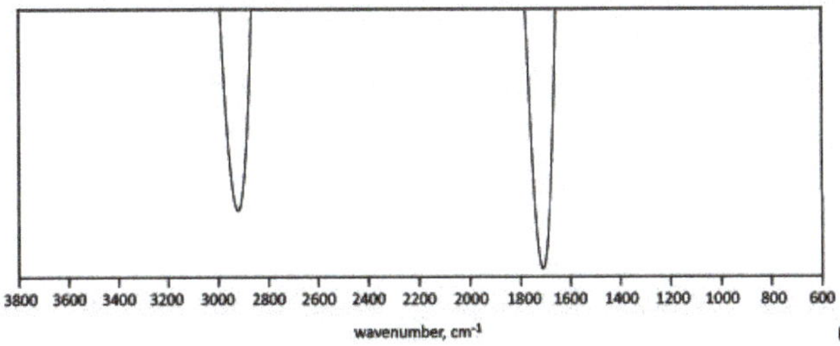

c)

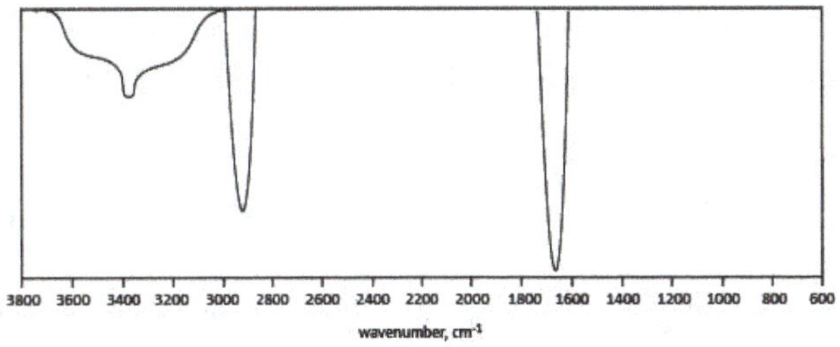

d)

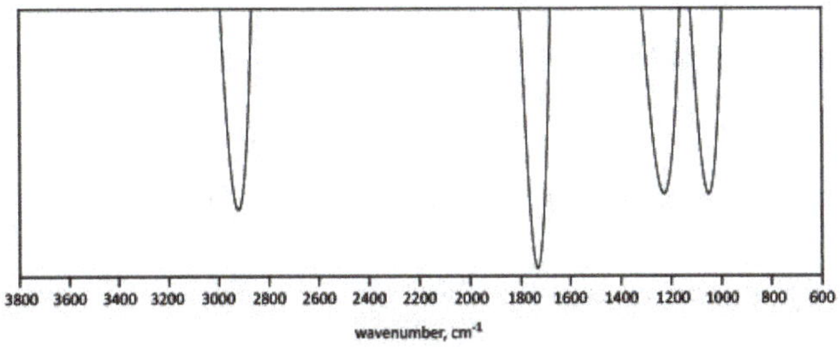

e)

16.12 Identify the following carbonyl compounds based on the available spectroscopic data.

a) 1.16 ppm 3H triplet, 2.38 ppm 2H quartet, 11.73 ppm 1H singlet
b) 7.5-7.9 ppm 5H multiplet, 10.02 ppm 1H singlet with the following IR spectrum:

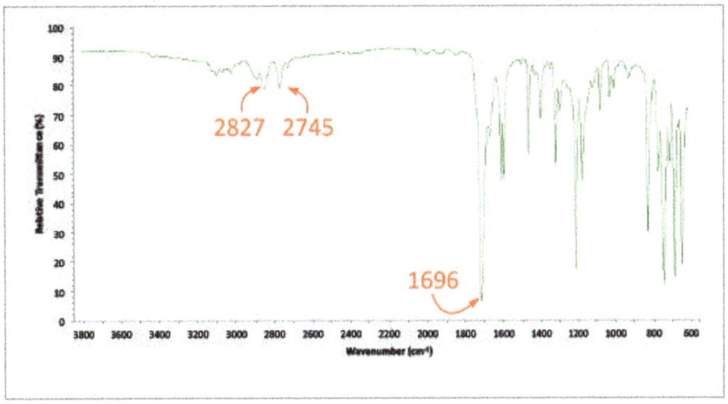

Carbonyl Compounds

16.13 Draw the products of the following oxidation reactions.

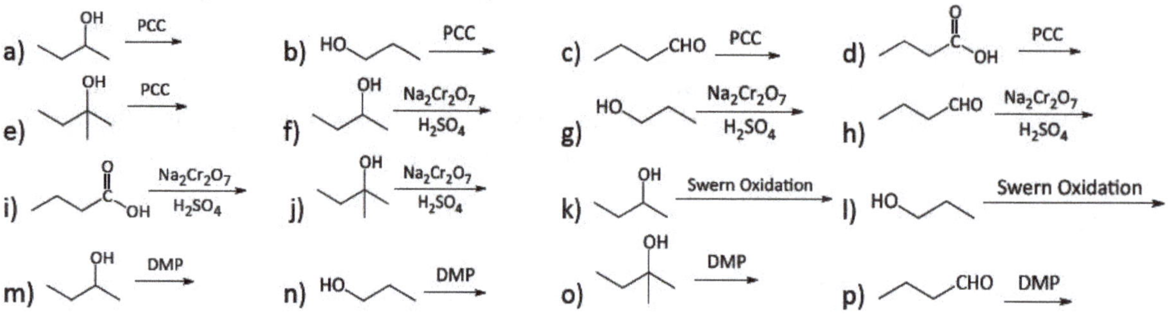

16.14 Identify an appropriate oxidizing agent to perform the following transformations.

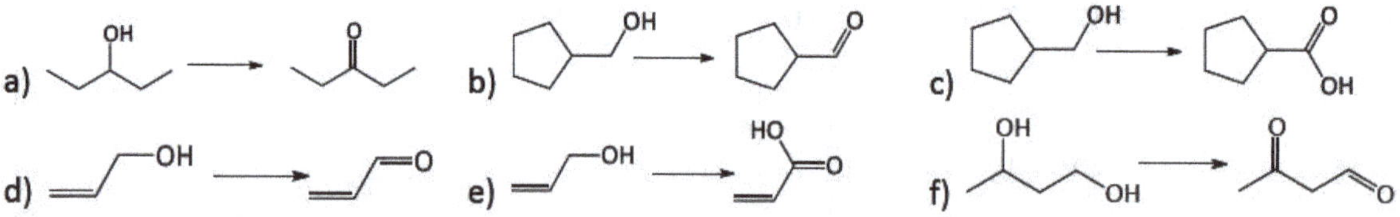

16.15 Show the products of 1-propyne with each of the following reagents.
a) $HgSO_4$, H_2SO_4 b) 1) Sia_2BH 2) H_2O_2, OH^- c) 1)O_3, 2) Me_2S d) warm, conc. basic $KMnO_4$

16.16 Identify the reagents necessary to perform the following reactions.
a) 2-butyne to acetic acid (two ways) b) 1-pentyne to 2-pentanone c) 1-pentyne to pentanal
d) 2-butyne to 2,3-butanedione

16.17 Show how you would make the following from an alkene or alkyne and other necessary reagents.
a) 2-butanone b) butyraldehyde (from an alkyne) c) propionic acid (two ways, from an alkene and an alkyne)
d) 3,4-heptanedione e) 2-butanone and butyric acid f) 2,6-heptanedione g) propanal and butanal

16.18 Use 1,3-dithiane to synthesize the following ketones and aldehydes.
a) octanal b) 3-heptanone c) methyl ethyl ketone d) $PhCH_2CH_2CHO$

16.19 Show how you would make the following from an appropriate carboxylic acid and other necessary reagents.
a) 2-butanone b) 3-hexanone c) ethyl phenyl ketone d) benzophenone (Ph(CO)Ph)

16.20 Show how you would make the following from an appropriate nitrile and other necessary reagents.
a) 2-butanone b) 3-hexanone c) ethyl phenyl ketone d) benzophenone (Ph(CO)Ph)
e) benzoic acid f) pentanoic acid

16.21 Show how you would make the following carboxylic acids from a Grignard reagent reacting with carbon dioxide. If the oxidation of an alkyl benzene can be used to synthesize the product, show that reaction as well.
a) 2-methylbutanoic acid b) benzoic acid c) 2-methylpropanoic acid d) hexanoic acid e) *p*-nitrobenzoic acid

16.22 There are two alkenes, each one contains only one double bond, that react with a hot, concentrated, basic solution of $KMnO_4$ to make acetic acid (ethanoic acid) as the only organic product. What are these two alkenes?

16.23 It is possible, through a series of reactions, to turn toluene ($PhCH_3$) into phenylacetic acid ($PhCH_2CO_2H$). Show one pathway using NBS and making a Grignard reagent. Can you find more than one way to do this?

16.24 a) Show how you would make 4-nonanone by using bromobutane as the only organic reagent. For one reaction, turn bromobutane into $CH_3CH_2CH_2$ CH_2CN. For the other, turn bromobutane and butyl Grignard. Then, perform any other necessary reactions. b) What would be the product of the acid hydrolysis of the pentane nitrile intermediate?

16.25 a) What is the product of 1-methyl-1-cyclopentanol reacted with concentrated sulfuric acid and heat? b) What is the product when the answer from a) is reacted with ozone followed by Me_2S. c) What is the product when the answer from part a) is reacted with OsO_4/H_2O_2, then HIO_4?

16.26 List three (or more) ways to synthesize pentanoic acid.

16.27 Provide the major organic products for the following reactions.

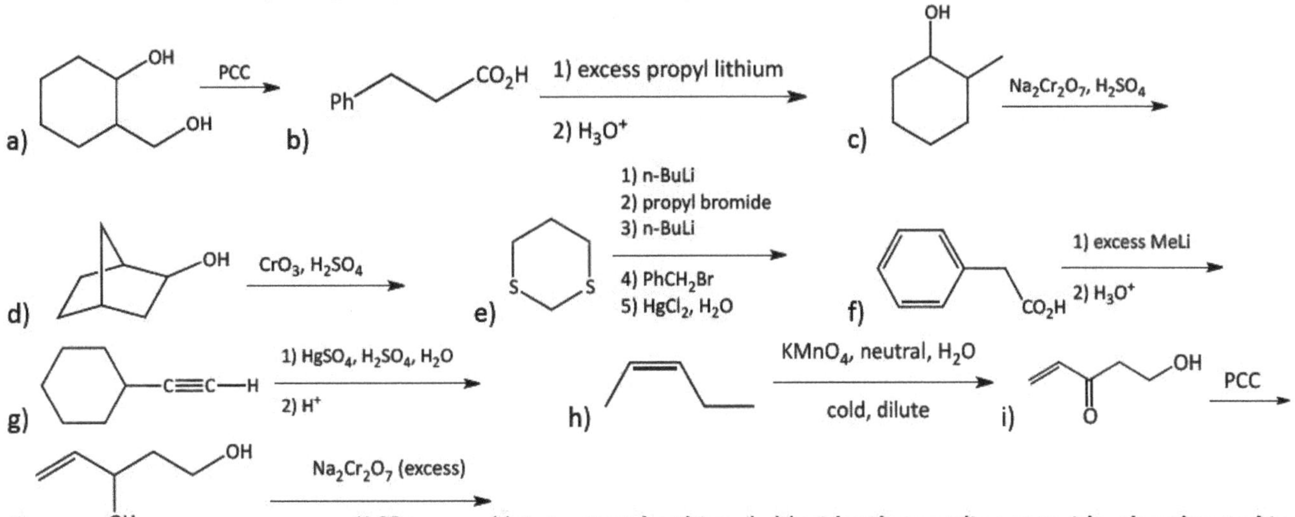

j) k) 1-pentanol + thionyl chloride, then sodium cyanide, then heated in H_3O^+

l) Swern oxidation of 1-butanol m) propanol + DMP n) heptanol + PCC

o) $PhCH_2COCl$ + 1) $LiAlH(OtBu)_3$ 2) H_3O^+ p) $PhCH_2COCl$ + $(CH_3CH_2)_2CuLi$

16.28 Fill in the blanks in the following reactions.

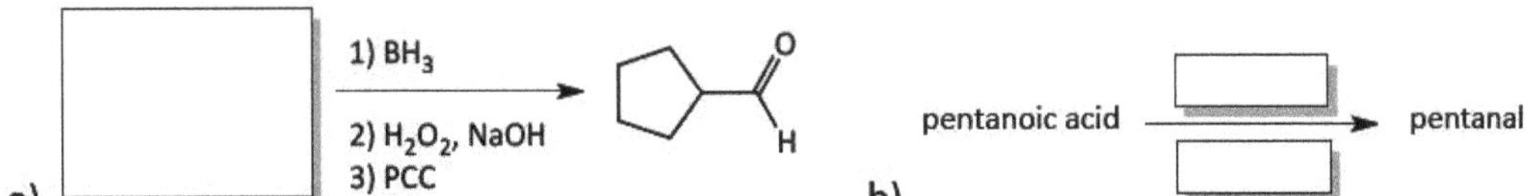

Carbonyl Compounds

16.29 **Describe the reagents that could be used to perform the following reactions. Sometimes, more than one reaction may be required.**

a) 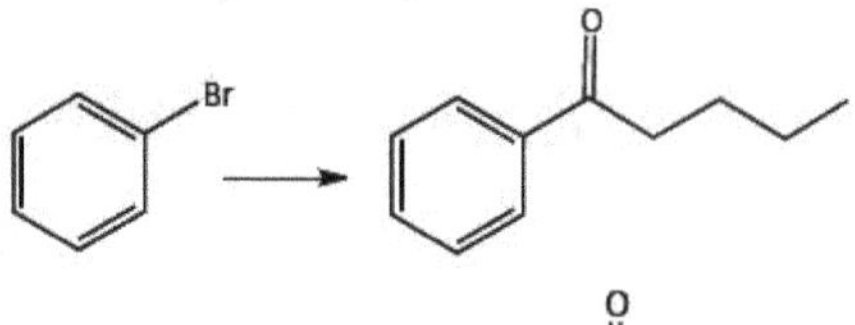 b) 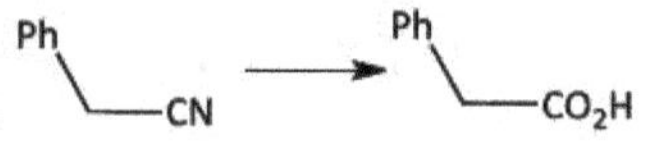 c) 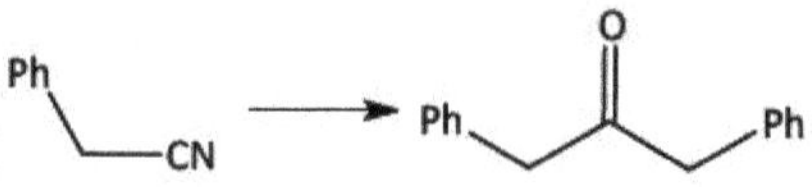

d) 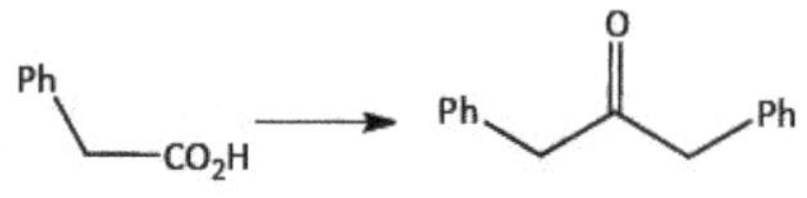 e) 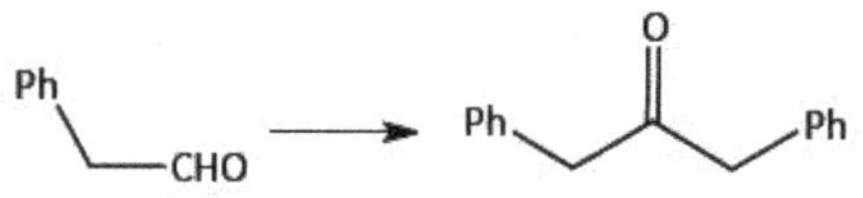 f) 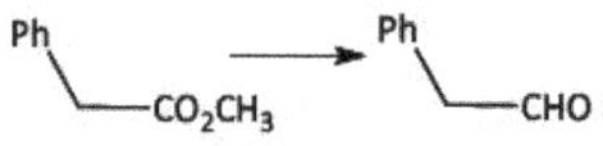

g) 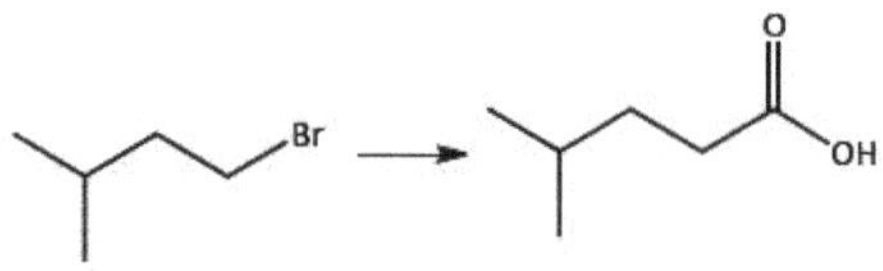 h) 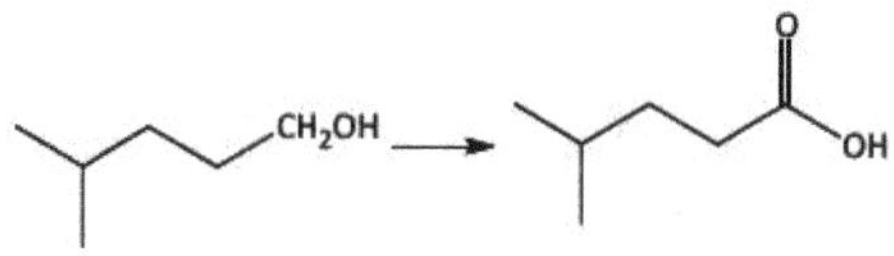

i)  j) 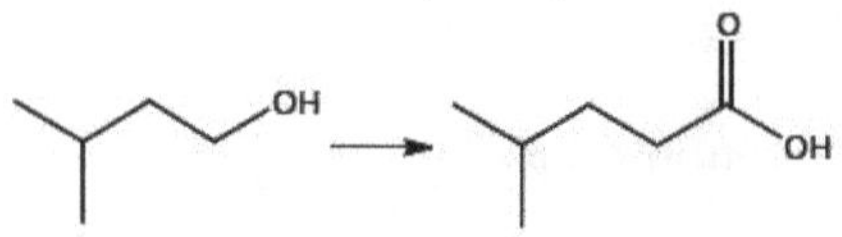

k) 2-butyne to acetic acid l) 2-butene to acetic acid m) bromobenzene to benzoic acid
n) benzaldehyde to benzoic acid o) propanoic acid to propenal

16.30 **Show how the following compounds can be synthesized from an appropriate carboxylic acid. For the aldehydes, provide two syntheses, one with an ester intermediate and one with an acid chloride intermediate.**

a) 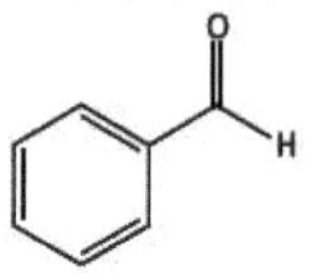 b) 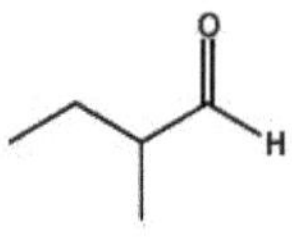 c) acetone d) 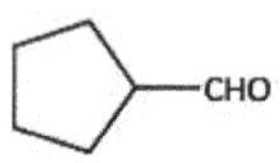 e) 3-hexanone

Chapter 17. Carbonyl Reactions

17.1 Reactions of Ketones and Aldehydes

In the previous chapter, we saw several ways to make carbonyl compounds. Let's take a closer look at how ketones and aldehydes react.

17.1.1 Ketone + primary amine makes imine

When a ketone or aldehyde reacts with a primary amine, a molecule of water can be expelled and an imine formed. Imines are also called a Schiff base.

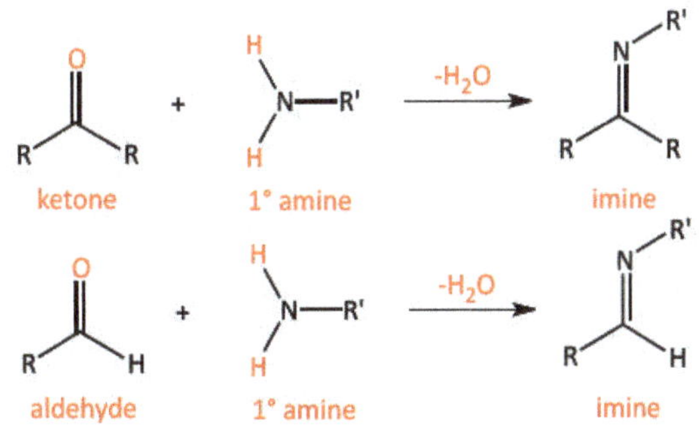

17.1.2 Wittig reaction – making alkene

Surprisingly, so far in organic chemistry class, we have learned very few ways to form carbon-carbon bonds. Georg Wittig won the 1979 Nobel Prize in Chemistry for his discovery of a reaction that forms alkene carbon-carbon double bonds from a ketone or aldehyde and a phosphine reagent.

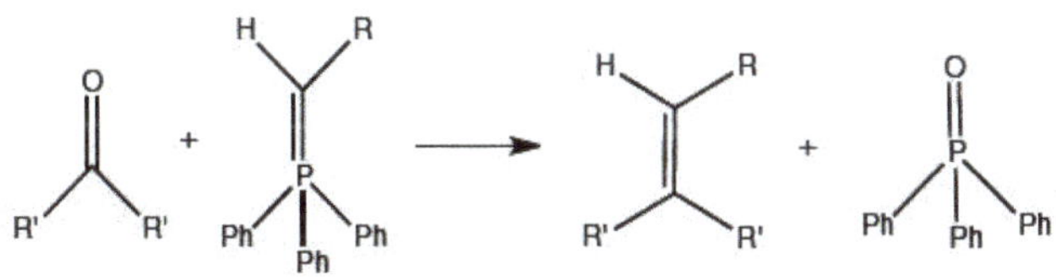

The phosphine for the Wittig reaction is formed by the reaction of triphenylphosphine with the appropriate alkyl halide followed by deprotonation with butyl lithium.

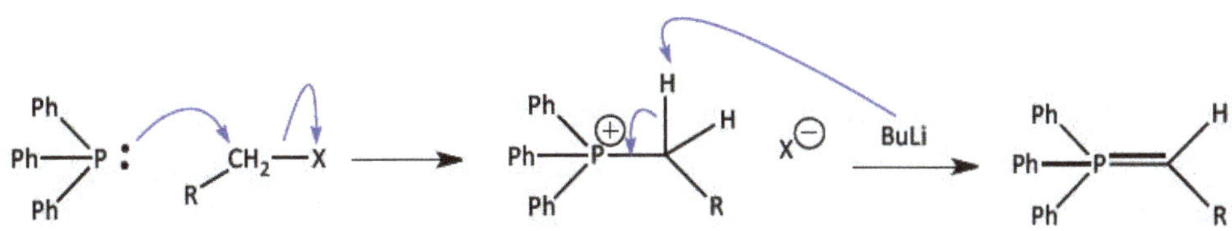

The phosphine reagent can be written with a phosphorus-carbon double bond, or it can be written with a positive and negative charge. A compound with a positive and negative charge separation is called an ylide.

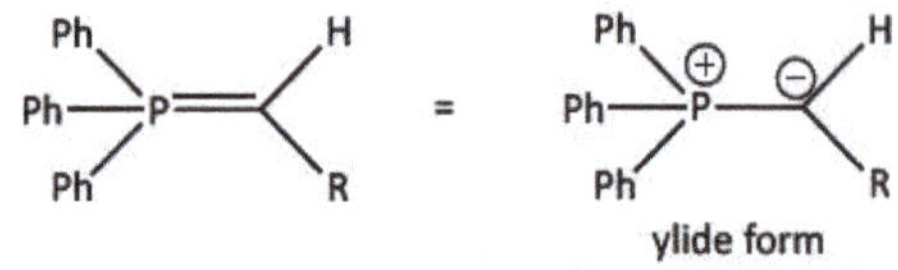

Many people prefer the ylide form, but I prefer the phosphorus-carbon double bond form, because then the mechanism is two arrows to form a square intermediate followed by two more arrows for form the final alkene and the very stable triphenylphosphine oxide. The driving force for this reaction is that phosphorus atoms love to bond to oxygen atoms. Overall, you can think of the reaction as removing the oxygen atom and the phosphine in the shaded box. You can simply cover the shaded box up with your hand and push together the two carbon atoms to make a new alkene.

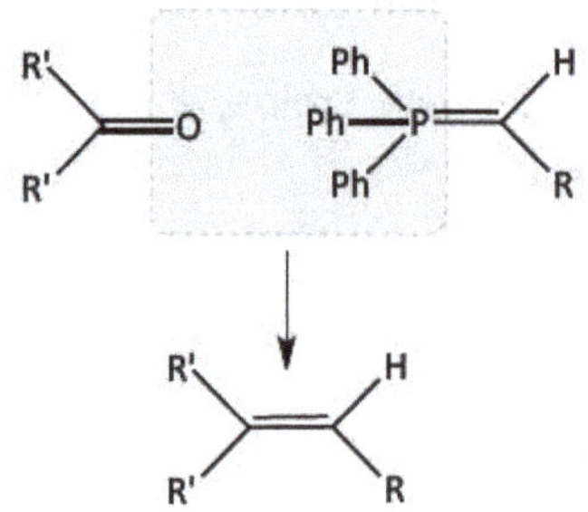

The mechanism is below.

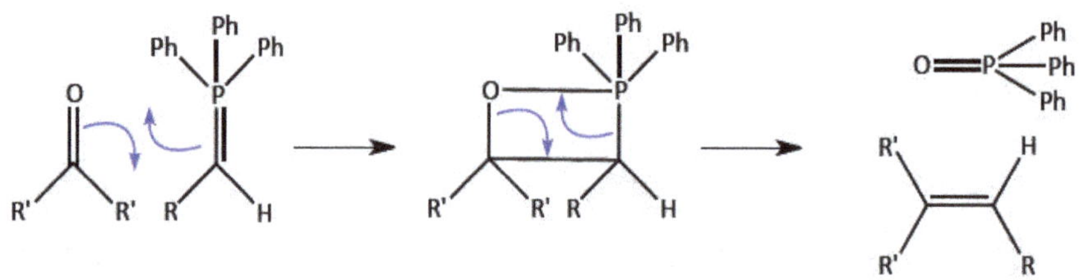

17.1 Fill in the blank boxes.

a) 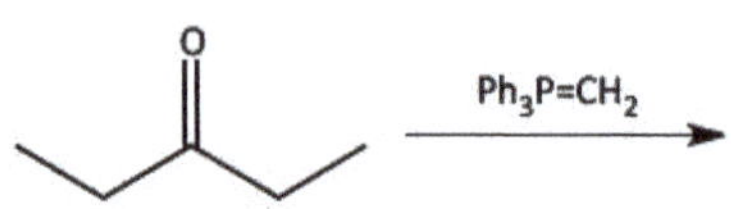

b) 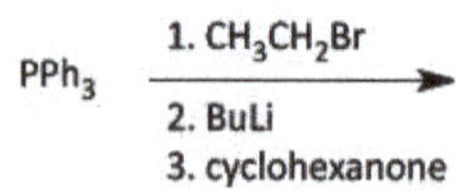

17.1.3 Acetals/ketals

Carbonyl compounds like aldehydes and ketones react with alcohols in acid to form acetals. Acetals are geminal diethers. Two ether groups are attached to the same carbon atom. A carbonyl consists of two carbon-oxygen bonds. The acetals that are formed also have two carbon-oxygen bonds that make two ether groups. The carbon groups that are attached to the oxygen atoms (R') come from the alcohol used as the solvent.

A compound is made halfway along the mechanism of turning an aldehyde into an acetal that is called a hemiacetal. The hemiacetal has an alcohol and an ether on it. The hydroxyl (OH) group of the hemiacetal further reacts to form an ether in the acetal.

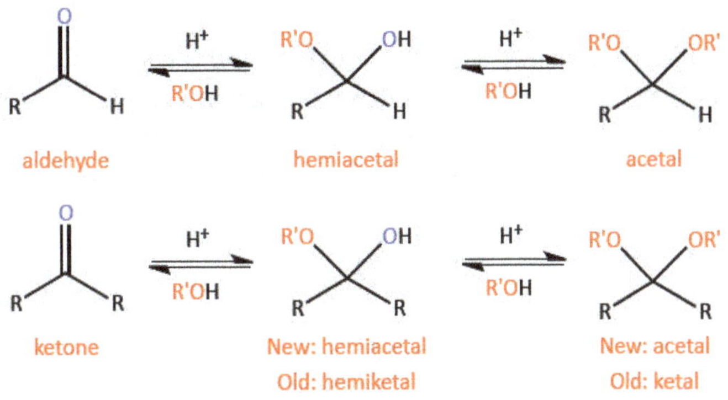

Figure 17.1 Carbonyls to acetals

All geminal diethers are now called acetals, but in older nomenclature, acetals that were formed from aldehydes were called acetals and those that were formed from ketones were called ketals. This is still often used.

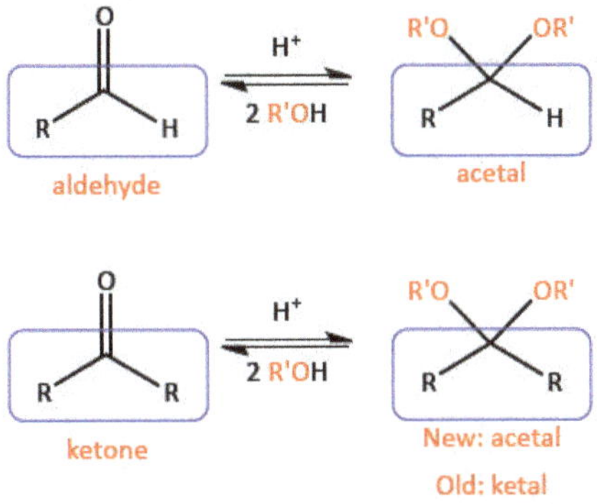

One thing I want you to stop and notice right now. The central carbon atom of the acetal, the one with two oxygen atoms attached to it, comes from the carbonyl compound used in the synthesis. This is easy to remember because a carbonyl has two carbon-oxygen bonds and the same acetal carbon has two carbon-oxygen bonds. The oxygen atoms and the carbon groups on the other side of them (the red groups above) come from the alcohols used. This little gem will help you solve problems later on.

The mechanism for acetal formation from a carbonyl is another example of "protonate and go to town." Protonate the carbonyl and then play around with the mechanism to form a hemiacetal and then an acetal.

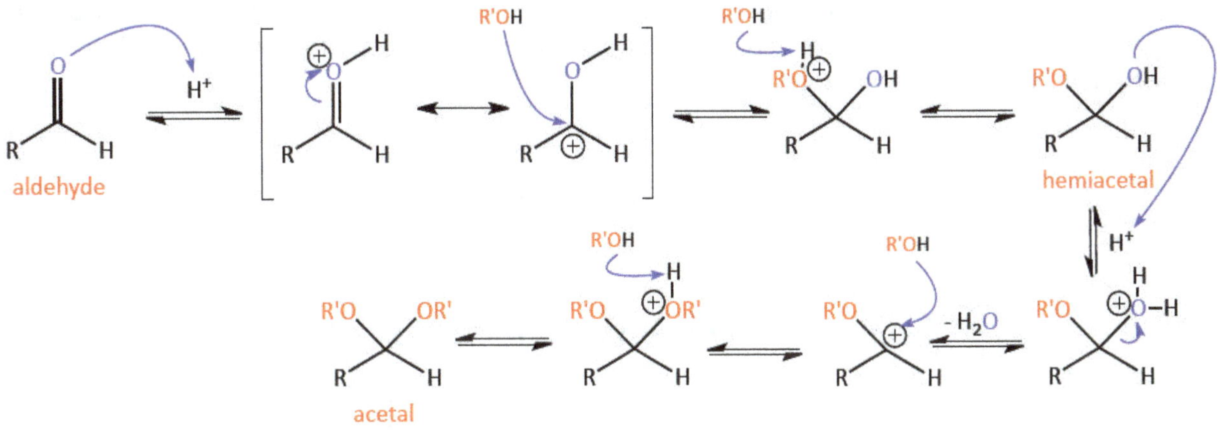

Carbonyl Reactions

Hydrolysis of acetal

The reverse mechanism turning an acetal into a carbonyl compound is called a hydrolysis. It involved the reaction of the acetal with water and acid. The steps of the mechanism are reversed, but one simply needs to remember that we need to protonate and go to town. Protonate the acetal, and play around with it until you make a hemiacetal. Then, protonate the ether of the hemiacetal and play around with it until you've formed the carbonyl compound.

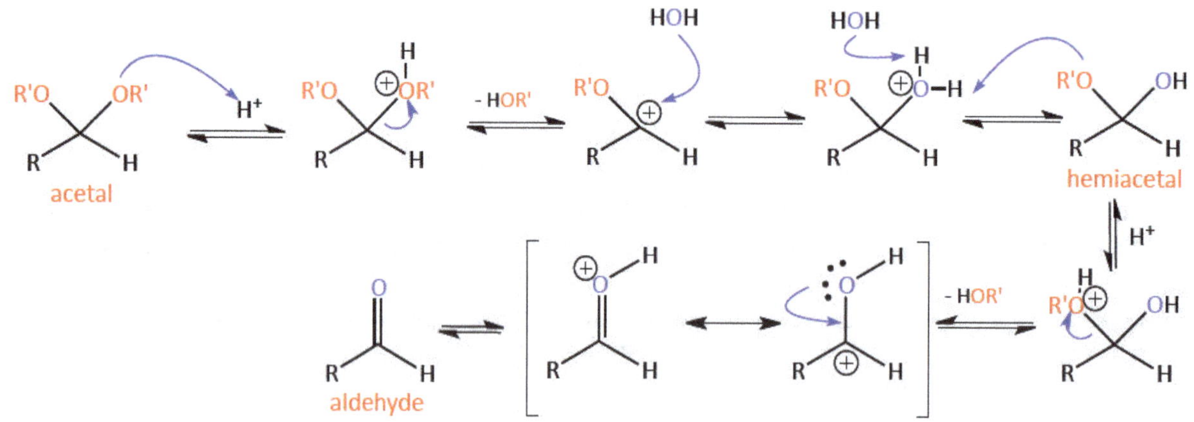

17.2 Fill in the blank boxes.

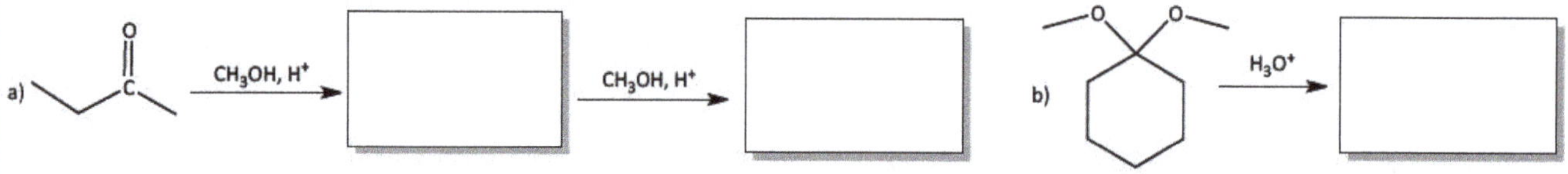

Protecting aldehydes with acetals

It is easier to form an acetal from an aldehyde than it is from a ketone. This can be used to our advantage. Let's say we have a compound that contains both an aldehyde and a ketone. If we want the ketone to react with a compound like $NaBH_4$, but not the aldehyde, we can "protect" the aldehyde and keep it from reacting. We can use an acetal as a protecting group. In this case, we can react our compound with ethylene glycol in acid. The aldehyde will preferentially react with the ethylene glycol (more than the ketone will) to form an acetal. $NaBH_4$ can then react with the ketone. Following the reaction with the ketone, the acetal can be deprotected with acid and water (a hydrolysis) to once again reform an aldehyde.

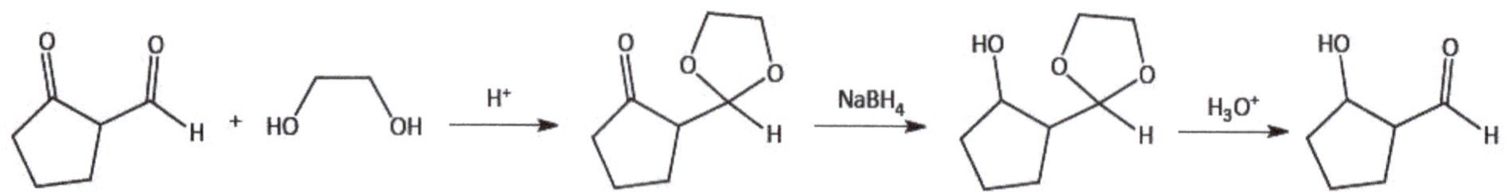

17.2 Reactions of Carboxylic acids and their derivatives

17.2.1 Deprotonation

Carboxylic acids are fairly acidic. A carboxylic acid can be deprotonated to form a carboxylate. This is because the conjugate base is quite stable. The negatively charged oxygen atom can be further stabilized through resonance.

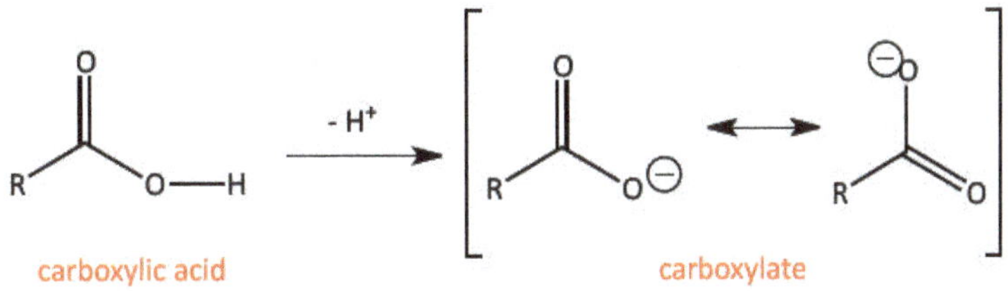

Since the carboxylates are so stable, carboxylic acids have a pKa around 5. Regular alcohols have a pKa of around 17. This makes carboxylic acids 1,000,000,000,000 (1 trillion or 10^{12}) times more acidic than alcohols.

Therefore, carboxylic acids react with bases to make carboxylate salts. For example, when carboxylic acids react with sodium hydroxide, or a very weak base like sodium bicarbonate ($NaHCO_3$) for that matter, they make the sodium salt of a carboxylic acid. These acid salts are water-soluble. Then, if the solution is acidified, the carboxylic acid can be reformed.

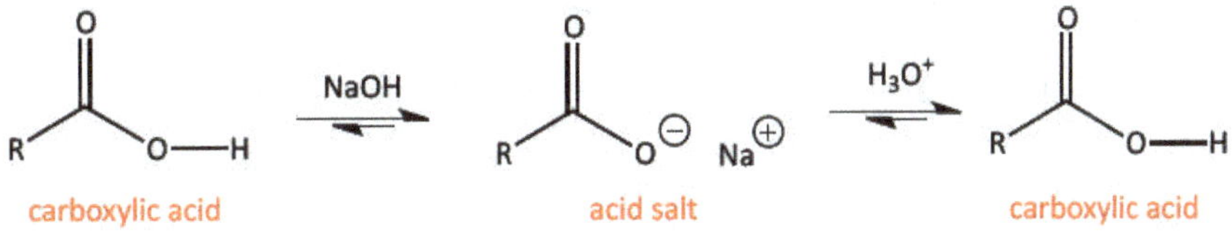

17.2.2 Fischer esterification (Review)

The Fischer Esterification is the acid catalyzed reaction of an alcohol and a carboxylic acid to form an ester. The Fischer esterification is a condensation reaction, meaning water is eliminated in the reaction. Overall, a carboxylic acid reacts with an alcohol in the presence of an acid catalyst. Water is eliminated and an ester is formed- hence the esterification part of the Fischer esterification name.

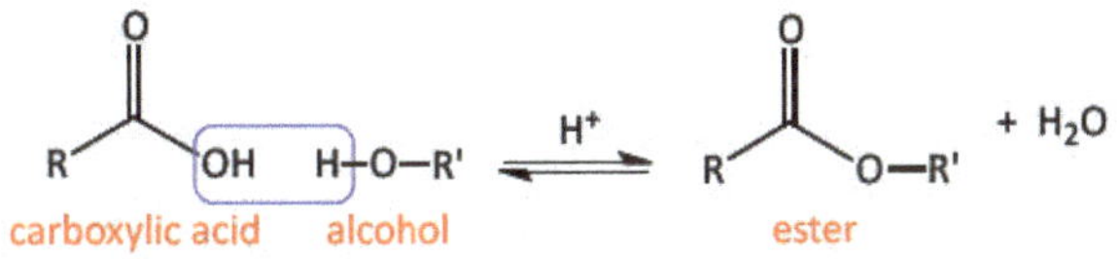

In order to drive the equilibrium of the Fischer Esterification to the ester product side, two things are commonly done. One trick is to run the reaction with an excess of starting materials. A large excess of either the alcohol or the carboxylic acid will work. The other trick is to remove the water product from the reaction using a dehydrating agent in the reaction flask that can absorb water, like molecular sieves. Also, water can be removed from a reaction using a Dean-Stark trap. This is a special attachment to a reflux apparatus that helps remove water from a reaction.

Once again, the mechanism for the esterification of an ester is a "protonate and go to town" mechanism. The carbonyl oxygen atom, with its negative lone pairs of electrons, is protonated as the mechanism's first step.

Carbonyl Reactions

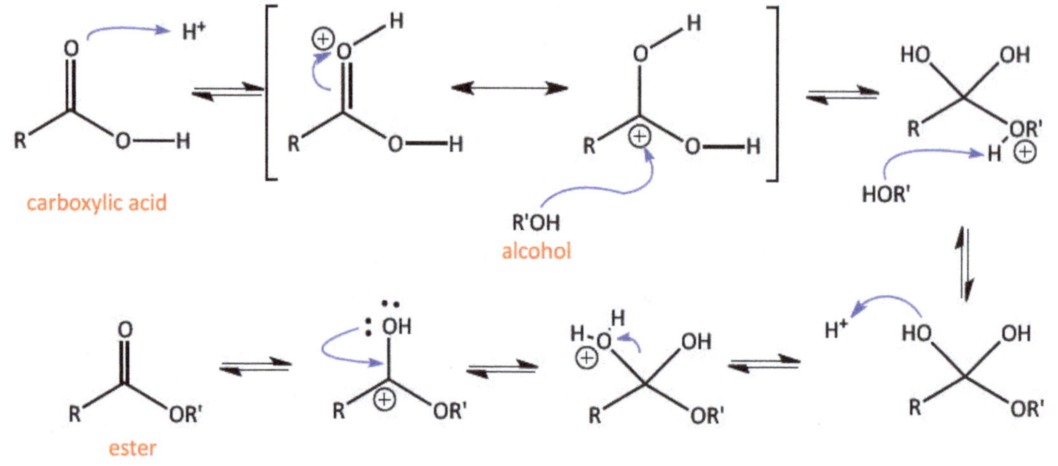

17.2.3 Carboxylic acid to acid chloride (Review)

We learned in the alcohol chapter that OH groups of alcohols can be transformed into chlorides by using thionyl chloride ($SOCl_2$). The same reaction can transform the OH group of a carboxylic acid into an acid chloride.

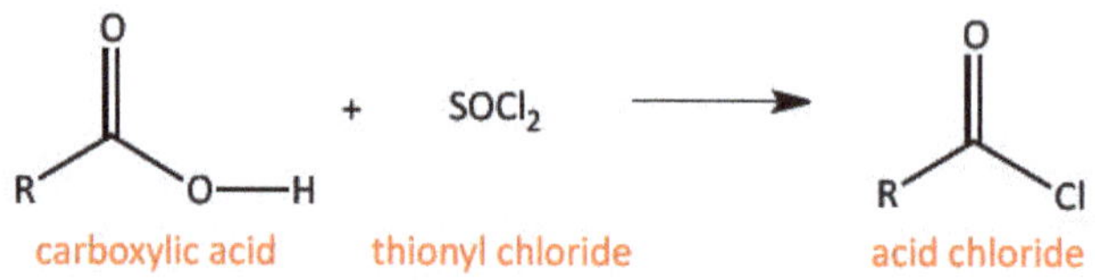

17.2.4 Acid chloride to ester (Review)

Another effective way to make an ester is to react and alcohol with an acid chloride. This is an exothermic reaction that is not an equilibrium reaction. The reaction usually proceeds well. Sometimes a base it put into the reaction flask to soak up the HCl that is produced. Acid chlorides can be made from reacting carboxylic acids with $SOCl_2$.

One drawback for this reaction is that it is more water sensitive than a Fischer Esterification. Water will quench the acid chloride and turn it into a carboxylic acid.

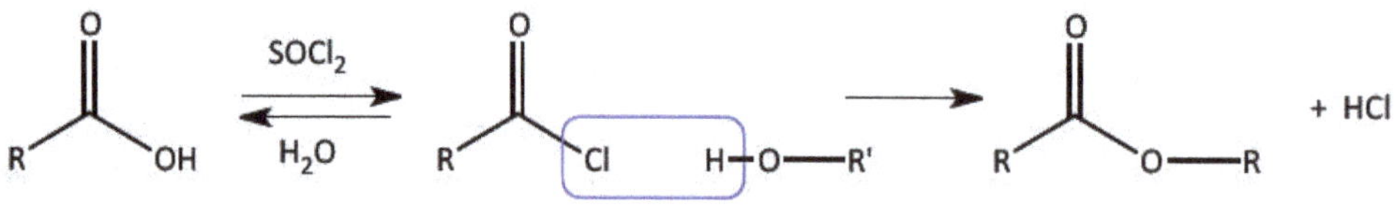

17.2.5 Acid chloride to amide

Acid chlorides can react with primary amines to form amides. HCl, hydrochloric acid, is released in this reaction. A base is needed to soak up or react with this acid. One easy way to accomplish this is to use two equivalents of amine. One reacts with the acid chloride to make an amide and the other equivalent of amine is used to react with the HCl that is produced.

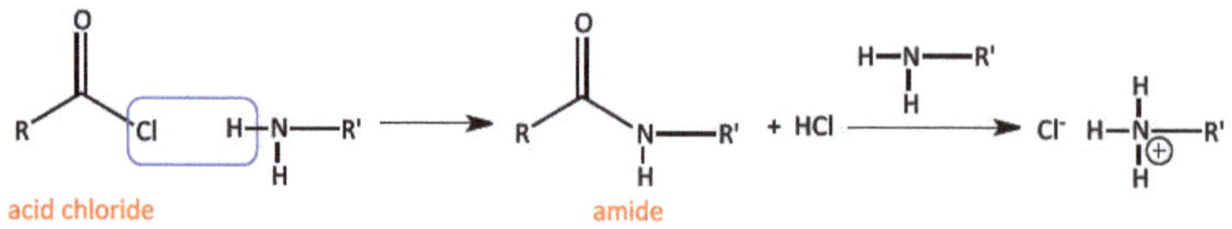

17.2.6 Carboxylic acid with 2 RLi (Review)

Carboxylic acids can react with two molecules of alkyl lithium to make ketones.

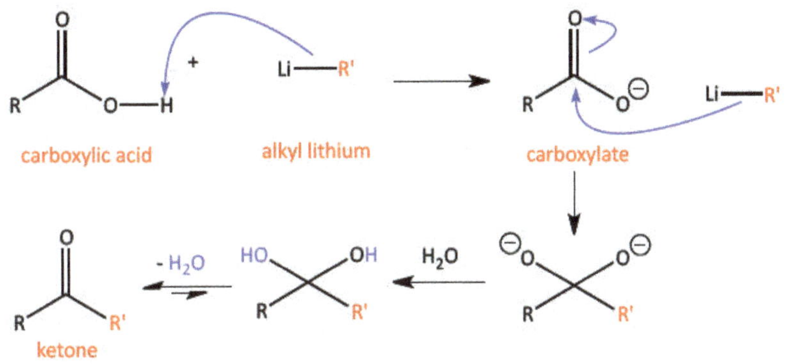

The first equivalent of alkyl lithium is simply used as a base to deprotonate the carboxylic acid to make a carboxylate. The second equivalent of alkyl lithium attacks the carbonyl carbon atom. The pi bond electrons go up onto the carbonyl oxygen atom to make a dianion. Following protonation, a geminal diol is formed. This geminal diol dehydrates to form a ketone.

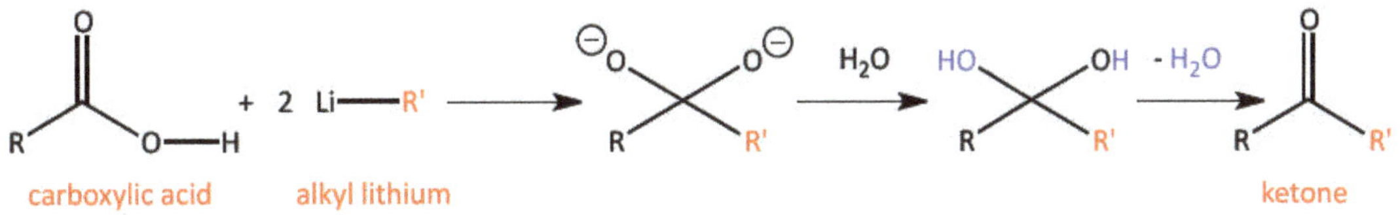

Since the first equivalent of alkyl lithium acts like a base, if it is expensive to buy or if it is valuable because it took time to synthesize, a cheaper and less valuable base can be used. Usually, this is lithium hydroxide (LiOH). LiOH can be used to deprotonate the carboxylic acid then the alkyl lithium can be added.

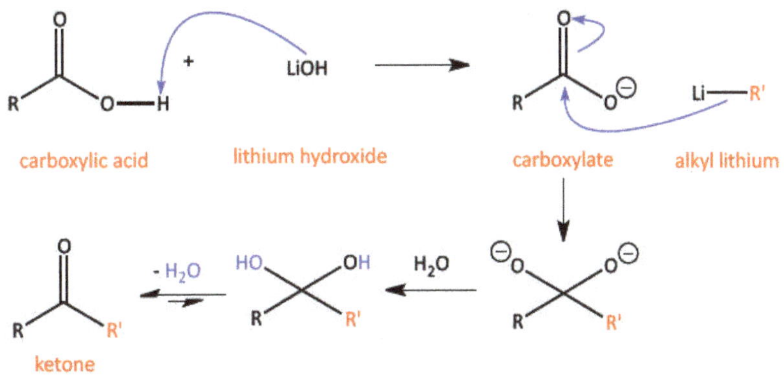

17.3 Carbonyl reductions (Review)

17.3.1 Reduction of aldehyde or ketone

In the overall reaction mechanism for the reduction of a ketone or an aldehyde, a hydride, attacks the partially positive carbon atom of a carbonyl.

The hydride attack of a ketone yields a secondary alcohol while the hydride attack of an aldehyde yields a primary ketone. But, there is no reason to try to memorize such a thing. Simply draw out the mechanism anytime you are asked such a thing and see what your product is. In the attack of the ketone, both R groups on the carbonyl are carbons and therefore terrible leaving groups. In the attack of the aldehyde, the R groups are carbon and hydrogen. These are bad leaving groups. So, after the attack of the nucleophile, the electrons stay up on the oxygen atom and do not come back down. Both the weaker $NaBH_4$ and the stronger $LiAlH_4$ can reduce aldehydes and ketones.

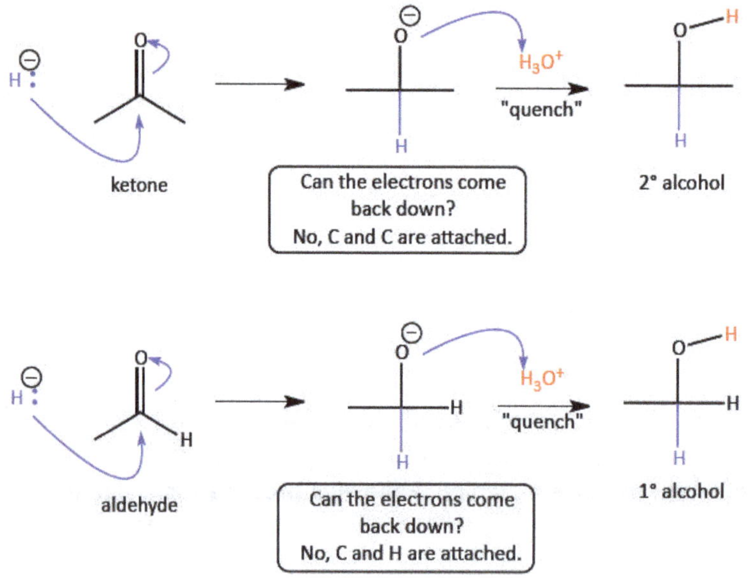

Figure 17.2 Reduction of aldehyde and ketone

17.3.2 Reduction of acid chloride to alcohol

An acid chloride can be reduced by a strong reducer like $LiAlH_4$ or even a weak reducing agent like $NaBH_4$. The hydride attacks the carbonyl carbon atom of the acid chloride. The electrons go up onto the oxygen atom of the carbonyl. There is a good leaving group, the chloride group. The electrons come back down and kick off the chloride group. A carbonyl is re-formed as an aldehyde. Aldehydes also react with $NaBH_4$ or LiAlH. The hydride attacks the carbonyl. The electrons go up onto the oxygen atom giving it a formal negative charge. We ask, "Can the electrons come back down?" In this case, they cannot because the aldehyde only has poor leaving groups, an H and an alkyl group.

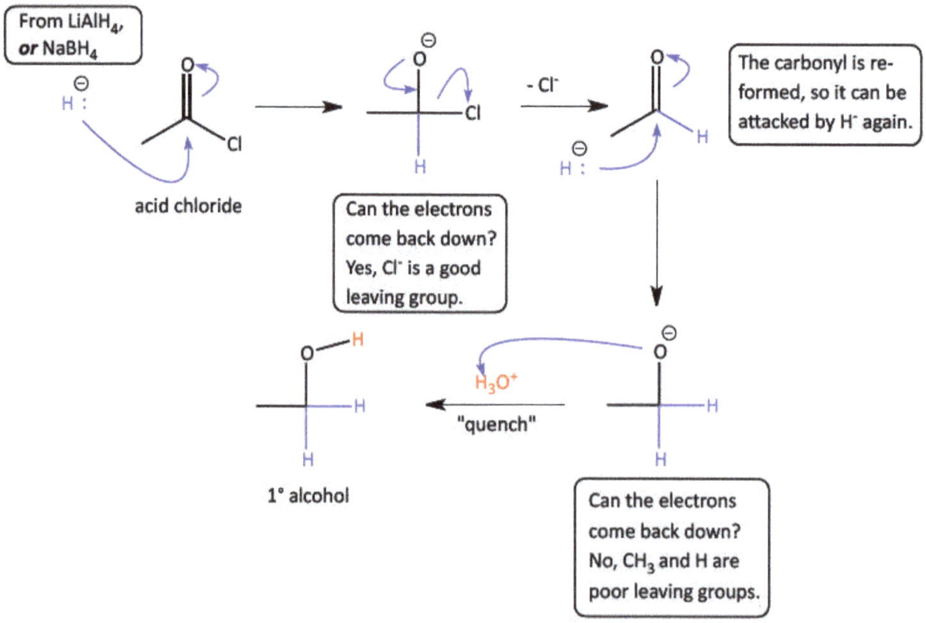

Figure 17.3 Reduction of acid chloride to alcohol

17.3.3 Reduction of ester or carboxylic acid to alcohol

Are other carbonyl compounds like esters or carboxylic acids any different? They are. We find that the lone pair of electrons on the extra oxygen atom donated electron density to the carbonyl carbon atom. Therefore, it is not as positively charged as it is in aldehydes or ketones. The end result of this is that a weak hydride sources like $NaBH_4$ is not strong enough to attack the carbonyl carbon of an ester or carboxylic acid. But, a very strong hydride source like $LiAlH_4$ *is* strong enough to attack the carbonyl carbon atom.

The hydride from the $LiAlH_4$ attacks the carbonyl carbon atom of the ester. The electrons go up onto the oxygen atom of the carbonyl. There is a good leaving group, the OR group. The electrons come back down and kick off the OR group. A carbonyl is reformed as an aldehyde. Aldehydes also react with $LiAlH_4$. The hydride attacks the carbonyl. The electrons go up onto the oxygen atom giving it a formal negative charge. We ask, "Can the electrons come back down?" In this case, they cannot because the aldehyde R groups are poor leaving groups, H and an alkyl group.

This alkoxide intermediate can be quenched with acid and water. This needs to be done carefully! $LiAlH_4$ is quite reactive with water and acid. The reaction mixture should be cooled down and the water added one drop at a time.

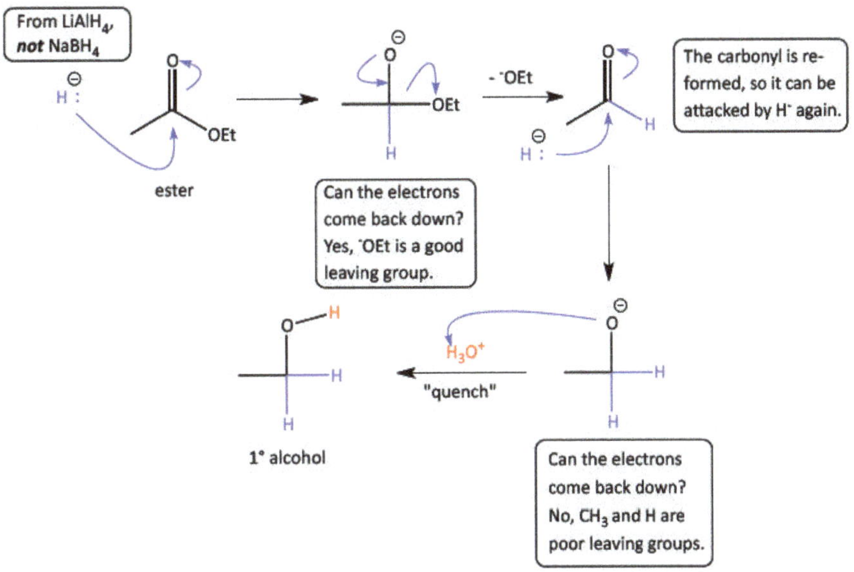

Figure 17.4 $LiAlH_4$ reduction of an ester to alcohol

Like esters, carboxylic acids are fully reduced to primary alcohols using $LiAlH_4$.

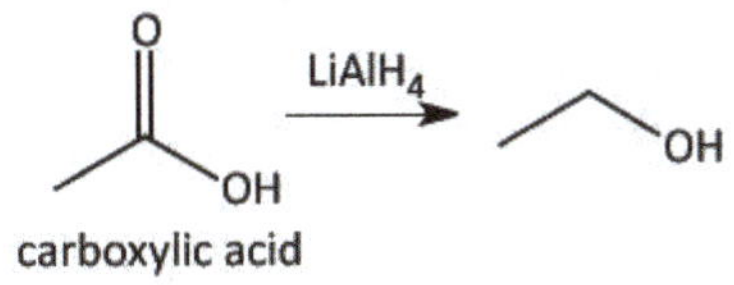

Figure 17.5 $LiAlH_4$ reduction of carboxylic acid to alcohol

Of course, the reduction of a carboxylic acid (three C-O bonds) to an alcohol (one C-O bond) goes through aldehyde (two C-O bonds).

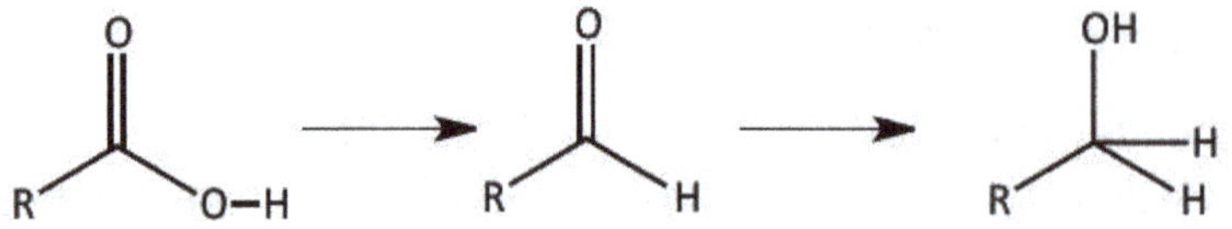

carboxylic acid
3 C-O bonds

aldehyde
2 C-O bonds

alcohol
1 C-O bond

If a strong reducing agent like $LiAlH_4$ reduces all the way to the primary alcohol, is there a way to stop halfway at the aldehyde?

17.3.4 Carboxylic acid reduction to aldehyde

The best way to stop the reduction of a carboxylic acid at the aldehyde stage to do that is to turn the carboxylic acid into an acid chloride with $SOCl_2$ followed by reduction with a gentler reducing agent. Instead of using the very strong $LiAlH_4$, we use lithium tris-t-butoxy aluminum hydride ($LiAl(O^tBu)_3H$). This reducing agent is bulkier and the oxygen atoms donate electron density to the aluminum atom making the hydride less reactive. The reduction therefore stops at the aldehyde because the aldehyde cannot be reduced by $LiAl(O^tBu)_3H$. This reagent is not strong enough to reduce a carboxylic acid, so we must first transform the carboxylic acid into an acid chloride.

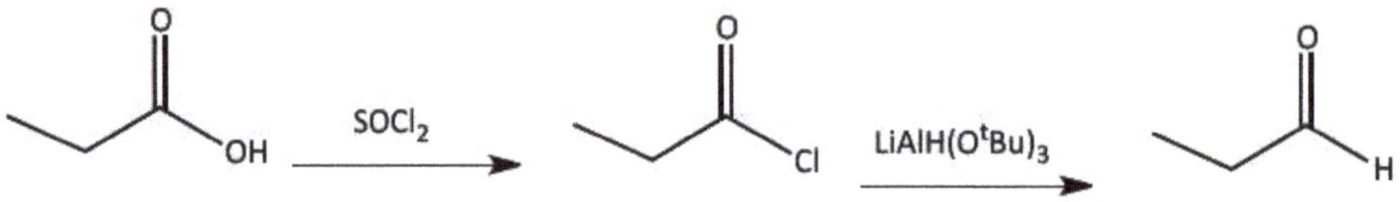

carboxylic acid

acid chloride

aldehyde

There is another method that is sometimes used. A carboxylic acid can also be turned into an ester followed by reduction of the ester with DIBAL-H to make an aldehyde.

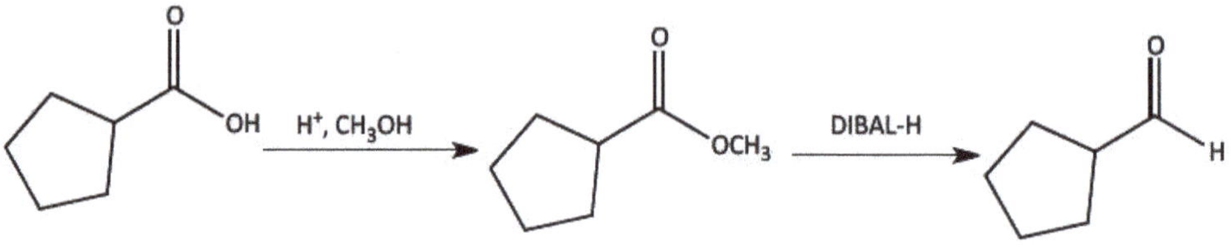

carboxylic acid

ester

aldehyde

17.3.5 Raney nickel, H_2 reductions

While the weaker $NaBH_4$ can reduce aldehydes, ketones, and acid chlorides while the much stronger $LiAlH_4$ can of course also reduce aldehydes, ketone as well as the more difficult to reduce esters and carboxylic acids. But, neither of these reducing agents can reduce alkenes to alkanes. But, H_2 and Raney nickel can. Raney nickel/H_2 performs a hydrogenation across double bonds by adding two hydrogen atoms across any pi bond, from a carbonyl (of an aldehyde or ketone) or an alkene, with syn stereochemistry.

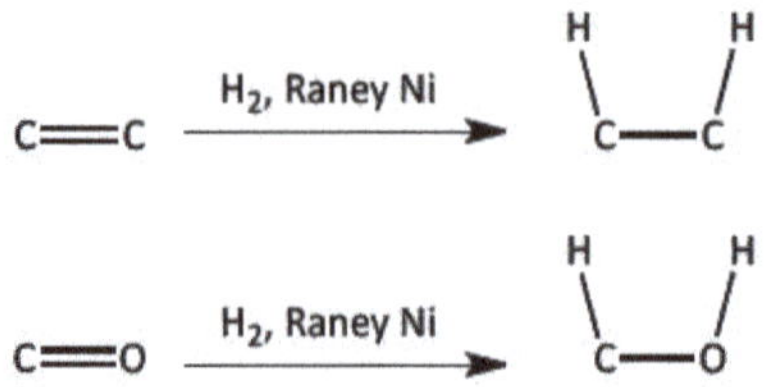

Figure 17.6 H_2 and Raney Nickel reaction with alkene and carbonyl

Notice that both the carbonyl *and* the alkene of cyclopentenone are hydrogenated when reacted with hydrogen and Raney nickel.

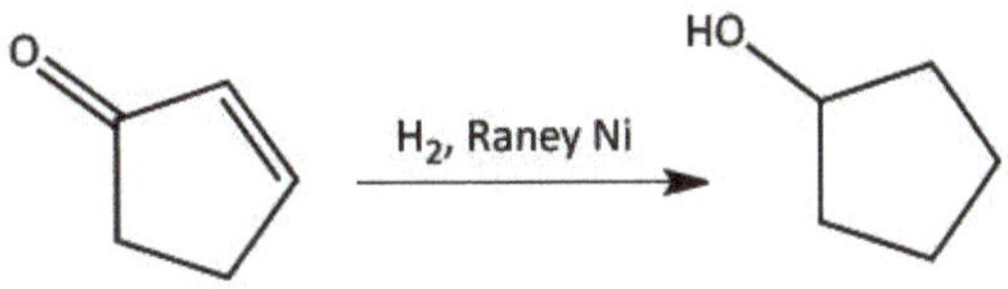

17.3 Fill in the blank boxes.

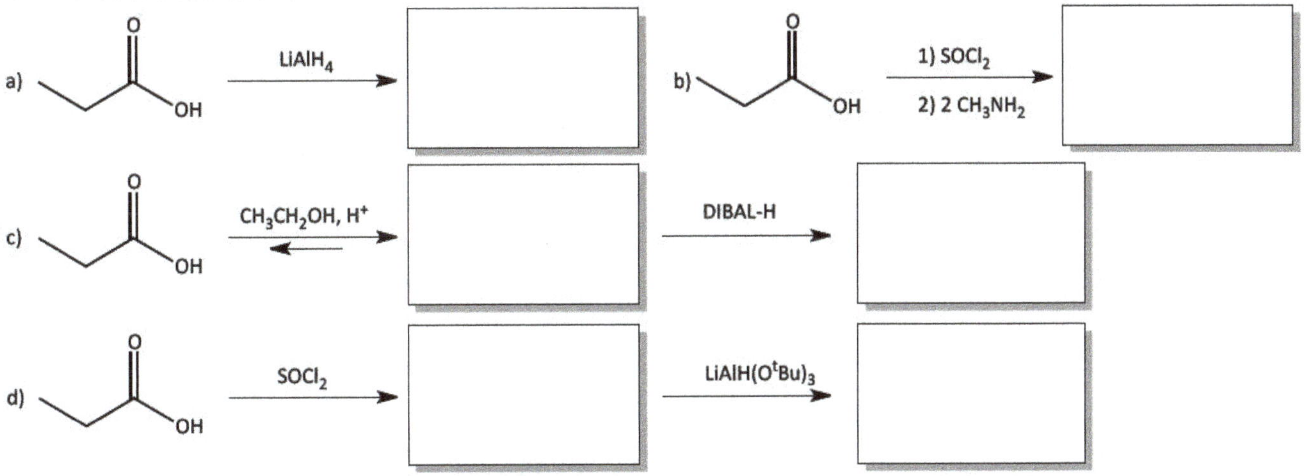

17.4 The nucleophilic attack on carbonyls (Review)

Let's review nucleophiles attacking carbonyl compounds. When a nucleophile, like a Grignard reagent or some other negatively charged species, attacks the partially positive carbon atom of the carbonyl. The electrons of the carbonyl pi bond go up on to the oxygen atom, giving it a formal negative charge. Every time this happens, a question needs to be asked. "Can the electrons come back down?"

What we are really asking with this question is whether the groups attached to the carbonyl are good leaving groups are not. If one of the groups is a good leaving group, like Cl or OR, the electrons come back down to reform the carbonyl pi bond. This would make five bonds to the carbonyl carbon, so the leaving group must leave. Acid chlorides and esters have good leaving groups on the carbonyl because Cl and O like the negative charge when they leave. Whenever the electrons come back down, a carbonyl compound is reformed. When a carbonyl is reformed, another nucleophile can attack, and we ask the same question again. Usually, at this point, the electrons cannot come back down because the remaining groups are all poor leaving groups.

Aldehydes and ketones do not have good leaving groups on the carbonyl carbon because H and alkyl groups are terrible leaving groups because H and C are not very electronegative and are too small to want the negative charge. In the case where there is no good leaving group on the carbonyl, the electrons stay up on the carbonyl oxygen as a negative charge.

Finally, a source of H^+, protons, must be added to the reaction. This is called quenching the reaction. Typically, a dilute aqueous acid is used as the proton source.

Carbonyl Reactions

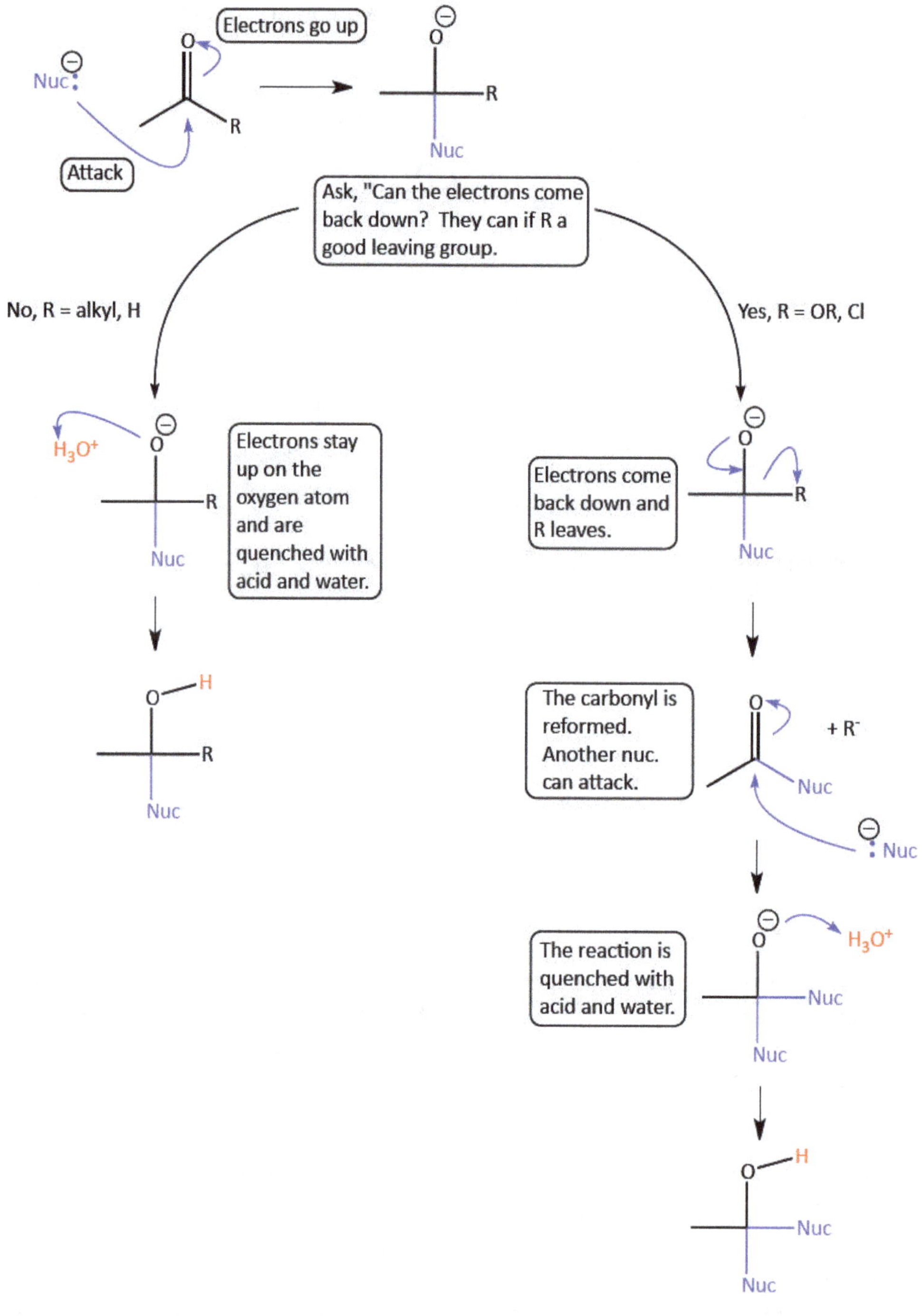

Figure 17.7 General mechanism for a nucleophilic attack of carbonyl compounds

17.4.1 Grignard attack of aldehydes and ketones (Review)

The carbanion of the Grignard reagent attacks the carbon atom of the carbonyl because the carbon atom has a slight positive charge compared to the more electronegative oxygen atom. The electrons go up onto the oxygen atom. When this happens, we ask ourselves, “Can the electrons come back down?” The answer to that question for aldehydes and ketones is “No” since alkyl groups and hydrogen atoms are poor leaving groups. The negative charge stays on the oxygen atom until it is quenched with acid and water to make an alcohol. For ketones, you can notice that a tertiary alcohol is formed. For aldehydes, a secondary alcohol is formed.

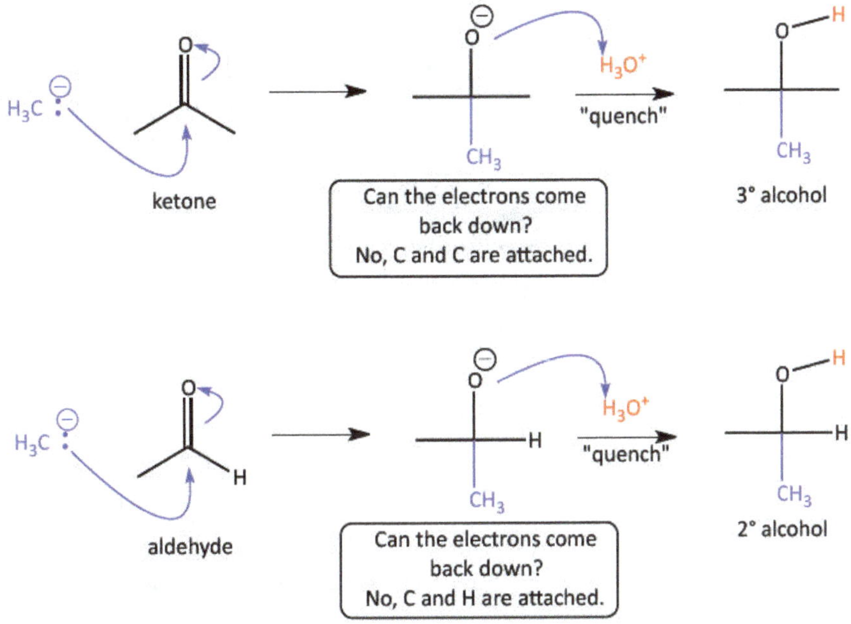

Figure 17.8 Grignard reagents attacking an aldehyde and ketone

17.4.2 Grignard attack of esters and acid chlorides (Review)

For esters and acid chlorides, the carbanion of the Grignard reagent again attacks the carbon atom of the carbonyl because the carbon atom has a slight positive charge compared to the more electronegative oxygen atom. The electrons go up onto the oxygen atom. When this happens, we ask ourselves, "Can the electrons come back down?" The answer to that question for aldehydes and ketones is "Yes" since chloride and $^{-}$O-R are pretty decent leaving groups. Therefore, the carbonyl can be reformed. The carbonyl can be attacked again by the Grignard reagent. The electrons go back up on the oxygen atom. Can they come back down? The answer is "No" because only alkyl groups remain, and they are poor leaving groups. The negative charge stays on the oxygen atom until it is quenched with acid and water to make an alcohol. Tertiary alcohols are then formed.

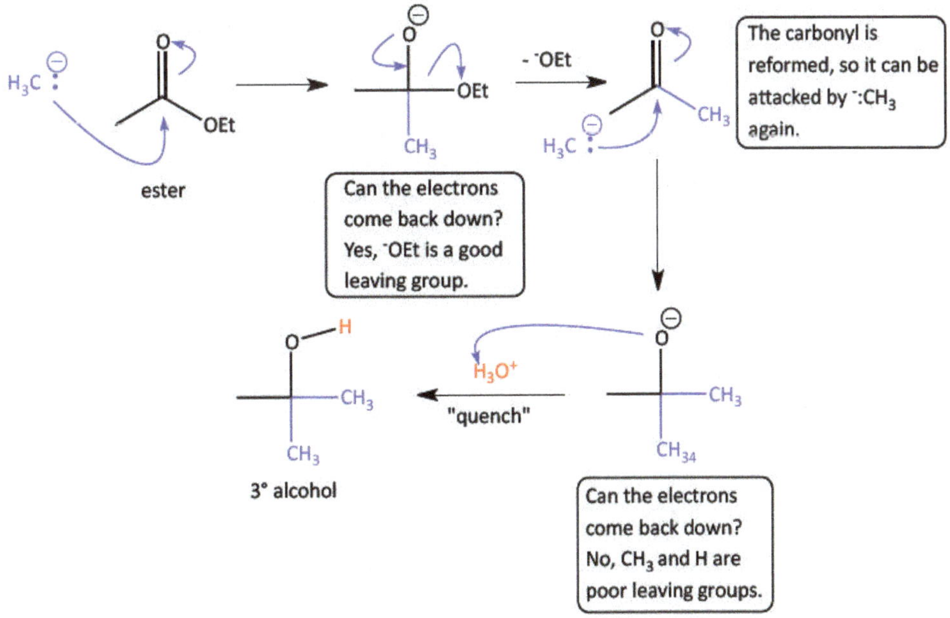

Figure 17.9 Grignard reagents reacting with an ester

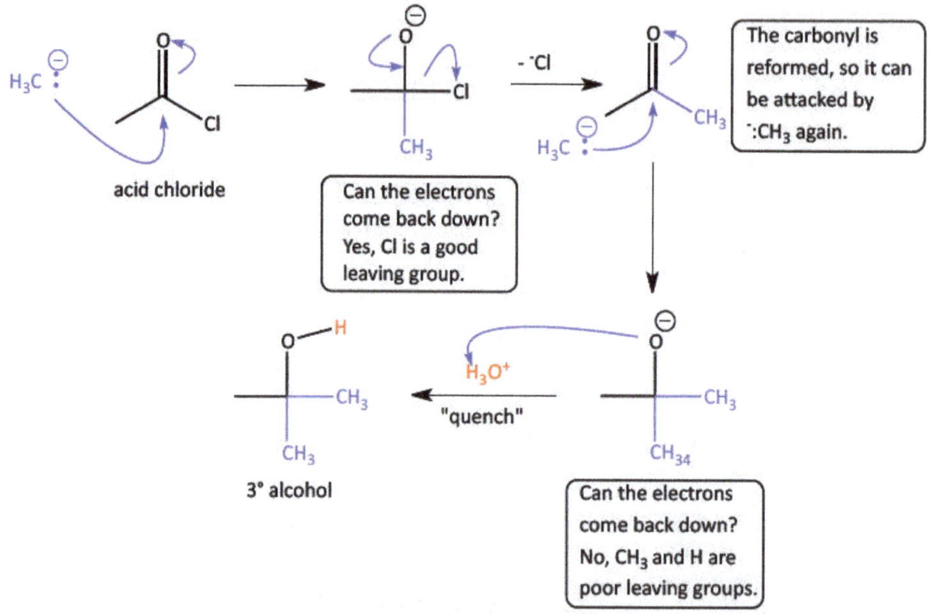

Figure 17.10 Grignard reagents reacting with an acid chloride

17.4.3 Reaction of a Grignard with a carboxylic acid (Review)

Look at this theoretical reaction of a Grignard reagent with a carboxylic acid. Do you foresee any problem with this reaction?

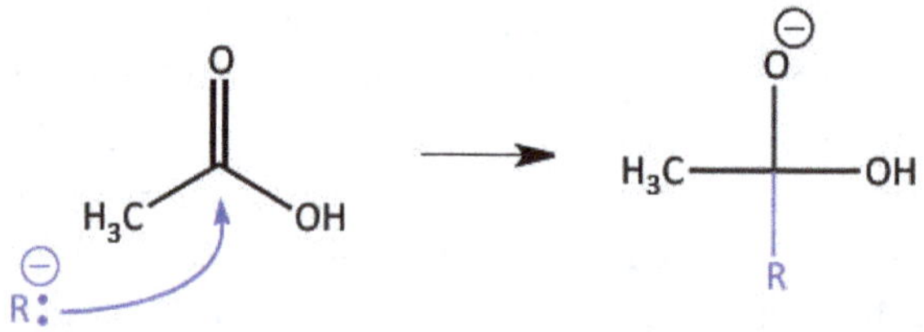

From what we just learned; the Grignard reagent will attack the proton on the oxygen making an alkane. This will effectively kill what we were trying to do in this reaction. The carbonyl will not be attacked.

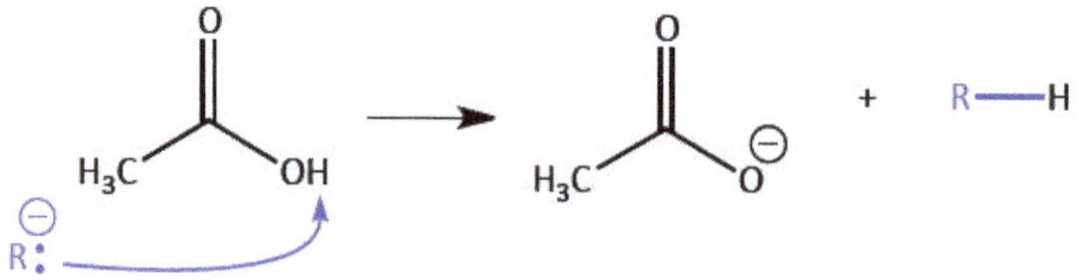

17.5 Reaction Summary

Ketones/aldehyde to imines

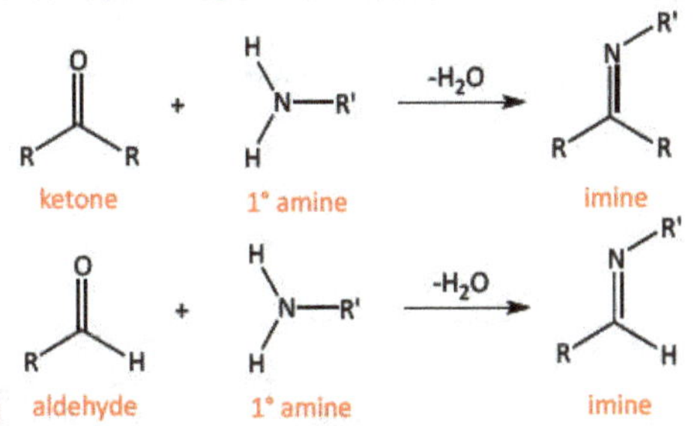

a)

Wittig reaction – Ketones to alkenes

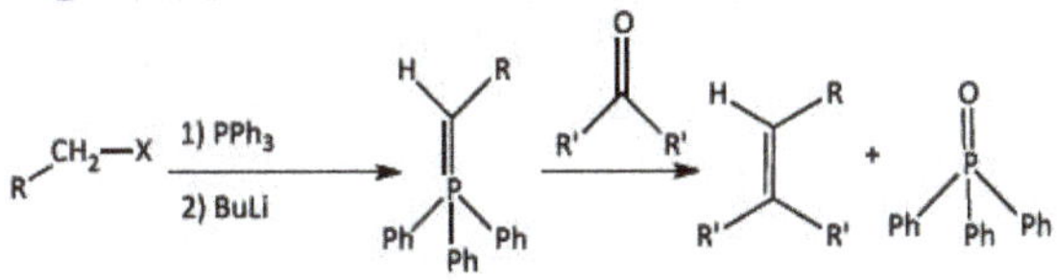

b)

Ketones/aldehydes to acetals/ketals

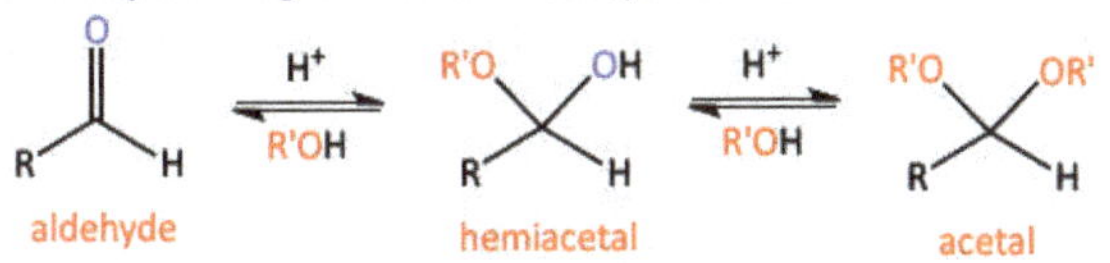

c)

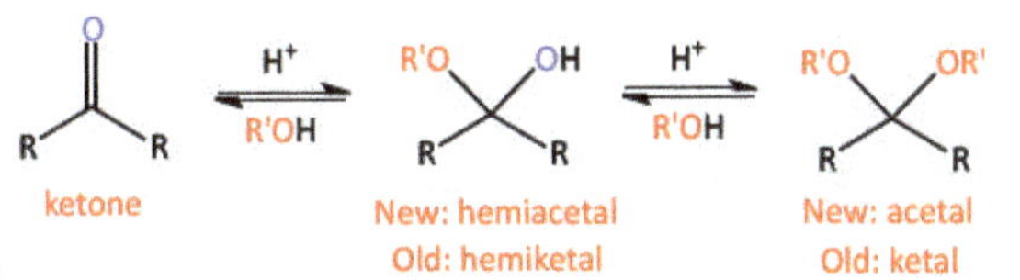

d)

Acetal protecting group

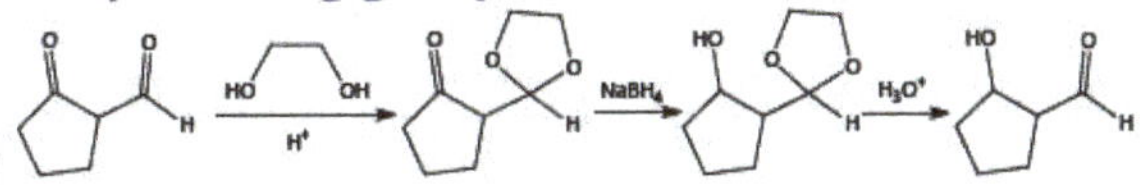

e)

Deprotonation or carboxylic acid

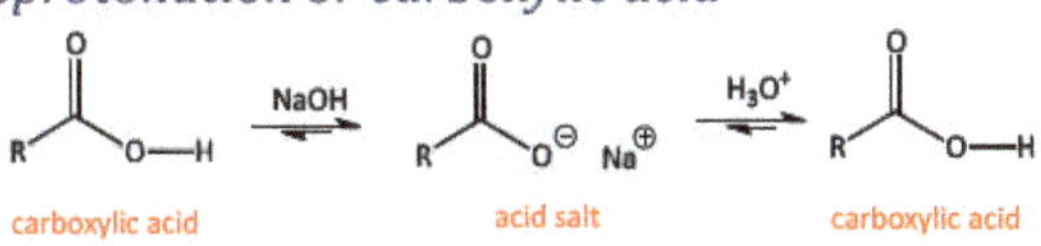

f)

Fischer Esterification – carboxylic acid to ester

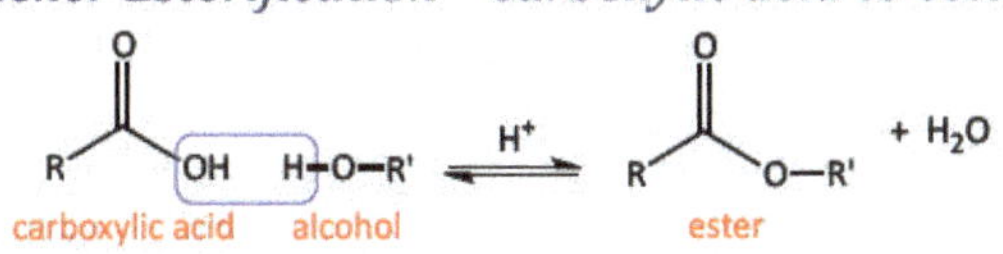

g)

Carboxylic acid to acid chloride to ester

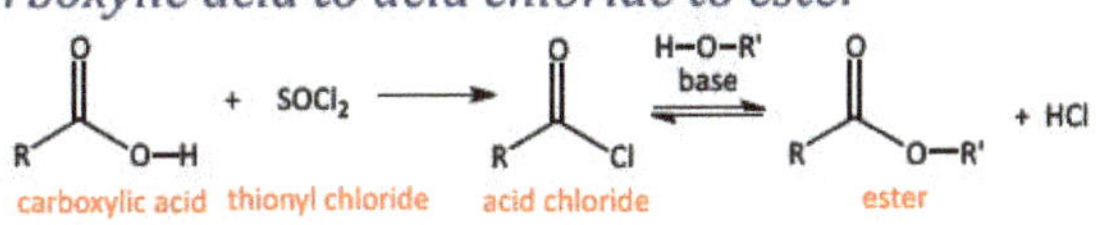

h)

Acid chloride to amide

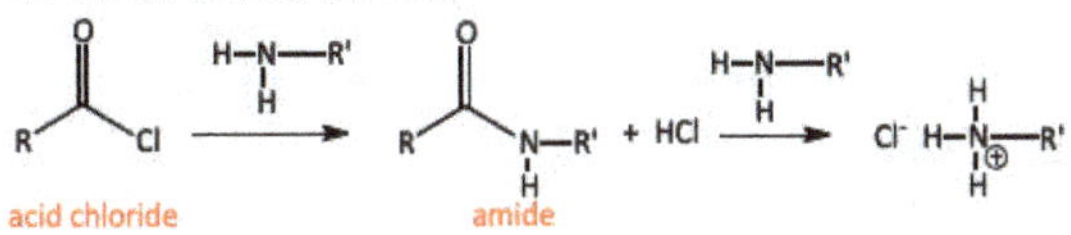

i)

Carboxylic acid to ketone

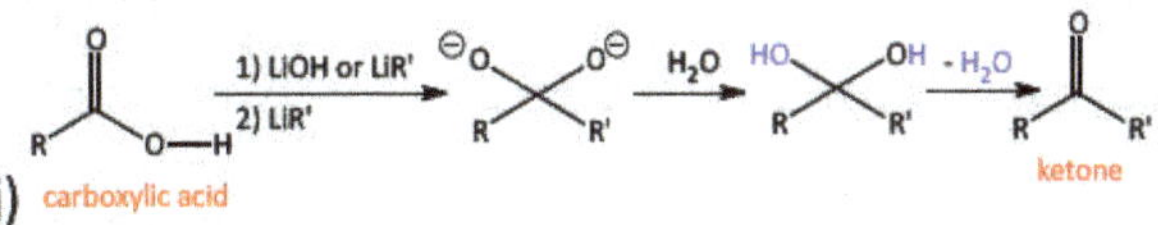

j)

Reductions of carbonyls

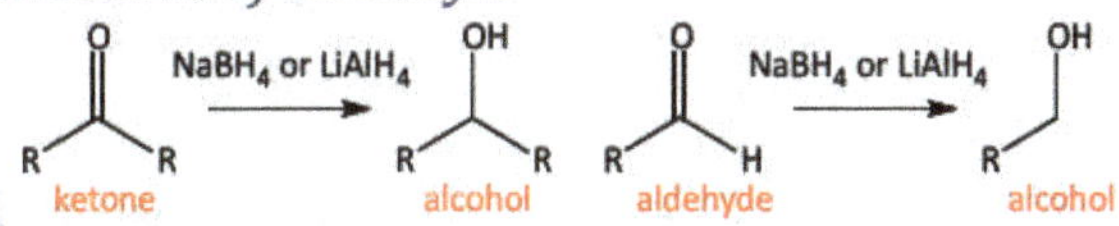

k)

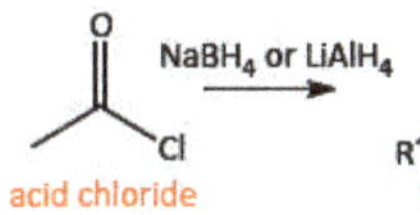

l)

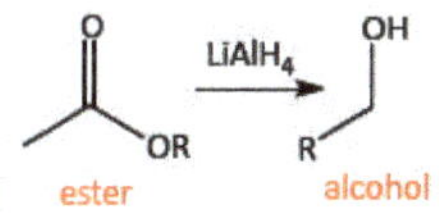

m)

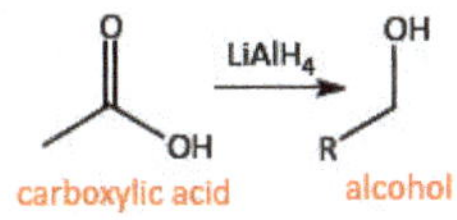

n)

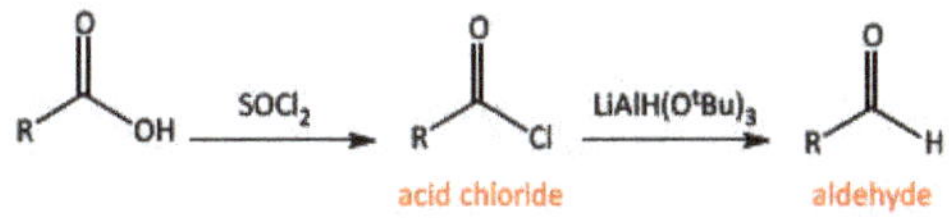

o)

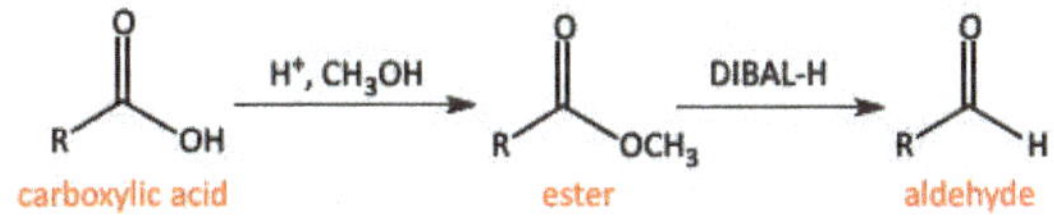

p)

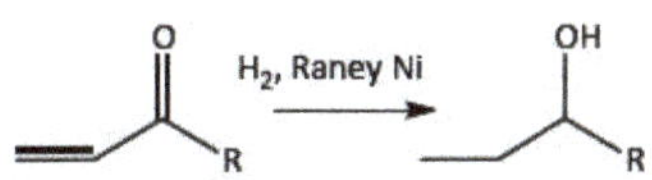

q)

Grignard reagent attacks of carbonyls

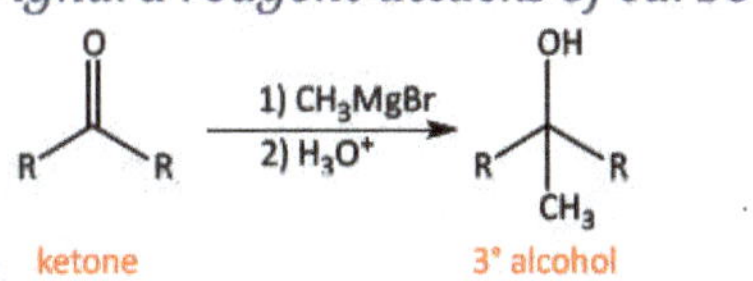

r)

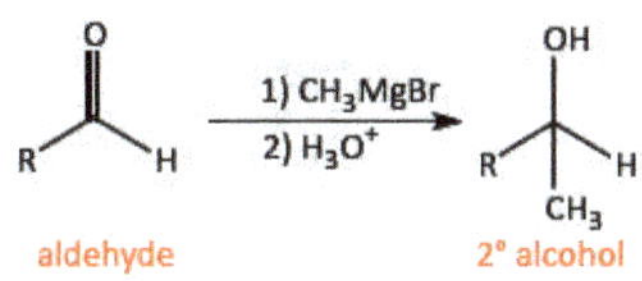

s)

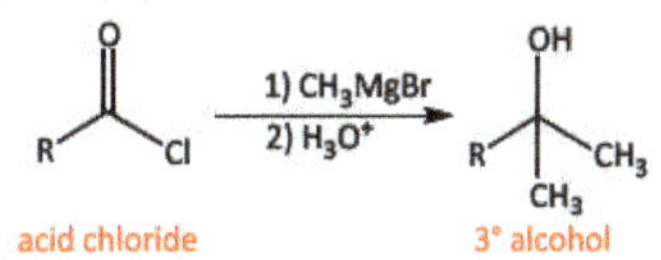

t)

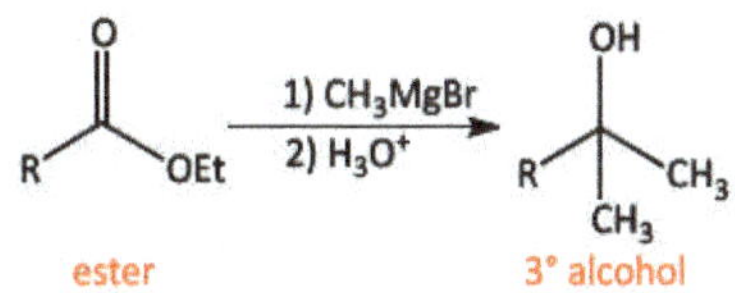

u)

17.6 Problems

17.4 Show how the following compounds can be synthesized from an appropriate carboxylic acid. For the aldehydes, provide two syntheses, one with an ester intermediate and one with an acid chloride intermediate.

a) 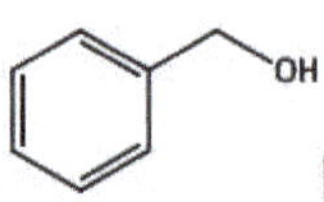 b) 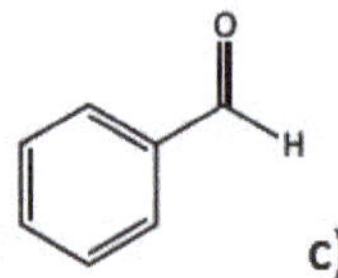 c)  d) 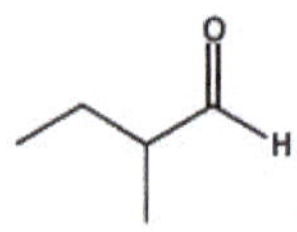 e) 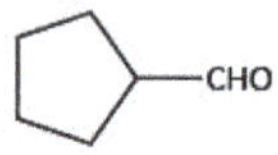 f) hexanol

17.5 Provide the major organic product that results when benzaldehyde, PhCHO, is treated with the following sequence of reagents: 1. CH_3MgBr; 2. H_3O^+; 3. $Na_2Cr_2O_7$, H_2SO_4.

17.6 Identify the carbonyl compound and amine that would be used to synthesize the following imines.

a) 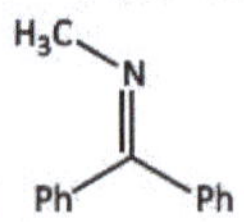 b) 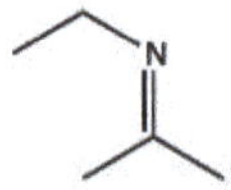 c) 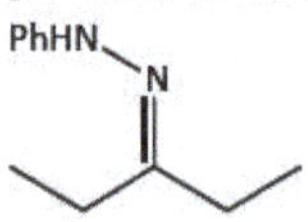 d) 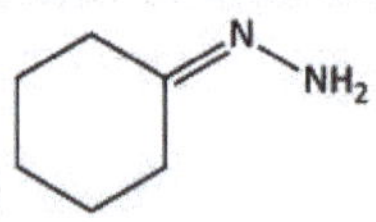 e) 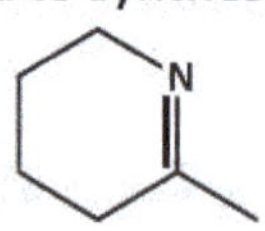

17.7 When 2-methylcyclopentanone reacts with ethylamine in acidic conditions, two different compounds are formed. These are geometric isomers of each other. Draw them.

17.8 Show how you would synthesize the following alkenes from an alkyl halide and aldehyde or ketone.

a) 2-methylpentene b) 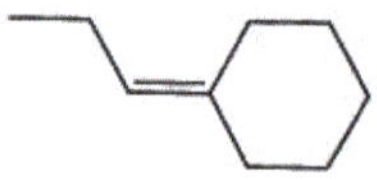 c) 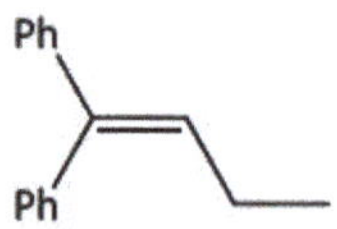 d) 3-heptene

17.9 D-glucose is a familiar monosaccharide. Its Fischer projection is shown below.

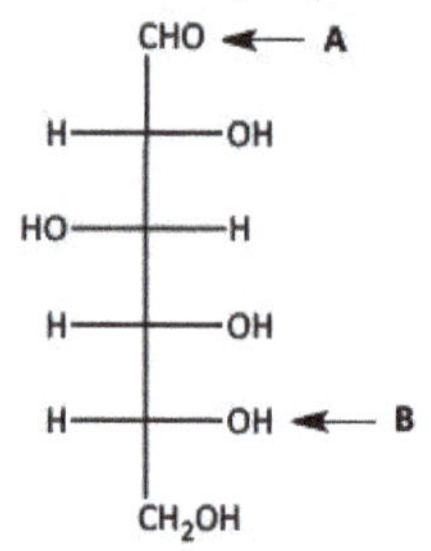

D-glucose

The aldehyde at position A reacts with the hydroxyl group at position B to give the ring structure shown here.

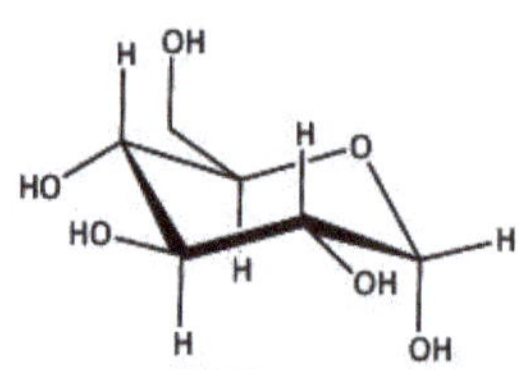

α-D-glucose

a) Circle the hemiacetal portion of the D-glucose ring shown above.

b) When a small amount of acid is present, the *hemiacetal* portion of the above molecule reacts with another alcohol to form what biochemists call a glycoside. Quite often this alcohol is one of the alcohol groups on another monosaccharide to form di-, tri-, or polysaccharides. In fact, glycosides formed by adding several glucose rings together make either starch or cellulose. But let's keep your task simple. *Show the glycoside formed from the reaction of α-D-glucose with methanol.*

17.10 **Identify the carbonyl compound and the alcohol(s) needed to make the following acetals or hemiacetals.**

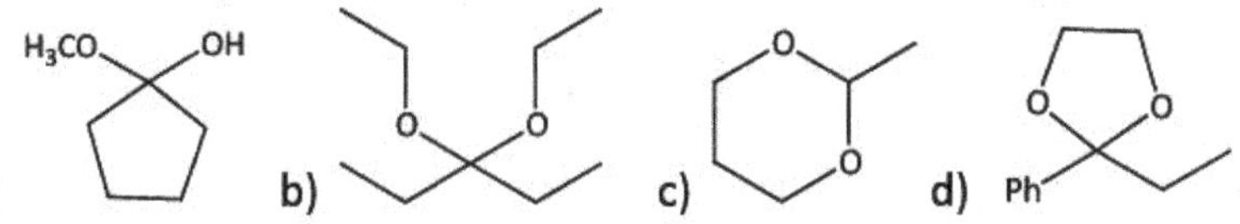

17.11 **Identify the products of the hydrolysis of the following compounds.**

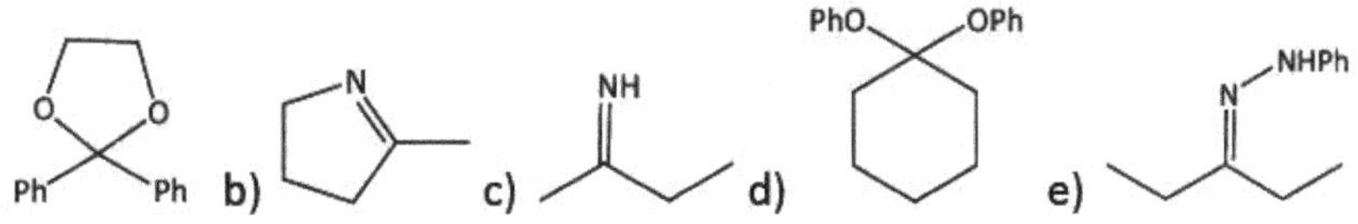

17.12 **Identify a ketone or aldehyde and a Grignard reagent that would make the following compounds.**
a)2-methyl-3-heptanol b) 1-phenylpropanol c) 1-ethylcyclohexanol d) 2-methyl-2-pentanol

17.13 **Show how to make the following compounds by using an excess of appropriate Grignard reagent and an acid chloride. Then show how to make it from a Grignard reagent and an ester.**
a) 1,1-diphenylpropanol b) 1,1-dicyclohexylethanol c) 3-ethyl-2,2-dimethylpentan-3-ol d) 2-methyl-2-pentaol

17.14 **Fill in the reagents or products (A-M) of the following reactions.**

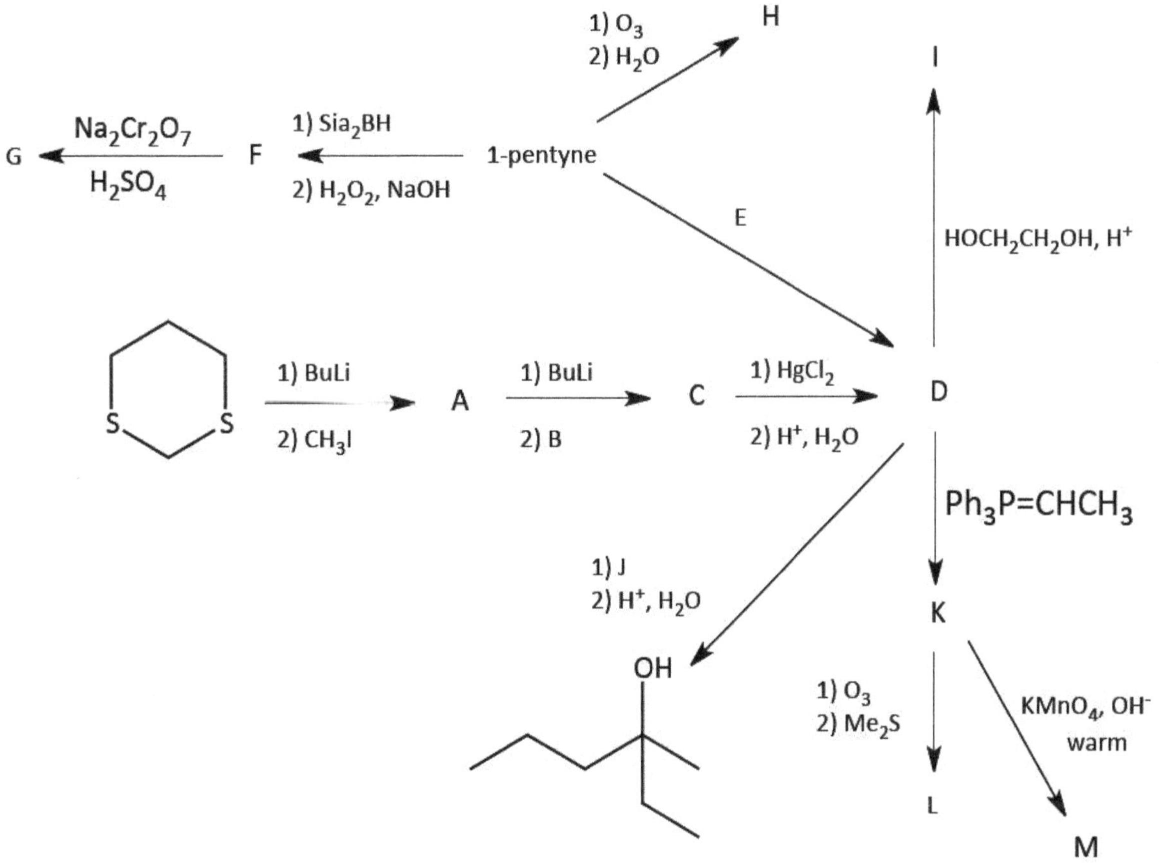

Carbonyl Reactions

17.15 Provide the major organic products for the following reactions.

a) PhCHO $\xrightarrow[\text{2) } H_3O^+]{\text{1) PhMgBr}}$

b) 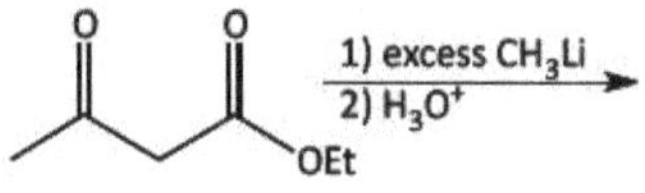

c) 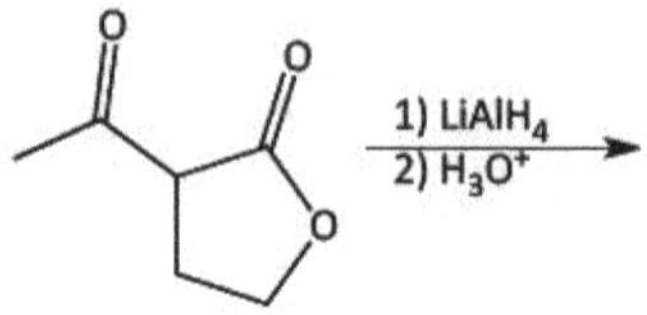

d) 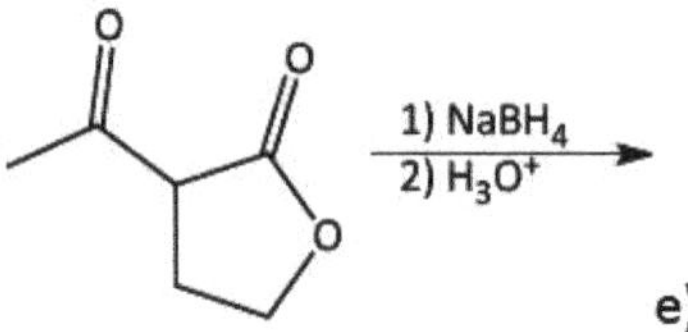

e) 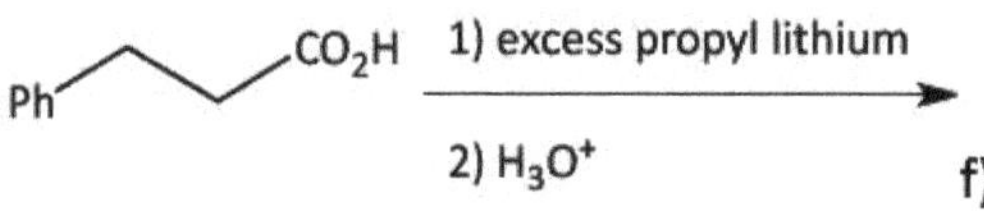

f) 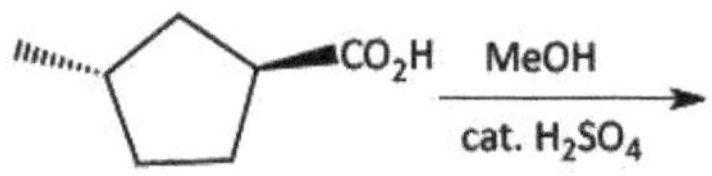

g) 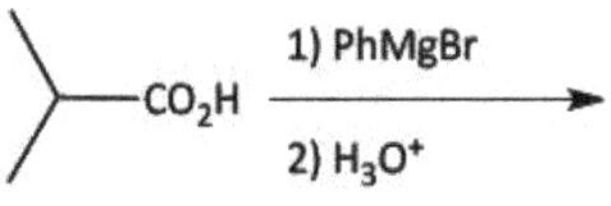

h) 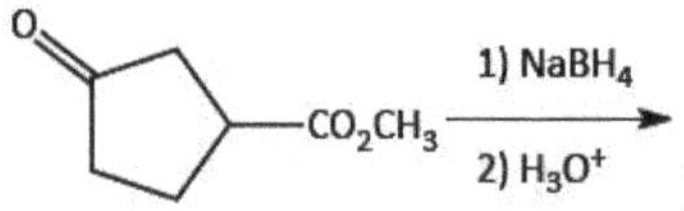

i) 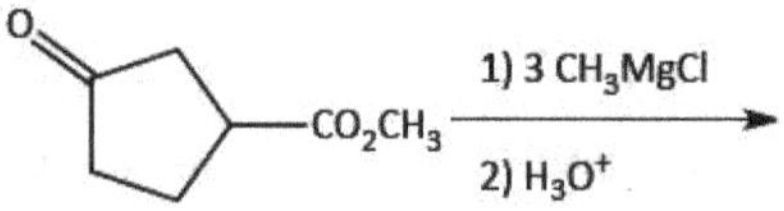

j) 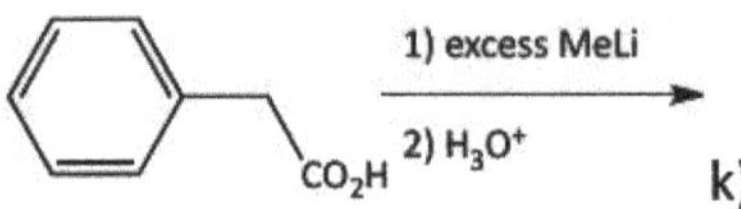

k) 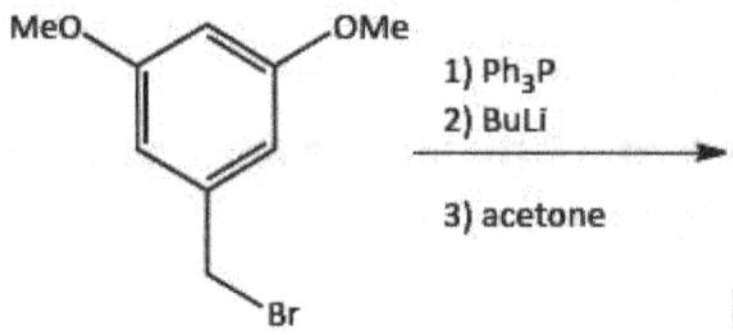

l) 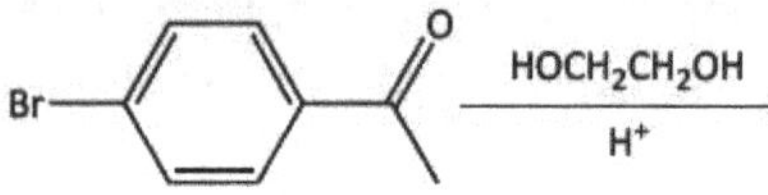

q)

m) 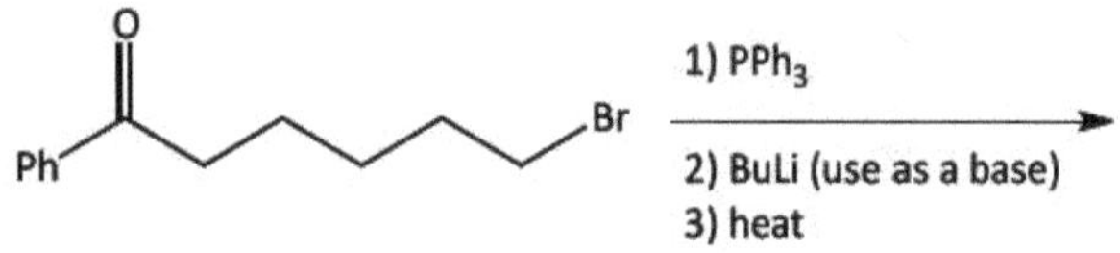

n) 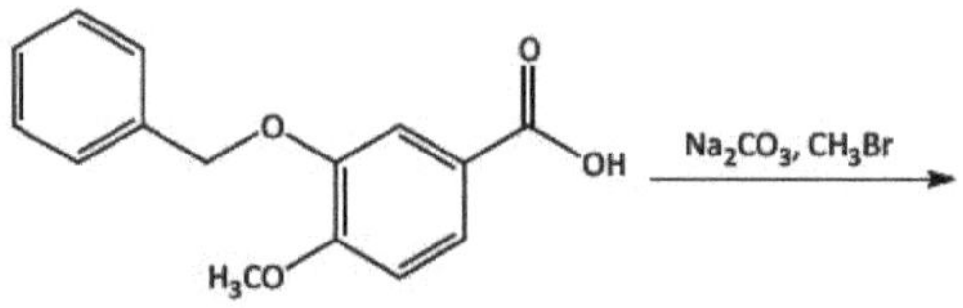

o) 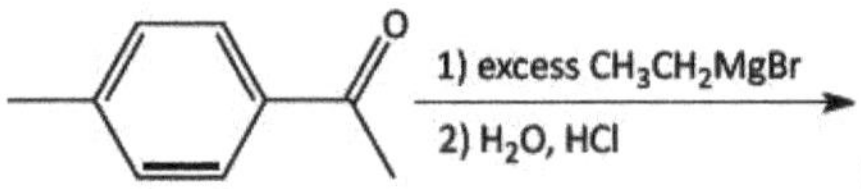

p) 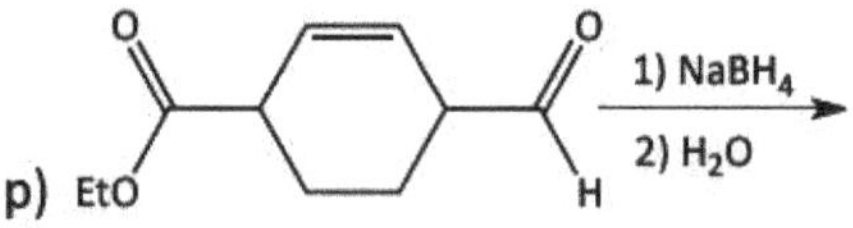

q) 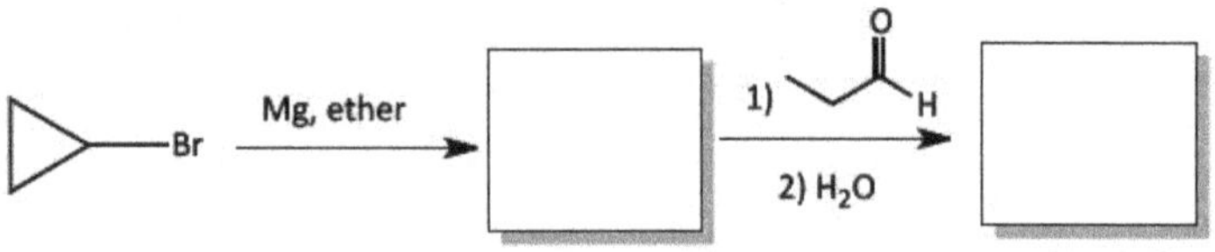

r) 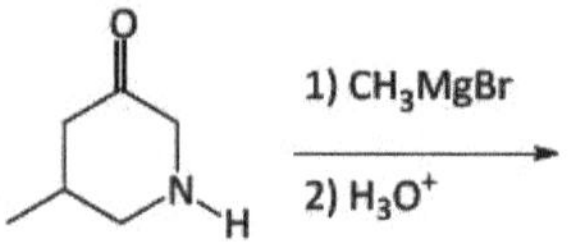

s) 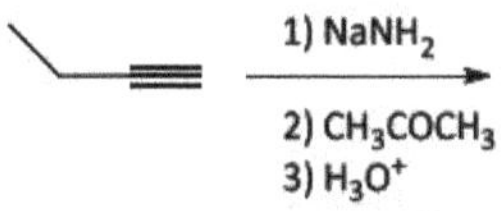

t) 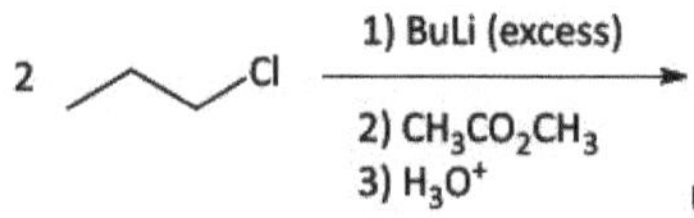

u) 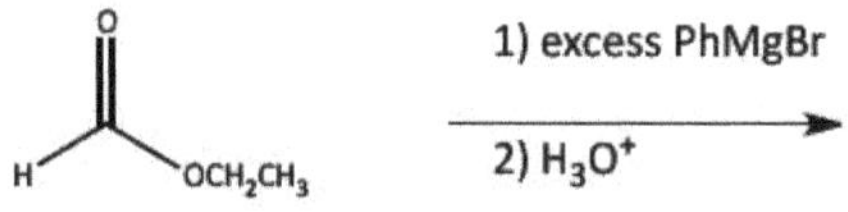

v) 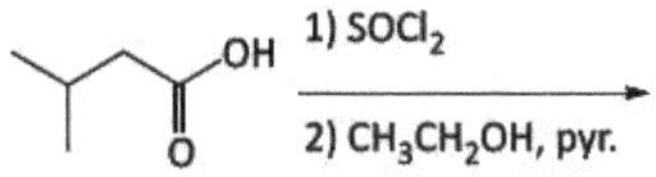

w) 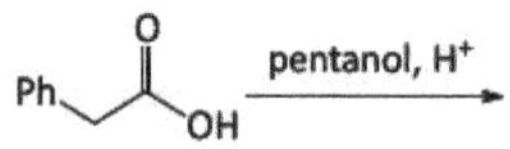

x) 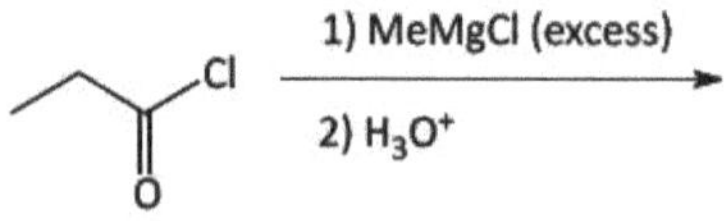

y) 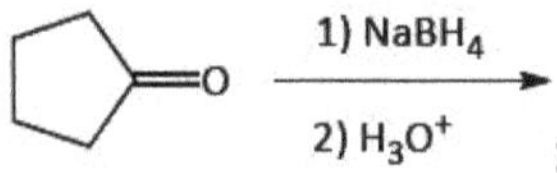

z) 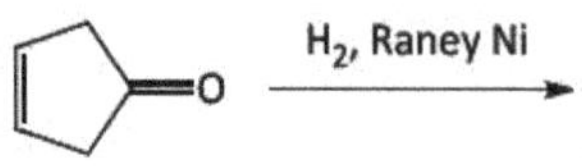

aa) $H_2C{=}CHCO_2H$ + 1) $LiAlH_4$, 2) H_3O^+

bb) $CH_3CO_2CH_2CH_2CHO$ + 1) $NaBH_4$, 2) H_3O^+ cc) $CH_3COCH_2CH{=}CHCH_2CO_2CH_3$ + 1) $LiAlH_4$, 2) H_3O^+

dd) CH_3CHO + 1) LAH, 2) H_3O^+ ee) $PhCH_2COCl$ + 1) $LiAlH_4$, 2) H_3O^+ ff) $PhCH_2CH_2COCl$ + 1) $LiH(OtBu)_3$, 2) H_3O^+

gg) $PhCOCl$ + $(CH_3)_2CuLi$

17.16 Draw the structures of A-P.

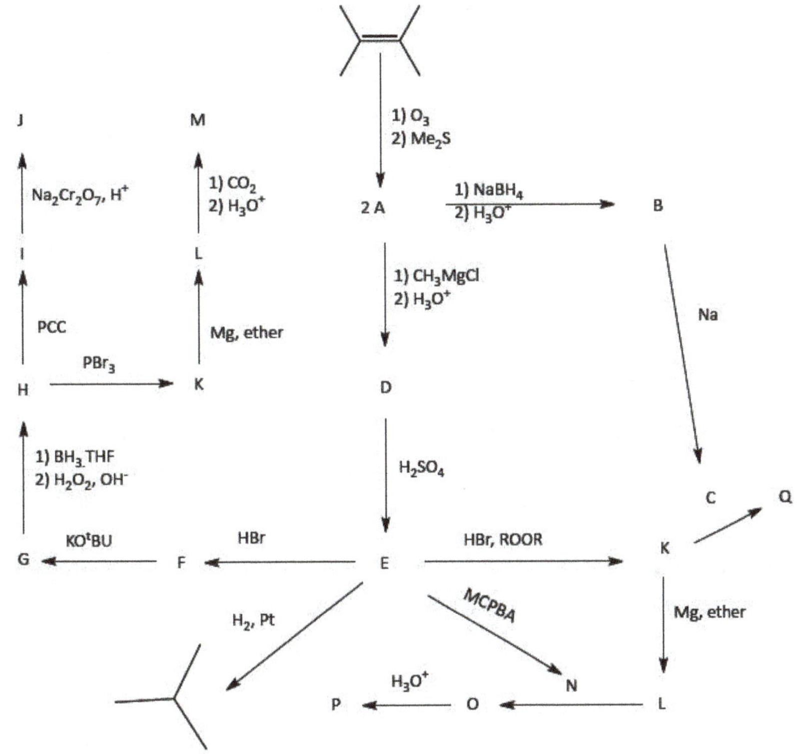

17.17 Show how you would make the following alcohol starting with bromobenzene and 1-butene. More than one reaction is needed.

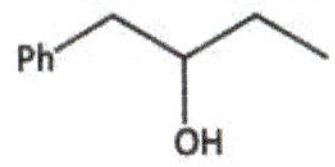

17.18 Show how you would make the following alcohol starting with 2-pentanol and acetylene (H-C≡C-H). More than one reaction is needed.

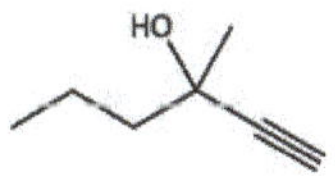

17.19 Show how you would make styrene starting with bromobenzene and ethyleneoxide. More than one reaction is needed.

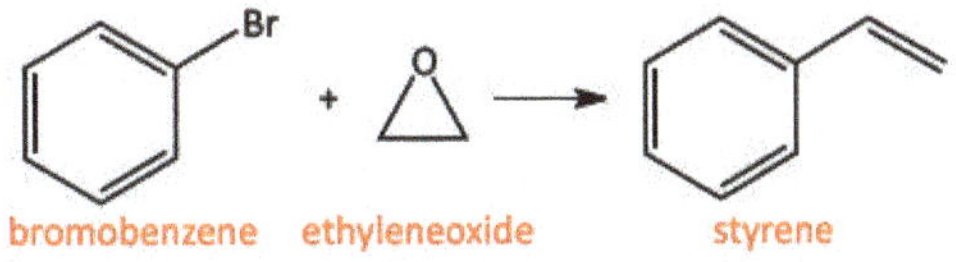

17.20 Show how you would make the following ketone starting with cyclopentylmethanol as your only organic reagent. More than one reaction is needed. You can use as more than one equivalent of cyclopentylmethanol if necessary.

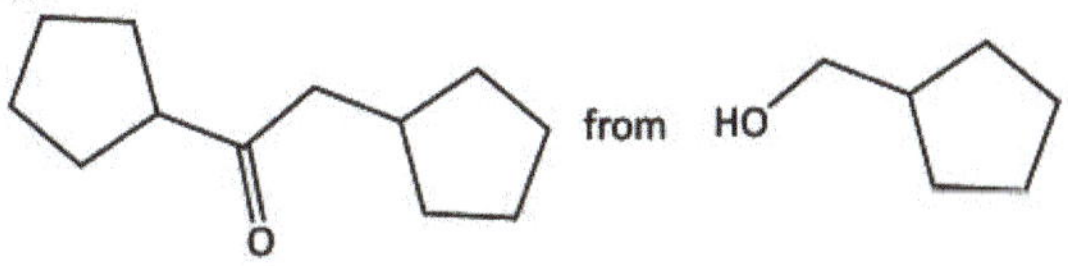

Carbonyl Reactions

17.21 Fill in the blanks in the following reactions.

a) 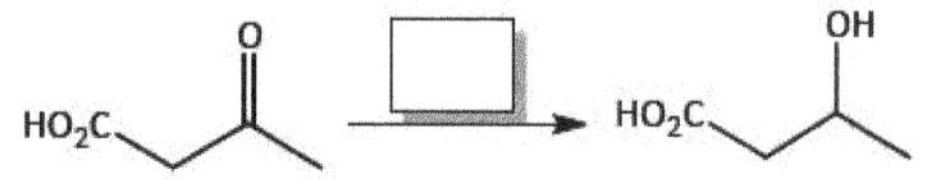

b) 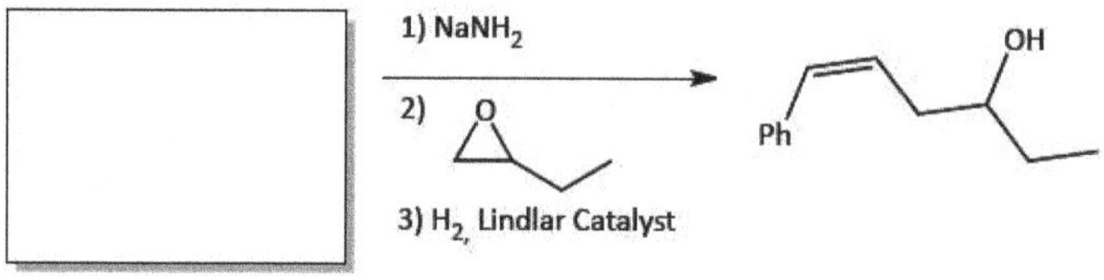

c) 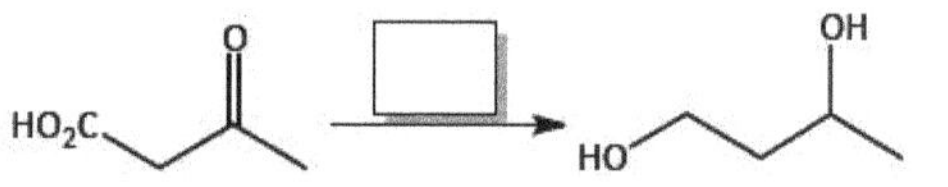

d) 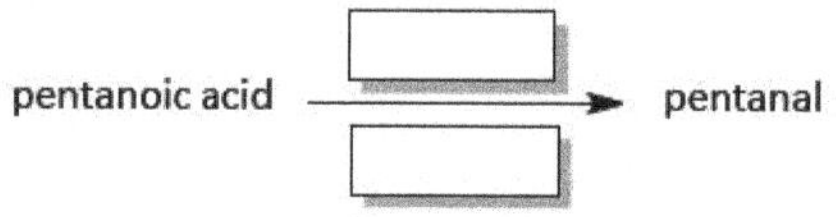

e) 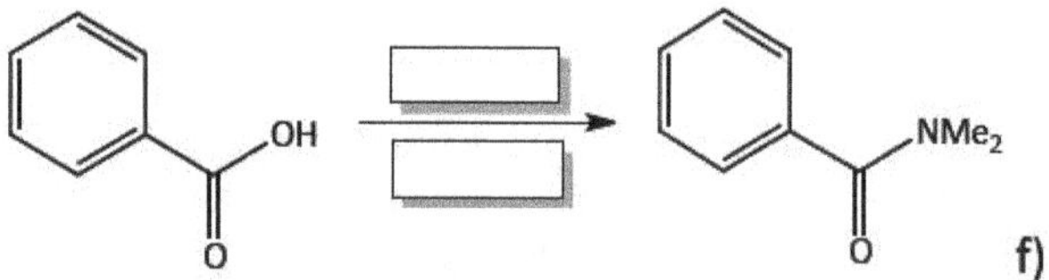

f) 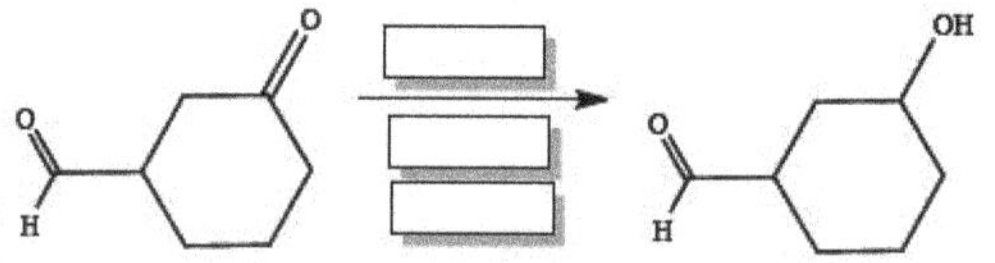

17.22 The mechanism for the reaction below begins the same way we have seen Grignard reagents attacking esters. Then, in an interesting twist, an ether is formed because of the bromide leaving group. Show the full mechanism for this reaction.

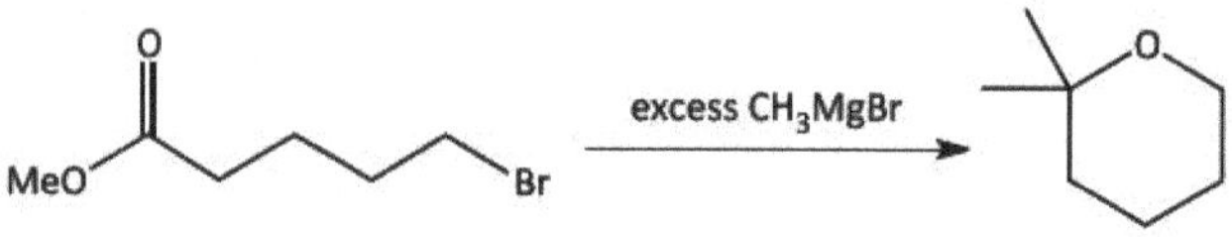

17.23 Why doesn't the following reaction work as written?

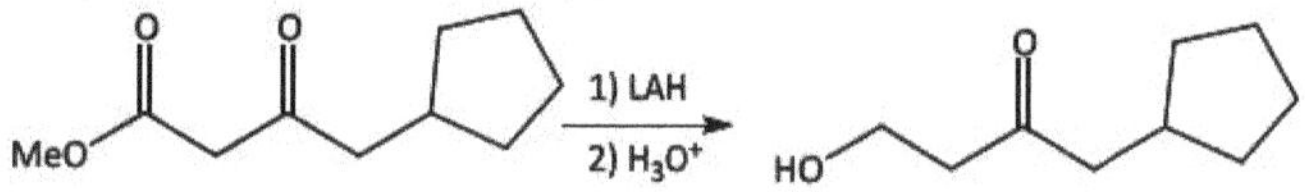

17.24 Describe the reagents that could be used to perform the following reactions. Sometimes, more than one reaction may be required.

a) 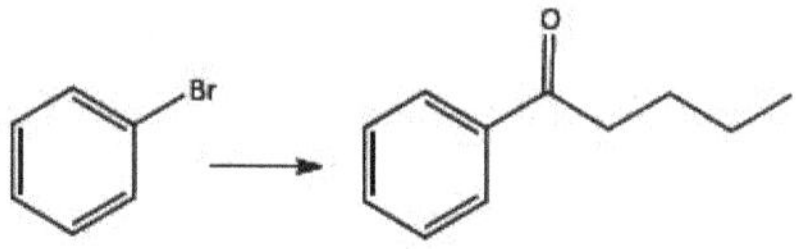

b) 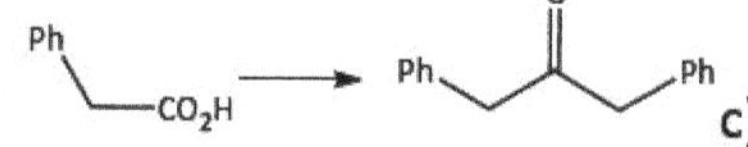

c) 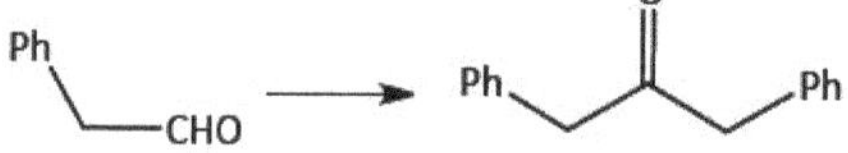

d) 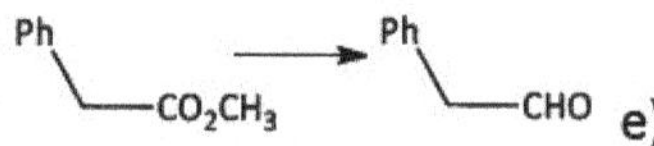

e) 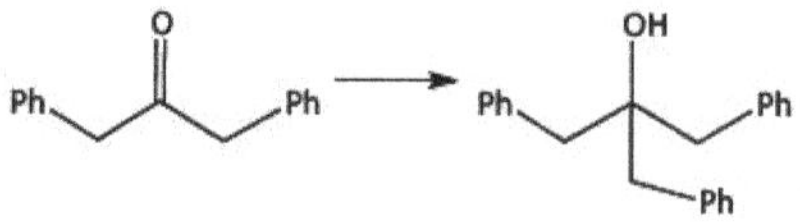

f) 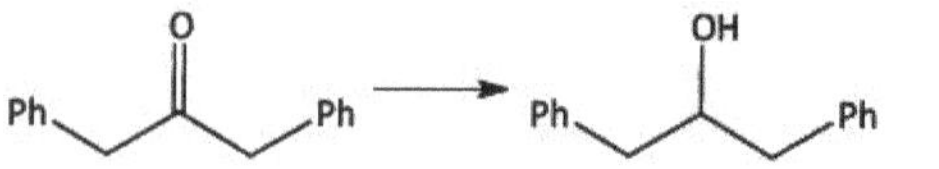

g) 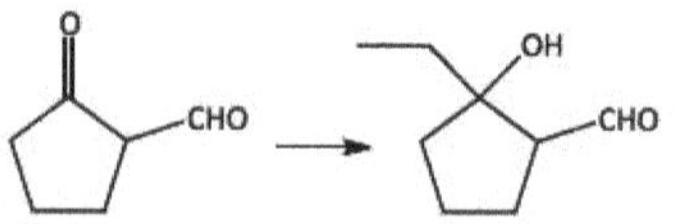

h) 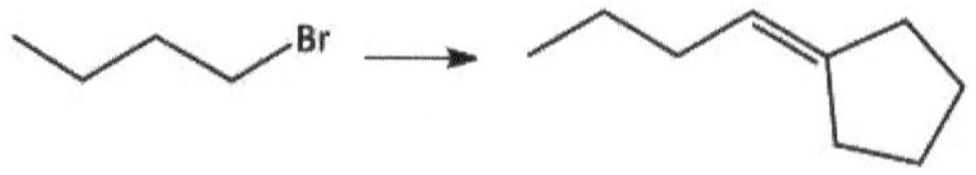

i) 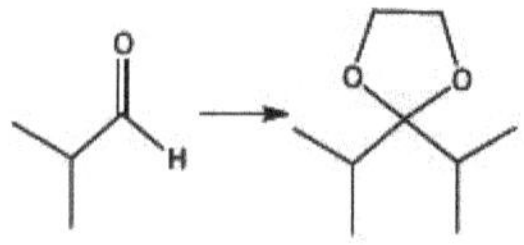

j) $PhCH_2CO_2H$ to $PhCH_2CO_2CH_2CH_3$ k) benzaldehyde to benzoic acid l) propanoic acid to propanal

m) $PhCH_2CO_2H$ to $PhCH_2CONHCH_2CH_3$

Chapter 18. Enolates

18.1 The Enolate

Protons that are on a carbon atom adjacent to a carbonyl, the alpha carbon atom, are quite acidic. This is because once the molecule is deprotonated, the resulting carbanion can be stabilized through resonance forms. The major resonance form is the one with the negative charge on the oxygen atom. That is more stable than the negative charge on the carbon atom. This negative species is called an enolate (pronounced een-ohl-ate) ion.

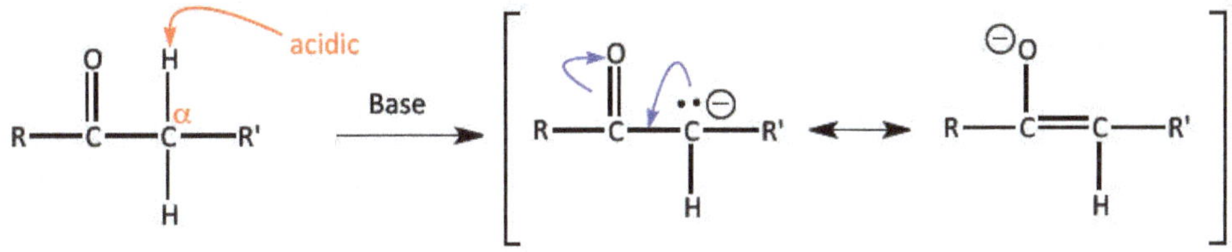

minor version   major version

Enolate ion

An enol is a compound with a carbon-carbon double bond and an alcohol group on that double bond. An enolate is an enol that is deprotonated making it have a negative charge. The negative charge is why it has the “ate” on the name.

Enol   Enolate

Even though the major form of the enolate is the one with the negative charge on the oxygen atom, I think it is beneficial to keep in mind the minor version, with the negative charge on the carbon atom, for the reactions in this chapter.

A good base like hydroxide (HO^-) is used to remove the alpha hydrogen atom. A bigger, bulkier base is also often used. The base of choice is one named LDA, lithium diisopropyl amide. This is a lithium salt of a negative nitrogen atom with two big, bulky isopropyl groups attached to it.

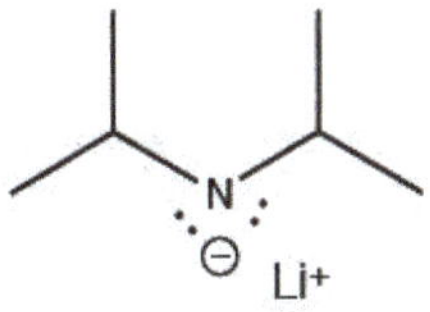

Figure 18.1 LDA, lithium diisopropyl amide

18.2 Enolate as nucleophile

Enolates are good nucleophiles. Once the negative carbon atom is formed, the enolate can attack halogens like Br_2 or Cl_2. The alpha carbon atom can be halogenated in this way.

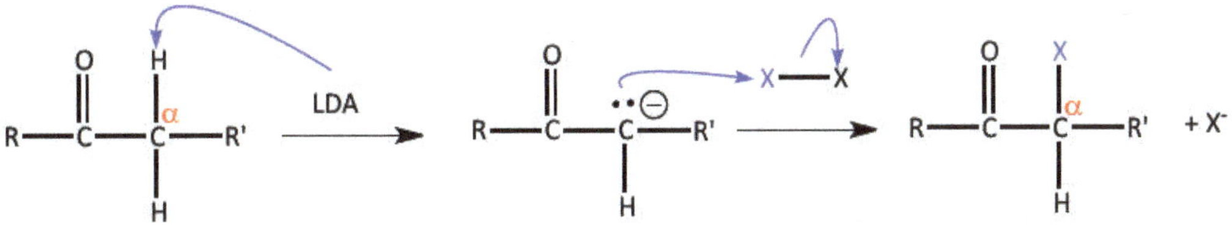

Enolates can also perform S_N2 reactions and attack alkyl halides.

Enolates

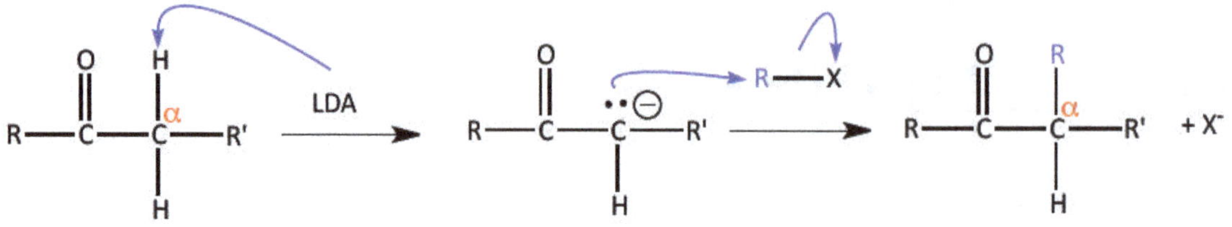

18.1 What are the products of the following reactions?

a) acetone + LDA followed by chloromethane b) cyclohexanone + LDA followed by Br_2 c) 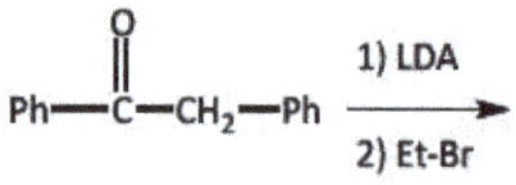

18.3 Aldol condensation

The aldol condensation is an important reaction. It is a reaction that helps form carbon-carbon bonds, which is extremely important in organic chemistry. In the aldol condensation, an enolate attacks a ketone or an aldehyde. Let's begin our study of the aldol condensation using acetone as our model molecule. Acetone is a dimethyl ketone that contains acidic, alpha hydrogen atoms.

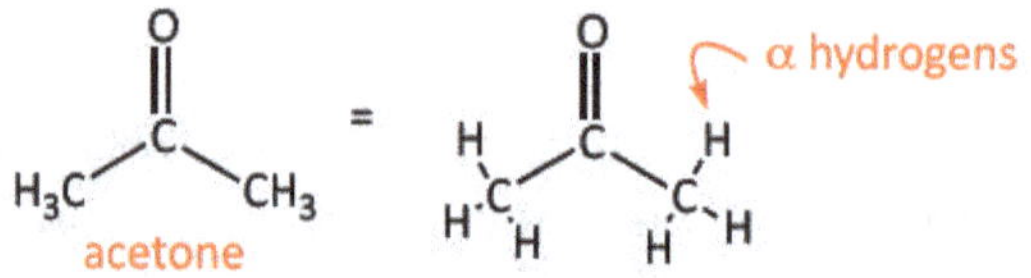

What would happen if we take a beaker full of acetone and add a base like hydroxide to it? The alpha hydrogen atoms of acetone can be deprotonated with the hydroxide to make an enolate. When this happens, the reactive, acetone enolate would be surrounded by gazillions of other molecules of acetone. The negative charge of the enolate can attack the carbonyl of the acetone, and the pi bond electrons can go up onto the oxygen atom to make a negative charge. The electrons cannot come back down because there is no good leaving group. This negatively charged species can be protonated, quenched, by acid and water to give an alcohol. Since a hydrogen atom and the hydroxyl group are on neighboring carbon atoms, a dehydration elimination can occur where water is given off and a conjugated double bond is formed. This dehydration is often exothermic, because a more stable, conjugated system is formed. A conjugated molecule is one that contains two double bonds separated by a single bond

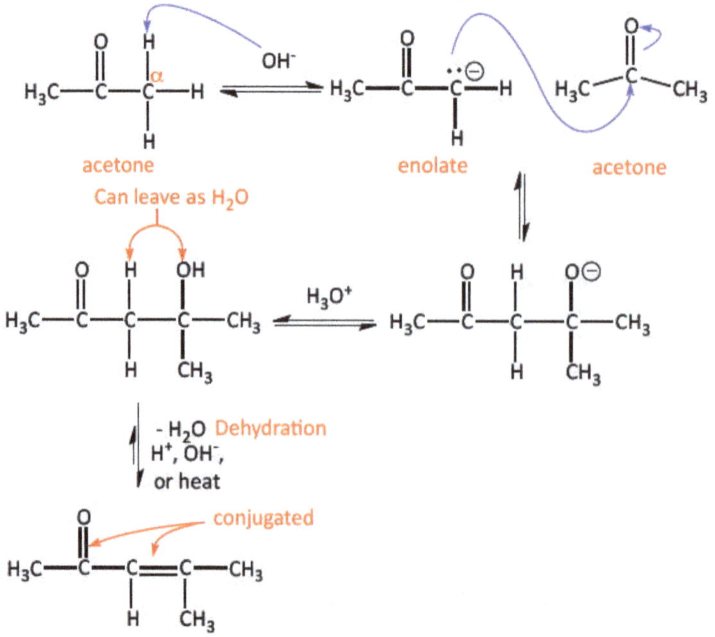

The aldol condensation is a reversible reaction. The reaction can go forward or backwards. If the water produced in the dehydration step of the reaction is removed, the equilibrium can be forced to the product side of the reaction.

18.3.1 To dehydrate or not

Not every aldol reaction results in the dehydration elimination.

1. If hydroxide is used as the base, the dehydration usually occurs.
2. If LDA is used as the base, the dehydration does not normally occur.
3. Under acidic conditions, the dehydration usually occurs.

18.2 The following compounds undergo aldol condensation reactions when they react with base. Show the products of the aldol condensation following dehydration.
a) propionaldehyde (propanal) b) acetophenone (methyl phenyl ketone) c) cyclohexanone

18.3.2 Crossed aldol

Many aldol condensations involve a molecule transforming into an enolate and then reacting with itself like in the acetone example above. But, that is not necessarily the case. It is possible for an enolate to react with a carbonyl compound that is different. When this happens, it is called a Crossed Aldol condensation. Let's begin our study of the crossed aldol condensation with an example. In this case, acetone is deprotonated to form an enolate followed by a reaction with benzaldehyde. The mechanism is pretty much the same as before. The negative lone pair of electrons attack the carbonyl of benzaldehyde. The electrons of the pi bond go up onto the oxygen atom to form a negative charge. The electrons cannot come back down because there is no good leaving group. The negative oxygen is protonated with acid and water followed by possible dehydration to form a conjugated compound.

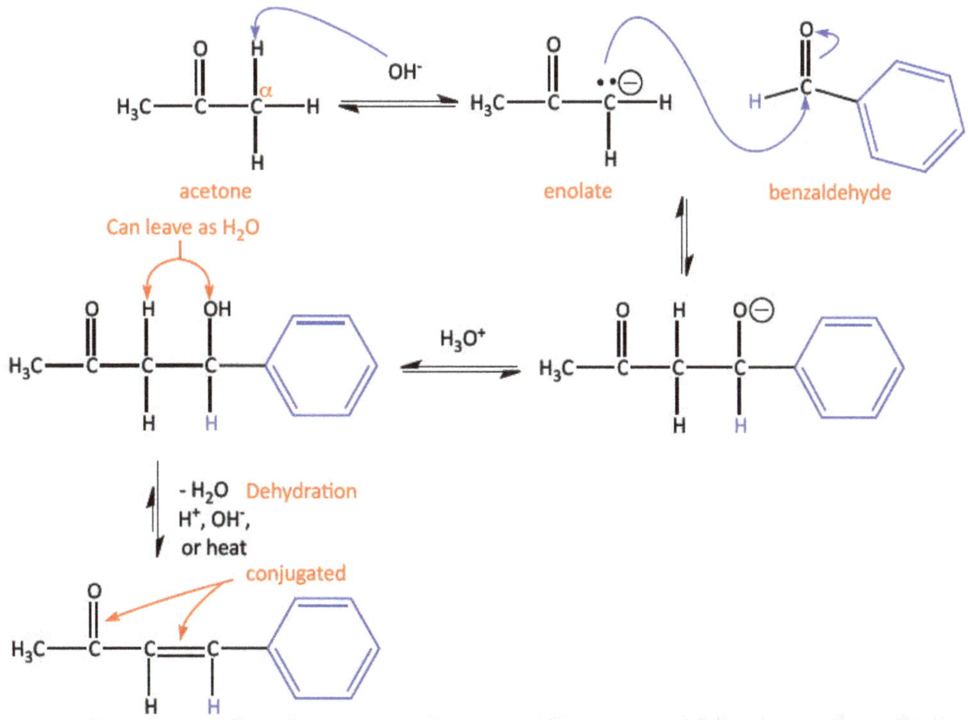

Once the acetone is deprotonated to form an enolate, any ketone or aldehyde can be added to the reaction. Benzaldehyde was chosen with a purpose. The second carbonyl molecule added to the reaction works much better if it does not have any alpha hydrogens on the molecule. The reason it is helpful that the second molecule does not have alpha hydrogen atoms is that we do not want it deprotonated to form an enolate. We want to be able to control which molecule makes the enolate. Notice, benzaldehyde does not have any alpha hydrogen atoms.

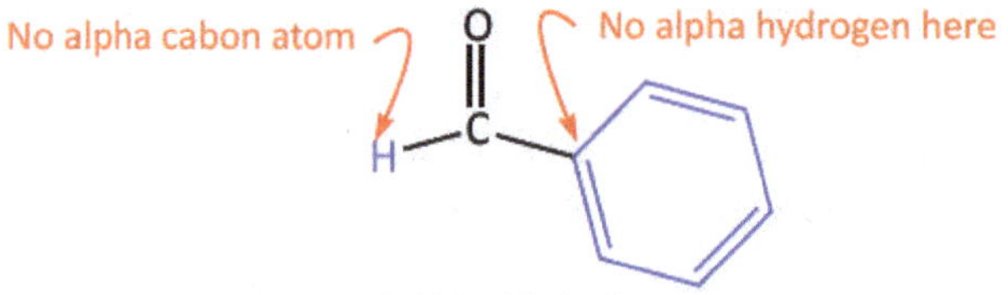

benzaldehyde

A second consideration is that if we want the enolate to attack benzaldehyde, we do not want the enolate to attack another molecule of acetone. How can we help ensure that will happen? The best way is to run the reaction with a large excess of benzaldehyde. This way, the enolate will be much more likely to bump into a benzaldehyde molecule in the reaction flask instead of an acetone molecule. Run the reaction with an excess of the one that cannot form an enolate.

It is helpful to pause and notice something helpful. Imagine the back of your thumb has written on it part of a water molecule (either the two hydrogen atoms or the one oxygen atom). Then, if you cover up part of the aldol product with your thumb, you can see the starting materials used in the aldol reaction. You place your thumb over half of the molecule at the double bond formed in the dehydration step.

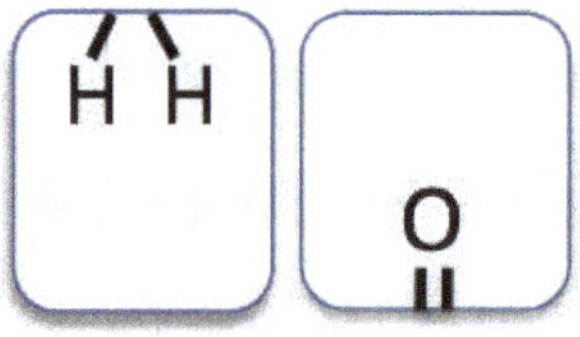

The blue square is your thumb.

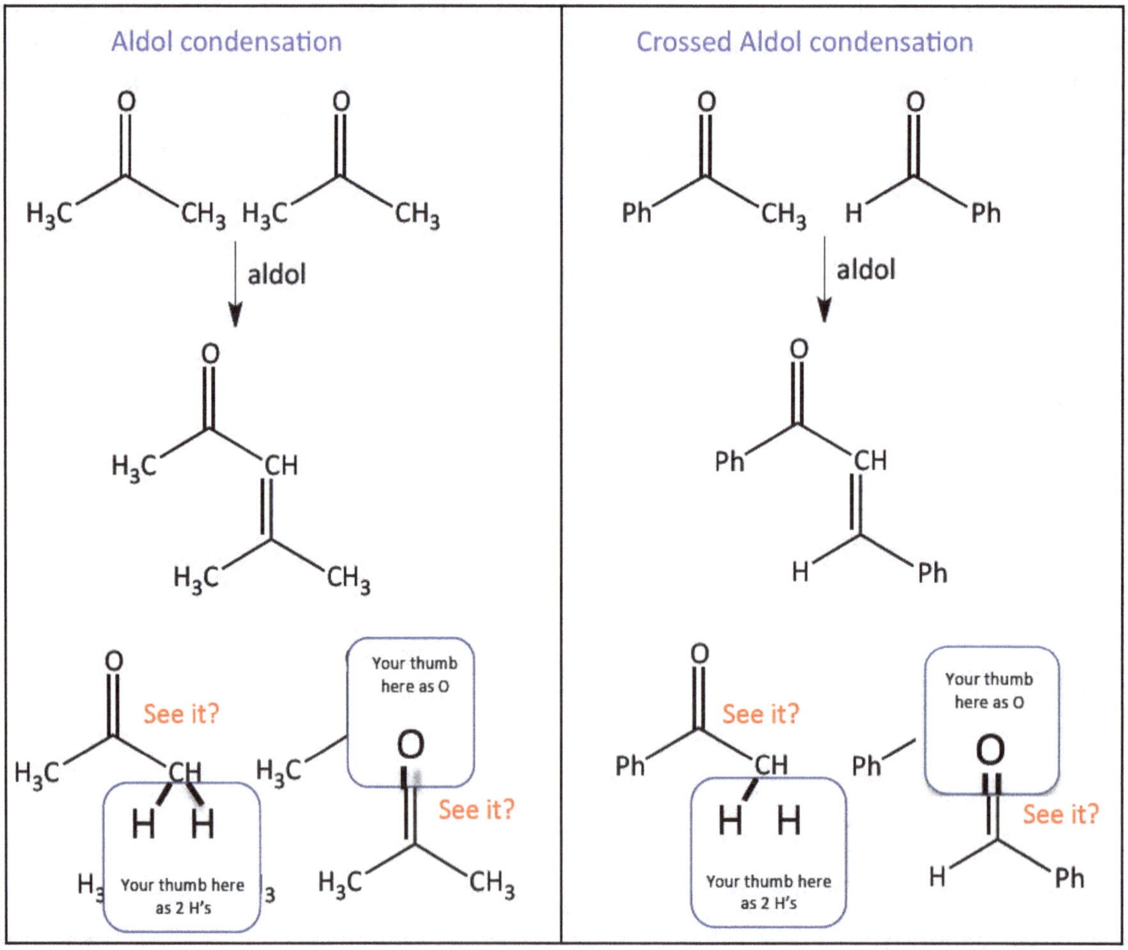

18.3 Draw the product of the following crossed aldol condensations – followed by dehydration.
   a) benzophenone (diphenyl ketone) and butyraldehyde (butanal)
   b) 2,2-dimethylpropanal and acetophenone (methyl phenyl ketone)

18.4 a) Jasmine aldehyde is a compound with a floral, jasmine scent used in the perfume industry. Draw the two compounds (both are carbonyl compounds) that would be used in a crossed aldol condensation to make jasmine aldehyde. (Hint: Use your thumb to cover up half of the molecule at a time at the C=C.)

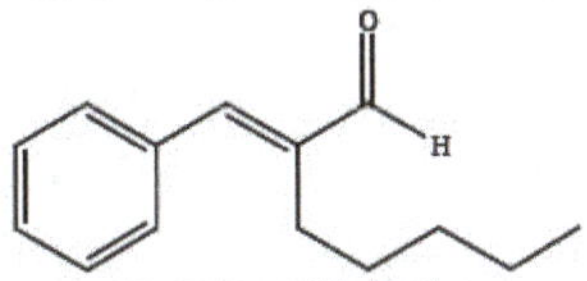

jasmine aldehyde

   b) When the two compounds in the solution for part a are used to make jasmine aldehyde, a second compound is also made. It is the product of an aldol condensation where one of the compounds reacts with itself. Draw this unwanted side-product of this reaction.

18.5 Show the mechanism for the following aldol condensation.

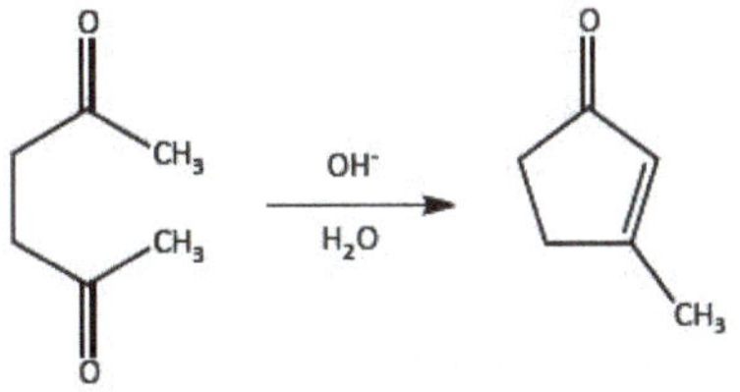

18.4 Claisen Condensation

A Claisen Condensation is very similar to an aldol condensation except esters are used instead of aldehydes and ketones. Esters can also have acidic alpha hydrogen atoms. These hydrogen atoms are not acidic as ketones or aldehydes, but they are still pretty acidic. Let's examine the Claisen Condensation using ethyl acetate as our example molecule with a base.

A base can deprotonate at the alpha spot of the ethyl acetate to make an enolate. The enolate can attack another molecule of ethyl acetate. The enolate attacks the carbonyl of ethyl acetate. The pi electrons of the carbonyl go up onto the oxygen atom, but they come back down because there is a good leaving group, the ethoxide, $^{-}$OEt. The electrons come back down from the oxygen atom to reform the carbonyl pi bond and the ethoxide, $^{-}$OEt, leaving group leaves. The final product of the Claisen Condensation is a β-dicarbonyl compound.

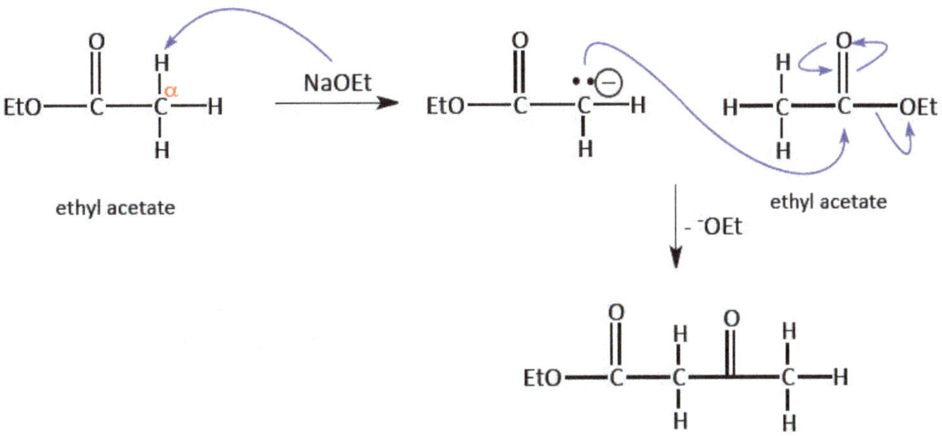

Enolates

Pay special attention to the base used, NaOEt, sodium ethoxide. Sodium ethoxide was carefully chosen so that the base used, $^{-}$OEt, is the same as the OEt that is on the ester of our final product. If this $^{-}$OEt attacks the carbonyl of the product, the electrons would go up, come back down, and the $^{-}$OEt leaving group would leave.

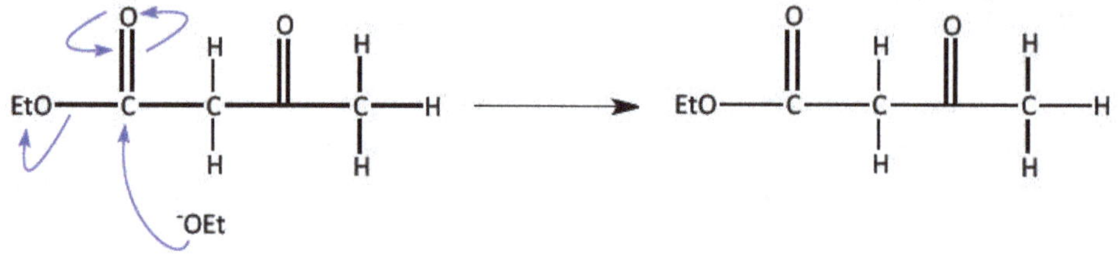

This would not be a problem for us, because we end up with the exact same product. But, if we used a different base, we could end up with an ester we didn't want! This careful selection of a base that keeps our final ester the same is a common trick in reactions involving esters.

18.6 Draw the products of the following Claisen Condensations.

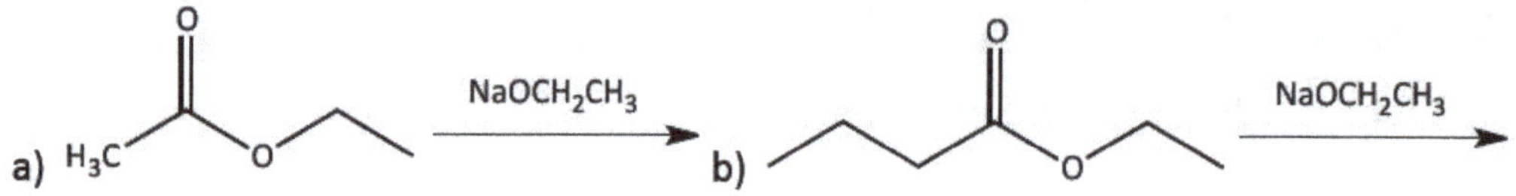

18.4.2 Dieckmann Condensation- A cyclic Claisen condensation

The Dieckmann Condensation is a cyclic Claisen condensation, meaning a ring is formed. For example, a chain with two esters attached can undergo a Dieckmann condensation.

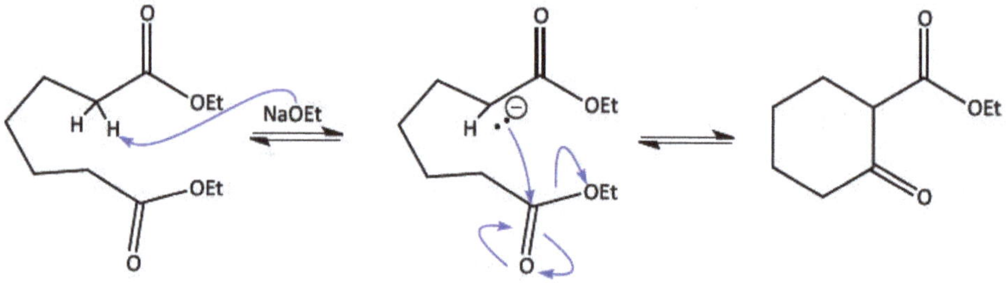

18.7 Draw the product when the following ester undergoes a Dieckmann Condensation.

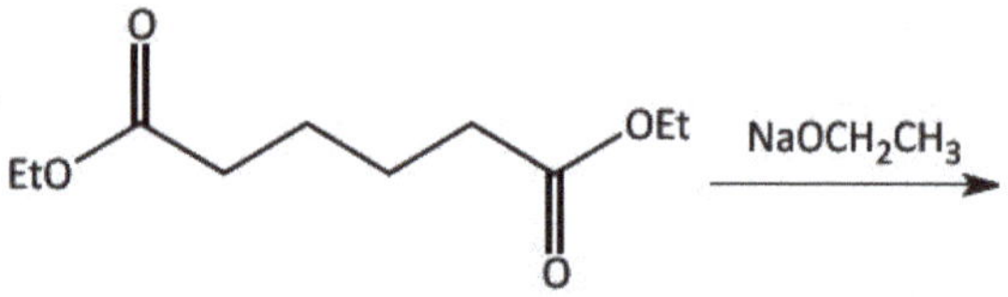

18.4.3 Crossed Claisen condensation

Much like an aldol condensation can undergo a crossed aldol with two different aldehydes or ketones, a crossed Claisen condensation is when two different esters are used. Again, it is useful if only one of the esters has alpha hydrogen atoms so only one enolate can be formed. In the following example, ethyl acetate and ethyl benzoate are mixed together with a base. Only the ethyl acetate has alpha hydrogen atoms, the ethyl benzoate does not. The ethyl acetate is the only one that can form an enolate.

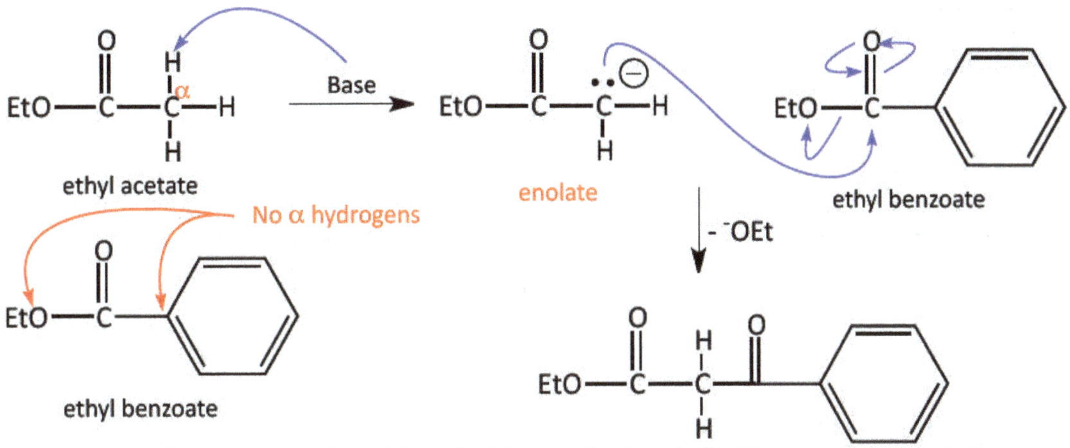

How can we help ensure that the enolate will react with a molecule of ethyl benzoate instead of another molecule of ethyl acetate? We need to run the reaction with a large excess of ethyl benzoate so the enolate is much more likely to encounter a molecule of ethyl benzoate in the flask instead of a molecule of ethyl acetate.

Ketones or aldehydes could also be the molecules that form an enolate that would attack an ester in a Claisen Condensation.

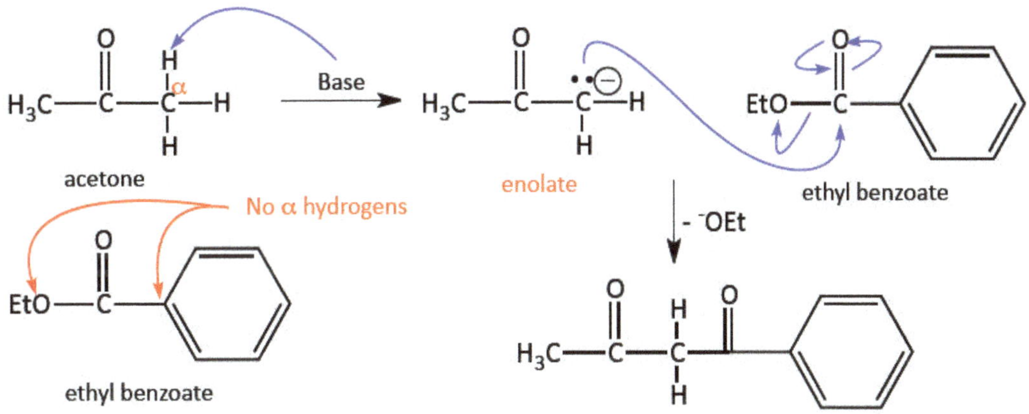

It is helpful to pause and notice something. Imagine the back of your thumb has written on it part of an alcohol (HOR) molecule (either the hydrogen atom or the alkoxy (OR) group). Then, if you cover up half of the Claisen Condensation product with your thumb, you can see the starting materials used in the Claisen reaction. You place your thumb over half of the molecule at the double bond formed in the dehydration step.

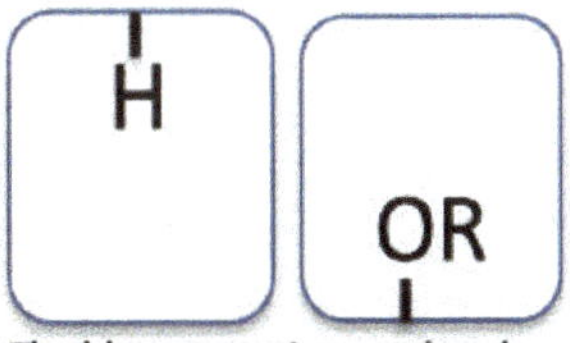

The blue square is your thumb.

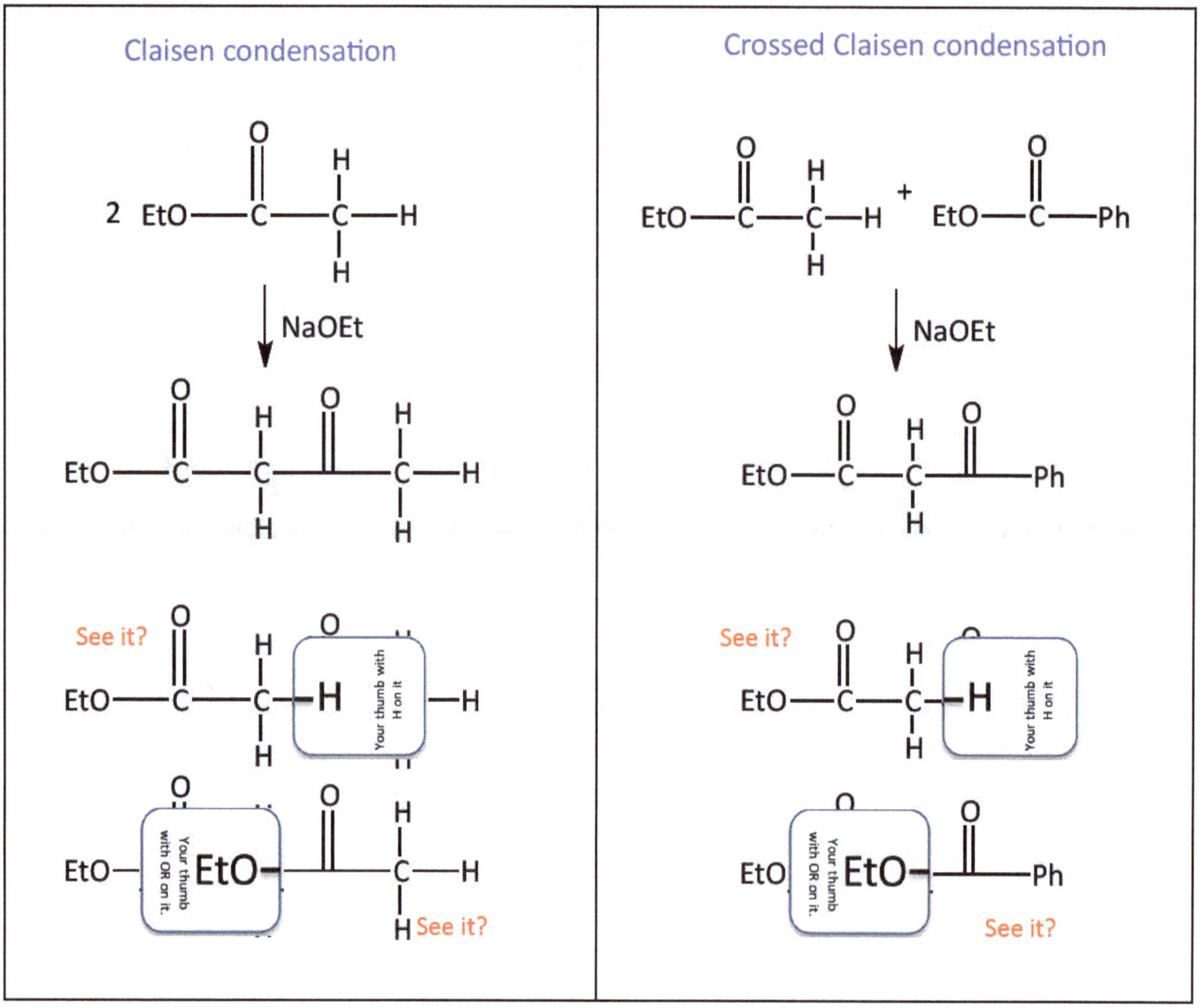

18.8 a) Draw the product when the following compounds undergo a Crossed Claisen reaction.

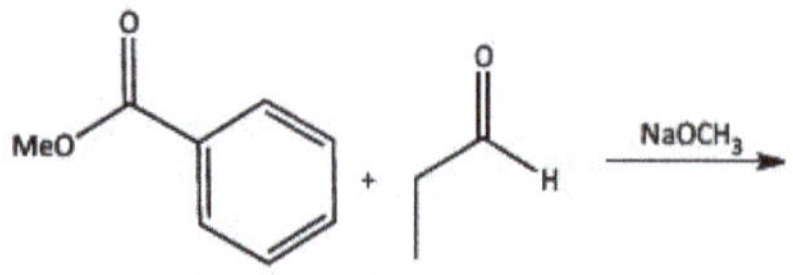

b) What two reagents would undergo a Crossed Claisen reaction to make the following compound?

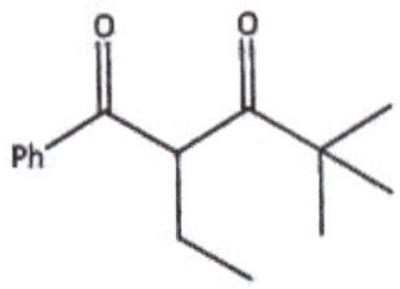

18.5 Acetoacetic ester synthesis

An *aceto-* (pronounced *uh-seat-oh*) group is a methyl next to a carbonyl. Acetic acid is a methyl on a carboxylic acid. Acetoacetic acid is the combination of the two. Acetoacetic ester is the ester made from acetoacetic acid.

Let's look more closely at the β-dicarbonyl compound, acetoacetic ester. There are two different types of acidic alpha hydrogen atoms. Which are the most acidic?

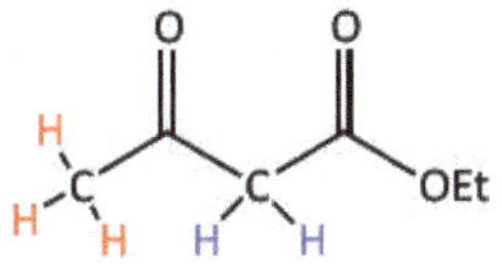

acetoacetic ester

If a red hydrogen atom from the methyl group is removed, the enolate conjugate base can be stabilized with one resonance form. But, the blue hydrogen atoms on the carbon atom between the two carbonyls are even more acidic since two resonance forms can be drawn for its enolate conjugate base.

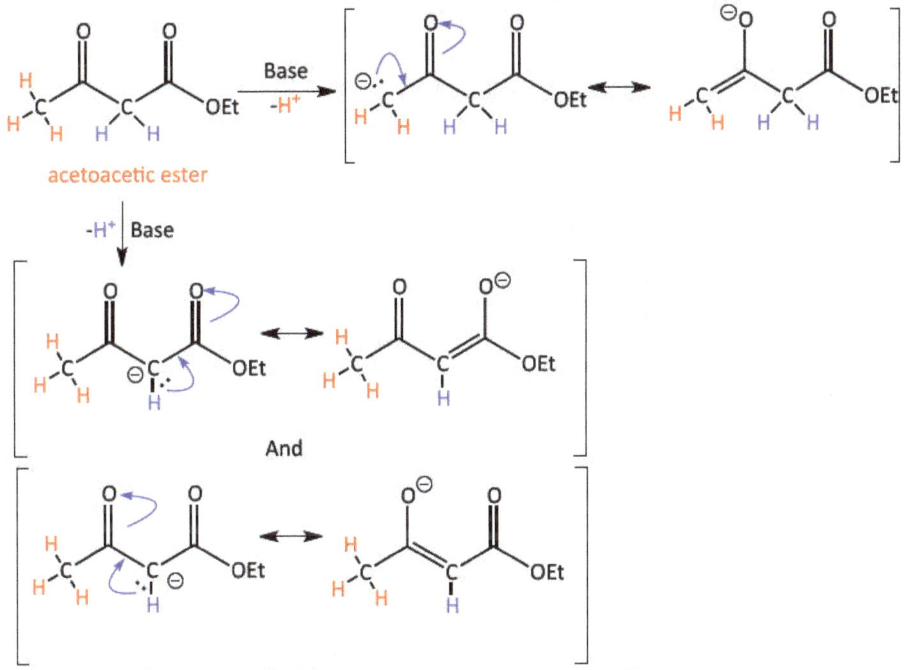

Whenever you encounter a choice of which protons to deprotonate, if you see these especially acidic protons between two carbonyl groups, deprotonate there.

In the acetoacetic ester synthesis, acetoacetic ester is first deprotonated between the carbonyl groups. The resulting enolate reacts with an alkyl halide via an S_N2 reaction to alkylate it. Once the ester is hydrolyzed with acid and heat, the esters turn into a carboxylic acid.

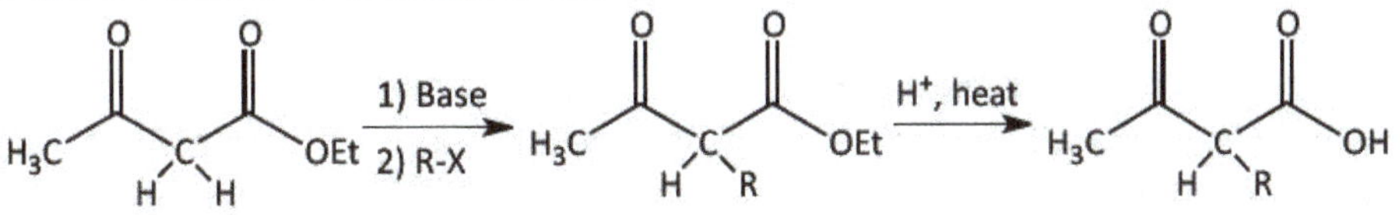

acetoacetic acid ester

A carboxylic acid that is beta to a carbonyl is interesting. It decarboxylates, or loses carbon dioxide.

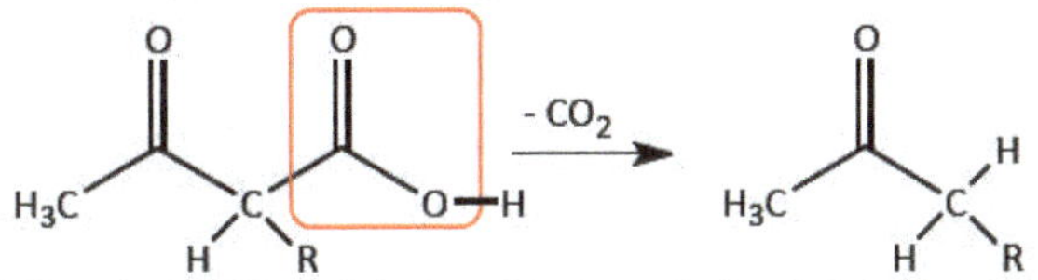

Let's redraw the carboxylic acid beta to a carbonyl in such a way as to show the mechanism of the decarboxylation. As the carbon dioxide is eliminated, an enol is formed. It tautomerizes under acidic conditions to a keto form. This ketone is a substituted acetone.

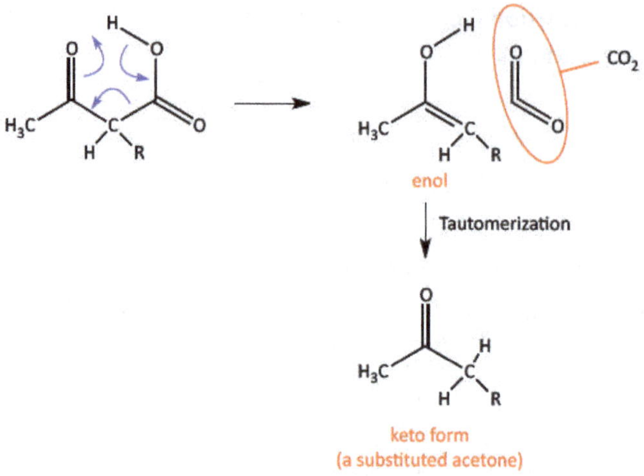

18.6 Malonic ester synthesis

The following dicarboxylic acid is called malonic acid. The ester it makes is called malonic ester.

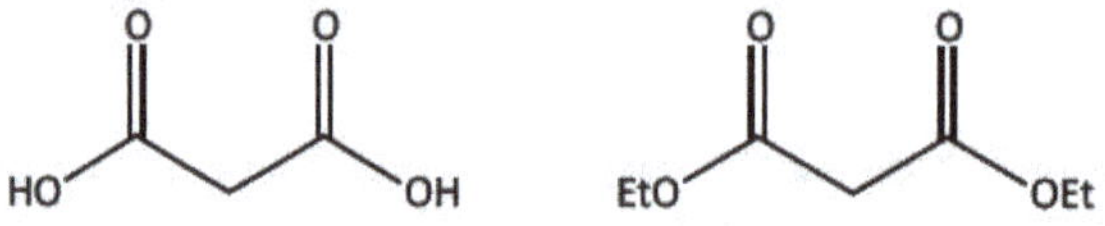

malonic acid malonic ester

In the malonic ester synthesis, acetoacetic acid ester is first deprotonated between the carbonyl groups. The resulting enolate reacts with an alkyl halide via an S_N2 reaction to alkylate it. Once the ester is hydrolyzed with acid and heat, the esters turn into a carboxylic acid.

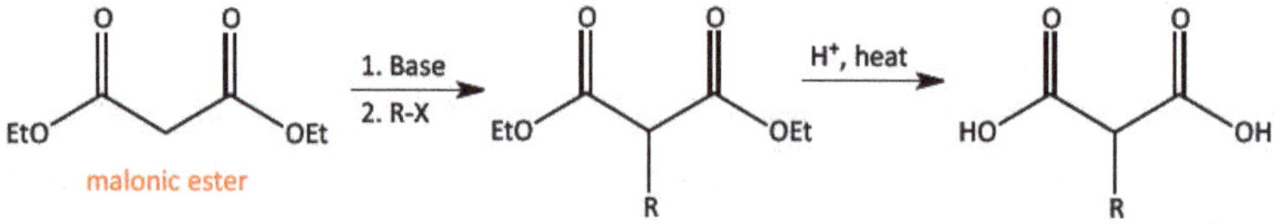

It decarboxylates to form a substituted acetic acid because the carboxylic acid is beta to another carbonyl group.

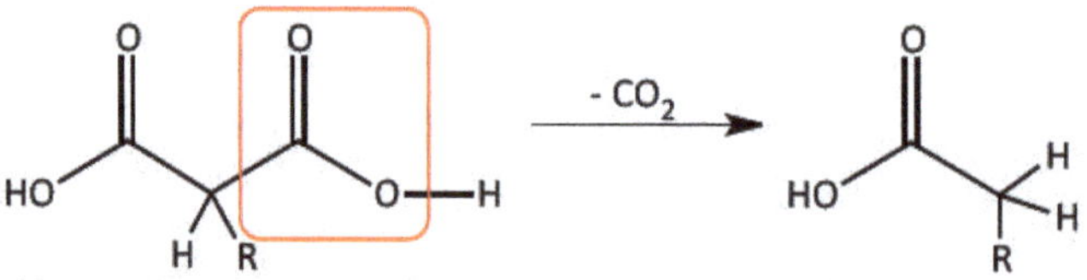

Take a moment and cover up the red square directly above with your thumb. When you cover up the CO_2 of the carboxylic acid (in the red square) with your thumb, you get the product on the right. Covering the carboxylic acid with your thumb is a quick way to envision the product you get following decarboxylation.

18.9 Draw the products of the following reactions.

a) 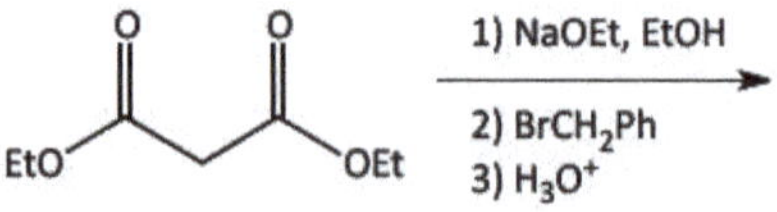

b) 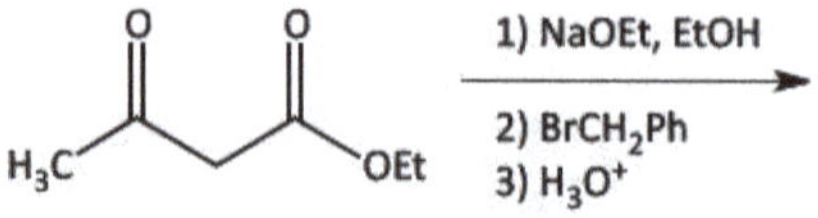

18.10 Circle all of the molecules in the following group that can decarboxylate.

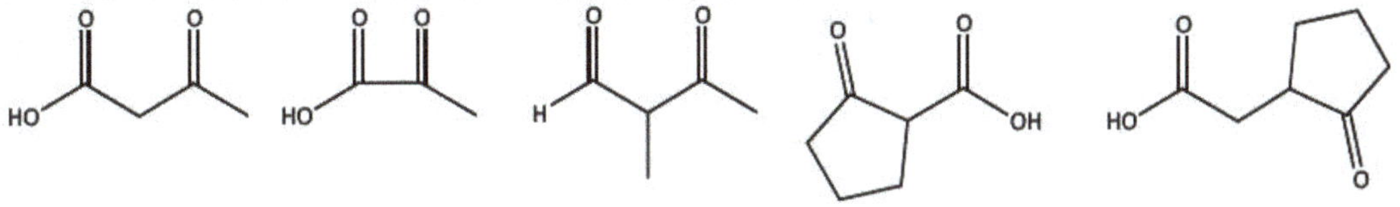

18.7 Michael addition

Conjugated carbonyl compounds are electrophilic. They can be attacked by nucleophiles in two ways. The carbonyl carbon can be attacked by a nucleophile in what is called 1,2-addition or the alkene can be attacked in what is called conjugate addition or 1,4-addition.

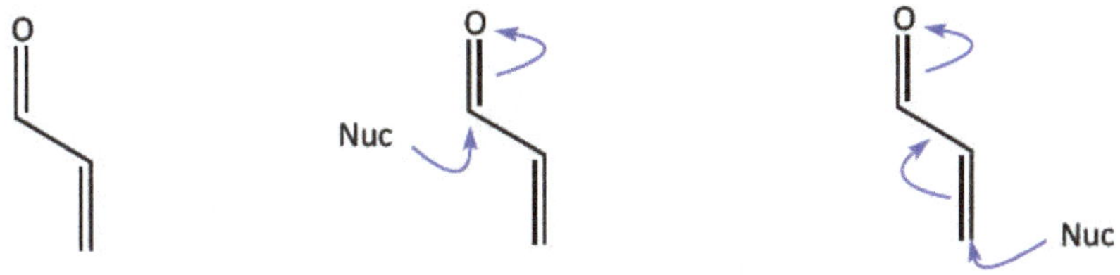

conjugated carbonyl　　1,2-addition　　1,4-addition or conjugate addition

If an enolate is a nucleophile, 1,4-addition occurs. This is called a Michael Addition. Here is one example:

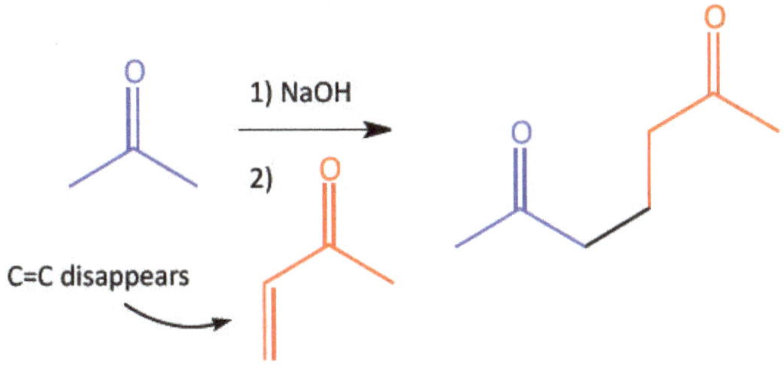

The enolates in a Michael Addition are called Michael donors and the conjugated carbonyl compounds are called Michael acceptors.

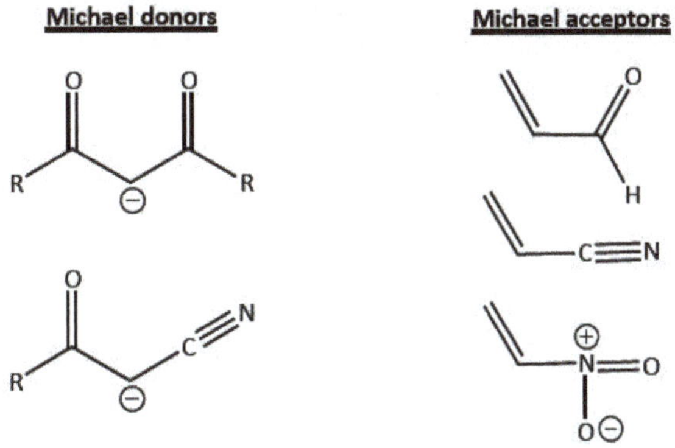

Figure 18.2 Michael donors and acceptors

Enolates
Another example:

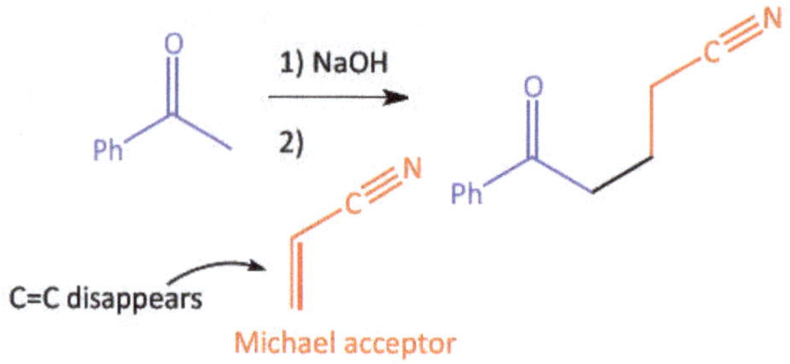

For the mechanism of the Michael Addition, an enolate attacks a conjugated carbonyl in a 1,4-addition. The electrons move all the way up to the oxygen atom making a negatively charged oxygen atom. The electrons come back down, and the pi bond grabs a proton from the solvent making a single bond.

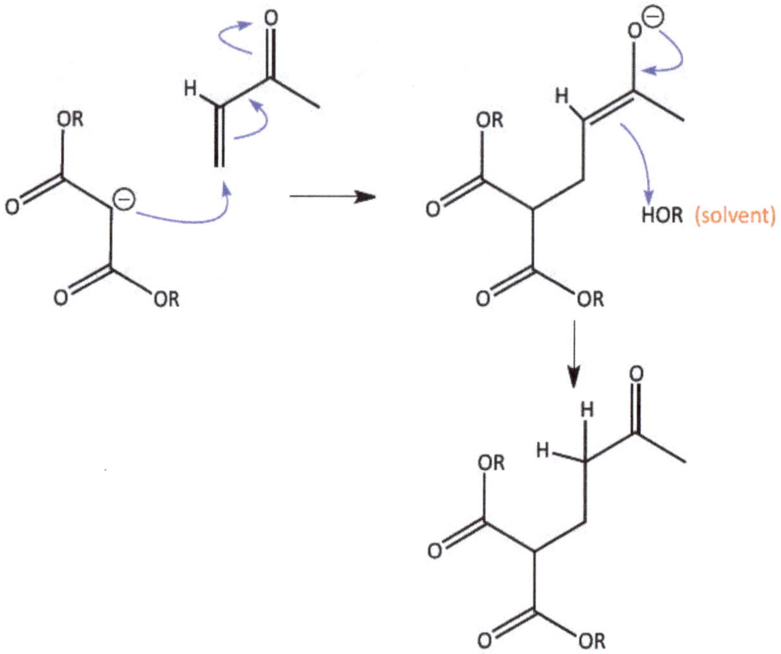

Often, the Michael Addition stops at this point. But, if an ester is used as the enolate, it can be hydrolyzed with acid and heat to make carboxylic acids. Since we now have a carboxylic acid beta to a carbonyl, this particular molecule can decarboxylate giving the final product.

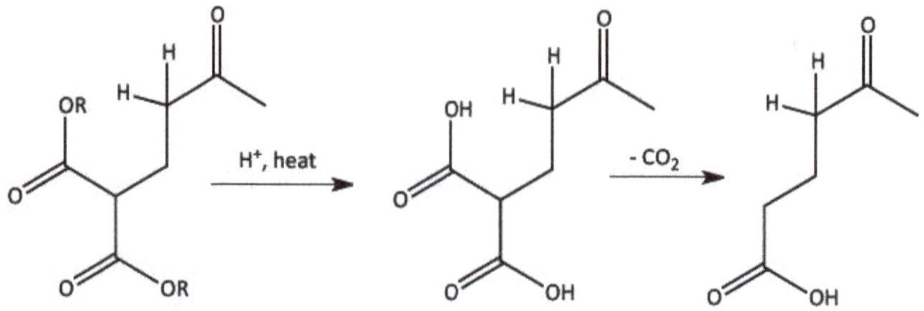

Overall:

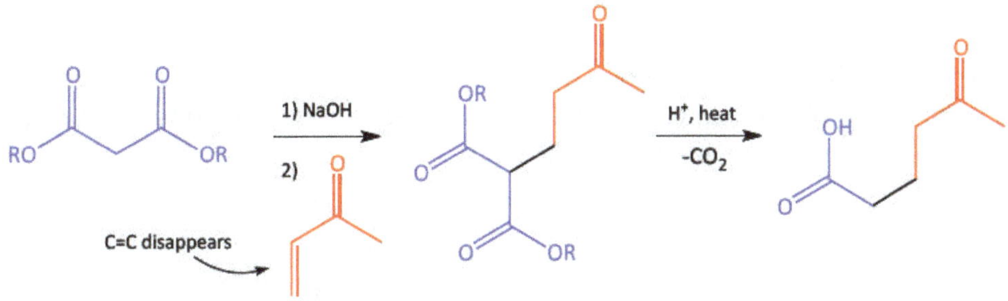

18.11 When cyclohexenone reacts with hydroxide base, it undergoes a Michael addition. The cylohexenone acts as both the Michael donor and Michael acceptor. Draw the product of this Michael addition.

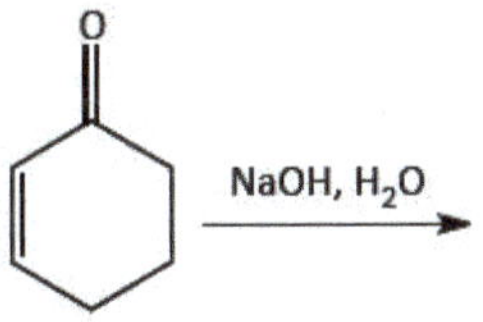

18.12 Draw the product of the following Michael addition reaction.

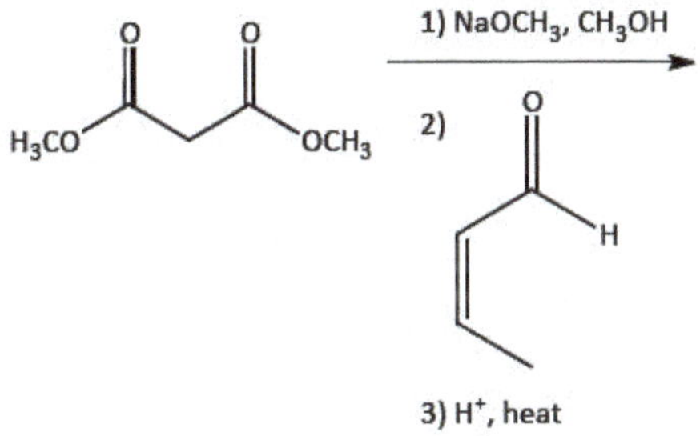

18.8 Robinson annulation

An annulation is a reaction that forms a ring or a cyclic compound. The Robinson annulation uses two reactions we have already learned, the Michael addition followed by an Aldol condensation to give a ring.
The first step of a Robinson annulation is a Michael addition.

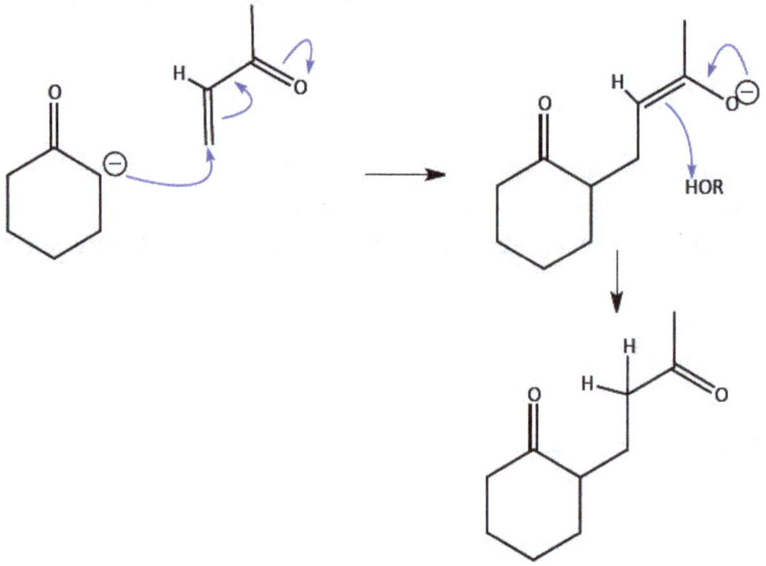

Following the Michael addition, an aldol condensation forms the ring. An alpha proton is removed and a carbonyl is attacked giving the ring. Following dehydration, a conjugated carbonyl ring system is produced.

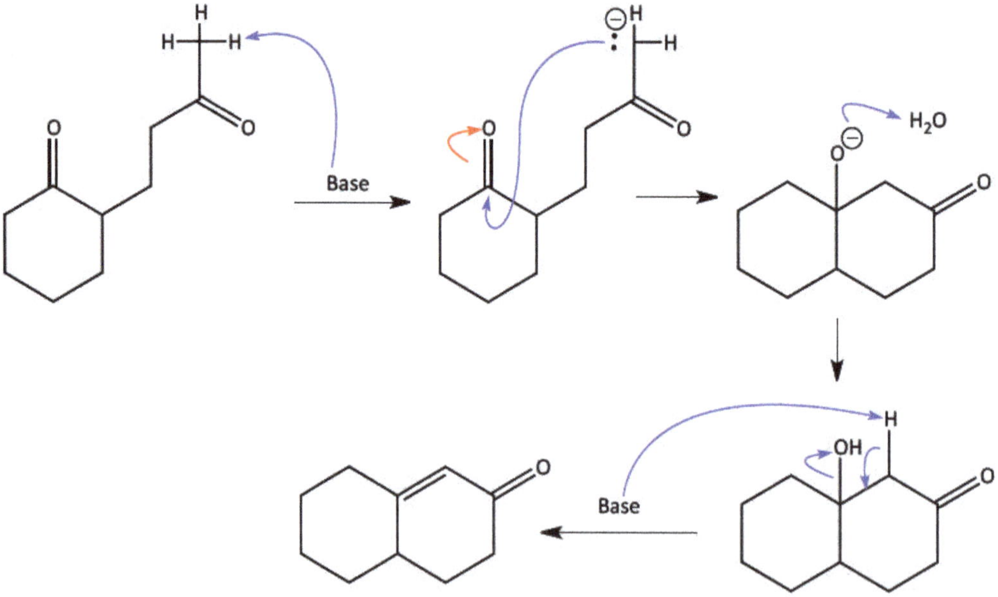

The Robinson annulation is an important reaction because the conjugated six-membered rings it can produce. These rings are are a part of steroids like testosterone.

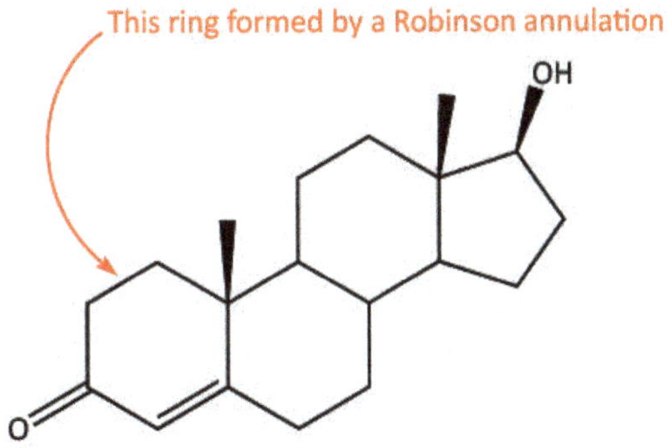

testosterone

18.9 Reaction Summary

Enolate formation

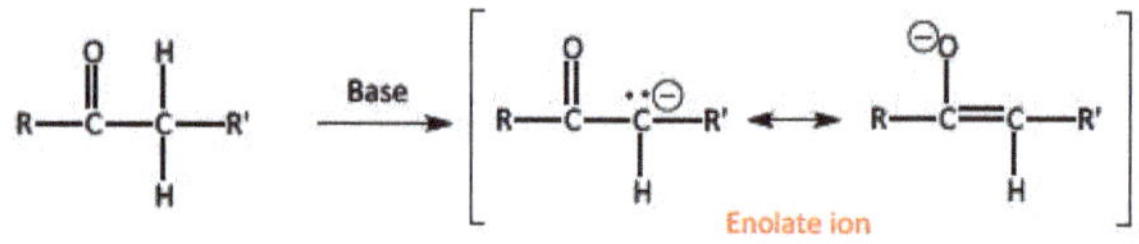

a)

Enolate halogenation

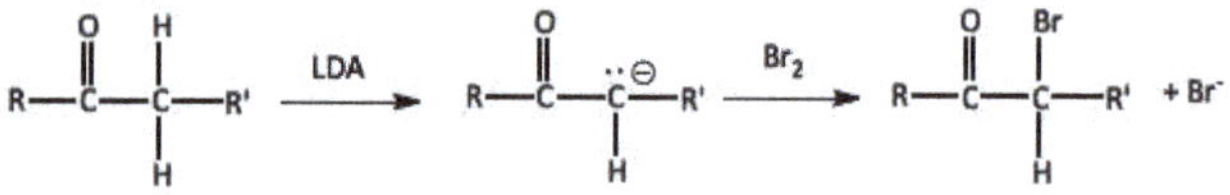

b)

Enolate S_N2 with alkyl halide

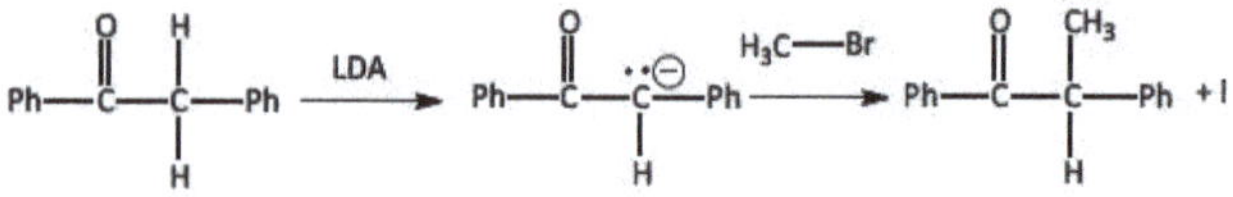

c)

Aldol condensation

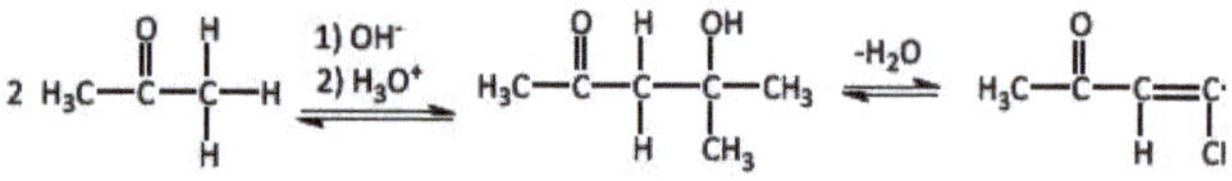

d)

Crossed Aldol condensation

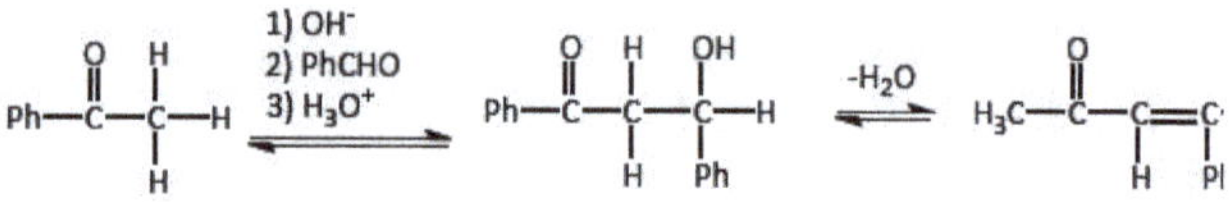

e)

Claisen condensation

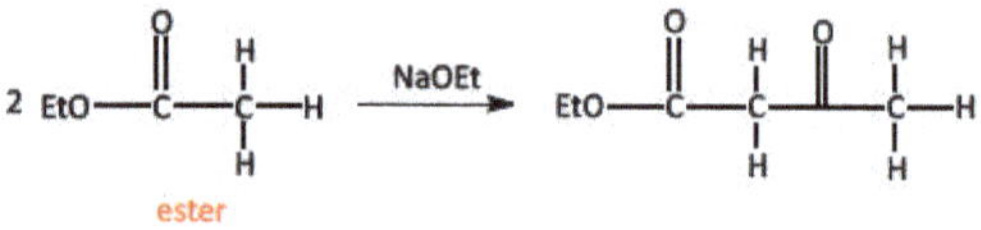

f)

Dieckmann condensation

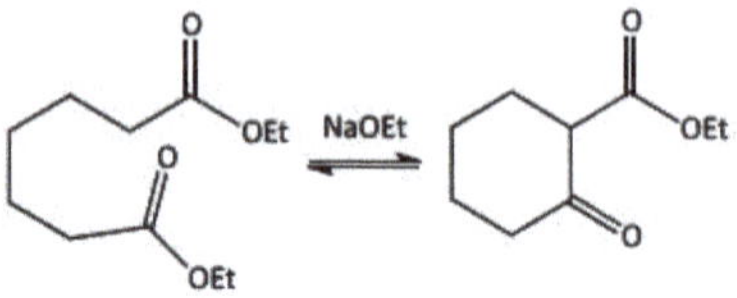

g)

Crossed Claisen condensation

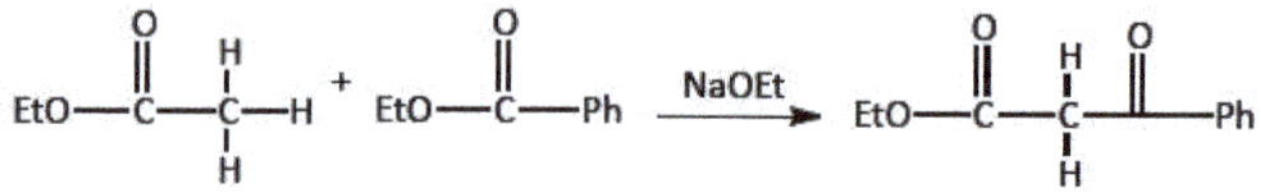

h)

Acetoacetic ester synthesis

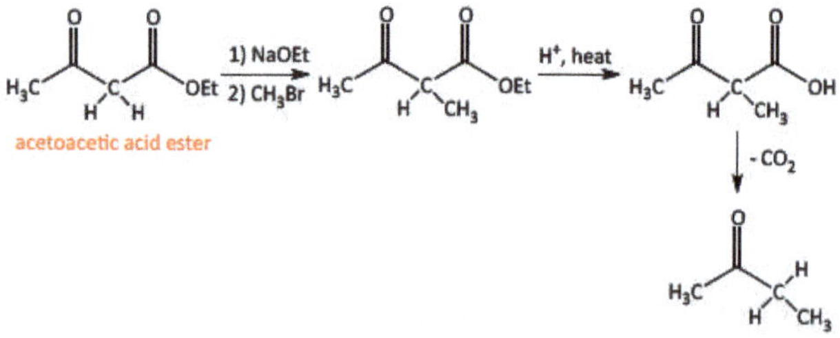

i)

Malonic ester synthesis

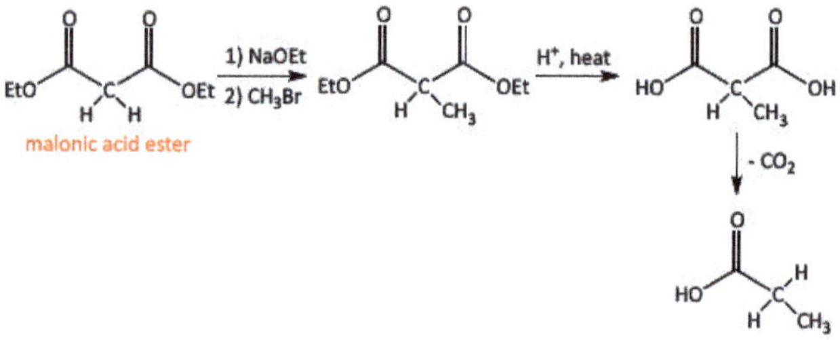

j)

Michael Addition

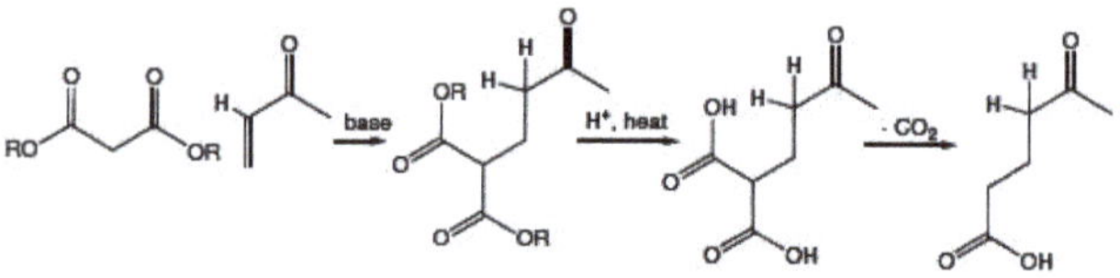

k)

Robinson annulation

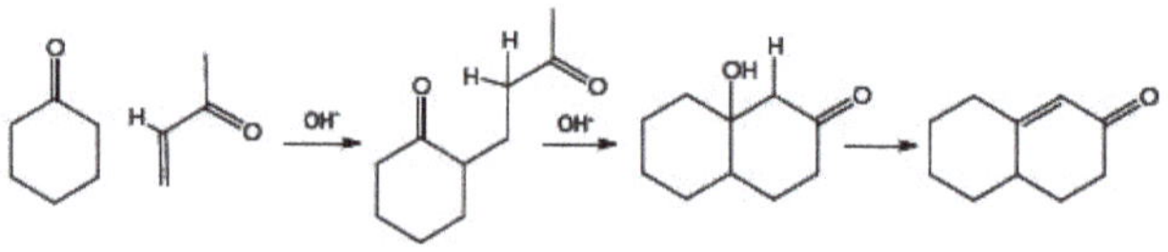

l)

18.10 Problems

18.13 What are the products of the following reactions?

a) 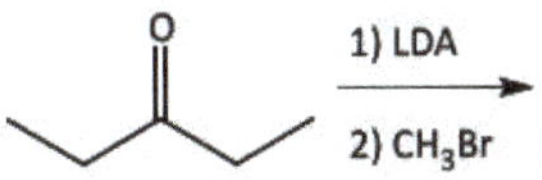 b) 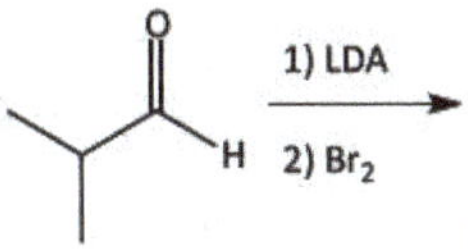 c) 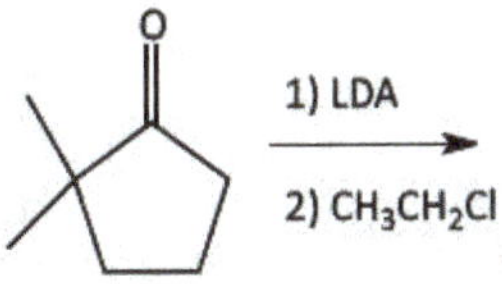 d) 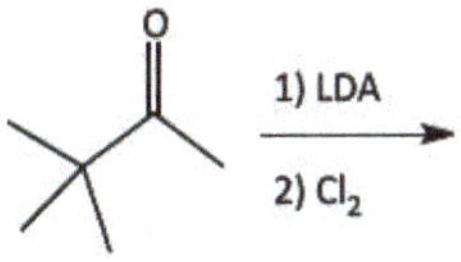

18.14 What are the products when the following reagents undergo aldol condensations following dehydration?

a) 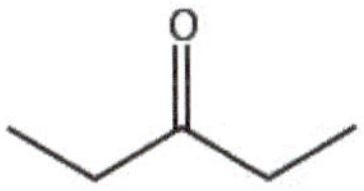 b) 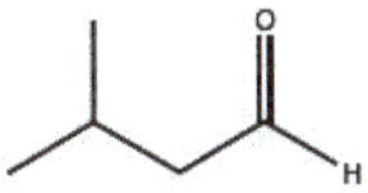 c) 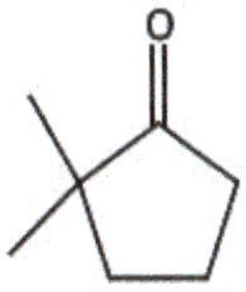 d) 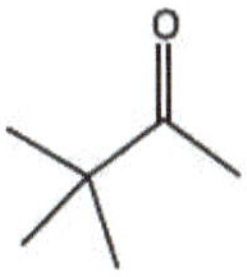 e) 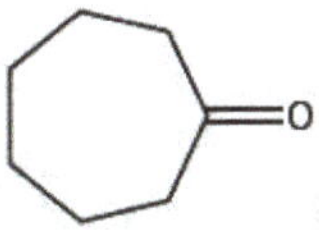 f) 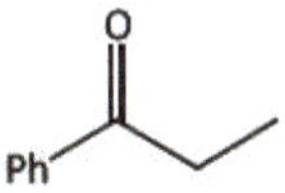

18.15 What molecule undergoes an aldol condensation to form the following products?

a) 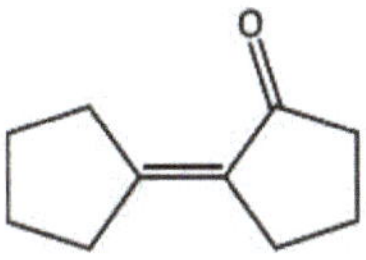 b) 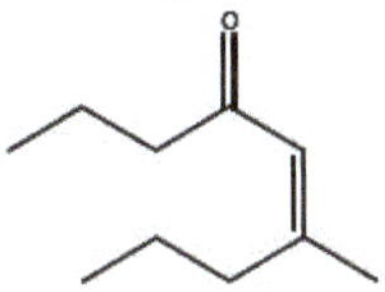 c) 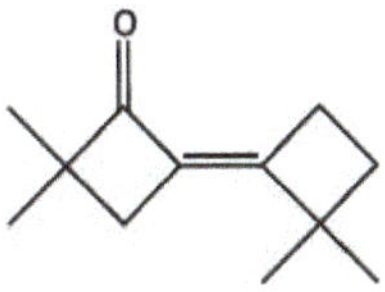 d) 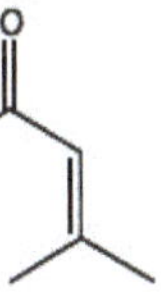

18.16 Provide the structure of the product of the crossed Aldol condensation between benzaldehyde and acetone.

18.17 a) Identify the two molecules that react in a crossed aldol reaction to produce the following? b) Identify which of the two molecules would be used in excess in the crossed aldol reaction.

a) 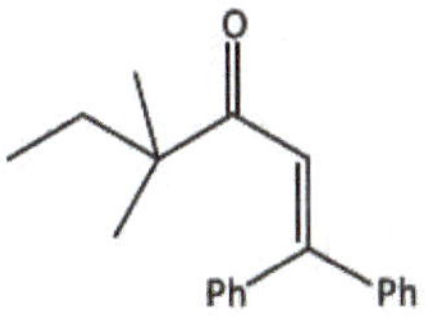 b) 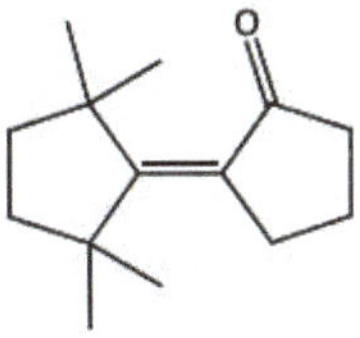 c) 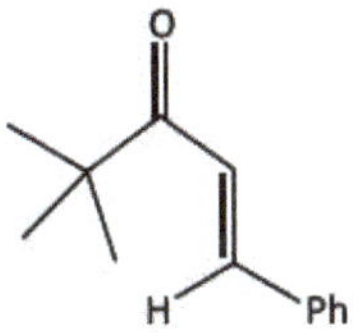 d) 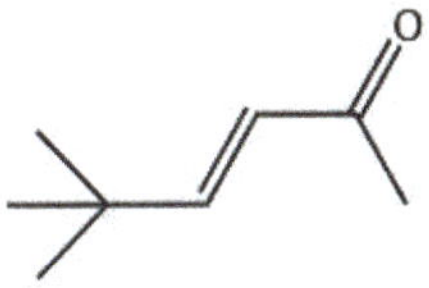

18.18 What are the products when the following reagents undergo Claisen condensations?

a) 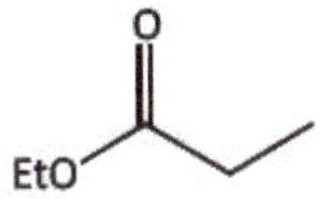 b) 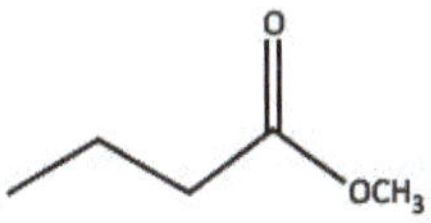 c) 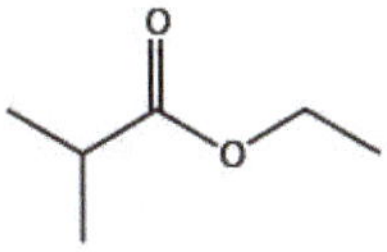 d) 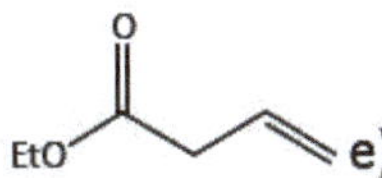 e) 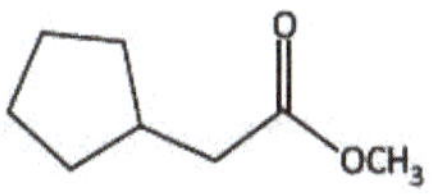

18.19 Provide the structure of the product of the crossed Claisen condensation between $PhCO_2CH_2CH_3$ and $CH_3CH_2CO_2CH_2CH_3$.

18.20 Provide the structure of the product of the following Dieckmann condensation.

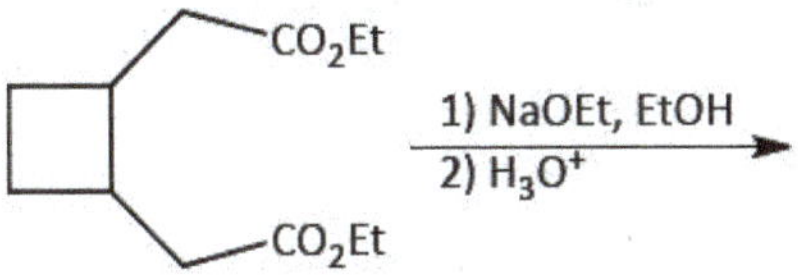

Enolates

18.21 What ester would undergo a Claisen condensation to make the following products?

a) 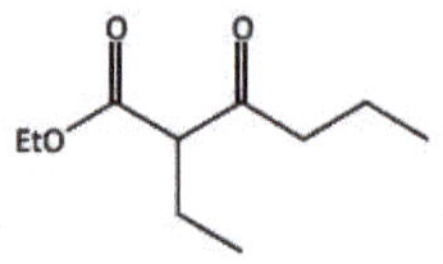

b) 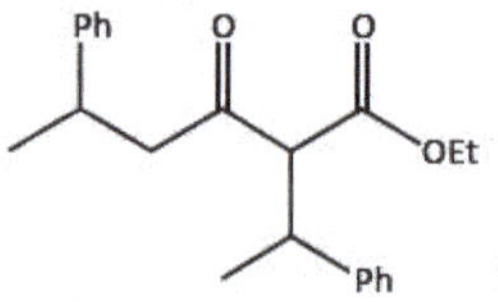

c) 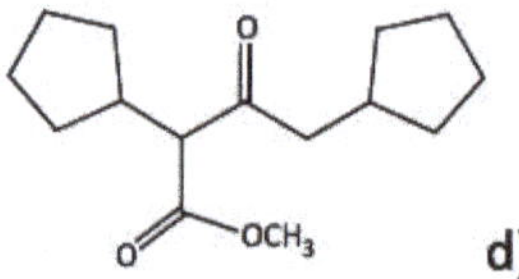

d) 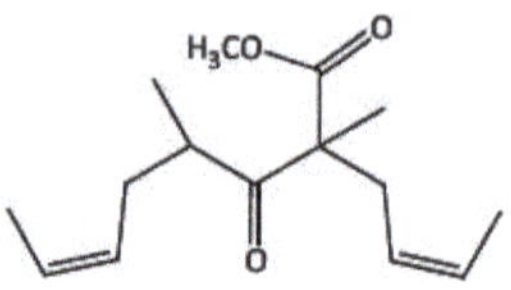

18.22 Show how the following compound could be synthesized from a Dieckmann condensation.

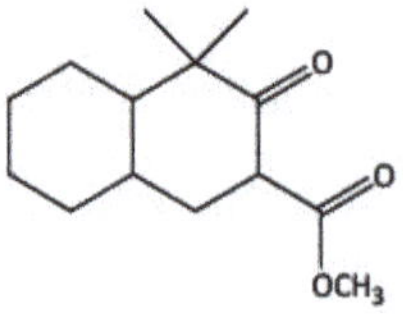

18.23 Show how you would use an aldol reaction to make the following.

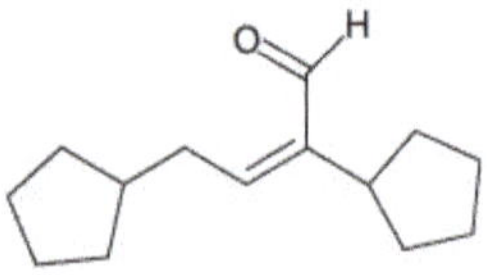

18.24 a) What reagents would you use to make the β-keto ester shown? It is the first part of the acetoacetic acid ester synthesis.

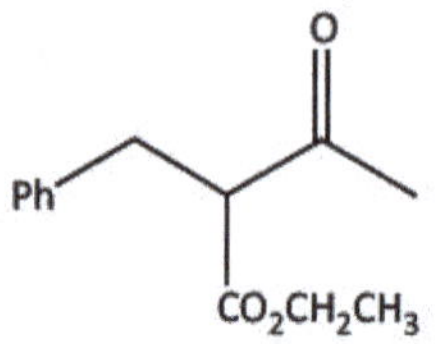

b) Show the ketone that would result from hydrolysis-decarboxylation of the following β-keto esters. (This is the final part of the acetoacetic ester synthesis).

18.25 Predict the products of the following aldol condensation. Show the products *both before and after dehydration.*

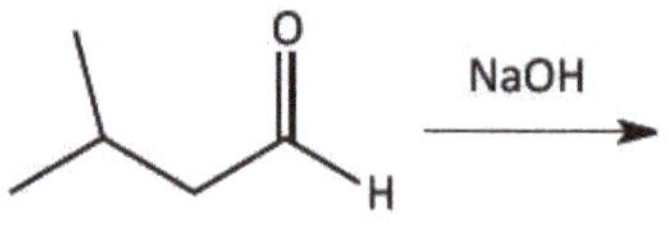

18.26 The following crossed aldol reaction is not a reasonable way to synthesize the product shown. Why?

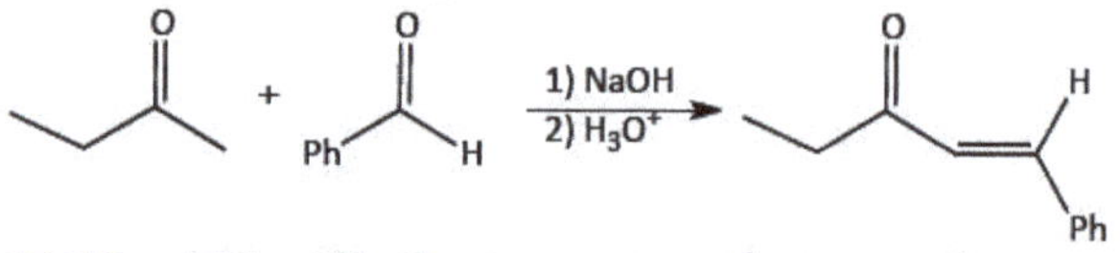

18.27 a) Identify the two esters that react in a crossed Claisen condensation to produce the following? b) Identify which of the two molecules would be used in excess in the crossed Claisen reaction.

a) 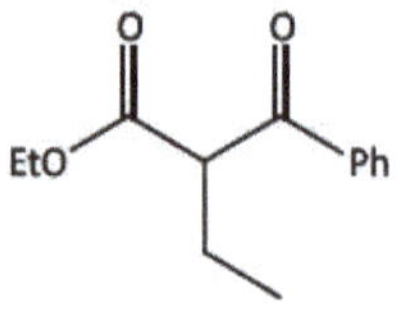

b) 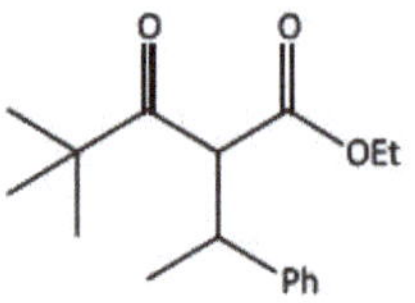

c) 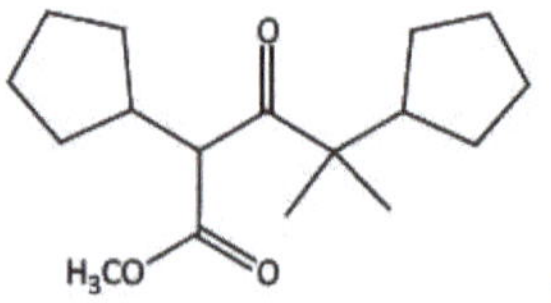

d) 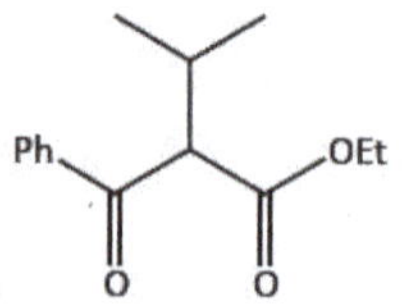

18.28 Draw the structures of the products of the following crossed Claisen condensations. Indicate if any pair of compounds would be a poor choice to perform a crossed Claisen condensation.

a) $PhCO_2CH_3 + PhCH_2CO_2CH_3$ b) $(CH_3)_3CCO_2CH_2CH_3 + CH_3CH_2CO_2CH_2CH_3$ c) $CH_3CH_2CO_2CH_3 + CH_3CO_2CH_3$

18.29 Provide the structures of the products of the following reactions.

a) 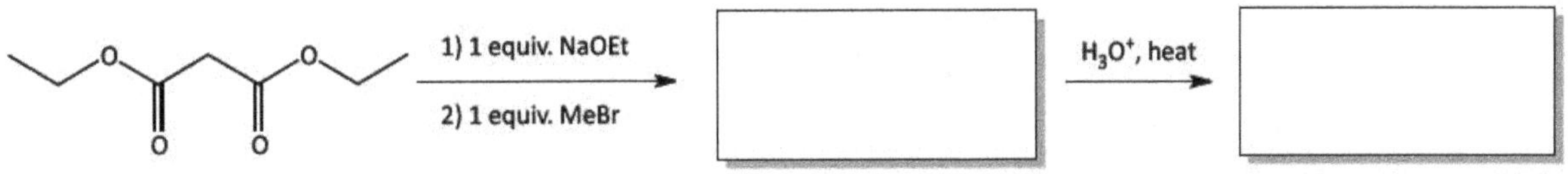

b) 

18.30 Show how you could use the acetoacetic ester synthesis or the malonic ester synthesis to make the following compounds.
a) 2-hexanone b) hexanoic acid c) 4-methyl-2-pentanone d) 4-methylpentanoic acid

18.31 Draw the product of the following compounds after they undergo hydrolysis and decarboxylation.

a) 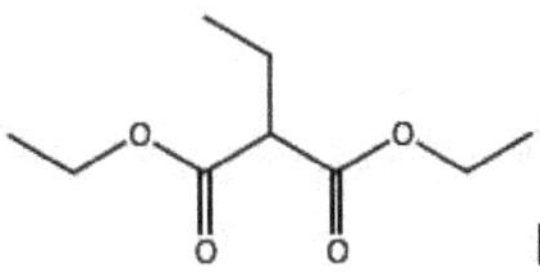

b) 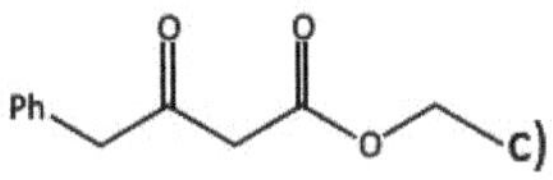

c) 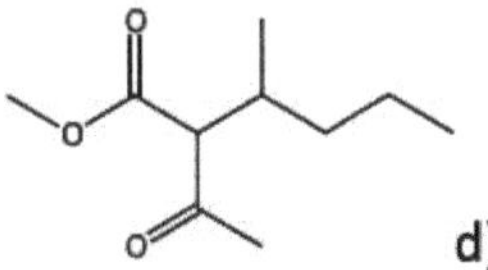

d) 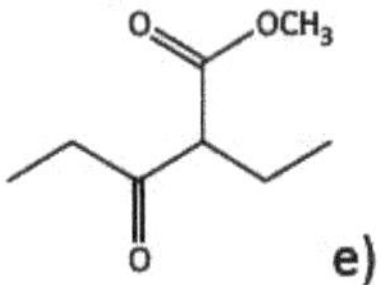

e) 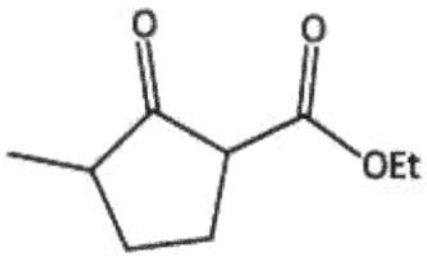

18.32 Draw the structures of the products of the following reactions.

a) 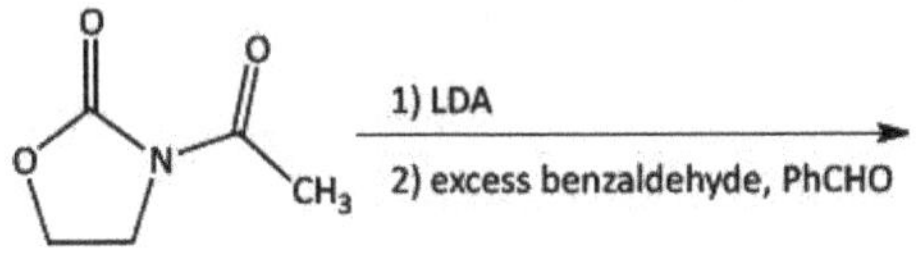

b) 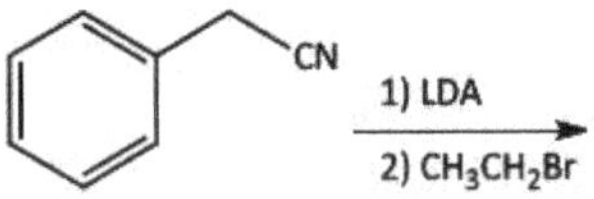

c) 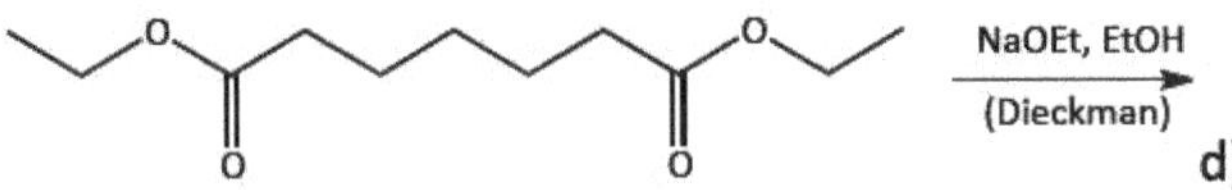

d) 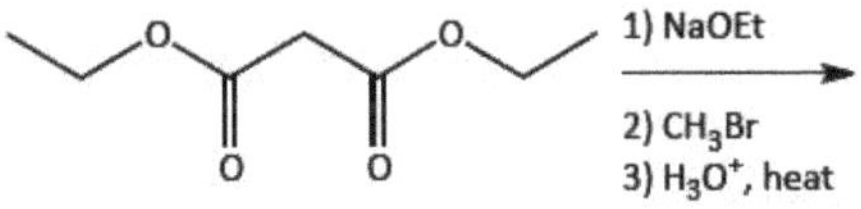

18.33 Show the mechanism of the following reaction.

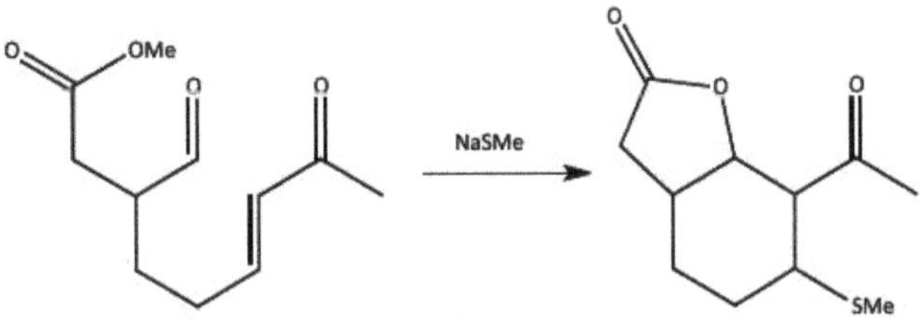

18.34 Show what starting material(s) you would use to get the following products. Use aldol or Claisen reactions.

a) 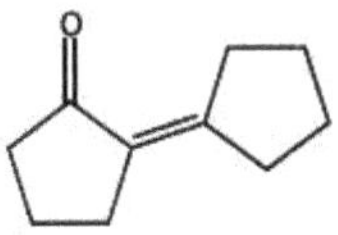

b) 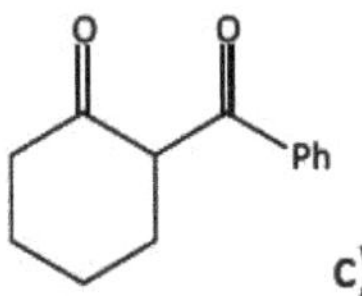

c) 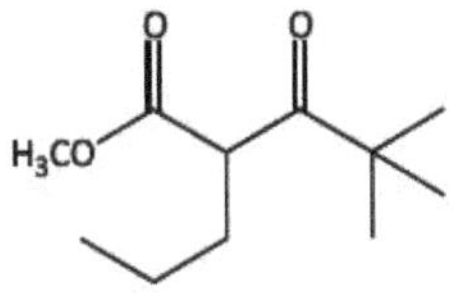

d) 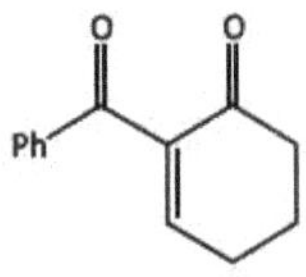

e) 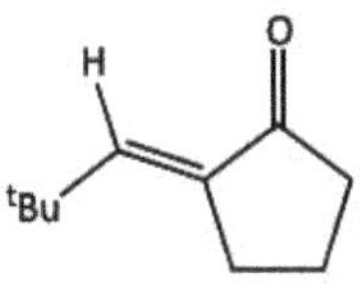

Enolates

18.35 Which of the following, when heated, can decarboxylate and lose CO_2? If it can decarboxylate, show the product that is formed.

a) 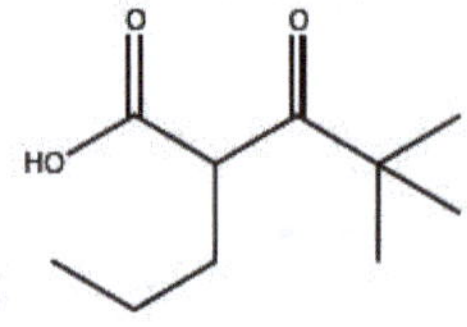 b) 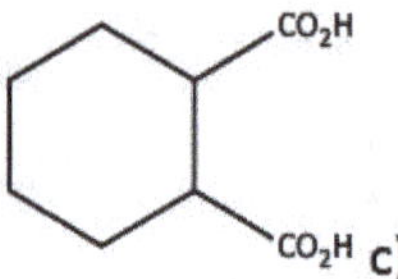 c) 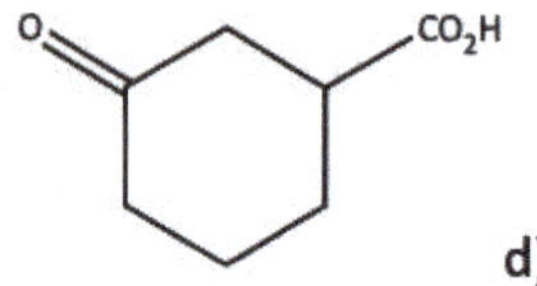 d) 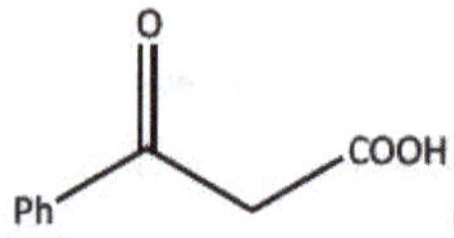 e) 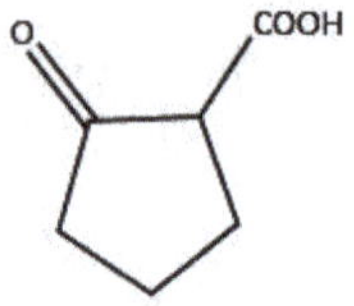

18.36 Show the products of the following Michael addition reactions.

a) 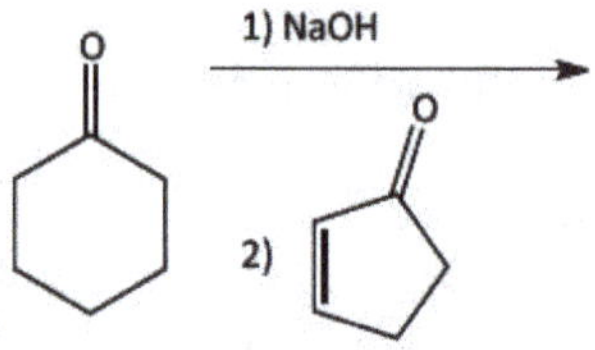 b) 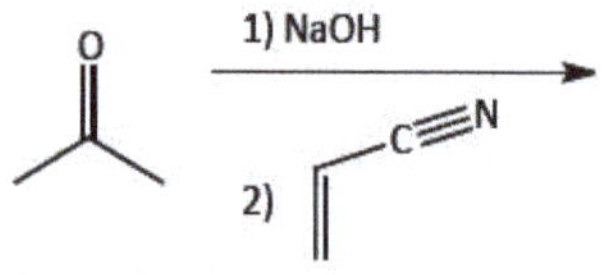 c) 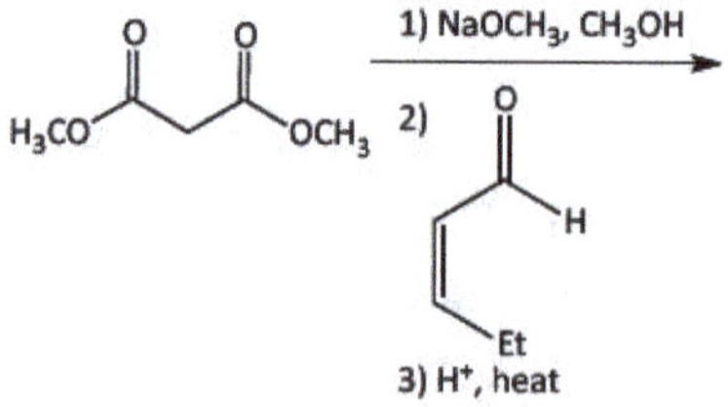

d) 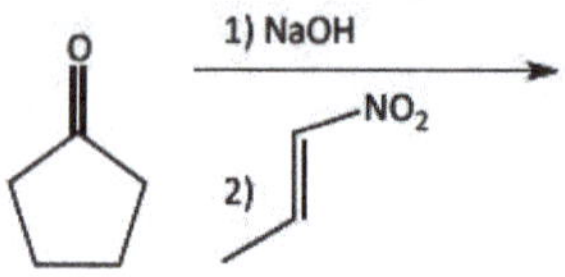 e) 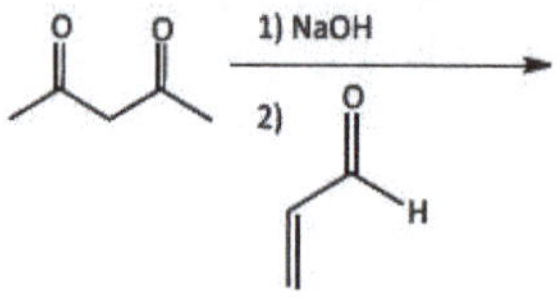 f) 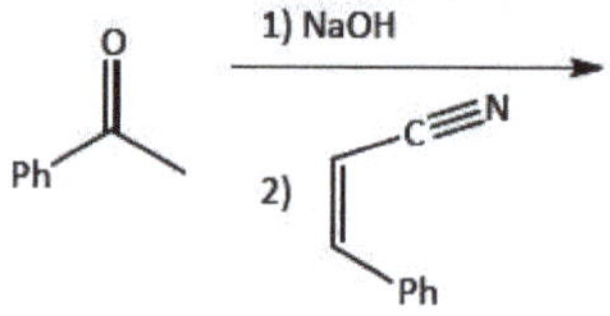

18.37 Show how the following compounds can be synthesized by a Michael addition reaction.

a) 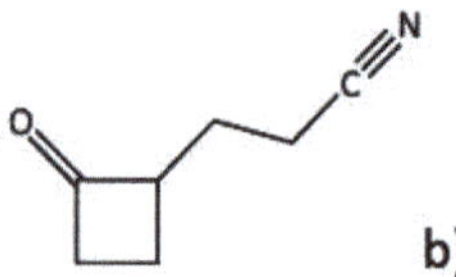 b) 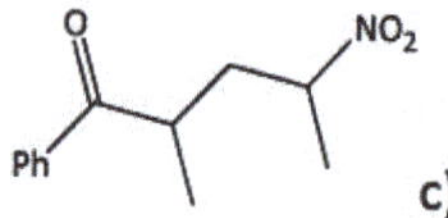 c) 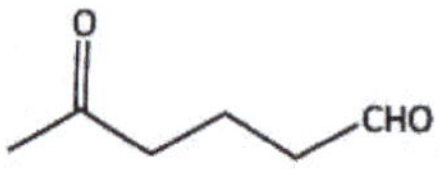 d) 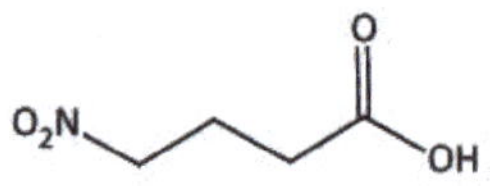 e) 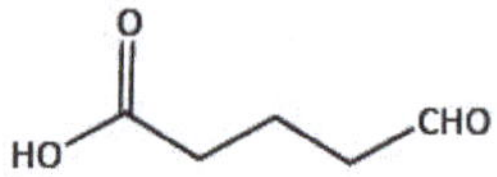

Chapter 19. Conjugated Alkenes

19.1 Color

Thankfully, we live in a world with color. Because of color, we can appreciate life and beauty more fully. I believe that understanding why things have color allows me to appreciate color even more deeply. In this chapter, we will be discussing why most things have color. First, we need to take a quick look at some simple color theory. Then, we will learn what types of molecules are involved in giving most organic pigments their colors.

Photo by jeanneg at Morguefile.com.

19.1.1 Nature of light

A light wave can be described by its wavelength (λ) and frequency (ν).

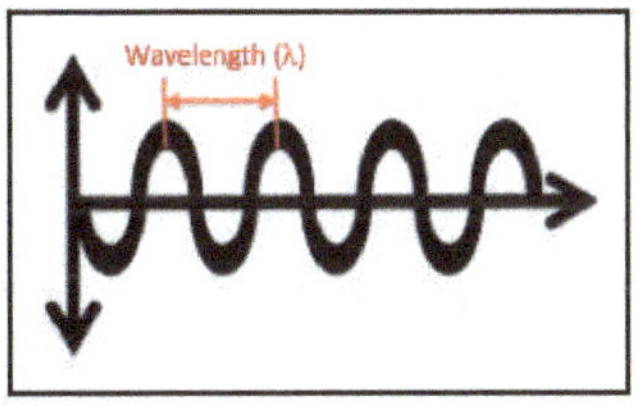

Figure 19.1 Wavelength

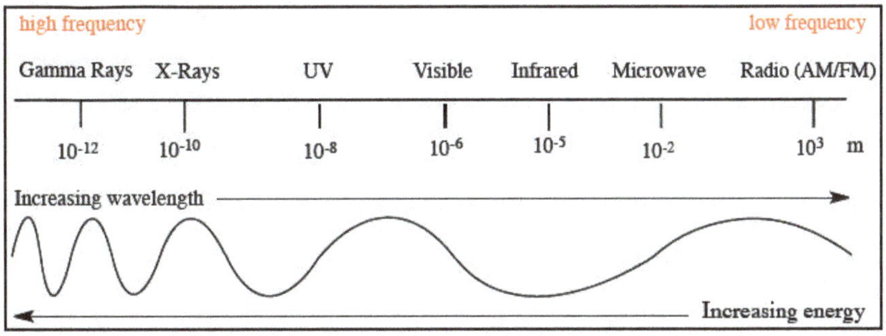

Figure 19.2 Various Wavelengths of the light spectrum

The wavelength of the light is the distance between the peaks of the waves. This is a distance, so it is usually measured in a unit like meters or nanometers. Since all of these types of light travels at the speed of light, we could also count how many waves cross a point in one second. This count of the number of waves in one second is called the frequency. It is written as a #/second. Sometimes the unit is written as s^{-1} which is also called Hz (Hertz). Wavelength and frequency are inversely related to each other. The longer the wave, the fewer can pass by in one second, so the lower the frequency. The product of the wavelength and the frequency is equal to the speed of light, c. The speed of light in a vacuum is $3X10^8$ m/s.

$$c = \lambda\nu$$

The energy of a light wave is directly proportional to the frequency. The constant of proportionality is Planck's constant ($h = 6.63X10^{-34}$Js). So, the higher the frequency of a light wave, the greater its energy is.

$$E = h\nu$$

19.1.2 Simple Color Theory

Sunlight is white light. It is a combination of all of the colors of the rainbow. We know this because the colors of white light can be separated if the sunlight passes through a prism or if it falls on a diffraction grating like a CD. So, when we are taking a hike along a trail and notice the colors of the beautiful flowers, we may wonder what gives them their color.

Photo by Alvimann at Morguefile.com

Photo by mzacha at Morguefile.com

Let's say we are looking at a daisy. Obviously, the flower itself is not generating its own light. The light that is entering our eyes and allowing use to see the daisy originated in the sun. This white light that contains all colors, hit the daisy, bounced off of it, and into our eyes. White is a combination of all colors. When the sunlight hits the white petals of the daisy, all of the colors of the white sunlight bounce off of the petals and into your eyes. This rebounding of all colors from the white petals makes them look white to you. The yellow center of the daisy is a different story.

Photo by DodgertonSkillhause at Morguefile.com

Photo by greyerbaby at Morguefile.com

The following, simplified color wheel roughly shows the colors of white light separated with the spectrum wrapped around so longest wavelength light, red, connect with the shortest wavelength light, violet. The color wheel is also arranged so that each color is directly opposite its complementary color.

For most things that have color, like the yellow center of the daisy, instead of all of the colors of light bouncing off of it, one of the colors is absorbed. It is absorbed because there is a chemical compound in it that absorbs one color. If one color from the white light is absorbed out of the white light, all of the other colors of the visible spectrum bounce off and enter your eye. This makes you see the complementary color of what is absorbed. For the yellow center of a daisy, there is a compound in it that absorbs light around 400-450 nm, which corresponds to blue/violet light. When blue/violet light is absorbed from the white light, the daisy center looks yellow to us. The compound that gives a tomato its red color absorbs light in the 440-500 nm range. This is an absorption in the blue and green part of the spectrum making the tomato their complementary color, red. Similarly, the compound responsible for giving blueberries their color absorbs around 530-600 nm of light, an orange color. The complementary color of the orange color that is absorbed from white light is the blue that we see. We will learn more about the compounds that absorb visible light in this chapter.

19.2 MO diagrams

19.2.1 Electron waves

The world really is strange. It is stranger than most people realize. Back in your Junior High Science class, you were probably told that light is both a particle and a wave. We call that the duality of light. This dual nature is not only true about light, but it is true about most of nature. This is the strangeness that we explore in a field called quantum mechanics.

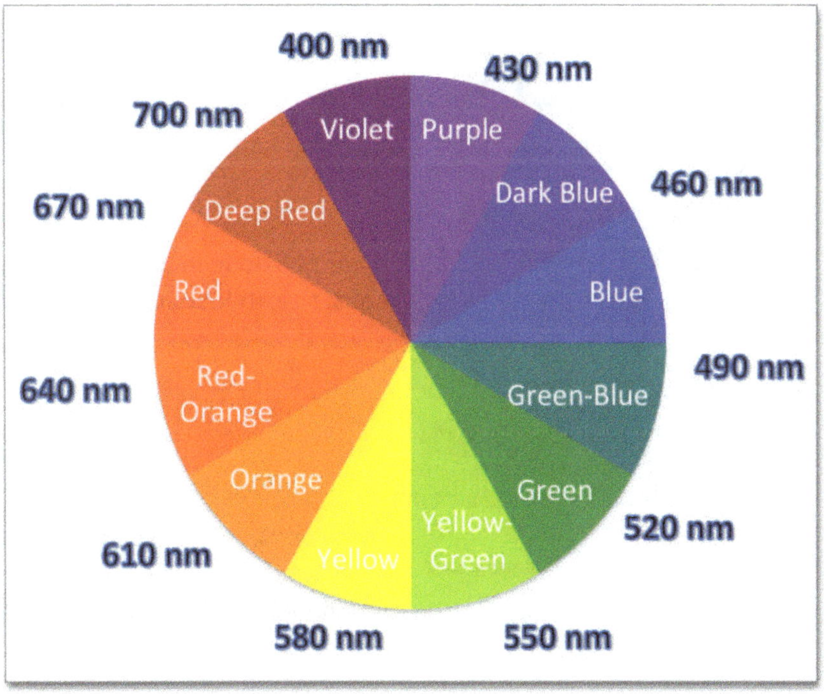

Figure 19.3 Color wheel with wavelengths

For now, I want us to focus on the wave-particle duality of electrons. One way to think about electrons is that they are little particles orbiting the nucleus. This is a fairly good model that many of us have in our minds, the Bohr planetary model. But, the interesting thing is that electrons are also waves! Let's focus on two very simple waves. Think of a guitar string. If I pluck a guitar string, it vibrates up and down.

If I combine these, the string vibrates so fast that it looks something like this.

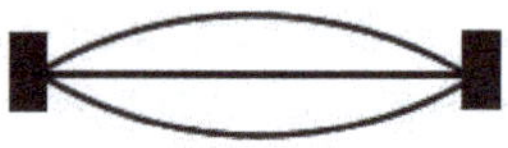

It has zero nodes in the center of it. This simple vibrational shape is spherical in nature and leads us to the shape of the simplest electron orbital, the s-orbital.

Figure 19.4 s-orbital

If I put my finger down in the middle of the guitar string and pluck it, I get a wave where when one side is up, the other side is down. A moment later, it switches which side is up and which is down.

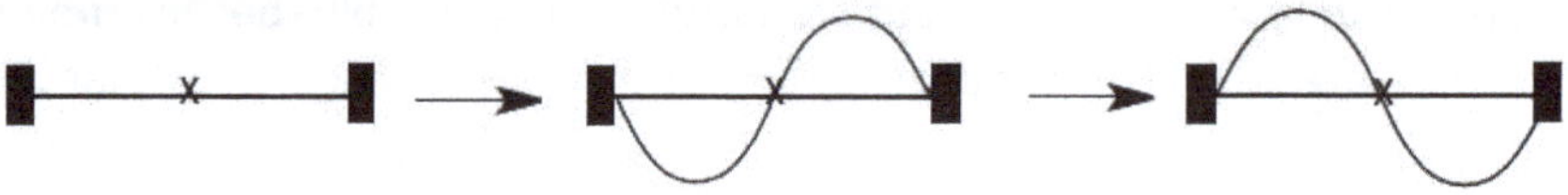

When the string vibrates fast, it looks something like this.

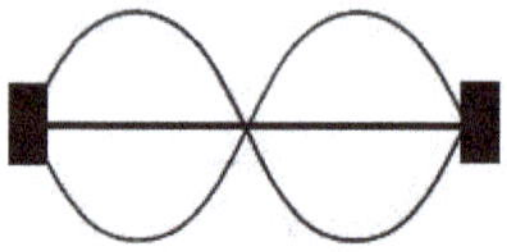

Photo by GaborfromHungary at Morguefile.com

It has one node in the center of it. This vibrational shape has two lobes of equal size. It leads us to the shape of another electron orbital, the p-orbital.

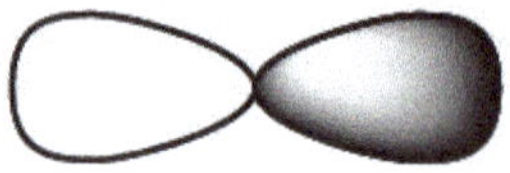

Figure 19.5 p-orbital

These orbital waves can be described with mathematical functions—just like all waves can. We call this mathematical function for the wave the wavefunction (Ψ). These mathematical functions can be added together or subtracted. Intuitively, you understand this about waves.

For example, a child on a swing set moves forward and backward. This motion is a wave-like motion. When children are learning to swing on a swing set, they are taught to kick their legs out when they are moving forward and bend at the knees and put their feet down when they are moving backwards. This motion of their feet is also a wave. They are taught to do this so that their feet wavefunction can be added to the wavefunction of the swing. This is *constructive interference* of these waves. This increases the amplitude of the wave, and the swing goes higher and higher.

Many children get confused when learning this, though. If instead they kicked their feet straight out when they were going backwards and bent their knees putting their feet down when they were going forwards, their feet wave would be opposite the swing's wave. This is *destructive interference*. Children who swing this way will not be swinging for long. When the wavefunctions are subtracted from each other, they get cancelled out.

Because they are waves, when electron orbitals meet and overlap, they can add constructively or destructively. This can happen when orbitals overlap on the same atom (to make hybrid orbitals) or when orbitals on two different atoms overlap when the atoms come together (to make molecular orbitals).

19.2.2 Sigma and Pi Bond review

When two atoms come together, the orbitals on the atoms can overlap. This overlapping of electron density between two atoms makes bonding molecular orbitals, also called bonds. There are two main types of bonds we will deal with in organic chemistry, sigma and pi bonds. A sigma bond is directly between two atomic nuclei. Or, a more technical way to describe it is the electron density lies on the internuclear axis. A pi bond has electron density above and below the internuclear axis.

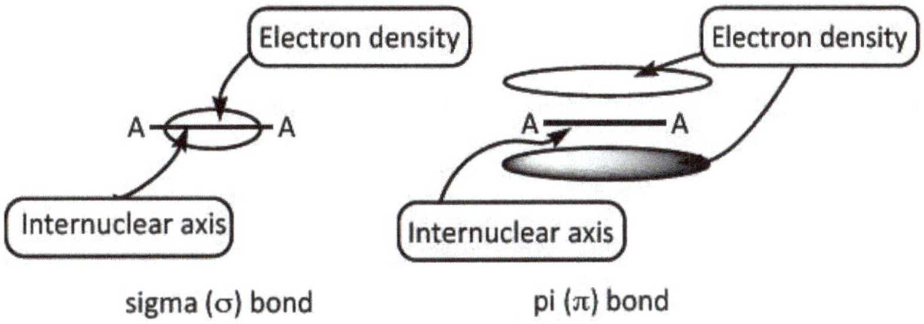

Figure 19.6 Sigma and pi-bonds

We've learned that hybridization is a mixing of atomic orbitals on one atom. We need to look at how two or more atoms come together to make bonds and form molecules. We mix hybrid and/or atomic orbitals to make these molecular orbitals. These bonds that we form between atoms are called molecular bonds.

Atomic orbitals are s and p orbitals on individual atoms.

Hybrid orbitals are sp, sp^2, sp^3 orbitals that are mixed from the s and p atomic orbitals <u>on the same atom</u>.

Molecular orbitals are a mixing of atomic and hybrid orbitals on atoms to make bonds <u>between atoms</u>. This is how we make molecules. But, interestingly, it is a little more complicated than that. These molecular orbitals can be spread out over more than two atoms. This may be a new concept for you. It will seem strange at first, but if you think about it enough, it will begin to make sense. Another strange feature of molecular orbitals is that for every bonding molecular orbital (also called a bond) that forms, an antibonding MO also forms. Sometimes an antibonding MO will have electrons in it. Sometimes it does not. When an antibonding MO has electrons in it, it is the opposite of a bond, it is a bond breaker. Again, this is a strange concept to you because it is new, but after a while, it begins to make sense.

We describe the energies of bonds and antibonding MOs using a diagram called a molecular orbital diagram (MO diagram). The MO diagram shows the energies of molecular bonds, molecular antibonding MO, and atomic orbitals, as well as which orbitals contain electrons. We will begin by showing some very simple molecules and describe their bonding using MO diagrams.

When making MO diagrams, there are a few rules we follow.

19.2.3 Rules for Molecular Orbital (MO) diagrams.

1. The number of atomic orbitals equals the number of molecular orbitals formed.

2. The number of bonding molecular orbitals equals the number of antibonding molecular orbitals.
3. The bonding molecular orbitals are lower in energy than the atomic orbitals. The antibonding molecular orbitals are higher in energy than the atomic orbitals. The amount of energy saved putting electrons in bonds is equal to the amount of energy lost putting electrons into antibonding MOs.
4. Any orbital, whether an atomic orbital or a molecular orbital can contain at most two electrons.

19.2.4 Dihydrogen Molecular Orbital Diagram

This is a hydrogen atom. It has a nucleus with one proton in it. But, for now, we are interested only in its electrons. It contains one electron in a 1s atomic orbital.

Hydrogen atom 1

This is another hydrogen atom with one electron in a 1s atomic orbital.

Hydrogen atom 2

What happens when these two hydrogen atoms approach each other to try to make the molecule H_2? There are two possibilities for how the two hydrogen atoms can interact. When the electrons of the hydrogen atoms (which are waves) begin to interact, they can constructively interfere with each other or destructively interfere with each other. When they constructively interfere, the electron waves add to each other and form a lot of electron density between the two hydrogen atoms, helping hold them together. We call this a bond. In fact, since this electron density is along the internuclear axis, it is called a sigma bond. When electrons from the two hydrogen atoms are out of phase with each other and destructively interfere with each other, the electron density is cancelled out between the two hydrogen atoms and the electron density is concentrated outside of the hydrogen nuclei. This is called an antibonding molecular orbital. This does not help hold the two hydrogen atoms together.

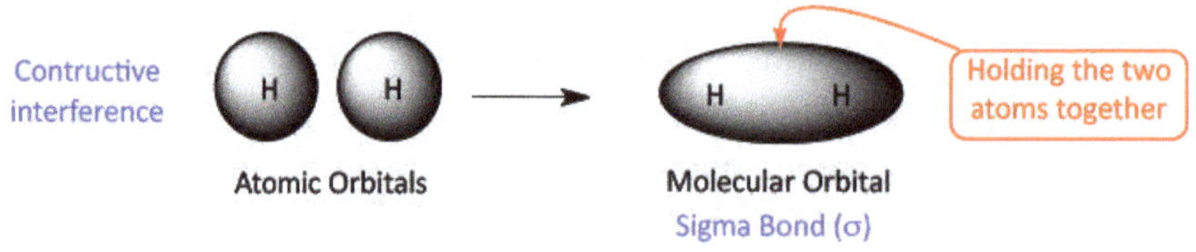

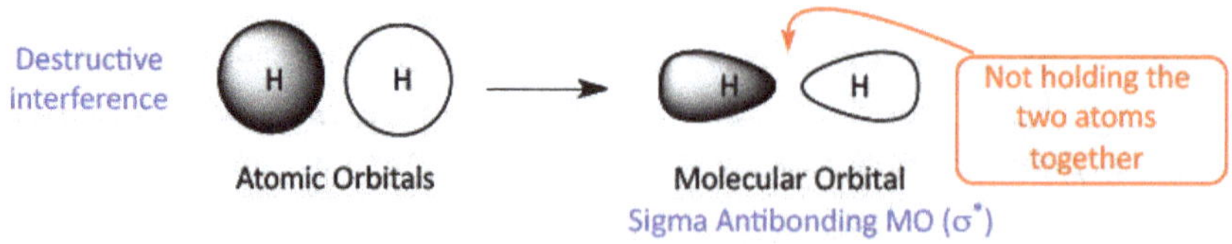

The sigma bond is lower in energy than the energy of the hydrogen atoms s-orbital atomic orbital. It is also lower in energy than the sigma antibonding MO. This leads to the following molecular orbital diagram for dihydrogen (H_2).

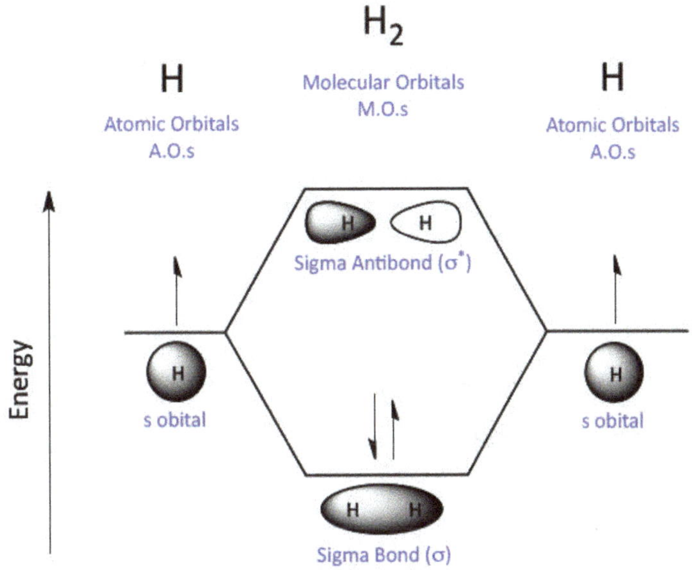

Figure 19.7 Molecular orbital diagram for dihydrogen

In the MO diagram, lower energy orbitals are lower in the diagram and higher energy orbitals are higher in the diagram. The atomic orbitals are on the left and right side of the diagram. In this case, the atomic orbitals are the 1s atomic orbitals of the two hydrogen atoms that come together. These atomic orbitals each have one fishhook arrow in them to describe the one electron in each of the 1s orbitals.

In the middle of the diagram are the molecular orbitals. The constructively overlapping sigma bond is lower in energy and the destructively overlapping sigma antibond is higher in energy. Antibonding orbitals are designated with a "*", so the antibonding orbital of H_2 is represented as σ*.

The total number of bonds between two atoms is called the bond order. A bond order of 1 is a single bond, a bond order of 2 is a double bond, a bond order of 3 is a triple bond.

Bond order (B.O) = Number of bonds – number of antibonding MOs

Since each hydrogen atom contained one electron in its atomic orbital, there is a total of two electrons the molecule will have. These electrons fill up the molecular orbitals starting at the lowest energy orbital and move up from there. So, the two electrons are placed into the sigma bond orbital. This means that for the molecule, two electrons are in the sigma bond making a single bond between the two hydrogen atoms. There are no more electrons, so the sigma antibond orbital (σ*) is empty.

Bond order = 1 full bond – 0 full antibonding MOs = 1

H_2 has a single bond between the two hydrogen atoms and can be symbolized as H-H.

19.2.5 Dihelium Molecular Orbital Diagram

What happens when two helium atoms come together to try to make a dihelium molecule, He_2? A helium atom contains two electrons in its 1s atomic orbital.

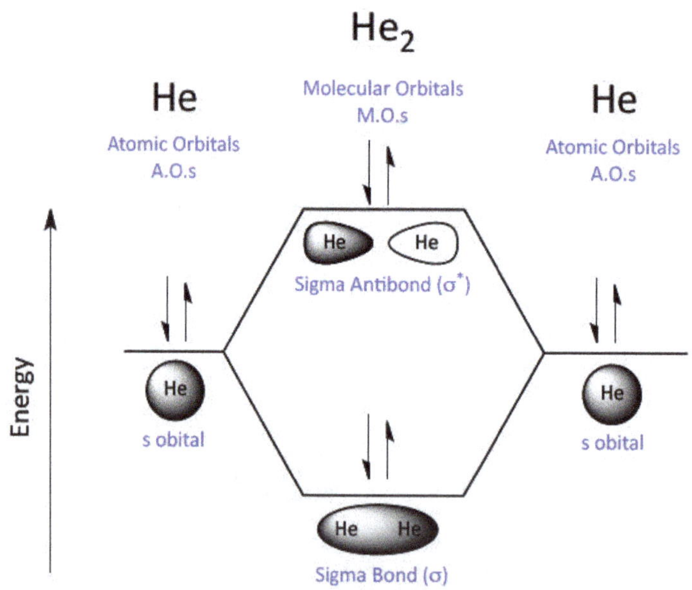

Figure 19.8 Molecular orbital diagram for dihelium

The helium atoms provide four electrons that must go into the molecular orbitals. Two electrons can go into the sigma bond. Each orbital, even molecular orbitals, can hold a maximum of two electrons in it. Therefore, the sigma bond is full with two electrons. The remaining two electrons must go into the antibonding σ^* orbital. When two electrons go into an antibonding orbital, a bond is broken.

Bond order = 1 full bond – 1 full antibonding MO = 0

He_2 has a bond order of 0. He_2 is not held together. The He atoms do not bond together. This is why He is a noble gas and does not form a diatomic molecule.

Now let's focus on pi bonding, in particular, conjugated double bonds.

19.3 Conjugation Introduction

In this chapter, we are looking to deeply understand conjugated alkenes. Conjugated double bonds or conjugated alkenes are those double bonds that are separated by only one single bond. Conjugated double bonds have a double-single-double pattern of carbon bonds.

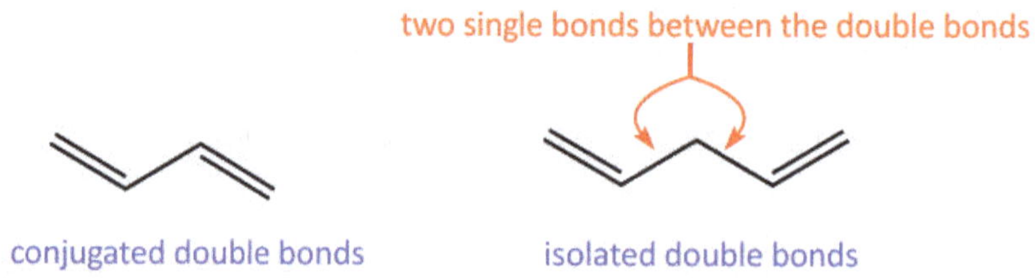

Conjugated double bonds are more stable, or lower in energy, than isolated double bonds. For example, 1,4-pentadiene, with isolated double bonds, has an energy of 57.4 kcal. The very similar, 1,3-pentadiene has an energy of 53.7 kcal. The only difference is 1,3-pentadiene has a conjugated pi system. This conjugation results in nearly 4 kcal of stability compared to the isolated double bonds.

1,4-pentadiene 1,3-pentadiene

57.4 kcal 53.7 kcal

This extra stability because of conjugation can be explained by looking at the pi-system molecular orbital (MO) diagrams. Let's begin our investigation of pi-system MO diagrams by first looking at the simplest case of pi-bonding, ethene. The double bond of ethene consists of one sigma-bond made from two overlapping sp^2 orbitals and one pi-bond with two overlapping p orbitals (in blue).

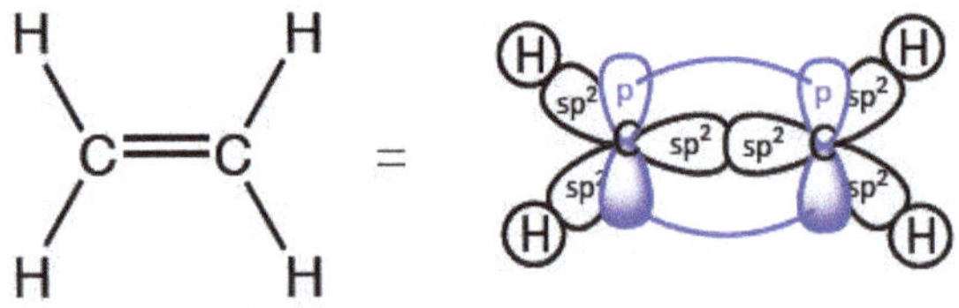

ethene

Now, let's look only at the pi bond of ethene and ignore the sigma bonds. The atomic p orbitals can overlap constructively to make a pi bond or destructively to make an antibonding π^*. The π designation is used because the electron density lies above and below the internuclear axis.

Contructive interference

Atomic Orbitals Molecular Orbital
Pi Bond (π)

Destructive interference

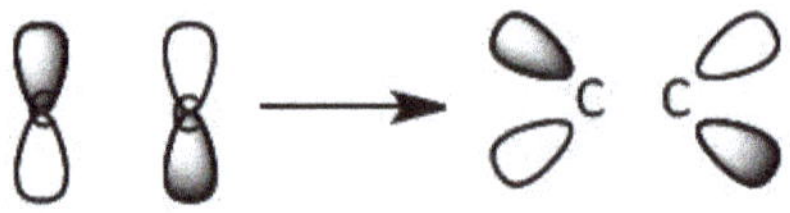

Atomic Orbitals Molecular Orbital
Pi Antibond (π^*)

We can describe the pi bonding for ethene using the following molecular orbital diagram.

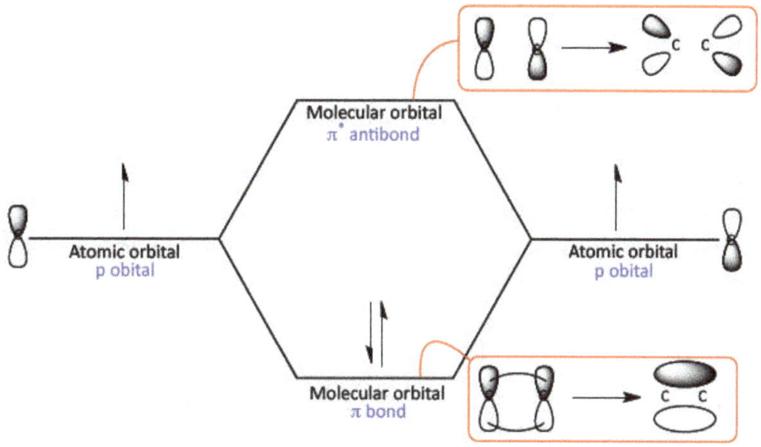

Figure 19.9 Molecular orbital diagram for ethene

Each p atomic orbital contains one electron. When combined, the bond order is 1 for this pi system. There is only one pi bond formed.

Bond order = 1 full bond – 0 full antibonding MOs = 1

19.1 Circle the following molecules that are conjugated alkenes.

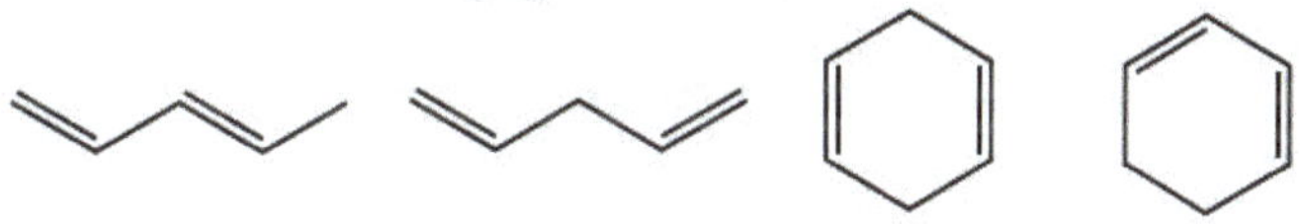

19.4 1,3-Butadiene

Butadiene is a simple, conjugated pi system.

$$H_2C{=}CH{-}CH{=}CH_2$$

Figure 19.10 1,3-Butadiene

To make the two double bonds, four carbon p-orbitals overlap. Each of the p-orbitals contains one electron each. Since these four p atomic orbitals are overlapping, it makes four molecular orbitals in the MO diagram. Since there are always an equal number of bonding and antibonding molecular orbitals, butadiene has two bonding molecular orbitals and two antibonding molecular orbitals.

To determine the shapes of the molecular orbitals, we overlap the four p-orbitals. Notice that in the lowest energy molecular orbital, there are no nodes. All of the p-orbitals overlap making one pi molecular orbital that spreads over the entire four carbon atoms. This is labeled π_1 in the MO diagram. There are two electrons in this orbital that are spread over the entire four carbon atoms.

The next higher energy molecular orbital, π_2, has one node in it, right in the center. The electron density of π_2 is located between carbon atoms C_1-C_2 and C_3-C_4. There are two electrons in this molecular orbital. This electron density is quite similar to the way we draw butadiene, with pi bonds drawn between carbons C_1-C_2 and C_3-C_4.

Between carbon atoms C_2-C_3, there is destructive overlapping of the p-orbitals. But, overall, since there are two bonding sites and one antibonding site, the molecular orbital is bonding.

The electrons in π_2 are the highest energy electrons in the molecule because they are in the highest occupied molecular orbital. This molecular orbital is given the acronym HOMO for **H**ighest **O**ccupied **M**olecular **O**rbital. These highest energy electrons are the most reactive electrons in butadiene.

The next higher molecular orbital is higher in energy than the four atomic p-orbitals. This molecular orbital, π_3*, contains two nodes. This orbital has some overlapping p-orbitals between carbon atoms C_2-C_3, but it is destructively overlapping in two places, between carbon atoms C_1-C_2 and C_3-C_4. Overall, this is more antibonding than it is bonding. This molecular orbital does not have any electros in it because the four p-orbitals each only had one electron for a total of four electrons that can be used in the molecular orbitals. All four were used in π_1 and π_2. Since π_3^* is the lowest molecular orbital without electrons in it, it is labeled with the acronym LUMO for Lowest **U**noccupied **M**olecular **O**rbital.

Finally, the highest molecular orbital of the conjugated pi system of butadiene is π_4*. In π_4*, all of the p-orbitals are out of phase with each other and destructively overlapping with each other. This gives three nodes for a totally antibonding orbital.

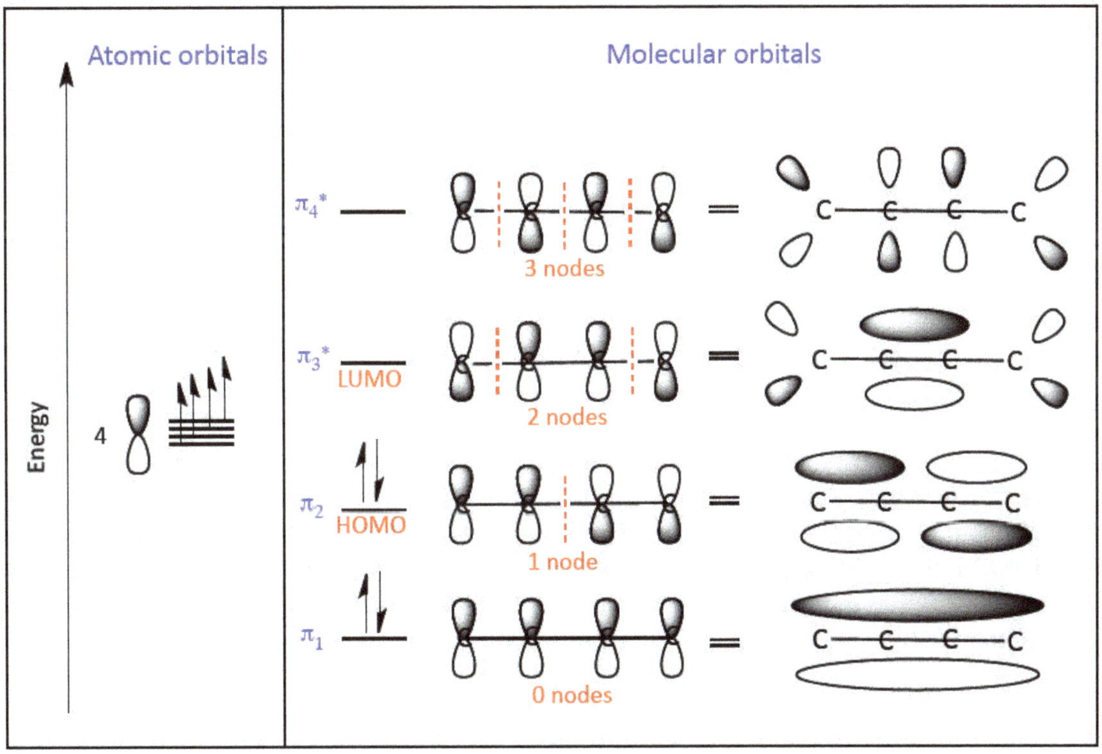

Figure 19.11 Molecular orbital diagram for 1,3-butadiene

The energy distance between the HOMO and LUMO molecular orbitals is of particular importance. This energy gap between the highest occupied molecular orbital and the lowest molecular orbital is called the HOMO-LUMO band gap or simply band gap. As the number of conjugated pi bonds increases in conjugated molecules, more molecular orbitals are needed to describe the conjugated system. This results in shrinking or smaller HOMO-LUMO band gaps energies.

19.2 Shade in the atomic p-orbitals to represent what would overlap to make the following molecular orbitals.

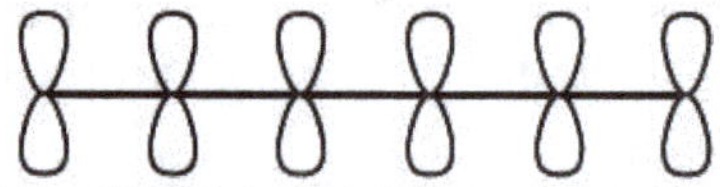
The HOMO of 1,3,5-hexatriene

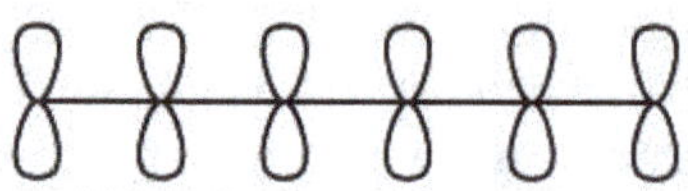
The LUMO of 1,3,5-hexatriene

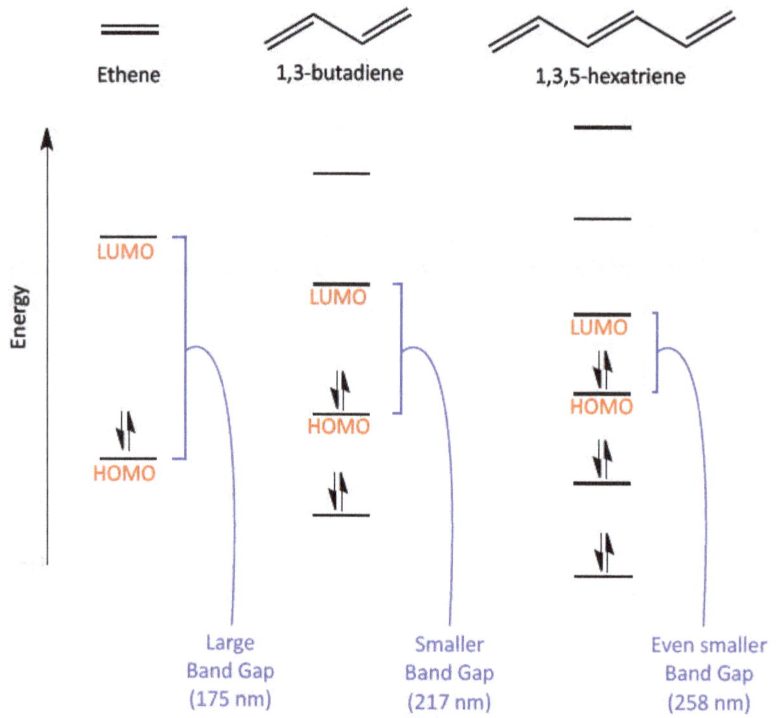

Figure 19.12 HOMO-LUMO band gap of conjugated compounds

For ethane, the HOMO-LUMO band gap is fairly large. In fact, the gap is of an energy that equals 175 nm of light. This is in the ultraviolet range. This is to say, if ultraviolet light of frequency 175 nm shines on ethane, ethane will absorb that light by promoting an electron from the HOMO to the LUMO. 1,3-butadiene, with more conjugation, has a lower HOMO-LUMO band gap. Ultraviolet light of 217 nm can promote an electron from its HOMO to its LUMO energy level. 1,3,5-hexadiene has an even smaller HOMO-LUMO band gap. It will absorb ultraviolet light with a wavelength of 258 nm. Notice, longer wavelength light equals lower energy.

All three of these compounds, ethane, 1,3-butadiene, and 1,3,5-hexatriene, all absorb light in the invisible, ultraviolet region. Because invisible light is being absorbed, all of these compounds are colorless. In order for a compound to have a visible color, the band gap needs to be sufficiently small to absorb wavelengths in the visible spectrum, between about 400-700 nm. The way organic molecules get this small of a band gap is to be very conjugated. Here are some examples of colorful pigments with their absorptions.

19.3 Circle the one molecule in the following grouping that absorbs the longest wavelength of light?

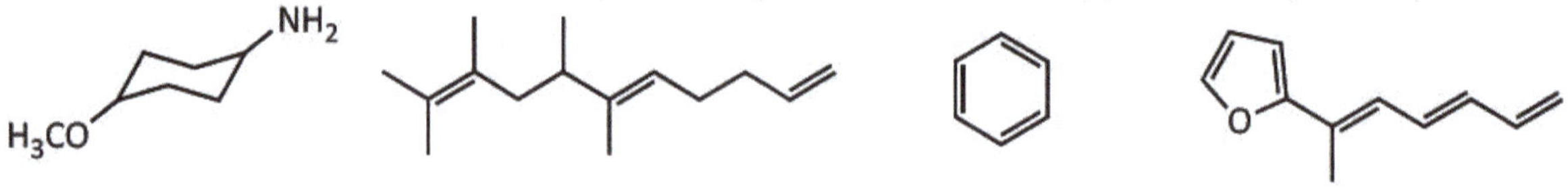

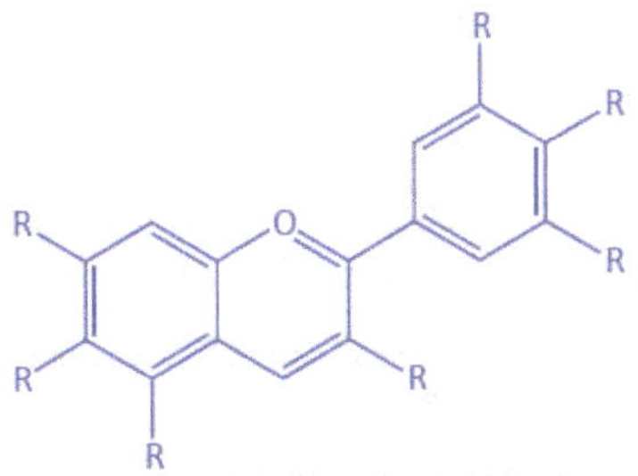

Anthrocyanins (in blueberries)
Absorbs 500-600 nm

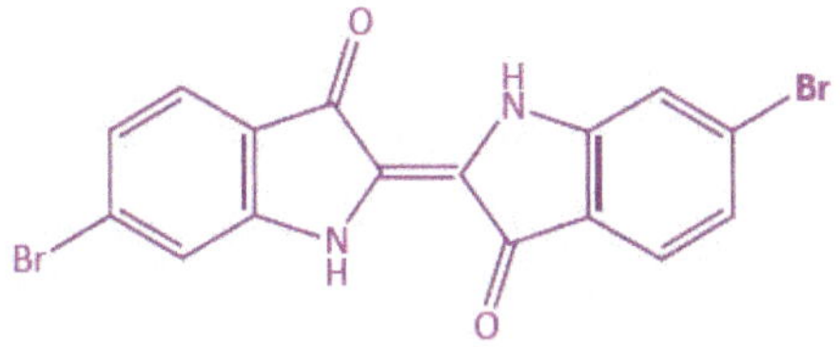

Tyrian Purple
Absorbs 530-600 nm

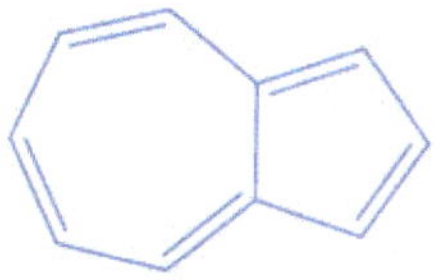

Azulene (from chamomile)
Absorbs 500-700 nm

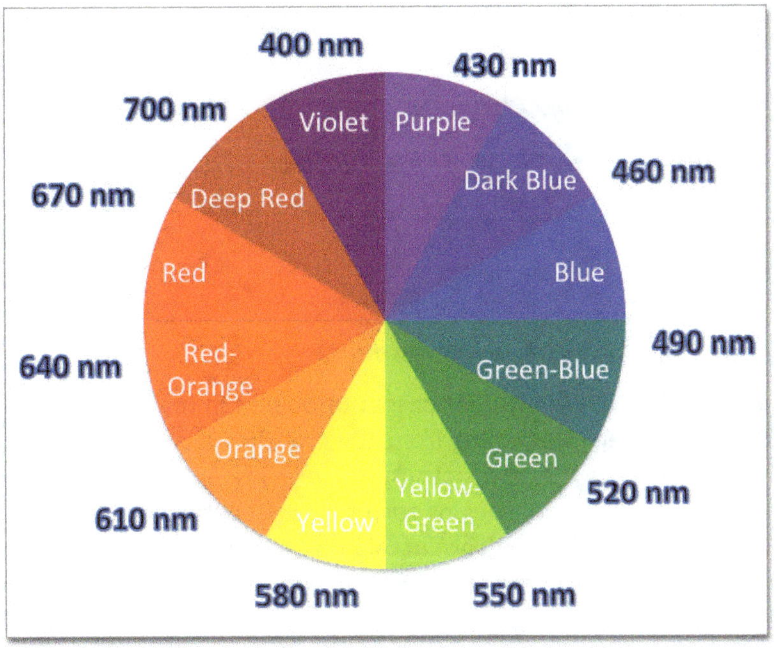

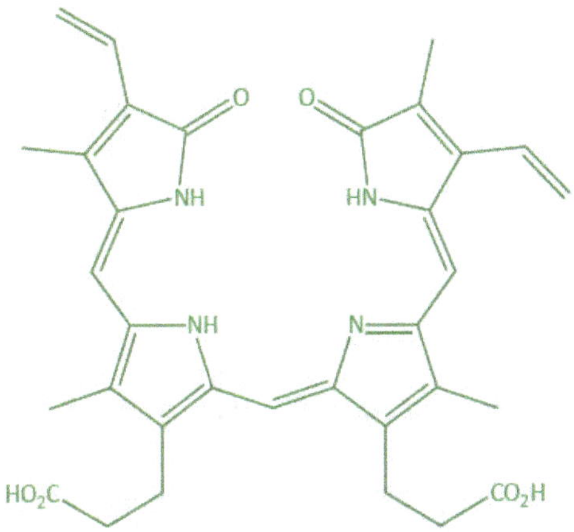

Green biliverdin (in bile)
Absorbs 600-700 nm

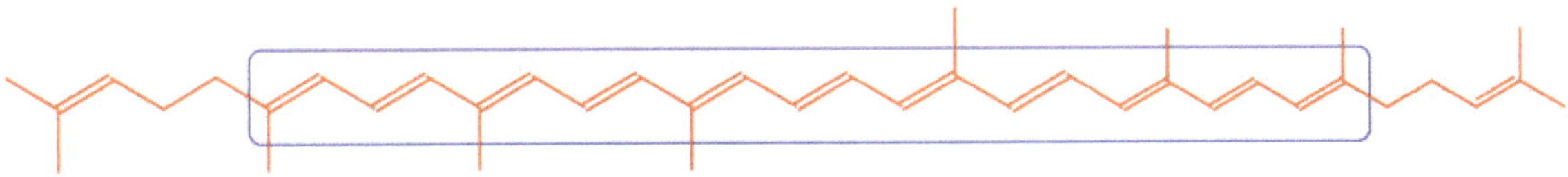

Lycopene (in tomatoes)
11 conjugated double bonds
Absorbs 450-500 nm

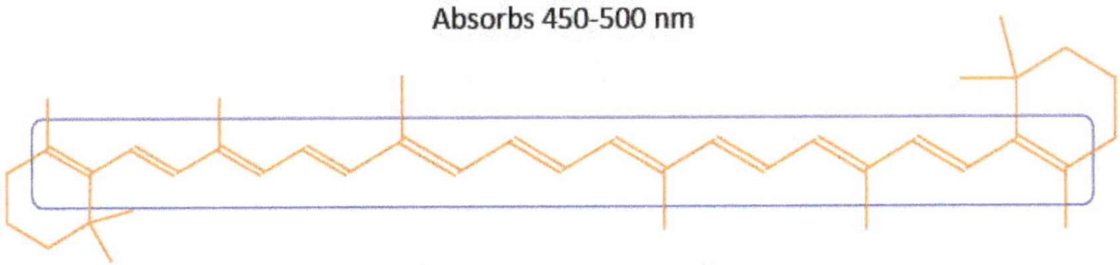

β-Carotene (in carrots)
11 conjugated double bonds
Absorbs 460-500 nm

Figure 19.13 Molecular structures of various conjugated compounds

19.5 UV-Visible Spectroscopy

We have seen that conjugated molecules can absorb ultraviolet and visible light. The process of measuring the absorption of ultraviolet and visible light is called UV-Visible (UV-Vis) spectroscopy. A UV-Vis spectrometer is an instrument that can shine specific wavelengths of ultraviolet or visible light through a sample. The sample is usually dissolved in a solvent and placed in a cuvette. A cuvette is a 1 cm X 1 cm sample holder. Cheap ones are made out of plastic or glass. But, glass and plastic absorb ultraviolet light. So, good cuvettes are made out of quartz and can allow ultraviolet light pass through them. A detector measures the amount of each wavelength of light that passes through the sample. If the sample begins to absorb the light, less passes through to the detector, and a peak is made. It shows an absorbance.

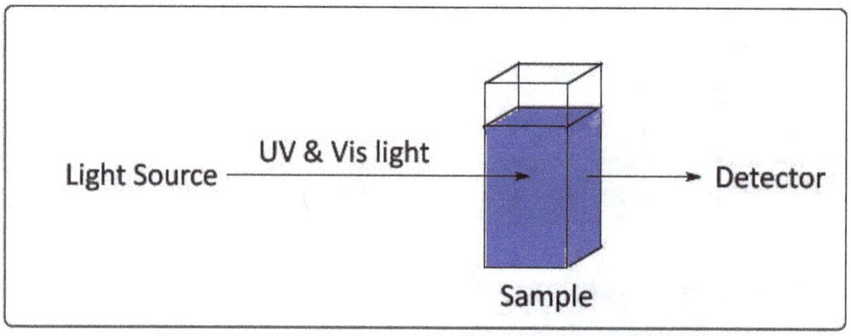

Figure 19.14 Simple concept for UV-Visible spectrometer

A UV-Vis spectrum usually plots absorbance vs. wavelength of light that passes through it in nm. Let's pretend that we are analyzing blue juice. The UV-Vis instrument begins scanning from 200 nm (the ultraviolet region) to 400 nm (violet light) on up to 800 nm (red light). Below is the UV-Vis spectrum. It shows that no light is absorbed between 200-500 nm (ultraviolet through purple). Then, the sample begins absorbing light between 500-700 nm. The light around 600 nm is the greatest absorbance with the highest spot at 605 nm. 605 nm is therefore called the λ_{max} (lambda max). This is the wavelength (λ) with the maximum absorbance. Since the juice absorbs light in the yellow-orange part of the spectrum, it looks blue.

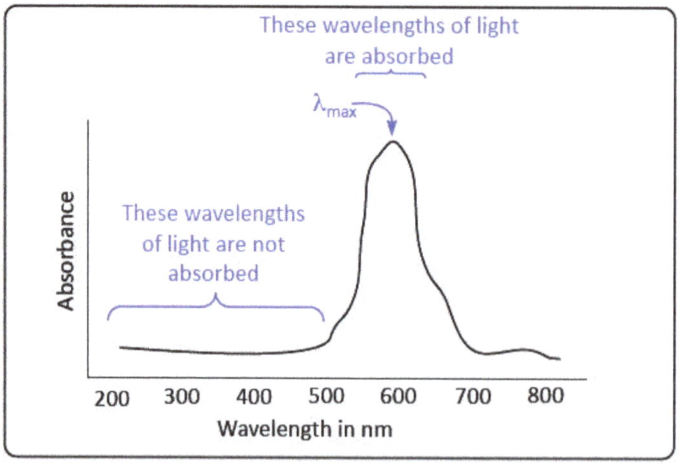

Figure 19.15 UV-Vis spectrum of blue juice

Even though a HOMO-LUMO gap is a very specific energy, UV-Vis peaks tend to be broad. This is because there are different rotational and vibrational states and energies for the molecule. This causes the energies of the HOMO and LUMO molecular orbitals to vary some.

Solved problem:

Given this UV-Vis spectrum, what is the energy of the HOMO-LUMO band gap for this blue juice?

The λ_{max} for the blue juice is 605 nm. This wavelength corresponds to a specific energy. We know that the energy of light is proportional to its frequency.

$$E = h\nu$$

We therefore need to know the frequency of this light given its wavelength. We know wavelength and frequency are related by the speed of light.

$$c = \lambda\nu$$

Using this equation, we solve for frequency. The units should be in *m/s* for the speed of light and in *m* for the wavelength.

$$\nu = \frac{c}{\lambda}$$

$$\nu = \frac{2.99X10^{8} m/s}{(605nm)\left(\frac{1\,m}{10^{9}nm}\right)} = 4.94X10^{14}\,s^{-1}$$

With the frequency, the energy can now be calculated. Plank's constant, h, is $6.63X10^{-34}$ Js.

$$E = h\nu$$

$$E = \left(6.63X10^{-34}Js\right)\left(4.94X10^{14}s^{-1}\right) = 3.27X10^{-19}J$$

This is the approximate energy of the HOMO-LUMO band gap.

19.6 Allyl groups

A propylene group is often called an allyl group.

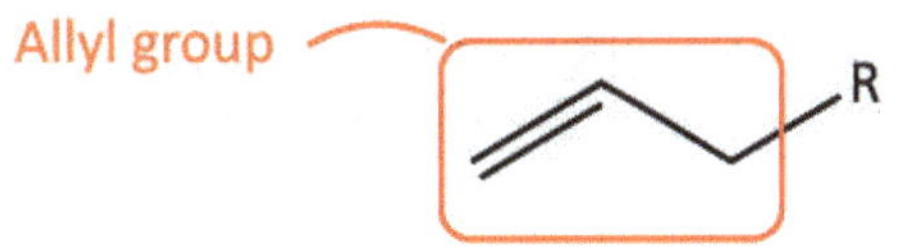

For instance, 3-bromopropene is commonly called allyl bromide.

The carbon position neighboring the double bond is called the allylic position.

19.7 Allyl carbocation, radical, and carbanion

This is an allyl carbocation.

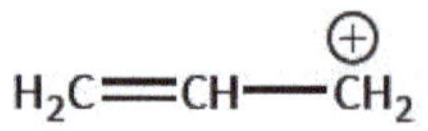

The allyl carbocation has three carbon atoms that are each sp^2 hybridized. They each contain one p-orbital for a total of three overlapping p-orbitals. Two p-orbitals are used to make the pi bond and one p-orbital is an empty p-orbital for the carbocation.

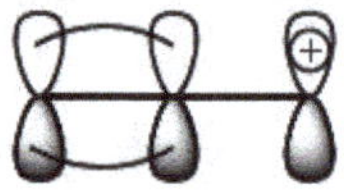

Two good resonance forms can be drawn for the allyl carbocation. What the allyl carbocation really looks like is a blend of these two resonance forms, the resonance hybrid. In the resonance forms, carbon atoms C_1 and C_3 have the positive charge on them. The hybrid shows carbon atoms C_1 and C_3 each have part of this positive charge or a partial positive charge, δ^+. The resonance forms indicate a double bond between carbon atoms C_1 and C_2 in the first resonance form and a single bond between carbon atoms C_1 and C_2 in the second resonance form. Therefore, there is the electron density of 1.5 bonds between carbon atoms C_1 and C_2. The same can be said between carbon atoms C_2 and C_3.

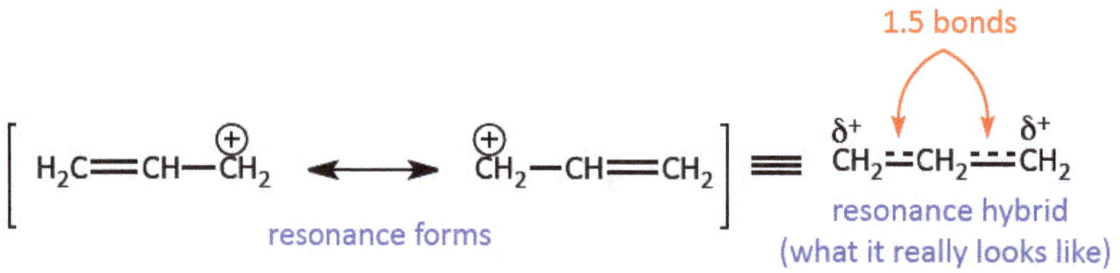

The resonance hybrid can be explained by carefully studying the molecular orbital diagram of the allyl carbocation's pi system. The allyl carbocation consists of three overlapping p atomic orbitals. These p-orbitals contain two electrons (two from the pi bond and zero from the carbocation). We indicate this on the atomic orbital, left side of the following diagram. Since there are three p atomic orbitals that we are overlapping, we must make three molecular orbitals. The lowest energy molecular orbital, π_1, contains zero nodes, the middle one, called nonbonding, contains one node, and the highest energy molecular orbital, π_2*, contains two nodes. Notice that the nodes are symmetrical.

There are always equal numbers of bonding and antibonding molecular orbitals. With three molecular orbitals for allyl carbocation, there is one bonding MO (π_1) and one antibonding MO (π_2*). The leftover MO is called nonbonding, and is always the middle energy. The nonbonding MO is the same energy as the original p atomic orbitals. With the one node for the nonbonding molecular orbital at the center carbon atom, we are left with two non-overlapping p-orbitals on carbon atoms C_1 and C_3. These are simply atomic p-orbitals. That is why it is the same energy and the p atomic orbitals and why we call this MO "nonbonding". The bonding MO and antibonding MO are symmetrically distributed about the nonbonding MO. The amount of energy saved by making the π_1 bonding MO is the same as the amount of energy lost in making a π_2*antibonding MO.

We place the two pi system electrons into the molecular orbitals starting with the lowest energy MOs. We can place at most two electrons in any orbital. Therefore, we end up with two electrons in the π_1 bonding molecular orbital. The π_1 MO has two electrons that are shared over carbon atoms C_1, C_2, and C_3. This is what the hybrid showed with 1.5 bonds between C_1 and C_2 and 1.5 bonds between C_2 and C_3. There are no electrons in the nonbonding MO. This nonbonding MO indicates that there are zero electrons in the p-orbitals on C_1 and C_3. This is shown in the partial positive charges on C_1 and C_3 of the resonance hybrid.

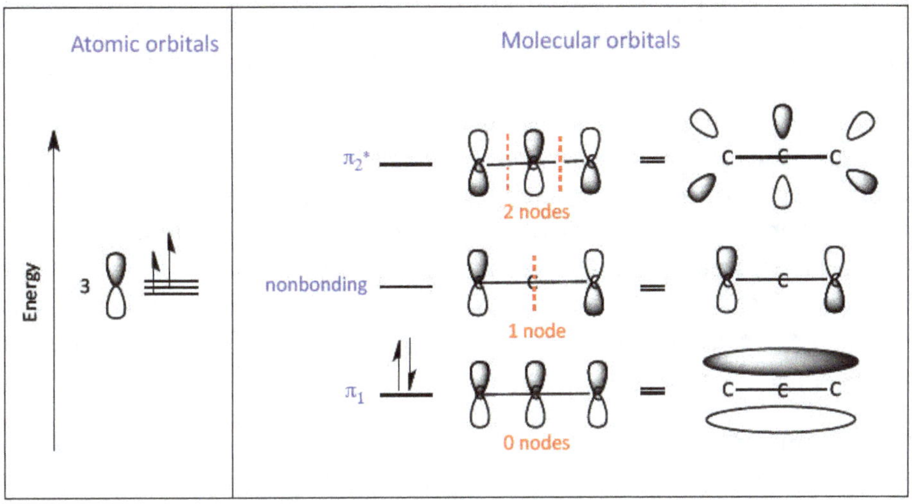

Figure 19.16 Allyl carbocation molecular orbital diagram

This is an allyl radical.

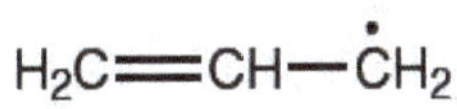

The allyl radical has three carbon atoms that are each sp^2 hybridized. They each contain one p-orbital for a total of three overlapping p-orbitals. Two p-orbitals are used to make the pi bond and one p-orbital has one electron in it for the radical.

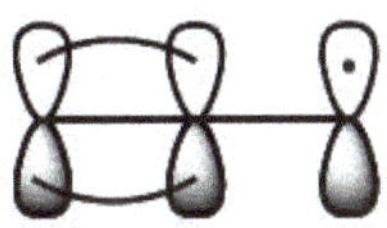

Two good resonance forms can be drawn for the allyl radical. What the allyl radical really looks like is a blend of these two resonance forms, the resonance hybrid. In the resonance forms, carbon atoms C_1 and C_3 have the free radical single electron on them. The hybrid shows carbon atoms C_1 and C_3 each have part of this radical or contain a partial radical, $\delta^{\cdot}$. The resonance forms indicate a double bond between carbon atoms C_1 and C_2 in the first resonance form and a single bond between carbon atoms C_1 and C_2 in the second resonance form. Therefore, there is the electron density of 1.5 bonds between carbon atoms C_1 and C_2. The same can be said between carbon atoms C_2 and C_3.

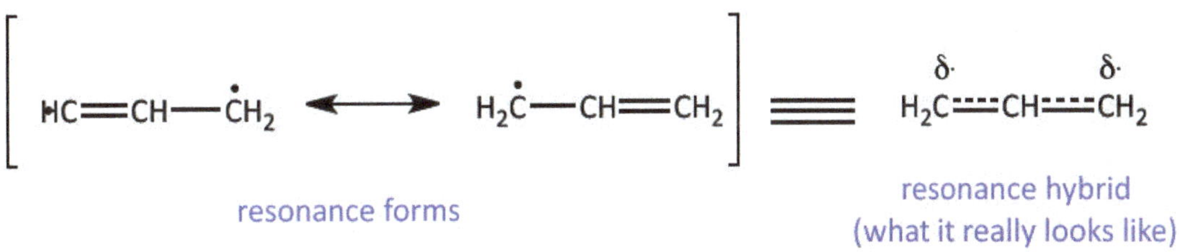

The resonance hybrid can be explained by carefully studying the molecular orbital diagram of the allyl radical's pi system. The allyl radical consists of three overlapping p atomic orbitals. These p-orbitals contain three electrons (two from the pi bond and one from the radical). We indicate this on the atomic orbital, left side of the following diagram. Since there are three p atomic orbitals that we are overlapping, we must make three molecular orbitals. The lowest energy molecular orbital, π_1, contains zero nodes, the middle one, called nonbonding, contains one node, and the highest energy molecular orbital, π_2*, contains two nodes. Notice that the nodes are symmetrical.

There are always equal numbers of bonding and antibonding molecular orbitals. With three molecular orbitals for allyl radical, there is one bonding MO (π_1) and one antibonding MO (π_2*). The leftover MO is called nonbonding, and is always the middle energy. The nonbonding MO is the same energy as the original p atomic orbitals. With the one node for the nonbonding molecular orbital at the center carbon atom, we are left with two non-overlapping p-orbitals on carbon atoms C_1 and C_3. These are simply atomic p-orbitals. That is why it is the same energy and the p atomic orbitals and why we call this MO "nonbonding". The bonding MO and antibonding MO are symmetrically distributed about the nonbonding MO. The amount of energy saved by making the π_1 bonding MO is the same as the amount of energy lost in making a π_2*antibonding MO.

We place the three pi system electrons into the molecular orbitals starting with the lowest energy MOs. We can place at most two electrons in any orbital. Therefore, we end up with two electrons in the π_1 bonding molecular orbital. We have one more electron that is placed into the nonbonding MO. The π_1 MO has two electrons that are shared over carbon atoms C_1, C_2, and C_3. This is what the hybrid showed with 1.5 bonds between C_1 and C_2 and 1.5 bonds between C_2 and C_3. There is one electron in the nonbonding MO. This nonbonding MO indicates that there is one electron shared by the p-orbitals on C_1 and C_3. This is shown in the partial radicals on C_1 and C_3 of the resonance hybrid. These p-orbitals are isolated and nonbonding. This is why this MO is called "nonbonding".

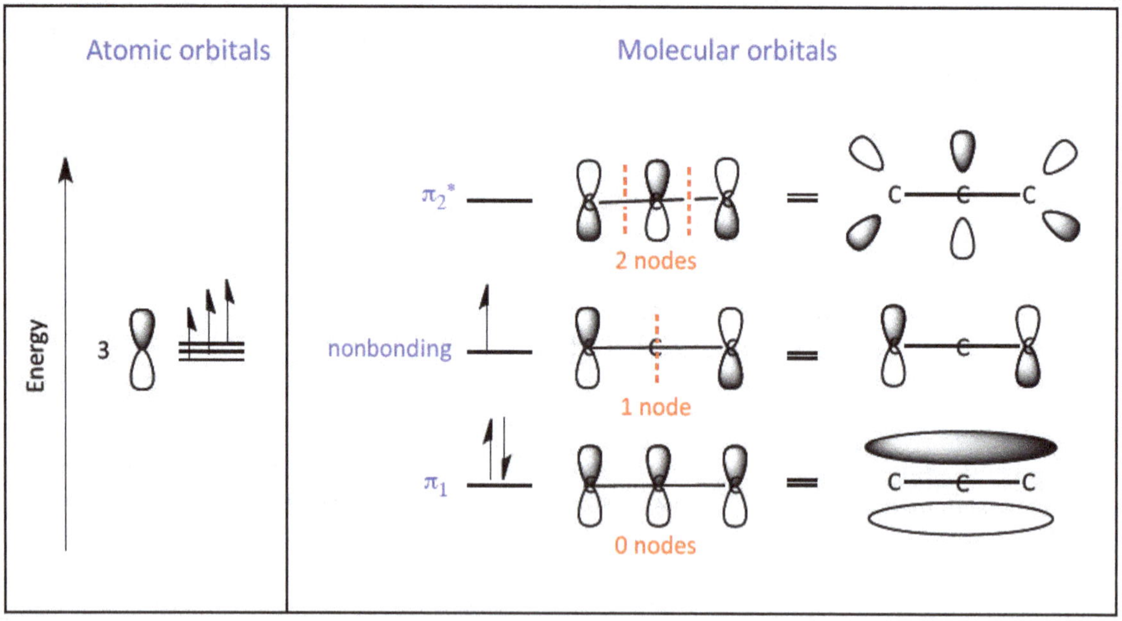

Figure 19.17 Allyl radical molecular orbital diagram

This is an allyl carbanion.

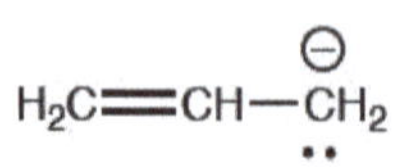

The allyl carbanion has three carbon atoms that are each sp^2 hybridized. They each contain one p-orbital for a total of three overlapping p-orbitals. Two p-orbitals are used to make the pi bond and one p-orbital has two electrons in it for the lone pair of electrons of the carbanion.

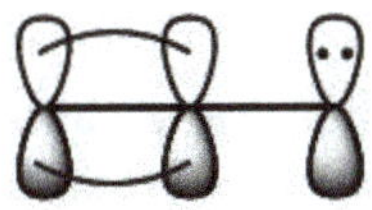

Two good resonance forms can be drawn for the allyl carbanion. What the allyl carbanion really looks like is a blend of these two resonance forms, the resonance hybrid. In the resonance forms, carbon atoms C_1 and C_3 have the lone pair of electrons (the negative charge) on them. The hybrid shows carbon atoms C_1 and C_3 each have part of this lone pair of electrons. There are also 1.5 bonds between carbon atoms C_1 and C_2 as well as between carbon atoms C_2 and C_3.

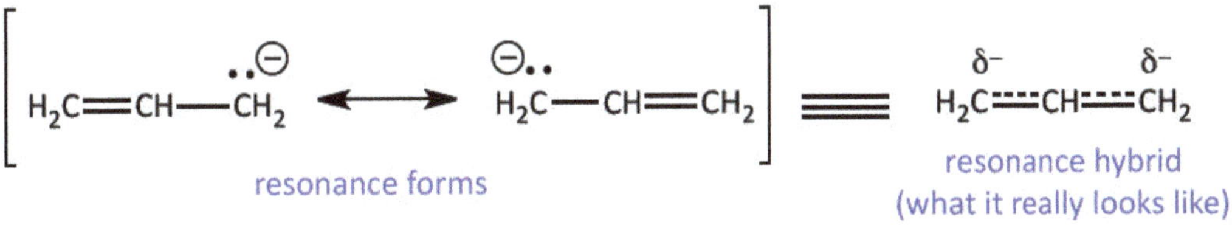

The MO diagram is very similar to the allyl free radical except there are a total of four pi electrons (two from the pi bond and two for the lone pair of electrons).

The MO diagram has two electrons in the nonbonding MO. This means the lone pair of electrons are shared over carbon atoms C_1 and C_3 as the resonance hybrid shows.

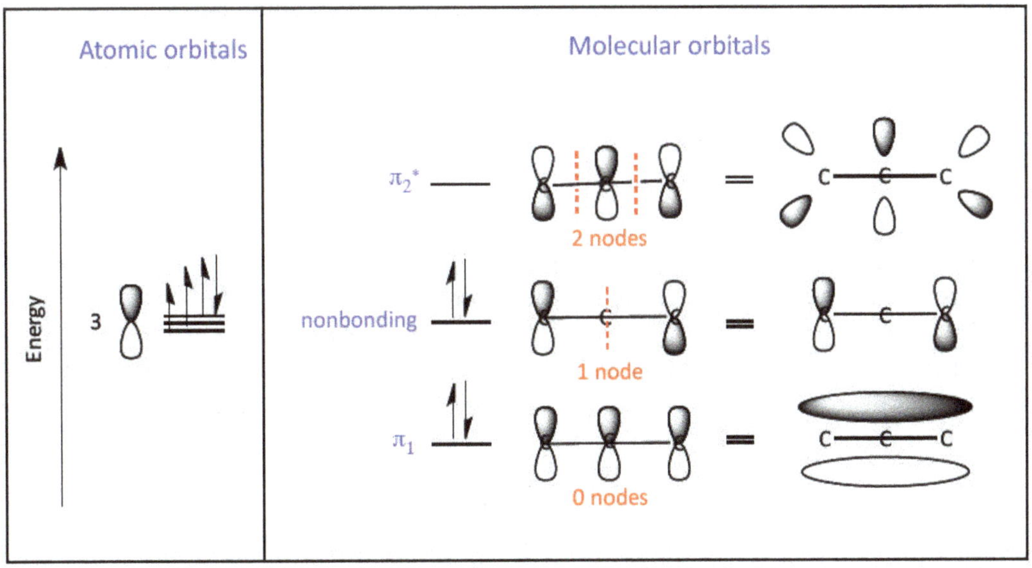

Figure 19.18 Allyl carbanion molecular orbital diagram

19.7.1 Stability of allylic cations

Allylic positions are important because they are especially reactive sites. For instance, when heat is applied to allyl bromide, or it is placed in an ionizing solvent, it can easily dissociate forming an allyl cation.

Conjugated Alkenes

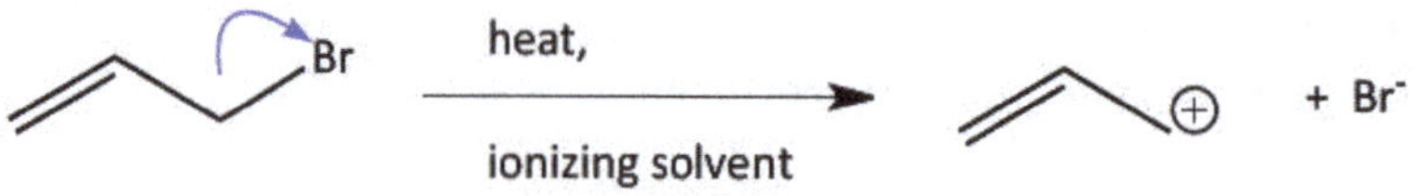

At first, this may seem surprising because a primary carbocation is formed, but the allyl cation is quite stable. In fact, its stability is similar to a secondary carbocation. We saw this in the previous section. The pi electrons of the allyl cation are stabilized, or lower in energy, because they end up in a bonding molecular orbital. Resonance forms can also be drawn for the allyl cation, spreading out the positive charge over two carbon atoms.

Another very stable position is next to a benzene ring with its double bonds.

19.7.2 Allylic Radicals

Like allyl carbocations, allyl radicals are also quite stable. They are resonance stabilized and the conjugated overlap allows for stabilized, lower energy pi molecular orbitals to be formed as we saw earlier.

When cyclohexene reacts with a bromine radical, the following product predominates.

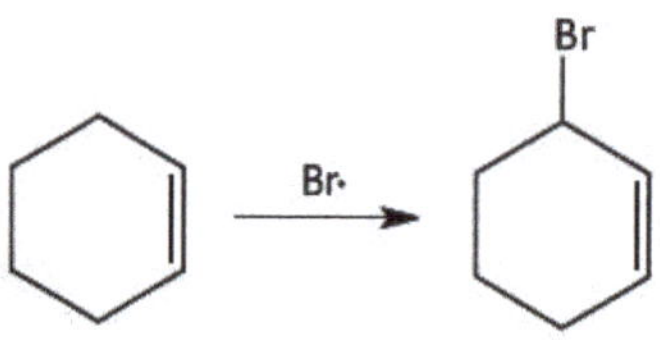

Let's analyze the hydrogen atoms that we could have replaced on cyclohexene. All of the hydrogen atoms on cyclohexene are secondary, yet their reactivity is very different. The hydrogen atoms on the double bond are called vinyl or vinylic hydrogen atoms. Hydrogen atoms on a carbon that neighbors the double bond are called allyl or allylic hydrogen atoms. This carbon is an allylic carbon atom. The other hydrogen atoms are regular alkyl hydrogen atoms. Allylic hydrogen atoms are the most reactive. Why is this since they are all secondary hydrogens?

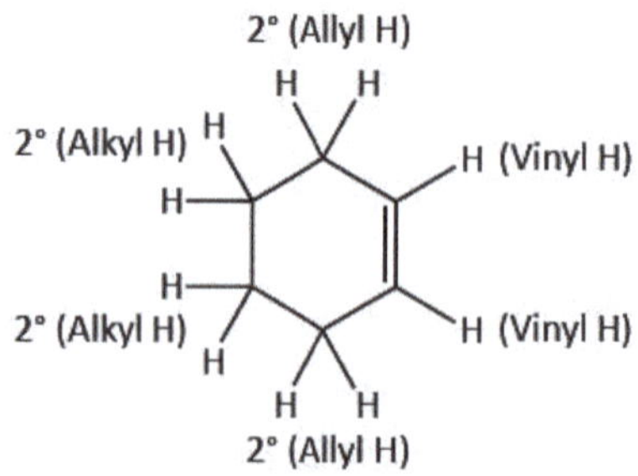

Figure 19.19 Vinyl vs. allyl vs. alkyl protons

If we look at the radical that is formed when the allylic hydrogen atom is removed, we find that resonance forms can be drawn for it. This greatly stabilizes the free radical.

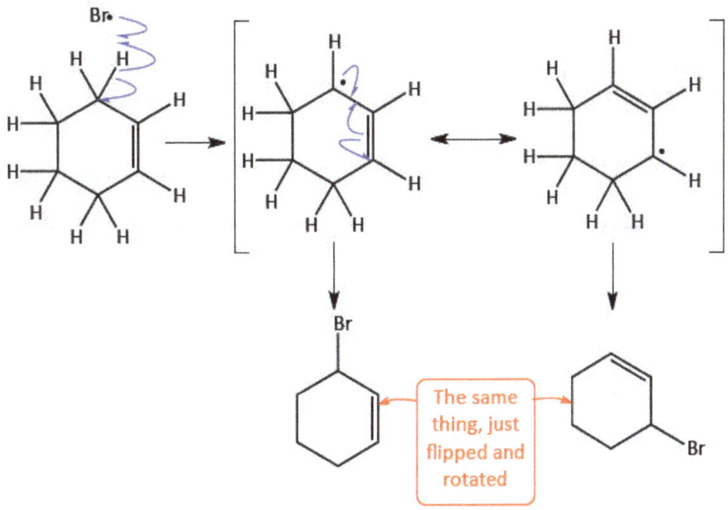

Figure 19.20 Free radical bromination at an allylic position

If a vinylic or alkyl hydrogen is removed, resonance forms cannot be drawn. Vinylic radicals are especially unstable. In a vinyl radical (much like in a vinyl cation), the electron deficient partially positive location is in an sp^2 hybrid orbital and held closer to the positive carbon nucleus than it is in an sp^3 hybrid orbital. This is a bad situation.

When we generate bromine radicals in the presence of a double bond, we need to be careful. If we did an initiation reaction with Br_2 and ultraviolet light like we have already learned, we would run into problems. The problem is that alkenes react with Br_2. We learned about this previously. Alkenes reacting with Br_2 can result in the dibromination of the double bond.

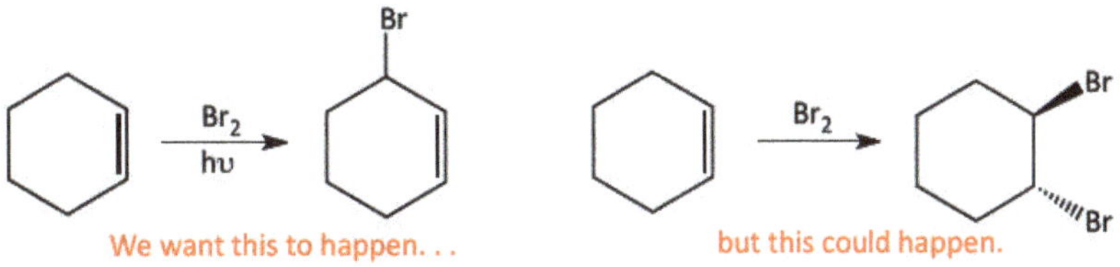

How do we avoid this problem? We need to make sure we don't have too much Br_2 present in our reaction mixture. The best way to do this is to perform this reaction with NBS (N-bromosuccinimide) and not Br_2. NBS generates Br_2 in small amounts so the dibromination of the double bond does not happen.

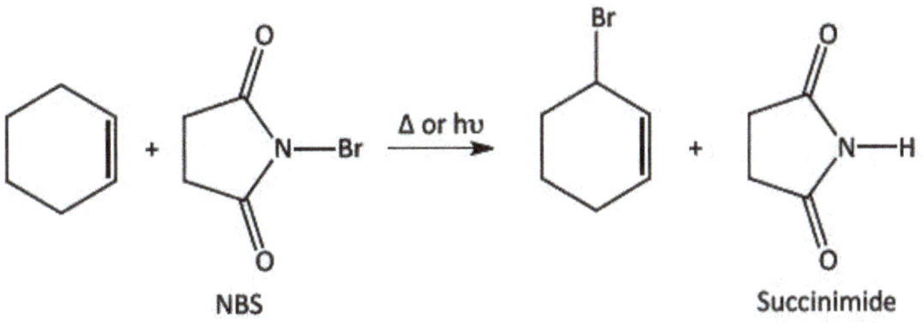

Now that you know about allylic bromination, predict the product of the following reaction.

19.4 Fill in the blanks.

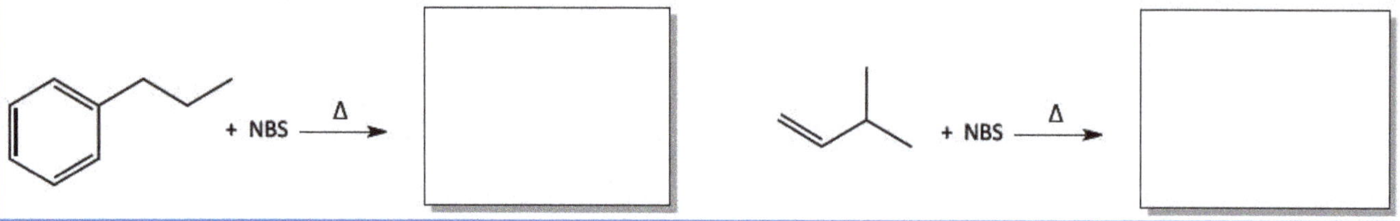

We call the carbon that neighbors a benzene ring a benzylic carbon atom. The hydrogen atoms on that carbon are benzylic hydrogen atoms.

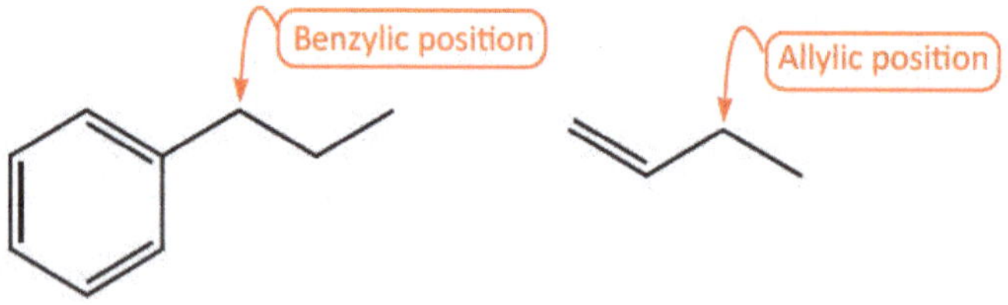

We can now include these resonance-stabilized positions in a rank of the stability of free radicals.

vinyl ≈ methyl < 1° < 2° < 3° < allylic ≈ benzylic

Least stable Most stable

When a free radical reaction occurs in a compound with a pi-bond, you need to always consider the various resonance forms. Sometimes, if the molecule is symmetrical, the products we get are the same, so it does not matter. We saw this in the bromination of cyclohexene above and in the following.

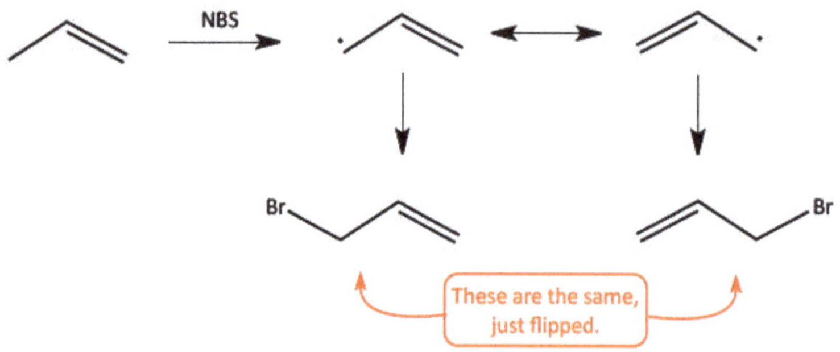

So, if we draw the overall reaction, we only need to draw one product.

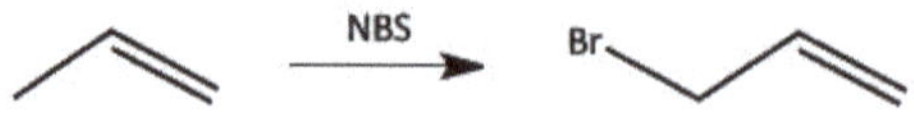

But, when the alkene is less symmetrical, it can lead to a mixture of products.

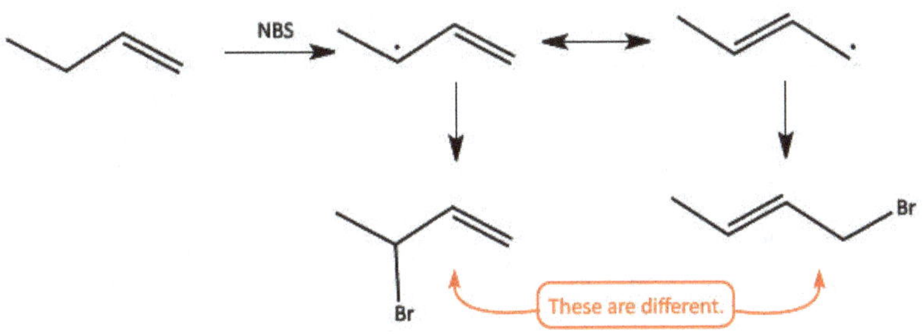

If we draw the overall reaction, we need to include both products, since they are different.

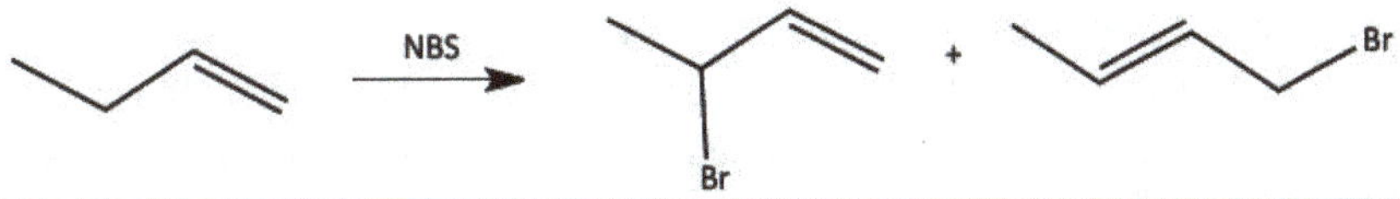

19.5 (a) Identify the most reactive hydrogen atom(s) in the following molecules if they undergo free radical reactions.

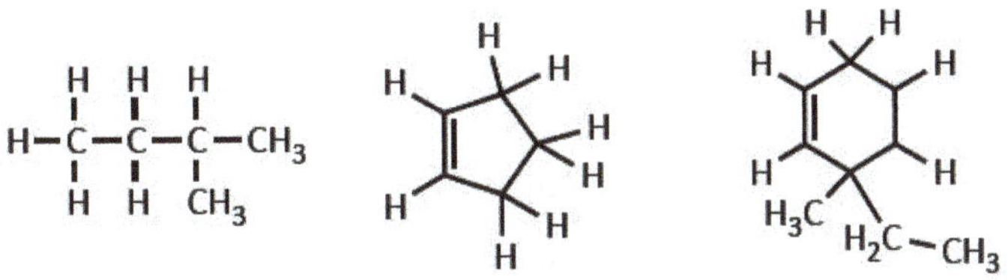

(b) For these molecules, draw the major monobrominated product(s) of each when they undergo a free radical bromination. Monobrominated means only one bromine atom is on the molecule.

19.7.3 S_N2 of allyl bromide

The S_N2, backside attack, reaction of a nucleophile with propyl bromide is a pretty good reaction since it is a primary halide. In the transition state, the attacked carbon atom is sp^2 hybridized. The nucleophile attacks a p-orbital as the leaving group leaves from the p-orbital.

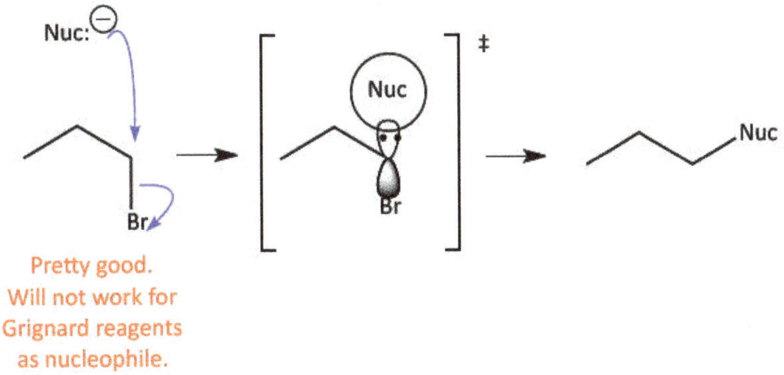

This reaction works quite well, but it does not work with Grignard reagents as the nucleophile.

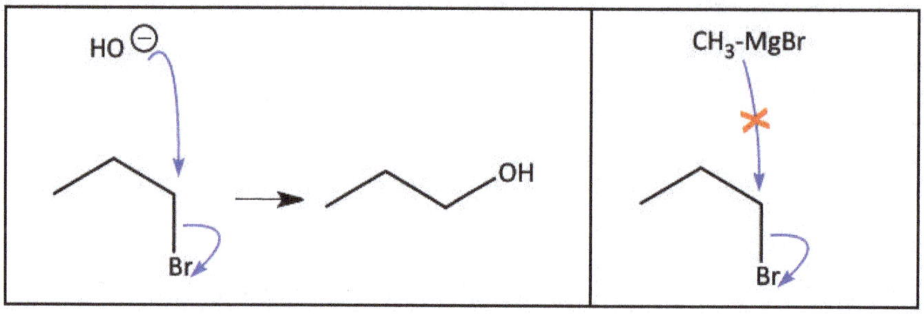

Allyl bromide can also do an S_N2 backside attack reaction. But, this S_N2 works *even better* than propyl bromide. Why would this be a more reactive substrate if it is a primary bromide like propyl bromide? The reason is, in the transition state, the p-orbital formed as the nucleophile attacks and the leaving group leaves can overlap with the p-orbitals of the pi bond and form an extra stable, lower energy, transition state. This is because it is conjugated. In fact, this substrate is so reactive that allyl bromide will do an S_N2 reaction with a Grignard reagent!

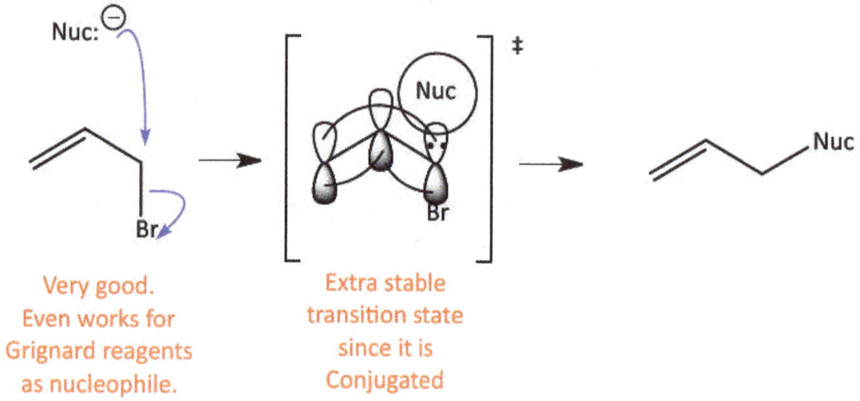

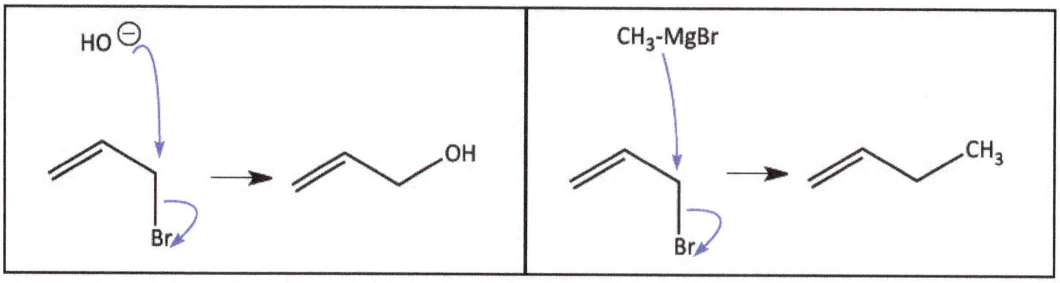

19.6 Draw the products of the following reactions.

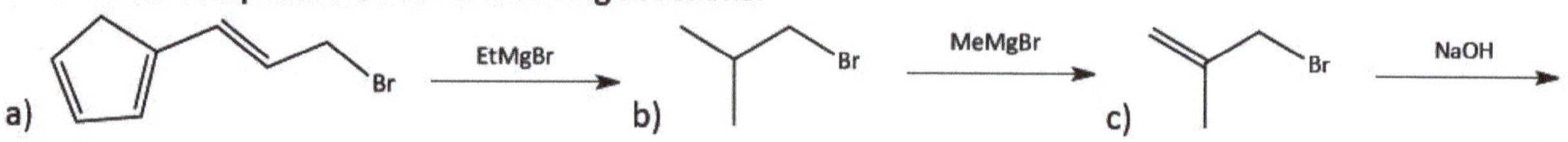

19.8 Kinetic vs. Thermodynamic reaction of HBr with 1,3-butadiene

The reaction of 1,3-butadiene with HBr is an interesting reaction to allow us to begin to consider kinetic and thermodynamic reaction conditions and how they influence the products of a reaction. A pi bond containing negative electrons attacks the proton of the HBr acid as Br^- leaves as a good leaving group. The hydrogen atom attached to carbon C_1 so the carbocation can form in the allylic position. This position is more stable because resonance forms can be drawn. When considering the two resonance forms, the secondary carbocation is more stable than the primary carbocation. We could also say the secondary carbocation is lower in energy than the primary carbocation.

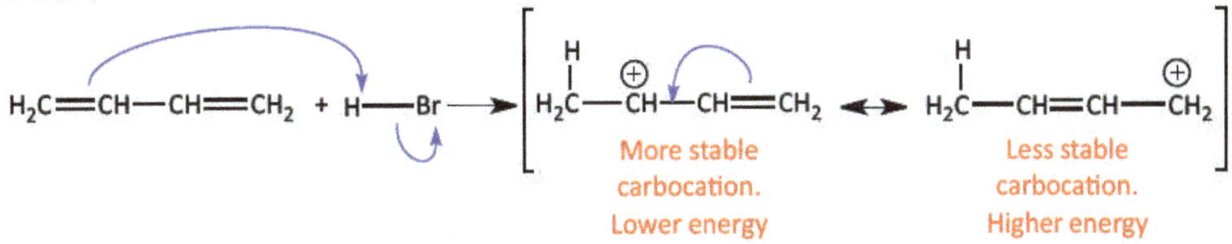

When the bromide attacks the carbocation, it can attack the carbocation of either resonance form. If it attacks the carbocation of the first resonance form, the H and Br end up on carbon atoms C_1 and C_2. This is called 1,2-addition. If the bromide attacks the carbocation of the second resonance form, the H and Br end up on carbon atoms C_1 and C_4. This is called 1,4-addition.

When comparing the stability of the two alkene final products, we find that the product of 1,4-addition is more stable than the 1,2-addition product according to Uncle Saytzeff because it contains a more substituted alkene. The 1,4-addition product has the double bond on the interior of the molecule with 2 carbon groups attached to it compared to the 1,2-addition product with the double bond on the end of the molecule with only one carbon group attached.

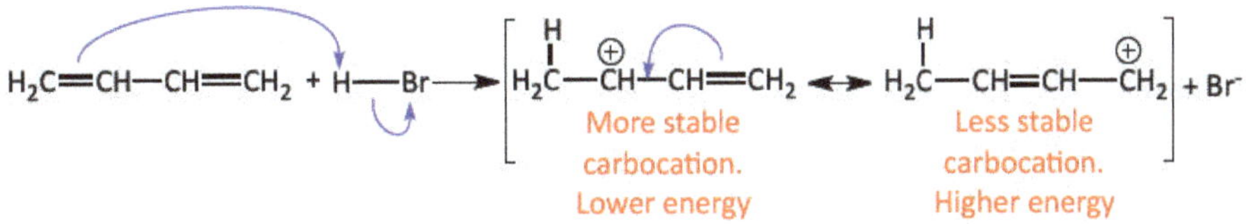

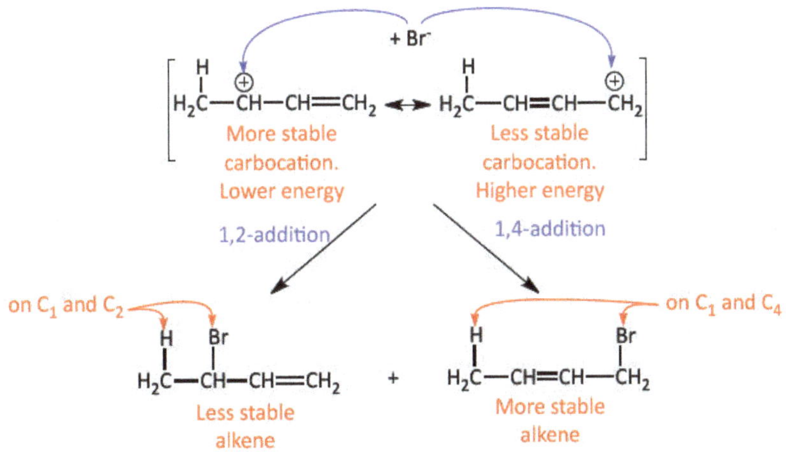

The reaction energy diagram for this would look something like this. Carefully analyze this to make sure you understand it. Notice that for the final products, the one for 1,2-addition is higher in energy than the product for 1,4-addition. This is because the 1,4-addition product is more stable because of Saytzeff's rule. Notice that the transition state going to the more stable 1,4-addition product is higher in energy than the transition state heading to the 1,2-addition product. This is because the higher energy transition state going to the 1,4-addition product more closely resembles the higher energy, less stable carbocation intermediate.

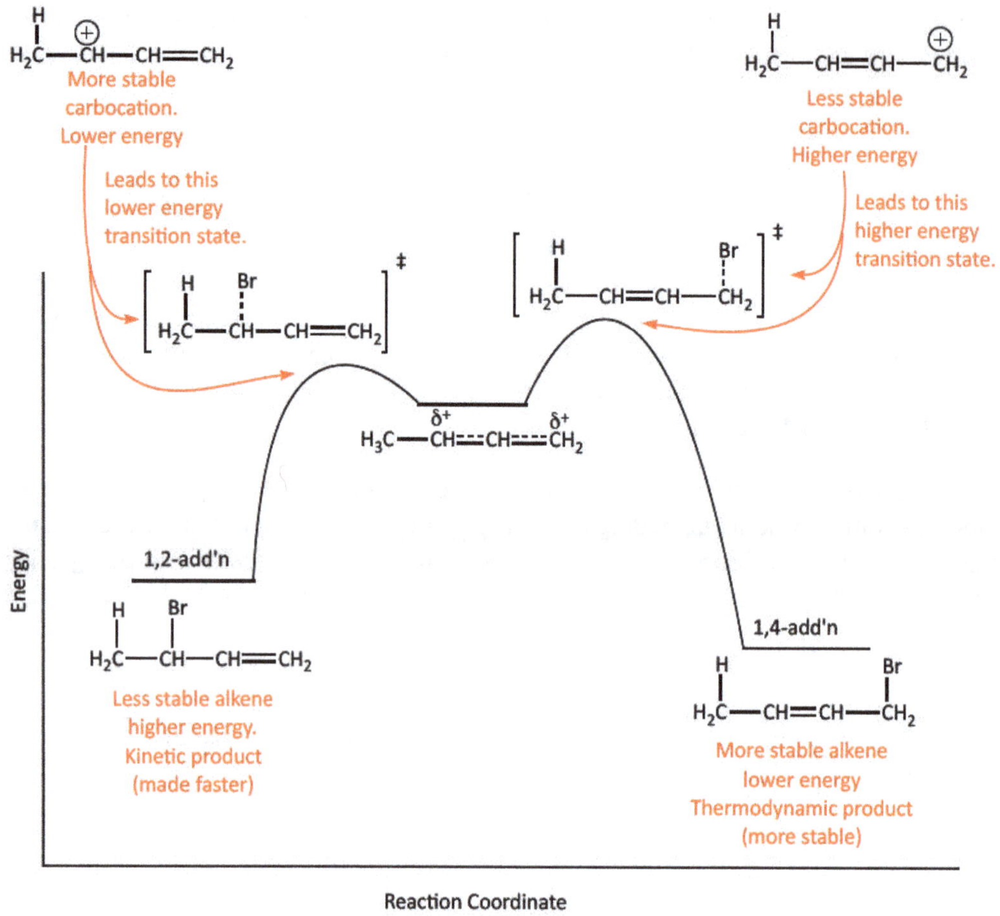

Figure 19.21 1,2 and 1,4-addition of HBr to 1,3-butadiene

The 1,2-addition product is formed faster in this reaction. This is because the transition state leading to the 1,2-addition product is lower in energy than the other transition state. Therefore, we call the 1,2-addition product the *kinetic product*. The 1,4-addition product is more stable than the 1,2-addition product. This is because the 1,4-addition product is lower in energy than the 1,2-addition product. The more stable product is called the *thermodynamic product*. Many times, the kinetic and thermodynamic products of a reaction are the same. This is one example of a reaction where the kinetic and thermodynamic products are different. What if we wanted more of the kinetic product in a reaction, or more of the thermodynamic product, is there a way to do that? Yes there is.

If we want more of the kinetic product, the reaction is performed under kinetic conditions or under kinetic control. This means the reaction is run at very cold temperatures. Usually, the reaction is run at -78°C. When dry ice is added to acetone it makes a slurry that is at -78°C. This is a very common cold temperature used in the chemistry lab. How does running the reaction at very cold temperatures help make more of the kinetic product? At very cold reaction temperatures, there is less energy in the reactants. If we're looking at the reaction coordinate diagram, we could see that there is enough energy to get over the lower energy transition state leading to the 1,2-addition, kinetic product, but not over the higher energy transition state leading to the 1,4-addition thermodynamic product. If this reaction is performed under kinetic conditions at -78°C, the 1,2-addition product is made in 80% yield.

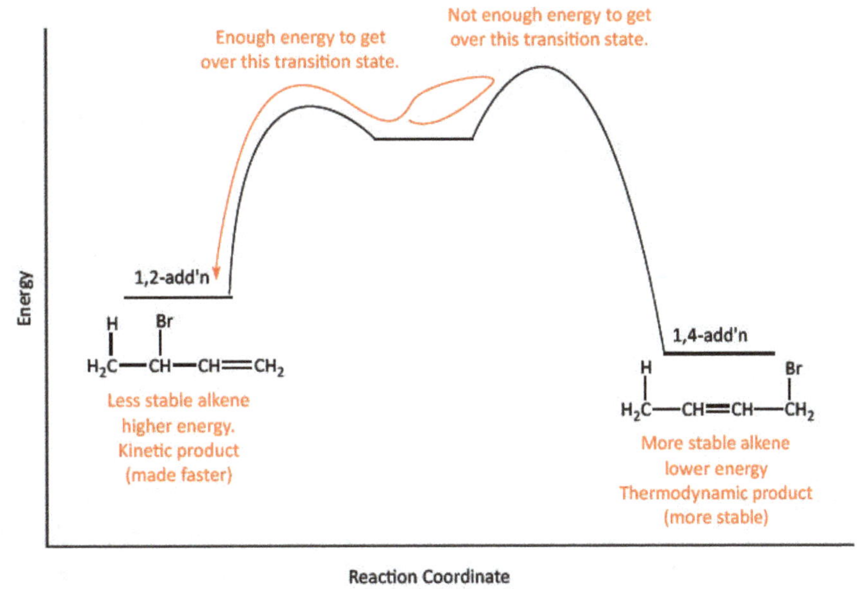

Figure 19.22 Kinetic conditions for HBr addition to 1,3-butadiene

If we want to make more of the thermodynamic product, we do the opposite. We heat up the reaction. If this reaction is performed at 40°C, we get the 1,4-addition product in 85% yield. At warmer temperatures, there is plenty of energy in the reaction to go over either transition state. We can make 1,2-addition product, and it can go back to the intermediate and then go on to make 1,4-addition product. In easily reversible reactions like this, the relative stability of the two products determines how much of each we get. Since the 1,4-addition product is more stable and lower in energy (Saytzeff's rule), we get more of the thermodynamic 1,4-addition product.

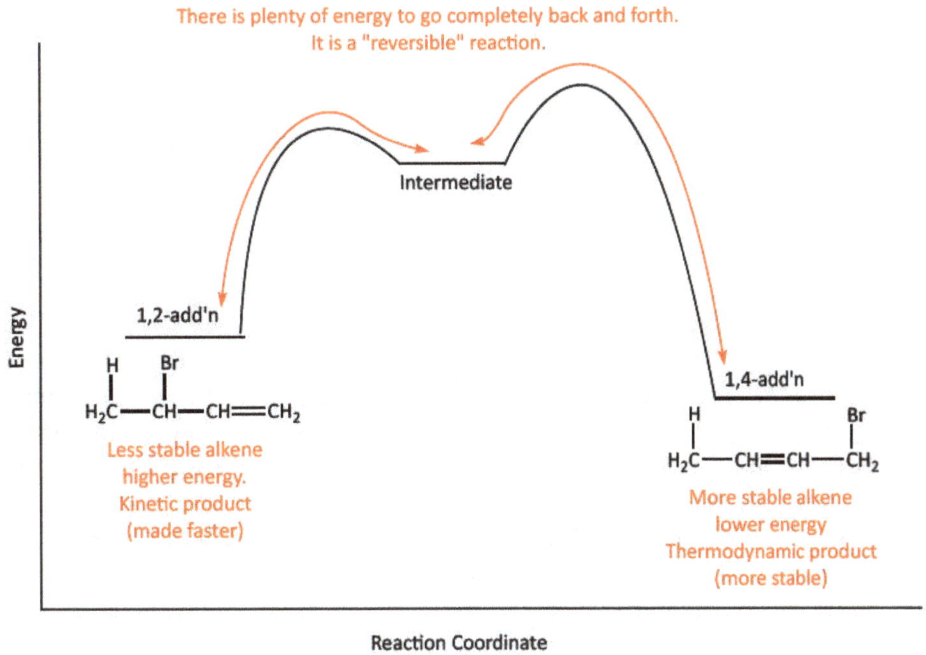

Figure 19.23 Thermodynamic conditions of HBr addition to 1,3-butadiene

19.7 Draw the product of the following reactions. Attempt to answer this by thinking through the mechanism and NOT looking at the information on the previous pages.

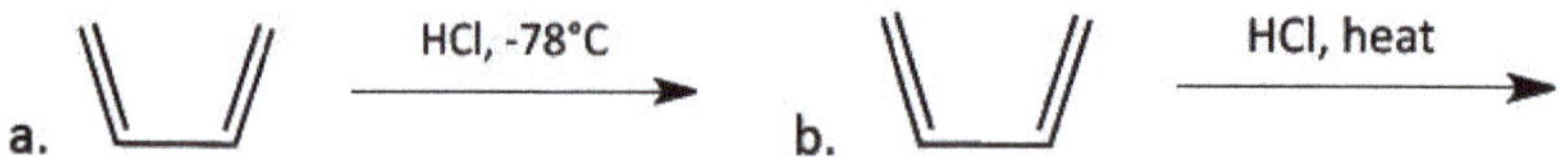

19.9 Diels-Alder Reaction

19.9.1 Overview

Otto Diels and Kurt Alder were awarded the 1950 Nobel Prize in Chemistry for their work on a very important reaction. This reaction, the Diels-Alder reaction, is a [4+2] cycloaddition. A [4+2] cycloaddition means four atoms react with two atoms to make a ring. In the Diels-Alder reaction, a conjugated diene reacts with an alkene that is commonly called a dienophile (lover of the diene) to make a 6-membered ring.

Remember that six-membered rings do not have much ring strain. Therefore, six-membered rings are very common in nature. We find them in many natural products and medicines. Therefore, the Diels-Alder reaction is a very commonly used reaction by synthetic organic chemists.

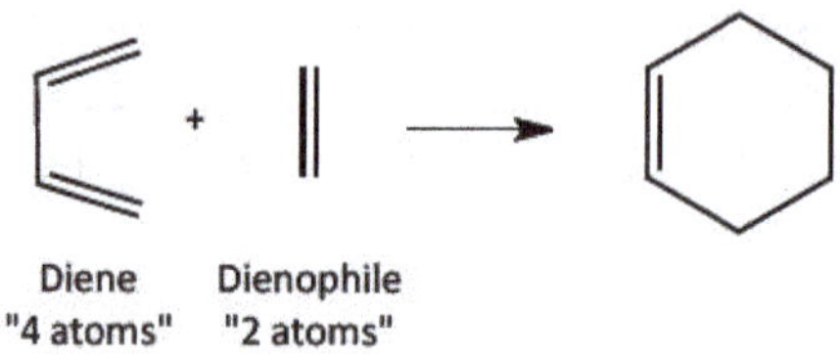

Figure 19.24 Diels-Alder reaction, a [4+2] cycloaddition

The Diels-Alder reaction has a concerted mechanism. This means all of the bonds are broken and made simultaneously. We usually draw the diene and dienophile lined up and then draw three arrows "around the horn".

The diene is usually electron rich. You will commonly find electron donating groups on the diene which push even more electron density into the conjugated diene. Electron donating groups that you will commonly see are alkyl groups, -OR, or –OH. Alkyl groups donate electron density through hyperconjugation and OR or OH donate electron density through resonance with their lone pairs of electrons. The dienophile prefers to be electron poor so it can accept the electrons from the electron rich diene. Therefore, the dienophile often has electron withdrawing groups on it. Good electron withdrawing groups are often carbonyl groups or cyano groups. They can pull electron density away from the dienophile through resonance.

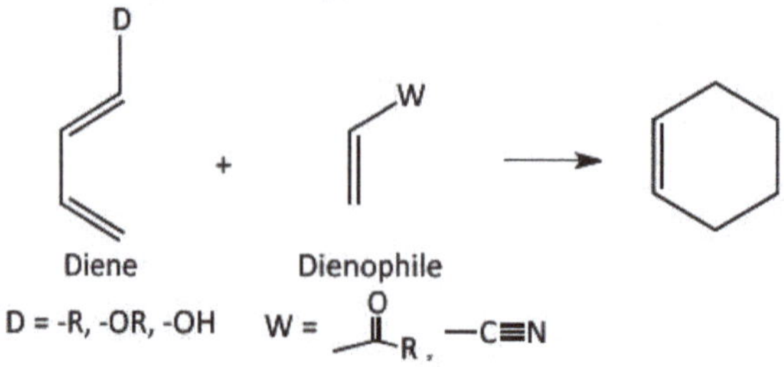

19.9.2 s-cis conformation

Both double bonds of the diene need to point towards the dienophile in order for the Diels-Alder reaction to proceed. This conformation of the diene is called the s-cis conformation. Previously, we learned that we used the prefix cis when two groups were on the same side of something that cannot rotate, like a double bond or a ring. We used trans when two groups were on the opposite side of a double bond or a ring. In the case of this diene, the two double bonds are on the same side of a single bond, or sigma bond. This is a different situation than when we used cis/trans before because it is possible to rotate completely around a single bond. Therefore, we designate these conformations to describe whether the double bonds are on the same side of the single bond, s-cis, or opposite sides of the single bond, s-trans.

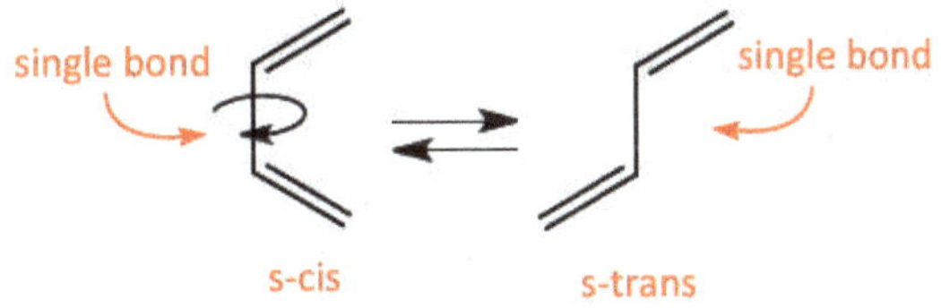

Figure 19.25 s-cis and s-trans conformation

The more easily the diene can get into an s-cis conformation, the more easily it can undergo a Diels-Alder reaction.

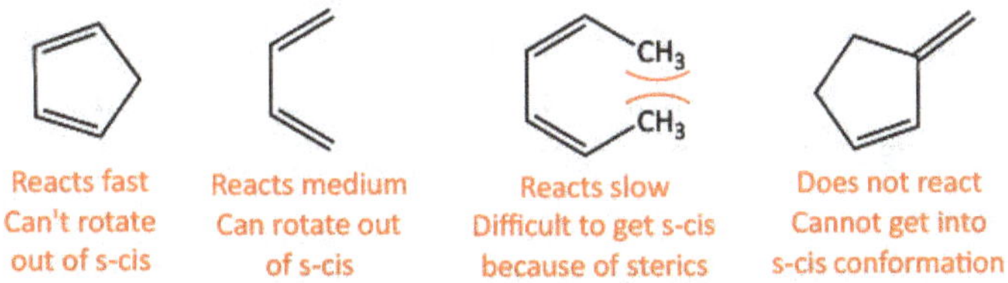

19.8 Rank the following dienes for their reactivity in a Diels-Alder reaction from most reactive (1) to least reactive (5).

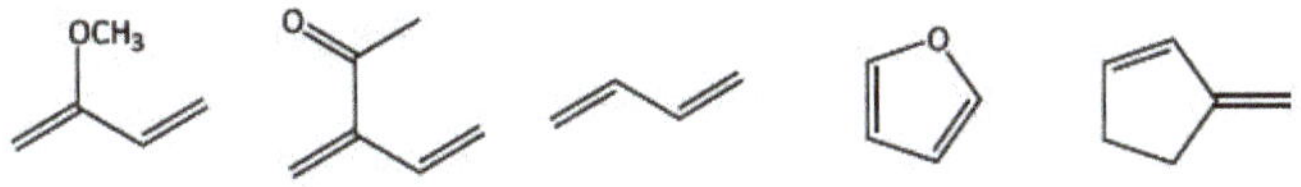

19.9.3 Orbital overlap

Below are the MO diagrams for the simplest diene (1,3-butadiene) and the simplest dienophile (ethane). When a reaction, like the Diels-Alder reaction, takes place, one of the reactants usually donates some electrons from a filled molecular orbital into an empty molecular orbital of the other reactant. In this case, the electron rich diene will donate a pair of electrons from its HOMO (π_2) into the LUMO of the dienophile (π_2*).

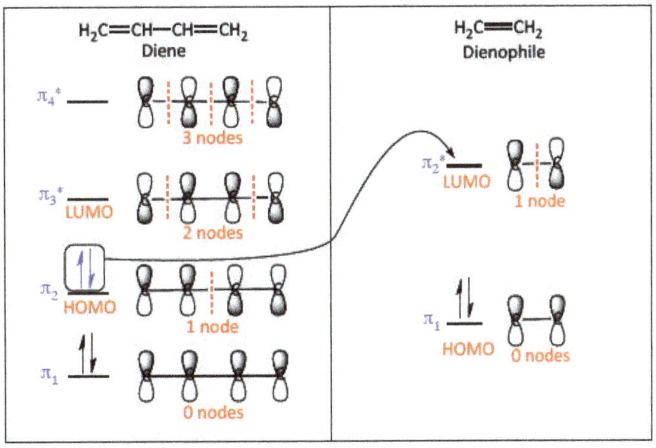

The following diagram shows the overlap of the molecular orbitals of the HOMO of the diene with the LUMO of the dienophile. Notice how the orbitals overlap perfectly. The constructively overlapping orbitals are in phase with each other.

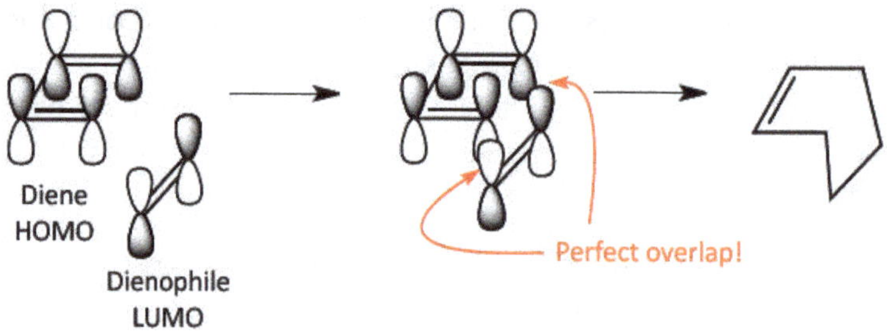

Figure 19.26 Orbital overlap for Diels-Alder reaction

Also, notice that the cyclohexene that is formed is not flat. It is puckered with an envelope shape.

19.9.4 Syn stereochemistry

If the dienophile has two groups on it that are cis, they end up cis on the cyclohexene ring formed. If the two groups are trans on the dienophile, they are trans on the cyclohexene ring. This is because the p-orbitals of the diene and dienophile must overlap and the mechanism for the reaction is concerted. There is no time for the groups to change their configuration.

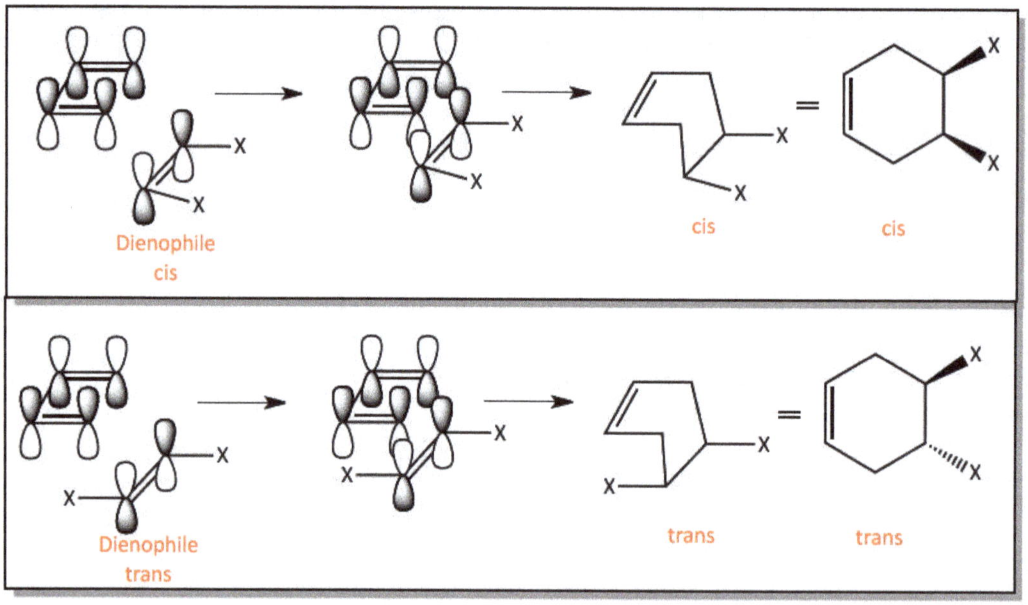

Figure 19.27 Syn-stereochemistry for Diels-Alder reaction

19.9 Draw the products of the following Diels-Alder reactions.

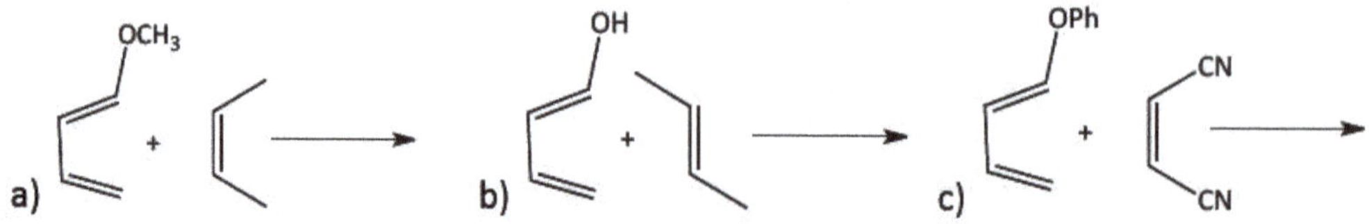

19.9.5 Endo rule

Similarly, if an electron-withdrawing group on the dienophile has a pi bond, it will overlap with the p-orbitals of carbon atoms C2 or C3 of the diene. This secondary overlap helps stabilize the transition state of the reaction. The result of this secondary overlap is the electron-withdrawing group ends up inside the pocket of the puckered envelope of the cyclohexene. This is called the Endo rule.

The groups that point into the puckered envelope fold of the cyclohexene are called the endo positions and those that point outside of the puckered fold are called the exo positions.

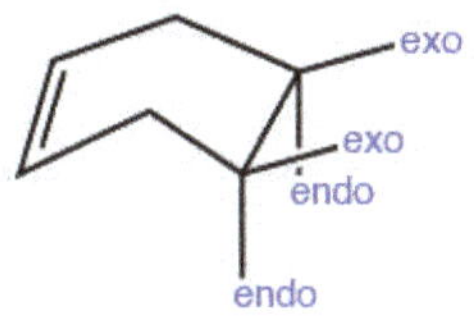

Figure 19.28 Endo vs. Exo positions

Let's look at the following Diels-Alder reaction.

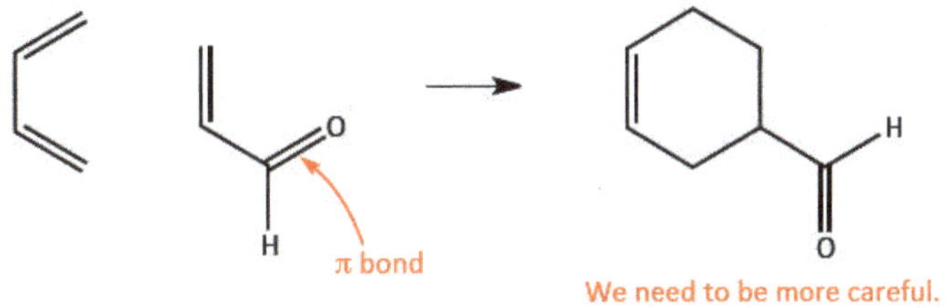

Because an aldehyde that contains a pi bond is the electron-withdrawing group on the dienophile, we need to be a little more careful and allow for the p-orbitals of the pi bond of the carbonyl to overlap with C_2 of the diene. In the end, we need to make sure that our carbonyl is in an endo position of our final product.

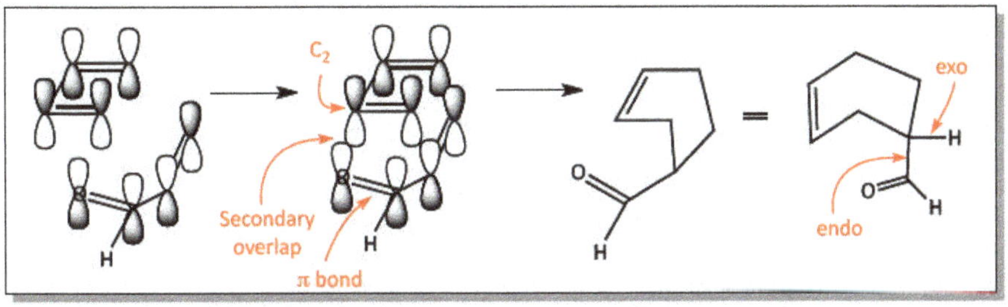

Figure 19.29 Orbital argument for Endo Rule

Similarly, in the following reaction, because the nitrile on the dienophile contains a pi bond, it will bend up in the endo position of the final product.

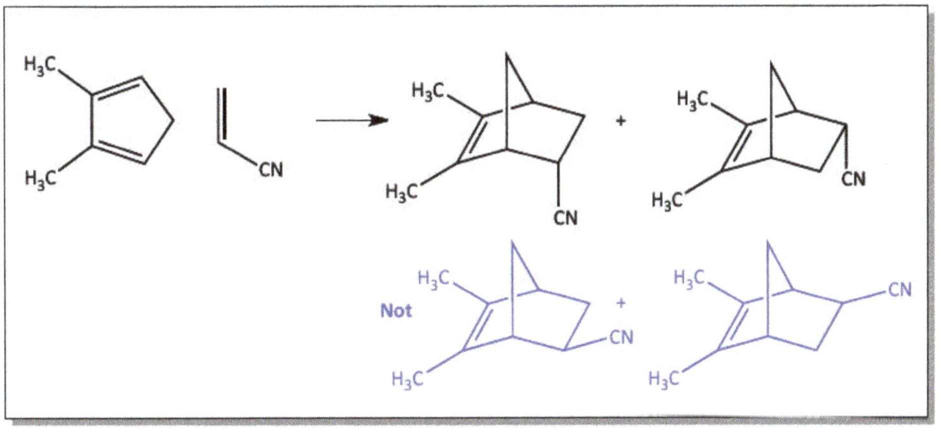

19.10 In the following Diels-Alder reactions, one hydrogen atom and one other group is missing in each of the final products. Draw in the missing hydrogen atoms and missing groups giving careful attention to the Endo Rule.

a)
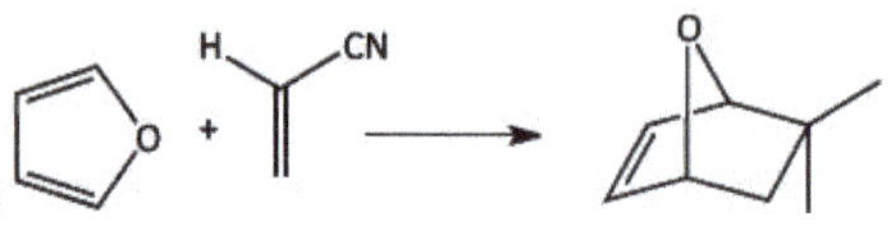

b)
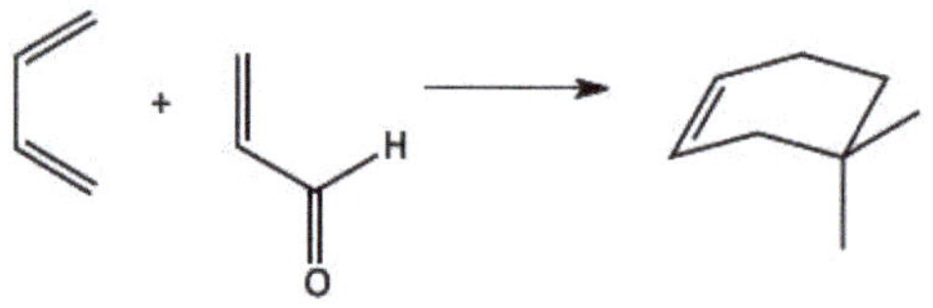

19.9.6 Diels-Alder regiochemistry

Finally, sometimes a decision needs to be made on the regiochemistry of the cyclohexene product. For instance, if 2-methoxy-1,3-butadiene reacts with cyanoethene, which product do we get? 1,3-addition or 1,4-addition? Here are the two possibilities.

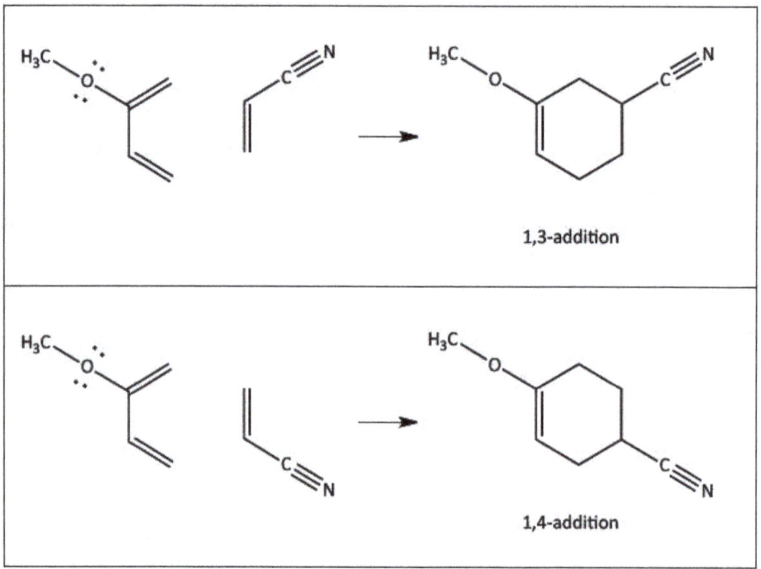

The best way to analyze the regiochemistry is to investigate where the most negative site is on the electron rich diene. We draw resonance forms to push the electrons from the electron-donating group out onto the end of the diene.

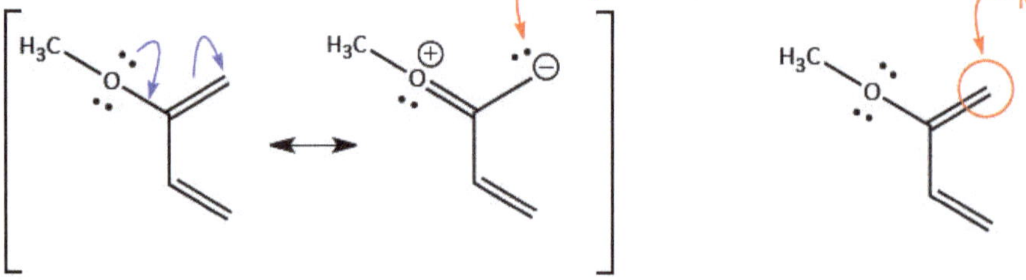

Which is the most negative site of the diene?

Then, we use resonance forms to determine the most positive site on the electron poor dienophile. We push the electrons away from the alkene onto the electron-withdrawing group.

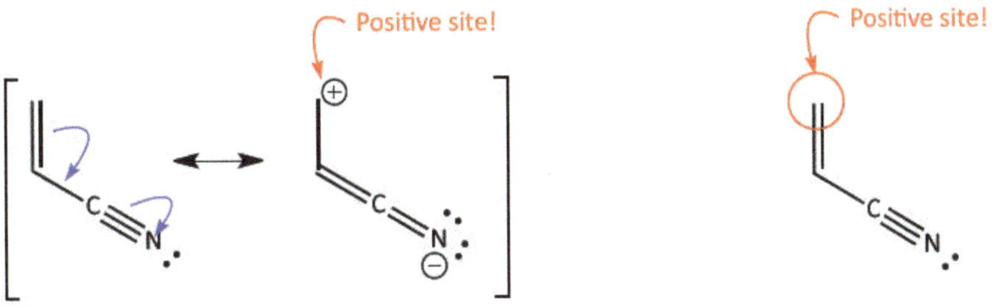

We then put the most negative site of the diene next to the most positive site of the dienophile and perform the Diels-Alder reaction. In this particular case, we see that the 1,4-addition product is made.

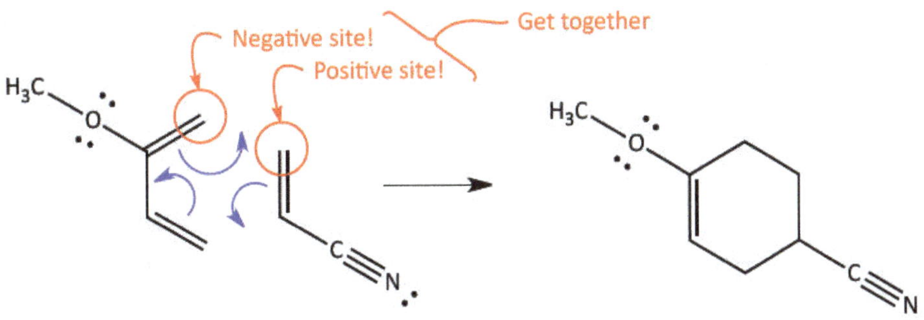

1,4-addition

In the reaction of 1-methoxy-1,3-butadiene with cyanoethene, 1,2- or 1,3-addition are both possible. Which is the correct product? Here are the two possibilities.

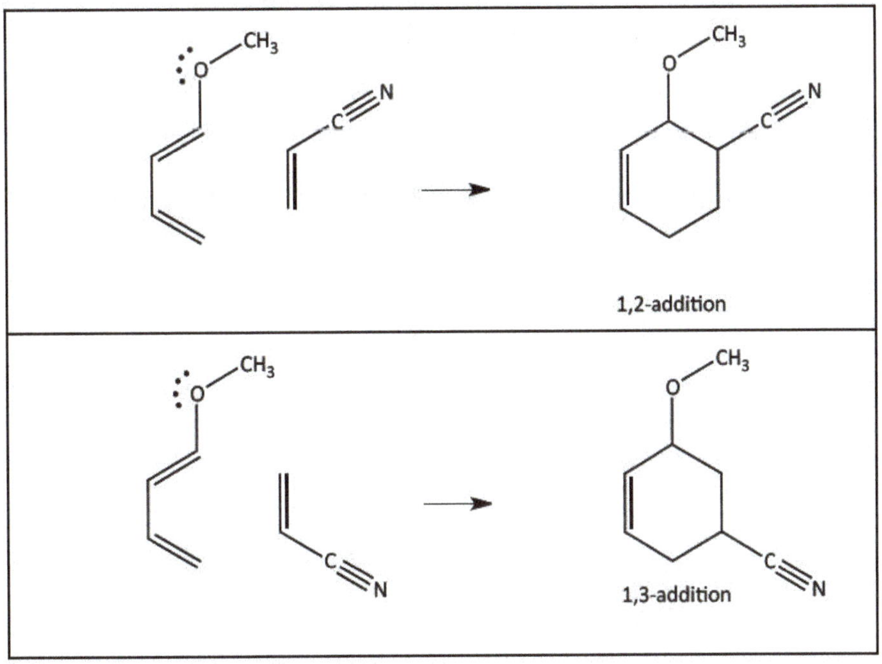

Conjugated Alkenes

Once again, we draw resonance forms to find the most negative site on the diene by pushing electrons from the methoxy electron-donating group to one of the edges of the diene.

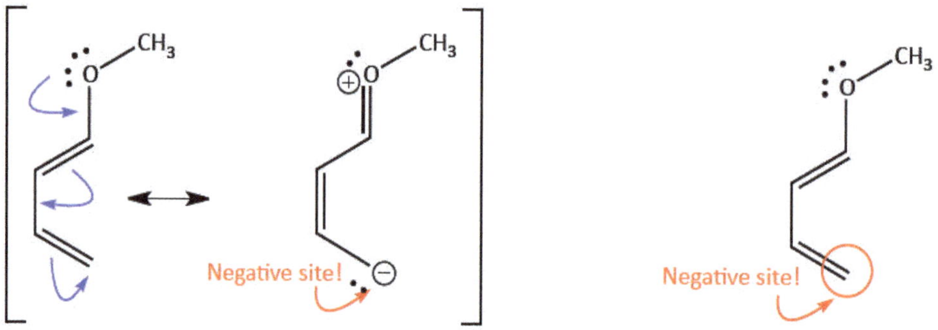

Which is the most negative site of the diene?

We then once again find the most positive site on the electron poor dienophile by finding resonance forms by pushing electrons from the alkene onto the electron-withdrawing group.

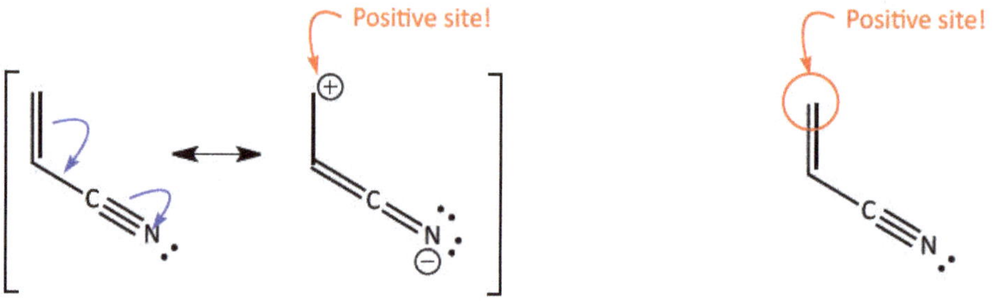

Which is the most positivesite of the dienophile?

We put the most negative site of the diene with the most positive site of the dienophile and perform the Diels-Alder reaction. The final product in this case is the 1,2-addition product. We find that 1,2 and 1,4 addition are common in Diels-Alder reactions, but the 1,3-addition product is not very common.

19.11 Draw the products of the following reactions giving careful attention to regiochemistry.

a) 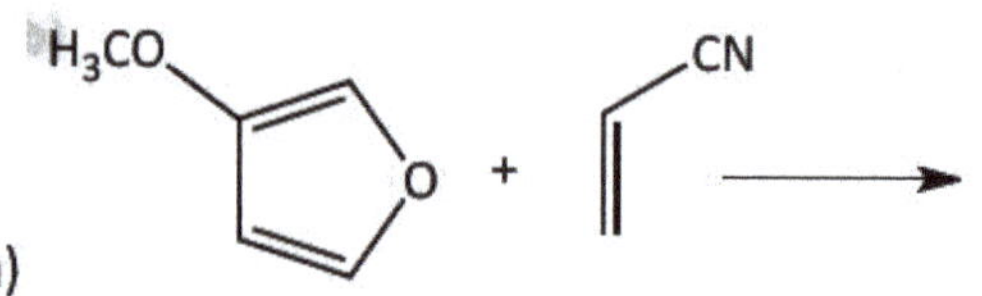

b) 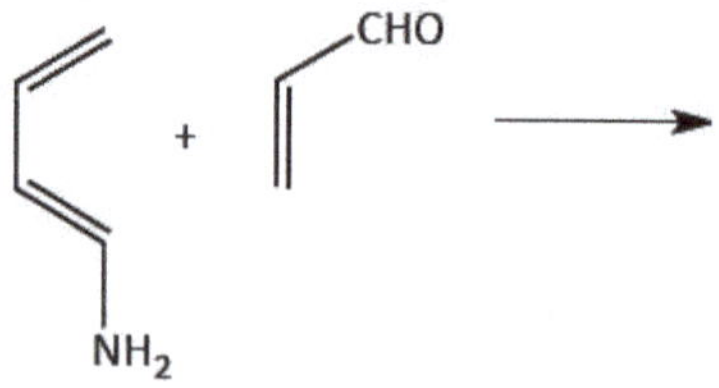

19.10 Reaction Summary

Allylic bromination

a) 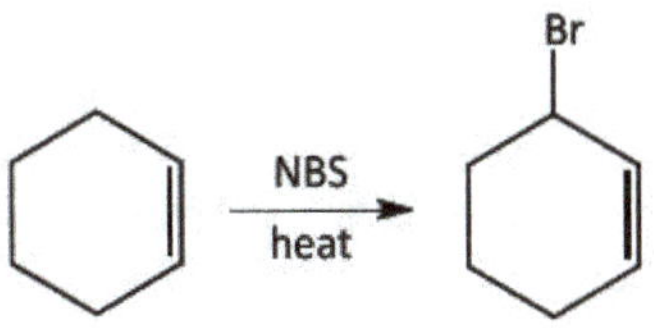

S_N2 of Allyl bromide

b) 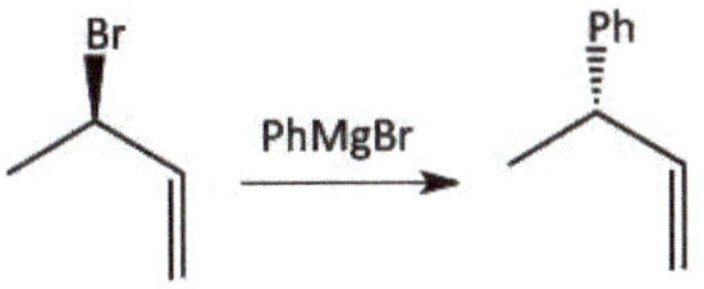

Hydrobromination of 1,3-butadiene

c) 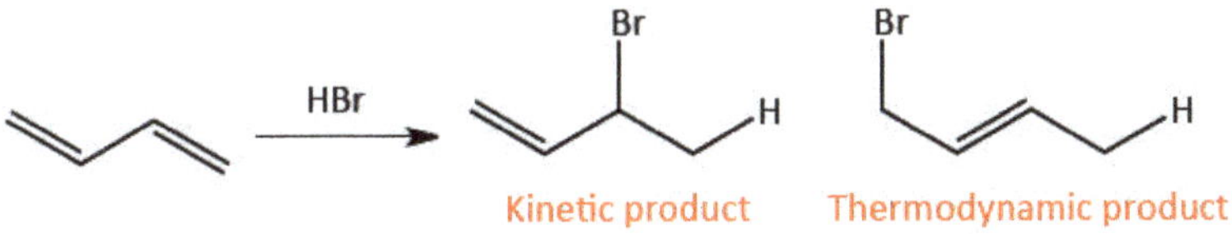

Diels-Alder reaction

d) 

Diels-Alder Endo Rule

e) 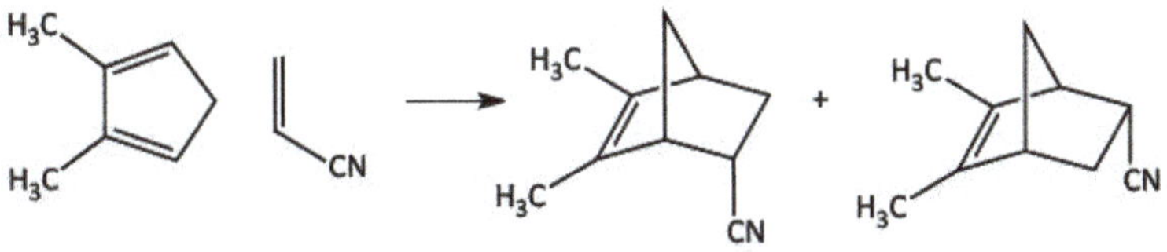

Diels-Alder Regiochemistry

f) 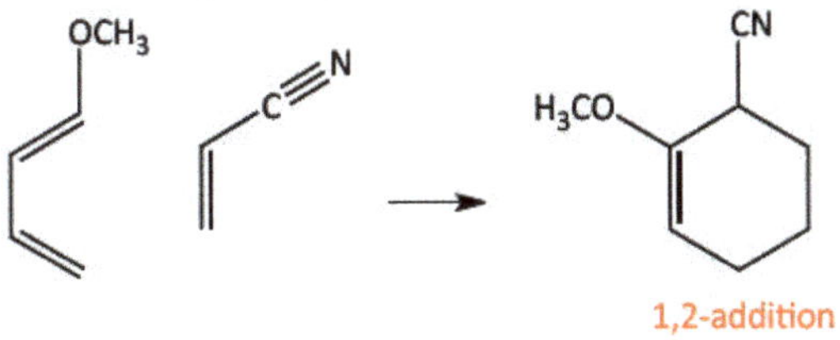

g) 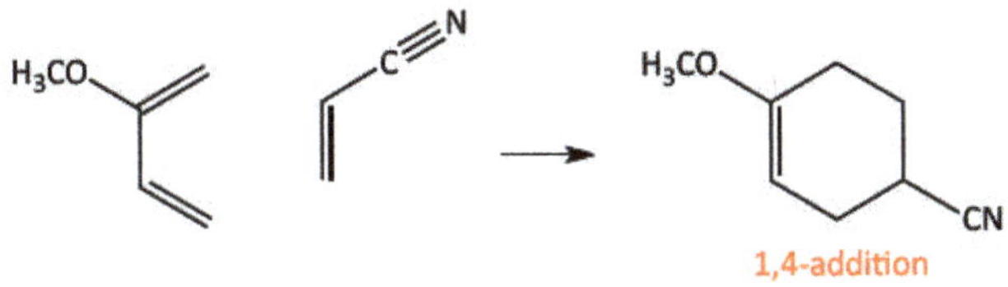

19.11 Problems

19.12 Draw the product when each of the following alkenes reacts with NBS?
a) cycloheptene b) ethylbenzene c) 1-pentene d) 4-methylcyclopentene

19.13 Draw the structure of the major product which results when 2,3-dimethyl-1,3-butadiene is treated with HBr under warm, thermodynamic conditions.

19.14 Draw the products when the following pairs of reagents undergo a Diels-Alder reaction.

a) 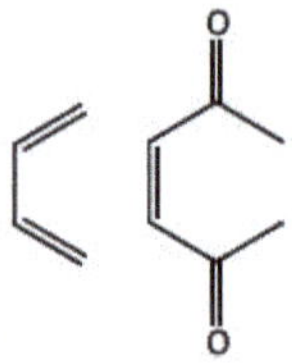

b) 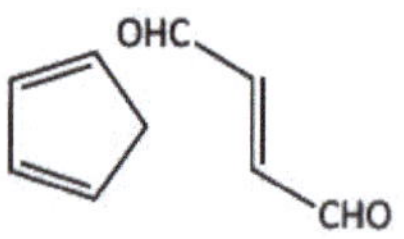

c) 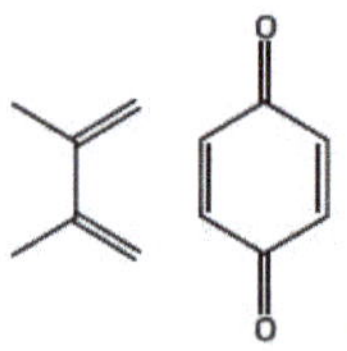

d) 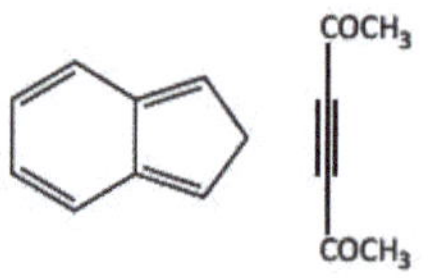

e) 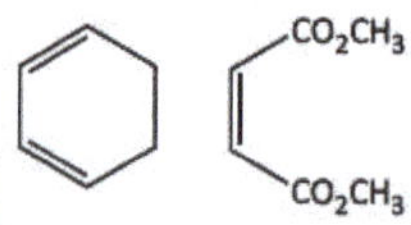

19.15 What diene and dienophile would react to give the product below?

a) 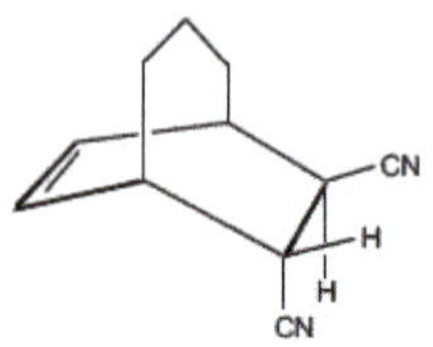

b) 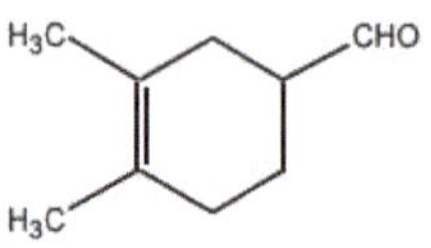

c) 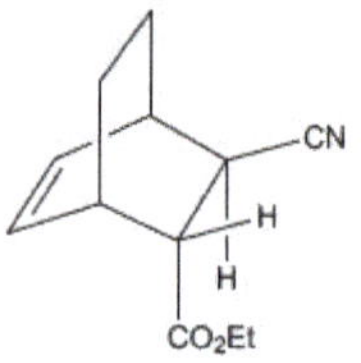

d) 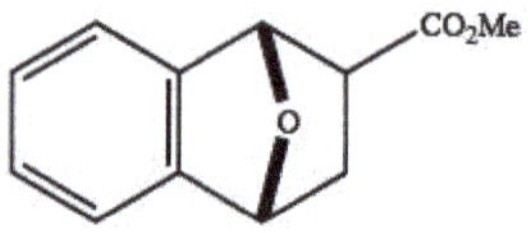

e) 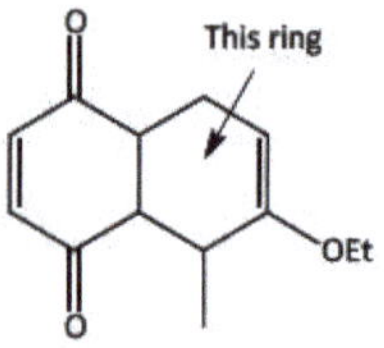

f) 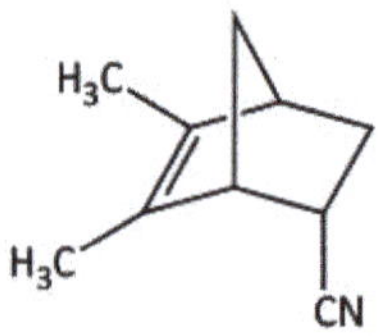

g) 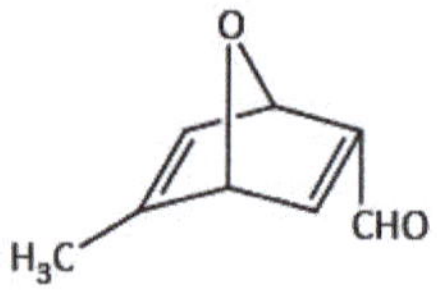

h) 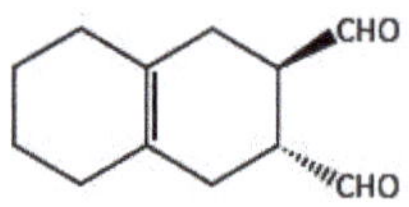

19.16 a) Draw the products when the following pairs of reagents undergo a Diels-Alder reaction. Sometimes, the reactants might need to be redrawn in a better orientation so you can better see how the reaction would proceed. b) Label each reagent for each reaction as either the diene or the dienophile.

a) 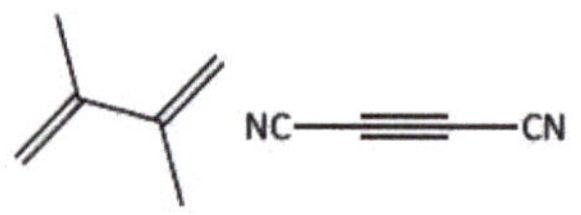

b) 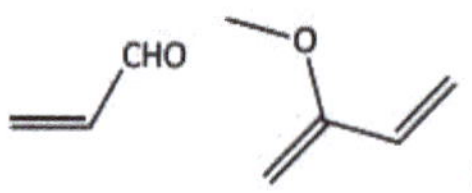

c) 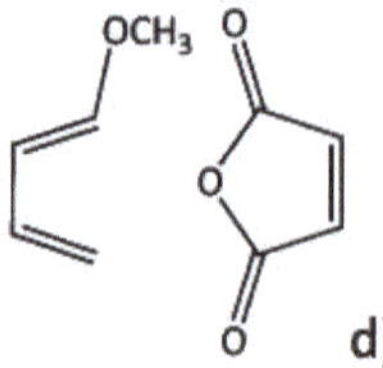

d) 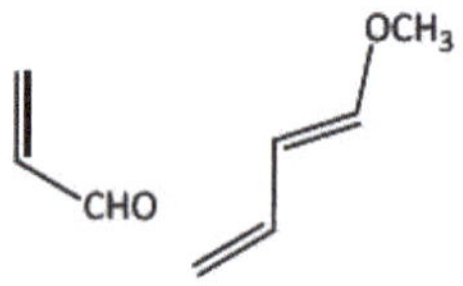

19.17 What is A, B, and C in the following reaction sequence?

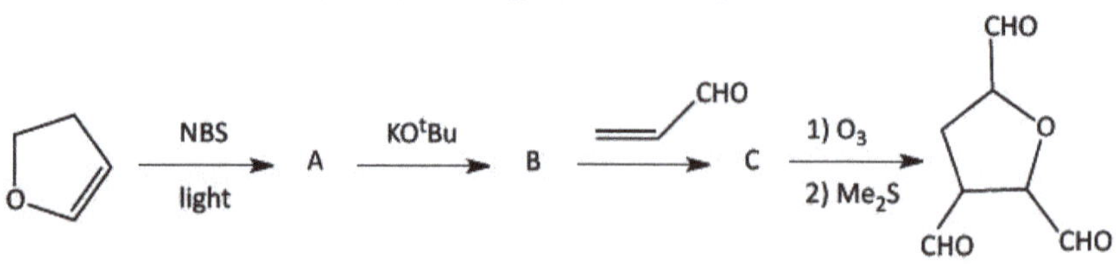

19.18 The pentadienyl anion, $H_2C{=}CH{-}CH{=}CH{-}CH_2{:}^-$, has its lone pair of electrons delocalized over three carbon atoms.

a) Use resonance forms to show which three carbon atoms contain part of the lone pair of electrons.
b) Draw the molecular orbital diagram of the pentadienyl system in order of increasing energy. Label them $\pi 1$, $\pi 2$, $\pi 3$, $\pi 4^*$, $\pi 5^*$. Fill in the pi electrons for each molecular orbital for this system in its ground state.
c) Sketch the molecular orbital for each energy level ($\pi 1$, $\pi 2$, $\pi 3$, $\pi 4^*$, $\pi 5^*$).
d) Label the HOMO, LUMO, and draw the nonbonding line.
e) Explain how your molecular orbital diagram agrees or disagrees with the resonance picture showing delocalization of the lone pair of electrons onto three carbon atoms.

19.19 Draw the product(s) of the following reactions.

a) 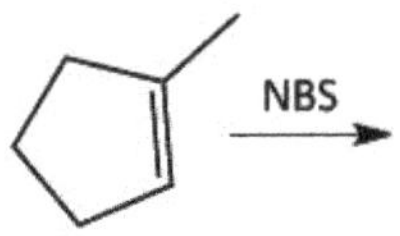

b) 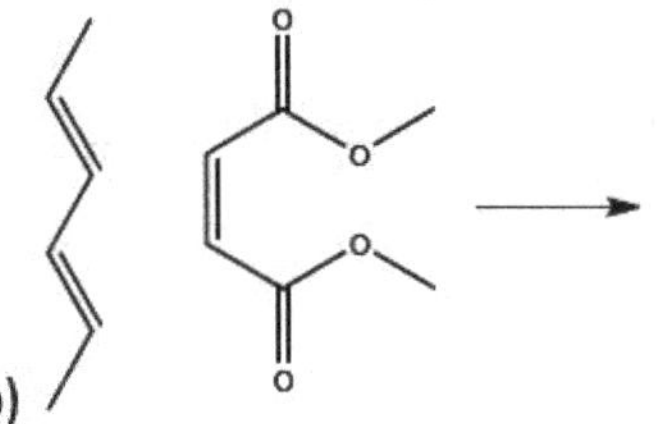

c) 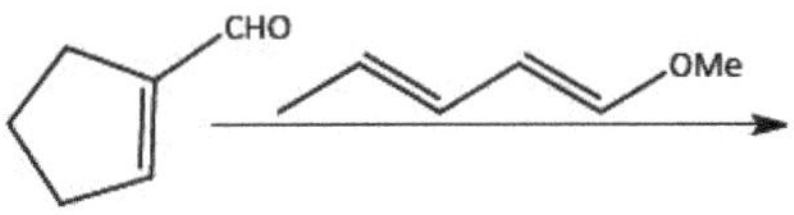

d) 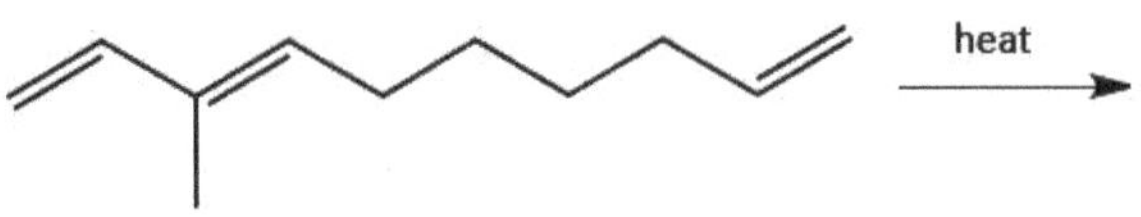

e) 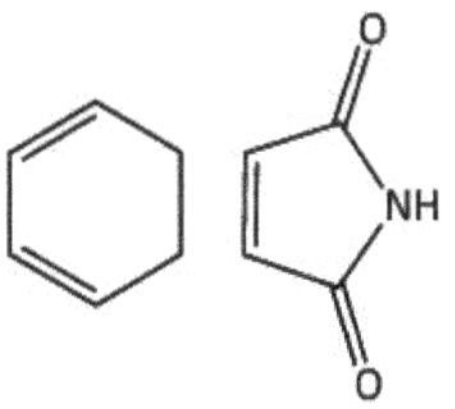

19.20 Show the step-by-step mechanism of the following reaction sequence.

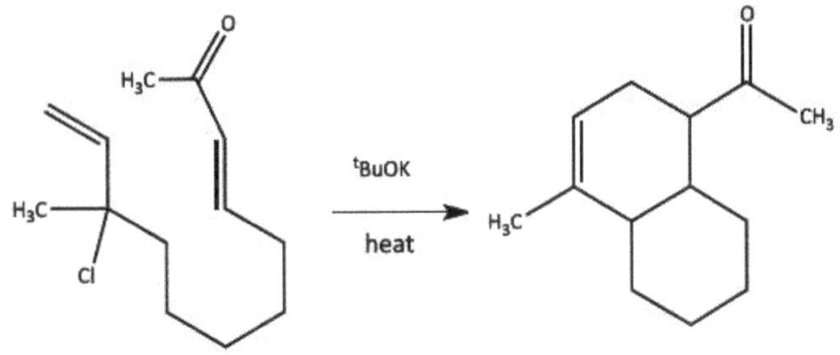

19.21 Show the step-by-step mechanism of the following reaction sequence.

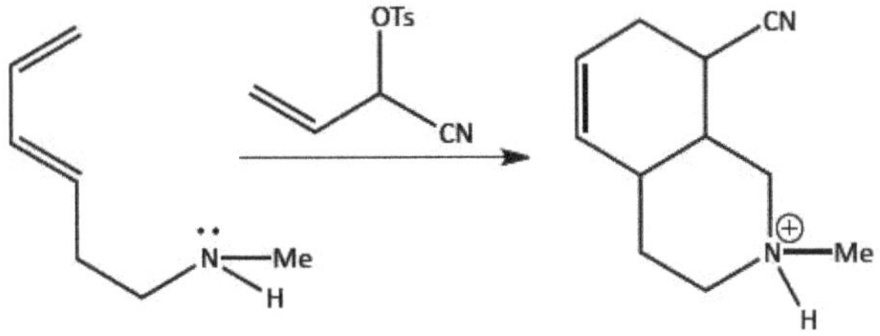

19.22 Show how you would synthesize the following compounds. You do not need to show mechanisms.

a) 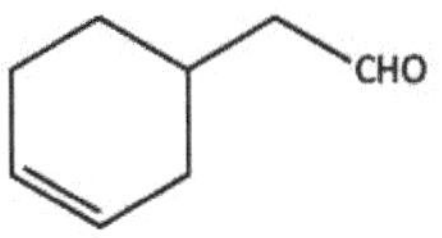

[Starting materials must contain four carbon atoms or less.]

b) 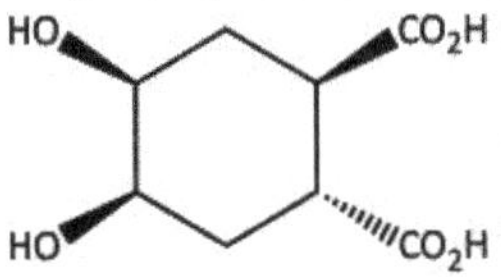

[Starting materials must contain four carbon atoms or less.]

19.23 The following S$_N$1, solvolysis, reaction results in two products. Show the mechanism for their formation.

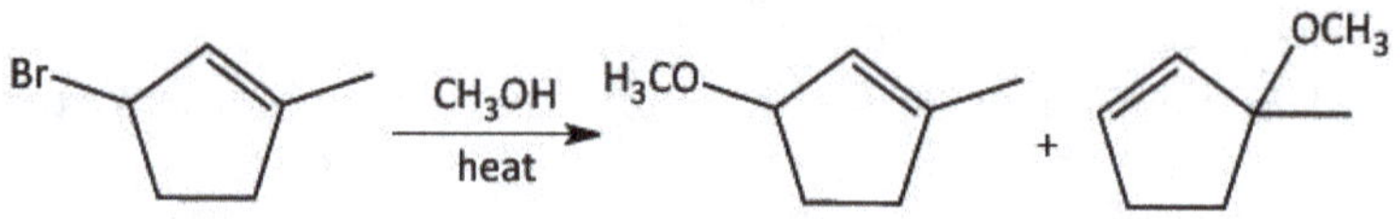

19.24 The following free-radical bromination results in two different products. Draw those products and show the mechanism for their formation.

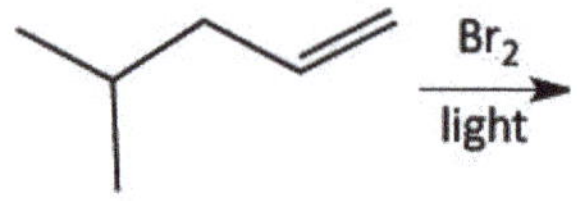

19.25 Draw good, reasonable resonance structures for the following intermediates.

a) 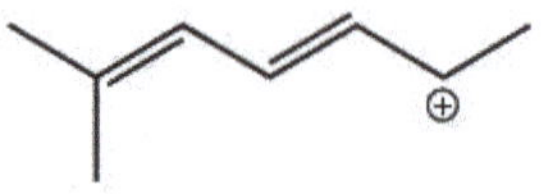 b) 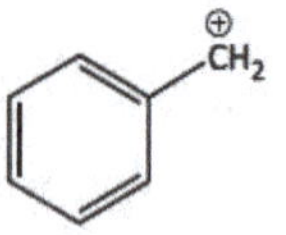 c) 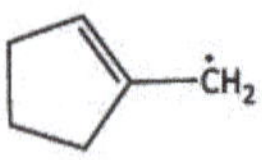 d) 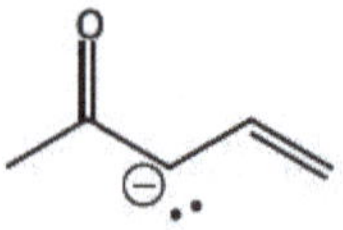 e) 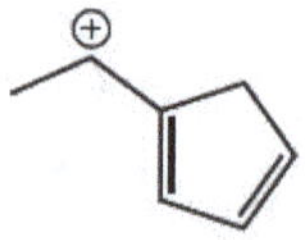

19.26 Draw the product(s) of the following reactions.

a) 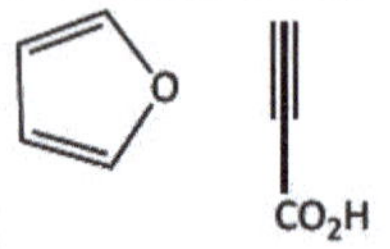 b) 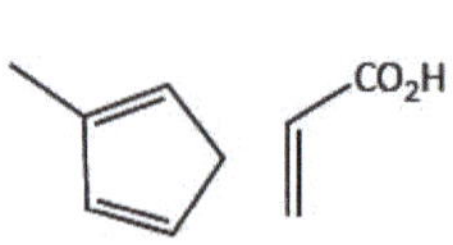 c) 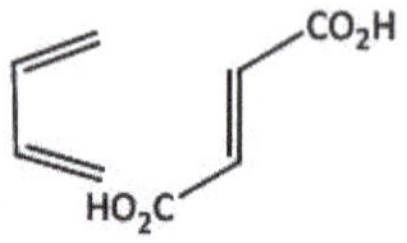 d) 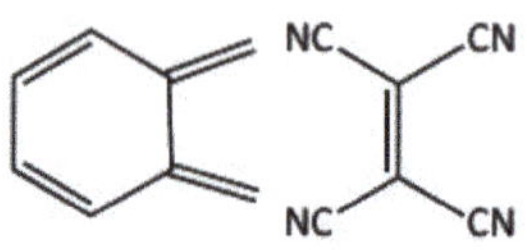

19.27 a) Show how you would make 1,3-cyclopentadiene from cyclopentane. b) In the lab, if you analyzed a flask full of 1,3-cyclopentadiene the day after you synthesized it, you would find the flask no longer contained 1,3-cyclopentadiene, but another compound. The 1,3-cyclopentadiene would react with itself via a Diels-Alder reaction. Show the mechanism for this reaction, and draw the structure of the new compound in the flask.

Chapter 20. Aromatic Compounds

20.1 Introduction

Benzene is a compound with six carbon atoms and six hydrogen atoms. It contains three pi bonds. Benzene and its derivatives are often called aromatic compounds. They were originally called this because of the odor these compounds had. But, more recently, the term "aromatic" has come to mean that these compounds are unusually stable.

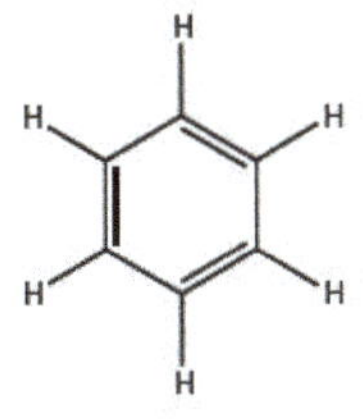

Benzene

When three pi bonds are drawn in the benzene ring of dimethylbenzene, it would incorrectly predict that two different isomers exist. One of the isomers would have a double bond between the methyl groups and the other would have a single bond between the methyl groups. In fact, there is only one type of dimethylbenzene. Somehow, both of these forms of dimethylbenzene are the same.

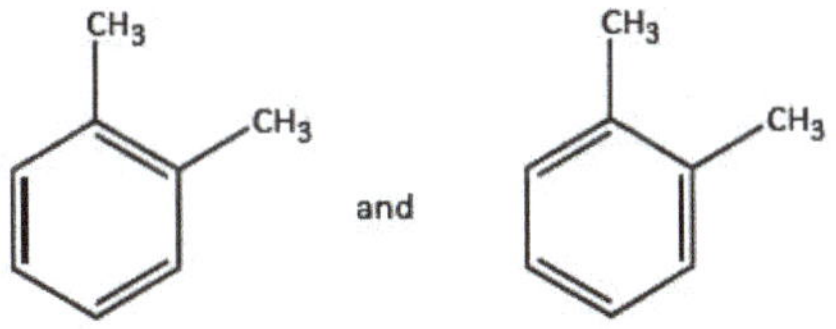

Dimethylbenzene

One way of thinking about how these two are the same is to consider them two resonance forms of each other. Each bond of the benzene ring is halfway between a double and single bond. We say that each of these bonds has a bond order of 1.5. The resonance hybrid of a benzene compound is sometimes drawn with a circle in the benzene ring. This is sometimes called the toilet bowl structure.

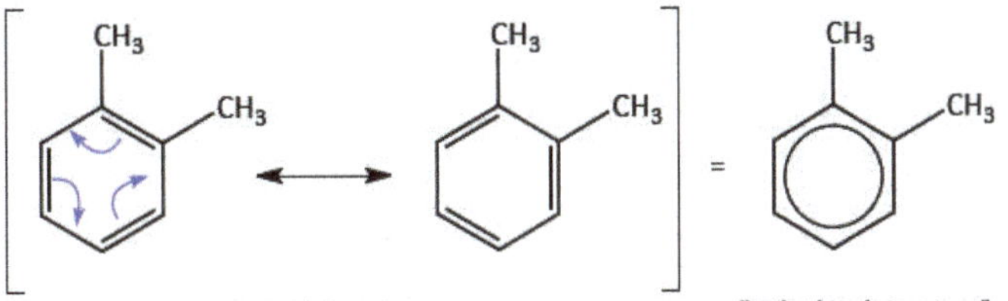

Dimethylbenzene "toilet bowl structure"

The toilet bowl structure is a pretty good approximation for the bonding in a benzene ring. With three pi bonds, each carbon atom of the benzene ring is sp^2 hybridized with an unhybridized p-orbital on it. These p-orbitals can all overlap causing equally sharing the pi electrons around the ring.

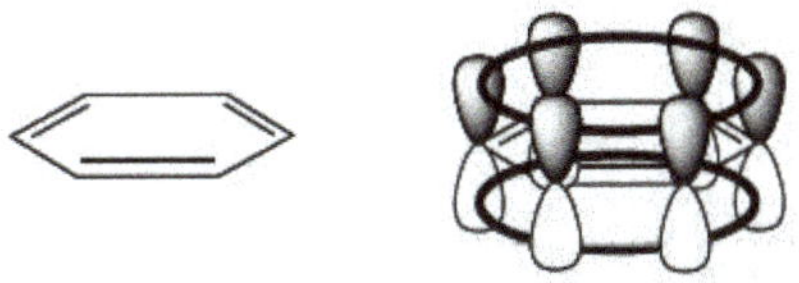

The unusual stability of benzene

The pi bonds in benzene do not react like normal pi bonds. Many of the alkene reactions we learned previously do not apply to benzene rings. For instance, the bromination of an alkene is simple, yet it does not occur with a benzene ring.

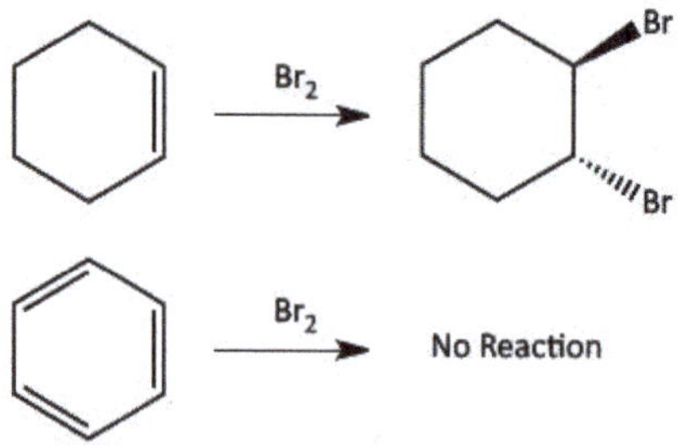

We see that the benzene ring is unusually stable. Its pi bonds do not behave like regular pi bonds. Why is benzene so stable? How can we explain that? At first glance, you might think benzene is so stable because we can draw resonance forms with it. But, this is incorrect. We can draw resonance forms with other similar compounds, yet they are not very stable.

20.2 Annulenes

Annulenes are cyclic hydrocarbons with alternating double and single bonds throughout its structure. Cyclobutadiene, benzene, and cyclooctatetraene are all annulenes.

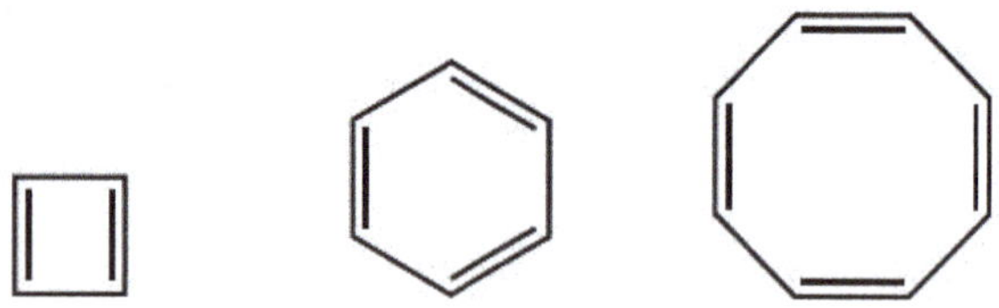

cyclobutadiene [4]annulene | benzene [6]annulene | cyclooctatetraene [8]annulene

Cyclobutadiene is very unstable. Benzene is very stable. Cyclooctatetraene has the stability of a normal annulene. All three of these annulenes have other good resonance forms. We clearly see that the unusual stability of benzene does not come from its ability to have another resonance form.

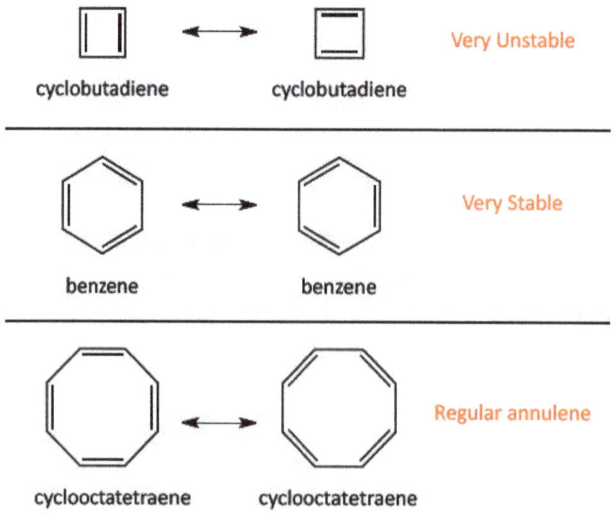

We need another theory beyond resonance forms to explain the unusual stability of benzene. We must use molecular orbital theory to explain it.

20.3 Molecular orbital diagrams of annulenes

20.3.1 MO diagram of benzene

Benzene is extra stable in its ring form. It is aromatic. The benzene pi system contains six sp^2 hybridized carbon atoms. Each carbon atom has an unhybridized p-orbital. These six p-orbitals can overlap. When they overlap, they can overlap constructively to make bonding sites or destructively to make antibonding sites.

The lowest energy is when all of the p-orbitals are in phase with each other. This totally overlapping set of orbitals has zero nodes. This makes a pi-bonding molecular orbital, π_1. There are two ways to overlap the p-orbitals to make one node, π_2 and π_3. There are two ways to overlap the p-orbitals to make two nodes, π_4^* and π_5^*. In these, there is more destructive overlap than constructive overlap, so these are antibonding molecular orbitals. We indicate they are antibonding by using *. Finally, there is one way to overlap p-orbitals where each p-orbital is out of phase with the next. This makes a totally anti-bonding pi orbital, π_6^*.

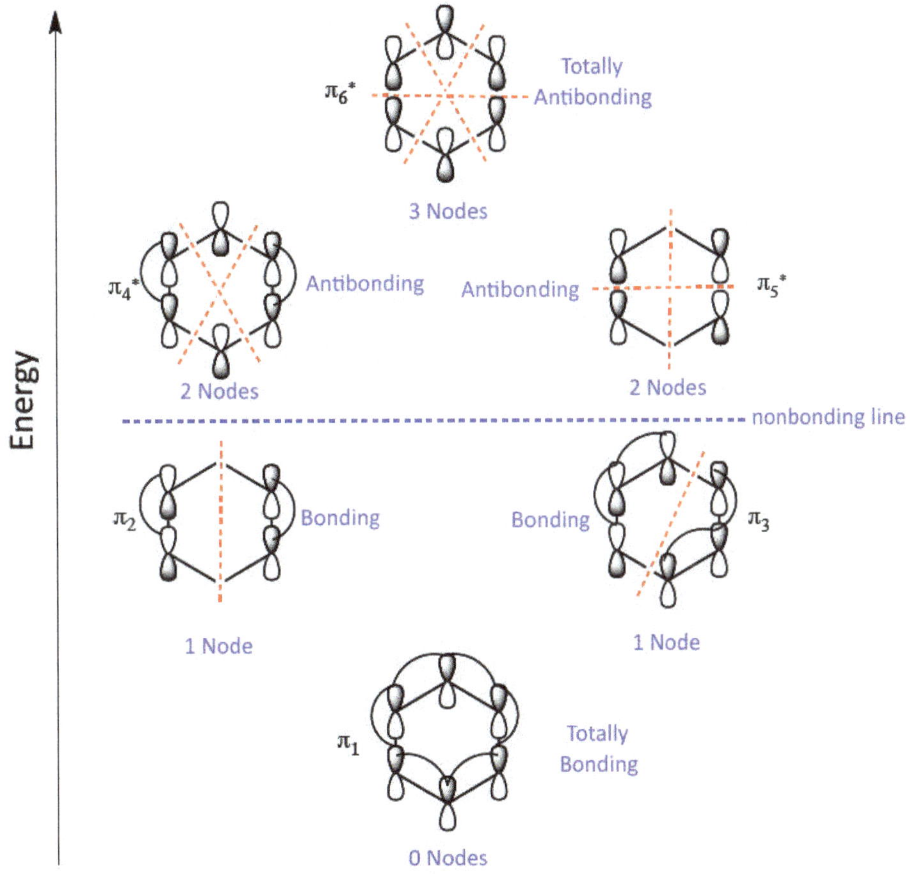

Once we know the ways the p-orbitals can overlap to form the various molecular orbitals, we can begin to draw a molecular orbital energy diagram. Benzene, with its three pi bonds, has six pi-system electrons. The molecular orbitals are filled in from the lowest energy up. This fills up π_1, π_2, and π_3 with a pair of electrons in each. Each of these molecular orbitals is a bonding molecular orbital. By sharing the pi electrons over more than two atoms in these molecular orbitals, we get even lower energy electrons. The amount of energy we save by dropping below the nonbonding line is all saved energy giving us extra stability for benzene.

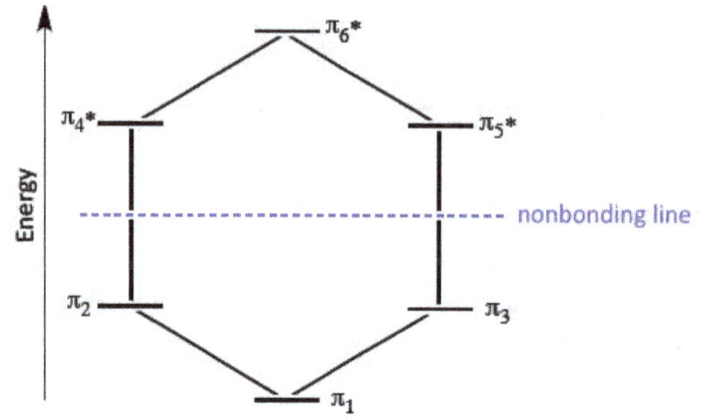

Overall, because benzene is more stable in its ring form than it is in its non-ring form, we say that benzene is aromatic.

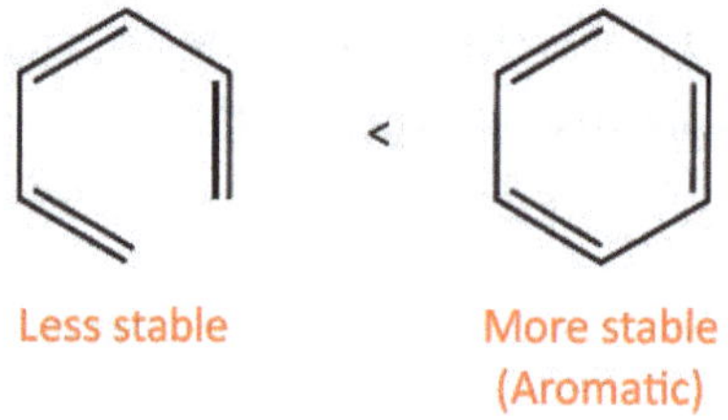

20.3.2 MO diagram of cyclobutadiene

Cyclobutadiene is especially unstable. We call it antiaromatic. Its molecular orbital diagram will help us understand why it is so reactive.

For cyclobutadiene, there are four sp^2 hybridized carbon atoms with an unhybridized p-orbital on each carbon atom. The lowest energy form for the overlapping p-orbitals is if they all overlap and make a totally bonding molecular orbital with zero nodes in it, π_1. There are two ways to overlap to make one node, π_2 and π_3. There is one way where the p-orbitals are all out of phase with each other to make two nodes and a totally antibonding molecular orbital, π_4^*.

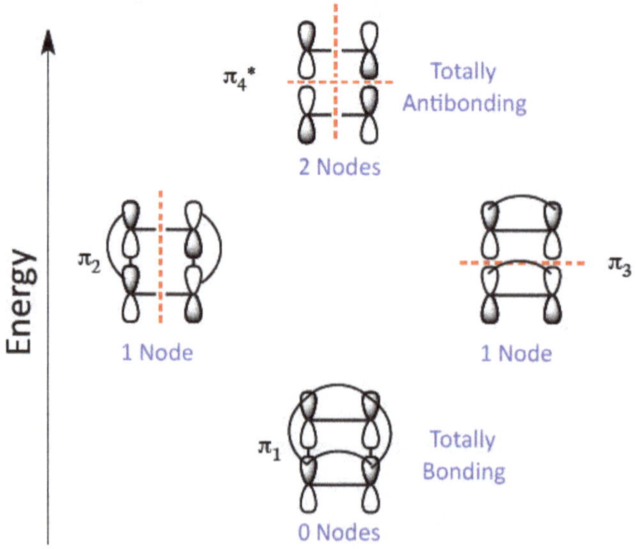

Notice that the molecular orbitals π_2 and π_3 are the same as the two resonance forms of the Lewis structures we draw for cyclobutadiene.

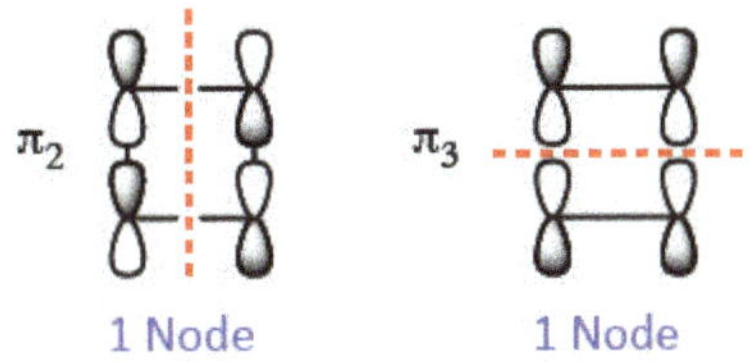

Once we know the ways the p-orbitals can overlap to form the various molecular orbitals, we can begin to draw a molecular orbital energy diagram. Cyclobutadiene, with its two pi bonds, has four pi-system electrons. The molecular orbitals are filled in from the lowest energy up. The molecular orbital π_1 is filled with a pair of electrons. Then, following Hund's rule where each orbital of the same energy is filled with one electron before the electrons are paired together, we get two electrons that are in separate orbitals, π_2 and π_3. This diradical (two unpaired electrons) is very reactive.

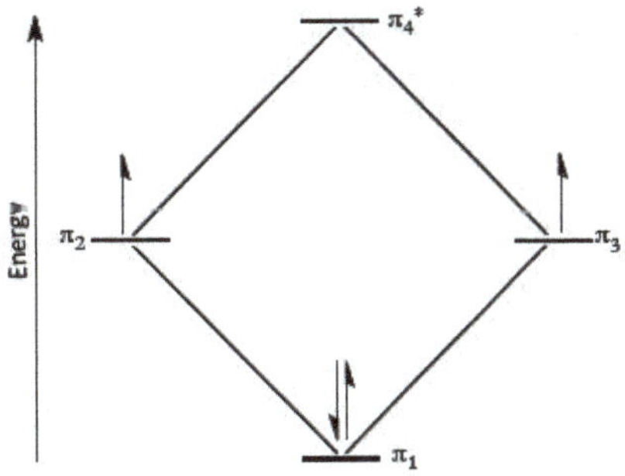

The molecule is unstable. Therefore, we say that cyclobutadiene is antiaromatic. Cyclobutadiene is less stable than its open ring counterpart.

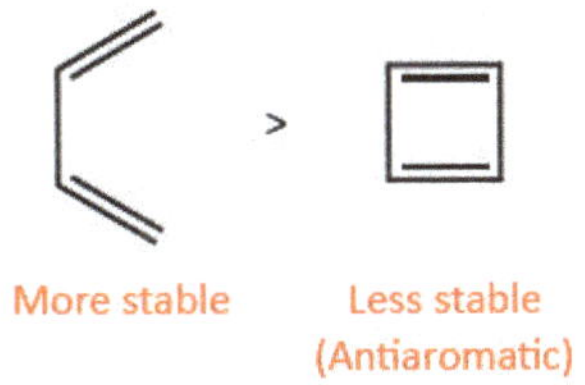

20.3.3 Counting Pi-System electrons

Let's practice counting pi electrons that are in the pi-system of these conjugated cyclic compounds. Each pi bond of the double bonds is obviously part of the pi system. There are two electrons for each of these pi bonds. If there is a carbocation in the system, there is an empty p-orbital there. This extends the conjugation of the system because this empty p-orbital can overlap with the other p-orbitals, but it donates zero electrons to the pi-system. Each carbanion is sp^2 hybridized with two electrons in the p-orbital adding to the electron count of the pi-system.

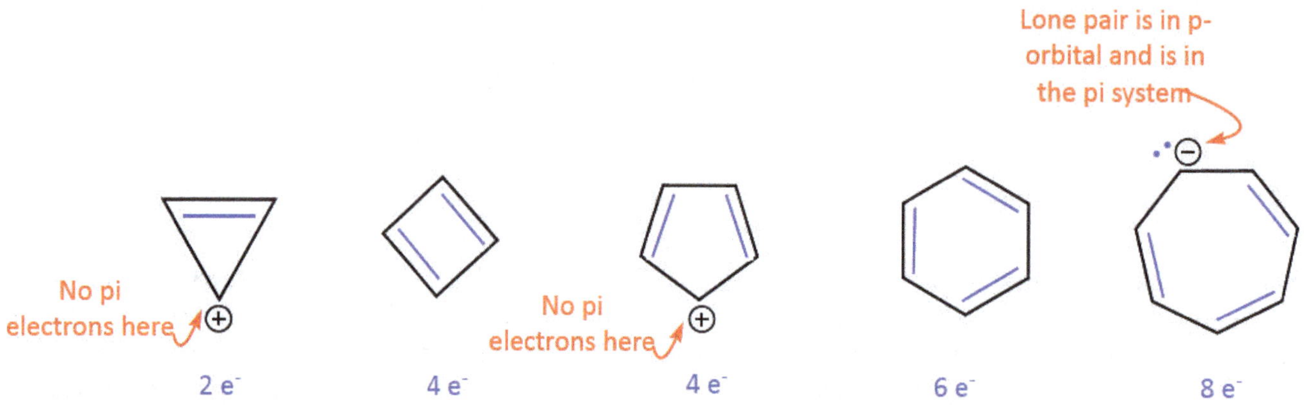

20.3.4 Polygon rule

The MO diagram of the pi system of a planar, totally conjugated ring system has the same polygon shape as the molecule does with the vertex (point) at the bottom of the diagram. We can remember the vertex is at the bottom because there is always only one lowest energy molecular orbital, the one where all of the p-orbitals overlap. There is always only one way to do this and have zero nodes.

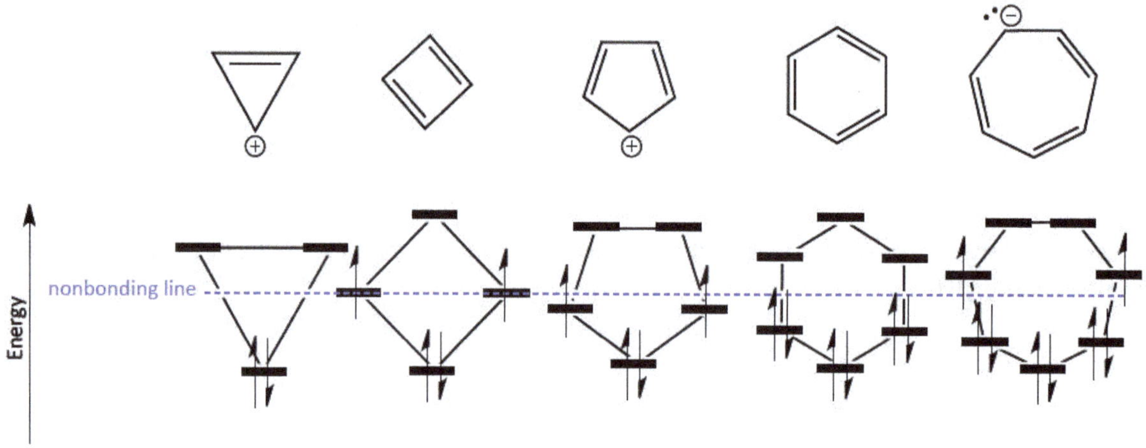

20.1 For the following MO diagrams of the compounds above, circle the word "aromatic" or "antiaromatic" to indicate whether the MO diagram represents an aromatic or an antiaromatic compound. Hint: Look at whether all of the electrons are paired up or if there are two unpaired electrons.

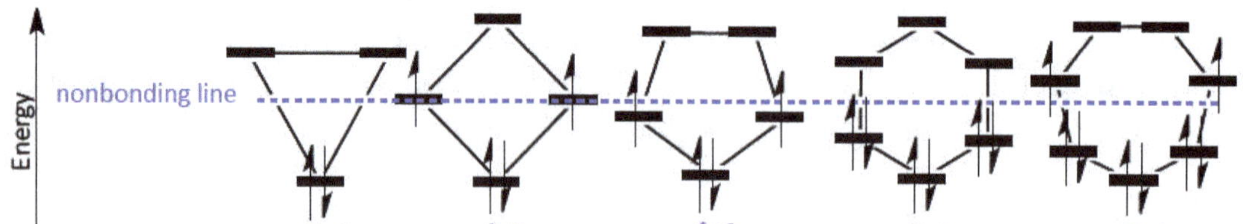

2 e⁻	4 e⁻	4 e⁻	6 e⁻	8 e⁻
a) aromatic or antiaromatic	b) aromatic or antiaromatic	c) aromatic or antiaromatic	d) aromatic or antiaromatic	e) aromatic or antiaromatic

20.2 Shade in the atomic p-orbitals to represent what would overlap to make the following molecular orbitals.

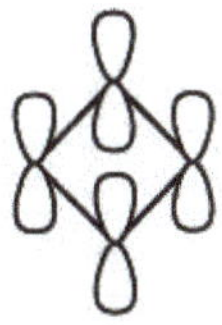

Lowest energy MO of cyclobutadiene

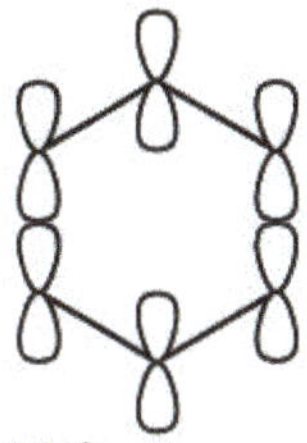

Highest energy MO of benzene

20.4 Aromatic, antiaromatic, nonaromatic

20.4.1 4N+2, Hückel's rule

In these flat, completely conjugated ring systems, we find that there is a pattern for which types of molecules are aromatic and which are antiaromatic.

2 e^- = aromatic

4 e^- = antiaromatic

6 e^- = aromatic

8 e^- = antiaromatic

We find that is the number of electrons in the pi system = 4N+2, where N is an integer, then the system is aromatic. If the number of electrons in the pi system = 4N, then the system is antiaromatic. This is Hückel's rule.

2 e^- = 4N+2 = 4(0)+2 = aromatic

4 e^- = 4N = 4(1) = antiaromatic

6 e^- = 4N+2 = 4(1)+2 = aromatic

8 e^- = 4N = 4(2) = antiaromatic

20.5 Identifying if aromatic, antiaromatic, or nonaromatic

Aromatic systems are cyclic, completely conjugated systems that are more stable than their open-ring counterparts. Antiaromatic systems are cyclic, completely conjugated systems that are less stable than their open-ring counterparts. In order for a system to be aromatic or antiaromatic, it MUST have the following characteristics.

1. It must be a cyclic compound. If it is not a ring, it is neither aromatic nor antiaromatic.
2. It must be completely conjugated at every atom in the ring. Each atom of the ring must have an unhybridized p-orbital that can overlap.
3. The ring must be a flat, planar structure. All of the p-orbitals must be able to overlap with each other so the pi-electrons can be shared.

If any of the three above considerations is not true for a compound, then the compound is a nonaromatic system.

20.3 Each of the following compounds is nonaromatic. Which of the previous rules, 1-3, does each compound break?

a)  b) 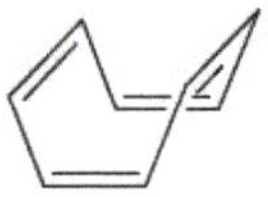 c) 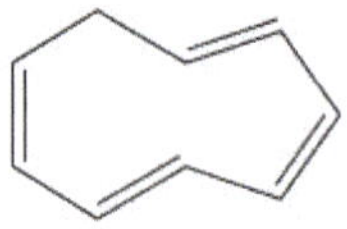

20.4 For the following compounds, determine whether the molecule is aromatic, antiaromatic, or nonaromatic. Assume the molecules are planer.

a) 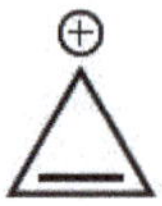 b) 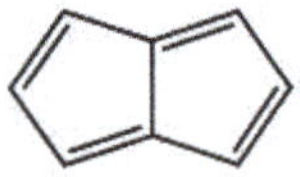 c) 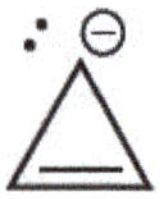 d) 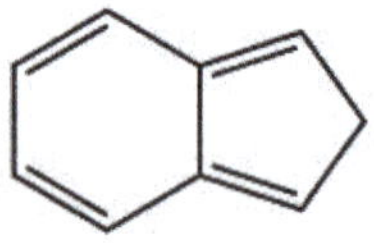 e) 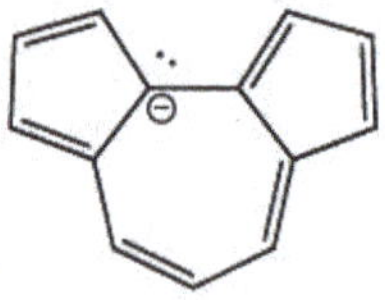

f) 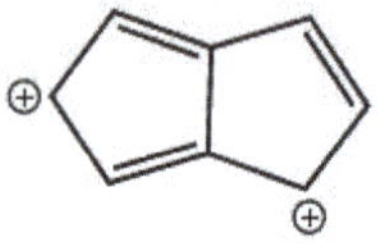 g) 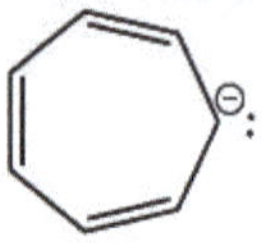 h) 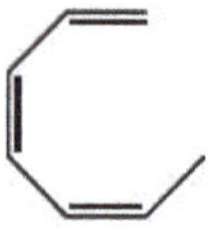 i) 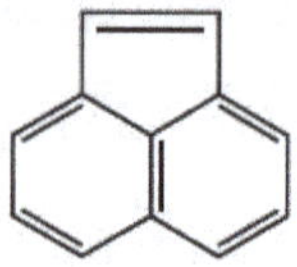

20.5 Predict whether the following compounds would be flat (planar) or not.

a)  b) 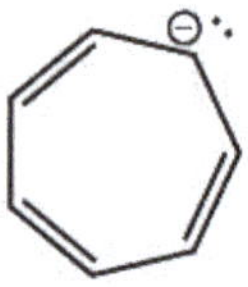 c) 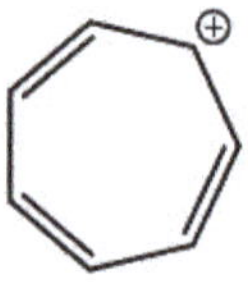 d) 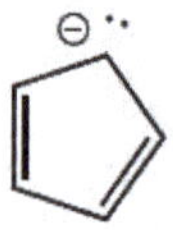 e) 

20.6 Aromatic or Antiaromatic Reaction Intermediates

Consider cyclopentadiene. Would we predict the protons on cyclopentadiene to be very acidic or not?

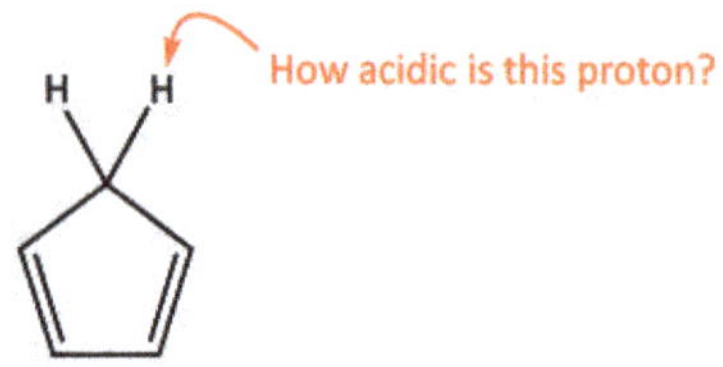

Like any time we consider the acidity of a proton, we look at the conjugate base once the proton is removed and see how stable that anion is. In this case, the cyclopentadienyl anion is a completely conjugated, flat ring system that contains six pi electrons in the pi system. This is an aromatic compound and is very stable. This is a species that would like to be made, so we know that the protons on cyclopentadiene are quite acidic. These protons have a pK_a of 16.

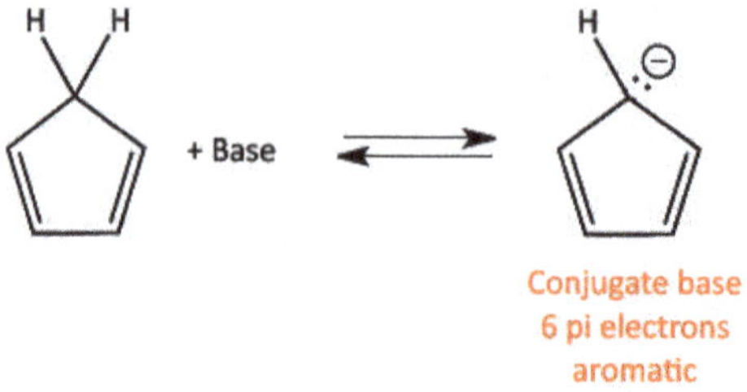

Let's predict the acidity of the protons of cyclopropene.

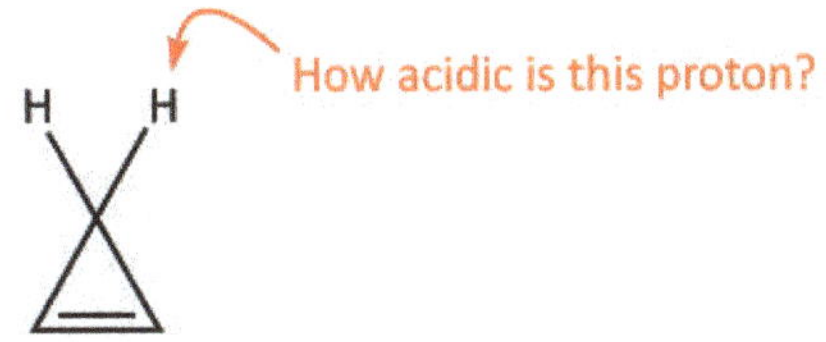

Once again, let's consider the stability of its conjugate base. In this case, the cyclopropenyl anion is a completely conjugated, flat ring system that contains four pi electrons in the pi system. This is an antiaromatic compound and is very unstable. This is a species that would not like to be made, so we know that the protons on cyclopropene are not very acidic. They have a pK_a of 61. This is 10^{45} times less acidic than cyclopentadiene.

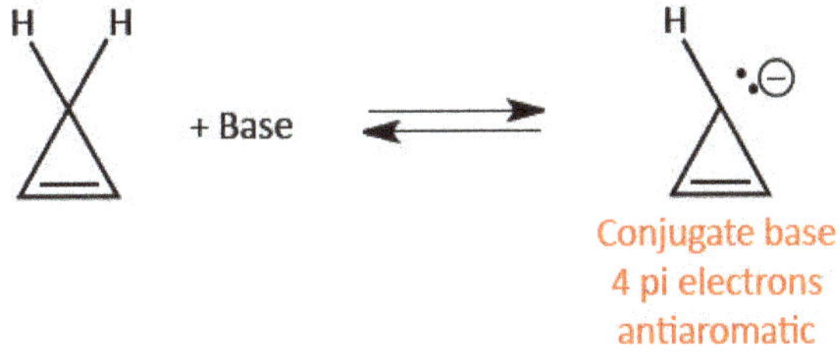

Two other reactions that give stable, aromatic species are the formation of a tropylium ion and the formation of a cyclooctatetraene dianion. A tropylium ion is a cycloheptatrienyl cation. It is a completely conjugated, flat ring system with six pi electrons in the pi system. It is aromatic. It can be formed by the acid dehydration of cycloheptatrienyl alcohol.

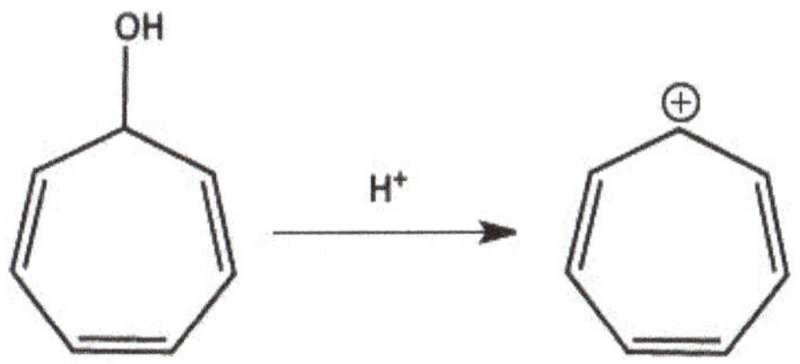

When cyclooctatetraene reacts with potassium metal, it can form a dianion. The cyclooctatetraenyl dianion is a fully conjugated, flat ring with ten pi electrons in the pi-system. This is a stable, aromatic compound.

The general rule is if we make a stable, aromatic compound as a product, this is good and the reaction will proceed easily. If we make an unstable, antiaromatic compound as a product, this is bad and the reaction will not proceed very well.

20.6 Predict whether or not the following reactions would proceed well.

a) 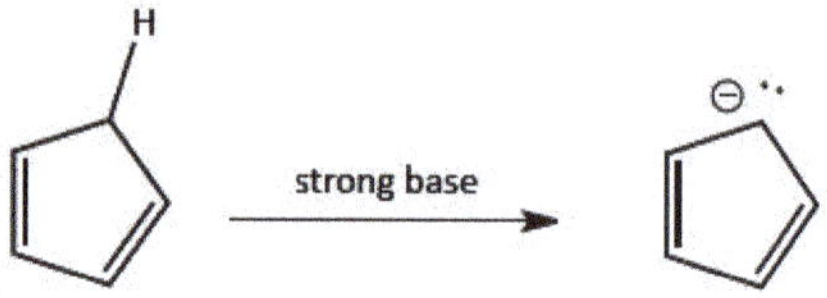

b) 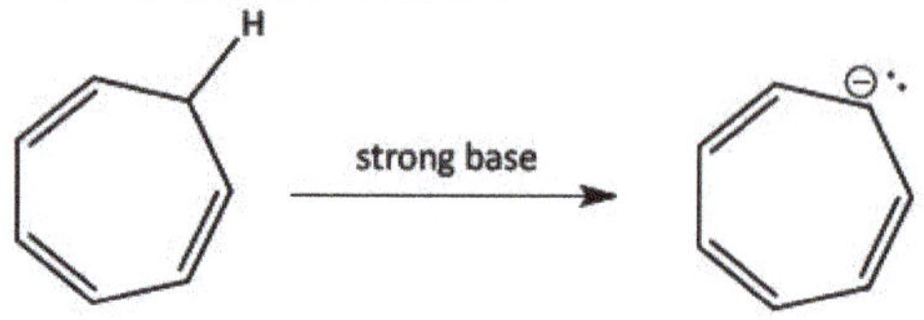

c) 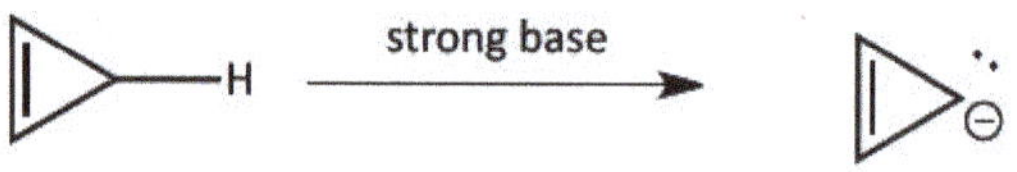

d) 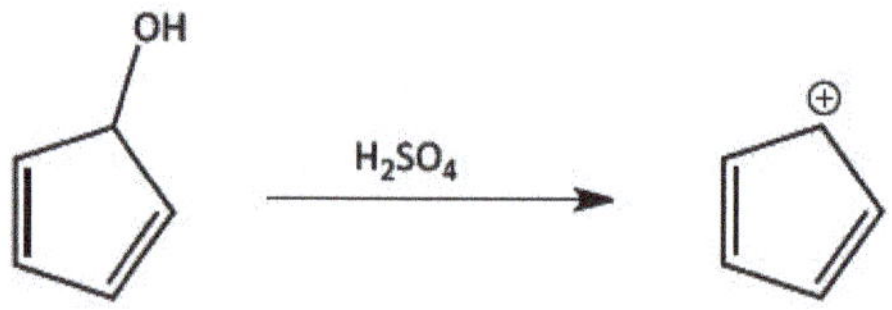

20.7 Aromatic bases, pyrrole, pyridine or Heterocyclic aromatic rings

Heteroatoms like N, S, or O can extend the conjugation of the pi-system of a flat, fully conjugated ring system. It is possible to have heterocyclic aromatic ring system. Some very simple heterocyclic aromatic compounds that you should know are:

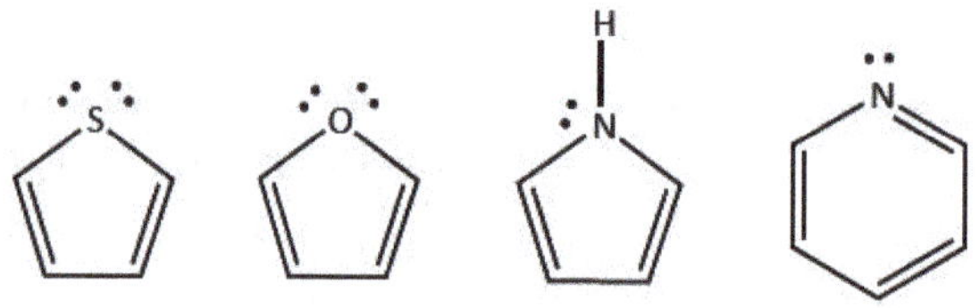

thiophene furan pyrrole pyridine

One of the lone pairs of electrons is joining in with the pi bonds to make an aromatic ring (6 pi electrons).

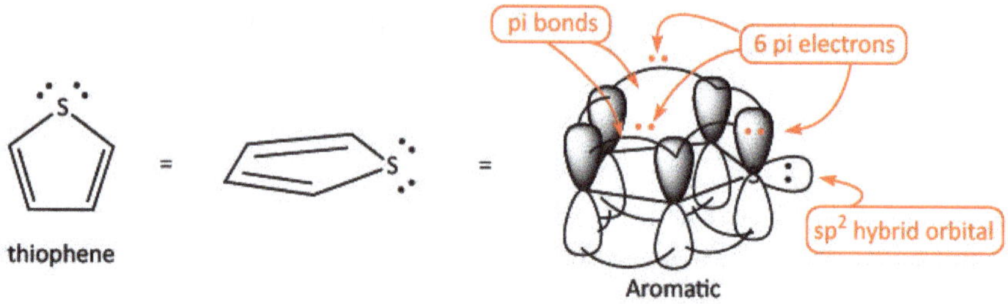

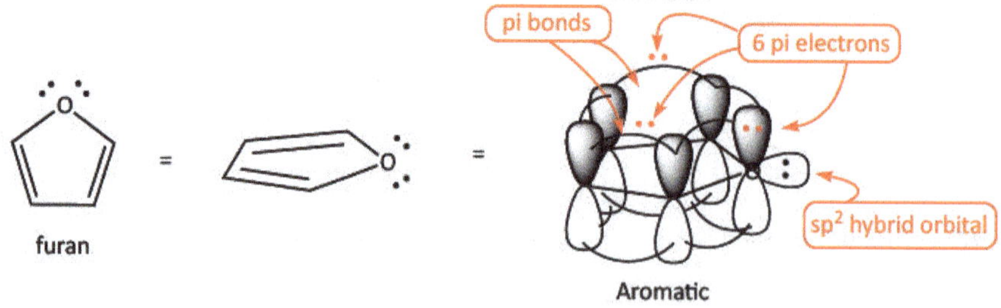

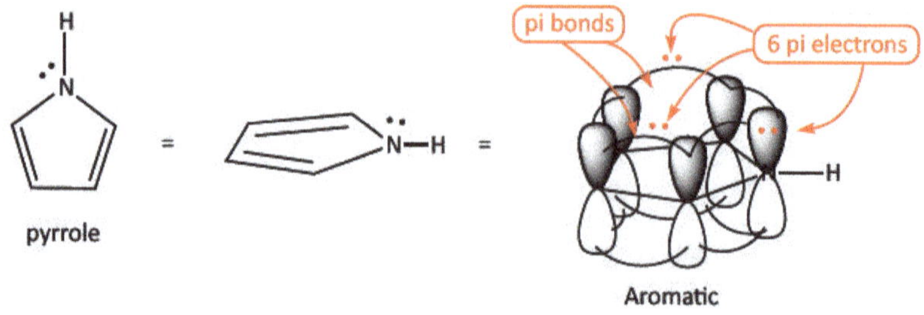

20.7.1 Basicity of nitrogen aromatics

The lone pair of electrons on the nitrogen atoms of most amines is usually pretty basic. Because of its partially negative charge, it can easily grab the H^+ of an acid.

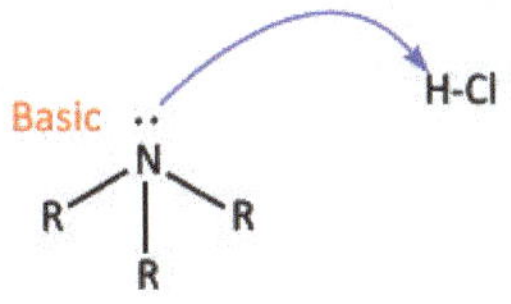

Regular amine

Even nitrogen-containing amine rings are often basic.

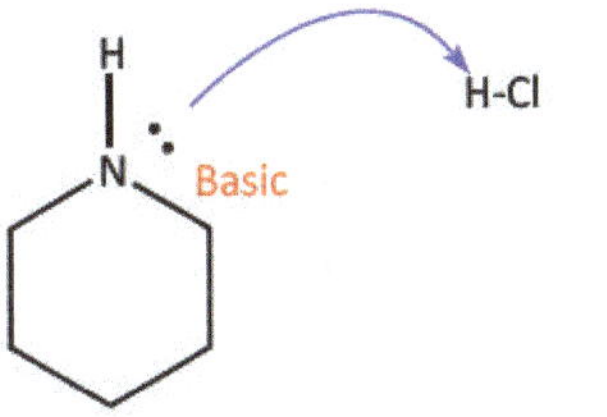

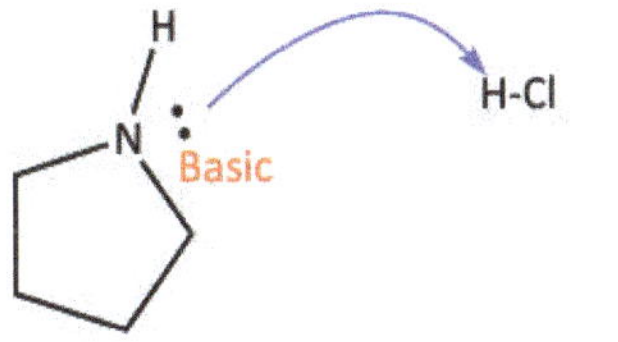

But, if we have a nitrogen *and* double bond-containing ring, the basicity can change. In these cases, we need to be a little more careful. For example, the nitrogen-containing ring, pyrrole, is not very basic. The ring has four pi-electrons from the pi bonds. It needs two more electrons to make the ring aromatic. It can get these two electrons into the pi system from the lone pair of electrons on the nitrogen atom. The nitrogen atom's lone pair of electrons enters a p-orbital of the pi system. Notice this makes the one electron pair on the nitrogen atom delocalized over the ring. Since it is being used in the pi system to make pyrrole aromatic, the negative pair of electrons is not available to go react with the H^+ of an acid, pyrrole is a very weak base. It has a pK_b of 13.6.

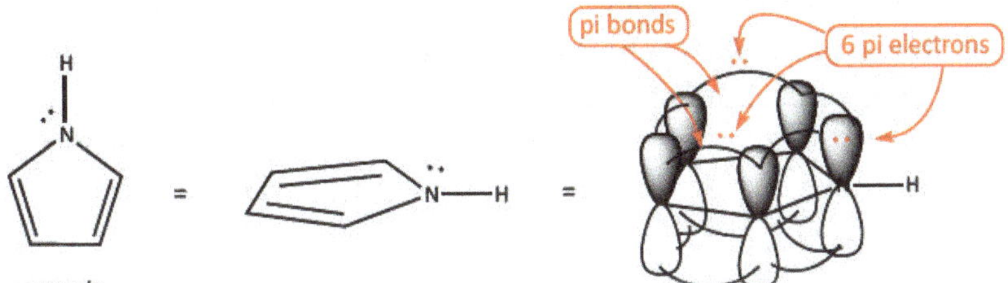

pyrrole

Aromatic

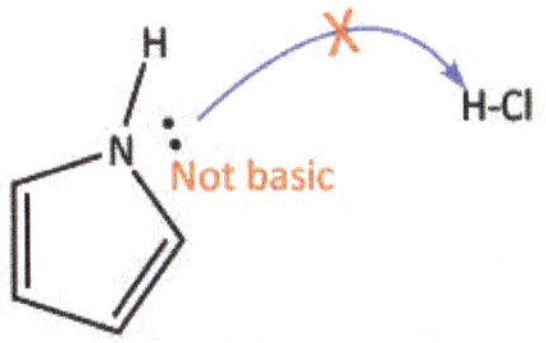

We compare that to the nitrogen and double bond-containing ring, pyridine, which is quite basic.

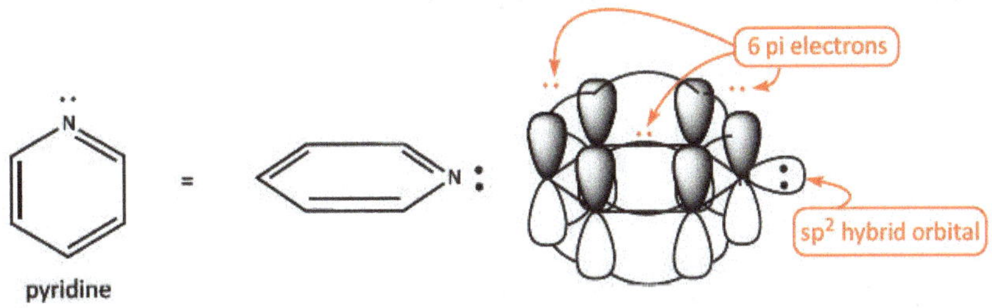

pyridine

The nonbonding lone pair of electrons on pyridine is *not* involved in the pi system. It is in an sp^2 hybrid orbital pointing outside the ring. The nitrogen atom's p-orbital that is part of the pi system is already filled with the pi electrons of the double bond attached to the nitrogen atom. The lone pair of electrons cannot go into that already filled pi system! Therefore, the nitrogen atom lone pair of electrons is localized and remains on the nitrogen atom in an sp^2 hybrid orbital. That electron pair does not spread out over the ring because it is not needed to make the pyridine molecule aromatic. Those electrons are available to go and grab and H^+ of an acid. Therefore, pyridine is quite basic. It has a pK_b of 8.8.

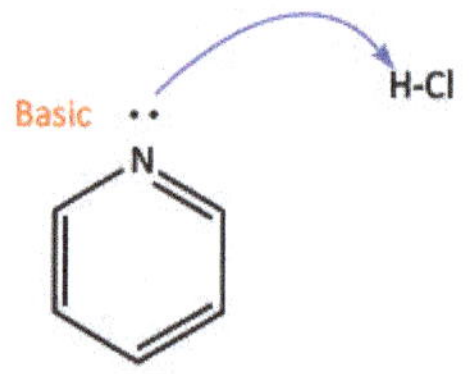

In order to help determine if a nitrogen atom's lone pair of electrons is basic or not, we look for a few things. If the nitrogen atom is not part of a ring or it is part of a ring that does not contain alkenes, it is probably pretty basic.

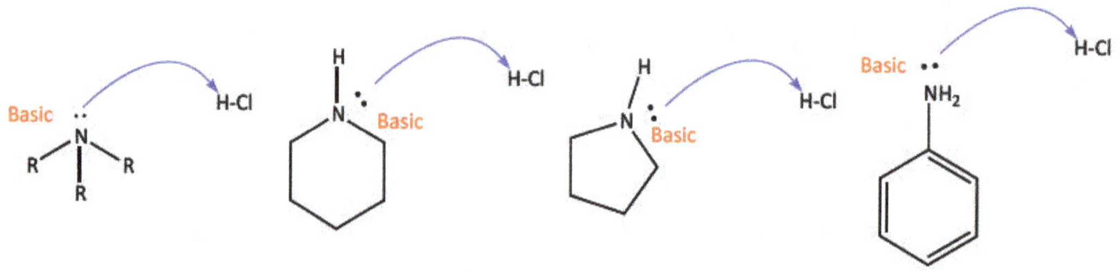

If the nitrogen atoms are inside of an alkene-containing double bond we need to be careful. But, if the nitrogen atom has a double bond attached to it, we can be confident that the p-orbital on that nitrogen atom would already be filled with the pi electrons of the double bond and the lone pair of electrons cannot go into the pi system. It will therefore be available to go react with an acid. That nitrogen atom would be basic. Pyrimidine has two basic nitrogen atoms because both nitrogen atoms have a double bond attached to it.

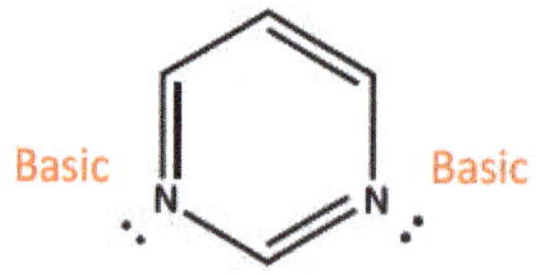

Pyrimidine

Imidazole is a nitrogen *and* double bond-containing ring. We need to be careful in our analysis. It has one basic nitrogen atom (with the double bond on the nitrogen). And, it has one nonbasic nitrogen atom (without a double bond on the nitrogen). The six electrons in the pi system to make this ring aromatic are the four from the two double bonds and the one from the nonbasic nitrogen atom.

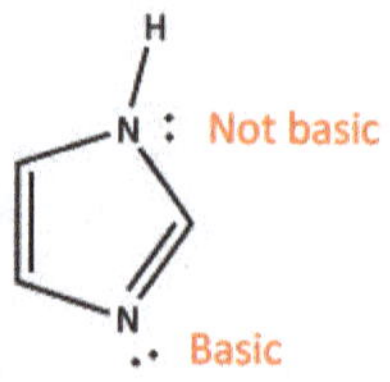

Imidazole

20.7 Label each of the following nitrogen atoms as basic or nonbasic.

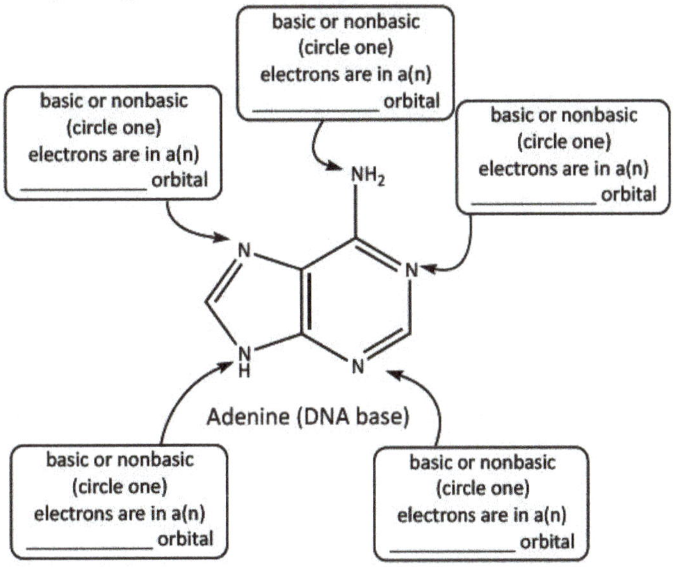

20.8 Monosubstituted Benzene Derivatives

There are several common names of benzene derivatives that you should know.

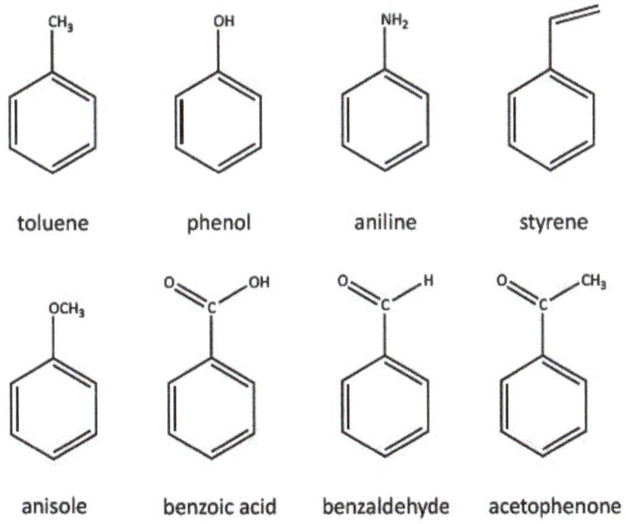

Figure 20.1 Monosubstituted Benzene Derivatives

20.9 Disubstituted Benzene Derivatives

20.9.1 Ortho, Meta, Para

When two groups are attached to a benzene ring, there is a common way to name them using the prefixes ortho, meta, and para. The way to remember the order of the prefixes is to remember "ROMP". Let's look at a generic R group on a phenyl ring. The carbon atom attached to the R group is called the ipso carbon atom. The carbon atom next door to the R group, the 2nd carbon atom, is the *ortho* carbon atom. The 3rd carbon atom is *meta*, and the 4th carbon atom is *para*. This spells "ROMP" around the ring.

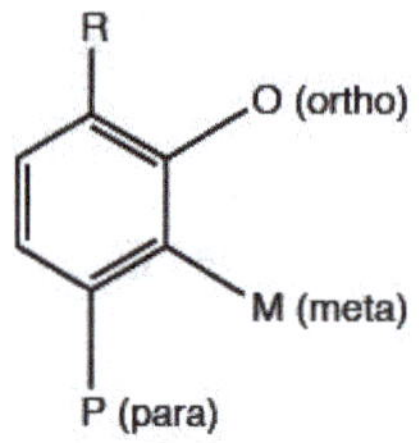

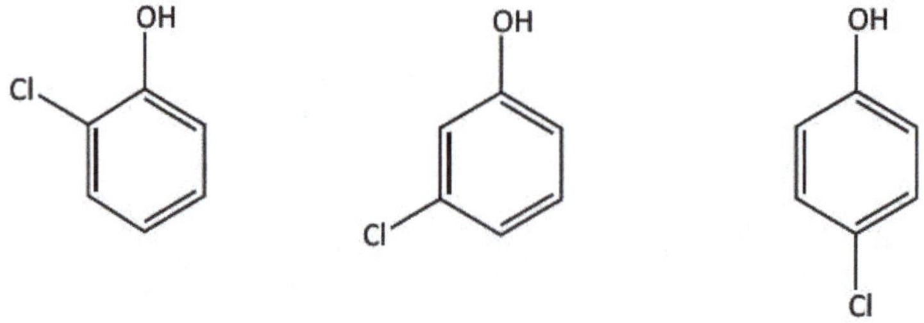

IUPAC	2-chlorophenol	3-chlorophenol	4-chlorophenol
Common	*ortho*-chlorophenol *o*-chlorophenol	*meta*-chlorophenol *m*-chlorophenol	*para*-chlorophenol *p*-chlorophenol

There are some common disubstituted benzenes that you should know, toluic acid, cresol, and xylene.

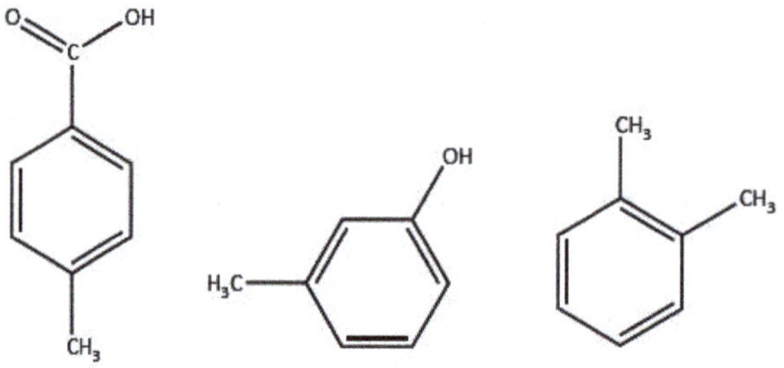

para-toluic acid
p-toluic acid

meta-cresol
m-cresol

ortho-xylene
o-xylene

20.9.2 Benzene groups as substituents

Benzene rings groups can be named as substituents. A benzene ring attachment is called a phenyl group. A group with a CH_2 followed by a benzene is called a benzyl group.

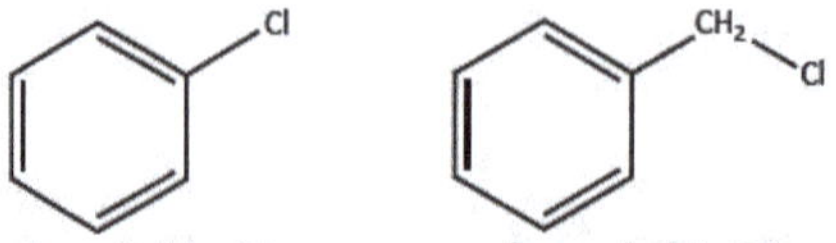

phenyl chloride   benzyl chloride

20.8 Name the following compounds using the common prefix (*ortho, meta, para*) when appropriate.

a) 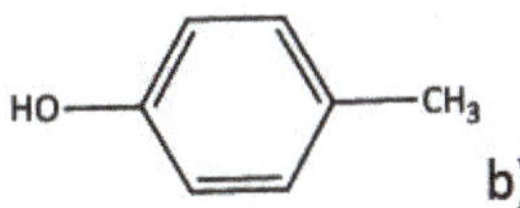

b) 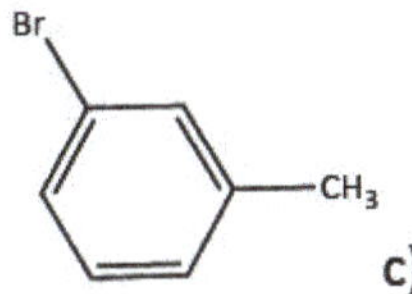

c) 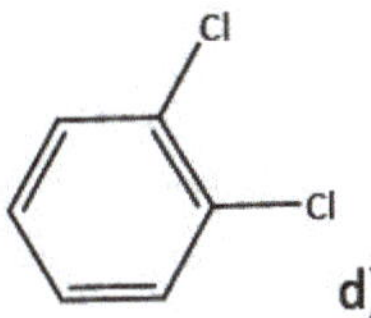

d) 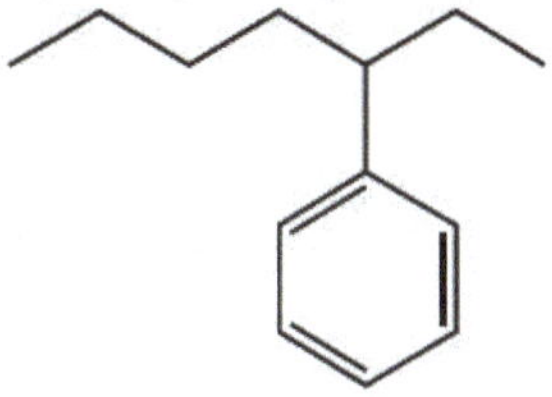

e) 

f) 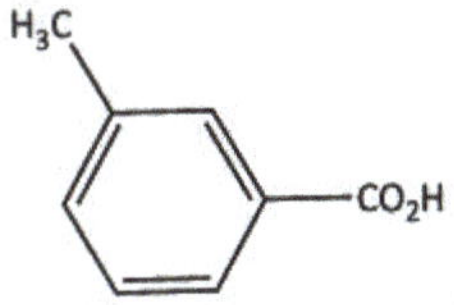

g) 

h) 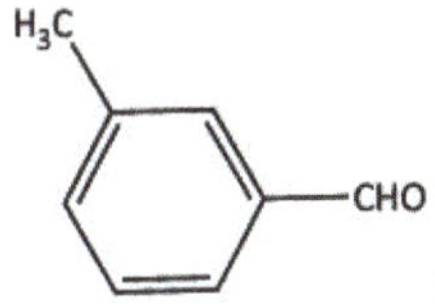

i) 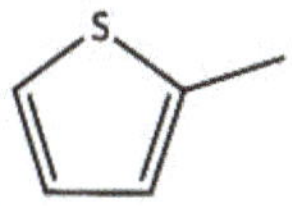

20.10 Polyaromatic hydrocarbons

Polyaromatic,hydrocarbons are aromatic compounds that are benzene rings fused together. Some simple ones are:

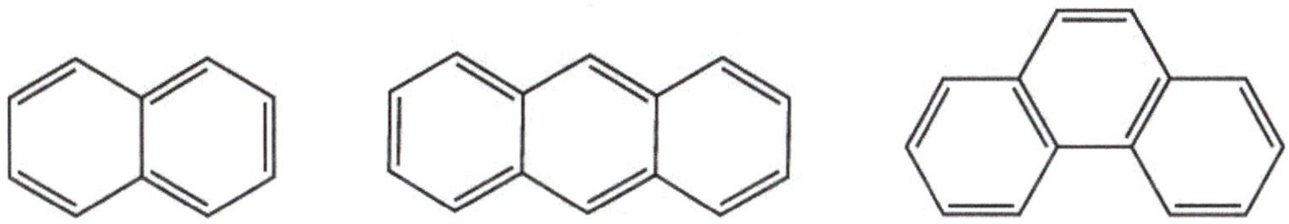

Figure 20.2 Polyaromatic Hydrocarbons (PAHs)

Naphthalene smells like mothballs. This is for good reason. Mothballs are made of naphthalene!

Larger fused benzene ring systems are often called polyaromatic hydrocarbons (PAHs). PAHs are implicated as cancer causing agents. PAHs are produced when substances like cigarettes or meat are burned. The tasty, burned edges of meat on the grill contain these. For that matter, the bits of charred material left on the grill contain PAHs. It is believed that PAHs actually become cancer-causing agents when they are oxidized into epoxides in the liver. These epoxides are electrophiles that can be attacked by nitrogen bases on our DNA. These PAHs then stick to our DNA causing a mutation.

20.11 Allotropes of Carbon

Imagine a fused ring system that is not only four or five benzenes fused together, but an infinite number of benzene rings fused together into an infinite place. Then, imagine several of these infinite planes stacked on top of each other. These planes would be made out of only carbon since it has no edges where the hydrogens can be. This stacked group of infinite planes of fused benzenes is called graphite.

Figure 20.3 Graphite

Each carbon atom of this infinite plane has an unhybridized p-orbital on it that can overlap with the others. Electrons are shared in the molecular orbitals formed by these p-orbitals giving a plane where electrons can freely move around. These p-orbitals cause the layers to loosely stack on each other. Here is a set of three benzenes stacked on top of each other.

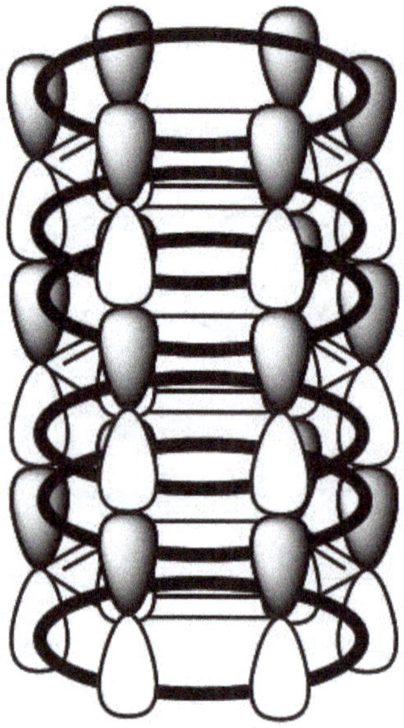

Figure 20.4 Pi stacking

Because of this loose stacking, graphite us used as pencil "lead". It is called pencil lead because long ago, lead metal was used to make the dark marks on paper. But, since lead is toxic, the nontoxic compound graphite is used. As the graphite is rubbed on paper, the bottom layer of graphite easily shaves off. The graphite is deposited one layer at a time onto the paper.

Graphite is one of the several orderly forms of pure carbon. These are called allotropes of carbon.

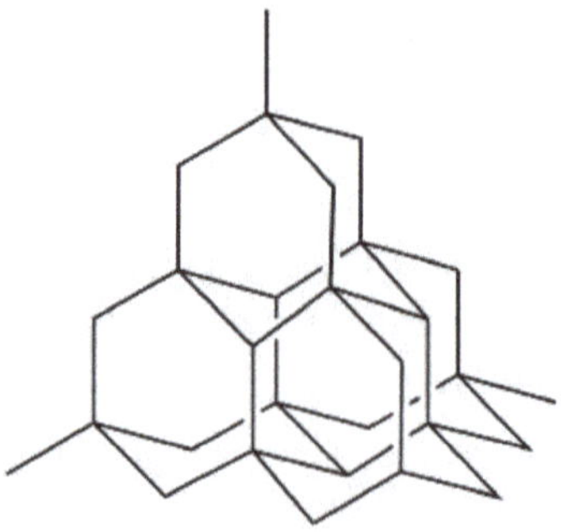

Figure 20.5 Structure of diamond

Another allotrope of carbon is diamond. The carbon atoms in diamond ar all sp^3 hybridized. A diamond structure is a series of cyclohexane chairs stacked on top of each other. Because each carbon atom in diamond is a perfect tetrahedral, diamond has tremendous strength. Of course, it is one of the hardest materials.

For many years, these were thought to be the only allotropes of carbon. But, in the 1980s, scientists discovered another allotrope of carbon. These are the fullerenes. Fullerenes are 5- and 6- membered aromatic rings arranged to form a "soccer ball" structure. The most common fullerene is C_{60}. This is a molecule that contains 60 carbon atoms and zero hydrogen atoms. It was named the Molecule of the Year in 1990 by the journal "Science". Yes, believe it, there is such a thing as the Molecule of the Year.

Figure 20.6 C_{60}, Buckyball

Even more recently, another form of all carbon is being developed, the nanotubes. You can think of carbon nanotubes as one layer of graphite wrapped into a cylinder. Nanotubes are incredibly strong. Nanotubes are nearly 120 times stronger than steel. One hope is to eventually make extremely strong, lightweight cables out of carbon nanotubes.

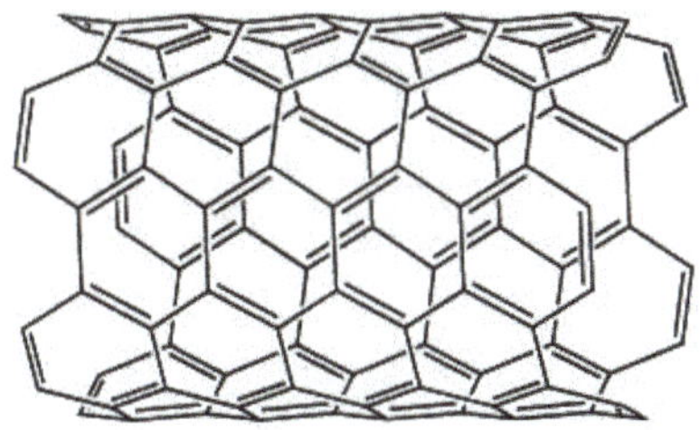

Figure 20.7 Nanotube

20.12 Phenyl spectroscopy

20.12.1 Infrared spectroscopy

Since phenyl groups and aromatic rings have sp^2 hybridized carbon atoms, the sp^2 C-H stretch appears just above 3000 cm-1. For phenyl rings, aromatic overtones (zig zags) show up at 1800-2000 cm-1.

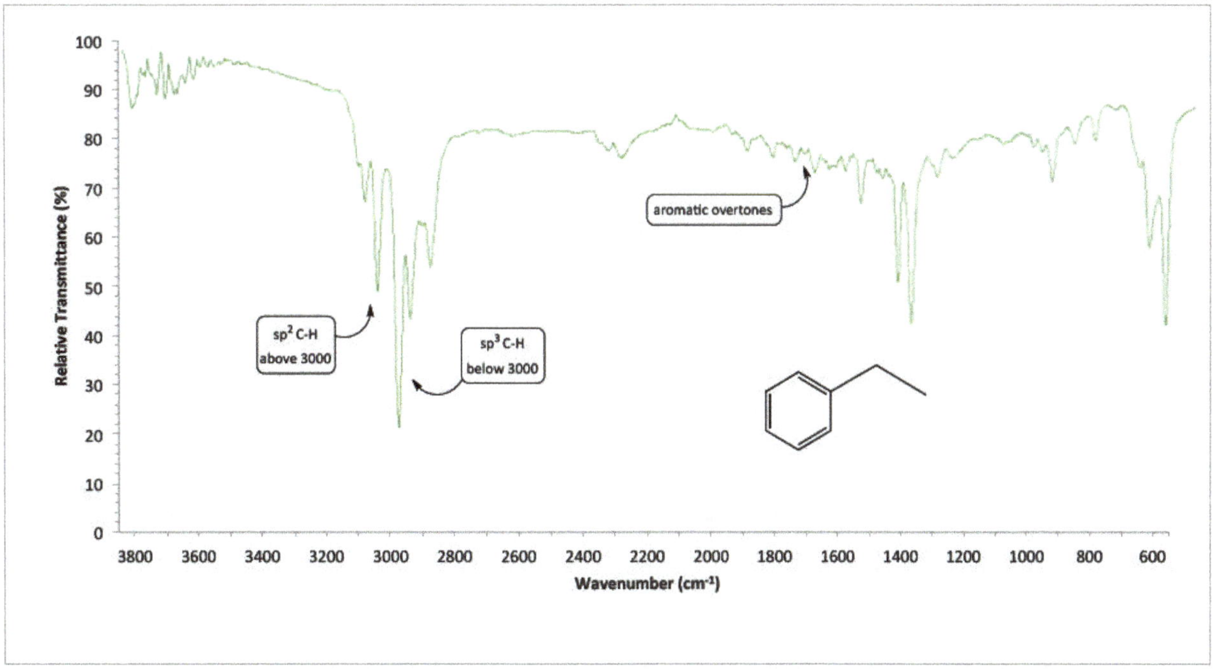

Figure 20.8 IR spectrum of ethyl benzene

20.12.2 NMR

In the ^1H NMR, aromatic hydrogen atoms appear around δ6- δ8.

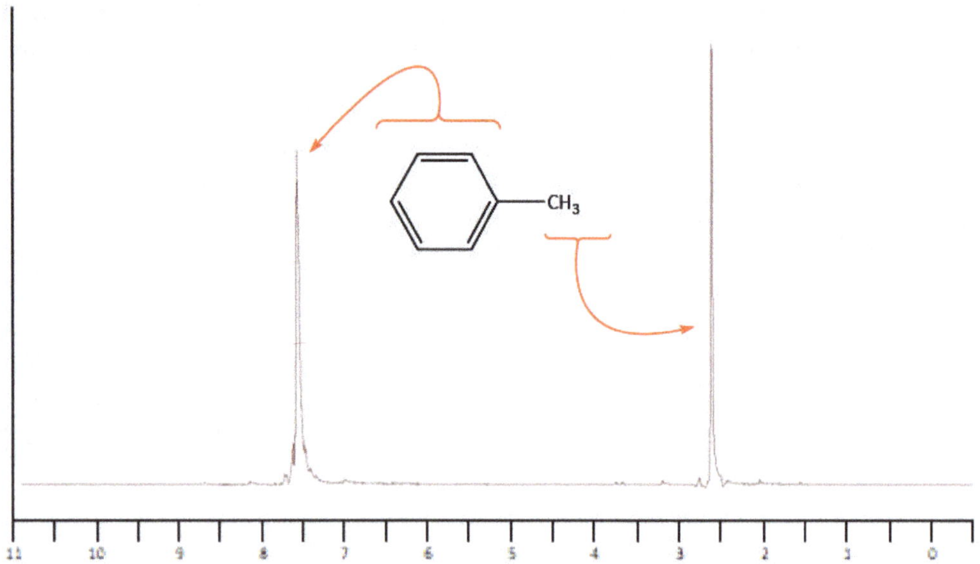

Figure 20.9 Proton NMR of toluene

In ^{13}C NMR, the sp^2 carbon atom peaks appear between δ100- δ150 where double bonded carbon atoms appear.

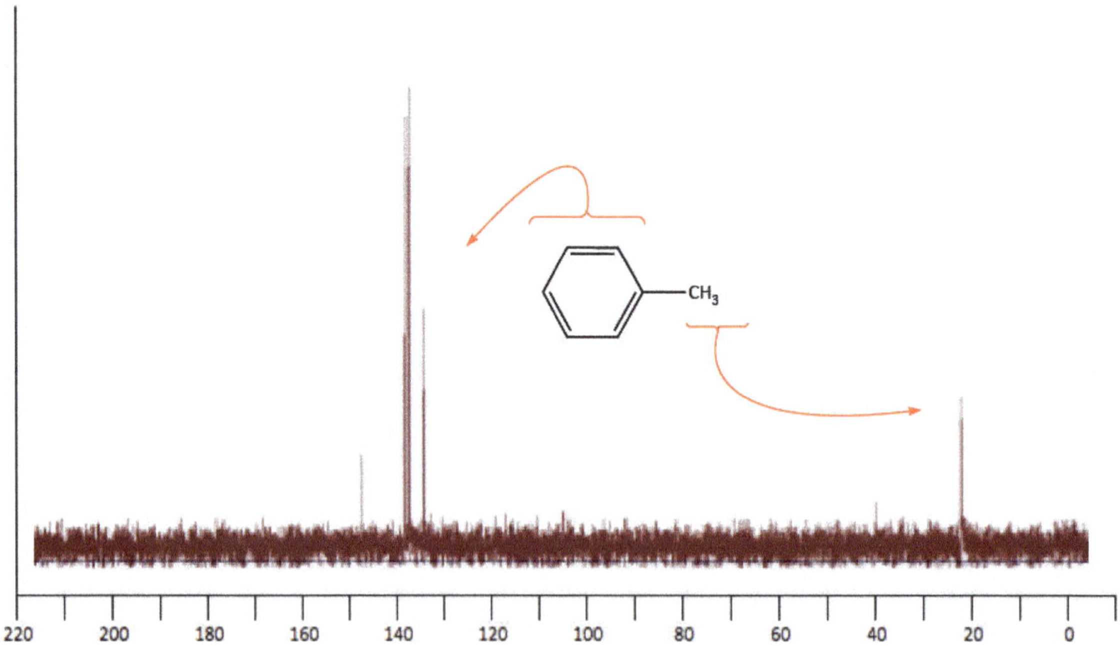

Figure 20.10 Carbon NMR of toluene

20.12.3 Mass Spectrometry

A very common cleavage in mass spectrometry is at the benzylic position to give a fragment of m/z = 91. We have seen earlier that benzylic cations are very stable because of resonance forms. In fact, benzylic cations can even rearrange to make an aromatic 7-membered ring cation called a tropylium ion.

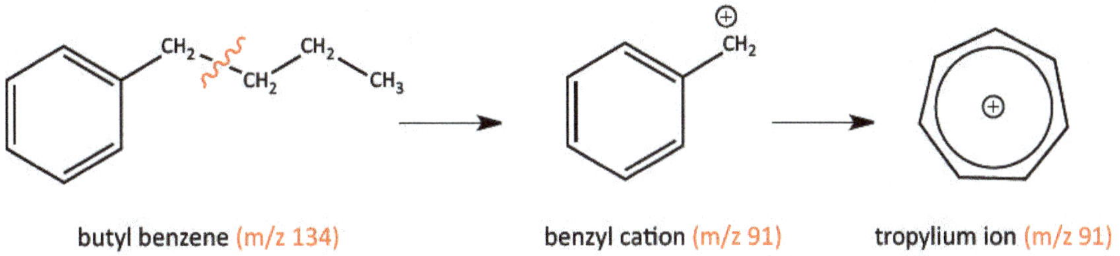

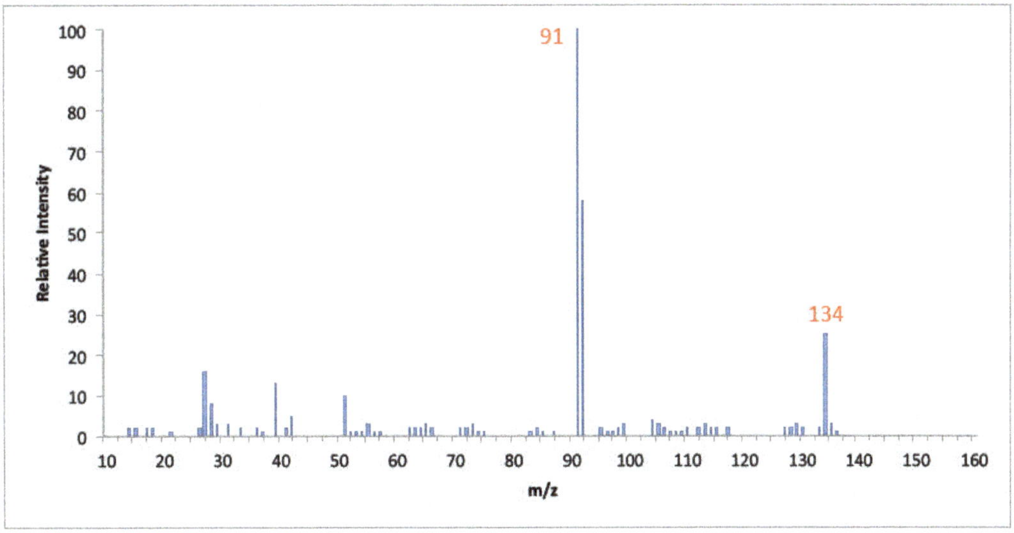

Figure 20.11 Mass Spectrum of n-butylbenzene

20.13 Problems

20.9 Assuming they are planar, label each of the following compounds/ions as aromatic, antiaromatic, or nonaromatic.

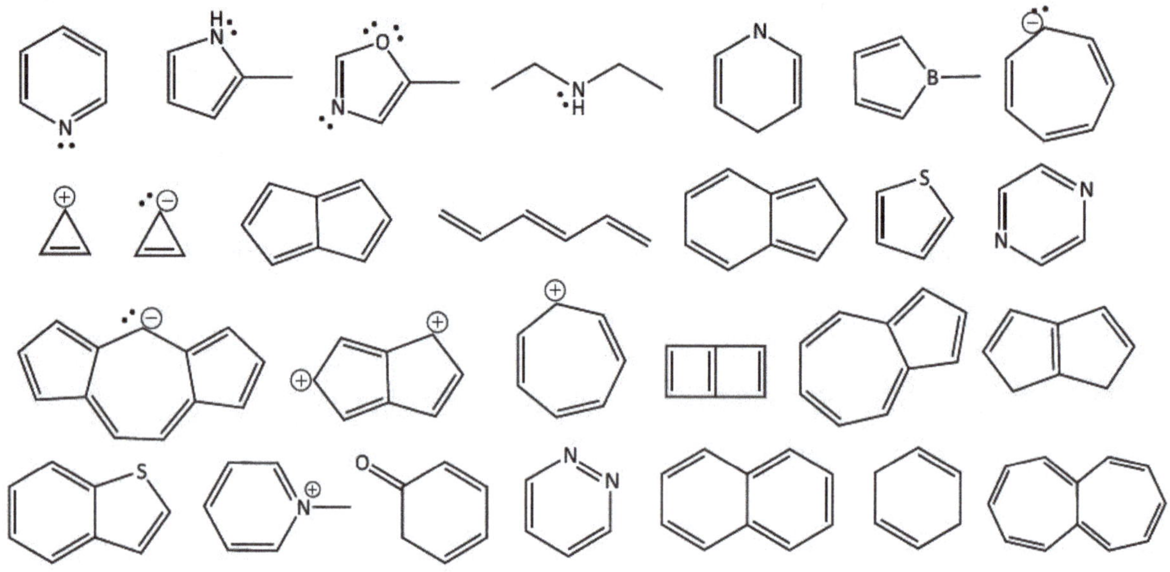

20.10 Is cycloheptatrienyl cation a planar ion? Explain.

20.11 Draw the following compounds.

a) *p*-xylene b) toluene c) *o*-cresol d) benzoic acid e) benzaldehyde f) aniline
g) *o*-ethoxytoluene h) *m*-nitroanisole i) *p*-nitroaniline j) 3-chlorostyrene k) 2,3-dimethylphenol
l) 3-aminobenzoic acid m) 2-methylpyrrole n) 3-ethylpyridine o) 2,3-dimethylthiophene

20.12 Identify each of the nitrogen atoms in the following compounds as basic or nonbasic.

a) 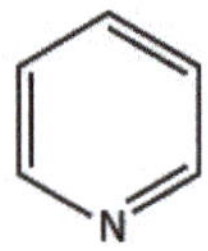 b) 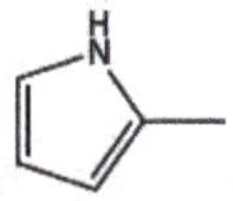 c) 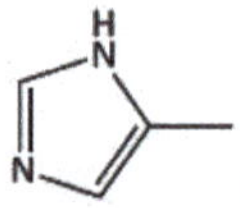 d) 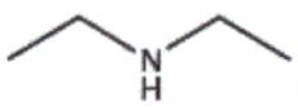 e) 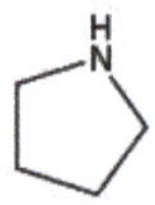 f) 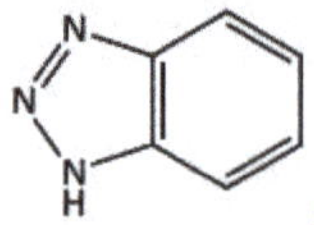 g) 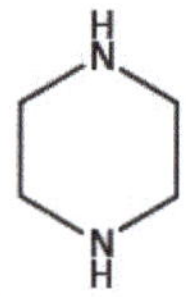

20.13 Use the polygon rule and draw the MO diagram for the cyclopentadienyl cation and cyclopentadienyl anion (including the pi electrons). a) Determine from the MO diagrams which is aromatic and which is anti-aromatic. b) Does this confirm what the 4N+2 rule would predict?

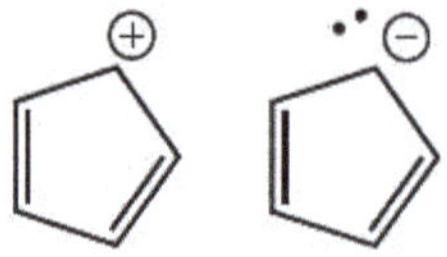

20.14 **In the following amino acids, identify the indicated nitrogen atoms as basic or nonbasic. Also, identify the type of orbital that contains that nitrogen atom's lone pair of electrons,**

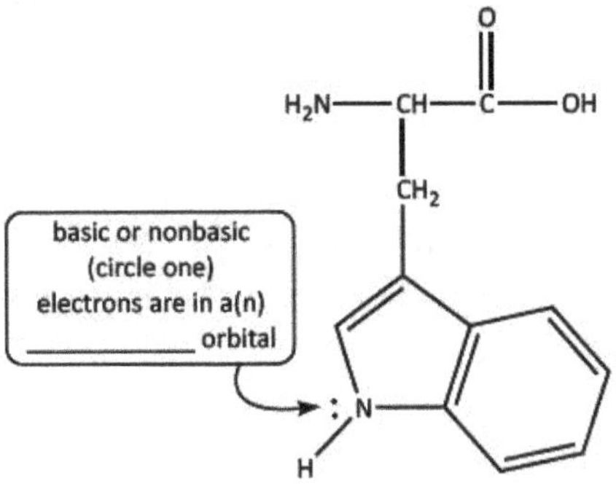

L-tryptophan (amino acid)

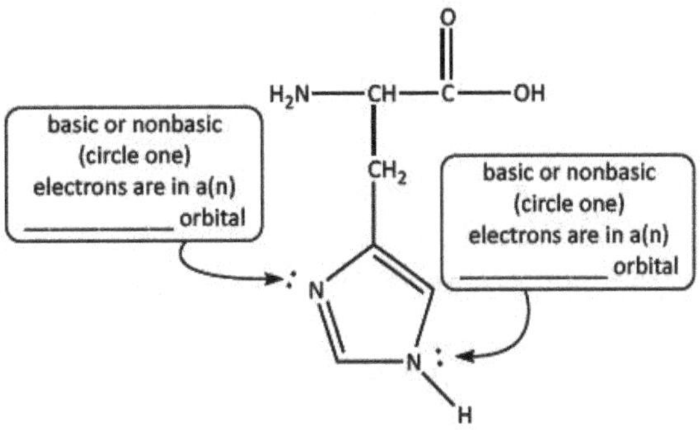

Histidine (amino acid)

20.15 **Use the polygon rule and draw the MO diagram for the cyclopropenyl cation and cyclopropenyl anion (including the pi electrons). a) Determine from the MO diagrams which is aromatic and which is anti-aromatic. b) Does this confirm what the 4N+2 rule would predict?**

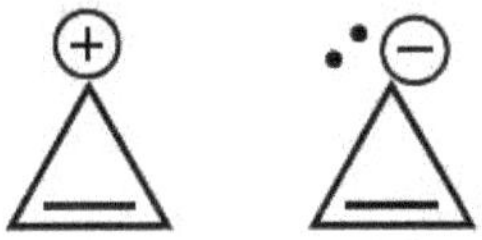

20.16 **Which of the two following compounds is the most acidic and why? Hint: As always, think about the stability of the conjugate base that is formed.**

20.17 **Vitamin B_1, thiamine, is shown below. Identify whether rings A and B are aromatic, nonaromatic, or antiaromatic. Then, identify whether the nitrogen amines (C-F) are basic or nonbasic.**

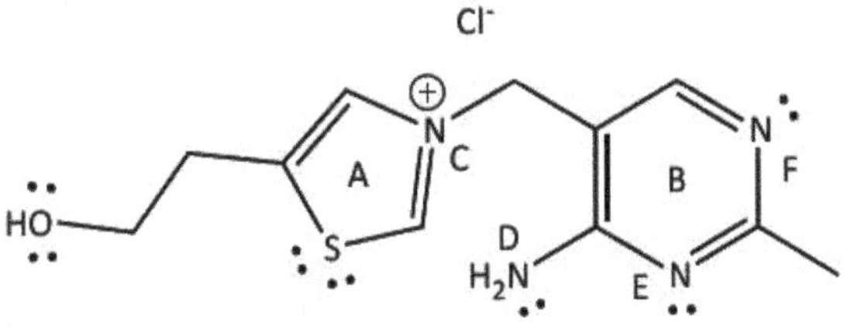

Vitamin B_1 (Thiamine)

Chapter 21. Aromatic Compound Reactions

21.1 Introduction

Aromatic rings like benzene are very stable. Therefore, they do not react easily. They will react with a very strong electrophile. In this chapter, we will see several reactions that occur on aromatic rings. These are very important reactions as aromatic rings are found in many natural products and medicines.

21.2 Electrophilic Aromatic Substitution

Aromatic rings usually do not react unless they are in the presence of a very positively charged species, an electrophile. But, if an electrophile has a strong enough positive charge, it can coax the negatively charged pi electrons of the aromatic ring to attack it. When the pi bond attacks the electrophile, the electrophile attaches to the benzene ring, and the ring loses its conjugation. It forms a non-aromatic species called a sigma complex. A base cleans up the reaction by pulling off a proton with the electrons in its bond reforming the pi bond of the aromatic ring. Overall, a hydrogen atom is replaced with an electrophile.

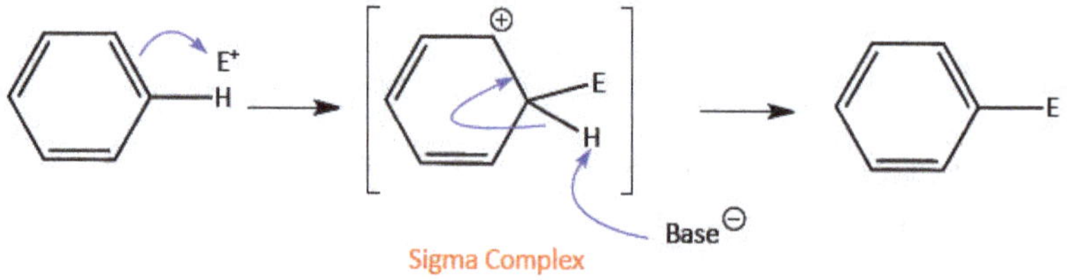

Let's look at a few of these reactions. You should immediately memorize these next few reactions. We will use them often.

21.2.1 Bromination of Benzene

We learned previously that pi bonds can be brominated using Br_2 to make anti-dibrominated compounds. But, we find that benzene cannot be brominated the same way. The stability of the aromatic benzene ring precludes that.

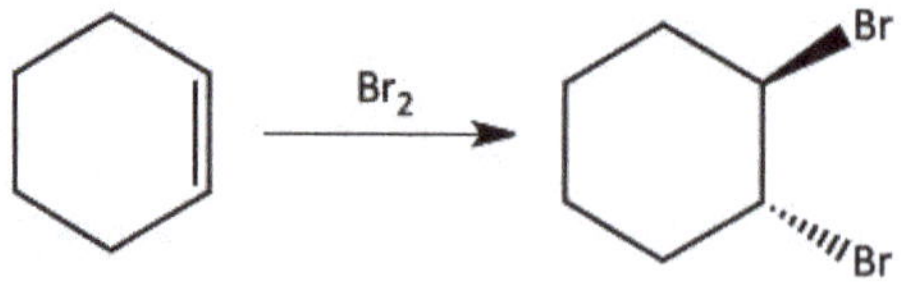

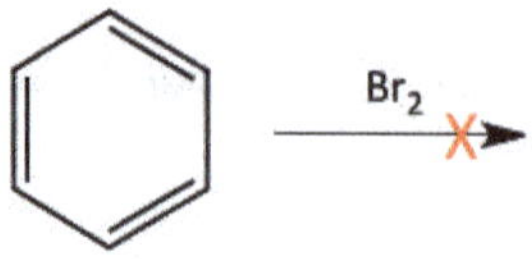

A stronger electrophile is needed in order to brominate benzene or other aromatic compounds. This is done by adding the Lewis acid, ferric bromide ($FeBr_3$), to bromine, Br_2. Br_2 donates a lone pair of electrons to the $FeBr_3$ Lewis acid. This makes the bromine positively charged and an electrophile.

This is a strong enough electrophile that the aromatic benzene ring will now donate electrons from a pi bond to it. We will not worry about the mechanism of this reaction, but you should memorize it.

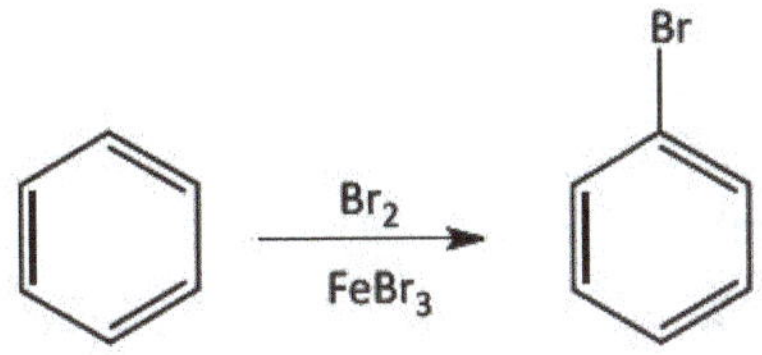

21.2.2 Chlorination of Benzene

The chlorination of benzene is similar to the bromination of benzene. On its own, chlorine is not a strong enough electrophile to react with an aromatic ring. A Lewis acid is needed. For the chlorination reaction, the Lewis acid is usually aluminum trichloride, $AlCl_3$. Memorize this reaction.

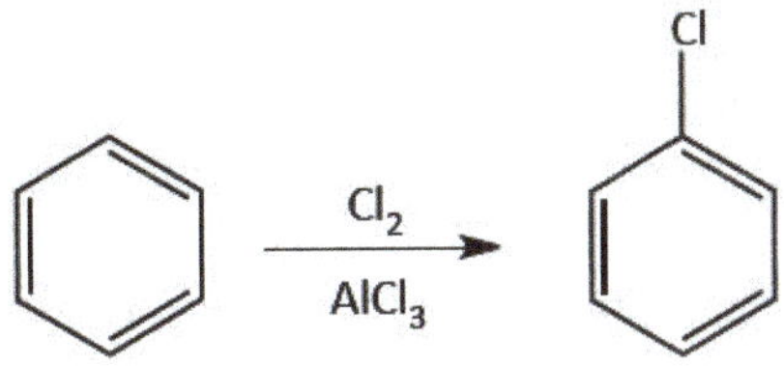

21.2.3 Iodination of Benzene

The iodination of benzene occurs in the presence of iodine and nitric acid.

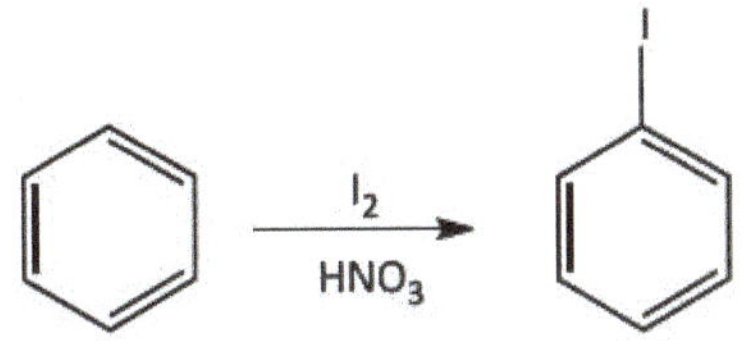

Iodine, I_2, is also not a strong enough electrophile on its own. Typically, nitric acid, HNO_3, is added to iodine to make the electrophilic iodonium ion.

$$H^+ + HNO_3 + I_2 \longrightarrow I^+ + NO_2 + H_2O$$

Iodonium ion

Once iodonium is produced, the iodination of an aromatic compound can occur. Memorize this reaction.

21.2.4 Nitration of Benzene

Benzene is nitrated in the presence of nitric and sulfuric acid.

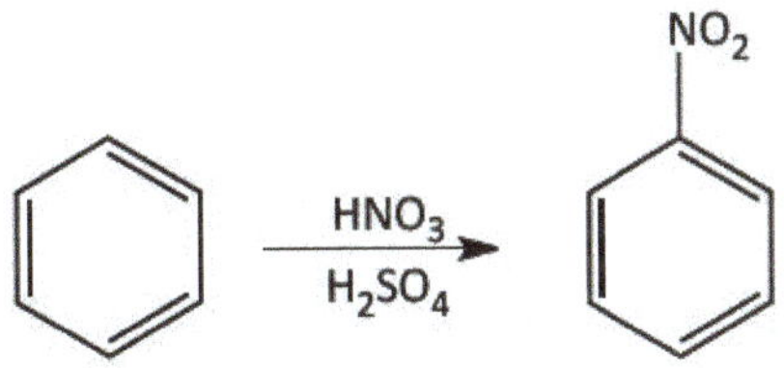

The mechanism to produce the electrophile in this reaction is interesting. When nitric acid and sulfuric acid react with each other, sulfuric acid acts like an acid and donated a proton while nitric acid acts like a base. Once the nitric acid is protonated, water leaves giving a nitronium ion. This nitronium ion is an excellent electrophile and can act with benzene producing a nitrobenzene.

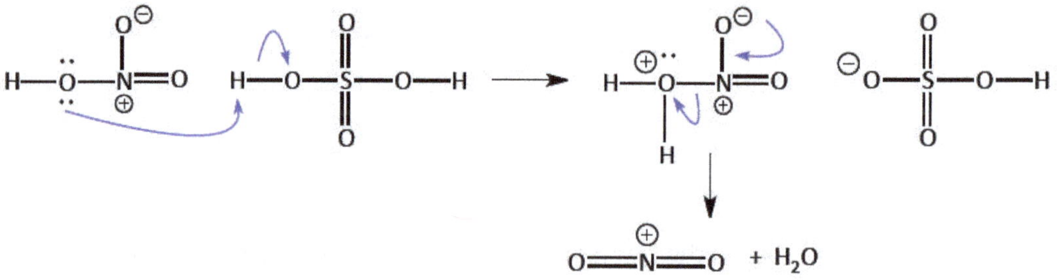

21.2.5 Sulfonation/Desulfonation of Benzene

This reaction is fun because it uses fuming sulfuric acid which just sounds menacing. Fuming sulfuric acid is sulfuric acid with sulfur trioxide gas dissolved in it. In the presence of fuming sulfuric acid, benzene can be sulfonated.

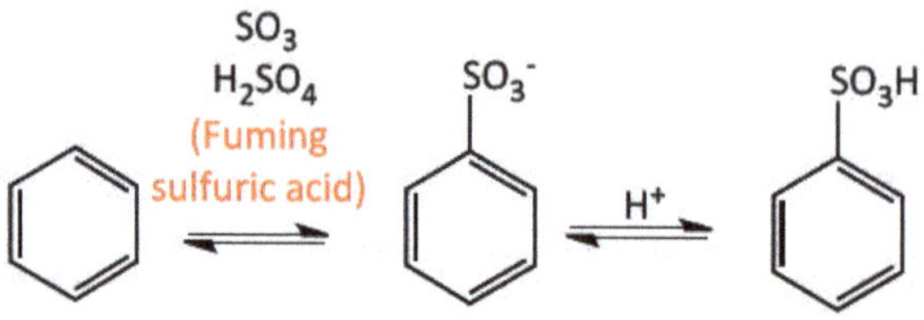

Sulfonated benzene can be desulfonated when it is heated with water and acid.

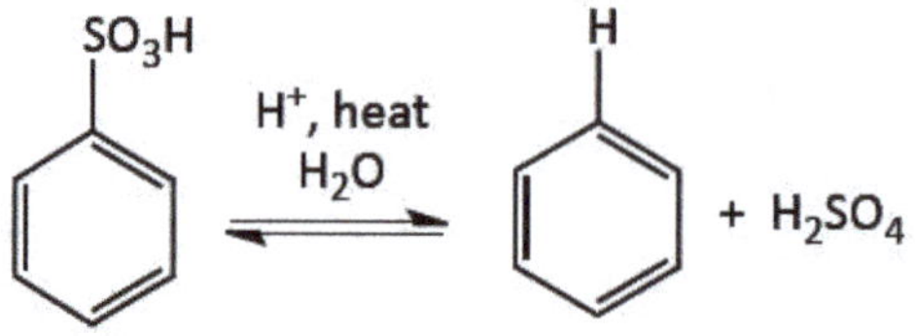

21.1 Write the reagents used to perform the following transformations.

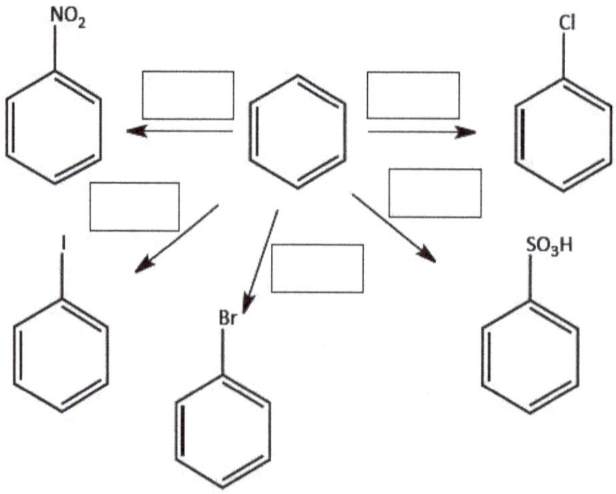

21.3 Activating groups

21.3.1 Alkyl groups

Activated

Let's look at the nitration of benzene compared to the nitration of toluene. Toluene is methyl benzene. We find that the nitration of toluene occurs faster than it does on benzene, 25 times faster. The methyl group makes the reaction go faster. We say that the methyl group, as well as other alkyl groups, is an activating group. Activating groups make aromatic reactions proceed faster.

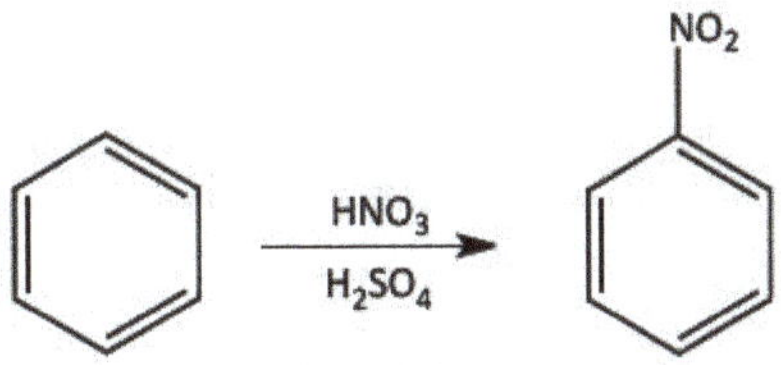

Nitration of benzene

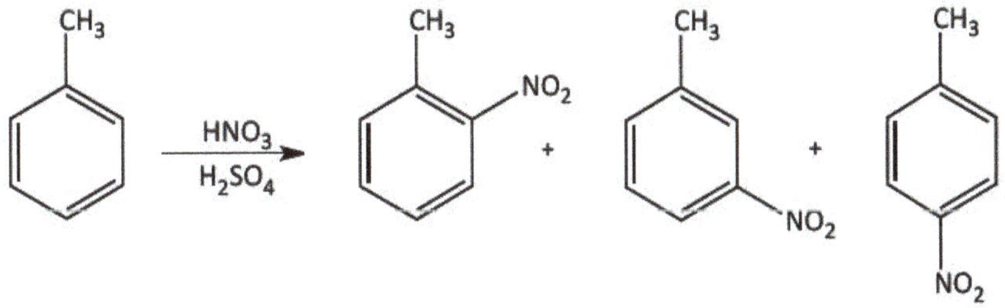

Nitration of toluene (occurs 25 times faster)

Ortho/Para Directing

When toluene is nitrated, the nitro group can go on ortho, meta, or para positions. If the process was totally random, we would expect ortho:meta:para isomers in a 2:2:1 ratio.

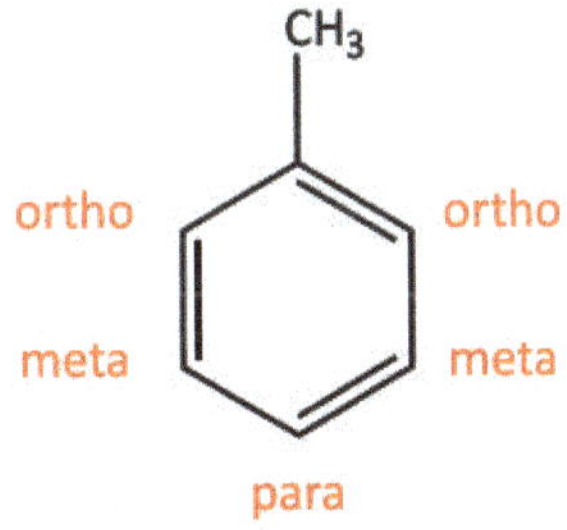

We find that much more of the ortho and para isomers are made with very little of the *meta* isomer being formed. We say that the methyl group, like other alkyl groups, is ortho-para directing.

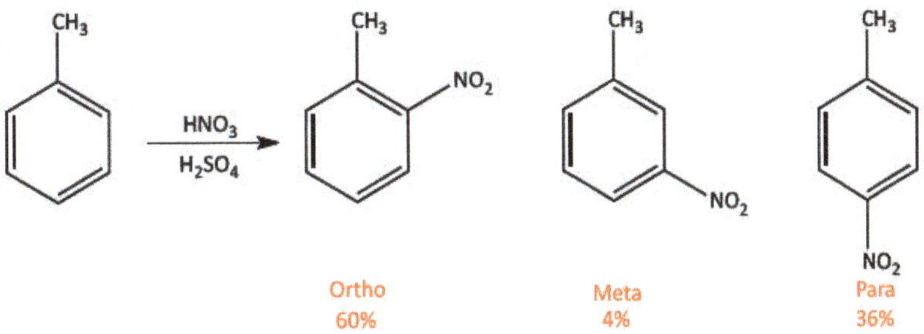

The reason methyl and other alkyl groups are activating (making the reaction go faster) and ortho-para directing can be explained by looking at the mechanism of the nitration. When benzene reacts with the nitronium ion, the sigma complexes are all secondary cations.

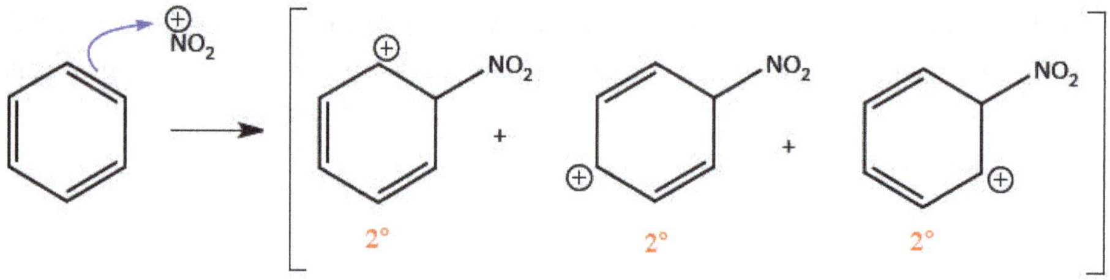

When toluene is nitrated, we get sigma complexes that are both tertiary and secondary carbocations. Because tertiary carbocations are lower in energy (more stable) than secondary carbocations, reactions that make tertiary carbocations proceed faster than those that only make secondary carbocations. This is why the toluene reactions are activated. Notice that only the nitration at the ortho and para sites produce tertiary carbocation sigma complexes. This is why methyl and other alkyl groups are ortho-para directing.

Ortho attack

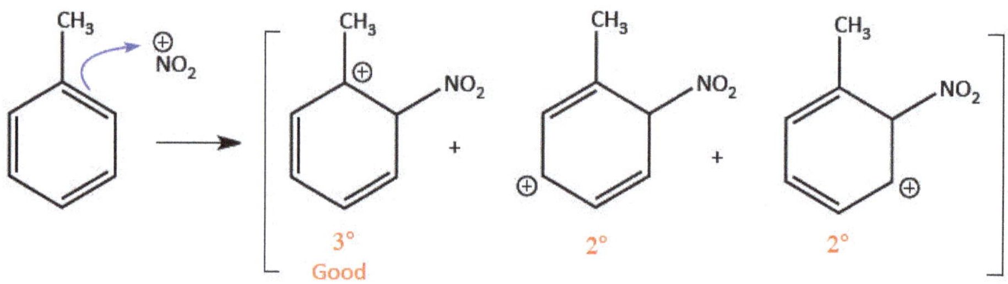

Nitration at ortho site gives tertiary carbocation

Meta attack

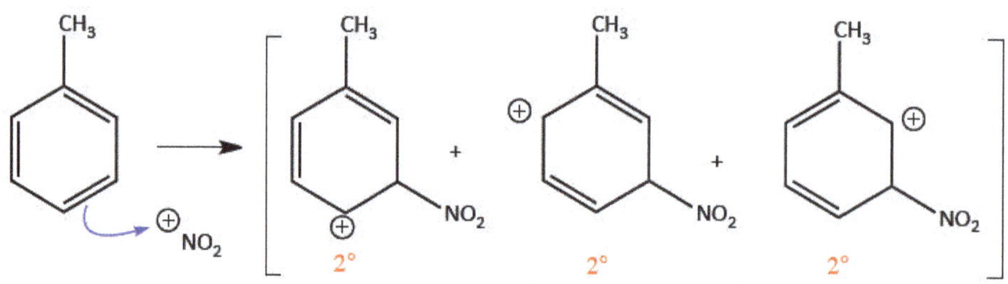

Nitration at meta site gives only secondary carbocations

Para attack

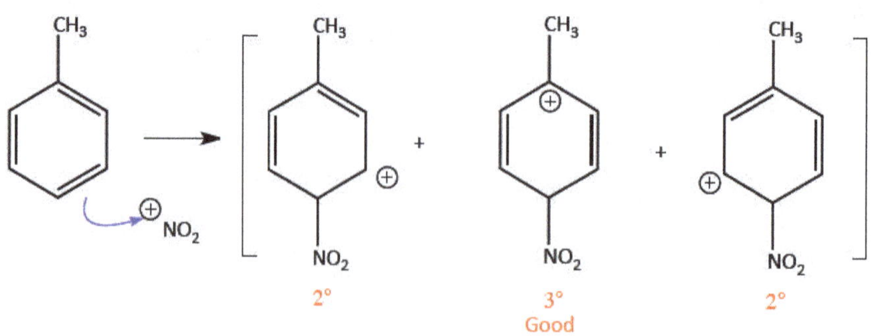

Nitration at para site gives a tertiary carbocation

21.3.2 Alkoxy and amino groups

Alkoxy (-OR) and amine ($-NR_2$) groups are even more activating than alkyl groups. The nitration of anisole (methoxy benzene) reacts 10,000 times faster than the nitration of benzene.

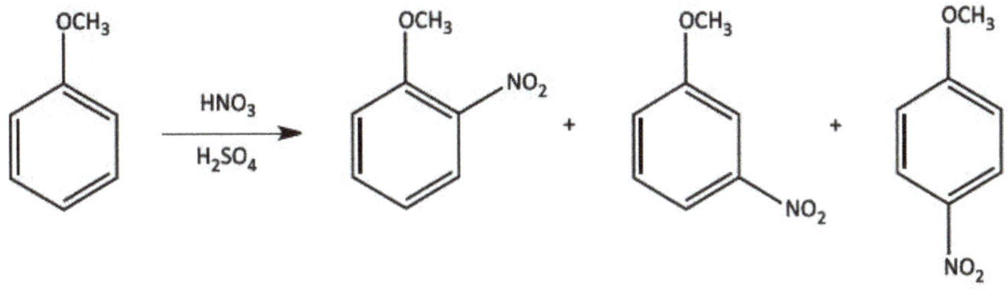

This fast nitration with alkoxy or amino groups is because carbocations that are formed in these sigma complexes are further stabilized through resonance. Also, when the nitration occurs at the ortho or para sites, this resonance-stabilized carbocation is formed but when the attack is at the meta site, no very good resonance stabilized sigma complexes are formed. For this reason, alkoxy and amino groups are ortho-para directing.

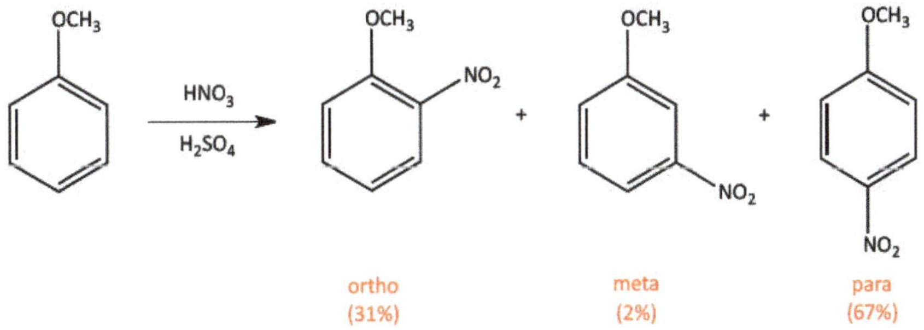

Ortho attack

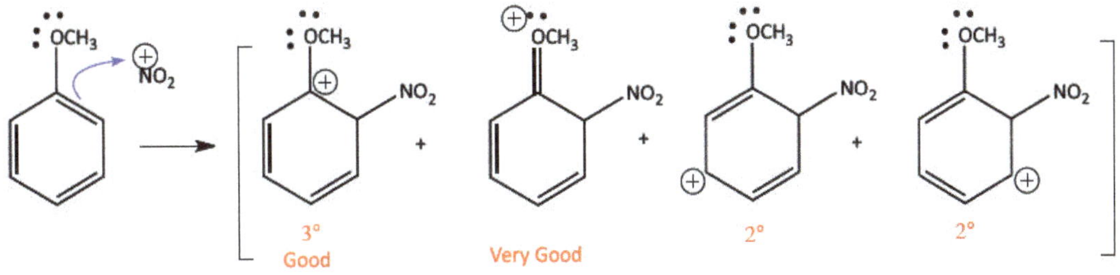

Meta attack

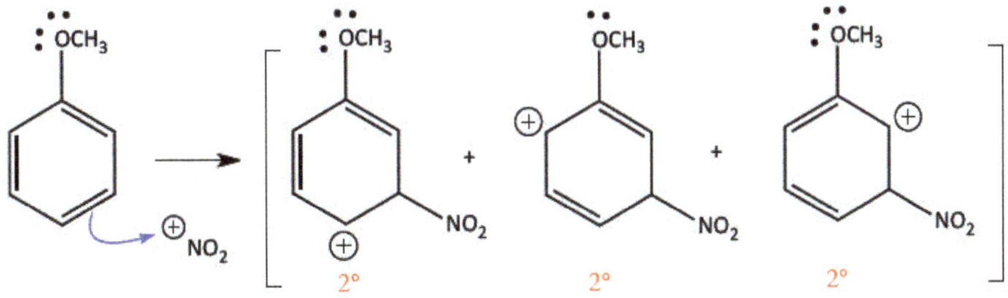

Para attack

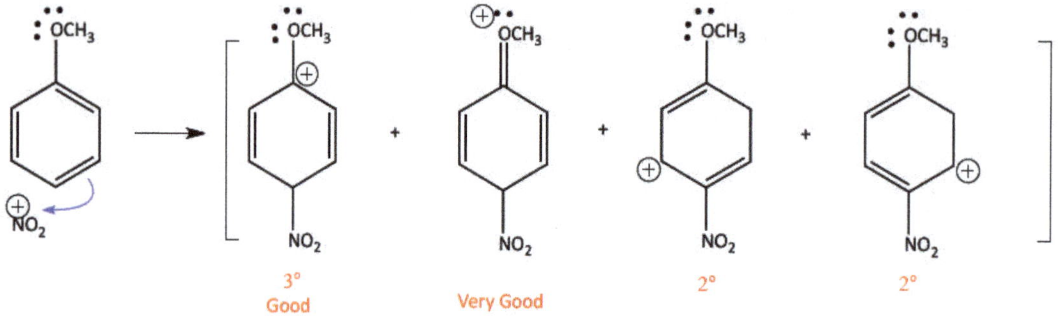

Alkoxy and amino groups are such excellent activating groups, that anisole and aniline can be brominated with bromine. The reaction doesn't stop after one bromination. But, all of the ortho and para positions can be brominated.

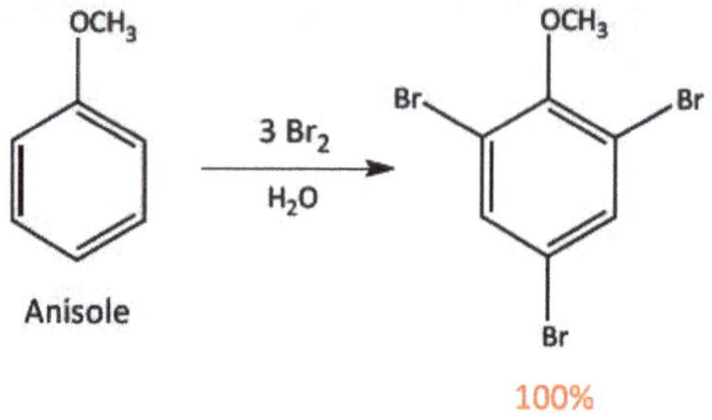

Bromination of anisole

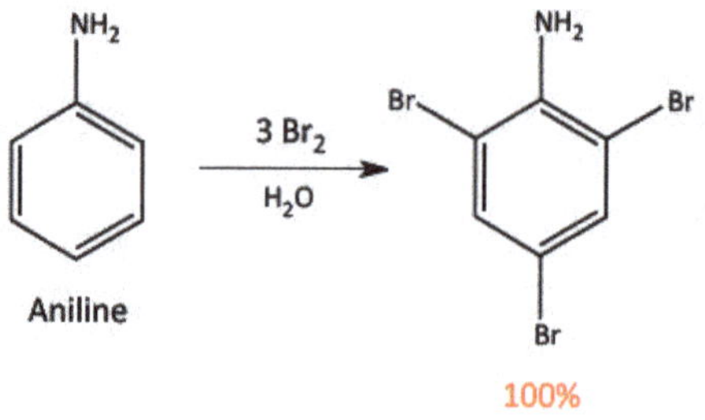

Bromination of aniline

The following chart shows the ranking of the activity of activating groups. The more electron donating the group, the more activating it is.

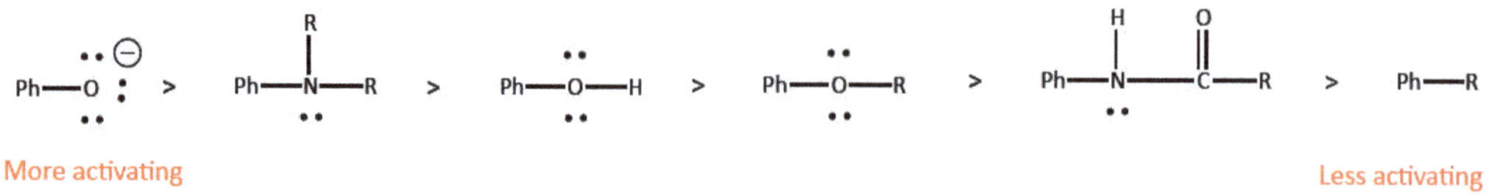

21.2 Draw the products of the following reactions.

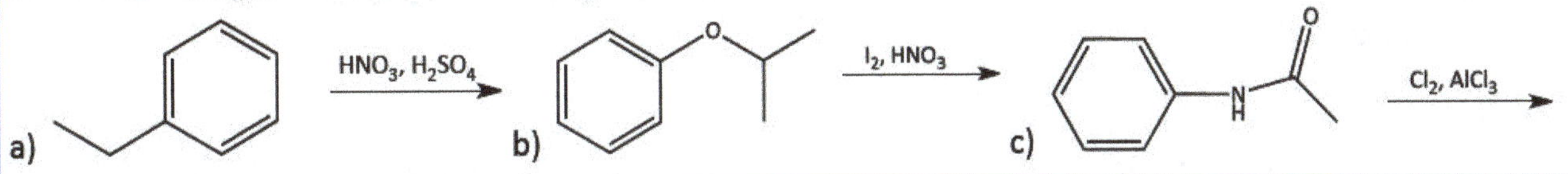

21.4 Deactivating Groups

Electron-donating groups are activating and ortho-para directing. If a group on an aromatic ring is an electron-withdrawing group, it has the opposite effect. Because electron-withdrawing groups suck electron density out of the aromatic ring, the ring is much slower to react. Therefore, an electron-withdrawing group slows down electrophilic aromatic reactions. A nitro group is an electron-withdrawing group. It is deactivating. While electron-donating activating groups make the ortho-para sites more reactive, electron-withdrawing deactivating groups make the ortho-para sites less reactive. Because the ortho-para sites are less reactive, we get more of the meta product. Deactivating groups are meta directors.

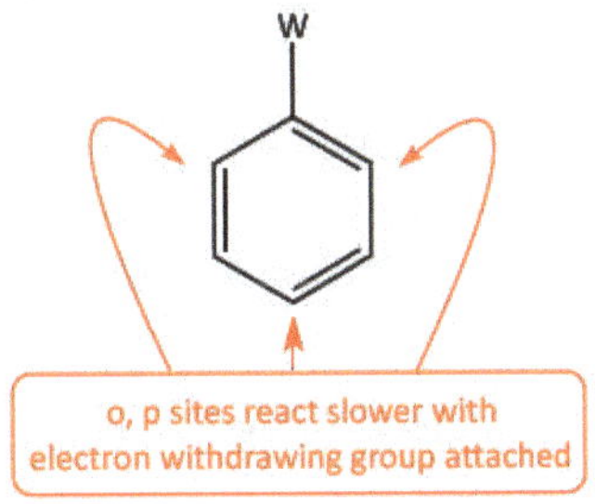

The nitration of nitrobenzene is 100,000 times slower than the nitration of benzene. It is deactivated. The nitration of nitrobenzene gives a majority of the *meta* product.

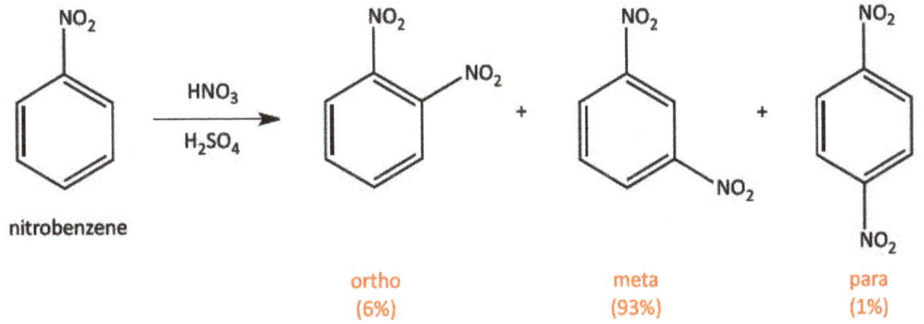

If we draw out an entire nitro group, it makes it easy to see why it is electron withdrawing. It has a positively charged nitrogen atom next to the benzene ring.

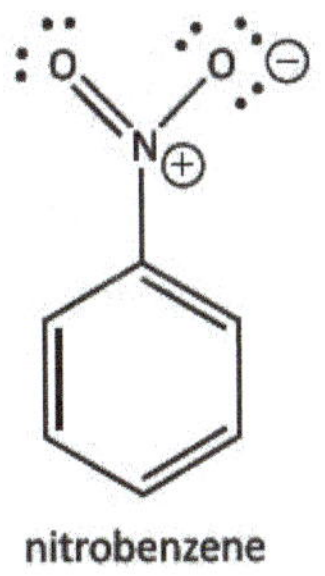

If we analyze the mechanism of electrophilic substitution on nitrobenzene, we can see why ortho and para substitution is less desirable. When ortho or para attacks happen, we get the bad situation where two positive charges next to each other.

Ortho attack

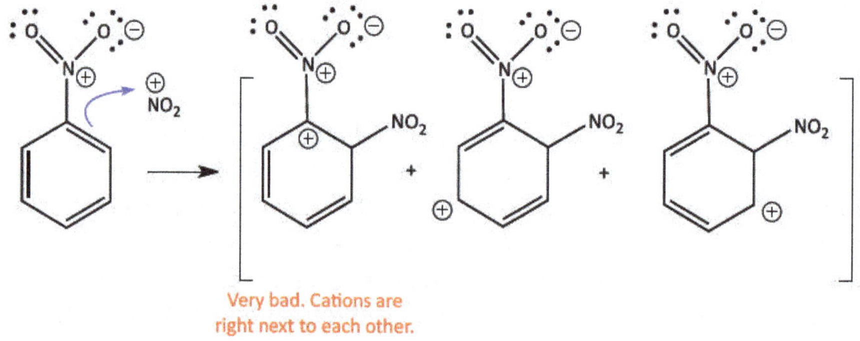

Meta attack

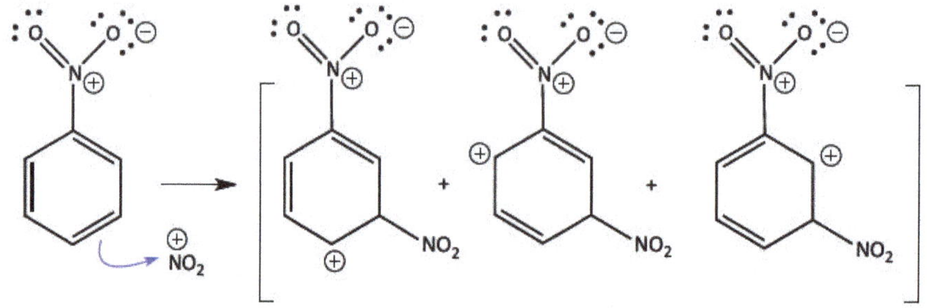

Para attack

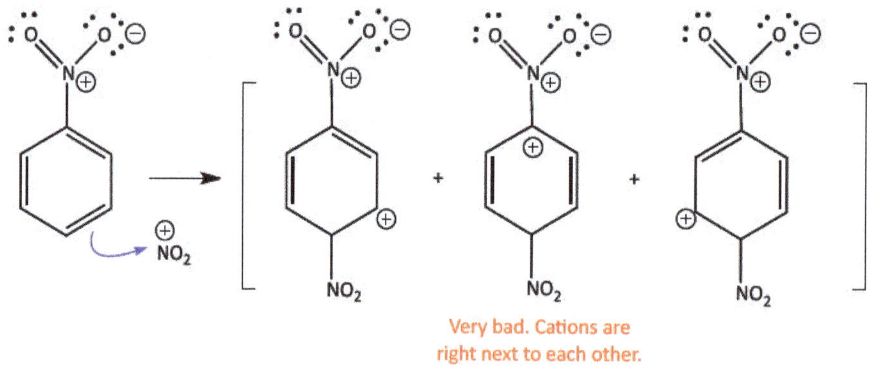

Carbonyl groups are also meta-directing deactivating groups. This is because the carbon atom of a carbonyl has a partially positive charge.

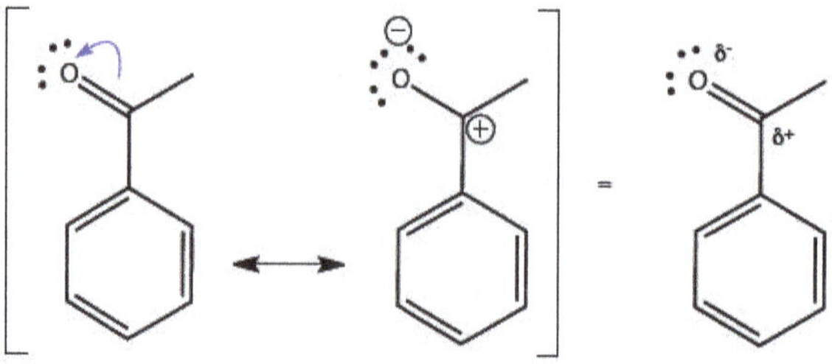

In summary, electron withdrawing, meta-directing, deactivating groups all carry at least a partial positive charge. The more positive the group is, the more deactivating it is.

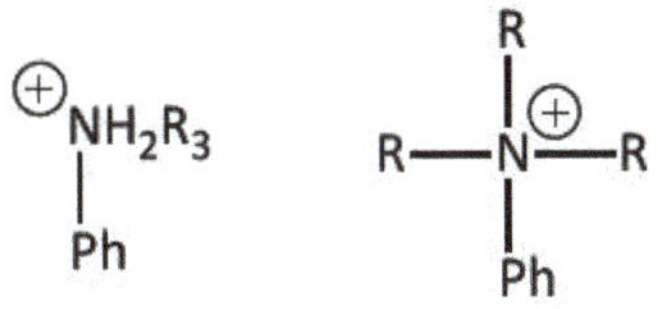

quarternary ammonium salt

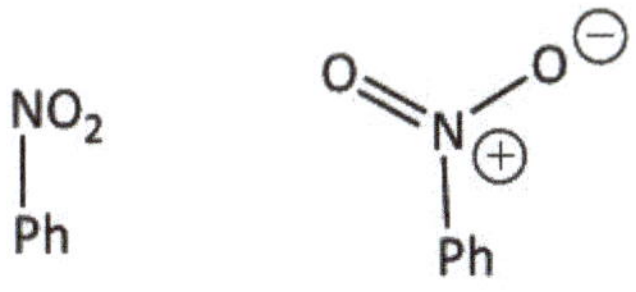

nitro

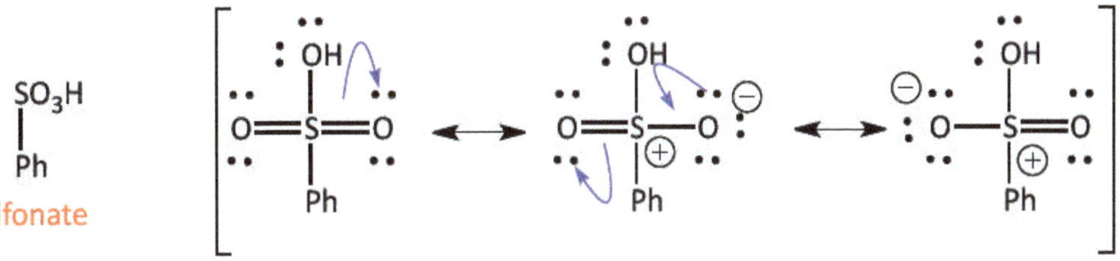

sulfonate

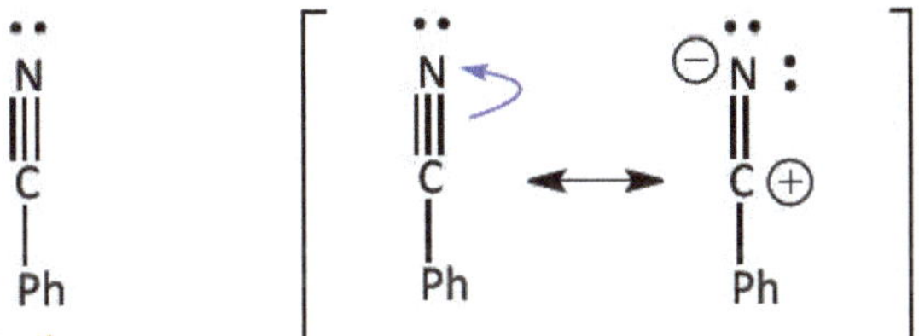

nitrile

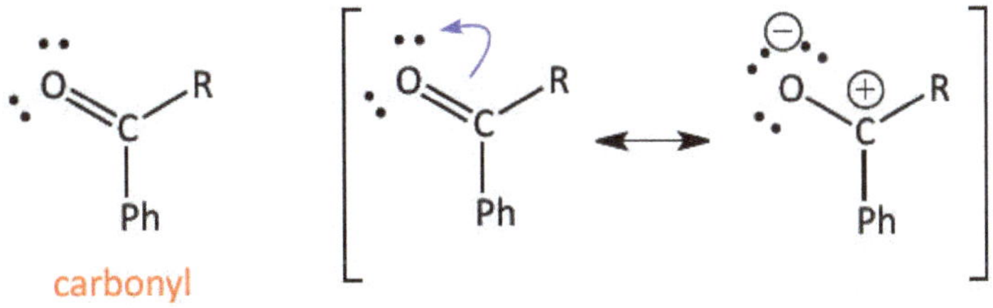

carbonyl

R = H aldehyde
R = OR ester
R = C ketone

21.5 Halide Group

Because halides are pretty electronegative, they suck electron density out of an aromatic ring. This causes the aromatic ring to react slowly in an aromatic electrophilic substitution reaction. The halide is therefore a deactivating group. But, halides are a little different than other deactivating groups. Most deactivating groups like carbonyl or nitro groups are meta-directing. Halides, because they have lone-pairs of electrons on them, can help stabilize a carbocation through resonance much like alkoxy (oxygen lone pair) or amino (nitrogen lone pair) groups did. Therefore, halides are not meta-directing, but they are ortho-para directing like electron-donating, activating groups are.

The nitration of chlorobenzene leads to mostly the ortho-para products because chloride is an o,p-director (even though the reaction is slow because chloride deactivates the ring).

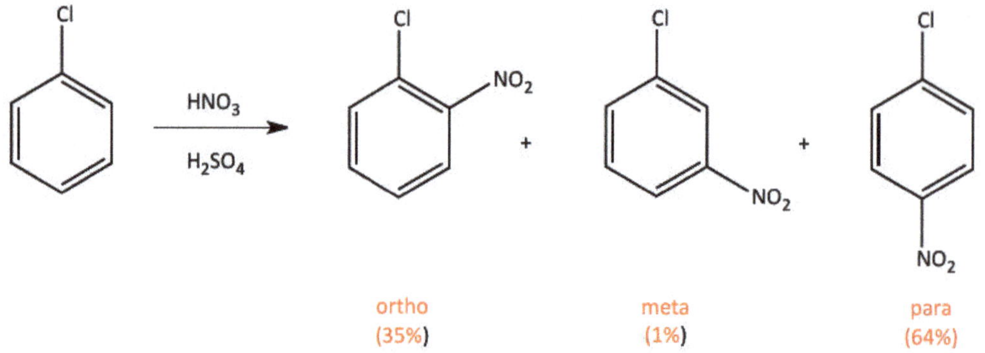

Summary

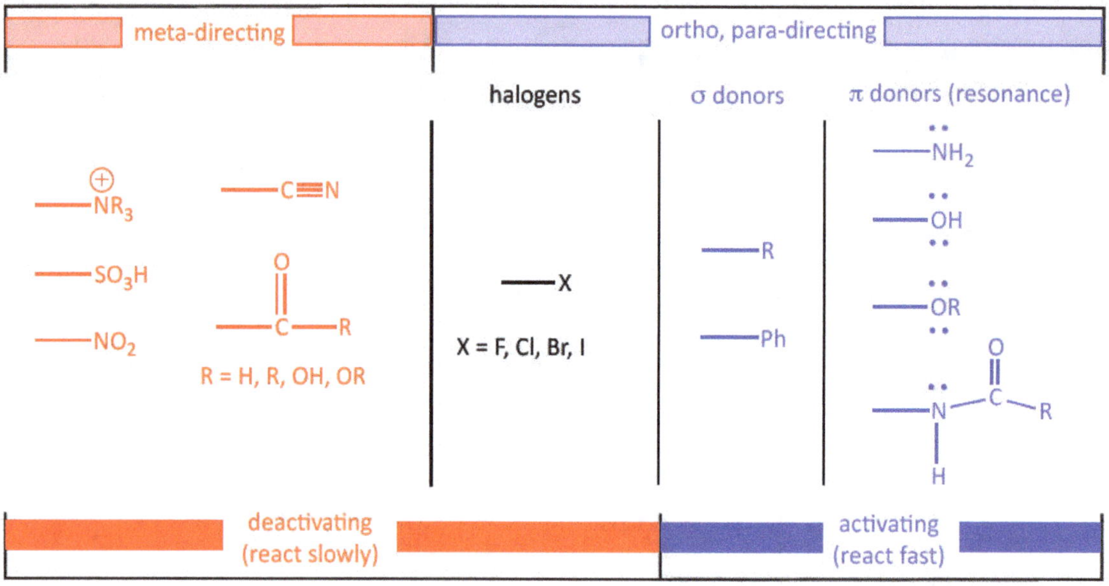

Figure 21.1 Summary of the Directing Effects of Substituents

We have seen previously that the amino group of aniline (aminobenzene) is a very activating, *ortho-para*, director. In fact, the bromination of aniline happens quite fast at all of the ortho and para sites giving 2,4,6-tribromoaniline.

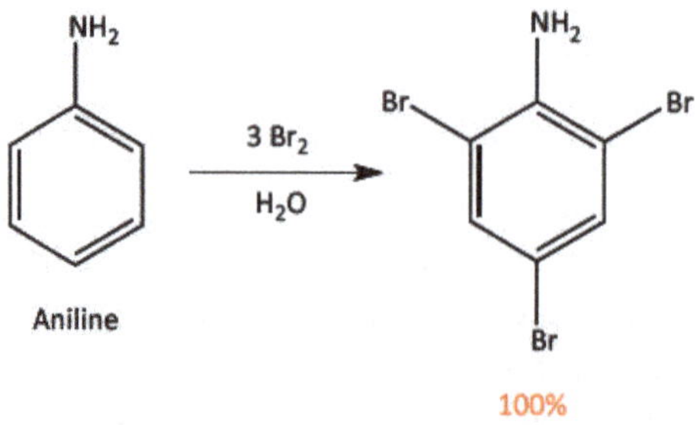

Interestingly, the nitration of aniline is quite different. When aniline reacts with nitric and sulfuric acid, the reaction is slow and gives primarily *meta*-nitroaniline. The reason this occurs is the nitrogen atom of the amino group of aniline is basic. It is protonated under the acidic conditions of the nitration to become a quaternary amino salt, $-NH_3^+$, a deactivating, *meta*-directing group.

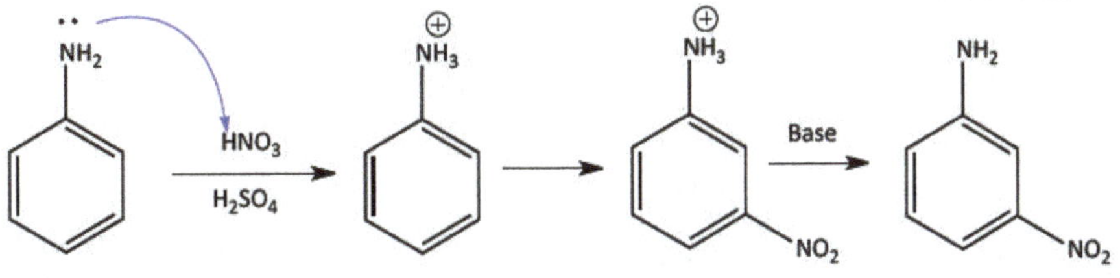

Aniline

21.3 Draw the products of the following reactions.

a) 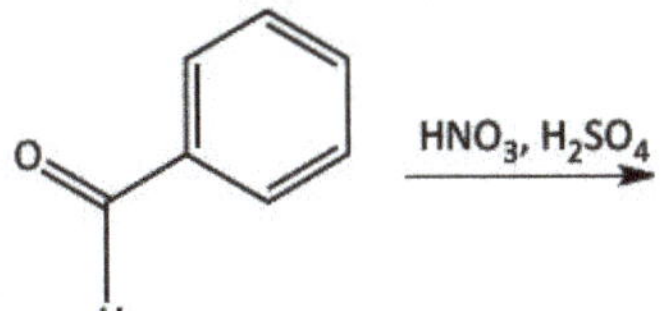

b) 

c) 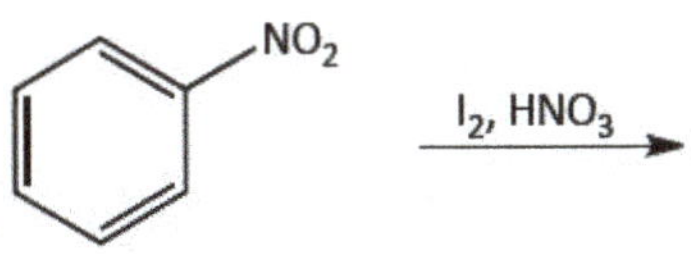

d) 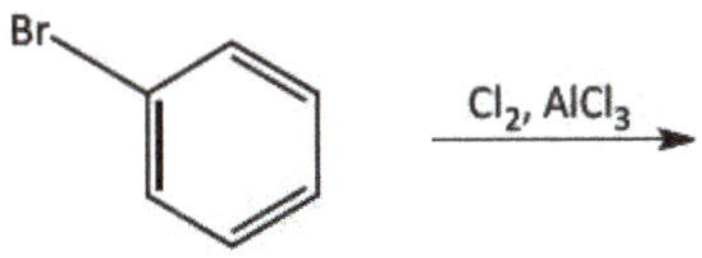

21.6 Disubstituted benzenes

A disubstituted benzene is when two groups are on a benzene. When both groups are activating, the benzene ring is activated. When both groups are deactivating, the ring is deactivated. When one group is activating and one is deactivating, it is difficult to tell.

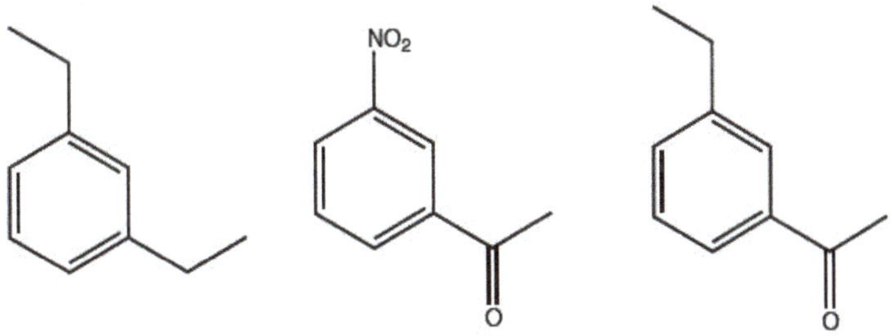

activated — deactivated — difficult to tell

When an electrophilic aromatic reaction occurs, analyze how each group directs the substitution. If the two groups agree where the substitution should occur then the substitution occurs there. If a site is sterically hindered, the substitution may not occur at that site. For *m*-ethyl methylbenzene, both alkyl groups are o, p-directing. Each alkyl group directs to the same locations. The red and blue arrows point to the same sites. The site between the two alkyl groups is a little crowded, so the substitution does not happen much there.

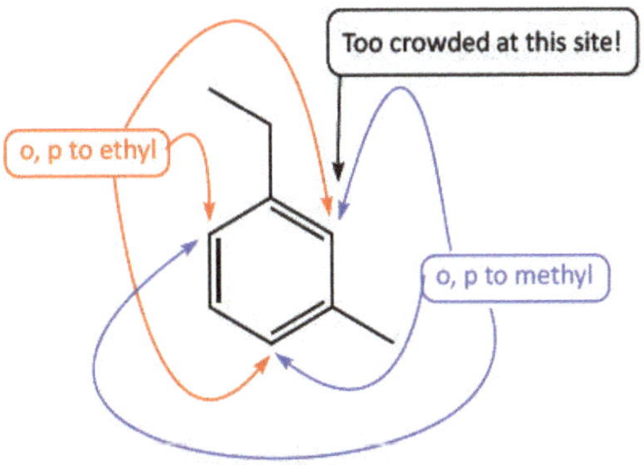

m-ethyl methyllbenzene

Aromatic Compound Reactions

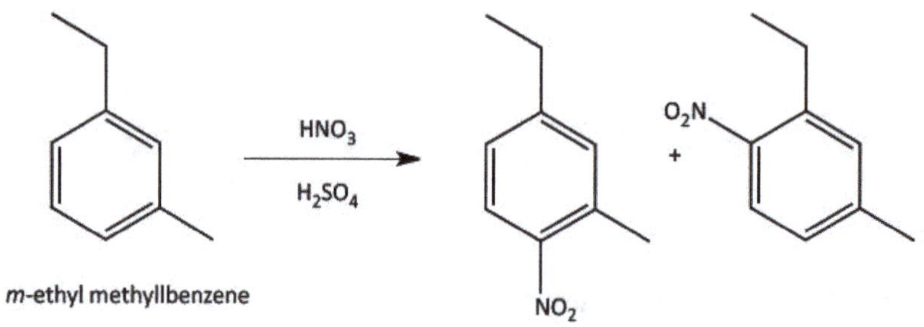

For *p*-methyl benzaldehyde, the methyl group is ortho directing. The carbonyl group is *meta* directing. These agree for the location of the substitution.

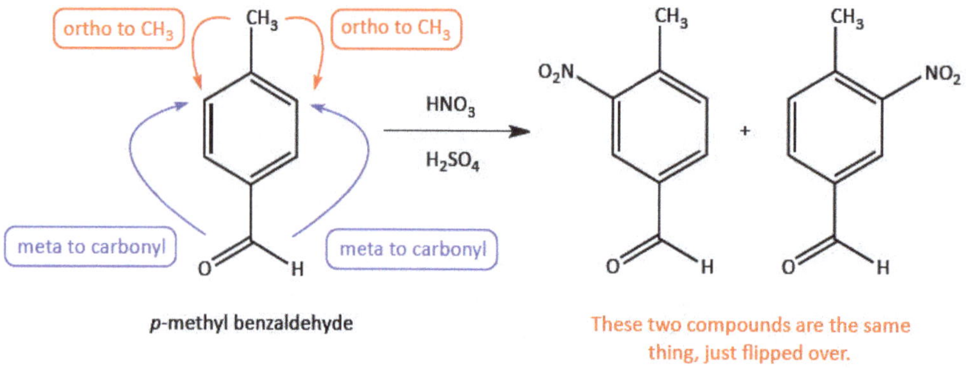

For *o*-xylene, each methyl group is o,p-directing, but they direct to different locations. A mixture of products is formed.

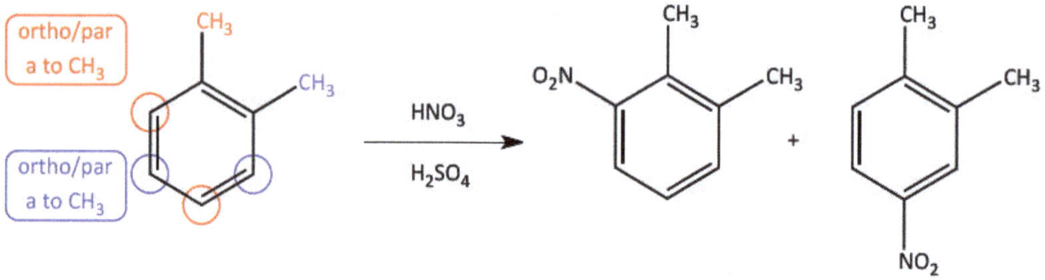

For *m*-nitroanisole, the methoxy is o,p-directing but the nitro is meta directing. These point to different locations. When there is a disagreement in locations, the group that is more activating wins because the reactions happen faster at those sites. In this case, methoxy (OCH_3) wins. Again, the site in between the two groups is pretty crowded, so the reaction does not happen there as much.

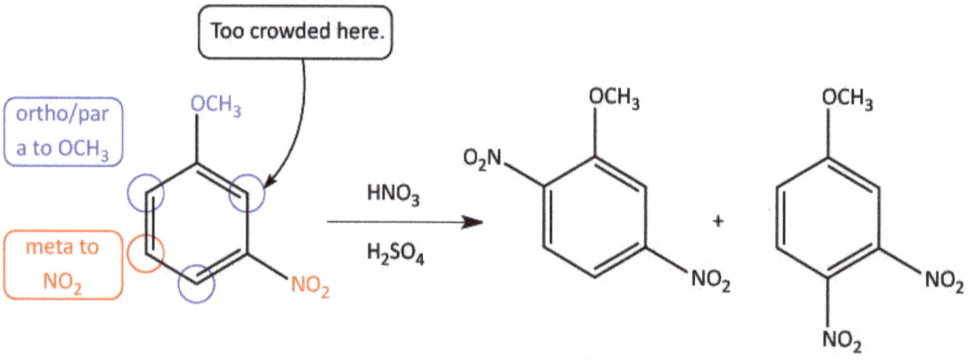

The strength of directing groups was summarized in the previous section. Pi donors (-OH, -OR, -NR_2) are more activating than sigma donors (alkyl groups). Halides are a little deactivating, but are not as deactivating as fully *meta*-directing deactivating groups like –NO_2, -CN, and carbonyls). Go back and study the directing group summary figure in the previous section.

21.4 Draw the products of the following reactions.

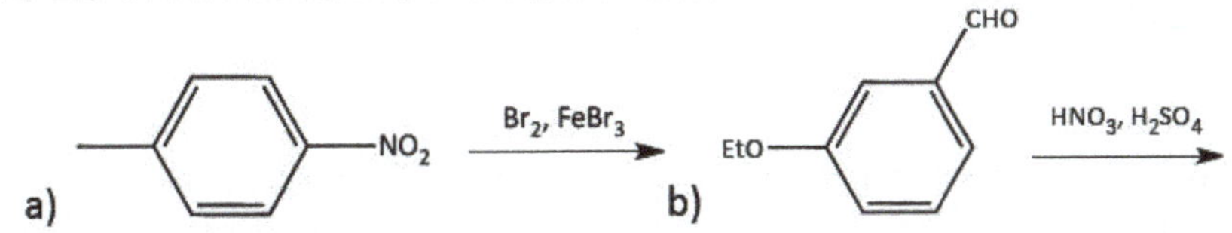

21.7 Friedel-Crafts

21.7.1 Friedel-Crafts alkylation

A Friedel-Crafts alkylation is a reaction where an alkyl group is added to an aromatic ring. In a Friedel-Crafts alkylation, an alkyl halide reacts with a Lewis acid like $AlCl_3$ or $FeBr_3$.

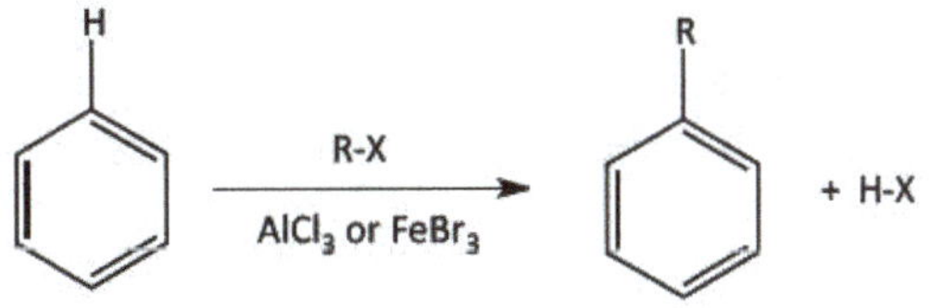

Below is an example Friedel-Crafts reaction.

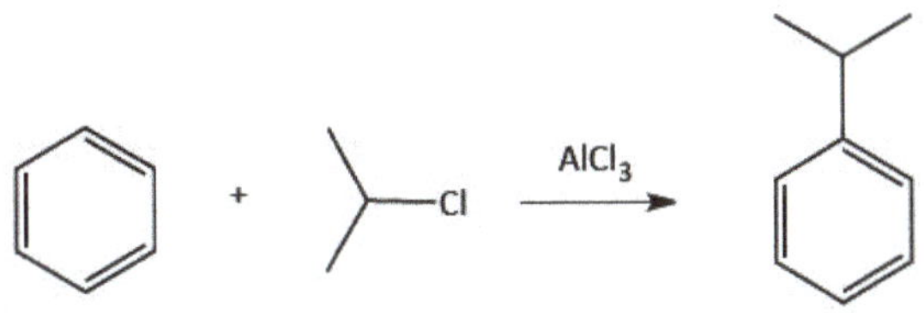

In the mechanism of the Friedel-Crafts reaction, the alkyl halide reacts with the Lewis acid to form a carbocation on the alkyl group. This positively charged alkyl group is a strong electrophile. Since a carbocation is formed, the Friedel-Crafts reaction works best on secondary or tertiary alkyl groups since they make the most stable carbocations.

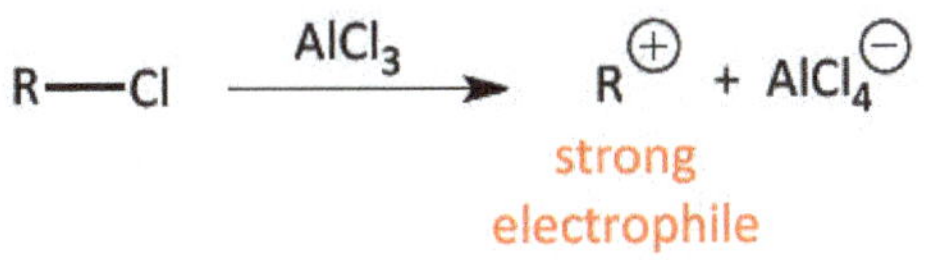

In the following example of a Friedel-Crafts reaction, the chloride of *tert*-butyl chloride is removed by the Lewis acid aluminum trichloride, $AlCl_3$, forming a stable tertiary carbocation.

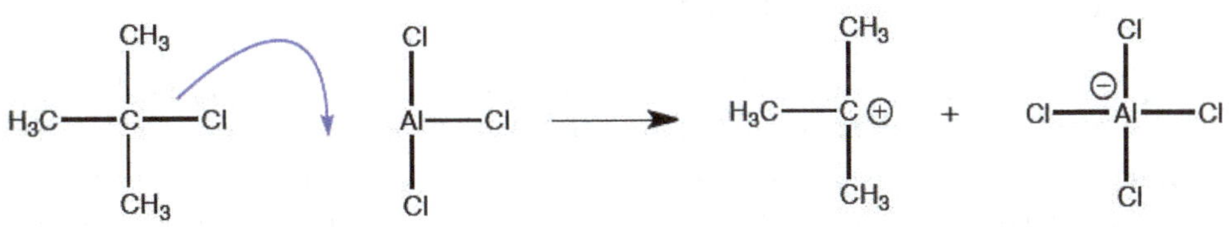

A pi bond of the benzene ring attacks the electrophilic carbocation forming a sigma complex.

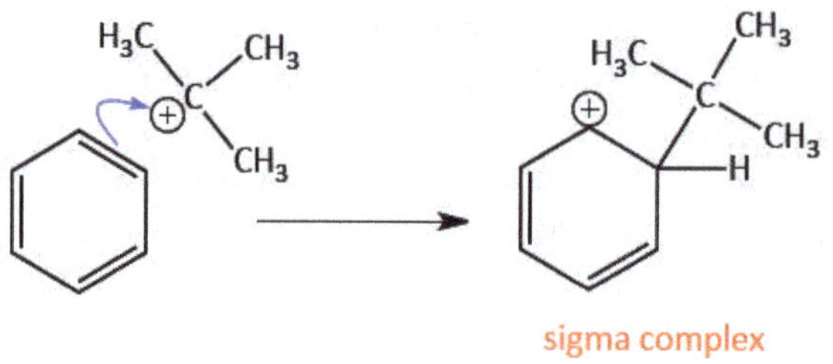

The sigma complex is deprotonated. The electrons return to reform a pi bond reforming the aromatic benzene ring. The final product is an alkyl benzene.

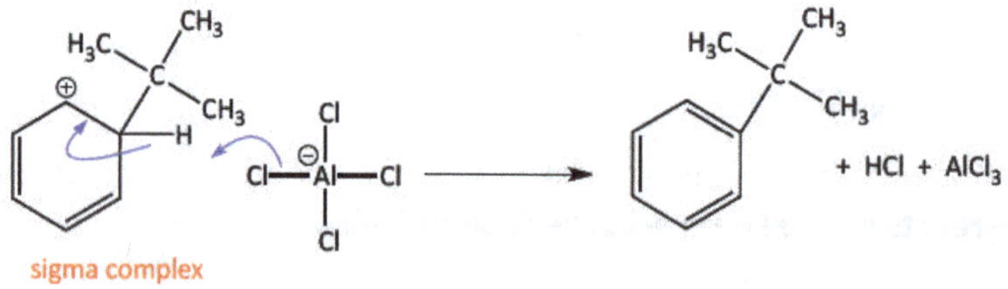

Even though secondary and tertiary carbocations are more stable, it is possible for a primary alkyl halide to do a Friedel-Crafts reaction even though primary carbocations do not want to form!

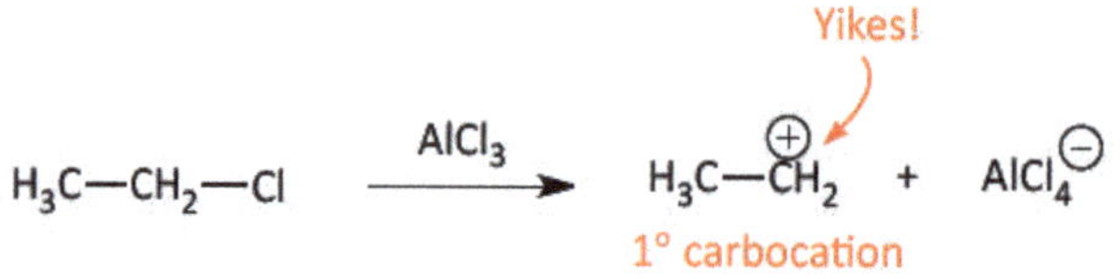

If a primary carbocation is going to be formed, the attack by the benzene happens at the same time as the chloride leaves.

In the following reaction, a primary ethyl group is added to benzene by the reaction of ethyl chloride with $AlCl_3$.

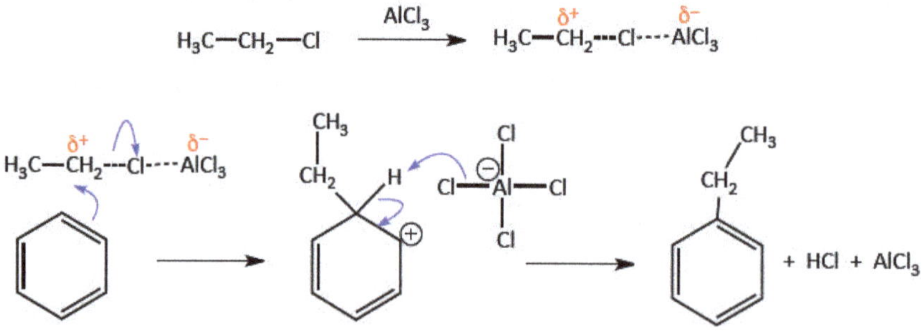

Limitations of Friedel-Crafts alkylation

There are three main limitations to the Friedel-Crafts alkylation. 1. Deactivated rings do not work. 2. Carbocations can rearrange. 3. Polyalkylations are possible.

1. *Deactivated rings do not work*. In order for a benzene ring to do a Friedel-Crafts alkylation, the ring needs a sufficient amount of electron density in it. Friedel-Crafts alkylations do not work on benzene rings that are more deactivated than halobenzenes. Benzene rings with nitro groups, cyano groups, or carbonyl groups attached will not work. For instance, nitrobenzene will not undergo a Friedel-Crafts alkylation.

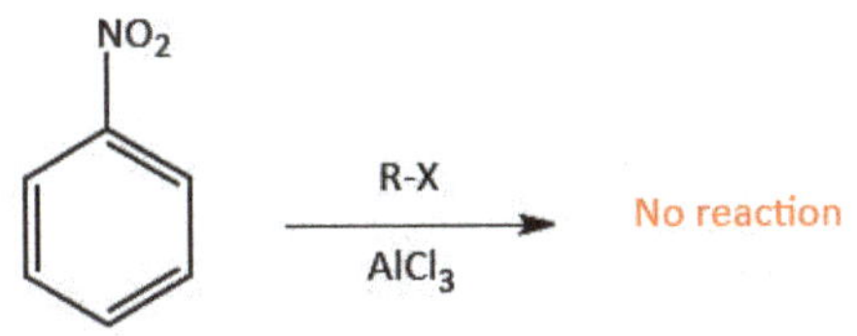

This is something to consider in some syntheses. For instance, let's say we wanted to make *p*-ethyl nitrobenzene from benzene. Two groups need added to the benzene, an ethyl and a nitro group.

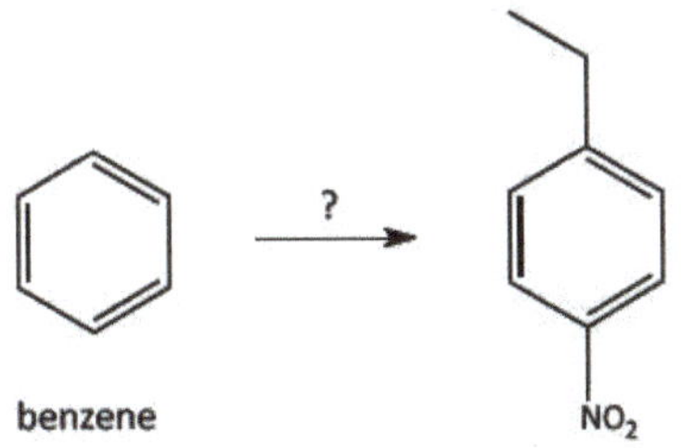

benzene

p-ethyl nitrobenzene

The order in which we add these groups is important. In reaction 1, we put the nitro group on first. This deactivates the ring and the subsequent Friedel-Crafts alkylation to add the ethyl group will not work. In reaction 2, we put the ethyl group on first with a Friedel-Crafts alkylation. The ethyl group is activating and ortho/para directing. The subsequent nitration easily works giving a mixture of the ortho and para products.

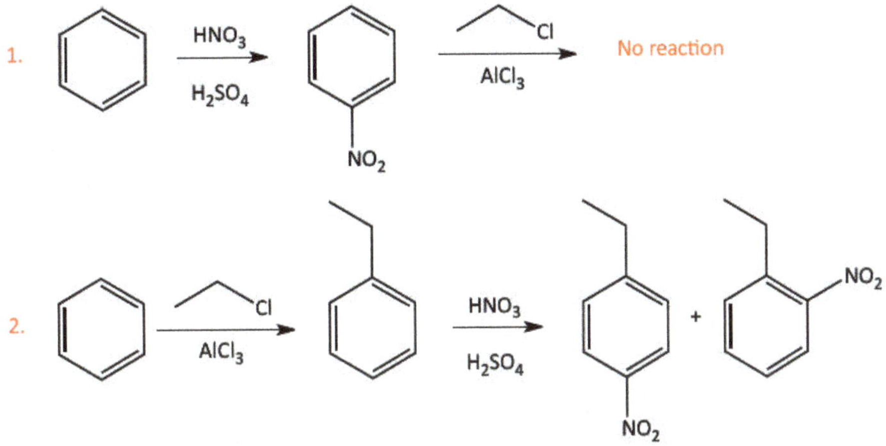

2. Carbocations can rearrange. In a Friedel-Crafts alkylation, the halogen of an alkyl halide is removed with $AlCl_3$ for form a carbocation. Any time a carbocation is formed, we need to watch out for rearrangements. For instance, in the following example, if we wanted to make propyl benzene, we might try to react benzene with propyl chloride and $AlCl_3$.

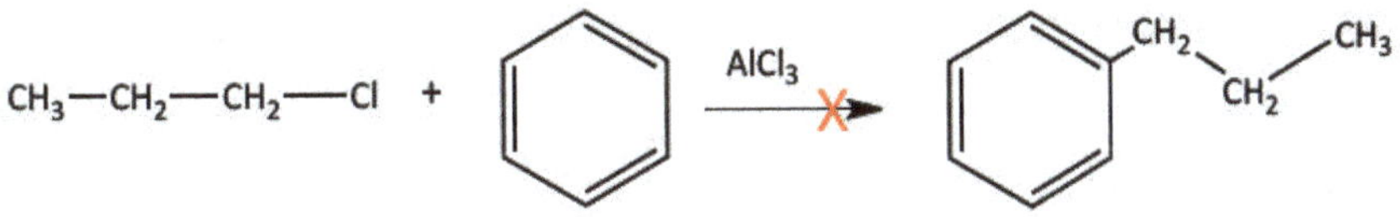

propyl benzene

This reaction does not work as intended. We find that propyl benzene is not made, but instead, isopropyl benzene is formed.

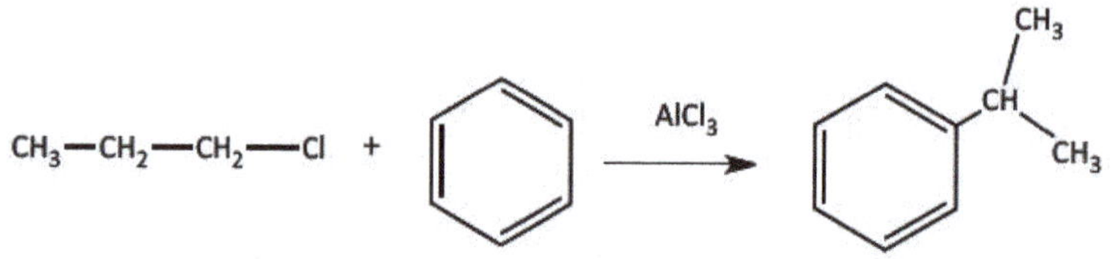

The reason isopropyl benzene is formed is the primary propyl cation rearranges. It does a 1,2-hydride shift to form a more stable secondary carbocation.

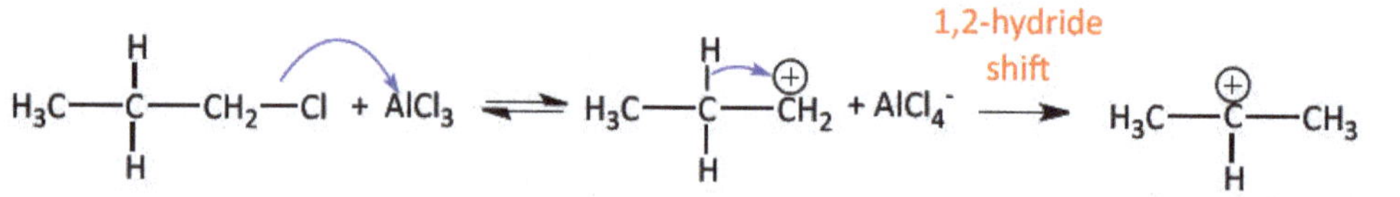

This secondary carbocation does the Friedel-Crafts alkylation with the benzene to form isopropyl benzene.

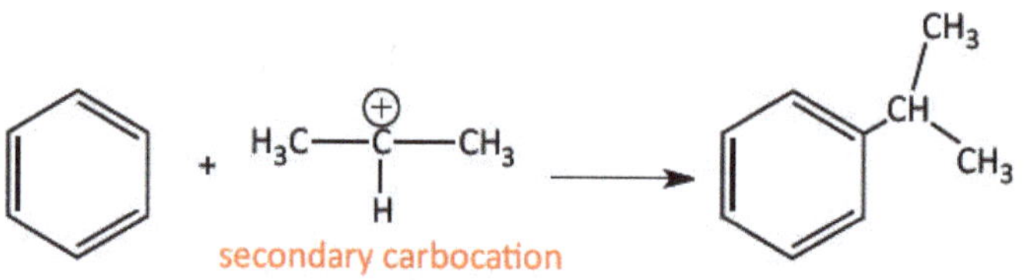

3. Polyalkylations are possible.

Because alkyl groups are activating groups, once one alkyl group is added to a benzene ring, it makes the benzene ring even more activated. The ring becomes even more reactive, so the ring can be alkylated again. Therefore, the ring can become di- and tri-alkylated.

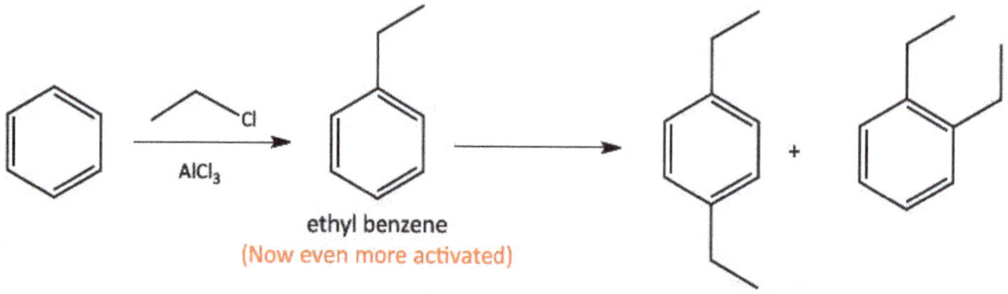

> 21.5 List the three limitations usually associated with the Friedel-Crafts alkylation reaction.

21.7.2 Friedel-Crafts acylation

An acyl group is a carbonyl ketone or aldehyde group.

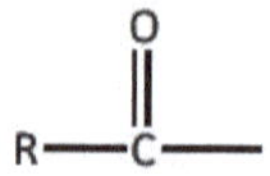

acyl group

An acyl group can be added to a benzene ring through a Friedel-Crafts acylation reaction.

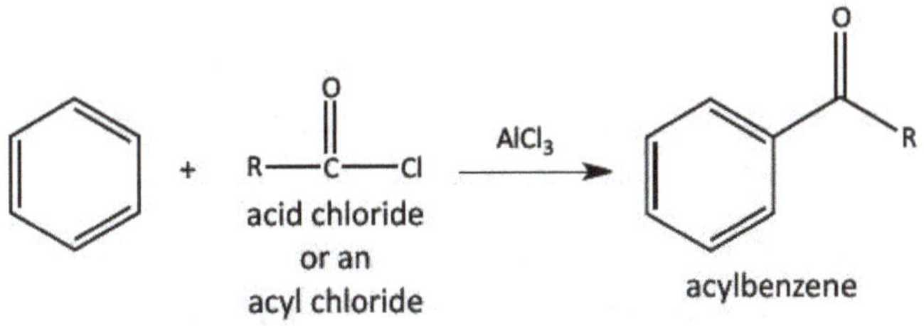

This is very similar to a Friedel-Crafts alkylation. A chloride is removed from an acid chloride with $AlCl_3$ to make a carbocation acylium ion.

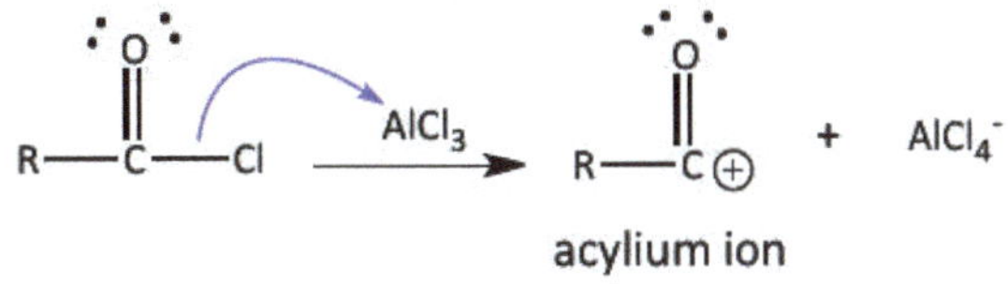

The acylium ion is a stable carbocation because it has a good resonance form. It is so stable; it does not rearrange.

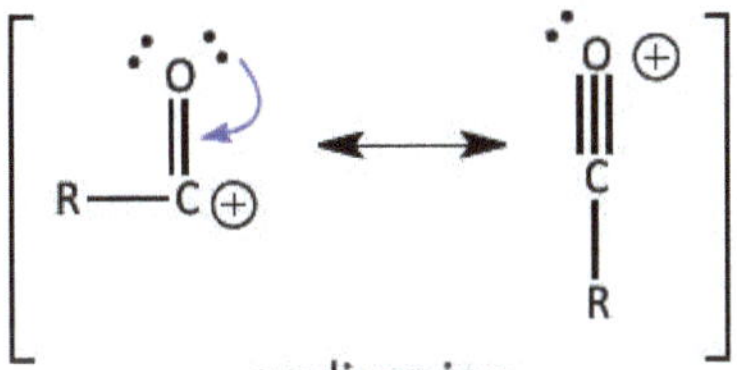

The acylium ion reacts with the benzene ring through an electrophilic aromatic substitution to make an acylbenzene.

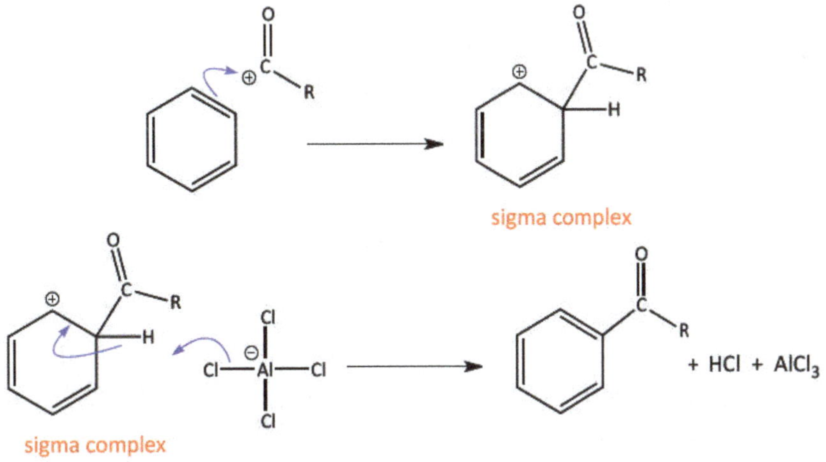

The acid chlorides used in Friedel-Crafts acylations can be synthesized from the appropriate carboxylic acid and thionyl chloride ($SOCl_2$).

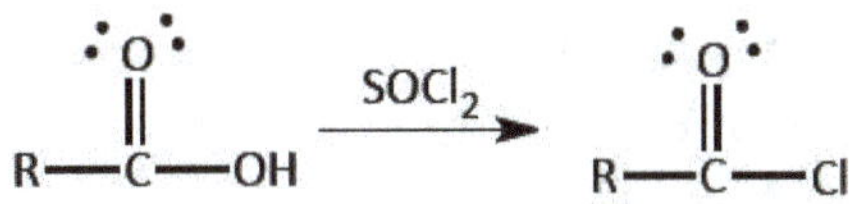

Let's consider the three limitations of Friedel-Crafts alkylations and see how Friedel-Crafts acylations compare.

1. *Deactivated rings do not work.* Once again, if the ring is deactivated more than halobenzenes, the Friedel-Crafts reaction does not work. This is the same for alkylations and acylations. Therefore, Friedel-Crafts reactions do not work on nitro, cyano, or acylbenzenes.

2. *Friedel-Crafts acylation carbocations do not rearrange.* In the Friedel-Crafts acylation, a resonance stabilized acylium ion is formed. It is so stable; rearrangements do not occur like they can in an alkylation.

3. *Friedel-Crafts acylations do not polyacylate.* When benzene is alkylated using a Friedel-Crafts alkylation, the alkyl benzene product that is made is more activated than the original benzene. This is because alkyl groups are activating groups. In a Friedel-Crafts acylation, the acylbenzene product formed is deactivated since carbonyl acyl groups are deactivating groups. Therefore, only one acyl group attached to the benzene ring. They do not polyacylate.

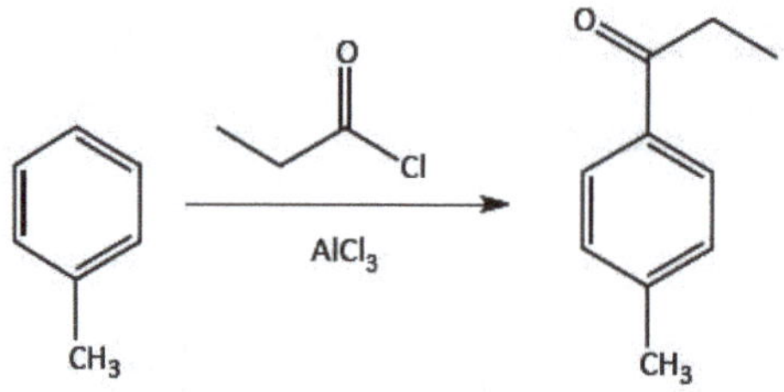

Figure 21.2 Example of Friedel-Crafts acylation

> 21.6 Why are Friedel-Crafts acylation reactions not prone to polyacylation?

21.7.3 Gatterman-Koch formylation

Consider if we wanted to make benzaldehyde from benzene using a Friedel-Crafts acylation. We would want to use a reagent called formyl chloride. A formyl group is an aldehyde group.

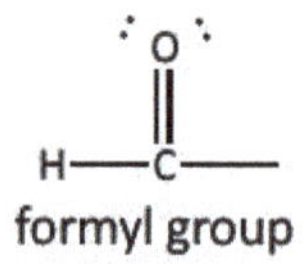

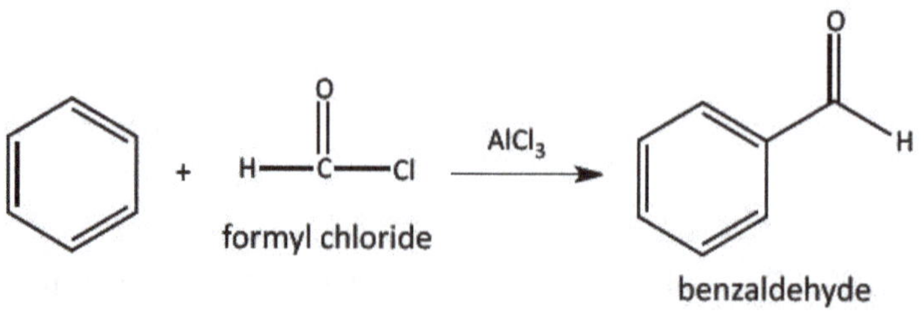

The problem with making benzaldehyde in this way is formyl chloride is unstable. We cannot buy a bottle of formyl chloride. Therefore, we must make formyl chloride when we need it. We make formyl chloride using carbon monoxide (CO) and hydrochloric acid (HCl). With the addition of the Lewis acid, $AlCl_3$, a formyl cation is formed.

Upon the reaction with benzene, benzaldehyde is formed. This reaction of making aromatic aldehydes is called a Gatterman-Koch formylation.

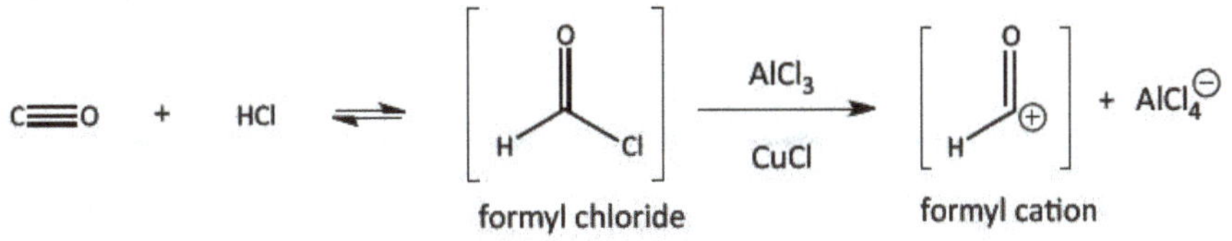

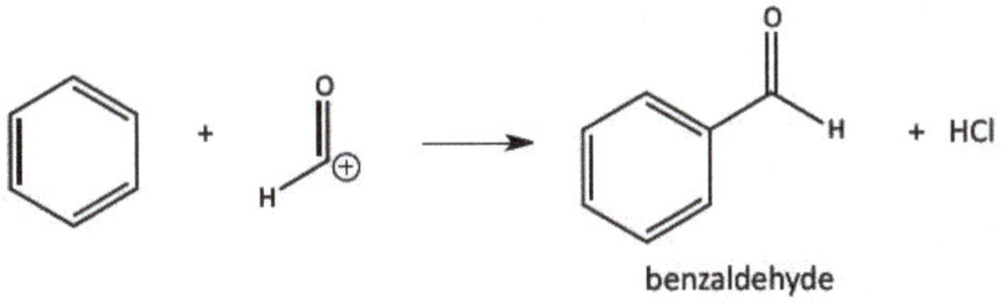

21.7 Draw the products of the following reactions.

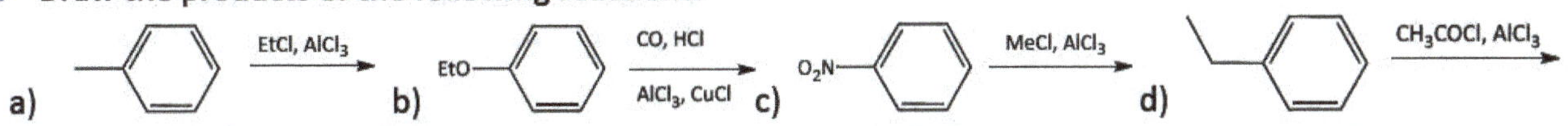

21.8 Clemmensen Reduction

Acylbenzenes can be reduced to alkyl benzenes using the Clemmensen Reduction. This is a reduction because a carbonyl with two carbon-oxygen bonds is reduced to an alkyl group with zero carbon-oxygen bonds.

Clemmensen Reduction reagents are a zinc/mercury amalgam in the presence of aqueous hydrochloric acid.

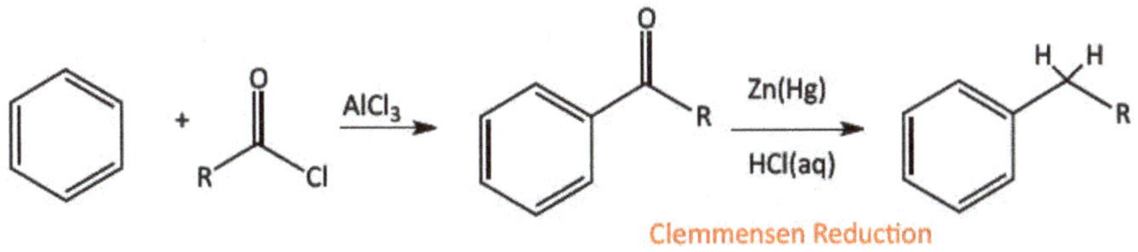

Think back to the difficulty we had making propyl benzene in a Friedel-Crafts alkylation. Instead of propyl benzene, we made isopropyl benzene because the alkyl carbocation rearranged. A combination of Friedel-Crafts acylation followed by Clemmensen Reduction allows us to make propylbenzene.

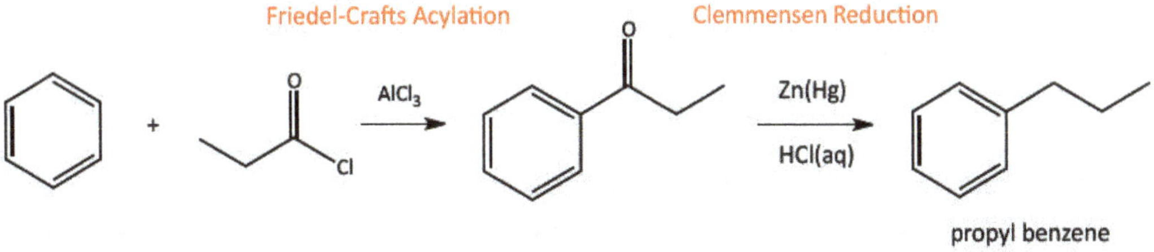

The reason this combination works is the acylium ion formed in the Friedel-Crafts acylation does not rearrange. The reduction of the acyl group gives us the alkyl group. This is the primary way we can make alkylbenzenes with primary alkyl groups.

21.8 A chemist attempted to make propyl benzene by performing a Friedel-Crafts alkylation on benzene using propyl chloride and $AlCl_3$. Unfortunately for the chemist, a rearrangement occurred and isopropyl benzene was made.

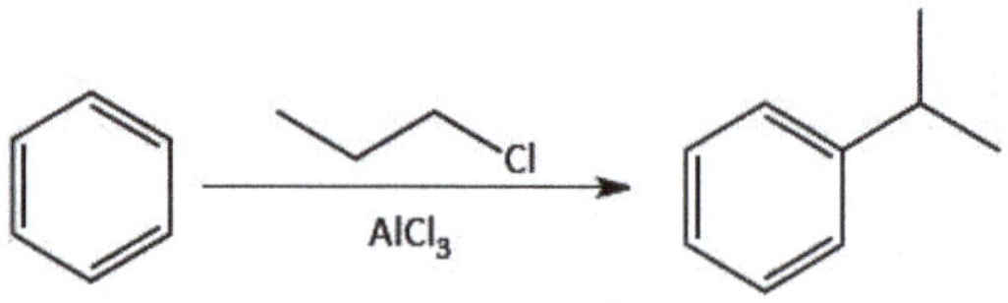

The chemist then decided to make propyl benzene from a Friedel-Crafts acylation followed by a Clemmensen Reduction. Show this synthesis and explain why this works better than the Friedel-Crafts alkylation did.

21.9 Sandmeyer reaction

21.9.1 Aniline synthesis to set up the Sandmeyer reaction

We have already learned how to nitrate an aromatic ring with nitric and sulfuric acid. A reduction of a nitro group on an aromatic ring to an amino group with fewer nitrogen-oxygen bonds is accomplished by reducing it with iron filings and hydrochloric acid followed by hydroxide. The meta-directing nitro group is now an o,p-directing amino group. Remember that aminobenzene is commonly called aniline.

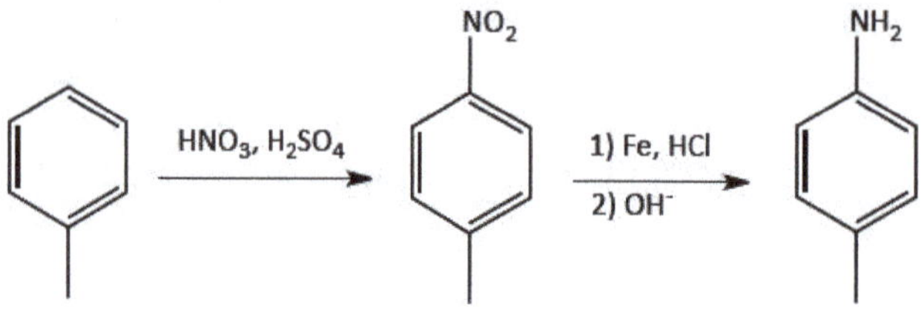

21.9 Show the two reaction steps needed to turn propyl benzene into 4-aminopropylbenzene.

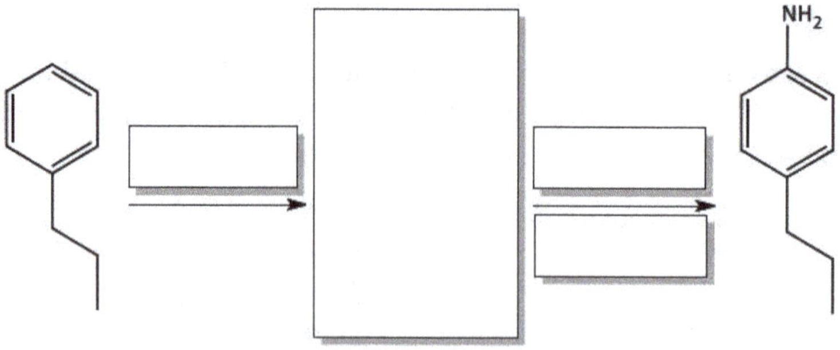

Amino groups on an aromatic ring can be turned into the world's greatest leaving group, N_2, by reacting it with sodium nitrite ($NaNO_2$) and HCl. Once this happens, the leaving group is so good, it is only stable below 10°C. It can be replaced with many different types of nucleophiles.

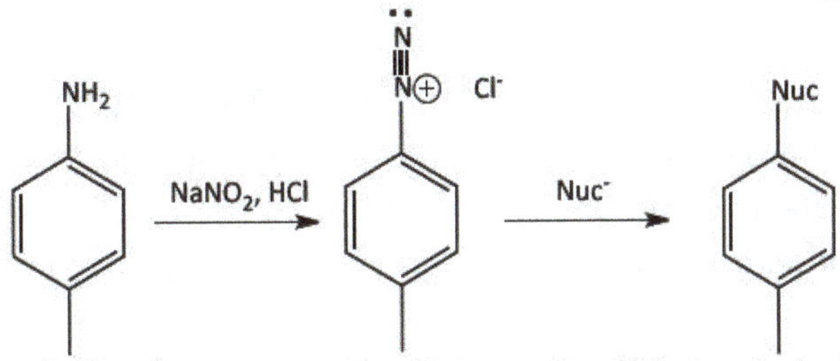

The substitution reaction is called a Sandmeyer reaction if the nucleophile is a Cu^{+} salt like cuprous cyanide (CuCN), cuprous chloride (CuCl), or cuprous bromide (CuBr).

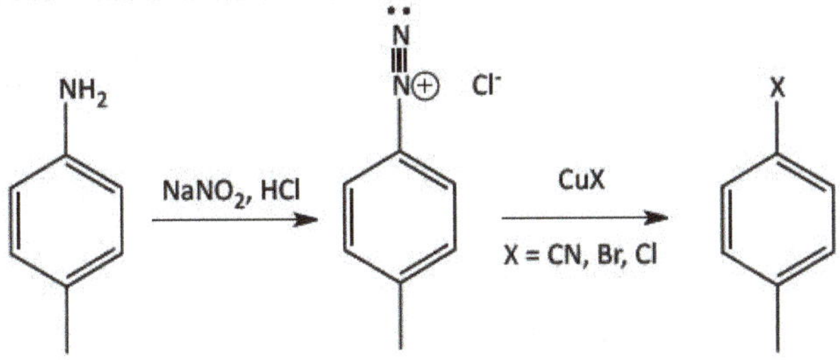

Figure 21.3 Sandmeyer Reaction

Many different nucleophilic substitutions can occur. With acid and water, we get a phenol. With HBF_4, we get an aromatic fluoride. With potassium iodide (KI), we get an aromatic iodide. With hypophosphorous acid (H_3PO_2), the N_2 is replaced with a proton to reform the benzene ring.

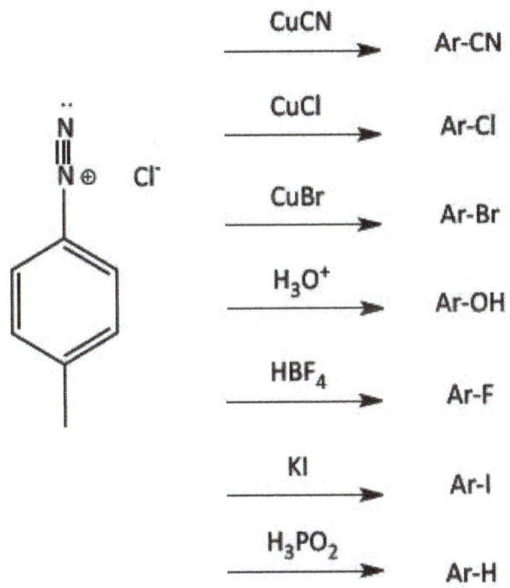

21.10 Write in the reagents to complete the following Sandmeyer and Sandmeyer-like reactions.

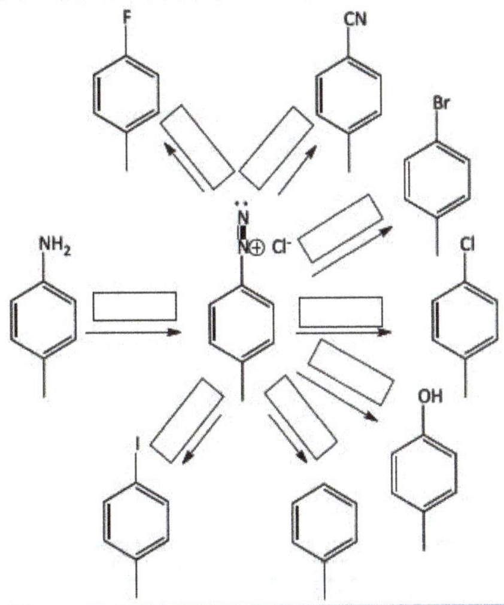

21.11 Draw in all reagents and products in the following reaction sequence. Pay careful attention if groups are ortho-para directing or meta-directing. Notice how groups sometimes direct to one location and then are changed into a different type of directing group.

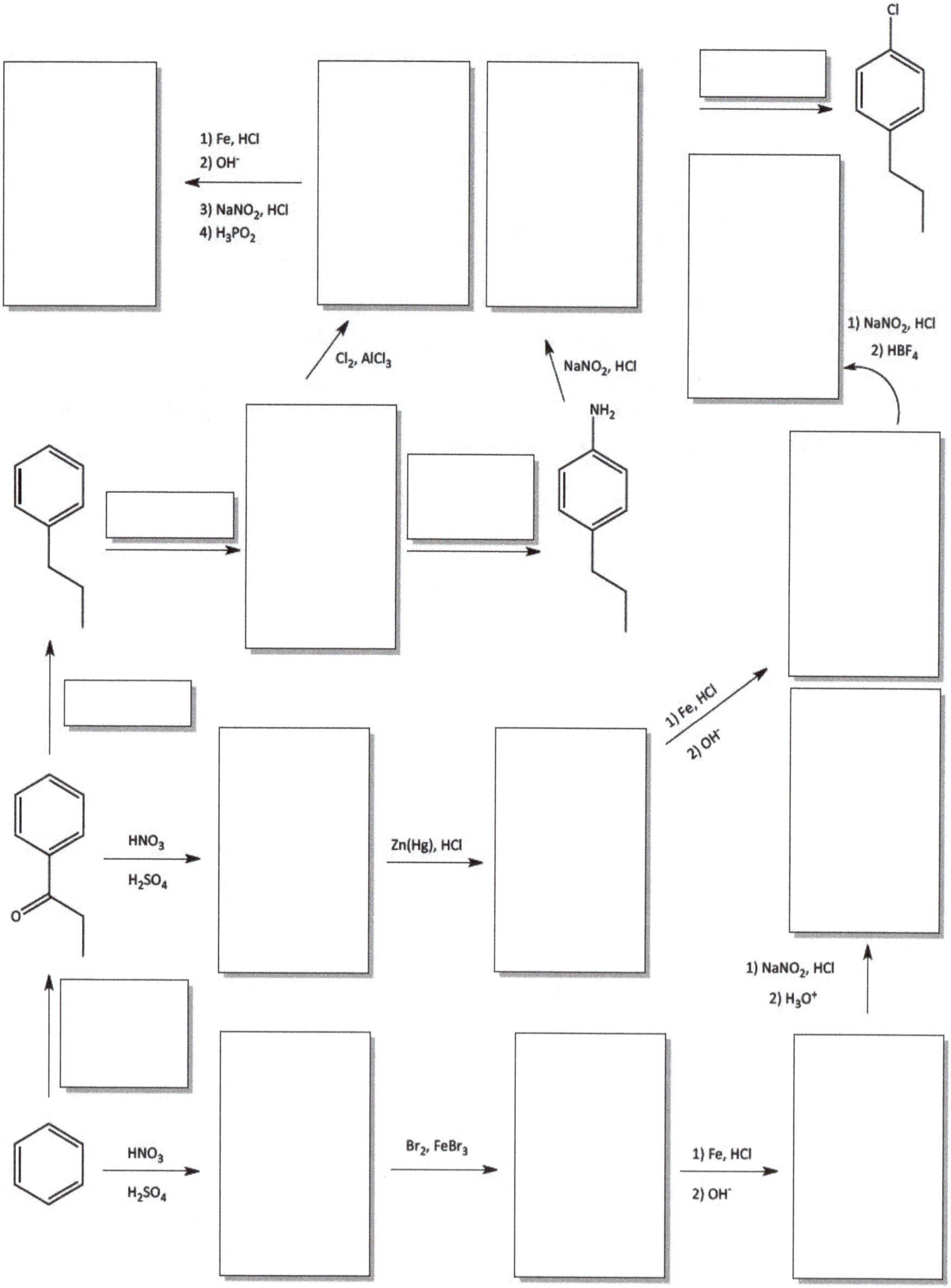

21.10 Friedel-Crafts on aniline

Suppose you wanted to make *o*- or *p*-methylaniline from aniline using a Friedel-Crafts alkylation. You might think you could react methyl chloride (CH_3Cl) with aluminum trichloride ($AlCl_3$) with aniline. You would find that this reaction does not work.

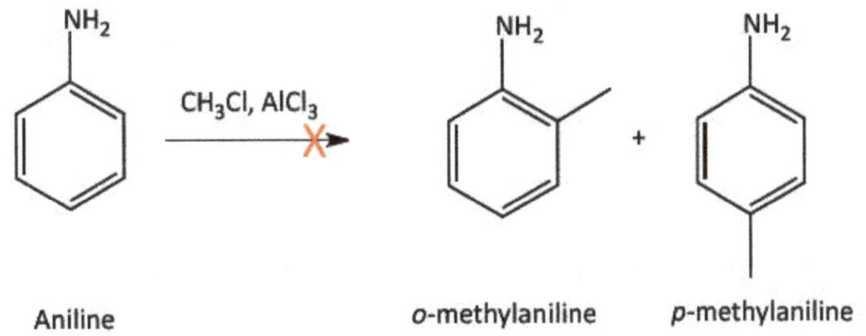

The reason it does not work is that $AlCl_3$ is a good Lewis acid and wants electrons. The amino NH_2- group of aniline has a negative, lone pair of electrons on it that can attack the $AlCl_3$ Lewis acid. This makes a positively charged quaternary ammonium salt that is a deactivating, meta-directing group. The reaction would, therefore, not proceed at all.

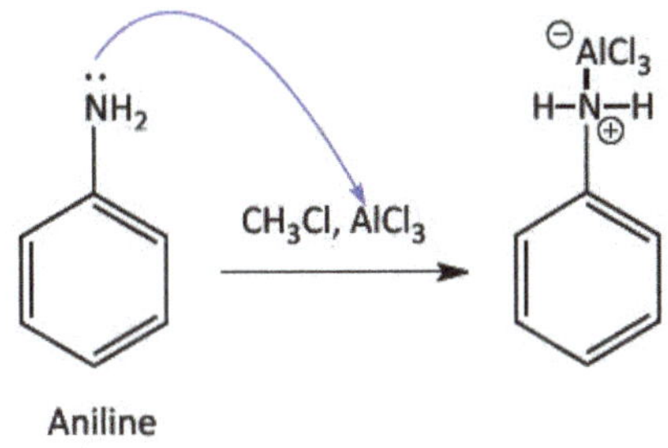

Like many times before in human history, when we hit a roadblock, we have figured out a way around it. In order to do a Friedel-Crafts alkylation on aniline to make *o*- or *p*-methylaniline, we need to turn the amino NH_2 group into something that is still electron donating and *o,p*-directing, but does not have such a reactive negative lone pair of electrons on it. The way we do that is we can turn the amino group into an amido group. To do this, we add an acid chloride like acetyl chloride to the aniline making N-phenylacetamide. The amido group has a lone pair of electrons on the nitrogen atom, but they are not as concentrated on the nitrogen atom as they were in aniline. The lone pair is spread out towards the oxygen atom of the amide through resonance. In fact, the electrons are now spread out enough that they will not react with the $AlCl_3$ Lewis acid. There is a lone pair of electrons on the nitrogen atom, so the amido group is still an *o,p*-director.

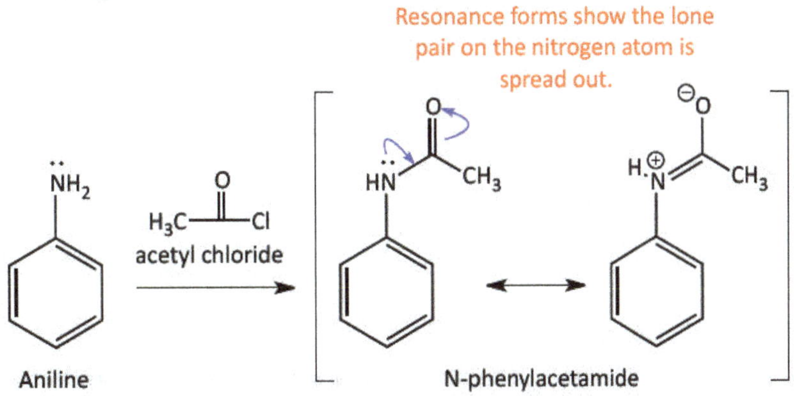

This amido compound can now undergo a Friedel-Crafts reaction.

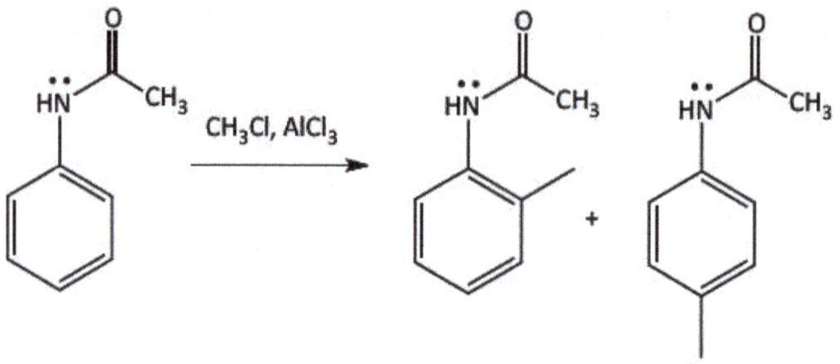

The amido groups can then be transformed back into amino groups through hydrolysis, or reacting with acid and water.

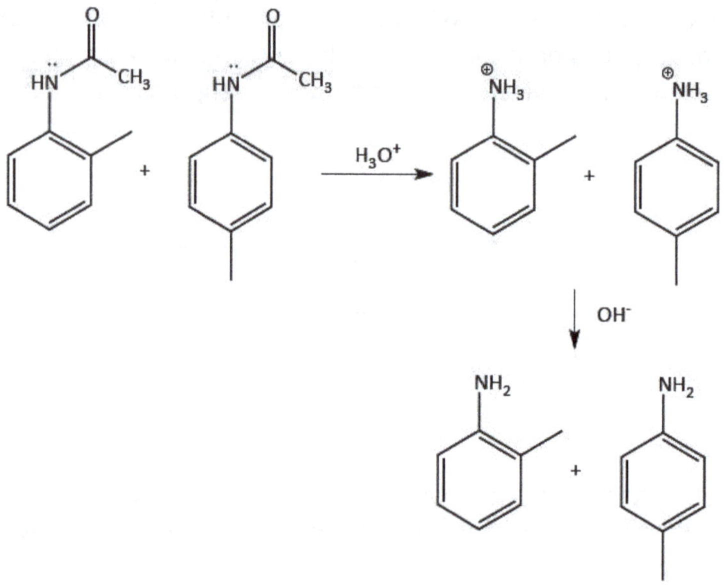

In this roundabout way, the Friedel-Crafts reaction of an amino benzene can occur. The amino group is transformed into an amido group, the Friedel-Crafts reaction occurs, then the amido group is hydrolyzed to an amino group once again.

21.12 a) Why is the following Friedel-Crafts reaction NOT a good way to make isopropyl aniline from aniline?

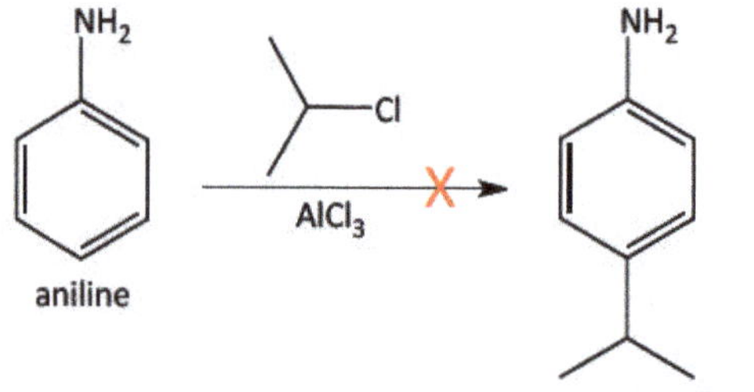

b) What is a better synthesis of isopropyl aniline?

21.11 Nucleophilic Aromatic Substitution

We are going to switch gears a bit. We have been looking at many electrophilic aromatic substitutions where the pi bonds of a benzene ring attack a positively charged species. Now, we are going to look at nucleophilic aromatic substitutions, where a negatively charged species attacks our benzene ring. We have seen nucleophilic substitutions before. For instance, the S_N2 mechanism is a nucleophilic substitution. We need a good attacking nucleophile with a large amount of electron density that attacks a substrate with a good leaving group.

This type of reaction normally does not work if the substrate is a halobenzene.

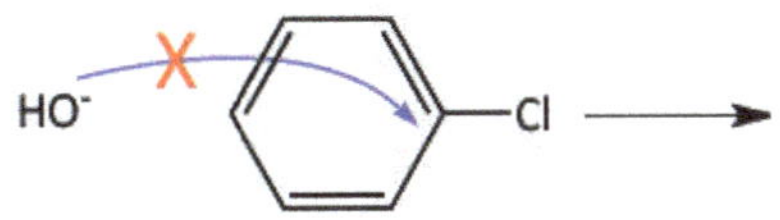

In order to get a halobenzene to do a nucleophilic substitution, we need to use a very strong nucleophile or make the benzene ring more reactive to the attack.

There are two main ways these substitutions occur. One is the Addition-Elimination mechanism and the other is the Benzyne mechanism.

21.11.1 Addition-Elimination (S_NAr)

One way to make a nucleophilic aromatic substitution occur on a benzene ring is to make the benzene ring very electron poor so the negative nucleophile will want to attack it. The benzene ring is electron poor when electron-withdrawing groups like nitro, cyano, or carbonyl groups are attached. The first part of the Addition-Elimination mechanism (also called an S_NAr mechanism) is obviously an addition. An electron rich nucleophile, in this case hydroxide (HO^-), attacks and adds to the benzene ring. It attacks the benzene ring where the nucleophile is attached. The electrons of the benzene ring are pushed over and "stored" next to the electron withdrawing nitro group. Resonance forms can be drawn to spread out this negative charge, helping it be more stable. Look carefully at the resonance forms. Notice when the electron-withdrawing nitro groups are in the *ortho* and *para* sites, they are right next to the negative lone pair of electrons. Electron withdrawing groups at the *ortho* and *para* positions are much better than if they are at the *meta* positions.

The addition

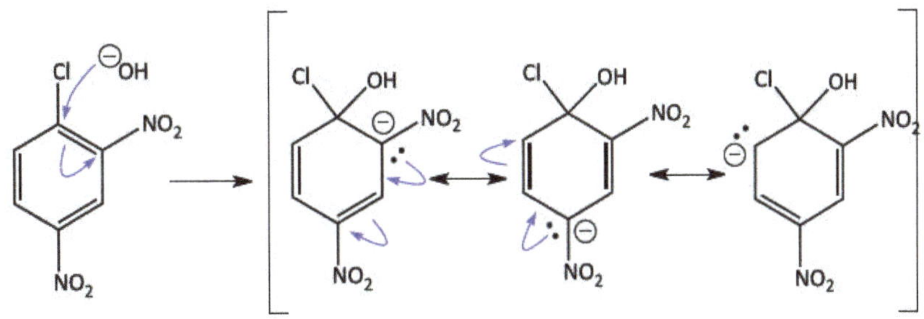

After the addition, the elimination part of the Addition-Elimination mechanism must occur. The lone pair of electrons that had been "stored" fall back to remake the benzene ring and kick off the halide leaving group.

The elimination

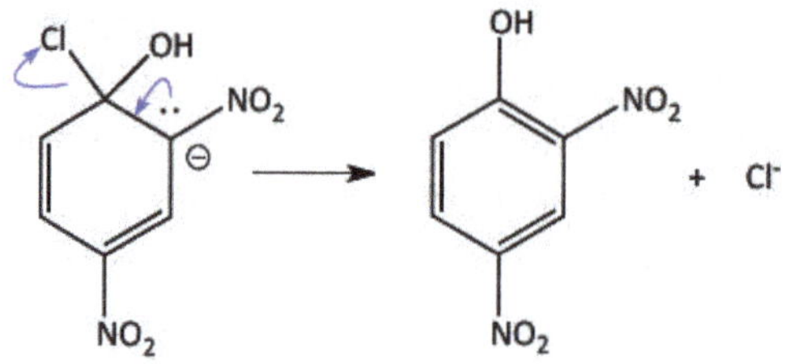

If the nucleophile is ammonia, NH_3, an aniline is formed. Notice that the final product is a benzene with an $-NH_2$ attached, not $-NH_3$. One proton is removed with another molecule of ammonia in order to make the neutral aniline compound.

Sometimes, even if the leaving group is not a great leaving group, the reaction can still proceed.

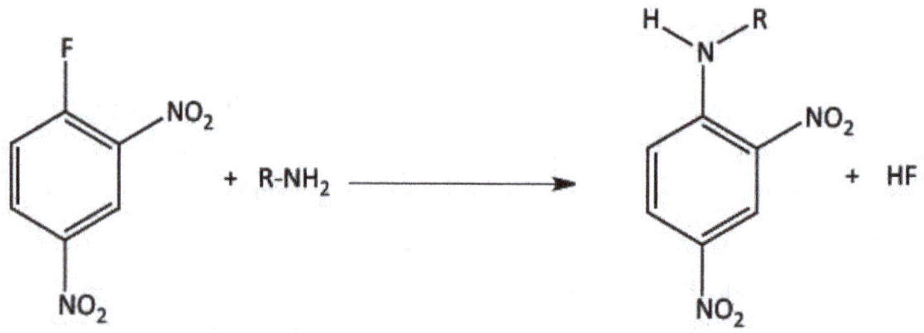

21.13 Which of the following aryl halides would undergo an addition-elimination reaction the best?

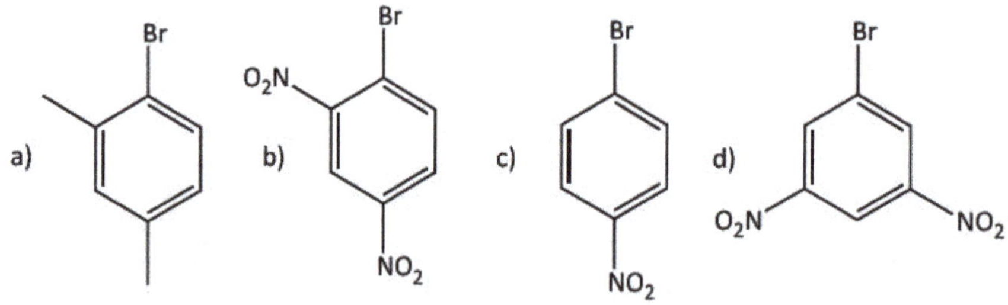

21.11.2 Benzyne

The other mechanism for nucleophilic aromatic substitutions is the benzyne mechanism. In the benzyne mechanism, the electron-withdrawing groups are usually absent on the benzene. Therefore, our benzene ring is not particularly electron poor. The only way to get the nucleophile to attack these benznenes is to use nucleophiles that are very electron rich and use harsh conditions, often heating the reactions to pretty high temperatures.

Chlorobenzene can react with the fairly aggressive hydroxide nucleophile, but the reaction must take place at 350°C.

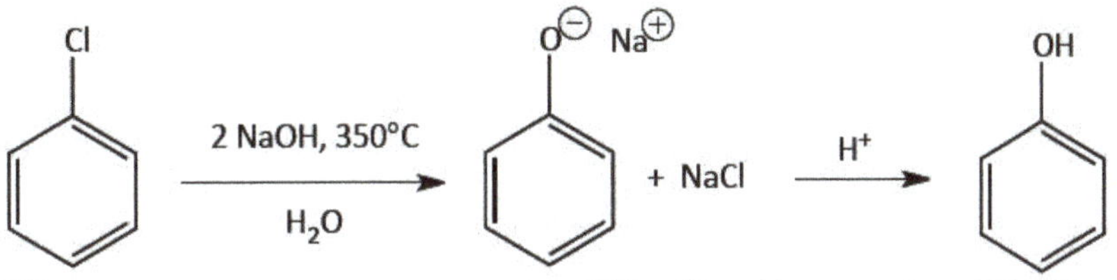

Sodium amide ($NaNH_2$) is an even more aggressive nucleophile. The nitrogen atom likes the negative charge even less than the oxygen atom of the hydroxide. Sodium amide can perform the aromatic nucleophilic substitution at lower temperatures.

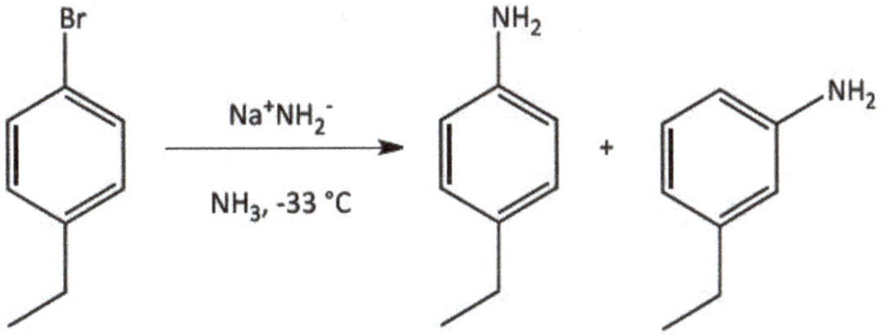

The mechanism for these reactions is called the Benzyne mechanism because it involves a benzene with a triple bond (alkyne) in it. In the Benzyne mechanism, the aggressive nucleophile performs an E2 type of elimination. It pulls off a proton beta to the leaving group. The electrons it leaves behind fall into the benzene ring forming a triple bond, the benzyne, as the leaving group leaves.

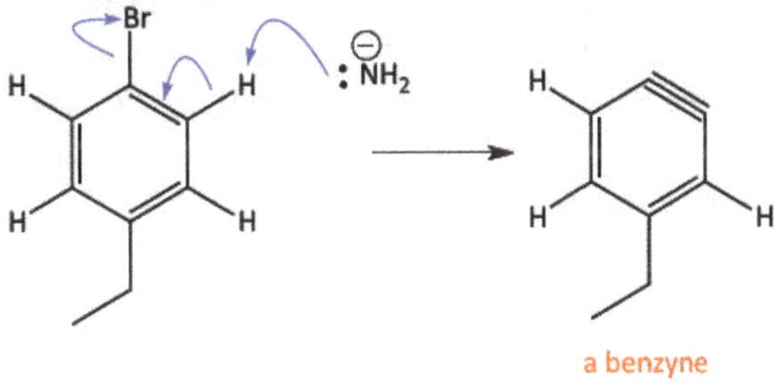

The aggressive nucleophile can then attack either side of the benzyne triple bond. It attacks and one pi bond from the triple bond goes onto the ring to form a carbanion. The carbanion is then protonated. Notice how either side of the benzyne triple bond can be attacked leading to a mixture of products that have the nucleophile in adjoining sites like *ortho* & *meta* or *meta* & *para* like this example.

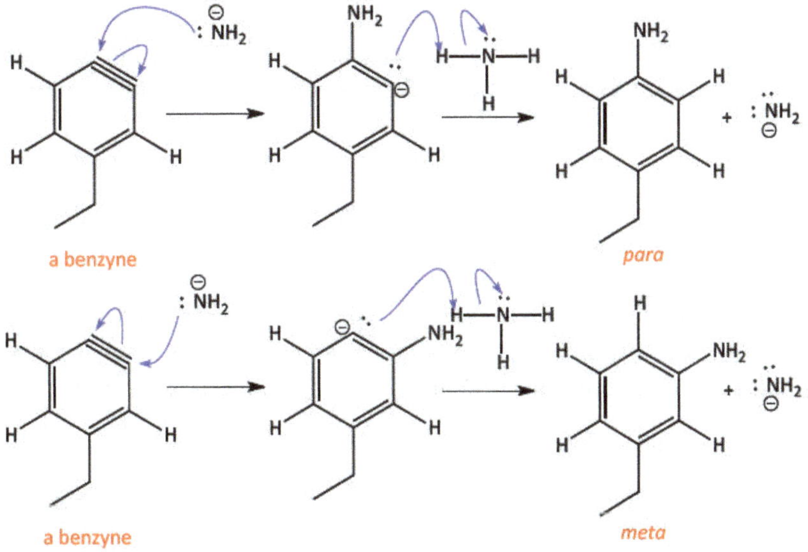

21.14 Classify the following reactions as a benzyne aromatic substitution, an addition-elimination substitution, or neither.

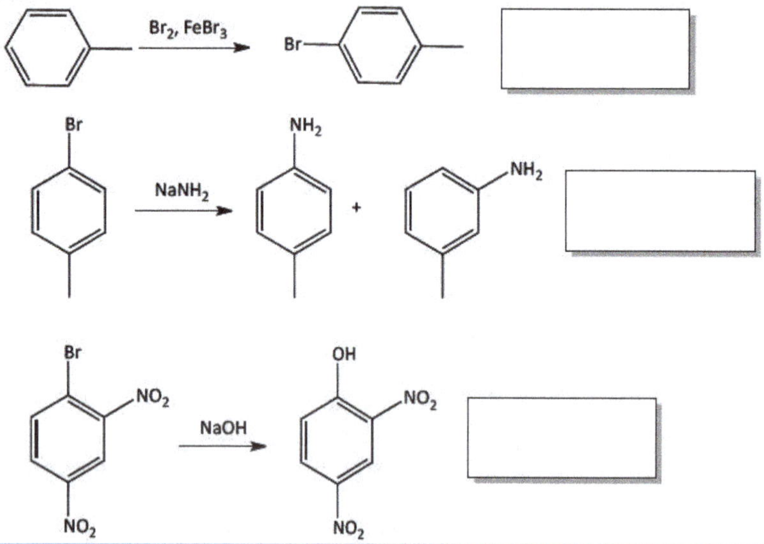

21.12 Gilman Reagents

One other way to do a nucleophilic substitution on an aromatic halide with a carbon group is to use a Gilman reagent. A Gilman reagent is a lithium organocuprate. It is so named because it contains lithium, organic groups, and copper. Many different halide compounds (alkyl, vinyl, or aryl) can be used. The halide can be converted into an organolithium reagent by the addition of two equivalents of lithium metal. The organolithium reagent can be converted into a Gilman reagent with cuprous iodide, CuI.

Halide to organolithium

$$R{-}X \quad + 2\,Li \longrightarrow R{-}Li \quad + LiX$$

Organolithium to Gilman reagent

$$2\ R{-}Li \quad + CuI \longrightarrow \underset{\text{Gilman reagent}}{R_2CuLi} \quad + LiI$$

The R groups in the Gilman reagent are attached to the positively charged copper and lithium ions. The R groups are therefore negatively charged and are nucleophilic. They can perform substitution reactions with alkyl and even aryl halides!

$$R'{-}X \quad + \underset{\text{Gilman reagent}}{R_2CuLi} \longrightarrow R'{-}R \quad + RCu \quad + LiX$$

Below are some examples of how this Gilman reagent coupling can be used. Notice how a carbon-carbon bond is formed in each of these. This is a little unusual compared to the previous ways we have learned to make carbon-carbon bonds. This substitution can occur on a double bond and even a benzene ring.

The substitution can occur on an alkyl halide.

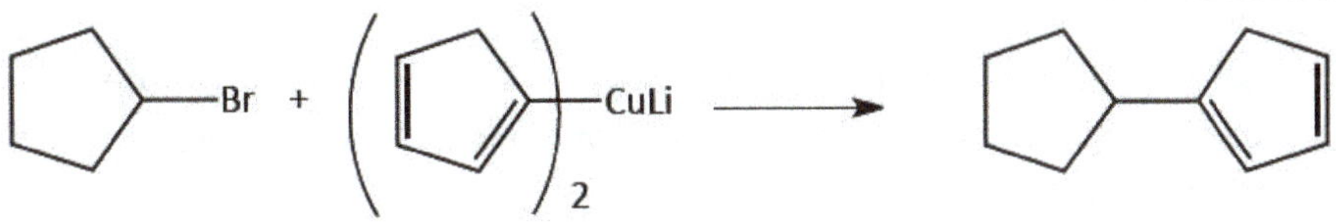

The substitution can occur on an aryl halide.

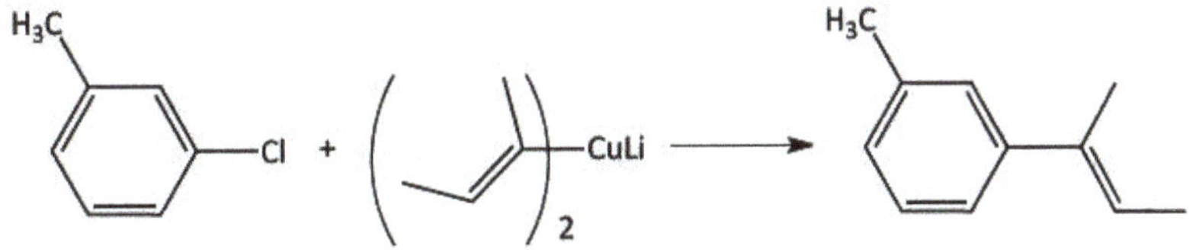

The substitution can occur on a vinyl halide.

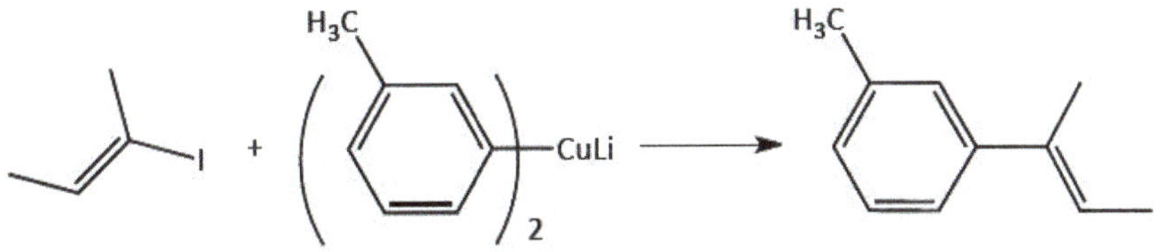

The substitution can occur on an acid chloride.

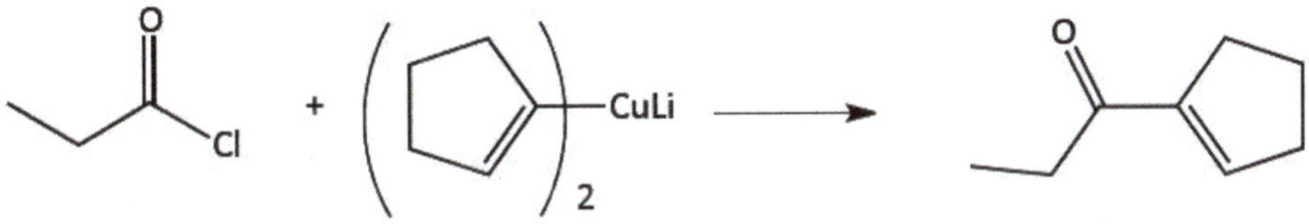

21.15 Give the products of the following reactions.

a) 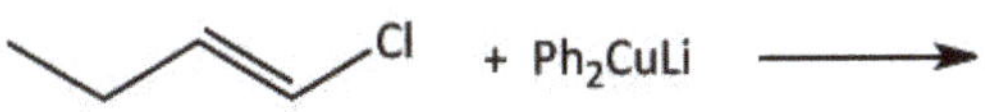

b) 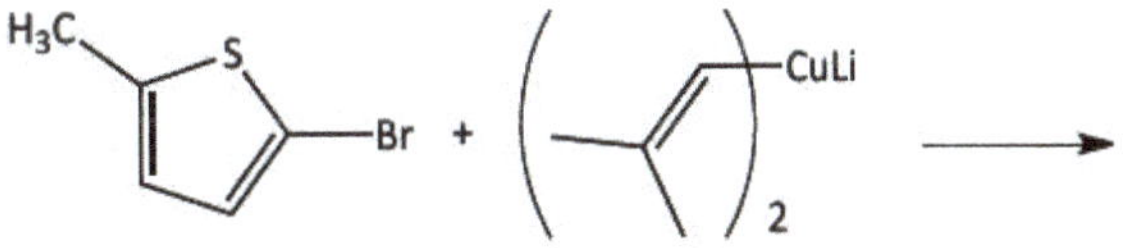

21.16 Give the Gilman reagent used in the following reactions.

a) 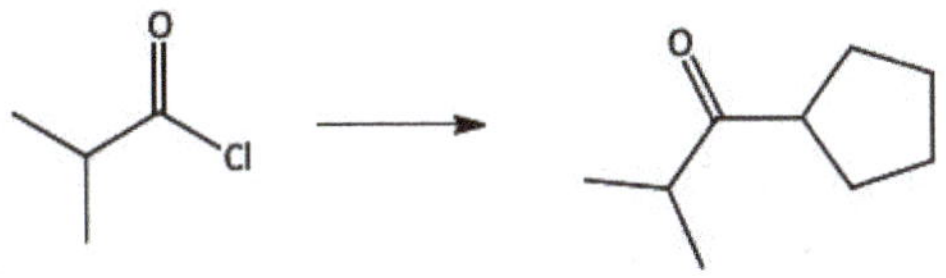

b) 

21.13 Addition Reactions

A benzene ring has three pi bonds in it. These pi bonds are usually quite unreactive because it takes all three pi bonds to make the benzene ring a very stable aromatic ring. But, sometimes under fairly harsh conditions, a molecule can add across one of the pi bonds. Once one pi bond reacts, the aromaticity is broken. This allows the remaining two pi bonds to react rather easily. Under enough heat and pressure, or under ultraviolet light, the pi bond of benzene can react with Cl_2 to be chlorinated. Once one pi bond is chlorinated, the other two pi bonds are chlorinated to make benzene hexachloride.

Chlorination of benzene

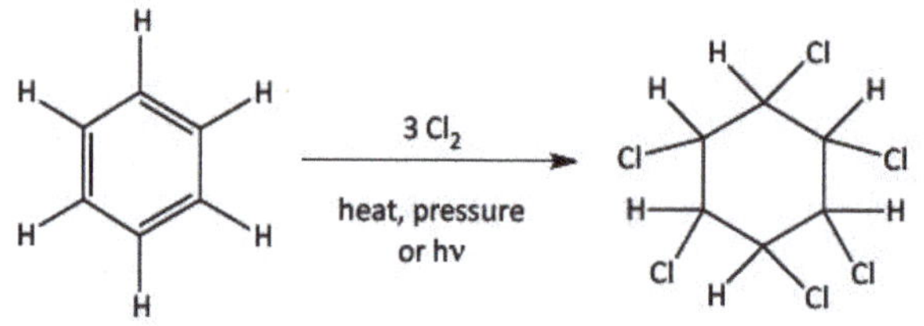

The pi bonds of benzene can be hydrogenated. Benzene can react with H_2 and a metal catalyst under high pressure to become fully hydrogenated making cyclohexane.

Hydrogenation of benzene

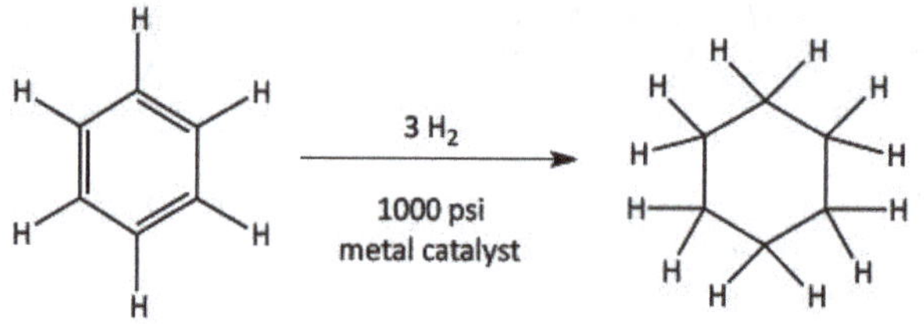

21.14 Birch Reduction

In the previous section we saw how benzene was fully hydrogenated with all three pi bonds reacting with hydrogen. It is possible for only two hydrogen atoms to be added to a benzene ring in a Birch reduction. If a benzene ring reacts with sodium and ammonia, the benzene can form a cyclohexa-1,4-diene.

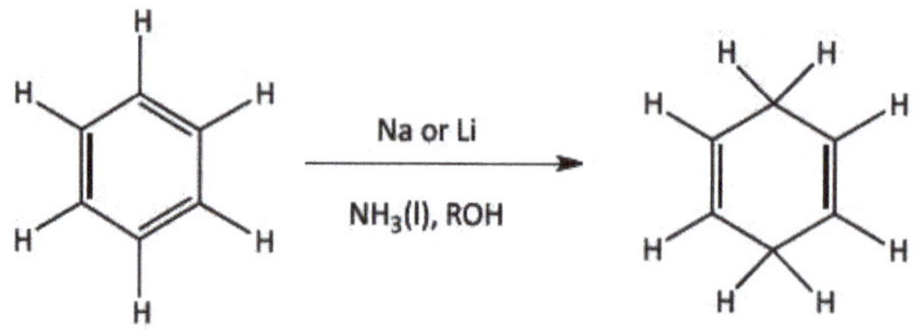

The Birch reduction is of course a reduction. Reductions need electrons. It would be nice if we could simply buy a bottle of electrons. Since we cannot, we need to make some. The reaction of sodium and ammonia is a good way to make a beaker full of electrons.

For this reaction, sodium or lithium metal is dissolved in liquid ammonia. Ammonia has a boiling point of -33°C. It is a gas at room temperature. In order to turn the gaseous ammonia into a liquid solvent, it is cooled in a special chamber using dry ice. Once the ammonia is a liquid, pieces of solid sodium metal are added. Once enough sodium is added, it turns a beautiful blue color. This blue color means we have electrons trapped in the ammonia solvent. These are called "solvated electrons" or "ammoniated electrons".

$$NH_3 + Na \longrightarrow Na^+ + NH_3 \cdot e^-$$

blue color,
solvated electron

How does this work? Sodium metal is an alkali metal. It wants to give up one electron to become Na^+. This is the source of the electrons. Ammonia has the perfect molecular shape to hold these electrons. Ammonia's trigonal pyramidal shape forms a pocket with the positively charged hydrogen atoms that can hold a negatively charged electron. These freed electrons can be thought of as "electrons in a bottle".

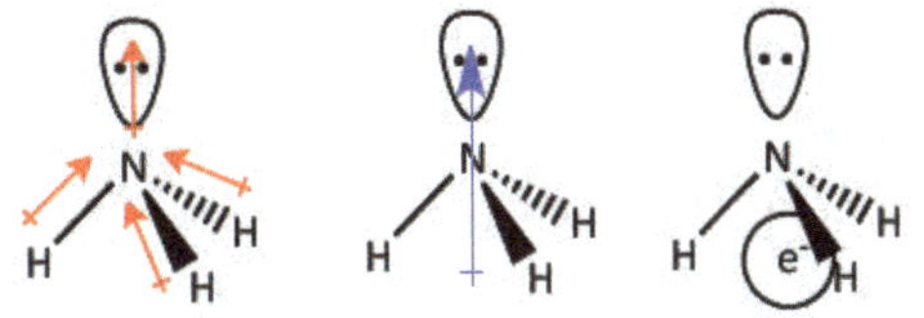

A negatively charged electron from the solvated electrons attacks the benzene ring. The electrons in the ring flow around to make a negatively charged carbanion as far away from the originally attacking electron as possible. The carbanion grabs a proton from the solvent.

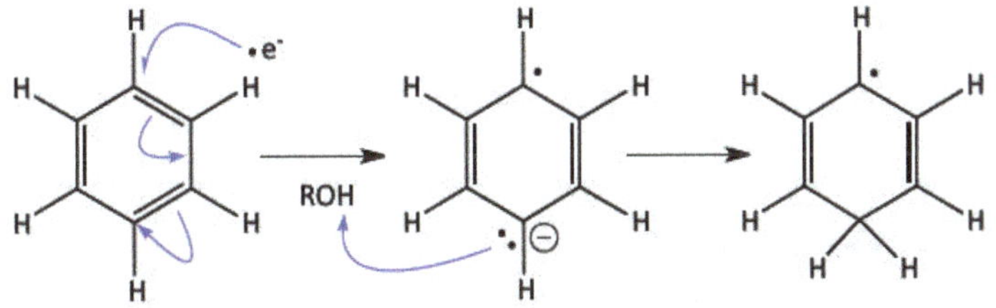

Another electron from the beaker full of electrons comes in to join its unpaired electron friend that is already on the ring forming a negatively charged carbanion. Once the carbanion is formed, it grabs a proton from the solvent making the cyclohexa-1,4-diene.

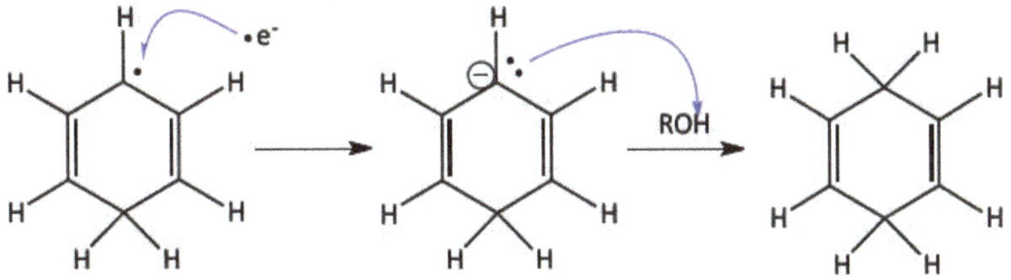

cyclohexa-1,4-diene

The two protons added to the benzene in a Birch reduction are on opposite sides of the ring. These two protons add where the negatively charged carbanions are formed. The best place for carbanions to form on a benzene ring would be where electron-withdrawing groups are attached like carbonyl, nitro, or cyano groups. Notice the hydrogen atom adds to the carbon atom of the benzene ring with the electron-withdrawing group attached.

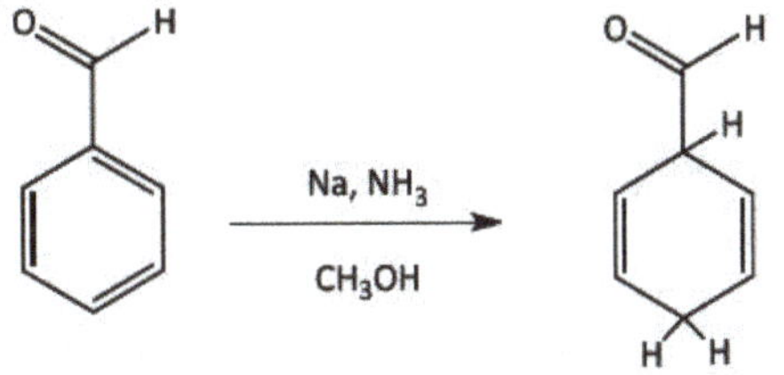

Carbanions would not want to form next to electron donating groups. So, the two hydrogen atoms added in a Birch reduction are *not* next to electron donating groups.

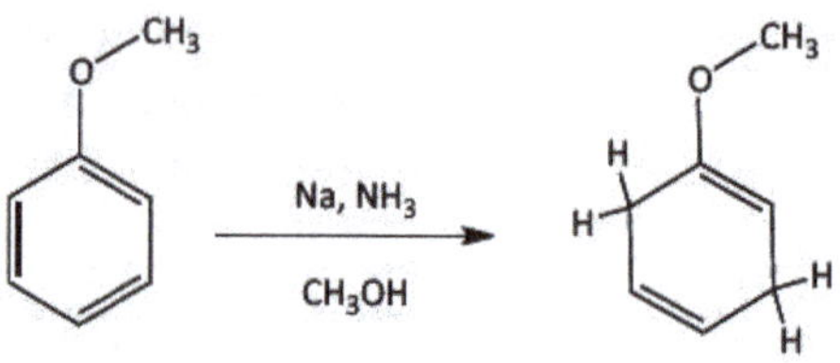

21.17 Draw the products of the following reactions.

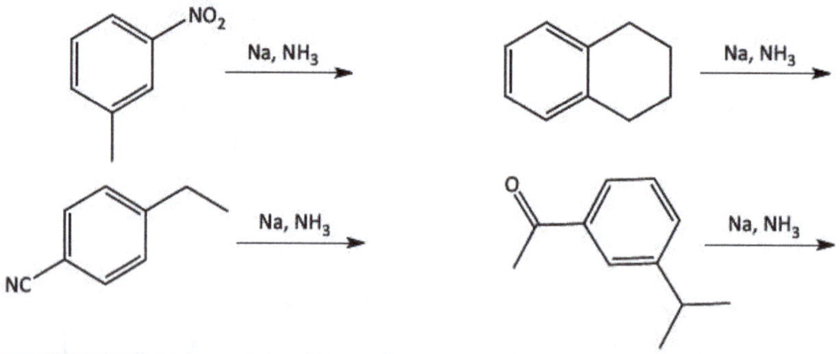

21.15 Oxidation of alkyl benzenes

Alkyl groups that are attached to benzene rings can be oxidized with potassium permanganate, $KMnO_4$, or a combination of sodium dichromate, $Na_2Cr_2O_7$, and H_2SO_4 to form carboxylic acids. This is a great way to form benzoic acid compounds.

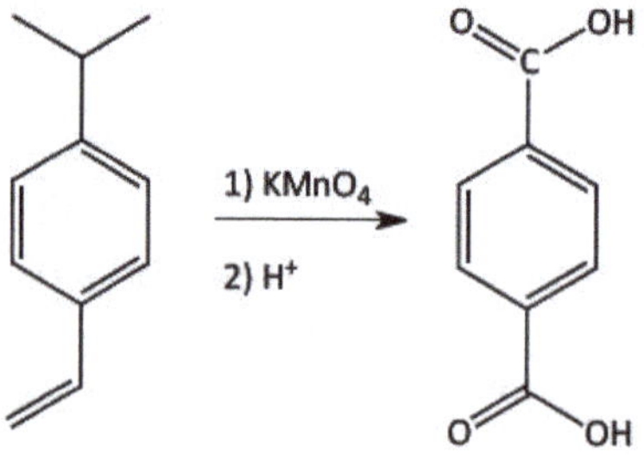

21.18 What is the product of the following reaction?

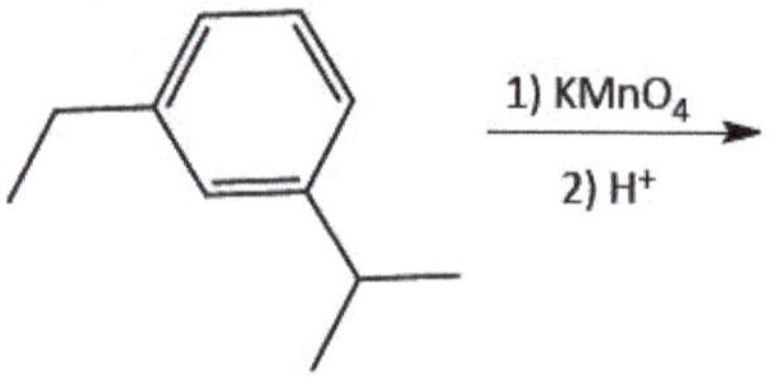

21.16 Halogenation of side chains

We have seen previously that benzylic positions are quite reactive to free radical halogenation because the radicals that form are stabilized through resonance.

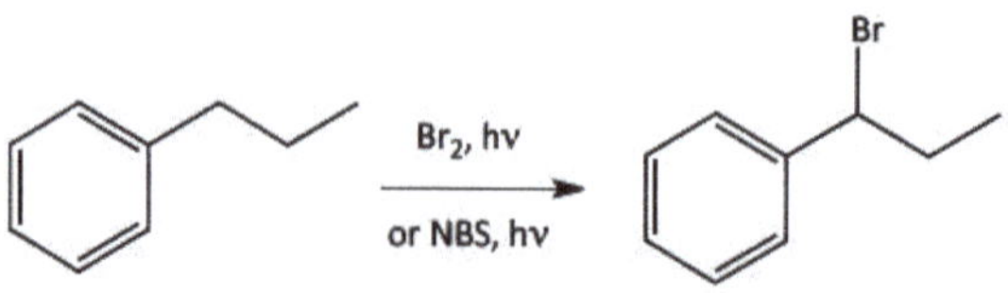

21.17 Oxidation of phenols

Phenols are easily oxidized to quinones with oxidizers like sodium dichromate.

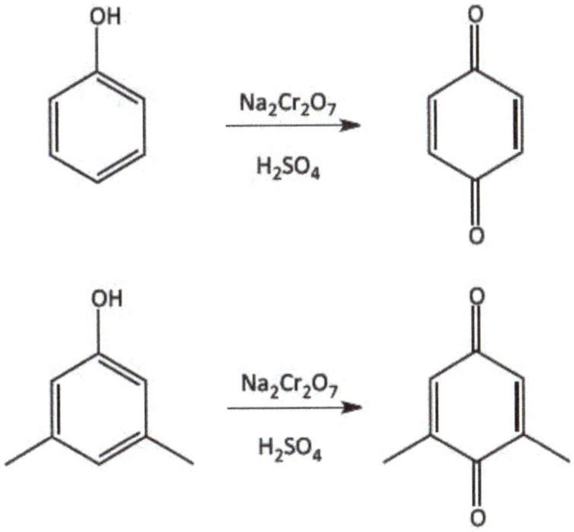

Because phenols are easily oxidized, they are sometimes used as food preservatives. They are added to our food so they are oxidized before our food is oxidized. This helps keep our food from spoiling as quickly. BHA is a preservative added to fats to help keep them from oxidizing.

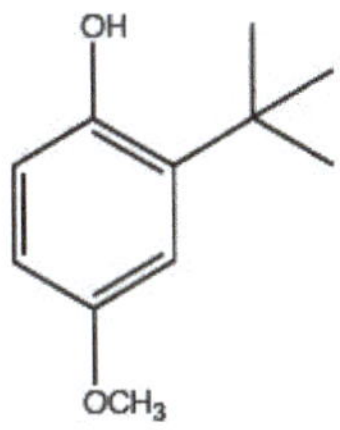

Figure 21.4 BHA structure

Foods that are high in antioxidants, like raspberries, often have polyphenols in them. The following is raspberry ellagitannin.

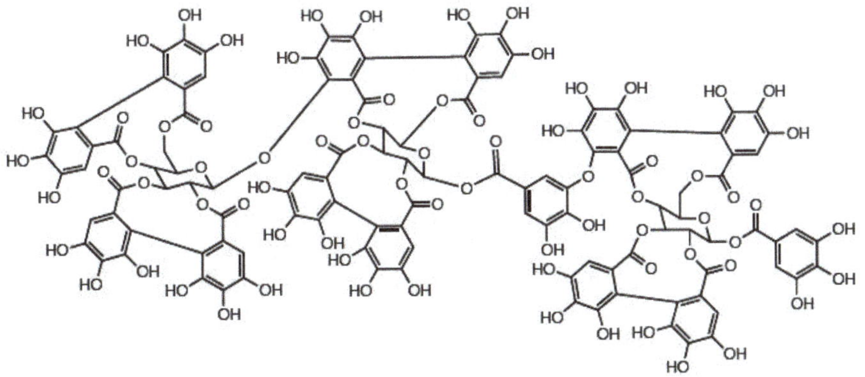

Photo by MaxStraeten at Morguefile.com

Figure 21.5 Raspberry Ellagitannin Structure

21.18 Reaction Summary

Bromination of benzene

a) 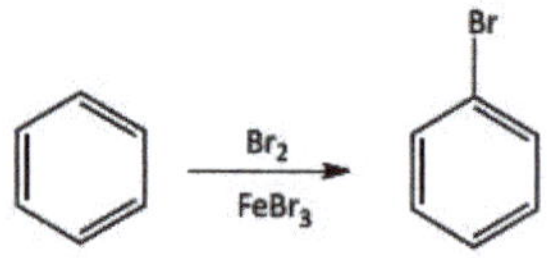

Chlorination of benzene

b) 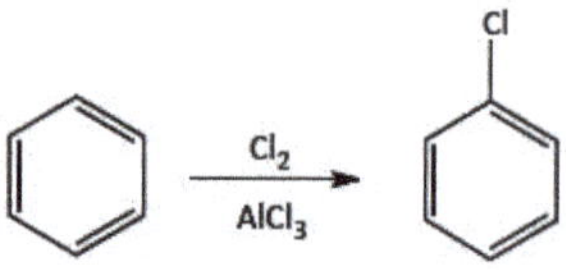

Iodination of benzene

c) 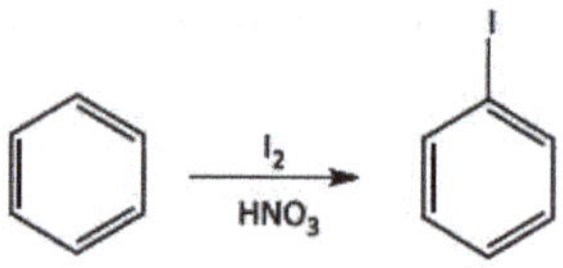

Nitration of benzene

d) 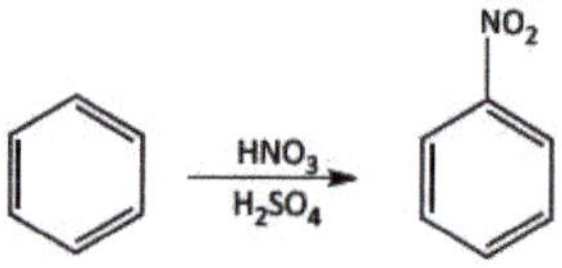

Sulfonation of benzene

e) 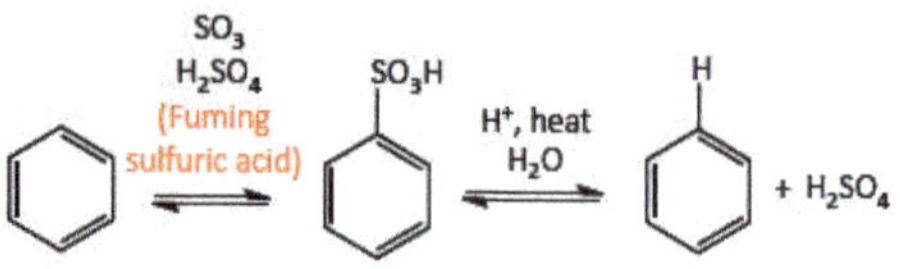

Ortho-para directors

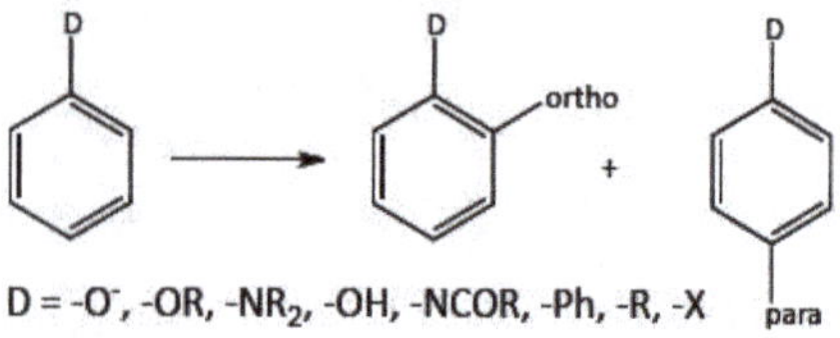

f) D = -O⁻, -OR, -NR$_2$, -OH, -NCOR, -Ph, -R, -X

Meta directors

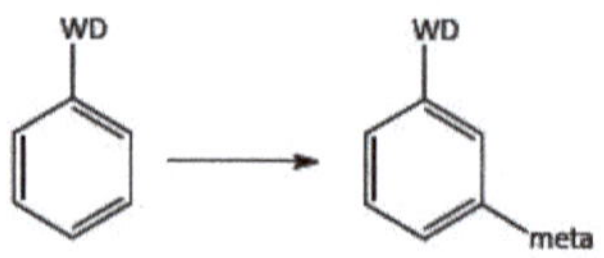

g) WD = -NR$_3^+$, -SO$_3$H, -NO$_2$, -CN, -Carbonyl

Friedel-Crafts alkylation

h) 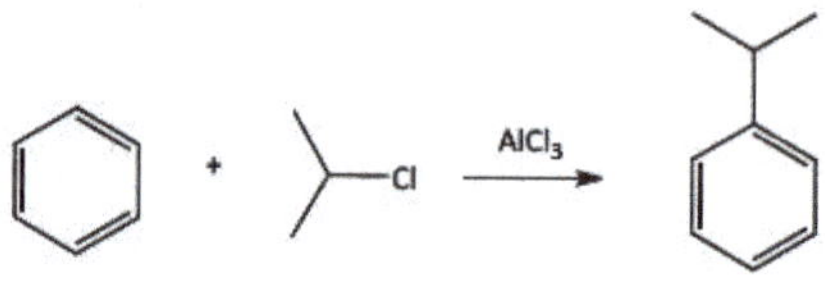

Friedel-Crafts Acylation

i) 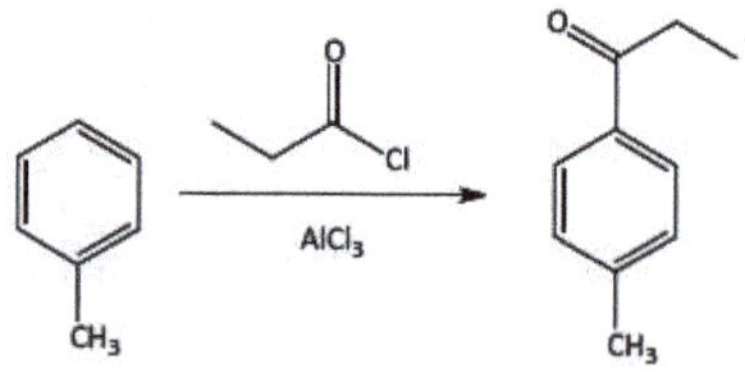

Gatterman-Koch formylation

j) 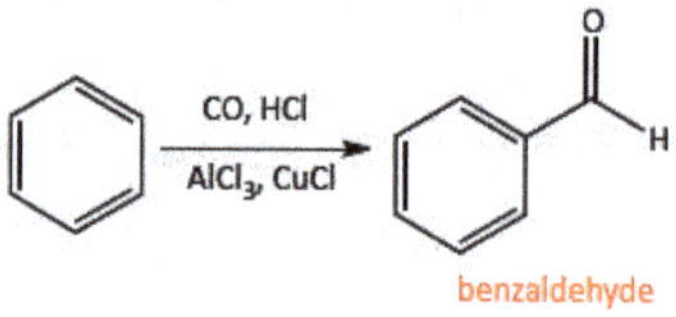

Clemmensen Reduction

k) 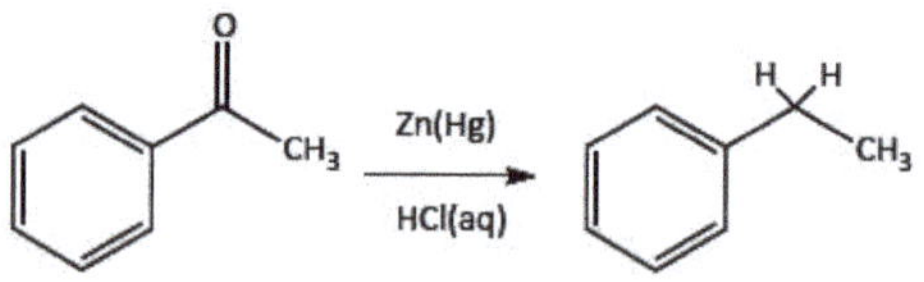

Aniline from nitrobenzene

l) 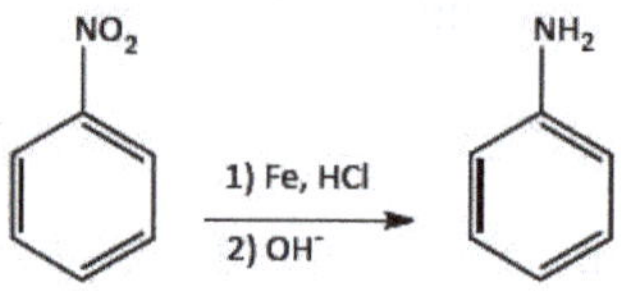

Sandmeyer reaction

m) 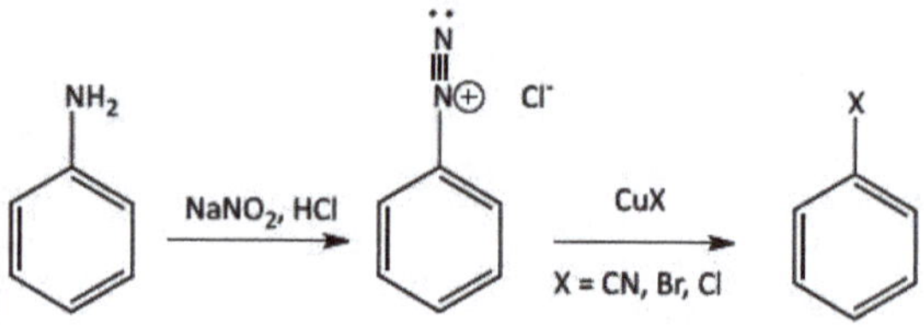

Sandmeyer-like reactions

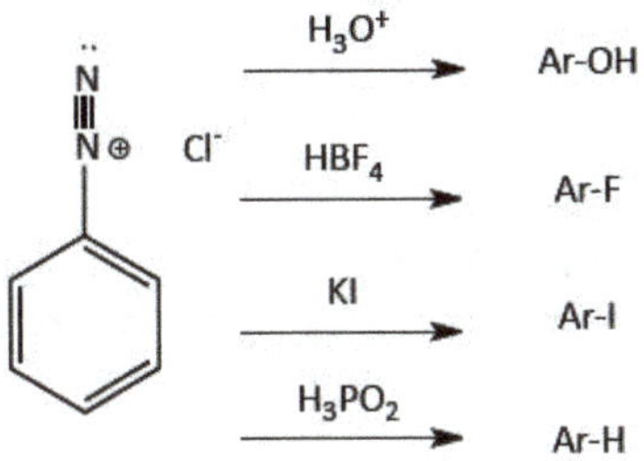

n)

Aniline to amide

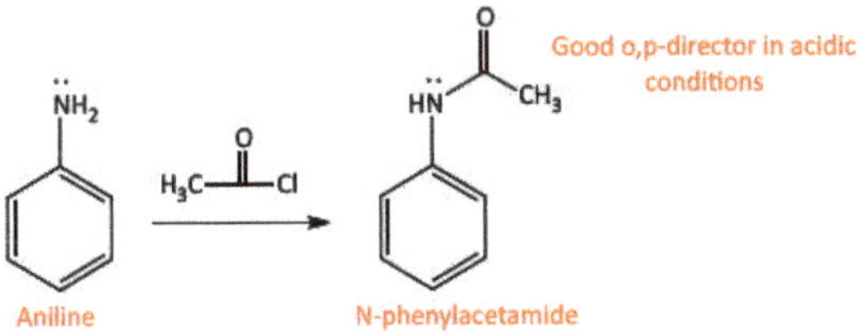

o)

Hydrolysis of amide to amine

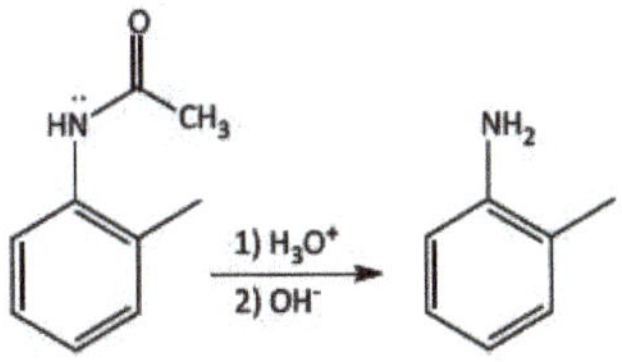

p)

Addition-Elimination Nucleophilic Aromatic Substitution

q)

Benzyne Nucleophilic Aromatic Substitution

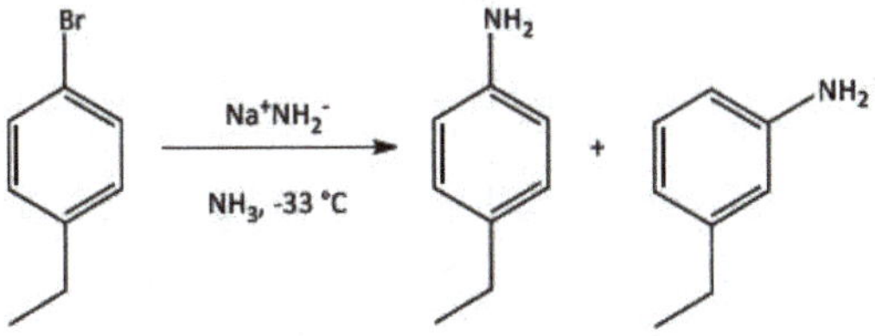

r)

Gilman Reagent Synthesis

s) $2\ R{-}Li + CuI \longrightarrow R_2CuLi + LiI$

Gilman reagent

Gilman Reagent reaction with alkyl halide

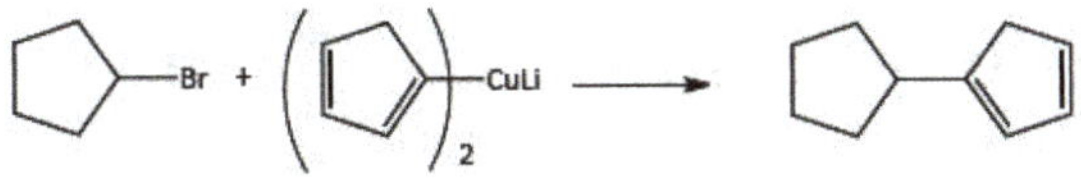

t)

Gilman Reagent reaction with aryl halide

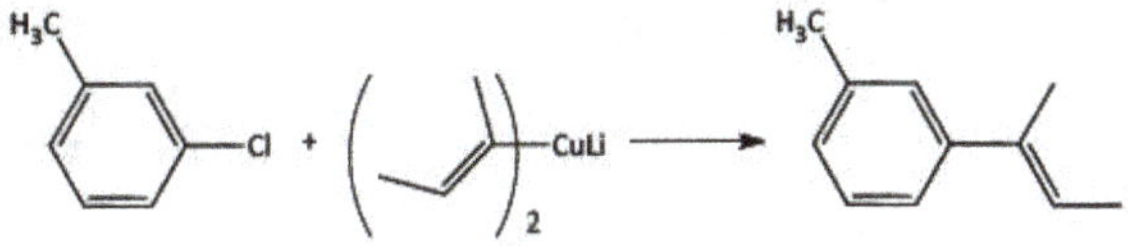

u)

Chlorination of benzene

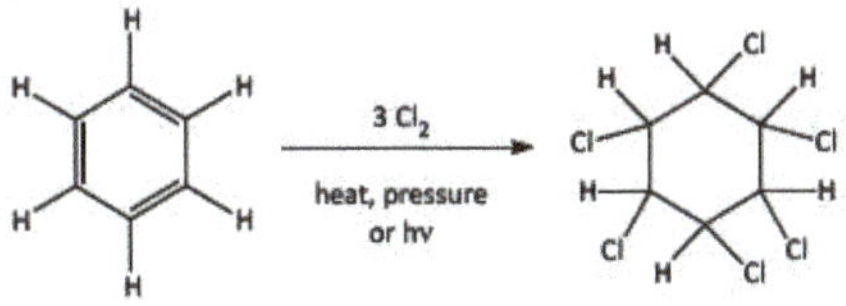

v)

Hydrogenation of benzene

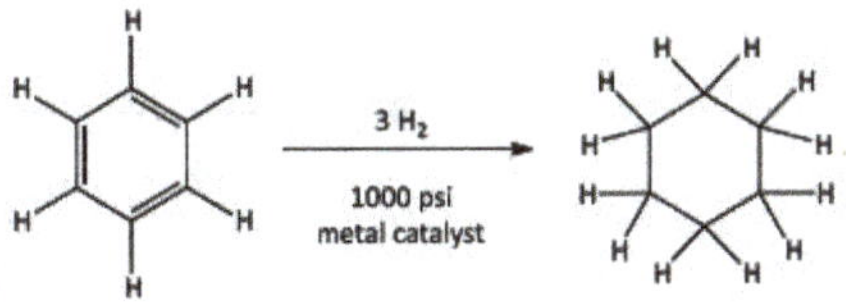

w)

Birch Reduction

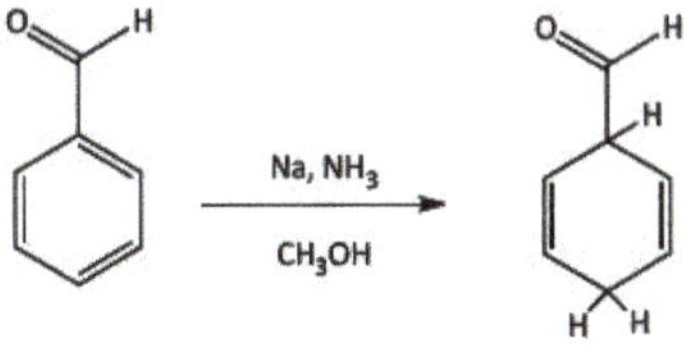

x)

Oxidation of alkyl benzene

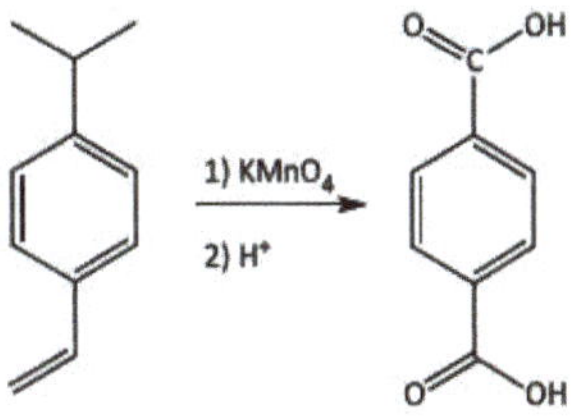

y)

Side-chain halogenation

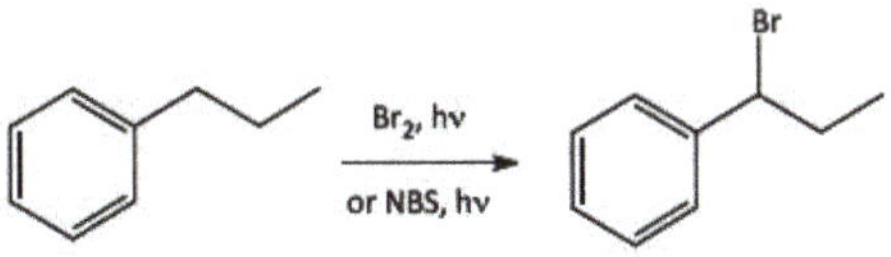

z)

Oxidation of phenol

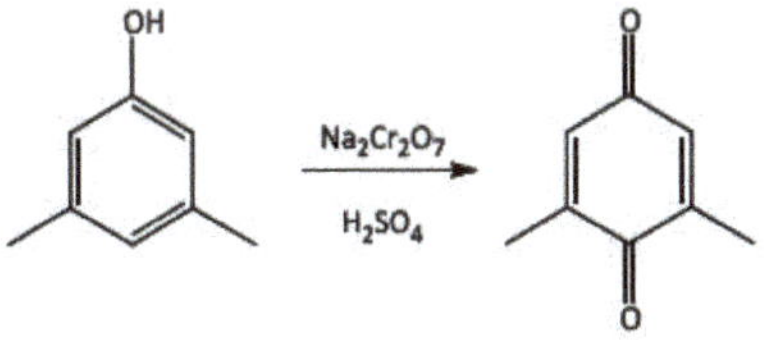

aa)

21.19 Problems

21.19 Show the product of each of the following compounds after adding a nitro group by reacting with nitric and sulfuric acid.
a) bromobenzene b) *o*-nitrochlorobenzene c) ethylbenzene d) benzaldehyde e) benzoic acid
f) benzonitrile (PhCN) g) *m*-nitroanisole h) *p*-bromobenzaldehyde i) *m*-ethoxybenzoic acid j) 3-ethylphenol

21.20 Fill in the reagents/products for the following reactions.

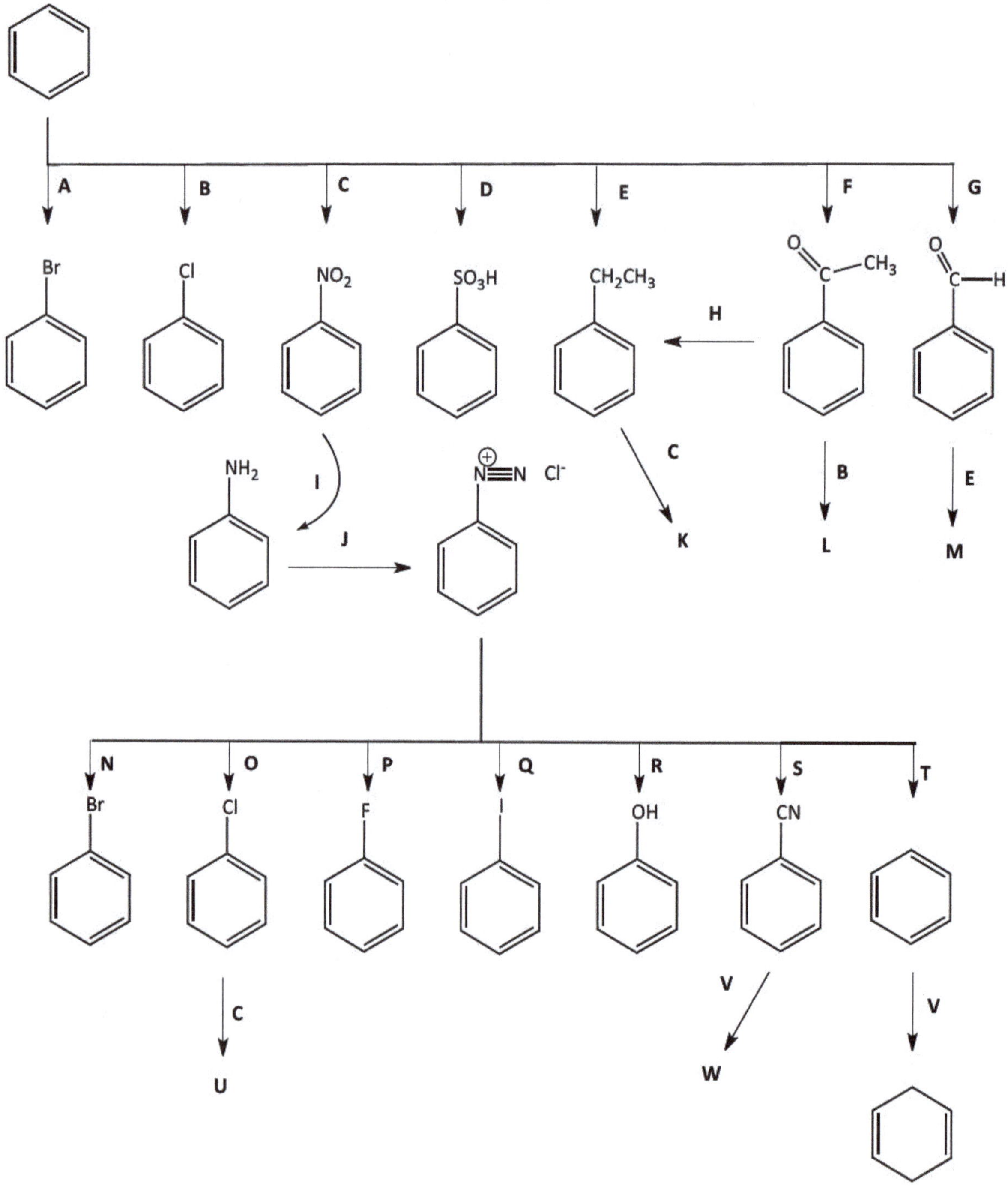

21.21 **Describe the orbitals that overlap and are involved in making the "triple" bond in benzyne. A sketch might be helpful.**

21.22 **Provide the structure of the major organic product(s) in the following reactions.**

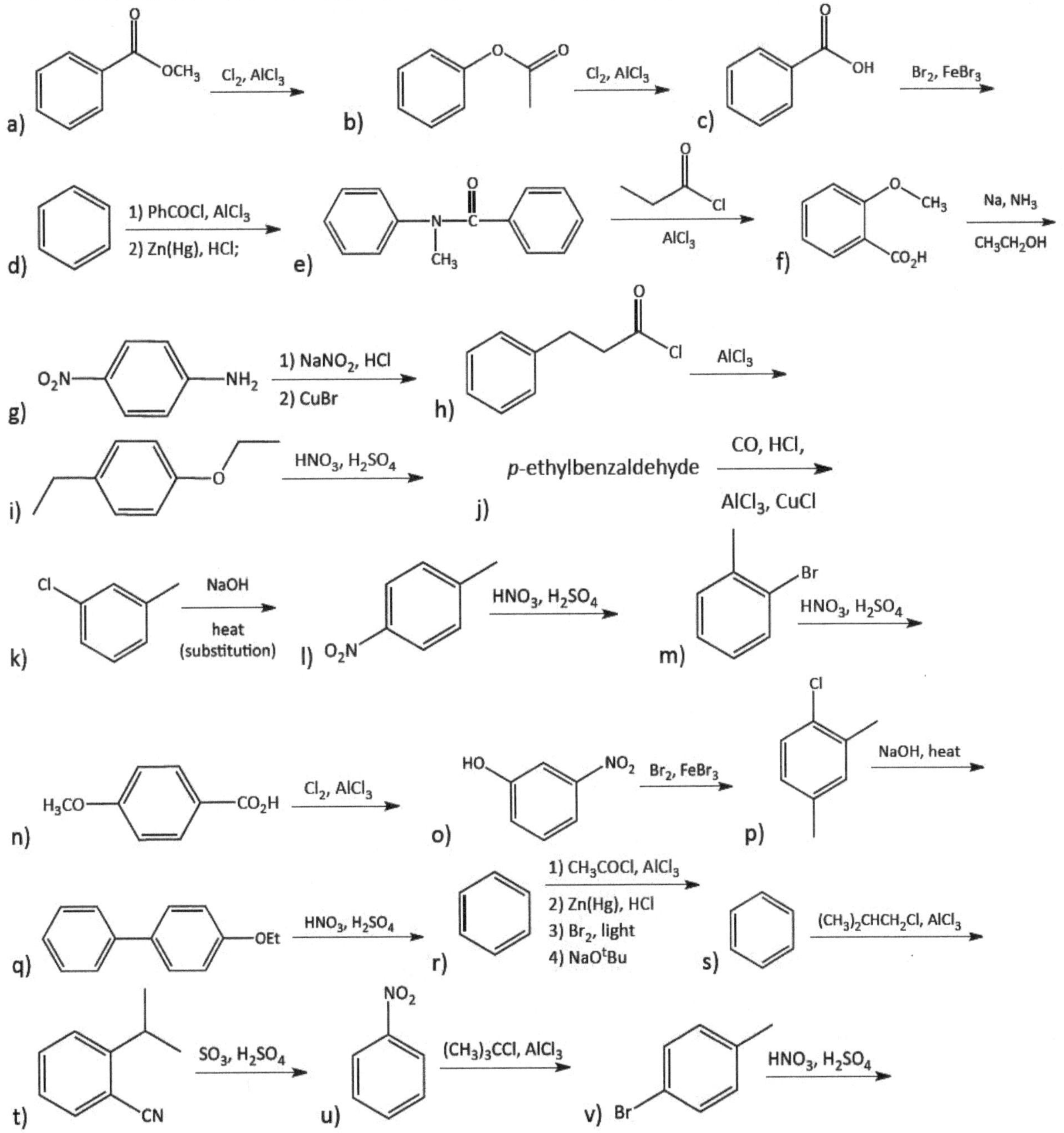

21.23 Provide the structure of the major organic product(s) in the following reactions.

a) 2,4-dinitrochlorobenzene + NaOEt b) 3-methylchlorobenzene + NaOH at 100°C
c) 1,3-dimethylbenzene (*m*-xylene) + Na/NH_3 d) *m*-xylene + concentrated potassium permanganate
e) chlorobenzene + fuming sulfuric acid f) 1-chloro-2-nitrobenzene + $NaNH_2$
g) anisole + sodium in liquid ammonia h) benzaldehyde + sodium in liquid ammonia

Aromatic Compound Reactions

21.24 Devise an efficient method to prepare the following compounds from benzene.
a) 4-bromobenzonitrile b) *m*-nitrobutylbenzene c) *p*-nitropropylbenzene d) *p*-nitrobenzoic acid
e) *p*-methylphenol (also called *p*-cresol) f) *p*-methylbenzaldehyde g) 3-bromoethylbenzene h) *m*-chloroaniline
i) *p*-bromotoluene j) phenol k) 1-phenyl-1-bromohexane l) 4-phenylheptan-4-ol
m) *o*-nitrobenzoic acid n) *p*-t-butylnitrobenzene o) n-butylbenzene

21.25 Show how you would use nitration, Friedel-Crafts alkylation, Friedel-Crafts acylation, Gatterman-Koch formylation, and/or Clemmensen reduction to make the following compounds from benzene.
a) ethylbenzene b) *p*-ethylbenzaldehyde c) benzophenone (PhCOPh) d) $PhCH_2Ph$ e) $(CH_3)_2CHCOPh$
f) *n*-pentylbenzene g) CH_3CH_2COPh h) *m*-nitrobenzaldehyde

21.26 Provide detailed, stepwise mechanisms for the following reactions.

a) 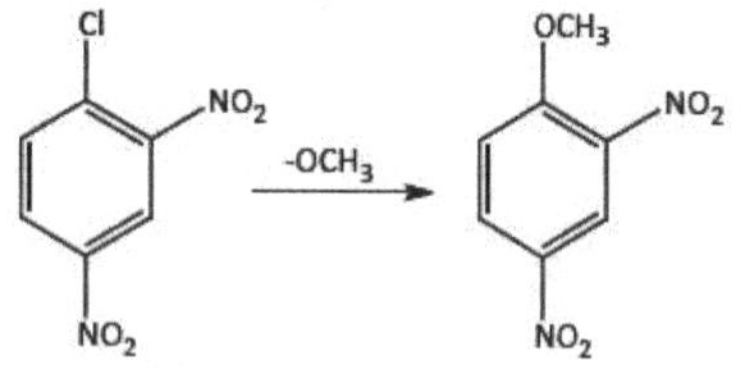

b) 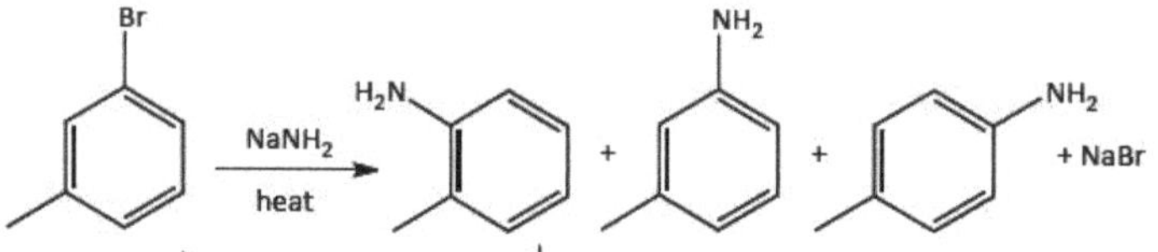

c) 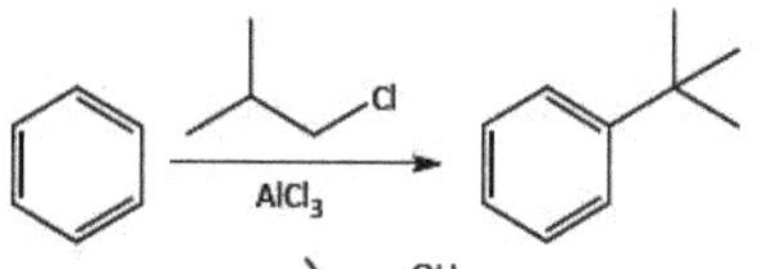

d) 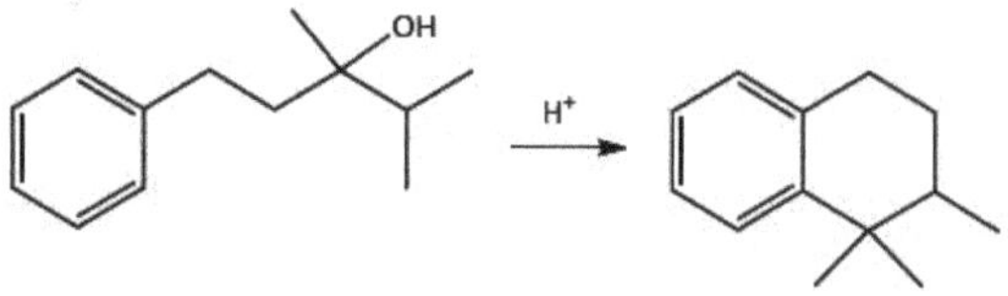

21.27 Draw the three major resonance structures of the sigma complex intermediate in the reaction of benzaldehyde with HNO_3/H_2SO_4 to yield *o*-nitrobenzaldehyde. Circle the resonance form that is *less* stable than the other two.

21.28 Predict the products of the following nucleophilic substitution reactions.
  a) 4-chloronitrobezene + sodium ethoxide ($NaOCH_2CH_3$)
  b) 2,4-dinitrobromobenzene + sodium hydroxide (NaOH), 100°C
  c) 2,4,6-trinitrobromobenzene + methylamine
  d) 4-bromotoluene + sodium hydroxide (NaOH), 350°C
  e) 2-chloroethylbenzene + sodium amide ($NaNH_2$)

21.29 Predict the products of the following reactions.
  a) toluene (methyl benzene) + propyl chloride/$AlCl_3$
  b) anisole (methoxy benzene) + isopropyl chloride/$AlCl_3$
  c) benzaldehyde + ethyl chloride/$AlCl_3$
  d) *n*-pentylbenzene + $CH_3CH_2COCl/AlCl_3$
  e) chlorobenzene + CO, HCl, $AlCl_3$, CuCl

21.30 Predict the products of ethyl benzene with the following:

a) $Cl_2/AlCl_3$
b) $CH_3CH_2CH_2COCl/AlCl_3$
c) HNO_3/H_2SO_4
d) Hot, concentrated $KMnO_4$
e) *n*-butyl chloride/$AlCl_3$
f) Br_2 and UV light
g) Na/NH_3
h) NBS
i) fuming sulfuric acid
j) 1) $Cl_2/AlCl_3$ 2) NaOH (350°C)
k) 3 Cl_2, heat, pressure
l) H_2, heat, pressure, metal catalyst
m) $Br_2/FeBr_3$

21.31 Devise an efficient method to prepare the following compounds from nitrobenzene.

a) iodobenzene
b) fluorobenzene
c) chlorobenzene
d) bromobezene
e) benzonitrile
f) phenol
g) 1,3,5-tribromobenzene

21.32 Predict the products of the following with hot, concentrated potassium permanganate, $KMnO_4$.
a) toluene b) *o*-xylene c) hexylbenzene d) *t*-butylbenzene

21.33 Show how to make the following compounds starting with benzene. Your reaction sequences should involve bromination or chlorination, making Grignard or Gilman reagents, and any other necessary reactions with a halide, carbonyl compound, or epoxide.

a) 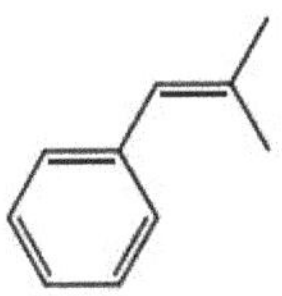

b) 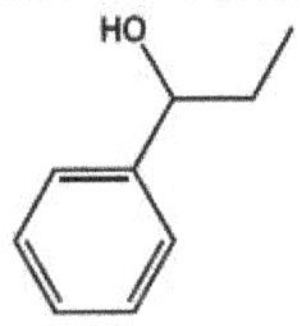

c) 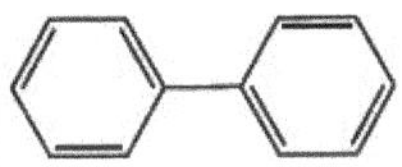

d) 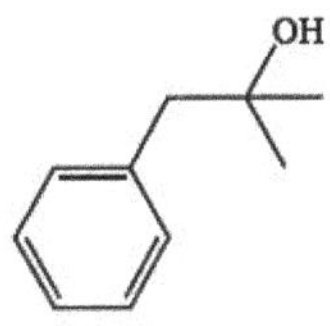

21.34 Show how to make the phenyl propyl ether starting with benzene. Your reaction sequence should involve bromination or chlorination, nucleophilic substitution with sodium hydroxide to form phenol, and any other necessary reactions.

21.35 Show how you would convert aniline to the following compounds:
a) chlorobenzene b) bromobenzene c) fluorobenzene d) benzonitrile (PhCN) e) benzene f) phenol g) iodobenzene h) *p*-methylaniline i) 1,3,5-tribromobenzene

21.36 Benzyne is a great dienophile in a Diels-Alder reaction. Show the Diels-Alder product when chlorobenzene reacts with sodium hydroxide at 350°C in the presence of furan.

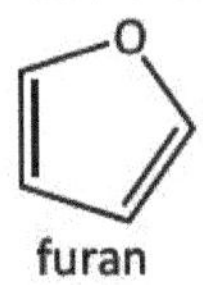

furan

21.37 Use benzene or toluene to make the following. You must first nitrate then do whatever else is necessary.
a) aniline b) *m*-chloroaniline c) *p*-bromobenzoic acid d) *para*-fluorotoluene e) 4-ethylbenzonitrile

21.38 Show the reagents that could be used to convert *o*-toluidine (*o*-methylaniline) to the following compounds.

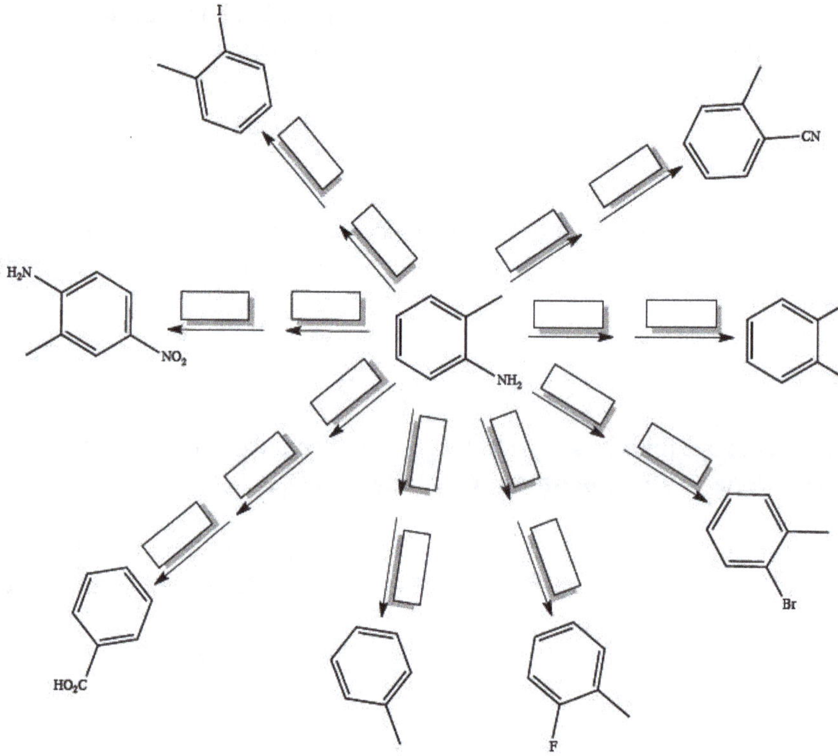

21.39 Provide a series of synthetic steps by which the following compound can be prepared from phenol.

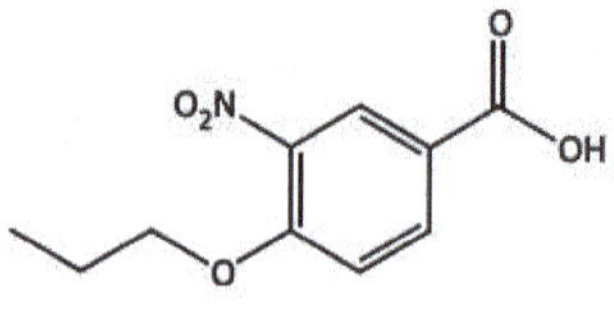

Chapter 22. Amines and Nitriles

22.1 Introduction

Nitrogen atoms with a zero formal charge have three bonds on them. Amines are organic compounds that have a nitrogen atom with three single bonds on it. Amines have one or more alkyl or aryl groups attached to this nitrogen atom. If one carbon-based group is attached to the nitrogen atom, it is called a primary amine. If two carbon-based groups are attached to the nitrogen atom, it is called a secondary amine. If three carbon-based groups are attached to a nitrogen atom, it is called a tertiary amine.

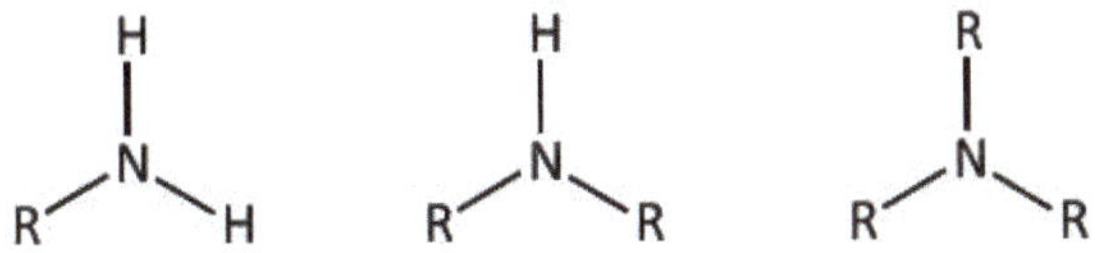

Primary, 1° amine   Secondary, 2° amine   Tertiary, 3° amine

Amines are found in many natural products and biologically important compounds.

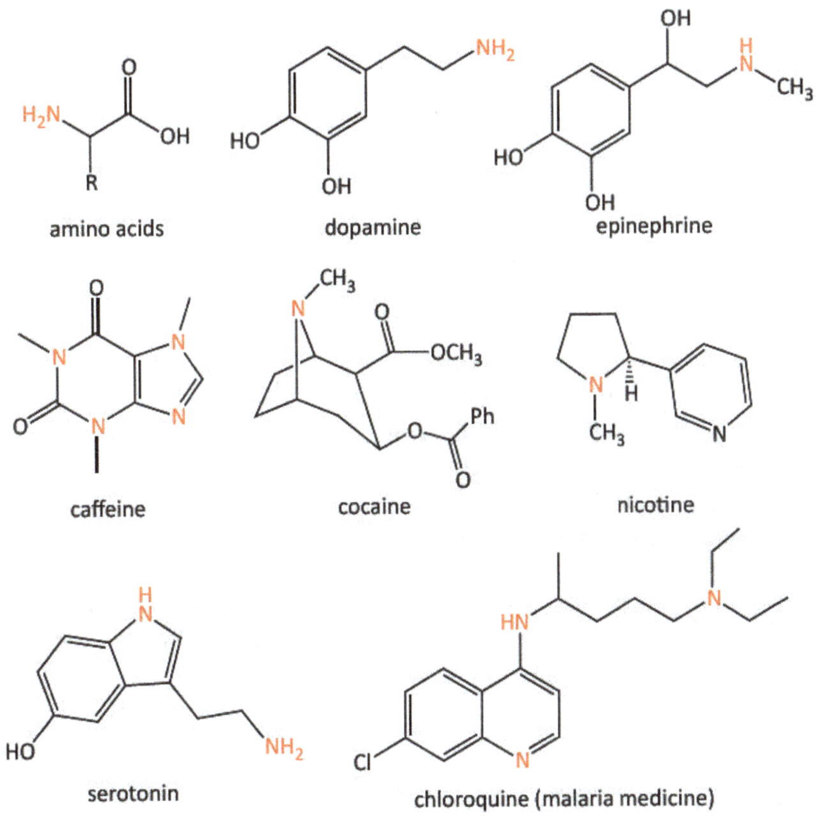

Figure 22.1 Biologically important amines

Bases are one of the building blocks of RNA and DNA. Each is made up of four different types of bases. These bases are amine rings called purines and pyrimidines. Both RNA and DNA contain the purine bases guanine (G), adenine (A). Both RNA and DNA contain the pyrimidine base cytosine (C). The fourth base of RNA and DNA is ever so slightly different between the two. RNA contains the pyrimidine base uracil (U), while DNA contains the pyrimidine base thymine (T). Uracil and thymine only differ by one methyl group attached to the ring.

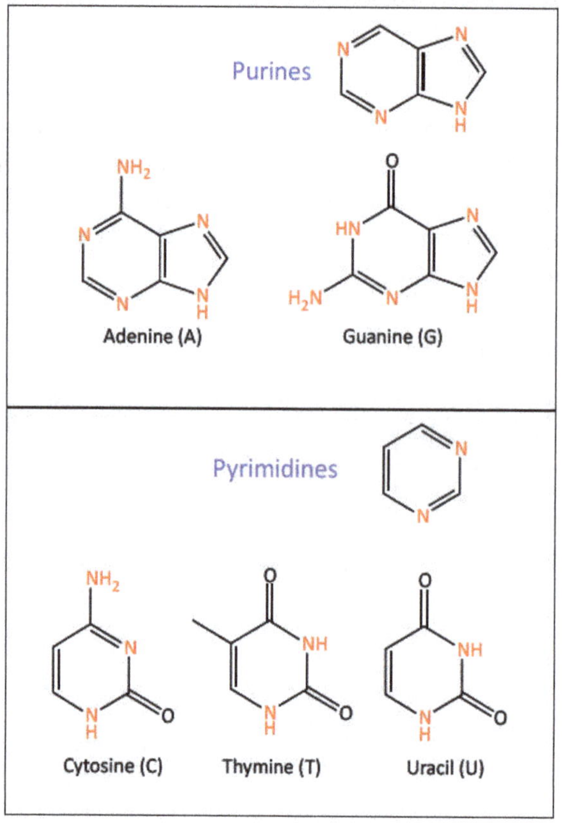

Figure 22.2 DNA/RNA Bases

Amino acids, peptides, proteins

Polymers are biological compounds responsible for many different functions. Proteins like collagen and keratin are structural proteins needed for our skin, tendons, hair, and fingernails. Enzymes are proteins that catalyze biological reactions. Hemoglobin is a protein that transports oxygen throughout your body. Proteins like insulin are hormones. Proteins can even be toxic, like snake venom.

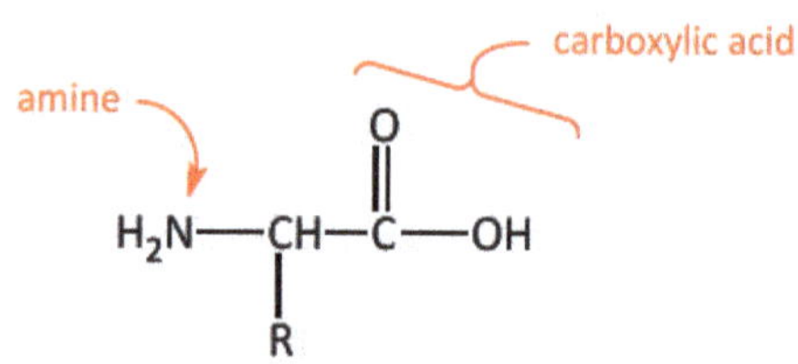

Figure 22.3 Amino acid

Amino acids are the building blocks proteins. There are twenty essential amino acids found in biochemistry. All amino acids have both an amine or *amino* end, a methane (CH) in the middle with some -R group attached to it, and a carboxylic *acid* end. That's why they are named amino acids.

Amino acids differ from each other by having different –R groups attached to them. Here are just a few of the twenty essential amino acids.

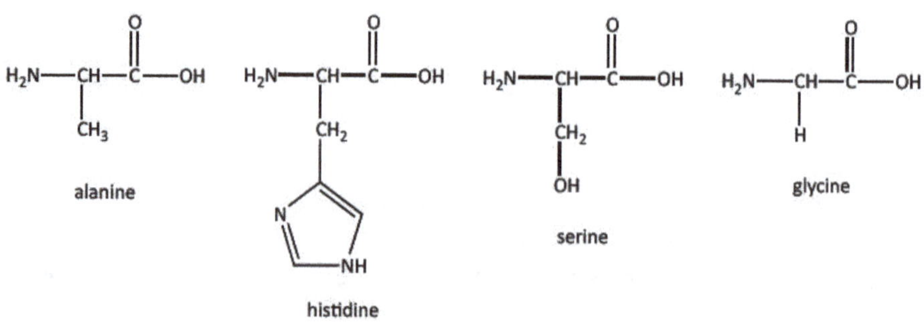

Several amino acids linked together are called peptides. A dipeptide is two amino acids linked together. Three linked together makes a tripeptide. An oligopeptide is a small chain of three to ten amino acids. A compound of many amino acids linked together into a polymer is called a polypeptide. Amino acids link together by amide bonds to form small chains called peptides.

dipeptide tripeptide

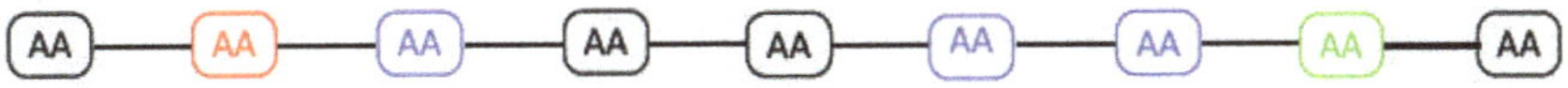

peptide

When peptides are folded into molecules with a specific shape it can have a biological function. This folded molecule made of peptides is called a protein.

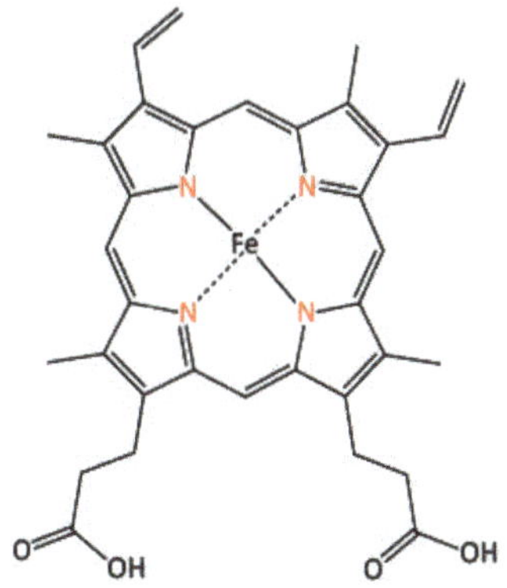

Figure 22.5 Heme

Porphyrin compounds

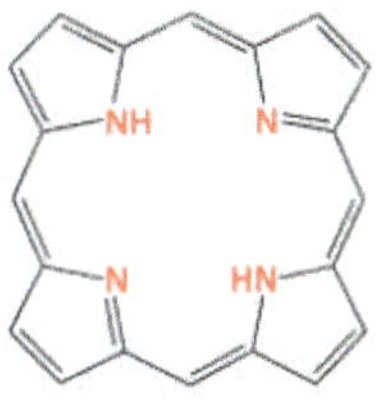

Figure 22.4 Porphyrin

Four pyrroles linked together by one-carbon units forming a conjugated ring are called porphyrins. Porphyrins are interesting subunits in biological compounds. The four nitrogen atoms with their negative lone pairs of electrons are perfectly situated to hold large, positively charged metal ions. The high degree of conjugation makes them absorb visible light and makes these compounds colorful. A porphyrin ring that holds an iron atom is the main active unit of heme in hemoglobin. This gives our red blood cells their bright red color.

Conjugated rings of four pyrroles with porphyrin-like structures are found in the biological compounds of vitamin B12 (cobalt metal and red color) and chlorophyll a (magnesium metal and green color).

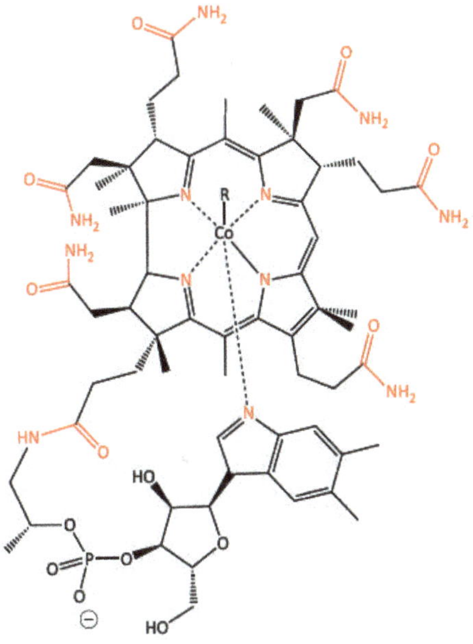

Figure 22.6 Vitamin B12

Vitamin B12 is an essential water-soluble vitamin for our diets. Because it contains cobalt, the group of compounds that have vitamin B12 activity are collectively called the cobalamins. Vitamin B12 is commonly found in animal products like fish, meat, eggs, and milk. Notice the amines of the pyrroles in the central ring as well as the amide groups along the periphery.

Chlorophyll a is found in the chloroplasts of plant cells and is crucial for photosynthesis. Of course, chlorophyll allows the plant to absorb sunlight converting carbon dioxide to oxygen while synthesizing carbohydrates like glucose.

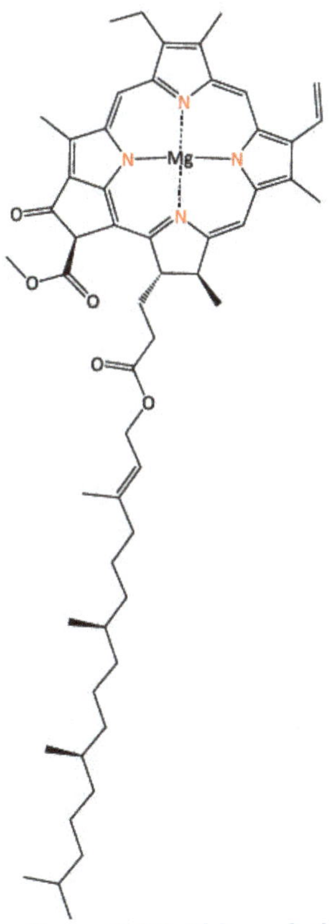

Figure 22.7 Chlorophyll a

Sulfanilamide

Para amino benzoic acid (PABA) is a molecule used to synthesize vitamin B9 (folic acid). Humans can obtain vitamin B9 through our diets, but bacteria cannot. Bacteria have an enzyme called dihydropteroate synthase (DHPS) which is necessary to convert PABA to vitamin B9. Sulfanilamide is the simplest sulfa drug (an antibiotic). Notice how sulfanilamide has a very similar structure to PABA. It acts as a competitive inhibitor to the DHPS enzyme. Since it has a similar shape, it overwhelms the enzyme and the enzyme is unable to convert PABA to vitamin B9. The bacteria therefore have trouble growing. Sulfanilamide is therefore bacteriostatic. It doesn't kill the bacteria. It only inhibits the bacteria's ability to replicate and grow.

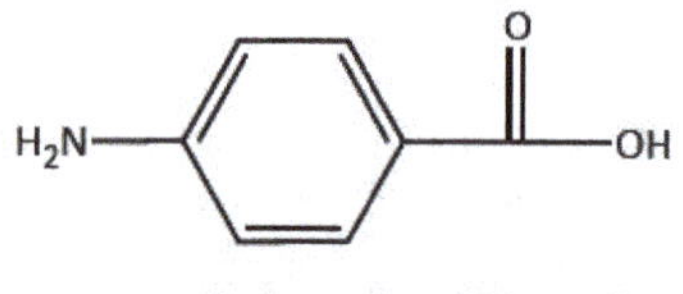

para aminobenzoic acid (PABA)

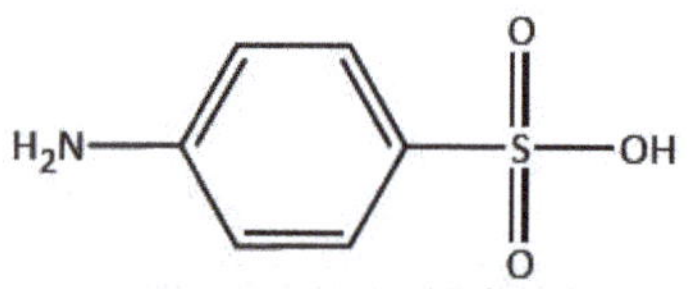

sulfanilamide (sulfa drug)

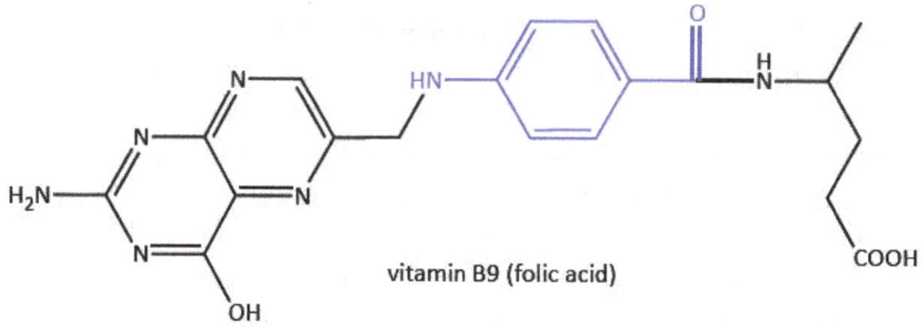

vitamin B9 (folic acid)

22.2 Nomenclature

22.2.1 Amines

IUPAC

To name an amine, the longest alkane chain is identified. The *–e* suffix is removed and is replaced with an *–amine* suffix.

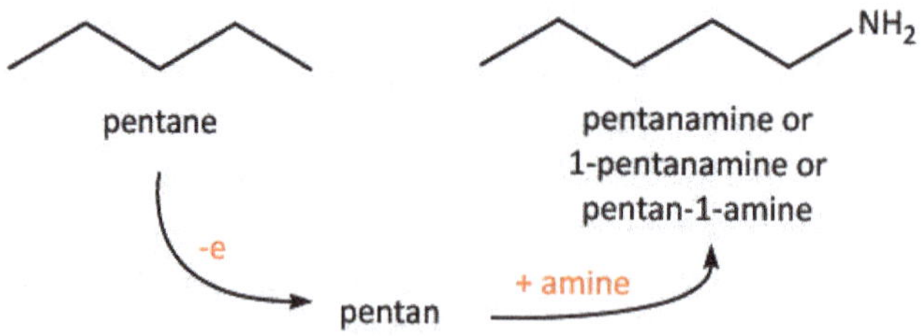

If the amino group is not on the end, #1, carbon atom, a number needs to be used to identify the location of the amino group. The carbon chain should be numbered starting at the end closest to the amino group.

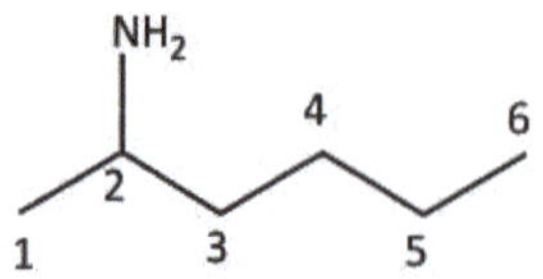

2-hexanamine or
hexan-2-amine

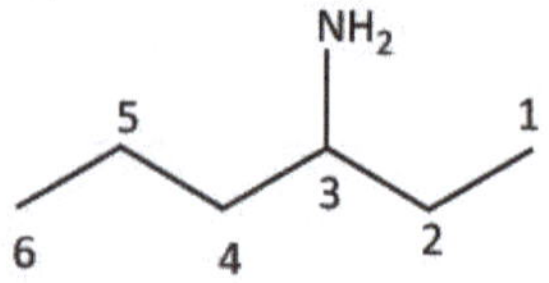

3-hexanamine or
hexan-3-amine

If it is secondary or tertiary, the prefix *N-* is used for each shorter substituent that is attached to the nitrogen (N) atom of the amine. The *N-* prefix tells us the shorter substituent is located on the nitrogen atom (N) instead of the longest, numbered carbon chain. The "N" prefix is written for each small alkyl substituent on the nitrogen atom.

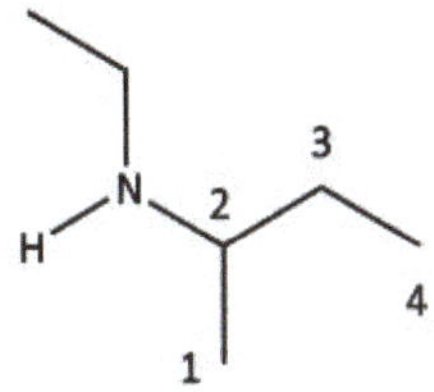

N-ethyl-2-butanamine or
N-ethyl-butan-2-amine

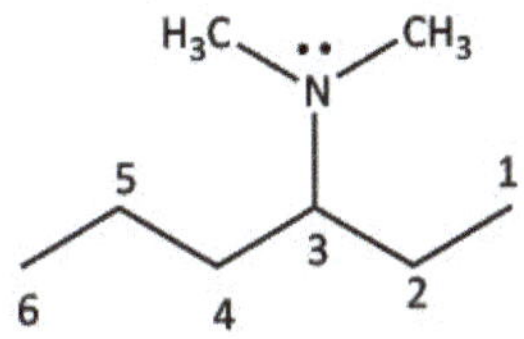

N,N-dimethyl-3-hexanamine or
N,N-dimethylhexan-3-amine

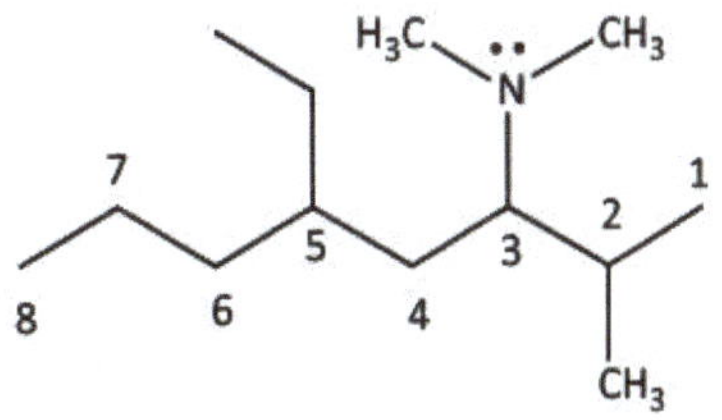

5-ethyl-2-N,N-trimethyl-3-octanamine or
5-ethyl-2-N,N-trimethyloctan-3-amine

Common names

Alkyl amines

Amines are commonly named by naming the alkyl groups on the nitrogen atom followed by the suffix *-amine*. If two or three identical groups are on the nitrogen atom, the prefixes *di-* or *tri-* are used.

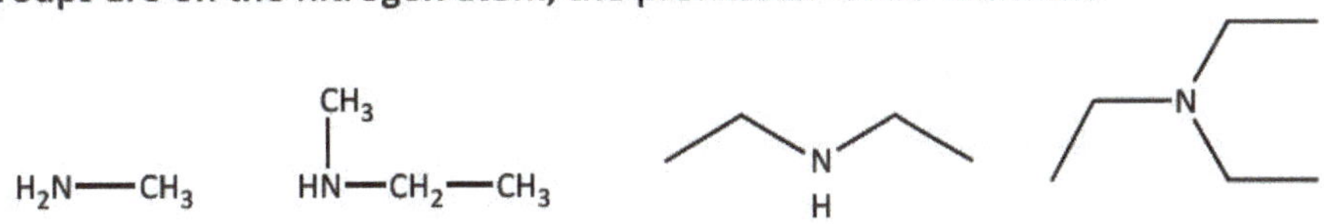

IUPAC	methanamine	N-methylethylamine	N-ethylethylamine	N,N-diethylethylamine
Common	methyl amine	ethylmethylamine	diethylamine	triethylamine

Aryl amines

The first aryl amines that you should memorize are aniline, pyrrole, and pyridine. Purine and pyrimidine are biologically important because they are the bases that we find in DNA. There are many more common names for amine-based rings that you will learn as you progress in your study of organic chemistry, but this is a good starting point.

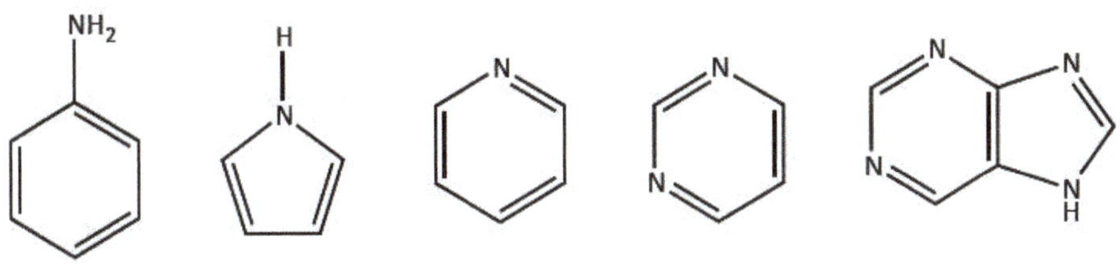

aniline pyrrole pyridine pyrimidine purine

As a substituent

If the molecule is more complex, an $-NH_2$ group can be named as an "amino" substituent.

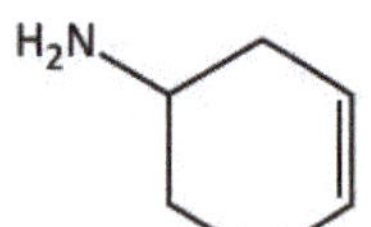

4-aminocyclohexene

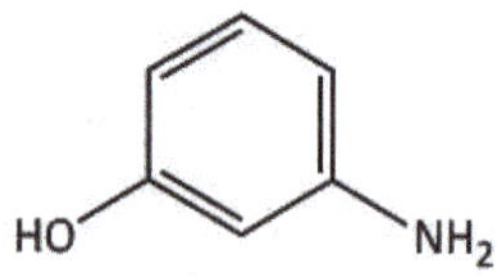

m-aminophenol

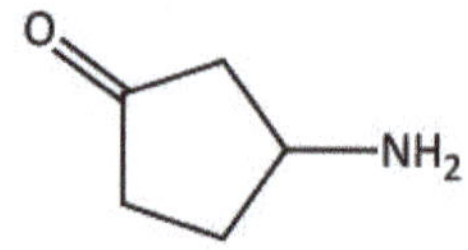

3-aminocyclohexanone

22.2.2 Nitriles

IUPAC

We name the parent compound by counting the number of carbon atoms of the longest chain (including the carbon atom of the C≡N). The numbering for the compound starts at this nitrile carbon atom. This is followed by the word *–nitrile.*

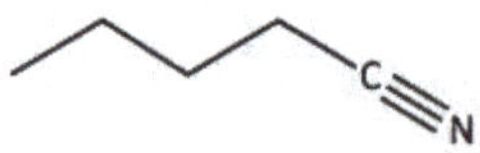

pentane nitrile

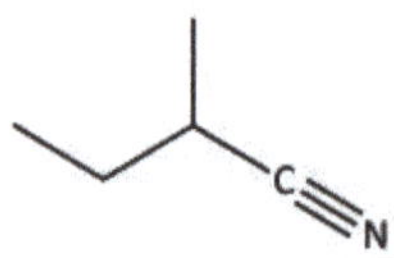

2-methylbutane nitrile

As a substituent

If the molecule is more complex, the -C≡N group can be named as a substituent, a "cyano" group.

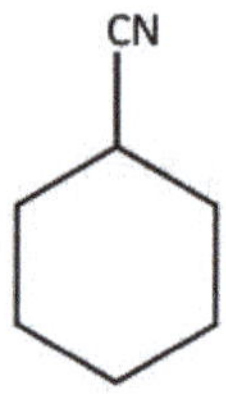

cyanocyclohexane

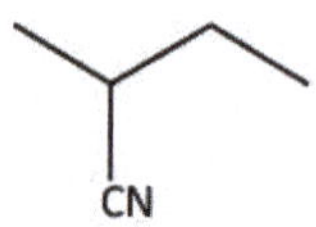

2-cyanobutane

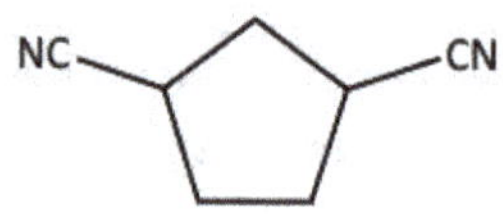

1,3-dicyanocyclopentane

Common names

Nitrile common names are derived from taking the common name of the corresponding carboxylic acid, dropping the *–ic acid* suffix and adding the suffix *–onitrile.* Notice the carbon of the carbonyl (with three C-O bonds) is the one that becomes the carbon of the nitrile (with three C-N bonds).

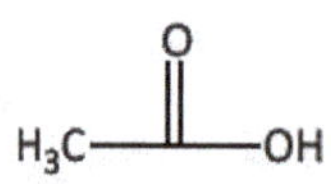

acetic acid

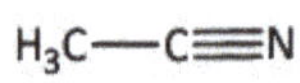

acetonitrile

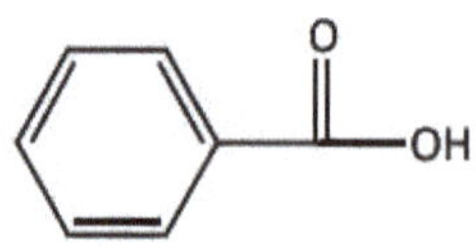

benzoic acid

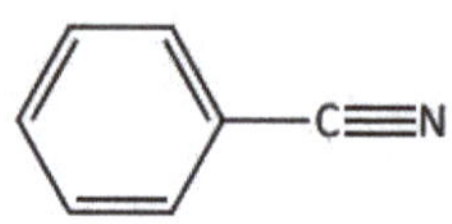

benzonitrile

22.1 Draw the following compounds.

a) 1,4-dicyanocyclohexane
b) 3-ethylheptanenitrile
c) 4-aminocyclohexanone
d) 3-ethylaniline
e) 2-methylpyridine
f) triethylamine
g) 2-octanamine
h) N-methyl-2-pentanamine

22.3 Spectroscopy

22.3.1 IR Spectroscopy

NH, NH_2

NH bond stretches all appear around 3300.

For benzylamine, there are sp^3 CH peaks below 3000 for the CH_2 and sp^2 CH peaks above 3000 for the aromatic ring. The aromatic overtones below 2000 confirm this. At 3300, we see a two-hump peak. This means there are two N-H bonds present (an NH_2).

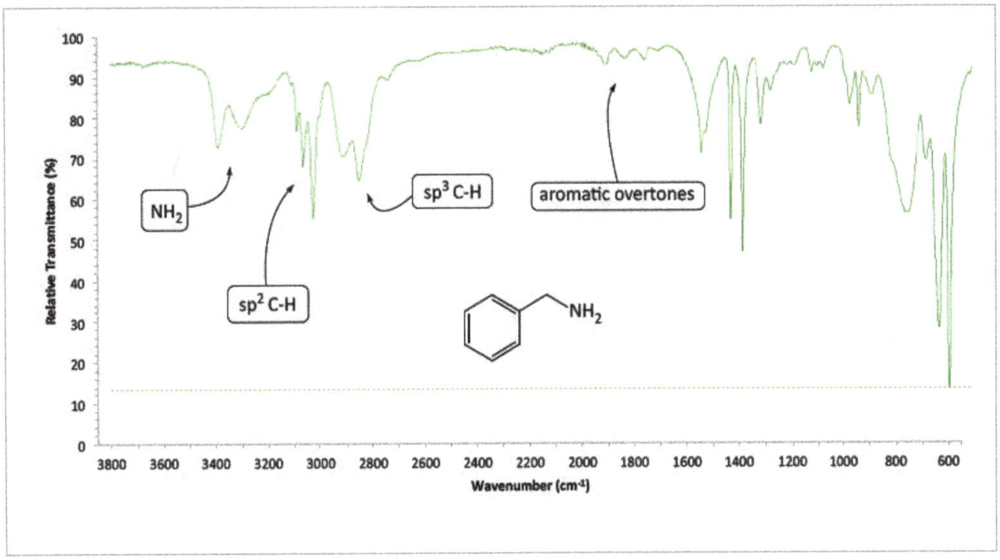

Figure 22.4 IR spectrum of benzylamine

For N-methylaniline, we see the sp^2 and sp^3 CH peaks surrounding 3000. We also see the aromatic overtones indicating aromatic ring. We do see a one-hump peak at 3300. We know this is an N-H stretch or an sp CH stretch. It is fairly sharp. An OH stretch is usually broader and NH_2 has two humps. If it was an sp CH stretch, we would also see the C≡C peak below 2200. Therefore, we know that we have an amine with one N-H bond.

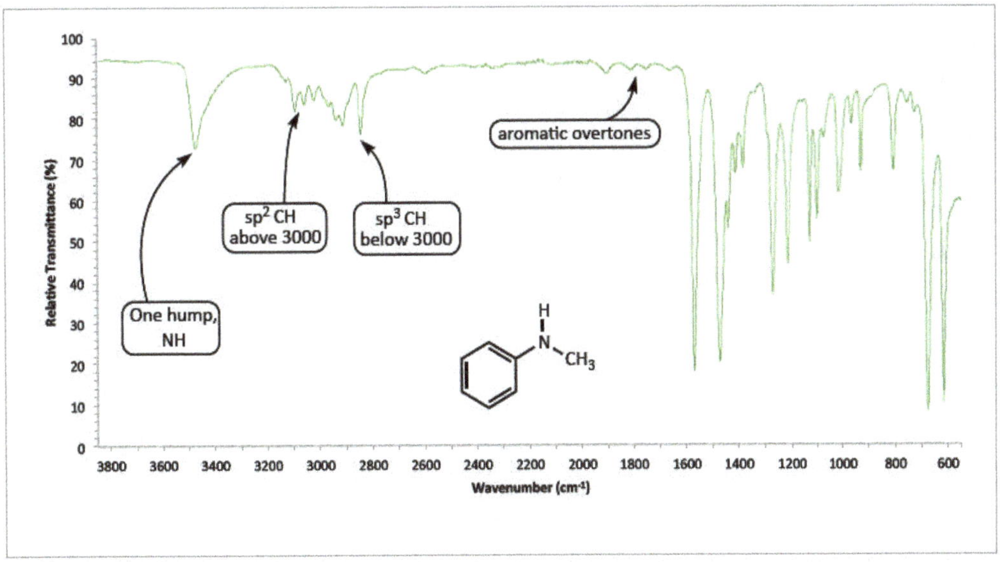

Figure 22.5 IR spectrum of N-methylaniline

Nitriles

In acetonitrile, the C≡N nitrile is the sharp peak just above 2200. C≡C IR peaks also show up near C≡N. We can tell this is not C≡C because C≡N shows up just above 2200 while C≡C is usually just below 2200. Additionally, we usually only see C≡C if it is at the end of a molecule (terminal alkyne). In that case, we also see the sp CH stretch at 3300, which is absent here.

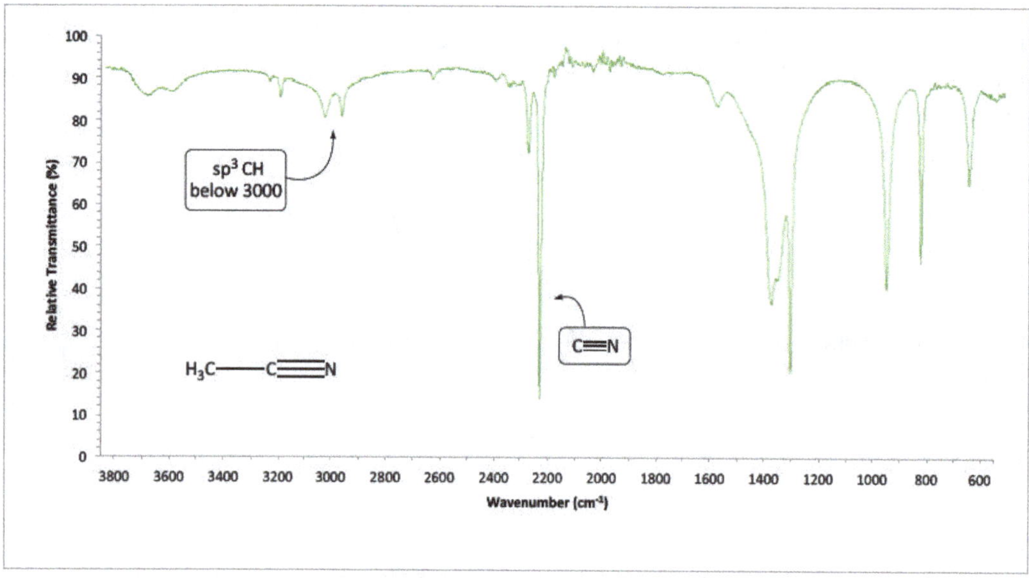

Figure 22.6 IR spectrum of acetonitrile

Amides

Amide carbonyl shifts are lower than normal carbonyl shifts around 1700. Amide carbonyl shifts are typically in the 1660s range. For acetamide, we see that the carbonyl peak is quite low in the 1660s indicating an amide. We look for secondary peaks at 3300 and find two N-H bonds. We also see sp_3 CH peaks below 3000. It is fairly small since there are only three C-H bonds.

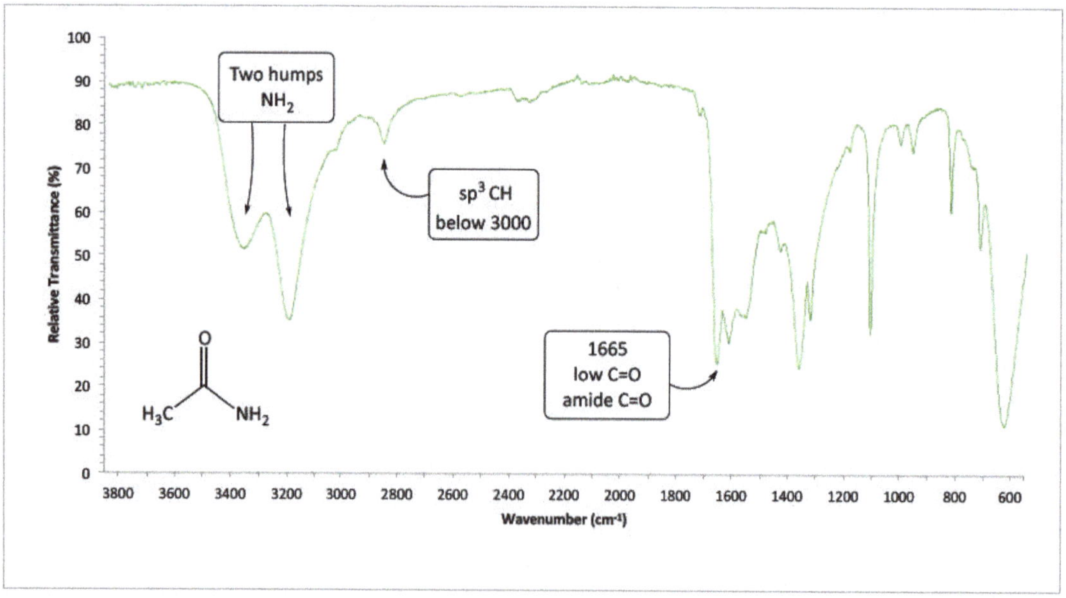

Figure 22.7 IR spectrum of acetamide

22.3.2 Mass Spectrometry

Nitrogen atoms do not have M+2 isotope peaks to help us identify them. But, they do give a clue when they are present. All of the M^+ peaks you have seen so far have something in common; they all have an even number m/z. This is remarkable. But, nitrogen compounds are different. If a molecule has an odd number of nitrogen atoms in it, it has an *odd* molecular mass. So, a typical nitrogen-containing compound with one nitrogen atom will be odd, but if there are two nitrogen atoms present, it is even. Below is the mass spectrum for methylamine, CH_3NH_2.

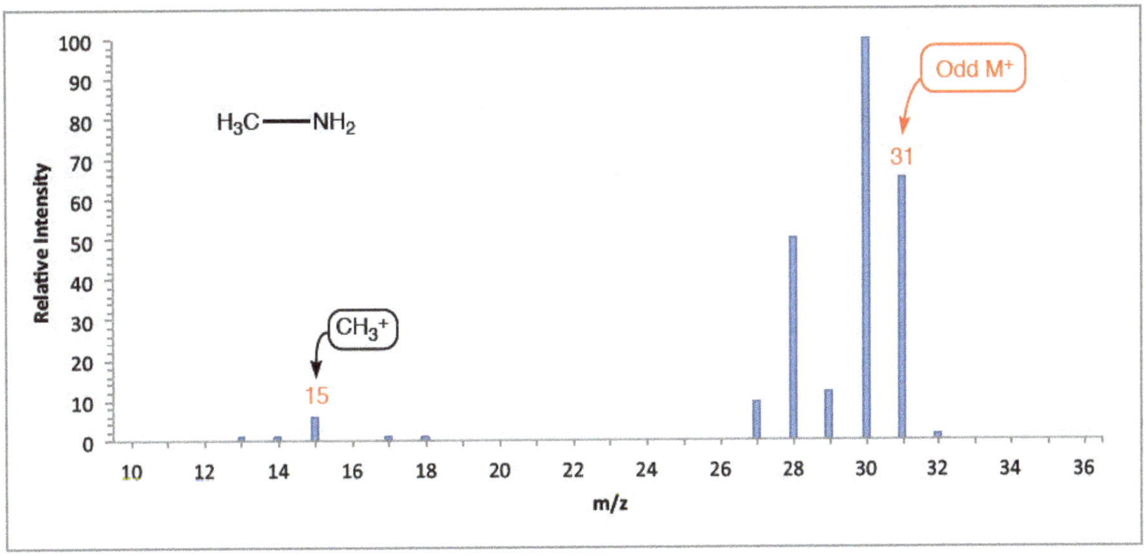

Figure 22.8 Mass Spectrum of methylamine

22.3.3 ¹H NMR Spectroscopy

Protons on oxygen or nitrogen atoms are a little strange. They can show up anywhere between 2-6 ppm. Because protons on OH and NH are involved in hydrogen bonding, their chemical shift can change depending on the temperature and concentration of the solution. They are not always broad peaks. Sometimes they stay on the molecule enough to couple.

Usually, these peaks are broad, hump-like peaks. This is because the hydrogen atom disassociates from the OH or NH bond. It falls off and returns to the molecule, so there usually is not time for it to couple with other protons.

One laboratory trick to help determine if a ^1H NMR peak is from an OH or NH proton is to add D_2O, heavy water, to the NMR sample. D_2O interacts with the OH or NH proton and the hydrogen atom and deuterium atoms switch to form O-D or N-D. The ^1H NMR peak then disappears confirming OH or NH. Protons on carbon atoms will not trade with deuterium of D_2O. A new peak will grow in the spectrum, though. HOD is formed in the reaction.

The protons on carbon atoms next door to a nitrile usually show up at 2-3 ppm, like when they are next door to a carbonyl. Protons on carbon atoms next door to the nitrogen atom of an amine usually show up at 3-4 ppm.

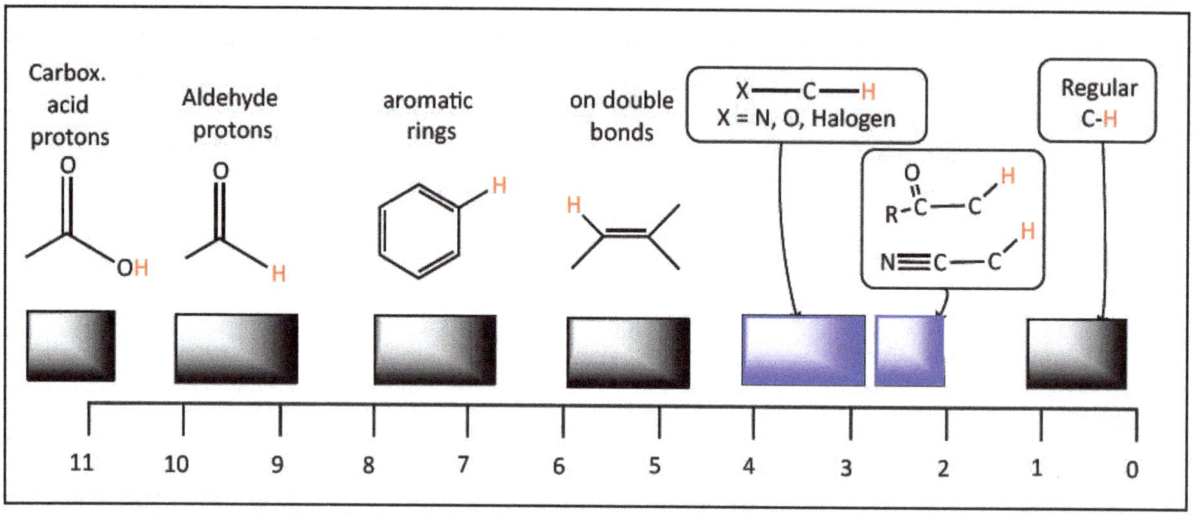

22.3.4 ^{13}C NMR Spectroscopy

Of course, nitrogen atoms do not show up in ^{13}C NMR spectra. But, carbon atoms next to nitrogen atoms show up between 50-100 ppm in a ^{13}C NMR spectrum. These carbon atoms are "next to something special". Nitrile carbon atoms typically show up 115-125 ppm.

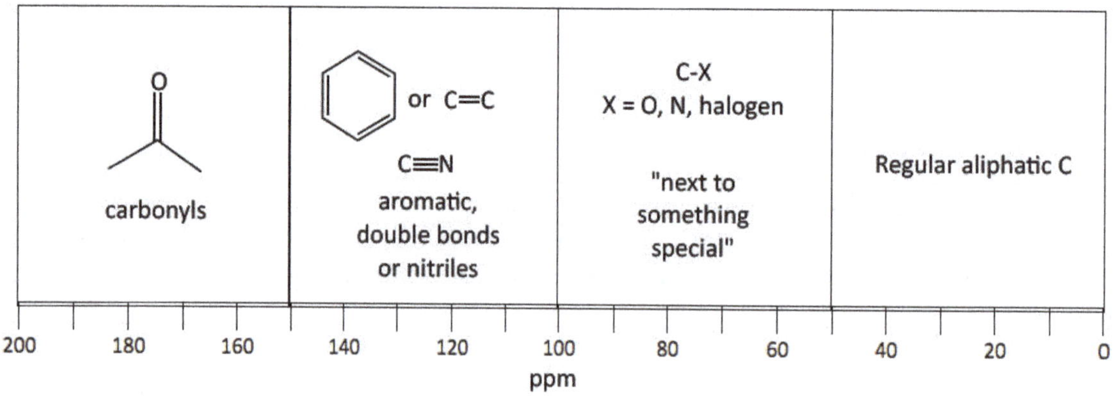

22.2 Fill in the blanks to complete the following statements about amine, nitrile, and amide spectroscopy.

a) If the M+ peak in a mass spectrum is ________, it tells us we have an odd number of nitrogen atoms in the compound.

b) If a compound is a primary amine and has an $-NH_2$ group, the IR spectrum should show a broad, double humped peak (rabbit ears) around ________ cm^{-1}.

c) If a compound is a nitrile, it should show a strong, sharp C≡N peak in the IR spectrum around ________ cm^{-1}.

d) If a compound is a secondary amine and has an –NH group, the IR spectrum should show a broad peak with one hump on it (a shark's tooth) around ________ cm^{-1}.

e) The C=O carbonyl peak for an amide shows up at a ___________ wavenumber than "regular" ketone or aldehyde C=O peaks around 1710 cm^{-1}. The carbonyl peak for an amide is closer to __________ cm^{-1}.

f) In the ^{13}C NMR spectrum, the carbon atom for the C=O of an amide appears in the ______________ ppm region. In the ^{13}C NMR spectrum, the carbon atom in the C≡N of a nitrile appears in the ____________ ppm region.

22.4 Amine basicity

Amines, with their negative lone pair of electrons, tend to be basic. This means the negative lone pair of electrons can grab a proton (H^+). This turns the amine into amine salt. Salts have both the positive cations and negative anions combined. Salts are soluble in water.

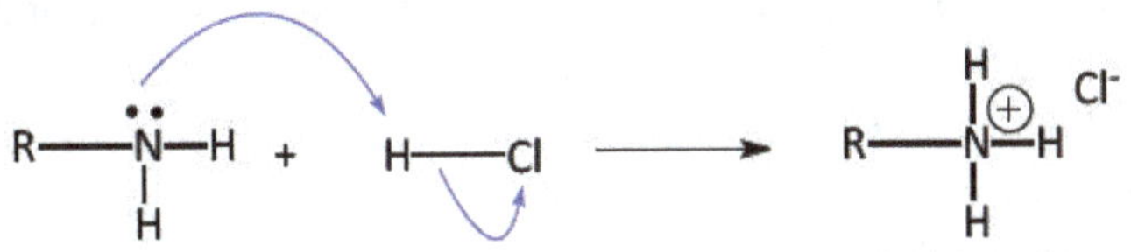

amine salt

H_2O soluble

In the lab, we know that amines will react with aqueous HCl to form a salt that can dissolve in water. If we have an unknown compound that we think might be an amine, we can test that hypothesis by trying to dissolve a small amound of our compound in 10% aqueous HCl. If our compound dissolves, we could determine that we have a basic compound – perhaps an amine.

In medicines, it is sometimes preferable to sell the amine as its water-soluble salt. For instance, Benadryl and Sudafed are the hydrochloride salts of the amines.

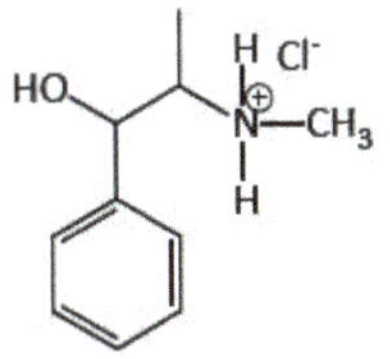

Ephedrine hydrochloride
Ephedrine HCl
Sudafed

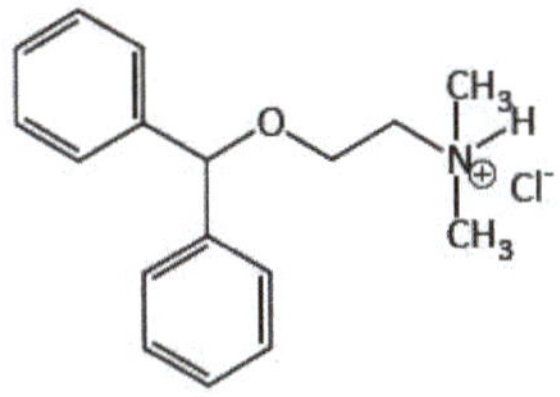

Diphenhydramine hydrochloride
Diphenhydramine HCl
Benadryl

An interesting example, although illicit, form of amines and amine salts is cocaine and crack. Crack (the free base) is formed when the hydrochloride salt powder form of cocaine is added to boiling water with sodium bicarbonate, $NaHCO_3$, (to deprotonate it). A rock forms. It is broken into smaller pieces and sold as crack.

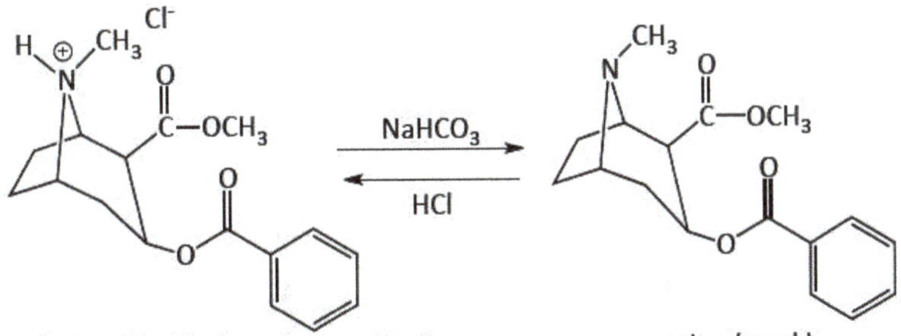

cocaine hydrochloride (powder cocaine) cocaine (crack)

If an acid like HCl is produced in a reaction, an amine base is often used to soak up the acid that is produced. Pyridine is a common base for this task. We have seen this before in the tosylation of an alcohol. Notice that the H from the alcohol and the Cl from TsCl are not in the product. The pyridine base soaked up the HCl that is produced in this reaction.

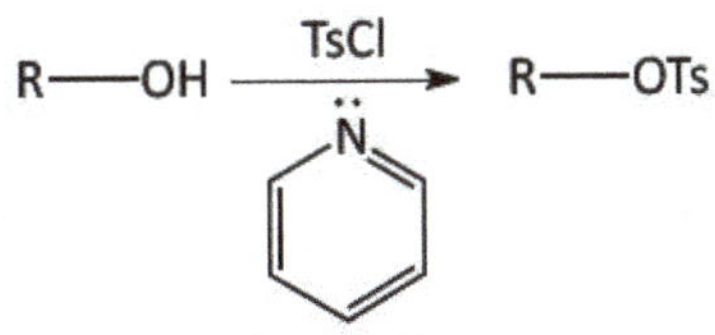

Pyridine

22.5 Amine & Nitrile Synthesis

22.5.1 S_N2 Synthesis

Amines and nitriles can be synthesized from an alkyl halide (or other good leaving group) with the appropriate amine or cyanide nucleophile. The same rules apply as always for an S_N2 backside attack. The reaction works best if the backside of the alkyl halide is not crowded or sterically hindered. Methyl and 1° alkyl halides work well, 2° alkyl halides work OK, but 3° alkyl halides do not work well. Any good leaving group on the substrate will work. The leaving group can be tosylates or halides like iodide, bromide, or chloride.

If the nucleophile is ammonia. . .

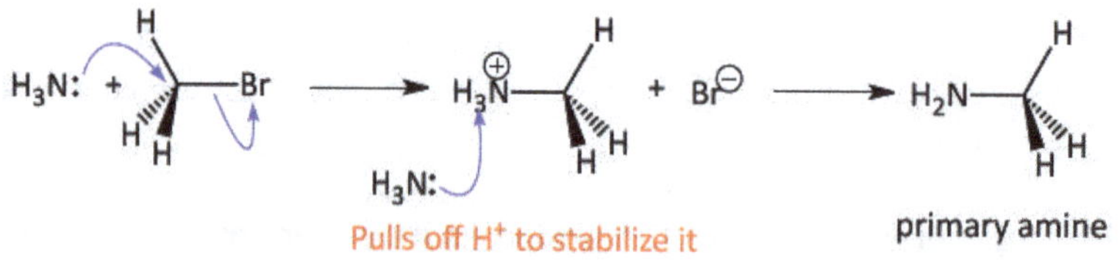

If the nucleophile is a primary amine. . .

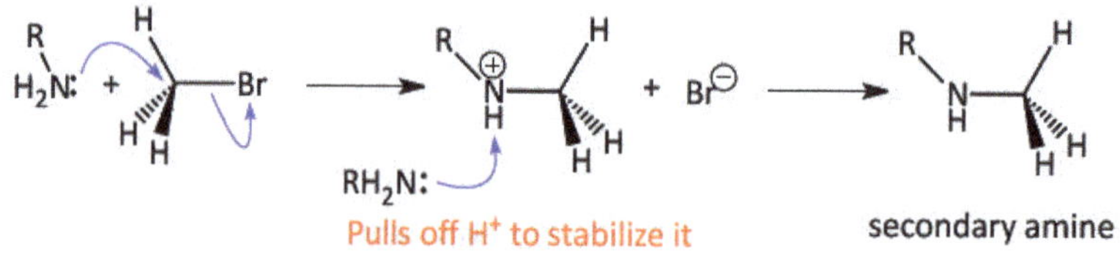

If the nucleophile is a secondary amine. . .

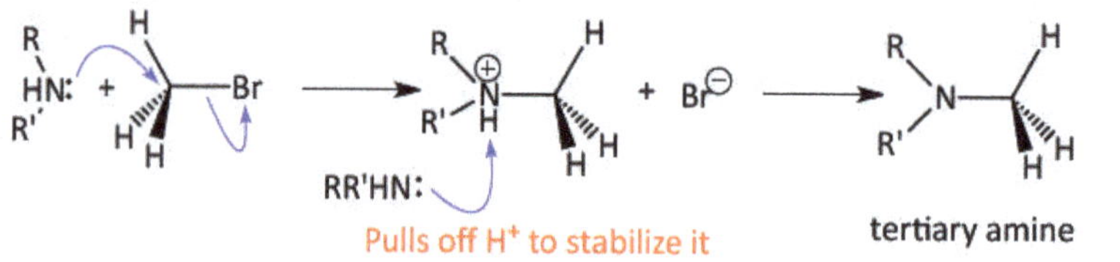

If the nucleophile is a cyanide. . .

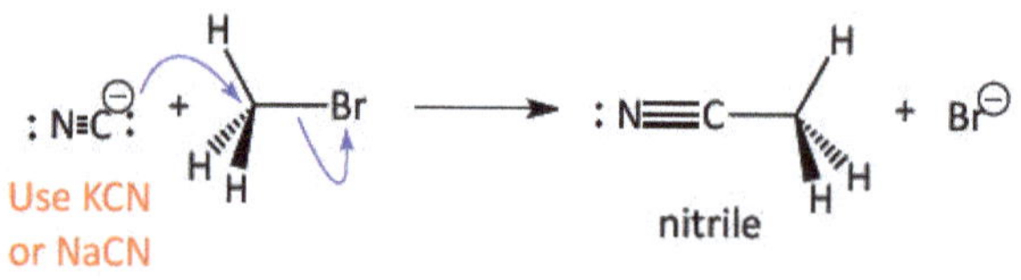

22.3 Draw the products of the following reactions.

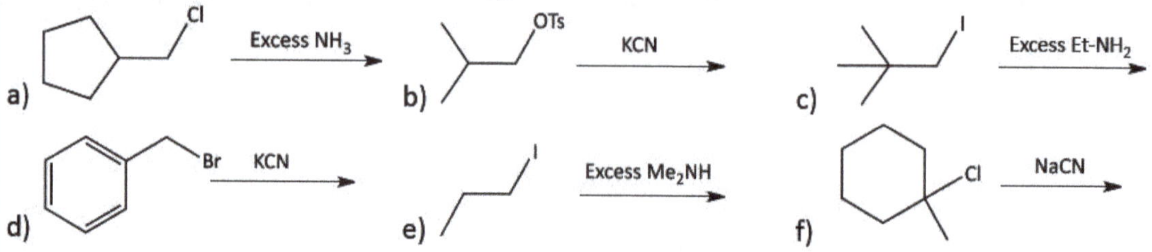

22.5.2 Reduction of nitriles

Nitriles can be reduced to primary amines. Notice that there are three carbon-nitrogen bonds in the nitrile. A good reducing agent like lithium aluminum hydride, $LiAlH_4$ (LAH) can reduce the cyano group to a primary amine making an $-NH_2$ group. The hydrides, H:⁻, can twice attack the carton atom of the nitrile. Following protonation of the doubly negative nitrogen atom, a primary amine is formed.

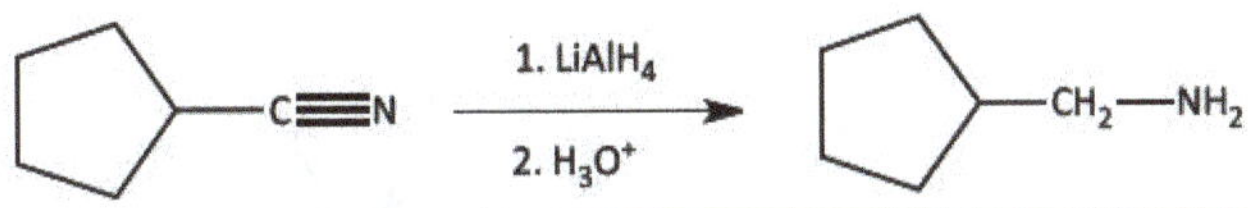

22.4 Draw the products of the following reactions.

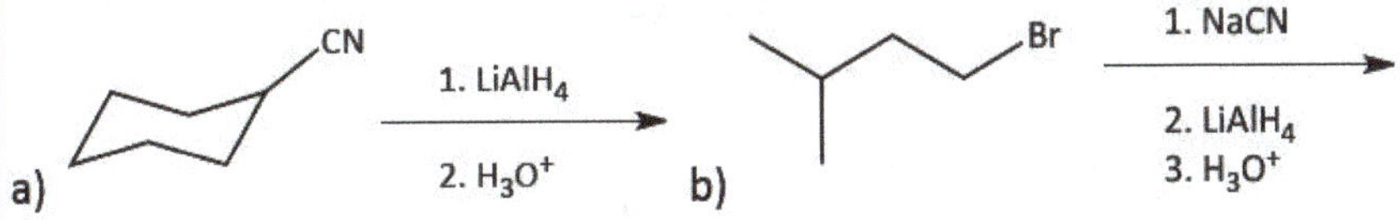

22.5.3 Reduction of imines

Much like nitriles, imines (C=N compounds) can be reduced with a good reducing agent like lithium aluminum hydride, $LiAlH_4$ (LAH). The imine is like the nitrile, but it is already half reduced to the final primary amine. Once the hydride has attacked the carbon atom of the imine, H:⁻, followed by quenching with acid/water, an amine is formed.

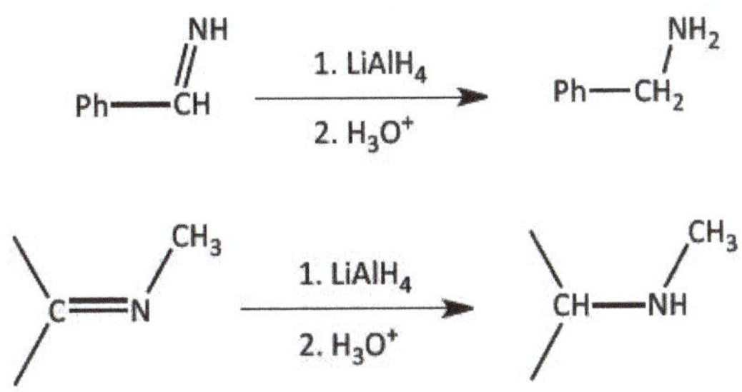

It is a good reminder that carbonyl (C=O) compounds can react with amines, lose water, and form imines.

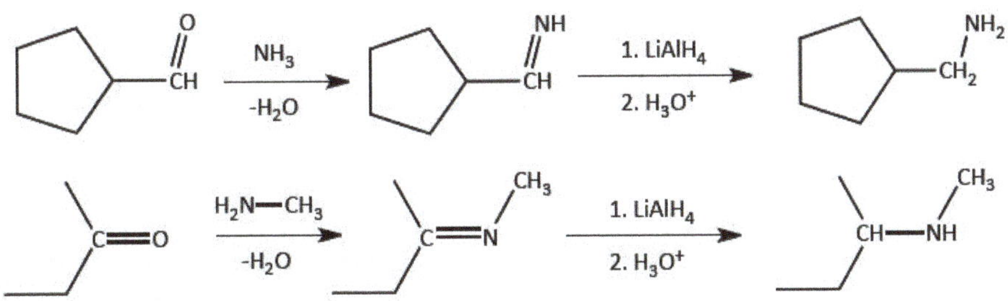

22.5 Draw the products of the following reactions.

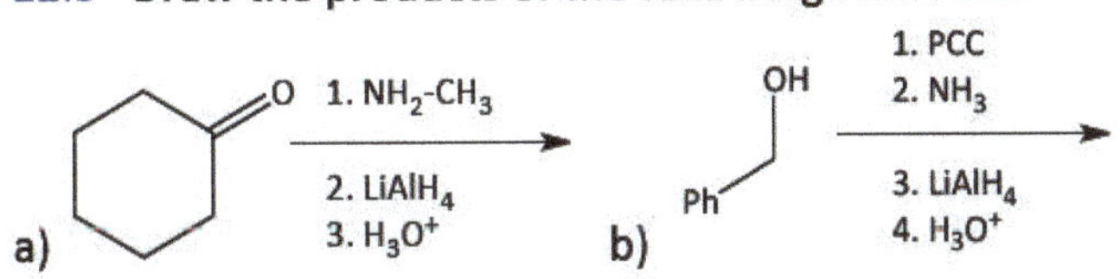

22.5.4 Reduction of amides

Amides can also be reduced to amines. The C=O carbonyl of the amide can be reduced (making fewer carbon-oxygen bonds) to a methylene $-CH_2-$ with a strong reducing agent like lithium aluminum hydride, $LiAlH_4$ (LAH) followed by quenching with acid water. Similarly, a hydrogenation using H_2 and Ni catalyst can perform this reduction.

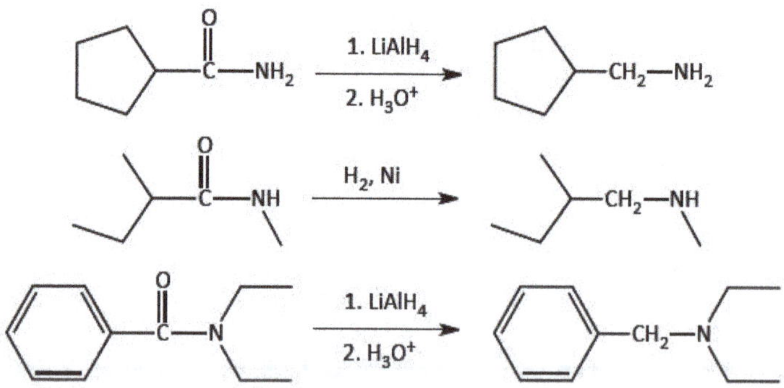

22.6 Draw the products of the following reactions.

a) 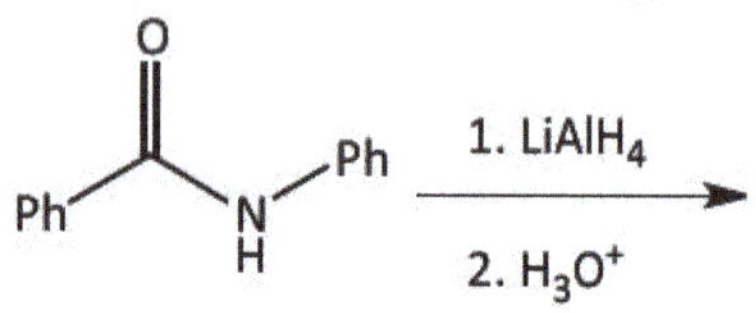

b) 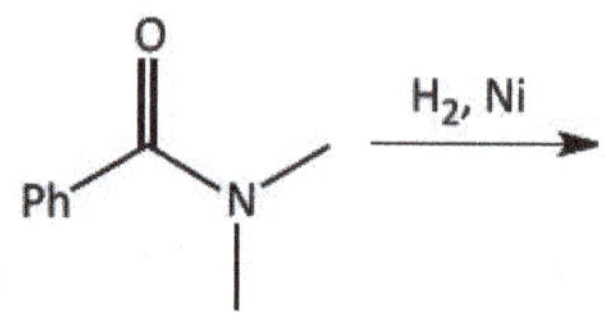

22.5.5 Reduction of nitrobenzene

Aniline synthesis to set up the Sandmeyer reaction

We have already learned how to nitrate an aromatic ring with nitric and sulfuric acid. A reduction of a nitro group on an aromatic ring to an amino group with fewer nitrogen-oxygen bonds is accomplished by reducing it with iron filings and hydrochloric acid followed by hydroxide. The meta-directing nitro group is now an o,p-directing amino group. Remember that aminobenzene is commonly called aniline. Memorize this Fe, HCl reagent, but realize that you might encounter other reductions in the future ($SnCl_2/HCl$ or H_2/Pt).

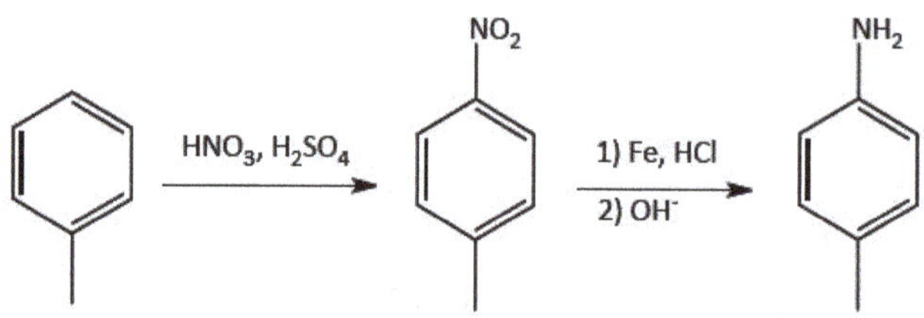

22.7 Show the two reaction steps needed to turn propyl benzene into 4-aminopropylbenzene.

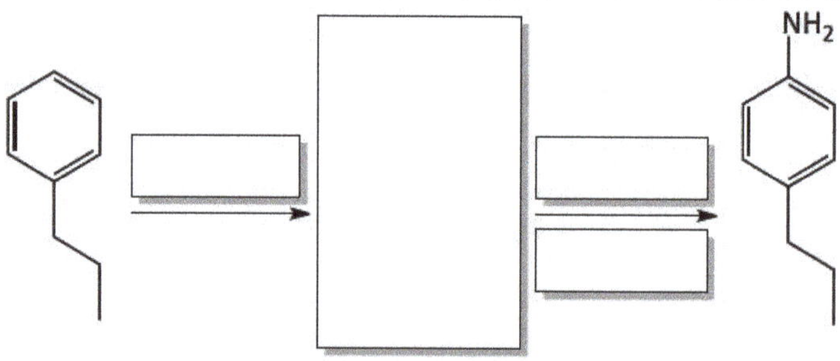

22.6 Amine & Nitrile Reactions

22.6.1 The Sandmeyer Reaction

Aniline (amino benzene) compounds are excellent precursors to make many different types of substituted benzene compounds. The amino group on an aromatic ring can be turned into the world's greatest leaving group, N_2, by reacting it with sodium nitrite ($NaNO_2$) and HCl. Once this happens, the N_2 leaving group is so good; this compound is only stable below 10°C. The N_2 leaving group can be replaced with many different types of nucleophiles.

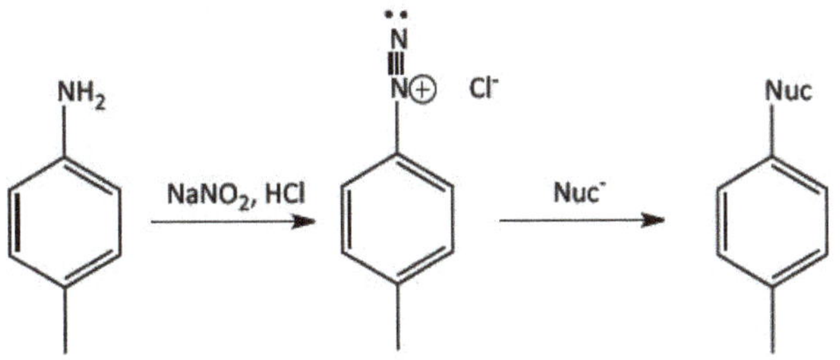

The substitution reaction is called a Sandmeyer reaction if the nucleophile is a Cu^+ salt like cuprous cyanide (CuCN), cuprous chloride (CuCl), or cuprous bromide (CuBr).

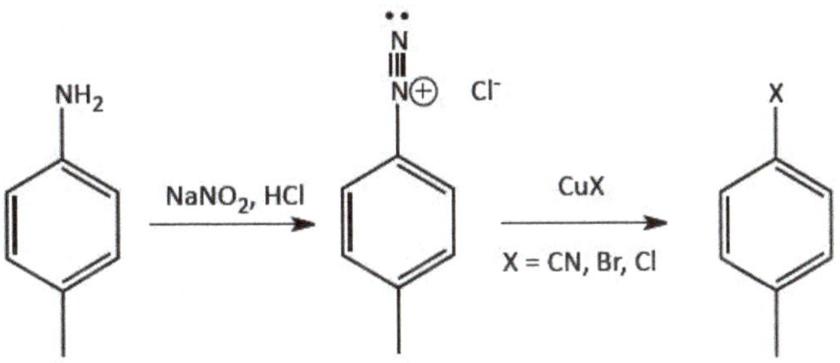

Many different nucleophilic substitutions can occur. With acid and water, we get a phenol. With HBF_4, we get an aromatic fluoride. With potassium iodide (KI), we get an aromatic iodide. With hypophosphorous acid (H_3PO_2), the N_2 is replaced with a proton to make a benzene ring. You should spend some time now memorizing these reagents and products.

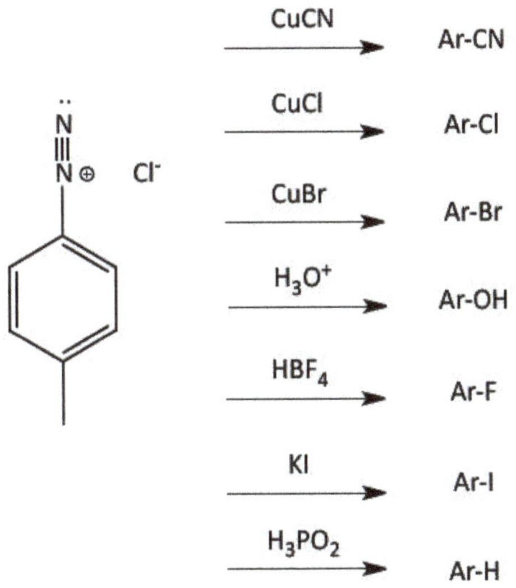

22.8 Draw the products of the following reagents with aniline (amino benzene).

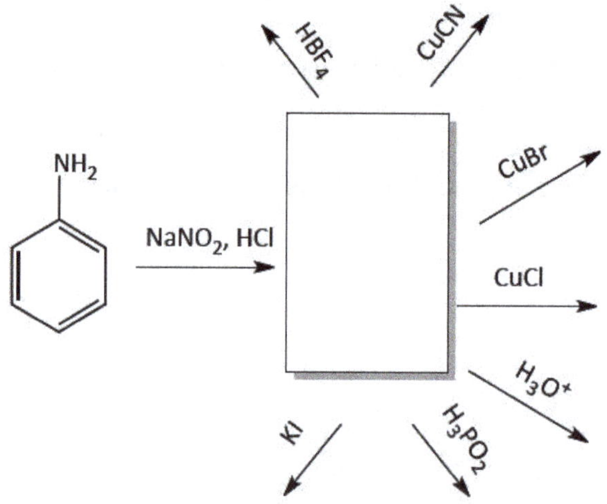

22.6.2 Acid chloride and amine to make amide

Perhaps the easiest way to make an amide is to react an amine with an acid chloride. The –Cl leaving group can be replaced with an amine. HCl, hydrochloric acid, is released in this reaction. That is why two equivalents of amine are used. One amine reacts with the acid chloride to make an amide and the other equivalent of amine reacts with the HCl that is produced. Acid chlorides can be easily prepared from carboxylic acids and thionyl chloride, $SOCl_2$.

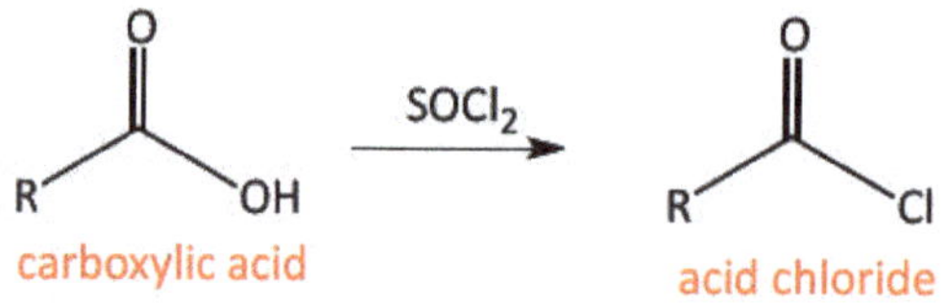

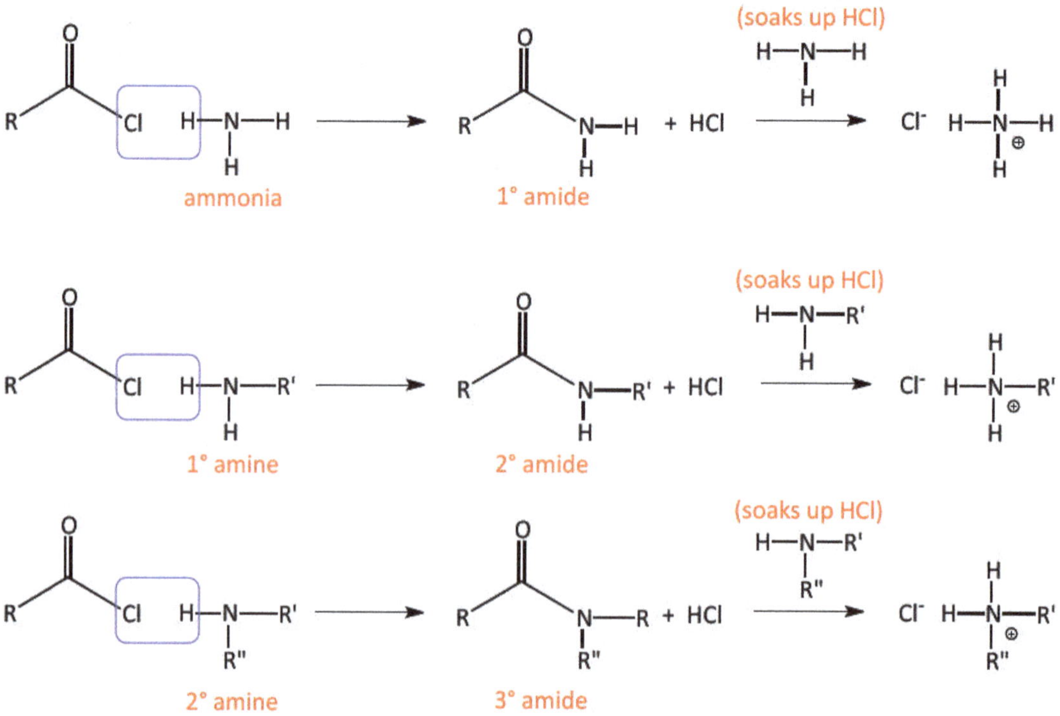

22.9 Draw in the appropriate products and reactants for the following reactions.

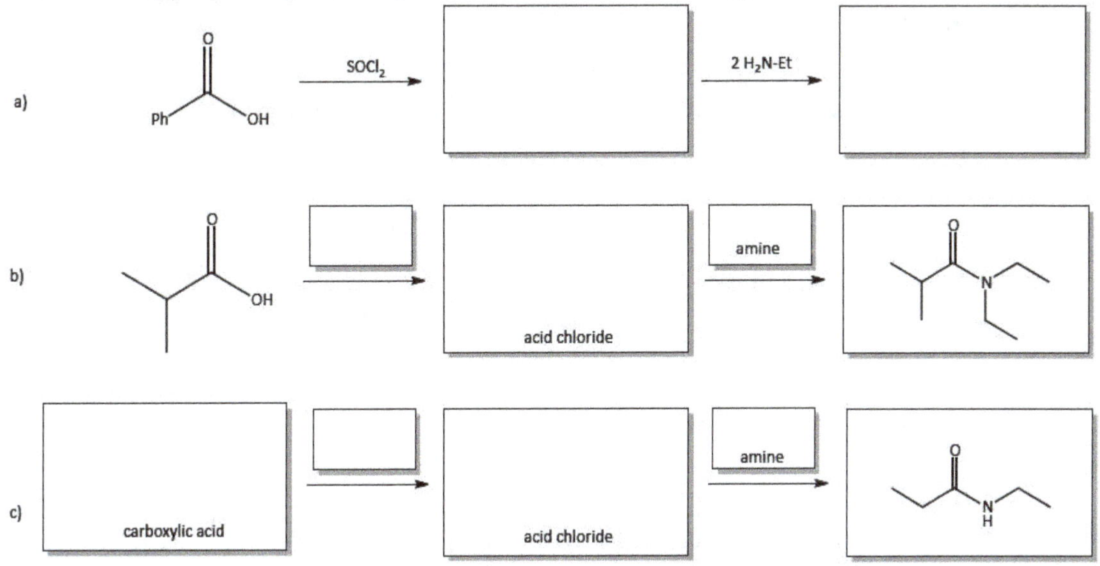

22.6.3 Carboxylic acid and amine to make amide

In the previous section, we first turned a carboxylic acid into an acid chloride to perform the quick, easy reaction with an amine to form an amide. If you don't want to take the extra time to first turn a carboxylic acid into an acid chloride, you can react the carboxylic acid directly with an amine, under heat, to form an amide. This reaction is not as easy as the reaction with an acid chloride, but it saves you one extra step in the synthesis. Water is produced in this reaction.

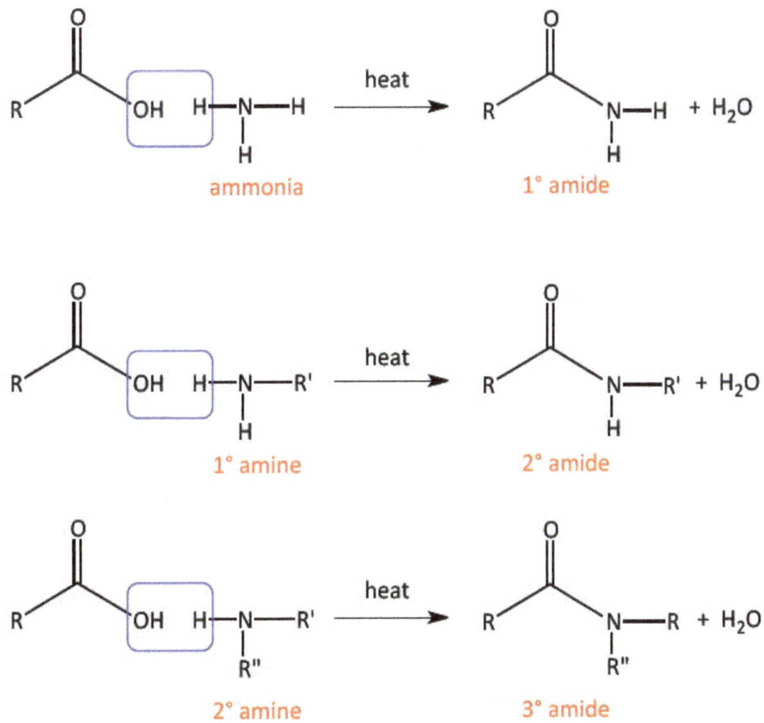

22.10 Draw in the appropriate products and reactants for the following reactions of carboxylic acids reacting with amines to make amides.

a) 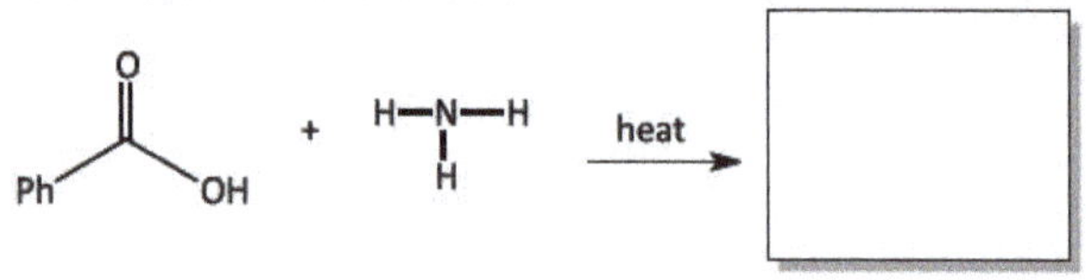

b) 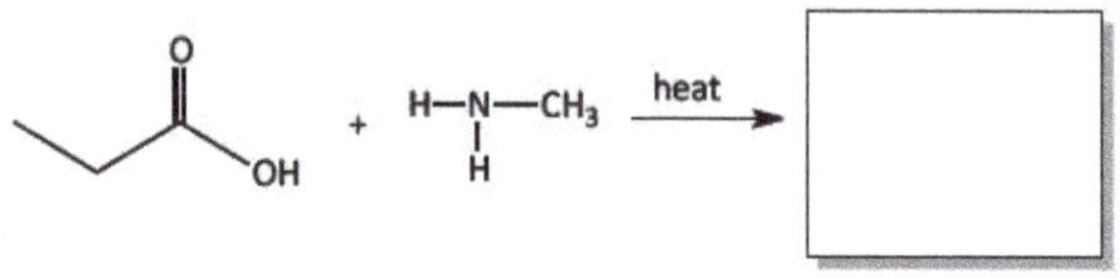

c) 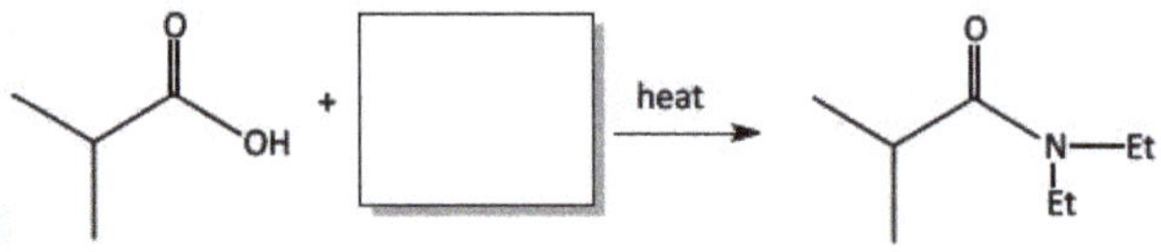

d) 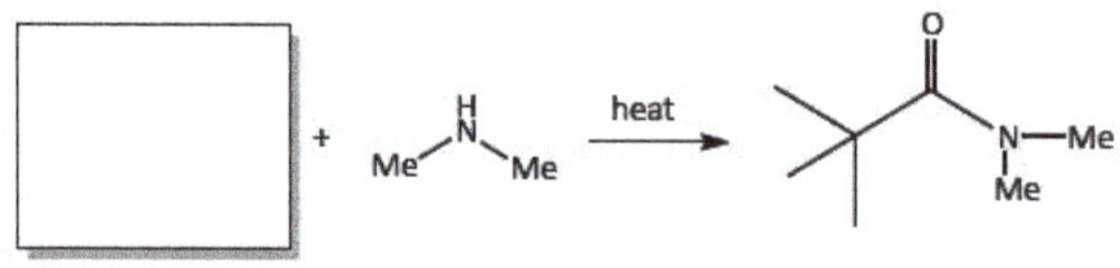

e) 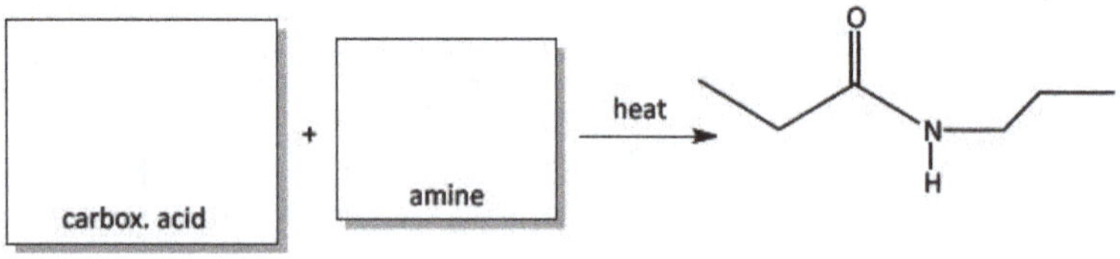

Linking amino acids to make peptides

This amide synthesis from the reaction of a carboxylic acid with an amine is an important biochemical reaction.

Remember, peptides are chunks of protein. They are long chains of amino acids linked together.

dipeptide tripeptide

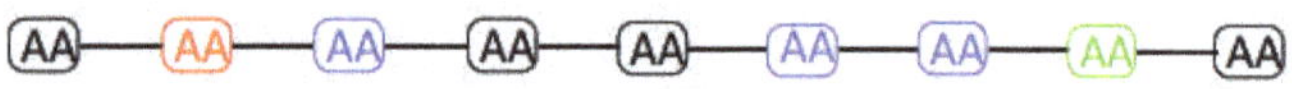

peptide

These amino acids are linked together by the reaction of an amino acid carboxylic acid reacting with another amino acid amino group to make an amide. In biochemistry, the bonds that link amino acids together in peptides are called peptide bonds or peptide linkages. In organic chemistry, the functional group made in this reaction is called an amide.

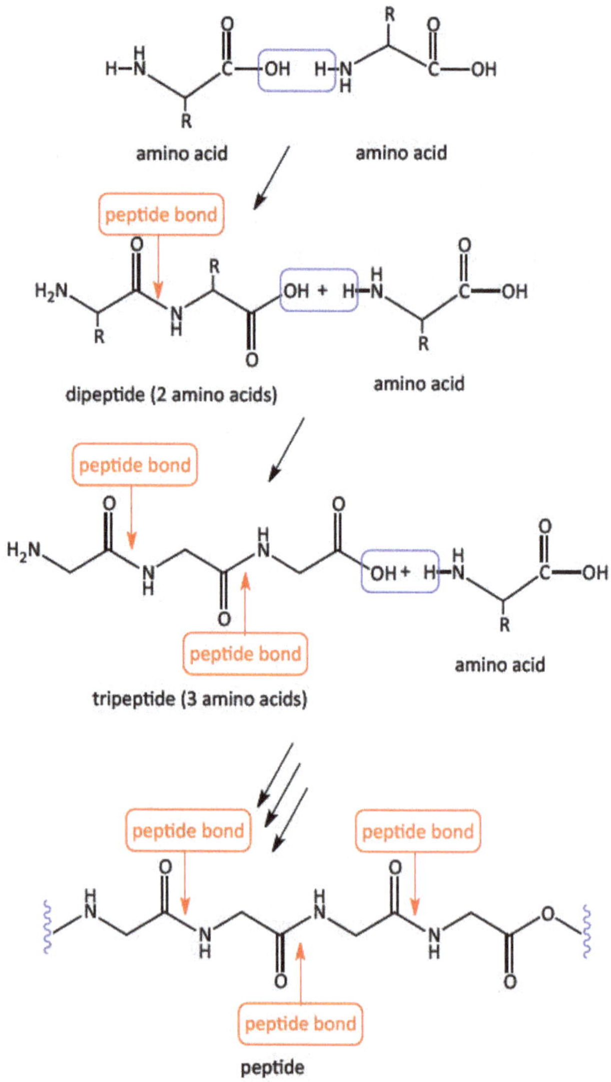

22.11 Draw the dipeptide made from two alanine amino acids.

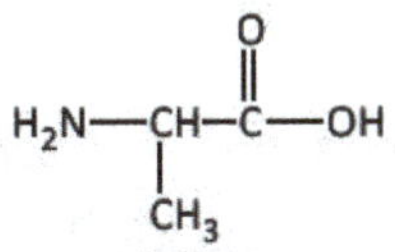

alanine

22.6.4 Ester and amine to make amide

We have seen that perhaps the easiest way to make an amide is with an acid chloride and an amine. The chloride is a great leaving group. Although it is a more difficult reaction, we can also make amides by reacting an amine with a carboxylic acid. The –OH is a worse leaving group than the chloride. Similarly, an ester can react with an amine to make an amide. The –OR alkoxy group is also not as good of a leaving group as the chloride, but the reaction does proceed with heat. Alcohol is produced in this reaction.

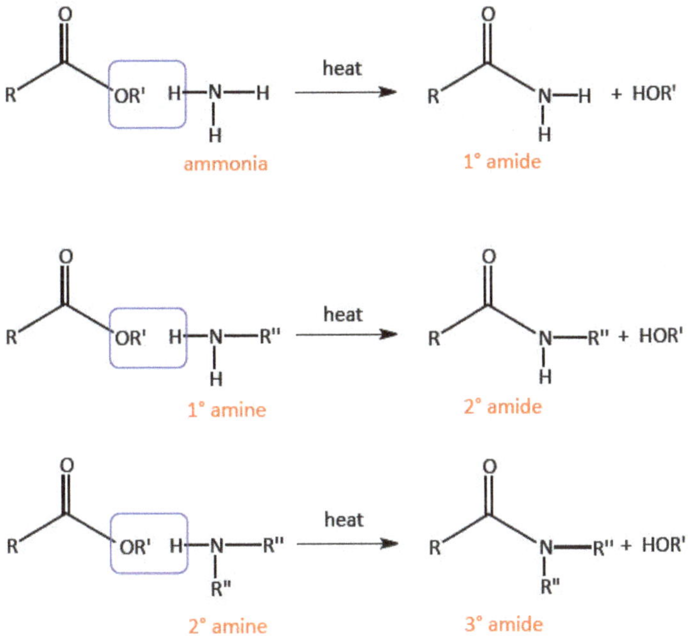

22.12 Draw in the appropriate products and reactants for the following reactions of esters reacting with amines to form amides.

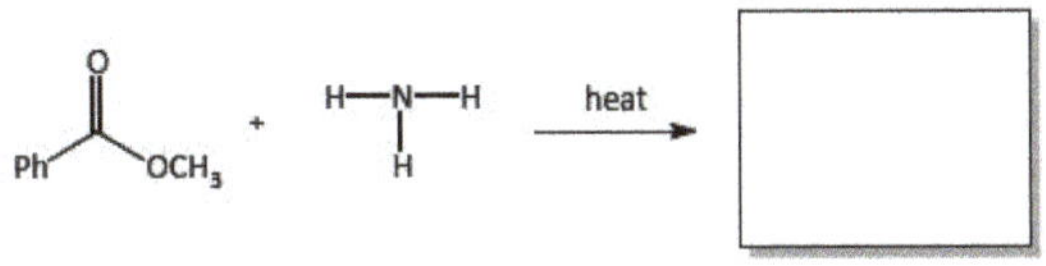

a)

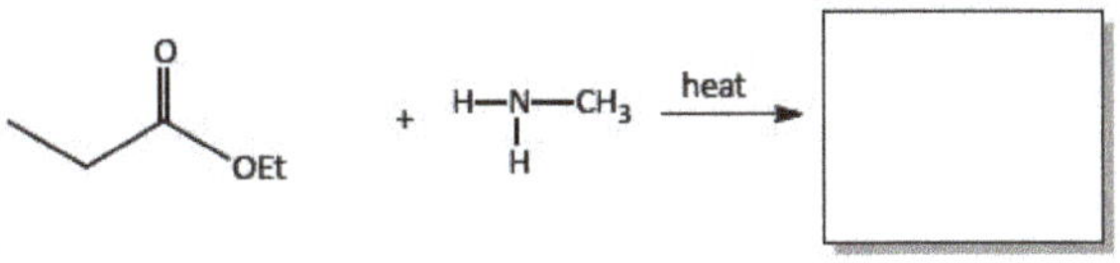

b)

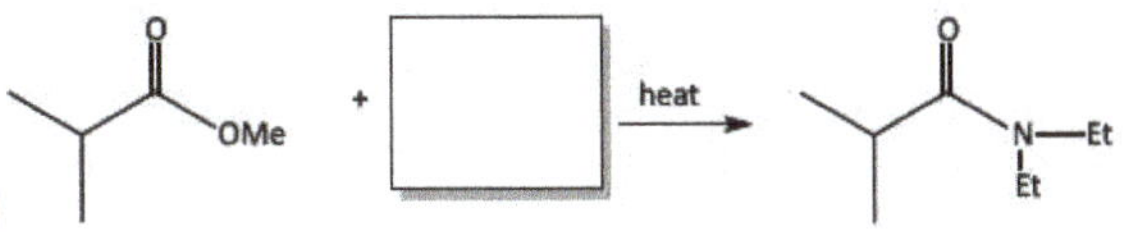

c)

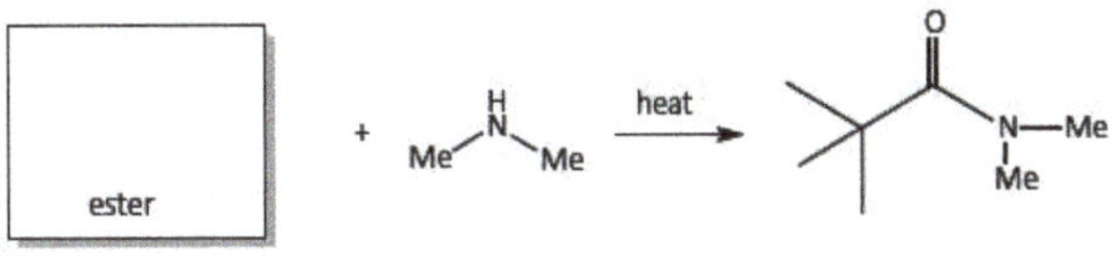

d)

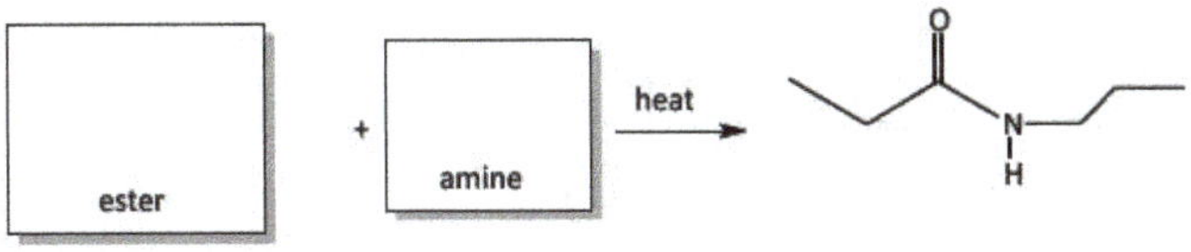

e)

22.6.5 Imine Hydrolysis

Imines have two carbon-nitrogen bonds. When imines are hydrolyzed (heated with acid and water), it forms ketones with two carbon-oxygen bonds. Notice the two carbon-nitrogen bonds form two carbon-oxygen bonds, the carbonyl, of the ketone.

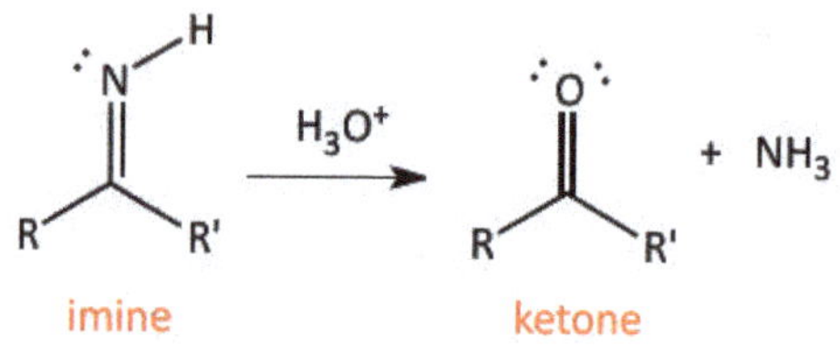

22.6.6 Ketones from Nitriles

The negative carbon atom of a Grignard reagent can attack the partially positive carbon atom of a nitrile. The electrons go up onto the nitrogen atom, until we quench the reaction with acid water to form an imine (C=N compound). We saw in the previous section that when imines are heated up with acid and water, they undergo hydrolysis to form ketones.

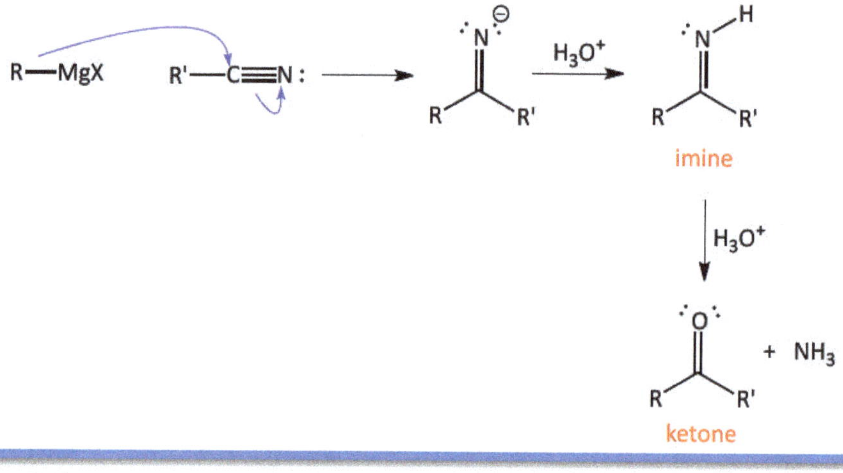

22.13 Fill in the blank boxes with the appropriate product or reagents.

a) 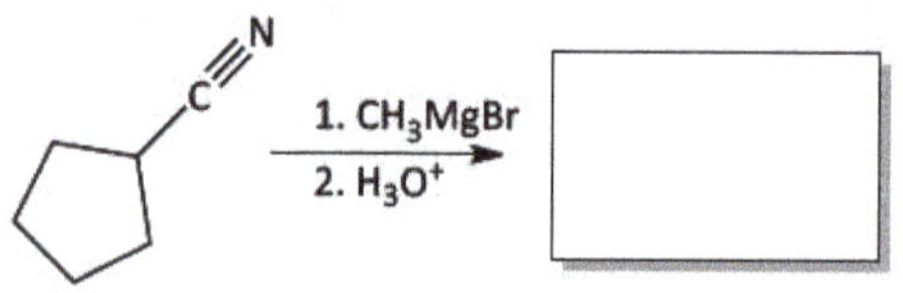

b) 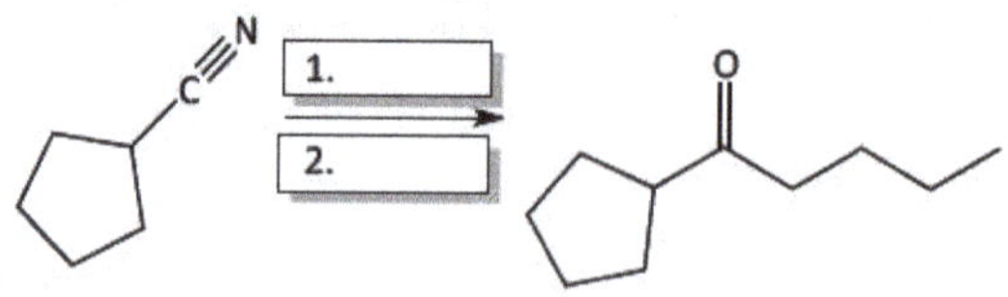

22.6.7 Hydrolysis of nitrile to carboxylic acid

Nitriles, with their three carbon-nitrogen bonds, can undergo hydrolysis to form carboxylic acids, which contain three carbon-oxygen bonds.

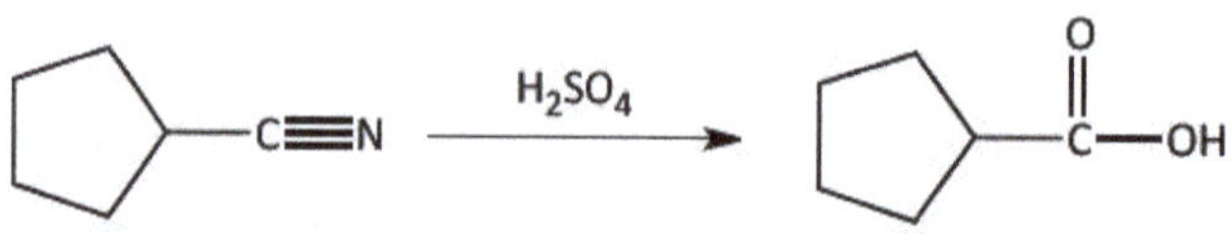

Nitrile synthesis review

Since the previous two sections were reactions of nitriles, we should pause briefly to remind ourselves of a few nitrile syntheses. Nitriles can be formed from the S_N2 reaction of cyanide (from a cyanide salt) with an alkyl halide.

On an aromatic ring, nitriles can be formed through the Sandmeyer reaction.

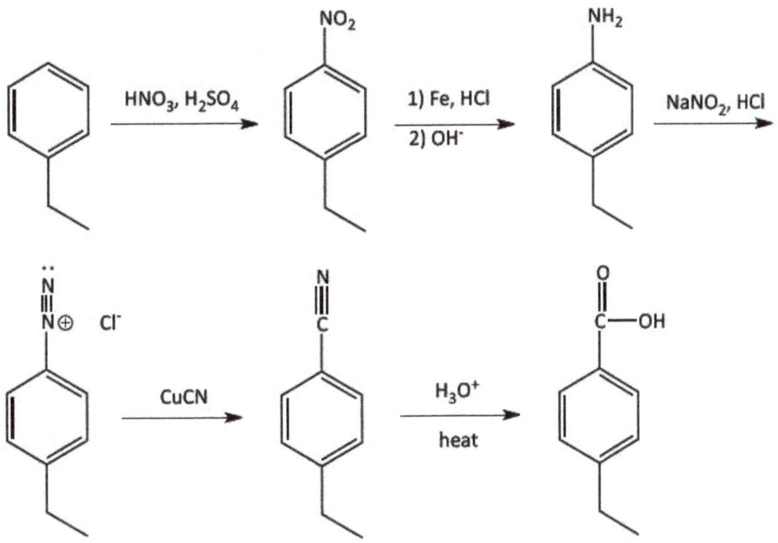

22.14 Fill in the blank boxes with the appropriate products or reagents.

a)

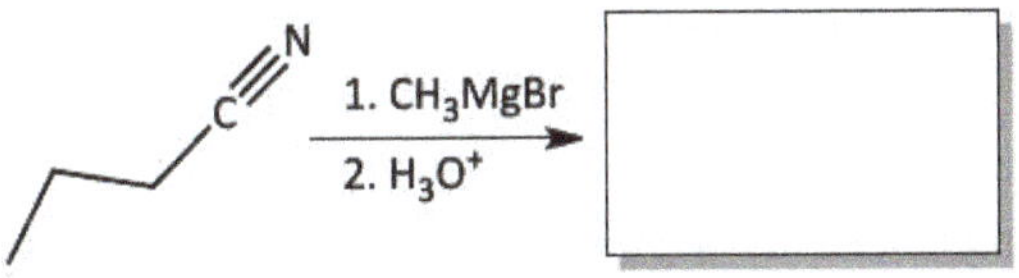

b)

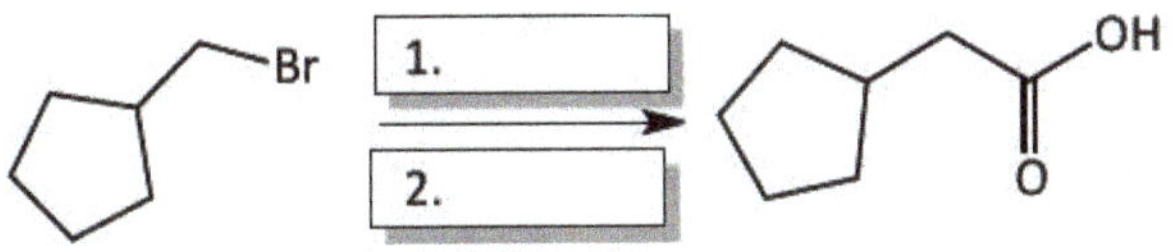

c)

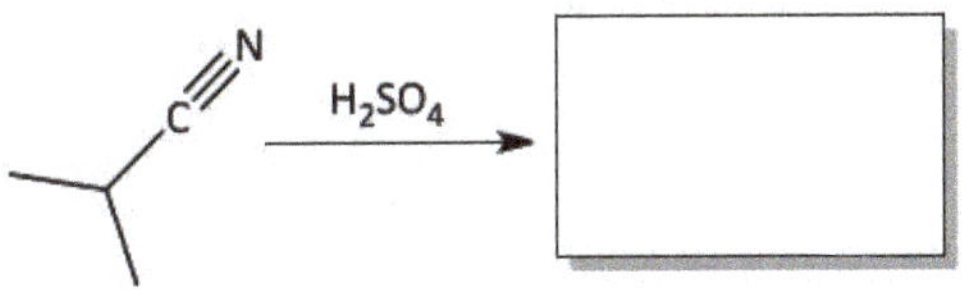

d)

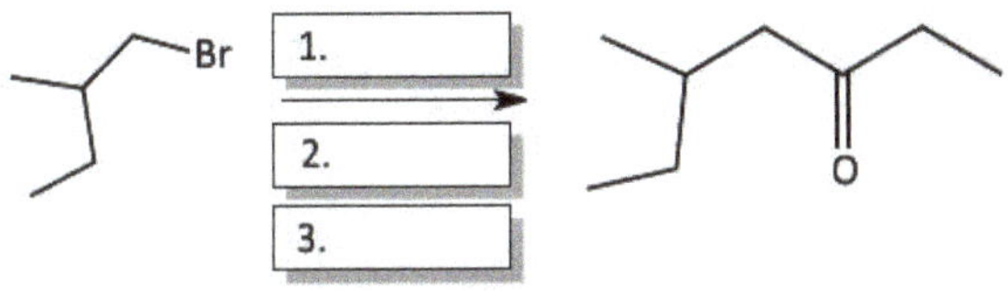

e)

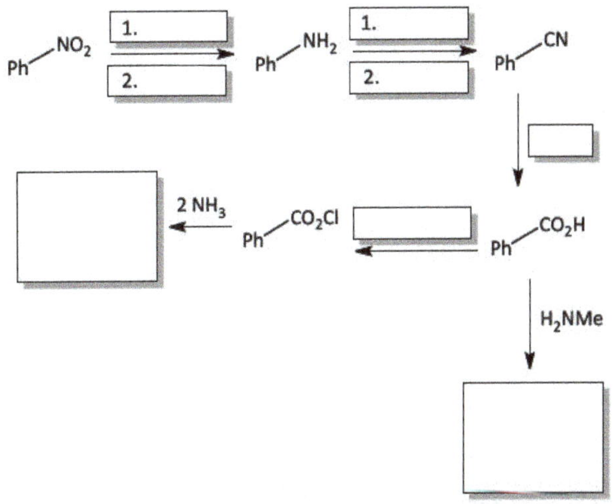

Reaction Summary

Amines as a base

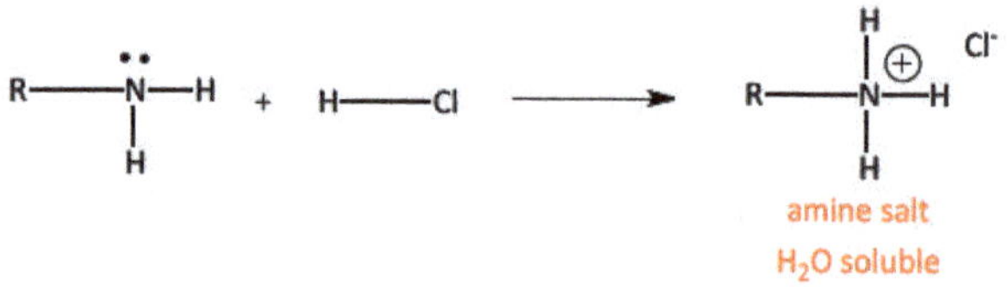

a)

SN2 Amine synthesis

b) 

SN2 Nitrile synthesis

c) 

Reduction of nitriles to amines

d) 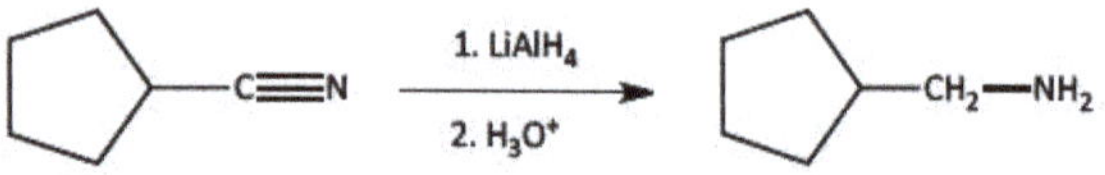

Reduction of imines to amines

e) 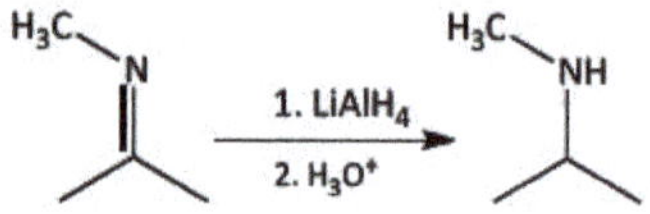

Reduction of amides to amines

f) 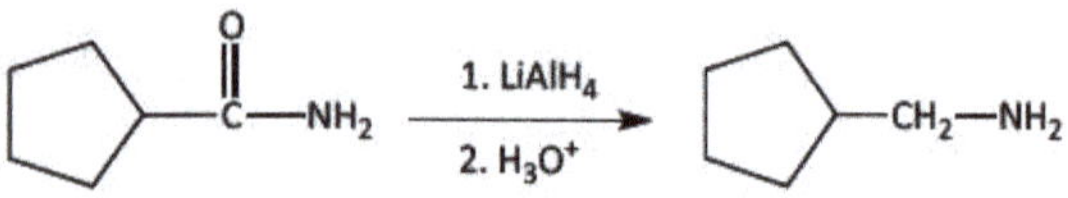

Reduction of nitrobenzene to aniline

g) 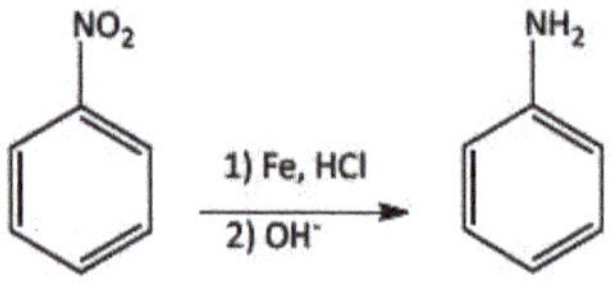

Sandmeyer reaction

h) 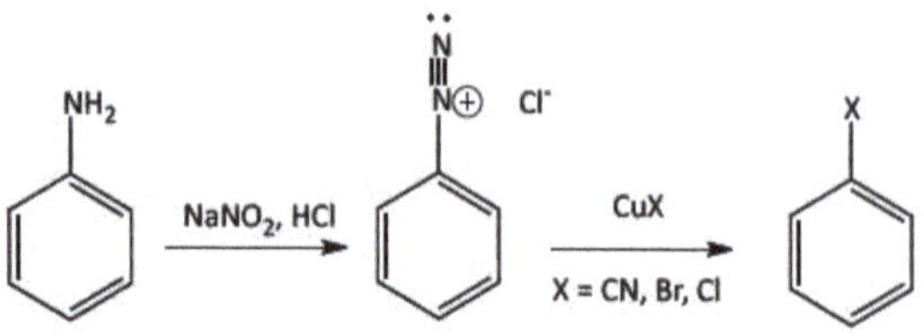

Sandmeyer-like reactions

i) 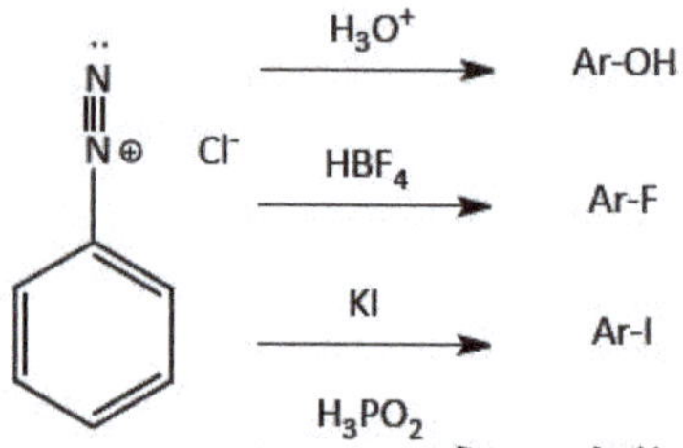

Acid chloride + amine to make amide

j) 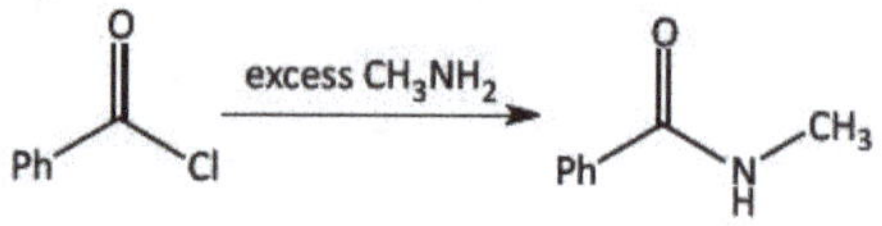

Carboxylic acid + amine to make amide

k) 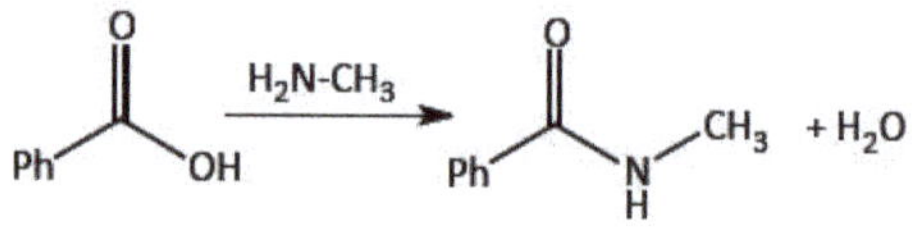

Peptide synthesis

l) 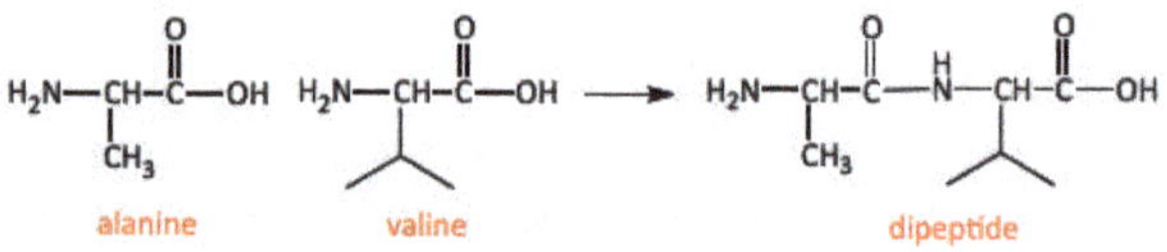

Ester + amine to make amide

m) 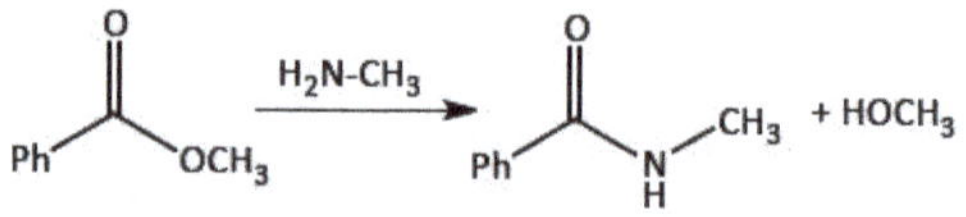

Imine hydrolysis to ketone

n) 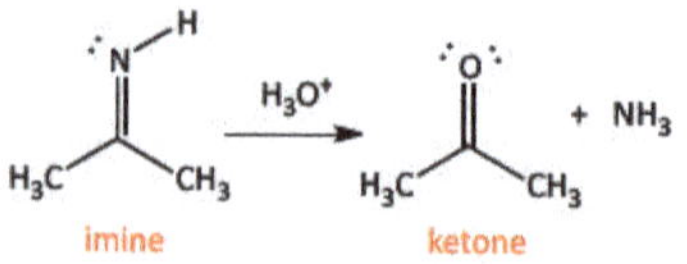

Nitriles to imines to ketones

o) 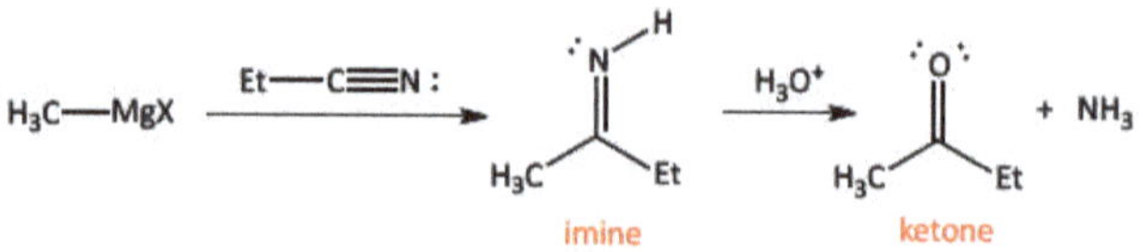

Hydrolysis of nitrile to carboxylic acid

p) 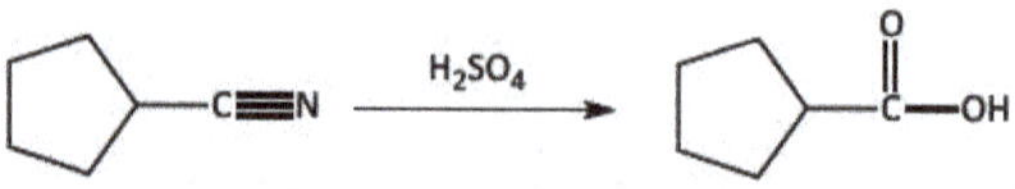

22.7 Problems

22.15 Draw the following compounds.
a) 1,4-dicyanocyclooctane b) 2-methylheptanenitrile c) 3-aminocyclopentanone d) *m*-nitroaniline
e) 3-methylpyridine f) diethylamine g) 3-octanamine h) N-methyl-2-butanamine

22.16 Name the following compounds.
a) PhCN b) $CH_3CH_2CH_2NH_2$ c) $CH_3CH_2CH_2NHCH_3$ d) CH_3CN e) $CH_3CH_2CH_2CN$

f) 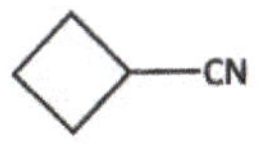

g) 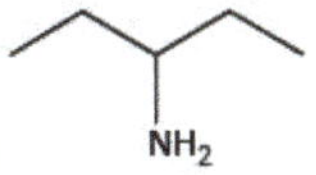

h) 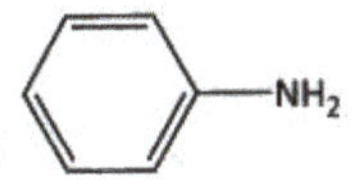

i) 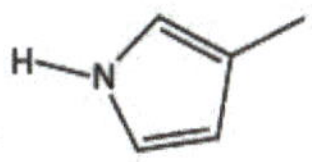

j) 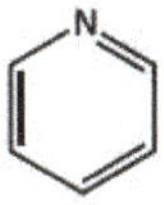

k) 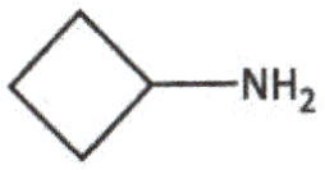

22.17 Identify the functional group(s) indicated in each generic IR spectrum between a primary amine, a secondary amine, a nitrile, and an amide. Additionally, which contains a phenyl ring?

a) 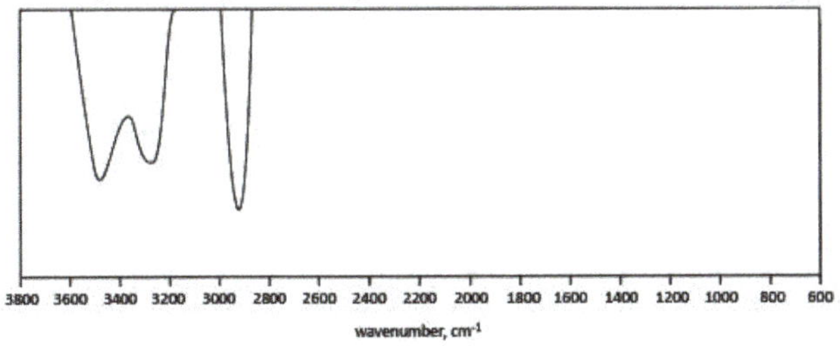

b) 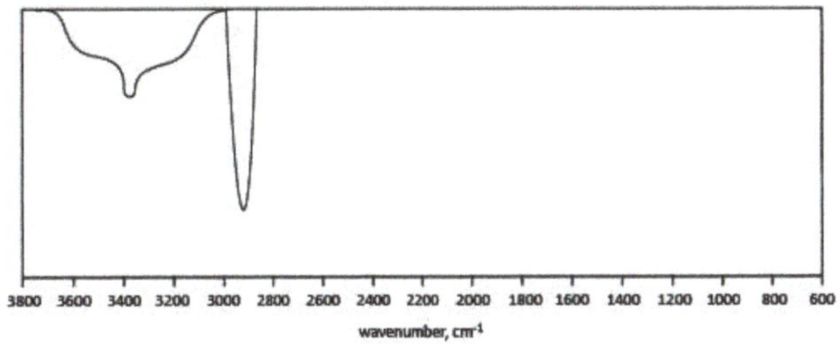

c) 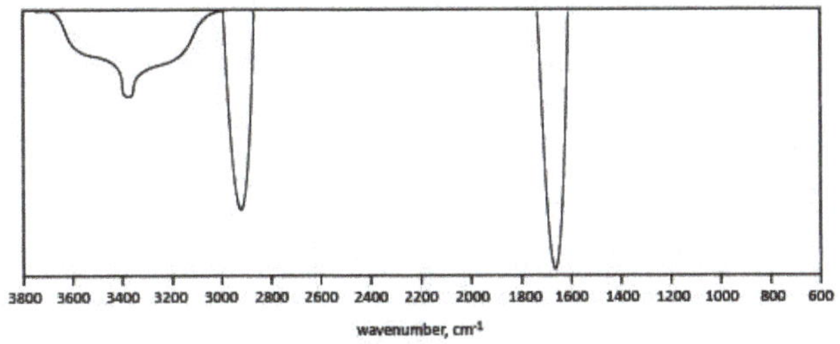

d) 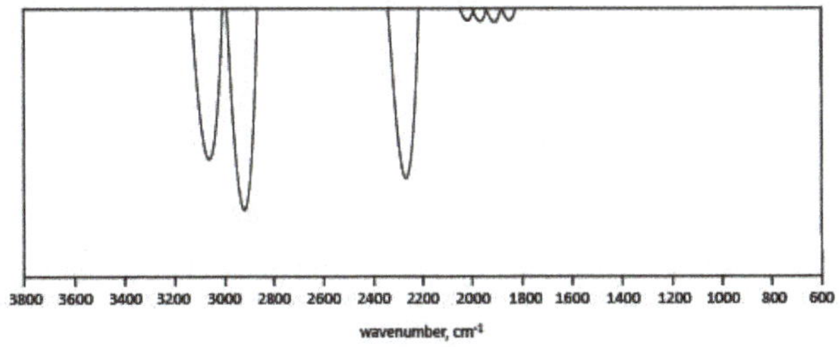

22.18 Describe how IR spectroscopy could be used to distinguish between triethylamine, diethylamine, and ethylamine.

22.19 What is the key feature of the M+ peak in a mass spectrum that indicates the compound has an odd number of nitrogen atoms in it?

22.20 Draw the products of the following reactions.

a) 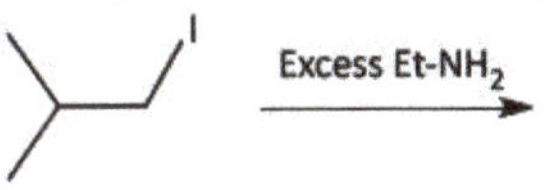

b) 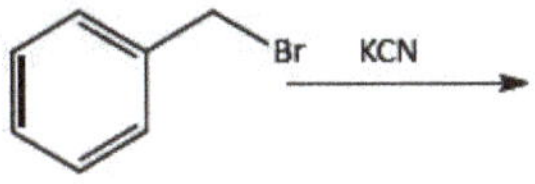

c) 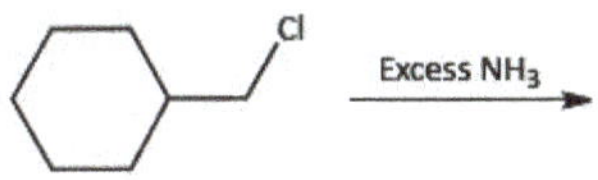

d) 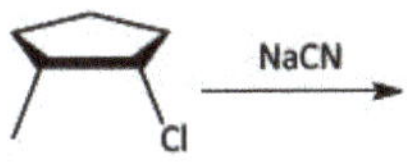

e) 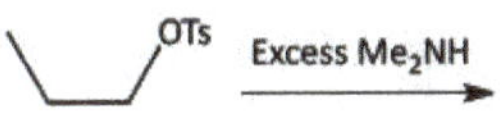

f) 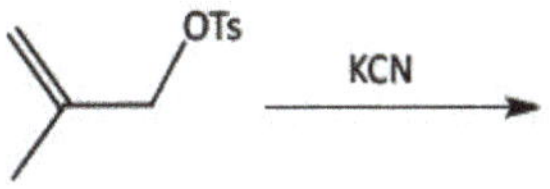

22.21 Provide an appropriate reagent to perform the following transformations.

a) 

b) bromopropane to N-ethylpropylamine c) $PhCH_2I$ to $PhCH_2NH_2$

22.22 Draw the products of the following reductions.
a) $PhCH_2CN$ + $LiAlH_4$ b) $PhCH_2CONH_2$ + LAH c) $PhCH_2C{=}NH$ + $LiAlH_4$ d) benzonitrile + $LiAlH_4$
e) PhCONHPh + $LiAlH_4$ f) *m*-nitroaniline + Fe/HCl followed by NaOH

22.23 Fill in the reagents to perform the following reactions.

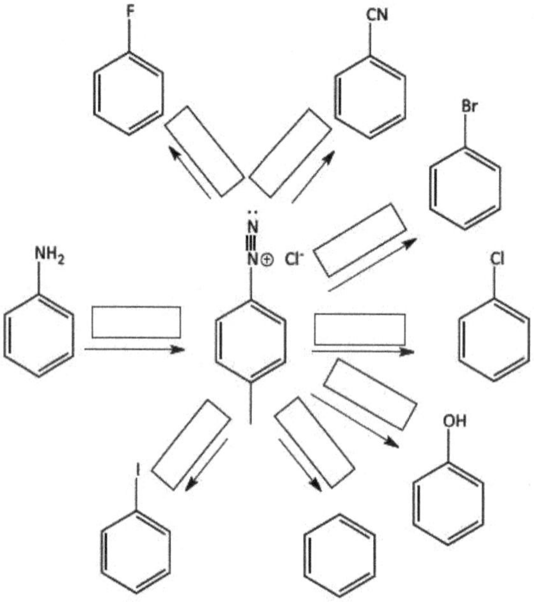

22.24 Draw the amides synthesized from the following reactions.
a) PhCOCl + excess ethylamine b) benzoic acid + methylamine c) propanoic acid + aniline
d) pentanoic acid + $SOCl_2$ followed by excess 2-aminobutane e) $CH_3CH{=}CHCH_2CO_2H$ + ammonia (NH_3)
f) butyric acid + thionyl chloride followed by excess propylamine g) $PhCH_2CO_2CH_3$ + H_2NCH_3

22.25 Draw appropriate carboxylic acids and amines that can synthesize the following amides.

a) 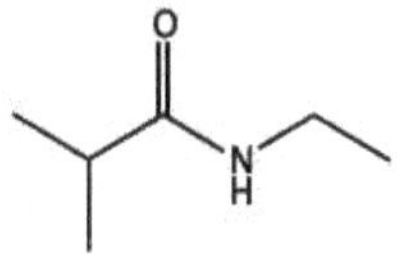

b) 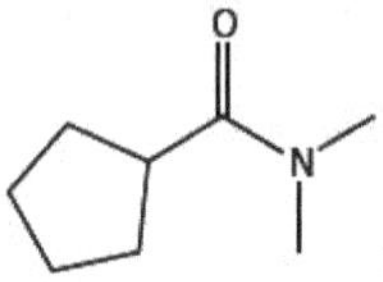

c) 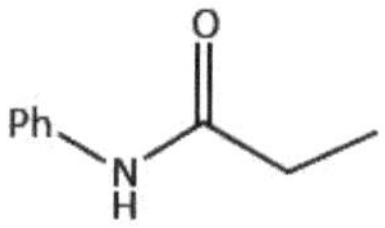

22.26 Provide appropriate acid chlorides and amines that can synthesize the following amides.

a) 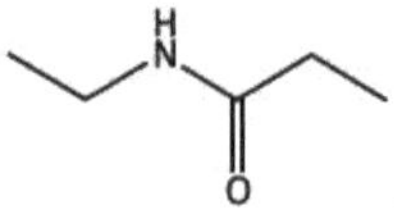

b) 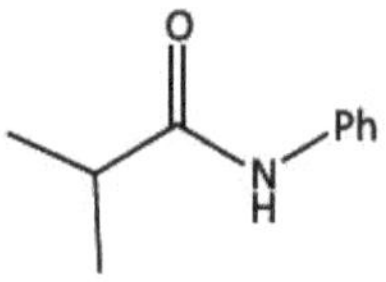

c) 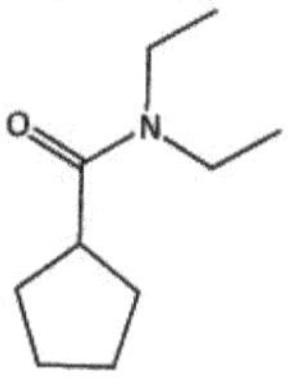

22.27 The following peptide was synthesized from several amino acids. a) How many? b) Draw out all of the amino acids used to make this peptide.

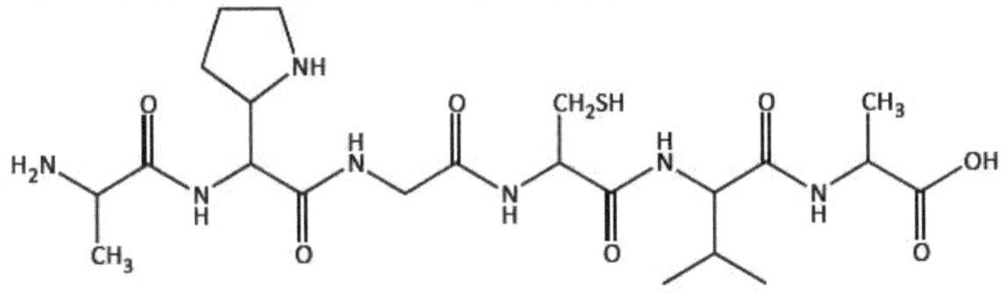

22.28 Provide the appropriate amines that could be used as reagents in the following reactions.

a) 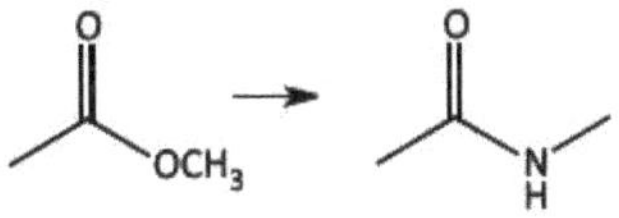 b) 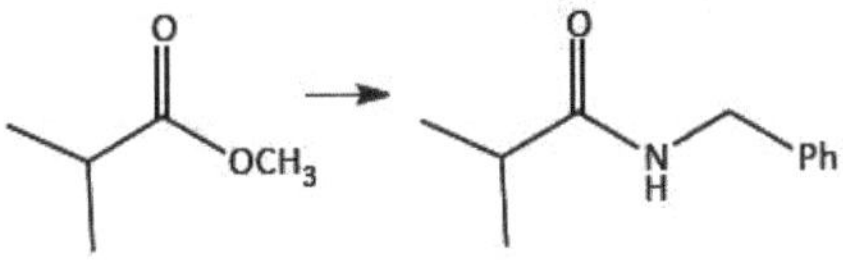 c) 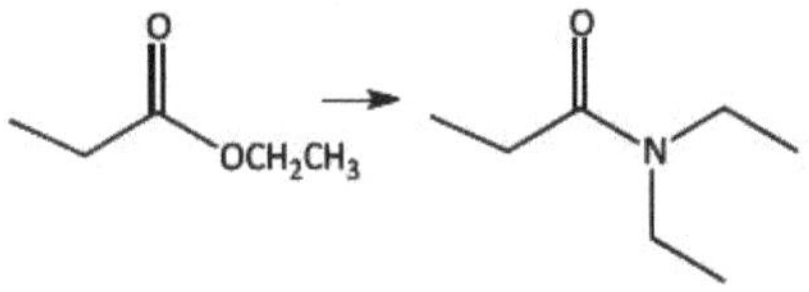

22.29 Draw the acid hydrolysis product of each of the following compounds.

a) 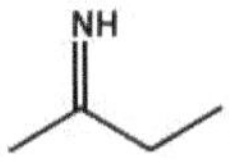 b) 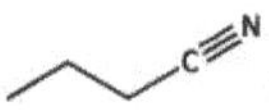 c) 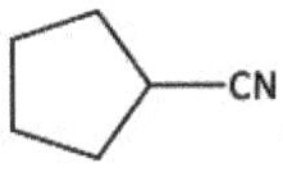 d) 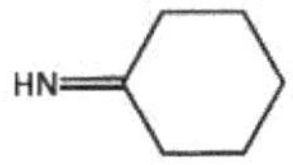 e) 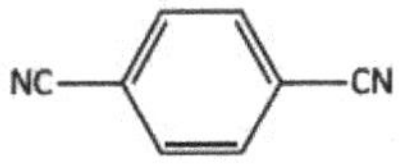 f) 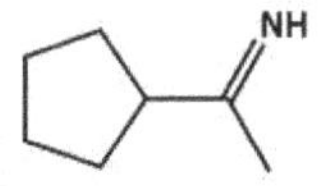

22.30 Draw the products of the following reactions.

a) 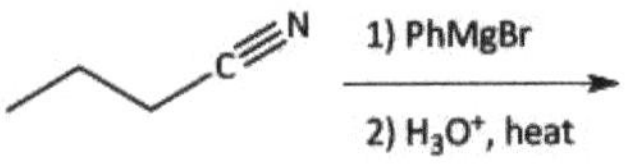 b) 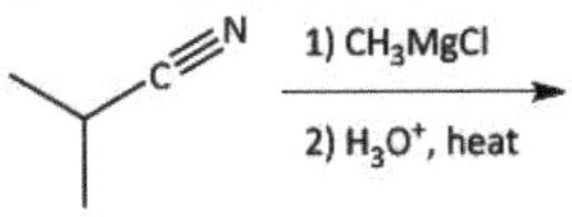 c) 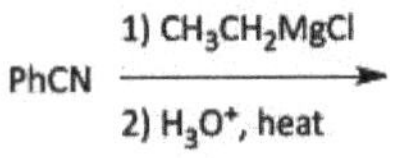

22.31 Show how you would synthesize the following products starting from a nitrile compound.
a) benzoic acid b) benzophenone (PhCOPh) c) 3-heptanone d) pentanamine e) 2-methylpentanoic acid f) cyclohexylmethylketone g) acetic acid h) $PhCH_2NH_2$ i) acetone j) butyric acid k) ethylamine

Answers to Exercises

Chapter 1

1.1

H He, LiBe BCNOFNe H He, LiBe BCNOFNe H He, LiBe BCNOFNe
H He, LiBe BCNOFNe H He, LiBe BCNOFNe H He, LiBe BCNOFNe

1.2

H																	He
Li	Be											B	C	N	O	F	Ne
Na	Mg											Al	Si	P	S	Cl	Ar
K	Ca	Sc	Ti	V	Cr	Mn	Fe	Co	Ni	Cu	Zn	Ga	Ge	As	Se	Br	Kr
Rb	Sr	Y	Zr	Nb	Mo	Tc	Ru	Rh	Pd	Ag	Cd	In	Sn	Sb	Te	I	Xe
Cs	Ba	Lu	Hf	Ta	W	Re	Os	Ir	Pt	Au	Hg	Tl	Pb	Bi	Po	At	Rn
Fr	Ra																

1.3 F Cl Br I At F Cl Br I At
F Cl Br I At F Cl Br I At
F Cl Br I At F Cl Br I At

1.4

Write the name of each element next to its atomic symbol.

H ___Hydrogen______________ He ______ Helium____________ Li ____ Lithium_______________
C ___Carbon________________ N ______ Nitrogen___________ O ___ Oxygen_______________
F ___ Fluorine________________ Cl _____ Chlorine____________ Br ___ Bromine______________

1.5

H																	He
Li	Be											B	C	N	O	F	Ne
Na	Mg											Al	Si	P	S	Cl	Ar
K	Ca	Sc	Ti	V	Cr	Mn	Fe	Co	Ni	Cu	Zn	Ga	Ge	As	Se	Br	Kr
Rb	Sr	Y	Zr	Nb	Mo	Tc	Ru	Rh	Pd	Ag	Cd	In	Sn	Sb	Te	I	Xe
Cs	Ba	Lu	Hf	Ta	W	Re	Os	Ir	Pt	Au	Hg	Tl	Pb	Bi	Po	At	Rn
Fr	Ra																

1.6

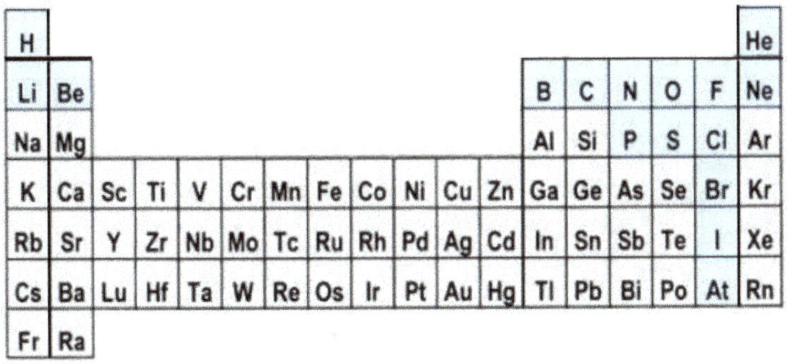

1.7 C, O

1.8

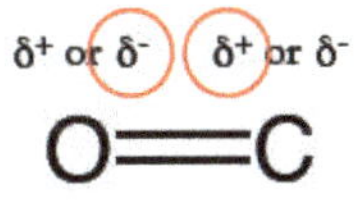

1.9

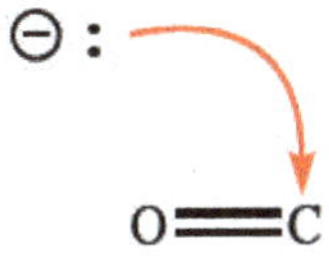

1.10

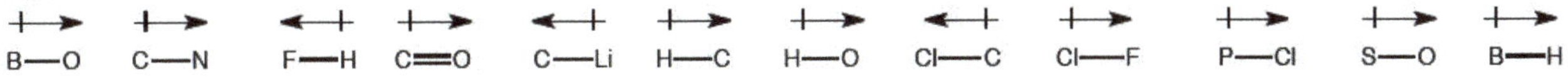

1.11 Li^- B^- C^- N^- O^- F^-

Li^-	B^-	C^-	N^-	O^-	F^-
6	5	4	3	2	1

1.12

a) **C** or O b) **Be** or F c) O or **N** d) N or **C**

1.13

a) Cl or **Br** b) **I** or Cl c) F or **Cl** d) **Br** or F e) **P** or N f) O or **S**

1.14

a) Cl^- or **Br^-** b) **I^-** or Cl^- c) F^- or **Cl^-**

1.15

a) N^- or **O^-** b) O^- or **S^-** c) O^- or **F^-**

Chapter 2

2.1

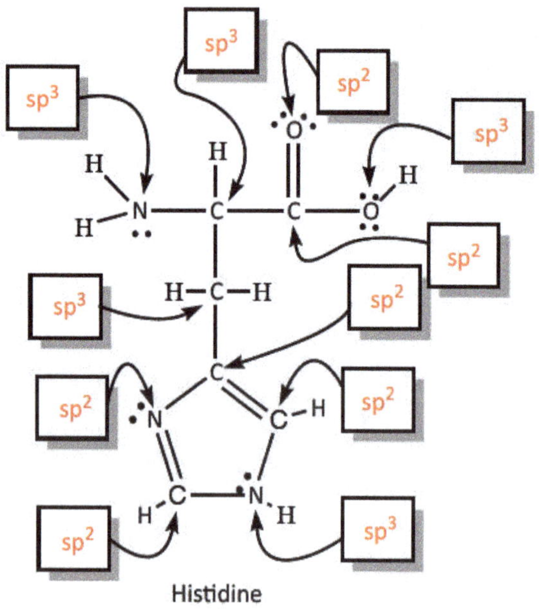

2.2

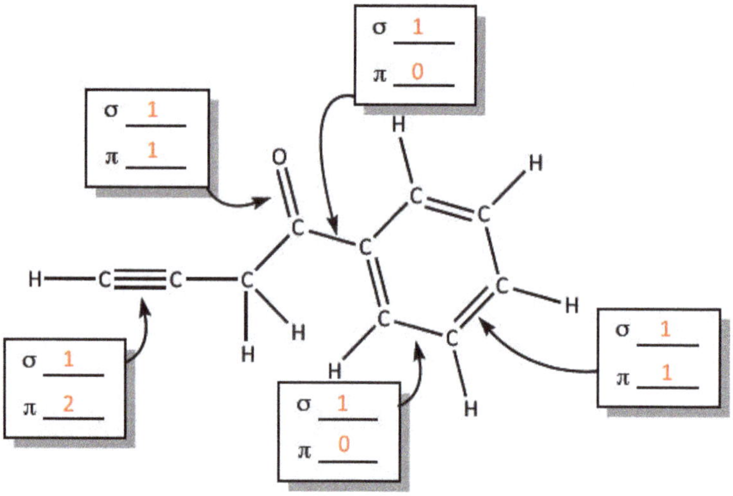

2.3

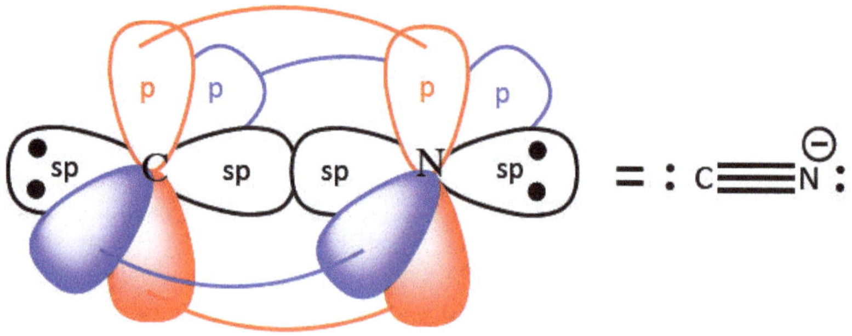

2.4

a)

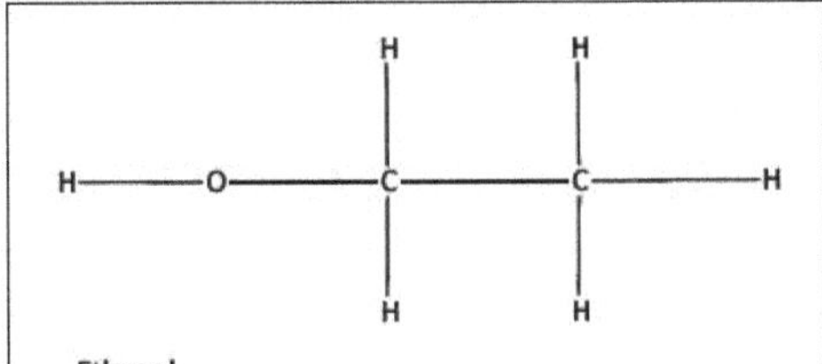

Ethanol

b)

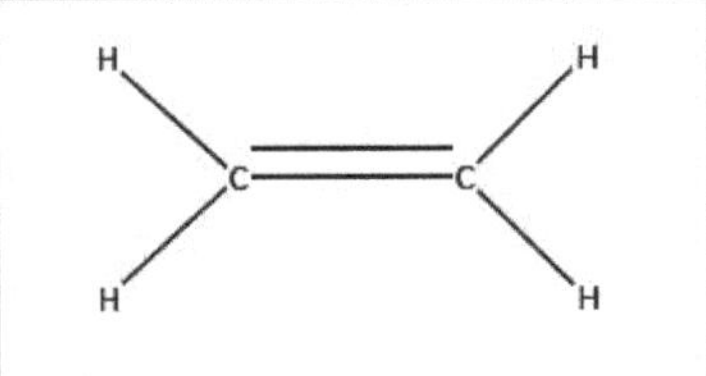

Ethene

c)

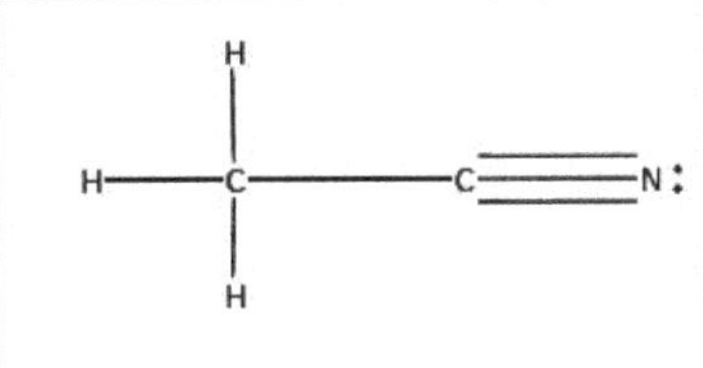

Acetonitrile

d)

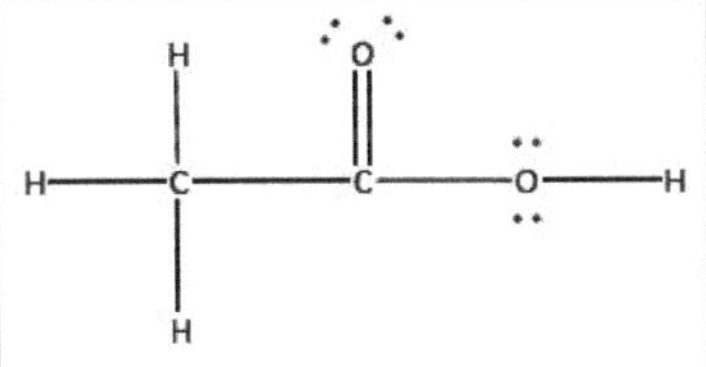

Acetic acid

e)

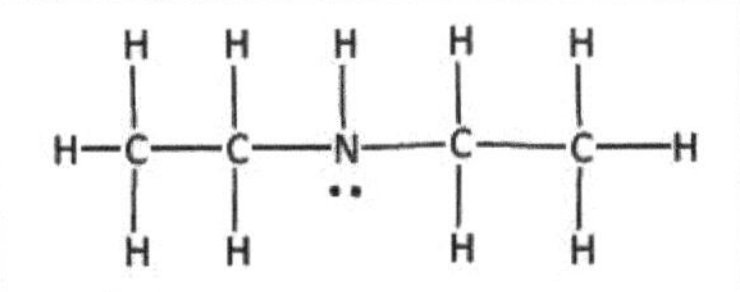

Diethylamine

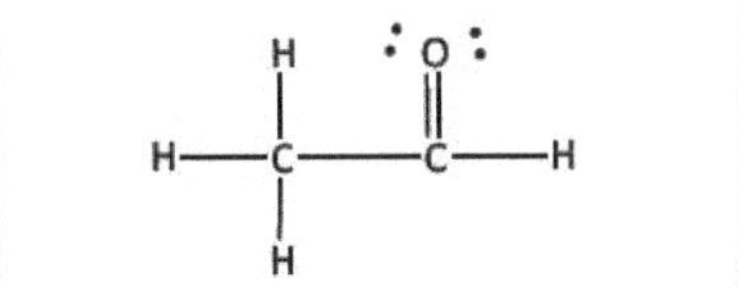

Acetaldehyde

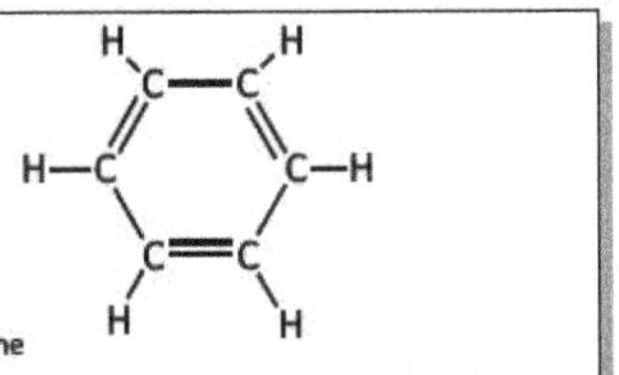

Benzene

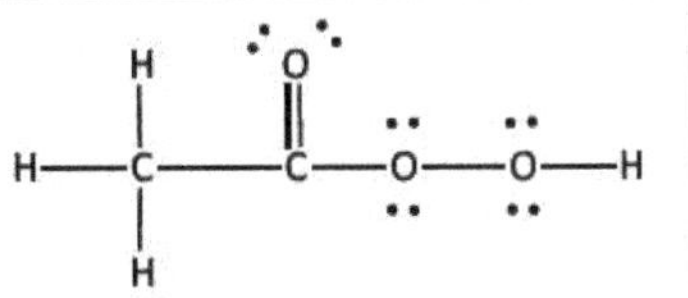

Peroxyacetic acid

The normal bonding pattern for oxygen atoms is two bonds and two lone pairs of electrons. So, if your answer for peroxyacetic acid had the incorrect bonding patterns for the oxygen atoms, don't worry about it. We'll get to it in the next section.

2. 5 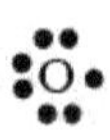 7, 6  5, 6

Answers to Exercises

2.6

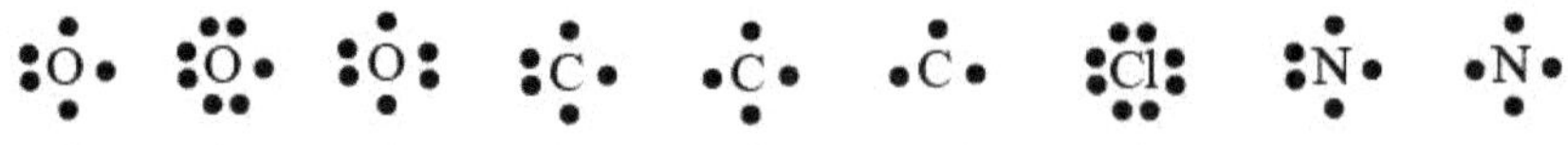

+1 -1 0 -1 0 +1 -1 0 +1

2.7

2.8

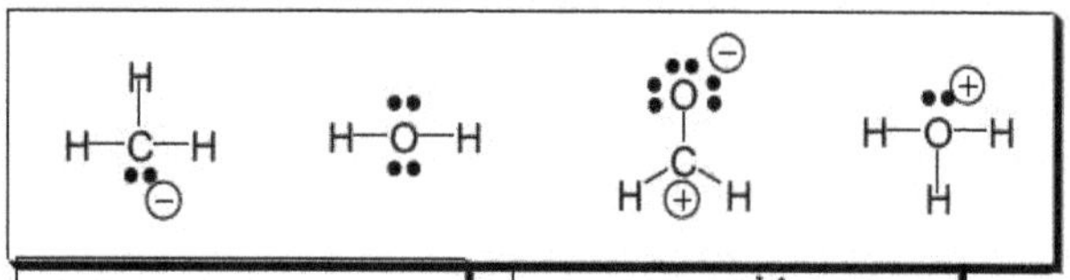

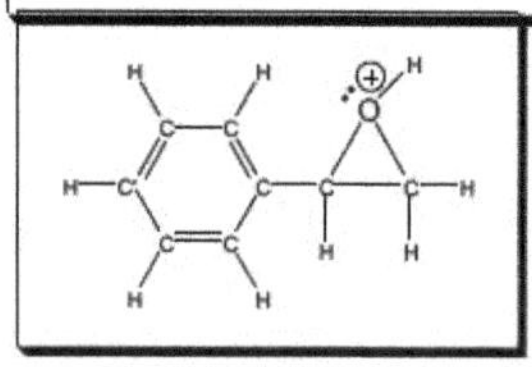

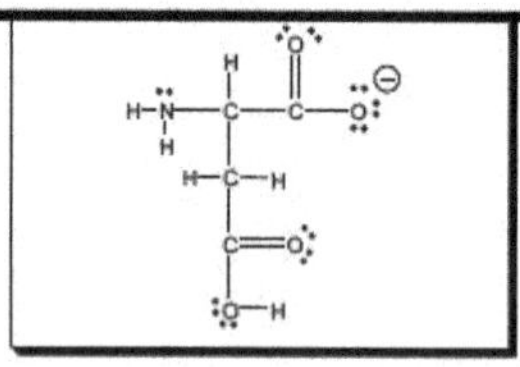

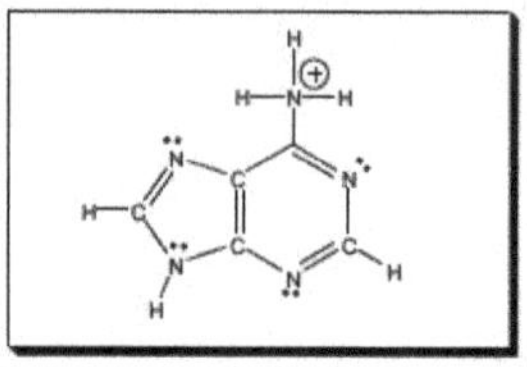

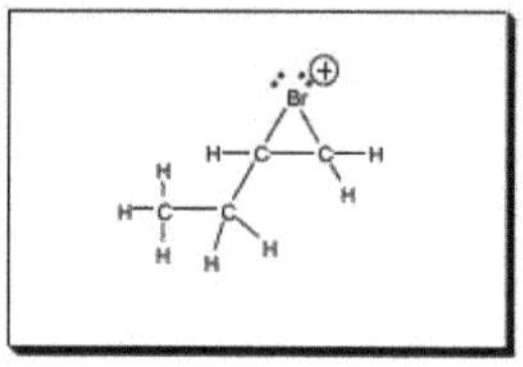

2.9

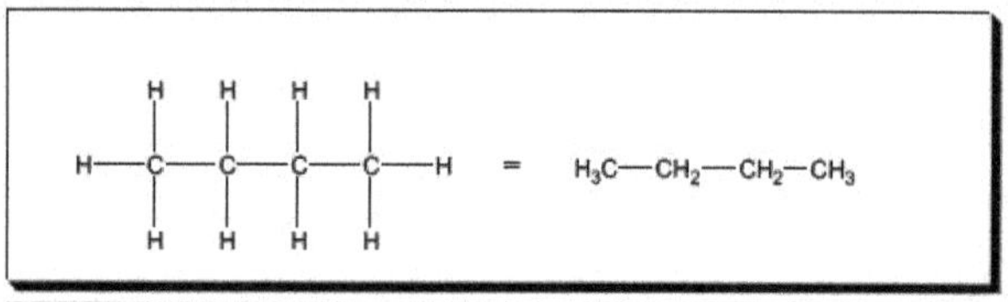

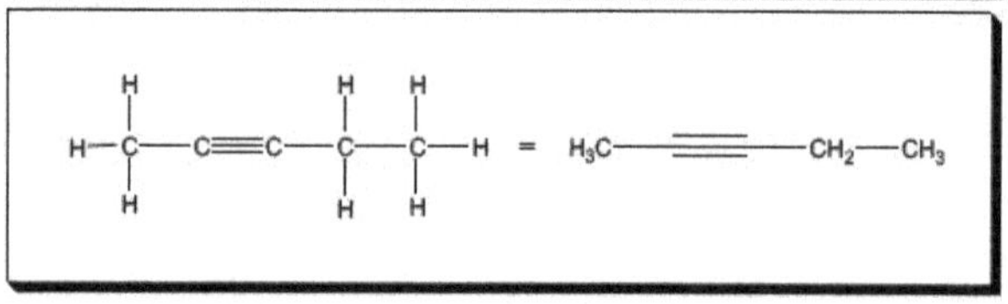

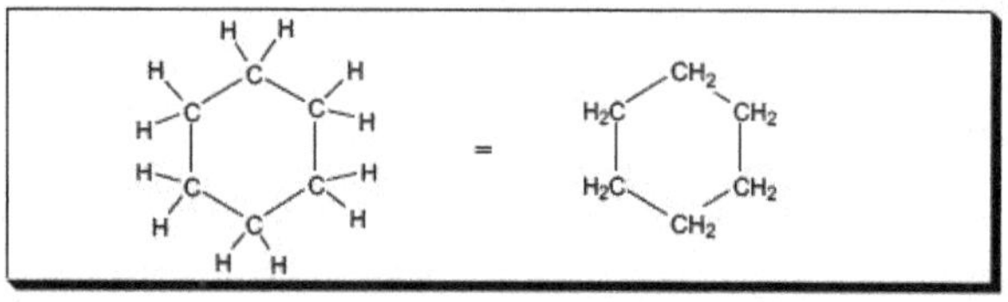

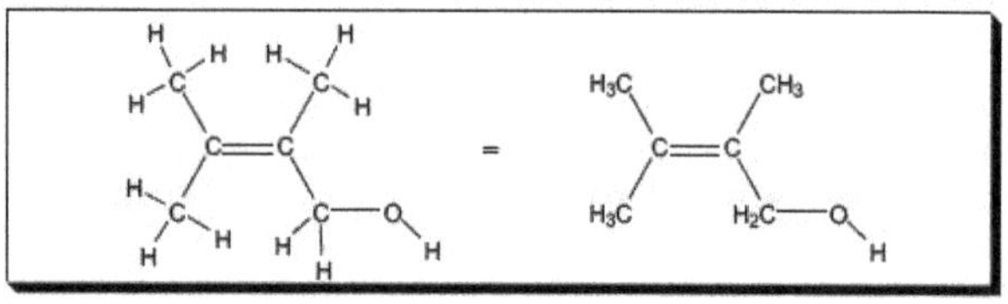

2.10 Draw the Lewis structure (every bond and atom) for the following molecules.

a) 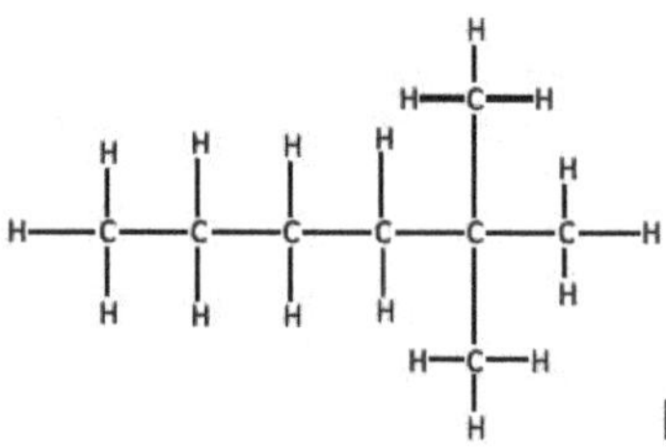

b) 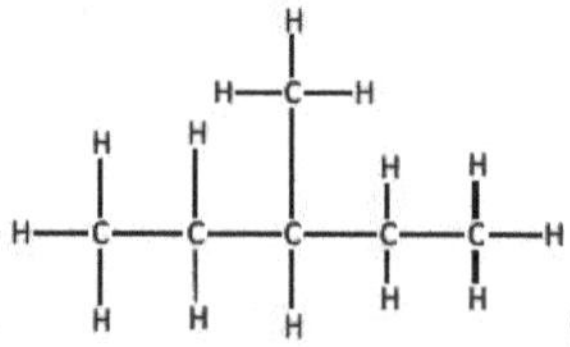

c) 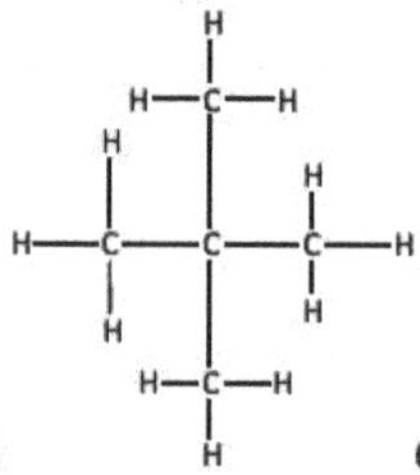

d) 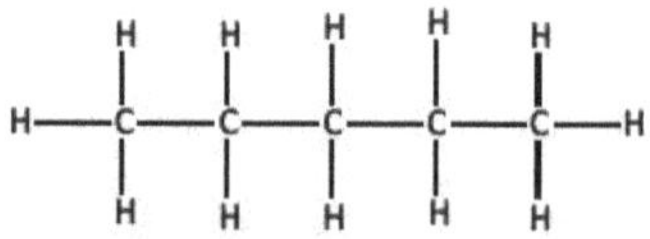

2.11

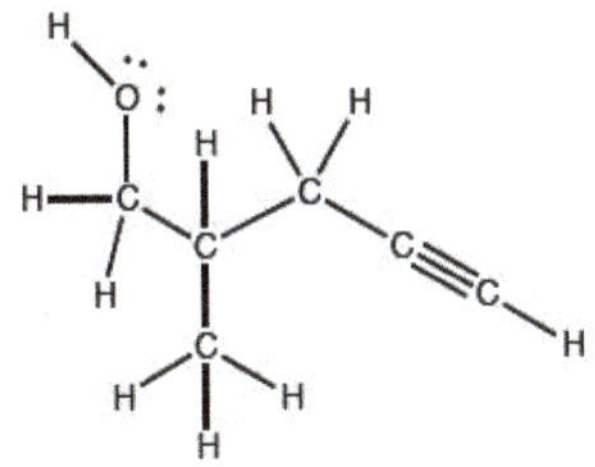

2.12

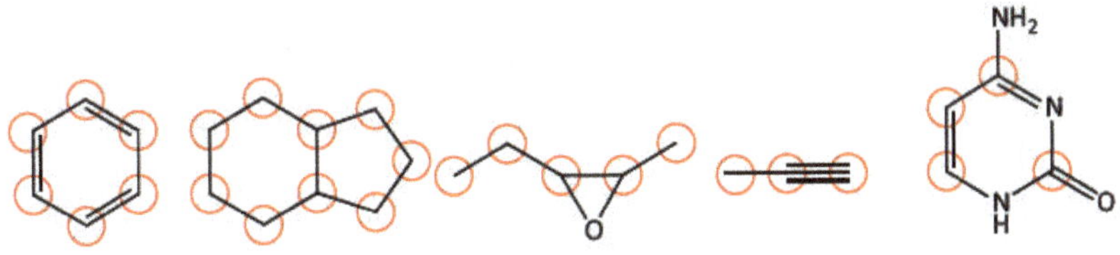

6 9 5 3 4

2.13

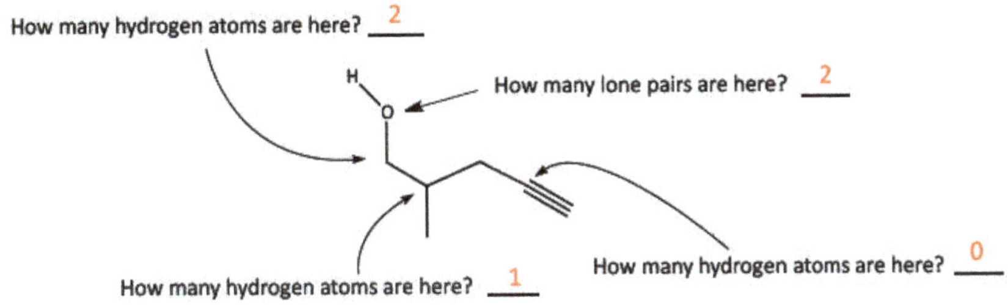

2.14

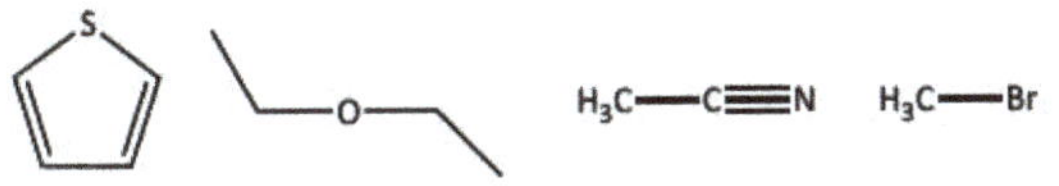

2 2 1 3

2.15

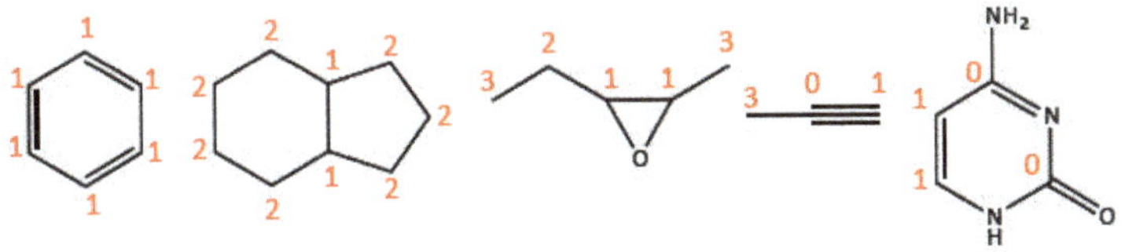

2.16

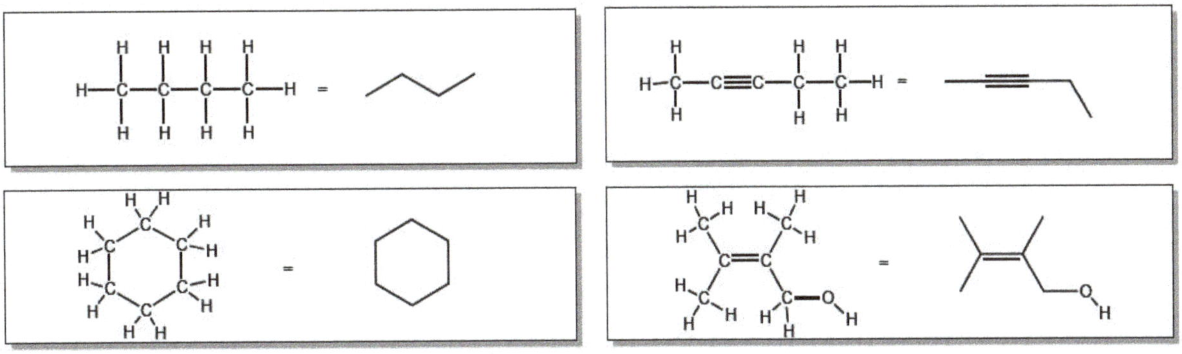

Answers to Exercises

2.17 State the geometric shape of the molecule with the following bond angles between atoms (assume no lone pairs of electrons).

a) 109° (tetrahedral) b) 120° (trigonal planar) c) 180° (linear)

2.18 Name the geometric shape of the following molecules.

a) CCl_4 (tetrahedral) b) PH_3 (trigonal pyramidal) c) $H_2C{=}O$ (trigonal planar) d) $O{=}C{=}O$ (linear)
e) F_2O (bent) f) $AlCl_3$ (trigonal planar)

2.19

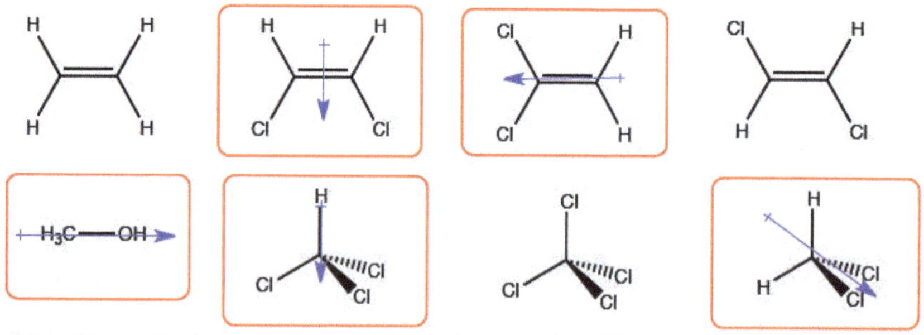

2.20 a) Cl_2 No polar bonds, nonpolar molecule b) HCl Yes polar bonds, polar molecule c) H_2O Yes polar bonds, polar molecule d) CH_4 Yes polar bonds, nonpolar molecule e) $CHBr_3$ Yes polar bonds, polar molecule f) CH_3OCH_3 Yes polar bonds, polar molecule (because it is bent)

2.21 Which compound is more soluble in water?

a) CH_3CH_2OH or $CH_3CH_2CH_2CH_3$
b) $CH_3CH_2OCH_2CH_3$ or $(CH_3)_2NH$

2.22

1 lone pair 0 N-H No H-bonding	1 lone pair 1 N-H Some H-bonding	1 lone pair 2 N-H More H-bonding

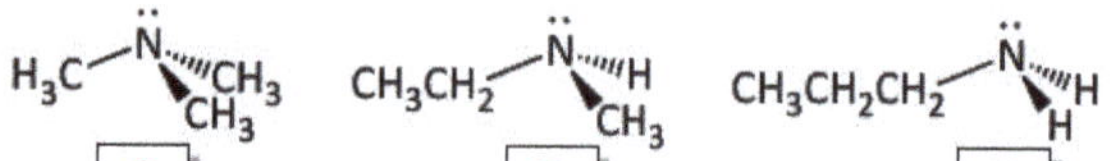

3	2	1
bp = 4°C	bp = 37°C	bp = 49°C

2.23

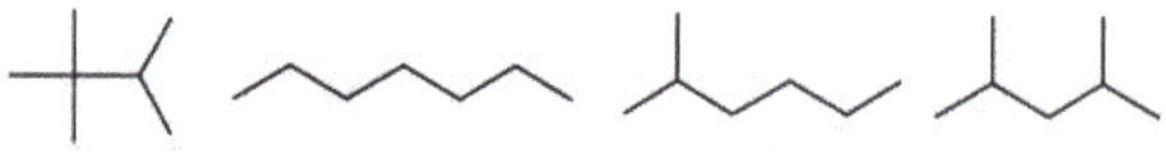

4 1 2 3

2.24

$H_3C—OH$	H_2O		
2	1	3	4
bp = 64°C	bp = 100°C	bp = 36°C	bp = 10°C
2 lone pairs 1 O-H Some H-bonding	2 lone pairs 2 O-H Most H-bonding	More surface area More London forces	More branching Less surface area Less London forces

2.25

pentane hexane heptane
octane nonane decane

2.26

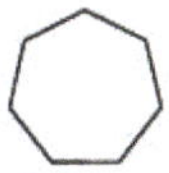
cycloheptane

cyclopropane

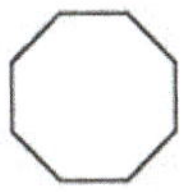
cyclooctane

2.27

$H_2C{=}CHCH_2CH_2CH_3$ $CH_3CH_2CH{=}CHCH_2CH_3$ $Cl_2C{=}CHCH_2CH_2CH_3$ $ClHC{=}CHCl$

2.28

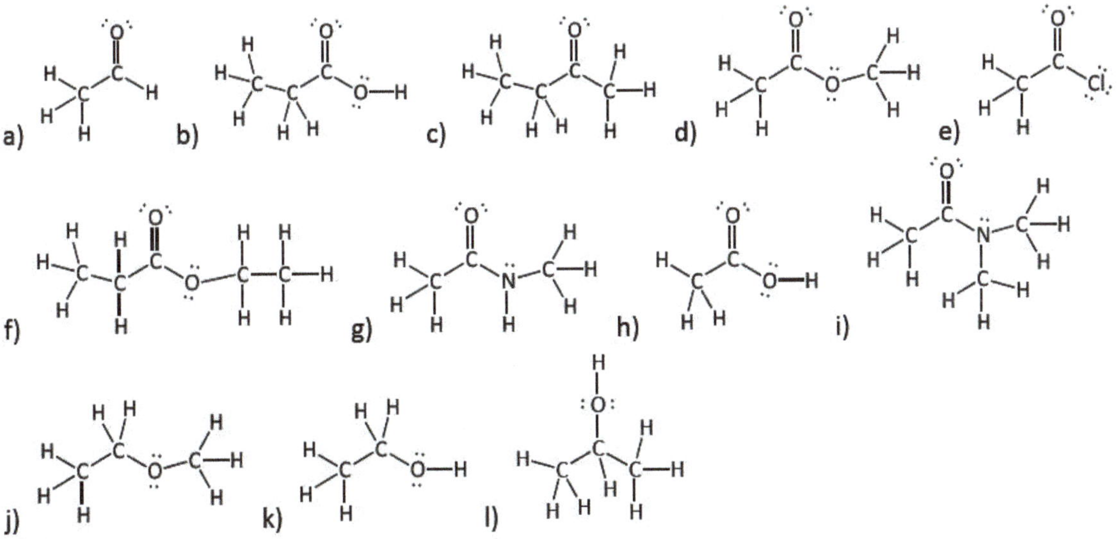

2.29

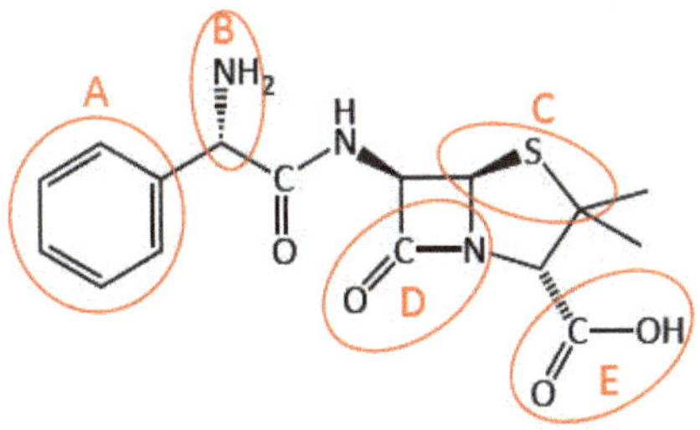

Amoxicillin (a penicillin antibiotic)

A phenyl
B amine
C thioether
D amide
E carboxylic acid

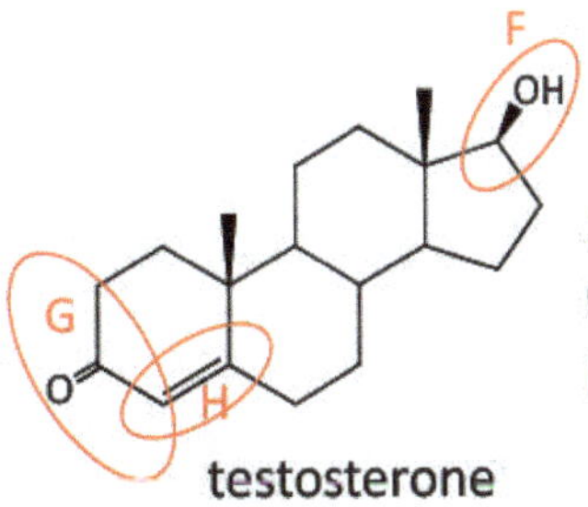

testosterone

F alcohol
G ketone
H alkene

Chapter 3

3.1 Identify the atom the following mass spectrum data indicates.

a) The M+2 peak is 1/3 the height of the M^+ peak. (Cl)

b) The M^+ peak is an odd number. (N)

c) The M+2 peak is nearly equal the height of the M^+ peak. (Br)

d) There is a large peak at 127 and M^+ - 127 is also prominent. (I)

e) The M+2 peak is slightly larger than usual, about 4% of the M^+ peak. (S)

3.2 How many degrees of unsaturation are in the compounds with the following formulas?

C_6H_6O __4___ C_2H_4ONCl __1___

C_3H_7OBr __0___ C_7H_8 __4___

3.3 No. Cyclopropane also has the formula C_3H_6.

3.4 Circle the molecule below if its bond that is marked red is and IR ACTIVE bond.

$CH_3–CH_3$ $CH_3CH_2–CH_3$ $HC≡CCH_3$ H–O–H

O=C=O C≡O H–C≡N H–C≡C–H

3.5

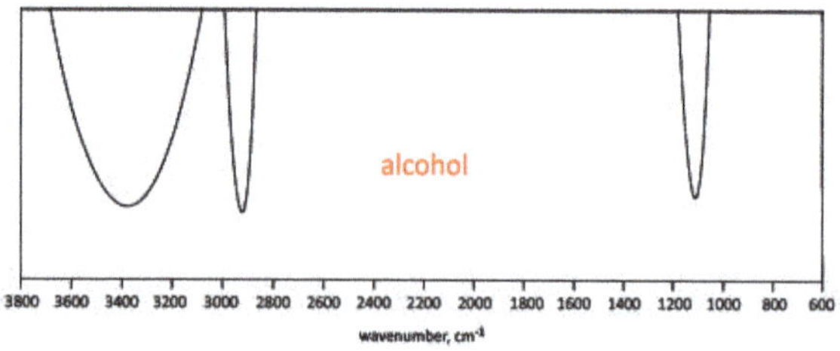

a)

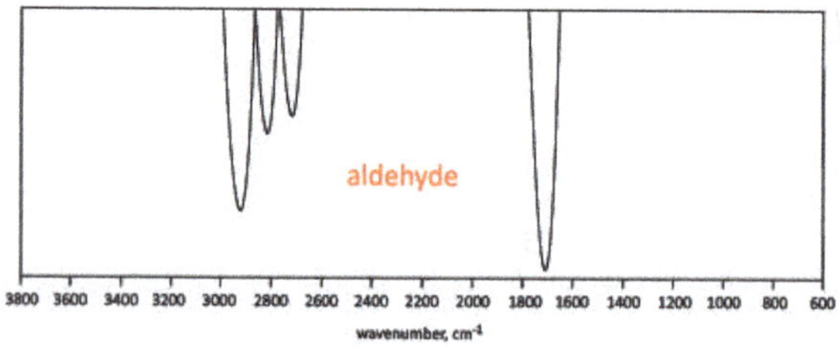

b)

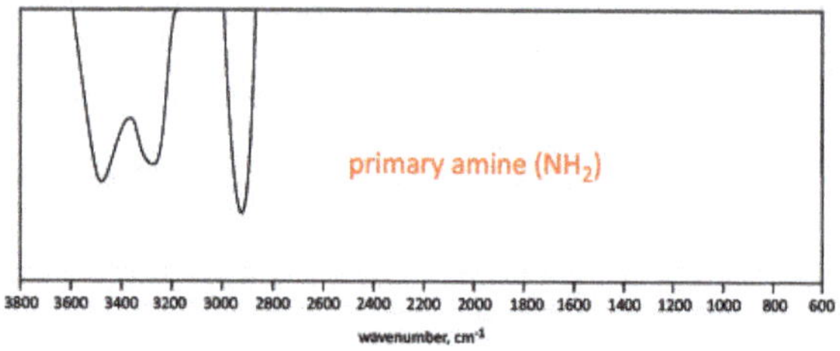

c)

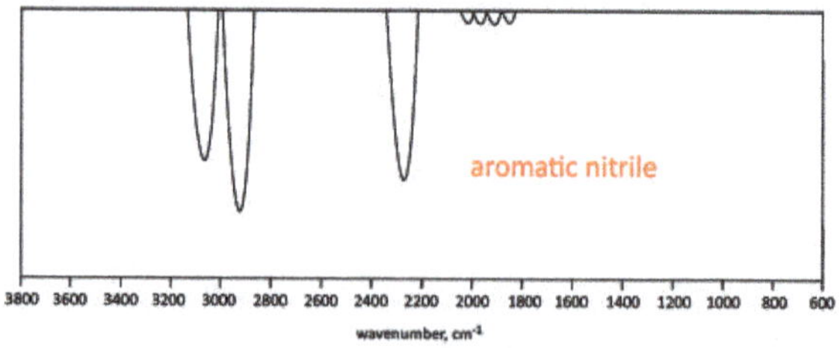

d)

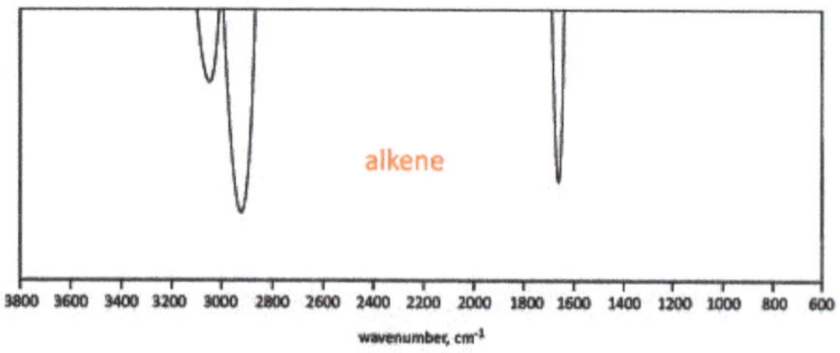

e)

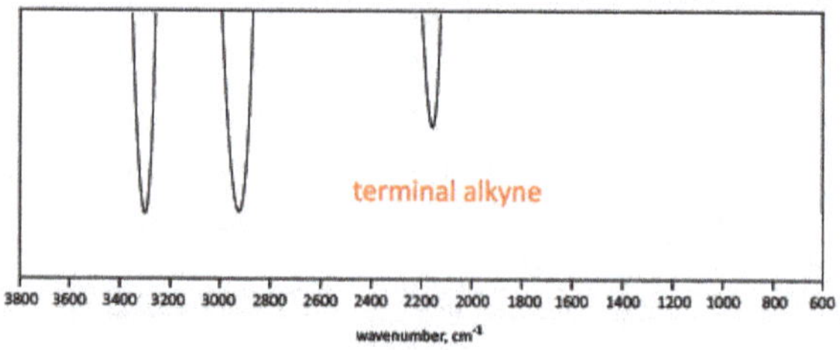

f)

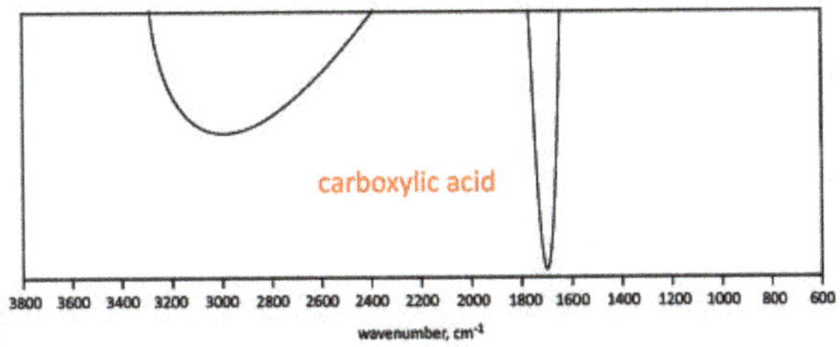

g)

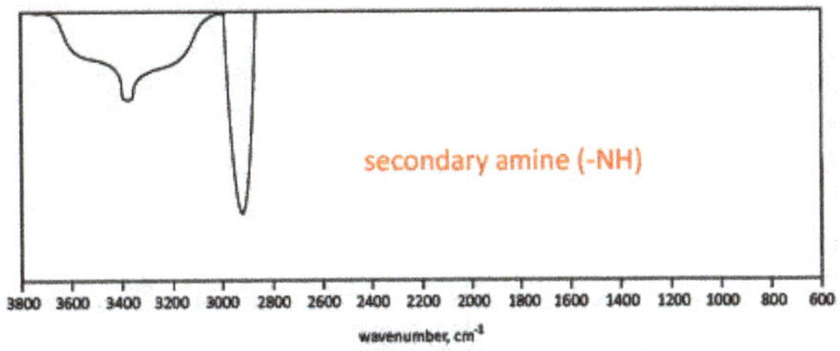

h)

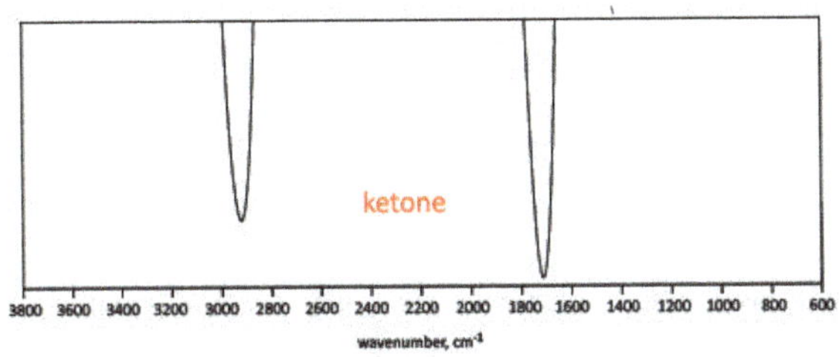

i)

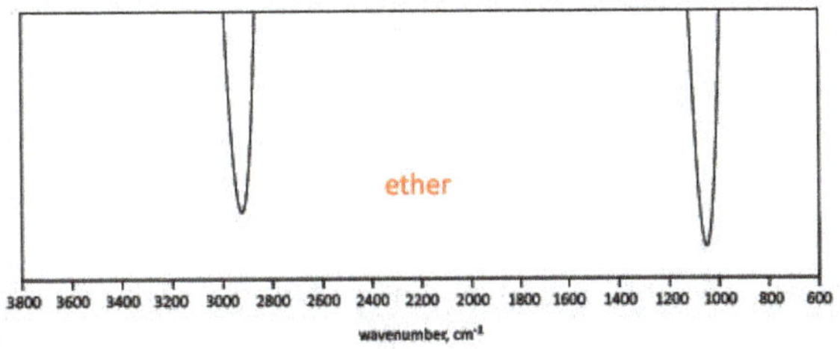

j)

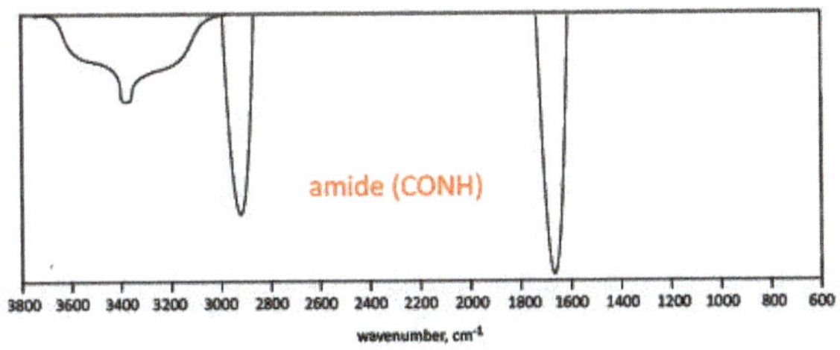

k)

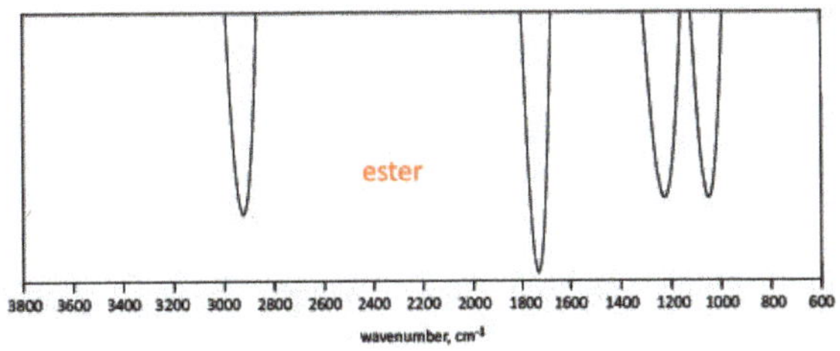

l)

3.6

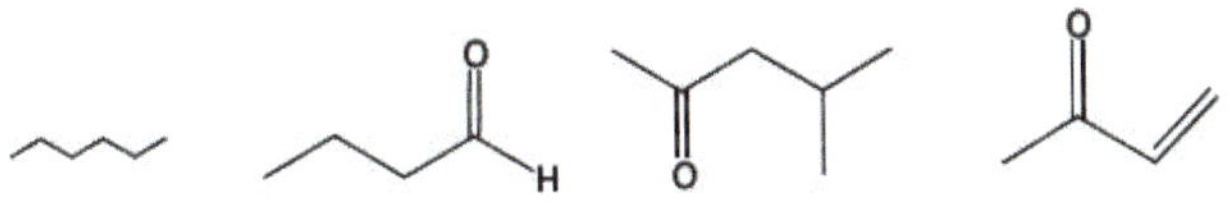

B D C A

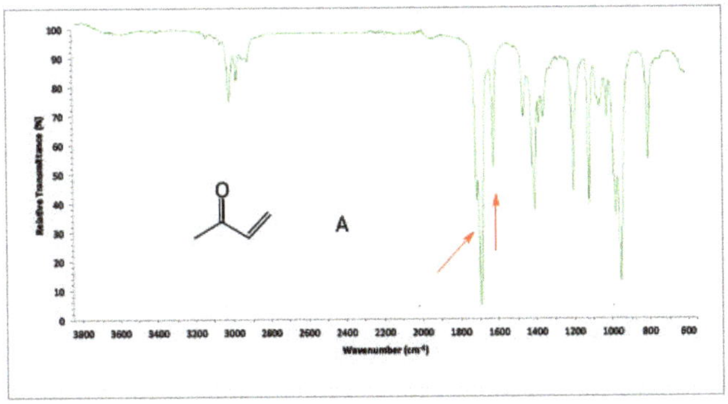

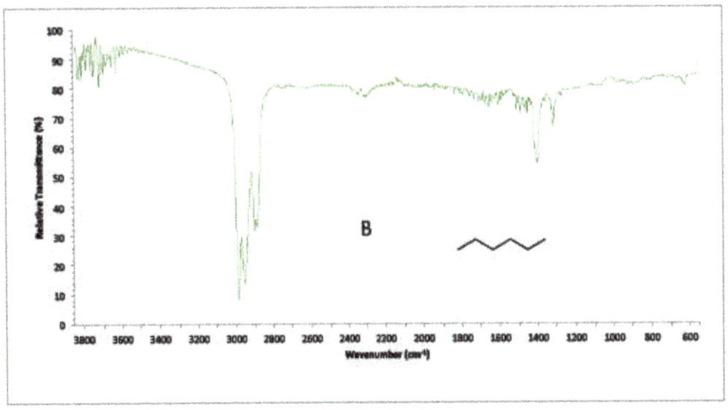

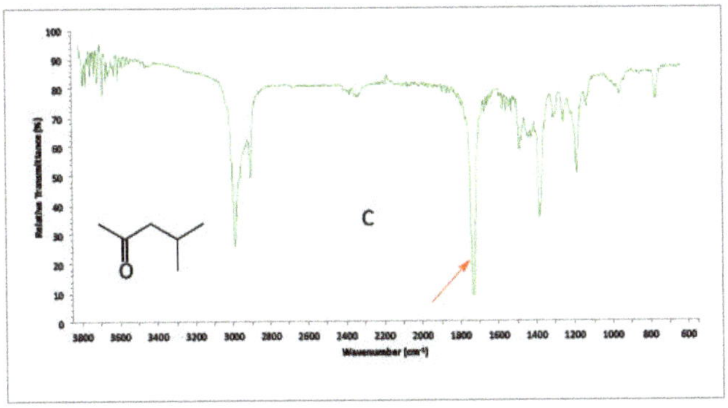

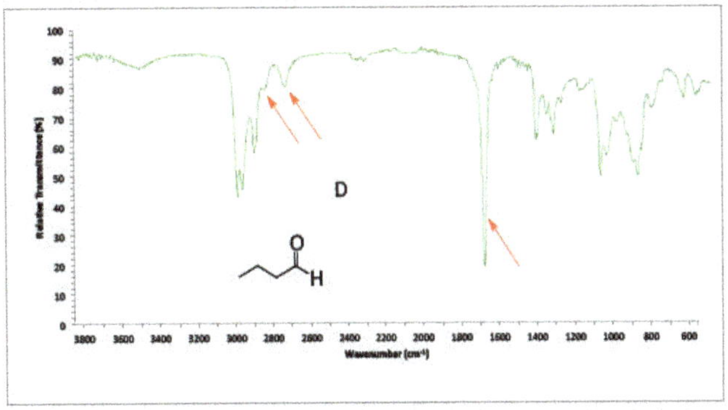

Answers to Exercises

3.7

C

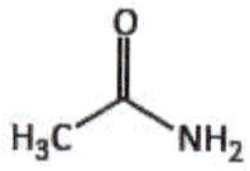

D

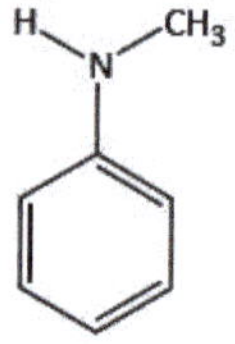

B

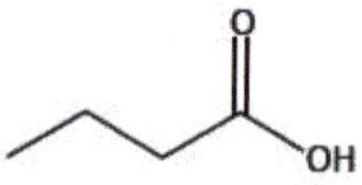

A

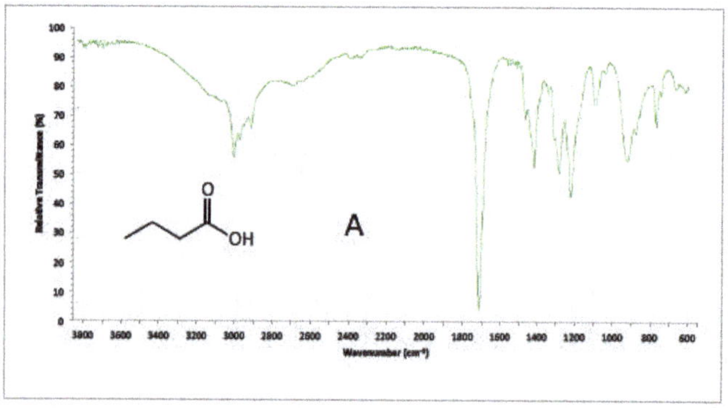

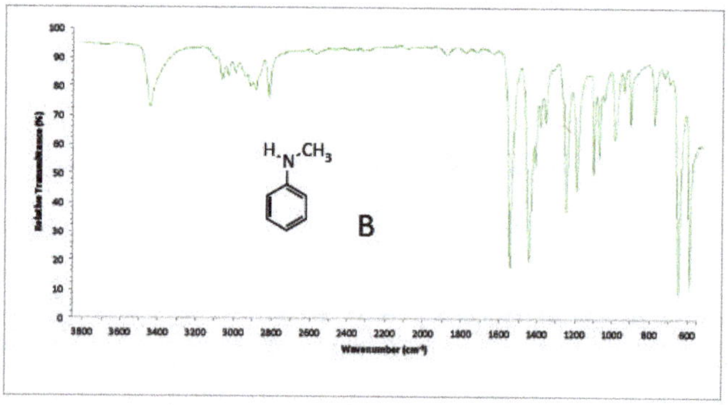

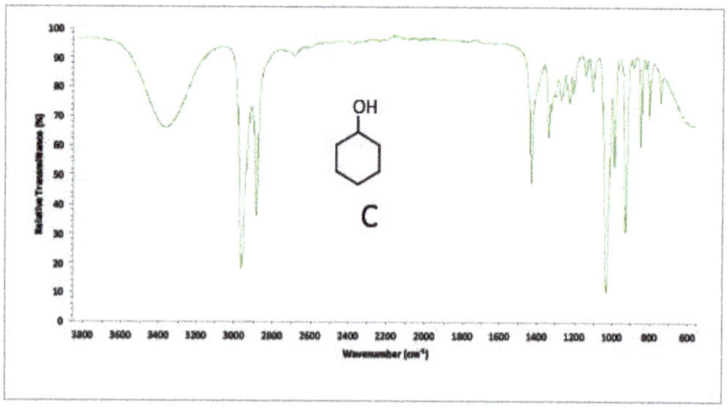

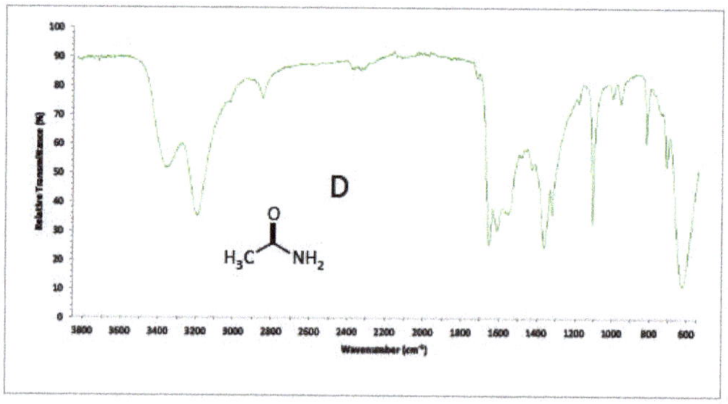

Chapter 4

4.1

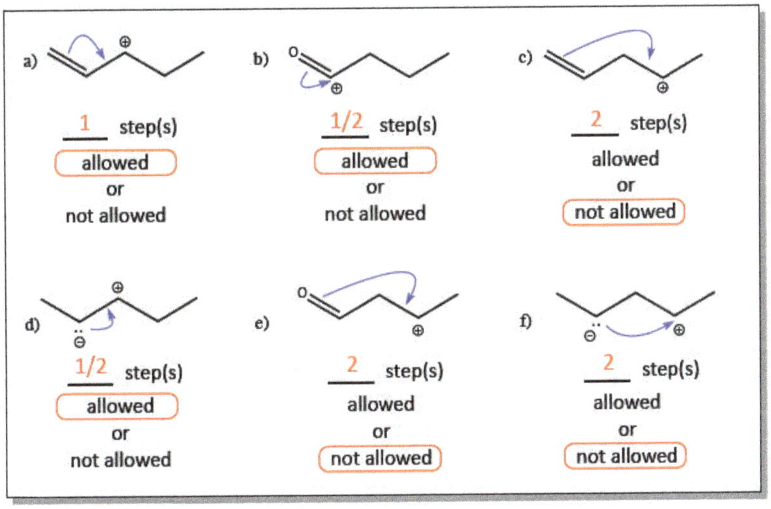

4.2

a) 

b) 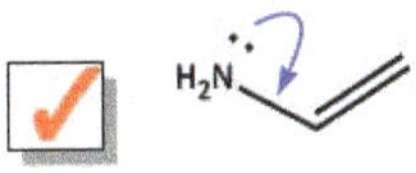

c) 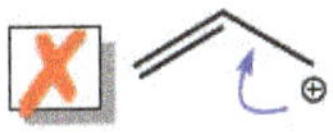

d) 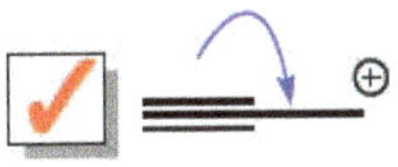

e) 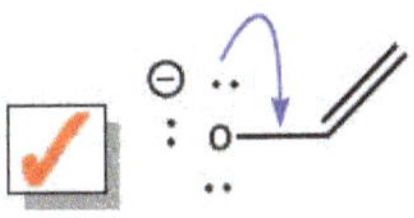

f) 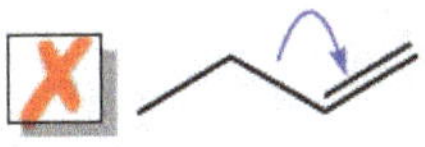

4.3

a) 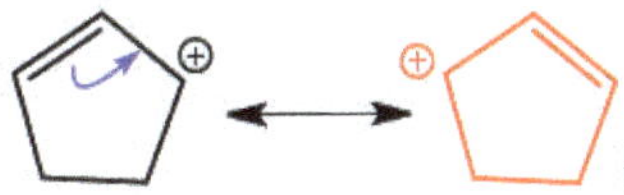

b) 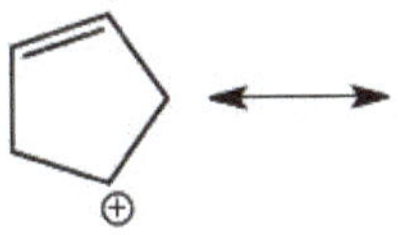

No resonance. Pi electrons must move 2 steps to get to cation.

4.4

a) 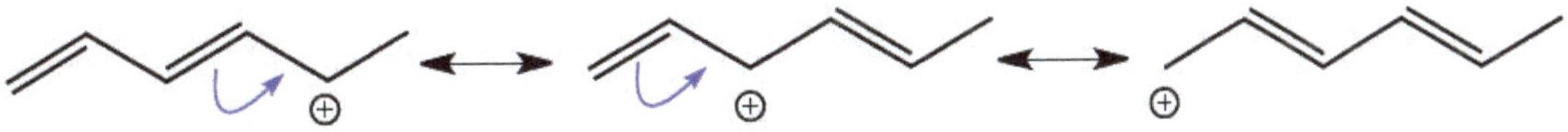

b) 

4.5

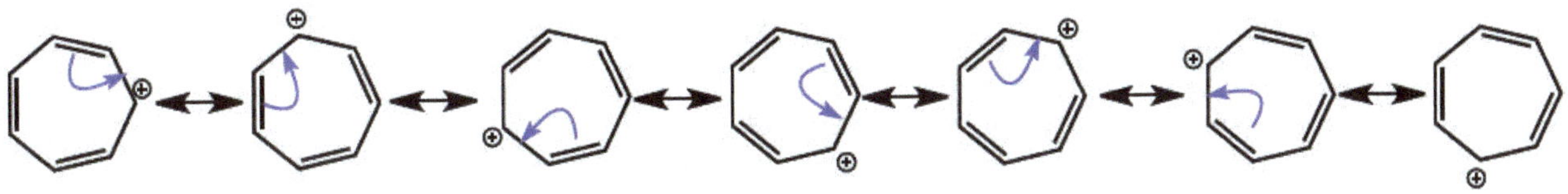

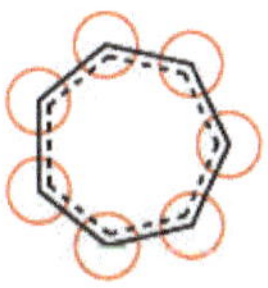

4.6

a)

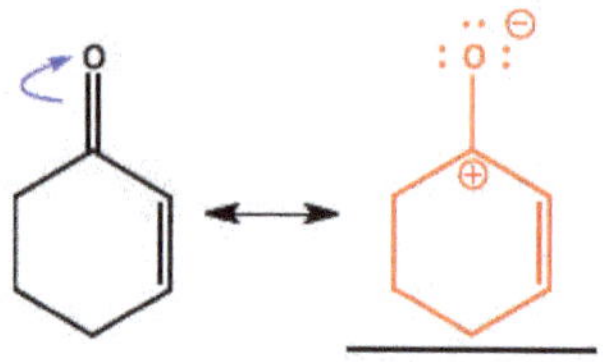

6a

Draw resonance structure

b)

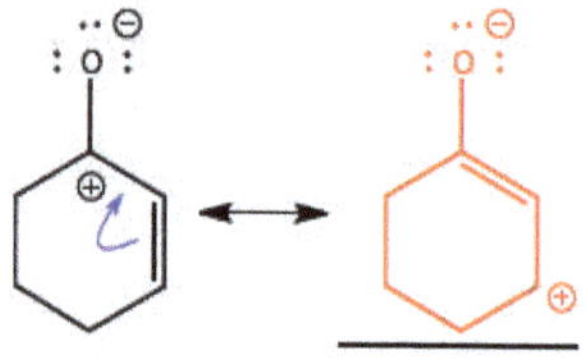

6a 6b

Draw resonance structure

c)

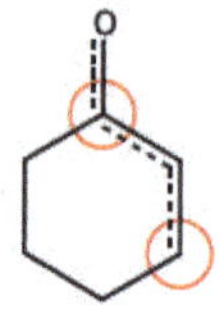

4.7

a)

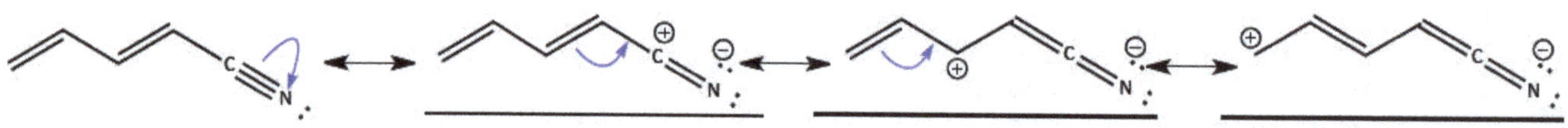

b)

4.8

a)

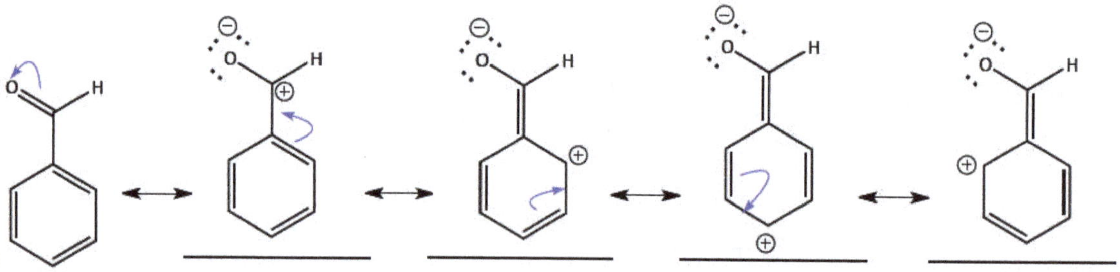

b)

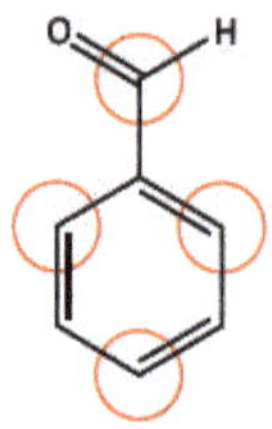

4.9

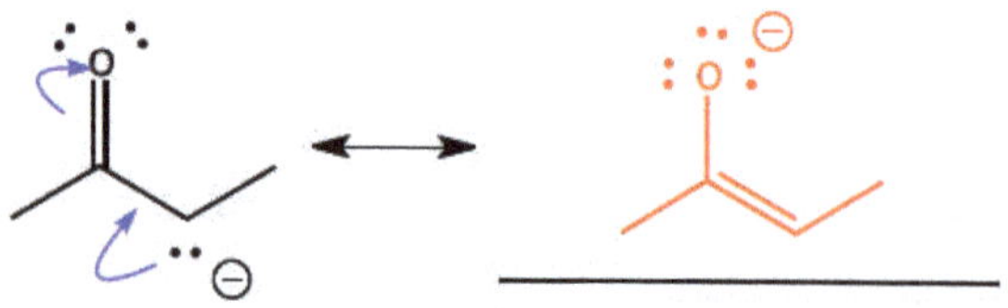

Draw one more arrow. Draw resonance structure.

4.10

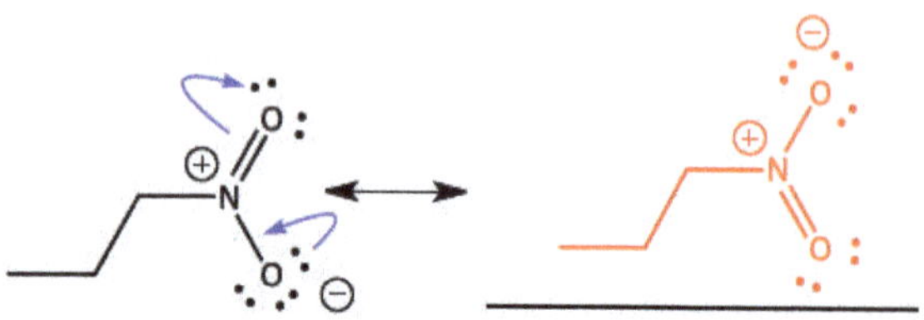

4.11

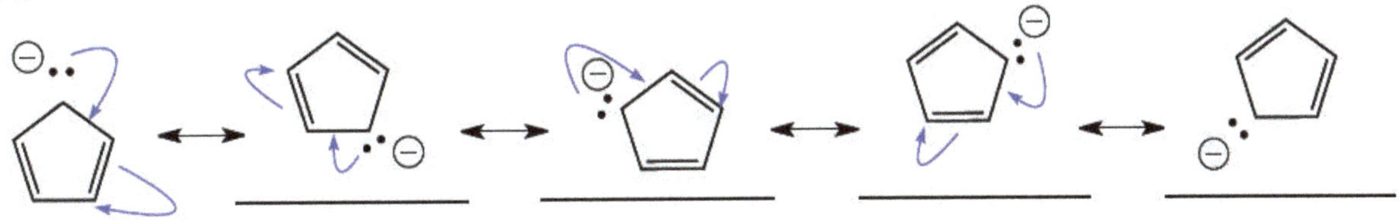

Draw resonance structure then draw two arrows

Draw resonance structure then draw two arrows

Draw resonance structure then draw two arrows

Draw resonance structure

4.12

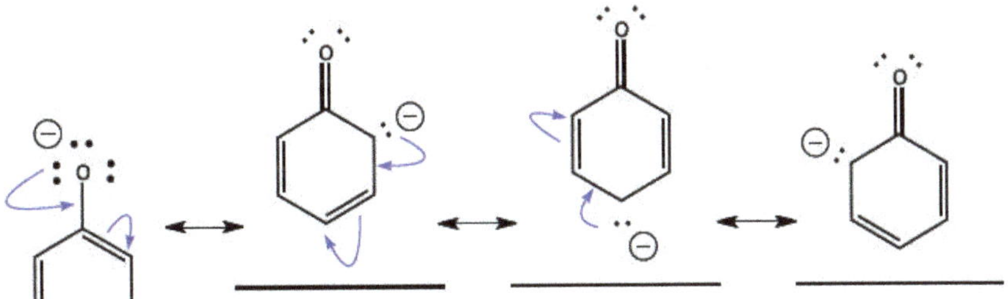

Draw resonance structure then draw two arrows

Draw resonance structure then draw two arrows

Draw resonance structure

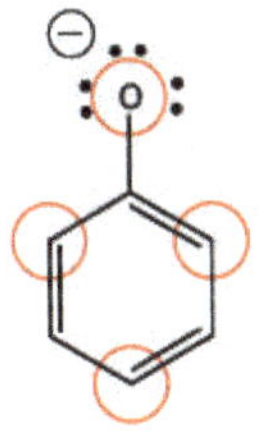

4.13

a)

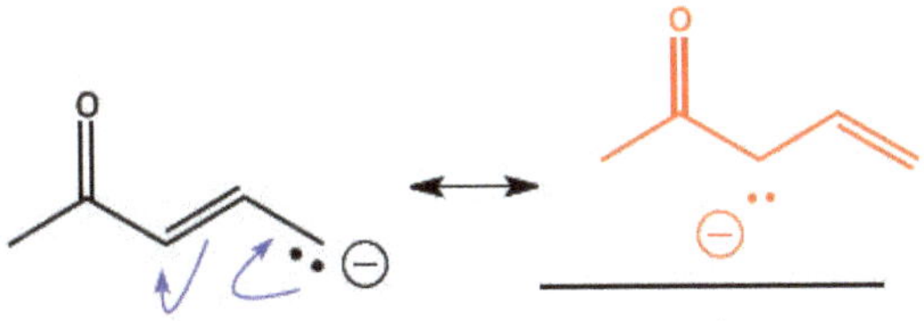

13a
Draw another arrow to move pi electrons 1/2 step

13b
Draw resonance structure

b)

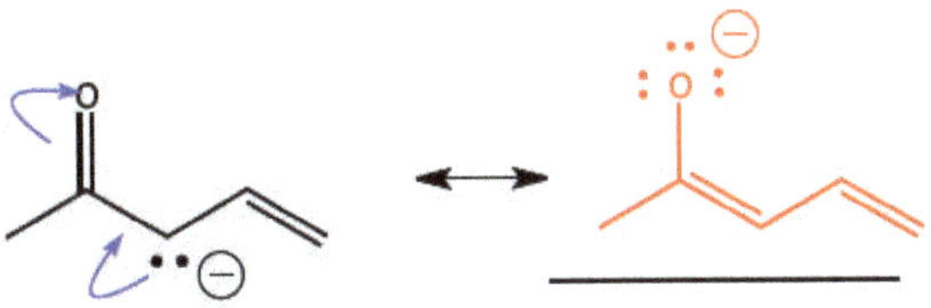

13b
Draw another arrow to move pi electrons 1/2 step

13c
Draw resonance structure

c)

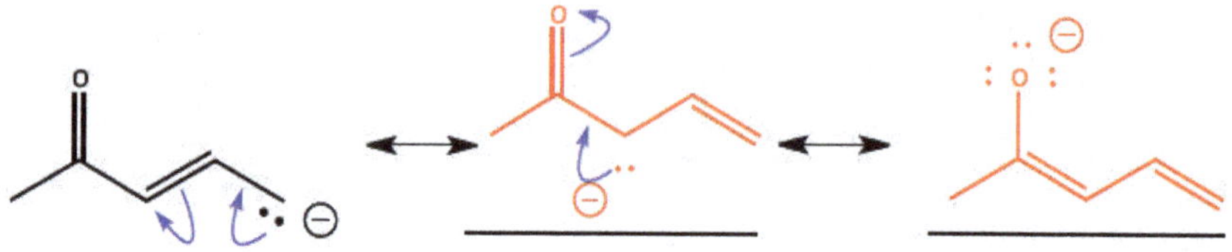

13a 13b 13c

d)

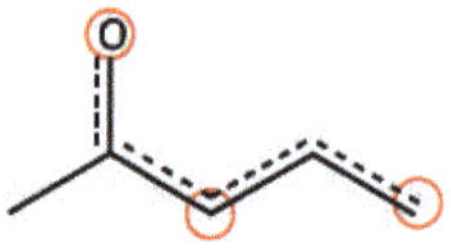

4.14

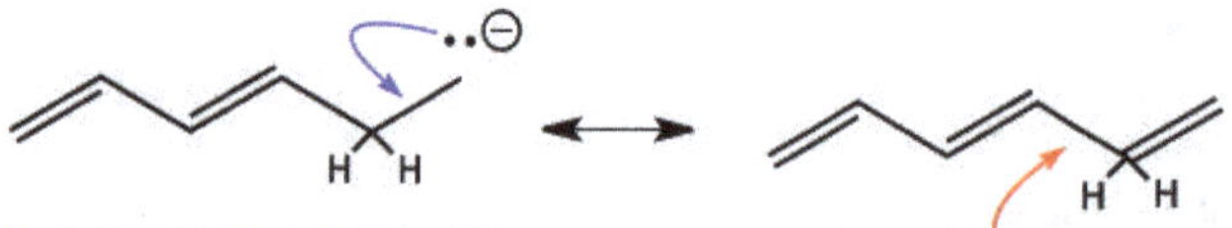

This move would make 5 bonds to carbon. It can't be fixed by taking electrons away on the other side.

Single bond here so can't take away electrons.

4.15

a)

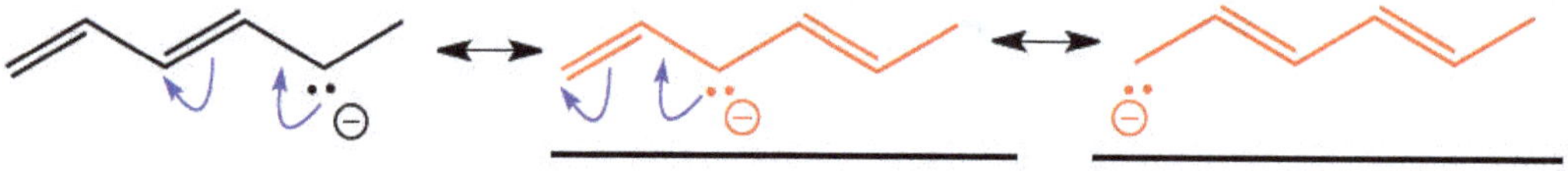

b)

Chapter 5

5.1

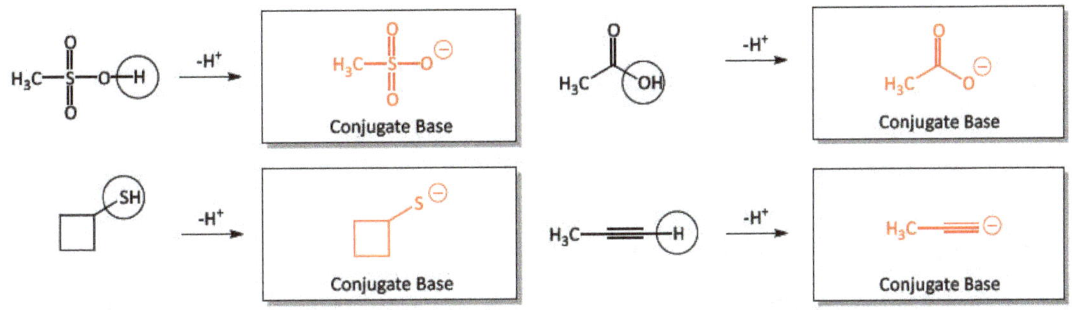

5.2

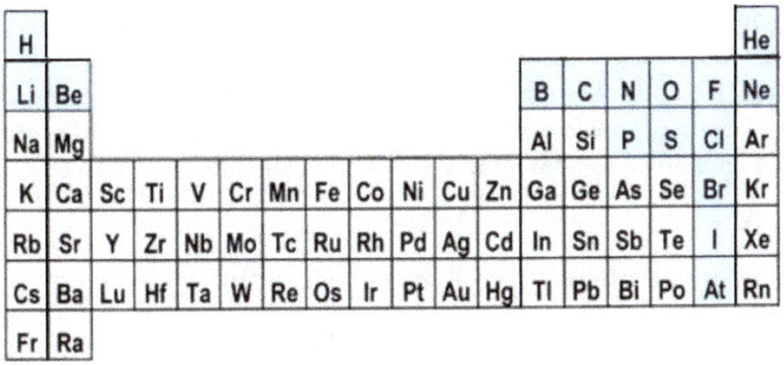

5.3

$R\text{-}NH_2$ vs. $R\text{-}CH_3$ H_2O vs. NH_3 $R\text{-}CH_3$ vs. $R\text{-}OH$

5.4

HBr vs. HI $CH_3\text{-}SH$ vs. $CH_3\text{-}OH$ HF vs. HCl

5.5

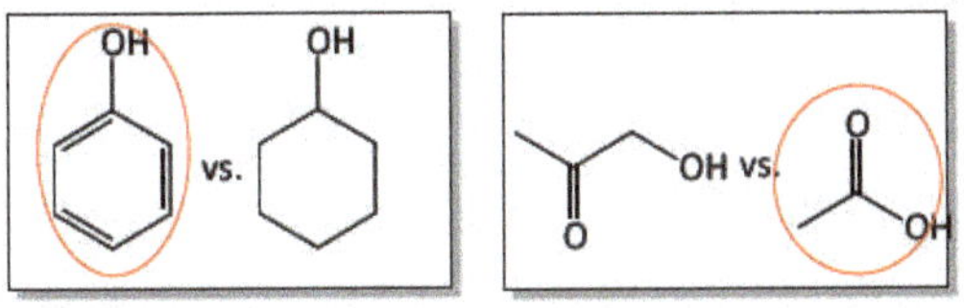

5.6

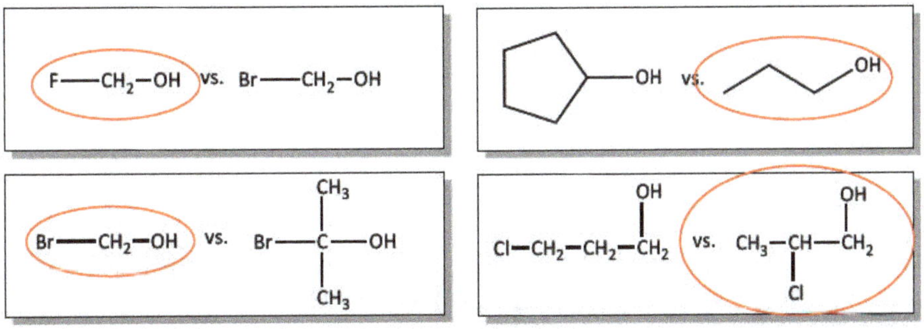

5.7

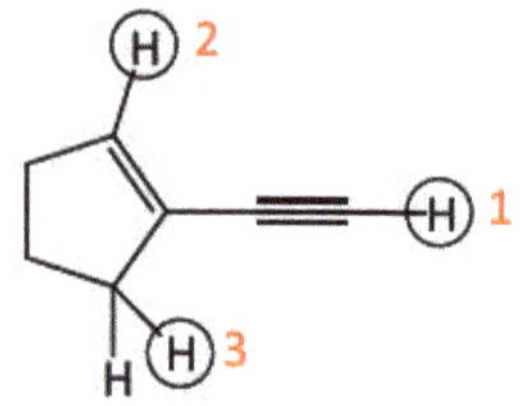

5.8

a) NH_3 and PH_3 b) F^- and Br^- c) NH_2^- and OH^- d) HS^- and Cl^- e) CH_3O^- and $(CH_3)_2CHO^-$

a) The lone pair of electrons are more spread out on the larger P atom than the N atom making the electron density more concentrated on the N and ready to go react.
b) The negative charge is spread out more on the Br- (larger) than the F-.
c) Both N and O are in the same row. O is more electronegative than N so is more stable with the negative charge making the NH_2^- the stronger base.
d) Same row. Cl is more electronegative and more stable with the negative charge than S.
e) The alkyl groups donate electron density towards the negative charge. Isopropoxide has more alkyl groups donating electron density making its negative oxygen less stable and more reactive, more basic.

5.9 a) CH_3NH_2 CH_3NH^- b) CH_3OH CH_3O^- c) HNO_3 NO_3^- d) Br^- none e) H_2O OH^-

5.10 Methyl mercaptan (CH_3SH) is a stronger acid than methanol (CH_3OH). Explain.
We look at the conjugate bases and see that CH_3S^- is more stable and happier with its negative charge than $CH3O^-$ since S is larger than O and can spread out the negative charge more. Therefore, methyl mercaptan is more likely to give up its proton making it the stronger acid.

5.11 The butyl anion ($CH_3CH_2CH_2CH_2:^-$) is a much stronger base than bromide (Br^-). Explain.
The negative charge on the carbon of the butyl anion is on a smaller, carbon atom. It is more localized. The negative charge is more spread out on the bromide ion since it is larger. This makes it more stable, happier. Therefore, the butyl anion is more likely to grab a proton, making it a stronger base.

Chapter 6

6.1

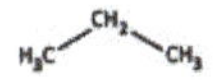 saturated

 unsaturated

 saturated

 unsaturated

6.2

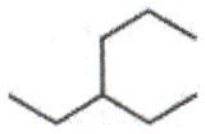
3-ethylhexane

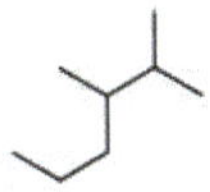
2,3-dimethylhexane

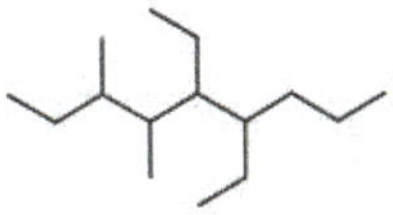
5,6-diethyl-3,4-dimethylnonane

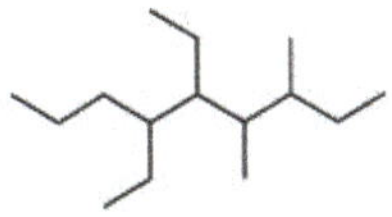
5,6-diethyl-3,4-dimethylnonane

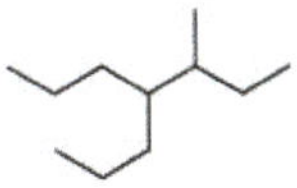
3-methyl-4-propylheptane

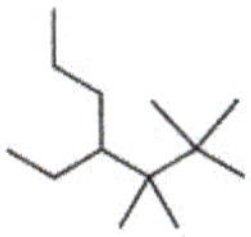
4-ethyl-2,2,3,3-tetramethylheptane

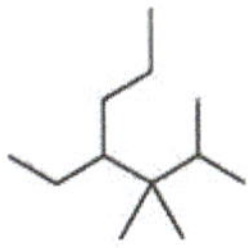
4-ethyl-2,3,3-trimethylheptane

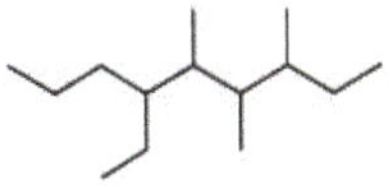
6-ethyl-3,4,5-trimethylnonane

6.3

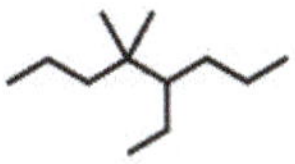
5-ethyl-4,4-dimethyloctane

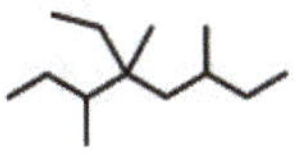
4-ethyl-3,4,6-trimethyloctane

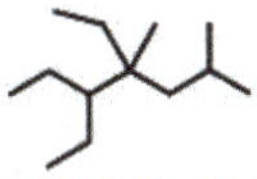
4,5-diethyl-2,4-dimethylheptane

6.4

a) 
4-isopropylheptane

b) 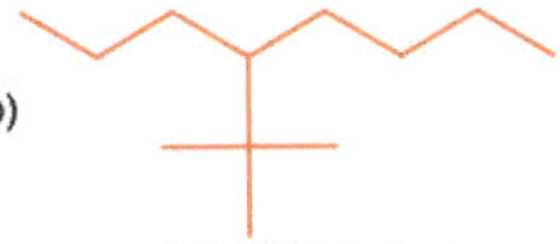
4-tert-butyloctane

c) 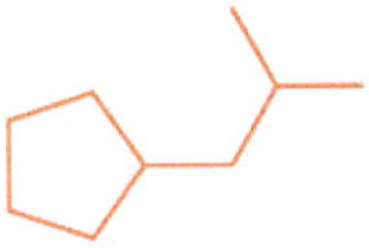
isobutylcyclopentane

d) 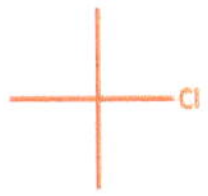

tert-butylchloride

6.5

a) 
4-isopropylheptane
or
4-(1-methyl ethyl)heptane

b) 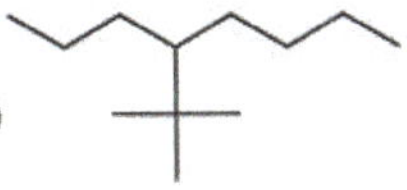
4-tert-butyloctane
or
4-(1,1-dimethyl ethyl)octane

6.6

a) 
4-(1-ethylpropyl)nonane

b) 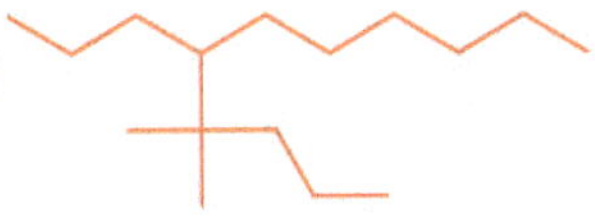
4-(1,1-dimethylbutyl)decane

6.7

a)  yes 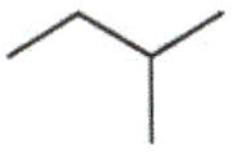

b)   no (just rotated)

c) 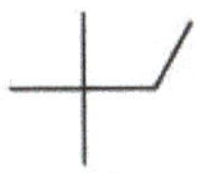 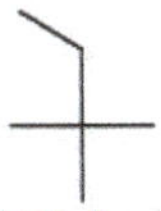 no (just rotated 90 degrees)

d)  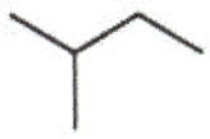 no (different formulas)

e)   yes

f)   no (different formulas)

Answers to Exercises

6.8

a) 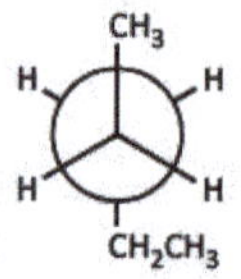

b) 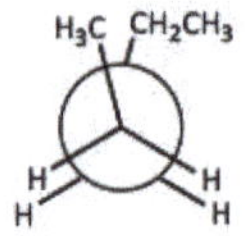

c) 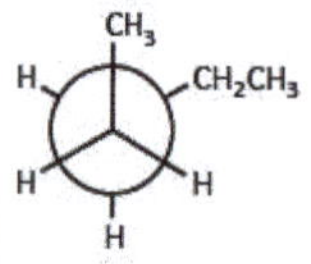

d) 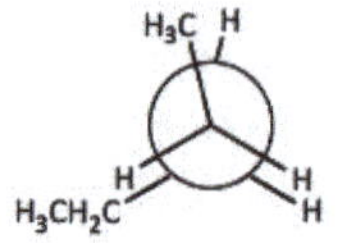

6.9

a) Anti b) Totally eclipsed c) Gauche d) Eclipsed

6.10

a < b < c < d   (a < c < d < b)   a < d < c < b   b < d < c < a   c < a < d < b

6.11

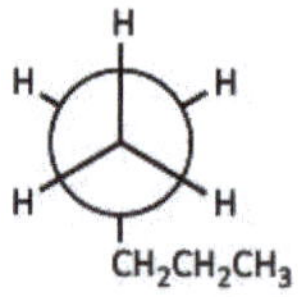

6.12

a)  conformations

b) 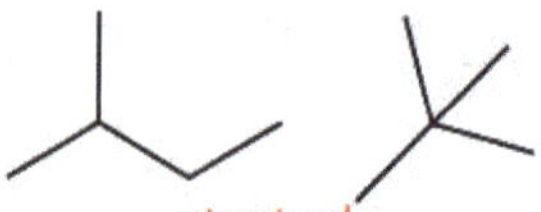 structural

c) 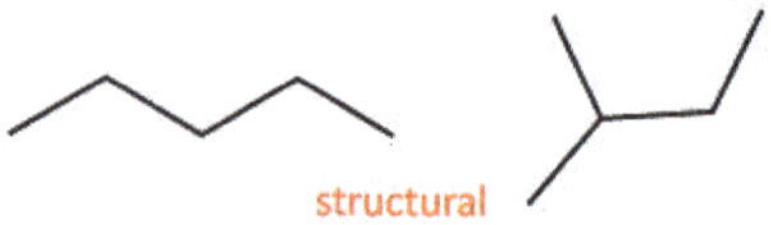 structural

d) 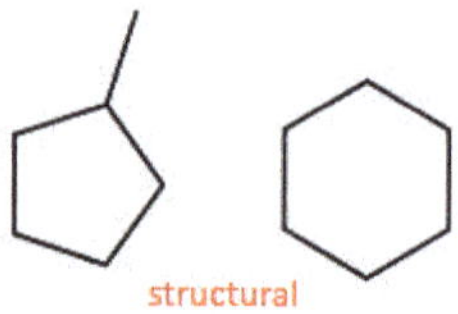 structural

e) 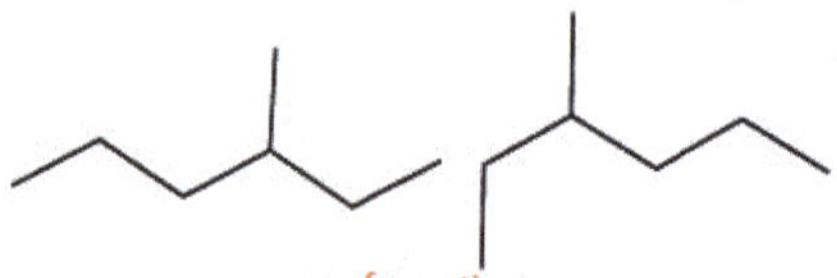 conformations

f)  structural

6.13

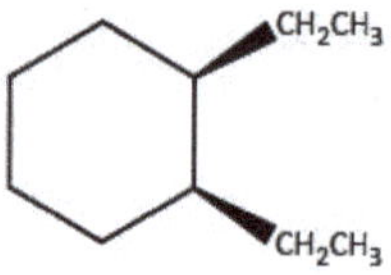

cis-1,2-diethylcyclohexane

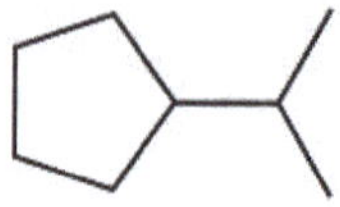

isopropylcyclopentane
1-isopropylcyclopentane
methylethylcyclopentane
1-(1-methylethyl)cyclopentane

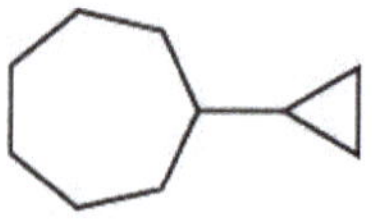

cyclopropylcycloheptane
1-cyclopropylcycloheptane

6.14

Draw the following compounds.

a) *trans*-1-ethyl-3-propylcyclohexane

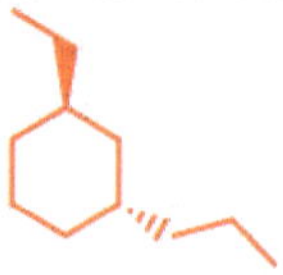

b) *cis*-1-propyl-3-(2,2-dimethylpropyl)cycloheptane

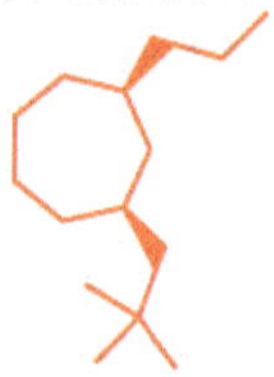

c) cyclopropylcyclobutane

d) *cis*-1,2-dichlorocyclopentane

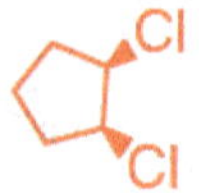

6.15

bicyclo[2.2.1]heptane

bicyclo[1.1.0]butane

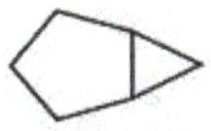
bicyclo[3.1.0]hexane

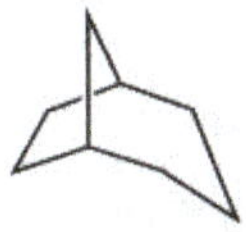
bicyclo[3.2.1]octane

6.16

bicyclo[2.1.1]hexane

bicyclo[2.1.0]pentane

bicyclo[2.2.2]octane

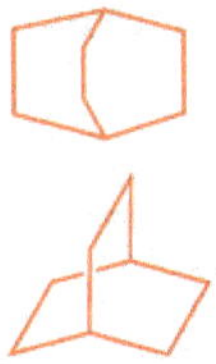

6.17

6.18

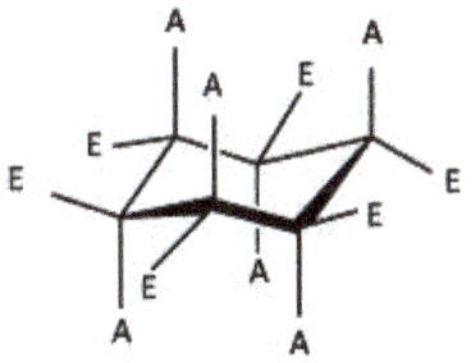

6.19

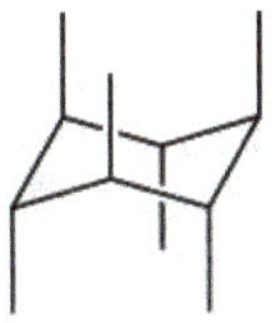

6.20

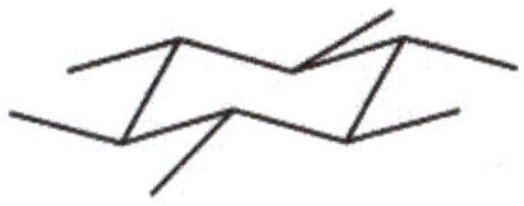

6.21

6.22

Your answer may look slightly different depending on where you chose to draw your ethyl group. Make sure it is equatorial in answer a and axial in answer b.

a) 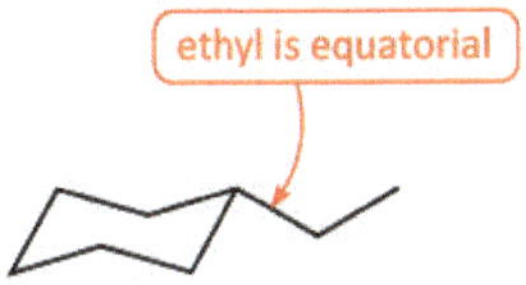

lower energy

b) 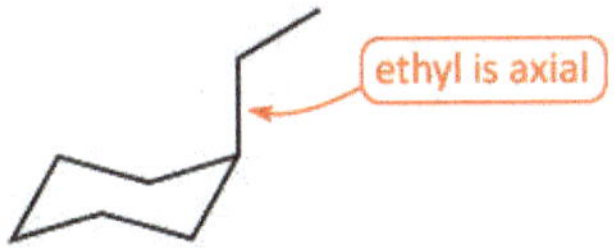

higher energy

6.23

a) 

equatorial

b) 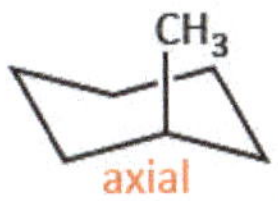

axial

c) 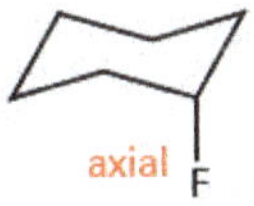

axial

d) 
equatorial

6.24

a) *trans*-1-ethyl-4-methylcyclohexane

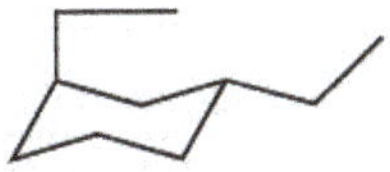
b) *trans*-1,3-diethylcyclohexane

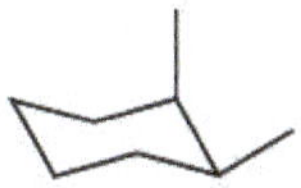
c) *cis*-1,2-dimethylcyclohexane

6.25

a. *trans*-1,2-diethylcyclohexane

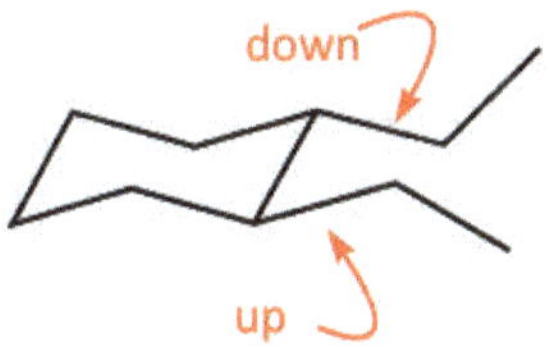

b. *cis*-1-(1,1-dimethylethyl)-4-propylcyclohexane

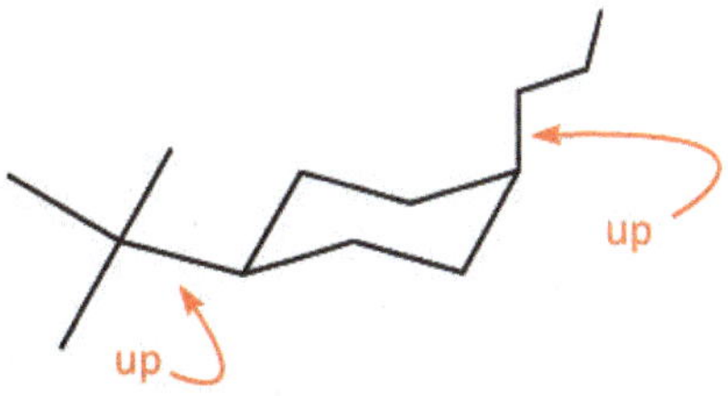

(big group equatorial)

Chapter 7

7.1 Are the following objects chiral or achiral?

a) a pair of scissors – Chiral (there are left and right handed scissors)

b) a screw – Chiral (because of the threads) c) a screwdriver - Achiral d) a hammer - Achiral

7.2

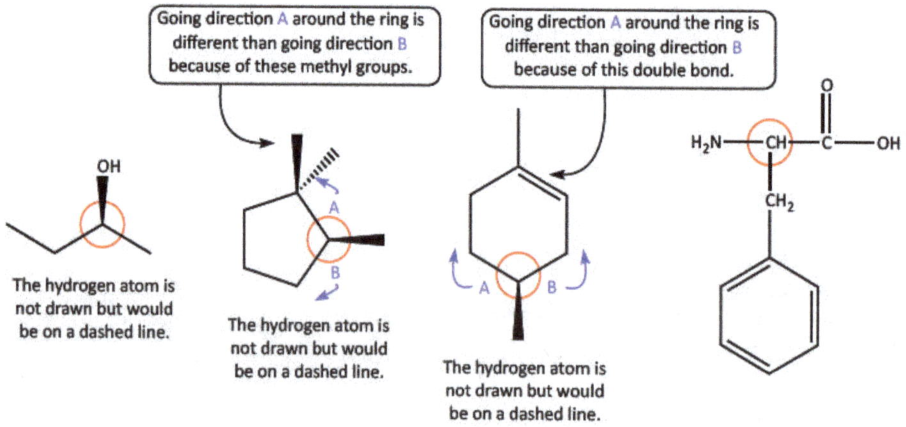

7.3

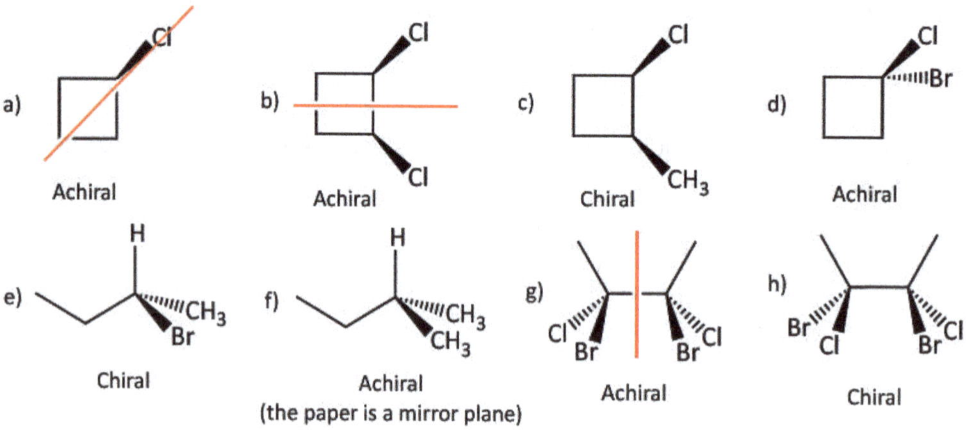

7.4

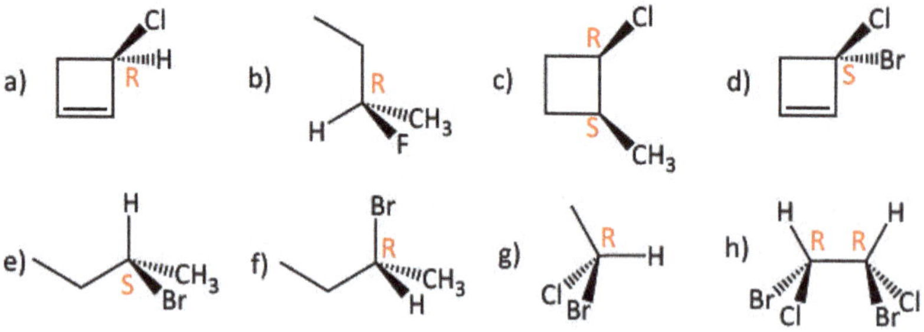

7.5

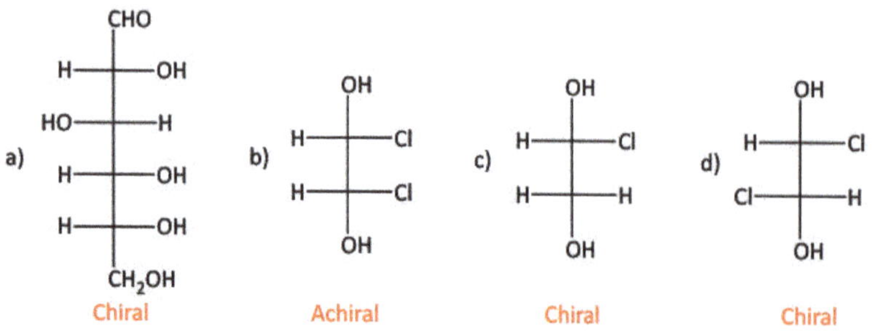

7.6

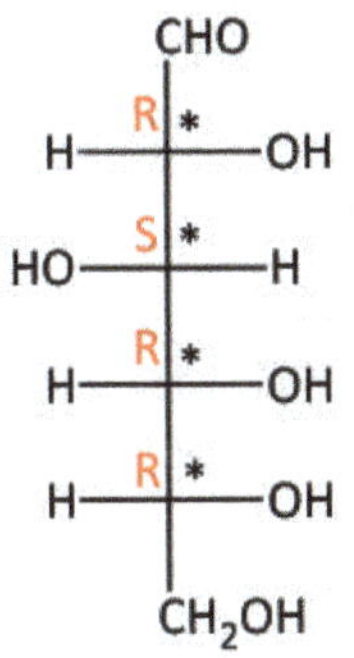

7.7

a)

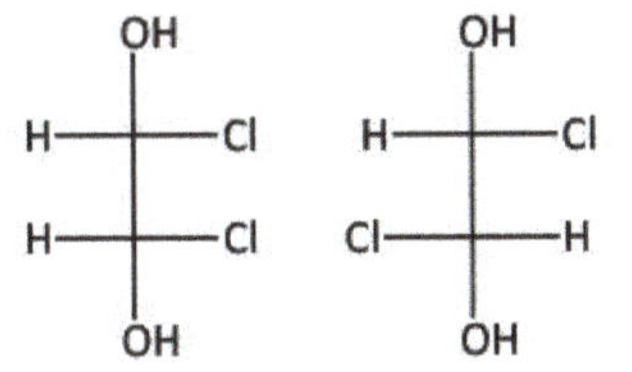

Diastereomers

b)

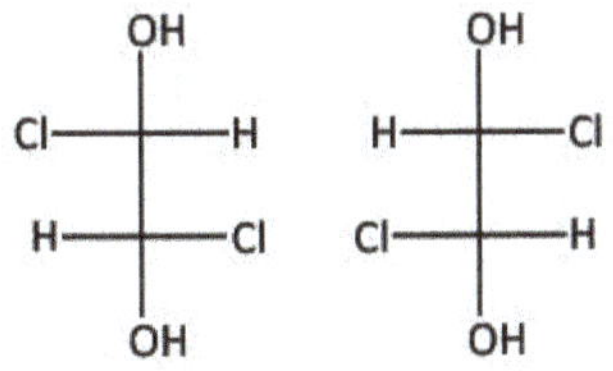

Enantiomers

c)

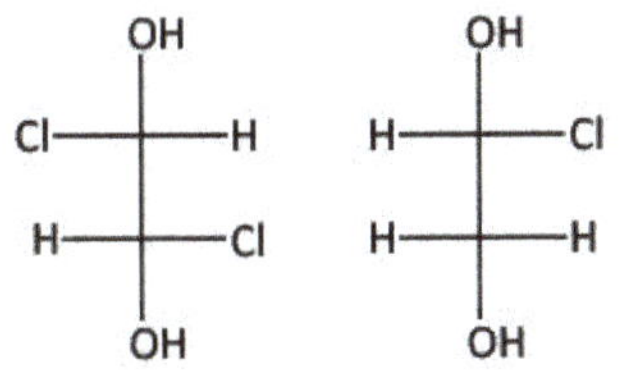

Different compounds

d)

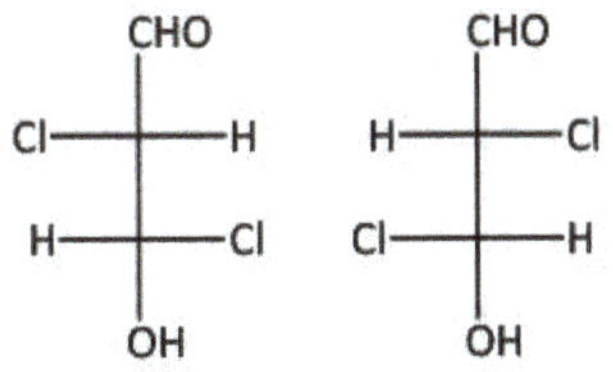

Enantiomers

7.8

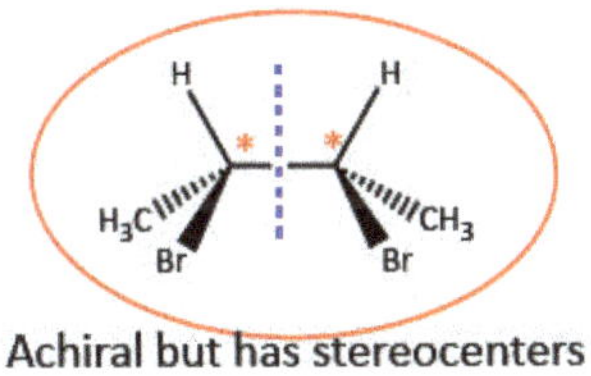

Achiral but has stereocenters
Meso

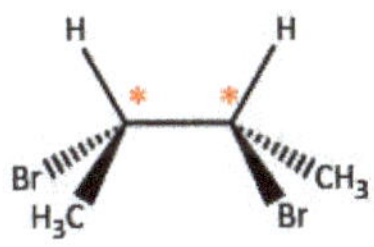

Chiral

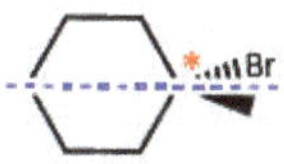

Achiral but has no stereocenters

Chiral

7.9.

a)

Not isomers

b)

structural isomers

c)

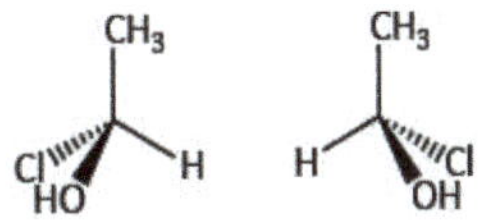

enatiomers

d)

somers
s

e)

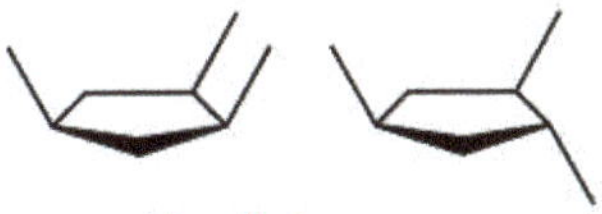

other diastereomers

f)

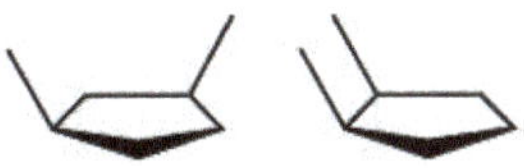

structural isomers

7. [illegible] tell from the (R) or (S) designation only. Just because the Cahn-Ingold-Prelog arrow is clockwise or [illegible] r (R), it has no bearing on if the compound is *d* or *l*. The experiment needs to be performed. (R)-2-[illegible] is levorotatory with a specific rotation of -23.1°.

7.11 60% e.e. of dextrorotatory.

7.12 20% e.e. of levorotatory

7.13 75% of *d*-enantiomer (25% of *l*-enantiomer)

Chapter 8

8.1

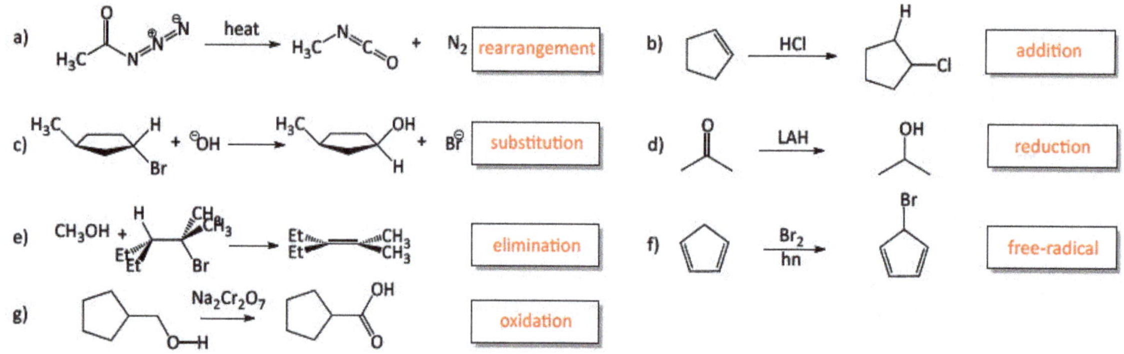

8.2

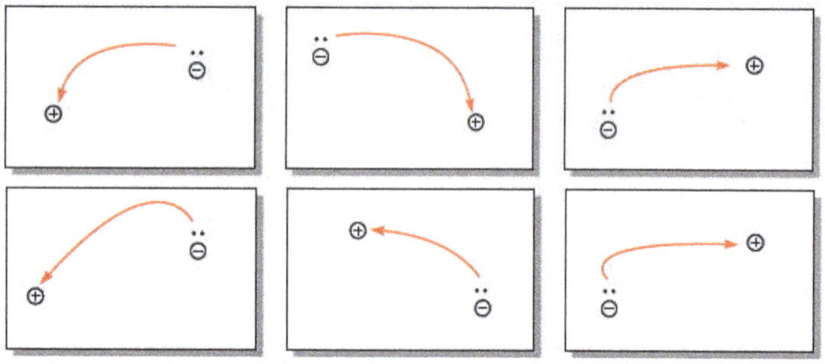

8.3

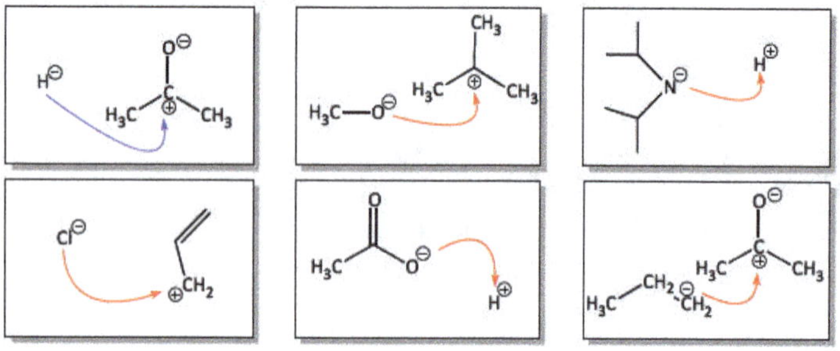

8.4

a) Nucleophiles
b) Electrophiles
c) Electrophiles

8.5

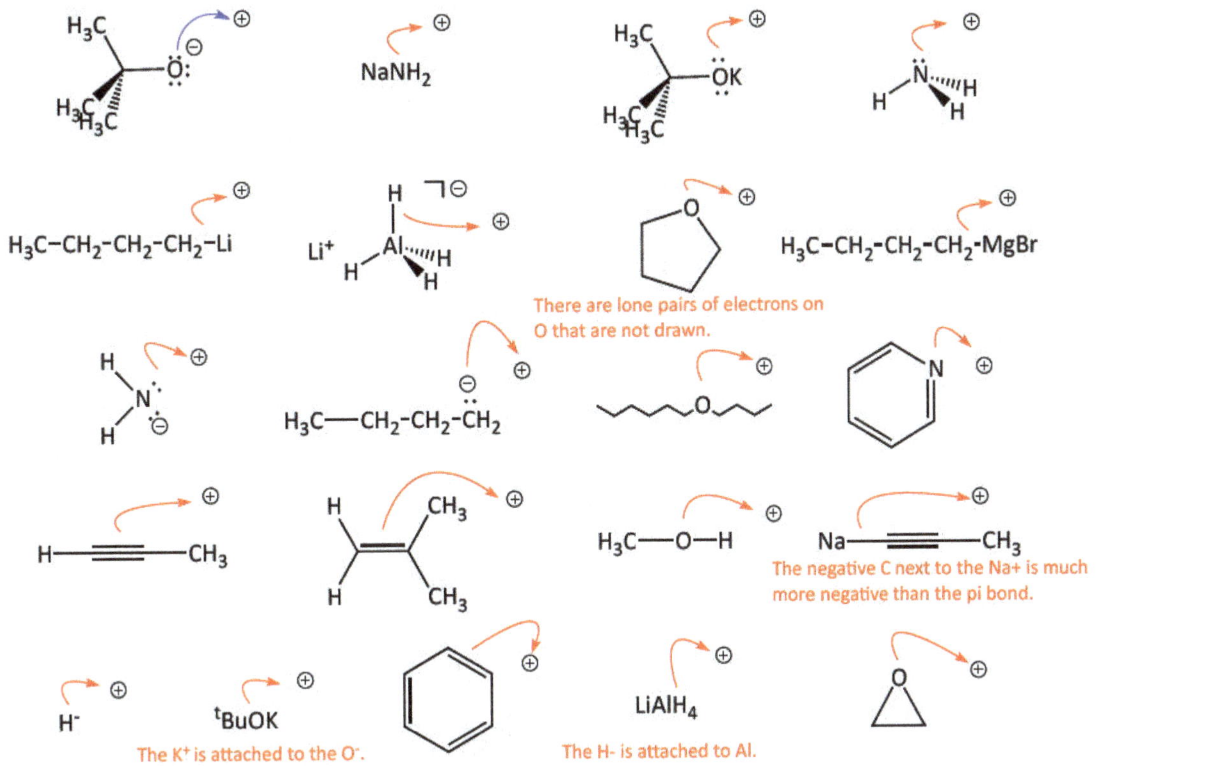

Notice in all of these, when there is a metal atom in the compound, the bond/atom next to it is the most negative site. Your arrow will start there! Pi bonds (double/triple bonds) are usually good locations for arrow beginnings, but not compared to spots next to very positive metal atoms!

8.6

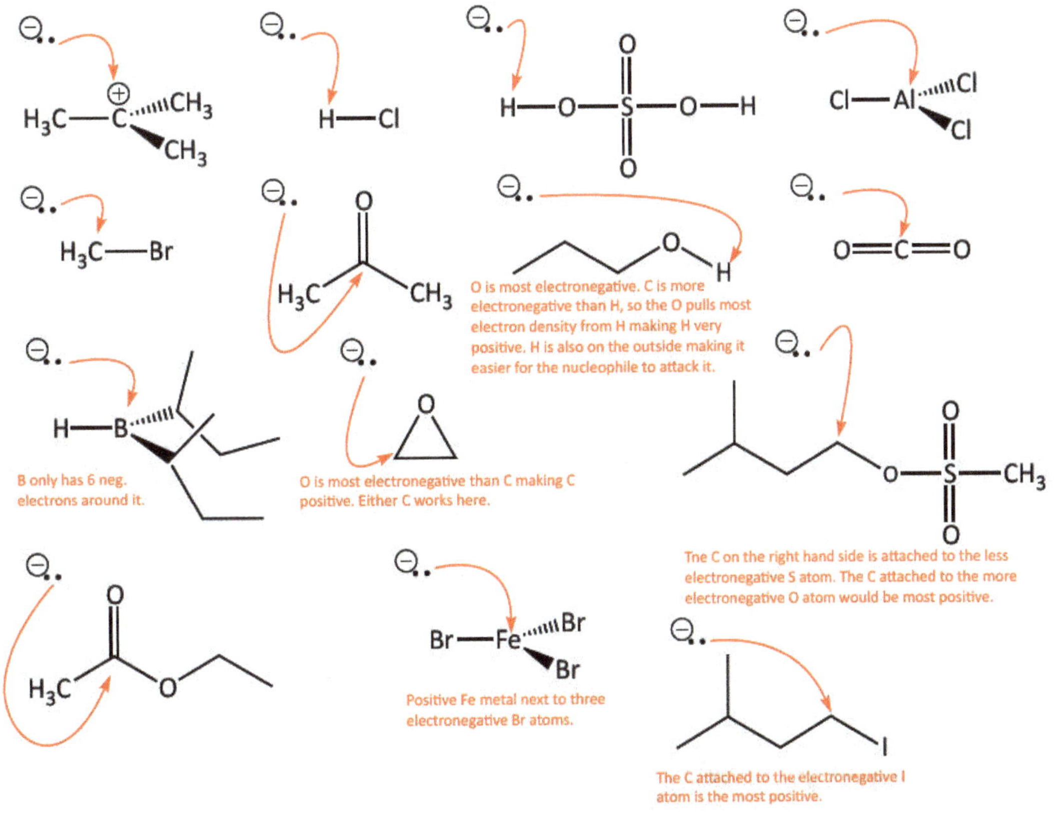

8.7

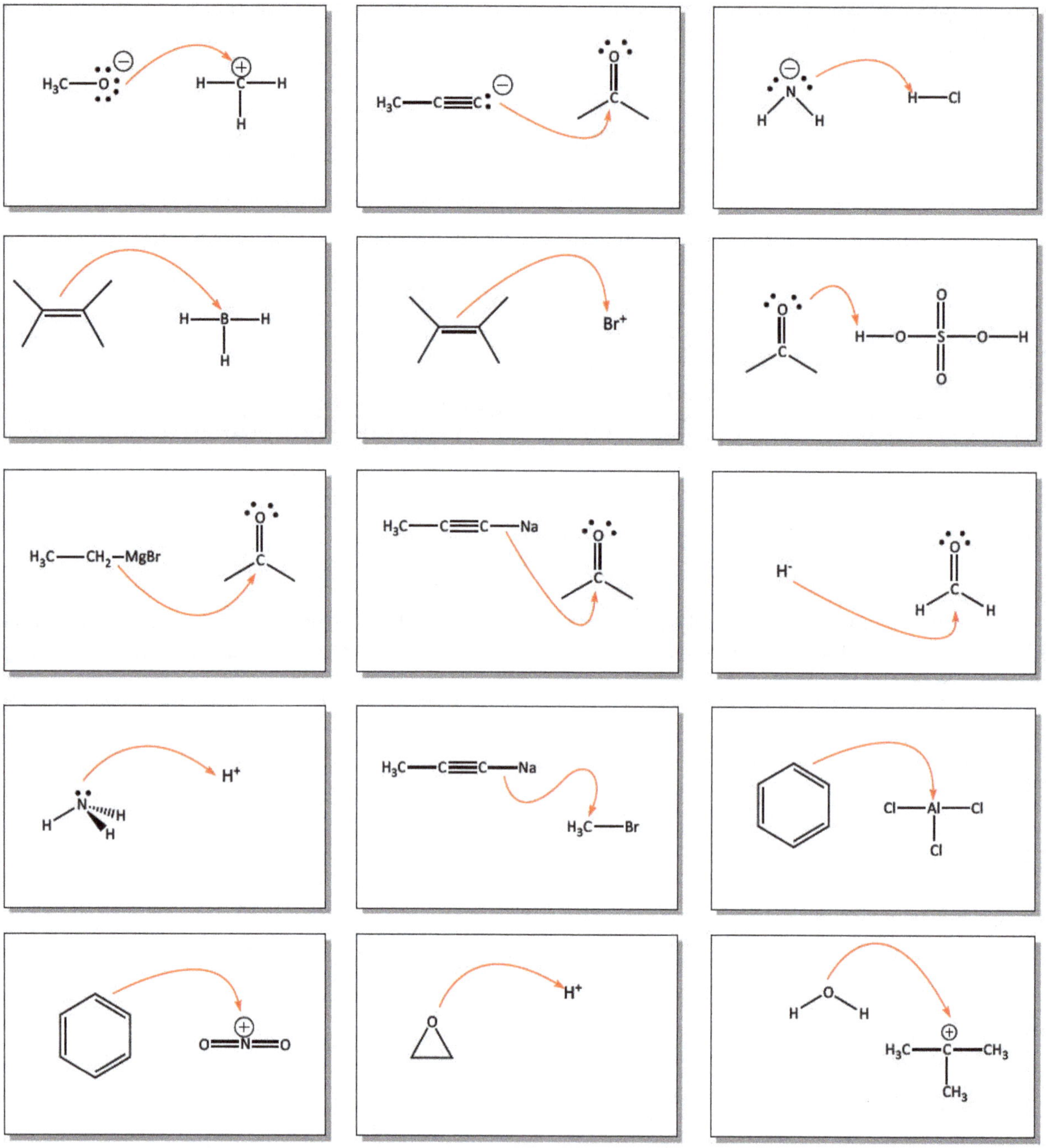

8.8

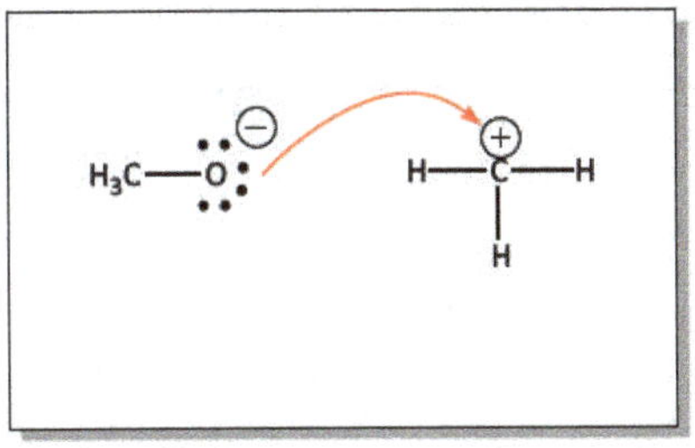
H3C—O
H—C—H
H

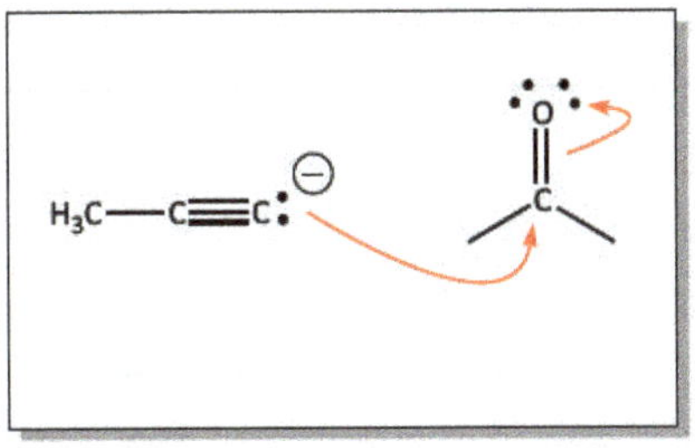
H3C—C≡C
O
C

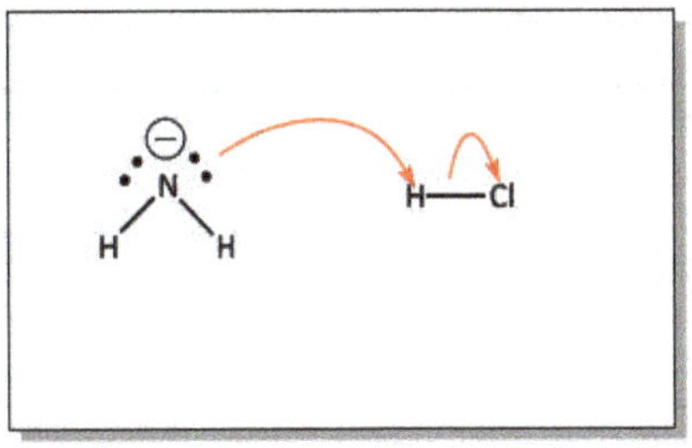
N
H H
H—Cl

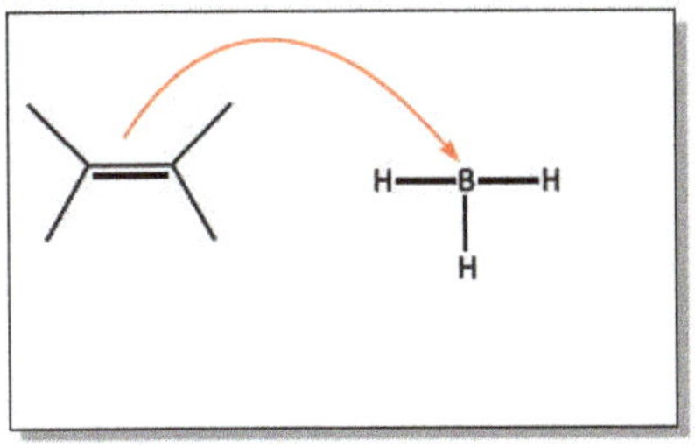
H—B—H
H

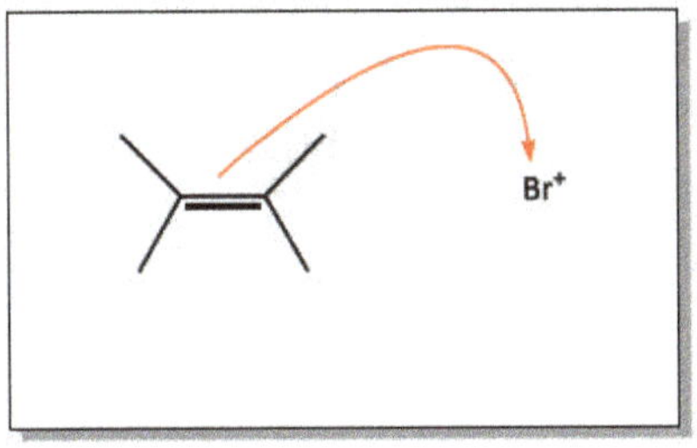
Br+

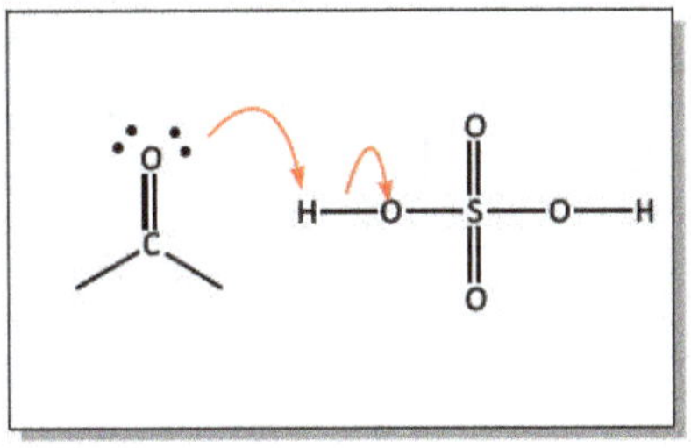
O
C
H—O—S—O—H
O
O

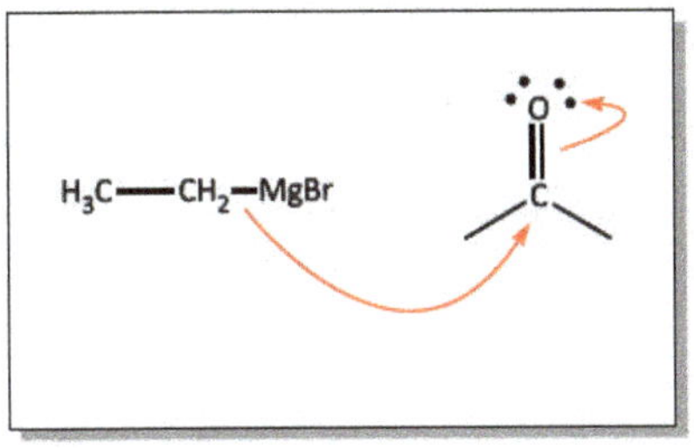
H3C—CH2—MgBr
O
C

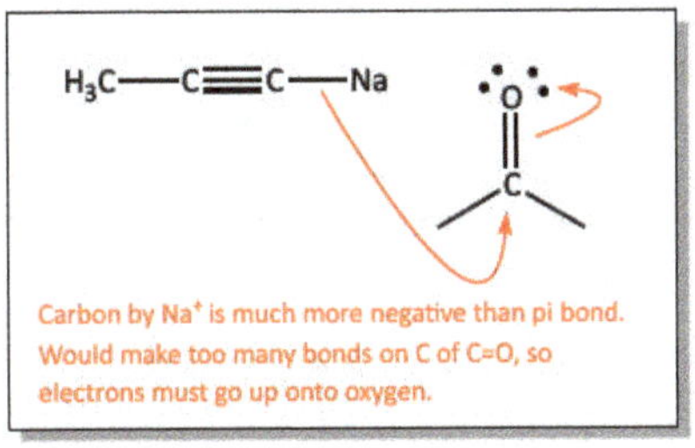
H3C—C≡C—Na
O
C
Carbon by Na+ is much more negative than pi bond.
Would make too many bonds on C of C=O, so
electrons must go up onto oxygen.

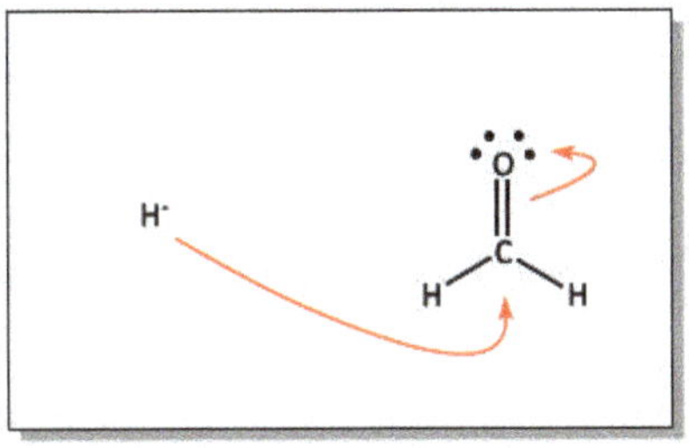
H-
O
C
H H

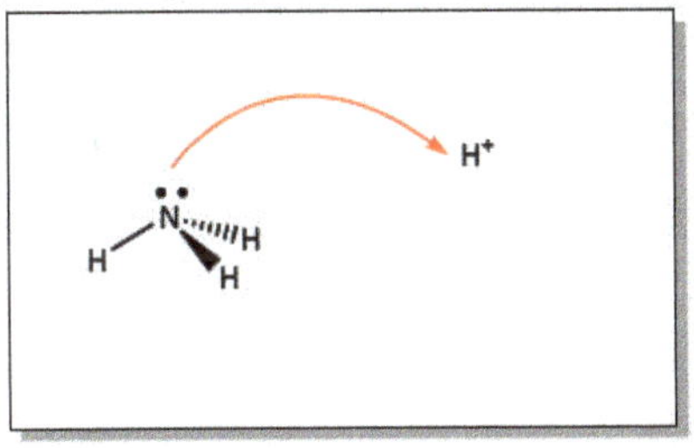
H+
N
H H H

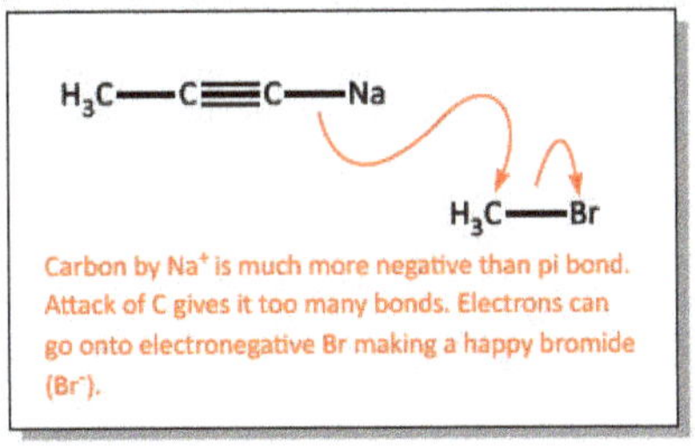
H3C—C≡C—Na
H3C—Br
Carbon by Na+ is much more negative than pi bond.
Attack of C gives it too many bonds. Electrons can
go onto electronegative Br making a happy bromide
(Br-).

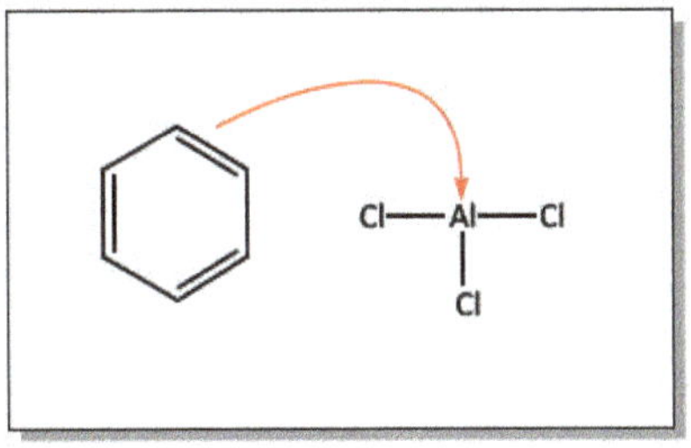
Cl—Al—Cl
Cl

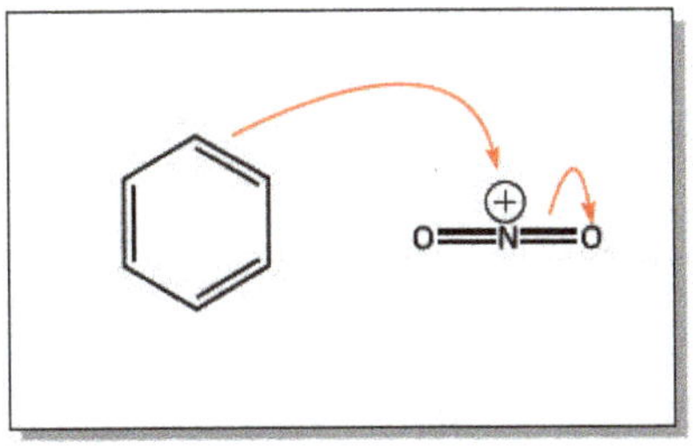
O=N=O

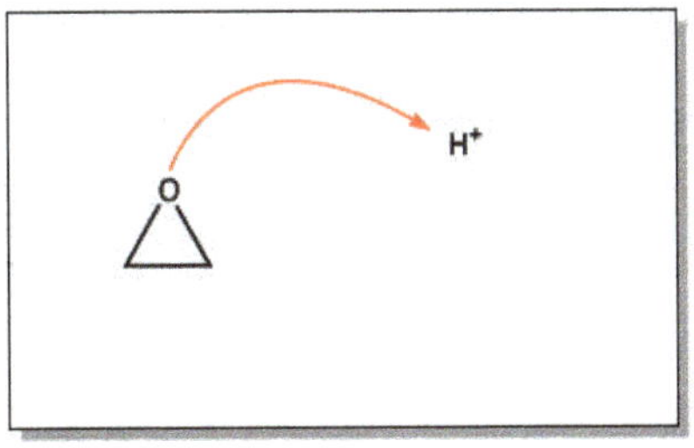
H+
O

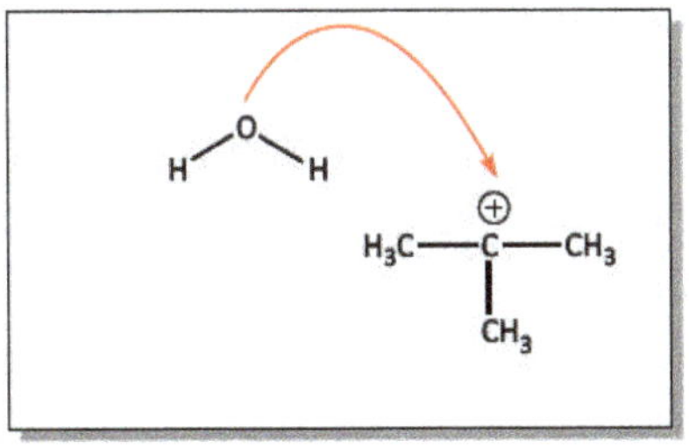
H O H
H3C—C—CH3
CH3

8.9

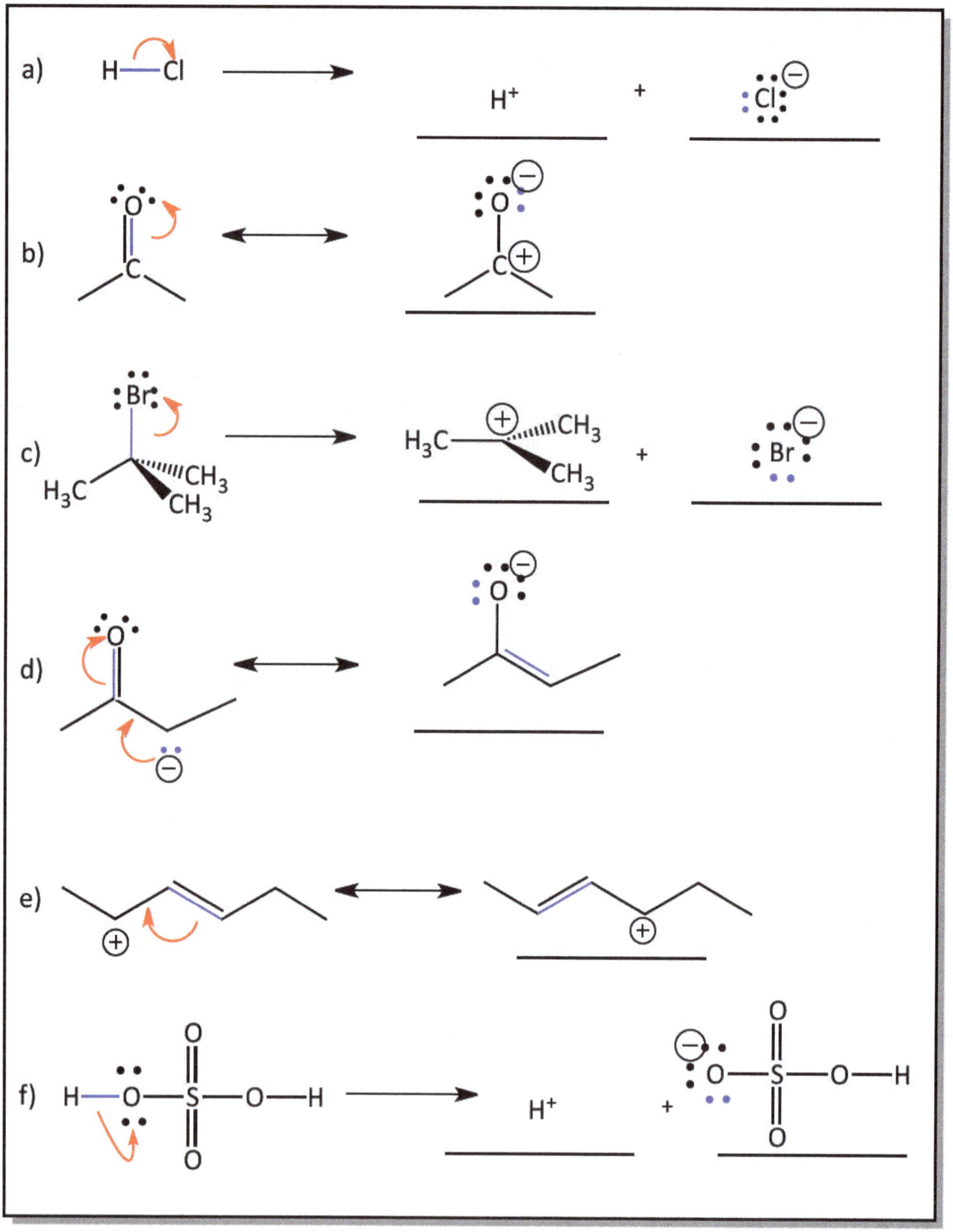
a)
H—Cl
H⁺
+
Cl
b)
O
C
O
C
c)
Br
H₃C
CH₃
CH₃
H₃C—
CH₃
CH₃
+
Br
d)
O
O
e)
f)
H—O—S—O—H
O
O
H⁺
+
O—S—O—H
O
O

8.10

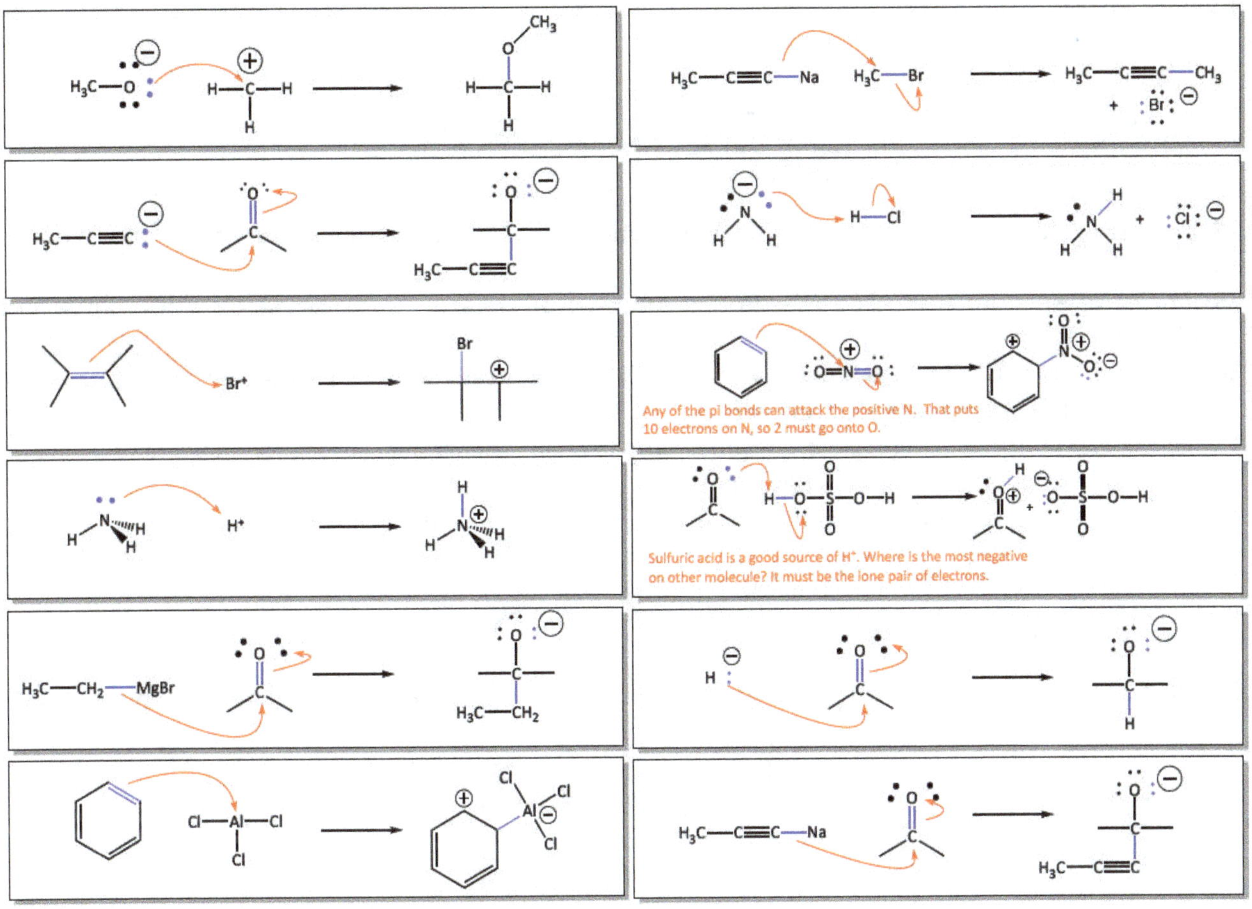

8.11 Predict if ΔH is positive or negative for the following situations.

a. In a reaction, one strong bond is broken and one weak bond is formed. __+_ ΔH

b. In a reaction, one weak bond is broken and one strong bond is formed. __-_ ΔH

8.12

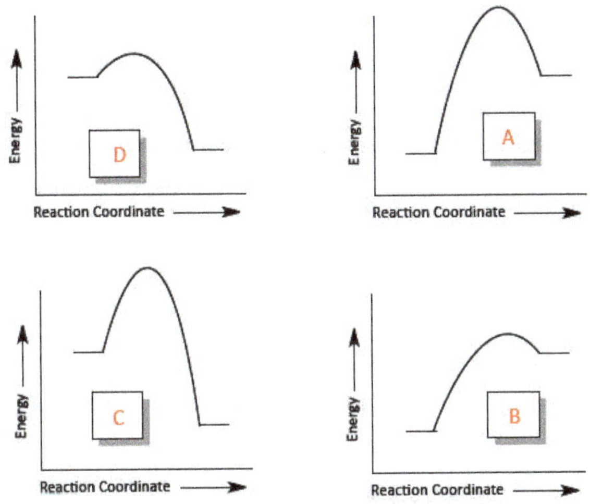

8.13

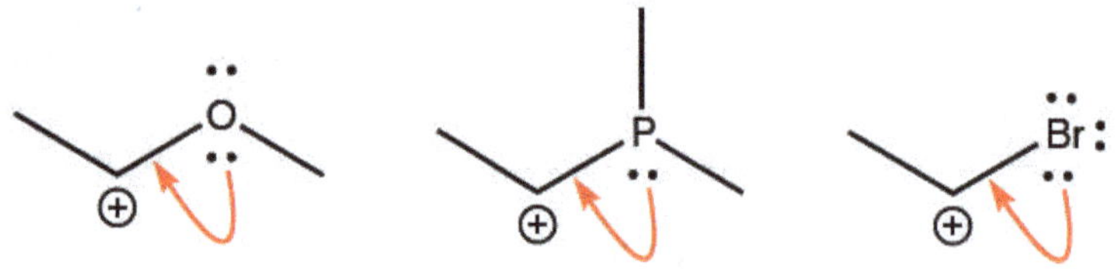

8.14 If negative electrons are pushed towards the positive p-orbital of a carbocation, does this help or hurt the carbocation? Helps

8.15 If negative electrons are sucked away from the positive p-orbital of a carbocation, does this help or hurt the carbocation? Hurts

8.16 The strength of an atom to pull negative electrons towards itself in a bond is called electronegativity. Between F and Br, which is more electronegative? F

8.17 Since F and Br are both fairly electronegative, do they help or hurt the carbocation they are on? Hurt

8.18 Which hurts the carbocation more, F or Br? F

8.19 Circle the structure below that is the most stable (or is hurt the least).

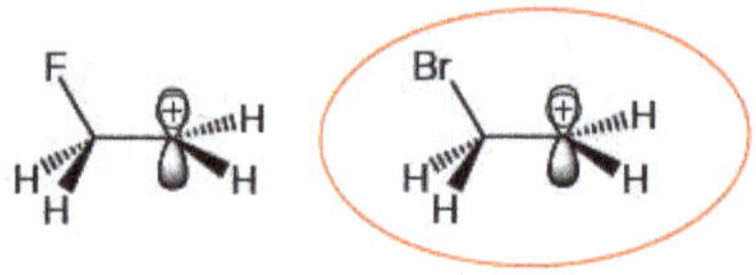

8.20 and 8.21 Draw a circle around the molecule that is hurt worse by the electronegative fluorine atom.

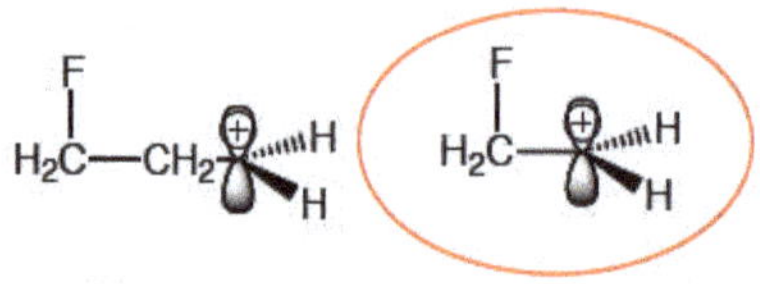

Distance matters. The fluorine being closer to the carbocations hurts worse than if it is farther away.

8.21 Now, think about which is the happiest, and draw a square around the most stable carbocation of the two.

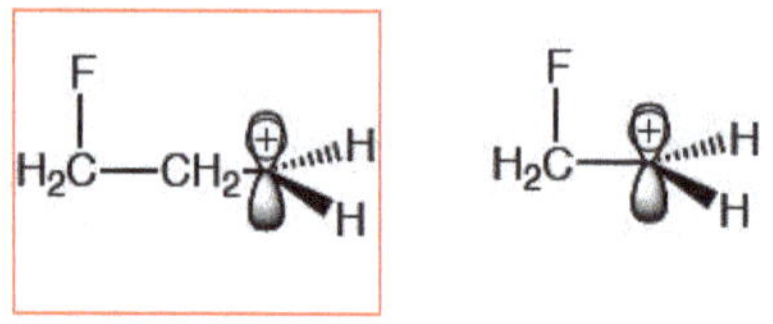

8.22

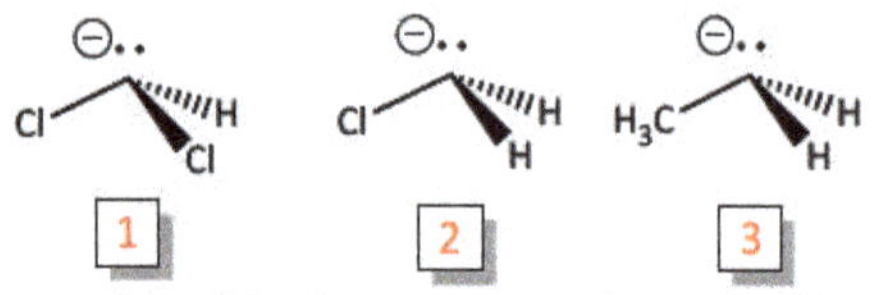

Electronegative chlorine atoms help carbanions by pulling electrons away through induction. The methyl group hurts the carbanion because it donates electron density towards the carbanion through hyperconjugation.

8.23

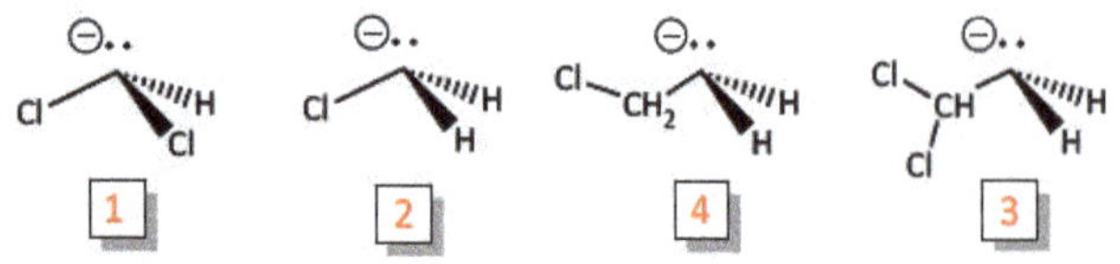

The closer the electronegative chlorine atoms are to the carbanion, and the more electronegative chlorine atoms there are, it helps the carbanion.

8.24 c) both electrophilic and nucleophilic

Chapter 9

9.1

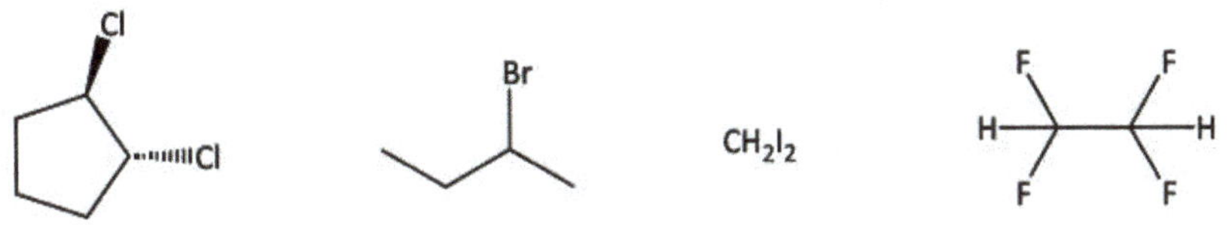

trans-1,2-dichlorocyclopentane 2-bromobutane diiodomethane 1,1,2,2-tetrafluoroethane

9.2

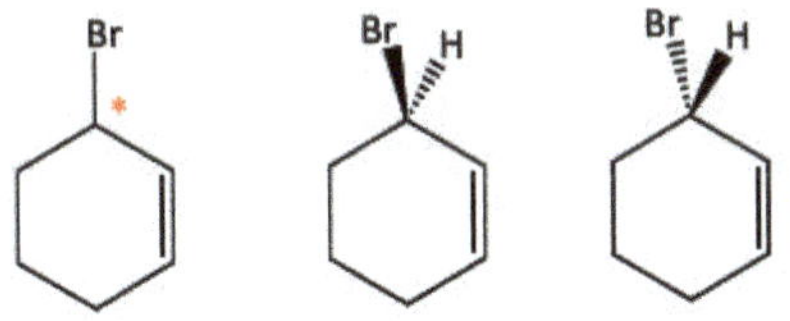

9.3

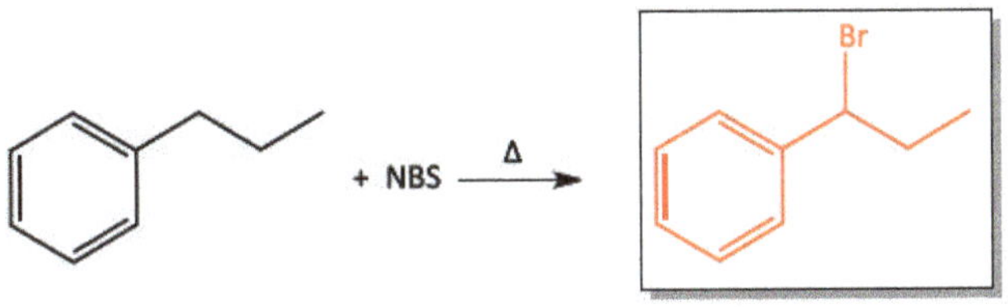

9.4

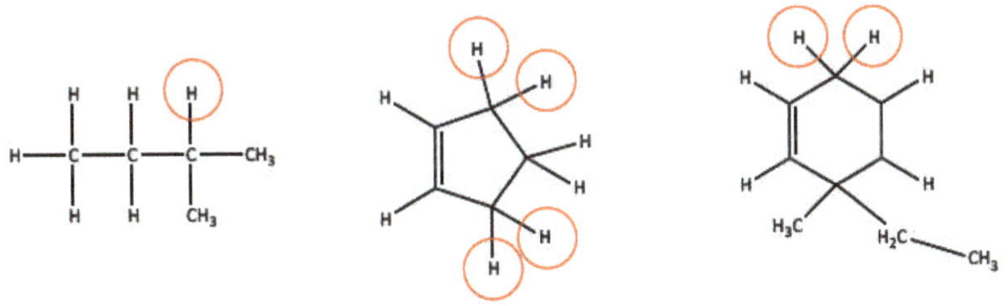

9.5

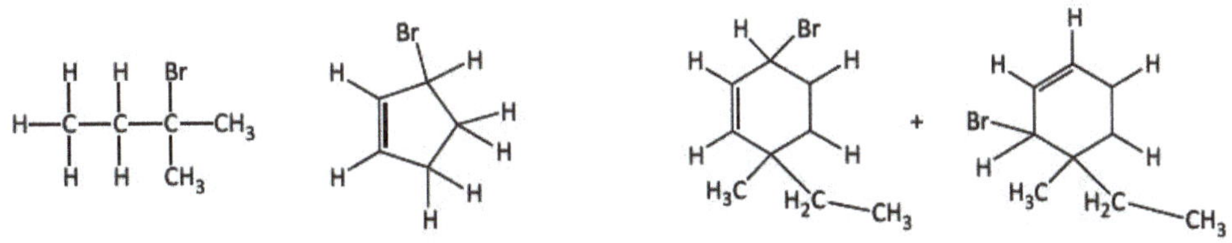

9.6

a. 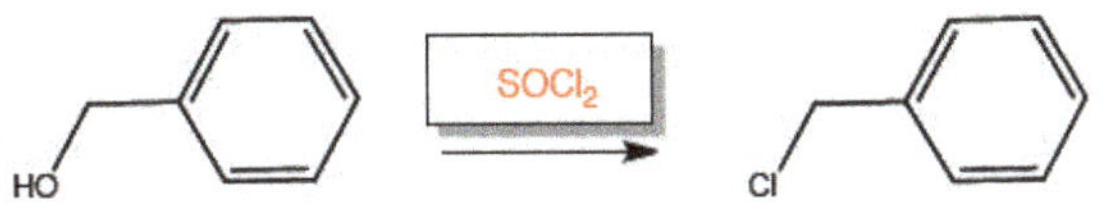

d. 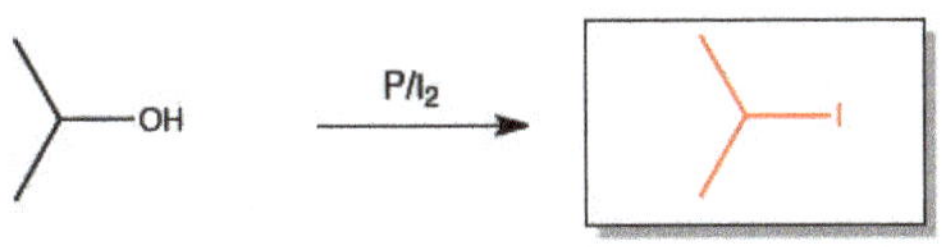

b. 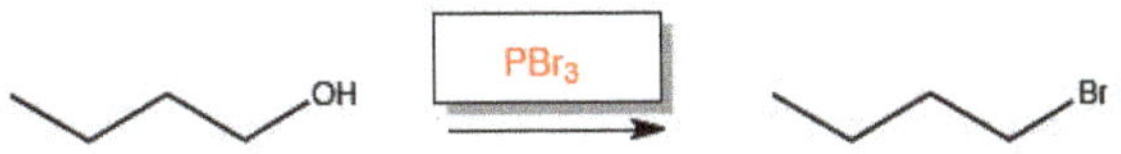

e. 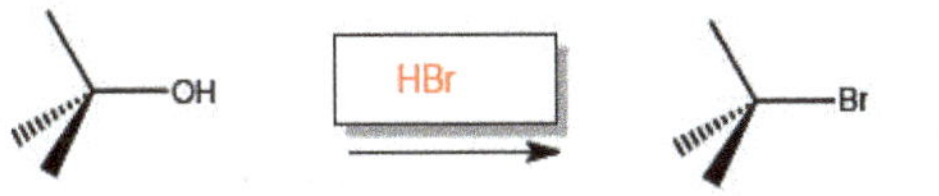

c. 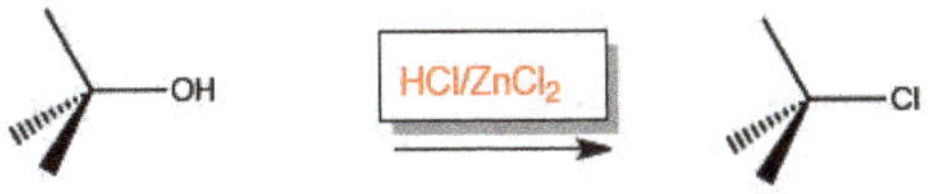

f. 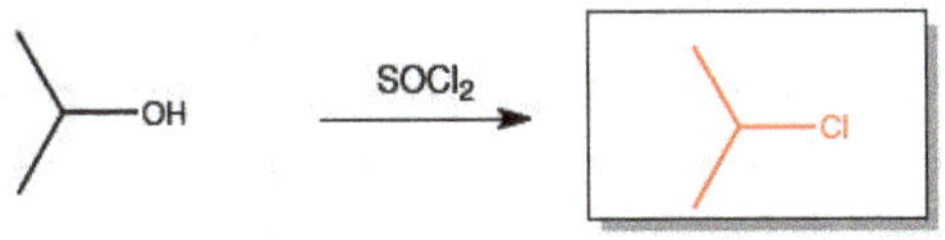

9.7

a) 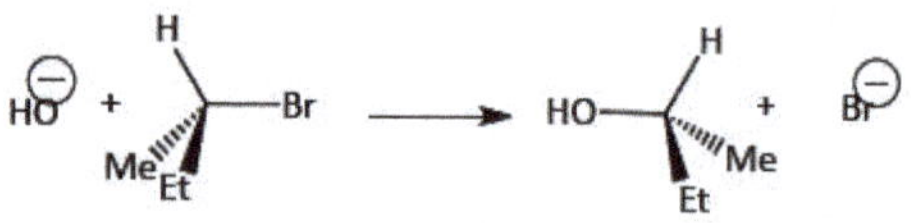

substitution

b) 

substitution

c) 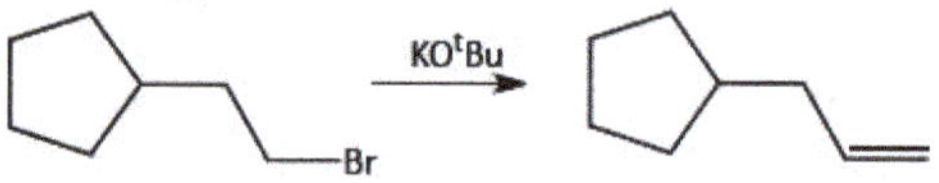

elimination

d) 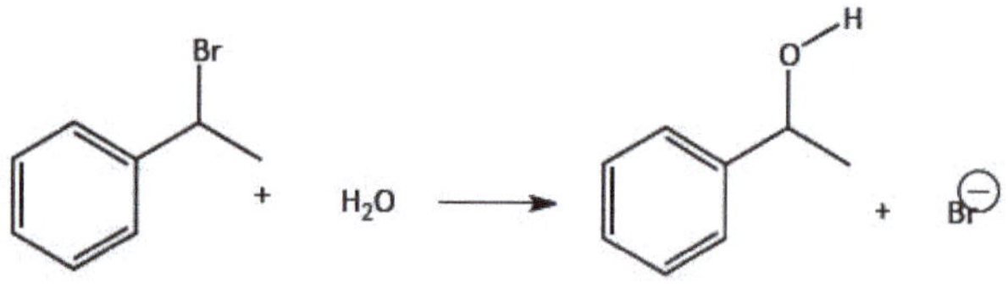

substitution

e) 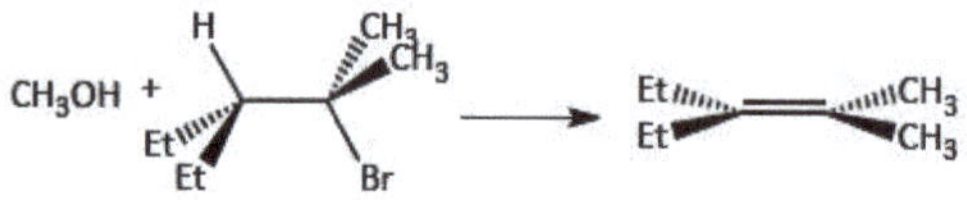

elimination

9.8

a) 

b) 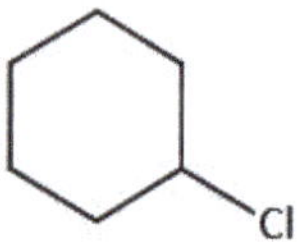

c) 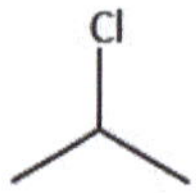

d) 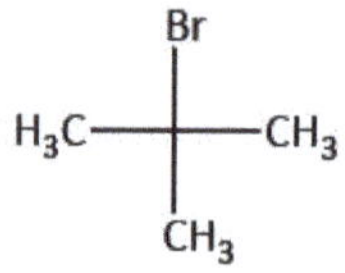

S_N2 Both Both S_N1

9.9

a) 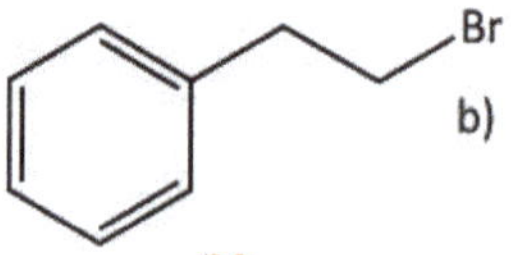

b) 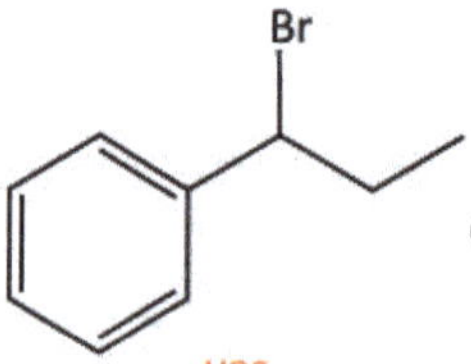

c) 

d) 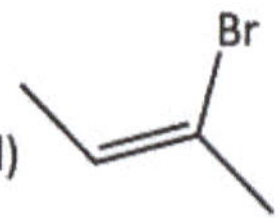

no yes yes no

9.10

a) $PhCH_2Cl$ b) 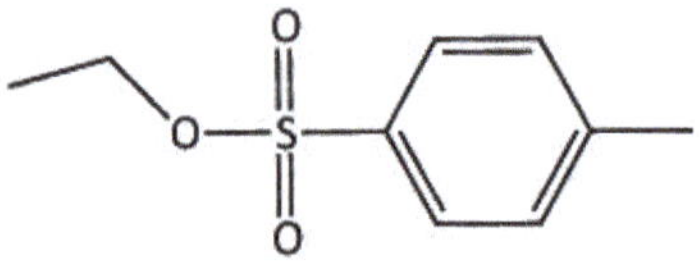

c) 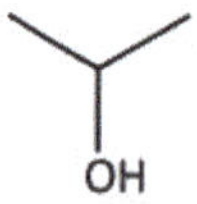

d) 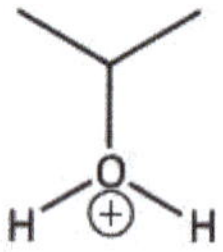

good good bad good

9.11

a) CH_3OH b) $(CH_3)_2N^-$ c) $CH_3CH_2CH_2CH_2^-$ d) CH_3CH_2SH e) H_2O

S_N1 S_N2 S_N2 S_N1 S_N1

9.12

CH_3CH_2OH does not react with NaBr, but adding H_2SO_4 forms CH_3CH_2Br. Explain

The OH group of CH_3CH_2OH (ethanol) is a bad leaving group. Therefore, the Br^- nucleophile of NaBr cannot attack the ethanol. Adding H_2SO_4 will protonate the OH group of the ethanol making it water, $-OH_2^+$. This water leaving group is a great leaving group which now allows the Br- nucleophile to attack the ethanol and the water leaving group leave making ethyl bromide, CH_3CH_2Br.

9.13
Even though it is a primary alkyl halide, neopentyl bromide is still a sterically hindered substrate preventing the backside attack S_N2 reaction from happening.

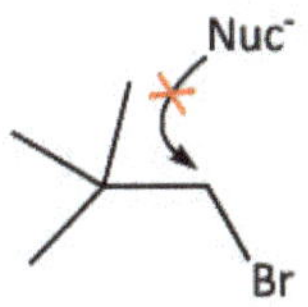

9.14

$CH_3O^- > OH^- > CH_3COO^- > H_2O$
Strongest............... ....weakest
Methoxide has electron donating methyl group on it making the negative oxygen atom very negative. The hydroxide oxygen has a negative charge on it, but the hydrogen atom doesn't donate in the electron density like the methyl does, so the oxygen is not as negative. The acetate has a resonance form making each oxygen atom carry half a negative charge. The water is not negatively charged at all so the lone pair of electrons will attack, but water is very stable and not eager to do so.

9.15

PH_3 is a stronger nucleophile than NH_3 because P is larger than N and is softer and more polarizable. It can donate its electron density into the substrate.

9.16 The primary alkyl halide, $CH_2{=}CHCH_2Cl$, undergoes an S_N1 reaction faster than the secondary alkyl halide, $(CH_3)_2CHCl$. Explain why.

Once the leaving group leaves the alkyl halide alkene, a resonance-stabilized carbocation is formed.

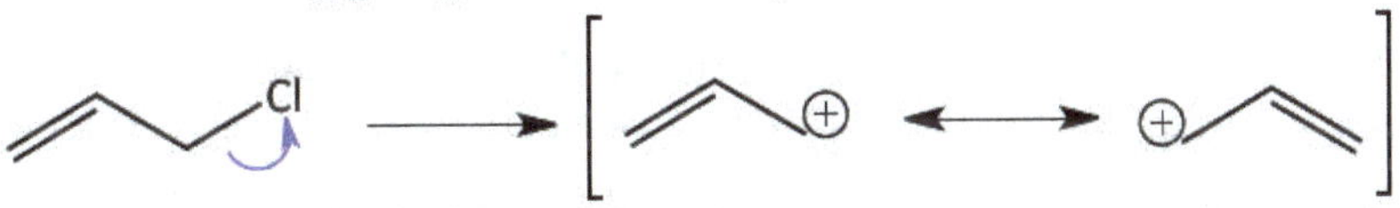

The secondary alkyl halide will make a secondary carbocation, not bad, but not as good as spreading out the positive charge over two carbon atoms.

9.17 Explain why $PhCH_2Cl$ is hydrolyzed (reacts with water) slowly to form $PhCH_2OH$, but forms the alcohol quickly if a small, catalytic amount of potassium iodide, KI, is added to the reaction.

The I- from KI can attack the $PhCH_2Cl$ by an SN_2 reaction to make $PhCH_2I$. I^- is a better leaving group than Cl^-. This makes the reaction with water to form $PhCH_2OH$ faster.

9.18 Give the organic product in the following substitution reactions.

a) $HS^- + CH_3CH_2CHBrCH_3 \rightarrow CH_3CH_2CHSHCH_3$
b) $I^- + CH_3CHBrCH_3 \rightarrow CH_3CHICH_3$
c) CH_3NH_2 + ethyl bromide → $CH_3CH_2NHCH_3$
d) ethyl bromide + PPh_3 → 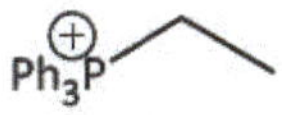

9.19 Because it is a bicyclic compound, it cannot do a Walden inversion for an S_N2 reaction. It is also sterically hindered and too crowded for a backside attack. Because of the bridges, it also has difficulty becoming a planar carbocation in an S_N1 reaction.

9.20

a) S_N2 Secondary alkyl halide substrate could go either way. Hydroxide nucleophile is aggressive.
b) S_N1 Tertiary alkyl halide is too crowded for S_N2 and will stabilize the carbocation of S_N1. Water is a weak nucleophile.
c) S_N1 OH is a poor leaving group. The acid can protonate it making water which is a good leaving group. It is tertiary which makes a good carbocation (S_N1). The chloride nucleophile from the HCl will then attack.
d) Neither, OCH_3 is not a good leaving group.

9.21

a) E1 b) E1 c) E2 d) E2

9.22

a) b) c)

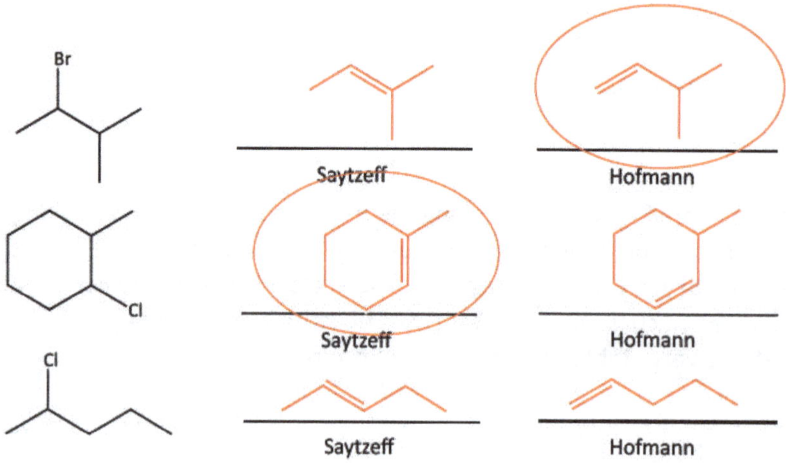

9.23

a) E2 b) E1 c) E1 d) Neither, poor leaving group

9.24

The first reaction goes S_N1. The water is a bad nucleophile and will wait around for the tertiary carbocation to form, then S_N1. Hydroxide is an aggressive nucleophile/base. It cannot S_N2 because the substrate is tertiary. It will then act like a base and do an E2 because it is not going to wait around for a carbocation to form.

9.25

a) propane, S_N2

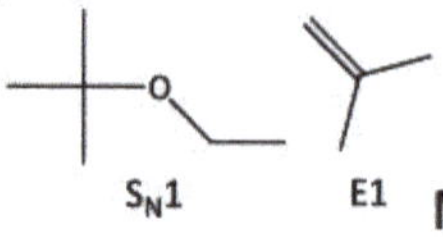

b) Maybe more of the E1 product because it is heated.

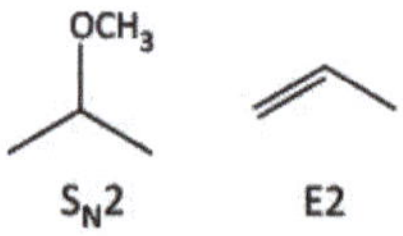

c) Probably a mixture of both. You could get more of the E2 product with a bigger, bulkier base like KOtBu.

d) $CH_3CH{=}CHCH_2\text{-}NH_2$, S_N2

e) $CH_2{=}C(CH_3)_2$, E2

Chapter 10

10.1 a) 2 b) 2 c) 0 d) 2 e) 4 f) 3

10.2

a) 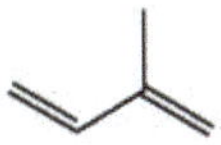 2-methyl-1,3-butadiene

b) 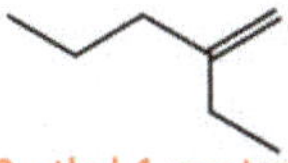 2-ethyl-1-pentene

c) 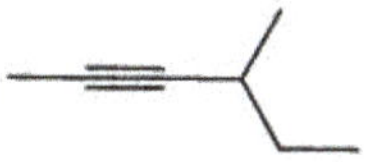 4-methyl-2-hexyne

10.3

a) 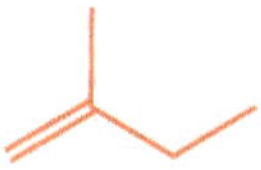 2-methyl-1-butene

b) 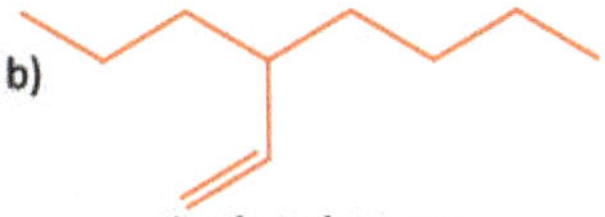 4-ethenyloctane

c)  1,5-hexadiene

d)  2-pentyne

10.4

a)  cis

b) 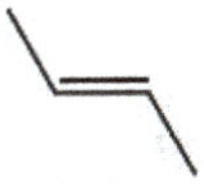 trans

c) 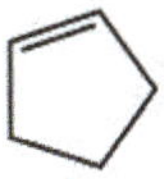 cis

d)  trans

10.5

a) 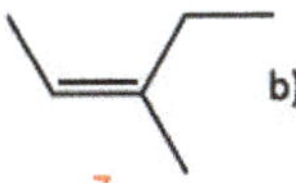 Z

b) 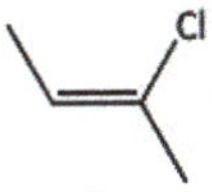 Z

c) 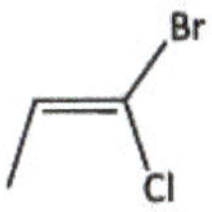 E

d) 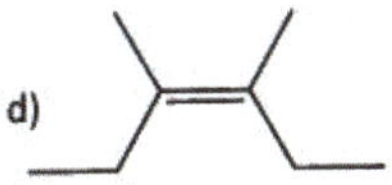 Z

10.6

a) 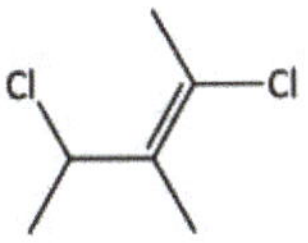
2,4-dichloro-3,4-dimethyl-(2E)-pentene
or
(E)-2,4-dichloro-3,4-dimethyl-2-pentene

b) 
cis-3-heptene
or
(Z)-3-heptene

c) 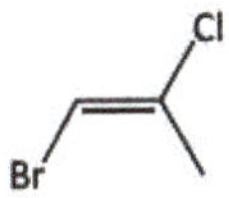
1-bromo-2-chloro-(1E)-propene
or
(E)-1-bromo-2-chloro-1-propene

d) 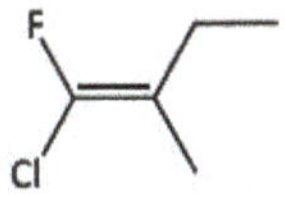
1-chloro-1-fluoro-2-methyl-(1E)-butene
or
(E)-1-chloro-1-fluoro-2-methyl-1-butene

10.7

a) 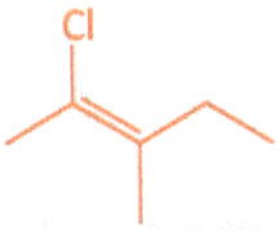 2-chloro-3-methyl-2Z-pentene

b)  *cis*-2-pentene

c)  *trans*-1,2-dibromoethene

d) 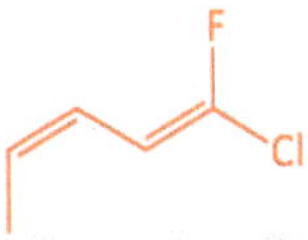 1-chloro-1-fluoro-(1E,3Z)-pentadiene

10.8

a)  no

b)  yes

c)  no

d)  yes

e) 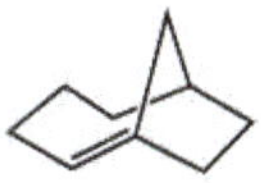 no

10.9

a) 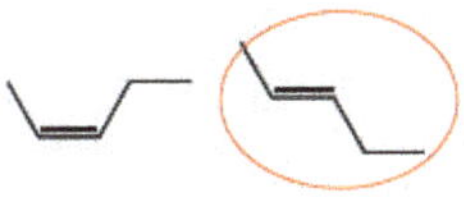

b) 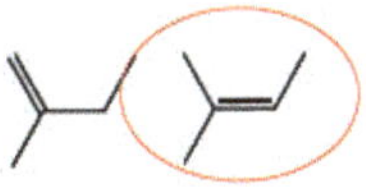

10.10

a) 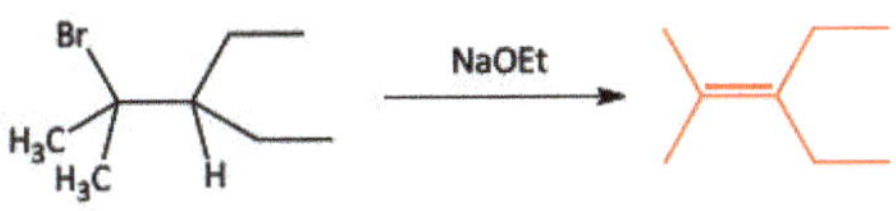

b) 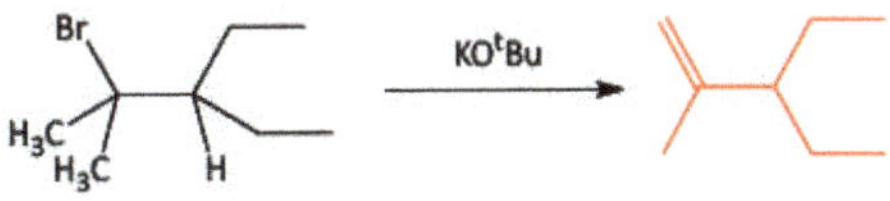

c) 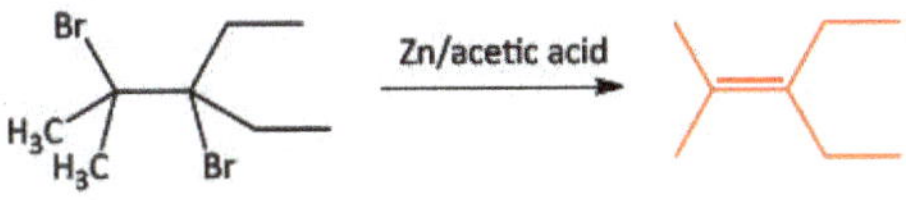

d) 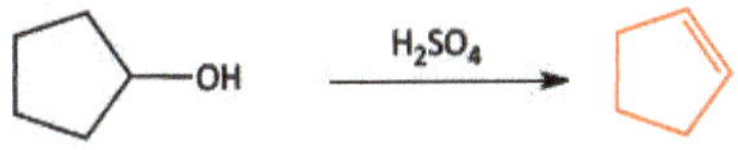

e) 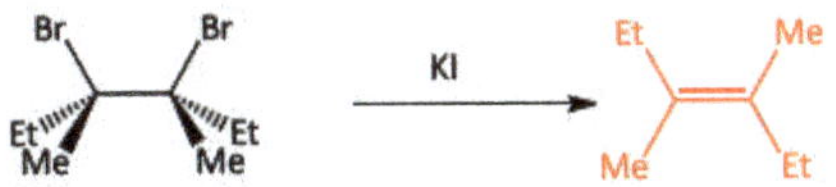

Chapter 11

11.1

11.2

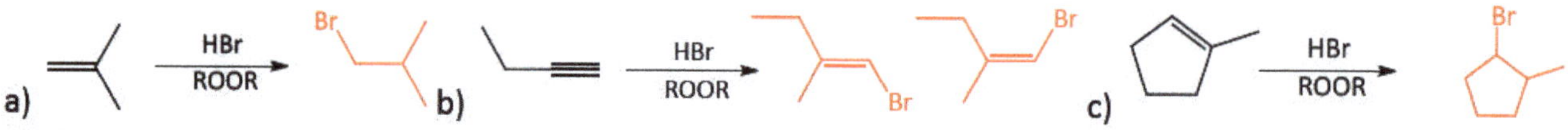

11.3

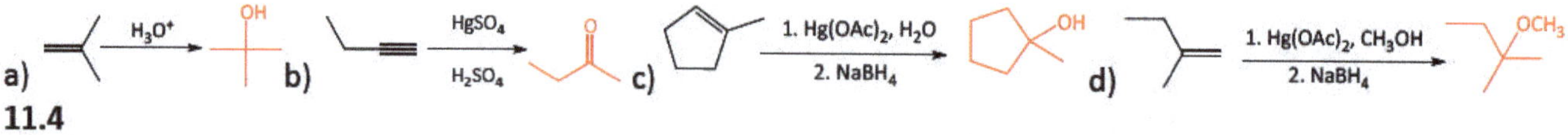

11.4

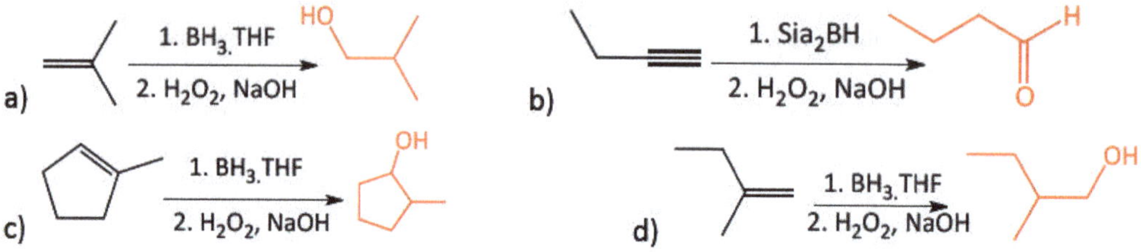

11.5

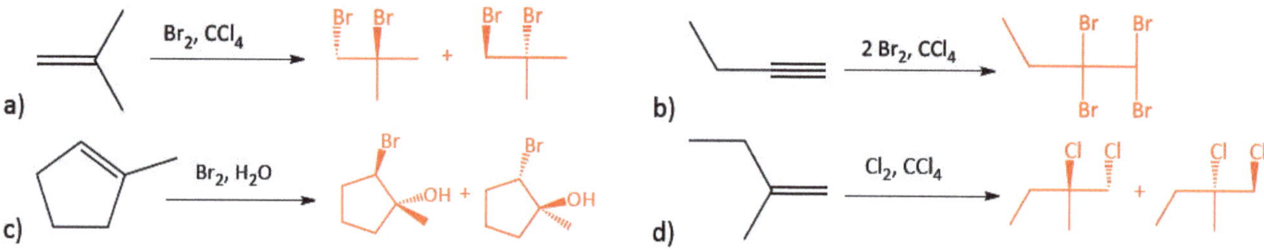

11.6

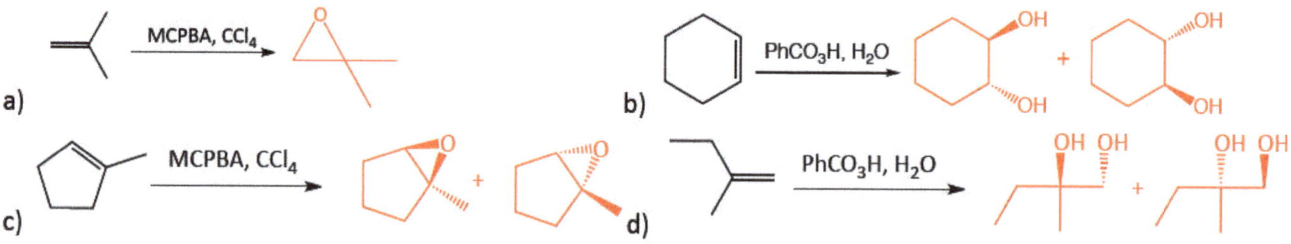

11.7

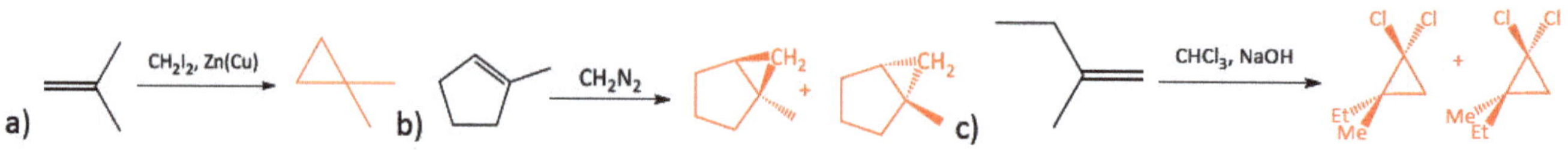

11.8

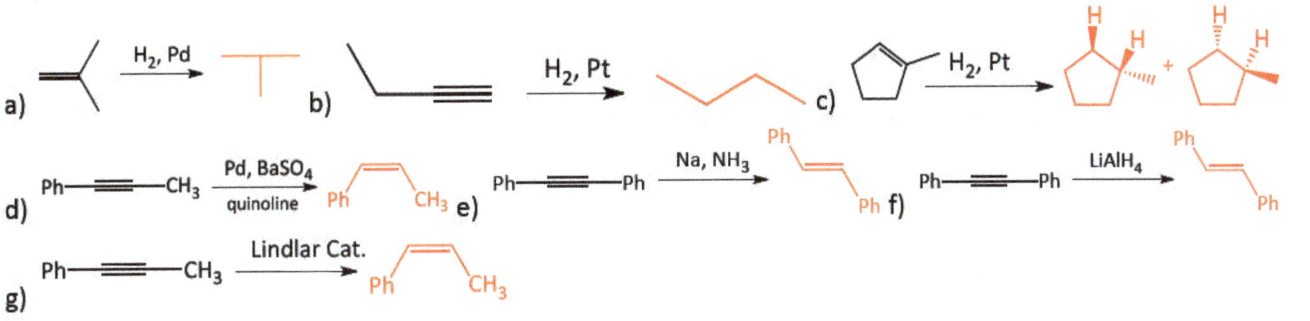

11.9

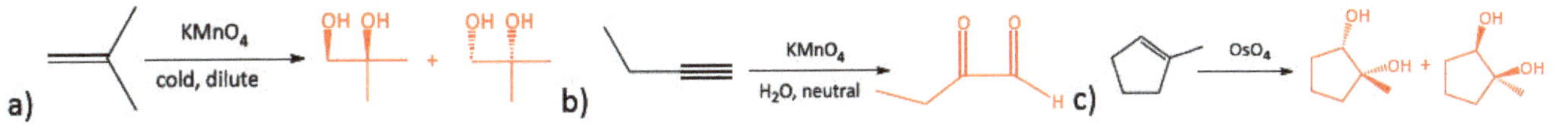

11.10

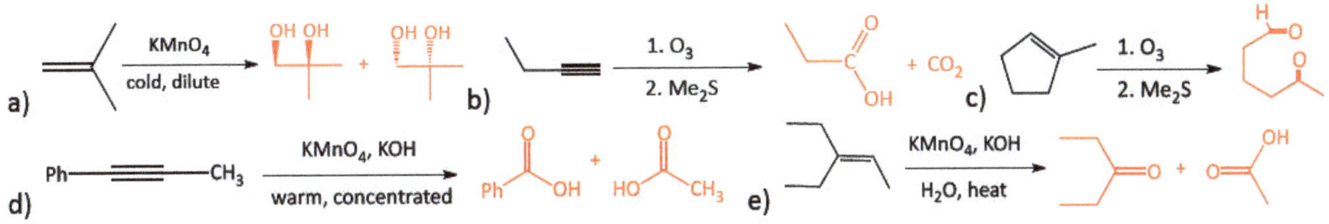

11.11

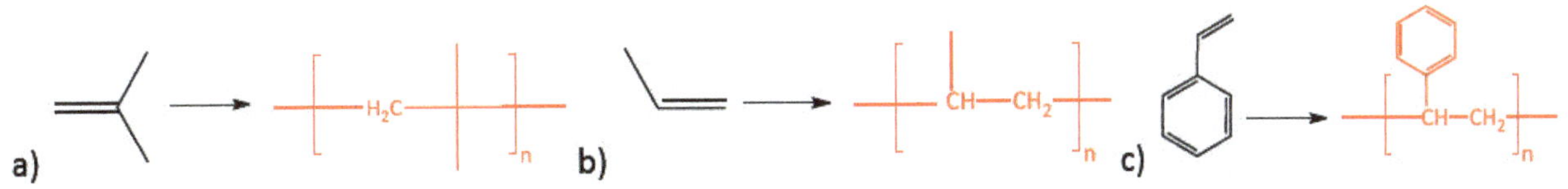

11.12

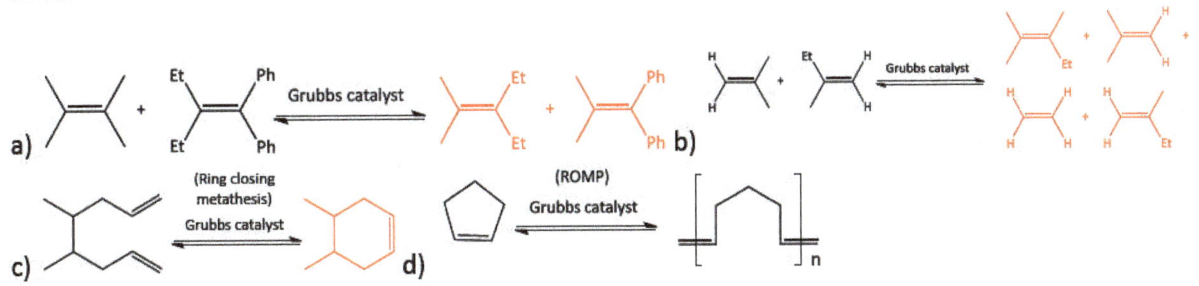

11.13

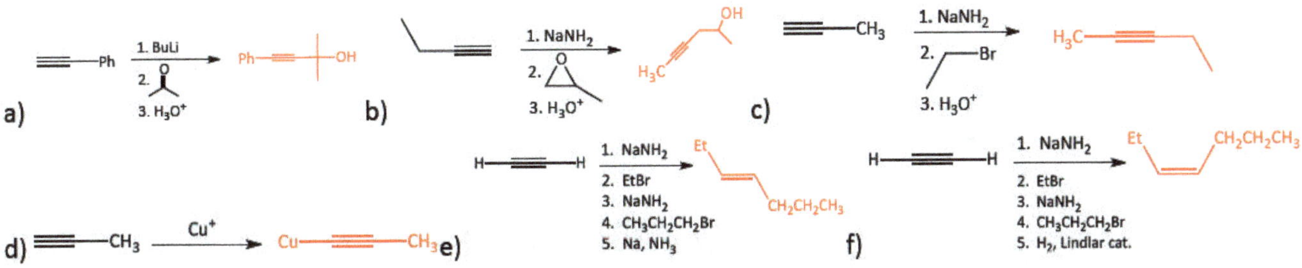

Chapter 12

12.1

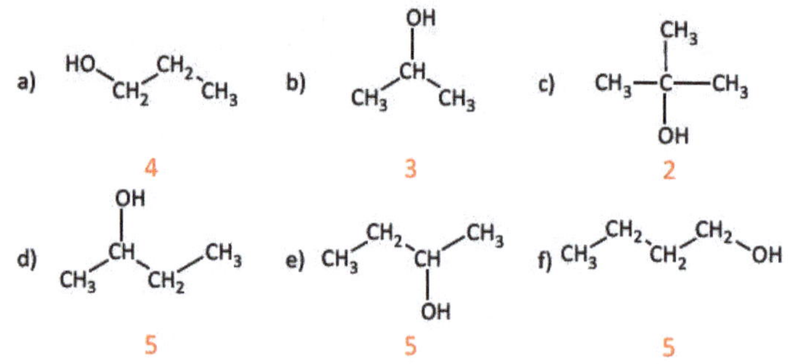

12.2

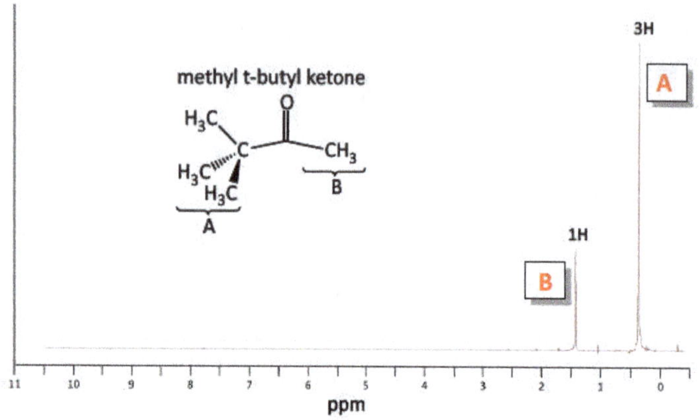

12.3

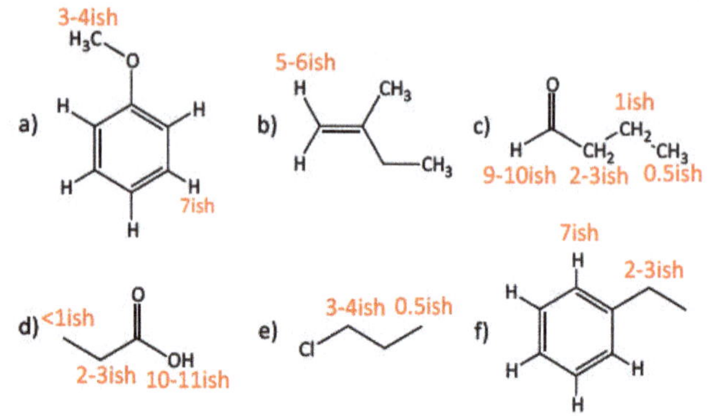

12.4

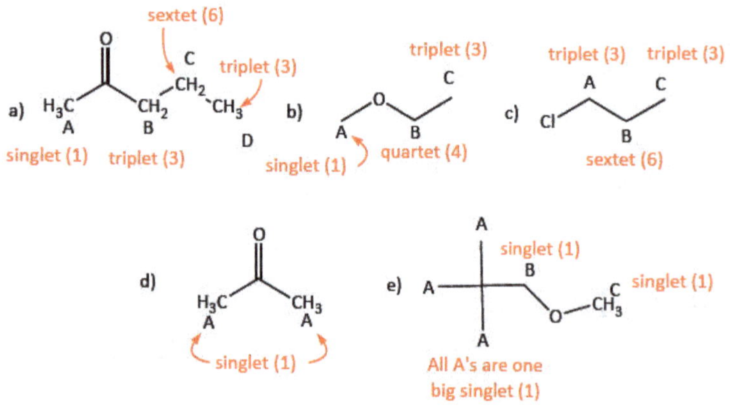

12.5 Two doublets. It does not have the 1:3:3:1 pattern that a quartet should have.

12.6

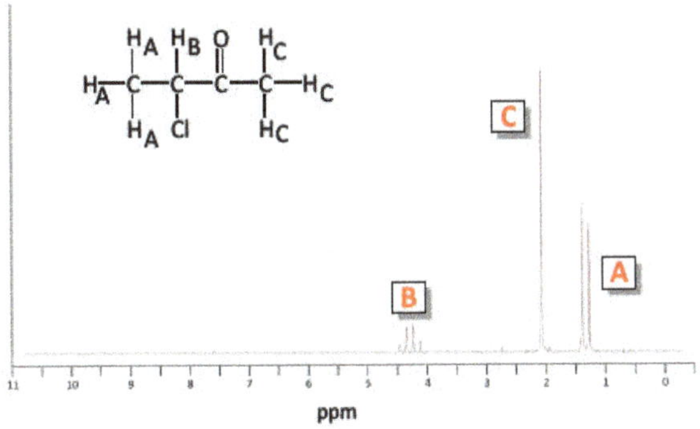

12.7

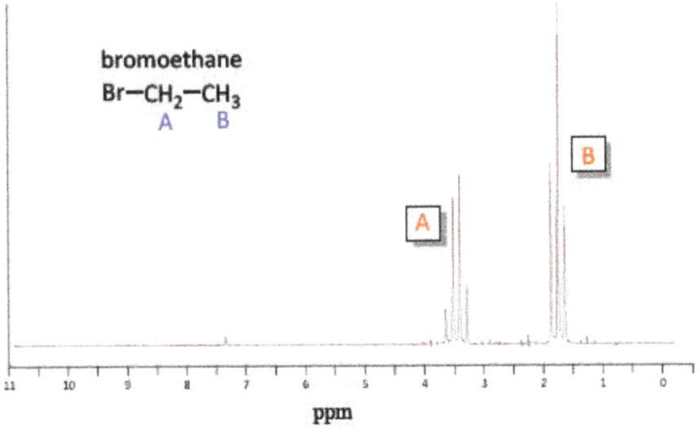

12.8 C

12.9

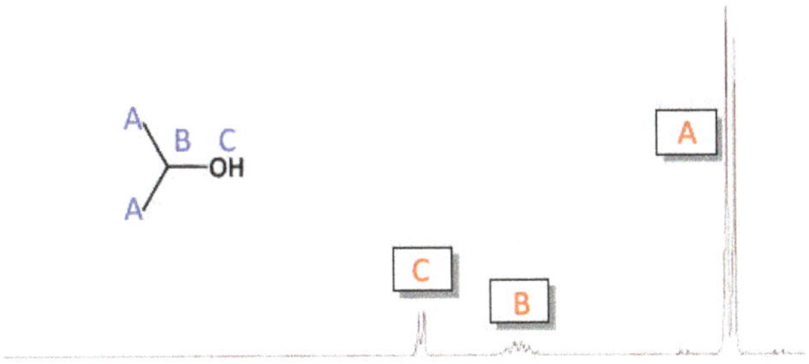

12.10

a) 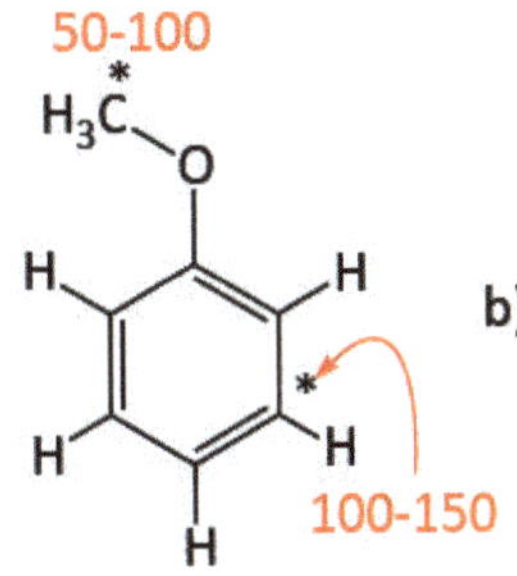

b) 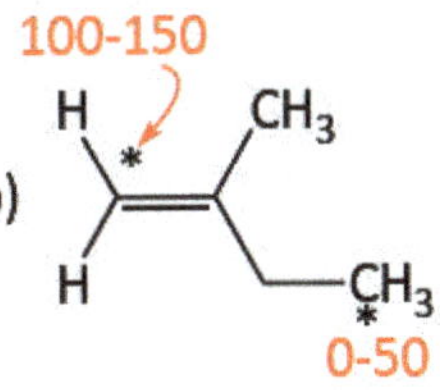

c) 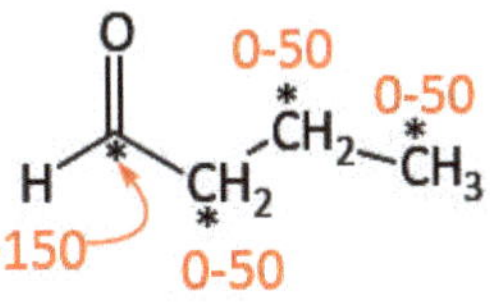

d) 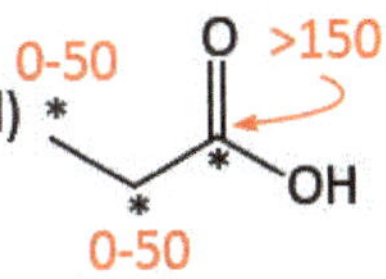

e) 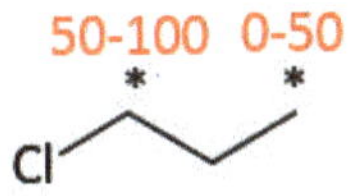

f) 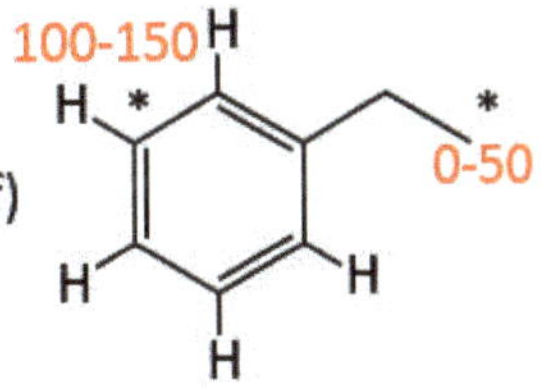

12.11 D. Notice the C=O peak and the symmetry with two types of CH_2 groups.

Chapter 13

13.1

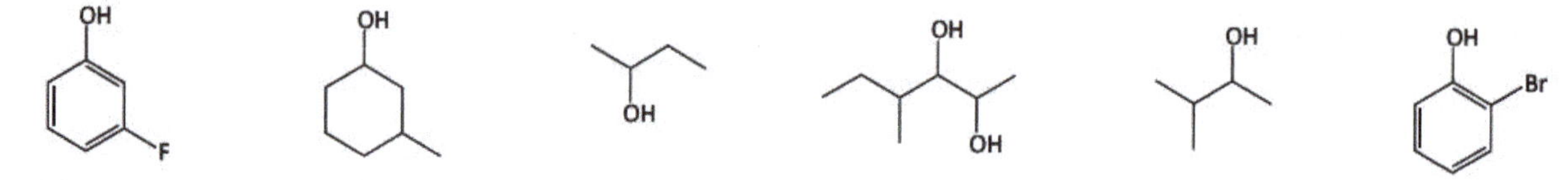

3-fluorophenol *meta*-fluorophenol | 3-methylcyclohexanol | butan-2-ol 2-butanol | 4-methylhexane-2,3-diol | 3-methylbutan-2-ol | 2-bromophenol *ortho*-bromophenol

13.2

A) 1. $Hg(OAc)_2$ or H^+, H_2O (H_3O^+) 2. $NaBH_4$ B) 1. BH_3.THF 2. H_2O_2, OH^- C) RCO_3H in H_2O

D) 1. OsO_4 or $KMnO_4$, OH^- (cold, dilute) 2. H_2O_2 E) 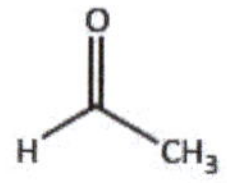

F) 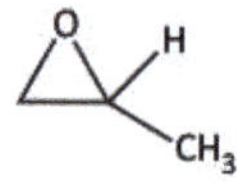

13.3

a) 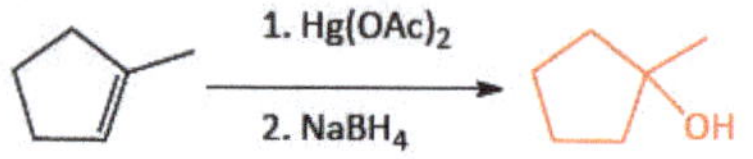

b) 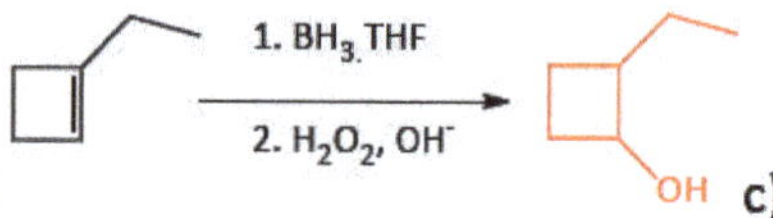

c) 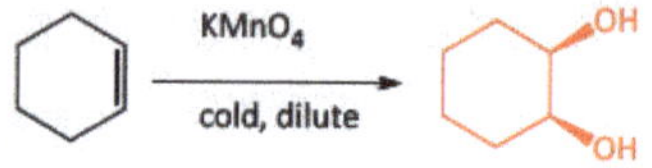

d) 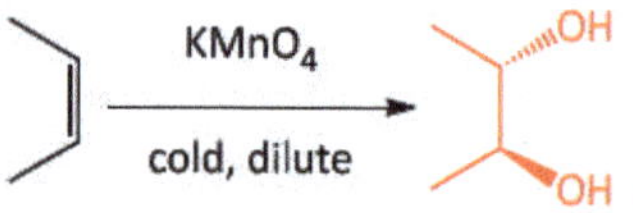

e) 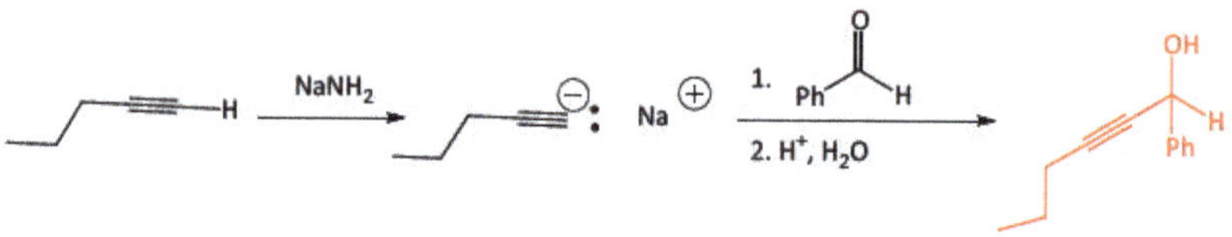

13.4

a) 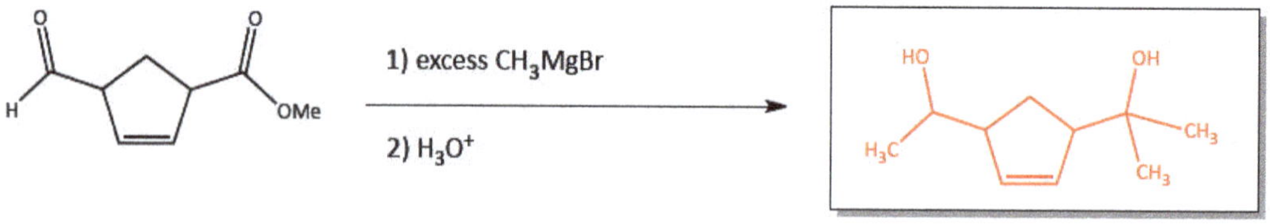

b) 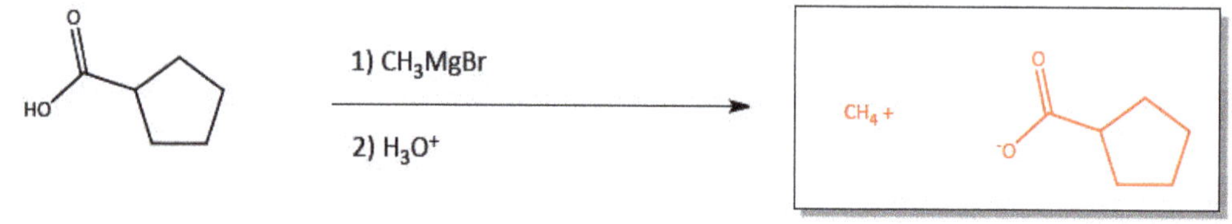

The carboxylic acid proton quenches the Grignard reagent.

c) 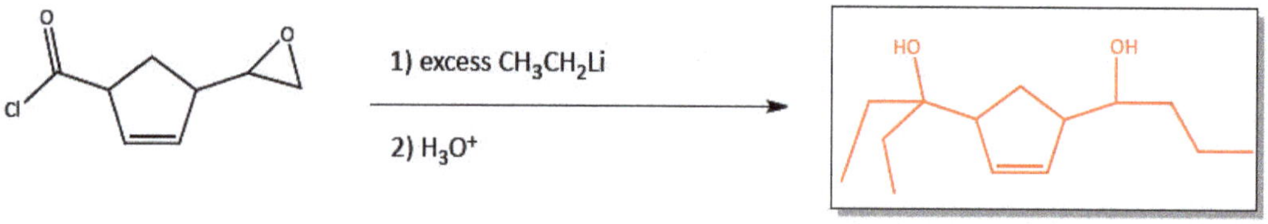

d) 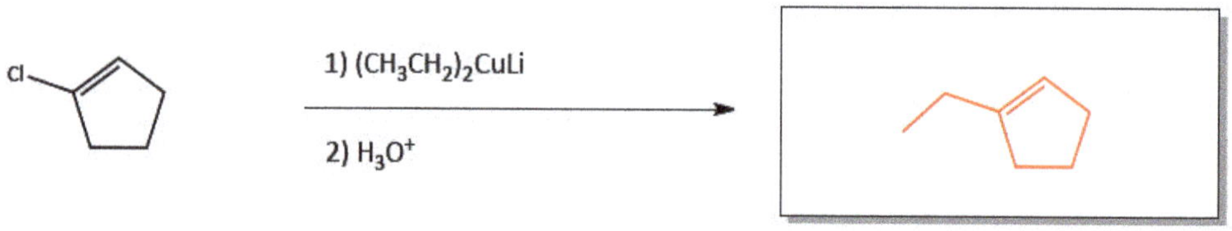

13.5

a) 

b) 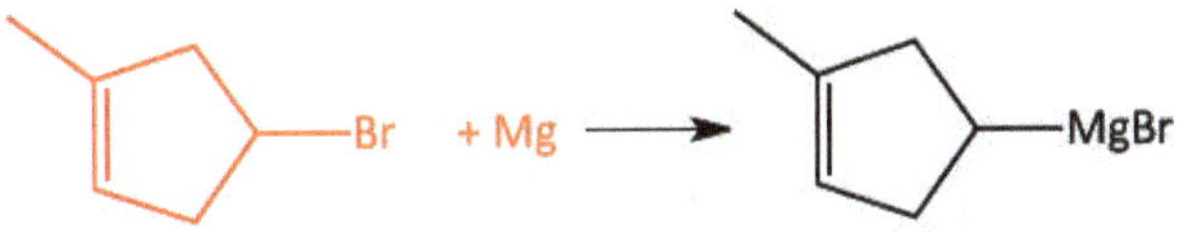

c) $2\ PhCl \xrightarrow{Li} 2\ PhLi \xrightarrow{CuI} Ph_2CuLi + LiI$

13.6

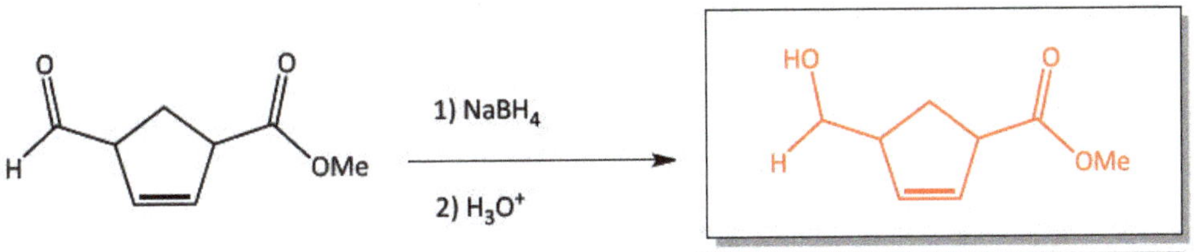

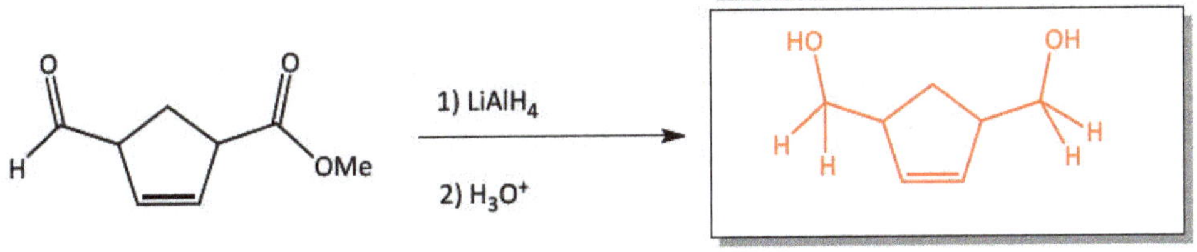

Chapter 14

14.1

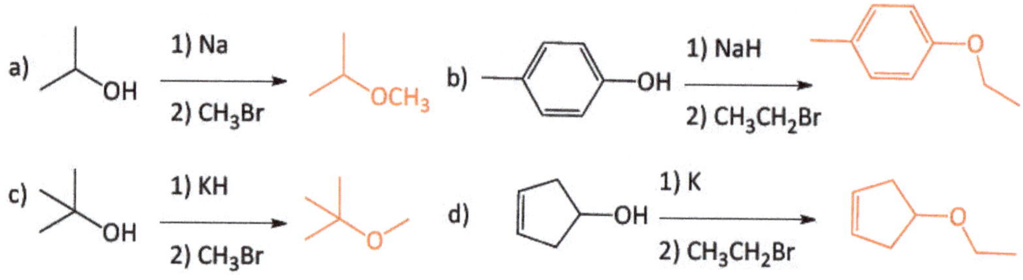

14.2 D

14.3

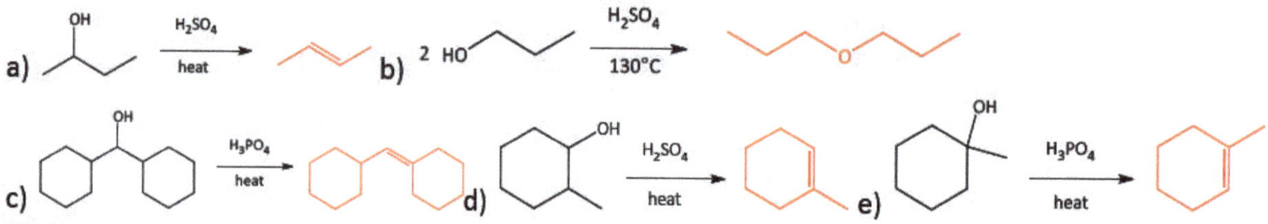

14.4

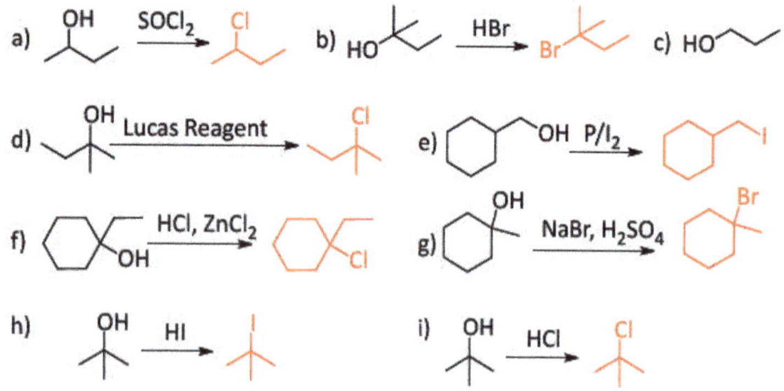

14.5

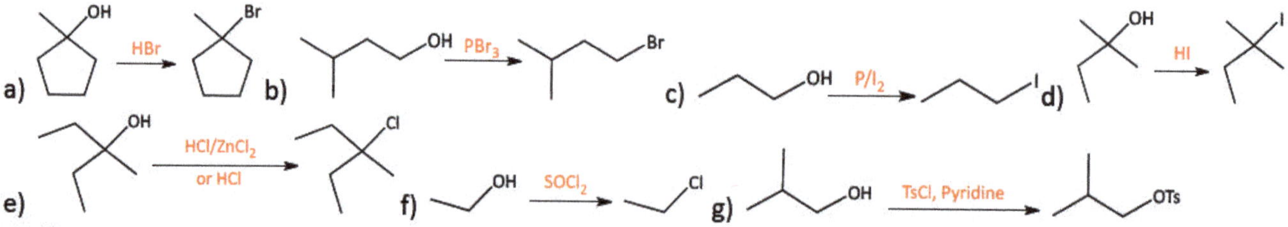

14.6

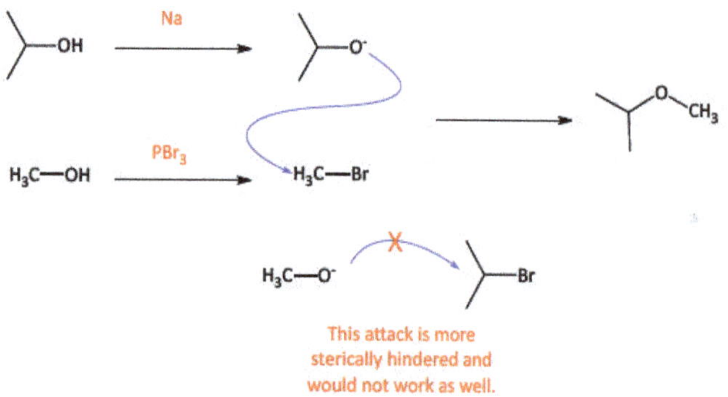

14.7

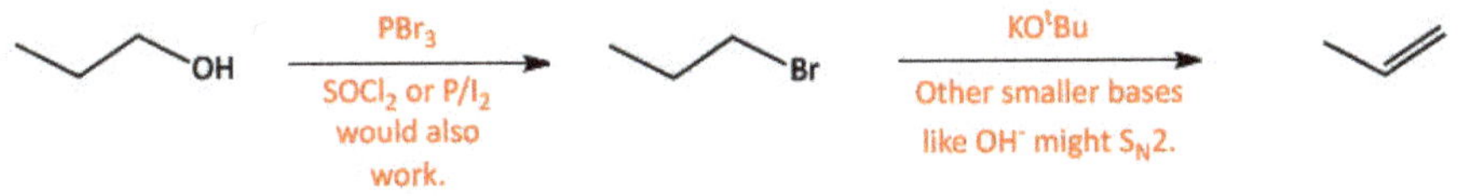

14.8

a)
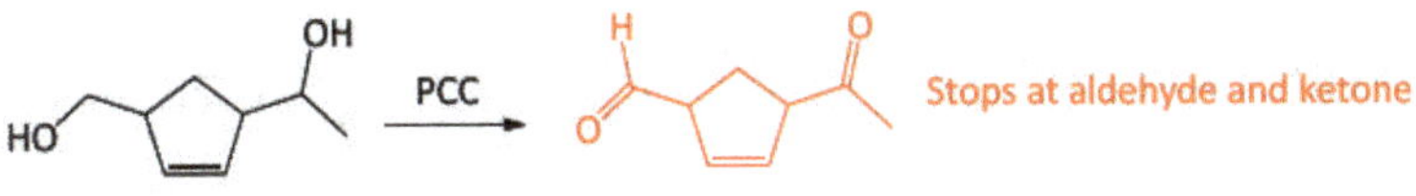

b)
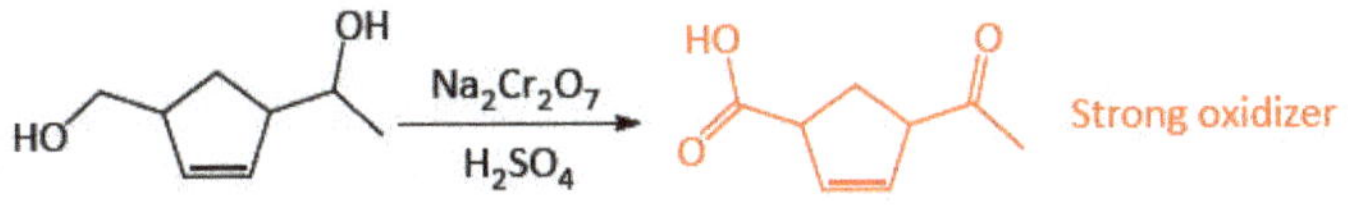

c)
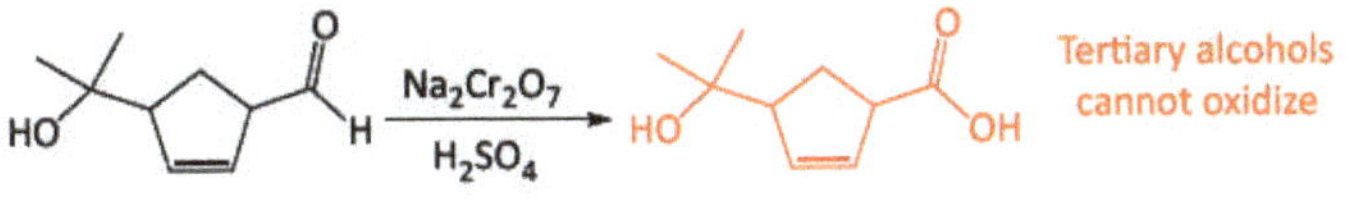

d)
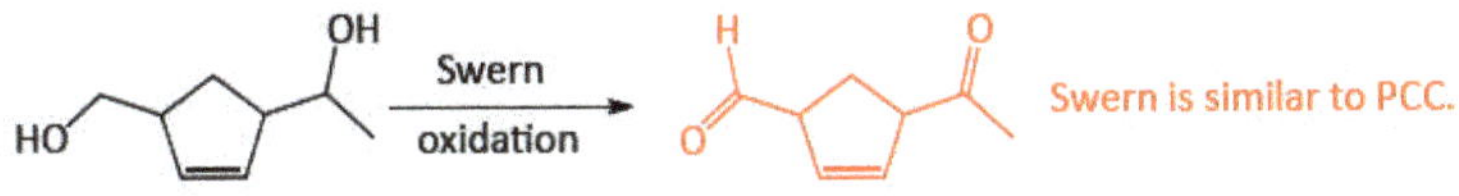

e)
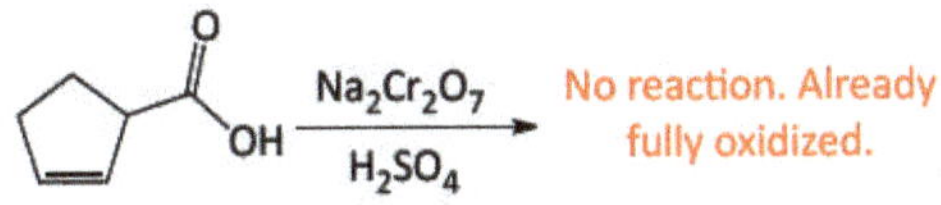

f)
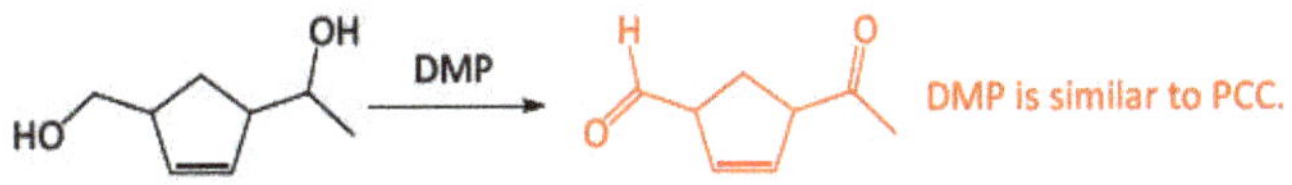

g)
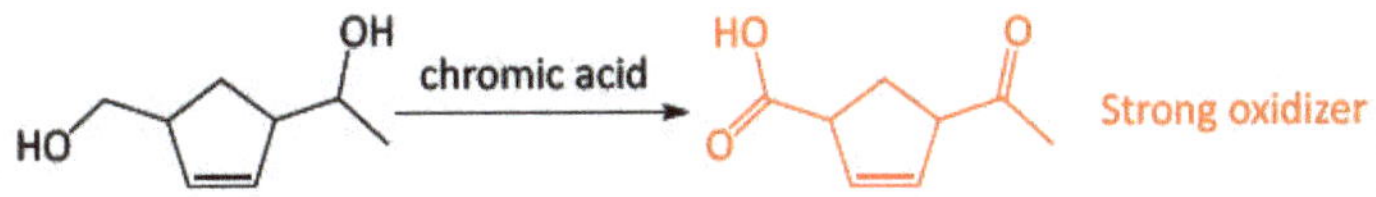

h)
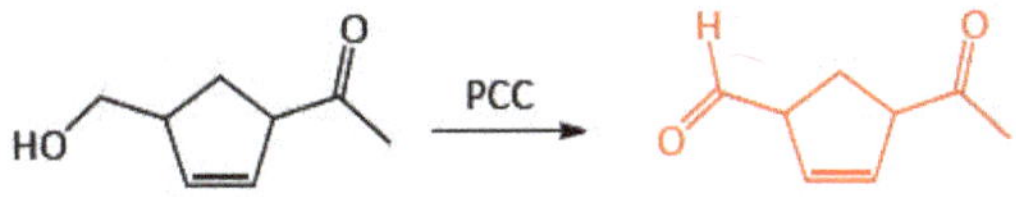

i)
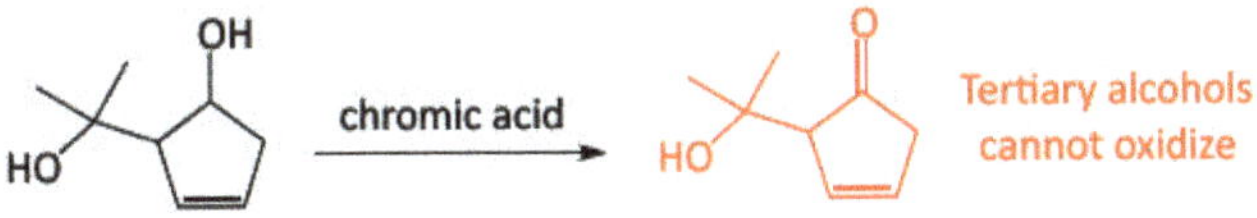

14.9

a)
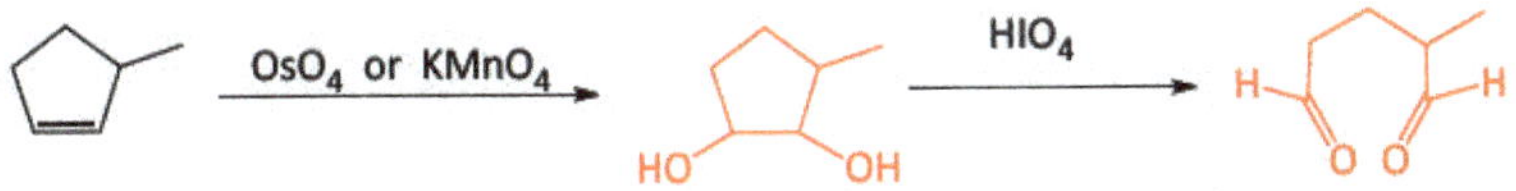

b)
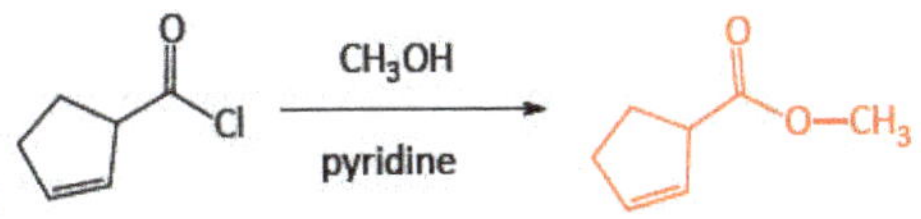

c)
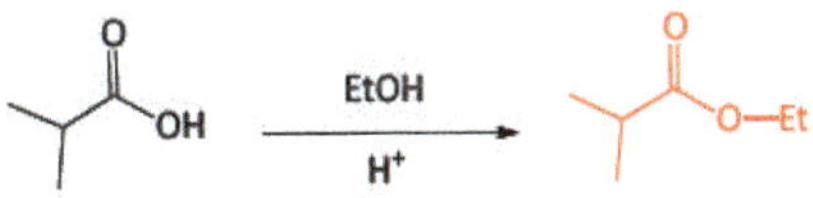

d)
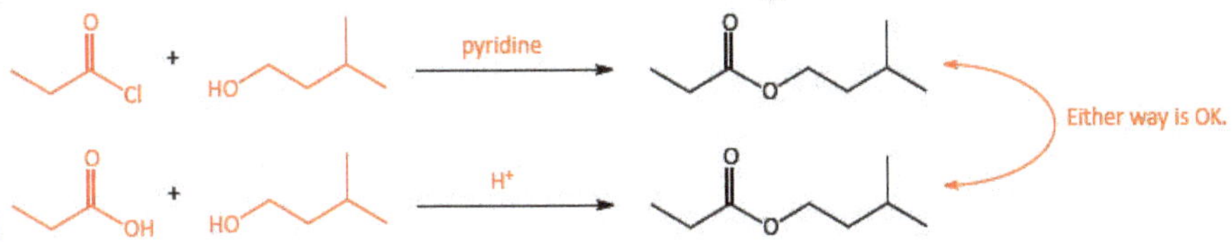

e)
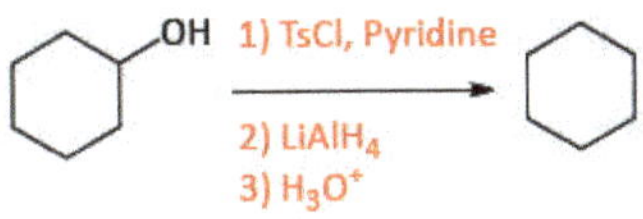

f)
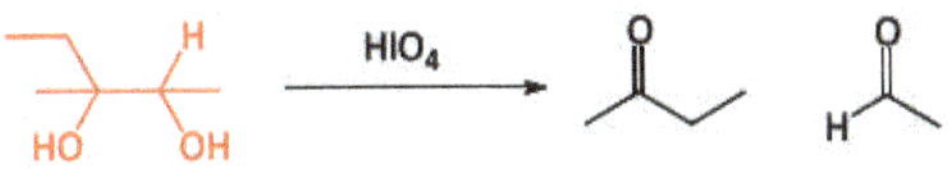

g)
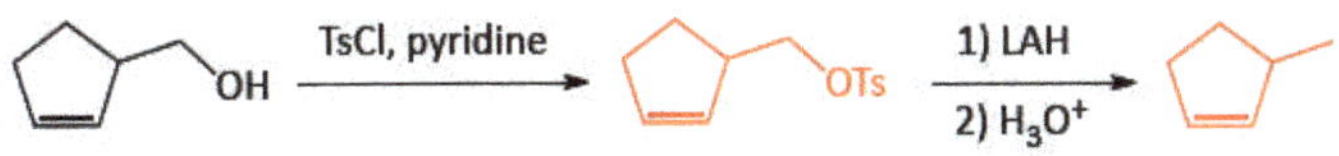

Chapter 15

15.1

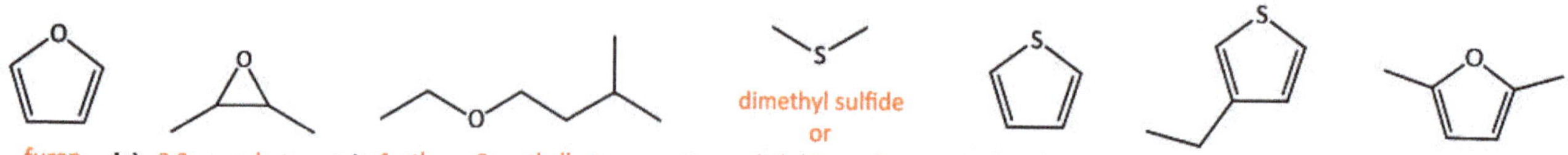

a) furan b) 2,3-epoxybutane c) 1-ethoxy-3-methylbutane d) dimethyl sulfide or methyl thiomethane e) thiophene f) 3-ethylthiophene g) 2,5-dimethylfuran

15.2

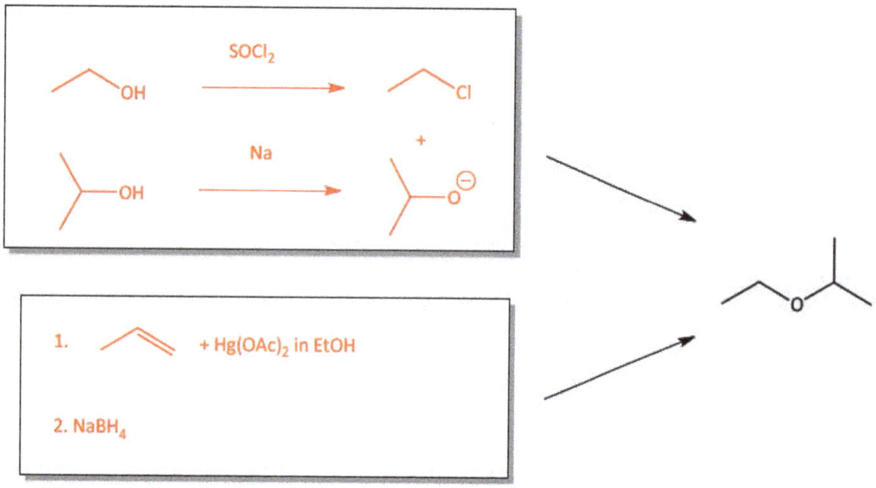

15.3

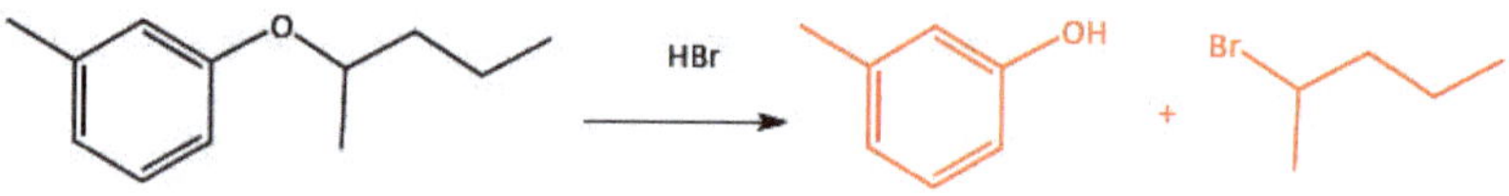

15.4

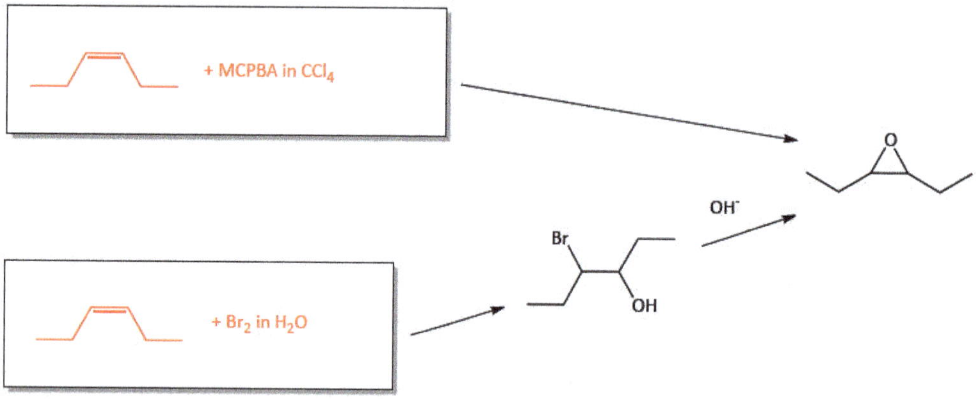

15.5

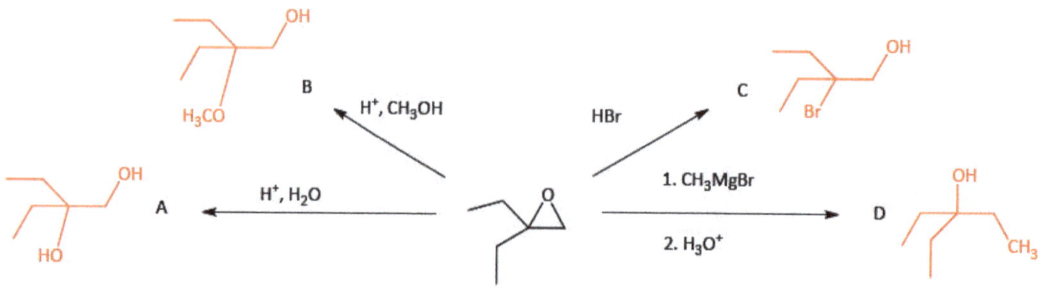

Chapter 16

16.1

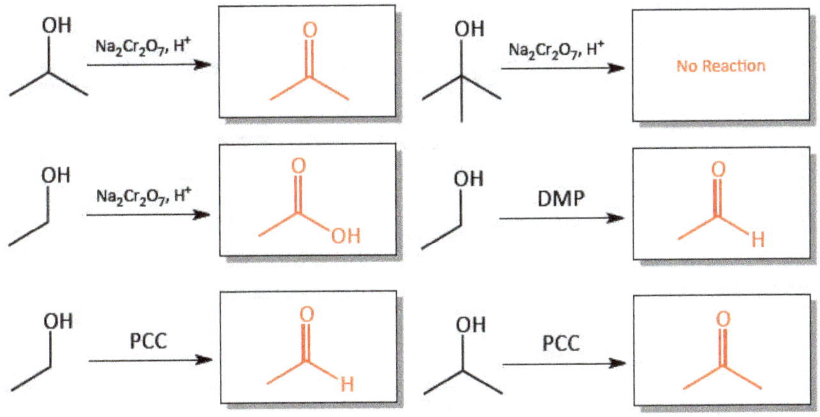

16.2

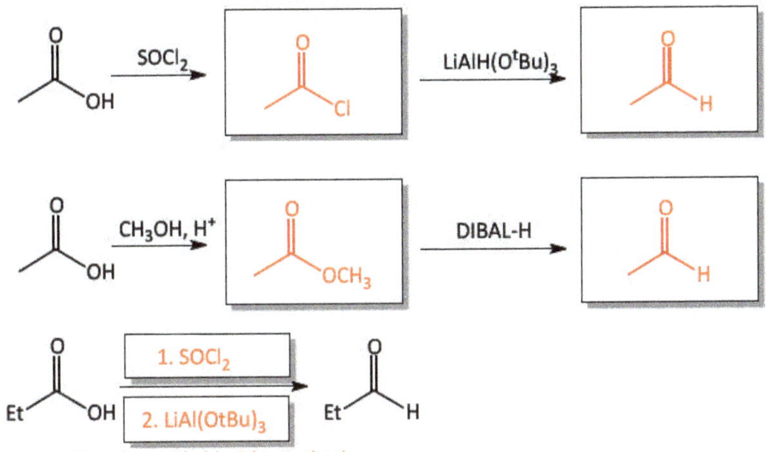

Use the acid chloride method

16.3

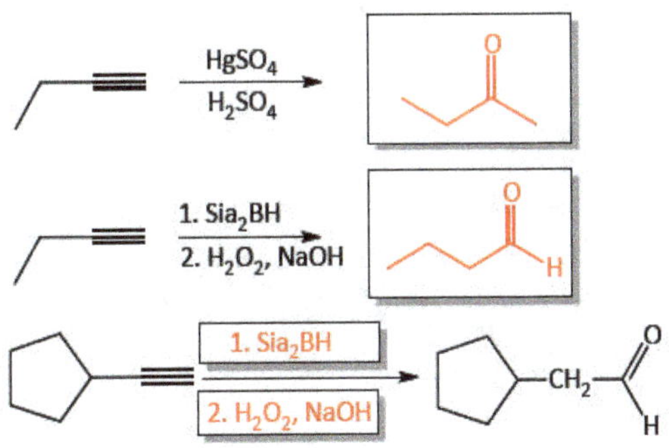

Answers to Exercises

16.4

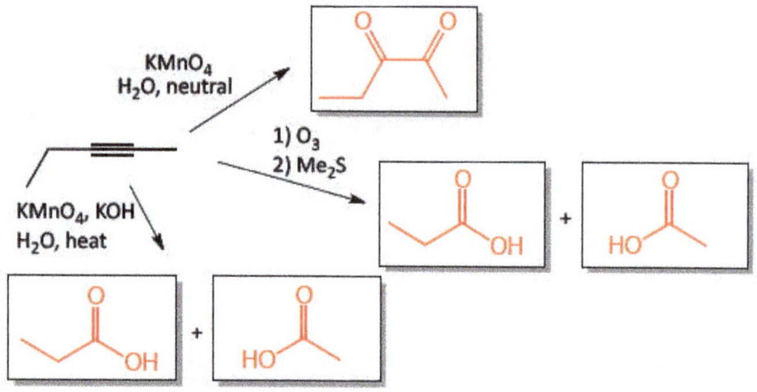

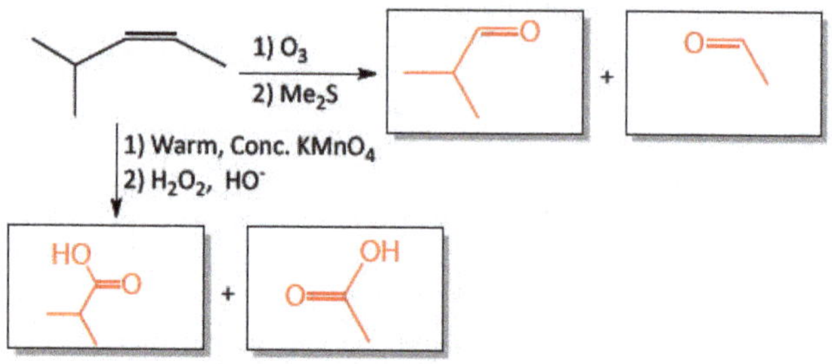

16.5

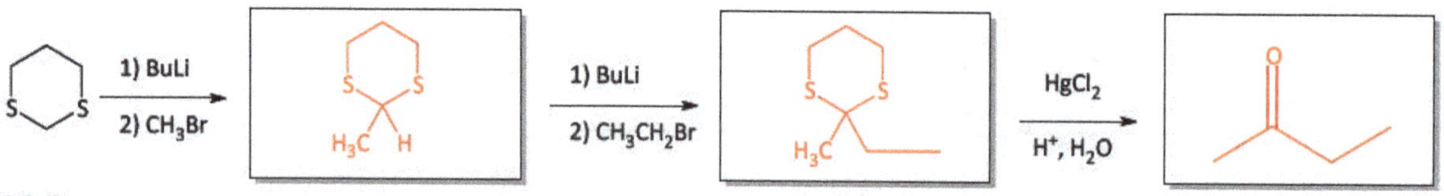

16.6

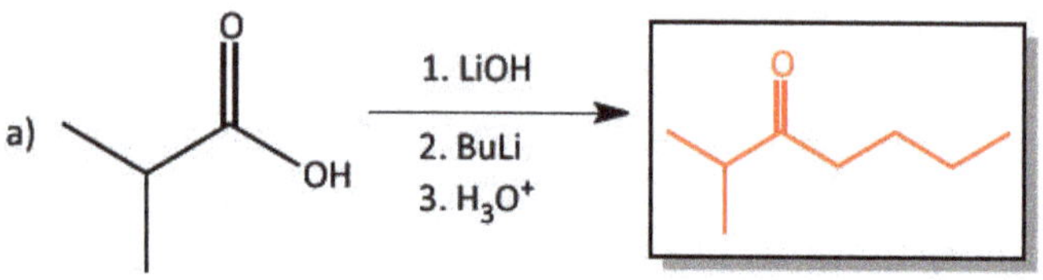

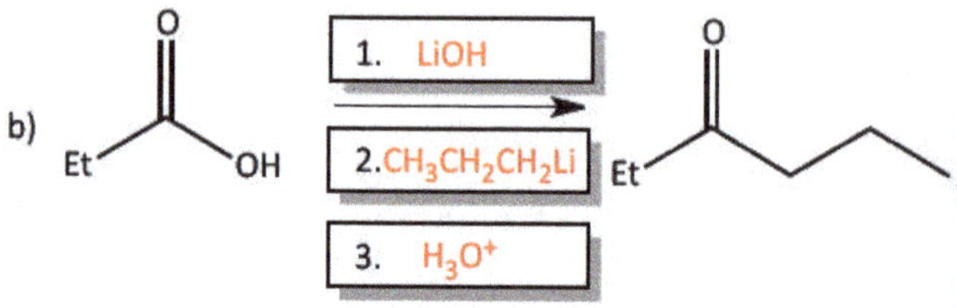

16.7

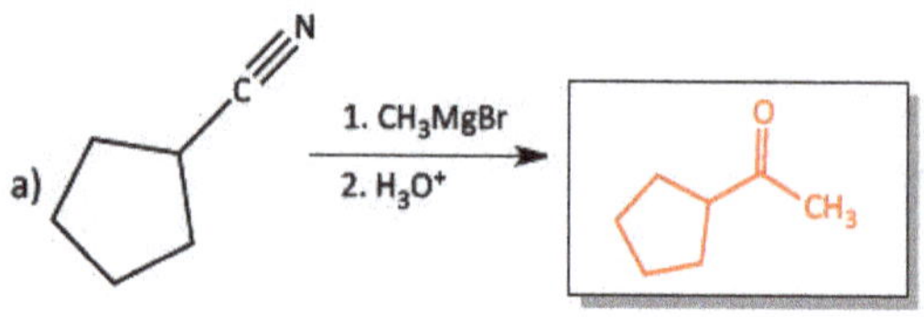

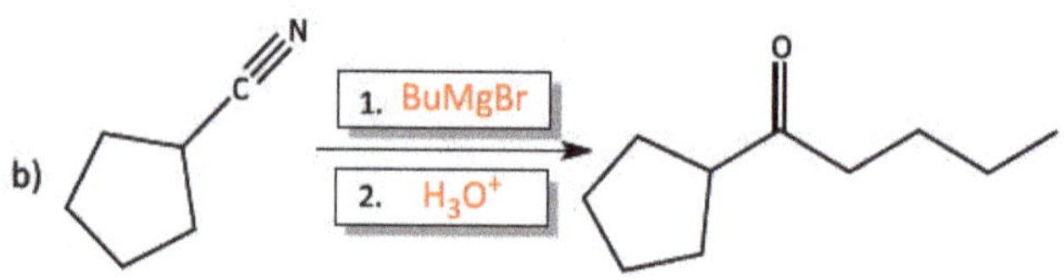

16.8

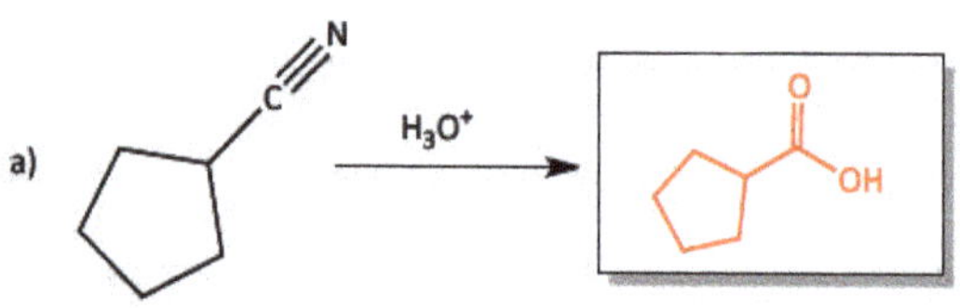

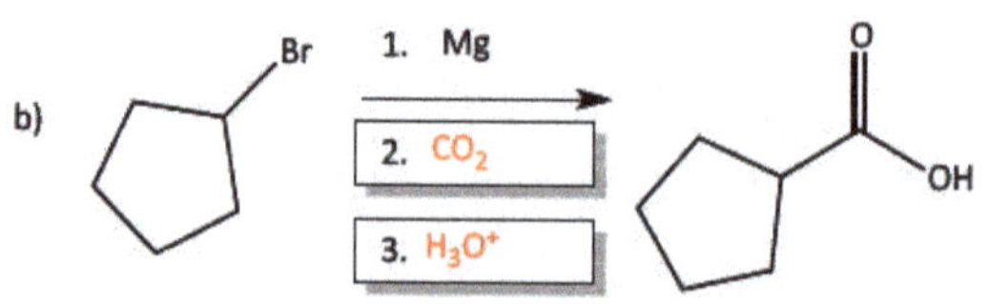

Chapter 17

17.1

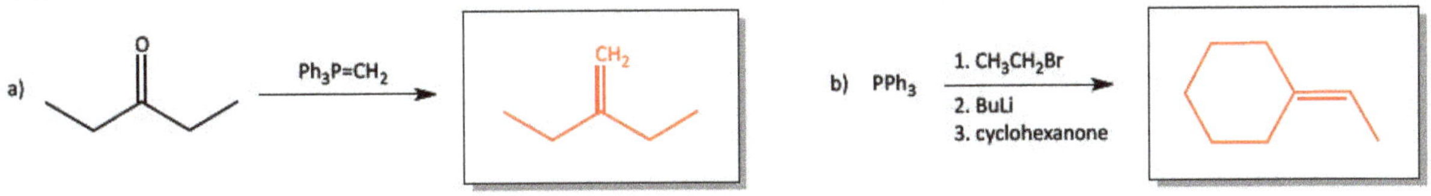

17.2

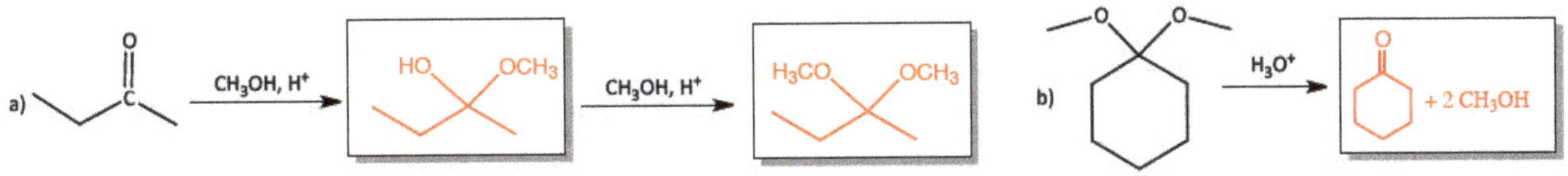

17.3

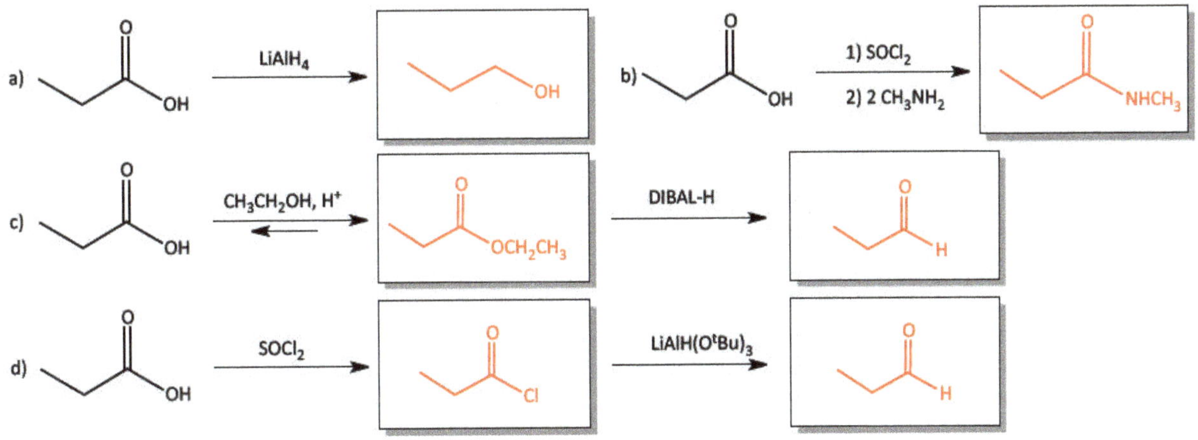

Chapter 18

18.1

a) 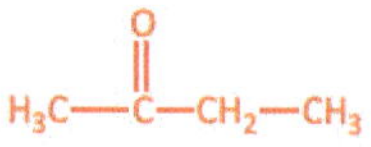

b) 

c) 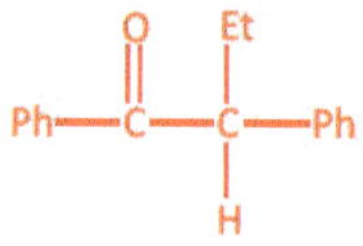

18.2

a) 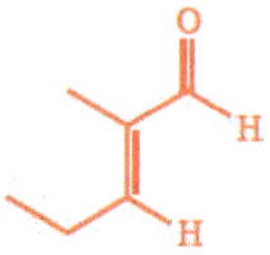

b) 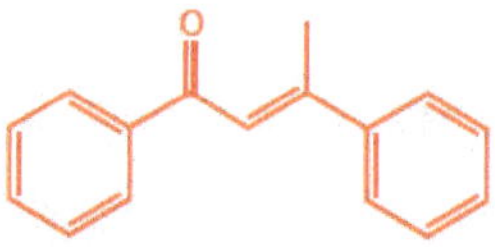

c) 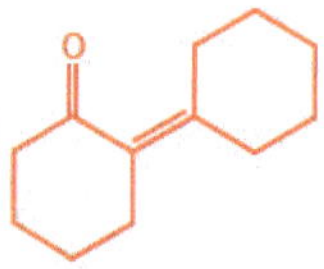

18.3

a) 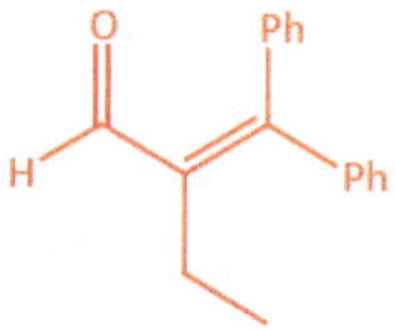

b) 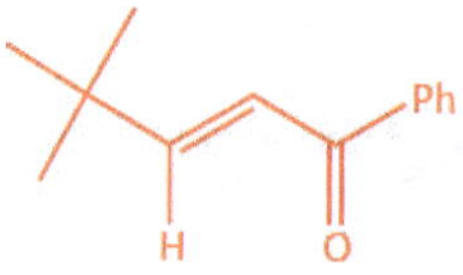

18.4

a)

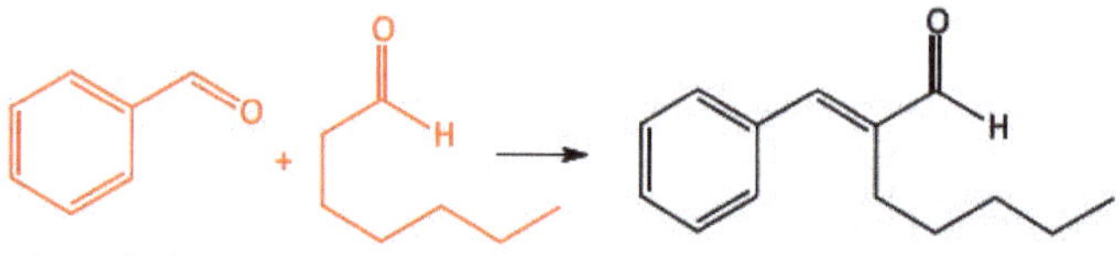

benzaldehyde + heptanal → jasmine aldehyde

b)

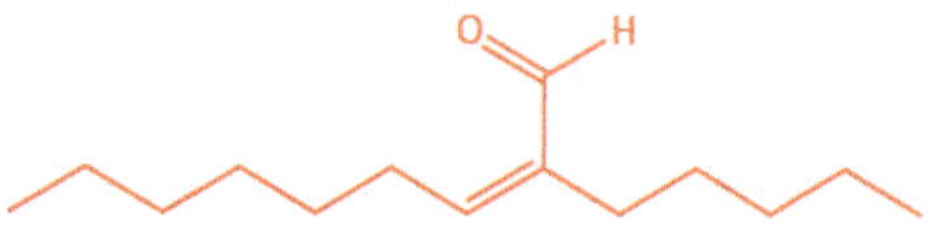

18.5

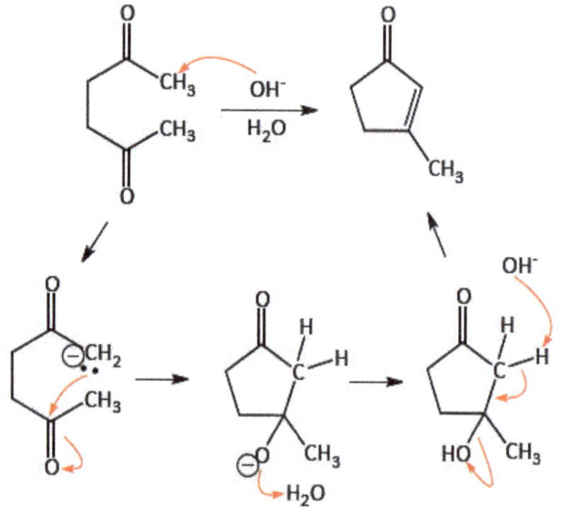

18.6

a) 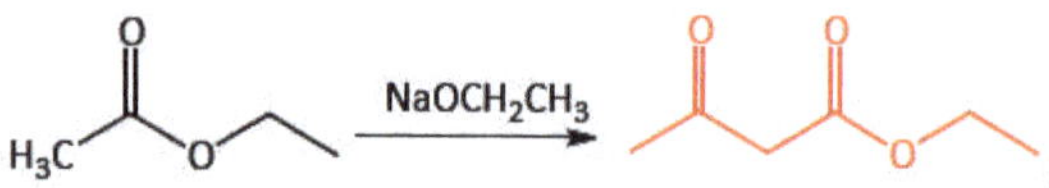

b) 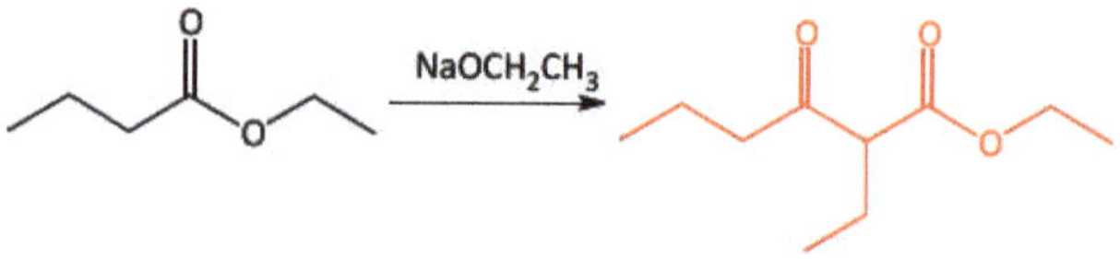

18.7

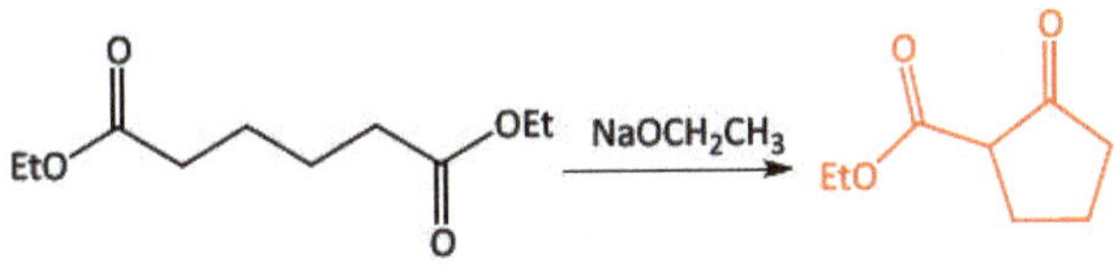

18.8

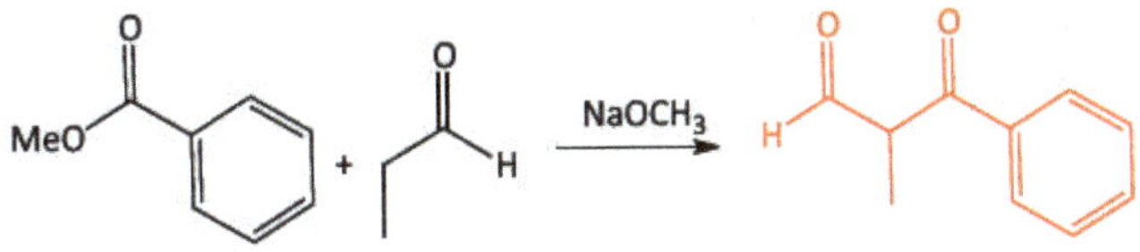

a)

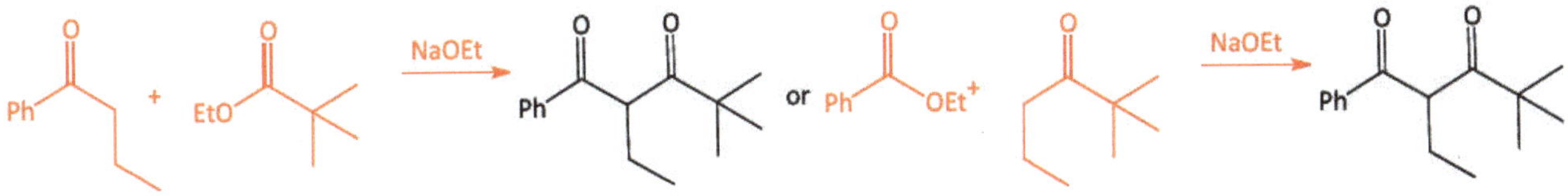

b)

18.9

a) 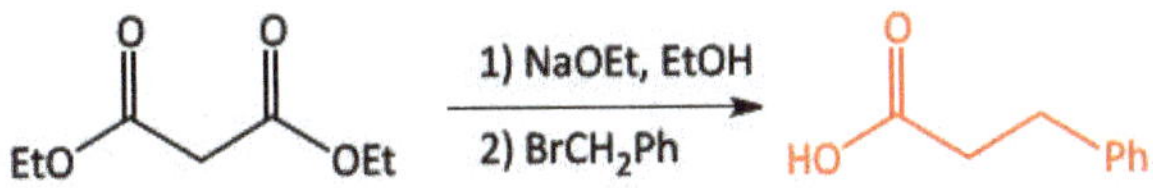 b) 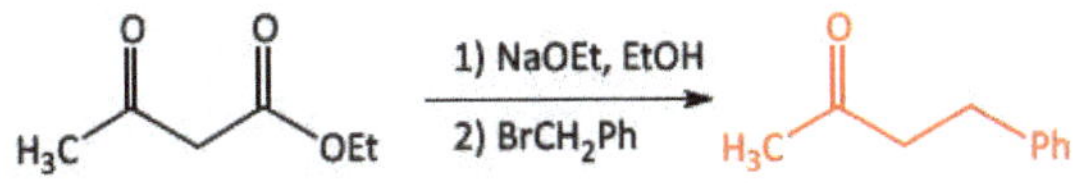

18.10

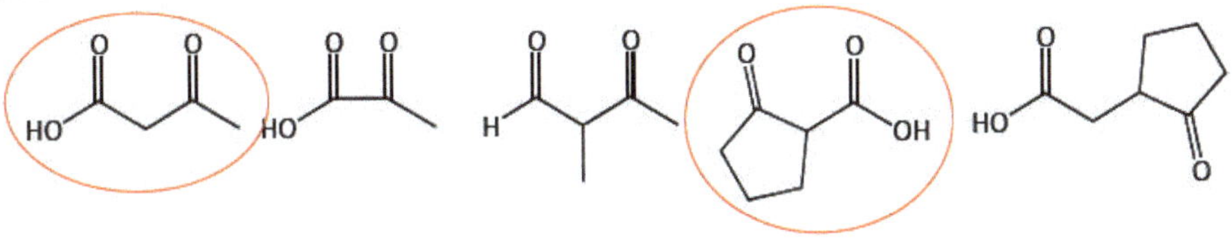

18.11

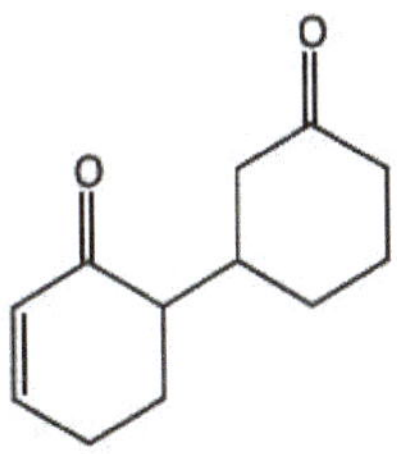

18.12

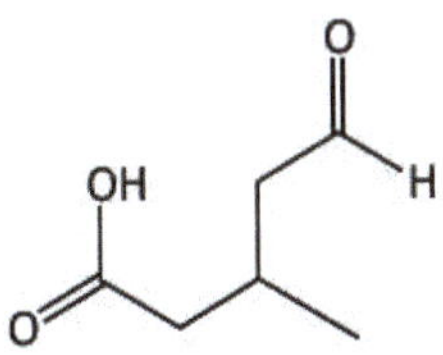

Chapter 19

19.1

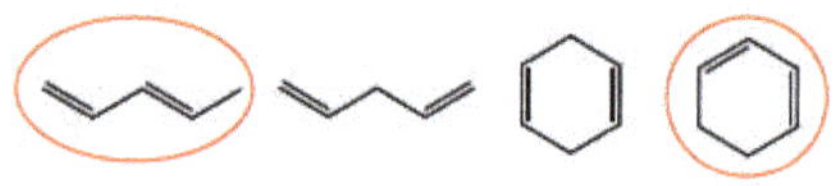

19.2

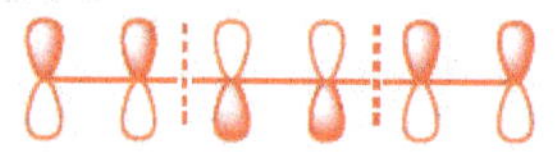

The HOMO of 1,3,5-hexatriene

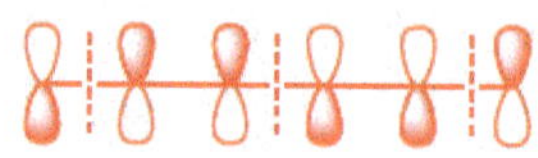

The LUMO of 1,3,5-hexatriene

19.3

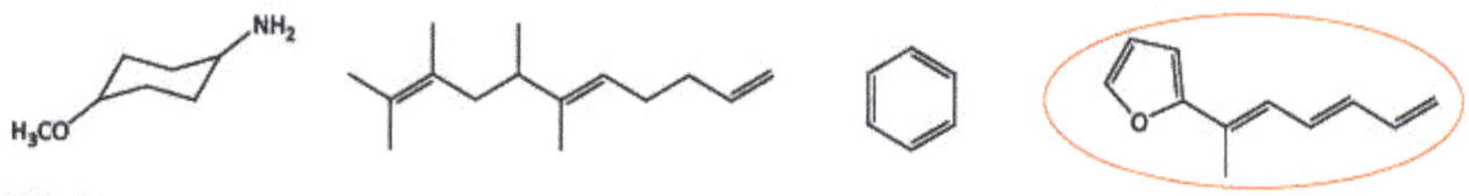

19.4

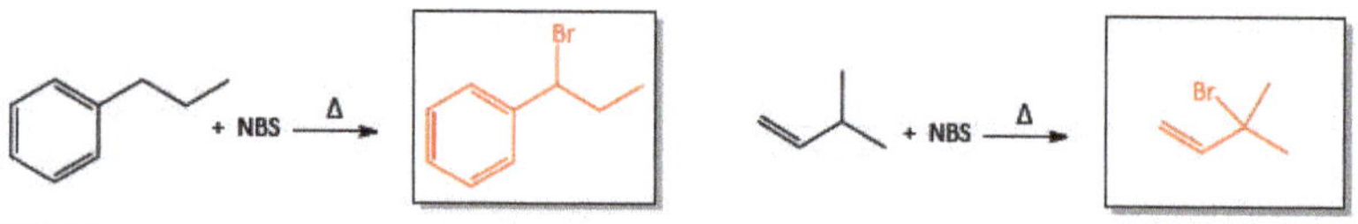

19.5

a)

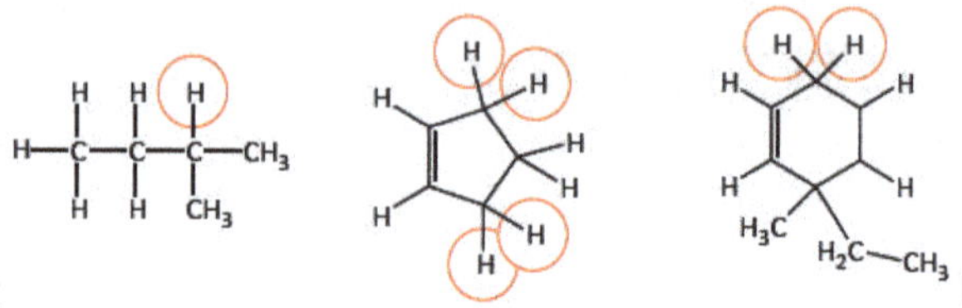

b)

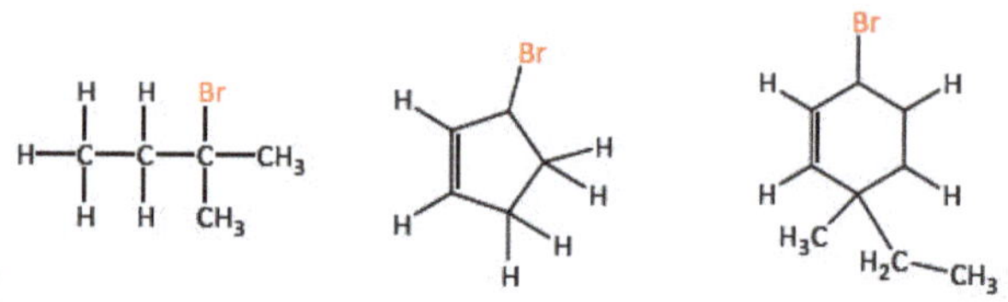

19.6

a)

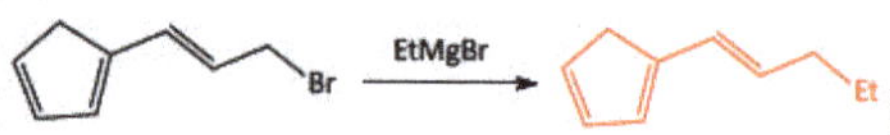

b)

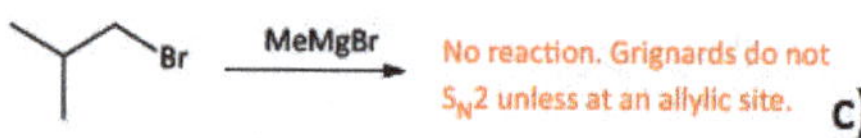

c)

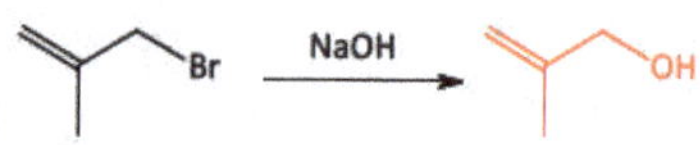

19.7

a)

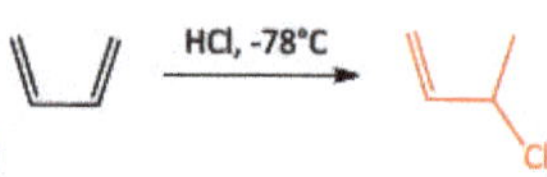

b)

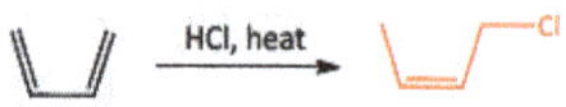

19.8

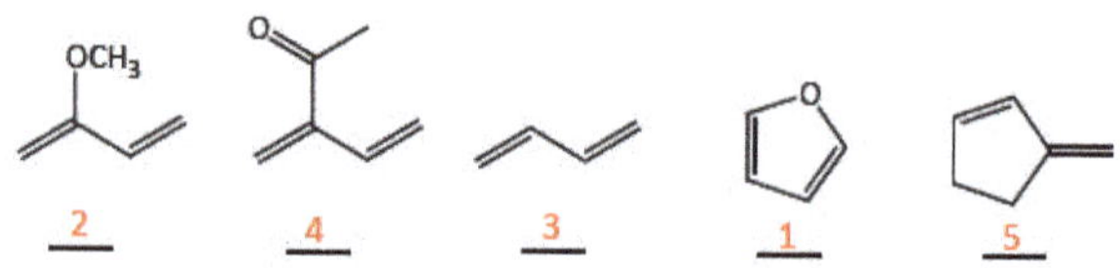

19.9

a)

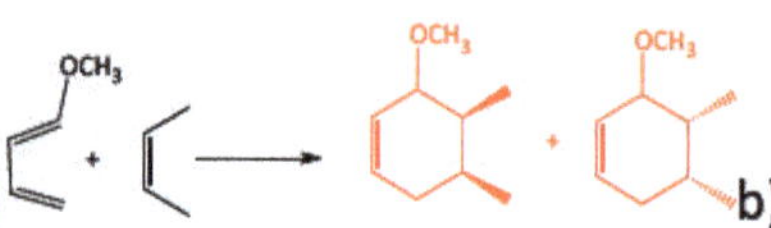

b)

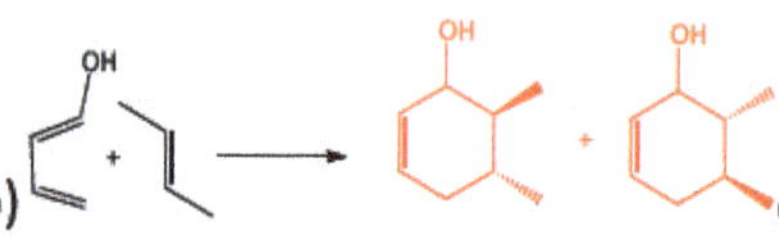

c)

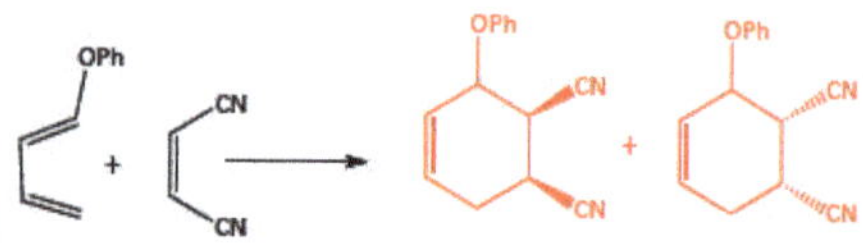

19.10

a)

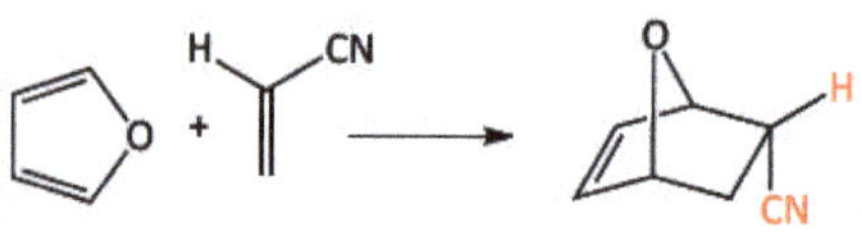

b)

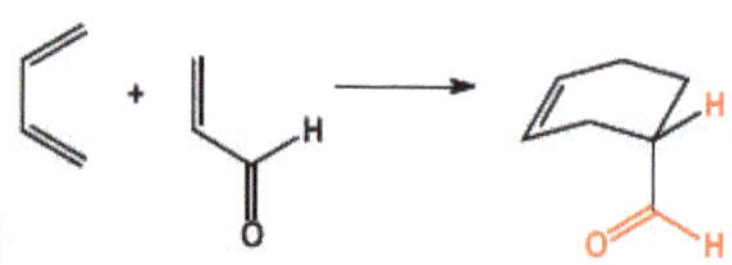

19.11

a)

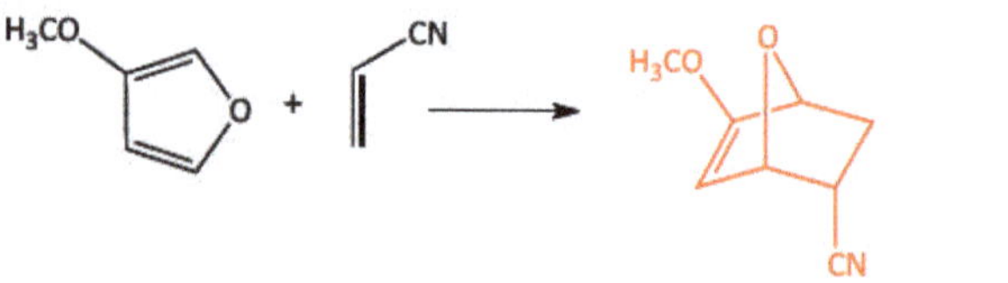

Notice, it is 1,4 and not 1,3.
Endo rule followed.

b)

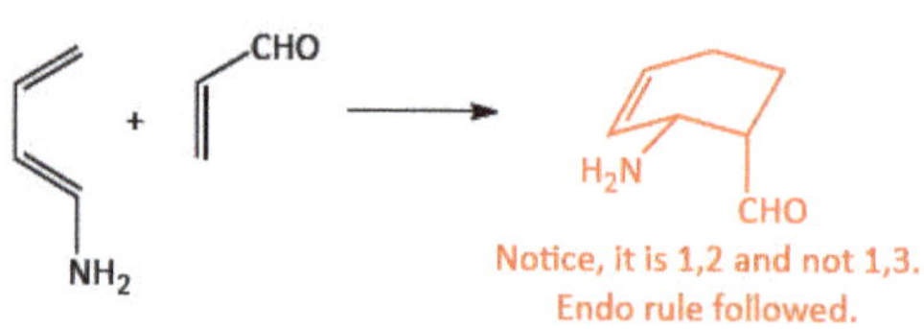

Notice, it is 1,2 and not 1,3.
Endo rule followed.

Chapter 20

20.1

a) aromatic b) antiaromatic c) antiaromatic d) aromatic e) antiaromatic

20.2

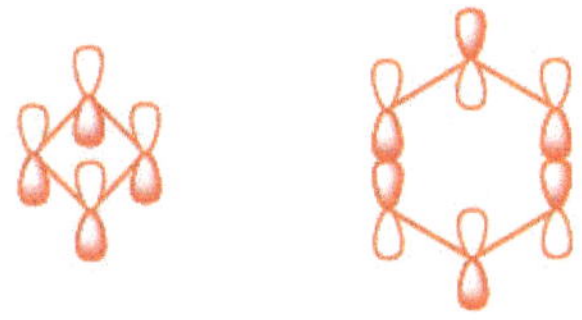

Lowest energy MO of cyclobutadiene   Highest energy MO of benzene

20.3

Each of the following compounds is nonaromatic. Which of the previous rules, 1-3, does each compound break?

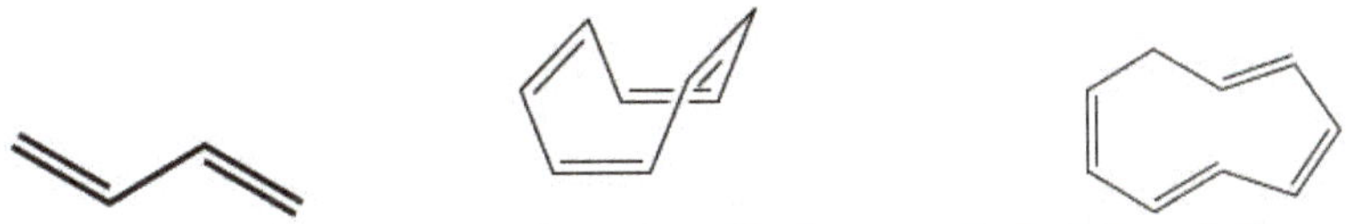

Rule 1 (not a ring)   Rule 3 (not flat)   Rule 2 (not fully conjugated)

20.4

a) aromatic b) antiaromatic c) antiaromatic d) nonaromatic e) aromatic f) aromatic g) antiaromatic
h) nonaromatic i) antiaromatic

20.5

a) no b) no (a and b would be antiaromatic if flat) c) yes d) yes e) yes (c, d, and e would be aromatic if flat.)

20.6

a) yes b) no c) no d) no

20.7

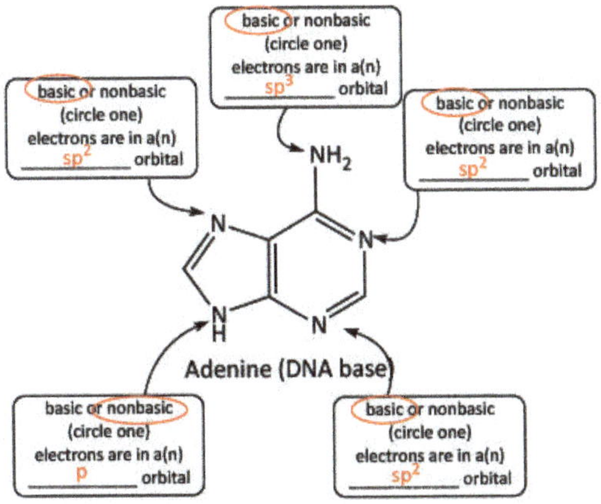

Adenine (DNA base)

20.8

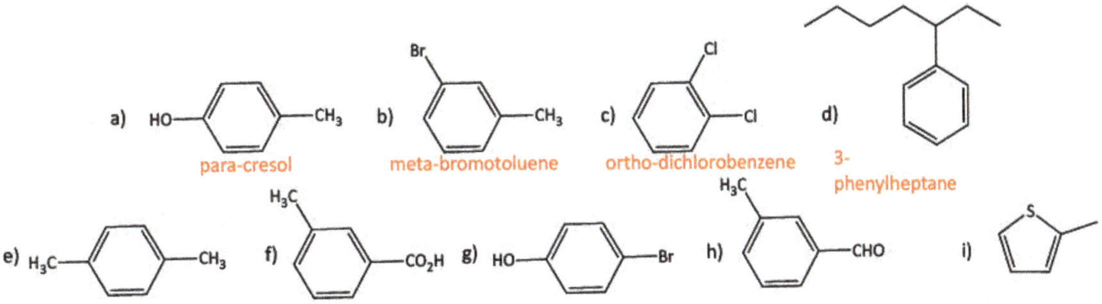

a) para-cresol b) meta-bromotoluene c) ortho-dichlorobenzene d) 3-phenylheptane
e) para-xylene f) meta-toluic acid g) para-bromophenol h) meta-methylbenzaldehyde i) 2-methylthiophene

Chapter 21

21.1

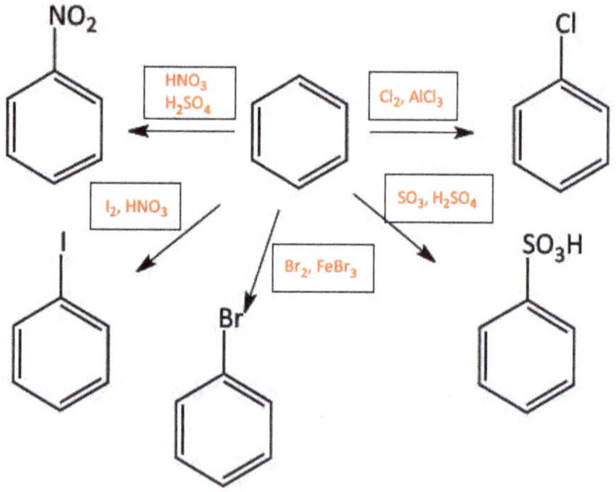

21.2

a) 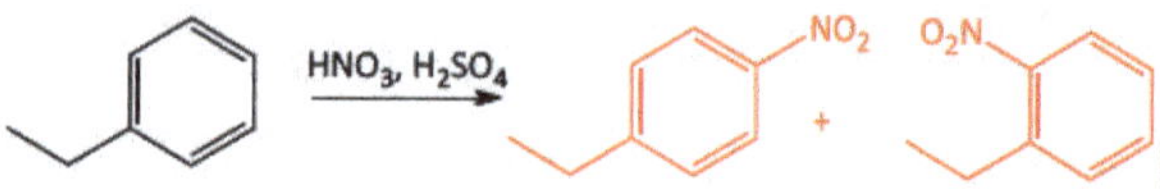

b) 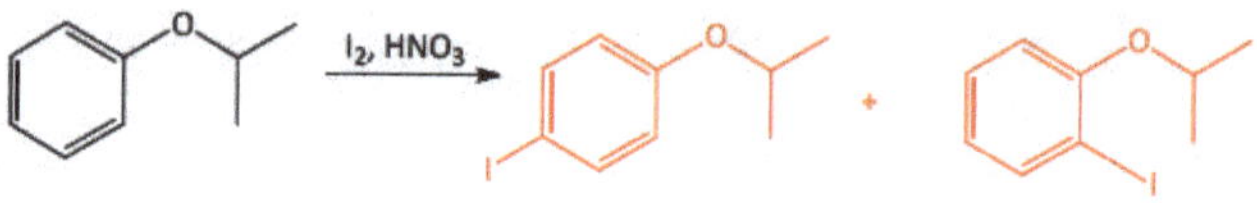

c) 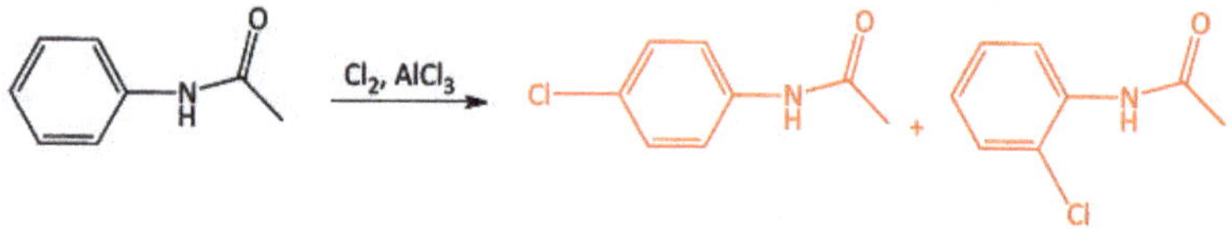

21.3

a) 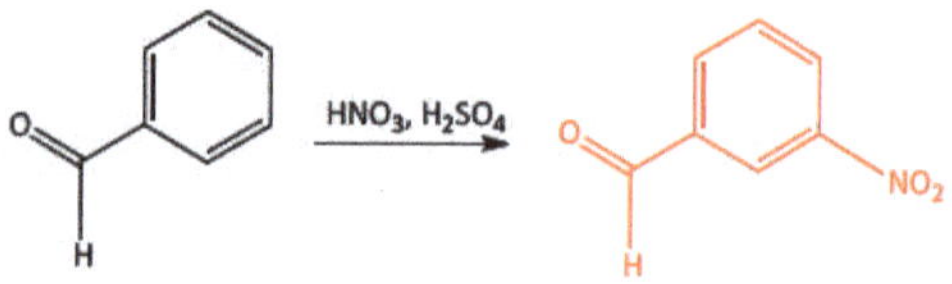

b) 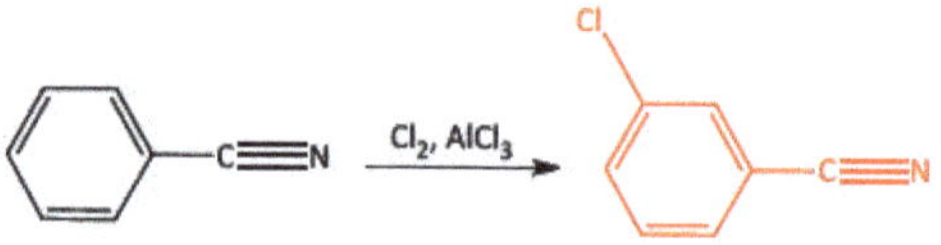

c) 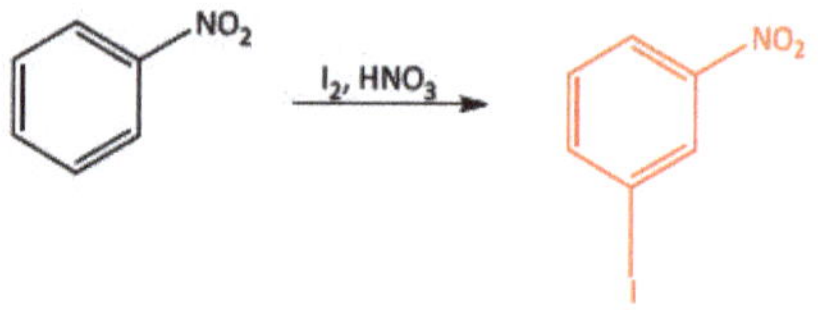

d) 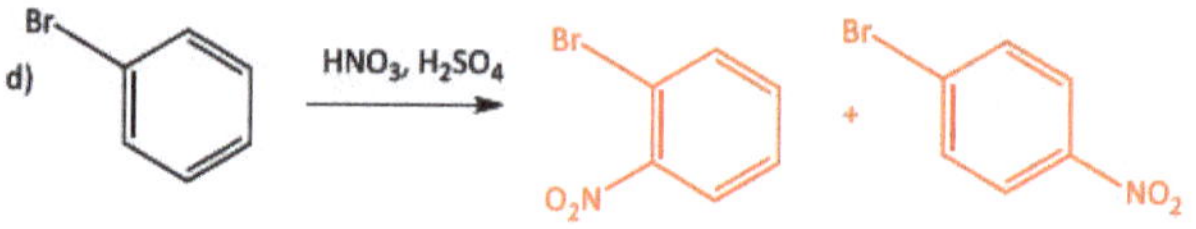

21.4

a) 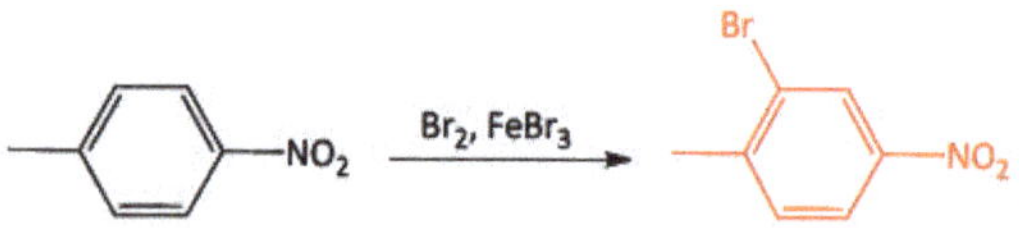

b) 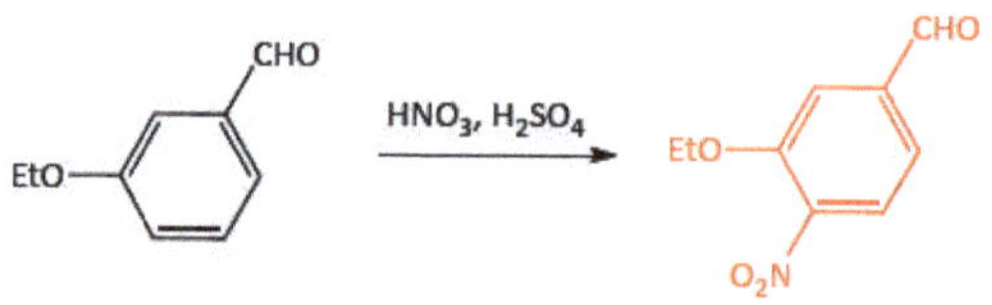

21.5 1. They can polyalkylate because after the first alkyl group goes on the ring, it becomes even more reactive because alkyl groups are activating. 2. They do not work on deactivated rings. 3. They can rearrange because they have a carbocation intermediate.

21.6 Once the first acyl group goes onto the ring; the ring becomes deactivated because carbonyl-containing groups are electron withdrawing.

21.7

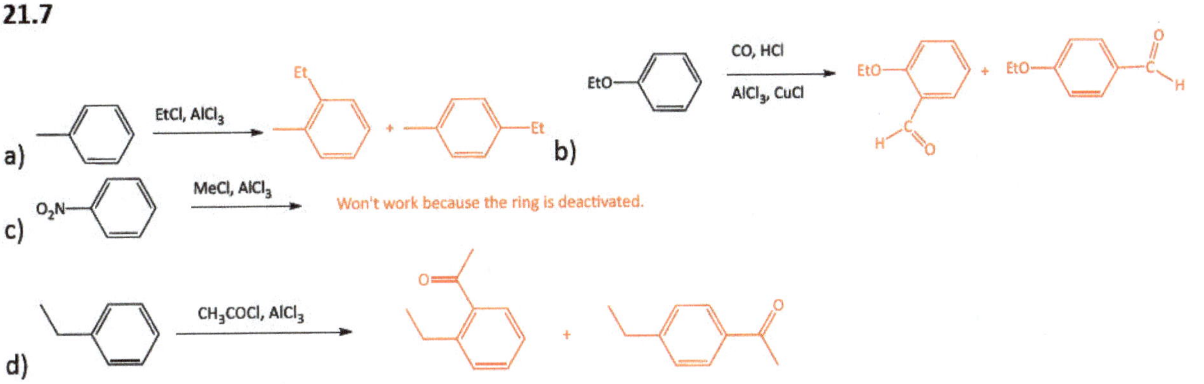

21.8

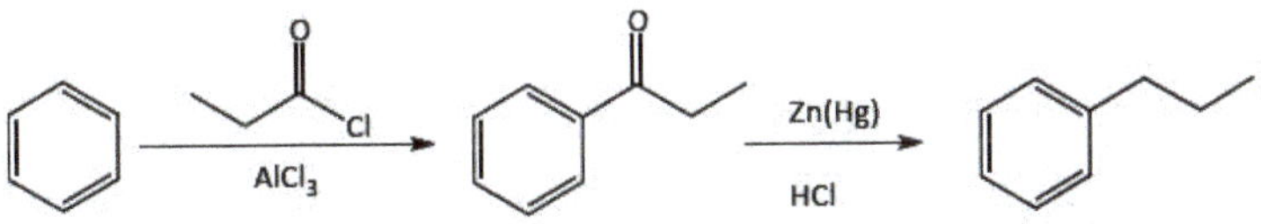

Friedel Crafts acylations cannot rearrange because the acylium carbocations are very stable because of resonance.

21.9

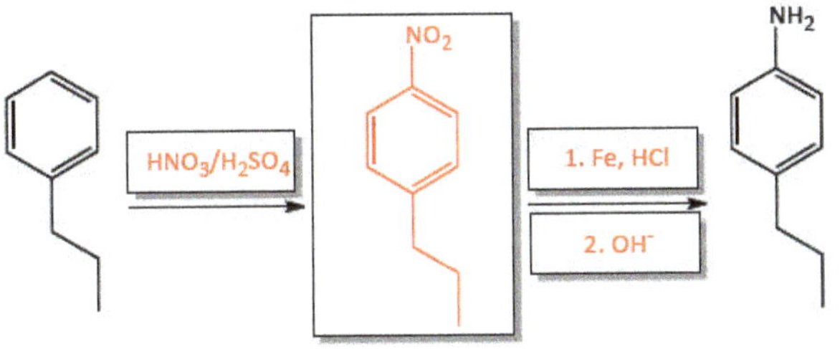

21.10

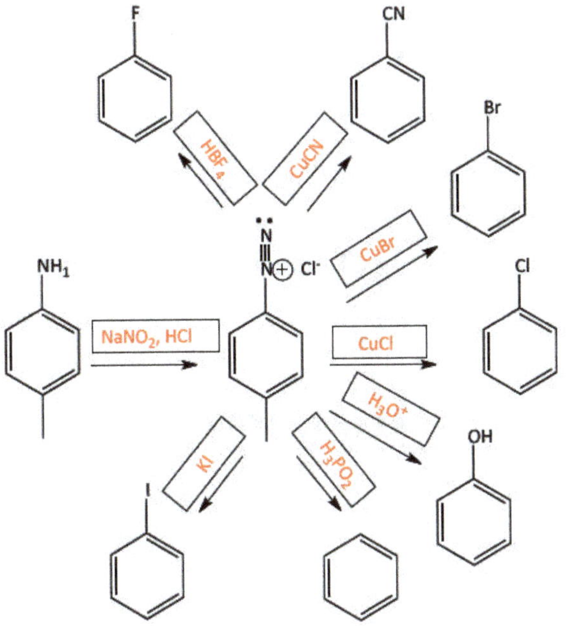

Answers to Exercises

21.11

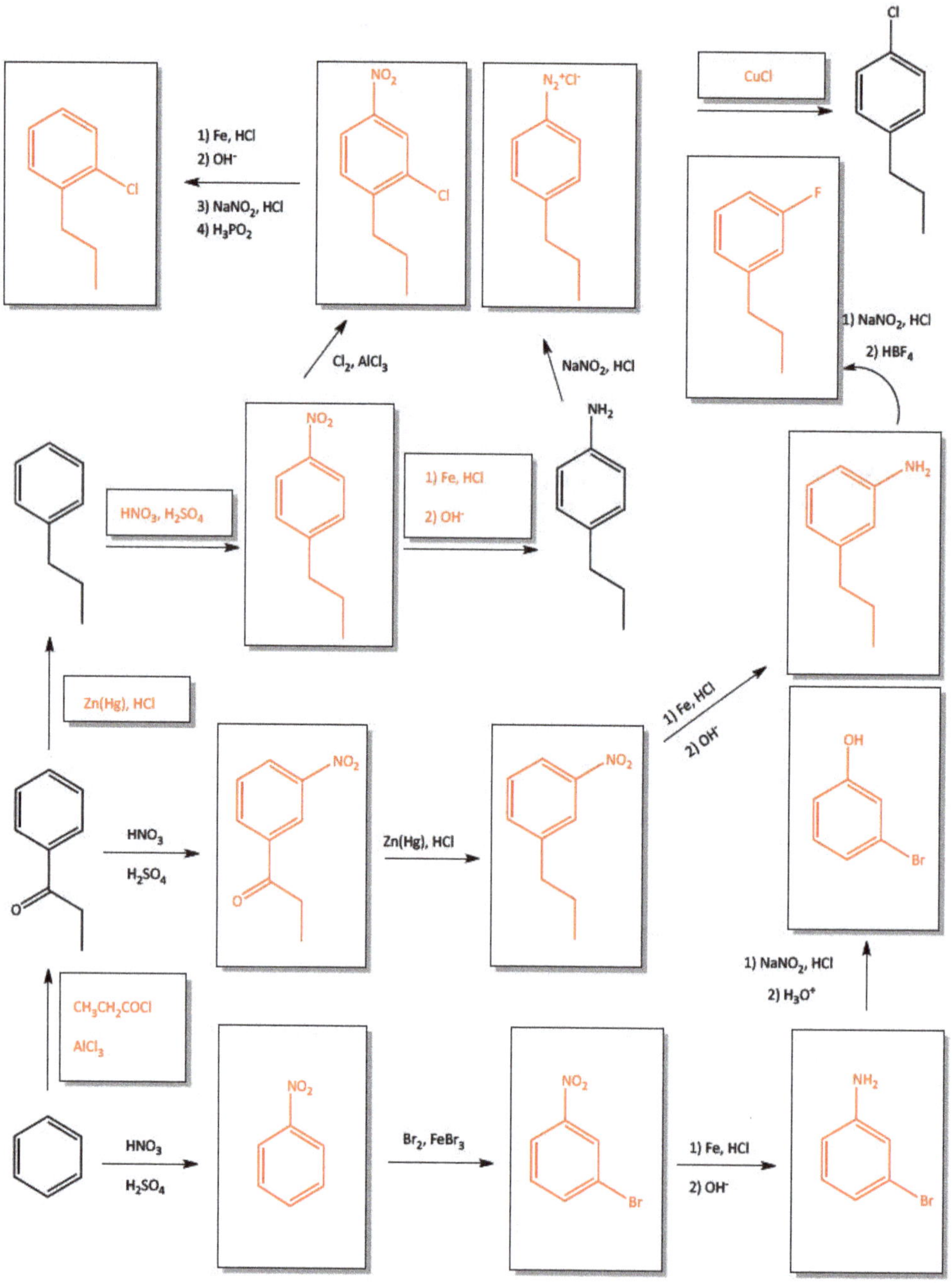

21.12

a) The lone pair of electrons on the amino group would react with the Lewis Acid, $AlCl_3$ instead of the $AlCl_3$ helping pull off the chloride of isopropyl chloride.

b)

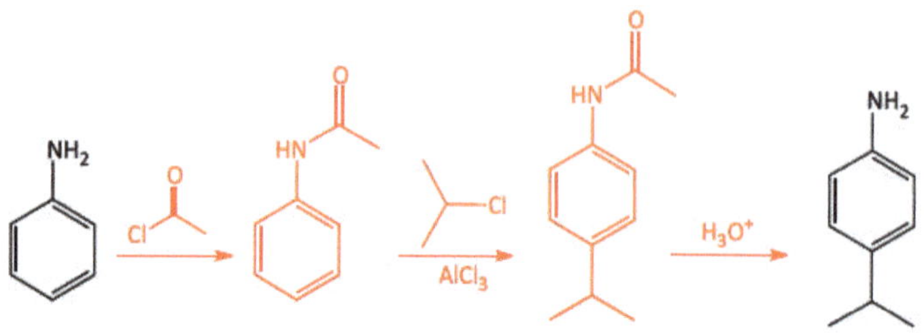

21.13 b It has the most electron withdrawing groups (-NO_2) in the ortho and para positions.

21.14

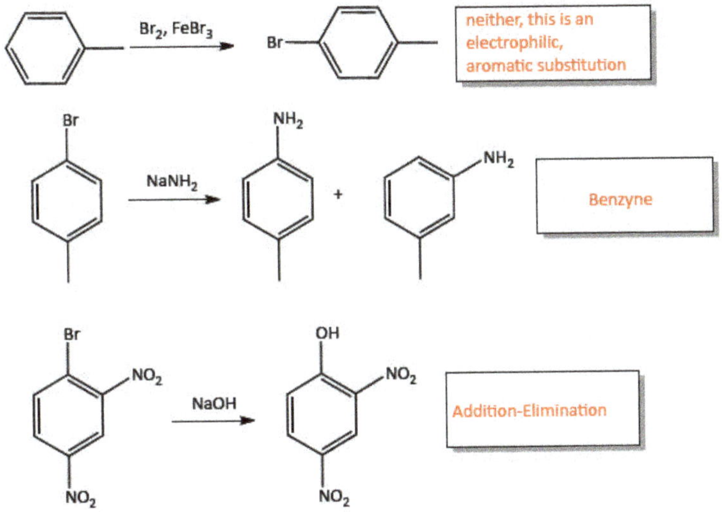

21.15

a) 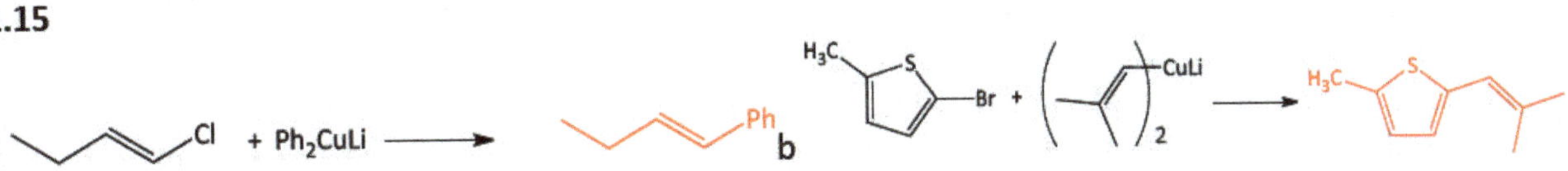

21.16

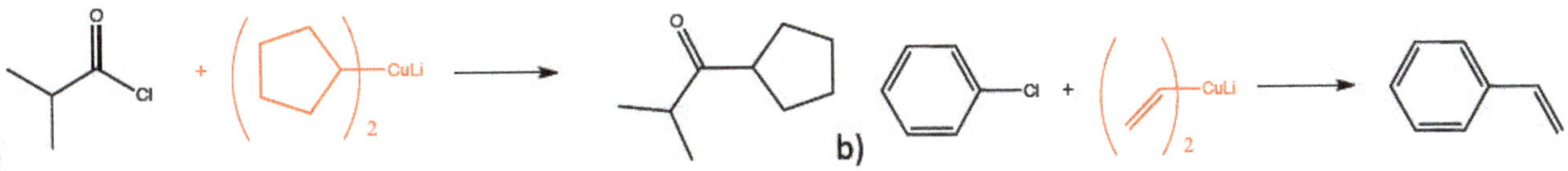

a)

21.17

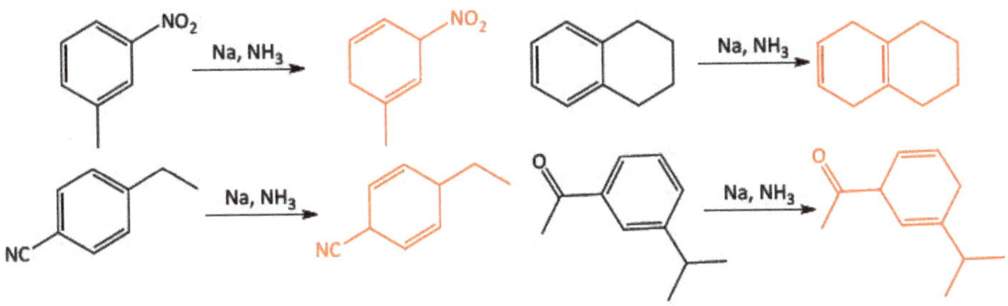

21.18

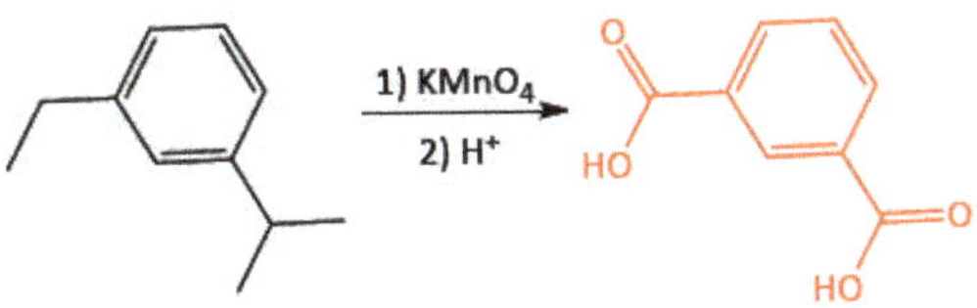

Chapter 22

22.1

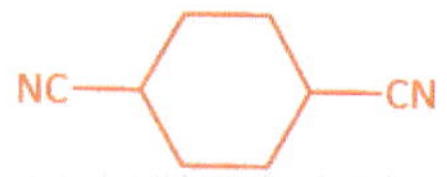
a) 1,4-dicyanocyclohexane

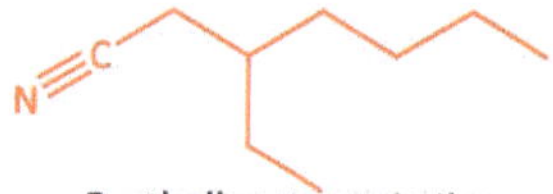
b) 3-ethylheptanenitrile

c) 4-aminocyclohexanone

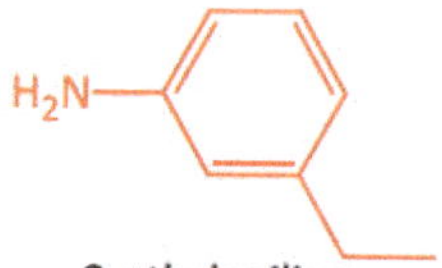
d) 3-ethylaniline

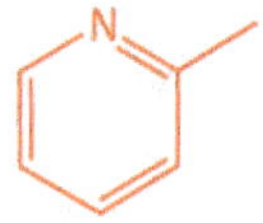
e) 2-methylpyridine

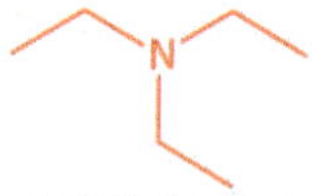
f) triethylamine

g) 2-octanamine

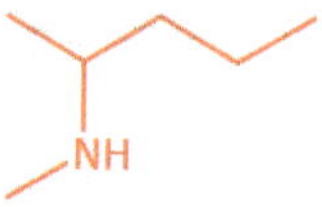
h) N-methyl-2-pentanamine

22.2

a) If the M^+ peak in a mass spectrum is odd, it tells us we have an odd number of nitrogen atoms in the compound.

b) If a compound is a primary amine and has an $-NH_2$ group, the IR spectrum should show a broad, double humped peak (rabbit ears) around 3200-3300 cm^{-1}.

c) If a compound is a nitrile, it should show a strong, sharp C≡N peak in the IR spectrum around 2200-2300 cm^{-1}.

d) If a compound is a secondary amine and has an –NH group, the IR spectrum should show a broad peak with one hump on it (a shark's tooth) around 3200-3300 cm^{-1}.

e) The C=O carbonyl peak for an amide shows up at a lower wavenumber than "regular" ketone or aldehyde C=O peaks around 1710 cm^{-1}. The carbonyl peak for an amide is closer to 1660 cm^{-1}.

f) In the ^{13}C NMR spectrum, the carbon atom for the C=O of an amide appears in the 150-200 ppm region. In the ^{13}C NMR spectrum, the carbon atom in the C≡N of a nitrile appears in the 100-150 ppm region.

22.3

a) 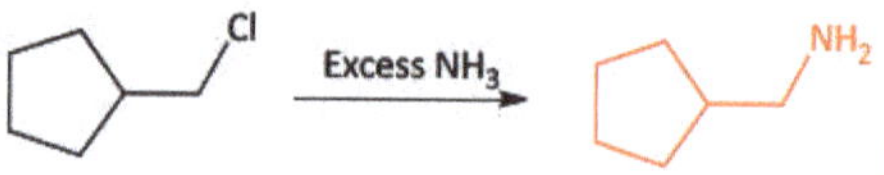

b) 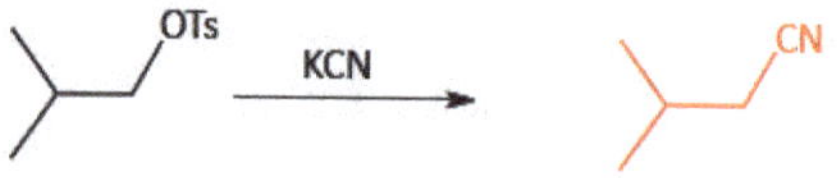

c) 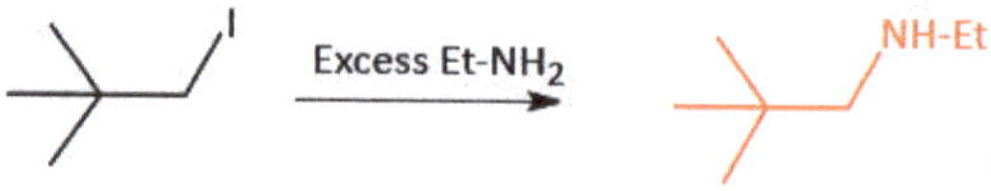

d) 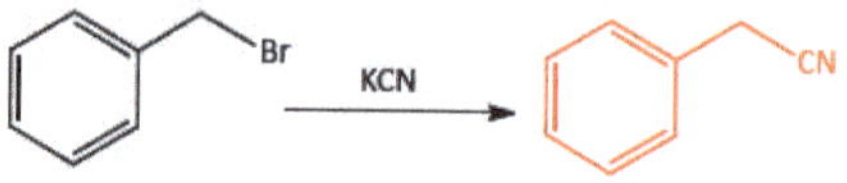

e) 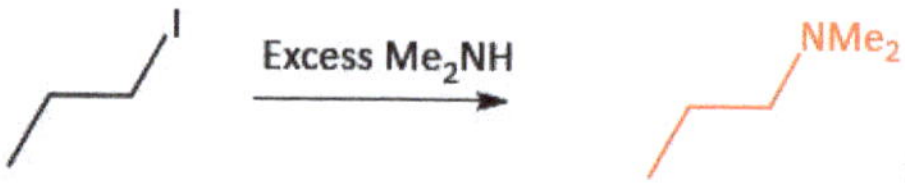

f) 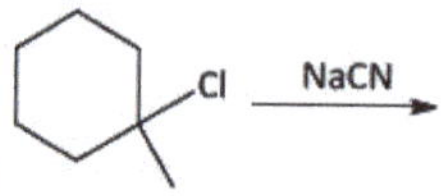
No reaction. Alkyl halide is tertiary.

22.4

a) 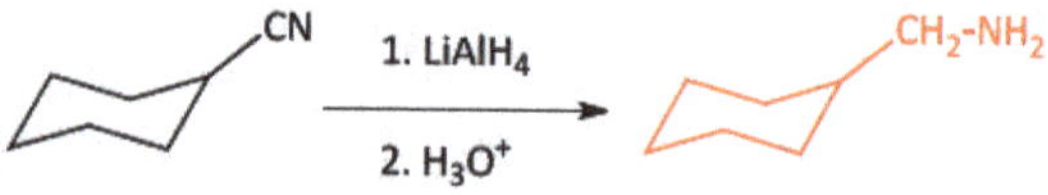

b) 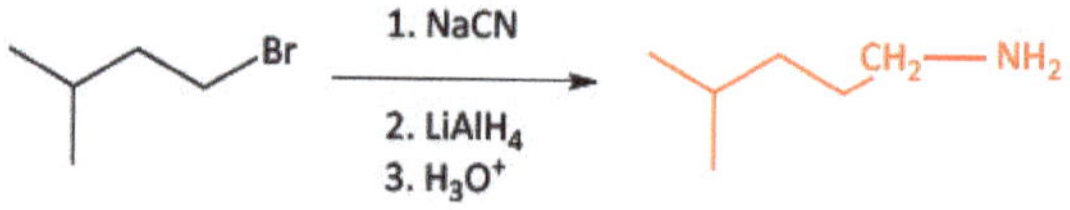

22.5

a) 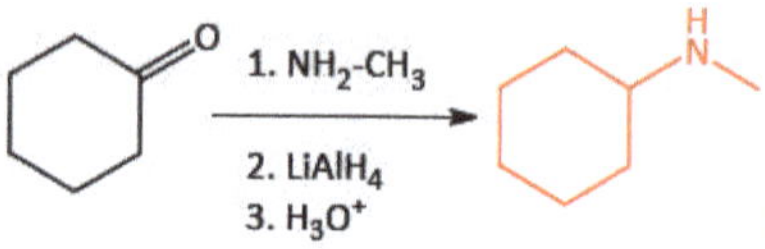

b) 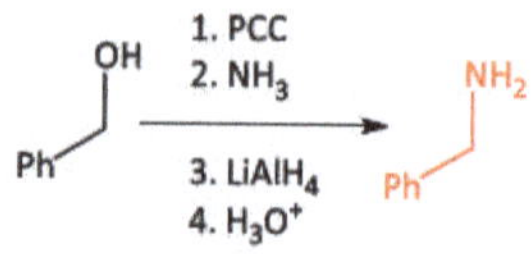

22.6

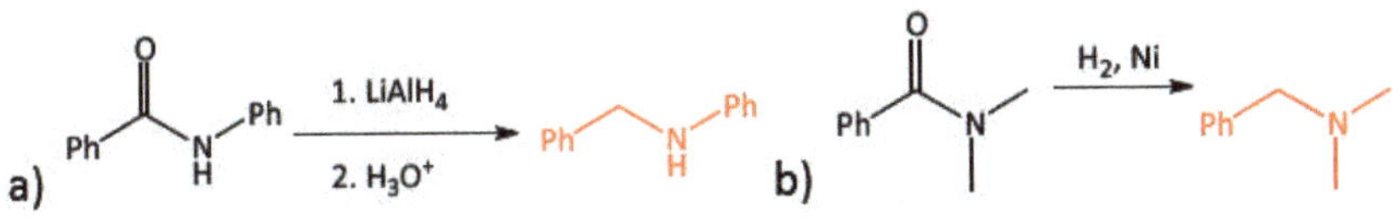

22.7

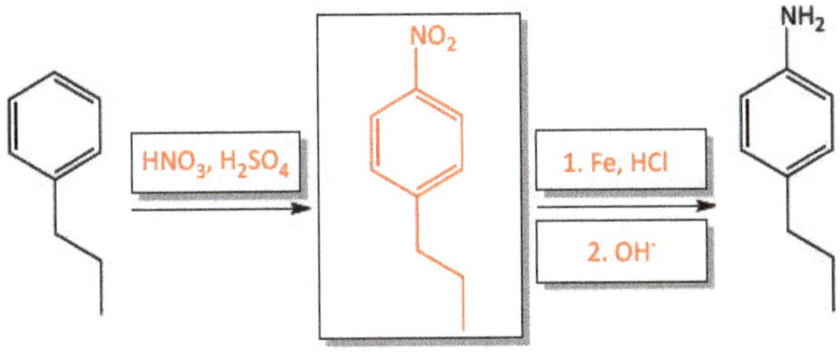

22.8

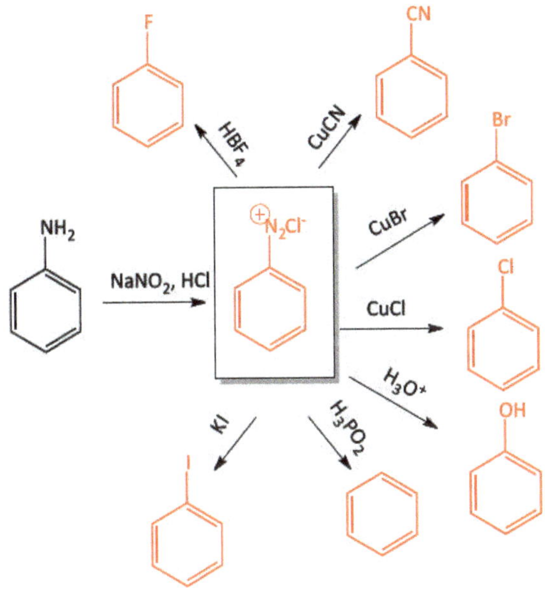

22.9

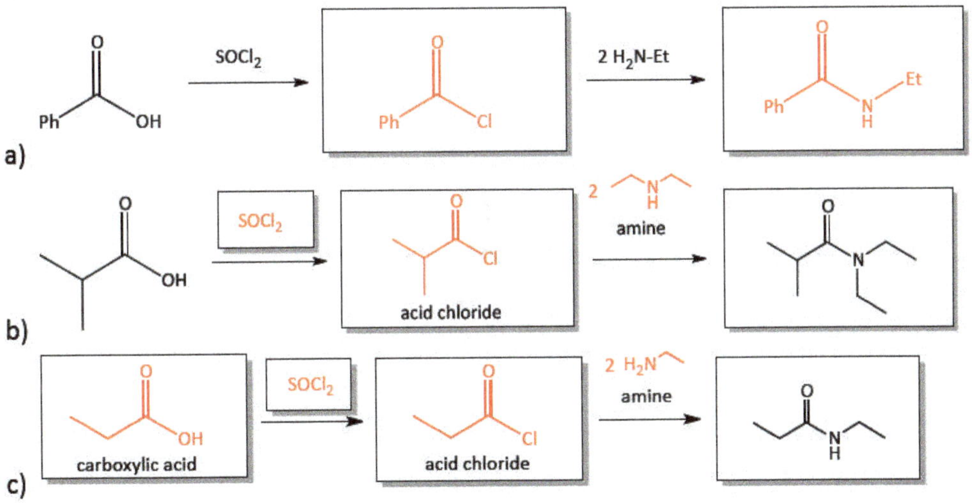

Answers to Exercises

22.10

a) 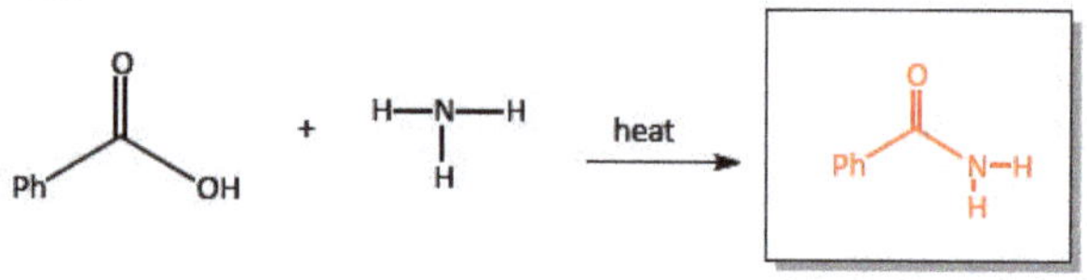

b) 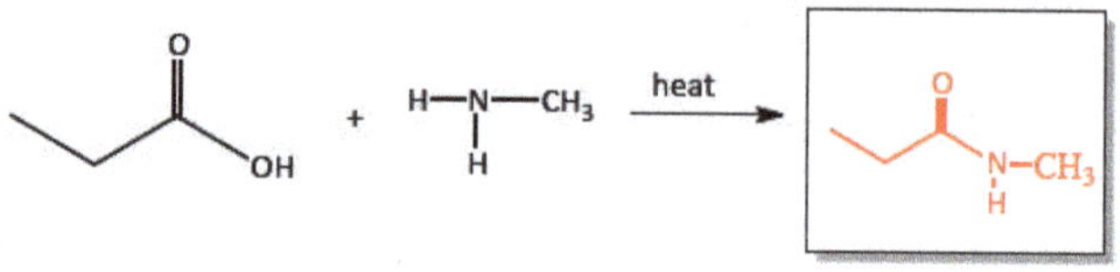

c) 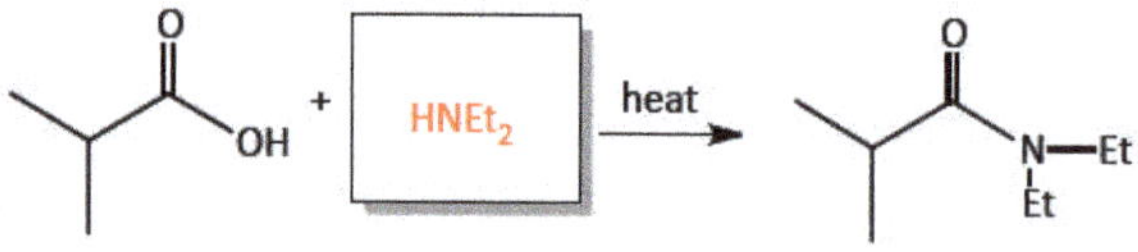

d) 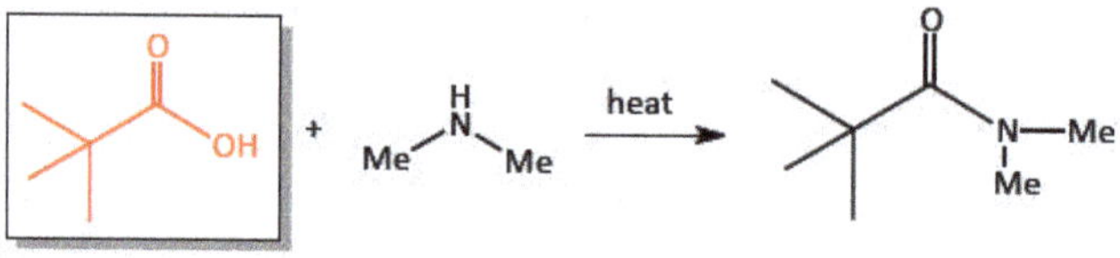

e) 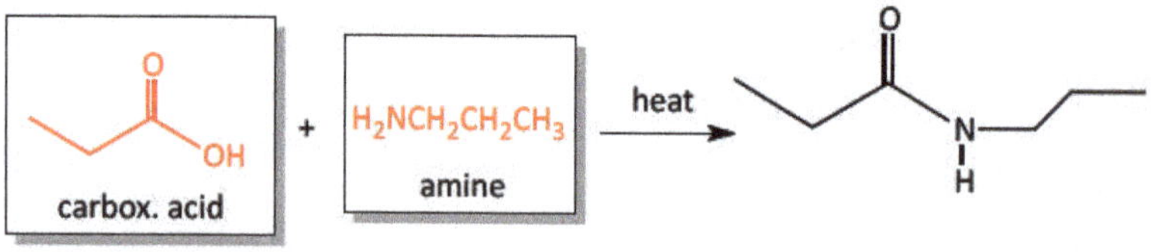

22.11

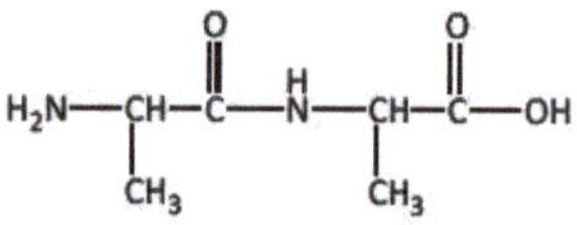

22.12

a) 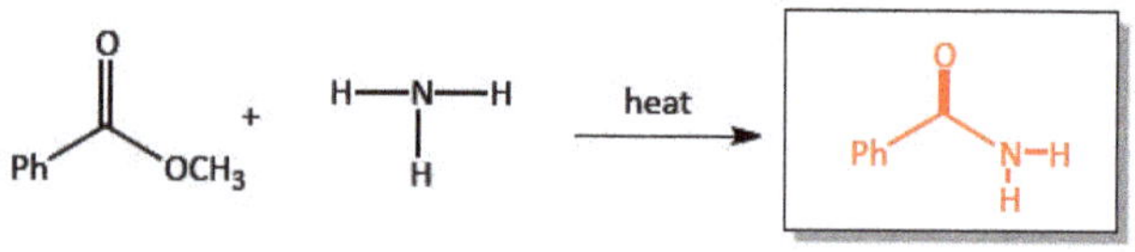

b) 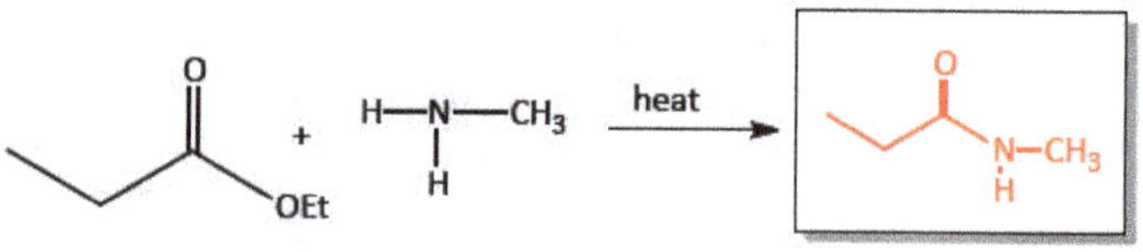

c) 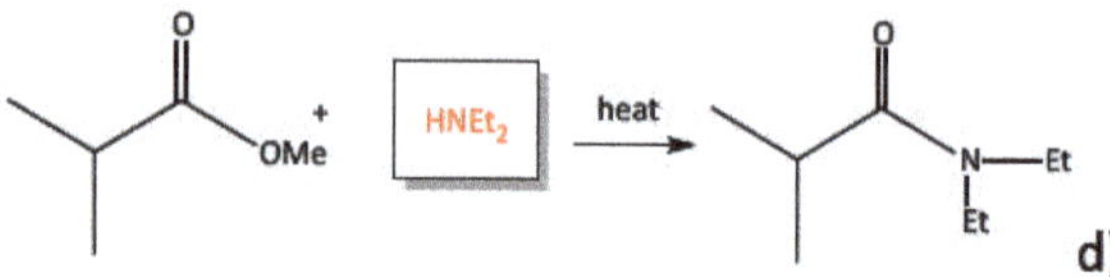

d) 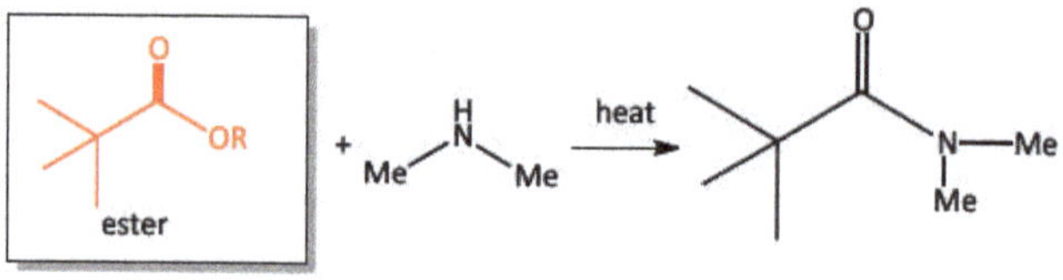

e) 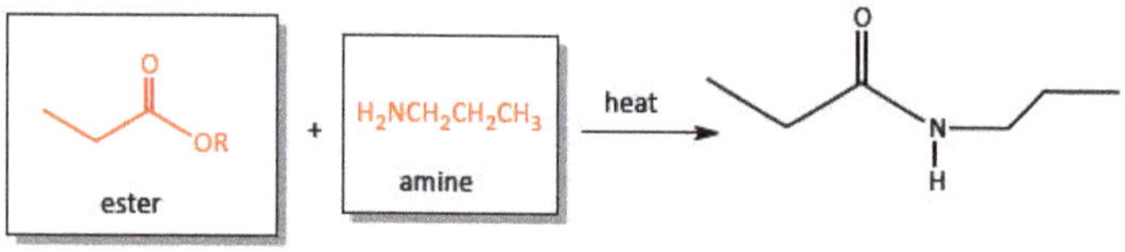

22.13

a) 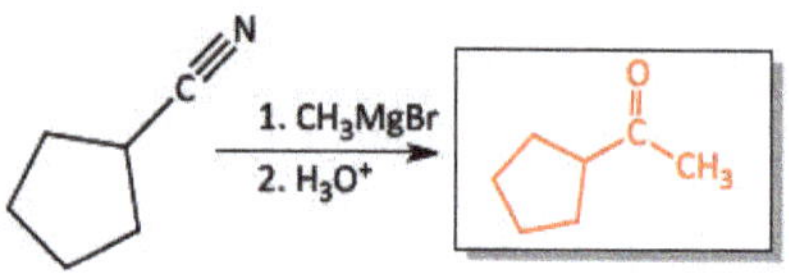

b) 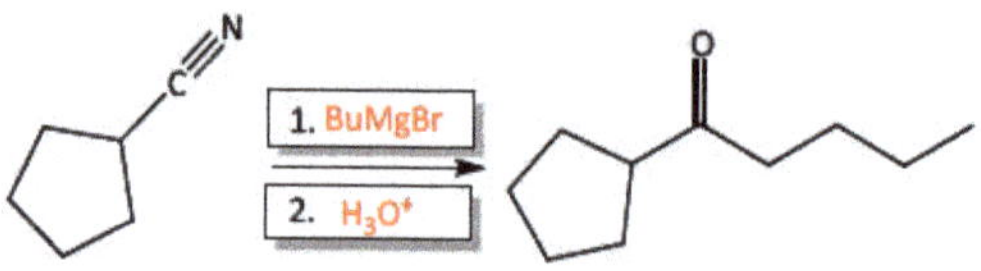

22.14

a) 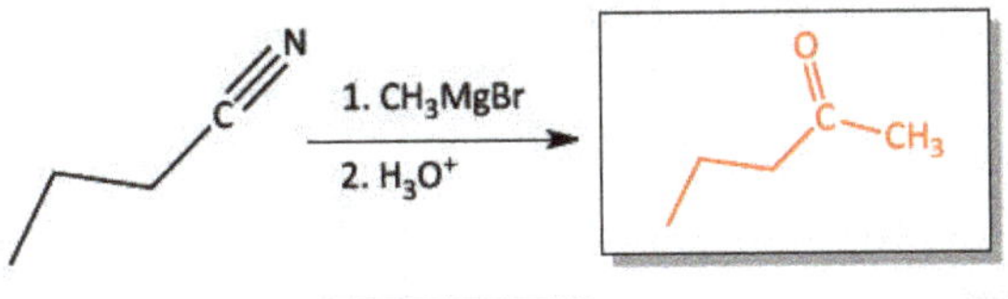

b) 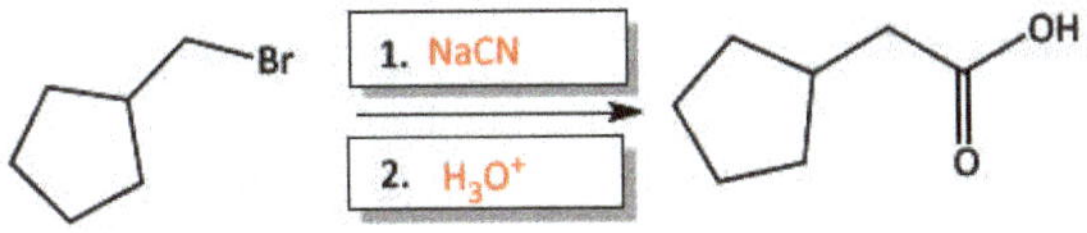

c) 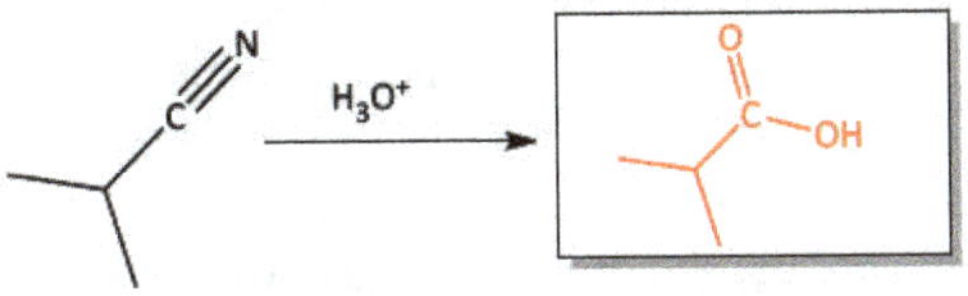

d) 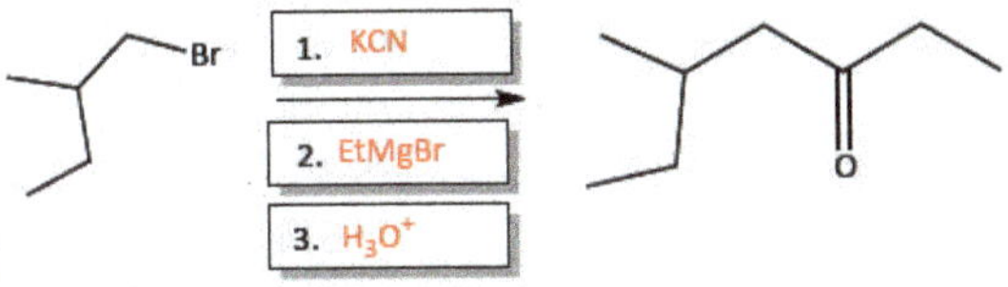

e) 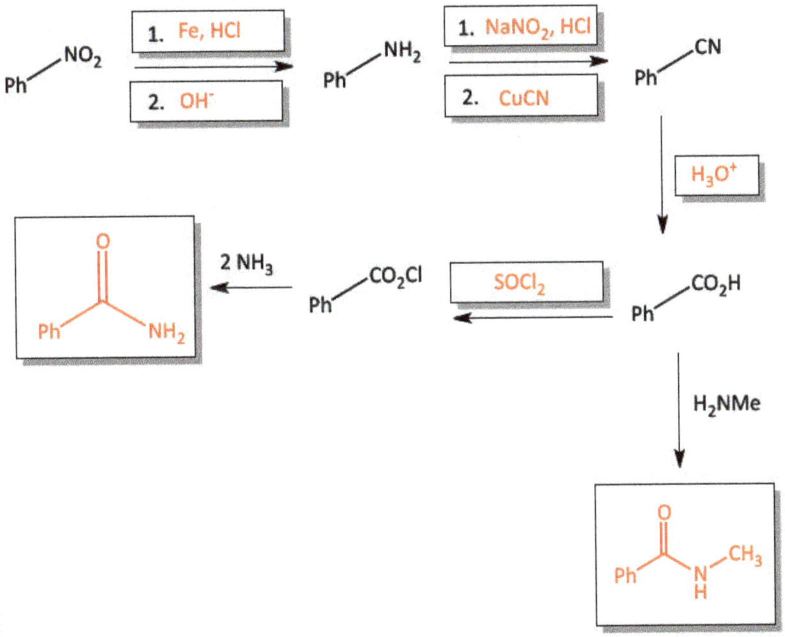

Index

D

E

P

Q

R

S

www.ingramcontent.com/pod-product-compliance
Lightning Source LLC
LaVergne TN
LVHW062151200926
848484LV00007B/1899
9781733039505